HANDBOOK of
WOOD TECHNOLOGY
and
HOUSE CONSTRUCTION

Staff of Research and Education Association

Research and Education Association
505 Eighth Avenue
New York, N. Y.   10018

HANDBOOK OF WOOD TECHNOLOGY
AND HOUSE CONSTRUCTION

Printed in the United States of America

Library of Congress Catalog Card Number 81-50949

International Standard Book Number 0-87891-529-X

PREFACE

The contents of this handbook have been selected to make it a comprehensive reference source for engineers and builders. The handbook includes technology sections that make it suitable for use in building complete houses, adding to existing houses, and home-improvement projects.

In this age of synthetic materials, wood remains as one of the most important construction materials. Wood is often found to be the most economical material for buildings, housing, support structures, furniture, and household appliances. Wood is easily shaped and wood members are easily linked. Special tools are rarely required in working with wood. When properly treated and maintained, wood can also be in service for very long periods of time.

To design and build with wood, it is often useful to know the properties of the different types of wood that are available. For this reason, Part I of the handbook begins with chapters on the structure and properties of woods and lumber that are commercially important to the United States. These chapters are followed by those that deal with fastening devices and gluing procedures. Plywoods and sandwich constructions are included, and the methods that may be used for preserving wood are extensively discussed.

Part II of the handbook deals with wood-frame house construction, which remains a most popular type of construction for building homes in the U. S. Separate chapters are devoted to excavations, foundations, and floors, in this part of the handbook. These are followed by wall, ceiling, and roof framing. Sheathing and coverings for framed walls, ceilings, and roofs are discussed in detail. Windows, doors, exterior trim and coverings, plumbing, heating, water supply, sewage disposal, electrical systems, and insulation are all described in a manner so as to be helpful to the practical designer and builder. Separate chapters in this part of the book are devoted moreover, to interior doors, cabinets, stairs, porches and garages, chimneys and fireplaces, driveways, gutter and sheet metal work. Protection against fire, termites, maintenance and repair are taught in individual chapters on these subjects.

As a result of evaluating many different designs and construction methods over a period of several years, the U. S. Dept. of Agriculture found particularly low-cost designs and constructions that are useful. These are presented in Part III of this handbook. These low-cost construction designs cover all exterior and interior aspects of all important parts connected with house construction.

A very extensive index makes it possible for users of the handbook to find topics quickly.

Much of the contents of this handbook represent the expertise of individuals in the U. S. Dept. of Agriculture and the Defense Department. The results of the vast research and development activities undertaken by these government agencies and others, have been incorporated in this handbook, and are gratefully acknowledged.

Max Fogiel, Ph. D.
Program Director

CONTENTS

PART I

WOOD AS A CONSTRUCTION MATERIAL

PART II

WOOD-FRAME HOUSE CONSTRUCTION

PART III

SPECIAL LOW COST METHODS IN DESIGN AND CONSTRUCTION

Part I

WOOD AS A CONSTRUCTION MATERIAL

Chapter 1

CHARACTERISTICS AND AVAILABILITY OF WOODS COMMERCIALLY IMPORTANT TO THE UNITED STATES

Through the ages the unique characteristics and comparative abundance of wood have made it a natural material for homes and other structures, furniture, tools, vehicles, and decorative objects. Today, for the same reasons, wood is prized for a multitude of uses.

All wood is composed of cellulose, lignin, ash-forming minerals, and extractives formed in a cellular structure. Variations in the characteristics and volume of the four components and differences in cellular structure result in some woods being heavy and some light, some stiff and some flexible, some hard and some soft. For a single species, the properties are relatively constant within limits; therefore, selection of wood by species alone may sometimes be adequate. However, to use wood to its best advantage and most effectively in engineering applications, the effect of specific characteristics or physical properties must be considered.

Historically, some woods have filled many purposes, while others which were not so readily available or so desirable qualitatively might serve only one or two needs. The tough, strong, and durable white oak, for example, was a highly prized wood for shipbuilding, bridges, cooperage, barn timbers, farm implements, railroad crossties, fenceposts, flooring, paneling, and other products. On the other hand, woods such as black walnut and cherry became primarily cabinet woods. Hickory was manufactured into tough, hard resilient striking-tool handles. Black locust was prized for barn timbers and treenails. What the early builder or craftsman learned by trial and error became the basis for the decision as to which species to use for a given purpose, and what characteristics to look for in selecting a tree for a given use. It was commonly accepted that wood from trees gown in certain locations under certain conditions was stronger, more durable, and more easily worked with tools, or finer grained than wood from trees in some other locations. Modern wood quality research has substantiated that location and growth conditions do significantly affect wood properties.

The gradual utilization of the virgin forests in the United States has reduced the available supply of large clear logs for lumber and veneer However, the importance of high quality logs has diminished as new concepts of wood use have been introduced. Second-growth timber (fig. 1–1), the balance of the old-growth forests, and imports continue to fill the needs for wood in the quality required. Wood is as valuable an engineering material as it ever was, and in many cases technological advances have made it even more useful.

The inherent factors which keep wood in the forefront of raw materials are many and varied, but one of the chief attributes is its availability in many species, sizes, shapes, and conditions to suit almost every demand. It has a high ratio of strength to weight and a remarkable record for durability and performance as a structural material. Dry wood has good insulating properties against heat, sound, and electricity. It tends to absorb and dissipate vibrations under some conditions of use, yet is an incomparable material for such musical instruments as violins. Because of grain patterns and colors, wood is inherently an esthetically pleasing material, and its appearance may be easily enhanced by stains, varnishes, lacquers, and other finishes. It is easily shaped with tools and fastened with adhesives, nails, screws, bolts, and dowels. When wood is damaged it is easily repaired, and wood structures are easily remodeled or altered. In addition, wood resists oxidation, acid, salt water, and other corrosive agents; has a high salvage value; has good shock resistance; takes treatments with preservatives and fire retardants; and combines with almost any other material for both functional and esthetic uses.

TIMBER RESOURCES AND WOOD USES

In the United States more than 100 woods are available to the prospective user, but it is very unlikely that all are available in any one locality. Commercially, there are about 60 native woods of major importance. Another 30 woods are commonly imported in the form of logs, cants, lumber, and veneer for industrial uses, the building trades, and the craftsman.

A continuing program of timber inventory is in effect in the United States through cooperation of Federal agencies and the States. As new information regarding timber resources becomes available it appears in State and Federal

Figure 1–1.—Reforested area on the Kaniksu National Forest in Idaho. Foreground is stocked with western larch and Douglas-fir reproduced naturally. The central area, edged by mature timber, is a field-planted western white pine plantation.

publications. One of the most valuable source books is "Timber Trends in the United States," Forest Service, U.S. Department of Agriculture Forest Resource Report No. 17.

The best source of current information on timber consumption, production, imports, and the demand and price situation is published periodically in a U.S. Department of Agriculture Miscellaneous Publication, entitled "The Demand and Price Situation for Forest Products." Both publications are available from the Superintendent of Documents, U.S. Government Printing Office, Washington, D.C. 20402.

HARDWOODS AND SOFTWOODS

Trees are divided into two broad classes, usually referred to as "hardwoods" and "softwoods." Some softwoods, however, are actually harder than some of the hardwoods, and some hardwoods are softer than softwoods. For ex-

ample, such softwoods as longleaf pine and Douglas-fir produce wood that is typically harder than the hardwoods basswood and aspen. Botanically, the softwoods are Gymnosperms, species that fall into a classification called conifers that have their seed exposed, usually in cones. Examples are the pines, spruces, redwoods, and junipers. The other broad classification, the Angiosperms, comprise the various orders of hardwoods. They have true flowers and broad leaves, and the seeds are enclosed in a fruit. United States softwoods have needlelike or scalelike leaves that, except for larches and baldcypress, remain on the trees throughout the year. The hardwoods, with a few exceptions, lose their leaves in fall or during the winter. Most of the imported woods, other than those from Canada, are hardwoods.

Major resources of softwood species are spread across the United States, except for the Great Plains where only small areas are for-

3

ested. Species are often loosely grouped in three general producing areas:

Western softwoods

Douglas-fir	Sitka spruce
Ponderosa pine	Idaho white pine
Western hemlock	Sugar pine
Western redcedar	Lodgepole pine
True firs	Port-Orford-cedar
Redwood	Incense-cedar
Engelmann spruce	Alaska-cedar
Western larch	

Northern softwoods

Eastern white pine
Red pine
Jack pine
Eastern hemlock
Balsam fir
Tamarack
Eastern spruces
Eastern redcedar
Northern white-cedar

Southern softwoods

Southern pine	Eastern redcedar
Baldcypress	Atlantic white-cedar

With some exceptions, most hardwoods occur east of the Great Plains area (fig. 1–2). The following classification is based on the principal producing region for each wood:

Southern hardwoods

Ash	Magnolia
Basswood	Soft maple
American beech	Red oak
Cottonwood	White oak
Elm	Sweetgum
Hackberry	American sycamore
Pecan hickory	Tupelo
True hickory	Black walnut
American holly	Black willow
Black locust	Yellow-poplar

Northern and Appalachian hardwoods

Ash	True hickory
Aspen	Black locust
Basswood	Hard maple
American beech	Soft maple
Birch	Red oak
Black cherry	White oak
American chestnut [1]	American sycamore
Cottonwood	Black walnut
Elm	Yellow-poplar
Hackberry	

Western hardwoods

Red alder	Bigleaf maple
Oregon ash	Paper birch
Aspen	Tanoak
Black cottonwood	

[1] American chestnut is no longer harvested as a living tree, but the lumber is still on the market as "wormy chestnut" and prices are quoted in the Hardwood Market Report.

Figure 1–2.—Mixed northern hardwoods on Ottawa National Forest in Michigan.

COMMERCIAL SOURCES OF WOOD PRODUCTS

Softwoods are available directly from the sawmill, wholesale and retail yards, or lumber brokers. Softwood lumber and plywood are used in construction for forms, scaffolding, framing, sheathing, flooring, ceiling, trim, paneling, cabinets, and many other building components. Softwoods may also appear in the form of shingles, sash, doors, and other millwork, in addition to some rough products such as round treated posts.

Hardwoods are used in construction for flooring, architectural woodwork, trim, and paneling. These items are usually available from lumberyards and building supply dealers. Most hardwood lumber and dimension are remanufactured into furniture, flooring, pallets, containers, dunnage, and blocking. Hardwood lumber and dimension are available directly from the manufacturer, through wholesalers and brokers, and in some retail yards.

Both softwood and hardwood forest products are distributed throughout the United States, although they tend to be more readily available in or near their area of origin. Local preferences and the availability of certain species may influence choice; but, a wide selection of woods is generally available for building construction, industrial uses, remanufacturing, and use by home craftsmen.

USE CLASSES AND TRENDS

Some of the many use classifications for wood are growing with the overall national economy, and others are holding about the same levels of production and consumption. The wood-based industries that are growing most vigorously convert wood to thin slices (veneer), particles (chips, flakes, etc.), or fiber pulps and reassemble the elements to produce plywood, numerous types of particleboard, paper, paperboard, and fiberboard products. Another growing wood industry specializes in producing laminated timbers. Annual production by the lumber industry has continued for several years at almost the same board footage. Some of the forest products industries, such as railroad crossties, cooperage, shingles, and shakes appear to have leveled off following a period of depressed production, and in some instances to be making modest increases in production.

COMMERCIAL SPECIES IN THE UNITED STATES

The following brief discussions of the principal localities of occurrence, characteristics, and uses of the main commercial species, or groups of species, will aid in selecting woods for specific purposes. More detailed information on the properties of these and other species is given in various tables throughout this handbook.

Certain uses listed under the individual species are no longer important. They have been included to provide some information on the historical and traditional uses of the species.

The common and botanical names given for the different species conform to the Forest Service official nomenclature for trees.

Hardwoods

Alder, Red

Red alder (*Alnus rubra*) grows along the Pacific coast between Alaska and California. It is used commercially along the coasts of Oregon and Washington and is the most abundant commercial hardwood species in these States.

The wood of red alder varies from almost white to pale pinkish brown and has no visible boundary between heartwood and sapwood. It is moderately light in weight, intermediate in most strength properties, but low in shock resistance. Red alder has relatively low shrinkage.

The principal use of red alder is for furniture, but it is also used for sash, doors, panel stock, and millwork.

Ash

Important species of ash are white ash (*Fraxinus americana*), green ash (*F. pennsylvanica*), blue ash (*F. quadrangulata*), black ash (*F. nigra*), pumpkin ash (*F. profunda*), and Oregon ash (*F. latifolia*). The first five of these species grow in the eastern half of the United States. Oregon ash grows along the Pacific coast.

Commercial white ash is a group of species that consists mostly of white ash and green ash, although blue ash is also included. Heartwood of commercial white ash is brown; the sapwood is light colored or nearly white. Second-growth trees have a large proportion of sapwood. Old-growth trees, which characteristically have little sapwood, are scarce.

Second-growth commercial white ash is particularly sought because of the inherent qualities of this wood; it is heavy, strong, hard, stiff, and has high resistance to shock. Because of these qualities such tough ash is used principally for handles, oars, vehicle parts, baseball bats, and other sporting and athletic goods. Some handle specifications call for not less than five or more than 17 growth rings per inch for handles of the best grade. The addition of a weight requirement of 43 or more pounds a cubic foot at 12 percent moisture content will assure excellent material.

Oregon ash has somewhat lower strength properties than white ash, but it is used locally for the same purposes.

Black ash is important commercially in the Lake States. The wood of black ash and pumpkin ash runs considerably lighter in weight than that of commercial white ash. Ash trees growing in southern river bottoms, especially in areas that are frequently flooded for long periods, produce buttresses that contain relatively lightweight and weak wood. Such wood is sometimes separated from tough ash when sold.

Ash wood of lighter weight, including black

ash, is sold as cabinet ash, and is suitable for cooperage, furniture, and shipping containers. Some ash is cut into veneer for furniture, paneling, wire-bound boxes.

Aspen

"Aspen" is a generally recognized name applied to bigtooth aspen (*Populus grandidentata*) and to quaking aspen (*P. tremuloides*). Aspen does not include balsam poplar (*P. balsamifera*) and the species of *Populus* that make up the group of cottonwoods. In lumber statistics of the U.S. Bureau of the Census, however, the term "cottonwood" includes all of the preceding species. Also, the lumber of aspens and cottonwood may be mixed in trade and sold either as poplar ("Popple") or cottonwood. The name "popple" or "poplar" should not be confused with yellow-poplar (*Liriodendron tulipifera*), also known in the trade as "poplar."

Aspen lumber is produced principally in the Northeastern and Lake States. There is some production in the Rocky Mountain States.

The heartwood of aspen is grayish white to light grayish brown. The sapwood is lighter colored and generally merges gradually into heartwood without being clearly marked. Aspen wood is usually straight grained with a fine, uniform texture. It is easily worked. Well-seasoned aspen lumber does not impart odor or flavor to foodstuffs.

The wood of aspen is lightweight and soft. It is low in strength, moderately stiff, moderately low in resistance to shock, and has a moderately high shrinkage.

Aspen is cut for lumber, pallets, boxes and crating, pulpwood, particleboard, excelsior, matches, veneer, and miscellaneous turned articles.

Basswood

American basswood (*Tilia americana*) is the most important of the several native basswood species; next in importance is white basswood (*T. heterophylla*). Other species occur only in very small quantities. Because of the similarity of the wood of the different species, no attempt is made to distinguish between them in lumber form. Other common names of basswood are linden, linn, and beetree.

Basswood grows in the eastern half of the United States from the Canadian provinces southward. Most basswood lumber comes from the Lake, Middle Atlantic, and Central States. In commercial usage, "white basswood" is used to specify white wood or sapwood of either species.

The heartwood of basswood is pale yellowish brown with occasional darker streaks. Basswood has wide, creamy-white or pale brown sapwood that merges gradually into the heartwood. When dry, the wood is without odor or taste. It is soft and light in weight, has fine, even texture, and is straight grained and easy to work with tools. Shrinkage in width and thickness during drying is rated as large; however, basswood seldom warps in use.

Basswood lumber is used mainly in venetian blinds, sash and door frames, molding, apiary supplies, woodenware, and boxes. Some basswood is cut for veneer, cooperage, excelsior, and pulpwood.

Beech, American

Only one species of beech, American beech (*Fagus grandifolia*), is native to the United States. It grows in the eastern one-third of the United States and adjacent Canadian provinces. Greatest production of beech lumber is in the Central and Middle Atlantic States.

Beechwood varies in color from nearly white sapwood to reddish-brown heartwood in some trees. Sometimes there is no clear line of demarcation between heartwood and sapwood. Sapwood may be 3 to 5 inches thick. The wood has little figure and is of close, uniform texture. It has no characteristic taste or odor.

The wood of beech is classed as heavy, hard, strong, high in resistance to shock, and highly adaptable for steam bending. Beech has large shrinkage and requires careful drying. It machines smoothly, is an excellent wood for turning, wears well, and is rather easily treated with preservatives.

Largest amounts of beech go into flooring, furniture, brush blocks, handles, veneer woodenware, containers, cooperage, and laundry appliances. When treated, it is suitable for railway ties.

Birch

The important species of birch are yellow birch (*Betula alleghaniensis*), sweet birch (*B. lenta*), and paper birch (*B. papyrifera*). Other birches of some commercial importance are river birch (*B. nigra*), gray birch (*B. populifolia*), and western paper birch (*B. papyrifera* var. *commutata*).

Yellow birch, sweet birch, and paper birch grow principally in the Northeastern and Lake States. Yellow and sweet birch also grow along the Appalachian Mountains to northern Georgia. They are the source of most birch lumber and veneer.

Yellow birch has white sapwood and light reddish-brown heartwood. Sweet birch has light-colored sapwood and dark brown heartwood tinged with red. Wood of yellow birch and sweet birch is heavy, hard, strong, and has good shock-resisting ability. The wood is fine and uniform in texture. Paper birch is lower in weight, softer, and lower in strength than yellow and sweet birch. Birch shrinks considerably during drying.

Yellow and sweet birch lumber and veneer go principally into the manufacture of furniture, boxes, baskets, crates, woodenware, cooperage, interior finish, and doors. Birch veneer goes into plywood used for flush doors, furniture, paneling, radio and television cabinets, aircraft, and other specialty uses. Paper birch is used for turned products, including spools, bobbins, small handles, and toys.

Buckeye

Buckeye consists of two species, yellow buckeye (*Aesculus octandra*) and Ohio buckeye (*A. glabra*). They range from the Appalachians of Pennsylvania, Virginia, and North Carolina westward to Kansas, Oklahoma, and Texas. Buckeye is not customarily separated from other species when manufactured into lumber and can be utilized for the same purposes as aspen, basswood, and sap yellow-poplar.

The white sapwood of buckeye merges gradually into the creamy or yellowish white of the heartwood. The wood is uniform in texture, generally straight-grained, light in weight, weak when used as a beam, soft, and low in shock resistance. It is rated low on machineability such as shaping, mortising, steam bending, boring, and turning.

Buckeye is suitable for pulping for paper and in lumber form has been used principally for furniture, boxes, and crates, food containers, woodenware, novelties, and planing mill products.

Butternut

Butternut (*Juglans cinerea*) is also called white walnut, American white walnut, and oilnut. It grows from southern New Brunswick and Maine, west to Minnesota. Its southern range extends into northeastern Arkansas and eastward to western North Carolina.

The narrow sapwood is nearly white, and the heartwood is a light brown, frequently modified by pinkish tones or darker brown streaks. The wood is moderately light in weight—about the same as eastern white pine—rather coarse-textured, moderately weak in bending and endwise compression, relatively low in stiffness, moderately soft, and moderately high in shock resistance. Butternut machines easily and finishes well. In many ways it resembles black walnut, but it does not have the strength or hardness. Principal uses are for lumber and veneer, which are further manufactured into furniture, cabinets, paneling, trim, and miscellaneous rough items.

Cherry, Black

Black cherry (*Prunus serotina*) is sometimes known as cherry, wild black cherry, wild cherry, or chokecherry. It is the only native species of the genus *Prunus* of commercial importance for lumber production. It occurs scatteringly from southeastern Canada throughout the eastern half of the United States. Production is centered chiefly in the Middle Atlantic States.

The heartwood of black cherry varies from light to dark reddish brown and has a distinctive luster. The sapwood is narrow in old trees and nearly white. The wood has a fairly uniform texture and very satisfactory machining properties. It is moderately heavy. Black cherry is strong, stiff, moderately hard, and has high shock resistance and moderately large shrinkage. After seasoning, it is very dimensionally stable in use.

Black cherry is used principally for furniture, fine veneer panels, architectural woodwork, and for backing blocks on which electrotype plates are mounted. Other uses include burial caskets, woodenware novelties, patterns, and paneling. It has proved satisfactory for gunstocks, but has a limited market for this purpose.

Chestnut, American

American chestnut (*Castanea dentata*) is known also as sweet chestnut. Before chestnut was attacked by a blight, it grew in commercial quantities from New England to northern Georgia. Practically all standing chestnut has been killed by blight, and supplies come from dead timber. There are still quantities of standing dead chestnut in the Appalachian Mountains, which may be available for some time because of the great natural resistance to decay of its heartwood.

The heartwood of chestnut is grayish brown or brown and becomes darker with age. The sapwood is very narrow and almost white. The wood is coarse in texture, and the growth rings are made conspicuous by several rows of large, distinct pores at the beginning of each year's growth. Chestnut wood is moderately light in weight. It is moderately hard, moderately low in strength, moderately low in resistance to shock, and low in stiffness. It seasons well and is easy to work with tools.

Chestnut was used for poles, railway ties, furniture, caskets, boxes, crates, and core stock for veneer panels. It appears most frequently now as "wormy chestnut" for paneling, trim, and picture frames, while a small amount is still used in rustic fences.

Cottonwood

Cottonwood includes several species of the genus *Populus*. Most important are eastern cottonwood (*P. deltoides* and varieties), also known as Carolina poplar and whitewood; swamp cottonwood (*P. heterophylla*), also known as cottonwood, river cottonwood, and swamp poplar; and black cottonwood (*P. trichocarpa*) and balsam poplar (*P. balsamifera*).

Eastern cottonwood and swamp cottonwood grow throughout the eastern half of the United States. Greatest production of lumber is in the Southern and Central States. Black cottonwood grows in the West Coast States and in western Montana, northern Idaho, and western Nevada. Balsam poplar grows from Alaska across Canada, and in the northern Great Lake states.

The heartwood of the three cottonwoods is grayish white to light brown. The sapwood is whitish and merges gradually with the heartwood. The wood is comparatively uniform in texture, and generally straight grained. It is odorless when well seasoned.

Eastern cottonwood is moderately low in bending and compressive strength, moderately limber, moderately soft, and moderately low in ability to resist shock. Black cottonwood is slightly below eastern cottonwood in most strength properties. Both eastern and black cottonwood have moderately large shrinkage. Some cottonwood is difficult to work with tools because of fuzzy surfaces. Tension wood is largely responsible for this characteristic.

Cottonwood is used principally for lumber, veneer, pulpwood, excelsior, and fuel. The lumber and veneer go largely into boxes, crates, baskets, and pallets.

Elm

Six species of elm grow in the eastern United States: American elm (*Ulmus americana*), slippery elm (*U. rubra*), rock elm (*U. thomasii*), winged elm (*U. alata*), cedar elm (*U. crassifolia*), and September elm (*U. serotina*). American elm is also known as white elm, water elm, and gray elm; slippery elm as red elm; rock elm as cork elm or hickory elm; winged elm as wahoo; cedar elm as red elm or basket elm; and September elm as red elm.

Supply of American elm is threatened by two diseases, Dutch Elm and phloem necrosis, which have killed hundreds of thousands of trees.

The sapwood of the elms is nearly white and the heartwood light brown, often tinged with red. The elms may be divided into two general classes, hard elm and soft elm, based on the weight and strength of the wood. Hard elm includes rock elm, winged elm, cedar elm, and September elm. American elm and slippery elm are the soft elms. Soft elm is moderately heavy, has high shock resistance, and is moderately hard and stiff. Hard elm species are somewhat heavier than soft elm. Elm has excellent bending qualities.

Production of elm lumber is chiefly in the Lake, Central and Southern States.

Elm lumber is used principally in boxes, baskets, crates, and slack barrels; furniture, agricultural supplies and implements; caskets and burial boxes, and vehicles. For some uses the hard elms are preferred. Elm veneer is used for furniture, fruit, vegetable, and cheese boxes, baskets, and decorative panels.

Hackberry

Hackberry (*Celtis occidentalis*) and sugarberry (*C. laevigata*) supply the lumber known in the trade as hackberry. Hackberry grows east of the Great Plains from Alabama, Georgia, Arkansas, and Oklahoma northward, except along the Canadian boundary. Sugarberry overlaps the southern part of the range of hackberry and grows throughout the Southern and South Atlantic States.

The sapwood of both species varies from pale yellow to greenish or grayish yellow. The heartwood is commonly darker. The wood resembles elm in structure.

Hackberry lumber is moderately heavy. It is moderately strong in bending, moderately weak in compression parallel to the grain, moderately hard to hard, high in shock resistance, but low in stiffness. It has moderately large to large shrinkage but keeps its shape well during seasoning.

Most hackberry is cut into lumber, with small amounts going into dimension stock and some into veneer. Most of it is used for furniture and some for containers.

Hickory, Pecan

Species of the pecan group include bitternut hickory (*Carya cordiformis*), pecan (*C. illinoensis*), water hickory (*C. aquatica*), and nutmeg hickory (*C. myristicaeformis*). Bitternut hickory grows throughout the eastern half of the United States. Pecan hickory grows from central Texas and Louisiana to Missouri and Indiana. Water hickory grows from Texas to South Carolina. Nutmeg hickory occurs principally in Texas and Louisiana.

The wood of pecan hickory resembles that of true hickory. It has white or nearly white sapwood, which is relatively wide, and somewhat darker heartwood. The wood is heavy and sometimes has very large shrinkage.

Heavy pecan hickory finds use in tool and implement handles and flooring. The lower grades are used in pallets. Many higher grade logs are sliced to provide veneer for furniture and decorative paneling.

Hickory, True

True hickories are found throughout most of the eastern half of the United States. The species most important commercially are shagbark (*Carya ovata*), pignut (*C. glabra*), shellbark (*C. laciniosa*), and mockernut (*C. tomentosa*).

The greatest commercial production of the true hickories for all uses is in the Middle Atlantic and Central States. The Southern and South Atlantic States produce nearly half of all hickory lumber.

The sapwood of hickory is white and usually quite thick, except in old, slowly growing trees. The heartwood is reddish. From the standpoint of strength, no distinction should be made between sapwood and heartwood having the same weight.

The wood of true hickory is exceptionally tough, heavy, hard, strong, and shrinks considerably in drying. For some purposes, both rings per inch and weight are limiting factors where strength is important.

The major use for hickory is for tool handles, which require high shock resistance. It is also used for ladder rungs, athletic goods, agricultural implements, dowels, gymnasium apparatus, poles, and furniture.

A considerable quantity of lower grade hickory is not suitable, because of knottiness or other growth features and low density, for the special uses of high-quality hickory. It appears particularly useful for pallets, blocking, and similar items. Hickory sawdust and chips and some solid wood is used by the major packing companies to flavor meat by smoking.

Holly, American

American holly (*Ilex opaca*) is sometimes called white holly, evergreen holly, and boxwood. The natural range of holly extends along the Atlantic coast, gulf coast, and Mississippi Valley.

Both heartwood and sapwood are white, the heartwood with an ivory cast. The wood has a uniform and compact texture; it is moderately low in strength when used as a beam or column and low in stiffness, but it is heavy and hard, and ranks high in shock resistance. It is readily penetrable to liquids and can be satisfactorily dyed. It works well, cuts smoothly, and is used principally for scientific and musical instruments, furniture inlays, and athletic goods.

Honeylocust

The wood of honeylocust (*Gleditsia triacanthos*) possesses many desirable qualities such as attractive figure and color, hardness, and strength, but is little used because of its scarcity. Although the natural range of honeylocust has been extended by planting, it is found most commonly in the eastern United States, except for New England and the South Atlantic and Gulf Coastal Plains.

The sapwood is generally wide and yellowish in contrast to the light red to reddish brown heartwood. It is very heavy, very hard, strong in bending, stiff, resistant to shock, and is durable when in contact with the ground. When available, it is restricted primarily to local uses, such as fence posts and lumber for general construction. Occasionally it will show up with other species in lumber for pallets and crating.

Locust, Black

Black locust (*Robinia pseudoacacia*) is sometimes called yellow locust, white locust, green locust, or post locust. This species grows from Pennsylvania along the Appalachian Mountains to northern Georgia. It is also native to a small area in northwestern Arkansas. The greatest production of black locust timber is in Tennessee, Kentucky, West Virginia, and Virginia.

Locust has narrow, creamy-white sapwood. The heartwood, when freshly cut, varies from greenish yellow to dark brown. Black locust is very heavy, very hard, very high in resistance to shock, and ranks very high in strength and stiffness. It has moderately small shrinkage. The heartwood has high decay resistance.

Black locust is used extensively for round, hewed, or split mine timbers and for fenceposts, poles, railroad ties, stakes, and fuel. An important product manufactured from black locust is insulator pins, a use for which the wood is well adapted because of its strength, decay resistance, and moderate shrinkage and swelling. Other uses are for rough construction, crating, ship treenails and mine equipment.

Magnolia

Three species comprise commercial magnolia—southern magnolia (*Magnolia grandiflora*), sweetbay (*M. virginiana*), and cucumbertree (*M. acuminata*). Other names for southern magnolia are evergreen magnolia, magnolia, big laurel, bull bay, and laurel bay. Sweetbay is sometimes called swamp magnolia, or more often simply magnolia.

The natural range of sweetbay extends along the Atlantic and gulf coasts from Long Island to Texas, and that of southern magnolia from North Carolina to Texas. Cucumbertree grows from the Appalachians to the Ozarks northward to Ohio. Louisiana leads in production of magnolia lumber.

The sapwood of southern magnolia is yellowish white, and the heartwood is light to dark brown with a tinge of yellow or green. The wood, which has close, uniform texture and is generally straight grained, closely resembles yellow-poplar. It is moderately heavy, moderately low in shrinkage, moderately low in bending and compressive strength, moderately hard and stiff, and moderately high in shock resistance. Sweetbay is reported to be much like southern magnolia. The wood of cucumbertree is similar to that of yellow-poplar, and cucumbertree growing in the yellow-poplar range is not separated from that species on the market.

Magnolia lumber is used principally in the manufacture of furniture, boxes, pallets, venetian blinds, sash, doors, veneer, and millwork.

Maple

Commercial species of maple in the United States include sugar maple (*Acer saccharum*), black maple (*A. nigrum*), silver maple (*A. saccharinum*), red maple (*A. rubrum*), boxelder (*A. negundo*), and bigleaf maple (*A. macrophyllum*). Sugar maple is also known as hard maple, rock maple, sugar tree, and black maple; black maple as hard maple, black sugar maple, and sugar maple; silver maple as white maple, river maple, water maple, and swamp maple; red maple as soft maple, water maple, scarlet maple, white maple, and swamp maple; boxelder as ash-leaved maple, three-leaved maple, and cut-leaved maple; and bigleaf maple as Oregon maple.

Maple lumber comes principally from the Middle Atlantic and Lake States, which together account for about two-thirds of the production.

The wood of sugar maple and black maple is known as hard maple; that of silver maple, red maple, and boxelder as soft maple. The sapwood of the maples is commonly white with a slight reddish-brown tinge. It is from 3 to 5 or more inches thick. Heartwood is usually light reddish brown, but sometimes is considerably darker. Hard maple has a fine, uniform texture. It is heavy, strong, stiff, hard, resistant to shock, and has large shrinkage. Sugar maple is generally straight grained but also occurs as "birdseye," "curley," and "fiddleback" grain. Soft maple is not so heavy as hard maple, but has been substituted for hard maple in the better grades, particularly for furniture.

Maple is used principally for lumber, veneer, crossties, and pulpwood. A large proportion is manufactured into flooring, furniture, boxes, pallets, and crates, shoe lasts, handles, woodenware, novelties, spools, and bobbins.

Oak (Red Oak Group)

Most red oak lumber and other products come from the Southern States, the southern mountain regions, the Atlantic Coastal Plains, and the Central States. The principal species are: Northern red oak (*Quercus rubra*), scarlet oak (*Q. coccinea*), Shumard oak (*Q. shumardii*), pin oak (*Q. palustris*), Nuttall oak (*Q. nuttallii*), black oak (*Q. velutina*), southern red oak (*Q. falcata*), cherrybark oak (*Q. falcata* var. *pagodaefolia*), water oak (*Q. nigra*), laurel oak (*Q. laurifolia*), and willow oak (*Q. phellos*).

The sapwood is nearly white and usually 1 to 2 inches thick. The heartwood is brown with a tinge of red. Sawed lumber of red oak cannot be separated by species on the basis of the characteristics of the wood alone. Red oak lumber can be separated from white oak by the number of pores in summerwood and because, as a rule, it lacks the membranous growth known as tyloses in the pores. The open pores of the red oaks make these species unsuitable for tight cooperage, unless the barrels are lined with sealer or plastic. Quarter-sawed lumber of the oaks is distinguished by the broad and conspicuous rays, which add to its attractiveness.

Wood of the red oaks is heavy. Rapidly grown second-growth oak is generally harder and tougher than finer textured old-growth timber. The red oaks have fairly large shrinkage in drying.

The red oaks are largely cut into lumber, railroad ties, mine timbers, fenceposts, veneer, pulpwood, and fuelwood. Ties, mine timbers, and fenceposts require preservative treatment for satisfactory service. Red oak lumber is remanufactured into flooring, furniture, general millwork, boxes, pallets and crates, agricultural implements, caskets, woodenware, and handles. It is also used in railroad cars and boats.

Oak (White Oak Group)

White oak lumber comes chiefly from the South, South Atlantic, and Central States, including the southern Appalachian area.

Principal species are white oak (*Quercus alba*), chestnut oak (*Q. prinus*), post oak (*Q. stellata*), overcup oak (*Q. lyrata*), swamp chestnut oak (*Q. michauxii*), bur oak (*Q. macrocarpa*), chinkapin oak (*Q. muehlenbergii*), swamp white oak (*Q. bicolor*), and live oak (*Q. virginiana*).

The heartwood of the white oaks is generally grayish brown, and the sapwood, which is from 1 to 2 or more inches thick, is nearly white. The pores of the heartwood of white oaks are usually plugged with the membranous growth known as tyloses. These tend to make

the wood impenetrable by liquids, and for this reason most white oaks are suitable for tight cooperage. Chestnut oak lacks tyloses in many of its pores.

The wood of white oak is heavy, averaging somewhat higher in weight than that of the red oaks. The heartwood has moderately good decay resistance.

White oaks are used for lumber, railroad ties, cooperage, mine timbers, fenceposts, veneer, fuelwood, and many other products. High-quality white oak is especially sought for tight cooperage. Live oak is considerably heavier and stronger than the other oaks, and was formerly used extensively for ship timbers. An important use of white oak is for planking and bent parts of ships and boats, heartwood often being specified because of its decay resistance. It is also used for flooring, pallets, agricultural implements, railroad cars, truck floors, furniture, doors, millwork, and many other items.

Sassafras

The range of sassafras (*Sassafras albidum*) covers most of the eastern half of the United States from southeastern Iowa and eastern Texas eastward.

The wood of sassafras is easily confused with black ash, which it resembles in color, grain, and texture. The sapwood is light yellow and the heartwood varies from dull grayish brown to dark brown, sometimes with a reddish tinge. The wood has an odor of sassafras on freshly cut surfaces.

Sassafras is moderately heavy, moderately hard, moderately weak in bending and endwise compression, quite high in shock resistance, and quite durable when exposed to conditions conducive to decay. It was highly prized by the Indians for dugout canoes, and some sassafras lumber is now used for small boats. Locally, it is used for fence posts and rails and general millwork, for foundation posts, and some wooden containers.

Sweetgum

Sweetgum (*Liquidambar styraciflua*) grows from southwestern Connecticut westward into Missouri and southward to the gulf. Lumber production is almost entirely from the Southern and South Atlantic States.

The lumber from sweetgum is usually divided into two classes—sap gum, the light-colored wood from the sapwood, and red gum, the reddish-brown heartwood.

11

Sweetgum has interlocked grain, a form of cross grain, and must be carefully dried. The interlocked grain causes a ribbon stripe, however, that is desirable for interior finish and furniture. The wood is rated as moderately heavy and hard. It is moderately strong, moderately stiff, and moderately high in shock resistance.

Sweetgum is used principally for lumber, veneer, plywood, slack cooperage, railroad ties, fuel, and pulpwood. The lumber goes principally into boxes and crates, furniture, radio and phonograph cabinets, interior trim, and millwork. Sweetgum veneer and plywood are used for boxes, pallets, crates, baskets, and interior woodwork.

Sycamore, American

American sycamore (*Platanus occidentalis*) is also known as sycamore and sometimes as buttonwood, buttonball tree, and planetree. Sycamore grows from Maine to Nebraska, southward to Texas, and eastward to Florida. In the production of sycamore lumber, the Central States rank first.

The heartwood of sycamore is reddish brown; sapwood is lighter in color and from 1½ to 3 inches thick. The wood has a fine texture and interlocked grain. It shrinks moderately in drying. Sycamore wood is moderately heavy, moderately hard, moderately stiff, moderately strong, and has good resistance to shock.

Sycamore is used principally for lumber, veneer, railroad ties, slack cooperage, fenceposts, and fuel. Sycamore lumber is used for furniture, boxes (particularly small food containers), pallets, flooring, handles, and butcher's blocks. Veneer is used for fruit and vegetable baskets, and some decorative panels and door skins.

Tanoak

In recent years tanoak (*Lithocarpus densiflorus*) has gained some importance commercially, primarily in California and Oregon. It is also known as tanbark-oak because at one time high-grade tannin in commercial quantities was obtained from the bark. This species is found in southwestern Oregon and south to Southern California, mostly near the coast but also in the Sierra Nevadas.

The sapwood of tanoak is light reddish brown when first cut and turns darker with age to become almost indistinguishable from the heartwood, which also ages to dark reddish brown. The wood is heavy, hard, and except for compression perpendicular to the grain has roughly the same strength properties as eastern white oak. Volumetric shrinkage during drying is more than for white oak, and it has a tendency to collapse during drying. It is quite susceptible to decay, but the sapwood takes preservatives easily. It has straight grain, machines and glues well, and takes staining readily.

Because of tanoak's hardness and abrasion resistance, it is an excellent wood for flooring in homes or commercial buildings. It is also suitable for industrial applications such as truck flooring. Tanoak treated with preservative has been used for railroad crossties. The wood has been manufactured into baseball bats with good results. It is also suitable for veneer, both decorative and industrial, and for high-quality furniture.

Tupelo

The tupelo group includes water tupelo (*Nyssa aquatica*), also known as tupelo gum, swamp tupelo, and gum; blacktupelo (*N. sylvatica*), also known as black gum; and sour gum; swamp tupelo (*N. sylvatica* var. *biflora*), also known as swamp blackgum, blackgum, tupelo gum, and sour gum; Ogeechee tupelo (*N. ogeche*), also known as sour tupelo, gopher plum, tupelo, and Ogeechee plum.

All except black tupelo grow principally in the southeastern United States. Black tupelo grows in the eastern United States from Maine to Texas and Missouri. About two-thirds of the production of tupelo lumber is from the Southern States.

Wood of the different tupelos is quite similar in appearance and properties. Heartwood is light brownish gray and merges gradually into the lighter colored sapwood, which is generally several inches wide. The wood has fine, uniform texture and interlocked grain. Tupelo wood is rated as moderately heavy. It is moderately strong, moderately hard and stiff, and moderately high in shock resistance. Buttresses of trees growing in swamps or flooded areas contain wood that is much lighter in weight than that from upper portions of the same trees. For some uses, as in the case of buttressed ash trees, this wood should be separated from the heavier wood to assure material of uniform strength. Because of interlocked grain, tupelo lumber requires care in drying.

Tupelo is cut principally for lumber, veneer, pulpwood, and some railroad ties and slack cooperage. Lumber goes into boxes, pallets, crates, baskets, and furniture.

Walnut, Black

Black walnut (*Juglans nigra*) is also known as American black walnut. Its natural range extends from Vermont to the Great Plains and southward into Louisiana and Texas. About three-quarters of the walnut timber is produced in the Central States.

The heartwood of black walnut varies from light to dark brown; the sapwood is nearly white and up to 3 inches wide in open-grown trees. Black walnut is normally straight grained, easily worked with tools, and stable in use. It is heavy, hard, strong, stiff, and has good resistance to shock. Black walnut wood is well suited for natural finishes.

The outstanding uses of black walnut are for furniture, architectural woodwork, and decorative panels. Other important uses are gunstocks, cabinets, and interior finish. It is used either as solid wood or as plywood.

Willow, Black

Black willow (*Salix nigra*) is the most important of the many willows that grow in the United States. It is the only one to supply lumber to the market under its own name.

Black willow is most heavily produced in the Mississippi Valley from Louisiana to southern Missouri and Illinois.

The heartwood of black willow is grayish brown or light reddish brown frequently containing darker streaks. The sapwood is whitish to creamy yellow. The wood of black willow is uniform in texture, with somewhat interlocked grain. The wood is light in weight. It has exceedingly low strength as a beam or post and is moderately soft and moderately high in shock resistance. It has moderately large shrinkage.

Willow is cut principally into lumber. Small amounts are used for slack cooperage, veneer, excelsior, charcoal, pulpwood, artificial limbs, and fenceposts. Black willow lumber is remanufactured principally into boxes, pallets, crates, caskets, and furniture. Willow lumber is suitable for roof and wall sheathing, subflooring, and studding.

Yellow-Poplar

Yellow-poplar (*Liriodendron tulipifera*) is also known as poplar, tulip poplar, tulipwood, and hickory poplar. Sapwood from yellow-poplar is sometimes called white poplar or whitewood.

Yellow-poplar grows from Connecticut and New York southward to Florida and westward to Missouri. The greatest commercial production of yellow-poplar lumber is in the South.

Yellow-poplar sapwood is white and frequently several inches thick. The heartwood is yellowish brown, sometimes streaked with purple, green, black, blue, or red. These colorations do not affect the physical properties of the wood. The wood is generally straight grained and comparatively uniform in texture. Old-growth timber is moderately light in weight and is reported as being moderately low in bending strength, moderately soft, and moderately low in shock resistance. It has moderately large shrinkage when dried from a green condition but is not difficult to season and stays in place well after seasoning.

Much of the second-growth yellow-poplar is heavier, harder, and stronger than old growth. Selected trees produce wood heavy enough for gunstocks. Lumber goes mostly into furniture, interior finish, siding, core stock for plywood, radio cabinets, and musical instruments, but use for core stock is decreasing as particleboard use increases. Yellow-poplar is frequently used for crossbands in plywood. Boxes, pallets, and crates are made from lower grade stock. Yellow-poplar plywood is used for finish, furniture, piano cases, and various other special products. Yellow-poplar is used also for pulpwood, excelsior, and slack-cooperage staves.

Lumber from the cucumbertree (*Magnolia acuminata*) sometimes may be included in shipments of yellow-poplar because of its similarity.

Softwoods

Alaska-Cedar

Alaska-cedar (*Chamaecyparis nootkatensis*) grows in the Pacific coast region of North America from southeastern Alaska southward through Washington to southern Oregon.

The heartwood of Alaska-cedar is bright, clear yellow. The sapwood is narrow, white to yellowish, and hardly distinguishable from the heartwood. The wood is fine textured and generally straight grained. It is moderately heavy, moderately strong and stiff, moderately hard, and moderately high in resistance to shock. Alaska-cedar shrinks little in drying, is stable in use after seasoning, and the heartwood is very resistant to decay. The wood has a mild, unpleasant odor.

Alaska-cedar is used for interior finish, furniture, small boats, cabinetwork, and novelties.

Baldcypress

Baldcypress (*Taxodium distichum*) is commonly known as cypress, also as southern cypress, red cypress, yellow cypress, and white cypress. Commercially, the terms "tidewater red cypress," "gulf cypress," "red cypress (coast type)," and "yellow cypress (inland type)" are frequently used.

About one-half of the cypress lumber comes from the Southern States and one-fourth from the South Atlantic States. It is not as readily available as formerly.

The sapwood of baldcypress is narrow and nearly white. The color of the heartwood varies widely, ranging from light yellowish brown to dark brownish red, brown, or chocolate. The wood is moderately heavy, moderately strong, and moderately hard, and the heartwood of old-growth timber is one of our most decay-resistant woods. Shrinkage is moderately small, but somewhat greater than that of cedar and less than that of southern pine.

Frequently the wood of certain cypress trees contains pockets or localized areas that have been attacked by a fungus. Such wood is known as pecky cypress. The decay caused by this fungus is arrested when the wood is cut into lumber and dried. Pecky cypress, therefore, is durable and useful where water tightness is unnecessary, and appearance is not important or a novel effect is desired. Examples of such usage are as paneling in restaurants, stores, and other buildings.

Cypress has been used principally for building construction, especially where resistance to decay is required. It was used for beams, posts, and other members in docks, warehouses, factories, bridges, and heavy construction.

It is well suited for siding and porch construction. It is also used for caskets, burial boxes, sash, doors, blinds, and general millwork, including interior trim and paneling. Other uses are in tanks, vats, ship and boat building, refrigerators, railroad-car construction, greenhouse construction, cooling towers, and stadium seats. It is also used for railroad ties, poles, piling, shingles, cooperage, and fenceposts.

Douglas-fir

Douglas-fir (*Pseudotsuga menziesii* and var. *glauca*) is also known locally as red fir, Douglas spruce, and yellow fir.

The range of Douglas-fir extends from the Rocky Mountains to the Pacific coast and from Mexico to central British Columbia. The Douglas-fir production comes from the Coast States of Oregon, Washington, and California and from the Rocky Mountain States.

Sapwood of Douglas-fir is narrow in old-growth trees but may be as much as 3 inches wide in second-growth trees of commercial size. Fairly young trees of moderate to rapid growth have reddish heartwood and are called red fir. Very narrow-ringed wood of old trees may be yellowish brown and is known on the market as yellow fir.

The wood of Douglas-fir varies widely in weight and strength. When lumber of high strength is needed for structural uses, selection can be improved by applying the density rule. This rule uses percentage of latewood and rate of growth as a basis.

Douglas-fir is used mostly for building and construction purposes in the form of lumber, timbers, piling, and plywood. Considerable quantities go into railroad ties, cooperage stock, mine timbers, poles, and fencing. Douglas-fir lumber is used in the manufacture of various products, including sash, doors, general millwork, railroad-car construction, boxes, pallets, and crates. Small amounts are used for flooring, furniture, ship and boat construction, wood pipe, and tanks. Douglas-fir plywood has found ever-increasing usefulness in construction, furniture, cabinets, and many other products.

Firs, True (Eastern Species)

Balsam fir (*Abies balsamea*) grows principally in New England, New York, Pennsylvania, and the Lake States. Fraser fir (*A. fraseri*) grows in the Appalachian Mountains of Virginia, North Carolina, and Tennessee.

The wood of the true firs, eastern as well as western species, is creamy white to pale brown. Heartwood and sapwood are generally indistinguishable. The similarity of wood structure in the true firs makes it impossible to distinguish the species by an examination of the wood alone.

Balsam fir is rated as light in weight, low in bending and compressive strength, moderately limber, soft, and low in resistance to shock.

The eastern firs are used mainly for pulpwood, although there is some lumber produced from them, especially in New England and the Lake States.

Firs, True (Western Species)

Six commercial species make up the western true firs: Subalpine fir (*Abies lasiocarpa*), California red fir (*A. magnifica*), grand fir (*A. grandis*), noble fir (*A. procera*), Pacific silver fir (*A. amabilis*), and white fir (*A. concolor*).

The western firs are light in weight, but, with the exception of subalpine fir, have somewhat higher strength properties than balsam fir. Shrinkage of the wood is rated from small to moderately large.

The western true firs are largely cut for lumber in Washington, Oregon, California, western Montana, and northern Idaho and marketed as white fir throughout the United States. Lumber of the western true firs goes principally into building construction, boxes and crates, planing-mill products, sash, doors, and general millwork. In house construction the lumber is used for framing, subflooring, and sheathing. Some western true fir lumber goes into boxes and crates. High-grade lumber from noble fir is used mainly for interior finish, moldings, siding, and sash and door stock. Some of the best material is suitable for aircraft construction. Other special and exacting uses of noble fir are for venetian blinds and ladder rails.

Hemlock, Eastern

Eastern hemlock (*Tsuga canadensis*) grows from New England to northern Alabama and Georgia, and in the Lake States. Other names are Canadian hemlock and hemlock spruce.

The production of hemlock lumber is divided fairly evenly between the New England States, the Middle Atlantic States, and the Lake States.

The heartwood of eastern hemlock is pale brown with a reddish hue. The sapwood is not distinctly separated from the heartwood but may be lighter in color. The wood is coarse and uneven in texture (old trees tend to have considerable shake); it is moderately light in weight, moderately hard, moderately low in strength, moderately limber, and moderately low in shock resistance.

Eastern hemlock is used principally for lumber and pulpwood. The lumber is used largely in building construction for framing, sheathing, subflooring, and roof boards, and in the manufacture of boxes, pallets, and crates.

Hemlock, Western

Western hemlock (*Tsuga heterophylla*) is also known by several other names, including west coast hemlock, hemlock spruce, western hemlock spruce, western hemlock fir, Prince Albert fir, gray fir, silver fir, and Alaska pine. It grows along the Pacific coast of Oregon and Washington and in the northern Rocky Mountains, north to Canada and Alaska.

A relative, mountain hemlock, *T. mertensiana*, inhabits mountainous country from central California to Alaska. It is treated as a separate species in assigning lumber properties.

The heartwood and sapwood of western hemlock are almost white with a purplish tinge. The sapwood, which is sometimes lighter in color, is generally not more than 1 inch thick. The wood contains small, sound, black knots that are usually tight and stay in place. Dark streaks often found in the lumber and caused by hemlock bark maggots as a rule do not reduce strength.

Western hemlock is moderately light in weight and moderate in strength. It is moderate in its hardness, stiffness, and shock resistance. It has moderately large shrinkage, about the same as Douglas-fir. Green hemlock lumber contains considerably more water than Douglas-fir, and requires longer kiln drying time.

Mountain hemlock has approximately the same density as western hemlock but is somewhat lower in bending strength and stiffness.

Western hemlock is used principally for pulpwood, lumber, and plywood. The lumber goes largely into building material, such as sheathing, siding, subflooring, joists, studding, planking, and rafters. Considerable quantities are used in the manufacture of boxes, pallets, crates, and flooring, and smaller amounts for refrigerators, furniture, and ladders.

Mountain hemlock serves some of the same uses as western hemlock although the quantity available is much lower.

Incense-Cedar

Incense-cedar (*Libocedrus decurrens*) grows in California and southwestern Oregon, and a little in Nevada. Most incense-cedar lumber comes from the northern half of California and the remainder from southern Oregon.

Sapwood of incense-cedar is white or cream colored, and the heartwood is light brown, often tinged with red. The wood has a fine, uniform texture and a spicy odor. Incense-cedar is light in weight, moderately low in strength, soft, low in shock resistance, and low in stiffness. It has small shrinkage and is easy to season with little checking or warping.

Incense-cedar is used principally for lumber and fenceposts. Nearly all the high-grade lumber is used for pencils and venetian blinds. Some is used for chests and toys. Much of the incense-cedar lumber is more or less pecky; that is, it contains pockets or areas of disintegrated wood caused by advanced stages of localized decay in the living tree. There is no further development of peck once the lumber is seasoned. This lumber is used locally for rough construction where cheapness and decay resistance are important. Because of its resistance to decay, incense-cedar is well suited for fenceposts. Other products are railroad ties, poles, and split shingles.

Larch, Western

Western larch (*Larix occidentalis*) grows in western Montana, northern Idaho, northeastern Oregon, and on the eastern slope of the Cascade Mountains in Washington. About two-thirds of the lumber of this species is produced in Idaho and Montana and one-third in Oregon and Washington.

The heartwood of western larch is yellowish brown and the sapwood yellowish white. The sapwood is generally not more than 1 inch thick. The wood is stiff, moderately strong and hard, moderately high in shock resistance, and moderately heavy. It has moderately large shrinkage. The wood is usually straight grained, splits easily, and is subject to ring shake. Knots are common but small and tight.

Western larch is used mainly in building construction for rough dimension, small timbers, planks and boards, and for railroad ties and mine timbers. It is used also for piles, poles, and posts. Some high-grade material is manufactured into interior finish, flooring, sash, and doors.

Pine, Eastern White

Eastern white pine (*Pinus strobus*) grows from Maine to northern Georgia and in the Lake States. It is also known as white pine, northern white pine, Weymouth pine, and soft pine.

About one-half the production of eastern white pine lumber occurs in the New England States, about one-third in the Lake States, and most of the remainder in the Middle Atlantic and South Atlantic States.

The heartwood of eastern white pine is light brown, often with a reddish tinge. It turns considerably darker on exposure. The wood has comparatively uniform texture, and is straight grained. It is easily kiln-dried, has small shrinkage, and ranks high in stability. It is also easy to work and can be readily glued.

Eastern white pine is light in weight, moderately soft, moderately low in strength, and low in resistance to shock.

Practically all eastern white pine is converted into lumber, which is put to a great variety of uses. A large proportion, which is mostly second-growth knotty lumber of the lower grades, goes into container and packaging applications. High-grade lumber goes into patterns for castings. Other important uses are sash, doors, furniture, trim, knotty paneling, finish, caskets and burial boxes, shade and map rollers, toys, and dairy and poultry supplies.

Pine, Jack

Jack pine (*Pinus banksiana*), sometimes known as scrub pine, gray pine, or black pine in the United States, grows naturally in the Lake States and in a few scattered areas in New England and northern New York. In lumber, jack pine is not separated from the other pines with which it grows, including red pine and eastern white pine.

The sapwood of jack pine is nearly white, the heartwood is light brown to orange. The sapwood may make up one-half or more of the volume of a tree. The wood has a rather coarse texture and is somewhat resinous. It is moderately light in weight, moderately low in bending strength and compressive strength, moderately low in shock resistance, and low in stiffness. It also has moderately small shrinkage. Lumber from jack pine is generally knotty.

Jack pine is used for pulpwood, box lumber, pallets, and fuel. Less important uses include railroad ties, mine timber, slack cooperage, poles, and posts.

Pine, Lodgepole

Lodgepole pine (*Pinus contorta*), also known as knotty pine, black pine, spruce pine, and jack pine, grows in the Rocky Mountain and Pacific coast regions as far northward as Alaska. The cut of this species comes largely from the central Rocky Mountain States; other producing regions are Idaho, Montana, Oregon, and Washington.

The heartwood of lodgepole pine varies from light yellow to light yellow-brown. The sapwood is yellow or nearly white. The wood is generally straight grained with narrow growth rings.

The wood is moderately light in weight, fairly easy to work, and has moderately large shrinkage. Lodgepole pine rates as moderately low in strength, moderately soft, moderately stiff, and moderately low in shock resistance.

Lodgepole pine is used for lumber, mine timbers, railroad ties, and poles. Less important uses include posts and fuel. It is being used in increasing amounts for framing, siding, finish, and flooring.

Pine, Pitch

Pitch pine (*Pinus rigida*) grows from Maine along the mountains to eastern Tennessee and northern Georgia. The heartwood is brownish red and resinous; the sapwood is thick and light yellow. The wood of pitch pine is medium heavy to heavy, medium strong, medium stiff, medium hard, and medium high in shock resistance. Its shrinkage is medium small to medium large. It is used for lumber, fuel, and pulpwood. Pitch pine lumber is classified as a "minor species" along with pond pine and Virginia pine in southern pine grading rules.

Pine, Pond

Pond pine (*Pinus serotina*) grows in the coast region from New Jersey to Florida. It occurs in small groups or singly, mixed with other pines on low flats. The wood is heavy, coarse-grained, and resinous, with dark, orange-colored heartwood and thick, pale yellow sapwood. At 12 percent moisture content it weighs about 38 pounds per cubic foot. Shrinkage is moderately large. The wood is moderately strong, stiff, medium hard, and medium high in shock resistance. It is used for general construction, railway ties, posts, and poles. As noted for pitch pine, the lumber of this species is graded as a "minor species" with pitch pine and Virginia pine.

Pine, Ponderosa

Ponderosa pine (*Pinus ponderosa*) is known also as pondosa pine, western soft pine, western pine, Califorina white pine, bull pine, and black jack. Jeffrey pine (*P. jeffreyi*), which grows in close association with ponderosa pine in California and Oregon, is usually marketed with ponderosa pine and sold under that name.

Major producing areas are in Oregon, Washington, and California. Other important producing areas are in Idaho and Montana; lesser amounts come from the southern Rocky Mountain region and the Black Hills of South Dakota and Wyoming.

Botanically, ponderosa pine belongs to the yellow pine group rather than the white pine group. A considerable proportion of the wood, however, is somewhat similar to the white pines in appearance and properties. The heartwood is light reddish brown, and the wide sapwood is nearly white to pale yellow.

The wood of the outer portions of ponderosa pine of sawtimber size is generally moderately light in weight, moderately low in strength, moderately soft, moderately stiff, and moderately low in shock resistance. It is generally straight grained and has moderately small shrinkage. It is quite uniform in texture and has little tendency to warp and twist.

Ponderosa pine is used mainly for lumber and to a lesser extent for piles, poles, posts, mine timbers, veneer, and ties. The clear wood goes into sash, doors, blinds, moldings, paneling, mantels, trim, and built-in cases and cabinets. Lower grade lumber is used for boxes and crates. Much of the lumber of intermediate or lower grades goes into sheathing, subflooring, and roof boards. Knotty ponderosa pine is used for interior finish. A considerable amount now goes into particleboard and pulp chips.

Pine, Red

Red pine (*Pinus resinosa*) is frequently called Norway pine. It is occasionally known as hard pine and pitch pine. This species grows in the New England States, New York, Pennsylvania, and the Lake States. In the past, lumber from red pine has been marketed with white pine without distinction as to species.

The heartwood of red pine varies from pale red to reddish brown. The sapwood is nearly white with a yellowish tinge, and is generally from 2 to 4 inches wide. The wood resembles the lighter weight wood of southern pine. Latewood is distinct in the growth rings.

Red pine is moderately heavy, moderately strong and stiff, moderately soft and moderately high in shock resistance. It is generally straight grained, not so uniform in texture as eastern white pine, and somewhat resinous. The wood has moderately large shrinkage but is not difficult to dry and stays in place well when seasoned.

Red pine is used principally for lumber and to a lesser extent for piles, poles, cabin logs, posts, pulpwood, and fuel. The wood is used for many of the purposes for which eastern white pine is used. It goes mostly into building construction, siding, flooring, sash, doors, blinds, general millwork, and boxes, pallets, and crates.

Pine, Southern

There are a number of species included in the group marketed as southern pine lumber. The most important, and their growth range, are:

(1) Longleaf pine (*Pinus palustris*), which grows from eastern North Carolina southward into Florida and westward into eastern Texas. (2) Shortleaf pine (*P. echinata*), which grows from southeastern New York and New Jersey southward to northern Florida and westward into eastern Texas and Oklahoma. (3) Loblolly pine (*P. taeda*), which grows from Maryland southward through the Atlantic Coastal Plain and Piedmont Plateau into Florida and westward into eastern Texas. (4) Slash pine (*P. elliottii*), which grows in Florida and the southern parts of South Carolina, Georgia, Alabama, Mississippi, and Louisiana east of the Mississippi River.

Lumber from any one or from any mixture of two or more of these species is classified as southern pine by the grading standards of the industry. These standards provide also for lumber that is produced from trees of the longleaf and slash pine species to be classified as longleaf pine if conforming to the growth-ring and latewood requirements of such standards. The lumber that is classified as longleaf in the domestic trade is known also as pitch pine in the export trade. Three southern pines—pitch pine, pond pine, and Virginia pine—are designated in published grading rules as "minor species," to distinguish them from the four principal species.

Southern pine lumber comes principally from the Southern and South Atlantic States. States that lead in production are Georgia, Alabama, North Carolina, Arkansas, and Louisiana.

The wood of the various southern pines is quite similar in appearance. The sapwood is yellowish white and heartwood reddish brown. The sapwood is usually wide in second-growth stands. Heartwood begins to form when the tree is about 20 years old. In old, slow-growth trees, sapwood may be only 1 to 2 inches in width.

Longleaf and slash pine are classed as heavy, strong, stiff, hard, and moderately high in shock resistance. Shortleaf and loblolly pine are usually somewhat lighter in weight than longleaf. All the southern pines have moderately large shrinkage but are stable when properly seasoned.

To obtain heavy, strong wood of the southern pines for structural purposes, a density rule has been written that specifies certain visual characteristics for structural timbers.

Dense southern pine is used extensively in construction of factories, warehouses, bridges, trestles, and docks in the form of stringers, beams, posts, joists, and piles. Lumber of lower density and strength finds many uses for building material, such as interior finish, sheathing, subflooring, and joists, and for boxes, pallets, and crates. Southern pine is used also for tight and slack cooperage. When used for railroad ties, piles, poles, and mine timbers, it is usually treated with preservatives. The manufacture of structural grade plywood from southern pine has become a major wood-using industry.

Pine, Spruce

Spruce pine (*Pinus glabra*), also known as cedar pine, poor pine, Walter pine, and bottom white pine, is found growing most commonly on low moist lands of the coastal regions of southeastern South Carolina, Georgia, Alabama, Mississippi, and Louisiana, and northern and northwestern Florida.

Heartwood is light brown, and the wide sapwood zone is nearly white. Spruce pine wood is lower in most strength values than the major southern pines. It compares favorably with white fir in important bending properties, in crushing strength perpendicular and parallel to the grain, and in hardness. It is similar to the denser species such as coast Douglas-fir and loblolly pine in shear parallel to the grain.

Until recent years the principal uses of spruce pine were locally for lumber, and for pulpwood and fuelwood. The lumber, which is classified as one of the minor southern pine species, reportedly was used for sash, doors, and interior finish because of its lower specific gravity and less marked distinction between earlywood and latewood. In recent years it has qualified for use in plywood.

Pine, Sugar

Sugar pine (*Pinus lambertiana*) is sometimes called California sugar pine. Most of the sugar pine lumber is produced in California and the remainder in southwestern Oregon.

The heartwood of sugar pine is buff or light brown, sometimes tinged with red. The sapwood is creamy white. The wood is straight grained, fairly uniform in texture, and easy to work with tools. It has very small shrinkage, is readily seasoned without warping or checking, and stays in place well. This species is light in weight, moderately low in strength, moderately soft, low in shock resistance, and low in stiffness.

Sugar pine is used almost entirely for lumber products. The largest amounts are used in boxes and crates, sash, doors, frames, blinds, general millwork, building construction, and foundry patterns. Like eastern white pine, sugar pine is suitable for use in nearly every part of a house because of the ease with which it can be cut, its ability to stay in place, and its good nailing properties.

Pine, Virginia

Virginia pine (*Pinus virginiana*), known also as Jersey pine and scrub pine, grows from New Jersey and Virginia throughout the Appalachian region to Georgia and the Ohio Valley. The heartwood is orange and the sapwood nearly white and relatively thick. The wood is rated as moderately heavy, moderately strong, moderately hard, and moderately stiff and has moderately large shrinkage and high shock resistance. It is used for lumber, railroad ties, mine props, pulpwood, and fuel. It is one of three southern pines to be classified as a "minor species" in the grading rules.

Pine, Western White

Western white pine (*Pinus monticola*) is also known as Idaho white pine or white pine. About four-fifths of the cut comes from Idaho (fig. 1–3) with the remainder mostly from Washington; small amounts are cut in Montana and Oregon.

Heartwood of western white pine is cream colored to light reddish brown and darkens on exposure. The sapwood is yellowish white

and generally from 1 to 3 inches wide. The wood is straight grained, easy to work, easily kiln-dried, and stable after seasoning.

This species is moderately light in weight, moderately low in strength, moderately soft, moderately stiff, moderately low in shock resistance, and has moderately large shrinkage.

Practically all western white pine is sawed into lumber and used mainly for building construction, matches, boxes, patterns, and millwork products, such as sash, frames, doors, and blinds. In building construction, boards of the lower grades are used for sheathing, knotty paneling, subflooring, and roof strips. High-grade material is made into siding of various kinds, exterior and interior trim, and finish. It has practically the same uses as eastern white pine and sugar pine.

Port-Orford-Cedar

Port-Orford-cedar (*Chamaecyparis lawsoniana*) is sometimes known as Lawson cypress, Oregon cedar, and white cedar. It grows along the Pacific coast from Coos Bay, Oreg., southward to California. It does not extend more than 40 miles inland.

The heartwood of Port-Orford-cedar is light yellow to pale brown in color. Sapwood is thin and hard to distinguish. The wood has fine texture, generally straight grain, and a pleasant spicy odor. It is moderately light in weight, stiff, moderately strong and hard, and moderately resistant to shock. Port-Orford-cedar heartwood is highly resistant to decay. The wood shrinks moderately, has little tendency to warp, and is stable after seasoning.

Some high-grade Port-Orford-cedar is used in the manufacture of battery separators and venetian-blind slats. Other uses are mothproof boxes, archery supplies, sash and door construction, stadium seats, flooring, interior finish, furniture, and boatbuilding.

Redcedar, Eastern

Eastern redcedar (*Juniperus virginiana*) grows throughout the eastern half of the United States, except in Maine, Florida, and a narrow strip along the gulf coast, and at the higher elevations in the Appalachian Mountain Range. Commercial production is principally in the southern Appalachian and Cumberland Mountain regions. Another species, southern redcedar (*J. silicicola*), grows over a limited area in the South Atlantic and Gulf Coastal Plains.

Figure 1–3.—Western white pine timber, mostly privately owned, viewed from Elk Butte in Clearwater National Forest in Idaho.

The heartwood of redcedar is bright red or dull red, and the thin sapwood is nearly white. The wood is moderately heavy, moderately low in strength, hard, and high in shock resistance, but low in stiffness. It has very small shrinkage and stays in place well after seasoning. The texture is fine and uniform. Grain is usually straight, except where deflected by knots, which are numerous. Eastern redcedar heartwood is very resistant to decay.

The greatest quantity of eastern redcedar is used for fenceposts. Lumber is manufactured into chests, wardrobes, and closet lining. Other uses include flooring, novelties, pencils, scientific instruments, and small boats. Southern redcedar is used for the same purposes.

Redcedar, Western

Western redcedar (*Thuja plicata*) grows in the Pacific Northwest and along the Pacific coast to Alaska. Western redcedar is also called canoe cedar, giant arborvitae, shinglewood, and Pacific redcedar. Western redcedar lumber is produced principally in Washington, followed by Oregon, Idaho, and Montana.

The heartwood of western redcedar is reddish or pinkish brown to dull brown and the sapwood nearly white. The sapwood is narrow, often not over 1 inch in width. The wood is generally straight grained and has a uniform but rather coarse texture. It has very small shrinkage. This species is light in weight, moderately soft, low in strength when used as a beam or posts, and low in shock resistance. Its heartwood is very resistant to decay.

Western redcedar is used principally for shingles, lumber, poles, posts, and piles. The lumber is used for exterior siding, interior finish, greenhouse construction, ship and boat building, boxes and crates, sash, doors, and millwork.

Redwood

Redwood (*Sequoia sempervirens*) is a very large tree growing on the coast of California. Another sequoia, gaint sequoia (*Sequoia gigantea*), grows in a limited area in the Sierra Nevada of California, but is used in very limited quantities. Other names for redwood are coast redwood, California redwood, and sequoia. Production of redwood lumber is limited to California, but a nationwide market exists.

The heartwood of redwood varies from a light cherry to a dark mahogany. The narrow sapwood is almost white. Typical old-growth redwood is moderately light in weight, moderately strong and stiff, and moderately hard. The wood is easy to work, generally straight grained, and shrinks and swells comparatively little. The heartwood has high decay resistance.

Most redwood lumber is used for building. It is remanufactured extensively into siding, sash, doors, blinds, finish, casket stock, and containers. Because of its durability, it is useful for cooling towers, tanks, silos, wood-stave pipe, and outdoor furniture. It is used in agriculture for buildings and equipment. Its use as timbers and large dimension in bridges and trestles is relatively minor. The wood splits readily and the manufacture of split products, such as posts and fence material, is an important business in the redwood area. Some redwood veneer is manufactured for decorative plywood.

Spruce, Eastern

The term "eastern spruce" includes three species, red (*Picea rubens*), white (*P. glauca*), and black (*P. mariana*). White spruce and black spruce grow principally in the Lake States and New England, and red spruce in New England and the Appalachian Mountains. All three species have about the same properties, and in commerce no distinction is made between them. The wood dries easily and is stable after drying, is moderately light in weight and easily worked, has moderate shrinkage, and is moderately strong, stiff, tough, and hard. The wood is light in color, and there is little difference between the heartwood and sapwood.

The largest use of eastern spruce is for pulpwood. It is also used for framing material, general millwork, boxes and crates, ladder rails, scaffold planks, and piano sounding boards.

Spruce, Engelmann

Engelmann spruce (*Picea engelmannii*) grows at high elevations in the Rocky Mountain region of the United States. This species is sometimes known by other names, such as white spruce, mountain spruce, Arizona spruce, silver spruce, and balsam. About two-thirds of the lumber is produced in the southern Rocky Mountain States. Most of the remainder comes from the northern Rocky Mountain States and Oregon.

The heartwood of Engelmann spruce is nearly white with a slight tinge of red. The sapwood varies from 3/4 inch to 2 inches in width and is often difficult to distinguish from heartwood. The wood has medium to fine texture and is without characteristic taste or odor. It is generally straight grained. Engelmann spruce is rated as light in weight. It is low in strength as a beam or post. It is limber, soft, low in shock resistance, and has moderately small shrinkage. The lumber typically contains numerous small knots.

Engelmann spruce is used principally for lumber and for mine timbers, railroad ties, and poles. It is used also in building construction in the form of dimension stock, flooring, sheathing, and studding. It has excellent properties for pulp and papermaking.

Spruce, Sitka

Sitka spruce (*Picea sitchensis*) is a tree of large size growing along the northwestern coast of North American from California to Alaska. It is generally known as Sitka spruce, although other names may be applied locally, such as yellow spruce, tideland spruce, western spruce, silver spruce, and west coast spruce. About two-thirds of the production of Sitka spruce lumber comes from Washington and one-third from Oregon.

The heartwood of Sitka spruce is a light pinkish brown. The sapwood is creamy white and shades gradually into the heartwood; it may be 3 to 6 inches wide or even wider in young trees. The wood has a comparatively fine, uniform texture, generally straight grain, and no distinct taste or odor. It is moderately light in weight, moderately low in bending and compressive strength, moderately stiff, moderately soft, and moderately low in resistance to shock. It has moderately small shrinkage. On the basis of weight, it rates high in strength properties and can be obtained in clear, straight grained pieces.

Sitka spruce is used principally for lumber, pulpwood, and cooperage. Boxes and crates account for a considerable amount of the remanufactured lumber. Other important uses are furniture, planing-mill products, sash, doors, blinds, millwork, and boats. Sitka spruce has been by far the most important wood for aircraft construction. Other specialty uses are ladder rails and sounding boards for pianos.

Tamarack

Tamarack (*Larix laricina*) is a small- to medium-sized tree with a straight, round, slightly tapered trunk. In the United States it grows from Maine to Minnesota, with the bulk of the stand in the Lake States. It was formerly used in considerable quantity for lumber, but in recent years production for that purpose has been small.

The heartwood of tamarack is yellowish brown to russet brown. The sapwood is whitish, generally less than an inch wide. The wood is coarse in texture, without odor or taste, and the transition from earlywood to latewood is abrupt. The wood is intermediate in weight and in most mechanical properties.

Tamarack is used principally for pulpwood, lumber, railroad ties, mine timbers, fuel, fenceposts, and poles. Lumber goes into framing material, tank construction, and boxes, pallets, and crates.

White-Cedar, Northern and Atlantic

Two species of white-cedar grow in the eastern part of the United States—northern white-cedar (*Thuja occidentalis*) and Atlantic white-cedar (*Chamaecyparis thyoides*). Northern white-cedar is also known as arborvitae, or simply cedar. Atlantic white-cedar is also known as juniper, southern white-cedar, swamp cedar, and boat cedar.

Northern white-cedar grows from Maine along the Appalachian Mountain Range and westward through the northern part of the Lake States. Atlantic white-cedar grows near the Atlantic coast from Maine to northern Florida and westward along the gulf coast to Louisiana. It is strictly a swamp tree.

Production of northern white-cedar lumber is probably greatest in Maine and the Lake States. Commercial production of Atlantic white-cedar centers in North Carolina and along the gulf coast.

The heartwood of white-cedar is light brown, and the sapwood is white or nearly so. The sapwood is usually thin. The word is light in weight, rather soft and low in strength, and low in shock resistance. It shrinks little in drying. It is easily worked, holds paint well, and the heartwood is highly resistant to decay. The two species are used for similar purposes, mostly for poles, ties, lumber, posts, and decorative fencing. White-cedar lumber is used principally where high degree of durability is needed, as in tanks and boats, and for woodenware.

IMPORTED WOODS

This section does not purport to discuss all of the woods that have been at one time or another imported into the United States. Only those species at present considered to be of commercial importance are included. The same species may be marketed in the United States under other common names.

Text information is necessarily brief, but when used in conjunction with the shrinkage and strength tables (ch. 3 and 4), a reasonably good picture may be obtained of a particular wood. The bibliography at the end of this chapter contains information on many species not described here.

Afrormosia (See Kokrodua)

Almon (See Lauans)

Andiroba

Because of the widespread distribution of andiroba (*Carapa guianensis*) in tropical America, the wood is known under a variety of names that include cedro macho, carapa, crabwood, and tangare. These names are also applied to the related species *Carapa nicaraguensis*, whose properties are generally inferior to those of *C. guianensis*.

The heartwood color varies from reddish brown to dark reddish brown. The texture (size of pores) is like that of mahogany (*Swietenia*). The grain is usually interlocked but is rated as easy to work, paint, and glue. The wood is rated as durable to very durable with respect to decay and insects. Andiroba is heavier than mahogany and accordingly is markedly superior in all static bending properties, compression parallel to the grain, hardness, shear, and toughness.

On the basis of its properties, andiroba appears to be suited for such uses as flooring, frame construction in the tropics, furniture and cabinetwork, millwork, and utility and decorative veneer and plywood.

Angelique

Angelique (*Dicorynia guianensis*), or basra locus, comes from French Guiana and Surinam and was previously identified under the name *D. paraensis*. Because of the variability in heartwood color between different trees, two forms are commonly recognized by producers. Heartwood that is russet colored when freshly cut, and becomes superficially dull brown with a purplish cast, is referred to as "gris." Heartwood that is more distinctly reddish and frequently shows wide bands of purplish color is called angelique rouge.

The texture is somewhat coarser than that of black walnut. The grain is generally straight or slightly interlocked. In strength, angelique is superior to teak and white oak, when either green or air dry, in all properties except tension perpendicular to grain. Angelique is rated as highly resistant to decay, and resistant to marine borer attack. Machining properties vary and may be due to differences in density, moisture content, and silica content. After the wood is thoroughly air dried or kiln dried, it can be worked effectively only with carbide-tipped tools.

The strength and durability of angelique make it especially suitable for heavy construction, harbor installations, bridges, heavy planking for pier and platform decking, and railroad bridge ties. The wood is particularly suitable for ship decking, planking, boat frames, and underwater members. It is currently being used in the United States for pier and dock fenders and flooring.

Apamate

Apamate (*Tabebuia rosea*) ranges from southern Mexico through Central America to Venezuela and Ecuador. The name roble is frequently applied to this species because of some fancied resemblance of the wood to that of oak (Quercus). Another common name for for apamate in Belize is mayflower.

The sapwood becomes a pale brown upon exposure. The heartwood varies through the browns, from a golden to a dark brown. Texture is medium, and grain is closely and narrowly interlocked. Heartwood is without distinctive odor or taste. The wood weighs about 38 pounds per cubic foot at 12 percent moisture content.

Apamate has excellent working properties in all machine operations. It finishes attractively in natural color and takes finishes with good results.

Apamate averages lighter in weight than the average of the American white oaks, but is comparable with respect to bending and compression parallel to grain. The white oaks are superior with respect to side hardness and shear.

The heartwood of apamate is generally rated as durable to very durable with respect to fungus attack; the darker colored and heavier

wood is regarded as more resistant than the lighter forms.

Within its region of growth, apamate is used extensively for furniture, interior trim, doors, flooring, boat building, ax handles, and general construction. The wood veneers well and produces an attractive paneling.

Apitong

Apitong is the most common structural timber of the Philippine Islands. The principal species are apitong (*Dipterocarpus grandiflorus*), panau (*D. gracilis*), and hagakhak (*D. warburgii*). All members of the genus are timber trees, and all are marketed under the name apitong. Other important species of the genus *Dipterocarpus* are marketed as keruing in Malaysia and Indonesia, yang in Thailand, and gurjun in India and Burma.

The wood is light to dark reddish brown in color, comparatively coarse to comparatively fine textured, straight grained or very nearly so, strong, hard, and heavy. The wood is characterized by the presence of resin ducts, which occur in short arcs as seen from end grain surfaces.

Although the heartwood is fairly resistant to decay and insect attack, the wood should be treated with preservatives when it is to be used in contact with the ground.

In machining research at the Forest Products Laboratory on apitong and the various species of "Philippine mahogany," apitong ranked appreciably above the average in all machining operations.

Apitong is used for heavy-duty purposes as well as for such items as mine guides, truck floors, chutes, flumes, agitators, pallets, and boardwalks.

Avodire

Avodire (*Turraeanthus africanus*) has rather extensive range from Sierra Leone westward to the Cameroons and southward to Zaire. It is a medium-sized tree of the rain forest in which it forms fairly dense but localized and discontinuous stands.

The wood is cream to pale yellow in color with a high natural luster and eventually darkens to a golden yellow. The grain is sometimes straight but more often is wavy or irregularly interlocked, which produces an unusual and attractive mottled figure when sliced or cut on the quarter.

Although its weight is only 85 percent that of English oak, avodire has almost identical strength properties except that it is lower in shock resistance and in shear. The wood works fairly easily with hand and machine tools and finishes well in most operations.

Figured material is usually converted into veneer for use in decorative work and it is this kind of material that is chiefly imported into the United States.

Bagtikan

The genus *Parashorea* consists of about seven species occurring in Southeast Asia. The principal species in the United States lumber trade is bagtikan (*P. plicata*) of the Philippines and Borneo. White seraya (*P. malagnonan*) from Sabah is also important. In the United States, bagtikan may be encountered under its usual common name or more frequently with the species comprising the light-red group of lauans. The heartwood is gray to straw colored or very pale brown and sometimes has a pinkish cast. It is not always clearly demarcated from the sapwood. The wood weighs about 34 pounds per cubic foot at 12 percent moisture content. The texture is similar to that of the light-red group of Philippine lauans. The grain is interlocked and shows a rather widely spaced stripe pattern on quartered surfaces.

With respect to strength, Philippine bagtikan exceeds the lauans in all properties. Its natural durability is very low and it is resistant or extremely resistant to preservative treatment.

The wood works fairly easily with hand and machine tools and has little blunting effect on tool cutting edges.

Bagtikan is used for many of the same purposes as the Philippine lauans, but in the solid form and in thin stock it is best utilized in the quartersawn condition to prevent excessive movement with changes in moisture conditions of service. It is perhaps most useful as a veneer for plywood purposes. In Britain it is best known as a decking timber which has been specially selected for this use in vessels.

Balsa

Balsa (*Ochroma pyramidale*) is widely distributed throughout tropical America from southern Mexico to southern Brazil and Bolivia, but Ecuador has been the principal area of growth since the wood gained commercial importance.

Balsa possesses several characteristics that make possible a wide variety of uses. It is the lightest and softest of all woods on the market. The lumber selected for use in the United States when dry weighs on the average of about 11 pounds per cubic foot and often as little as 6 pounds. Because of its light weight and exceedingly porous composition, balsa is highly efficient in uses where buoyancy, insulation against heat and cold, or absorption of sound and vibration are important considerations.

The wood is readily recognized by its light weight, white to very pale gray color, and its unique "velvety" feel.

The principal uses of balsa are in life-saving equipment, floats, rafts, core stock, insulation, cushioning, sound modifiers, models, and novelties. Balsa is imported in larger volume than most of the foreign woods entering the United States.

Banak

More than 40 species of *Virola* occur in tropical America, but only three species supply the bulk of the timber known as banak. These are: *V. koschnyi* of Central America, and *V. surinamensis* and *V. sebifera* of northern South America.

The heartwood is usually pinkish brown or grayish brown in color and is not differentiated from the sapwood. The wood is straight grained and is of a medium to coarse texture.

The various species are nonresistant to decay and insect attack but can be readily treated with preservatives. Their machining properties are very good, but fuzzing and grain tearing are to be expected when zones of tension wood are present. The wood finishes readily and is easily glued. It is rated as a first-class veneer species. Its strength properties are similar to yellow-poplar.

Banak is considered as a general utility wood in both lumber and plywood form.

Basra Locus (See Angelique)

Benge

Although benge (*Guibourtia arnoldiana*) and ehie (*G. ehie*) belong to the same botanical genus, they differ rather markedly with respect to their color. The heartwood of benge is a yellow-brown to medium brown with gray to almost black striping. Ehie heartwood tends to be a more golden brown and is striped as in benge. Ehie appears to be the more attractive of the two species.

The technical aspects of these species have not been investigated, but both are moderately hard and heavy. Benge is fine textured and in this respect similar to birch; the texture of ehie is somewhat coarser. Both are straight grained or have slightly interlocked grain.

These woods are as yet little known in the United States, but would provide both veneer and lumber for decorative purposes and furniture manufacture.

Capirona

A genus of about five species found throughout most of Latin America. The species best known in the United States and particularly in the archery field is degame (*Calycophyllum candidissimum*), which was imported in the past in some quantities from Cuba. Capirona (*C. spruceanum*) of the Amazon Basin is a much larger tree and occurs in considerably greater abundance than degame.

The heartwood of degame ranges from a light brown to gray, while that of capirona has a distinct yellowish cast. The texture is fine and uniform. The grain is usually straight or infrequently shows a shallow interlocking, which may produce a narrow and indistinct stripe on quartered faces. The luster is medium and the wood is without odor and taste. The wood weighs about 50 pounds per cubic foot.

Natural durability is low when the wood is used under conditions favorable to stain, decay, and insect attack.

In strength, degame is above the average for woods of similar density. Tests show degame superior to persimmon (*Diospyros virginiana*) in all respects but hardness. Limited tests on hardness of capirona from Peru gave higher values of side hardness than those of degame or persimmon.

Degame is moderately difficult to machine because of its density and hardness, although it produces no appreciable dulling effect on cutting tools. Machined surfaces are very smooth.

Degame and capirona are little used in the United States at the present time, but the characteristics of the wood should make it particularly adaptable for shuttles, picker sticks, and other textile industry items in which resilience and strength are required. It should find application for many of the same purposes as hard maple and yellow birch.

Carapa (See Andiroba)

Cativo

Cativo (*Prioria copaifera*) is one of the few tropical American species that occur in abundance and often in nearly pure stands. Commercial stands are found in Nicaragua, Costa Rica, Panama, and Colombia. The sapwood is usually thick, and in trees up to 30 inches in diameter the heartwood may be only 7 inches in diameter. The sapwood that is utilized commercially may be a very pale pinkish color or may be distinctly reddish. The grain is straight and the texture of the wood is uniform, comparable to that of mahogany. Figure on flat-sawn surfaces is rather subdued and results from the exposure of the narrow bands of parenchyma tissue. Odor and taste are not distinctive, and the luster is low.

The wood can be seasoned rapidly and easily with very little degrade. The dimensional stability of the wood is very good; it is practically equal to that of mahogany. Cativo is classed as a nondurable wood with respect to decay and insects. Cativo may contain appreciable quantities of gum, which may interfere with finishes. In wood that has been properly seasoned, however, the gum presents no difficulties.

The tendency of the wood to bleed resinous material in use and in warping of narrow cuttings kept this species in disfavor for many years. Improved drying and finishing techniques have materially reduced the prominence of these inherent characteristics, and the uses for this wood are rapidly increasing. Considerable quantities are used for interior trim, and resin-stabilized veneer has become an important pattern material, particularly in the automotive industry. Cativo is widely used for furniture and cabinet parts, lumber core for plywood, picture frames, edge banding for doors, and bases for piano keyboards.

Cedro (See Spanish-Cedar)

Cedro Macho (See Andiroba)

Cocal (See Sande)

Courbaril

The genus *Hymenaea* consists of about 30 species occurring in the West Indies and from southern Mexico, through Central America, into the Amazon Basin of South America. The best known and most important species is *H. courbaril*, which occurs throughout the range of the genus.

Courbaril sapwood is gray-white and usually

quite wide. The heartwood is sharply differentiated and varies through shades of brown to an occasional purplish cast. The texture is medium. Grain is interlocked, and luster is fairly high. The heartwood is without distinctive odor or taste. The wood weighs about 50 pounds per cubic foot at 12 percent moisture content.

The strength properties of courbaril are quite high but very similar to those of shagbark hickory, species of lower specific gravity.

In decay resistance, courbaril is rated as very durable to durable.

Courbaril can be finished smoothly, and it turns and glues well. It compares favorably with white oak in steam-bending behavior.

Courbaril has been little utilized in the United States, but should find application for a number of uses. Its high shock resistance recommends it for certain types of sporting equipment, or as a substitute for ash in handle stock. It promises to be a suitable substitute for white oak in steam-bent boat parts. It makes an attractive veneer and should also find application in the solid form for furniture. The thick sapwood would provide a excellent source of blond wood.

Crabwood (See Andiroba)

Cuangare (See Virola)

Degame (See Capirona)

Ehie (See Benge)

Encino (See Oak)

Freijo (See Laurel)

Gola

Gola (*Tetraberlinia tubmaniana*) is known presently only from Liberia. The names "African pine" and "Liberian pine" have been applied to this species, but because it is a hardwood and not a pine, these names are most inappropriate and very misleading.

The heartwood is light reddish brown and is distinct from the lighter colored sapwood, which may be up to 2 inches thick. The wood is moderately coarse textured. Luster is medium. Grain is interlocked, showing a narrow stripe pattern on quartered surfaces. The wood weighs about 39 pounds per cubic foot at 12 percent moisture content.

Tests made at the Forest Products Laboratory indicate no potential difficulties in the machining of gola. It also peels and slices very well.

Gola is a very recent newcomer to the timber market and its potential has yet to be developed. Its workability and relatively light color should permit utilization in both the solid and veneer form for both utility and decorative purposes.

Goncalo Alves

The major and early imports of goncalo alves (*Astronium graveolens & fraxinifolium*) have been from Brazil. These species range from southern Mexico, through Central America into the Amazon Basin.

The heartwood ranges from various shades of brown to red with narrow to wide, irregular stripes of dark brown or nearly black. The sapwood is grayish white and sharply demarcated from the heartwood. The texture is medium and uniform. Grain is variable from straight to interlocked and wavy. The wood is very heavy and averages about 63 pounds per cubic foot at 12 percent moisture content.

It turns readily, finishes very smoothly, and takes a high natural polish. The heartwood is highly resistant to moisture absorption and the pigmented areas, because of their high density, may present some difficulties in gluing.

The heartwood is rated as very durable with respect to fungus attack.

The high density of the wood is accompanied by equally high strength values, which are considerably higher in most respects than those of any well known U.S. species. It is not expected, however, that goncalo alves will be imported for purposes where strength is an important criterion.

In the United States the greatest value of goncalo alves is in its use for specialty items such as archery bows, billiard cue butts, brush backs, cutlery handles, and for fine and attractive products of turnery or carving.

Greenheart

Greenheart (*Ocotea rodiaei*) is essentially a Guyana tree although small stands also occur in Surinam. The heartwood varies in color from light to dark olive-green or nearly black. The texture is fine and uniform.

Greenheart is stronger and stiffer than white oak and generally more difficult to work with tools because of its high density. The heartwood is rated as very resistant to decay and termites. It also is very resistant to marine

26

borers in temperate waters but much less so in warm tropical waters.

Greenheart is used principally where strength and resistance to wear are required. Uses include ship and dock building, lock gates, wharves, piers, jetties, engine bearers, planking, flooring, bridges, and trestles.

Curjun (See Apitong)

Ilomba

Ilomba (*Pycnanthus angolensis*) is a tree of the rain forest and ranges from Guinea and Sierra Leone through west tropical Africa to Uganda and Angola. This species is also referred to in the literature under the synonymous name *Pycnanthus kombo*.

The wood is a grayish white to pinkish brown and in some trees may be a uniform light brown. There is generally no distinction between heartwood and sapwood. The texture is moderately coarse and even. Luster is low. Grain is generally straight. The wood weighs about 32 pounds per cubic foot at 12 percent moisture content. This species is generally similar to banak (*Virola*), but is somewhat coarser textured.

The wood is rated as perishable, but permeable to preservative treatment. The general characteristics of ilomba would suggest similarity to banak in working properties, seasoning, finishing, and utilization.

This species has been utilized in the United States only in the form of plywood for general utility purposes.

Ipe (See Lapacho)

Jacaranda (See Rosewood, Brazilian)

Jarrah

Jarrah (*Eucalyptus marginata*) is native to the coastal belt of southwestern Australia and one of the principal timbers of the sawmill industry.

The heartwood is a uniform pinkish to dark red, often a rich, dark red mahogany hue, turning to a deep brownish red with age and exposure to light. The sapwood is pale in color and usually very narrow in old trees. The texture is even and moderately coarse. The grain, though usually straight, is frequently interlocked or wavy. The wood weighs about 44 pounds per cubic foot at 12 percent moisture content. The common defects of jarrah include gum veins or pockets which, in extreme instances, separate the log into concentric shells.

Jarrah is a heavy, hard timber possessing correspondingly high strength properties. It is resistant to attack by termites and rated as very durable with respect to fungus attack. The heartwood is rated as extremely resistant to preservative treatment.

Jarrah is fairly hard to work in machines and difficult to cut with hand tools.

Jarrah is used for decking and underframing of piers, jetties, and bridges, and also for piling and fenders in dock and harbor installations. As a flooring timber it has a high resistance, but is inclined to splinter under heavy traffic.

Jelutong

Jelutong (*Dyera costulata*) is an important species in Malaya where it is best known for its latex production rather than its timber.

The wood is white or straw-colored and there is no differentiation between heartwood and sapwood. The texture is moderately fine and even. The grain is straight, and luster is low. The wood weighs about 29 pounds per cubic foot at 12 percent moisture content.

The wood is reported to be very easy to season with little tendency to split or warp, but staining may cause trouble.

It is easy to work in all operations, finishes well, and can be glued satisfactorily.

The wood is rated as nondurable, but readily permeable to preservatives.

Jelutong would make an excellent core stock if it were economically feasible to fill the latex channels which radiate outward in the stem at the branch whorls. Because of its low density and ease of working, it is well suited for sculpture and pattern. Jelutong is essentially a "short-cutting" species, because the wood between the channels is remarkably free of other defects.

Kapur

The genus *Dryobalanops* comprises some nine species distributed over parts of Malaya, Sumatra, and Borneo, including North Borneo and Sarawak. For the export trade, however, the species are combined under the name kapur.

The heartwood is light reddish brown, clearly demarcated from the pale colored sapwood. The wood is fairly coarse textured but uniform. In general appearance the wood resembles that of apitong and keruing, but on the whole it is straighter grained and not quite so coarse in texture. The Malayan timber averages about 48 pounds per cubic foot at 12 percent moisture content.

Strength property values available for *D. lanceolata* show it to be on a par with apitong or keruing of similar specific gravity.

The heartwood is rated as very durable and extremely resistant to preservative treatment.

The wood works with moderate ease in most hand and machine operations. A good surface is obtainable from the various machining operations, but there is a tendency toward "raised grain" if dull cutters are used. It takes nails and screws satisfactorily.

The wood provides good and very durable construction timbers and is suitable for all the purposes for which apitong and keruing are used in the United States.

Karri

Karri (*Eucalyptus diversicolor*) is a very large tree limited to Western Australia, occurring in the southwestern portion of the state.

Karri resembles jarrah (*E. marginata*) in structure and general appearance. It is usually paler in color, and, on the average, slightly heavier (57 lb. per cu. ft. at 12 pct. m.c.).

The heartwood is rated as moderately durable and extremely resistant to preservative treatment.

Karri is a heavy hardwood possessing mechanical properties of a correspondingly high order.

The wood is fairly hard to work in machines and difficult to cut with hand tools. It is generally more resistant to cutting than jarrah and has slightly more dulling effect on tool edges.

It is inferior to jarrah for underground use and waterworks, but where flexural strength is required, such as in bridges, floors, rafters, and beams, it is an excellent timber. Karri is popular in the heavy construction field because of its strength and availability in large sizes and long lengths that are free of defects.

Keruing (See Apitong)

Khaya

The bulk of the khaya or "African Mahogany"[2] shipped from west central Africa is *Khaya ivorensis*, which is the most widely distributed and most plentiful species of the genus

[2] Forest Service nomenclature restricts the name mahogany to the species belonging to the botanical genus *Swietenia*.

found in the coastal belt of the so-called closed or high forest. The closely allied species, *Khaya anthotheca*, has a more restricted range and is found farther inland in regions of lower rainfall but well within the area now being worked for the export trade.

The heartwood varies from a pale pink to a dark reddish brown. The grain is interlocked, and the texture is equal to that of mahogany (*Swietenia*). The wood is very well known in the United States and large quantities are imported annually. The wood is easy to season, machines and finishes well. In decay resistance, it is generally rated below American mahogany.

Principal uses include furniture, interior finish, boat construction, and veneer.

Kokrodua

Kokrodua (*Pericopsis elata*) is the vernacular name used in Ghana. It is also known as afrormosia, its former generic name.

This large West African tree shows promise of becoming a substitute for teak (*Tectona grandis*). The heartwood is fine textured, with straight to interlocked grain. The wood is brownish yellow with darker streaks, moderately hard and heavy, weighing about 44 pounds per cubic foot at 15 percent moisture content. The wood strongly resembles teak in appearance but lacks the oily nature of teak and is finer textured.

The wood seasons readily with little degrade and has good dimensional stability. It is somewhat heavier than teak and stronger. The heartwood appears to be highly resistant to decay and should prove extremely durable under adverse conditions. The wood can undoubtedly be used for the same purposes as teak, such as boat construction, interior trim, and decorative veneer.

Korina (See Limba)

Krabak (See Mersawa)

Lapacho

The lapacho group or series of the genus *Tabebuia* consists of about 20 species of trees and occurs in practically every Latin American country except Chile. Another commonly used name is ipe.

The sapwood is relatively thick, yellowish gray or gray brown and sharply differentiated from the heartwood, which is a light to dark

28

olive brown. The texture is fine. Grain is closely and narrowly interlocked. Luster is medium. The wood is very heavy and averages about 64 pounds per cubic foot at 12 percent moisture content. Thoroughly air-dried specimens of heartwood generally sink in water.

Lapacho is moderately difficult to machine because of its high density and hardness. Glassy smooth surfaces can be readily produced.

Being a very heavy wood, lapacho is also very strong in all properties and in the air-dry condition is comparable to greenheart.

Lapacho is highly resistant to decay and insects, including both subterranean and dry-wood termites. It is, however, susceptible to marine borer attack. The heartwood is impermeable, but the sapwood can be readily treated with preservatives.

Lapacho is used almost exclusively for heavy duty and durable construction. Because of its hardness (two to three times that of oak or apitong) and very good dimensional stability, it would be particularly well suited for heavy duty flooring in trucks and box cars.

Lauans

The term "lauan" or "Philippine mahogany" is applied commercially to Philippine woods belonging to three genera—*Shorea*, *Parashorea*, and *Pentacme*. These woods are usually grouped by the United States trade into "dark red Philippine mahogany" and "light red Philippine mahogany." The species found in these two groups and their heartwood color are:

"Dark red Philippine mahogany"

Red lauan, *Shorea negrosensis*	Dark reddish-brown to brick red
Tanguile, *Shorea polysperma*	Red to reddish-brown
Tianong, *Shorea agsaboensis*	Light red to light reddish-brown

"Light red Philippine mahogany"

Almon, *Shorea almon*	Light red to pinkish
Bagtikan, *Parashorea plicata*	Grayish-brown
Mayapis, *Shorea squamata*	Light red to reddish-brown
White lauan, *Pentacme contorta*	Grayish to very light red

The species within each group are shipped interchangeably when purchased in the form of lumber. Mayapis of the light red group is quite variable with respect to color and frequently shows exudations of resin. For this reason, some purchasers of "Philippine mahogany'" specify that mayapis be excluded from their shipments.

"Philippine mahoganies" as a whole have a coarser texture than mahogany or the "African mahoganies" and do not have the dark colored deposits in the pores. Forest Products Laboratory studies showed that the average decay resistance was greater for mahogany than for either the "African mahoganies" or the "Philippine mahoganies." The resistance of "African mahogany" was of the moderate type and seemed no greater than that of some of the "Philippine mahoganies." Among the Philippine species, the woods classified as "dark red Philippine mahogany" usually were more resistant than the woods belonging to the light red group.

In machining trials made at the Laboratory, the Philippine species appeared to be about equal with the better of the hardwoods found in the United States. Tanguile was consistently better than average in all or most of the tests. Mayapis, almon, and white lauan were consistently below average in all or most of the trials. Red lauan and bagtikan were intermediate. All of the species showed interlocked grain.

The shrinkage and swelling characteristics of the Philippine species are comparable to those found in the oaks and maples of the United States.

Principal uses include interior trim, paneling, flush doors, plywood, cabinets, furniture, siding, and boat construction. The use of the woods of the dark red group for boatbuilding in the United States exceeds in quantity that of any foreign wood.

Laurel

The genus *Cordia* contains numerous species, but only a relatively small number are trees of commercial size. The three most important species are *Cordia alliodora* (laurel) with the most extensive range from the West Indies and Mexico southward to northern Bolivia and eastern Peru; *C. goeldiana* (freijo) of the Amazon Basin; and *C. trichotoma* (peterebi) of southeastern Brazil, northern Argentina, and adjacent Paraguay.

The heartwood is light to medium brown, plain or frequently with a pigment figure outlining the growth ring pattern. Sapwood is generally distinct and of a yellowish or very light brown color. Grain is generally straight or shallowly interlocked. Texture is medium and uniform. The wood is variable in weight, but in the same density range as mahogany and cedro.

The wood saws and machines easily with good to excellent results in all operations. It is reported to glue readily and holds its place well when manufactured.

The *Cordia* woods are rated as moderately durable to durable when used in contact with the ground, and are rated slightly above mahogany with respect to resistance against drywood termites. The darker colored wood is reputed to be more durable with respect to decay and have better termite resistance than the lighter colored material.

The strength properties of these *Cordias* are generally on a par with those of mahogany and cedro.

Because of their ease of working, good durability, low shrinkage, and attractiveness, the woods are used extensively within their areas of growth for furniture, cabinetwork, general construction, boat construction and many other uses. The characteristics of the wood should qualify it for use in the United States for many of the same purposes as mahogany and cedro.

Lignum Vitae

Lignum vitae (*Guaiacum offinale*) native to the West Indies, northern Venezuela, northern Colombia, and Panama, was for a great many years the only species used on a large scale. With the near exhaustion of commercial-size timbers of *G. officinale* the principal species of commerce is now *G. sanctum*. The latter species occupies the same range as *G. officinale*, but is more extensive and includes the Pacific side of Central America as well as southern Mexico and southern Florida.

Lignum vitae is one of the heaviest and hardest woods on the market. The wood is characterized by its unique green color and oily or waxy feel. The wood has a fine, uniform texture and closely interlocked grain. Its resin content may constitute up to about one-fourth of the air-dry weight of the heartwood.

Lignum vitae wood is used chiefly for bearing or bushing blocks for the lining of stern tubes of steamship propeller shafts. The great strength and tenacity of lignum vitae, combined with the self-lubricating properties that are due to the high resin content, make it especially adaptable for underwater use. It is also used for such articles as mallets, pulley sheaves, caster wheels, stencil and chisel block, various turned articles, and brush backs.

Vera or verawood (*Bulnesia arborea*) of Colombia and Venezuela is sometimes substituted for lignum vitae; however, vera is not suitable for underwater bearings.

Limba

Abundant supplies of limba (*Terminalia superba*) occur in west central Africa and the Congo region.

The wood varies in color from a gray-white to creamy brown and may contain dark streaks, which are valued for special purposes. The light colored wood is considered an important asset for the manufacture of blond furniture. The wood is generally straight grained and of uniform but coarse texture.

The wood is easy to season and the shrinkage is reported to be rather small. Limba is not resistant to decay, insects, or termites. It is easy to work with all types of tools and is veneered without difficulty.

Principal uses include interior trim, paneling, and furniture. Selected limba plywood is sold in the United States under the copyrighted name, "korina."

Lupuna

Lupuna (*Ceiba samauma*) is a very large tree found in the Amazon Basin. In the Peruvian-Amazon region it is known as lupuna. In its Brazilian range it is known as samauma. The wood is white or grayish to very pale reddish, very soft and light, weighing about 25 pounds per cubic foot air dry. The wood is coarse textured and has a dull luster. It is nondurable with respect to decay and insect attack. Some of the veneer used for plywood cores has been known to give off highly objectionable odor when subjected to high humidity.

The wood is available in large sizes, and its low density combined with a rather high degree of dimensional stability make it ideally suited for pattern and core stock.

The mechanical properties have not been investigated.

Mahogany

Mahogany (*Swietenia macrophylla*) ranges from southern Mexico through Central America into South America as far south as Bolivia. Mexico, Belize, and Nicaragua furnish about 70 percent of the mahogany imported into the United States.

The heartwood varies from a pale to a dark reddish brown. The grain is generally straighter than that of "African mahogany;" however, a wide variety of grain patterns are obtained from this species.

Among the properties that mahogany possesses to a high degree are dimensional stability, fine finishing qualities, and ease of working with tools. The wood is without odor or taste. It weighs about 32 pounds per cubic foot at 12 percent moisture content.

The principal uses for mahogany are furniture, models and patterns, boat construction, radio and television cabinets, caskets, interior trim, paneling, precision instruments, and many other uses where an attractive and dimensionally stable wood is required.

Mahogany, African (See Khaya)

Mahogany, Philippine (See Lauans)

Mayapis (See Lauans)

Mayflower (See Apamate)

Meranti

The trade name meranti covers a number of closely related species of *Shorea* from which light or only moderately heavy timber is produced. This timber is imported from Malaysia and Indonesia. On the Malay Peninsula this timber is commonly classified for export either as light red or dark red meranti. Each of these color varieties is the product of several species of *Shorea*. Meranti exported from Sarawak and various parts of Indonesia is generally similar to the Malayan timber. Meranti corresponds roughly to seraya from North Borneo and lauan from the Philippines, which are names used for the lighter types of *Shorea* and allied genera.

Meranti shows considerable variation in color, weight, texture, and related properties, according to the species. The grain tends to be slightly interlocked so that quartered stock shows a broad stripe figure. The texture is moderately coarse but even. Resin ducts with or without white contents occur in long tangential lines on the end surfaces of the wood, but the wood is not resinous like some of the keruing species. Wood from near the center of the log is apt to be weak and brittle.

Light red meranti is classed as a light-weight utility hardwood and comprises those species yielding a red or reddish but not a dark red timber. The actual color of the heartwood varies from pale pink to light reddish brown. The weight of the wood may vary over a rather wide range from 25 to 44 pounds per cubic foot in the seasoned condition.

Dark red meranti is darker in color than ordinary red meranti and appreciably heavier, weighing on the average about 43 pounds per cubic foot seasoned. This color variation is the product of a more limited number of species and consequently tends to be more uniform in character than light red meranti. Because of the number of species contributing to the production of meranti, appreciable variation may be encountered with respect to mechanical and physical properties, durability, and working characteristics.

The wood is used in both plywood and solid form for much the same purposes as the Philippine lauans.

Mersawa

Mersawa is one of the *Anisoptera*, a genus of about 15 species distributed from the Philippine Islands and Malaysia to East Pakistan. Names applied to the timber vary with the source and three names are generally encountered in the lumber trade: *krabak* (Thailand), *mersawa* (Malaysia), and *palosapis* (Philippines).

The *Anisoptera* species produce wood of light color and moderately coarse texture. The heartwood when freshly sawn is pale yellow or yellowish brown and darkens on exposure. Some timber may show a pinkish cast or pink streaks, but these eventually disappear on exposure. The wood weighs about 39 pounds per cubic foot in the seasoned condition at 12 percent moisture content and about 59 pounds when green.

The sapwood is susceptible to attack by powderpost beetles and the heartwood is not resistant to termites. With respect to fungus resistance, the heartwood is rated as moderately resistant and should not be used under conditions favoring decay. The heartwood does not absorb preservative solutions readily.

The wood machines readily, but because of the presence of silica, the dulling effect on the cutting edges of ordinary tools is severe and is very troublesome with saws.

It appears probable that the major volume of these timbers will be used in the form of plywood, because conversion in this form presents considerably less difficulty than lumber production.

Nogal, Tropical Walnut

Nogal or tropical walnut includes two species, *Juglans neotropica* of the eastern slope of the Andes and *J. olanchana* of northern Central America. There is widespread interest in walnut from sources where lumber costs are decidedly lower than in the United States, but unfortunately little technical information is available regarding these species.

The wood of the tropical species is generally darker than that of typical American black walnut, and the texture (pore size) is somewhat coarser. From the limited number of specimens available for examination, nogal also appears to be somewhat lighter in weight than U.S. black walnut. Logs frequently show streaks of lighter color in the heartwood and this characteristic has caused some concern about the potential utility of the tropical wood. Other features that are mentioned whenever tropical walnut is under discussion are the extreme slowness with which the wood dries and the dull yellowish-green coloring of the inner portion of some boards that occurs during drying. It has been stated that lumber to ⅝-inch thickness can be dried at the same rate as American black walnut, but thicker stock takes an appreciably longer period of time, and in the thicker stock the wood is prone to collapse and honeycomb.

Tropical walnut peels and slices readily, but the veneer is said to dry more slowly than American black walnut. Tension wood and compression failures have been observed in a number of specimens, and these invariably came from the central core of the tree.

It appears that these species require rather intensive study, particularly with respect to seasoning and machining, in order to ascertain their true potential.

Oak

The oaks (*Quercus* spp.) are abundantly represented in Mexico and Central America with about 150 species, which are nearly equally divided between red and white oak groups. Mexico is represented with over 100 species and Guatemala with about 25; the numbers diminish southward to Colombia, which has two species. The usual Spanish name applied is encino or roble and no distinction is made in the use of these names.

The wood of the various species is in most cases heavier than the species of the United States.

Strength data are available for only four species and the values obtained fall between those of white oak and the southern live oak or are equal to those of the latter. The average specific gravity for these species is 0.72 based on volume when green and weight ovendry, with an observed maximum average for one species from Guatemala of 0.86.

Utilization of the tropical oaks is very limited at present due to difficulties encountered in the drying of the wood. The major volume is used in the form of charcoal.

Obeche

Obeche (*Triplochiton scleroxylon*) trees of west central Africa reach heights of 150 feet or more and diameters of up to 5 feet. The trunk is usually free of branches for considerable heights so that clear lumber of considerable size is obtainable.

The wood is creamy white to pale yellow with little or no difference between the sapwood and heartwood. It is fairly soft, of uniform texture, and the grain is straight or more often interlocked. The wood weighs about 24 pounds per cubic foot in the air-dry condition.

The wood seasons readily with little degrade. It is not resistant to decay, and the sapwood blue stains readily unless appropriate precautions are taken after the trees are felled as well as after they have been converted into lumber.

The wood is easy to work and machine, veneers and glues well, and takes nails and screws without splitting.

The characteristics of this species make it especially suitable for veneer and core stock.

This species is also called samba and wawa.

Okoume

The natural distribution of okoume (*Aucoumea klaineana*) is rather restricted and is found only in west central Africa and Guinea. This species has been popular in European markets for many years, but its extensive use in the United States is rather recent. When first introduced in volume in the plywood and door fields, its acceptance was phenomenal because it provided attractive appearance at moderate cost. The wood has a salmon-pink color with a uniform texture and high luster. The texture is slightly coarser than that of birch. Okoume offers unusual flexibility in both working and finishing because the color, which is of medium intensity, permits toning to either lighter or darker shades.

In this country it is used for decorative plywood paneling, general utility plywood, and for doors. Its use as solid lumber has been hampered because special saws and planer knives are required to effectively machine this species because of the silica content of the wood.

Palosapis (See Mersawa)

"Parana Pine"

The wood that is commonly called "Parana pine" (Araucaria angustifolia) is not a true pine. It is a softwood that comes from southeastern Brazil and adjacent areas of Paraguay and Argentina.

"Parana pine" has many desirable characteristics. It is available in large sizes of clear boards with uniform texture. The small pinhead knots (leaf traces) that appear on flat-sawn surfaces and the light brown or reddish-brown heartwood, which is frequently streaked with red, provide desirable figured effects for matching in paneling and interior finishes. The growth rings are fairly distinct and more nearly like those of white pine (Pinus strobus) rather than those of the yellow pines. The wood has relatively straight grain, takes paint well, glues easily, and is free from resin ducts, pitch pockets, and streaks.

The strength values of this species compare favorably with those softwood species of similar density found in the United States and, in some cases, approach the strength values of species with greater specific gravity. It is especially good in shearing strength, hardness, and nail-holding ability, but notably deficient in strength in compression across the grain.

Some tendency towards splitting of kiln-dried "Parana pine" and warping of seasoned and ripped lumber is caused by the presence of compression wood, an abnormal type of wood structure with intrinsically large shrinkage along the grain. Boards containing compression wood should be excluded from exacting uses. The principal uses of "Parana pine" include framing lumber, interior trim, sash and door stock, furniture, case goods, and veneer.

This species is known in Brazil as pinheiro do Parana or pinho do Parana.

Pau Marfim

The growing range of Pau marfim (Balfourodendron riedelianum) is rather limited, extending from the State of Sao Paulo, Brazil, into Paraguay and the provinces of Corrientes and Missiones of northern Argentina. In Brazil it is generally known as pau marfim and in Argentina and Paraguay as guatambu.

In color and general appearance the wood is very similar to birch or hard maple sapwood. Although growth rings are present, they do not show as distinctly as in birch and maple. The wood is straight grained, easy to work and finish but is not considered to be resistant to decay. There is no apparent difference in color between heartwood and sapwood.

The average specific gravity of pau marfim is about 0.63 based on the volume when green and weight when ovendry. On the basis of its specific gravity, its strength values would be above those of hard maple which has an average specific gravity of 0.56.

In the areas of growth it is used for much the same purposes as our native hard maple and birch. Pau marfim was introduced to the U.S. market in the late 1960's and has been very well received and is especially esteemed for turned items.

Peroba de Campos

Peroba de campos (Paratecoma peroba) occurs in the coastal forests of eastern Brazil ranging from Bahia to Rio de Janeiro. It is the only species in the genus.

The heartwood is variable in color, but generally is in shades of brown with tendencies toward casts of olive and reddish color. The sapwood is a yellowish gray, clearly defined from the heartwood. The texture is relatively fine and approximates that of birch. The wood averages about 47 pounds per cubic foot at 12 percent moisture content.

The wood machines easily, but when smooth surfaces are required particular care must be taken in planing to prevent excessive grain tearing of quartered surfaces because of the presence of interlocked or irregular grain. There is some evidence that the fine dust arising from machining operations may produce allergic responses in certain individuals.

Peroba de campos is heavier than teak or white oak and is proportionately stronger than either of these species.

The heartwood is rated as very durable with respect to fungus attack and is rated as resistant to preservative treatment.

In Brazil, the wood is used in the manufacture of fine furniture, flooring, and decorative paneling. The principal use in the United Sates is in shipbuilding, where it serves as an alternate for white oak for all purposes except bent members. The wood is classified as a poor "bender."

Peterebi (See Laurel)

Pine, Caribbean

Caribbean pine (*Pinus caribaea*) occurs along the Caribbean side of Central America from Belize to northeastern Nicaragua. It is also native to the Bahamas and Cuba. It is primarily a tree of the lower elevations.

The heartwood is a golden brown to red brown and distinct from the sapwood which is 1 to 2 inches in thickness and a light yellow. The wood has a strong resinous odor and a greasy feel. The wood averages about 51 pounds per cubic foot at 12 percent moisture content.

The lumber can be kiln dried satisfactorily using the same schedule as that for ocote pine.

Caribbean pine is easy to work in all machining operations but the high resin content may necessitate occasional stoppages to permit removal of accumulated resin from the equipment.

Caribbean pine is an appreciably heavier wood than slash pine (*P. elliottii*), but the mechanical properties of these two species are rather similar.

Caribbean pine is used for the same purposes as the southern pines of the United States.

Pine, Ocote

Ocote pine (*Pinus oocarpa*) is a species of the higher elevations and occurs from northwestern Mexico southward through Guatemala into Nicaragua. The largest and most extensive stands occur in northern Nicaragua and Honduras.

The sapwood is a pale yellowish brown and generally up to 3 inches in thickness. The heartwood is a light reddish brown. Grain is straight. Luster is medium. The wood has a resinous odor, and weighs about 41 pounds per cubic foot at 12 percent moisture content.

The strength properties of ocote pine are comparable in most respects with those of longleaf pine (*P. palustris*).

Decay resistance studies show ocote pine heartwood to be very durable with respect to attack by a white-rot fungus and moderately durable with respect to brown rot.

Ocote pine is comparable to the southern pines in workability and machining characteristics.

Ocote pine is a general construction timber and is suited for the same uses as the southern pines.

Primavera

The natural distribution of primavera (*Cybistax donnell-smithii*) is restricted to southwestern Mexico, the Pacific coast of Guatemala and El Salvador, and north central Honduras.

Primavera is regarded as one of the primary light-colored woods, but its use was limited because of its rather restricted range and the relative scarcity of wild trees within its natural growing area.

Plantations now coming into production have increased the availability of this species and provided a more constant source of supply. The quality of the plantation-grown wood is equal in all respects to that obtained from wild trees.

The heartwood is whitish to straw-yellow and in some logs may be tinted with pale brown or pinkish streaks. The wood has a very high luster.

Primavera produces a wide variety of figure patterns.

The shrinkage properties are very good, and the wood shows a high degree of dimensional stability. Although the wood has considerable grain variation, it machines remarkably well. With respect to decay resistance it is rated as durable to very durable.

The dimensional stability, ease of working, and pleasing appearance recommend primavera for solid furniture, paneling, interior trim, and special exterior uses.

Ramin

Ramin (*Gonystylus bancanus*) is one of the very few moderately heavy woods that are classified as a "blond" wood. This species is native to southeast Asia from the Malay Peninsula to Sumatra and Borneo.

The wood is a uniform pale straw or yellowish to whitish in color. The grain is straight or shallowly interlocked. The texture is moderately fine, similar to that of mahogany (*Swietenia*), and even. The wood is without figure or luster. Ramin is moderately hard and heavy, weighing about 42 pounds per cubic foot in the air-dry condition. The wood is easy to work, finishes well, and glues satisfactorily.

With respect to natural durability ramin is rated as perishable, but it is permeable with regard to preservative treatment.

Ramin has been used in the United States in the form of plywood for doors and in the solid form for interior trim.

Red lauan (See Lauans)

Roble (See Oak)

Rosewood, Brazilian

Brazilian rosewood or jacaranda (*Dalbergia nigra*) occurs in the eastern forests of the State of Bahia to Rio de Janeiro. Having been exploited for a long period of time it is, at present, nowhere abundant.

The wood of commerce is very variable with respect to color, ranging through shades of brown, red, and violet and is irregularly and conspicuously streaked with black. Many kinds are distinguished locally on the basis of prevailing color. The texture is coarse, and the grain is generally straight. Heartwood has an oily or waxy appearance and feel. The odor is fragrant and distinctive. The wood is hard and heavy; thoroughly air-dried wood is just barely floatable in water.

The strength properties have not been determined, but for the purposes for which Brazilian rosewood is utilized they are more than adequate. In hardness, for example, it exceeds by far any of the native hardwood species used in the furniture and veneer field.

The wood machines and veneers well. It can be glued satisfactorily, providing the necessary precautions are taken to ensure good glue bonds as with other woods in this density class.

Brazilian rosewood has an excellent reputation for durability with respect to fungus and insect attack, including termites, although the wood is not used for purposes where these would present a problem.

Brazilian rosewood is used primarily in the form of veneer for decorative plywood. Limited quantities are used in the solid form for specialty items such as cutlery handles, brush backs, billiard cue butts, and fancy articles of turnery.

Rosewood, Indian

Indian rosewood (*Dalbergia latifolia*) is native to most provinces of India except in the northwest.

The heartwood is a dark purplish brown with denser blackish streaks terminating the growth zones and giving rise to an attractive figure on flat-sawn surfaces. The average weight is about 53 pounds per cubic foot at 12 percent moisture content. The texture is uniform and moderately coarse. The wood of this species is quite similar in appearance to that of the Brazilian and Honduras rosewood. The timber is said to kiln dry well, but rather slowly, and the color is said to improve during drying.

Indian rosewood is a heavy timber with high strength properties and is particularly hard for its weight after being thoroughly seasoned.

The wood is moderately hard to work with handtools and offers a fair resistance in machine operations. Lumber containing calcareous deposits tends to blunt tools rapidly. The wood turns well and has high screw-holding properties. Filling of the pores is desirable if a very smooth surface is required for certain purposes.

Indian rosewood is essentially a decorative wood for high-class furniture and cabinetwork. In the United States it is used primarily in the form of veneer.

Samauma (See Lupuna)

Samba (See Obeche)

Sande

Practically all of the exportation of sande (*Brosimum* spp.-utile group) is from Pacific Ecuador and Colombia. It is also known as cocal.

The sapwood and heartwood show no distinction, being a uniform yellowish white to yellowish brown or light brown. The pores are moderately coarse and evenly distributed. The grain is straight to widely and shallowly interlocked. In many respects, sande has much the same appearance as white seraya (*Parashorea malaanonan*) from Sabah. The wood averages about 33 pounds per cubic foot at 12 percent moisture content.

The wood is nondurable with respect to stain, decay, and insect attack and care must be exercised to prevent degrade from these agents.

Strength data for sande are too limited to permit comparison with woods of similar density such as banak, although it is suspected that they would be rather similar.

Normal wood of sande machines easily, takes stains, and finishes readily, and presents no gluing problems. Sande should find utilization for many of the same purposes as banak, and with the current demand for molding species, it should assist in relieving the ever-increasing wood demand of this industry.

Santa Maria

Santa Maria (*Calophyllum brasiliense*) ranges from the West Indies to southern Mexico and southward through Central America into northern South America.

The heartwood is pinkish to brick red or rich reddish brown and marked by a fine and slightly darker striping on flat-sawn surfaces. The sapwood is lighter in color and generally distinct from the heartwood. Texture is medium and fairly uniform. Luster is medium. The heartwood is rather similar in appearance to the red lauan of the Philippines. The wood averages about 38 pounds per cubic foot at 12 percent moisture content.

The wood is moderately easy to work and good surfaces can be obtained when attention is paid to machining operations.

Santa Maria is in the density class of hard maple and its strength properties are generally similar, with the exception of hardness, in which property hard maple is superior to Santa Maria.

The heartwood is generally rated as moderately durable to very durable in contact with the ground, but apparently has little resistance against termites and marine borers.

The inherent natural durability, color, and figure on the quarter suggest utilization as face veneer for plywood in boat construction. It also offers possibilities for use in flooring, furniture, cabinetwork, millwork, and decorative plywood.

Sapele

Sapele (*Entandrophragma cylindricum*) is a large African rain forest tree ranging from Sierra Leone to Angola and eastward through the Congo to Uganda.

The heartwood ranges in color from that of mahogany to a dark reddish or purplish brown. The lighter colored and distinct sapwood may be up to 4 inches thick. Texture is finer than that of mahogany. Grain is interlocked and produces a narrow and uniform stripe pattern on quartered surfaces. The wood averages about 39 pounds per cubic foot at 12 percent moisture content.

Sapele has the same average density as white oak, and its mechanical properties are in general higher than those of white oak.

The wood works fairly easily with machine tools, although interlocked grain offers difficulties in planing and molding. Sapele finishes and glues well.

The heartwood is rated as moderately durable and as resistant to preservative treatment.

Sapele is used extensively, primarily in the form of veneer for decorative plywood.

Spanish-Cedar

Spanish-cedar or cedro (*Cedrela* spp.) comprises a group of about seven species that are widely distributed in tropical America from southern Mexico to northern Argentina. The wood is more or less distinctly ring porous, and the heartwood varies from light reddish brown to dark reddish brown. The heartwood is characterized by its distinctive cedarlike odor.

The wood seasons readily. It is not high in strength but is roughly rated to be similar to Central American mahogany in most properties except in hardness and compression perpendicular to the grain where mahogany is definitely superior. It is considered decay resistant and works and glues well.

Spanish-cedar is used locally for all purposes where an easily worked, light but strong, straight grained, and durable wood is required. Spanish-cedar and mahogany are the classic timbers of Latin America.

Tangare (See Andiroba)

Tanguile (See Lauans)

Teak

Teak (*Tectona grandis*) occurs in commercial quantities in India, Burma, Thailand, Laos, Cambodia, North and South Vietnam, and the East Indies. Numerous plantations have been developed within its natural range and tropical areas of Latin America and Africa, and many of these are now producing timber.

The heartwood varies from a yellow-brown to a rich brown. It has a coarse texture, is usually straight grained, and has a distinctly oily feel. The heartwood has excellent dimensional stability and possesses a very high degree of natural durability.

Although not generally used in the United States where strength is of prime importance, the values for teak are generally on a par with those of our native oaks.

Teak generally works with moderate ease with hand and machine tools. Because of the presence of silica, its dulling effect on tools is sometimes considerable. Finishing and gluing are satisfactory although pretreatment may be necessary to ensure good bonding of finishes and glues.

Intrinsically, teak is one of the most valuable of all woods, but its use is limited by scarcity and high cost. Teak is unique in that it does not cause rust or corrosion when in contact with metal; hence, it is extremely useful in the shipbuilding industry. It is currently used in the construction of expensive boats, furniture, flooring, decorative objects, and veneer for decorative plywood.

Tianong (See Lauans)

Verawood (See Lignum Vitae)

Virola

Virola is the common name being currently applied to the wood of two or more species of *Dialyanthera* originating in the Pacific forests of Colombia and Ecuador. The local name for this wood is cuangare, and would be preferred for common usage because the common name "virola" is frequently confused with the botanical genus *Virola*.

The wood is a pale pinkish brown with a high luster. There is no sharp demarcation between heartwood and sapwood. The wood is generally straight grained, easy to work, holds nails well, and finishes smoothly. The texture is quite similar to that of okoume.

Virola is a relatively low-density wood. On the dry-weight basis, it is equal to that of alder, aspen, and basswood. Shrinkage properties of virola are about the same as those of sugar maple. The wood is rated very low with respect to natural durability; hence it is best suited for use under interior conditions.

Currently the wood is being used for paneling, interior trim, and core stock.

The mechanical properties have not been investigated.

Walnut, European

Although generally referred to as European walnut or by its country of origin, walnut (*Juglans regia*) is a native of western and central Asia, extending to China and northern India. Trees are grown in commercial quantities mainly in Turkey, Italy, France, and Yugoslavia.

Walnut is variable in color, with a grayish-brown background, marked with irregular dark-colored streaks. The figure, which is due to the infiltration of coloring matter, is sometimes accentuated by the naturally wavy grain. The highly figured veneers used in cabinetmaking and decorative paneling are obtained from the stumps, burls, and crotches of a relatively small percentage of the trees. The wood weighs about 40 pounds per cubic foot in the air-dry condition.

The product of any one locality may vary considerably in color, figure, and texture, but the selected export timber generally shows certain typical characteristics. French walnut is typically paler and grayer than English walnut, while the Italian wood is characterized by its elaborate figure and dark, streaky coloration. Because of the ease of machining, finishing, and gluing, walnut is used extensively in veneer form as well as in the solid form for furniture, paneling, and decorative objects. It and American black walnut are the classic woods for rifle stocks.

Walnut, tropical (See Nogal)

Wawa (See Obeche)

White lauan (See Lauans)

White seraya (See Bagtikan)

Yang (See Apitong)

BIBLIOGRAPHY

Bellosillo, S. B., and Miciano, R. J.
    1959. Progress report on the survey of mechanical
           properties of Philippine woods. The Lum-
           berman. Philippines.

British Forest Products Research Laboratory
    1956. A handbook of hardwoods. H. M. Stationery
           Office, 269 pp. London.

————
    1960. The strength properties of timber. Forest
           Prod. Res. Bull. 50. H. M. Stationery
           Office, 34 pp. London.

Brown, H. P., Panshin, A. J., and Forsaith, C. C.
    1949. Textbook of wood technology. Vol. I. Struc-
           ture, identification, defects, and uses of the
           commercial woods of the United States.
           652 pp. New York.

Kukachka, B. F.
    1970. Properties of imported tropical woods.
           USDA Forest Serv. Res. Pap. FPL 125.
           Forest Prod. Lab., Madison, Wis.

Markwardt, L. J.
    1930. Comparative strength properties of woods
           grown in the United States. U.S. Dep.
           Agr. Tech. Bull. 158, 39 pp.

Record, S. J., and Hess, R. W.
    1949. Timbers of the new world. Yale Univ. Press,
           640 pp.

Wangaard, F. F.
    1950. Mechanical properties of wood. 377 pp. New
           York.

Chapter 2

STRUCTURE OF WOOD

The fibrous nature of wood strongly influences how it is used. Specifically, wood is composed mostly of hollow, elongate, spindle-shaped cells that are arranged parallel to each other along the trunk of a tree. When lumber is cut from the tree, the characteristics of these fibrous cells and their arrangement affects such properties as strength and shrinkage, as well as grain pattern of the wood.

A brief description of some elements of anatomical structure are given in this chapter.

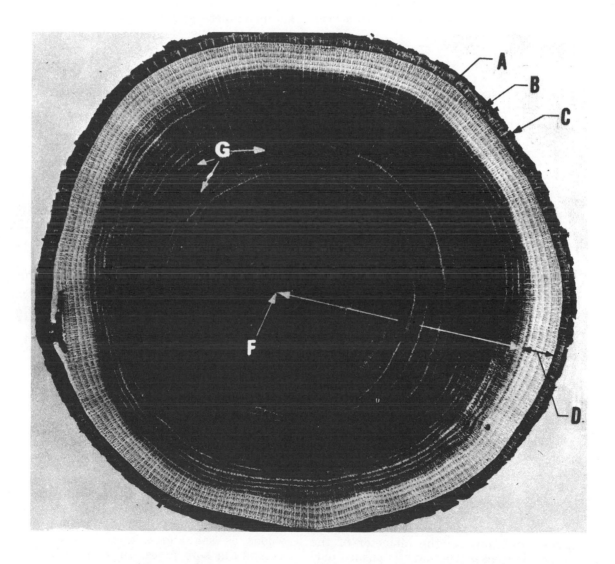

Figure 2–1.—Cross section of a white oak tree trunk: A, Cambium layer (microscopic) is inside inner bark and forms wood and bark cells. B, Inner bark is moist, soft, and contains living tissue. Carries prepared food from leaves to all growing parts of tree. C, Outer bark containing corky layers is composed of dry dead tissue. Gives general protection against external injuries. Inner and outer bark are separated by a bark cambium. D, Sapwood, which contains both living and dead tissues, is the light-colored wood beneath the bark. Carries sap from roots to leaves. E, Heartwood (inactive) is formed by a gradual change in the sapwood. F, Pith is the soft tissue about which the first wood growth takes place in the newly formed twigs. G, Wood rays connect the various layers from pith to bark for storage and transfer of food.

BARK, WOOD, AND PITH

A cross section of a tree (fig. 2–1) shows the following well-defined features in succession from the outside to the center: (1) Bark, which may be divided into the outer, corky, dead part that varies greatly in thickness with different species and with age of trees, and the thin, inner living part; (2) wood, which in merchantable trees of most species is clearly differentiated into sapwood and heartwood; and (3) the pith, indicated by a small central core, often darker in color, which represents primary growth formed when woody stems or branches elongate.

Most branches originate at the pith, and their bases are intergrown with the wood of the trunk as long as they are alive. These living branch bases constitute intergrown knots. After the branches die, their bases continue to be surrounded by the wood of the growing trunk. Such enclosed portions of dead branches constitute the loose or encased knots. After the dead branches drop off, the dead stubs become overgrown and, subsequently, clear wood is formed. In a tree, the part containing intergrown knots comprises a cylinder, extending the entire length of the tree; the part containing loose knots forms a hollow cylinder, extending from the ground to the base of the green crown. Clear wood constitutes an outer cylinder covering overgrown branch ends. In second-growth trees, the clear zone and even the zone of loose knots may be absent.

GROWTH RINGS

Between the bark and the wood is a layer of thin-walled living cells called the cambium, invisible without a microscope, in which most growth in thickness of bark and wood arises by cell division. No growth in either diameter or length takes place in wood already formed; new growth is purely the addition of new cells, not the further development of old ones. New wood cells are formed on the inside and new bark cells on the outside of the cambium. As the diameter of the woody trunk increases, the bark is pushed outward, and the outer bark layers become stretched, cracked, and ridged in patterns often characteristic of a species. A bark cambium forms from living cells and this tissue separates the outer bark from the inner bark.

With most species in temperate climates, there is sufficient difference between the wood

formed early and that formed late in a growing season to produce well-marked annual growth rings. The age of a tree at the stump or the age at any cross section of the trunk may be determined by counting these rings (fig. 2–2). If the growth of trees in diameter is interrupted by drought or defoliation by insects, more than one ring may be formed in the same season. In such an event, the inner rings usually do not have sharply defined boundaries and are termed false rings. Trees that have only very small crowns or that have accidentally lost most of their foliage may form only an incomplete growth layer, sometimes called a discontinuous ring, until the crown is restored.

Figure 2–2.—Cross section of a ponderosa pine log showing growth rings: Light bands are earlywood, dark bands are latewood. An annual ring is composed of the earlywood ring and the latewood ring outside it.

Growth rings are most readily seen in species with sharp contrast between earlywood and latewood, such as the native ring-porous hardwoods as ash and oak, and most softwoods except the soft pines. In some other species, such as water tupelo, sweetgum, and soft maple, differentiation of early and late growth is slight, and the annual growth rings are difficult to recognize. In some tropical regions, growth may be practically continuous throughout the year, and no well-defined annual rings are formed.

EARLYWOOD AND LATEWOOD

The inner part of the growth ring formed first in the growing season is called earlywood or springwood, and the outer part formed later in the growing season, latewood or summerwood. Actual time of formation of these two parts of a ring may vary with environmental and weather conditions. Earlywood is characterized by cells having relatively large cavities and thin walls. Latewood cells have smaller cavities and thicker walls. The transition from earlywood to latewood may be gradual or abrupt, depending on the kind of wood and the growing conditions at the time it was formed. In some species, such as the maples, gums, and yellow-poplar, there is little difference in the appearance of the earlywood and latewood parts of a growth ring.

When growth rings are prominent, as in most softwoods and the ring-porous hardwoods, earlywood differs markedly from latewood in physical properties. Earlywood is lighter in weight, softer, and weaker than latewood; it shrinks less across and more lengthwise along the grain of the wood. Because of the greater density of latewood, the proportion of latewood is sometimes used to judge the quality or strength of wood. This method is useful with such species as the southern pines, Douglas-fir, and the ring-porous hardwoods—ash, hickory, and oak.

SAPWOOD AND HEARTWOOD

Sapwood is located next to the cambium. It contains only a few living cells and functions primarily in the storage of food and the mechanical transport of sap.

The sapwood layer may vary in thickness and in the number of growth rings contained in it. Sapwood commonly ranges from 1½ to 2 inches in radial thickness. In certain species, such as catalpa and black locust, the sapwood contains very few growth rings and sometimes does not exceed one-half inch in thickness. The maples, hickories, ashes, some of the southern pines, and ponderosa pine may have sapwood 3 to 6 inches or more in thickness, especially in second-growth trees. As a rule, the more vigorously growing trees of a species have wider sapwood layers. Many second-growth trees of merchantable size consist mostly of sapwood.

Heartwood consists of inactive cells that have been slightly changed, both chemically and physically, from the cells of the inner sapwood rings. In this condition these cells cease to conduct sap.

The cell cavities of heartwood also may contain deposits of various materials which frequently give much darker color to the heartwood. All heartwood, however, is not dark colored. Species in which heartwood does not darken to a great extent include the spruces (except Sitka spruce), hemlock, the true firs, Port-Orford-cedar, basswood, cottonwood, and buckeye. The infiltrations or materials deposited in the cells of heartwood usually make the wood more durable when used in exposed situations. Unless treated, all sapwood is nondurable when exposed to conditions that favor decay.

In some species, such as the ashes, hickories, and certain oaks, the pores become plugged to a greater or lesser degree with ingrowths, known as tyloses, before the change to heartwood is completed. Heartwood having pores tightly plugged by tyloses, as in white oak, is suitable for tight cooperage.

Heartwood has a higher extractives content than sapwood, and because of this, exhibits a higher specific gravity. For most species the difference is so small as to be quite unimportant. The weight and strength of wood are influenced more by growth conditions of the trees at the time the wood is formed than they are by the change from sapwood to heartwood. In some instances, as in redwood, western redcedar, and black locust, considerable amounts of infiltrated material may somewhat increase the weight of the wood and its resistance to crushing.

WOOD CELLS

Wood cells that make up the structural elements of wood are of various sizes and shapes and are quite firmly grown together. Dry wood cells may be empty or partly filled with deposits, such as gums and resins, or with tyloses. A majority of cells are considerably elongated and pointed at the ends; they are customarily called fibers or tracheids. The length of wood fibers is highly variable within a tree and among species. Hardwood fibers average about one twenty-fifth of an inch in length (1 mm.); softwood fibers (called tracheids) range from one-eighth to one-third of an inch in length (3 to 8 mm.).

In addition to their fibers, hardwoods have cells of relatively large diameter known as vessels. These form the main arteries in the movement of sap. Softwoods do not contain special vessels for conducting sap longitudinally in the tree; this function is performed by the tracheids.

Both hardwoods and softwoods have cells (usually grouped into structures) that are oriented horizontally in the direction from the pith toward the bark. These structures conduct sap radially across the grain and are called rays or wood rays. The rays are most easily seen on quartersawed surfaces. They vary greatly in size in different species. In oaks and sycamores, the rays are conspicuous and add to the decorative features of the wood.

Wood also has other cells, known as longitudinal, or axial, parenchyma cells, that function mainly for the storage of food.

CHEMICAL COMPOSITION OF WOOD

Dry wood is made up chiefly of the following substances, listed in decreasing order of amounts present: Cellulose, lignin, hemicelluloses, extractives, and ash-forming minerals.

Cellulose, the major constituent, comprises approximately 50 percent of wood substance by weight. It is a high-molecular weight linear polymer that, on chemical degradation by mineral acids, yields the simple sugar glucose as the sole product. During growth of the tree, the linear cellulose molecules are arranged into highly ordered strands called fibrils, which in turn are organized into the larger structural elements comprising the cell wall of wood fibers. The intimate physical, and perhaps partially chemical, association of cellulose with lignin and the hemicelluloses imparts to wood its useful physical properties. Delignified wood fibers have great commercial value when reconstituted into paper. Moreover, they may be chemically altered to form synthetic textiles, films, lacquers, and explosives.

Lignin comprises 23 to 33 percent of softwoods, but only 16 to 25 percent of hardwoods. It occurs in the wood largely as an intercellular material. Like cellulose, it has a macromolecular chemical structure, but its three-dimensional network is far more complex and not yet completely worked out. As a chemical, lignin is an intractable, insoluble material, probably bonded at least loosely to the cellulose. To remove it from the wood on a commercial scale requires vigorous reagents, high temperatures, and high pressures. Such conditions greatly modify the lignin molecule, producing a complex mixture of high-molecular-weight phenolic compounds.

To the paper industry, lignin is difficult to solubilize and is a sometimes troublesome by-product. Theoretically, it might be converted to a variety of chemical products but, practically, a large percentage of the lignin removed from wood during pulping operations is burned for heat and recovery of pulping chemicals. One sizable commercial use for lignin is in the formulation of drilling muds, used in the drilling of oil wells, where its dispersant and metal-combining properties are valuable. It has found use also in rubber compounding and as an air-entraining agent in concrete mixes. Lesser amounts are processed to yield vanillin for flavoring purposes and to product solvents such as dimethyl sulfide and dimethyl sulfoxide.

The hemicelluloses are intimately associated with cellulose in nature and, like cellulose, are polymeric units built up from simple sugar moleclues. Unlike cellulose, however, the hemicelluloses yield more than one type of sugar on acid cleavage. Also, the relative amounts of these sugars vary markedly with species. Hardwoods contain an average of 20 to 30 percent hemicelluloses with xylose as the major sugar. Lesser amounts of arabinose, mannose, and a sugar acid are also attached to the main polymer chain. Softwoods contain an average of 15 to 20 percent hemicelluloses, with mannose as the main sugar unit. Xylose, arabinose, and the sugar acid are again present at lower levels. The hemicelluloses play an important role in fiber-to-fiber bonding in the papermaking process. The component sugars of hemicellulose are of potential interest for conversion into chemical products.

Unlike the major constituents just discussed. the extractives are not part of the wood structure. However, they do contribute to such properties of wood as color, odor, taste, decay resistance, strength, density, hygroscopicity, and flammability. They include tannins and other poly-phenolics, coloring matters, essential oils, fats, resins, waxes, gums, starch, and simple metabolic intermediates. They can be removed from wood by extraction with such inert neutral solvents as water, alcohol, acetone, benzene, and ether. In quantity, the extractives may range from roughly 5 to 30 percent, depending on such factors as species, growth conditions, and time of year the tree is cut.

Ash-forming minerals comprise from 0.1 to 3 percent of wood substance, although considerably higher values are occasionally reported. Calcium, potassium, phosphate, and silica are common constituents. Due to the uniform distribution of these inorganic materials throughout the wood, ash often retains the microstructural pattern of wood.

A significant dollar value of nonfibrous products is produced from wood including naval stores, pulp byproducts, vanillin, ethyl alcohol, charcoal, extractives, and bark products.

IDENTIFICATION

Many species of wood have unique physical, mechanical, or chemical properties. Efficient utilization dictates that species should be matched to use requirements through an understanding of properties. This requires identification of the species in wood form, independent of bark, foliage, and other characteristics of the tree.

Field identification can often be made on the basis of readily visible characteristics such as color, presence of pitch, or grain pattern. Sometimes odor, density, or splitting tendency is helpful. Where more positive identification is required, a laboratory investigation of the microscopic anatomy of the wood can be made. Detailed descriptions of identifying characteristics are given in texts such as "Textbook of Wood Technology" by Panshin and de Zeeuw.

BIBLIOGRAPHY

Bratt, L. C.
  1965. Trends in the production of silvichemicals in the United States and abroad. Tappi 48(7): 46A–49A. Tech. Assoc. Pulp and Paper Indus.

Brauns, F. E., and Brauns, D. A.
  1960. The chemistry of lignin—supplement volume, 804 pp. Academic Press.

Browning, B. L.
  1963. The chemistry of wood. 689 pp. Interscience Publishers, N.Y.

Freudenberg, K.
  1965. Lignin: Its constitution and formation from p-hydroxy-cinnamyl alcohols. Sci. 148: 595–600.

Hamilton, J. K., and Thompson, N. S.
  1959. A comparison of the carbohydrates of hardwoods and softwoods. Tappi 42: 752–760. Tech. Assoc. Pulp and Paper Indus.

Ott, E., Spurlin, H. M., and Grafflin, M. W.
  1954. Cellulose and cellulose derivatives. Volume V. Parts I, II, and III (1955) of High Polymers. 1601 pp. Interscience Publishers, N.Y.

Panshin, A. J., and de Zeeuw, C.
  1970. Textbook of wood technology. Volume 1. 3d edition. McGraw-Hill.

Wise, L. E., and Jahn, E. C.
  1952. Wood chemistry. Volumes I and II, 1259 pp. Reinhold.

Chapter 3

PHYSICAL PROPERTIES OF WOOD

The versatility of wood is demonstrated by a wide variety of products. This variety is a result of a spectrum of desirable physical characteristics or properties that is available among the many species of wood. Often more than one property of wood is important to an end product. For example, to select a species for a product, the value of appearance-type properties such as texture, grain pattern, or color may be evaluated against the influence of characteristics such as machinability, stability, or decay resistance. This chapter discusses the physical properties most often of interest in the design of wood products.

Some physical properties discussed and tabulated are influenced by species as well as variables like moisture content; other properties tend to be more independent of species. The thoroughness of sampling and the degree of variability influences the confidence with which species-dependent properties are known. In this chapter an effort is made to indicate either the general or specific nature of the properties tabulated.

APPEARANCE

Grain and Texture

The terms "grain" and "texture" are commonly used rather loosely in connection with wood. Grain is often used in reference to annual rings, as in fine grain and coarse grain, but it is also employed to indicate the direction of fibers, as in straight grain, spiral grain, and curly grain. Grain, as a synonym for fiber direction, is discussed in more detail relative to mechanical properties in chapter 4. Wood finishers refer to woods as open grained and close grained as terms reflecting the relative size of the pores, which determines whether the surface needs a filler. Texture is often used synonymously with grain, but usually it refers to the finer structure of wood rather than to annual rings. When the words "grain" or "texture" are used in connection with wood, the meaning intended should be made perfectly clear (see glossary).

Plainsawed and Quartersawed Lumber

Lumber can be cut from a log in two distinct ways: Tangent to the annual rings,

producing "plainsawed" lumber in hardwoods and "flat-grained" or "slash-grained" lumber in softwoods; and radially to the rings or parallel to the rays, producing "quartersawed" lumber in hardwoods and "edge-grained" or "vertical-grained" lumber in softwoods (fig. 3–1). Usually quartersawed or edge-grained lumber is not cut strictly parallel with the rays; and often in plainsawed boards the surfaces next to the edges are far from being tangent to the rings. In commercial practice, lumber with rings at angles of 45° to 90° with the wide surface is called quartersawed, and lumber with rings at angles of 0° to 45° with the wide surface is called plainsawed. Hardwood lumber in which annual rings make angles of 30° to 60° with the wide faces is sometimes called "bastard sawn."

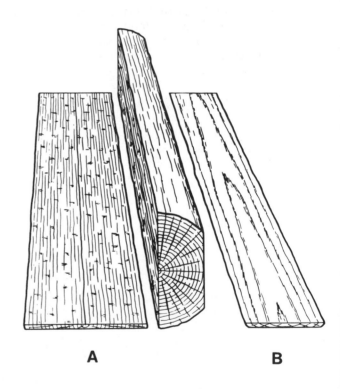

Figure 3–1.—Quartersawed A and plainsawed B boards cut from a log.

For many purposes either plainsawed or quartersawed lumber is satisfactory. Each type has certain advantages, however, that may by important in a particular use. Some of the advantages are given in table 3–1.

Table 3–1—*Some advantages of plainsawed and quartersawed lumber*

Plainsawed	Quartersawed
Figure patterns resulting from the annual rings and some other types of figure are brought out more conspicuously by plainsawing.	Quartersawed lumber shrinks and swells less in width.
Round or oval knots that may occur in plainsawed boards affect the surface appearance less than spike knots that may occur in quartersawed boards. Also, a board with a round or oval knot is not as weak as a board with a spike knot.	It twists and cups less. It surface-checks and splits less in seasoning and in use. Raised grain caused by separation in the annual rings does not become so pronounced.
Shakes and pitch pockets, when present, extend through fewer boards.	It wears more evenly. Types of figure due to pronounced rays, interlocked grain, and wavy grain are brought out more conspicuously.
It is less susceptible to collapse in drying. It shrinks and swells less in thickness.	It does not allow liquids to pass into or through it so readily in some species.
It may cost less because it is easier to obtain.	It holds paint better in some species. The sapwood appearing in boards is at the edges and its width is limited according to the width of the sapwood in the log.

Decorative Features of Common Woods

The decorative value of wood depends upon its color, figure, luster, and the way in which it bleaches or takes fillers, stains, and transparent finishes. Because of the combinations of color and the multiplicity of shades found in wood, it is impossible to give detailed descriptions of colors of the various kinds. Sapwood of most species, however, is light in color, and in some species it is practically white. White sapwood of certain species, such as maple, may be preferred to the heartwood for specific uses. In some species, such as hemlock, spruce, the true firs, basswood, cottonwood, and beech, there is typically little or no difference in color between sapwood and heartwood, but in most species heartwood is darker and fairly uniform in color. Table 3–2 describes in a general way the color of heartwood of the more common kinds of woods.

In plainsawed boards and rotary-cut veneer, the annual growth rings frequently form ellipses and parabolas that make striking figures, especially when the rings are irregular in width and outline on the cut surface. On quartersawed surfaces, these rings form stripes, which are not especially ornamental unless they are irregular in width and direction. The relatively large rays, sometimes referred to as flecks, form a conspicuous figure in quartersawed oak and sycamore. With interlocked grain, which slopes in alternate directions in successive layers from the center of the tree outward, quartersawed surfaces show a ribbon effect, either because of the difference in reflection of light from successive layers when the wood has a natural luster or because cross grain of varying degree absorbs stains unevenly. Much of this type of figure is lost in plainsawed lumber.

In open-grained hardwoods, the appearance of both plainsawed and quartersawed lumber can be varied greatly by the use of fillers of different colors. In softwoods, the annual growth layers can be made to stand out more by applying a stain.

Knots, pin wormholes, bird pecks, decay in isolated pockets, birdseye, mineral streaks, swirls in grain, and ingrown bark are decorative in some species when the wood is carefully selected for a particular architectural treatment.

MOISTURE CONTENT

Moisture content of wood is defined as the weight of water in wood expressed as a fraction, usually as a percentage, of the weight of ovendry wood. Weight, shrinkage, strength, and other properties depend upon moisture content of wood.

In trees, moisture content may range from about 30 percent to more than 200 percent of the weight of wood substance. Moisture content of the sapwood portion is usually high. Heartwood moisture content may be much less than sapwood moisture content in some species, greater in others. Table 3–3 gives some moisture content values for heartwood and sapwood of some domestic species. These values are considered typical, but there is considerable variation within and between trees. Variability of moisture content exists even within individual boards cut from the same tree. Information on heartwood and sapwood moisture content is not available for imported species.

Table 3-2—Color and figure of common kinds of domestic wood

Species	Color of dry heartwood [1]	Type of figure in—	
		Plainsawed lumber or rotary-cut veneer	Quartersawed lumber or quarter-sliced veneer
HARDWOODS			
Alder, red	Pale pinkish brown	Faint growth ring	Scattered large flakes, sometimes entirely absent.
Ash:			
Black	Moderately dark grayish brown	Conspicuous growth ring; occasional burl.	Distinct, not conspicuous growth-ring stripe; occasional burl.
Oregon	Grayish brown, sometimes with reddish tinge.	do	Do.
White	do	do	Do.
Aspen	Light brown	Faint growth ring	None.
Basswood	Creamy white to creamy brown, sometimes reddish.	do	Do.
Beech, American	White with reddish tinge to reddish brown.	do	Numerous small flakes up to ⅛ inch in height.
Birch:			
Paper	Light brown	do	None.
Sweet	Dark reddish brown	Distinct, not conspicuous growth ring; occasionally wavy.	Occasionally wavy.
Yellow	Reddish brown	do	Do.
Butternut	Light chestnut brown with occasional reddish tinge or streaks.	Faint growth ring	None.
Cherry, black	Light to dark reddish brown	Faint growth ring; occasional burl	Occasional burl.
Chestnut, American	Grayish brown	Conspicuous growth ring	Distinct, not conspicuous growth-ring stripe.
Cottonwood	Grayish white to light grayish brown	Faint growth ring	None.
Elm:			
American and rock	Light grayish brown, usually with reddish tinge.	Distinct, not conspicuous with fine wavy pattern within each growth ring.	Faint growth-ring stripe.
Slippery	Dark brown with shades of red	Conspicuous growth ring with fine pattern within each growth ring.	Distinct, not conspicuous growth-ring stripe.
Hackberry	Light yellowish or greenish gray	Conspicuous growth ring	Faint growth-ring stripe.
Hickory	Reddish brown	Distinct, not conspicuous growth ring	Do.
Honeylocust	Cherry red	Conspicuous growth ring	Distinct, not conspicuous growth-ring stripe.
Locust, black	Golden brown, sometimes with tinge of green.	do	Do.
Magnolia	Light to dark yellowish brown with greenish or purplish tinge.	Faint growth ring	None.
Maple:			
Black, bigleaf, red, silver, and sugar.	Light reddish brown	Faint growth ring, occasionally birds-eye, curly, and wavy.	Occasionally curly and wavy.
Oak:			
All red oaks	Grayish brown usually with fleshy tinge.	Conspicuous growth ring	Pronounced flake; distinct, not conspicuous growth-ring stripe.
All white oaks	Grayish brown, rarely with fleshy tinge.	do	Do.
Sweetgum	Reddish brown	Faint growth ring; occasional irregular streaks.	Distinct, not pronounced ribbon; occasional streak.

Species	Color[1]	Figure (flat-sawn surface)	Figure (quarter-sawn surface)
Sycamore	Flesh brown	Faint growth ring	Numerous pronounced flakes up to ¼ inch in height.
Tupelo:			
Black and water	Pale to moderately dark brownish gray.	do	Distinct, not pronounced ribbon.
Walnut, black	Chocolate brown occasionally with darker, sometimes purplish streaks.	Distinct, not conspicuous growth ring; occasionally wavy, curly, burl, and other types.	Distinct, not conspicuous growth-ring stripe; occasionally wavy, curly, burl, crotch, and other types.
Yellow-poplar	Light to dark yellowish brown with greenish or purplish tinge.	Faint growth ring	None.

SOFTWOODS

Species	Color[1]	Figure (flat-sawn surface)	Figure (quarter-sawn surface)
Baldcypress	Light yellowish brown to reddish brown.	Conspicuous irregular growth ring	Distinct, not conspicuous growth-ring stripe.
Cedar:			
Alaska-	Yellow	Faint growth ring	None.
Atlantic white-	Light brown with reddish tinge	Distinct, not conspicuous growth ring	Do.
Eastern redcedar	Brick red to deep reddish brown	Occasionally streaks of white sapwood alternating with heartwood.	Occasionally streaks of white sapwood alternating with heartwood.
Incense-	Reddish brown	Faint growth ring	Faint growth-ring stripe.
Northern white-	Light to dark brown	do	Do.
Port-Orford-	Light yellow to pale brown	do	None.
Western redcedar	Reddish brown	Distinct, not conspicuous growth ring	Faint growth-ring stripe.
Douglas-fir	Orange red to red; sometimes yellow	Conspicuous growth ring	Distinct, not conspicuous growth-ring stripe.
Fir:			
Balsam	Nearly white	Distinct, not conspicuous growth ring	Faint growth-ring stripe.
White	Nearly white to pale reddish brown	Conspicuous growth ring	Distinct, not conspicuous growth-ring stripe.
Hemlock:			
Eastern	Light reddish brown	Distinct, not conspicuous growth ring	Faint growth-ring stripe.
Western	do	do	Do.
Larch, western	Russet to reddish brown	Conspicuous growth ring	Distinct, not conspicuous growth-ring stripe.
Pine:			
Eastern white	Cream to light reddish brown	Faint growth ring	None.
Lodgepole	Light reddish brown	Distinct, not conspicuous growth ring; faint "pocked" appearance.	None.
Ponderosa	Orange to reddish brown	Distinct not conspicuous growth ring	Faint growth-ring stripe.
Red	do	do	Do.
Southern: Longleaf, loblolly, shortleaf, and slash	do	Conspicuous growth ring	Distinct, not conspicuous growth-ring stripe.
Sugar	Light creamy brown	Faint growth ring	None.
Western white	Cream to light reddish brown	do	None.
Redwood	Cherry to deep reddish brown	Distinct, not conspicuous growth ring; occasionally wavy and burl.	Faint growth-ring stripe; occasionally wavy and burl.
Spruce:			
Black, Engelmann, red, white.	Nearly white	Faint growth ring	None.
Sitka	Light reddish brown	Distinct, not conspicuous growth ring	Faint growth-ring stripe.
Tamarack	Russet brown	Conspicuous growth ring	Distinct, not conspicuous growth-ring stripe.

[1] The sapwood of all species is light in color or virtually white unless discolored by fungus or chemical stains.

Table 3–3.—*Average moisture content of green wood, by species*

Species	Moisture content [1]		Species	Moisture content [1]	
	Heart-wood *Pct.*	Sap-wood *Pct.*		Heart-wood *Pct.*	Sap-wood *Pct.*
HARDWOODS			SOFTWOODS		
Alder, red	---	97	Baldcypress	121	171
Apple	81	74	Cedar:		
Ash:			Alaska-	32	166
Black	95	---	Eastern redcedar	33	---
Green	---	58	Incense-	40	213
White	46	44	Port Orford-	50	98
Aspen	95	113	Western redcedar	58	249
Basswood, American	81	133	Douglas-fir:		
Beech, American	55	72	Coast type	37	115
Birch:			Fir:		
Paper	89	72	Grand	91	136
Sweet	75	70	Noble	34	115
Yellow	74	72	Pacific silver	55	164
Cherry, black	58	---	White	98	160
Chestnut, American	120	---	Hemlock:		
Cottonwood, black	162	146	Eastern	97	119
Elm:			Western	85	170
American	95	92	Larch, western	54	110
Cedar	66	61	Pine:		
Rock	44	57	Loblolly	33	110
Hackberry	61	65	Lodgepole	41	120
Hickory, pecan:			Longleaf	31	106
Bitternut	80	54	Ponderosa	40	148
Water	97	62	Red	32	134
Hickory, true:			Shortleaf	32	122
Mockernut	70	52	Sugar	98	219
Pignut	71	49	Western white	62	148
Red	69	52	Redwood, old-growth	86	210
Sand	68	50	Spruce:		
Magnolia	80	104	Eastern	34	128
Maple:			Engelmann	51	173
Silver	58	97	Sitka	41	142
Sugar	65	72	Tamarack	49	---
Oak:					
California black	76	75			
Northern red	80	69			
Southern red	83	75			
Water	81	81			
White	64	78			
Willow	82	74			
Sweetgum	79	137			
Sycamore, American	114	130			
Tupelo:					
Black	87	115			
Swamp	101	108			
Water	150	116			
Walnut, black	90	73			
Yellow-poplar	83	106			

[1] Based on weight when ovendry.

Green Wood and Fiber Saturation Point

Moisture can exist in wood as water or water vapor in cell lumens (cavities) and as water "bound" chemically within cell walls. Green wood often is defined as wood in which the cell walls are completely saturated with water; however, green wood usually contains additional water in the lumens. The moisture content at which cell walls are completely saturated (all "bound" water) but no water exists in cell cavities is called the "fiber satura-tion point." The fiber saturation point of wood averages about 30 percent moisture content, but individual species and individual pieces of wood may vary by several percentage points from that value.

The fiber saturation point often also is considered as that moisture content below which the physical and mechanical properties of wood begin to change as a function of moisture content.

Equilibrium Moisture Content

The moisture content of wood below the fiber saturation point or "green" condition is a function of both relative humidity and temperature of the surrounding air. The equilibrium moisture content is defined as that moisture content at which the wood is neither gaining nor losing moisture; an equilibrium condition has been reached. The relationship between equilibrium moisture content, relative humidity, and temperature is shown in table 3–4. This table illustrates that below the fiber saturation point wood will attain a moisture content in equilibrium with widely differing atmospheric conditions. For most practical purposes the values in table 3–4 may be applied to wood of any species.

Wood in service usually is exposed to both long-term (seasonal) and short-term (such as daily) changes in the relative humidity and temperature of the surrounding air. Thus, wood virtually always is undergoing at least slight changes in moisture content. These changes usually are gradual and short-term fluctuations tend to influence only the wood surface. Moisture content changes may be retarded by protective coatings, such as varnish, lacquer, or paint. The practical objective of all wood seasoning, handling, and storing methods should be to minimize moisture content changes in wood in service. Favored procedures are those that bring the wood to a moisture content corresponding to the average atmospheric conditions to which it will be exposed (see ch. 14 and 16).

SHRINKAGE

Wood is dimensionally stable when the moisture content is above the fiber saturation point. Wood changes dimension as it gains or loses moisture below that point. It shrinks when losing moisture from the cell walls and swells when gaining moisture in the cell walls. This shrinking and swelling may result in warping, checking, splitting, or performance problems that detract from its usefulness. It is therefore important that these phenomena be understood and considered when they may affect a product in which wood is used.

Wood is an anisotropic material in shrinkage characteristics. It shrinks most in the direction of the annual growth rings (tangentially), about one-half as much across the rings (radially), and only slightly along the grain (longitudinally). The combined effects of radial and tangential shrinkage can distort the shape of wood pieces because of the difference in shrinkage and the curvature of annual rings. Figure 3–2 illustrates the major types of distortion due to these effects.

Transverse and Volumetric Shrinkage

Data have been collected to represent the average radial, tangential, and volumetric shrinkage of numerous domestic species by methods described in ASTM Designation D143. These shrinkage values, expressed as a percentage of the green dimension, are summarized in table 3–5. Shrinkage values collected from the world literature for selected imported species are summarized in table 3–6.

The shrinkage of wood is affected by a number of variables. In general, greater shrinkage is associated with greater density. The size and shape of a piece of wood may also affect shrinkage, as may the temperature and rate of drying for some species. Transverse and volumetric shrinkage variability can be expressed by a coefficient of variation of approximately 15 percent, based on a study in which 50 species were represented.

Longitudinal Shrinkage

Longitudinal shrinkage of wood (shrinkage parallel to the grain) is generally quite small. Average values for green to ovendry shrinkage are between 0.1 and 0.2 percent for most species of wood. Certain atypical types of wood, however, exhibit excessive longitudinal shrinkage, and these should be avoided in uses where longitudinal stability is important. Reaction wood, whether compression wood in softwoods or tension wood in hardwoods, tends to shrink excessively along the grain. Wood from near the center of trees (juvenile wood) of some species also shrinks excessively lengthwise. Wood with cross grain exhibits increased shrinkage along the longitudinal axis of the piece.

Reaction wood exhibiting excessive longitudinal shrinkage may occur in the same board with normal wood. The presence of this type of wood, as well as cross grain, can cause serious warping such as bow, crook, or twist, and cross breaks may develop in the zones of high shrinkage.

Table 3–4—Moisture content of wood in equilibrium with stated dry-bulb temperature and relative humidity

Temperature dry-bulb, °F.	Relative humidity, percent																			
	5	10	15	20	25	30	35	40	45	50	55	60	65	70	75	80	85	90	95	98
30	1.4	2.6	3.7	4.6	5.5	6.3	7.1	7.9	8.7	9.5	10.4	11.3	12.4	13.5	14.9	16.5	18.5	21.0	24.3	26.9
40	1.4	2.6	3.7	4.6	5.5	6.3	7.1	7.9	8.7	9.5	10.4	11.3	12.3	13.5	14.9	16.5	18.5	21.0	24.3	26.9
50	1.4	2.6	3.6	4.6	5.5	6.3	7.1	7.9	8.7	9.5	10.3	11.2	12.3	13.4	14.8	16.4	18.4	20.9	24.3	26.9
60	1.3	2.5	3.6	4.6	5.4	6.2	7.0	7.8	8.6	9.4	10.2	11.1	12.1	13.3	14.6	16.2	18.2	20.7	24.1	26.8
70	1.3	2.5	3.5	4.5	5.4	6.2	6.9	7.7	8.5	9.2	10.1	11.0	12.0	13.1	14.4	16.0	17.9	20.5	23.9	26.6
80	1.3	2.4	3.5	4.4	5.3	6.1	6.8	7.6	8.3	9.1	9.9	10.8	11.7	12.9	14.2	15.7	17.7	20.2	23.6	26.3
90	1.2	2.3	3.4	4.3	5.1	5.9	6.7	7.4	8.1	8.9	9.7	10.5	11.5	12.6	13.9	15.4	17.3	19.8	23.3	26.0
100	1.2	2.3	3.3	4.2	5.0	5.8	6.5	7.2	7.9	8.7	9.5	10.3	11.2	12.3	13.6	15.1	17.0	19.5	22.9	25.6
110	1.1	2.2	3.2	4.0	4.9	5.6	6.3	7.0	7.7	8.4	9.2	10.0	11.0	12.0	13.2	14.7	16.6	19.1	22.4	25.2
120	1.1	2.1	3.0	3.9	4.7	5.4	6.1	6.8	7.5	8.2	8.9	9.7	10.6	11.7	12.9	14.4	16.2	18.6	22.0	24.7
130	1.0	2.0	2.9	3.7	4.5	5.2	5.9	6.6	7.2	7.9	8.7	9.4	10.3	11.3	12.5	14.0	15.8	18.2	21.5	24.2
140	.9	1.9	2.8	3.6	4.3	5.0	5.7	6.3	7.0	7.7	8.4	9.1	10.0	11.0	12.1	13.6	15.3	17.7	21.0	23.7
150	.9	1.8	2.6	3.4	4.1	4.8	5.5	6.1	6.7	7.4	8.1	8.8	9.7	10.6	11.8	13.1	14.9	17.2	20.4	23.1
160	.8	1.6	2.4	3.2	3.9	4.6	5.2	5.8	6.4	7.1	7.8	8.5	9.3	10.3	11.4	12.7	14.4	16.7	19.9	22.5
170	.7	1.5	2.3	3.0	3.7	4.3	4.9	5.6	6.2	6.8	7.4	8.2	9.0	9.9	11.0	12.3	14.0	16.2	19.3	21.9
180	.7	1.4	2.1	2.8	3.5	4.1	4.7	5.3	5.9	6.5	7.1	7.8	8.6	9.5	10.5	11.8	13.5	15.7	18.7	21.3
190	.6	1.3	1.9	2.6	3.2	3.8	4.4	5.0	5.5	6.1	6.8	7.5	8.2	9.1	10.1	11.4	13.0	15.1	18.1	20.7
200	.5	1.1	1.7	2.4	3.0	3.5	4.1	4.6	5.2	5.8	6.4	7.1	7.8	8.7	9.7	10.9	12.5	14.6	17.5	20.0
210	.5	1.0	1.6	2.1	2.7	3.2	3.8	4.3	4.9	5.4	6.0	6.7	7.4	8.3	9.2	10.4	12.0	14.0	16.9	19.3

Table 3–5.—*Shrinkage values of domestic woods*

Species	Shrinkage from green to ovendry moisture content [1]			Species	Shrinkage from green to ovendry moisture content [1]		
	Radial	Tangential	Volumetric		Radial	Tangential	Volumetric
	Pct.	*Pct.*	*Pct.*		*Pct.*	*Pct.*	*Pct.*

HARDWOODS

Species	Radial	Tangential	Volumetric	Species	Radial	Tangential	Volumetric
Alder, red	4.4	7.3	12.6	Honeylocust	4.2	6.6	10.8
Ash:				Locust, black	4.6	7.2	10.2
Black	5.0	7.8	15.2	Madrone, Pacific	5.6	12.4	18.1
Blue	3.9	6.5	11.7	Magnolia:			
Green	4.6	7.1	12.5	Cucumbertree	5.2	8.8	13.6
Oregon	4.1	8.1	13.2	Southern	5.4	6.6	12.3
Pumpkin	3.7	6.3	12.0	Sweetbay	4.7	8.3	12.9
White	4.9	7.8	13.3	Maple:			
Aspen:				Bigleaf	3.7	7.1	11.6
Bigtooth	3.3	7.9	11.8	Black	4.8	9.3	14.0
Quaking	3.5	6.7	11.5	Red	4.0	8.2	12.6
Basswood,				Silver	3.0	7.2	12.0
American	6.6	9.3	15.8	Striped	3.2	8.6	12.3
Beech, American	5.5	11.9	17.2	Sugar	4.8	9.9	14.7
Birch:				Oak, red:			
Alaska paper	6.5	9.9	16.7	Black	4.4	11.1	15.1
Gray	5.2	....	14.7	Laurel	4.0	9.9	19.0
Paper	6.3	8.6	16.2	Northern red	4.0	8.6	13.7
River	4.7	9.2	13.5	Pin	4.3	9.5	14.5
Sweet	6.5	9.0	15.6	Scarlet	4.4	10.8	14.7
Yellow	7.3	9.5	16.8	Southern red	4.7	11.3	16.1
Buckeye, yellow	3.6	8.1	12.5	Water	4.4	9.8	16.1
Butternut	3.4	6.4	10.6	Willow	5.0	9.6	18.9
Cherry, black	3.7	7.1	11.5	Oak, white:			
Chestnut,				Bur	4.4	8.8	12.7
American	3.4	6.7	11.6	Chestnut	5.3	10.8	16.4
Cottonwood:				Live	6.6	9.5	14.7
Balsam poplar	3.0	7.1	10.5	Overcup	5.3	12.7	16.0
Black	3.6	8.6	12.4	Post	5.4	9.8	16.2
Eastern	3.9	9.2	13.9	Swamp			
Elm:				chestnut	5.2	10.8	16.4
American	4.2	7.2	14.6	White	5.6	10.5	16.3
Cedar	4.7	10.2	15.4	Persimmon,			
Rock	4.8	8.1	14.9	common	7.9	11.2	19.1
Slippery	4.9	8.9	13.8	Sassafras	4.0	6.2	10.3
Winged	5.3	11.6	17.7	Sweetgum	5.3	10.2	15.8
Hackberry	4.8	8.9	13.8	Sycamore,			
Hickory, Pecan	4.9	8.9	13.6	American	5.0	8.4	14.1
Hickory, True:				Tanoak	4.9	11.7	17.3
Mockernut	7.7	11.0	17.8	Tupelo:			
Pignut	7.2	11.5	17.9	Black	5.1	8.7	14.4
Shagbark	7.0	10.5	16.7	Water	4.2	7.6	12.5
Shellbark	7.6	12.6	19.2	Walnut, black	5.5	7.8	12.8
Holly, American	4.8	9.9	16.9	Willow, black	3.3	8.7	13.9
				Yellow-poplar	4.6	8.2	12.7

SOFTWOODS

Species	Radial	Tangential	Volumetric	Species	Radial	Tangential	Volumetric
Baldcypress	3.8	6.2	10.5	Fir:			
Cedar:				Balsam	2.9	6.9	11.2
Alaska-	2.8	6.0	9.2	California red	4.5	7.9	11.4
Atlantic white-	2.9	5.4	8.8	Grand	3.4	7.5	11.0
Eastern				Noble	4.3	8.3	12.4
redcedar	3.1	4.7	7.8	Pacific silver	4.4	9.2	13.0
Incense-	3.3	5.2	7.7	Subalpine	2.6	7.4	9.4
Northern				White	3.3	7.0	9.8
white-	2.2	4.9	7.2	Hemlock:			
Port-Orford-	4.6	6.9	10.1	Eastern	3.0	6.8	9.7
Western				Mountain	4.4	7.1	11.1
redcedar	2.4	5.0	6.8	Western	4.2	7.8	12.4
Douglas-fir: [2]				Larch, western	4.5	9.1	14.0
Coast	4.8	7.6	12.4	Pine:			
Interior north	3.8	6.9	10.7	Eastern white	2.1	6.1	8.2
Interior west	4.8	7.5	11.8	Jack	3.7	6.6	10.3

51

Table 3-5.—*Shrinkage values of domestic woods*—continued

Species	Shrinkage from green to ovendry moisture content [1]			Species	Shrinkage from green to ovendry moisture content [1]		
	Radial	Tangential	Volumetric		Radial	Tangential	Volumetric
	Pct.	*Pct.*	*Pct.*		*Pct.*	*Pct.*	*Pct.*
SOFTWOODS—Continued							
Pine (cont.)				Spruce:			
Loblolly	4.8	7.4	12.3	Black	4.1	6.8	11.3
Lodgepole	4.3	6.7	11.1	Engelmann	3.8	7.1	11.0
Longleaf	5.1	7.5	12.2	Red	3.8	7.8	11.8
Pitch	4.0	7.1	10.9	Sitka	4.3	7.5	11.5
Pond	5.1	7.1	11.2	Tamarack	3.7	7.4	13.6
Ponderosa	3.9	6.2	9.7				
Red	3.8	7.2	11.3				
Shortleaf	4.6	7.7	12.3				
Slash	5.4	7.6	12.1				
Sugar	2.9	5.6	7.9				
Virginia	4.2	7.2	11.9				
Western white	4.1	7.4	11.8				
Redwood:							
Old-growth	2.6	4.4	6.8				
Young-growth	2.2	4.9	7.0				

[1] Expressed as a percentage of the green dimension.
[2] Coast Douglas-fir is defined as Douglas-fir growing in the States of Oregon and Washington west of the summit of the Cascade Mountains. Interior West includes the State of California and all counties in Oregon and Washington east of but adjacent to the Cascade summit. Interior North includes the remainder of Oregon and Washington and the States of Idaho, Montana, and Wyoming.

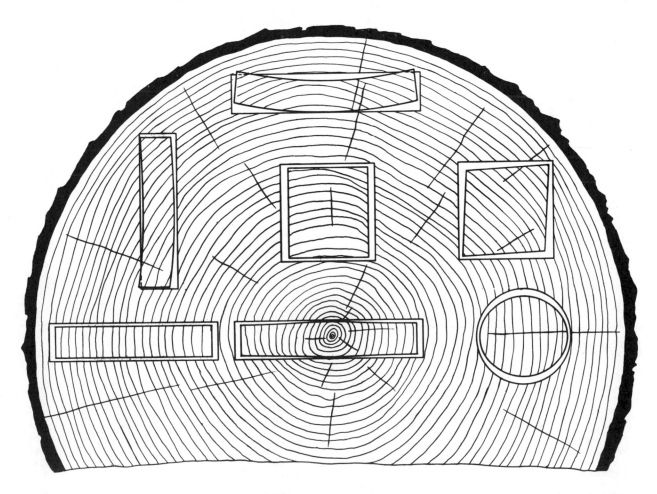

Figure 3–2.—Characteristic shrinkage and distortion of flats, squares, and rounds as affected by the direction of the annual rings. Tangential shrinkage is about twice as great as radial.

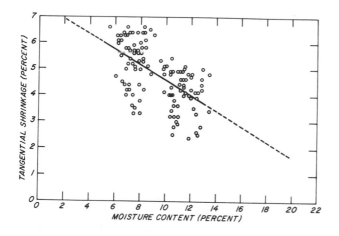

Figure 3–3.—An illustration of variation in individual tangential shrinkage values of several boards of Douglas-fir from one locality, dried from green condition.

Moisture-Shrinkage Relationship

The shrinkage of a small piece of wood normally begins at about the fiber saturation point and continues in a fairly linear manner until the wood is completely dry. However, in the normal drying of lumber or other large pieces of wood the surface of the wood dries first. When the surface gets below the fiber saturation point, it begins to shrink. Meanwhile, the interior may still be quite wet. The exact form of the moisture content-shrinkage curve depends on several variables, principally size and shape of the piece, species of wood, and drying conditions employed.

Considerable variation in shrinkage occurs for any species. Figure 3–3 is a plot of shrinkage data for boards, $\frac{7}{8}$ by $5\frac{1}{2}$ inches in cross section, of Douglas-fir from one locality when dried under mild conditions from green to near equilibrium at 65° F. and 30 percent relative humidity. The figure shows that it is impossible to predict accurately the shrinkage of an individual piece of wood; the average shrinkage of a quantity of pieces is more predictable.

If the shrinkage-moisture content relationship is not known for a particular product and drying condition, the data in tables 3–5 and 3–6 can be used to estimate shrinkage from the green condition to any moisture content:

$$S_m = S_o \left(\frac{30 - m}{30} \right)$$

where S_m is shrinkage (in percent) from the green condition to moisture content m (below 30 pct.) and S_o is total shrinkage (in percent) from table 3–5 or 3–6. If the moisture content at which shrinkage from the green condition begins is known to vary from 30 percent for a species, the shrinkage estimate can be improved by replacing 30 in the equation with the appropriate moisture content.

TABLE 3–6.—*Shrinkage for some woods imported into the United States* [1]

Species	Shrinkage from green to ovendry moisture content [2]		Species	Shrinkage from green to ovendry moisture content [2]	
	Radial	Tangential		Radial	Tangential
	Pct.	*Pct.*		*Pct.*	*Pct.*
Andiroba (*Carapa guianensis*)	4.0	7.8	Mahogany (*Swietenia macrophylla*)	3.7	5.1
Angelique (*Dicorynia guianensis*)	5.2	8.8	Nogal (*Juglans* spp.)	2.8	5.5
Apitong (*Dipterocarpus* spp.)	5.2	10.9	Obeche (*Triplochiton scleroxylon*)	3.1	5.3
Avodire (*Turraeanthus africanus*)	3.7	6.5	Okoume (*Aucoumea klaineanv*)	5.6	6.1
Balsa (*Ochroma pyramidale*)	3.0	7.6	Parana pine (*Araucaria angustifolia*)	4.0	7.9
Banak (*Virola surinamensis*)	4.6	8.8	Primavera (*Cybistax donnell-smithii*)	3.1	5.2
Cativo (*Prioria copaifera*)	2.3	5.3	Ramin (*Gonystylus* spp.)	3.9	8.7
Greenheart (*Ocotea rodiaei*)	8.2	9.0	Santa Maria (*Calophyllum brasiliense*)	5.4	7.9
Ishpingo (*Amburana acreana*)	2.7	4.4	Spanish-cedar (*Cedrela* spp.)	4.1	6.3
Khaya (*Khaya* spp.)	4.1	5.8	Teak (*Tectona grandis*)	2.2	4.0
Kokrodua (*Pericopsis elata*)	3.2	6.3	Virola (*Dialyanthera* spp.)	5.3	9.6
Lauan (*Shorea* spp.)	3.8	8.0	Walnut, European (*Juglans regia*)	4.3	6.4
Limba (*Terminalia superba*)	4.4	5.4			
Lupuna (*Ceiba samauma*)	3.5	6.3			

[1] Shrinkage values in this table were obtained from world literature and may not represent a true species average.

[2] Expressed as a percentage of the green dimension.

S_o may be an appropriate value of radial, tangential, or volumetric shrinkage. Tangential values should be used for estimating width shrinkage of flat-sawed material; radial values for quartersawed material. For mixed or unknown ring orientations, the tangential values are suggested. Individual pieces will vary from predicted shrinkage values. As noted previously, shrinkage variability is characterized by a coefficient of variation of approximately 15 percent. Chapter 14 contains further discussion of shrinkage-moisture content relations.

DENSITY–SPECIFIC GRAVITY–WEIGHT

Two primary sources of variation affect the weight of wood products. One is the density of the basic wood structure; the other is the variable moisture content. A third source, minerals and extractable substances, has a marked effect only on a limited number of species. The density of wood, exclusive of water, varies greatly both within and between species. While the density of most species falls between about 20 and 45 pounds-mass per cubic foot, the range of densities actually extends from about 10 pounds-mass per cubic foot for balsa to over 65 pounds-mass per cubic foot for some other imported woods. A coefficient of variation of about 10 percent is considered suitable for describing the variability of density within common domestic species.

Wood is used in a wide range of conditions and thus has a wide range of moisture contents in use. Since moisture makes up part of the weight of each product in use, the density must reflect this fact. This has resulted in the density of wood often being determined and reported on a moisture content-in-use condition.

The calculated density of wood, including the water contained in it, is usually based on average species characteristics. This value should always be considered an approximation because of the natural variation in anatomy, moisture content, and the ratio of heartwood to sapwood that occurs. Nevertheless, this determination of density usually is sufficiently accurate to permit proper utilization of wood products where weight is important. Such applications range from estimation of structural loads to the calculating of approximate shipping weights.

To standardize comparisons of species or products and estimations of product weights, specific gravity is used as a standard reference basis, rather than density. The traditional definition of specific gravity is the ratio of the density of the wood to the density of water at a specified reference temperature (often 4° C. where the density of water is 1.0000 g. per cc.). To reduce any confusion introduced by the variable moisture content the specific gravity of wood usually is based on the ovendry weight and the volume at some specified moisture content. A coefficient of variation of about 10 percent describes the variability inherent in many common domestic species.

In research activities specific gravity may be reported on the basis of both weight and volume ovendry. For engineering work, the basis commonly is ovendry weight and volume at the moisture content of test or use. Often the moisture content of use is taken as 12 percent. Some specific gravity data are reported in table 4–2, chapter 4, on this basis.

If the specific gravity of wood is known, based on ovendry weight and volume at a specified moisture content, the specific gravity at any other moisture content between 0 and 30 percent can be approximated from figure 3–4. This figure adjusts for average shrinkage and swelling that affects the volume of the wood. The specific gravity of wood based on ovendry weight does not change at moisture contents above approximately 30 percent (the approximate fiber saturation point). To use figure 3–4 locate the point corresponding to the known specific gravity on the vertical axis and the specified moisture content on the horizontal axis. From this point, move left or right parallel to the inclined lines until vertically above the target moisture content. Then read the new specific gravity corresponding to this point at the left-hand side of the graph.

With a knowledge of the specific gravity at the moisture content of interest, the density of wood including water at that moisture content can be read directly from table 3–7. [1] For example, to estimate the density of white ash at 12 percent moisture content, consult table 4–2 in chapter 4. The average green specific gravity for the species is 0.55. Using figure 3–4, the 0.55 green specific gravity curve is found to intersect with the vertical 12 percent moisture content line at a point corresponding

[1] Table 3–7 is repeated as A–3–7 in metric (SI) units at the end of this chapter.

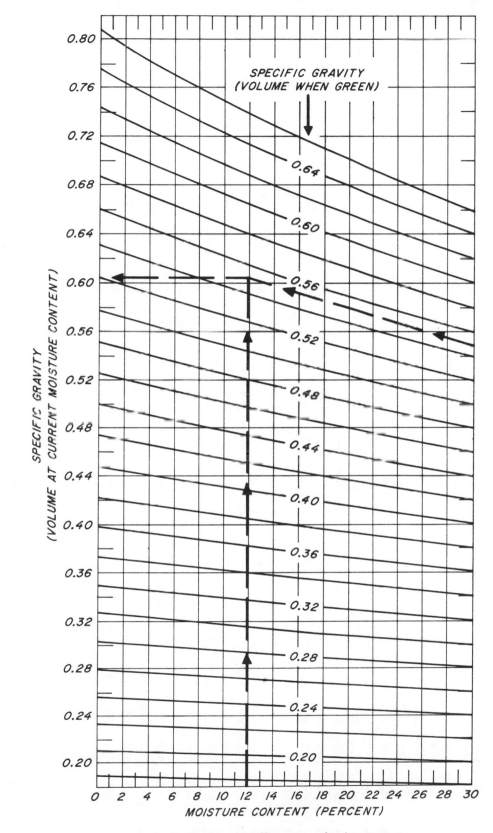

Figure 3-4.—Relation of specific gravity and moisture content.

55

Table 3–7—Density of wood as a function of specific gravity and moisture control

Moisture content of wood (%)	Density in pounds-mass per cubic foot when the specific gravity [1] is—																				
	0.30	0.32	0.34	0.36	0.38	0.40	0.42	0.44	0.46	0.48	0.50	0.52	0.54	0.56	0.58	0.60	0.62	0.64	0.66	0.68	0.70
0 --	18.7	20.0	21.2	22.5	23.7	25.0	26.2	27.5	28.7	30.0	31.2	32.4	33.7	34.9	36.2	37.4	38.7	39.9	41.2	42.4	43.7
4 --	19.5	20.8	22.1	23.4	24.7	26.0	27.2	28.6	29.8	31.2	32.4	33.7	35.0	36.3	37.6	38.9	40.2	41.5	42.8	44.1	45.4
8 --	20.2	21.6	22.9	24.3	25.6	27.0	28.3	29.6	31.0	32.3	33.7	35.0	36.4	37.7	39.1	40.4	41.8	43.1	44.5	45.8	47.2
12 --	21.0	22.4	23.8	25.2	26.6	28.0	29.4	30.8	32.2	33.5	34.9	36.3	37.7	39.1	40.5	41.9	43.3	44.7	46.1	47.5	48.9
16 --	21.7	23.2	24.6	26.0	27.5	29.0	30.4	31.8	33.3	34.7	36.2	37.6	39.1	40.5	42.0	43.4	44.9	46.3	47.8	49.2	50.7
20 --	22.5	24.0	25.5	27.0	28.4	30.0	31.4	32.9	34.4	35.9	37.4	38.9	40.4	41.9	43.4	44.9	46.4	47.9	49.4	50.9	52.4
24 --	23.2	24.8	26.3	27.8	29.4	31.0	32.5	34.0	35.6	37.1	38.7	40.2	41.8	43.3	44.9	46.4	48.0	49.5	51.1	52.6	54.2
28 --	24.0	25.6	27.2	28.8	30.4	31.9	33.5	35.1	36.7	38.3	39.9	41.5	43.1	44.7	46.3	47.9	49.5	51.1	52.7	54.3	55.9
32 --	24.7	26.4	28.0	29.7	31.3	32.9	34.6	36.2	37.9	39.5	41.2	42.8	44.5	46.1	47.8	49.4	51.1	52.7	54.4	56.0	57.7
36 --	25.5	27.2	28.9	30.6	32.2	33.9	35.6	37.3	39.0	40.7	42.4	44.1	45.8	47.5	49.2	50.9	52.6	54.3	56.0	57.7	59.4
40 --	26.2	28.0	29.7	31.4	33.2	34.9	36.7	38.4	40.2	41.9	43.7	45.4	47.2	48.9	50.7	52.4	54.2	55.9	57.7	59.4	61.2
44 --	27.0	28.8	30.6	32.3	34.1	35.9	37.7	39.5	41.3	43.1	44.9	46.7	48.5	50.3	52.1	53.9	55.7	57.5	59.3	61.1	62.9
48 --	27.7	29.6	31.4	33.2	35.1	36.9	38.8	40.6	42.5	44.3	46.2	48.0	49.9	51.7	53.6	55.4	57.3	59.1	61.0	62.8	64.6
52 --	28.5	30.4	32.2	34.1	36.0	37.9	39.8	41.7	43.6	45.5	47.4	49.3	51.2	53.1	55.0	56.9	58.8	60.7	62.6	64.5	66.4
56 --	29.2	31.2	33.1	35.0	37.0	38.9	40.9	42.8	44.8	46.7	48.7	50.6	52.6	54.5	56.5	58.4	60.4	62.3	64.2	66.2	68.1
60 --	30.0	31.9	33.9	35.9	37.9	39.9	41.9	43.9	45.9	47.9	49.9	51.9	53.9	55.9	57.9	59.9	61.9	63.9	65.9	67.9	69.9
64 --	30.7	32.7	34.8	36.8	38.9	40.9	43.0	45.0	47.1	49.1	51.2	53.2	55.3	57.3	59.4	61.4	63.4	65.5	67.5	69.6	71.6
68 --	31.4	33.5	35.6	37.7	39.8	41.9	44.0	46.1	48.2	50.3	52.4	54.5	56.6	58.7	60.8	62.9	65.0	67.1	69.2	71.3	73.4
72 --	32.2	34.3	36.5	38.6	40.8	42.9	45.1	47.2	49.4	51.5	53.7	55.8	58.0	60.1	62.3	64.4	66.5	68.7	70.8	73.0	75.1
76 --	32.9	35.1	37.3	39.5	41.7	43.9	46.1	48.3	50.5	52.7	54.9	57.1	59.3	61.5	63.7	65.9	68.1	70.3	72.5	74.7	76.9
80 --	33.7	35.9	38.2	40.4	42.7	44.9	47.2	49.4	51.7	53.9	56.2	58.4	60.7	62.9	65.1	67.4	69.6	71.9	74.1	76.4	78.6
84 --	34.4	36.7	39.0	41.3	43.6	45.9	48.2	50.5	52.8	55.1	57.4	59.7	62.0	64.3	66.6	68.9	71.2	73.5	75.8	78.1	80.4
88 --	35.2	37.5	39.9	42.2	44.6	46.9	49.3	51.6	54.0	56.3	58.7	61.0	63.3	65.7	68.0	70.4	72.7	75.1	77.4	79.8	82.1
92 --	35.9	38.3	40.7	43.1	45.5	47.9	50.3	52.7	55.1	57.5	59.9	62.3	64.7	67.1	69.5	71.9	74.3	76.7	79.1	81.5	83.9
96 --	36.7	39.1	41.6	44.0	46.5	48.9	51.4	53.8	56.3	58.7	61.2	63.6	66.0	68.5	70.9	73.4	75.8	78.3	80.7	83.2	85.6
100 --	37.4	39.9	42.4	44.9	47.4	49.9	52.4	54.9	57.4	59.9	62.4	64.9	67.4	69.9	72.4	74.9	77.4	79.9	82.4	84.9	87.4
110 --	39.3	41.9	44.6	47.2	49.8	52.4	55.0	57.7	60.3	62.9	65.5	68.1	70.8	73.4	76.0	78.6	81.2	83.9	86.5	89.1	91.7
120 --	41.2	43.9	46.7	49.4	52.2	54.9	57.7	60.4	63.1	65.9	68.6	71.4	74.1	76.9	79.6	82.4	85.1	87.9	90.6	93.4	96.1
130 --	43.1	45.9	48.8	51.7	54.5	57.4	60.3	63.1	66.0	68.9	71.8	74.6	77.5	80.4	83.2	86.1	89.0	91.9	94.7	97.6	100.5
140 --	44.9	47.9	50.9	53.9	56.9	59.9	62.9	65.9	68.9	71.9	74.9	77.9	80.9	83.9	86.9	89.9	92.9	95.8	98.8	101.8	104.8
150 --	46.8	49.9	53.0	56.2	59.3	62.4	65.5	68.6	71.8	74.9	78.0	81.1	84.2	87.4	90.5	93.6	96.7	99.8	103.0	106.1	109.2

[1] Based on mass when ovendry and volume at tabulated moisture content.

to a specific gravity of 0.605 based on ovendry weight and volume at 12 percent moisture content (see dashed lines in fig. 3–4). Table 3–7 then can be used to convert the specific gravity of 0.605 to a density of 42 pounds-mass per cubic foot.

WORKING QUALITIES

The ease of working wood with hand tools generally varies directly with the specific gravity of the wood. The lower the specific gravity, the easier it is to cut the wood with a sharp tool. Tables 4–2 and 4–3 list the specific gravity values for various native and imported species. These specific gravity figures can be used as a general guide to the ease of working with hand tools.

A wood species that is easy to cut does not necessarily develop a smooth surface when it is machined. Consequently, tests have been made with many United States hardwoods to evaluate them for machining properties. Results of these evaluations are given in table 3–8.

Machining evaluations are not available for many imported woods. However, it is known that three major factors other than density may affect production of smooth surfaces during wood machining. These factors are: Interlocked and variable grain; hard mineral deposits; and reaction wood, particularly tension wood in hardwoods. Interlocked grain is characteristic of the majority of tropical species and presents difficulty in planing quartered surfaces unless attention is paid to feed rate, cutting angles, and sharpness of knives. Hard deposits such as calcium carbonate and silica may have a pronounced dulling effect on all cutting edges. This dulling effect becomes more pronounced as the wood is dried to the usual inservice requirements. Tension wood may cause fibrous and fuzzy surfaces. It can be very troublesome in species of lower density. Reaction wood may also be responsible for the pinching effect on saws due to stress relief. The pinching may result in burning and dulling of the sawteeth. Table 3–9 lists some of the imported species that have irregular grain, hard deposits, or tension wood.

Table 3–8.—*Some machining and related properties of selected domestic hardwoods*

Kind of wood [1]	Planing—perfect pieces	Shaping—good to excellent pieces	Turning—fair to excellent pieces	Boring—good to excellent pieces	Mortising—fair to excellent pieces	Sanding—good to excellent pieces	Steam bending—unbroken pieces	Nail splitting—pieces free from complete splits	Screw splitting—pieces free from complete splits
	Pct.	Pct.	Pct.	Pct.	Pct.	Pct.	Pct.	Pct.	Pct.
Alder, red	61	20	88	64	52				
Ash	75	55	79	94	58	75	67	65	71
Aspen	26	7	65	78	60				
Basswood	64	10	68	76	51	17	2	79	68
Beech	83	24	90	99	92	49	75	42	58
Birch	63	57	80	97	97	34	72	32	48
Birch, paper	47	22							
Cherry, black	80	80	88	100	100				
Chestnut	74	28	87	91	70	64	56	66	60
Cottonwood	21	3	70	70	52	19	44	82	78
Elm, soft	33	13	65	94	75	66	74	80	74
Hackberry	74	10	77	99	72		94	63	63
Hickory	76	20	84	100	98	80	76	35	63
Magnolia	65	27	79	71	32	37	85	73	76
Maple, bigleaf	52	56	80	100	80				
Maple, hard	54	72	82	99	95	38	57	27	52
Maple, soft	41	25	76	80	34	37	59	58	61
Oak, red	91	28	84	99	95	81	86	66	78
Oak, white	87	35	85	95	99	83	91	69	74
Pecan	88	40	89	100	98		78	47	69
Sweetgum	51	28	86	92	58	23	67	69	69
Sycamore	22	12	85	98	96	21	29	79	74
Tanoak	80	39	81	100	100				
Tupelo, water	55	52	79	62	33	34	46	64	63
Tupelo, black	48	32	75	82	24	21	42	65	63
Walnut, black	62	34	91	100	98		78	50	59
Willow	52	5	58	71	24	24	73	89	62
Yellow-poplar	70	13	81	87	63	19	58	77	67

[1] Commercial lumber nomenclature.

Table 3–9.—*Some characteristics of imported woods that may effect machining.*

Irregular and interlocked grain	Hard mineral deposits (silica or calcium carbonate)	Reaction wood (tension wood)
Apamate	Angelique	Andiroba
Apitong	Apitong	Banak
Avodire	Kapur	Cativo
Capirona	Okoume	Khaya
Courbaril	Palosapis	Lupuna
Gola	Rosewood,	Mahogany
Goncalo alves	Indian	Nogal
Ishpingo	Teak	Sande
Jarrah		Spanish-cedar
Kapur		
Karri		
Khaya		
Kokrodua		
Lapacho		
Laurel		
Lignum vitae		
Limba		
Meranti		
Obeche		
Okoume		
Palosapis		
Peroba de campos		
Primavera		
Rosewood, Indian		
Santa Maria		
Sapele		

WEATHERING

Without protective treatment, freshly cut wood exposed to the weather changes materially in color. Other changes due to weathering include warping, loss of some surface fibers, and surface roughening and checking. The effects of weathering on wood may be desirable or undesirable, depending on the requirements for the particular wood product. The time required to reach the fully weathered appearance depends on the severity of the exposure to sun and rain. Once weathered, wood remains nearly unaltered in appearance.

The color of wood is affected very soon on exposure to weather. With continued exposure all woods turn gray; however, only the wood at or near the exposed surfaces is noticeably affected. This very thin gray layer is composed chiefly of partially degraded cellulose fibers and micro-organisms. Further weathering causes fibers to be lost from the surface but the process is so slow that only about ¼ inch is lost in a century.

In the weathering process, chemical degradation is influenced greatly by the wavelength of light. The most severe effects are produced by exposure to ultraviolet light. As cycles of wetting and drying take place, most woods develop physical changes such as checks

or cracks that are easily visible. Moderate to low density woods acquire fewer checks than do high density woods. Vertical-grain boards check less than flat-grain boards.

As a result of weathering, boards tend to warp (particularly cup) and pull out their fastenings. The cupping tendency varies with the density, width, and thickness of a board. The greater the density and the greater the width in proportion to the thickness, the greater is the tendency to cup. Warping also is more pronounced in flat-grain boards than in vertical-grain boards. For best cup resistance, the width of a board should not exceed eight times its thickness.

Biological attack of a wood surface by micro-organisms is recognized as a contributing factor to color changes. When weathered wood has an unsightly dark gray and blotchy appearance, it is due to dark-colored fungal spores and mycelium on the wood surface. The formation of a clean, light gray, silvery sheen on weathered wood occurs most frequently where micro-organism growth is inhibited by a hot, arid climate or a salt atmosphere in coastal regions.

The contact of fasteners and other metallic products with the weathering wood surface is a source of color, often undesirable if a natural color is desired. Chapter 16 discusses this effect in more detail.

Details of treatments to preserve the natural color, retard biological attack, or impart additional color to the wood, are covered in chapters 16, 17, and 18.

DECAY RESISTANCE

Wood kept constantly dry does not decay. Further, if it is kept continuously submerged in water even for long periods of time, it is not decayed significantly by the common decay fungi regardless of the wood species or the presence of sapwood. Bacteria and certain soft-rot fungi can attack submerged wood but the resulting deterioration is very slow. A large proportion of wood in use is kept so dry at all times that it lasts indefinitely. Moisture and temperature, which vary greatly with local conditions, are the principal factors affecting rate of decay. When exposed to conditions that favor decay, wood deteriorates more rapidly in warm, humid areas than in cool or dry areas. High altitudes, as a rule, are less favorable to decay than low altitudes because the average temperatures are lower and the growing seasons for fungi, which cause decay, are shorter.

The heartwoods of some common native species of wood have varying degrees of natural decay resistance. Untreated sapwood of substantially all species has low resistance to decay and usually has a short service life under decay-producing conditions. The decay resistance of heartwood is greatly affected by differences in the preservative qualities of the wood extractives, the attacking fungus, and the conditions of exposure. Considerable difference in service life may be obtained from pieces of wood cut from the same species, or even from the same tree, and used under apparently similar conditions. There are further complications because, in a few species, such as the spruces and the true firs (not Douglas-fir), heartwood and sapwood are so similar in color that they cannot be easily distinguished. Marketable sizes of some species such as southern pine and baldcypress are becoming largely second growth and contain a high percentage of sapwood.

Table 3–10.—*Grouping of some domestic woods according to heartwood decay*

Resistant or very resistant	Moderately resistant	Slightly or nonresistant
Baldcypress (old growth)[1]	Baldcypress (young growth)[1]	Alder
Catalpa	Douglas-fir	Ashes
Cedars	Honeylocust	Aspens
Cherry, black	Larch, western	Basswood
Chestnut	Oak, swamp chestnut	Beech
Cypress, Arizona	Pine, eastern white[1]	Birches
Junipers	Southern pine:	Buckeye
Locust, black[2]	Longleaf[1]	Butternut
Mesquite	Slash[1]	Cottonwood
Mulberry, red[2]	Tamarack	Elms
Oak:		Hackberry
Bur		Hemlocks
Chestnut		Hickories
Gambel		Magnolia
Oregon white		Maples
Post		Oak (red and black species)
White		Pines (other than longleaf, slash, and eastern white)
Osage orange[2]		Poplars
Redwood		Spruces
Sassafras		Sweetgum
Walnut, black		True firs (western and eastern)
Yew, Pacific[2]		Willows
		Yellow-poplar

[1] The southern and eastern pines and baldcypress are now largely second growth with a large proportion of sapwood. Consequently, substantial quantities of heartwood lumber of these species are not available.
[2] These woods have exceptionally high decay resistance.

Precise ratings of decay resistance of heartwood of different species are not possible because of differences within species and the variety of service conditions to which wood is exposed. However, broad groupings of many of the native species, based on service records, laboratory tests, and general experience, are helpful in choosing heartwood for use under conditions favorable to decay. Table 3–10 shows such groupings for some domestic woods, according to their average heartwood decay resistance, and table 3–11 gives similar groupings for some imported woods. The extent of variations in decay resistance of individual trees or wood samples of a species is much greater for most of the more resistant species than for the slightly or nonresistant species.

Where decay hazards exist, heartwood of species in the resistant or very resistant category generally gives satisfactory service, but heartwood of species in the other two categories will usually require some form of preservative treatment. For mild decay conditions, a simple preservative treatment—such as a short soak in preservative after all cutting and boring operations are complete—will be adequate for wood low in decay resistance. For more severe decay hazards, pressure treatments are often required; even the very decay-resistant species may require preservative treatment for important structural or other uses where failure would endanger life or require expensive repairs. Preservative treatments and methods are discussed in chapter 18.

CHEMICAL RESISTANCE

Wood is highly resistant to many chemicals. In the chemical processing industry, it is the preferred material for numerous applications, such as various types of tanks and other containers, and for structures adjacent to or housing chemical equipment. Wood is widely used in cooling towers where the hot water to be cooled contains boiler conditioning chemicals as well as dissolved chlorine for algae suppression. It is also used in the fabrication of buildings for bulk chemical storage where the wood may be in direct contact with chemicals.

Wood owes its extensive use in chemical processing operations largely to its superiority over cast iron and ordinary steel in resistance to mild acids and solutions of acidic salts. While iron is superior to untreated wood in resistance to alkaline solutions, wood may be

Table 3–11.—*Grouping of some woods imported into the United States according to approximate relative heartwood decay resistance*

Resistant or very resistant	Moderately resistant	Slightly or nonresistant
Angelique	Andiroba [1]	Balsa
Apamate	Apitong [1]	Banak
Brazilian	Avodire	Cativo
rosewood	Capirona	Ceiba
Caribbean pine	European	Jelutong
Courbaril	walnut	Limba
Encino	Gola	Lupuna
Goncalo alves	Khaya	Mahogany,
Greenheart	Laurel	Philippine:
Guijo	Mahogany,	Mayapis
Iroko	Philippine:	White lauan
Jarrah	Almon	Obeche
Kapur	Bagtikan	Parana pine
Karri	Red lauan	Ramin
Kokrodua	Tanguile	Sande
(Afrormosia)	Ocote pine	Virola
Lapacho	Palosapis	
Lignum vitae	Sapele	
Mahogany,		
American		
Meranti [1]		
Peroba de campos		
Primavera		
Santa Maria		
Spanish-cedar		
Teak		

[1] More than 1 species included, some of which may vary in resistance from that indicated.

treated to greatly enhance its durability in this respect.

In general, heartwood is more resistant to chemical attack than sapwood, basically because heartwood is more resistant to penetration by liquids. The heartwoods of cypress, southern pine, Douglas-fir, and redwood are preferred for water tanks. Heartwoods of the first three of these species are preferred where resistance to chemical attack is an important factor. These four species combine moderate to high resistance to water penetration with moderate to high resistance to chemical attack and decay.

Chemical solutions may affect wood by two general types of action. The first is an almost completely reversible effect involving swelling of the wood structure. The second type of action is irreversible and involves permanent changes in the wood structure due to alteration of one or more of its chemical constituents.

In the first type, liquids such as water, alcohols, and some other organic liquids swell the wood with no degradation of the wood structure. Removal of the swelling liquid allows the wood to return to its original condition. Petroleum oils and creosote do not swell wood.

The second type of action causes permanent changes due to hydrolysis of cellulose and hemicelluloses by acids or acidic salts, oxidation of wood substance by oxidizing agents, or delignification and solution of hemicelluloses by alkalies or alkaline salt solutions. Experience and available data indicate species and conditions where wood is equal or superior to other materials in resisting degradative actions of chemicals. In general, heartwood of such species as cypress, Douglas-fir, southern pine, redwood, maple, and white oak is quite resistant to attack by dilute mineral and organic acids. Oxidizing acids, such as nitric acid, have a greater degradative action than nonoxidizing acids. Alkaline solutions are more destructive than acidic solutions, and hardwoods are more susceptible to attack by both acids and alkalies than softwoods.

Highly acidic salts tend to hydrolyze wood when present in high concentrations. Even relatively low concentrations of such salts have shown signs that the salt may migrate to the surface of railroad ties that are occasionally wet and dried in a hot, arid region. This migration, combined with the high concentrations of salt relative to the small amount of water present, causes an acidic condition sufficient to make wood brittle.

Iron salts, which develop at points of contact with tie plates, bolts, and the like, have a degradative action on wood, especially in the presence of moisture. In addition, iron salts probably precipitate toxic extractives and thus lower the natural decay resistance of wood. The softening and discoloration of wood around corroded iron fastenings is a commonly observed phenomenon; it is especially pronounced in acidic woods, such as oak, and in woods such as redwood which contain considerable tannin and related compounds. The oxide layer formed on iron is transformed through reaction with wood acids into soluble iron salts which not only degrade the surrounding wood but probably catalyze the further corrosion of the metal. The action is accelerated by moisture; oxygen may also play an important role in the process. This effect is not encountered with well-dried wood used in dry locations. Under damp use conditions, it can be avoided or minimized by using corrosion-resistant fastenings.

Many substances have been employed as impregnants to enhance the natural resistance of wood to chemical degradation. One of the more economical treatments involves pressure impregnation with a viscous coke-oven coal tar to retard liquid penetration. Acid resistance of wood is increased by impregnation with phenolic resin solutions followed by appropriate drying and curing. Treatment with furfuryl alcohol has been used to increase resistance to alkaline solutions. A newer development involves massive impregnation with a monomeric resin, such as methyl methacrylate, followed by polymerization. Chapters 16, 17, and 18 discuss coatings and finishes, other chemical treatments, and preservation.

THERMAL PROPERTIES

Four important thermal properties of wood are (1) thermal conductivity, (2) specific heat, (3) thermal diffusivity, and (4) coefficient of thermal expansion. Thermal conductivity is a measure of the rate of heat flow through materials subjected to a temperature gradient. Specific heat of a material is the ratio of the heat capacity of the material to the heat capacity of water; the heat capacity of a material is the thermal energy required to produce one unit change of temperature in one unit mass. Thermal diffusivity is a measure of how quickly a material can absorb heat from its surroundings; it is the ratio of thermal conductivity to the product of density and specific heat. The coefficient of thermal expansion is a measure of the change of dimension caused by temperature change.

Thermal Conductivity

The thermal conductivity of common structural woods is a small fraction of the conductivity of metals with which it often is mated in construction. It is about two to four times that of common insulating material. For example, structural softwood lumber has a conductivity of about 0.8 British thermal units per inch per hour per square foot per degree Fahrenheit (Btu · in/hr · deg F) compared with 1390 for aluminum, 320 for steel, 8 for concrete, 5 for glass, 3 for plaster, and 0.3 for mineral wool.

The thermal conductivity of wood is affected by a number of basic factors: (1) density, (2) moisture content, (3) extractive content, (4) grain direction, and (5) structural irregularities such as checks and knots. It is nearly the same in the radial and tangential direction with respect to the growth rings but is 2.0 to 2.8 times greater parallel to the grain than in either the radial or tangential direc-

tions. It increases as the density, moisture content, or extractive content of the wood increases.

Figure 3–5 shows the average thermal conductivity perpendicular to the grain as related to wood density and moisture content up to approximately 40 percent moisture content. This chart is a plot of the empirical equation:

$$k = S(1.39 + 0.028M) + 0.165$$

where k is thermal conductivity in Btu · in./hr · deg F, S is specific gravity based on volume at current moisture content and weight when ovendry, and M is the moisture content in percent of dry weight. For wood at a moisture content of 40 percent or greater, the following equation has been applied:

$$k = S(1.39 + 0.038M) + 0.165$$

The equations presented were derived by averaging the results of studies on a variety of species. Individual wood specimen conductivity will vary from these predicted values because of the five variability sources noted above.

Specific Heat

The specific heat of wood depends on the temperature and moisture content of the wood but is practically independent of density or species. Specific heat of dry wood is approximately related to temperature t, in °F. by

$$\text{Specific heat} = 0.25 + 0.0006t$$

When wood contains water, the specific heat is increased because the specific heat of water is larger than that of dry wood. The apparent specific heat of moist wood, however, is larger than would be expected from a simple sum of the separate effects of wood and water. The additional apparent specific heat is due to thermal energy absorbed by the wood-water bonds. As the temperature increases the apparent specific heat increases because the energy of absorption of wood increases with temperature.

If the specific heat of water is considered to be unity, the specific heat of moist wood is given by:

$$\text{Specific heat} = \frac{M + c_o}{1 + M} + A$$

where M is the fractional moisture content of the wood, c_o is the specific heat of dry wood,

61

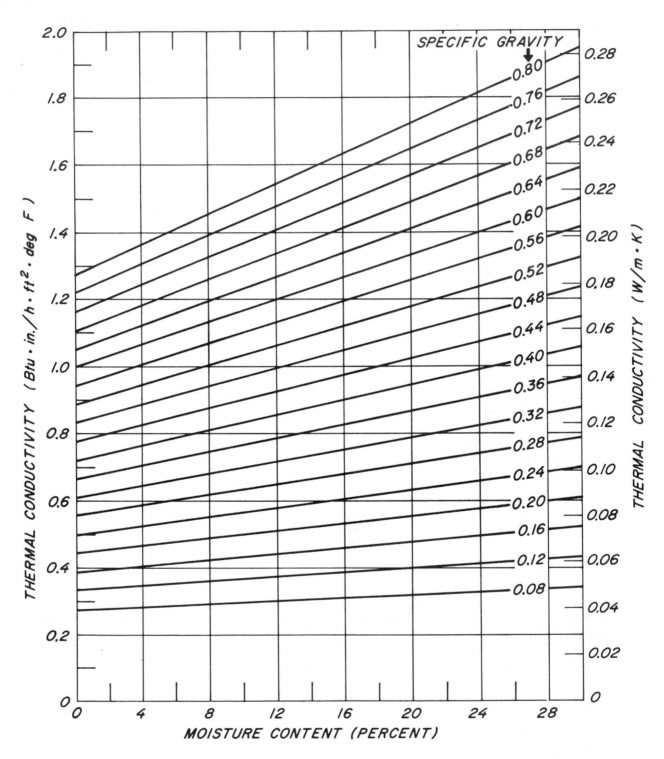

Figure 3–5.—Computed thermal conductivity of wood perpendicular to grain as related to moisture content and specific gravity.

and A is the additional specific heat due to the wood-water bond energy. A increases with increasing temperature. For wood at 10 percent moisture content, A ranges from about 0.02 at 85° F. to about 0.04 at 140° F. A ranges from about 0.04 at 85° F. to about 0.09 at 140° F. for wood at about 30 percent moisture content.

Thermal Diffusivity

Because of the small thermal conductivity and moderate density and specific heat of wood, the thermal diffusivity of wood is much smaller than that of other structural materials such as metals, brick, and stone. A typical value for wood is 0.00025 inch² per second compared to 0.02 inch² per second for steel, and 0.001 inch² per second for mineral wool. For this reason wood does not feel extremely hot or cold to the touch as do some other materials.

Few investigators have measured the diffusivity of wood directly. Since diffusivity is defined as the ratio of conductivity to the product of specific heat and density, conclusions regarding its variation with temperature and density often are based on calculating the effect of these variables on specific heat and conductivity.

All investigations illustrate that diffusivity is influenced slightly by both specific gravity and moisture content in an inverse fashion. The diffusivity increases approximately 0.0001 inch² per second over a decreasing specific gravity range of 0.65 to 0.30. Calculations suggest the effect of moisture is to increase diffusivity by about 0.00004 inch² per second as the moisture content is reduced from 12 to 0 percent.

Coefficient of Thermal Expansion

The thermal expansion coefficients of completely dry wood are positive in all directions— that is, wood expands on heating and contracts on cooling. Only limited research has been carried out to explore the influence of wood property variability on thermal expansion. The linear expansion coefficient of ovendry wood parallel to the grain appears to be independent of specific gravity and species. In tests of both hardwoods and softwoods, the parallel-to-the-grain values have ranged from about 0.0000017 to 0.0000025 per degree Fahrenheit.

The linear expansion coefficients across the grain (radial and tangential) are proportional to wood density. These coefficients range from about five to over ten times greater than the parallel-to-the-grain coefficients and thus are of more practical interest. The radial and tangential thermal expansion coefficients for ovendry wood, α_r and α_t, can be approximated by the following equations, over an ovendry specific gravity range of about 0.1 to 0.8:

$$\alpha_r = [(32)(\text{specific gravity}) + 9.9][10^{-6}] \text{ per °F.}$$

$$\alpha_t = [(33)(\text{specific gravity}) + 18.4][10^{-6}] \text{ per °F.}$$

Thermal expansion coefficients can be considered independent of temperature over the temperature range of $-60°$ to $+130°$ F.

Wood that contains moisture reacts to varying temperature differently than does dry wood. When moist wood is heated, it tends to expand because of normal thermal expansion and to shrink because of loss in moisture content. Unless the wood is very dry initially (perhaps 3 or 4 pct. M.C. or less), the shrinkage due to moisture loss on heating will be greater than the thermal expansion, so the net dimensional change on heating will be negative. Wood at intermediate moisture levels (about 8 to 20 pct.) will expand when first heated, then gradually shrink to a volume smaller than the initial volume, as the wood gradually loses water while in the heated condition.

Even in the longitudinal (grain) direction, where dimensional change due to moisture change is very small, such changes will still predominate over corresponding dimensional changes due to thermal expansion unless the wood is very dry initially. For wood at usual moisture levels, net dimensional changes will generally be negative after prolonged heating.

ELECTRICAL PROPERTIES

The most important electrical properties of wood are (1) conductivity, (2) dielectric constant, and (3) dielectric power factor.

The conductivity of a material determines the current that will flow when the material is placed under a given voltage gradient. The dielectric constant of a nonconducting material determines the amount of electric potential energy, in the form of induced polarization, that is stored in a given volume of the material when that material is placed in an electric field. The power factor of a nonconducting material determines the fraction of stored energy that is dissipated as heat when the material experiences a complete polarize-depolarize cycle.

Examples of industrial wood processes and applications in which electrical properties of wood are important include crossarms and poles for high-voltage powerlines, linemen's tools, and the heat-curing of adhesives in wood products by high-frequency electric fields. Moisture meters for wood utilize the relation between electrical properties and moisture content to estimate the moisture content.

Electrical Conductivity

The electrical conductivity of wood varies slightly with applied voltage and approximately doubles for each temperature increase of 10° C. The electrical conductivity of wood or its reciprocal, resistivity, varies greatly with moisture content, especially below the fiber saturation point. As the moisture content of wood increases from near zero to fiber saturation, the electrical conductivity increases (resistivity decreases) by 10^{10} to 10^{13} times. The resistivity is about 10^{14} to 10^{16} ohm-meters for ovendry wood and 10^3 to 10^4 ohm-meters for wood at fiber saturation. As the moisture content increases from fiber saturation to complete saturation of the wood structure, the further increase in conductivity is smaller and erratic, generally amounting to less than a hundredfold.

Figure 3–6 illustrates the change in resistance along the grain with moisture content, based on tests of many domestic species. Variability between test specimens is illustrated by the shaded area. Ninety percent of the experimental data points fall within this area.

The resistance values were obtained using a standard moisture meter electrode at 80° F. Conductivity is greater along the grain than across the grain and slightly greater in the radial direction than in the tangential direction. Relative conductivities in the longitudinal, radial, and tangential directions are in the approximate ratio of 1.0:0.55:0.50.

When wood contains abnormal quantities of water-soluble salts or other electrolytic substances, such as from preservative or fire-retardant treatment or prolonged contact with seawater, the electrical conductivity may be substantially increased. The increase is small when the moisture content of the wood is less than about 8 percent but becomes large rapidly as the moisture content exceeds 10 or 12 percent.

Dielectric Constant

The dielectric constant is the ratio of the dielectric permittivity of the material to that of free space; it is essentially a measure of the potential energy per unit volume stored in the material in the form of electric polarization when the material is in a given electric field. As measured by practical tests, the dielectric constant of a material is the ratio of the capacitance of a capacitor using the material as the dielectric, to the capacitance of the same capacitor using free space as the dielectric.

The dielectric constant of ovendry wood ranges from about 2 to 5 at room temperature, and decreases slowly but steadily with increasing frequency of the applied electric field. It increases as either temperature or moisture content increase, with a moderate positive interaction between temperature and moisture. There is an intense negative interaction between moisture and frequency: At 20 Hz the dielectric constant may range from 4 for dry wood to near 1,000,000 for wet wood; at 1 KHz, from 4 when dry to 5,000 wet; and at 1 MHz from 3 when dry to wet. The dielectric constant is larger for polarization parallel to the grain than for across the grain.

Dielectric Power Factor

When a nonconductor is placed in an electric field, it absorbs and stores potential energy. The amount of energy stored per unit volume depends upon the dielectric constant and the magnitude of the applied field. An ideal dielectric releases all of this energy to

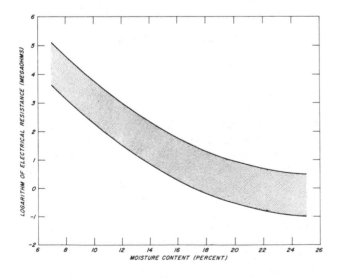

Figure 3–6.—Change in electrical resistance of wood with varying moisture content levels for many United States species. Ninety percent of test values are represented by the shaded area.

the external electric circuit when the field is removed, but practical dielectrics dissipate some of the energy as heat. The power factor is a measure of that portion of the stored energy converted to heat. Power factor values always fall between zero and unity. When the power factor does not exceed about 0.1, the fraction of the stored energy that is lost in one charge-discharge cycle is approximately equal to 2π times the power factor of the dielectric; for larger power factors, this fraction is approximated simply by the power factor itself.

The power factor of wood is large compared to inert plastic insulating materials; some materials, for example some formulations of rubber, have equally large power factors. The power factor of wood varies from about 0.01 for dry low-density woods to as large as 0.95 for dense woods at high moisture levels. It is usually, but not always, greater along the grain than across the grain.

The power factor of wood is affected by several factors, including frequency, moisture content, and temperature. These factors interact in complex ways to cause the power factor to have maximum and minimum values at various combinations of these factors.

COEFFICIENT OF FRICTION

The coefficient of friction depends on the moisture content of the wood and surface roughness. It varies little with species except for those species that contain abundant oily or waxy extractives, such as lignum vitae.

Coefficients of static friction for wood on unpolished steel have been reported to be approximately 0.70 for dry wood and 0.40 for green wood. Corresponding values for lignum vitae on unpolished steel are 0.20 and 0.34. Coefficients of static friction for smooth wood on smooth wood are 0.60 for dry wood and 0.83 for green wood.

Coefficients of sliding friction differ from those for static friction, and depend on the rate of relative movement between the rubbing parts. Coefficients for wood on steel of 0.70 for dry wood and 0.15 for green wood have been obtained at a relative movement of 4 meters per second.

NUCLEAR RADIATION

Radiation passing through matter is reduced in intensity according to the relationship

$$I = I_o\epsilon^{-\mu x}$$

where I is the reduced intensity of the beam at a depth of x in the material, I_o is the incident intensity of a beam of radiation, and μ, the linear absorption coefficient of the material, is the fraction of energy removed from the beam per unit depth traversed. Where the density is a factor of interest in energy absorption, the linear absorption coefficient is divided by the density of the material to derive the mass absorption coefficient. The absorption coefficient of a material varies with type and energy of radiation.

The linear absorption coefficient for gamma (γ) radiation of wood is known to vary directly with moisture content and density; inversely with the γ ray energy. As an example, the radiation of ovendry yellow-poplar with 0.047 MEV γ rays yielded linear absorption coefficients ranging from about 0.065 to about 0.11 per centimeter over the ovendry specific gravity range of about 0.33 to 0.62. An increase in the linear absorption coefficient of about 0.01 per centimeter occurs with an increase in moisture content from ovendry to fiber saturation. Absorption of γ rays in wood is of practical interest, in part, for measurement of the density of wood.

The interaction of wood with beta (β) radiation is similar in character to that with γ radiation, except that the absorption coefficients are larger. The linear absorption coefficient of wood with a specific gravity of 0.5 for a 0.5 MEV β ray is about 3.0 per centimeter. The result of the larger coefficient is that even very thin wood products are virtually opaque to β rays.

The interaction of neutrons with wood is of interest because wood and the water it contains are compounds of hydrogen, and hydrogen has a relatively large probability of interaction with neutrons. High energy level neutrons lose energy much more quickly through interaction with hydrogen than with other elements found in wood. The lower energy neutrons that result from this interaction thus are a measure of the hydrogen density of the specimen. Measurement of the lower energy level neutrons can be related to the moisture content of the wood.

When neutrons interact with wood, an additional result is the production of radioactive isotopes of the elements present in the wood. The radioisotopes produced can be identified by the type, energy, and half-life of their emissions, and the specific activity of each indicates amount of isotope present. This procedure, called neutron activation, provides a sensitive, nondestructive method of analysis for trace elements.

In the discussions above, moderate radiation levels that leave the wood physically unchanged have been assumed. Very large doses of γ rays or neutrons can cause substantial degradation of wood. The effect of large radiation doses on the mechanical properties of wood is discussed in chapter 4.

BIBLIOGRAPHY

American Society for Testing and Materials
  Standard methods for testing small clear specimens of timber. D 143 (see current edition.) Philadelphia, Pa.
James, W.
  1968. Effect of temperature on the readings of electric moisture meters. Forest Prod. J. 18(10): 23–31.
——, and Hamill, D.
  1965. Dielectric properties of Douglas-fir measured at microwave frequencies. Forest Prod. J. 15(2): 57.
Kleuters, W.
  1964. Determining local density of wood by beta ray method. Forest Prod. J. 14(9): 414.
Kukachka, B. F.
  1970. Properties of imported tropical woods. USDA Forest Serv. Res. Pap. FPL 125. Forest Prod. Lab., Madison, Wis.
Loos, Wesley
  1961. Gamma ray absorption and moisture content and density. Forest Prod. J. 11(3): 145.
Lynn, R.
  1967. Review of dielectric properties of wood and cellulose. Forest Prod. J. 17(7): 61.
McKenzie, W. M., and Karpovich, H.
  1968. Frictional behavior of wood. Wood Sci. and Tech. 2(2): 138.
MacLean, J. D.
  1941. Thermal conductivity of wood. Trans. Amer. Soc. Heat. Ventil. Eng. 47: 323.
Nanassy, A.
  1964. Electric polarization measurements on yellow birch. Can. J. Phys. 42(6): 1270.
Panshin, A. J., and deZeeuw, C.
  1970. Textbook of wood technology. Vol. 1. 3rd Edition. McGraw-Hill.
Wangaard, Frederick F.
  1969. Heat transmissivity of southern pine wood, plywood, fiberboard, and particleboard. Wood Sci. 2(1): 54.
Wong, Phillip T.Y.
  1964. Thermal conductivity and diffusivity of partially charred wood. Forest Prod. J. 14(5): 195.

Table A–3–7—Density of wood as a function of specific gravity and moisture content

Moisture content of wood (percent)	Density in kilograms per cubic meter when the specific gravity [1] is—																				
	0.30	0.32	0.34	0.36	0.38	0.40	0.42	0.44	0.46	0.48	0.50	0.52	0.54	0.56	0.58	0.60	0.62	0.64	0.66	0.68	0.70
0	300	320	340	360	380	400	420	440	460	480	500	519	540	559	580	599	620	639	660	679	700
4	312	333	354	375	396	416	436	458	477	500	519	540	561	581	602	623	644	665	686	706	727
8	324	346	367	389	410	432	453	474	497	517	540	561	583	604	626	647	670	690	713	734	756
12	336	359	381	404	426	449	471	493	516	537	559	581	604	626	649	671	694	716	738	761	783
16	348	372	394	416	440	465	487	509	533	556	580	602	626	649	673	695	719	742	766	788	812
20	360	384	408	432	455	481	503	527	551	575	599	623	647	671	695	719	743	767	791	815	839
24	372	397	421	445	471	497	521	545	570	594	620	644	670	694	719	743	769	793	819	843	868
28	384	410	436	461	487	511	537	562	588	614	639	665	690	716	742	767	793	819	844	870	895
32	396	423	449	476	501	527	554	580	607	633	660	686	713	738	766	791	819	844	871	897	924
36	408	436	463	490	516	543	570	597	625	652	679	706	734	761	788	815	843	870	897	924	951
40	420	449	476	503	532	559	588	615	644	671	700	727	756	783	812	839	868	895	924	951	980
44	432	461	490	517	546	575	604	633	662	691	719	748	777	806	835	863	892	921	950	979	1,010
48	444	474	503	532	562	591	622	650	681	710	740	769	799	828	859	887	918	947	977	1,010	1,030
52	457	487	516	546	577	607	638	668	698	729	759	790	820	851	881	911	942	972	1,000	1,030	1,060
56	468	500	530	561	593	623	655	686	718	748	780	811	843	873	905	935	968	998	1,030	1,060	1,090
60	481	511	543	575	607	639	671	703	735	767	799	831	863	895	927	960	992	1,020	1,060	1,090	1,120
64	492	524	557	589	623	655	689	721	754	786	820	852	886	918	951	984	1,020	1,050	1,080	1,110	1,150
68	508	537	570	604	638	671	705	738	772	806	839	873	907	940	974	1,010	1,040	1,070	1,110	1,140	1,180
72	516	549	585	618	654	687	722	756	791	825	860	894	929	963	998	1,030	1,070	1,100	1,130	1,170	1,200
76	527	562	597	633	668	703	738	774	809	844	879	915	950	985	1,020	1,060	1,090	1,130	1,160	1,200	1,230
80	540	575	612	647	684	719	756	791	828	863	900	935	972	1,010	1,040	1,080	1,110	1,150	1,190	1,220	1,260
84	551	588	625	662	698	735	772	809	846	883	919	956	993	1,030	1,070	1,100	1,140	1,180	1,210	1,250	1,290
88	564	601	639	676	714	751	790	827	865	902	940	977	1,010	1,050	1,090	1,130	1,160	1,200	1,240	1,280	1,320
92	575	614	652	690	729	767	806	844	883	921	960	998	1,040	1,070	1,110	1,150	1,190	1,230	1,270	1,310	1,340
96	588	626	666	705	745	783	823	862	902	940	980	1,020	1,060	1,100	1,140	1,180	1,210	1,250	1,290	1,330	1,370
100	599	639	679	719	759	799	839	879	919	960	1,000	1,040	1,080	1,120	1,160	1,200	1,240	1,280	1,320	1,360	1,400
110	630	671	714	756	798	839	881	924	966	1,010	1,050	1,090	1,130	1,180	1,220	1,260	1,300	1,340	1,390	1,430	1,470
120	660	703	748	791	836	879	924	968	1,010	1,060	1,100	1,140	1,190	1,230	1,280	1,320	1,360	1,410	1,450	1,500	1,540
130	690	735	782	828	873	919	966	1,010	1,060	1,100	1,150	1,190	1,240	1,290	1,330	1,380	1,420	1,470	1,520	1,560	1,610
140	719	767	815	863	911	960	1,010	1,060	1,100	1,150	1,200	1,250	1,300	1,340	1,390	1,440	1,490	1,530	1,580	1,631	1,679
150	750	799	849	900	950	1,000	1,050	1,100	1,150	1,200	1,250	1,300	1,350	1,400	1,450	1,500	1,550	1,599	1,650	1,700	1,749

[1] Based on mass when ovendry and volume at tabulated moisture content.

Chapter 4
MECHANICAL PROPERTIES OF WOOD

Mechanical properties discussed in this chapter have been obtained from tests of small pieces of wood termed "clear" and "straight grained" because they did not contain characteristics such as knots, cross grain, checks, and splits. These test pieces do contain wood structure characteristics such as growth rings that occur in consistent patterns within the piece. Clear wood specimens are usually considered "homogeneous" in wood mechanics.

Many of the mechanical properties of wood tabulated in this chapter were derived from extensive sampling and analysis procedures. These properties often are represented as the average mechanical properties of the species and are used to derive allowable properties for design. A number of other properties, particularly those less common and those for imported species, often are based on a more limited number of specimens not subject to the same sampling and analysis procedures. The appropriateness of these latter properties to represent the average properties of a species often is uncertain; nevertheless, they illustrate important wood behavior and provide guidance for wood design.

Variability, or variation in properties, is common to all materials. Since wood is a natural material and the tree is subject to numerous constantly changing influences (such as moisture, soil conditions, and growing space), wood properties vary considerably even in clear material. This chapter provides information where possible on the nature and magnitude of property variability.

ORTHOTROPIC NATURE OF WOOD

Wood may be described as an orthotropic material; that is, it has unique and independent mechanical properties in the directions of three mutually perpendicular axes—longitudinal, radial, and tangential. The longitudinal axis (L) is parallel to the fiber (grain); the radial axis (R) is normal to the growth rings (perpendicular to the grain in the radial direction); and the tangential axis (T) is perpendicular to the grain but tangent to the growth rings. These axes are shown in figure 4–1.

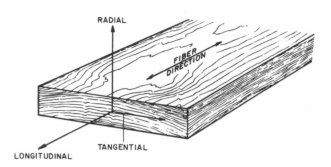

Figure 4–1.—The three principal axes of wood with respect to grain direction and growth rings.

ELASTIC PROPERTIES OF CLEAR WOOD

Twelve constants (nine are independent) are needed to describe the elastic behavior of wood: Three moduli of elasticity, E, three moduli of rigidity, G, and six Poisson's ratios, μ. The moduli of elasticity and Poisson's ratios are related by expressions of the form

$$\frac{\mu_{ij}}{E_i} = \frac{\mu_{ji}}{E_j}, i \neq j; i,j = L, R, T$$

General relations between stress and strain for a homogeneous, orthotropic material can be found in texts on anisotropic elasticity.

Modulus of Elasticity

The three moduli of elasticity denoted by E_L, E_R, and E_T are, respectively, the elastic moduli along longitudinal, radial, and tangential axes of wood. These moduli are usually obtained from compression tests; however, data for E_R and E_T are not extensive. Values of E_R and E_T for samples from a few species are presented in table 4–1 as ratios with E_L. These ratios, as well as the three elastic constants themselves, vary within and between species and with moisture content and specific gravity.

Often E_L determined from bending, rather than from an axial test, is the only E available. Average values of E_L obtained from bending tests are given in tables 4–2, 4–3, and 4–4. A representative coefficient of variation of E_L determined with bending tests for clear wood is reported in table 4–5. E_L as tabulated includes an effect of shear deflection. E_L from bending can be increased by 10 percent to approximately remove this effect. This adjusted bending E_L can be used to obtain E_R and E_T from table 4–1.

Modulus of Rigidity

The three moduli of rigidity denoted by G_{LR}, G_{LT}, G_{RT} are the elastic constants in the LR, LT, and RT planes, respectively. For example, G_{LR} is the modulus of rigidity based on shear strain in the LR plane and shear stresses in the LT and RT planes. Values of shear moduli for samples of a few species expressed as ratios with E_L are given in table 4–1. As with moduli of elasticity, the moduli of rigidity vary within and between species and with moisture content and specific gravity.

Poisson's Ratio

The six Poisson's ratios are denoted by μ_{LR}, μ_{RL}, μ_{LT}, μ_{TL}, μ_{RT}, μ_R. The first letter of the subscript refers to direction of applied stress and the second letter refers to direction of lateral deformation. For example, μ_{LR} is the Poisson's ratio for deformation along the radial axis caused by stress along the longitudinal axis. Values of Poisson's ratios for samples of a few species are given in table 4–1. Poisson's ratios vary within and between species and are affected slightly by moisture content.

STRENGTH PROPERTIES OF CLEAR STRAIGHT-GRAINED WOOD

Common Properties

Strength values most commonly measured and represented as "strength properties" for design include the modulus of rupture in bending, the maximum stress in compression parallel to the grain, compression strength perpendicular to the grain, and shear strength parallel to the grain. Additional measurements often made include work to maximum load in bending, impact bending strength, tensile strength perpendicular to the grain, and hardness. These properties, grouped according to the broad forest tree categories of hardwood and softwood (not correlated with hardness or softness), are given in tables 4–2, 4–3, and 4–4 for many of the commercially important species. Coefficients of variation for these properties from a limited sampling of specimens are reported in table 4–5.

The modulus of rupture in bending reflects the maximum load-carrying capacity of the member and is proportional to the maximum moment borne by the specimen. The work to maximum load is a measure of the energy absorbed by the specimen as it is slowly loaded to failure. On the other hand, the impact bending height of drop is related to the energy absorption due to a rapid or falling load. Hardness is the load required to embed a 0.444-inch ball to one-half its diameter in a direction perpendicular to the grain.

Less Common Properties

Strength properties less commonly measured in clear wood include tensile strength parallel to the grain, torsion, toughness, creep, rolling shear, and fatigue resistance.

Tensile Strength Parallel to Grain

Relatively few data are available on the tensile strength of various species parallel to grain. In the absence of sufficient tension test data, the modulus of rupture values are sometimes substituted for tensile strength of small, clear, straight-grained pieces of wood. The modulus of rupture is considered to be a low or conservative estimate of tensile strength for these specimens. Chapter 6 should be consulted for discussion of the tensile properties of commercial structural members. Table 4–6 lists average tensile strength values for a limited number of specimens of a few species.

Torsion

For solid wood members, the torsional shear strength often is taken as the shear strength parallel to the grain. Two-thirds of this value often is used as the torsional shear stress at the proportional limit.

Toughness

Toughness represents the energy required to rapidly cause complete failure in a centrally loaded bending specimen. Table 4–7 gives average toughness values for samples of a few hardwood and softwood species. Table 4–5 records the average coefficient of variation for toughness as determined from approximately 50 species.

Fatigue Strength

The resistance of wood to fatigue is sometimes an important consideration in design. Tests indicate that wood, like many fibrous materials, is less sensitive to repeated loads than are more crystalline structural materials, such as metals. In proportion to ultimate strength values, the fatigue strength of wood is higher than for some of the metals. A brief résumé of the results of several fatigue studies is given in table 4–8. Interpretation of fatigue data, and a discussion of fatigue as a function of the service environment, are included later in this chapter.

Rolling Shear Strength

The term "rolling shear" describes the shear strength of wood where the shearing force is in a longitudinal-transverse plane and perpendicular to the grain. Test procedures for rolling shear in solid wood are of recent origin; few test values have been reported. In limited tests, rolling shear strengths were 10 to 20 percent of the parallel-to-grain shear values. Rolling shear values were about the same in the longitudinal-radial and the longitudinal-tangential planes.

VIBRATION PROPERTIES

The vibration properties of primary interest in structural materials are the speed of sound and the damping capacity or internal friction.

Speed of Sound

The speed of sound in a structural material varies directly with the square root of the modulus of elasticity and inversely with the square root of the density. For example, a parallel-to-grain value for speed of sound of 150,000 inches per second corresponds to a modulus of elasticity of about 1,800,000 p.s.i., and a density of 30 pounds per cubic foot. The speed of sound in wood varies strongly with grain angle since the transverse modulus of elasticity may be as small as 1/20 of the longitudinal value. Thus, the speed of sound across the grain is about one-fifth to one-third of the longitudinal value.

The speed of sound decreases with increasing temperature or moisture content in proportion to the influence of these variables on the modulus of elasticity and density. The speed of sound decreases slightly with increasing frequency and amplitude of vibration, although for most common applications this effect is too small to be significant. There is no recognized independent effect of species on the speed of sound. Variability in the speed of sound in wood is directly related to the variability of modulus of elasticity and density.

Internal Friction

When solid material is strained, some mechanical energy is dissipated as heat. Internal friction is the term used to denote the mechanism that causes this energy dissipation. The internal friction mechanism in wood is a complex function of temperature and moisture content. At normal ambient temperatures the internal friction generally increases as the moisture content increases, up to the fiber saturation point. At room temperature internal friction is a minimum at about 6 to 8 percent moisture content. Below room temperature the minimum occurs at a higher moisture content; above room temperature it occurs at a lower moisture content. The parallel-to-grain internal friction of wood under normal use conditions of moisture content and temperature is approximately 10 times that of structural metals, explaining in part why wood structures damp vibration more quickly than metal structures of similar design.

SUMMARY TABLES ON MECHANICAL PROPERTIES OF CLEAR STRAIGHT-GRAINED WOOD

The mechanical properties listed in tables 4–1 through 4–7 are based on a variety of sampling methods. Generally, the greatest amount of sampling is represented in tables 4–2, 4–3, and 4–4. The values in table 4–2 are averages derived for a number of species grown in the United States. The table value is intended to estimate the average clear wood property of the species. Many of the values were obtained from test specimens taken at heights between 8 and 16 feet above the stump of the tree. Values reported in table 4–3 represent average clear wood properties of species grown in Canada and commonly imported into the United States.

Methods of data collection and analysis have changed over the years that the data in tables 4–2 and 4–3 have been collected. In addition, the character of some forests changes with time. Thus, when these data are used as a basis for critical applications such as stress grades of lumber, the current appropriateness of the data should be reviewed.

Values reported in table 4–4 were collected from the world literature; thus, the appropriateness of these properties to represent a species is not known. The properties reported in tables 4–1, 4–6, 4–7, and 4–8 are not intended to represent species characteristics in the broad sense; they suggest the relative influence of species and other specimen parameters on the mechanical behavior recorded.

Variability in properties can be important in both production and consumption of wood products. Often the fact that a piece may be stronger, harder, or stiffer than the average is of less concern to the user than if it is weaker; however, this may not be true if lightweight material is selected for a specific purpose or if harder or tougher material is hard to work. It is desirable, therefore, that some indication of the spread of property values be given. Average coefficients of variation for many mechanical properties are presented in table 4–5.

The mechanical properties reported in the tables are significantly affected by the moisture content of the specimens at the time of test. Some tables include properties evaluated at differing moisture levels; these moisture levels are reported. As indicated in the tables, many of the dry test data have been adjusted to a common moisture content base of 12 percent. The differences in properties displayed in the tables as a result of differing moisture levels are not necessarily consistent for larger wood pieces such as lumber. Guidelines for adjusting clear wood properties to arrive at allowable properties for lumber are discussed in chapter 6, "Lumber Stress Grades and Allowable Properties."

Specific gravity is reported in many of the tables because it often is used as an index of clear wood properties. The specific gravity values given in tables 4–2 and 4–3 represent the estimated average clear wood specific gravity of the species. In the other tables, the specific gravity represents only the specimens tested. The variability of specific gravity, represented by the coefficient of variation derived from tests on 50 species, is included in table 4–5.

Mechanical and physical properties as measured and reported often reflect not only the characteristics of the wood but also the influence of the shape and size of test specimen and the mode of test. The methods of test used to establish properties in tables 4–2, 4–3, 4–6, and 4–7 are based on standard procedures, ASTM Designation D 143. The methods of test for properties presented in other tables are reported in the bibliography at the end of this chapter.

Names of species listed in the tables conform to standard nomenclature of the U.S. Forest Service. Other common names may be used locally, and frequently one common name is applied to several species.

Tables 4–2, 4–3, and 4–4 are repeated in metric (SI) units in appendix 4–1 at the end of this chapter.

Table 4-1.—*Elastic constants of various species*

Species	Approximate specific gravity [1]	Approximate moisture content (pct.)	Modulus of elasticity ratios		Ratio of modulus of rigidity to modulus of elasticity			Poisson's ratios					
			E_T/E_L	E_R/E_L	G_{LR}/E_L	G_{LT}/E_L	G_{RT}/E_L	μ_{LR}	μ_{LT}	μ_{RT}	μ_{RL}	μ_{TR}	μ_{TL}
Balsa ---------	0.13	9	0.015	0.046	0.054	0.037	0.005	0.229	0.488	0.665	0.217	0.011	0.007
Birch, yellow ------	.64	13	.050	.078	.074	.068	.017	.426	.451	.697	.447	.033	.023
Douglas-fir -------	.50	12	.050	.068	.064	.078	.007	.292	.449	.390	.287	.020	.022
Spruce, Sitka ------	.38	12	.043	.078	.064	.061	.003	.372	.467	.435	.240	.029	.020
Sweetgum ---------	.53	11	.050	.115	.089	.061	.021	.325	.403	.682	.297	.037	.020
Walnut, black ------	.59	11	.056	.106	.085	.062	.021	.495	.632	.718	.379	.052	.035
Yellow-poplar ------	.38	11	.043	.092	.075	.069	.011	.318	.392	.703	.329	.029	.017

[1] Based on ovendry weight and volume at the moisture content shown.

Table 4-2.—Mechanical properties[1] of some commercially important woods grown in the United States

Common names of species	Specific gravity	Static bending			Impact bending—height of drop causing complete failure	Compression parallel to grain—maximum crushing strength	Compression perpendicular to grain—fiber stress at proportional limit	Shear parallel to grain—maximum shearing strength	Tension perpendicular to grain—maximum tensile strength	Side hardness—load perpendicular to grain
		Modulus of rupture	Modulus of elasticity[2]	Work to maximum load						
		P.s.i.	Million p.s.i.	In.-lb. per cu. in.	In.	P.s.i.	P.s.i.	P.s.i.	P.s.i.	Lb.
					HARDWOODS					
Alder, red	0.37	6,500	1.17	8.0	22	2,960	250	770	390	440
	.41	9,800	1.38	8.4	20	5,820	440	1,080	420	590
Ash:										
Black	.45	6,000	1.04	12.1	33	2,300	350	860	490	520
	.49	12,600	1.60	14.9	35	5,970	760	1,570	700	850
Blue	.53	9,600	1.24	14.7	---	4,180	810	1,540	---	---
	.58	13,800	1.40	14.4	---	6,980	1,420	2,030	---	---
Green	.53	9,500	1.40	11.8	35	4,200	730	1,260	590	870
	.56	14,100	1.66	13.4	32	7,080	1,310	1,910	700	1,200
Oregon	.50	7,600	1.13	12.2	39	3,510	530	1,190	590	720
	.55	12,700	1.36	14.4	33	6,040	1,250	1,790	720	1,160
White	.55	9,600	1.44	16.6	33	3,990	670	1,380	590	960
	.60	15,400	1.74	17.6	43	7,410	1,160	1,950	940	1,320
Aspen:										
Bigtooth	.36	5,400	1.12	5.7	---	2,500	210	730	---	---
	.39	9,100	1.43	7.7	---	5,300	450	1,080	---	---
Quaking	.35	5,100	.86	6.4	22	2,140	180	660	230	300
	.38	8,400	1.18	7.6	21	4,250	370	850	260	350
Basswood, American	.32	5,000	1.04	5.3	16	2,220	170	600	280	250
	.37	8,700	1.46	7.2	15	4,730	370	990	350	410
Beech, American	.56	8,600	1.38	11.9	43	3,550	540	1,290	720	850
	.64	14,900	1.72	15.1	41	7,300	1,010	2,010	1,010	1,300
Birch:										
Paper	.48	6,400	1.17	16.2	49	2,360	270	840	380	560
	.55	12,300	1.59	16.0	34	5,690	600	1,210	---	910
Sweet	.60	9,400	1.65	15.7	43	3,740	470	1,240	430	970
	.65	16,900	2.17	18.0	47	8,540	1,080	2,240	950	1,470
Yellow	.55	8,300	1.50	16.1	43	3,380	430	1,110	430	780
	.62	16,600	2.01	20.8	55	8,170	970	1,880	920	1,260

73

Table 4-2.—*Mechanical properties* [1] *of some commercially important woods grown in the United States*—continued

Common names of species	Specific gravity	Static bending — Modulus of rupture	Static bending — Modulus of elasticity [2]	Work to maximum load	Impact bending — height of drop causing complete failure	Compression parallel to grain — maximum crushing strength	Compression perpendicular to grain — fiber stress at proportional limit	Shear parallel to grain — maximum shearing strength	Tension perpendicular to grain — maximum tensile strength	Side hardness — load perpendicular to grain
		P.s.i.	*Million p.s.i.*	*In.-lb. per cu. in.*	*In.*	*P.s.i.*	*P.s.i.*	*P.s.i.*	*P.s.i.*	*Lb.*
HARDWOODS—continued										
Butternut	.36	5,400	0.97	8.2	24	2,420	220	760	430	390
	.38	8,100	1.18	8.2	24	5,110	460	1,170	440	490
Cherry, black	.47	8,000	1.31	12.8	33	3,540	360	1,130	570	660
	.50	12,300	1.49	11.4	29	7,110	690	1,700	560	950
Chestnut, American	.40	5,600	.93	7.0	24	2,470	310	800	440	420
	.43	8,600	1.23	6.5	19	5,320	620	1,080	460	540
Cottonwood:										
Balsam poplar	.31	3,900	.75	4.2	---	1,690	140	500	---	---
	.34	6,800	1.10	5.0	---	4,020	300	790	---	---
Black	.31	4,900	1.08	5.0	20	2,200	160	610	270	250
	.35	8,500	1.27	6.7	22	4,500	300	1,040	330	350
Eastern	.37	5,300	1.01	7.3	21	2,280	200	680	410	340
	.40	8,500	1.37	7.4	20	4,910	380	930	580	430
Elm:										
American	.46	7,200	1.11	11.8	38	2,910	360	1,000	590	620
	.50	11,800	1.34	13.0	39	5,520	690	1,510	660	830
Rock	.57	9,500	1.19	19.8	54	3,780	610	1,270	---	940
	.63	14,800	1.54	19.2	56	7,050	1,230	1,920	---	1,320
Slippery	.48	8,000	1.23	15.4	47	3,320	420	1,110	640	660
	.53	13,000	1.49	16.9	45	6,360	820	1,630	530	860
Hackberry	.49	6,500	.95	14.5	48	2,650	400	1,070	630	700
	.53	11,000	1.19	12.8	43	5,440	890	1,590	580	880
Hickory, pecan:										
Bitternut	.60	10,300	1.40	20.0	66	4,570	800	1,240	---	---
	.66	17,100	1.79	18.2	66	9,040	1,680	---	---	---
Nutmeg	.56	9,100	1.29	22.8	54	3,980	760	1,030	---	---
	.60	16,600	1.70	25.1	---	6,910	1,570	---	---	---

Species										
Pecan	.60 / .66	9,800 / 13,700	1.37 / 1.73	14.6 / 13.3	53 / 44	3,990 / 7,850	780 / 1,720	1,480 / 2,080	680 / ---	1,310 / 1,820
Water	.61 / .62	10,700 / 17,800	1.56 / 2.02	18.8 / 19.3	56 / 53	4,660 / 8,600	880 / 1,550	1,440 / ---	--- / ---	--- / ---
Hickory, true:										
Mockernut	.64 / .72	11,100 / 19,200	1.57 / 2.22	26.1 / 22.6	88 / 77	4,480 / 8,940	810 / 1,730	1,280 / 1,740	--- / ---	--- / ---
Pignut	.66 / .75	11,700 / 20,100	1.65 / 2.26	31.7 / 30.4	89 / 74	4,810 / 9,190	920 / 1,980	1,370 / 2,150	--- / ---	--- / ---
Shagbark	.64 / .72	11,000 / 20,200	1.57 / 2.16	23.7 / 25.8	74 / 67	4,580 / 9,210	840 / 1,760	1,520 / 2,430	--- / ---	--- / ---
Shellbark	.62 / .69	10,500 / 18,100	1.34 / 1.89	29.9 / 23.6	104 / 88	3,920 / 8,000	810 / 1,800	1,190 / 2,110	--- / ---	--- / ---
Honeylocust	.60 / ---	10,200 / 14,700	1.29 / 1.63	12.6 / 13.3	47 / 47	4,420 / 7,500	1,150 / 1,840	1,660 / 2,250	930 / 900	1,390 / 1,580
Locust, black	.66 / .69	13,800 / 19,400	1.85 / 2.05	15.4 / 18.4	44 / 57	6,800 / 10,180	1,160 / 1,830	1,760 / 2,480	770 / 640	1,570 / 1,700
Magnolia:										
Cucumbertree	.44 / .48	7,400 / 12,300	1.56 / 1.82	10.0 / 12.2	30 / 35	3,140 / 6,310	330 / 570	990 / 1,340	440 / 660	520 / 700
Southern	.46 / .50	6,800 / 11,200	1.11 / 1.40	15.4 / 12.8	54 / 29	2,700 / 5,460	460 / 860	1,040 / 1,530	610 / 740	740 / 1,020
Maple:										
Bigleaf	.44 / .48	7,400 / 10,700	1.10 / 1.45	8.7 / 7.8	23 / 28	3,240 / 5,950	450 / 750	1,110 / 1,730	600 / 540	620 / 850
Black	.52 / .57	7,900 / 13,300	1.33 / 1.62	12.8 / 12.5	48 / 40	3,270 / 6,680	600 / 1,020	1,130 / 1,820	720 / 670	840 / 1,180
Red	.49 / .54	7,700 / 13,400	1.39 / 1.64	11.4 / 12.5	32 / 32	3,280 / 6,540	400 / 1,000	1,150 / 1,850	--- / ---	700 / 950
Silver	.44 / .47	5,800 / 8,900	.94 / 1.14	11.0 / 8.3	29 / 25	2,490 / 5,220	370 / 740	1,050 / 1,480	560 / 500	590 / 700
Sugar	.56 / .63	9,400 / 15,800	1.55 / 1.83	13.3 / 16.5	40 / 39	4,020 / 7,830	640 / 1,470	1,460 / 2,330	--- / ---	970 / 1,450
Oak, red:										
Black	.56 / .61	8,200 / 13,900	1.18 / 1.64	12.2 / 13.7	40 / 41	3,470 / 6,520	710 / 930	1,220 / 1,910	--- / ---	1,060 / 1,210
Cherrybark	.61 / .68	10,800 / 18,100	1.79 / 2.28	14.7 / 18.3	54 / 49	4,620 / 8,740	760 / 1,250	1,320 / 2,000	800 / 840	1,240 / 1,480

| Common names of species | Specific gravity | Static bending | | | Impact bending—height of drop causing complete failure | Compression parallel to grain—maximum crushing strength | Compression perpendicular to grain—fiber stress at proportional limit | Shear parallel to grain—maximum shearing strength | Tension perpendicular to grain—maximum tensile strength | Side hardness—load perpendicular to grain |
| | | Modulus of rupture | Modulus of elasticity[2] | Work to maximum load | | | | | | |
		P.s.i.	*Million p.s.i.*	*In.-lb. per cu. in.*	*In.*	*P.s.i.*	*P.s.i.*	*P.s.i.*	*P.s.i.*	*Lb.*
HARDWOODS—continued										
Oak, red (Cont.)										
Laurel	.56	7,900	1.39	11.2	39	3,170	570	1,180	770	1,000
	.63	12,600	1.69	11.8	39	6,980	1,060	1,830	790	1,210
Northern red	.56	8,300	1.35	13.2	44	3,440	610	1,210	750	1,000
	.63	14,300	1.82	14.5	43	6,760	1,010	1,780	800	1,290
Pin	.58	8,300	1.32	14.0	48	3,680	720	1,290	800	1,070
	.63	14,000	1.73	14.8	45	6,820	1,020	2,080	1,050	1,510
Scarlet	.60	10,400	1.48	15.0	54	4,090	830	1,410	700	1,200
	.67	17,400	1.91	20.5	53	8,330	1,120	1,890	870	1,400
Southern red	.52	6,900	1.14	8.0	29	3,030	550	930	480	860
	.59	10,900	1.49	9.4	26	6,090	870	1,390	510	1,060
Water	.56	8,900	1.55	11.1	39	3,740	620	1,240	820	1,010
	.63	15,400	2.02	21.5	44	6,770	1,020	2,020	920	1,190
Willow	.56	7,400	1.29	8.8	35	3,000	610	1,180	760	980
	.69	14,500	1.90	14.6	42	7,040	1,130	1,650	---	1,460
Oak, white:										
Bur	.58	7,200	.88	10.7	44	3,290	680	1,350	800	1,110
	.64	10,300	1.03	9.8	29	6,060	1,200	1,820	680	1,370
Chestnut	.57	8,000	1.37	9.4	35	3,520	530	1,210	690	890
	.66	13,300	1.59	11.0	40	6,830	840	1,490	---	1,130
Live	.80	11,900	1.58	12.3	---	5,430	2,040	2,210	---	---
	.88	18,400	1.98	18.9	---	8,900	2,840	2,660	---	---
Overcup	.57	8,000	1.15	12.6	44	3,370	540	1,320	730	960
	.63	12,600	1.42	15.7	38	6,200	810	2,000	940	1,190
Post	.60	8,100	1.09	11.0	44	3,480	860	1,280	790	1,130
	.67	13,200	1.51	13.2	46	6,600	1,430	1,840	780	1,360
Swamp chestnut	.60	8,500	1.35	12.8	45	3,540	570	1,260	670	1,110
	.67	13,900	1.77	12.0	41	7,270	1,110	1,990	690	1,240

Species									
Swampy white	.64 / .72	9,900 / 17,700	1.59 / 2.05	14.5 / 19.2	50 / 49	4,360 / 8,600	760 / 1,190	1,300 / 2,000	860 / 830	1,160 / 1,620
White	.60 / .68	8,300 / 15,200	1.25 / 1.78	11.6 / 14.8	42 / 57	3,560 / 7,440	670 / 1,070	1,250 / 2,000	770 / 800	1,060 / 1,360
Sassafras	0.42 / .46	6,000 / 9,000	0.91 / 1.12	7.1 / 8.7	--- / ---	2,730 / 4,760	370 / 850	950 / 1,240	--- / ---	--- / ---
Sweetgum	.46 / .52	7,100 / 12,500	1.20 / 1.64	10.1 / 11.9	36 / 32	3,040 / 6,320	370 / 620	990 / 1,600	540 / 760	600 / 850
Sycamore, American	.46 / .49	6,500 / 10,000	1.06 / 1.42	7.5 / 8.5	26 / 25	2,920 / 5,380	360 / 700	1,000 / 1,470	630 / 720	610 / 770
Tanoak	.58 / ---	10,500 / ---	1.55 / ---	13.4 / ---	--- / ---	4,650 / ---	--- / ---	--- / ---	--- / ---	--- / ---
Tupelo: Black	.46 / .50	7,000 / 9,600	1.03 / 1.20	8.0 / 6.2	30 / 22	3,040 / 5,520	480 / 930	1,100 / 1,340	570 / 500	640 / 810
Water	.46 / .50	7,300 / 9,600	1.05 / 1.26	8.3 / 6.9	30 / 23	3,370 / 5,920	480 / 870	1,190 / 1,590	600 / 700	710 / 880
Walnut, black	.51 / .55	9,500 / 14,600	1.42 / 1.68	14.6 / 10.7	37 / 34	4,300 / 7,580	490 / 1,010	1,220 / 1,370	570 / 690	900 / 1,010
Willow, black	.36 / .39	4,800 / 7,800	.79 / 1.01	11.0 / 8.8	--- / ---	2,040 / 4,100	180 / 430	680 / 1,250	--- / ---	--- / ---
Yellow-poplar	.40 / .42	6,000 / 10,100	1.22 / 1.58	7.5 / 8.8	26 / 24	2,660 / 5,540	270 / 500	790 / 1,190	510 / 540	440 / 540

SOFTWOODS

Species									
Baldcypress	.42 / .46	6,600 / 10,600	1.18 / 1.44	6.6 / 8.2	25 / 24	3,580 / 6,360	400 / 730	810 / 1,000	300 / 270	390 / 510
Cedar: Alaska	.42 / .44	6,400 / 11,100	1.14 / 1.42	9.2 / 10.4	27 / 29	3,050 / 6,310	350 / 620	840 / 1,130	330 / 360	440 / 580
Atlantic white	.31 / .32	4,700 / 6,800	.75 / .93	5.9 / 4.1	18 / 13	2,390 / 4,700	240 / 410	690 / 800	180 / 220	290 / 350
Eastern redcedar	.44 / .47	7,000 / 8,800	.65 / .88	15.0 / 8.3	35 / 22	3,570 / 6,020	700 / 920	1,010 / ---	330 / ---	650 / 900
Incense	.35 / .37	6,200 / 8,000	.84 / 1.04	6.4 / 5.4	17 / 17	3,150 / 5,200	370 / 590	830 / 880	280 / 270	390 / 470
Northern white	.29 / .31	4,200 / 6,500	.64 / .80	5.7 / 4.8	15 / 12	1,990 / 3,960	230 / 310	620 / 850	240 / 240	230 / 320

Table 4-2.—Mechanical properties [1] of some commercially important woods grown in the United States—continued

Common names of species	Specific gravity	Static bending			Impact bending—height of drop causing complete failure	Compression parallel to grain—maximum crushing strength	Compression perpendicular to grain—fiber stress at proportional limit	Shear parallel to grain—maximum shearing strength	Tension perpendicular to grain—maximum tensile strength	Side hardness—load perpendicular to grain
		Modulus of rupture	Modulus of elasticity [2]	Work to maximum load						
		P.s.i.	Million p.s.i.	In.-lb. per cu. in.	In.	P.s.i.	P.s.i.	P.s.i.	P.s.i.	Lb.
SOFTWOODS—continued										
Cedar (Cont.)										
Port-Orford	.39	6,600	1.30	7.4	21	3,140	300	840	180	380
	.43	12,700	1.70	9.1	28	6,250	720	1,370	400	630
Western redcedar	.31	5,200	.94	5.0	17	2,770	240	770	230	260
	.32	7,500	1.11	5.8	17	4,560	460	990	220	350
Douglas-fir [3]:										
Coast	.45	7,700	1.56	7.6	26	3,780	380	900	300	500
	.48	12,400	1.95	9.9	31	7,240	800	1,130	340	710
Interior West	.46	7,700	1.51	7.2	26	3,870	420	940	290	510
	.50	12,600	1.82	10.6	32	7,440	760	1,290	350	660
Interior North	.45	7,400	1.41	8.1	22	3,470	360	950	340	420
	.48	13,100	1.79	10.5	26	6,900	770	1,400	390	600
Interior South	.43	6,800	1.16	8.0	15	3,110	340	950	250	360
	.46	11,900	1.49	9.0	20	6,220	740	1,510	330	510
Fir:										
Balsam	.34	4,900	.96	4.7	16	2,400	170	610	180	290
	.36	7,600	1.23	5.1	20	4,530	300	710	180	400
California red	.36	5,800	1.17	6.4	21	2,760	330	770	380	360
	.38	10,400	1.49	8.9	24	5,470	610	1,050	390	500
Grand	.35	5,800	1.25	5.6	22	2,940	270	740	240	360
	.37	8,800	1.57	7.5	28	5,290	500	910	240	490
Noble	.37	6,200	1.38	6.0	19	3,010	270	800	230	290
	.39	10,700	1.72	8.8	23	6,100	520	1,050	220	410
Pacific silver	.40	6,400	1.42	6.0	21	3,140	220	750	240	310
	.43	10,600	1.72	9.3	24	6,530	450	1,180	---	430
Subalpine	.31	4,900	1.05	---	---	2,300	190	700	---	260
	.32	8,600	1.29	---	---	4,860	390	1,070	---	350
White	.37	5,900	1.16	5.6	22	2,900	280	760	300	340
	.39	9,800	1.49	7.2	20	5,810	530	1,100	300	480

78

Species		Specific gravity	Modulus of rupture	Modulus of elasticity	Work to maximum load	Impact bending	Compression parallel	Compression perpendicular	Shear parallel	Tension perpendicular	Side hardness
Hemlock:											
Eastern	Green	.38	6,400	1.07	6.7	21	3,080	360	850	230	400
	Dry	.40	8,900	1.20	6.8	21	5,410	650	1,060	---	500
Mountain	Green	.42	6,300	1.04	11.0	32	2,880	370	930	330	470
	Dry	.45	11,500	1.33	10.4	32	6,440	860	1,540	---	680
Western	Green	.42	6,600	1.31	6.9	22	3,360	280	860	290	410
	Dry	.45	11,300	1.64	8.3	23	7,110	550	1,250	340	540
Larch, western	Green	.48	4,900	.96	10.3	29	3,760	400	870	330	510
	Dry	.52	13,100	1.87	12.6	35	7,640	930	1,360	430	830
Pine:											
Eastern white	Green	.34	4,900	.99	5.2	17	2,440	220	680	250	290
	Dry	.35	8,600	1.24	6.8	18	4,800	440	900	310	380
Jack	Green	.40	6,000	1.07	7.2	26	2,950	300	750	360	400
	Dry	.43	9,900	1.35	8.3	27	5,660	580	1,170	420	570
Loblolly	Green	.47	7,300	1.40	8.2	30	3,510	390	860	260	450
	Dry	.51	12,800	1.79	10.4	30	7,130	790	1,390	470	690
Lodgepole	Green	.38	5,500	1.08	5.6	20	2,610	250	680	220	330
	Dry	.41	9,400	1.34	6.8	20	5,370	610	880	290	480
Longleaf	Green	.54	8,500	1.59	8.9	35	4,320	480	1,040	330	590
	Dry	.59	14,500	1.98	11.8	34	8,470	960	1,510	470	870
Pitch	Green	.47	6,800	1.20	9.2	---	2,950	---	860	---	---
	Dry	.52	10,800	1.43	9.2	---	5,940	---	1,360	---	---
Pond	Green	.51	7,400	1.28	7.5	---	3,660	440	940	---	---
	Dry	.56	11,600	1.75	8.6	---	7,540	910	1,380	---	---
Ponderosa	Green	.38	5,100	1.00	5.2	21	2,450	280	700	310	320
	Dry	.40	9,400	1.29	7.1	19	5,320	580	1,130	420	460
Red	Green	.41	5,800	1.28	6.1	26	2,730	260	690	300	340
	Dry	.46	11,000	1.63	9.9	26	6,070	600	1,210	460	560
Sand	Green	.46	7,500	1.02	9.6	---	3,440	450	1,140	---	450
	Dry	.48	11,600	1.41	9.6	---	6,920	836	---	---	836
Shortleaf	Green	.47	7,400	1.39	8.2	30	3,530	350	910	320	440
	Dry	.51	13,100	1.75	11.0	33	7,270	820	1,390	470	690
Slash	Green	.54	8,700	1.53	9.6	---	3,820	530	960	---	---
	Dry	.59	16,300	1.98	13.2	---	8,140	1,020	1,680	---	---
Spruce	Green	.41	6,000	1.00	---	---	2,840	280	900	---	450
	Dry	.44	10,400	1.23	---	---	5,650	730	1,490	---	660
Sugar	Green	.34	4,900	1.03	5.4	17	2,460	210	720	270	270
	Dry	.36	8,200	1.19	5.5	18	4,460	500	1,130	350	380

Table 4-2.—*Mechanical properties ¹ of some commercially important woods grown in the United States*—continued

Common names of species	Specific gravity	Static bending			Impact bending—height of drop causing complete failure	Compression parallel to grain—maximum crushing strength	Compression perpendicular to grain—fiber stress at proportional limit	Shear parallel to grain—maximum shearing strength	Tension perpendicular to grain—maximum tensile strength	Side hardness—load perpendicular to grain
		Modulus of rupture	Modulus of elasticity ²	Work to maximum load						
		P.s.i.	Million p.s.i.	In.-lb. per cu. in.	In.	P.s.i.	P.s.i.	P.s.i.	P.s.i.	Lb.
SOFTWOODS—continued										
Pine (Cont.)										
Virginia	.45	7,300	1.22	10.9	34	3,420	390	890	400	540
	.48	13,000	1.52	13.7	32	6,710	910	1,350	380	740
Western white	.35	4,700	1.19	5.0	19	2,430	190	680	260	260
	.38	9,700	1.46	8.8	23	5,040	470	1,040	---	420
Redwood:										
Old-growth	.38	7,500	1.18	7.4	21	4,200	420	800	260	410
	.40	10,000	1.34	6.9	19	6,150	700	940	240	480
Young-growth	.34	5,900	.96	5.7	16	3,110	270	890	300	350
	.35	7,900	1.10	5.2	15	5,220	520	1,110	250	420
Spruce:										
Black	.38	5,400	1.06	7.4	24	2,570	140	660	100	370
	.40	10,300	1.53	10.5	23	5,320	530	1,030	---	520
Engelmann	.33	4,700	1.03	5.1	16	2,180	200	640	240	260
	.35	9,300	1.30	6.4	18	4,480	410	1,200	350	390
Red	.38	5,800	1.19	6.9	18	2,650	280	760	220	350
	.41	10,200	1.52	8.4	25	5,890	470	1,080	350	490
Sitka	.37	5,700	1.23	6.3	24	2,670	280	760	250	350
	.40	10,200	1.57	9.4	25	5,610	580	1,150	370	510
White	.37	5,600	1.07	6.0	22	2,570	240	690	220	320
	.40	9,800	1.34	7.7	20	5,470	460	1,080	360	480
Tamarack	.49	7,200	1.24	7.2	28	3,480	390	860	260	380
	.53	11,600	1.64	7.1	23	7,160	800	1,280	400	590

¹ Results of tests on small, clear straight-grained specimens. [Values in the first line for each species are from tests of green material; those in the second line are adjusted to 12 pct. moisture content.]Specific gravity is based on weight when ovendry and volume when green or at 12 pct. moisture content.

² Modulus of elasticity measured from a simply supported, center-loaded beam, on a span-depth ratio of 14/1. The modulus can be corrected for the effect of shear deflection by increasing it 10 pct.

³ Coast Douglas-fir is defined as Douglas-fir growing in the States of Oregon and Washington west of the summit of the Cascade Mountains. Interior West includes the State of California and all counties in Oregon and Washington east of but adjacent to the Cascade summit. Interior North includes the remainder of Oregon and Washington and the States of Idaho, Montana, and Wyoming. Interior South is made up of Utah, Colorado, Arizona, and New Mexico.

Table 4-3.—*Mechanical properties of some commercially important woods grown in Canada and imported into the United States* [1,2]

Common names of species	Specific gravity	Static bending		Compression parallel to grain —maximum crushing strength	Compression perpendicular to grain —fiber stress at protional limit	Shear parallel to grain —maximum shearing strength
		Modulus of rupture	Modulus of elasticity			
		P.s.i.	Million P.s.i.	P.s.i.	P.s.i.	P.s.i.
HARDWOODS						
Aspen:						
Quaking	0.37	5,500	1.31	2,350	200	720
		9,800	1.63	5,260	510	980
Big-toothed	.39	5,300	1.08	2,390	210	790
		9,500	1.26	4,760	470	1,100
Cottonwood:						
Balsam, poplar	.37	5,000	1.15	2,110	180	670
		10,100	1.67	5,020	420	890
Black	.30	4,100	.97	1,860	100	560
		7,100	1.28	4,020	260	860
Eastern	.35	4,700	.87	1,970	210	770
		7,500	1.13	3,840	470	1,160
SOFTWOODS						
Cedar:						
Alaska-	.42	6,600	1.34	3,240	350	880
		11,600	1.59	6,640	690	1,340
Northern white-	.30	3,900	.52	1,890	200	660
		6,100	.63	3,590	390	1,000
Western redcedar	.31	5,300	1.05	2,780	280	700
		7,800	1.19	4,290	500	810
Douglas-fir	.45	7,500	1.61	3,610	460	920
		12,800	1.97	7,260	870	1,380
Fir:						
Subalpine	.33	5,200	1.26	2,500	260	680
		8,200	1.48	5,280	540	980
Pacific silver	.36	5,500	1.35	2,770	230	710
		10,000	1.64	5,930	520	1,190
Balsam	.34	5,300	1.13	2,440	240	680
		8,500	1.40	4,980	460	910
Hemlock:						
Eastern	.40	6,800	1.27	3,430	400	910
		9,700	1.41	5,970	630	1,260
Western	.41	7,000	1.48	3,580	370	750
		11,800	1.79	6,770	660	940
Larch, western	.55	8,700	1.65	4,420	520	920
		15,500	2.08	8,840	1,060	1,340

Table 4–3.—*Mechanical properties of some commercially important woods grown in Canada and imported into the United States* [1,2]—Continued

Common names of species	Specific gravity	Static bending		Compression parallel to grain—maximum crushing strength	Compression perpendicular to grain—fiber stress at proportional limit	Shear parallel to grain—maximum shearing strength
		Modulus of rupture	Modulus of elasticity			
		P.s.i.	Million P.s.i.	P.s.i.	P.s.i.	P.s.i.
SOFTWOODS—continued						
Pine:						
Eastern white	.36	5,100	1.18	2,590	240	640
		9,500	1.36	5,230	490	880
Jack	.42	6,300	1.17	2,950	340	820
		11,300	1.48	5,870	830	1,190
Lodgepole	.40	5,600	1.27	2,860	280	720
		11,000	1.58	6,260	530	1,240
Ponderosa	.44	5,700	1.13	2,840	350	720
		10,600	1.38	6,130	760	1,020
Red	.39	5,000	1.07	2,370	280	710
		10,100	1.38	5,500	720	1,090
Western white	.36	4,800	1.19	2,520	240	650
		9,300	1.46	5,240	470	920
Spruce:						
Black	.41	5,900	1.32	2,760	300	800
		11,400	1.52	6,040	620	1,250
Engelmann	.38	5,700	1.25	2,810	270	700
		10,100	1.55	6,150	540	1,100
Red	.38	5,900	1.32	2,810	270	810
		10,300	1.60	5,590	550	1,330
Sitka	.35	5,400	1.37	2,560	290	630
		10,100	1.63	5,480	590	980
White	.35	5,100	1.15	2,470	240	670
		9,100	1.45	5,360	500	980
Tamarack	.48	6,800	1.24	3,130	410	920
		11,000	1.36	6,510	900	1,300

[1] Results of tests on small, clear, straight-grained specimens. Property values based on American Society for Testing and Materials Standard D 2555–70, "Standard methods for establishing clear wood values." Information on additional properties can be obtained from Department of Forestry, Canada, Publication No. 1104.

[2] The values in the first line for each species are from tests of green material; those in the second line are adjusted from the green condition to 12 pct. moisture content using dry to green clear wood property ratios as reported in ASTM D 2555–70. Specific gravity is based on weight when ovendry and volume when green.

Table 4-4.—Mechanical properties [1,2] of some woods imported into the United States

Common and botanical names of species	Moisture content Pct.	Specific gravity [3]	Static bending Modulus of rupture P.s.i.	Static bending Modulus of elasticity [4] Million p.s.i.	Static bending Work to maximum load In.-lb. per cu. in.	Compression parallel to grain—maximum crushing strength P.s.i.	Shear parallel to grain—maximum shearing strength P.s.i.	Side hardness—load perpendicular to grain Lb.	Sample Number of trees	Sample Origin [5]
Andiroba (Carapa guianensis)	Green	0.56	11,100	1.56	11.4	4,930	1,320	1,060	2	BR
	12	----	15,600	1.85	13.4	7,900	1,680	1,220	2	BR
Andiroba (C. nicaraguensis)	13	.45	----	----	----	6,240	----	1,240	3	EC
Angelique (Dicorynia guianensis)	Green	.60	11,400	1.84	12.0	5,590	1,340	1,100	2	SU
	12	----	17,400	2.19	15.2	8,770	1,660	1,290	2	SU
Apamate (Tabebuia rosea)	Green	.51	10,600	1.47	11.2	4,930	1,240	890	10	CS
	12	----	13,800	1.60	12.5	7,340	1,450	960	9	CS
Apitong (Dipterocarpus spp.)	Green	.59	9,200	1.79	----	4,410	1,040	800	57	PH
	12	----	15,200	2.35	----	8,540	1,690	1,200	53	PH
Avodire (Turraeanthus africanus)	12	.51	12,700	1.48	9.4	7,180	2,040	1,080	3	AF
Balsa (Ochroma pyramidale)	12	.17	2,800	.55	----	1,700	300	100	(6)	EC
Banak (Virola koschnyi)	Green	.44	6,200	1.47	5.3	3,050	660	440	8	CA
	12	----	10,800	1.72	8.1	5,720	1,300	640	8	CA
Banak (V. surinamensis)	Green	.42	5,600	1.64	4.1	2,390	720	320	2	BR
	12	----	10,900	2.04	10.0	5,140	980	510	2	BR
Capirona (Calycophyllum candidissimum)	Green	.67	14,300	1.93	18.6	6,200	1,660	1,630	2	VE
	12	----	22,300	2.27	27.0	9,670	2,120	1,940	2	VE
Capirona (C. spruceanum)	14	.85	----	----	----	9,280	----	2,550	1	PE
Cativo (Prioria copaifera)	Green	.40	5,900	.95	5.4	2,590	860	450	4	PA
	12	----	3,700	1.15	7.2	4,490	1,040	610	4	PA
Courbaril (Hymenaea courbaril)	Green	.72	12,900	1.82	15.7	5,800	1,770	2,030	9	CS
	12	----	13,400	2.17	17.6	9,680	2,470	2,440	9	CS
Gola (Tetraberlinia tubmaniana)	14	.66	15,700	2.21	----	9,010	----	----	11	AF

Table 4-4.—Mechanical properties [1,2] of some woods imported into the United States—continued

Common and botanical names of species	Moisture content	Specific gravity [3]	Static bending			Compression parallel to grain—maximum crushing strength	Shear parallel to grain—maximum shearing strength	Side hardness—load perpendicular to grain	Sample	
			Modulus of rupture	Modulus of elasticity [4]	Work to maximum load				Number of trees	Origin [5]
	Pct.		P.s.i.	Million p.s.i.	In.–lb. per cu. in.	P.s.i.	P.s.i.	Lb.		
Goncalo alves (Astronium graveolens)	Green	.86	12,400	1.90	7.4	6,880	1,840	1,990	4	CS
	12	------	17,100	2.17	10.4	10,560	2,060	2,230	4	CS
Greenheart (Ocotea rodiaei)	Green	.83	19,400	2.98	13.0	10,360	1,480	2,190	5	GY
	14	.93	25,500	3.70	22.0	13,040	1,830	2,630	1	GY
Ilomba (Pycnanthus angolensis)	12	.44	8,900	1.75	------	5,510	------	750	(7)	AF
Jarrah (Eucalyptus marginata)	Green	.67	9,900	1.48	------	5,190	1,325	1,285	28	AU
	12	------	16,200	1.88	------	8,870	2,185	1,915	28	AU
Jelutong (Dyera costulata)	Green	.36	5,600	1.16	5.6	3,050	760	330	3	AS
	16	.38	7,300	1.18	6.4	3,920	840	390	3	AS
Kapur (Dryobalanops lanceolata)	Green	.64	12,200	1.70	12.8	5,970	1,040	980	5	AS
	12	------	17,400	2.02	15.5	9,700	1,710	1,230	5	AS
Karri (Eucalyptus diversicolor)	Green	.70	10,600	2.07	------	5,250	1,340	1,360	26	AU
	12	------	19,200	2.76	------	10,400	2,140	2,030	21	AU
Kerving (Dipterocarpus spp.)	Green	.67	11,900	2.44	9.2	6,230	1,160	1,110	21	MI
	16	.69	14,500	2.63	13.3	8,000	1,360	1,160	11	MI
Khaya (Khaya anthotheca)	Green	.47	7,800	1.18	9.2	3,770	1,090	730	9	AF
	12	------	11,500	1.41	9.8	6,300	1,700	900	9	AF
Khaya (K. ivorensis)	Green	.43	7,400	1.16	8.3	3,500	930	640	11	AF
	12	------	10,700	1.39	8.3	6,460	1,500	830	11	AF
Kokrodua (Pericopsis elata)	Green	.66	14,800	1.77	19.5	7,490	1,670	1,600	6	AF
	12	------	18,400	1.94	18.5	9,940	2,090	1,560	6	AF
Lapacho (Tabebuia heterotricha)	Green	.80	20,100	2.12	27.3	7,680	2,140	2,530	3	PA
	12	------	22,600	2.32	26.0	10,930	2,280	3,010	3	PA
Lapacho (T. serratifolia)	Green	.92	22,800	3.06	25.6	10,660	2,050	2,970	3	SM
	12	------	26,300	3.31	23.0	13,420	2,070	3,670	3	SM

Lauan:
Dark red:

Common and botanical names	Moisture	Sp. gr.								
Red lauan (Shorea negrosensis)	Green	.44	7,700	1.38	3,700	---	930	570	15	AS
	12	---	11,300	1.63	5,890	---	1,220	680	15	AS
Tanguile (S. polysperma)	Green	.46	8,300	1.54	3,940	---	940	620	19	PH
	12	---	12,900	1.81	6,580	---	1,290	770	17	PH
Light red:										
Almon (S. almon)	Green	.41	7,500	1.44	3,750	---	840	500	12	PH
	12	.44	11,300	1.67	5,750	---	1,090	590	12	PH
Bagtikan (Parashorea plicata)	Green	.48	8,800	1.47	4,360	---	990	700	32	AS
	12	---	12,600	1.73	6,850	---	1,300	810	32	AS
Mayapis (Shorea squamata)	Green	.41	7,300	1.40	3,470	---	770	480	14	AS
	12	---	11,100	1.66	5,620	---	1,090	590	12	AS
White lauan (Pentacme contorta)	Green	.43	7,500	1.38	3,700	---	910	580	19	AS
	12	---	11,700	1.69	6,070	---	1,200	700	18	AS
Laurel (Cordia alliodora)	Green	.44	8,800	1.26	4,000	---	1,130	790	13	CA
	12	---	12,100	1.49	6,280	---	1,220	790	13	CA
Lignumvitae (Guaiacum sanctum)	12	1.09	11,400					4,500	(8)	SM
Limba (Terminalia superba)	12	.49	11,500	1.64	5,290		1,010	680	1	PE
Lupuna (Ceiba samauma)	13	.54						740		
Mahogany (Swietenia macrophylla)	Green	.45	9,300	1.28	4,510	9.6	1,310	700	77	CS
	12	---	11,600	1.51	6,630	7.9	1,290	810	77	CS
Meranti, red (Shorea dasyphylla)	Green	.43	8,600	1.50	4,450	8.8		560	2	AS
	12	---	12,100	1.63	6,970	11.7		630	2	AS
Oak (Quercus costaricensis)	12	.68	17,600	2.64		16.8		1,570	2	CR
Oak (Q. eugeniaefolia)	12	.75	16,400	2.84		14.1		2,170	1	CR
Obeche (Triplochiton scleroxylon)	Green	.33	5,100	.71	2,570	6.2	670	420	2	AF
	12	---	7,500	.86	3,930	6.9	990	430	2	AF
Okoume (Aucoumea klaineana)	12	.37	7,300	1.14	3,900	---		380	(9)	AF
Palosapis (Anisoptera spp.)	Green	.51	7,500	1.43	3,780	9.8	1,000	810	18	AS
	12	---	12,800	1.82	6,630	12.2	1,410	920	16	AS
"Parana pine" (Araucaria angustifolia)	Green	.46	7,100	1.35	4,000	---	970	560	(10)	SM
	12	---	13,500	1.62	7,650	---	1,730	780	(10)	SM
Pau marfim (Balfourodendron riedelianum)	Green	.73	14,400		6,100	---	1,890	1,530	5	BR
	15	---	18,900		8,200	---			5	BR

Table 4-4.—*Mechanical properties* [1,2] *of some woods imported into the United States*—continued

Common and botanical names of species	Moisture content	Specific gravity [3]	Static bending			Compression parallel to grain—maximum crushing strength	Shear parallel to grain—maximum shearing strength	Side hardness—load perpendicular to grain	Sample	
			Modulus of rupture	Modulus of elasticity [4]	Work to maximum load				Number of trees	Origin [5]
	Pct.		P.s.i.	Million p.s.i.	In.-lb. per cu. in.	P.s.i.	P.s.i.	Lb.		
Peroba de campos (*Paratecoma peroba*)	12	.75	15,400	1.76	10.2	8,920	2,140	1,600	(11)	BR
Pine, Caribbean (*Pinus caribaea*)	Green	.68	10,000	1.69	12.0	4,780	1,200	820	19	CA
	12	---	15,200	2.03	15.3	8,000	1,870	1,150	14	CA
Pine, ocote (*P. oocarpa*)	Green	.55	8,000	1.74	6.9	3,690	1,040	580	3	HO
	12	---	14,900	2.25	10.9	7,680	1,720	910	3	HO
Primavera (*Cybistax donnell-smithii*)	Green	.39	7,700	.98	6.9	3,630	1,050	660	4	HO
	12	---	10,900	1.22	10.3	6,140	1,710	700	4	HO
Ramin (*Gonystylus bancanus*)	Green	.59	9,800	1.57	9.0	5,395	994	640	9	AS
	12	---	18,400	2.17	17.0	10,080	1,514	1,300	9	AS
Rosewood, Indian (*Dalbergia latifolia*)	Green	.75	9,200	1.19	11.6	4,530	1,400	1,270	5	AS
	12	---	16,900	1.78	13.1	9,220	2,090	2,630	5	AS
Sande (*Brosimum utile*)	12	.44	---	---	---	6,310	---	500	3	EC
Santa Maria (*Calophyllum brasiliense*)	Green	.54	10,500	1.57	10.6	5,160	1,290	1,010	18	CA
	12	---	14,800	1.82	13.2	8,060	1,910	1,210	18	CA
Sapele (*Entandrophragma cylindricum*)	Green	.60	10,200	1.49	10.5	5,011	1,250	1,020	5	AF
	12	---	15,300	1.82	15.7	8,160	2,288	1,510	5	AF
Spanish-cedar (*Cedrela angustifolia*)	Green	.38	6,700	1.17	7.4	3,100	790	450	2	BR
	12	---	11,300	1.42	12.5	6,010	1,200	570	2	BR
Spanish-cedar (*C. oaxacensis*)	Green	.41	7,500	1.31	7.1	3,370	990	550	3	PA
	12	---	11,500	1.44	9.4	6,210	1,100	600	3	PA
Spanish-cedar (*C. odorata*) {(*Nicaragua*) (*Guatemala*) (*Nicaragua*)}	Green	.34	5,200	.87	7.4	2,760	720	350	1	NI
	Green	.43	9,500	1.48				620	1	GU
	12	.36	7,900	1.01	5.6	4,450		500	1	NI
Teak (*Tectona grandis*)	Green	.57	11,000	1.51	10.8	5,470	1,290	1,070	134	IN
	12	.62	13,300	1.39	10.3	6,770	1,600	1,110	3	HO
	12	.63	12,800	1.59	10.1	7,110	1,480	1,030	56	IN

"Virola" (Dialyanthera otoba)	12	------	------	------	------	------	------	300	------	10	AS
Walnut, European (Juglans regia) {	Green	8,710	.47	1.31	10.4	4,010	1,060	670	10	AS
{	8	13,090	------	1.54	9.8	7,320	1,320	860	10	AS

[1] Results of tests on small, clear, straight-grained specimens. Property values were taken from world literature (not obtained from experiments conducted at the U.S. Forest Products Laboratory). Other species may be reported in the world literature, as well as additional data on many of these species.

[2] Some property values have been adjusted to 12 pct. moisture content; others are based on moisture content at time of test.

[3] Specific gravity based on weight when ovendry and volume at moisture content indicated.

[4] Modulus of elasticity measured from a simply supported, center loaded beam, on a span-depth ratio of 14/1. The modulus can be corrected for the effect of shear deflection by increasing it 10 pct.

[5] Key to code letters: AF, Africa; AS, Southeast Asia; AU, Australia; BR, Brazil; CA, Central America; CH, Chile; CR, Costa Rica; CS, Central and South America; EC, Ecuador; GU, Guatemala; GY, Guyana (British Guiana); HO, Honduras; IN, India; MI, Malaysia—Indonesia; NI, Nicaragua; PA, Panama; PE, Peru; PH, Philippine Islands; SM, South American; SU, Surinam; and VE, Venezuela.

[6] 1,500 board feet.
[7] 1 bolt.
[8] 195 tests.
[9] 21 tests.
[10] 26 planks.
[11] 11 planks.

Table 4–5.—*Average coefficient of variation for some mechanical properties of clear wood*

Property	Coefficient of variation [1]
	Pct.
Static bending:	
Fiber stress at proportional limit	22
Modulus of rupture	16
Modulus of elasticity	22
Work to maximum load	34
Impact bending, height of drop causing complete failure	25
Compression parallel to grain:	
Fiber stress at proportional limit	24
Maximum crushing strength	18
Compression perpendicular to grain, fiber stress at proportional limit	28
Shear parallel to grain, maximum shearing strength	14
Tension perpendicular to grain, maximum tensile strength	25
Hardness:	
Perpendicular to grain	20
Toughness	34
Specific gravity	10

[1] Values given are based on results of tests of green wood from approximately 50 species. Values for wood adjusted to 12 pct. moisture content may be assumed to be approximately of the same magnitude.

Table 4–6.—*Average parallel-to-grain tensile strength for specimens of some species of wood*[1]

Species	Specific gravity	Tensile strength
		P.s.i.
HARDWOODS		
Elm:		
Cedar	0.59	17,500
	.64	20,200
Winged	.68	27,000
Oak, overcup	.57	11,300
	.63	14,700
Sweetgum	.46	13,600
	.52	17,300
Willow, black	.37	10,600
	.41	15,800
Yellow-poplar	.42	15,900
	.46	22,400
SOFTWOODS		
Douglas-fir, interior north	.46	15,600
	....	18,900
Fir:		
California red	.37	11,300
	.39	13,100
Pacific silver	.36	13,800
	.37	15,700
Larch, western	.51	16,200
	.55	19,400
Pine:		
Eastern white	.34	10,600
	.35	11,300
Virginia	.45	13,700
	.48	15,000
Red	.42	15,300
Spruce, Engelmann	.32	12,300
	.34	13,000

[1] Results of tests on small, clear, straight-grained specimens. The values in the first line for each species are from tests of green material; those in the second line are from tests of dry material with the properties adjusted to 12 pct. moisture content. Specific gravity values are not from the tension specimens but others representing the species shipment. Specific gravity is based on weight when ovendry and volume when green and an adjustment to approximately 12 pct. moisture content.

Table 4–7.—*Average toughness values* [1] *for samples of a few species of wood*

Species	Moisture content	Specific gravity [2]	Toughness [3]	
			Radial	Tangential
	Pct.		In.-lb.	In.-lb.
HARDWOODS				
Birch, yellow	12	0.65	500	620
Hickory:				
(Mockernut, pignut, sand) { Green	Green	.64	700	720
	12	.71	620	660
Maple, sugar	14	.64	370	360
Oak, red:				
Pin	12	.64	430	430
Scarlet	11	.66	510	440
Oak, white:				
Overcup { Green	Green	.56	730	680
	13	.62	340	310
Sweetgum { Green	Green	.48	340	330
	13	.51	260	260
Willow, black { Green	Green	.38	310	360
	11	.40	210	230
Yellow-poplar { Green	Green	.43	320	300
	12	.45	220	210
SOFTWOODS				
Cedar:				
Alaska-	10	.48	210	230
Western redcedar	9	.33	90	130
Douglas-fir:				
Coast { Green	Green	.44	210	360
	12	.47	200	360
Interior West { Green	Green	.48	200	300
	13	.51	210	340
Interior North { Green	Green	.43	170	240
	14	.46	160	250
Interior South { Green	Green	.38	130	180
	14	.40	120	180
Fir:				
California red { Green	Green	.36	130	180
	12	.39	120	170
Noble { Green	Green	.36	------	240
	12	.39	------	220
Pacific silver { Green	Green	.37	150	230
	13	.40	170	260
White { Green	Green	.36	140	220
	13	.38	130	200
Hemlock:				
Mountain { Green	Green	.41	250	280
	14	.44	140	170
Western { Green	Green	.38	150	170
	12	.41	140	210
Larch, western { Green	Green	.51	270	400
	12	.55	210	340

Table 4-7.—*Average toughness values* [1] *for samples of a few species of wood*—continued

Species	Moisture content	Specific gravity [2]	Toughness [3]	
			Radial	Tangential
	Pct.		In.-lb.	In.-lb.

SOFTWOODS—Continued

Species	Moisture content	Specific gravity [2]	Radial	Tangential
Pine:				
Eastern white	Green	.33	120	160
	12	.34	110	120
Jack	Green	.41	200	380
	12	.42	140	240
Loblolly	Green	.48	310	380
	12	.51	160	260
Lodgepole	Green	.38	160	210
Ponderosa	Green	.38	190	270
	11	.43	150	190
Red	Green	.40	210	350
	12	.43	160	290
Shortleaf	Green	.47	290	400
	13	.50	150	230
Slash	Green	.55	350	450
	12	.59	210	320
Virginia	Green	.45	340	470
	12	.49	170	250
Redwood:				
Old-growth	Green	.39	110	200
	11	.39	90	140
Young-growth	Green	.33	110	140
	12	.34	90	110
Spruce, Engelmann	Green	.34	150	190
	12	.35	110	180

[1] Results of tests on small, clear, straight-grained specimens.
[2] Based on ovendry weight and volume at moisture content of test.
[3] Properties based on specimen size of 2 cm. square by 28 cm. long; radial indicates load applied to radial face and tangential indicates load applied to tangential face of specimens.

Table 4-8.—*A summary of reported results of fatigue studies* [1]

Loading	Conditions	Range ratio (minimum stress ÷ maximum stress)	Fatigue life (million cycles)	Fatigue strength (percent of strength from static test)
Tension parallel to grain	Clear, air dry	0.10	30	50
Cantilever bending	Clear, air dry, solid wood	−1.00	30	30
Simple beam bending	Clear, green	.10	30	60
Rotational bending	Clear, air dry	−1.00	30	28

[1] Results from Forest Products Laboratory studies except for rotational bending results (from Fuller, F. B., and Oberg, T. T. 1943. Fatigue Characteristics of Natural and Resin-Impregnated Compressed Laminated Woods. J. Aero. Sci. 10(3): 81–85.)

INFLUENCE OF GROWTH CHARACTERISTICS ON THE PROPERTIES OF CLEAR STRAIGHT-GRAINED WOOD

Clear straight-grained wood is used for determining fundamental mechanical properties; however, because of natural growth characteristics of trees, wood products vary in specific gravity, may contain cross grain, or have knots and localized slope of grain. In addition, natural defects such as pitch pockets may occur due to biological or climatic elements acting on the living tree. These wood characteristics must be taken into account in assessing actual properties or estimating actual performance of wood products.

Specific Gravity

The substance of which wood is composed is actually heavier than water, its specific gravity being about 1.5 regardless of the species of wood. In spite of this fact, the dry wood of most species floats in water, and it is thus evident that part of the volume of a piece of wood is occupied by cell cavities and pores.

Variations in the size of these openings and in the thickness of the cell walls cause some species to have more wood substance per unit volume than others and therefore to have higher specific gravity. Specific gravity thus is an excellent index of the amount of wood substance a piece of dry wood contains; it is a good index of mechanical properties so long as the wood is clear, straight grained, and free from defects. It should be noted, however, that specific gravity values also reflect the presence of gums, resins, and extractives, which contribute little to mechanical properties.

The relationships between specific gravity and various other properties have been expressed for clear straight-grained wood as power functions, based on average results of strength tests of more than 160 species. These relationships, given in table 4–9, are only approximate. For any single species, more consistently accurate relationships can be obtained from specific test results.

Table 4–9.—*Functions relating mechanical properties to specific gravity of clear, straight-grained wood*

Property	Specific gravity-strength relation [1]	
	Green wood	Air-dry wood (12 pct. moisture content)
Static bending:		
Fiber stress at proportional limit _____p.s.i.	$10,200G^{1.25}$	$16,700G^{1.25}$
Modulus of elasticity _____million p.s.i.	$2.36G$	$2.80G$
Modulus of rupture _____p.s.i.	$17,600G^{1.25}$	$25,700G^{1.25}$
Work to maximum load _____in.-lb. per cu. in.	$35.6G^{1.75}$	$32.4G^{1.75}$
Total work _____in.-lb. per cu. in.	$103G^2$	$72.7G^2$
Impact bending, height of drop causing complete failure _____in.	$114G^{1.75}$	$94.6G^{1.75}$
Compression parallel to grain:		
Fiber stress at proportional limit _____p.s.i.	$5,250G$	$8,750G$
Modulus of elasticity _____million p.s.i.	$2.91G$	$3.38G$
Maximum crushing strength _____p.s.i.	$6,730G$	$12,200G$
Compression perpendicular to grain, fiber stress at proportional limit _____p.s.i.	$3,000G^{2.25}$	$4,630G^{2.25}$
Hardness:		
End _____lb.	$3,740G^{2.25}$	$4,800G^{2.25}$
Side _____lb.	$3,420G^{2.25}$	$3,770G^{2.25}$

[1] The properties and values should be read as equations; for example: modulus of rupture for green wood = $17,600G^{1.25}$, where G represents the specific gravity of ovendry wood, based on the volume at the moisture condition indicated.

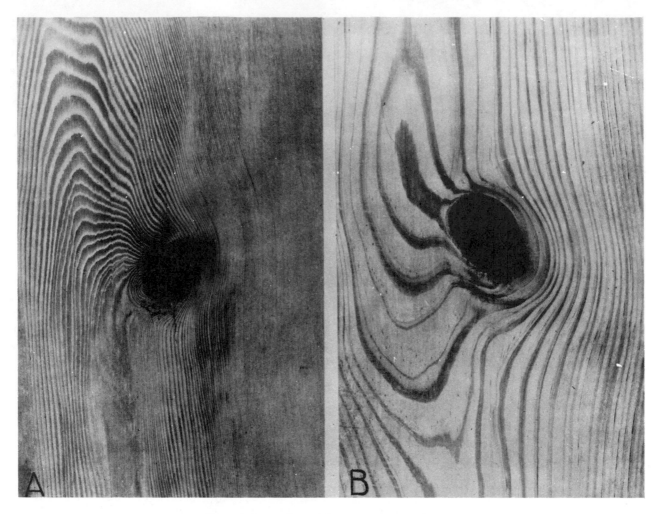

Figure 4–2.—*A*, Encased knot; *B*, intergrown knot.

Knots

A knot is that portion of a branch which has become incorporated in the bole of the tree. The influence of a knot on mechanical properties of a product is due to the interruption of continuity and change in direction of wood fibers. The influence of knots depends on their size, location, shape, soundness, attendant local slope of grain, and the type of stress to which they are subjected.

The shape (form) of a knot appearing on a sawed surface depends upon the direction of the exposing cut. When a branch is sawed through at right angles to its length, a nearly round knot results; when cut diagonally, an oval knot; and when sawed lengthwise, a spike knot.

Knots are further classified as intergrown or encased (fig. 4–2). As long as a limb remains alive, there is continuous growth at the junction of the limb and the trunk of the tree, and the resulting knot is called intergrown. After the branch has died, additional growth on the trunk encloses the dead limb, and an encased knot results; fibers of the trunk are not continuous with fibers of encased knots. Encased knots and knotholes tend to be accompanied by less cross grain than are intergrown knots and are, therefore, generally less serious with regard to some mechanical properties.

Knots decrease most mechanical properties because (1) the clear wood is displaced by the knot, (2) the fibers around the knot are distorted causing cross grain, (3) the discontinuity of wood fiber leads to stress concentrations, and (4) checking often occurs around knots in drying. Conversely, knots actually increase hardness and strength in compression perpendicular to the grain and are objectionable in regard to these properties only in that they cause nonuniform wear or nonuniform stress distributions at contact surfaces.

Wood members loaded uniformly in tension are usually more seriously affected by knots than if loaded in other ways. In some structural members, the effect of knots on strength depends not only on knot size but also on the location of the knot. For example, in a simply supported beam, knots on the lower side are placed in tension, those on the upper side in compression, and those at or near the neutral axis in horizontal shear. A knot has a marked effect on the maximum load a beam will sustain when on the tension side at the point of maximum stress; knots on the compression side are somewhat less serious.

In long columns, knots are important in that they affect stiffness. In short or intermediate columns, the reduction in strength caused by knots is approximately proportional to the size of the knot; however, large knots have a somewhat greater relative effect than do small knots.

Knots in round timbers, such as poles and piles, have less effect on strength than knots in sawed timbers. Although the grain is irregular around knots in both forms of timber, its angle with the surface is less in naturally round than in sawed timber.

Fatigue strength of clear wood is reduced by knots. The effect of knots, as well as an additive effect of knots and slope of grain on fatigue strength is discussed under the section on Time-Fatigue.

The effects of knots in structural lumber are discussed in chapter 6.

Fiber and Ring Orientation

In some wood product applications, the directions of important stresses may not coincide with the natural axes of fiber orientation in the wood. This may occur by choice in design, by the way the wood was removed from the log, or because of grain irregularities that occurred during growth.

Elastic properties in directions other than along the natural axes can be obtained from elastic theory. Strength properties in directions ranging from parallel to perpendicular to the fibers can be approximated using a Hankinson-type formula:

$$N = \frac{PQ}{P \sin^n \theta + Q \cos^n \theta}$$

in which N represents the strength property at an angle θ from the fiber direction, Q is the strength across the grain, P is the strength parallel to the grain, and n is an empirically determined constant. The formula has been used for modulus of elasticity as well as strength properties. Values of n and associated ratios of Q/P have been tabulated below from available literature:

Property	n	Q/P
Tensile strength	1.5–2	0.04–0.07
Compressive strength	2–2.5	0.03–0.4
Bending strength	1.5–2	0.04–0.1
Modulus of elasticity	2	0.04–0.12
Toughness	1.5–2	0.06–0.1

A Hankinson-type formula can be graphically depicted as a function of Q/P and n. Figure 4–3 shows the strength in any direction expressed as a fraction of the strength parallel to the fiber direction, plotted against angle to the fiber direction θ. The plot is for a range of values of Q/P and n.

The term "slope of grain" relates the fiber direction to the edges of a piece. Slope of grain is usually expressed by the ratio between a 1-inch deviation of the grain from the edge or long axis of the piece and the distance in inches within which this deviation occurs ($\tan \theta$). Table 4–10 gives the effect of grain slope on some properties of wood, as determined from tests. The values in table 4–10 for modulus of rupture fall very close to the curve in figure 4–3 for $Q/P = 0.1$ and $n = 1.5$. Similarly, the impact bending values fall close to the curve for $Q/P = 0.05$ and $n = 1.5$; and for compression, $Q/P = 0.1$, $n = 2.5$.

The term "cross grain" indicates the condition measured by slope of grain. Two important forms of cross grain are spiral grain and diagonal grain (fig. 4–4). Other types are wavy, dipped, interlocked, and curly grain.

Table 4–10.—*Strength of wood members with various grain slopes compared to strength of a straight-grained member, expressed as percentages*

Maximum slope of grain in member	Modulus of rupture	Impact bending— height of drop causing complete failure (50-lb. hammer)	Compression parallel to grain— maximum crushing strength
	Pct.	*Pct.*	*Pct.*
Straight-grained	100	100	100
1 in 25	96	95	100
1 in 20	93	90	100
1 in 15	89	81	100
1 in 10	81	62	99
1 in 5	55	36	93

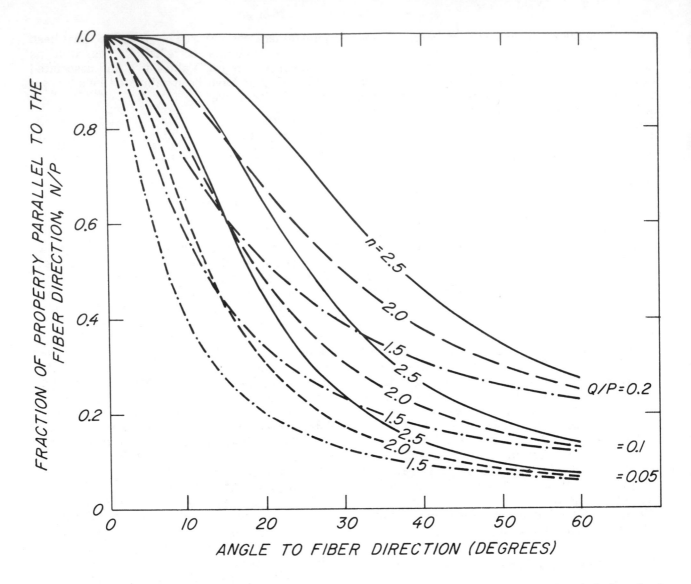

Figure 4–3.—Effect of grain angle on mechanical property of clear wood according to a Hankinson-type formula. Q/P is the ratio of the mechanical property across the grain (Q) to that parallel to the grain (P); n is an empirically determined constant.

Spiral grain in a tree is caused by fibers growing in a winding or spiral course about the bole of the tree instead of in a vertical course. In sawn products spiral grain can be defined as fibers lying in the tangential plane of the growth rings, not parallel to the longitudinal axis of the product (see fig. 4–4B for a simple case). Spiral grain often is not readily detected by ordinary visual inspection in sawn products. The best test for spiral grain is to split a sample section from the piece in the radial direction. A nondestructive method of determining the presence of spiral grain is to note the alinement of pores, rays, and resin ducts on the flat-grain face. Drying checks on a flat-sawn surface follow the fibers and indicate the fiber slope.

Diagonal grain describes cross grain caused by growth rings not parallel to one or both surfaces of the sawn piece. Diagonal grain is produced by sawing parallel to the axis (pith) of the tree in a log having pronounced taper, and is a common occurrence in sawing crooked or swelled logs.

Cross grain can be quite localized as a result of the disturbance of growth patterns by a branch. This condition, termed "local slope of grain," may be present even though the branch (knot) may have been removed in a sawing operation. Often the degree of local cross grain may be difficult to determine.

Any form of cross grain can have a serious effect on mechanical properties or machining characteristics. Spiral and diagonal grain com-

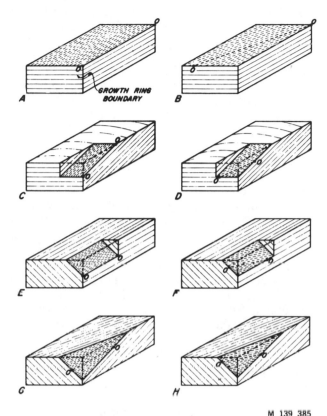

Figure 4-4.—Schematic views of wood specimens containing straight grain and cross grain to illustrate the relationship of fiber orientation (0–0) to the axes of the piece. Specimens A through D have radial and tangential surfaces; E through H do not. A and E contain no cross grain. B, D, F, and H have spiral grain. C, D, G, and H have diagonal grain.

bine to produce a more complex cross grain. To determine net cross grain, regardless of origin, fiber slopes on contiguous surfaces of a piece must be measured and combined. The combined slope of grain is determined by taking the square root of the sum of the squares of the two slopes. For example, assume the spiral grain slope on the flat grain surface of figure 4–4D is 1 in 12 and the diagonal grain slope is 1 in 18. The combined slope is

$$\sqrt{\left(\frac{1}{18}\right)^2 + \left(\frac{1}{12}\right)^2} = \frac{1}{10} \text{ or a slope of 1 in 10}$$

Stresses perpendicular to the fiber (grain) direction may be at any angle from 0° (T) to 90° (R) to the growth rings (fig. 4–5). Perpendicular-to-grain properties depend somewhat upon orientation of annual rings with respect to the direction of stress. Compression perpendicular-to-grain values in table 4–2 are derived from tests in which the load is applied parallel to the growth rings (T-direction); tension perpendicular-to-grain val-

ues are an average of an equal number of specimens with 0° and 90° growth ring orientation. Modulus of elasticity is least for the 45° orientation, intermediate at 0°, and highest at 90 to the growth rings. Proportional limit in compression increases gradually as the ring angle increases from 0° (T) to 90° (R), the total increase amounting to about one-third. Similar observations have been made in tension. For some softwoods in compression the values at 0° and 90° are about the same, with the value at 45° about two-thirds of the others.

Reaction Wood

Abnormal woody tissue is frequently associated with leaning boles and crooked limbs of both conifers and hardwoods. It is generally believed that it is formed as a natural response of the tree to return its limbs or bole to a more normal position, hence the term "reaction wood." In softwoods, the abnormal tissue is called "compression wood." It is common to all softwood species and is found on the lower side of the limb or inclined bole. In hardwoods, the abnormal tissue is known as "tension wood;" it is located on the upper side of the inclined member, although in some instances it is distributed irregularly around the cross section. Reaction wood is more prevalent in some species than in others.

Many of the anatomical, chemical, physical, and mechanical properties of reaction wood differ distinctly from those of normal wood. Perhaps most evident is the increase in the density over that of normal wood. The specific gravity of compression wood is frequently 30 to 40 percent greater than normal, while tension wood commonly ranges between 5 and 10

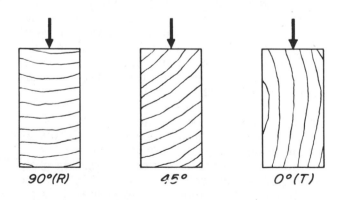

90°(R)   45°   0°(T)

Figure 4–5.—The direction of load in relation to the direction of the annual growth rings: 90° or perpendicular (R); 45°; 0° or parallel (T).

Figure 4-6.—The darker areas shown within the rectangle are compression wood.

about the pith and by the large proportion of summerwood at the point of greatest eccentricity (fig. 4-6). It is more difficult to detect in lumber; however, it is usually somewhat darker because of the greater proportion of summerwood and frequently has a relatively lifeless appearance, especially in woods which normally have an abrupt transition from springwood to summerwood (fig. 4-7). Because it is more opaque than normal wood, intermediate stages of compression wood can be detected by transmitted light through thin cross sections.

Tension wood is more difficult to detect than compression wood. However, eccentric growth as seen on the transverse section frequently indicates its presence. Also, the tough tension wood fibers resist being cut cleanly and result in a woolly condition on the surfaces of sawn boards, especially when surfaced in the green condition (fig. 4-8). In some species, tension wood may show up on a smooth surface as areas of contrasting colors. Examples of this are the silvery appearance of tension wood in sugar maple and the darker color of tension wood in mahogany.

Compression Failures

Excessive bending of standing trees from wind or snow, felling trees across boulders, logs, or irregularities in the ground, or the rough handling of logs or lumber may produce excessive compression stresses along the grain that cause minute compression failures. In some instances, such failures are visible on the surface of a board as minute lines or zones formed by the crumpling or buckling of the cells (fig. 4-9A), although usually they appear only as white lines or may even be invisible to the naked eye. Their presence frequently is

percent greater but may be as much as 30 percent greater.

Compression and tension wood undergo excessive longitudinal shrinkage when subjected to moisture loss reaching below the fiber saturation point. Longitudinal shrinkage ranges to 10 times normal in compression wood and perhaps five times normal or more in tension wood. When present in the same board with normal wood, unequal longitudinal shrinkage causes internal stresses that result in warping. This warp sometimes occurs in rough lumber but more often in planed, ripped, or resawed lumber. Fortunately, the most serious effects occurring in pronounced compression wood can be detected by ordinary visual examination, as compared to borderline forms that merge with normal wood and frequently are only detected by microscopic examination.

Reaction wood, particularly compression wood in the green condition, may be somewhat stronger than normal wood. However, when compared to normal wood of comparable specific gravity, the reaction wood is definitely weaker. Possible exceptions to this are compression parallel-to-grain properties of compression wood and impact bending properties of tension wood.

Because of its abnormal properties, it is frequently desirable to eliminate reaction wood from raw material. In logs, compression wood may be characterized by eccentric growth

Figure 4-7.—Sitka spruce boards containing normal wood (left) and compression wood (right).

indicated by fiber breakage on end grain (fig. 4–9B). Compression failures should not be confused with compression wood.

Products containing visible compression failures can have seriously low strength properties, especially in tensile strength and shock resistance. Tensile strength of wood containing compression failures has been found to be as low as one-third of the strength of matched clear wood. Even small compression failures, visible only under the microscope, seriously reduce strength and cause brittle fracture. Because of the low strength asssociated with compression failures, many safety codes require certain structural members, such as ladder rails and scaffold planks, to be entirely free of them.

Compression failures are often difficult to detect with the unaided eye, and special efforts including optimum lighting are required to aid detection.

Pitch Pockets

A pitch pocket is a well-defined opening that contains free resin. It extends parallel to the annual rings, and is almost flat on the pith side and curved on the bark side. Pitch pockets are confined to such species as the

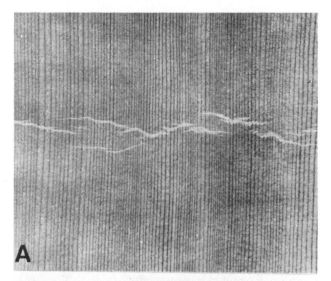

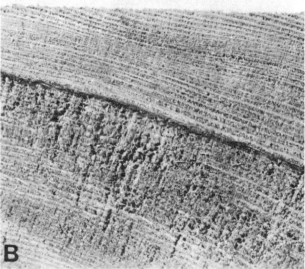

Figure 4–9.—*A.* Compression failure is shown by the irregular lines across the grain. *B.* End-grain surfaces of surfaces of spruce lumber show fiber breakage caused by compression failures below the dark line.

pines, spruces, Douglas-fir, tamarack, and western larch.

The effect of pitch pockets on strength depends upon their number, size, and location in the piece. A large number of pitch pockets indicates a lack of bond between annual growth layers, and a piece containing them should be inspected for shake or separations along the grain.

Bird Peck

Maple, hickory, white ash, and a number of other species are often damaged by small holes made by woodpeckers. These holes or bird

Figure 4–8.—Projecting tension wood fibers on the sawn surface of a mahogany board.

pecks are often placed in horizontal rows, sometimes encircling the tree, and a brown or black discoloration known as a mineral streak originates there. Holes for tapping maple trees are also a source of mineral streaks. The streaks are caused by oxidation and other chemical changes in the wood.

Bird pecks and mineral streaks are not generally important in regard to strength, although they do impair the appearance of the wood. However, if several bird pecks occur in a row across the outer surface of a piece of wood that is to be used in a bent product, such as a handle, the holes can appreciably weaken the product.

Extractives

Many species of wood contain extraneous materials or extractives that can be removed by solvents that do not degrade the cellulosic/lignin structure of the wood. These extractives are especially abundant in species such as larch, redwood, western redcedar, and black locust.

A small increase in modulus of rupture and in strength in compression parallel to grain has been measured for some species containing extractives. The extent to which the extractives influence the strength is apparently a function of the amount of extractives, the moisture content of the piece, and the mechanical property under consideration.

Timber From Live Versus Dead Trees

Timber from trees killed by insects, blight, wind, or fire may be as good for any structural purpose as that from live trees, provided further insect attack, staining, decay, or seasoning degrade has not occurred. In considering the subject, it may be useful to remember that the heartwood of a living tree is entirely dead, and in the sapwood only a comparatively few cells are living. Therefore, most wood is dead when cut, regardless of whether the tree itself is living or not. However, if a tree stands on the stump too long after its death, the sapwood is likely to decay or to be attacked severely by wood-boring insects, and in time the heartwood will be similarly affected. Such deterioration occurs also in logs that have been cut from live trees and improperly cared for afterwards. Because of variations in climatic and local weather conditions and in other factors that affect deterioration, the length of the period during which dead timber may stand or lie in the forest without serious deterioration varies.

Tests on wood from trees that had stood as long as 15 years after being killed by fire demonstrated that this wood was as sound and as strong as wood from live trees. Also, logs of some of the more durable species have had thoroughly sound heartwood after lying on the ground in the forest for many years.

On the other hand, decay may cause great loss of strength within a very brief time, both in trees standing dead on the stump and in logs cut from live trees and allowed to lie on the ground. The important consideration is not whether the trees from which timber products are cut are alive or dead, but whether the products themselves are free from decay or other defects that would render them unsuitable for use.

EFFECT OF MANUFACTURING AND SERVICE ENVIRONMENT ON MECHANICAL PROPERTIES OF CLEAR STRAIGHT-GRAINED WOOD

Moisture Content—Drying

Many mechanical properties are affected by changes in moisture content below the fiber saturation point. Most properties reported in tables 4–2, 4–3, and 4–4 increase with decrease in moisture content. The relation that describes these clear wood property changes in the vicinity of 70° F. is:

$$P = P_{12} \left(\frac{P_{12}}{P_g} \right)^{-\left(\frac{M-12}{M_p -12} \right)}$$

where P is the property and M the moisture content in percent. M_p is the moisture content at which property changes due to drying are first observed. This moisture content is slightly less than the fiber saturation point. (Table 4–11 gives values of M_p for a few species; for other species, $M_p = 25$ may be assumed.)

P_{12} is the property value at 12 percent moisture content, and P_g (green condition) is the property value for all moisture contents greater than M_p. Average property values of P_{12} and P_g are given for many species in tables 4–2, 4–3, and 4–4.

The formula for moisture content adjustment is not recommended for work to maximum load, impact bending, and tension perpendicular. These properties are known to be erratic in

Table 4–11.—*Moisture content at which proper-ties change due to drying for selected species*

Species	M_p
	Pct.
Ash white	24
Birch, yellow	27
Chestnut, American	24
Douglas-fir	24
Hemlock, western	28
Larch, western	28
Pine, loblolly	21
Pine, longleaf	21
Pine, red	24
Redwood	21
Spruce, red	27
Spruce, Sitka	27
Tamarack	24

their response to moisture content change, often apparently as a function of species.

The formula can be used to estimate a property at any moisture content below M_p from the species data given. For example, suppose the modulus of rupture of white ash at 8 percent moisture content is wanted. Using information from table 4–2:

$$P_8 = (15,400) \left(\frac{15,400}{9,600} \right)^{-\left(\frac{-4}{12} \right)}$$

$$P_8 = 18,020 \text{ p.s.i.}$$

The increase in mechanical properties discussed above assume small, clear specimens in a drying process in which no deterioration of the product (degrade) occurs. The property changes applied to large wood specimens such as lumber are discussed in chapter 6.

Drying degrade can take several forms. Perhaps the most common degrade is surface and end checking. Checks most often limit mechanical properties. Some loss of strength, especially in shock resistance, may occur when wood is dried at excessively high temperatures, although visible signs of degrade may not be present.

Further information is included in chapter 14.

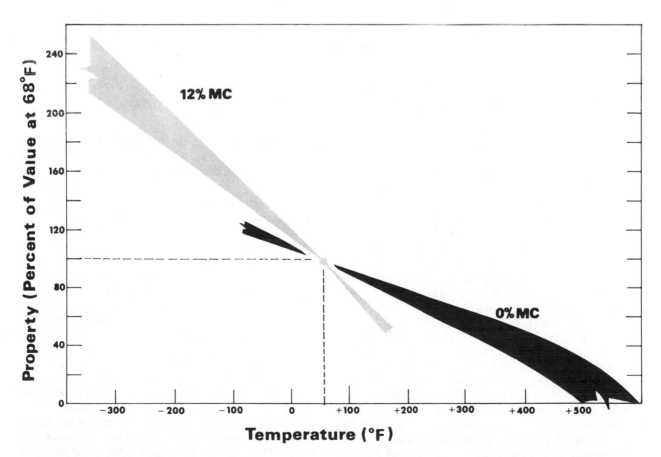

Figure 4–10.—The immediate effect of temperature on strength properties, expressed as percent of value at 68° F. Trends illustrated are composites from studies on three strength properties—modulus of rupture in bending, tensile strength perpendicular to grain, and compressive strength parallel to grain—as examined by several investigators. Variability in reported results is illustrated by the width of the bands.

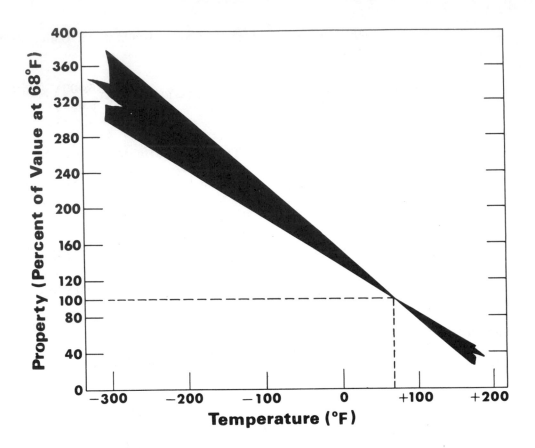

Figure 4–11.—The immediate effect of temperature on the proportional limit in compression perpendicular to grain at approximately 12 percent moisture content relative to the value at 68° F. Variability in the reported results of several investigators is illustrated by the width of the band.

Temperature

In general, the mechanical properties of wood decrease when heated and increase when cooled. This effect is immediate, but prolonged exposure at high temperature causes an irreversible decrease in properties.

At a constant moisture content and below about 400° F. mechanical properties are essentially linearly related to temperature. The change in properties that occurs when wood is quickly heated or cooled and then tested at that condition is termed an "immediate effect." At temperatures below 200° F. the immediate effect is essentially reversible; that is, the property will return to the value at the original temperature if the temperature change is rapid.

Figure 4–10 illustrates the immediate effect of temperature on strength, based on a composite of tests for modulus of rupture in bending, tensile strength perpendicular to the grain, and compressive strength parallel to the grain relative to 68° F. Figure 4–11 gives similar information for proportional limit in compression perpendicular to grain. Figure

4–12 illustrates the immediate effect of temperature on the modulus of elasticity as a composite of measurements made in bending, in tension parallel to grain, and in compression parallel to grain. Figures 4–10 through 4–12 represent an interpretation of data from a limited number of investigators. The width of the band illustrates variability between and within reported results. The influence of moisture content is illustrated where data are available.

In addition to the reversible effect of temperature on wood, there is an irreversible effect at elevated temperature. This permanent effect is one of degradation of wood substance, which results in loss of weight and strength. Quantitatively, the loss depends on factors which include moisture content, heating medium, temperature, exposure period, and, to some extent, species and size of piece involved.

The decrease of modulus of rupture due to heating in steam and in water is shown as a function of temperature and heating time in figure 4–13, based on tests of Douglas-fir and Sitka spruce. From the same studies work to maximum load was more greatly affected than modulus of rupture by heating in water (fig. 4–14). The effect of oven heating (wood at 0 pct. moisture content) on the modulus of rupture and modulus of elasticity is shown in figures 4–15 and 4–16, respectively, as derived from tests on four softwoods and two hardwoods. Note that the permanent property losses discussed above are based on tests conducted after the specimens have been cooled to near 75° F. and conditioned to the range of 7 to 12 percent moisture content. If tested hot, presumably immediate and permanent effects would be additive. The extent of property loss is considered greater for hardwoods than for softwoods.

Repeated exposure to elevated temperature has a cumulative effect on wood properties. For example, at a given temperature the property loss will be about the same after six exposure periods of 1 month each as it would after a single 6-month exposure period.

The shape and size of wood pieces are im-

portant in analyzing the influence of temperature. If the exposure is for only a short time, so that the inner parts of a large piece do not reach the temperature of the surrounding medium, the immediate effect on strength of inner parts will be less than for outer parts. The type of loading must be considered however. If the member is to be stressed in bending, the outer fibers of a piece are subjected to the greatest load and will ordinarily govern the ultimate strength of the piece; hence, under this loading condition, the fact that the inner part is at a lower temperature may be of little significance.

For extended, noncyclic exposures, it can be assumed that the entire piece reaches the temperature of the heating medium and will, therefore, be subject to permanent strength losses throughout the volume of the piece, regardless of size and mode of stress application. However, wood often will not reach the daily extremes in temperature of the air around

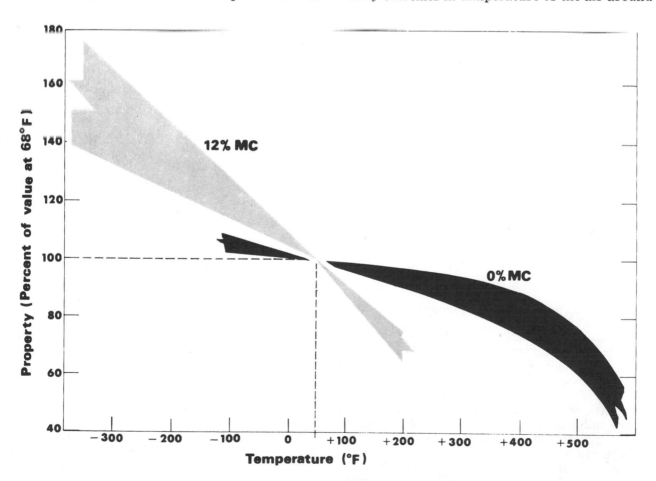

Figure 4–12.—The immediate effect of temperature on the modulus of elasticity, relative to the value at 68° F. The plot is a composite of studies on the modulus as measured in bending, in tension parallel to the grain, and in compression parallel to grain by several investigators. Variability in reported results is illustrated by the width of the bands.

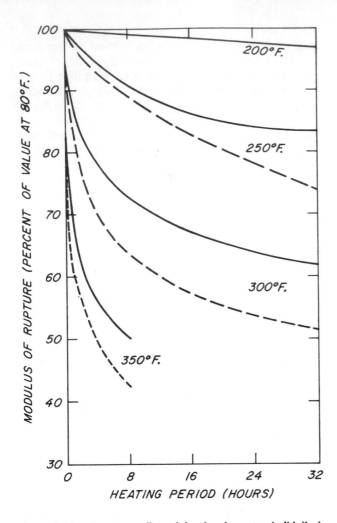

Figure 4–13.—Permanent effect of heating in water (solid line) and in steam (dashed line) on the modulus of rupture. Data based on tests of Douglas-fir and Sitka spruce.

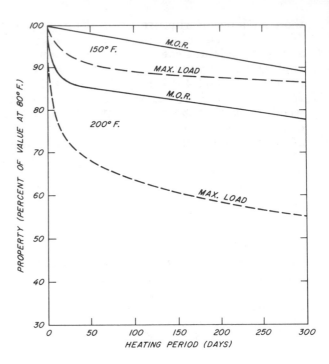

Figure 4–14.—Permanent effect of heating in water on work to maximum load and on modulus of rupture. Data based on tests of Douglas-fir and Sitka spruce.

it in ordinary construction; thus, long-term effects should be based on the accumulated temperature experience of critical structural parts.

Time

Time—Creep/Relaxation

When first loaded, a wood member deforms elastically. If the load is maintained, additional time-dependent deformation occurs. This is called creep. Even at very low stresses, creep takes place and can continue over a period of years. For suitably high loads, failure will eventually occur. This failure phenomenon, termed "duration of load," is discussed in the next section.

At typical design levels the additional deformation due to creep may approximately equal the initial, instantaneous elastic deformation if the environmental conditions are not

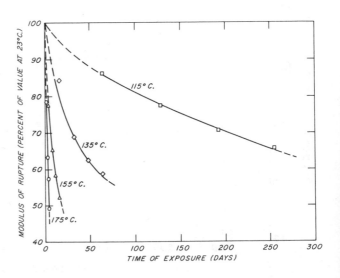

Figure 4–15.—Permanent effect of oven heating at four temperatures on the modulus of rupture, based on four softwood and two hardwood species.

102

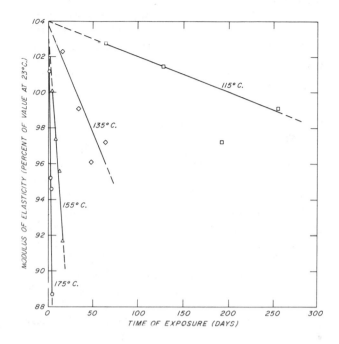

Figure 4—16.—Permanent effect of oven heating at four temperatures on modulus of elasticity, based on four softwood and two hardwood species.

changed. For illustration, a creep curve based on one species at several stress levels is shown in figure 4–17. It suggests that creep is greater under higher stresses than lower ones.

Ordinary climatic variations in temperature and humidity will cause creep to increase. An increase of about 50° F. in temperature can cause a two- to three-fold increase in creep. Green wood may creep four to six times the initial deformation as it dries under load.

Unloading the member results in an immediate and complete disappearance of the original elastic deformation as well as a delayed recovery of approximately one half of the creep deformation. Fluctuations in temperature and humidity increase the magnitude of the recovered deformation.

Creep at low stress levels is similar in bending, tension, or compression parallel to grain. It is reported to be somewhat less in tension than in bending or compression under varying moisture conditions. Creep across the grain is qualitatively similar to, but likely to be greater than, creep parallel to the grain. The creep behavior of all species studied for creep properties is approximately the same.

If, instead of controlling load or stress, a constant deformation is imposed and maintained on a wood member, the initial stress re-

laxes at a decreasing rate to about 60 to 70 percent of its original value within a few months. This reduction of stress with time is commonly termed "relaxation."

In limited bending tests carried out between approximately 65° F. and 120° F. over 2 to 3 months, the curve of stress vs. time that expresses relaxation is approximately the mirror image of the creep curve (deformation vs. time). These tests were carried out at stresses up to about 50 percent of the bending strength of the wood. As with creep, relaxation is markedly affected by fluctuations in temperature and humidity.

Time—Duration of Stress

The duration of stress, or the time during which a load acts on a wood member, is an important factor in determining the load that a member can safely carry. For members that continuously carry loads for long periods of time, the load required to produce failure is much less than that determined from the strength properties in tables 4–2, 4–3, and 4–4. For example, a wood member under the continuous action of bending stress for 10 years will carry only about 60 percent of the load required to produce failure in the same specimen loaded in a standard bending strength test of only a few minutes' duration.

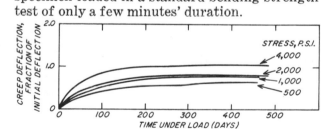

Figure 4–17.—An illustration of creep as influenced by four levels of stress. (Adapted from Kingston.)

Conversely, if the duration of stress is very short, the load-carrying capacity may be considerably higher than that determined from strength properties given in the tables. The load required to produce failure in a wood member in 1 second is approximately 25 percent higher than that obtained in ASTM standard strength tests. As an approximate indication of relation of strength to duration of stress, the strength may be said to increase or decrease 7 to 8 percent as the duration of stress is respectively decreased or increased by a factor of 10. The duration of stress is one of the factors used in establishing safe allowable stresses for structural timbers; this aspect is discussed in chapter 6.

Time—Fatigue

Fatigue in engineering material is defined as the progressive damage and failure that occurs when a structure or part is subjected to repeated loads of a magnitude smaller than the static strength.

Fatigue should be considered in wood design when repetitions of design stress or near-design stress are expected to be more than 100,000 cycles during the normal life of a structure. In many design considerations, the fatigue criteria specified will be: (1) Fatigue life (the number of stress cycles to be sustained), (2) range ratio (the ratio of minimum to maximum stress), and (3) the type of loading expected (tension parallel, tension perpendicular, or flexure). The problem will be to determine the greatest stress that can be sustained (proportioned to fatigue strength). If an indefinite fatigue life is specified, the fatigue limit stress should not be exceeded. This is the stress below which a material can be presumed to endure an infinite number of stress cycles.

The repetition of fatigue stresses can take different forms of range ratio, as from zero to some specified stress, or from some specified stress value to a higher stress in the same direction. They may be partially or completely reversed. Partially reversed stresses occur where repeated stresses are not of the same magnitude in the two directions; fully reversed stresses are of equal magnitude, as in compression and tension or as in positive and negative shear. Fully reversed stressing is the most severe loading condition. The range ratio is expressed as −1 for fully reversed loading, as a negative decimal for partially reversed loading, and as a positive decimal for repeated loading. Fatigue strength, fatigue life, and range ratio are reported for several types of loading in table 4–8.

Tests demonstrate that specimens of clear, straight-grained wood subjected to 2 million cycles of bending will have 60 percent of the strength of similar specimens tested under static conditions. In similar fatigue tests, specimens with small knots (ranging from 50 to 90 pct. in estimated strength ratio) had 50 percent of the static strength of clear, straight-grained wood; specimens with 1:12 slope of grain had 45 percent of the static strength value. When knots and slope of grain were both present, the specimens had approximately 30 percent of the static strength figure. This can be illustrated as follows:

Strength	Specimen	Index of Strength (Pct.)
Static	Clear, straight grained	100
Fatigue (2 x 10^6 cycles)	Clear, straight grained	60
Do	Small knots, straight grained	50
Do	Clear, 1:12 slope of grain	45
Do	Small knots, 1:12 slope of grain	30

Time—Age

In relatively dry and moderate temperature conditions where wood is protected from deteriorating influences such as decay, the mechanical properties of wood show little change with time. Test results for very old timbers suggest that significant losses in strength only occur after several centuries of normal aging conditions. The soundness of centuries-old wood in some standing trees (redwood, for example) also attests to the durability of wood.

Chemicals

The effect of chemicals on mechanical properties depends on the specific type of chemical. Nonswelling liquids, such as petroleum oils and creosote, have no appreciable effect on properties. Properties are lowered in the presence of water, alcohol, or other wood-swelling organic liquids even though these liquids do not chemically degrade the wood substance. The loss in properties largely depends on amount of swelling and this loss is regained upon removal of the swelling liquid. Liquid ammonia markedly reduces the strength and stiffness of wood, but most of the reduction is regained upon removal of the ammonia.

Chemical solutions that decompose wood substance have a permanent effect on strength. The following generalizations summarize the effect of chemicals: (1) Some species are quite resistant to attack by dilute mineral and organic acids, (2) oxidizing acids such as nitric acid degrade wood more than nonoxidizing acids, (3) alkaline solutions are more destructive than acidic solutions, and (4) hardwoods are more susceptible to attack by both acids and alkalies than are softwoods. Because both species and application are extremely important, reference to industrial sources with a specific history of use is recommended where possible. For example, large cypress tanks have survived long continuous use where exposure conditions involved mixed acids at the boiling point.

Wood products sometimes are treated with preservative or fire-retarding salts, usually in water solution, to impart resistance to decay or fire. Such products generally are kiln dried after treatment. Mechanical properties are essentially unchanged by preservative treatment.

Properties are, however, affected to some extent by the combined effects of fire-retardant chemicals, treatment methods, and kiln drying. A variety of fire-retardant treatments have been studied. Collectively the studies indicate modulus of rupture, work to maximum load, and toughness are reduced by varying amounts depending on species and type of fire retardant. Work to maximum load and toughness are most affected, with reductions of as much as 45 percent. A reduction in modulus of rupture of as much as 20 percent has been observed; a design reduction of 10 percent is frequently used. Stiffness is not appreciably affected by fire-retardant treatments.

Wood is also sometimes impregnated with monomers, such as methyl methacrylate, which are subsequently polymerized. Many of the properties of the resulting composite are higher than those of the original wood, generally as a result of filling the void spaces in the wood structure with plastic. The polymerization process and both the chemical nature and quantity of monomers are variables that influence composite properties.

A general discussion of the resistance of wood to chemical degradation is given in chapter 3.

Nuclear Radiation

Very large doses of gamma rays or neutrons can cause substantial degradation of wood. In general, irradiation with gamma rays in doses up to about 1 megarad has little effect on the strength properties of wood. As dosage increases above 1 megarad, tensile strength parallel to grain and toughness decrease. At a dosage of 300 megarads, tensile strength is reduced about 90 percent. Gamma rays also affect compressive strength parallel to grain above 1 megarad, but strength losses with further dosage are less than for tensile strength. Only about one-third of the compressive strength is lost when the total dose is 300 megarads. Effects of gamma rays on bending and shear strength are intermediate between the effects on tensile and compressive strength.

Molding and Staining Fungi

Molding and staining fungi do not seriously affect most mechanical properties of wood because they feed upon substance within the structural cell wall rather than on the structural wall itself. Specific gravity may be reduced by from 1 to 2 percent, while most of the strength properties are reduced by a comparable or only slightly greater extent. Toughness or shock resistance, however, may be reduced by up to 30 percent. The duration of infection and the species of fungi involved are important factors in determining the extent of weakening.

Although molds and stains themselves often do not have a major effect on the strength of wood products, conditions that favor the development of these organisms are likewise ideal for the growth of wood-destroying (decay) fungi, which can greatly reduce mechanical properties (see ch. 17).

Decay

Unlike the molding and staining fungi, the wood-destroying (decay) fungi seriously reduce strength. Even apparently sound wood adjacent to obviously decayed parts may contain hard-to-detect, early (incipient) decay that is decidedly weakening, especially in shock resistance.

All wood-destroying fungi do not affect wood in the same way. The fungi that cause an easily recognized pitting of the wood, for example, may be less injurious to strength than those that, in the early stages, give a slight discoloration of the wood as the only visible effect.

No method is known for estimating the amount of reduction in strength from the appearance of decayed wood. Therefore, when strength is an important consideration, the safe procedure is to discard every piece that contains even a small amount of decay. An exception may be pieces in which decay occurs in a knot but does not extend into the surrounding wood.

Insect Damage

Insect damage may occur in standing trees, logs, and unseasoned or seasoned lumber. Damage in the standing tree is difficult to control, but otherwise insect damage can be largely eliminated by proper control methods.

Insect holes are generally classified as pinholes, grub holes, and powderpost holes. The powderpost larvae, by their irregular burrows, may destroy most of the interior of a piece, while the surface shows only small holes, and the strength of the piece may be reduced virtually to zero.

No method is known for estimating the amount of reduction in strength from the appearance of insect-damaged wood, and, when strength is an important consideration, the safe procedure is to eliminate pieces containing insect holes.

BIBLIOGRAPHY

American Society for Testing and Materials
  Standard methods for testing small clear specimens of timber. D 143. (See current edition.) Philadelphia, Pa.
Bendtsen, B. A., Freese, Frank, and Ethington, R. L.
  1970. Methods for sampling clear, straight-grained wood from the forest. Forest Prod. J. 20(11): 38–47.
Boller, K. H.
  1954. Wood at low temperatures. Modern Packaging 27(9).
Coffey, D. J.
  1962. Effects of knots and holes on the fatigue strength of quarter-scale timber bridge stringers. M.S. Thesis, Civil Eng., Univ. of Wisconsin.
Comben, A. J.
  1964. The effect of low temperatures on the strength and elastic properties of timber. Jour. Inst. Wood Sci. 13: 44–55. (Nov.)
Ellwood, E. L.
  1954. Properties of American beech in tension and compression perpendicular to the grain and their relation to drying. Yale Univ. Bull. 61.
Fukuyama, M., and Takemura, T.
  1962. The effects of temperature on compressive properties perpendicular to grain of wood. Jour. Jap. Wood Res. Soc. 8(4).
———, and Takemura, T.
  1962. The effect of temperature on tensile properties perpendicular to grain of wood. Jour. Jap. Wood Res. Soc. 8(5).
Gerhards, C. C.
  1968. Effects of type of testing equipment and specimen size on toughness of wood. USDA Forest Serv. Res. Pap. FPL 97. Forest Prod. Lab., Madison, Wis.
Hearmon, R. F. S.
  n.d. Applied anisotropic elasticity. Pergamon Press, London.
Kingston, R. S. T.
  1962. Creep, relaxation, and failure of wood. Res. App. in Ind. 15(4).
Kollmann, Franz F. P., and Côté, Wilfred A. Jr.
  1968. Principles of wood science and technology. Springer Verlag, New York.

Kukachka, B. F.
  1970. Properties of imported tropical woods. USDA Forest Serv. Res. Pap. FPL 125. Forest Prod. Lab., Madison, Wis.
Little, E. L. Jr.
  1953. Checklist of native and naturalized trees of the United States. USDA Agr. Handbook 41. 472 pp.
MacLean, J. D.
  1953. Effect of steaming on the strength of wood. Amer. Wood-Preservers Assoc. 49: 88–112.

———
  1954. Effect of heating in water on the strength properties of wood. Amer. Wood-Preservers Assoc. 50: 253–281.
Millett, M. A., and Gerhards, C. C.
  1972. Accelerated aging: Residual weight and flexural properties of wood heated in air at 115° to 175° C. Wood Sci. 4(4).
Munthe, B. P., and Ethington, R. L.
  1968. Method for evaluating shear properties of wood. USDA Forest Serv. Res. Note FPL–0195. Forest Prod. Lab., Madison, Wis.
Pillow, M. Y.
  1949. Studies of compression failures and their detection in ladder rails. Forest Prod. Lab. Rep. D 1733.
Schaffer, E. L.
  1970. Elevated temperature effect on the longitudinal mechanical properties of wood. Ph.D. thesis. Univ. of Wis.
Sulzberger, P. H.
  1953. The effect of temperature on the strength of wood, plywood and glued joints. Asst. Aero. Res. Cons. Comm. Rep. ACA–46.
U.S. Department of Defense
  1951. Design of wood aircraft structures. ANC–18 Bull. (Issued by Subcommittee) on Air-Force-Navy-Civil Aircraft. Design Criteria Aircraft Comm.) 2nd ed. Munitions Bd. Aircraft Comm. 234 pp., illus.
Wangaard, F. F.
  1966. Resistance of wood to chemical degradation. Forest Prod. J. 16(2): 53–64.
Youngs, R. L.
  1957. Mechanical properties of red oak related to drying. Forest Prod. J. 7(10): 315–324.

APPENDIX

In this appendix, tables 4–2, 4–3, and 4–4 have been rewritten in SI or metric units, and thus appear as tables A–4–2, A–4–3, and A–4–4. The units used follow the ASTM Metric Practice Guide, E 380–72, where the newton is the basic measure of force, the pascal that of stress, the meter that of length, and the joule that of energy.

Table A-4-2.—Strength properties [1,2] of some commercially important woods grown in the United States

Common names of species	Specific gravity [3]	Static bending Modulus of— Rupture (Kilopascals)	Static bending Modulus of— Elasticity [4] (Megapascals)	Work to maximum load (Kilojoules per cu. m.)	Impact bending—height of drop causing complete failure (Mm.)	Compression parallel to grain—maximum crushing strength (Kilopascals)	Compression perpendicular to grain—fiber stress at proportional limit (Kilopascals)	Shear parallel to grain—maximum shearing strength (Kilopascals)	Tension perpendicular to grain—maximum tensile strength (Kilopascals)	Side hardness—load perpendicular to grain (Newtons)
HARDWOODS										
Alder, red	0.37	45,000	8,100	55	560	20,400	1,700	5,300	2,700	2,000
	.41	68,000	9,500	58	510	40,100	3,000	7,400	2,900	2,600
Ash:										
Black	.45	41,000	7,200	83	840	15,900	2,400	5,900	3,400	2,300
	.49	87,000	11,000	103	890	41,200	5,200	10,800	4,800	3,800
Blue	.53	66,000	8,500	101	----	24,800	5,600	10,600	----	----
	.58	95,000	9,700	99	----	48,100	9,800	14,000	----	----
Green	.53	66,000	9,700	81	890	29,000	5,000	8,700	4,100	3,900
	.56	97,000	11,400	92	810	48,800	9,000	13,200	4,800	5,300
Oregon	.50	52,000	7,800	84	990	24,200	3,700	8,200	4,100	3,500
	.55	88,000	9,400	99	840	41,600	8,600	12,300	5,000	5,200
White	.55	66,000	9,900	114	970	27,500	4,600	9,500	4,100	4,300
	.60	106,000	12,000	121	1090	51,100	8,000	13,400	6,500	5,900
Aspen:										
Bigtooth	.36	37,000	7,700	39	----	17,200	1,400	5,000	----	----
	.39	63,000	9,900	53	----	36,500	3,100	7,400	----	----
Quaking	.35	35,000	5,900	44	560	14,800	1,200	4,600	1,600	1,300
	.38	58,000	8,100	52	530	29,300	2,600	5,900	1,800	1,600
Basswood, American	.32	34,000	7,200	37	410	15,300	1,200	4,100	1,900	1,100
	.37	60,000	10,100	50	410	32,600	2,600	6,800	2,400	1,800
Beech, American	.56	59,000	9,500	82	1090	24,500	3,700	8,900	5,000	3,800
	.64	103,000	11,900	104	1040	50,300	7,000	13,900	7,000	5,800
Birch:										
Paper	.48	44,000	8,100	112	1240	16,300	1,900	5,800	2,600	2,500
	.55	85,000	11,000	110	860	39,200	4,100	8,300	----	4,000
Sweet	.60	65,000	11,400	108	1220	25,800	3,200	8,500	3,000	4,300
	.65	117,000	15,000	124	1190	58,900	7,400	15,400	6,600	6,500
Yellow	.55	57,000	10,300	111	1220	23,300	3,000	7,700	3,000	3,500
	.62	114,000	13,900	143	1400	56,300	6,700	13,000	6,300	5,600

Table A-4-2.—*Strength properties* [1,2] *of some commercially important woods grown in the United States*—continued

Common names of species	Specific gravity [3]	Static bending — Modulus of — Rupture (Kilopascals)	Static bending — Modulus of — Elasticity [4] (Megapascals)	Work to maximum load (Kilojoules per cu. m.)	Impact bending—height of drop causing complete failure (Mm.)	Compression parallel to grain—maximum crushing strength (Kilopascals)	Compression perpendicular to grain—fiber stress at proportional limit (Kilopascals)	Shear parallel to grain—maximum shearing strength (Kilopascals)	Tension perpendicular to grain—maximum tensile strength (Kilopascals)	Side hardness—load perpendicular to grain (Newtons)
HARDWOODS—continued										
Butternut	.36	37,000	6,700	57	610	16,700	1,500	5,200	3,000	1,700
	.38	56,000	8,100	57	610	35,200	3,200	8,100	3,000	2,200
Cherry, black	.47	55,000	9,000	88	840	24,400	2,500	7,800	3,900	2,900
	.50	85,000	10,300	79	740	49,000	4,800	11,700	3,900	4,200
Chestnut, American	.40	39,000	6,400	48	610	17,000	2,100	5,500	3,000	1,900
	.43	59,000	8,500	45	480	36,700	4,300	7,400	3,200	2,400
Cottonwood: Balsam poplar	.31	27,000	5,200	29	- - -	11,700	1,000	3,400	- - -	- - -
	.34	47,000	7,600	34	- - -	27,700	2,100	5,400	- - -	- - -
Black	.31	34,000	7,400	34	510	15,200	1,100	4,200	1,900	1,100
	.35	59,000	8,800	46	560	31,000	2,100	7,200	2,300	1,600
Eastern	.37	37,000	7,000	50	530	15,700	1,400	4,700	2,800	1,500
	.40	59,000	9,400	51	510	33,900	2,600	6,400	4,000	1,900
Elm: American	.46	50,000	7,700	81	970	20,100	2,500	6,900	4,100	2,800
	.50	81,000	9,200	90	990	38,100	4,800	10,400	4,600	3,700
Rock	.57	66,000	8,200	137	1370	26,100	4,200	8,800	- - -	- - -
	.63	102,000	10,600	132	1420	48,600	8,500	13,200	- - -	- - -
Slippery	.48	55,000	8,500	106	1190	22,900	2,900	7,700	4,400	2,900
	.53	90,000	10,300	117	1140	43,900	5,700	11,200	3,700	3,800
Hackberry	.49	45,000	6,600	100	1220	18,300	2,800	7,400	4,300	3,100
	.53	76,000	8,200	88	1090	37,500	6,100	11,000	4,000	3,900
Hickory, pecan: Bitternut	.60	71,000	9,700	138	1680	31,500	5,500	8,500	- - -	- - -
	.66	118,000	12,300	125	1680	62,300	11,600	- - -	- - -	- - -
Nutmeg	.56	63,000	8,900	157	1370	27,400	5,200	7,100	- - -	- - -
	.60	114,000	11,700	173	- - -	47,600	10,800	- - -	- - -	- - -

Species	Sp. gr.									
Pecan	.60	68,000	9,400	101	1350	27,500	5,400	10,200	4,700	5,800
	.66	94,000	11,900	95	1120	54,100	11,900	14,300	---	8,100
Water	.61	74,000	10,800	130	1420	32,100	6,100	9,900	---	---
	.62	123,000	13,900	133	1350	59,300	10,700	---	---	---
Hickory, true: Mockernut	.64	77,000	10,800	180	2240	30,900	5,600	8,800	---	---
	.72	132,000	15,300	156	1960	61,600	11,900	12,000	---	---
Pignut	.66	81,000	11,400	219	2260	33,200	6,300	9,400	---	---
	.75	139,000	15,600	210	1880	63,400	13,700	14,800	---	---
Shagbark	.64	76,000	10,800	163	1880	31,600	5,800	10,500	---	---
	.72	139,000	14,900	178	1700	63,500	12,100	16,800	---	---
Shellbark	.62	72,000	9,200	206	2640	27,000	5,600	8,200	---	---
	.69	125,000	13,000	163	2240	55,200	12,400	14,500	---	---
Honeylocust	.60	70,000	8,900	87	1190	30,500	7,900	11,400	6,400	6,200
	---	101,000	11,200	92	1190	51,700	12,700	15,500	6,200	7,000
Locust, black	.66	95,000	12,800	106	1120	46,900	8,000	12,100	5,300	7,000
	.69	134,000	14,100	127	1450	70,200	12,600	17,100	4,400	7,600
Magnolia: Cucumber tree	.44	51,000	10,800	69	760	21,600	2,300	6,800	3,000	2,300
	.48	85,000	12,500	84	890	43,500	3,900	9,200	4,600	3,100
Southern	.46	47,000	7,700	106	1370	18,600	3,200	7,200	4,200	3,300
	.50	77,000	9,700	88	740	37,600	5,900	10,500	5,100	4,500
Maple: Bigleaf	.44	51,000	7,600	60	580	22,300	3,100	7,700	4,100	2,800
	.48	74,000	10,000	54	710	41,000	5,200	11,900	3,700	3,800
Black	.52	54,000	9,200	88	1220	22,500	4,100	7,800	5,000	3,700
	.57	92,000	11,200	86	1020	46,100	7,000	12,500	4,600	5,200
Red	.49	53,000	9,600	79	810	22,600	2,800	7,900	---	3,100
	.54	92,000	11,300	86	810	45,100	6,900	12,800	---	4,200
Silver	.44	40,000	6,500	76	740	17,200	2,600	7,200	3,900	2,600
	.47	61,000	7,900	57	640	36,000	5,100	10,200	3,400	3,100
Sugar	.56	65,000	10,700	92	1020	27,700	4,400	10,100	---	4,300
	.63	109,000	12,600	114	990	54,000	10,100	16,100	---	6,400
Oak, red: Black	.56	57,000	8,100	84	1020	23,900	4,900	8,400	---	4,700
	.61	96,000	11,300	94	1040	45,000	6,400	13,200	---	5,400
Cherrybark	.61	74,000	12,300	101	1370	31,900	5,200	9,100	5,500	5,500
	.68	125,000	15,700	126	1240	60,300	8,600	13,800	5,800	6,600

109

Table A-4-2.—Strength properties [1,2] of some commercially important woods grown in the United States—continued

Common names of species	Specific gravity [3]	Static bending Modulus of Rupture (Kilopascals)	Static bending Modulus of Elasticity [4] (Megapascals)	Work to maximum load (Kilojoules per cu. m.)	Impact bending—height of drop causing complete failure (Mm.)	Compression parallel to grain—maximum crushing strength (Kilopascals)	Compression perpendicular to grain—fiber stress at proportional limit (Kilopascals)	Shear parallel to grain—maximum shearing strength (Kilopascals)	Tension perpendicular to grain—maximum tensile strength (Kilopascals)	Side hardness—load perpendicular to grain (Newtons)
HARDWOODS—continued										
Oak, red (cont.)										
Laurel	.56	54,000	9,600	77	990	21,900	3,900	8,100	5,300	4,400
	.63	87,000	11,700	81	990	48,100	7,300	12,600	5,400	5,400
Northern red	.56	57,000	9,300	91	1120	23,700	4,200	8,300	5,200	4,400
	.63	99,000	12,500	100	1090	46,600	7,000	12,300	5,500	5,700
Pin	.58	57,000	9,100	97	1220	25,400	5,000	8,900	5,500	4,800
	.63	97,000	11,900	102	1140	47,000	7,000	14,300	7,200	6,700
Scarlet	.60	72,000	10,200	103	1370	28,200	5,700	9,700	4,800	5,300
	.67	120,000	13,200	141	1350	57,400	7,700	13,000	6,000	6,200
Southern red	.52	48,000	7,900	55	740	20,900	3,800	6,400	3,300	3,800
	.59	75,000	10,300	65	660	42,000	6,000	9,600	3,500	4,700
Water	.56	61,000	10,700	77	990	25,800	4,300	8,500	5,700	4,500
	.63	106,000	13,900	148	1120	46,700	7,000	13,900	6,300	5,300
Willow	.56	51,000	8,900	61	890	20,700	4,200	8,100	5,200	4,400
	.69	100,000	13,100	101	1070	48,500	7,800	11,400	---	6,500
Oak, white:										
Bur	.58	50,000	6,100	74	1120	22,700	4,700	9,300	5,500	4,900
	.64	71,000	7,100	68	740	41,800	8,300	12,500	4,700	6,100
Chestnut	.57	55,000	9,400	65	890	24,300	3,700	8,300	4,800	4,000
	.66	92,000	11,000	76	1020	47,100	5,800	10,300	---	5,000
Live	.80	82,000	10,900	85	---	37,400	14,100	15,200	---	---
	.88	127,000	13,700	130	---	61,400	19,600	18,300	---	---
Overcup	.57	55,000	7,900	87	1120	23,200	3,700	9,100	5,000	4,300
	.63	87,000	9,800	108	970	42,700	5,600	13,800	6,500	5,300
Post	.60	56,000	7,500	76	1120	24,000	5,900	8,800	5,400	5,000
	.67	91,000	10,400	91	1170	45,300	9,900	12,700	5,400	6,000
Swamp chestnut	.60	59,000	9,300	88	1140	24,400	3,900	8,700	4,600	4,900
	.67	96,000	12,200	83	1040	50,100	7,700	13,700	4,800	5,500

Species	Specific gravity	(col 2)	(col 3)	(col 4)	(col 5)	(col 6)	(col 7)	(col 8)	(col 9)	(col 10)
Swamp white	.64 / .72	68,000 / 122,000	11,000 / 14,100	100 / 132	1270 / 1240	30,100 / 59,300	5,200 / 8,200	9,000 / 13,800	5,900 / 5,700	5,200 / 7,200
White	.60 / .68	57,000 / 105,000	8,600 / 12,300	80 / 102	1070 / 940	24,500 / 51,300	4,600 / 7,400	8,600 / 13,800	5,300 / 5,500	4,700 / 6,000
Sassafras	.42 / .46	41,000 / 62,000	6,300 / 7,700	49 / 60	--- / ---	18,800 / 32,800	2,600 / 5,900	6,600 / 8,500	--- / ---	--- / ---
Sweetgum	.46 / .52	49,000 / 86,000	8,300 / 11,300	70 / 82	910 / 810	21,000 / 43,600	2,600 / 4,300	6,800 / 11,000	3,700 / 5,200	2,700 / 3,800
Sycamore, American	.46 / .49	45,000 / 69,000	7,300 / 9,800	52 / 59	660 / 660	20,100 / 37,100	2,500 / 4,800	6,900 / 10,100	4,300 / 5,000	2,700 / 3,400
Tanoak	.58	72,000	10,700	92	---	32,100	---	---	---	---
Tupelo: Black	.46 / .50	48,000 / 66,000	7,100 / 8,300	55 / 43	760 / 560	21,000 / 38,100	3,300 / 6,400	7,600 / 9,200	3,900 / 3,400	2,800 / 3,600
Water	.46 / .50	50,000 / 66,000	7,200 / 8,700	57 / 48	760 / 580	23,200 / 40,800	3,300 / 6,000	8,200 / 11,000	4,100 / 4,800	3,200 / 3,900
Walnut, black	.51 / .55	66,000 / 101,000	9,800 / 11,600	101 / 74	940 / 860	29,600 / 52,300	3,400 / 7,000	8,400 / 9,400	3,900 / 4,800	4,000 / 4,500
Willow, black	.36 / .39	33,000 / 54,000	5,400 / 7,000	76 / 61	--- / ---	14,100 / 28,300	1,200 / 3,000	4,700 / 8,600	--- / ---	--- / ---
Yellow-poplar	.40 / .42	41,000 / 70,000	8,400 / 10,900	52 / 61	660 / 610	18,300 / 38,200	1,900 / 3,400	5,400 / 8,200	3,500 / 3,700	2,000 / 2,400
SOFTWOODS										
Baldcypress	.42 / .46	46,000 / 73,000	8,100 / 9,900	46 / 57	640 / 610	24,700 / 43,900	2,800 / 5,000	5,600 / 6,900	2,100 / 1,900	1,700 / 2,300
Cedar: Alaska-	.42 / .44	44,000 / 77,000	7,900 / 9,800	63 / 72	690 / 740	21,000 / 43,500	2,400 / 4,300	5,800 / 7,800	2,300 / 2,500	2,000 / 2,600
Atlantic white-	.31 / .32	32,000 / 47,000	5,200 / 6,400	41 / 28	460 / 330	16,500 / 32,400	1,700 / 2,800	4,800 / 5,500	1,200 / 1,500	1,300 / 1,600
Eastern redcedar	.44 / .47	48,000 / 61,000	4,500 / 6,100	103 / 57	890 / 560	24,600 / 41,500	4,800 / 6,300	7,000 / ---	2,300 / ---	2,900 / 4,000
Incense-	.35 / .37	43,000 / 55,000	5,800 / 7,200	44 / 37	430 / 430	21,700 / 35,900	2,600 / 4,100	5,700 / 6,100	1,900 / 1,900	1,700 / 2,100
Northern white-	.29 / .31	29,000 / 45,000	4,400 / 5,500	39 / 33	380 / 300	13,700 / 27,300	1,600 / 2,100	4,300 / 5,900	1,700 / 1,700	1,000 / 1,400

Table A-4.2.—*Strength properties* [1,2] *of some commercially important woods grown in the United States*—continued

Common names of species	Specific gravity [3]	Static bending — Modulus of — Rupture (Kilopascals)	Static bending — Modulus of — Elasticity [4] (Megapascals)	Work to maximum load (Kilojoules per cu. m.)	Impact bending—height of drop causing complete failure (Mm.)	Compression parallel to grain—maximum crushing strength (Kilopascals)	Compression perpendicular to grain—fiber stress at proportional limit (Kilopascals)	Shear parallel to grain—maximum shearing strength (Kilopascals)	Tension perpendicular to grain—maximum tensile strength (Kilopascals)	Side hardness—load perpendicular to grain (Newtons)
SOFTWOODS—continued										
Cedars (cont.)										
Port-Orford-	.39	46,000	9,000	51	530	21,600	2,100	5,800	1,200	1,700
	.43	88,000	11,700	63	710	43,100	5,000	9,400	2,800	2,800
Western redcedar	.31	35,900	6,500	34	430	19,100	1,700	5,300	1,600	1,200
	.32	51,700	7,700	40	430	31,400	3,200	6,800	1,500	1,600
Douglas-fir: [5]										
Coast	.45	53,000	10,800	52	660	26,100	2,600	6,200	2,100	2,200
	.48	85,000	13,400	68	790	49,900	5,500	7,800	2,300	3,200
Interior West	.46	53,000	10,400	50	660	26,700	2,900	6,500	2,000	2,300
	.50	87,000	12,500	73	810	51,300	5,200	8,900	2,400	2,900
Interior North	.45	51,000	9,700	56	560	23,900	2,500	6,600	2,300	1,900
	.48	90,000	12,300	72	660	47,600	5,300	9,700	2,700	2,700
Interior South	.43	47,000	8,000	55	380	21,400	2,300	6,600	1,700	1,600
	.46	82,000	10,300	62	510	42,900	5,100	10,400	2,300	2,300
Fir:										
Balsam	.34	34,000	6,600	32	410	16,500	1,200	4,200	1,200	1,300
	.36	52,000	8,500	35	510	31,200	2,100	4,900	1,200	1,800
California red	.36	40,000	8,100	44	530	19,000	2,300	5,300	2,600	1,600
	.38	72,000	10,300	61	610	37,700	4,200	7,200	2,700	2,200
Grand	.35	40,000	8,600	39	560	20,300	1,900	5,100	1,700	1,600
	.37	61,000	10,800	52	710	36,500	3,400	6,300	1,700	2,200
Noble	.37	43,000	9,500	41	480	20,800	1,900	5,500	1,600	1,300
	.39	74,000	11,900	61	580	42,100	3,600	7,200	1,500	1,800
Pacific silver	.40	44,000	9,800	41	530	21,600	1,500	5,200	1,700	1,400
	.43	73,000	11,900	64	610	45,000	3,100	8,100	---	1,900
Subalpine	.31	34,000	7,200	---	---	15,900	1,300	4,800	---	1,200
	.32	59,000	8,900	---	---	33,500	2,700	7,400	---	1,600
White	.37	41,000	8,000	39	560	20,000	1,900	5,200	2,100	1,500
	.39	68,000	10,300	50	510	40,100	3,700	7,600	2,100	2,100

Species	Sp. gr.									
Hemlock:										
Eastern	.38	44,000	7,400	46	530	21,200	2,500	5,900	1,600	1,800
	.40	61,000	8,300	47	530	37,300	4,500	7,300	---	2,200
Mountain	.42	43,000	7,200	76	810	19,900	2,600	6,400	2,300	2,100
	.45	79,000	9,200	72	810	44,400	5,900	10,600	---	3,000
Western	.42	46,000	9,000	48	560	23,200	1,900	5,900	2,000	1,800
	.45	78,000	11,300	57	580	49,000	3,800	8,600	2,300	2,400
Larch, western	.48	53,000	10,100	71	740	25,900	2,800	6,000	2,300	2,300
	.52	90,000	12,900	87	890	52,700	6,400	9,400	3,000	3,700
Pine:										
Eastern white	.34	34,000	6,800	36	430	16,800	1,500	4,700	1,700	1,300
	.35	59,000	8,500	47	460	33,100	3,000	6,200	2,100	1,700
Jack	.40	41,000	7,400	50	660	20,300	2,100	5,200	2,500	1,800
	.43	68,000	9,300	57	690	39,000	4,000	8,100	2,900	2,500
Loblolly	.47	50,000	9,700	57	760	24,200	2,700	5,900	1,800	2,000
	.51	88,000	12,300	72	760	49,200	5,400	9,600	3,200	3,100
Lodgepole	.38	38,000	7,400	39	510	18,000	1,700	4,700	1,500	1,500
	.41	65,000	9,200	47	510	37,000	4,200	6,100	2,000	2,100
Longleaf	.54	59,000	11,000	61	890	29,800	3,300	7,200	2,300	2,600
	.59	100,000	13,700	81	860	58,400	6,600	10,400	3,200	3,900
Pitch	.47	47,000	8,300	63	---	20,300	---	5,900	---	---
	.52	74,000	9,900	63	---	41,000	---	9,400	---	---
Pond	.51	51,000	8,800	52	---	25,200	3,000	6,500	---	---
	.56	80,000	12,100	59	---	52,000	6,300	9,500	---	---
Ponderosa	.38	35,000	6,900	36	530	16,900	1,900	4,800	2,100	1,400
	.40	65,000	8,900	49	480	36,700	4,000	7,800	2,900	2,000
Red	.41	40,000	8,800	42	660	18,800	1,800	4,800	2,100	1,500
	.46	76,000	11,200	68	660	41,900	4,100	8,400	3,200	2,500
Sand	.46	52,000	7,000	66	---	23,700	3,100	7,900	---	---
	.48	80,000	9,700	66	---	47,700	5,800	---	---	---
Shortleaf	.47	51,000	9,600	57	760	24,300	2,400	6,300	2,200	2,000
	.51	90,000	12,100	76	840	50,100	5,700	9,600	3,200	3,100
Slash	.54	60,000	10,500	66	---	26,300	3,700	6,600	---	---
	.59	112,000	13,700	91	---	56,100	7,000	11,600	---	---
Spruce	.41	41,000	6,900	---	---	19,600	1,900	6,200	---	2,000
	.44	72,000	8,500	---	---	39,000	5,000	10,300	---	2,900
Sugar	.34	34,000	7,100	37	430	17,000	1,400	5,000	1,900	1,200
	.36	57,000	8,200	38	460	30,800	3,400	7,800	2,400	1,700

Table A-4.—*Strength properties* [1,2] *of some commercially important woods grown in the United States*—continued

Common names of species	Specific gravity [3]	Static bending		Work to maximum load	Impact bending—height of drop causing complete failure	Compression parallel to grain—maximum crushing strength	Compression perpendicular to grain—fiber stress at proportional limit	Shear parallel to grain—maximum shearing strength	Tension perpendicular to grain—maximum tensile strength	Side hardness—load perpendicular to grain
		Modulus of—								
		Rupture	Elasticity [4]							
		Kilopascals	Megapascals	Kilojoules per cu. m.	Mm.	Kilopascals	Kilopascals	Kilopascals	Kilopascals	Newtons
SOFTWOODS—continued										
Pine (Cont.)										
Virginia	.45	50,000	8,400	75	860	23,600	2,700	6,100	2,800	2,400
	.48	90,000	10,500	94	810	46,300	6,300	9,300	2,600	3,300
Western white	.35	32,000	8,200	34	480	16,800	1,300	4,700	1,800	1,200
	.38	67,000	10,100	61	580	34,700	3,200	7,200	- - - -	1,900
Redwood:										
Old-growth	.38	52,000	8,100	51	530	29,000	2,900	5,500	1,800	1,800
	.40	69,000	9,200	48	480	42,400	4,800	6,500	1,700	2,100
Young-growth	.34	41,000	6,600	39	410	21,400	1,900	6,100	2,100	1,600
	.35	54,000	7,600	36	380	36,000	3,600	7,600	1,700	1,900
Spruce:										
Black	.38	37,000	7,300	51	610	17,700	1,000	4,600	700	1,600
	.40	71,000	10,500	72	580	36,700	3,700	7,100	- - - -	2,300
Engelmann	.33	32,000	7,100	35	410	15,000	1,400	4,400	1,700	1,150
	.35	64,000	8,900	44	460	30,900	2,800	8,300	2,400	1,750
Red	.38	40,000	8,200	48	460	18,300	1,900	5,200	1,500	1,600
	.41	70,000	10,500	58	640	40,600	3,200	7,400	2,400	2,200
Sitka	.37	39,000	8,500	43	610	18,400	1,900	5,200	1,700	1,600
	.40	70,000	10,800	65	640	38,700	4,000	7,900	2,600	2,300
White	.37	39,000	7,400	41	560	17,700	1,700	4,800	1,500	1,400
	.40	68,000	9,200	53	510	37,700	3,200	7,400	2,500	2,100
Tamarack	.49	50,000	8,500	50	710	24,000	2,700	5,900	1,800	1,700
	.53	80,000	11,300	49	580	49,400	5,500	8,800	2,800	2,600

[1] Results of tests on small clear specimens in the green and air-dry condition, converted to metric units directly from table 4–2.

[2] Values in the first line for each species are from tests of green material; those in the second line are adjusted to 12 pct. moisture content.

[3] Specific gravity is based on weight when ovendry and volume when green or at 12 pct. moisture content.

[4] Modulus of elasticity measured from a simply supported, center-loaded beam, on a span-depth ratio of 14/1. The modulus can be corrected for the effect of shear deflection by increasing it 10 pct.

[5] Coast Douglas-fir is defined as Douglas-fir growing in the States of Oregon and Washington west of the summit of the Cascade Mountains. Interior West includes the State of California and all counties in Oregon and Washington east of but adjacent to the Cascade summit. Interior North includes the remainder of Oregon and Washington and the States of Idaho, Montana, and Wyoming. Interior South is made up of Utah, Colorado, Arizona, and New Mexico.

114

Table A–4–3.—*Mechanical properties of some commercially important woods grown in Canada and imported into the United States* [1]

Common names of species	Specific gravity	Static bending		Compression parallel to grain— maximum crushing strength	Compression perpendicular to grain— fiber stress at proportional limit	Shear parallel to grain— maximum shearing strength
		Modulus of rupture	Modulus of elasticity			
		Kilo-pascals	Mega-pascals	Kilo-pascals	Kilo-pascals	Kilo-pascals
HARDWOODS						
Aspen:						
Quaking	0.37	38,000	9,000	16,200	1,400	5,000
		68,000	11,200	36,300	3,500	6,800
Big-toothed	.39	36,000	7,400	16,500	1,400	5,400
		66,000	8,700	32,800	3,200	7,600
Cottonwood:						
Black	.30	28,000	6,700	12,800	700	3,900
		49,000	8,800	27,700	1,800	5,900
Eastern	.35	32,000	6,000	13,600	1,400	5,300
		52,000	7,800	26,500	3,200	8,000
Balsam, poplar	.37	34,000	7,900	14,600	1,200	4,600
		70,000	11,500	34,600	2,900	6,100
SOFTWOODS						
Cedar:						
Alaska-	.42	46,000	9,200	22,300	2,400	6,100
		80,000	11,000	45,800	4,800	9,200
Northern white-	.30	27,000	3,600	13,000	1,400	4,600
		42,000	4,300	24,800	2,700	6,900
Western redcedar	.31	36,000	7,200	19,200	1,900	4,800
		54,000	8,200	29,600	3,400	5,600
Douglas-fir	.45	52,000	11,100	24,900	3,200	6,300
		88,000	13,600	50,000	6,000	9,500
Fir:						
Subalpine	.33	36,000	8,700	17,200	1,800	4,700
		56,000	10,200	36,400	3,700	6,800
Pacific silver	.36	38,000	9,300	19,100	1,600	4,900
		69,000	11,300	40,900	3,600	7,500
Balsam	.34	36,000	7,800	16,800	1,600	4,700
		59,000	9,600	34,300	3,200	6,300
Hemlock:						
Eastern	.40	47,000	8,800	23,600	2,800	6,300
		67,000	9,700	41,200	4,300	8,700
Western	.41	48,000	10,200	24,700	2,600	5,200
		81,000	12,300	46,700	4,600	6,500
Larch, western	.55	60,000	11,400	30,500	3,600	6,300
		107,000	14,300	61,000	7,300	9,200

Common names of species	Specific gravity	Static bending		Compression parallel to grain—maximum crushing strength	Compression perpendicular to grain—fiber stress at proportional limit	Shear parallel to grain—maximum shearing strength
		Modulus of rupture	Modulus of elasticity			
		Kilo-pascals	Mega-pascals	Kilo-pascals	Kilo-pascals	Kilo-pascals

SOFTWOODS—continued

Common names of species	Specific gravity	Modulus of rupture	Modulus of elasticity	Compression parallel	Compression perpendicular	Shear parallel
Pine:						
Eastern white	.36	35,000	8,100	17,900	1,600	4,400
		66,000	9,400	36,000	3,400	6,100
Jack	.42	43,000	8,100	20,300	2,300	5,600
		78,000	10,200	40,500	5,700	8,200
Lodgepole	.40	39,000	8,800	19,700	1,900	5,000
		76,000	10,900	43,200	3,600	8,500
Ponderosa	.44	39,000	7,800	19,600	2,400	5,000
		73,000	9,500	42,300	5,200	7,000
Red	.39	34,000	7,400	16,300	1,900	4,900
		70,000	9,500	37,900	5,000	7,500
Western white	.36	33,000	8,200	17,400	1,600	4,500
		64,100	10,100	36,100	3,200	6,300
Spruce:						
Black	.41	41,000	9,100	19,000	2,100	5,500
		79,000	10,500	41,600	4,300	8,600
Engelmann	.38	39,000	8,600	19,400	1,900	4,800
		70,000	10,700	42,400	3,700	7,600
Red	.38	41,000	9,100	19,400	1,900	5,600
		71,000	11,000	38,500	3,800	9,200
Sitka	.35	37,000	9,400	17,600	2,000	4,300
		70,000	11,200	37,800	4,100	6,800
White	.35	35,000	7,900	17,000	1,600	4,600
		63,000	10,000	37,000	3,400	6,800
Tamarack	.48	47,000	8,600	21,600	2,800	6,300
		76,000	9,400	44,900	6,200	9,000

[1] Results of tests on small, clear, straight-grained specimens. Property values based on American Society for Testing and Materials Standard D 2555–70, "Standard methods for establishing clear wood values." Information on additional properties can be obtained from Department of Forestry, Canada, Publication No. 1104.

[2] The values in the first line for each species are from tests of green material; those in the second line are adjusted from the green condition to 12 pct. moisture content using dry to green clear wood property ratios as reported in ASTM D 2555–70. Specific gravity is based on weight when ovendry and volume when green.

Table A–4.—*Mechanical properties* [1,2] *of some woods imported into the United States*

Common and botanical names of species	Moisture content	Specific gravity [3]	Static bending			Compression parallel to grain—maximum crushing strength	Shear parallel to grain—maximum shearing strength	Side hardness—load perpendicular to grain	Sample	
			Modulus of rupture	Modulus of elasticity [4]	Work to maximum load				No. of trees	Origin [5]
	Pct.		*Kilopascals*	*Megapascals*	*Kilojoules per cu. m.*	*Kilopascals*	*Kilopascals*	*Newtons*		
Andiroba (*Carapa guianensis*)	Green	0.56	76,600	10,800	79	34,000	9,100	4,700	2	BR
	12	---	107,700	12,800	92	54,500	11,600	5,400	2	BR
Andiroba (*C. nicaraguensis*)	13	.45	---	---	---	43,000	---	5,500	3	EC
Angelique (*Dicorynia guianensis*)	Green	.60	78,700	12,700	83	38,600	9,200	4,900	2	SU
		---	119,900	15,100	105	60,500	11,400	5,700	2	SU
Apamate (*Tabebuia rosea*)	Green	.51	73,400	10,100	77	34,000	8,500	4,000	10	CS
	12	---	95,000	11,000	86	50,600	10,000	4,300	9	CS
Apitong (*Dipterocarpus* spp.)	Green	.59	63,600	12,300	---	30,400	7,200	3,600	57	PH
	12	---	111,800	16,200	---	58,900	11,700	5,300	53	PH
Avodire (*Turraeanthus africanus*)	12	.51	87,800	10,200	65	49,500	14,100	4,800	3	AF
Balsa (*Ochroma pyramidale*)	12	.17	19,300	3,800	---	11,700	2,100	400	(6)	EC
Banak (*Virola koschnyi*)	Green	.44	42,700	10,100	37	21,000	4,600	2,000	8	CA
	12	---	74,500	11,900	56	39,400	9,000	2,800	8	CA
Banak (*V. surinamensis*)	Green	.42	38,600	11,300	28	16,500	5,000	1,400	2	BR
	12	---	75,500	14,100	69	35,400	6,800	2,300	2	BR
Capirona (*Calycophyllum candidissimum*)	Green	.67	98,500	13,300	128	42,700	11,400	7,300	2	VE
	12	---	153,800	15,700	186	66,700	14,600	8,600	2	VE
Capirona (*C. spruceanum*)	14	.85	---	---	---	64,000	---	11,300	1	PE
Cativo (*Prioria copaifera*)	Green	.40	40,900	6,600	37	17,900	5,900	2,000	4	PA
	12	---	60,200	7,900	50	31,000	7,200	2,700	4	PA
Courbaril (*Hymenaea courbaril*)	Green	.72	89,300	12,500	108	40,000	12,200	9,000	9	CS
	12	---	133,800	15,000	121	66,700	17,000	10,900	9	CS
Gola (*Tetraberlinia tubmaniana*)	14	.66	115,500	15,200	---	62,100	---	---	11	AF

117

Table A–4.—Mechanical properties [1,2] of some woods imported into the United States —continued

Common and botanical names of species	Moisture content (Pct.)	Specific gravity [3]	Static bending — Modulus of rupture (Kilopascals)	Static bending — Modulus of elasticity [4] (Megapascals)	Static bending — Work to maximum load (Kilojoules per cu. m.)	Compression parallel to grain—maximum crushing strength (Kilopascals)	Shear parallel to grain—maximum shearing strength (Kilopascals)	Side hardness—load perpendicular to grain (Newtons)	Sample — No. of trees	Sample — Origin [5]
Goncalo alves (Astronium graveolens)	Green	.86	85,500	13,100	51	47,400	12,700	8,900	4	CS
	12	---	117,700	15,000	72	72,800	14,200	9,900	4	CS
Greenheart (Ocotea rodiaei)	Green	.83	133,800	20,600	90	71,400	10,200	9,700	5	GY
	14	.93	175,800	25,500	152	89,900	12,600	11,700	1	GY
Ilomba (Pycnanthus angolensis)	12	.44	61,400	12,100	---	38,000	---	3,300	(7)	AF
Jarrah (Eucalyptus marginata)	Green	.67	68,100	10,200	---	35,800	9,100	5,700	28	AU
	12	---	111,700	13,000	---	61,200	15,100	8,500	28	AU
Jelutong (Dyera costulata)	Green	.36	38,600	8,000	39	21,000	5,200	1,500	3	AS
	16	.38	50,300	8,100	44	27,000	5,800	1,700	3	AS
Kapur (Dryobalanops lanceolata)	Green	.64	83,800	11,700	88	41,200	7,200	4,400	5	AS
	12	---	120,000	13,900	107	66,900	11,800	5,500	5	AS
Karri (Eucalyptus diversicolor)	Green	.70	73,100	14,300	---	36,200	9,200	6,100	26	AU
	12	---	132,400	19,000	---	71,700	14,700	9,000	21	AU
Kerving (Dipterocarpus supp.)	Green	.67	81,800	16,900	63	42,900	8,000	4,900	21	MI
	12	.69	99,900	18,100	92	55,100	9,400	5,200	11	MI
Khaya (Khaya anthotheca)	Green	.47	53,800	8,100	63	26,000	7,500	3,200	9	AF
	12	---	79,000	9,700	68	43,400	11,700	4,000	9	AF
Khaya (K. ivorensis)	Green	.43	51,000	8,000	57	24,100	6,400	2,800	11	AF
	12	---	73,800	9,600	57	44,500	10,300	3,700	11	AF
Kokrodua (Pericopsis elata)	Green	.66	102,200	12,200	134	51,600	11,500	7,100	6	AF
	12	---	127,100	13,400	128	68,500	14,400	6,900	6	AF
Lapacho (Tabebuia heterotricha)	Green	.80	138,400	14,600	188	53,000	14,800	11,300	3	PA
	12	---	156,000	16,000	179	75,400	15,700	13,400	3	PA
Lapacho (T. serratifolia)	Green	.92	157,500	21,100	177	73,500	14,100	13,200	3	SM
	12	---	181,400	22,800	159	92,500	14,300	16,300	3	SM

Lauan:
  Dark red:

Species	Moisture	Sp. gr.							No.	Region
Red lauan (*Shorea negrosensis*)	Green	.44	53,100	9,500	---	25,500	6,400	2,500	15	AS
	12	---	77,900	11,200	---	40,600	8,400	3,000	15	AS
Tanguile (*S. polysperma*)	Green	.46	57,200	10,600	---	27,200	6,500	2,800	19	PH
	12	---	88,900	12,500	---	45,400	8,900	3,400	17	PH
Light red:										
Almon (*Shorea almon*)	Green	.41	51,700	9,900	---	25,900	5,800	2,200	12	PH
	12	.44	77,900	11,500	---	39,600	7,500	2,600	12	PH
Bagtikan (*Parashorea plicata*)	Green	.48	60,700	10,100	---	30,100	6,800	3,100	32	AS
	12	---	86,600	11,900	---	47,300	9,000	3,600	32	AS
Mayapis (*Shorea squamata*)	Green	.41	50,300	9,700	---	23,900	5,300	2,100	14	AS
	12	---	76,500	11,400	---	38,700	7,500	2,600	12	AS
White lauan (*Pentacme contorta*)	Green	.43	51,700	9,500	---	25,500	6,300	2,600	19	AS
	12	---	80,700	11,700	---	41,900	8,300	3,100	18	AS
Laurel (*Cordia alliodora*)	Green	.44	61,000	8,700	---	27,600	7,800	3,500	13	CA
	12	---	83,200	10,300	---	42,300	8,400	3,500	13	CA
Lignumvitae (*Guaiacum sanctum*)	12	1.09	78,600	---	---	78,600	---	20,000	---	AF
Limba (*Terminalia superba*)	12	.49	79,300	8,800	---	36,500	7,000	3,000	(8)	AF
Lupuna (*Ceiba samauma*)	13	.54	---	---	---	---	---	3,300	1	PE
Mahogany (*Swietenia macrophylla*)	Green	.45	64,000	11,300	66	31,100	9,000	3,100	77	CS
	12	---	80,300	10,400	54	45,700	8,900	3,600	77	CS
Meranti, red (*Shorea dasphylla*)	Green	.43	59,600	10,300	61	30,700	---	2,500	2	AS
	12	---	83,200	11,200	81	48,100	---	2,800	2	AS
Oak (*Quercus costaricensis*)	12	.68	121,100	13,200	116	---	---	7,000	2	CR
Oak (*Q. eugeniaefolia*)	12	.75	113,100	19,600	97	---	---	9,700	1	CR
Obeche (*Triplochiton scleroxylon*)	Green	.33	35,400	4,900	43	17,700	4,600	1,900	2	AF
	12	---	51,700	5,900	48	27,100	6,800	1,900	2	AF
Okoume (*Aucoumea klaineana*)	12	.37	50,700	7,900	---	26,900	6,800	1,700	(9)	AF
Palosapis (*Anisoptera* spp.)	Green	.51	52,100	9,900	---	26,100	6,900	3,600	18	AS
	12	---	88,100	12,500	---	45,700	9,700	4,100	16	AS
"Parana pine" (*Araucaria angustifolia*)	Green	.46	49,200	9,300	68	27,600	6,700	2,500	(10)	SM
	12	---	93,100	11,200	84	52,700	11,900	3,500	(10)	SM
Pau marfim (*Balfourodendron riedelianum*)	Green	.73	99,300	13,000	---	42,100	13,000	6,800	5	BR
	15	---	130,300	---	---	56,500	---	---	5	BR

Table A–4.—*Mechanical properties* [1,2] *of some woods imported into the United States* —continued

Common and botanical names of species	Moisture content	Specific gravity [3]	Static bending			Compression parallel to grain—maximum crushing strength	Shear parallel to grain—maximum shearing strength	Side hardness—load perpendicular to grain	Sample	
			Modulus of rupture	Modulus of elasticity [4]	Work to maximum load				No. of trees	Origin [5]
	Pct.		Kilopascals	Megapascals	Kilojoules per cu. m.	Kilopascals	Kilopascals	Newtons		
Peroba de campos (*Paratecoma peroba*)	12	.75	106,200	12,100	70	61,500	14,800	7,100	(11)	BR
Pine, Caribbean (*Pinus caribaea*)	Green	.68	68,800	11,700	83	33,000	8,300	3,600	19	CA
	12	---	105,000	14,000	105	55,200	12,900	5,100	14	CA
Pine, ocote (*P. oocarpa*)	Green	.55	55,000	12,000	48	25,400	7,200	2,600	3	HO
	12	---	102,600	15,500	75	53,000	11,900	4,000	3	HO
Primavera (*Cybistax donnell-smithii*)	Green	.39	53,200	6,800	48	25,000	7,200	2,900	4	HO
	12	---	75,200	8,400	71	42,300	11,800	3,100	4	HO
Ramin (*Gonystylus bancanus*)	Green	.59	67,500	10,800	62	37,200	6,900	2,800	9	AS
	12	---	127,100	15,000	117	69,500	10,400	5,800	9	AS
Rosewood, Indian (*Dalbergia latifolia*)	Green	.75	63,400	8,200	80	31,200	9,700	5,600	5	AS
	12	---	116,700	12,300	90	63,600	14,400	11,700	5	AS
Sande (*Brosimum utile*)	12	.44	---	---	---	43,500	---	2,200	3	EC
Santa Maria (*Calophyllum brasiliense*)	Green	.54	72,200	10,800	73	35,600	8,900	4,500	18	CA
	12	---	101,800	12,500	91	55,600	13,200	5,400	18	CA
Sapele (*Entandrophragma cylindricum*)	Green	.60	70,100	10,300	72	34,500	8,600	4,500	5	AF
	12	---	105,500	12,500	108	56,300	15,800	6,700	5	AF
Spanish-cedar (*Cedrela angustifolia*)	Green	.38	44,400	8,100	51	21,400	5,400	2,000	2	BR
	12	---	77,900	9,800	86	41,400	8,300	2,500	2	BR
Spanish-cedar (*C. oaxacensis*)	Green	.41	51,800	9,000	49	23,200	6,800	2,400	3	PA
	12	---	79,500	9,900	65	42,800	7,600	2,700	3	PA
Spanish-cedar (*C. odorata*) (Nicaragua)	Green	.34	35,900	6,000	51	19,000	5,000	1,600	1	NI
(Guatemala)	Green	.43	65,500	10,200	---	---	---	2,800	1	GU
(Guatemala)	12	.36	54,200	7,000	39	30,700	---	2,200	1	NI
Teak (*Tectona grandis*) (India)	Green	.57	75,700	10,400	75	37,700	8,900	4,800	134	IN
(Honduras)	12	.62	91,800	9,600	71	46,900	11,000	4,900	356	HO
(India)	12	.63	88,100	11,000	70	49,000	10,200	4,600	---	IN

"Virola" (Dialyanthera otoba) -------------	12	.34	----	----	----	----	----	2,100	----	1	EC
Walnut, European (Juglans regia) --------- { Green		.47	60,100	9,000	72	27,700	7,300	3,000	10	AS
8		----	90,300	10,600	68	50,500	9,100	3,800	10	AS

[1] Results of tests on small clear, straight-grained specimens. Property values were taken from world literature (not obtained from experiments conducted at the U.S. Forest Products Laboratory). Other species may be reported in the world literature as well as additional data on many of these species.

[2] Some property values have been adjusted to 12 pct. moisture content; others are based on moisture content at time of test.

[3] Specific gravity based on weight when ovendry and volume at moisture content indicated.

[4] Modulus of elasticity measured from a simply supported, center loaded beam. The modulus can be corrected for the effect of shear deflection by increasing it 10 pct.

[5] Key to code letters: AF, Africa; AS, Southeast Asia; AU, Australia; BR, Brazil; CA, Central America; CH, Chile; CR, Costa Rica; CS, Central and South America; EC, Equador; GU, Guatemala; GY, Guyana (British Guana); HO, Honduras; MI, Malaysia-Indonesia; IN, India; NI, Nicaragua; PA, Panama; PE, Peru; PH, Philippine Islands; SM, South America; SU, Surinam; and VE, Venezuela.

[6] 1,500 bd. ft.
[7] 1 bolt.
[8] 195 tests.
[9] 21 tests.
[10] 26 planks.
[11] 11 planks.

Clear wood strength values and standard deviations for several species of wood (Unseasoned).

Species	Modulus of Rupture and Tension Parallel		Modulus of Elasticity		Compression Parallel to Grain		Shear Strength		Compression Perpendicular at Proportional Limit		Specific Gravity[1]	
	Avg. psi	Standard Deviation psi	Avg. 1000 psi	Standard Deviation 1000 psi	Avg. psi	Standard Deviation psi	Avg. psi	Standard Deviation psi	Avg. psi	Standard Deviation psi	Avg.	Standard Deviation
Douglas fir												
Coast	7665	1317	1560	315	3784	734	904	131	382	107	0.45	0.057
Interior West	7713	1322	1513	324	3872	799	936	137	418	117	0.46	0.058
Interior North	7438	1163	1409	274	3469	602	947	126	356	100	0.45	0.049
Interior South	6784	908	1162	200	3113	489	953	153	337	94	0.43	0.045
Southern pine												
Longleaf	8670	1387	1598	352	4300	774	1037	145	479	134	0.54	0.054
Slash	8570	1371	1588	349	4210	758	958	134	529	148	0.54	0.054
Loblolly	7340	1174	1406	309	3490	628	850	119	389	109	0.47	0.047
Shortleaf	7300	1168	1391	306	3430	619	851	119	353	99	0.46	0.046
Western hemlock	6637	1088	1307	258	3364	615	864	105	282	79	0.42	0.053
Western larch	7652	1001	1458	249	3756	564	869	85	399	112	0.48	0.048

[1]Based on volume when green and weight at 12 percent moisture content.

Chapter 5

COMMERCIAL LUMBER

A log when sawed yields lumber of varying quality. The objective of grading is to enable a user to buy the quality that best suits his purpose. This is accomplished by dividing the lumber from the log into use categories, each having an appropriate range in quality.

Except as noted later, the grade of a piece of lumber is based on the number, character, and location of features that may lower the strength, durability, or utility value of the lumber. Among the more common visual features are knots, checks, pitch pockets, shake, and stain, some of which are a natural part of the tree. Some grades are free or practically free from these features. Other grades comprising the great bulk of lumber contain fairly numerous knots and other features that may affect quality. With proper grading, lumber containing these features is entirely satisfactory for many uses.

The principal grading operation for most lumber takes place at the sawmill. Establishment of grading procedures is largely the responsibility of manufacturing associations. Because of the wide variety of wood species, industrial practices, and customer needs, different lumber grading practices coexist. The grading practices of most interest are considered in the sections that follow, under the major categories of Hardwood Lumber and Softwood Lumber.

HARDWOOD LUMBER

Hardwood lumber is graded according to three basic marketing categories: Factory Lumber, Dimension Parts, and Finished Market Products. Both factory lumber and dimension parts are intended to serve the industrial customer; the important difference is that the factory lumber grades reflect the proportion of a piece that can be cut into useful smaller pieces while the dimension grades are based on use of the entire piece. Finished market products are graded for their unique end use with little or no remanufacture. Examples of finished products include molding, stair treads, and hardwood flooring.

Factory Grades

The rules adopted by the National Hardwood Lumber Association are considered standard in grading hardwood lumber for cutting into smaller pieces to make furniture or other fabricated products. In these rules the grade of a piece of hardwood lumber is determined by the proportion of a piece that can be cut into a certain number of smaller pieces of material generally clear on one side and not smaller than a specified size. In other words, the grade classification is based upon the amount of usable lumber in the piece rather than upon the number or size of growth features that characterize softwood grades. This usable material, commonly termed 'cuttings," must have one face clear and the reverse face sound, which means free from such things as wane, rot, pith, and shake that materially impair the strength of the cutting. The lowest cutting grades require only that the cuttings be sound.

Cutting Grades

The highest cutting grade is termed "Firsts" and the next grade "Seconds." First and Seconds are nearly always combined in one grade and referred to as "FAS." The third grade is termed "Selects" followed by No. 1 Common, No. 2 Common, Sound Wormy, No. 3A Common, and No. 3B Common. A description of the standard hardwood cutting grades is given in table 5–1. This table illustrates, for example, that Firsts call for pieces that will allow 91⅔ percent of their surface measure to be cut into clear face material. Not more than 8⅓ percent of each piece can be wasted in making the required cuttings. In general the minimum acceptable length, width surface measure, and percent of piece that must work into a cutting decreases with decreasing grade. The grade of hardwood lumber called "Sound Wormy" has the same requirements as No. 1 Common and Better except that wormholes and limited sound knots and other imperfections are allowed in the cuttings. Figure 5–1 is an illustration of grading for cuttings.

Table 5-1.—*Standard hardwood cutting grades* [1]

Grade and lengths allowed (feet)	Widths allowed	Surface measure of pieces	Amount of each piece that must work into clear-face cuttings	Maximum cuttings allowed	Minimum size of cuttings required
	In.	*Sq. ft.*	*Pct.*	*Number*	
Firsts: [2]					
8 to 16 (will admit 30 percent of 8- to 11-foot, ½ of which may be 8- and 9-foot.)	6+	4 to 9	91⅔	1	4 inches by 5 feet, or 3 inches by 7 feet
		10 to 14	91⅔	2	
		15+	91⅔	3	
Seconds: [2]					
8 to 16 (will admit 30 percent of 8- to 11-foot, ½ of which may be 8- and 9-foot).	6+	4 and 5	83⅓	1	Do.
		6 and 7	83⅓	1	
		6 and 7	91⅔	2	
		8 to 11	83⅓	2	
		8 to 11	91⅔	3	
		12 to 15	83½	3	
		12 to 15	91⅔	4	
		16+	83⅓	4	
Selects:					
6 to 16 (will admit 30 percent of 6- to 11-foot, ⅙ of which may be 6- and 7-foot).	4+	2 and 3	91⅔	1	Do.
		4+	(*3*)		
No. 1 Common:					
4 to 16 (will admit 10 percent of 4- to 7-foot, ½ of which may be 4- and 5-foot).	3+	1	100	0	4 inches by 2 feet, or 3 inches by 3 feet
		2	75	1	
		3 and 4	66⅔	1	
		3 and 4	75	2	
		5 to 7	66⅔	2	
		5 to 7	75	3	
		8 to 10	66⅔	3	
		11 to 13	66⅔	4	
		14+	66⅔	5	
No. 2 Common:					
4 to 16 (will admit 30 percent of 4- to 7-foot, ⅓ of which may be 4- and 5-foot).	3+	1	66⅔	1	3 inches by 2 feet
		2 and 3	50	1	
		2 and 3	66⅔	2	
		4 and 5	50	2	
		4 and 5	66⅔	3	
		6 and 7	50	3	
		6 and 7	66⅔	4	
		8 and 9	50	4	
		10 and 11	50	5	
		12 and 13	50	6	
		14+	50	7	
No. 3A Common:					
4 to 16 (will admit 50 percent of 4- to 7-foot, ½ of which may be 4- and 5-foot).	3+	1+	[4] 33⅓	(*5*)	Do.
No. 3B Common:					
4 to 16 (will admit 50 percent of 4- to 7-foot, ½ of which may be 4- and 5-foot).	3+	1+	[6] 25	(*5*)	1½ inches by 2 feet

[1] Inspection to be made on the poorer side of the piece, except in Selects.

[2] Firsts and Seconds are combined as 1 grade (FAS). The percentage of Firsts required in the combined grade varies from 20 to 40 percent, depending on the species.

[3] Same as Seconds with reverse side of board not below No. 1 Common or reverse side of cuttings sound.

[4] This grade also admits pieces that grade not below No. 2 Common on the good face and have the reverse face sound.

[5] Unlimited.

[6] The cuttings must be sound; clear face not required.

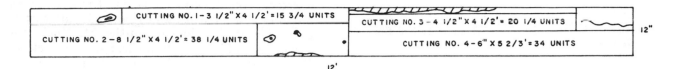

| | | CUTTING NO.1-3 1/2"X 4 1/2'=15 3/4 UNITS | | CUTTING NO. 3 - 4 1/2" X 4 1/2'= 20 1/4 UNITS | | 12" |
| CUTTING NO. 2 - 8 1/2" X 4 1/2'= 38 1/4 UNITS | | | | CUTTING NO. 4 - 6" X 5 2/3'= 34 UNITS | | |

12'

1. Determine Surface Measure (S.M.) using lumber scale stick or from formula:

$$\frac{\text{Width in inches x length in feet}}{12} = \frac{12" \times 12'}{12}$$

= 12 sq. ft. S.M.

2. No. 1 Common is assumed grade of board. Percent of clear-cutting area required for No. 1 Common—66-2/3% or 8/12.

3. Determine maximum number of cuttings permitted.

For No. 1 Common grade (S.M. + 1) ÷ 3

$$= \frac{(12 + 1)}{3} = \frac{13}{3} = 4 \text{ cuttings.}$$

4. Determine minimum size of cuttings.

For No. 1 Common grade 4" x 2' or 3" x 3'.

5. Determine clear-face cutting units needed.

For No. 1 Common grade S.M. x 8 = 12 x 8 = 96 units.

6. Determine total area of permitted clear-face cutting in units.

Width in inches and fractions of inches x length in feet and fractions of feet.

Cutting #1—3-1/2" x 4-1/2' = 15-3/4 units
Cutting #2—8-1/2" x 4-1/2' = 38   units
Cutting #3—4-1/2" x 4-1/2' = 20-1/4 units
Cutting #4—6"   x 5-2/3' = 34   units

Total Units   108

Units required for No. 1 Common—96.

7. Conclusion: Board meets requirements for No. 1 Common grade.

M 139 690

Figure 5–1.—An example of hardwood grading for cuttings using a No. 1 Common factory grade.

This brief summary of the factory grades should not be regarded as a complete set of grading rules, as numerous details, exceptions, and special rules for certain species are not included. The complete official rules of the National Hardwood Lumber Association should be followed as the only full description of existing grades.

Cutting Sizes

Standard lengths

Standard lengths are 4, 5, 6, 7, 8, 9, 10, 11, 12, 13, 14, 15, and 16 feet, but not more than 50 percent of odd lengths are allowed in any single shipment.

Standard thickness

Standard thicknesses for hardwood lumber, rough and surfaced (S2S), are given in table 5–2. The thickness of S1S lumber is subject to contract agreement.

Standard widths

Hardwood lumber is usually manufactured to random width. The hardwood factory grades do not specify standard widths; however, the grades do specify minimum widths for each grade as follows:

Firs ts _____ 6 inches
Seconds _____ 6 inches
Selects _____ 4 inches
Nos. 1, 2, 3A, 3B Common _____ 3 inches

If width is specified by purchase agreement, S1E or S2E lumber is ⅜ inch scant of nominal in lumber less than 8 inches wide and ½ inch scant in lumber 8 inches and wider.

Table 5–2.—*Standard thickness for rough and surfaced (S2S) hardwood lumber*

Rough	Surfaced	Rough	Surfaced
In.	*In.*	*In.*	*In.*
⅜	3/16	2 ½	2 ¼
½	5/16	3	2 ¾
⅝	7/16	3 ½	3 ¼
¾	9/16	4	3 ¾
1	13/16	4 ½	(¹)
1 ¼	1 1/16	5	(¹)
1 ½	1 5/16	5 ½	(¹)
1 ¾	1 ½	6	(¹)
2	1 ¾		

¹ Finished size not specified in rules. Thickness subject to special contract.

Alder, red
Ash
Aspen
Basswood
Beech
Birch
Boxelder
Buckeye
Butternut
Cedar, aromatic red
Cedar, Spanish-
Cherry
Chestnut
Cottonwood
Cypress
Elm:
  Rock (or cork)
  Soft
Gum:
  Black
  Red and sap
  Tupelo
Hackberry
Hardwoods (Philippine)

Hardwoods (Tropical
  American other than
  mahogany and Spanish
  cedar)
Hickory
Locust
Magnolia
Mahogany
  African
  Cuban and San
    Dominican
  Philippine
Maple:
  Hard (or sugar)
  Soft
  Pacific Coast
Oak:
  Red
  White
Pecan
Poplar
Sycamore
Walnut
Willow

Dimension Parts

Hardwood dimension parts are generally graded under the rules of the Hardwood Dimension Manufacturers Association. Dimension signifies primarily that the stock is processed so it can be used virtually in the sizes provided.

Hardwood dimension rules encompass three classes of material: Solid dimension flat stock, kiln-dried dimension flat stock, and solid dimension squares. Each class may be rough, semifabricated, or fabricated. Rough dimension blanks are usually kiln dried and are supplied sawn and ripped to size. Surfaced or semifabricated stock has been further processed by gluing, surfacing, tenoning, etc. Fabricated stock has been completely processed for the end use. Solid dimension flat stock has five grades: Clear—two faces, clear—one face, paint, core, and sound. Squares have three grades if rough (clear, select, sound) and four if surfaced (clear, select, paint, second).

Finished Market Products

Some hardwood lumber products are graded in relatively finished form, with little or no further processing anticipated. Flooring is probably the highest volume finished product.

Other examples are lath, siding, ties, planks, carstock, construction boards, timbers, trim, molding, stair treads, and risers. Grading rules promulgated for flooring anticipate final consumer use and are summarized in this section. Details on grades of other finished products are found in appropriate association grading rules.

Hardwood flooring generally is graded under the rules of the Maple Flooring Manufacturers Association and the rules of the National Oak Flooring Manufacturers Association. Tongued-and-grooved and end-matched hardwood flooring is commonly furnished. Square edge and square end strip flooring is also available as well as parquet flooring suitable for laying on a mastic base or on an ordinary subfloor.

The Maple Flooring Manufacturers Association grading rules cover flooring manufactured from hard maple, beech, and birch. Each species is graded into four categories—First grade, Second grade, Third grade, and Fourth grade. Combination grades of Second and Better and Third and Better are sometimes specified. There are also three special grades— Selected First grade light northern hard maple, Selected First grade amber northern hard maple, Selected First grade red (produced from northern beech or birch) which are made up of special stock selected for color.

First grade flooring must have one face practically free from all imperfections. Variations in the natural color of the wood are allowed. Second grade flooring admits tight, sound knots and other slight imperfections but must lay without waste. Third grade flooring has few restrictions as to imperfections in the grain but must permit proper laying and provide a good, serviceable floor.

The standard thickness of maple, beech, and birch flooring is $\frac{25}{32}$ inch. Face widths are $1\frac{1}{2}$, 2, $2\frac{1}{4}$, and $3\frac{1}{4}$ inches. Standard lengths are 2 feet and longer in First and Second grade flooring and $1\frac{1}{4}$ feet and longer in Third grade flooring.

The grading rules of the National Oak Flooring Manufacturers Association mainly cover quartersawed and plain-sawed oak flooring. Quartersawed flooring has two grades—Clear and Select. Plainsawed flooring has four grades—Clear, Select, No. 1 Common, and No. 2 Common. The Clear grade in both plainsawed and quartersawed flooring must have the face free from surface imperfections except for three-eighths inch of bright sap. Color is not

---

[1] Species names are those used in grading rules of the National Hardwood Lumber Association. Two woods—cedar (eastern redcedar, known as aromatic red) and cypress (baldcypress)—are not hardwoods. Cypress lumber has a different set of grading rules from those used for the hardwoods.

Table 5–3.—*Nomenclature for some types of hardwood lumber*

Commercial name for lumber	Official common tree name	Botanical name
Alder, red	Red alder	*Alnus rubra*
Ash:		
Black	Black ash	*Fraxinus nigra*
Oregon	Oregon ash	*F. latifolia*
White	Blue ash	*F. quadrangulata*
	Green ash	*F. pennsylvanica*
	White ash	*F. americana*
Aspen (popple)	Bigtooth aspen	*Populus grandidentata*
	Quaking aspen	*P. tremuloides*
Basswood	American basswood	*Tilia americana*
	White basswood	*T. heterophylla*
Beech	Beech	*Fagus grandifolia*
Birch	Gray birch	*Betula populifolia*
	Paper birch	*B. papyrifera*
	River birch	*B. nigra*
	Sweet birch	*B. lenta*
	Yellow birch	*B. alleghaniensis*
Box elder	Boxelder	*Acer negundo*
Buckeye	Ohio buckeye	*Aesculus glabra*
	Yellow buckeye	*A. octandra*
Butternut	Butternut	*Juglans cinerea*
Cherry	Black cherry	*Prunus serotina*
Chestnut	Chestnut	*Castanea dentata*
Cottonwood	Balsam poplar	*Populus balsamifera*
	Eastern cottonwood	*P. deltoides*
	Plains cottonwood	*P. sargentii*
Cucumber	Cucumbertree	*Magnolia acuminata*
Dogwood	Flowering dogwood	*Cornus florida*
	Pacific dogwood	*C. nuttallii*
Elm:		
Rock	Cedar elm	*Ulmus crassifolia*
	Rock elm	*U. thomasii*
	September elm	*U. serotina*
	Winged elm	*U. alata*
Soft	American elm	*U. americana*
	Slippery elm	*U. rubra*
Gum	Sweetgum	*Liquidambar styraciflua*
Hackberry	Hackberry	*Celtis occidentalis*
	Sugarberry	*C. laevigata*
Hickory	Mockernut hickory	*Carya tomentosa*
	Pignut hickory	*C. glabra*
	Shagbark hickory	*C. ovata*
	Shellbark hickory	*C. laciniosa*
Holly	American holly	*Ilex opaca*
Ironwood	Eastern hophornbeam	*Ostrya virginiana*
Locust	Black locust	*Robinia pseudoacacia*
	Honeylocust	*Gleditsia triacanthos*
Madrone	Pacific madrone	*Arbutus menziesii*
Magnolia	Southern magnolia	*Magnolia grandiflora*
	Sweetbay	*M. virginiana*

Table 5-3.—*Nomenclature for some types of hardwood lumber*—continued

Commercial name for lumber	Official common tree name	Botanical name
Maple:		
Hard	Black maple	*Acer nigrum*
	Sugar maple	*A. saccharum*
Oregon	Big leaf maple	*A. macrophyllum*
Soft	Red maple	*A. rubrum*
	Silver maple	*A. saccharinum*
Oak:		
Red	Black oak	*Quercus velutina*
	Blackjack oak	*Q. marilandica*
	California black oak	*Q. kelloggi*
	Cherrybark oak	*Q. falcata var. pagodaefolia*
	Laurel oak	*Q. laurifolia*
	Northern pin oak	*Q. ellipsoidalis*
	Northern red oak	*Q. rubra*
	Nuttall oak	*Q. nuttallii*
	Pin oak	*Q. palustris*
	Scarlet oak	*Q. coccinea*
	Shumard oak	*Q. shumardii*
	Southern red oak	*Q. falcata*
	Turkey oak	*Q. laevis*
	Willow oak	*Q. phellos*
White	Arizona white oak	*Q. arizonica*
	Blue oak	*Q. douglasii*
	Bur oak	*Q. macrocarpa*
	California white oak	*Q. lobata*
	Chestnut oak	*Q. primus*
	Chinkapin oak	*Q. muehlenbergii*
	Emory oak	*Q. emoryi*
	Gambel oak	*Q. gambelii*
	Mexican blue oak	*Q. oblongifolia*
	Live oak	*Q. virginiana*
	Oregon white oak	*Q. garryana*
	Overcup oak	*Q. lyrata*
	Post oak	*Q. stellata*
	Swamp chestnut oak	*Q. michauxii*
	Swamp white oak	*Q. bicolor*
	White oak	*Q. alba*
Oregon myrtle	California-laurel	*Umbellularia californica*
Osage orange (bois d'arc)	Osage-orange	*Maclura pomifera*
Pecan	Bitternut hickory	*Carya cordiformis*
	Nutmeg hickory	*C. myristicaeformis*
	Water hickory	*C. aquatica*
	Pecan	*C. illinoensis*
Persimmon	Common persimmon	*Diospyros virginiana*
Poplar	Yellow-poplar	*Liriodendron tulipifera*
Sassafras	Sassafras	*Sassafras albidum*
Sycamore	American sycamore	*Platanus occidentalis*
Tupelo	Black tupelo	*Nyssa sylvatica*
	Ogeechee tupelo	*N. ogeche*
	Water tupelo	*N. aquatica*
Walnut	Black walnut	*Juglans nigra*
Willow	Black willow	*Salix nigra*
	Peachleaf willow	*S. amygdaloides*

considered in the Clear grade. Select flooring (plain-sawed or quartersawed) may contain sap and will admit a few features such as pin wormholes and small tight knots. No. 1 Common plain-sawed flooring must contain material that will make a sound floor without cutting. No. 2 Common may contain grain and surface imperfections of all kinds but must provide a serviceable floor.

Standard thicknesses of oak flooring are $25/32$, ½, and ⅜ inch. Standard face widths are 1½, 2, 2¼, and 3¼ inches. Lengths in upper grades are 2 feet and up with a required average of 4½ feet in a shipment. In the lower grades lengths are 1¼ feet and up with a required average of 2½ or 3 feet per shipment.

A voluntary commercial standard (CS 56) has been in effect for oak flooring since 1936, being revised periodically.

The rules of the National Oak Flooring Manufacturers Association also include specifications for flooring of pecan, hard maple, beech, and birch. The grades of pecan flooring are: First grade, practically clear but unselected for color; First grade red, practically clear with an all-heartwood face; First grade white, practically clear with an all-bright sapwood face; Second grade, admits sound tight knots, pin wormholes, streak, and slight machining imperfections; Second grade red, similar to Second grade but must have a heartwood face; Third grade, must make a sound floor without cutting; and Fourth grade, must provide a serviceable floor. The standard sizes for pecan flooring are the same as those for oak flooring.

The National Oak Flooring Manufacturers Association rules for hard maple, beech, and birch flooring are the same as those of the Maple Flooring Manufacturers Association.

Hardwood Lumber Species

The names used by the trade to describe commercial lumber in the United States are not always the same as the names of trees adopted as official by the USDA Forest Service. Table 5–3 shows the common trade name, the USDA Forest Service tree name, and the botanical name. Table 5–4 lists United States agencies and associations that prepare rules for and supervise grading of hardwoods.

Table 5–4.—*Hardwood grading associations in United States*

Name and Address	Species Covered by Grading Rules
National Hardwood Lumber Association 59 East Van Buren Street Chicago, Illinois 60605	Hardwoods (furniture cuttings, construction lumber, siding, panels)
Hardwood Dimension Manufacturers Association 3813 Hillsboro Road Nashville, Tennessee 37215	Hardwoods (hardwood furniture dimension, squares, laminated stock, interior trim, stair treads and risers)
Maple Flooring Manufacturers Association 424 Washington Avenue, Suite 104 Oshkosh, Wisconsin 54901	Maple, beech, birch (flooring)
National Oak Flooring Manufacturers Association 814 Sterick Building Memphis, Tennessee 38103	Oak, pecan, beech, birch, and hard maple (flooring)
Northern Hardwood and Pine Manufacturers Association Suite 207, Northern Building Green Bay, Wisconsin 54301	Aspen (construction lumber—see discussion Softwood Lumber Grading)

SOFTWOOD LUMBER

Softwood lumber for many years has demonstrated the versatility of wood by serving as a primary raw material for construction and manufacture. In this role it has been produced in a wide variety of products from many different species. The first industry-sponsored grading rules (product descriptions) for softwoods were established before 1900 and were comparatively simple because the sawmills marketed their lumber locally and grades had only local significance. As new timber sources were developed and lumber was transported to distant points, each producing region continued to establish its own grading rules, so lumber from various regions differed in size, grade name, and permitted grade characteristics. When different species were graded under different rules and competed in the chief consuming areas, confusion and dissatisfaction were inevitable.

To eliminate unnecessary differences in the grading rules of softwood lumber and to improve and simplify these rules, a number of conferences were organized from 1919 to 1925 by the U.S. Department of Commerce. These were attended by representatives of lumber manufacturers, distributors, wholesalers, retailers, engineers, architects, and contractors. The result was a relative standardization of sizes, definitions, and procedures for deriving properties, formulated as a voluntary American Lumber Standard. This standard has been modified several times since. The current edition of the standard is issued in pamphlet form as the American Softwood Lumber Standard PS 20–70.

Softwood lumber is classified for market use by form of manufacture, species, and grade. For many products the American Softwood Lumber Standard serves as a basic reference. For specific information on other products, reference must be made to industry marketing aids, trade journals, and grade rules. The following sections outline general classifications of softwood lumber.

Softwood Lumber Grades

Softwood lumber grades can be considered in the context of two major categories of use: (1) Construction and (2) remanufacture. Construction relates principally to lumber expected to function as graded and sized after primary processing (sawing and planing). Remanufacture refers to lumber that will undergo a number of further manufacturing steps and reach the consumer in a significantly different form.

Lumber for Construction

The grading requirements of construction lumber are related specifically to the major construction uses intended and little or no further grading occurs once the piece leaves the sawmill. Construction lumber can be placed in three general categories—stress-graded, nonstress-graded, and appearance lumber. Stress-graded and nonstress-graded lumber are employed where the structural integrity of the piece is the primary requirement. Appearance lumber, as categorized here, encompasses those lumber products in which appearance is of primary importance; structural integrity, while sometimes important, is a secondary feature.

Stress-graded lumber

Almost all softwood lumber nominally 2 to 4 inches thick is stress graded under the national grading rules promulgated within the American Softwood Lumber Standard. For lumber of this kind there is a single set of grade names and descriptions used throughout the United States. Other stress-graded products include timbers, posts, stringers, beams, decking, and some boards. Stress grades and the National Grading Rule are discussed in chapter 6.

Nonstress-graded lumber

Traditionally, much of the lumber intended for general building purposes with little or no remanufacture has not been assigned allowable properties (stress graded). This category of lumber has been referred to as yard lumber; however, the assignment of allowable properties to an increasing number of former "yard" items has diluted the meaning of the term yard lumber.

In nonstress-graded structural lumber, the section properties (shape, size) of the pieces combine with the visual grade requirements to provide the degree of structural integrity intended. Typical nonstress-graded items include boards, lath, battens, crossarms, planks, and foundation stock.

Boards, sometimes referred to as "commons," are one of the more important nonstress-graded products. Common grades of boards are suitable for construction and general utility purposes. They are separated into three to five different grades depending upon the species and lumber manufacturing association involved. Grades may be described by number (No. 1, No. 2) or by descriptive terms (Construction, Standard).

Since there are differences in the inherent properties of the various species and in corresponding names, the grades for different species are not always interchangeable in use. First-grade boards are usually graded primarily for serviceability, but appearance is also considered. This grade is used for such purposes as siding, cornice, shelving, and paneling. Features such as knots and knotholes are permitted to be larger and more frequent as the grade level becomes lower. Second- and third-grade boards are often used together for such purposes as subfloors, roof and wall sheathing, and rough concrete work. Fourth-grade boards are not selected for appearance but for adequate strength. They are used for roof and wall sheathing, subfloor, and rough concrete form work.

Grading provisions for other nonstress-graded products vary by species, product, and grading association. Lath, for example, is available generally in two grades, No. 1 and No. 2; one grade of batten is listed in one grade rule and six in another. For detailed descriptions it is necessary to consult the appropriate grade rule for these products.

Appearance lumber

Appearance lumber often is nonstress-graded but forms a separate category because of the distinct importance of appearance in the grading process. This category of construction lumber includes most lumber worked to a pattern. Secondary manufacture on these items is usually restricted to onsite fitting such as cutting to length and mitering. There is an increasing trend toward prefinishing many items. The appearance category of lumber includes trim, siding, flooring, ceiling, paneling, casing, base, stepping, and finish boards. Finish boards are commonly used for shelving and built-in cabinetwork.

Most appearance lumber grades are described by letters and combinations of letters (B&BTR, C&BTR, D). (See Standard Lumber Abbreviations at the end of this chapter for

definitions of letter grades.) Appearance grades are also often known as "Select" grades. Descriptive terms such as "prime" and "clear" are applied to a limited number of species. The specification FG (flat grain), VG (vertical grain), or MG (mixed grain) is offered as a purchase option for some appearance lumber products. In cedar and redwood, where there is a pronounced difference in color between heartwood and sapwood and heartwood has high natural resistance to decay, grades of heartwood are denoted as "heart." In some species and products two, or at most three, grades are available. A typical example is casing and base in the grades of C&BTR and D in some species and in B&BTR, C,C&BTR, and D in other species. Although several grades may be described in grade rules, often fewer are offered on the retail market.

Grade B&BTR allows a few small imperfections, mainly in the form of minor skips in manufacture, small checks or stains due to seasoning, and, depending on the species, small pitch areas, pin knots, or the like. Since appearance grades emphasize the quality of one face, the reverse side may be lower in quality. In construction, grade C&BTR is the grade combination most commonly available. It is used for high-quality interior and exterior trim, paneling, and cabinetwork, especially where these are to receive a natural finish. It is the principal grade used for flooring in homes, offices, and public buildings. In industrial uses it meets the special requirements for large sized, practically clear stock.

The number and size of imperfections permitted increases as the grades drop from B&BTR to D and E. Appearance grades are not uniform across species and products, however, and official grade rules must be used for detailed reference. C is used for many of the same purposes as B&BTR, often where the best paint finish is desired. Grade D allows larger and more numerous surface imperfections that do not detract from the appearance of the finish when painted. Grade D is used in finish construction for many of the same uses as C. It is also adaptable to industrial uses requiring short-length clear lumber.

Lumber for Remanufacture

A wide variety of species, grades, and sizes of softwood lumber is supplied to industrial accounts for cutting to specific smaller sizes which become integral parts of other products. In this secondary manufacturing process, grade descriptions, sizes, and often the entire appearance of the wood piece are changed. Thus the role of the grading process for these remanufacture items is to describe as accurately as possible the yield to be obtained in the subsequent cutting operation. Typical of lumber for secondary manufacture are the factory grades, industrial clears, box lumber, molding stock, and ladder stock. The variety of species available for these purposes has led to a variety of grade names and grade definitions. The following section briefly outlines some of the more common classifications. For details, reference must be made to industry sources. Availability and grade designation often vary by region and species.

Factory (Shop) grades.

Traditionally softwood lumber used for cuttings has been termed Factory or Shop. This lumber forms the basic raw material for many secondary manufacturing operations. Some grading associations refer to cutting grades as Factory while others refer to Shop. All impose a somewhat similar nomenclature in the grade structure. Factory Select and Select Shop are typical high grades, followed by No. 1, No. 2, and No. 3 Shop. Door cuttings and sash cuttings are specialized grade categories under which grade levels of No. 1, No. 2, and No. 3 Cuttings and No. 1, No. 2, and No. 3 Shop are applied.

Grade characteristics of cuttings are influenced by the width, length, and thickness of the basic piece and are based on the amount of high-quality material that can be removed by cutting. Typically, a Select Shop would be required to contain either (a) 70 percent of cuttings of specified size, clear on both sides or (b) 70 percent cuttings of different size equal to a B&BTR Finish grade on one side. No. 1 Shop would be required to have 50 percent of (a) or (b); No. 2 Shop would be required to have $33\frac{1}{3}$ percent. Because of different characteristics assigned to grades with similar nomenclature, grades labeled Factory or Shop must be referenced to the appropriate industry source.

Industrial clears

These grades are used for cabinet stock, door stock, and other product components where excellent appearance, mechanical and physical properties, and finishing characteristics are important. The principal grades are B&BTR, C, and D. Grading is based primarily on the best face, although the influence of edge characteristics is important and varies depending upon piece width and thickness. In redwood this grade may include an "all heart" requirement for decay resistance in manufacture of cooling towers, tanks, pipe, and similar products.

Molding, ladder, pole, tank, and pencil stock

Within producing regions, grading rules delineate the requirements for a variety of lumber items oriented to specific consumer products. Custom and the characteristics of the wood supply lead to different grade descriptions and terminology. For example, in West Coast species, the ladder industry can choose from one "ladder and pole stock" grade plus two ladder rail grades and one ladder rail stock grade. In southern pine, ladder stock is available as Select and Industrial. Molding stock, tank stock, pole stock, stave stock, stadium seat stock, box lumber, and pencil stock are other typical industrial grades oriented to the final product. Some have only one grade level; a few offer two or three levels. Special features of these grades may include a restriction on sapwood related to desired decay resistance, specific requirements for slope of grain and growth ring orientation for high-stress use such as ladders, and particular cutting requirements as in pencil stock. All references to these grades should be made directly to current lumber association grading rules.

Structural laminations

Structural laminating grades describe the characteristics used to segregate lumber for structural glued-laminated timbers. Three typical basic categories, L1, L2, and L3, exist with additional provisions for "Dense." The grade characteristics permitted are based on anticipated performance as a portion of the laminated product; however, allowable properties are not assigned separately to the laminating grades.

Softwood Lumber Manufacture

Size

Lumber length is recorded in actual dimensions while width and thickness are traditionally recorded in "nominal" dimensions—the actual dimension being somewhat less.

Softwood lumber is manufactured in length multiples of 1 foot as specified in various grading rules. In practice, 2-foot multiples (in even numbers) are the rule for most construction lumber. Width of softwood lumber varies, commonly from 2 to 16 inch nominal. The thickness of lumber can be generally categorized as follows:

Boards.—lumber less than 2 inches in nominal thickness.

Dimension.—lumber from 2 inches to, but not including, 5 inches in nominal thickness.

Timbers.—lumber 5 or more inches in nominal thickness in the least dimension.

To standardize and clarify nominal-actual sizes the American Lumber Standard specifies thickness and width for lumber that falls under the standard.

The standard sizes for stress-graded and nonstress-graded construction lumber are given in table 5–5. Timbers are usually surfaced while green and only green sizes are given. Dimension and boards may be surfaced green or dry at the prerogative of the manufacturer; therefore, both green and dry standard sizes are given. The sizes are such that a piece of green lumber, surfaced to the standard green size, will shrink to approximately the standard dry size as it dries down to about 15 percent moisture content. The American Lumber Standard definition of dry is a moisture content of 19 percent or less. Many types of lumber are dried before surfacing and only dry sizes for these products are given in the standard.

Lumber for remanufacture is offered in specified sizes to fit end product requirements. Factory grades for general cuttings (Shop) are offered in thicknesses from less than 1 inch to over 3 inches depending on species. Thicknesses of door and sash cuttings start at 1⅜ inches. Cuttings are various lengths and widths. Laminating stock sometimes is offered oversize, compared to standard dimension sizes, to permit resurfacing prior to laminating. Industrial Clears can be offered rough or surfaced in a variety of sizes starting from less than 2 inches thick and as narrow as 3 inches. Sizes

Table 5–5.—*American Standard lumber sizes for stress-graded and non-stress-graded lumber for construction* [1]

Item	Thickness			Face width		
	Nominal	Minimum dressed		Nominal	Minimum dressed	
		Dry	Green		Dry	Green
	In.	*In.*	*In.*	*In.*	*In.*	*In.*
Boards	1	¾	25/32	2	1½	1 9/16
	1¼	1	1 1/32	3	2½	2 9/16
	1½	1¼	1 9/32	4	3½	3 9/16
				5	4½	4⅝
				6	5½	5⅝
				7	6½	6⅝
				8	7¼	7½
				9	8¼	8½
				10	9¼	9½
				11	10¼	10½
				12	11¼	11½
				14	13¼	13½
				16	15¼	15½
Dimension	2	1½	1 9/16	2	1½	1 9/16
	2½	2	2 1/16	3	2½	2 9/16
	3	2½	2 9/16	4	3½	3 9/16
	3½	3	3 1/16	5	4½	4⅝
	4	3½	3 9/16	6	5½	5⅝
	4½	4	4 1/16	8	7¼	7½
				10	9¼	9½
				12	11¼	11½
				14	13¼	13½
				16	15¼	15½
Timbers	5 and greater		½ less than nominal	5 and greater		½ less than nominal

[1] Nominal sizes in the table are used for convenience. No inference should be drawn that they represent actual sizes.

for special product grades such as molding stock and ladder stock are specified in appropriate grading rules or handled by purchase agreements.

Surfacing

Lumber can be produced either rough or surfaced (dressed). Rough lumber has surface imperfections caused by the primary sawing operations. It may be oversize by variable amounts in both thickness and width, depending upon the type of sawmill equipment. Rough lumber serves as a raw material for further manufacture and also for some decorative purposes. A rough sawn surface is common in post and timber products. Because of surface roughness, grading of rough lumber generally is difficult.

Surfaced lumber has been planed or sanded on one side (S1S), two sides (S2S), one edge (S1E), two edges (S2E), or combinations of sides and edges (S1S1E, S2S1E, S1S2E, or S4S). Surfacing may be done to attain smoothness or uniformity of size or both.

A number of surfaced lumber imperfections or blemishes are classified as "manufacturing imperfections" or "mismanufacture." For example, chipped and torn grain are irregularities of the surface where the particles of the surface have been torn out by the surfacing operation. Chipped grain is a "barely perceptible" characteristic, while torn grain is classified by depth. Raised grain, skip, machine burn and gouge, chip marks, and wavy dressing are other defined manufacturing imperfections. Manufacturing imperfections are defined in the American Lumber Standard and further detailed in the grade rules. Classifications of manufacturing imperfections (combinations of the imperfections allowed in the rules) are established in the rules as STANDARD "A", STANDARD "B", etc. For example, STANDARD "A" admits very light torn grain, occasional slight chip marks, and very slight knife marks. These classifications are used as part of the grade description of some lumber products to specify the allowable surfacing quality.

Patterns

Lumber which, in addition to being surfaced, has been matched, shiplapped, or otherwise patterned is often classed as "worked lumber." Figure 5–2 shows typical patterns of lumber.

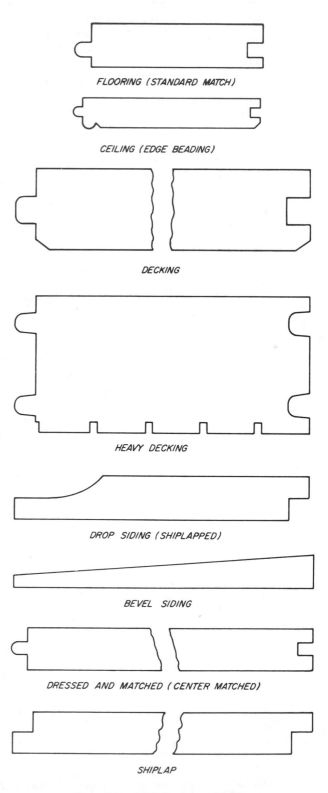

FLOORING (STANDARD MATCH)

CEILING (EDGE BEADING)

DECKING

HEAVY DECKING

DROP SIDING (SHIPLAPPED)

BEVEL SIDING

DRESSED AND MATCHED (CENTER MATCHED)

SHIPLAP

Figure 5–2.—Typical patterns of lumber.

Table 5–6.—*Nomenclature of commercial softwood lumber*

Standard lumber name under American Lumber Standards	Official Forest Service tree name used in this handbook	Botanical name
Cedar:		
Alaska	Alaska-cedar	*Chamaecyparis nootkatensis*
Eastern red	Eastern redcedar	*Juniperus virginiana*
Incense	Incense-cedar	*Libocedrus decurrens*
Northern white	Northern white-cedar	*Thuja occidentalis*
Port Orford	Port-Orford-cedar	*Chamaecyparis lawsoniana*
Southern white	Atlantic white-cedar	*C. thyoides*
Western red	Western redcedar	*Thuja plicata*
Cypress, red (coast type), yellow (inland type), white (inland type)	Baldcypress	*Taxodium distichum*
Douglas-fir	Douglas-fir	*Pseudotsuga menziesii*
Fir:		
Balsam	Balsam fir	*Abies balsamea*
	Fraser fir	*A. fraseri*
Noble	Noble fir	*A. procera*
White	California red fir	*A. magnifica*
	Grand fir	*A. grandis*
	Pacific silver fir	*A. amabilis*
	Subalpine fir	*A. lasiocarpa*
	White fir	*A. concolor*
Hemlock:		
Eastern	Eastern hemlock	*Tsuga canadensis*
Mountain	Mountain hemlock	*T. mertensiana*
West Coast	Western hemlock	*T. heterophylla*
Juniper, western	Alligator juniper	*Juniperus deppeana*
	Rocky Mountain juniper	*J. scopulorum*
	Utah juniper	*J. osteosperma*
	Western juniper	*J. occidentalis*
Larch, western	Western larch	*Larix occidentalis*
Pine:		
Idaho white	Western white pine	*Pinus monticola*
Jack	Jack pine	*P. banksiana*
Lodgepole	Lodgepole pine	*P. contorta*
Longleaf yellow [1]	Longleaf pine	*P. palustris*
	Slash pine	*P. elliottii*
Northern white	Eastern white pine	*P. strobus*
Norway	Red pine	*P. resinosa*
Southern yellow	Longleaf pine	*P. palustris*
	Shortleaf pine	*P. echinata*
	Loblolly pine	*P. taeda*
	Slash pine	*P. elliottii*
	Pitch pine	*P. rigida*
	Virginia pine	*P. virginiana*
Sugar	Sugar pine	*P. lambertiana*
Redwood	Redwood	*Sequoia sempervirens*
Spruce:		
Eastern	Black spruce	*Picea mariana*
	Red spruce	*P. rubens*
	White spruce	*P. glauca*
Engelmann	Blue spruce	*P. pungens*
	Engelmann spruce	*P. engelmannii*
Sitka	Sitka spruce	*P. sitchensis*
Tamarack	Tamarack	*Larix laricina*
Yew, Pacific	Pacific yew	*Taxus brevifolia*

[1] The commercial requirements for longleaf yellow pine lumber are that not only must it be produced from the species *Pinus elliottii* and *P. palustris*, but each piece must average either on 1 end or the other not less than 6 annual rings per inch and not less than ⅓ summerwood. Longleaf yellow pine lumber is sometimes designated as pitch pine in the export trade.

With softwood flooring, "standard match" means that the upper lip of the groove is thicker than the lower. The thickness of the lower lip is the same for all standard thicknesses of flooring and hence the difference between upper and lower lips becomes more pronounced in the greater thicknesses. Ceiling is usually machined with a "V" while partition usually has a bead and "V" but may be patterned on both sides. Decking is available in a variety of patterns including grooved V-joint and striated. It is available in panel form as well as in single pieces. Drop siding probably is made in more patterns than any other product except molding. Some siding patterns are shiplapped, others are tongued-and-grooved. Bevel siding is made by resawing ¼- or ⅝-inch lumber on an angle. Square-edged lumber in boards, timbers, or dimension forms only rectangles of different dimensions. Dressed and matched (D&M) boards have the tongue and groove in the center, making the pieces center matched. Dressed and matched boards are considered preferable to shiplap boards for some uses.

Softwood Lumber Species

The names of lumber as adopted by the trade as American Standard may vary from the names of trees adopted as official by the USDA Forest Service. Table 5–6 shows the American Lumber Standard commercial name for lumber, the USDA Forest Service tree name, and the botanical name. Some softwood species are marketed primarily in combinations. Designations such as Southern Pine and Hem-Fir represent typical combinations. The grading organizations listed in table 5–7 should be contacted for questions regarding combination names and species not listed in table 5–6. Further discussion of species grouping is contained in chapter 6.

Table 5–7.—*Organizations promulgating softwood grades*

Name and Address	Species Covered by Grading Rules
National Hardwood Lumber Association 59 East Van Buren Street Chicago, Illinois 60605	Baldcypress, eastern redcedar
Northeastern Lumber Manufacturers Association, Inc. 13 South Street Glens Falls, New York 12801	Balsam fir, eastern white pine, red pine, eastern hemlock, black spruce, white spruce, red spruce, pitch pine, tamarack, jack pine, northern white cedar
Northern Hardwood and Pine Manufacturers Association Suite 207, Northern Building Green Bay, Wisconsin 54301	Bigtooth aspen, quaking aspen, eastern white pine, red pine, jack pine, black spruce, white spruce, red spruce, balsam fir, eastern hemlock, tamarack
Red Cedar Shingle & Handsplit Shake Bureau 5510 White Building Seattle, Washington 98101	Western redcedar (shingles and shakes)
Redwood Inspection Service 617 Montgomery Street San Francisco, California 94111	Redwood
Southern Cypress Manufacturers Association P.O. Box 5816 Jacksonville, Florida 32207	Baldcypress
Southern Pine Inspection Bureau Box 846 Pensacola, Florida 32502	Longleaf pine, slash pine, shortleaf pine, loblolly pine, Virginia pine, pond pine, pitch pine
West Coast Lumber Inspection Bureau Box 25406 1750 SW. Skyline Boulevard Portland, Oregon 97225	Douglas-fir, western hemlock, western redcedar, incense-cedar, Port-Orford-cedar, Alaska-cedar, western true firs, mountain hemlock, Sitka spruce
Western Wood Products Association 700 Yeon Building Portland, Oregon 97204	Ponderosa pine, western white pine, Douglas-fir, sugar pine, western true firs, western larch, Engelmann spruce, incense-cedar, western hemlock, lodgepole pine, western redcedar, mountain hemlock, red alder

Softwood Lumber Grading

Most lumber is graded under the supervision of inspection bureaus and grading agencies. These organizations supervise lumber mill grading, and provide reinspection services to resolve disputes concerning lumber shipments. Some of the agencies also author grading rules which reflect the species and products in the geographic regions they represent.[2] Many of the grading rules and procedures fall under the American Softwood Lumber Standard. This can be an important consideration because it provides for recognized uniform grading procedures. Names and addresses of rules-writing organizations in the United States, and the species for which they are concerned, are given in table 5–7. Canadian softwood lumber imported into the United States is graded by inspection agencies in Canada. Names and addresses of Canadian grading agencies may be obtained from the Canadian Lumber Standards Administration Board, 1460–1055 West Hastings Street, Vancouver 1, B.C., Canada.

PURCHASING LUMBER

After primary manufacture, most lumber products are marketed through wholesalers to remanufacture plants or to retail outlets. Because of the extremely wide variety of lumber products, wholesaling is very specialized with some organizations dealing only with a limited number of species or products. Where the primary manufacturer can readily identify the customers, direct sales may be made. Examples are manufacturer sales to large retail chains, contractors, and to truss rafter fabricators. There is an increasing trend to direct sales, particularly for mobile and modular housing.

Lumber Distribution

Large primary manufacturers and wholesale organizations set up distribution yards in lumber-consuming areas to more effectively distribute both hardwood and softwood products. Retail yards draw inventory from distribution yards and, in wood-producing areas, from local lumber producers. Few lumber products are readily available at the retail level in the wide

---

[2] A limited number of hardwoods are also being graded under the provisions of the standards used for grading softwoods. These hardwoods include aspen and red alder.

range of grades and species suggested by the grade rules.

Transportation is a vital factor in lumber distribution. On the eastern seaboard of the United States, lumber from the Pacific Coast is readily available because of low-cost water transportation via the Panama Canal. Often the lumber shipped by water is green because weight is not a major factor in this type of shipping. On the other hand, lumber reaching the East Coast from the Pacific Coast by rail is largely kiln dried because rail shipping rates are based on weight. A shorter rail haul places southern and northeastern species in a favorable shipping cost position in this same market.

Changing transportation costs have influenced shifts in market distribution of species and products. Trucks have become a major factor in lumber transport for regional remanufacture plants, for retail supply from distribution yards, and for much construction lumber distribution where the distance from primary manufacture to customer is within an approximate 1,500-mile radius. The development of foreign hardwood and softwood manufacturing and the availability of water transport has brought foreign lumber products to the United States market, particularly in coastal areas.

Retail Yard Inventory

The small retail yards throughout the United States carry softwoods required for ordinary construction purposes and often small stocks of one or two hardwoods in the grades suitable for finishing or cabinetwork. Special orders must be made for other hardwoods. Trim items such as molding in either softwood or hardwood are available cut to size and standard pattern. Cabinets are usually made by millwork plants ready for installation and many common styles and sizes are carried or cataloged by the modern retail yard. Hardwood flooring is available to the buyer only in standard patterns. Some retail yards may carry specialty stress grades of lumber such as structural light framing for truss rafter fabrication.

The assortment of species in general construction items carried by retail yards depends largely upon geographic location, and both transportation costs and tradition are important factors. Retail yards within, or close to, a major lumber-producing region may therefore emphasize the local timber. For example, a local retail yard on the coast in the Pacific Northwest may stock only green Douglas-fir and cedar in dimension grades, dry pine and hemlock in boards and molding, plus assorted specialty items such as redwood posts, cedar shingles and shakes, and rough cedar siding. The only hardwoods carried may be walnut and "Philippine mahogany." [3] Retail yards farther from a major softwood supply, such as in the Midwest on the other hand, may draw from several species-growing areas and may stock spruce or southern pine. Being located in a major hardwood production area, these yards will stock, or have available to them, a different and wider variety of hardwoods.

Geography has less influence where consumer demands are more specific. For example, where long length construction lumber (20 to 26 ft.) is required, west coast species often are marketed because the size of the trees in several of the species makes long lengths a practical market item. As another example, ease of treatability makes treated southern pine construction lumber available in a wide geographic area.

Some lumber grades and sizes serve a variety of construction needs. Some species or species groups are available at the retail level only in grade groups. Typical are house framing grades such as joist and plank which are often sold as No. 2 and Better (2&BTR). The percentage of each grade in a grouping is part of the purchase agreement between the primary lumber manufacturer and the wholesaler; however, this ratio may be altered at the retail level by sorting. Where grade grouping is the practice, a requirement for a specific grade such as No. 1 at the retail level will require sorting or special purchase. Grade grouping occurs for reasons of tradition and of efficiency in distribution.

Another important factor in retail yard inventory is that not all grades, sizes, and species described by the grade rules are produced and not all those produced are distributed uniformly to all marketing areas. Regional consumer interest, building code requirements, and transportation costs influence distribution patterns. Often small retail yards will stock only a limited number of species and grades. Large yards, on the other hand, may cater to particular construction industry needs and carry more dry dimension grades along with

clears, finish, and decking. The effect of these variable retail practices is that the grades, sizes, and species outlined in the grade rules must be examined to determine what actually is available. A brief description of lumber products commonly carried by retail yards follows:

Stress-Graded Lumber for Construction

Dimension is the principal stress-graded lumber item available in a retail yard. It is primarily framing lumber for joists, rafters, and studs. Strength, stiffness, and uniformity of size are essential requirements. Dimension is stocked in all yards, frequently in only one or two of the general purpose construction woods such as pine, fir, hemlock, or spruce. Two by six, 2 by 8, and 2 by 10 dimension are found in grades of Select Structural, No. 1, No. 2, and No. 3; often in combinations of No. 2&BTR or possibly No. 3&BTR. In 2 by 4, the grades available would normally be Construction and Standard, sold as Standard and Better (STD&BTR), Utility and Better (UTIL&BTR), or Stud, in lengths of 10 feet and shorter.

Dimension is often found in nominal 2-, 4-, 6-, 8-, 10-, or 12-inch widths and 8- to 18-foot lengths in multiples of 2 feet. Dimension formed by structural end-jointing procedures may be found. Dimension thicker than 2 inches and longer than 18 feet is not available in large quantity.

Other stress-graded products generally present are posts and timbers, with some beams and stringers also possibly in stock. Typical stress grades in these products are Select Structural and No. 1 Structural in Douglas-fir and No. 1SR and No. 2SR in southern pine.

Nonstress-Graded Lumber for Construction

Boards are the most common nonstress-graded general purpose construction lumber in the retail yard. Boards are stocked in one or more species, usually in nominal 1-inch thickness. Standard nominal widths are 2, 3, 4, 6, 8, 10, and 12 inches. Grades most generally available in retail yards are No. 1, No. 2, and No. 3 (or Construction, Standard, and Utility). These will often be combined in grade groups. Boards are sold square edged, dressed and matched (tongued and grooved) or with a shiplapped joint. Boards formed by end-jointing of shorter sections may form an appreciable portion of the inventory.

---

[3] Common market name encompassing many species including tanguile, red lauan, and white lauan.

Appearance Lumber

Completion of a construction project usually depends on a variety of lumber items available in finished or semifinished form. The following items often may be stocked in only a few species, finishes, or in limited sizes depending on the yards.

Finish

Finish boards usually are available in a local yard in one or two species principally in grade C&BTR. Redwood and cedar have different grade designations. Grades such as Clear Heart, A, or B are used in cedar; Clear All Heart, Clear, and Select are typical redwood grades. Finish boards are usually a nominal 1 inch thick, dressed two sides to ¾ inch. The widths usually stocked are nominal 2 to 12 inches in even-numbered inches.

Siding

Siding, as the name implies, is intended specifically to cover exterior walls. Beveled siding is ordinarily stocked only in white pine, ponderosa pine, western redcedar, cypress, or redwood. Drop siding, also known as rustic siding or barn siding, is usually stocked in the same species as beveled siding. Siding may be stocked as B&BTR or C&BTR except in cedar where Clear, A, and B may be available and redwood where Clear All Heart and Clear will be found. Vertical grain (VG) is sometimes a part of the grade designation. Drop siding sometimes is stocked also in sound knotted C and D grades of southern pine, Douglas-fir, and hemlock. Drop siding may be dressed, matched, or shiplapped.

Flooring

Flooring is made chiefly from hardwoods such as oak and maple, and the harder softwood species, such as Douglas-fir, western larch, and southern pine. Often at least one softwood and one hardwood are stocked. Flooring is usually nominal 1 inch thick dressed to $\frac{25}{32}$ inch, and 3- and 4-inch nominal width. Thicker flooring is available for heavy-duty floors both in hardwoods and softwoods. Thinner flooring is available in hardwoods, especially for recovering old floors. Vertical and flat grain (also called quartersawed and plain-sawed) flooring is manufactured from both softwoods and hardwoods. Vertical-grained flooring shrinks and swells less than flat-grained flooring, is more uniform in texture, wears more uniformly, and the joints do not open as much.

Softwood flooring is usually available in B and Better grade, C Select, or D Select. The chief grades in maple are Clear No. 1 and No. 2. The grades in quartersawed oak are Clear and Select, and in plain-sawed Clear, Select, and No. 1 Common. Quartersawed hardwood flooring has the same advantages as vertical-grained softwood flooring. In addition, the silver or flaked grain of quartersawed flooring is frequently preferred to the figure of plain-sawed flooring. Beech, birch, and walnut and mahogany (for fancy parquet flooring) are also occasionally used.

Casing and base

Casing and base are standard items in the more important softwoods and are stocked by most yards in at least one species. The chief grade, B and Better, is designed to meet the requirements of interior trim for dwellings. Many casing and base patterns are dressed to $\frac{11}{16}$ by $2\frac{1}{4}$; other sizes used include $\frac{9}{16}$ by 3, $3\frac{1}{4}$, and $3\frac{1}{2}$. Hardwoods for the same purposes, such as oak and birch, may be carried in stock in the retail yard or may be obtained on special order.

Shingles and shakes

Shingles usually available are sawn from western redcedar, northern white-cedar, and redwood. The shingle grades are: Western redcedar, No. 1, No. 2, No. 3; northern white-cedar, Extra, Clear, 2nd Clear, Clear Wall, Utility; redwood, No. 1, No. 2 VG, and No. 2 MG.

Shingles that are all heartwood give greater resistance to decay than do shingles that contain sapwood. Edge-grained shingles are less likely to warp than flat-grained shingles; thick-butted shingles less likely than thin shingles; and narrow shingles less likely than wide shingles. The standard thicknesses of shingles are described as ⅖, ⅖, ¼, and ⅖ (four shingles to 2 in. of butt thickness, five shingles to 2¼ in. of butt thickness, and five shingles to 2 in. of butt thickness). Lengths may be 16, 18, or 24 inches. Random widths and specified widths ("dimension" shingles) are available in western redcedar, redwood, and cypress.

Shingles are usually packed four bundles to the square. A square of shingles will cover 100 square feet of roof area when the shingles are applied at standard weather exposures.

Shakes are handsplit or handsplit and resawn from western redcedar. Shakes are of a single grade and must be 100 percent clear, graded from the split face in the case of handsplit and resawn material. Handsplit shakes are graded from the best face. Shakes must be 100 percent heartwood free of bark and sapwood. The standard thickness of shakes ranges from ⅜ to 1¼ inches. Lengths are 18 and 24 inches, and a 15-inch "Starter-Finish Course" length.

Important Purchase Considerations

The following outline lists some of the points to consider when ordering lumber or timbers.

1. *Quantity.*—Feet, board measure, number of pieces of definite size and length. Consider that the board measure depends on the thickness and width nomenclature used and that the interpretation of this must be clearly delineated. In other words, nominal or actual, pattern size, etc., must be considered.

2. *Size.*—Thickness in inches—nominal and actual if surfaced on faces. Width in inches—nominal and also actual if surfaced on edges. Length in feet—may be nominal average length, limiting length, or a single uniform length. Often a trade designation, "random" length, is used to denote a nonspecified assortment of lengths. Note that such an assortment should contain critical lengths as well as a range. The limits allowed in making the assortment "random" can be established at the time of purchase.

3. *Grade.*—As indicated in grading rules of lumber manufacturing associations. Some grade combinations (B&BTR) are official grades; other [Standard and Better (STD&BTR) light framing, for example] are grade combinations and subject to purchase agreement. A typical assortment is 75 percent Construction and 25 percent Standard, sold under the label STD&BTR. In softwood, each piece of such lumber typically is stamped with its grade, a name or number identifying the producing mill, the dryness at the time of surfacing, and a symbol identifying the inspection agency supervising the grading inspection. The grade designation stamped on a piece indicates the quality at the time the piece was graded. Subsequent exposure to unfavorable storage conditions, improper drying, or careless handling may cause the material to fall below its original grade.

Note that working or rerunning a graded product to a pattern may result in changing or invalidating the original grade. The purchase specification should be clear regarding regrading or acceptance of worked lumber. In softwood lumber, grades for dry lumber generally are determined after kiln drying and surfacing. This practice is not general for hardwood factory lumber, however, where the grade is generally based on grade and size prior to kiln drying.

4. *Species or groupings of wood.*—Douglas-fir, cypress, Hem-Fir, etc. Some species have been grouped for marketing convenience; others are traded under a variety of names. Be sure the species or species group is correctly and clearly depicted on the purchase specification.

5. *Product.*—Flooring, siding, timbers, boards, etc. Nomenclature varies by species, region, and grading association. To be certain the nomenclature is correct for the product, refer to the grading rule by number and paragraph.

6. *Condition of seasoning.*—Air dry, kiln dry, etc. Softwood lumber dried to 19 percent moisture content or less (S–DRY) is defined as dry by the American Lumber Standard. Other degrees of dryness are partially air dried (PAD), green (S–GRN), and 15 percent max. (KD in southern pine). There are several specified levels of moisture content for redwood. If the moisture requirement is critical, the levels and determination of moisture content must be specified.

7. *Surfacing and working.*—Rough (unplaned), dressed (surfaced), or patterned stock. Specify condition. If surfaced, indicate S4S, S1S1E, etc. If patterned, list pattern number with reference to the appropriate grade rules.

8. *Grading rules.*—Official grading agency name, product identification, paragraph number or page number or both, date of rules or official rule volume (rule No. 16, for example).

9. *Manufacturer.*—Name of manufacturer or trade name of specific product or both. Most lumber products are sold without reference to a specific manufacturer. If proprietary names or quality features of a manufacturer are required, this must be stipulated clearly on the purchase agreement.

10. *Reinspection.*—Procedures for resolution of purchase disputes. The American Lumber Standard (ALS) provides for procedures to be followed in resolution of manufacturer-wholesaler-consumer conflicts over quality or quantity of softwood lumber graded under ALS jurisdiction. The dispute may be resolved by

reinspecting the shipment. Time limits, liability, costs, and complaint procedures are outlined in the grade rules of both softwood and hardwood agencies under which the disputed shipment was graded and purchased.

STANDARD LUMBER ABBREVIATIONS

The following standard lumber abbreviations are commonly used in contracts and other documents for purchase and sale of lumber.

AAR	Association of American Railroads
AD	air dried
ADF	after deducting freight
ALS	American Lumber Standard
AST	antistain treated. At ship tackle (western softwoods)
AV or avg	average
AW&L	all widths and lengths
B1S	see EB1S, CB1S, and E&CB1S
B2S	see EB2S, CB2S, and E&CB2S
B&B, B&BTR	B and Better
B&S	beams and stringers
BD	board
BD FT	board feet
BDL	bundle
BEV	bevel or beveled
BH	boxed heart
B/L, BL	bill of lading
BM	board measure
BSND	bright sapwood no defect
BTR	better
c	allowable stress in compression in pounds per square inch
CB	center beaded
CB1S	center bead on one side
CB2S	center bead on two sides
CC	cubical content
cft or cu. ft.	cubic foot or feet
CF	cost and freight
CIF	cost, insurance, and freight
CIFE	cost, insurance, freight, and exchange
CG2E	center groove on two edges
C/L	carload
CLG	ceiling
CLR	clear
CM	center matched
Com	common
CS	calking seam
CSG	casing
CV	center V
CV1S	center V on one side
CV2S	center V on two sides
DB Clg	double beaded ceiling (E&CB1S)
DB Part	double beaded partition (E&CB2S)
DET	double end trimmed
DF	Douglas-fir
DIM	dimension
DKG	decking
D/S, DS, D/Sdg	drop siding
D1S, D2S	See S1S and S2S
D&M	dressed and matched
D&CM	dressed and center matched
D&SM	dressed and standard matched
D2S&CM	dressed two sides and center matched
D2S&SM	dressed two sides and standard matched
E	edge
EB1S	edge bead one side
EB2S, SB2S	edge bead on two sides
EE	eased edges

EG	edge (vertical or rift) grain
EM	end matched
EV1S, SV1S	edge V one side
EV2S, SV2S	edge V two sides
E&CB1S	edge and center bead one side
E&CB2S, DB2S, BC&2S	edge and center bead two sides
E&CV1S, DV1S, V&CV1S	edge and center V one side
E&CV2S, DV2S, V&CV2S	edge and center V two sides
f	allowable stress in bending in pounds per square inch
FA	facial area
Fac	factory
FAS	free alongside (vessel)
FAS	Firsts and Seconds
FBM, Ft. BM	feet board measure
FG	flat or slash grain
FJ	finger joint. End-jointed lumber using a finger joint configuration
FLG, Flg	flooring
FOB	free on board (named point)
FOHC	free of heart center
FOK	free of knots
FRT, Frt	freight
FT, ft	foot or feet
FT. SM	feet surface measure
G	girth
GM	grade marked
G/R	grooved roofing
HB, H. B.	hollow back
HEM	hemlock
Hrt	heart
H&M	hit and miss
H or M	hit or miss
IN, in.	inch or inches
Ind	industrial
J&P	joists and planks
JTD	jointed
KD	kiln dried
LBR, Lbr	lumber
LCL	less than carload
LGR	longer
LGTH	length
Lft, Lf	lineal foot or feet
LIN, Lin	lineal
LL	longleaf
LNG, Lng	lining
M	thousand
MBM, MBF, M. BM	thousand (feet) board measure
MC, M.C.	moisture content
MERCH, Merch	merchantable
MG	medium grain or mixed grain
MLDG, Mldg	molding
Mft	thousand feet
MSR	machine stress rated

N	nosed	SM	surface measure
NBM	net board measure	Specs	specifications
No.	number	SQ	square
N1E or N2E	nosed one or two edges	SQRS	squares
		SR	stress rated
Ord	order	STD, Std	standard
PAD	partially air dry	Std. lgths.	standard lengths
PAR, Par	paragraph	SSND	sap stain no defect (stained)
PART, Part	partition	STK	stock
PAT, Pat	pattern	STPG	stepping
Pcs.	pieces	STR, STRUCT	structural
PE	plain end	SYP	southern yellow pine
PET	precision end trimmed	S&E	side and edge (surfaced on)
P&T	posts and timbers	S1E	surfaced one edge
P1S, P2S	see S1S and S2S	S2E	surfaced two edges
RDM	random	S1S	surfaced one side
REG, Reg	regular	S2S	surfaced two sides
Rfg.	roofing	S4S	surfaced four sides
RGH, Rgh	rough	S1S&CM	surfaced one side and center matched
R/L, RL	random lengths	S2S&CM	surfaced two sides and center matched
R/W, RW	random widths	S4S&CS	surfaced four sides and calking seam
RES	resawn	S1S1E	surfaced one side, one edge
SB1S	single bead one side	S1S2E	surfaced one side, two edges
SDG, Sdg	siding	S2S1E	surfaced two sides, one edge
S–DRY	surfaced dry. Lumber 19 percent moisture content or less per American Lumber Standard for softwood	S2S&SL	surfaced two sides and shiplapped
		S2S&SM	surfaced two sides and standard matched
SE	square edge	t	allowable stress in tension in pounds per square inch
SEL, Sel	select or select grade	TBR	timber
SE&S	square edge and sound	T&G	tongued and grooved
SG	slash or flat grain	VG	vertical (edge) grain
S–GRN	surfaced green. Lumber unseasoned, in excess of 19 percent moisture content per American Lumber Standard for softwood	V1S	see EV1S, CV1S, and E&CV1S
		V2S	see EV2S, CV2S, and E&CV2S
		WCH	west coast hemlock
		WDR, wdr	wider
SGSSND	Sapwood, gum spots and streaks, no defect	WHAD	worm holes a defect
		WHND	worm holes no defect
SIT. SPR	Sitka spruce	WT	weight
S/L, SL, S/Lap	shiplap	WTH	width
		WRD	western redcedar
STD. M	standard matched	YP	yellow pine

BIBLIOGRAPHY

U.S. Department of Commerce
  Strip oak flooring. Comm. Stand. CS 56. (See current edition.)

———

  American softwood lumber standard. Prod. Stand. PS 20. (See current edition.)

Chapter 6

LUMBER STRESS GRADES AND ALLOWABLE PROPERTIES

Lumber of any species and size, as it is sawed from the log, is quite variable in its mechanical properties. Pieces may differ in strength by several hundred percent. For simplicity and economy in use, pieces of lumber of similar mechanical properties can be placed in a single class called a stress grade.

A stress grade is characterized by:

1. One or more sorting criteria
2. A set of allowable properties for engineering design
3. A unique grade name

This chapter discusses sorting criteria for two stress grading methods, along with the philosophy of how allowable properties are derived. The allowable properties depend upon the particular sorting criteria and on additional factors that are independent of the sorting criteria. Allowable properties are different from, and usually much lower than, the properties of clear, straight-grained wood tabulated in chapter 4.

From one to six allowable properties are associated with a stress grade—modulus of elasticity and stresses in tension and compression parallel to the grain, in compression perpendicular to the grain, in shear parallel to the grain, and in extreme fiber in bending. As with any structural material, the strength properties used to derive the five allowable stresses must be inferred or measured nondestructively to avoid damage to pieces of lumber. Any nondestructive test provides both sorting criteria and a means of calculating appropriate mechanical properties.

The philosophies contained in this chapter are used by a number of organizations to develop commercial stress grades. The exact procedures they use and the resulting allowable stresses are not detailed in the Wood Handbook, but reference to them is given.

DEVELOPING VISUAL GRADES

Visual Sorting Criteria

Visual grading is the oldest stress grading method. It is based on the premise that mechanical properties of lumber differ from mechanical properties of clear wood because of characteristics that can be seen and judged by eye. These visual characteristics are used to sort the lumber into stress grades. The following are major visual sorting criteria:

Density

Strength is related to the weight or the density of clear wood. Properties assigned to lumber are sometimes modified by using the rate of growth and the percentage of latewood as measures of density. Selection for rate of growth required that the number of annual rings per inch be within a specified range. It is possible to eliminate some very low strength pieces from a grade by excluding those that are exceptionally light in weight.

Decay

Decay in most forms should be severely restricted or prohibited in stress grades because its extent is difficult to determine and its effect on strength is often greater than visual observation would indicate. Limited decay of the pocket type (e.g., *Fomes pini*) can be permitted to some degree in stress grades, as can decay that occurs in knots but does not extend into the surrounding wood.

Heartwood and Sapwood

Heartwood and sapwood of the same species have equal mechanical properties, and no heartwood requirement need be made in stress grading. Since heartwood of some species is more resistant to decay than sapwood, heartwood may be required if the untreated wood is to be exposed to a decay hazard. On the other hand, sapwood takes preservative treatment more readily and should not be limited in lumber that is to be treated.

Slope of Grain

In zones of cross grain, the direction of the wood fibers is not parallel to the edges of the lumber. Cross grain reduces the mechanical properties of lumber. Severely cross-grained pieces are also undesirable because they tend to warp with changes in moisture content. Stresses caused by shrinkage during drying are greater in structural lumber than in small, clear specimens and are increased in zones of sloping or distorted grain. To provide a margin of safety, the reduction of strength due to cross grain in visually graded structural lumber should be about twice the reduction observed in tests of small, clear specimens that contain similar cross grain.

Knots

Knots interrupt the direction of grain and cause localized cross grain with steep slopes. Intergrown or live knots resist some kinds of stress, but encased knots or knotholes transmit little or no stress. On the other hand, distortion of grain is greater around an intergrown knot than around an encased or dead knot. As a result, overall strength effects are roughly equalized, and often no distinction is made in stress grading between live knots, dead knots, and knotholes.

The zone of distorted grain (cross grain) around a knot has less "parallel to piece" stiffness than straight-grained wood; thus, localized areas of low stiffness are often associated with knots. Such zones generally comprise only a minor part of the total volume of a piece of lumber, however, and overall piece stiffness reflects the character of all parts.

The presence of a knot in a piece modifies some of the clear wood strength properties more than it affects the overall stiffness. The effect of a knot on strength depends approximately on the proportion of the cross section of the piece of lumber occupied by the knot, upon knot location, and upon the distribution of stress in the piece. Limits on knot sizes are therefore made in relation to the width of and location on the face in which the knot appears. Compression members are stressed about equally throughout, and no limitation related to location of knots is imposed. In tension, knots along the edge of a member produce an eccentricity that induces bending stresses, and should therefore be more restricted than knots away from the edge. In structural members subjected to bending, stresses are greater in the middle part of the length and are greater at the top and bottom edges than at midheight. These facts can be recognized in differing limitations on the sizes of knots in different locations.

Knots in glued-laminated structural members are not continuous as in sawed structural lumber, and different methods are used for evaluating their effect on strength (see ch. 10).

Shake

Shake in members subjected to bending reduces the resistance to shear and therefore is limited most closely in those parts of a bending member where shear stresses are highest. In members subjected only to tension or compression, shake does not greatly affect strength; it may be limited because of appearance and because it permits entrance of moisture that results in decay.

Checks and Splits

While shake indicates a weakness of fiber bond that is presumed to extend lengthwise without limit, checks and splits are rated only by the area of actual opening. An end split is considered equal to an end check that extends through the full thickness of the piece. The effects of checks and splits upon strength and the principles of their limitation are the same as for shake.

Wane

Requirements of appearance, fabrication, or the need for ample bearing or nailing surfaces generally impose stricter limitations on wane than does strength. Wane is therefore limited in structural lumber on those bases.

Pitch Pockets

Pitch pockets ordinarily have so little effect on structural lumber that they can be disregarded in stress grading if they are small and limited in number. The presence of a large number of pitch pockets, however, may indicate shake or weakness of bond between annual rings.

Deriving Properties for Visually Graded Lumber

The derivation of mechanical properties of visually graded lumber is based on clear wood properties and on the lumber characteristics allowed by the visual sorting criteria. The influence of the sorting criteria is handled with "strength ratios" for the strength properties of wood and with "quality factors" for the modulus of elasticity.

From piece to piece, there is variation both in the clear wood properties and in the occurrence of the property-modifying characteristics. The influence of this variability on lumber properites is handled differently for strength than for modulus of elasticity.

Once the clear wood properties have been modified for the influence of sorting criteria and variability, additional modifications for size, moisture content, and load duration are applied. The composite of these adjustments is an "allowable property," to be discussed in more detail later in the chapter.

Strength Properties

Each strength property of a piece of lumber is derived from the product of the clear wood strength for the species and the limiting strength ratio. The strength ratio is the hypothetical ratio of the strength of a piece of lumber with visible strength-reducing characteristics to its strength if those characteristics were absent.

The true strength ratio of a piece of lumber is never known and must be estimated. The strength ratio assigned to a lumber characteristic, therefore, serves as a predictor of lumber strength. Strength ratio usually is expressed in percent, ranging from zero to 100, although it may be greater than 100 when related to growth rate and percentage of latewood.

Estimated strength ratios for cross grain and density have been obtained empirically; strength ratios for other wood characteristics have been derived theoretically. For example, to account for the weakening effect of knots, the assumption is made that the knot is effectively a hole through the piece, reducing the cross section as shown in figure 6–1. The ratio of the moment a beam with the reduced cross section will carry to the moment of the beam without the knot is

$$SR = \left(1 - \frac{k}{h}\right)^2$$

where SR is strength ratio, k is the knot size, and h is the width of the face containing the knot. This is the basic expression for the effect of a knot at the edge of the vertical face of a beam that is deflected vertically. Figure 6–2 shows how strength ratio changes with knot size in the formula.

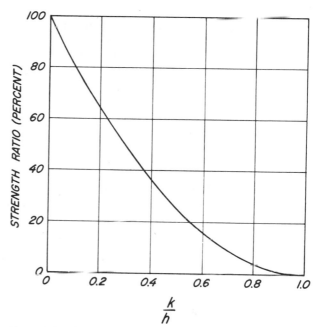

Figure 6–2.—A relation between strength ratio and size of edge knot expressed as a fraction of face width.

Strength ratios for all knots, shakes, checks, and splits are derived using similar concepts. Strength ratio formulas are given in American Society for Testing and Materials Designation D 245. The same reference contains rules for measuring the various growth characteristics.

An individual piece of lumber often will have several characteristics that can affect any particular strength property. Only the characteristic that gives the lowest strength ratio is used to derive the estimated piece strength. A visual stress grade contains lumber ranging from pieces having the minimum strength ratio permitted in the grade to pieces having the minimum for the next higher grade.

The range of strength ratios in a grade, and the natural variation in clear wood strength, give rise to variation in strength between pieces in the grade. To account for this variation, and provide for safety in design, it is intended that any strength property associated with a grade be less than the actual strength of at least 95 percent of the pieces in the grade. In visual grading according to ASTM Designation D 245, this is handled by using a near-minimum clear wood strength value, and multiplying it by the minimum strength ratio permitted in the grade to obtain the grade strength property. The near-minimum value is called the 5 percent exclusion limit. ASTM Designation D 2555 provides clear wood strength data and gives a

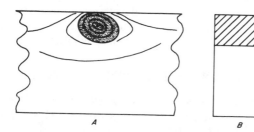

Figure 6–1.—An edge knot in lumber, A, and the assumed loss of cross section, B.

method for estimating the 5 percent exclusion limit.

Suppose a typical 5 percent exclusion limit for the clear wood bending strength for a species in the green condition is 7,000 p.s.i. Suppose also that among the characteristics allowed in a grade of lumber, one characteristic (a knot, for example) provides the lowest strength ratio in bending—assumed in this example as 40 percent. Using these numbers, the bending strength for the grade can be obtained by multiplying the strength ratio (0.40) by 7,000 p.s.i., equalling 2,800 p.s.i. This is shown in figure 6–3. The bending strength in the green condition of 95 percent of the pieces in this species that have a strength ratio of 40 percent is expected to be 2,800 p.s.i. or more. Similar procedures are followed for other strength properties, using the appropriate clear wood property value and strength ratio. As noted, additional multiplying factors then are applied to produce allowable properties for design, as summarized later in this chapter.

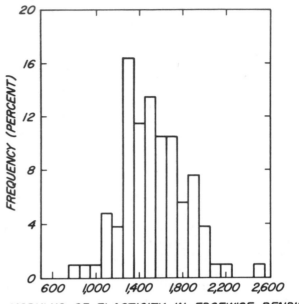

Figure 6—4.—Histogram of modulus of elasticity observed in a single visual grade. From pieces selected over a broad geographical range.

by lumber grade. This procedure is outlined in ASTM D 245.

For example, assume a clear wood average modulus of elasticity of 1.8 million p.s.i. for the example shown earlier. The limiting bending strength ratio was 40 percent. ASTM D 245 assigns a quality multiplying factor of 0.80 for lumber with this bending strength ratio. The modulus of elasticity for that grade would be the product of the clear wood modulus and the quality factor; i.e., 1.44 million p.s.i.

Actual modulus of elasticity of individual pieces of the grade varies from the mean assumed for design. Figure 6–4 shows a typical histogram of the modulus of elasticity, measured within a single visual grade, sampled from a wide selection of sources. Small individual lots of lumber can be expected to deviate from the distribution shown by the histogram. The additional multiplying factors used to derive final design values of modulus of elasticity are discussed later in this chapter.

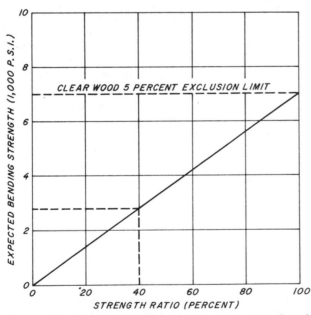

Figure 6–3.—An example of the relation between strength and strength ratio.

Modulus of Elasticity

In visual grading, the modulus of elasticity assigned is an estimate of the mean modulus of the lumber grade. The average modulus of elasticity for clear wood of the species, as recorded in ASTM D 2555, is used as a base. The clear wood average is multiplied by empirically derived "quality factors" to represent the reduction in modulus of elasticity that occurs

DEVELOPING MECHANICAL GRADES

Mechanical stress grading is based on an observed relation between modulus of elasticity and bending strength, tensile strength, or compressive strength parallel to the grain. The modulus of elasticity of lumber thus is the sorting criterion used in this method of grading. Mechanical devices operating up to relatively high rates of speed measure the modulus of elasticity or stiffness for a series of stress grades.

Mechanical Sorting Criteria

The modulus of elasticity used as a sorting criterion for mechanical properties of lumber can be measured in a variety of ways. Usually the apparent modulus or a stiffness-related deflection is the actual measurement made. Because lumber is heterogeneous, the apparent modulus of elasticity depends upon span, orientation (edgewise or flatwise in bending), mode of test (static or dynamic), and method of loading (tension, bending, concentrated, uniform, etc.). Any of the apparent moduli can be used, so long as the grading machine is properly calibrated to give the appropriate

strength property. Most grading machines in the United States are designed to detect the lowest flatwise bending stiffness that occurs in any approximate 4-foot span.

Deriving Properties of Mechanically Graded Lumber

A stress grade derived for mechanically graded lumber relates allowable strength in bending and in compression and tension parallel to grain to the modulus of elasticity levels by which the grade is identified. This relationship between properties is chosen so that 95 percent of the pieces encountered will be at least as strong as indicated by the grading process. Figure 6–5 shows an example of a relation between bending strength, and the modulus of elasticity measured in flatwise bending over an 84-inch span.

As in visual grading, the modulus of elastic-

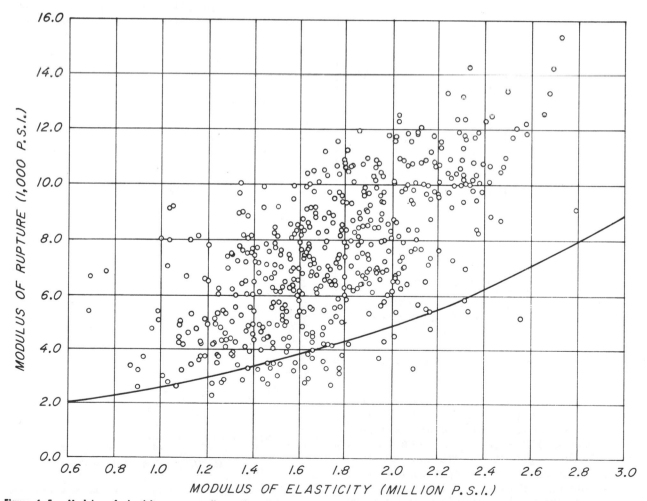

Figure 6–5.—Modulus of elasticity as a predictor of modulus of rupture. The line is a 5 percent exclusion line; the points are test results.

ity assigned to a grade is intended to be an average value for the grade. However, because the basis for mechanical stress grading is the sorting of lumber by modulus of elasticity classes, machines can be adjusted so the modulus for a grade varies less in a mechanical stress grade than in a visual stress grade. Figure 6–6 presents a typical histogram of the dispersion of modulus of elasticity within a single grade, obtained by mechanical stress rating. The characteristics of small lots of lumber can be expected to deviate from such a histogram.

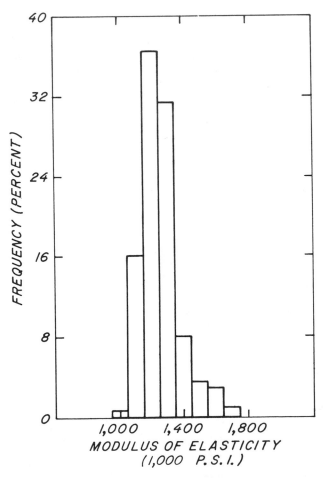

Figure 6–6.—Histogram of modulus of elasticity observed in a single machine stress grade.

Strengths in shear parallel to the grain and in compression perpendicular to the grain have not been shown to be well related to modulus of elasticity. Therefore, in mechanical stress grading these properties are handled in relationship to clear wood properties and lumber visual characteristics.

Most commercial mechanical grading practice in the United States combines a modified

strength ratio concept with the modulus of elasticity as a predictor of grade properties. Emphasis is placed on edge defects when deriving the limitations to visual characteristics of the grade. It has been shown that strength ratio and modulus of elasticity used together provide a somewhat more efficient strength prediction than either by itself.

In a fashion similar to visual grading, the properties derived from mechanical sorting criteria may be further modified for design use by consideration of moisture content and load duration.

ADJUSTING PROPERTIES FOR DESIGN USE

The mechanical properties associated with lumber quality, for the stress grading methods, are adjusted to give allowable unit stresses and an allowable modulus of elasticity suitable for most engineering uses.

A composite adjustment factor is applied to each strength property to adjust for an assumed 10-year duration of full design load. The composite factor includes a safety factor

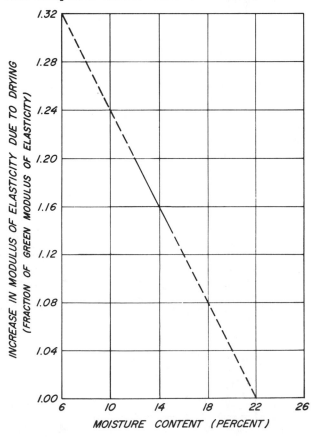

Figure 6–7.—Modulus of elasticity as a function of moisture content for lumber. Solid line represents range of experimental data on which graph is based.

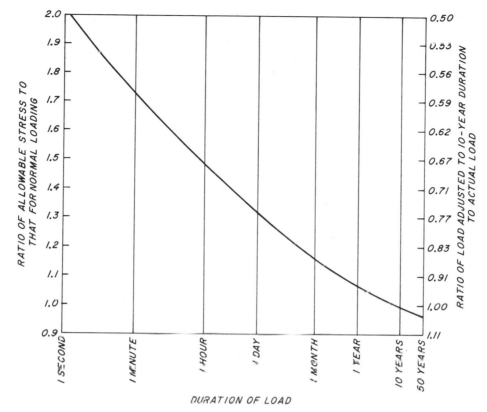

Figure 6-8.—Relation of strength to duration of load.

Size Factor

In bending, a size effect causes small members to have a greater unit strength than large members. If bending strength is known for one size of lumber, it can be approximated for another size by using formulas given in chapter 8. These formulas are used in the development of visual grades to convert from the sizes of standard small, clear specimens to lumber sizes.

Moisture Adjustments

For lumber 4 inches thick or less that has been dried, properties are related to moisture content. As an example, the relationship for modulus of elasticity is shown in figure 6-7, where the modulus at any moisture content is expressed as an increase above the modulus for green lumber.

of about 1.3. Additional adjustments are often made for size and moisture content. Discussion of these adjustment factors follows; specific adjustments are given in ASTM Designation D 245.

For lumber thicker than 4 inches, often no adjustment for moisture content is made; most allowable properties are assigned on the basis of wood in the green condition. Lumber in large sizes is usually put in place without drying.

Duration of Load

If loading will not be for a 10-year period (called normal loading), allowable stress can be adjusted using figure 6-8. There is some evidence that an intermittent load causes a cumulative effect on strength, and that the total duration should be considered in establishing the duration of load effect.

In many design circumstances there are several loads on the structure, some acting simultaneously and each with a different duration. Each increment of time during which the total load is constant should be treated separately, and the most severe condition governs the design. Either the allowable stress or the total design load (but not both) can be adjusted using figure 6-8.

For example, suppose a structure is expected to support a load of 100 pounds per square foot (p.s.f.) off and on for a cumulative duration of 1 year. Also, it is expected to support its own dead load of 20 p.s.f. for the anticipated 50-year life of the structure. The adjustments to be made to arrive at a design load are listed below. The more critical design load is 112 p.s.f., and this load and the allowable stress for lumber based on normal loading would be used to pick members of suitable size. In this case, it was convenient to adjust the loads on the structure, although the same result can be obtained by adjusting the allowable stress.

Time	Total load	Load adjustment	Design load
(Yr.)	(P.s.f.)		(P.s.f.)
1	$100 + 20 = 120$	0.93	112
50	20	1.04	21

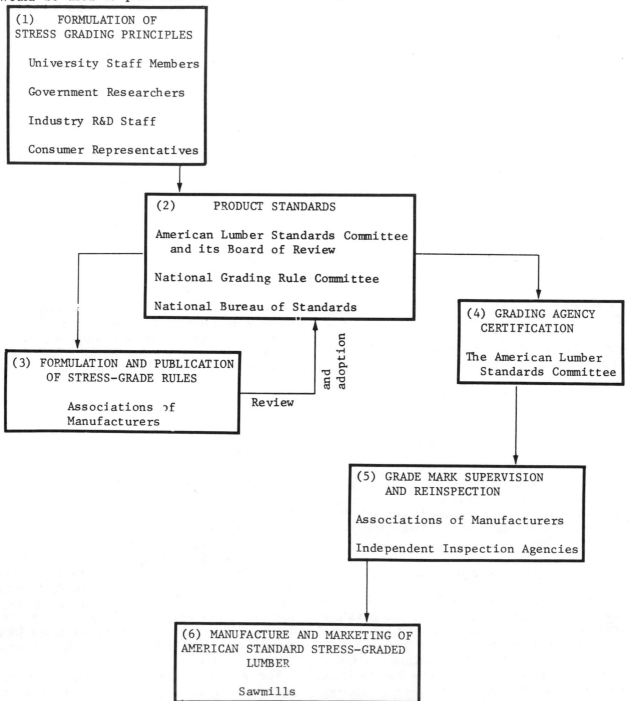

(1)   FORMULATION OF STRESS GRADING PRINCIPLES

University Staff Members

Government Researchers

Industry R&D Staff

Consumer Representatives

(2)   PRODUCT STANDARDS

American Lumber Standards Committee and its Board of Review

National Grading Rule Committee

National Bureau of Standards

(3) FORMULATION AND PUBLICATION OF STRESS-GRADE RULES

Associations of Manufacturers

Review

and adoption

(4) GRADING AGENCY CERTIFICATION

The American Lumber Standards Committee

(5) GRADE MARK SUPERVISION AND REINSPECTION

Associations of Manufacturers

Independent Inspection Agencies

(6) MANUFACTURE AND MARKETING OF AMERICAN STANDARD STRESS-GRADED LUMBER

Sawmills

Figure 6-9.—Voluntary system of responsibilities for stress grading under the American Softwood Lumber Standard.

Cumulative effects of repeated loads of the very short durations shown in figure 6–8 are called fatigue. For comments on fatigue, see chapter 4.

PRACTICE OF STRESS GRADING

An orderly, voluntary, but circuitous, system of responsibilities has evolved in the United States for the development, manufacture, and merchandising of most stress-graded lumber. The system is shown schematically in figure 6–9. Stress-grading principles are developed from research findings and engineering concepts, often within committees and subcommittees of the American Society for Testing and Materials.

The National Bureau of Standards cooperates with lumber producers, distributors, and users through an American Lumber Standards Committee (ALSC) to assemble a voluntary softwood standard of manufacture, called the American Softwood Lumber Standard (see ch. 5). The American Lumber Standard and the related National Grading Rule prescribe the ways in which stress-grading principles can be used to formulate grading rules said to be American Standard.

Organizations of lumber manufacturers publish grading rule books containing stress-grade descriptions. If an organization wants its published rules to be American Standard, it submits them to the ALSC's Board of Review for review of conformance with the American Softwood Lumber Standard (ALS).

Organizations that write grading rules, as well as independent agencies, can be certified by ALSC to issue grade marks corresponding to published stress-grade rules and provide grade marking supervision and reinspection services to individual lumber manufacturers. The performance of these organizations is then under the scrutiny of the Board of Review of the ALSC.

Most commercial softwoods are stress graded under standard practice in the United States. The principles of stress grading also can be applied to hardwoods; several are graded under provisions of the ALS.

Lumber found in the marketplace may be stress graded by methods approved by the American Lumber Standards Committee, by some other grading rule, or it may not be stress graded. Stress grades that meet the requirements of the voluntary American Lumber Standard are developed by the principles that have been described in this chapter, and only these stress grades are discussed here.

Stress grading under the auspices of the American Lumber Standards Committee is applied to many of the sizes and several of the patterns of lumber meeting the provisions of the American Lumber Standard. A majority of stress-graded lumber is dimension, however, and a uniform procedure, the National Grading Rule, is used for writing grading rules for this size lumber. Grade rules for other sizes may vary by grading agencies or species.

National Grading Rule

The American Softwood Lumber Standard, PS 20–70, provides for a National Grading Rule for lumber from 2 up to, but not including, 5 inches in nominal thickness (dimension lumber). All American Standard lumber in that size range is required to conform to the

Table 6–1.—*Visual grades described in the National Grading Rule* [1]

Lumber classification	Grade name	Bending strength ratio
		Pct.
Light framing (2 to 4 in. thick, 4 in. wide)[2]	Construction	34
	Standard	19
	Utility	9
Structural light framing (2 to 4 in. thick, 2 to 4 in. wide)	Select structural	67
	1	55
	2	45
	3	26
Studs (2 to 4 in. thick, 2 to 4 in. wide)	Stud	26
Structural joists and planks (2 to 4 in. thick, 6 in. and wider)	Select structural	65
	1	55
	2	45
	3	26
Appearance framing (2 to 4 in. thick, 2 to 4 in. wide)	Appearance	55

[1] Sizes shown are nominal.
[2] Widths narrower than 4 in. may have different strength ratio than shown. Contact rules-writing agencies for additional information.

National Grading Rule, except for special products such as scaffold planks.

The National Grading Rule establishes the lumber classifications and grade names for visually stress-graded lumber shown in table 6–1. The approximate minimum bending strength ratio is also shown to provide a comparative index of quality. The corresponding visual descriptions of the grades can be found in the grading rule books of most of the softwood rule-writing agencies listed in chapter 5. Grades of lumber that meet these requirements should have about the same appearance regardless of species. They will not have the same allowable properties. The allowable properties for each species and grade are given in the appropriate rule books and in the National Design Specification.

The National Grading Rule also establishes some limitations on sizes of permissible edge knots and other visual characteristics for American Standard lumber that is graded by a combination of mechanical and visual methods.

Grouping of Species

Some species have always been grouped together and the lumber from them treated as equivalent. This usually has been done for species that have about the same mechanical properties, for the wood of two or more species very similar in appearance, or for marketing convenience. For visual stress grades, ASTM D 2555 contains some rules for calculating clear wood properties for groups of species. The properties assigned to a group by such a procedure often will not be identical with any of the species that make up the group. The group will display a unique identity with nomenclature approved by the American Lumber Standards Committee. The grading association under whose auspices the lumber was graded should be contacted if the identities, properties, and characteristics of individual species of the group are desired.

In the case of mechanical stress grading, the inspection agency that supervises the grading certifies by test that the allowable properties in the grading rule are appropriate for the species or species group and the grading process.

BIBLIOGRAPHY

American Society for Testing and Materials
    Standard methods for establishing structural grades for visually graded lumber. D 245. (See current edition.) Philadelphia, Pa.

———

    Standard methods for establishing clear wood strength values. D 2555. (See current edition.) Philadelphia, Pa.

Galligan, W. L., and Snodgrass, D. V.
    1970. Machine stress rated lumber: Challenge to design. Forest Prod. J. 20(9): 63–69.

National Forest Products Association
    National design specification for stress-grade lumber and its fastenings. (See current edition.) Washington, D.C.

U.S. Department of Commerce
    American softwood lumber standard. Prod. Stand. PS 20. (See current edition.)

Chapter 7

FASTENINGS

The strength and stability of any structure depend heavily on the fastenings that hold its parts together. One prime advantage of wood as a structural material is the ease with which wood structural parts can be joined together with a wide variety of fastenings—nails, spikes, screws, bolts, lag screws, drift pins, staples, and metal connectors of various types. For utmost rigidity, strength, and service, each type of fastening requires joint designs adapted to the strength properties of wood along and across the grain, and to dimensional changes that may occur with changes in moisture content.

NAILS

Nails are the most common mechanical fastenings used in construction. There are many types, sizes, and forms of nails (fig. 7–1). The formulas presented in this chapter for loads apply for bright, smooth, common steel wire nails driven into wood when there is no visible splitting. For nails other than common wire nails, the loads can be adjusted by factors given later in the chapter.

Nails in use resist either withdrawal loads or lateral loads, or a combination of the two. Both withdrawal and lateral resistance are affected by the wood, the nail, and the condition of use. In general, however, any variation in these factors has a more pronounced effect on withdrawal resistance than on lateral resistance. The serviceability of joints with nails laterally loaded is not greatly dependent upon withdrawal resistance unless large joint distortion can be tolerated. Nails, as well as such other driven fasteners as staples and T-nails, should be used so that they are loaded laterally (perpendicular to the shank or direction of driving) rather than being loaded in direct withdrawal.

Withdrawal Resistance

The resistance of a nail shank to direct withdrawal from a piece of wood depends on the density of the wood, the diameter of the nail, and the depth of penetration. The surface condition of the nail at the time of driving also influences the initial withdrawal resistance.

For bright, common wire nails driven into the side grain of seasoned wood or unseasoned wood that remained wet, the results of many tests have shown that the maximum withdrawal load is given by the empirical formula:

$$p = 7,850 G^{5/2} DL \qquad (7-1)$$

where p is the maximum load in pounds; L, the depth, in inches, of penetration of the nail in the member holding the nail point; G, the specific gravity of the wood based on ovendry weight and volume at 12 percent moisture content (see table 4–2, ch. 4); and D, the diameter of the nail in inches.

The diameters of various penny or gage sizes of bright common nails are given in table 7–1. Bright box nails are generally of the same length but slightly smaller diameter (table 7–2), while cement-coated nails such as coolers, sinkers, and coated box nails are slightly shorter ($1/8$ in.) and of smaller diameter for the same penny size than common nails. Annularly and helically threaded nails generally have smaller diameters than common nails for the same penny size (table 7–3). The loads expressed by equation 7–1 represent average

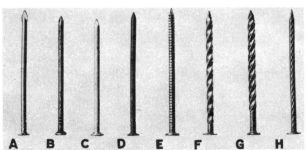

Figure 7–1.—Various types of nails: *A*, bright, smooth wire nail; *B*, cement-coated; *C*, zinc-coated; *D*, chemically etched; *E*, annularly threaded; *F*, helically threaded; *G*, helically threaded and barbed; and *H*, barbed.

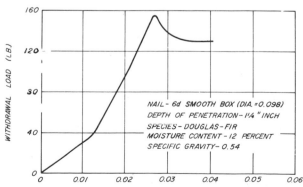

Figure 7–2.—Typical load-displacement curve for direct withdrawal of a nail.

data. A typical load-displacement curve for nail withdrawal (fig. 7–2) shows that maximum load occurs at relatively small values of displacement.

Although the formula for nail-withdrawal resistance indicates that the dense, heavy woods offer greater resistance to nail withdrawal than the lighter weight ones, lighter species should not be disqualified for uses requiring high resistance to withdrawal. As a rule, the less dense species do not split so readily as the denser ones and thus offer an opportunity for increasing the diameter, length, and number of the nails to compensate for the wood's lower resistance to nail withdrawal.

Table 7–1.—*Sizes of bright, common wire nails*

Size	Gage	Length	Diameter
		In.	*In.*
6d	11½	2	0.113
8d	10¼	2½	.131
10d	9	3	.148
12d	9	3¼	.148
16d	8	3½	.162
20d	6	4	.192
30d	5	4½	.207
40d	4	5	.225
50d	3	5½	.244
60d	2	6	.262

Table 7–2.—*Sizes of smooth box nails*

Size	Gage	Length	Diameter
		In.	*In.*
3d	14½	1¼	0.076
4d	14	1½	.080
5d	14	1¾	.080
6d	12½	2	.098
7d	12½	2¼	.098
8d	11½	2½	.113
10d	10½	3	.128
16d	10	3½	.135
20d	9	4	.148

Table 7–3.—*Sizes of helically and annularly threaded nails*

Size	Length	Diameter
	In.	*In.*
6d	2	0.120
8d	2½	.120
10d	3	.135
12d	3¼	.135
16d	3½	.148
20d	4	.177
30d	4½	.177
40d	5	.177
50d	5½	.177
60d	6	.177
70d	7	.207
80d	8	.207
90d	9	.207

The withdrawal resistance of nail shanks is greatly affected by such factors as type of nail point, type of shank, time the nail remains in the wood, surface coatings, and moisture content changes in the wood.

Effect of Seasoning

With practically all species, nails driven into green wood and pulled before any seasoning takes place offer about the same withdrawal resistance as nails driven into seasoned wood and pulled soon after driving. If, however, common smooth-shank nails are driven into green wood that is allowed to season, or into seasoned wood that is subjected to cycles of wetting and drying before the nails are pulled, they lose a major part of their initial withdrawal resistance. The withdrawal resistance for nails driven into wood that is subjected to changes in moisture content may be as low as 25 percent of the values for nails tested soon after driving. On the other hand if the wood fibers are affected or the nail corrodes under some conditions of moisture variation and time, withdrawal resistance is erratic; resistance may be regained or even increased over the immediate withdrawal resistance. However, such sustained performance should not be relied on in the design of a nailed joint.

In seasoned wood that is not subjected to appreciable moisture content changes, the withdrawal resistance of nails may also diminish with lapse of time. Under all these conditions of use, the withdrawal resistance of nails differs among species as well as within individual species, making it difficult to evaluate their behavior.

Effect of Nail Form

The surface condition of nails is frequently modified during the manufacturing process to improve withdrawal resistance. Such modification is usually done by surface coating, surface roughening, or mechanical deformation of the shank. Other factors that affect the surface condition of the nail are the oil film remaining on the shank after manufacture or corrosion resulting from storage under adverse conditions; but these factors are so variable that their influence on withdrawal resistance cannot be adequately evaluated.

A common surface treatment for nails is the so-called cement coating. Cement coatings, contrary to what the name implies, do not include cement as an ingredient; they generally are a composition of resin applied to the nail to increase the resistance to withdrawal by increasing the friction between the nail and the wood. If properly applied, they increase the resistance of nails to withdrawal immediately after the nails are driven into the softer woods. In the denser woods (such as hard maple, birch, or oak), however, cement-coated nails have practically no advantage over plain nails, because most of the coating is removed in driving. Some of the coating may also be removed in the cleat or facing member before the nail penetrates the foundation member.

Good-quality cement coatings are uniform, not sticky to the touch, and cannot be rubbed off easily. Different techniques of applying the cement coating and variations in its ingredients may cause large differences in the relative resistance to withdrawal of different lots of cement-coated nails. Some nails may show only a slight initial advantage over plain nails. The increase in withdrawal resistance of cement-coated nails is not permanent but drops off about one-half after a month or so for the softer woods. Cement-coated nails are used primarily in construction of boxes, crates, and other containers usually built for rough handling and relatively short service.

Nails that have special coatings, such as zinc, are intended primarily for uses where corrosion and staining are important factors in permanence and appearance. If the zinc coating is evenly applied, withdrawal resistance may be increased, but extreme irregularities of the coating may actually reduce it. The advantage that zinc-coated nails with a uniform coating may have over plain nails in resistance to initial withdrawal is usually reduced by repeated cycles of wetting and drying.

Nails have also been made with plastic coatings. The usefulness and characteristics of these coatings are influenced by the quality and type of coating, the effectiveness of the bond between the coating and base fastener, and the effectiveness of the bond between the coating and wood fibers. Some plastic coatings appear to resist corrosion or improve resistance to withdrawal, while others offer little improvement.

Fasteners with properly applied nylon coating tend to retain their initial resistance to withdrawal, as compared to other coatings which exhibit a marked decrease in withdrawal resistance within the first month after driving.

A chemically etched nail has somewhat higher withdrawal resistance than some coated nails, as the minutely pitted surface is an integral part of the nail shank. Under impact loading, however, the withdrawal resistance of etched nails is little different from that of plain or cement-coated nails under various moisture conditions.

Sand-blasted nails perform in much the same manner as do chemically etched nails.

Nail shanks may be varied from a smooth, circular form to give an increase in surface area without an increase in nail weight. Special nails with barbed, helically or annularly threaded, and other irregular shanks (fig. 7–1) are offered commercially.

The form and magnitude of the deformations along the shank influence the performance of the nails in the various wood species. The withdrawal resistance of these nails, except some types of barbed nails, is generally somewhat greater than that of common wire nails of the same diameter, in wood remaining at a uniform moisture content. For instance, annular shank nails have about 40 percent greater withdrawal resistance. Under conditions involving changes in the moisture content of the wood, however, some special nail forms provide considerably greater withdrawal resistance than the common wire nail—about four times greater for annular and helical shank nails of the same diameter. This is especially true of nails driven into green wood that subsequently dries. In general, annularly threaded nails sustain larger withdrawal loads, and helically threaded nails sustain greater impact withdrawal work values than the other nail forms.

A smooth, round shank nail with a long, sharp point will usually have a higher withdrawal resistance, particularly in the softer woods, than the common wire nail (which usually has a diamond point). Sharp points, however, accentuate splitting of certain species, which may reduce withdrawal resistance. A blunt or flat point without taper reduces splitting, but its destruction of the wood fibers when driven reduces withdrawal resistance to less than that of the common wire nail. A nail tapered at the end and terminating in a blunt point will cause less splitting. In the heavier woods, such a tapered, blunt-pointed nail will provide about the same withdrawal resistance, but in the less dense woods, its resistance to withdrawal is lower than the common nail.

Nailhead classifications include flat, oval, countersunk, deep-countersunk, and brad. Nails with all types of heads, except the deep-countersunk, brad, and some of the thin flathead nails, are sufficiently strong to withstand the force required to pull them from most woods in direct withdrawal. The deep-countersunk and brad nails are usually driven below the wood surface and are not intended to carry large withdrawal loads. In general, the thickness and diameter of the heads of the common wire nails increase as the size of the nail increases.

The development of some pneumatically operated portable nailers has introduced special headed nails such as T-nails and nails with a segment of the head cut off. Although the resistance of these heads to pulling through the wood might be less than for conventional nailheads, the performance of the modified heads appears adequate. It is preferable that the T-head be oriented so that the head is perpendicular to the grain of the adjoining wood.

Nails of copper alloys, aluminum alloys, stainless steel, and other alloys are used mainly where corrosion or staining is an important factor for appearance or permanence. Specially hardened nails are also frequently used where driving conditions are difficult, or to obtain improved performance, such as in pallet assembly. Sometimes even the mechanically deformed shank nails are given heat treatments. Hardened nails are brittle and care should be exercised to avoid injuries from fragments of nails broken during driving.

In general, the withdrawal resistance of copper and other alloy nails is somewhat comparable to that of common steel wire nails when pulled soon after driving.

Driving and Clinching

The resistance of nails to withdrawal is generally greatest when they are driven perpendicular to the grain of the wood. When the nail is driven parallel to the wood fibers—that is, into the end of the piece—withdrawal resistance in the softer woods drops to 75 or even 50 percent of the resistance obtained when the nail is driven perpendicular to the grain. The difference between side- and end-grain withdrawal loads is less for dense woods than for softer woods. With most species, the ratio between the end- and side-grain withdrawal loads of nails pulled after a time interval, or after moisture content changes have occurred, is usually somewhat higher than that of nails pulled immediately after driving.

Toenailing, a common method of joining wood framework, involves slant driving a nail or group of nails through the end or edge of an attached member and into a main member. Toenailing requires greater skill in assembly than does ordinary end nailing but provides joints of greater strength and stability. Tests show that the maximum strength of toenailed joints under lateral and uplift loads is obtained by (1) using the largest nail that will not cause excessive splitting, (2) allowing an end distance (distance from the end of the attached member to the point of initial nail entry) of approximately one-third the length of the nail, (3) driving the nail at a slope of 30° with the attached member, and (4) burying the full shank of the nail but avoiding excessive mutilation of the wood from hammer blows.

In tests of stud-to-sill assemblies with the number and size of nails frequently used in toenailed and end-nailed joints, a joint toenailed with four eightpenny common nails was superior to a joint end nailed with two sixteenpenny common nails. With such woods as Douglas-fir, toenailing with tenpenny common nails gave greater joint strength than the commonly used eightpenny nails.

The results of withdrawal tests with multiple nail joints in which the piece attached is pulled directly away from the main member show that slant driving is usually superior to straight driving when nails are driven into dry wood and pulled immediately, and decidedly superior when nails are driven into green or partially dry wood that is allowed to season for a month or more. However, the loss in depth of penetration due to slant driving may, in some types of joints, offset the advantages of slant nailing. Cross slant driving of groups of nails is usually somewhat more effective than parallel slant driving.

Nails driven into lead holes with a diameter slightly smaller than the nail have somewhat higher withdrawal resistance than nails driven without lead holes. Lead holes also prevent or reduce splitting of the wood, particularly for dense species.

The withdrawal resistance of smooth-shank, clinched nails is considerably higher than that of unclinched nails. The ratio between the loads for clinched and unclinched nails varies enormously, depending upon the moisture content of the wood when the nail is driven and withdrawn, the species of wood, the size of nail, and the direction of clinch with respect to the grain of the wood.

In dry or green wood, a clinched nail provides from 45 to 170 percent more withdrawal resistance than an unclinched nail when withdrawn soon after driving. In green wood that seasons after a nail is driven, a clinched nail gives from 250 to 460 percent greater withdrawal resistance than an unclinched nail. However, this improved strength of the clinched- over the unclinched-nail joint does not justify the use of green lumber, because the joints may loosen as the lumber seasons. Furthermore, laboratory tests were made with single nails, and the effects of drying, such as warping, twisting, and splitting, may reduce the efficiency of a joint that has more than one nail. Clinching of nails is generally confined to such construction as boxes and crates and other container applications.

Nails clinched across the grain have approximately 20 percent more resistance to withdrawal than nails clinched along the grain.

Plywood

The nailing characteristics of plywood are not greatly different from those of solid wood except for plywood's greater resistance to splitting when nails are driven near an edge. The nail shank withdrawal resistance of plywood is from 15 to 30 percent less than that of solid wood of the same thickness. The reason is that fiber distortion is less uniform in plywood than in solid wood. For plywood less than one-half inch thick, the high splitting resistance tends to offset the lower withdrawal resistance as compared to solid wood. The withdrawal resistance per inch of penetration decreases with increase in the number of plies. The direction of the grain of the face ply has little influence on the withdrawal resistance along the end or edge of a piece of plywood. The direction of the grain of the face ply may influence the pullthrough resistance of staples or nails with severely modified heads, such as T-heads. Fastener design information for plywood is available from the American Plywood Association.

Allowable Loads

The preceding discussion has dealt with maximum withdrawal loads obtained in short-time test conditions. For design, these loads must be reduced to account for variability and duration of load effects. A value of one-sixth the maximum load has usually been accepted as the allowable load for longtime loading conditions. For normal duration of load, this value may be increased by 10 percent.

Lateral Resistance

Test loads at joint slips of 0.015 inch for bright, common wire nails in lateral resistance driven into the side grain (perpendicular to the wood fibers) of seasoned wood were found to be expressed by the following empirical formula:

$$p = KD^{3/2} \qquad (7\text{--}2)$$

where p is the lateral load in pounds per nail at a joint slip of 0.015 inch (approximately proportional limit load); K is a coefficient; and D is the diameter of the nail in inches. Values of the coefficient K are listed in table 7–4 for ranges of specific gravity of hardwoods and softwoods.

The ultimate lateral nail loads for softwoods may approach $3\frac{1}{2}$ times the loads expressed by the formula, and for hardwoods they may be seven times as great. The joint slip at maximum load, however, is over 20 times 0.015 inch. This is demonstrated by the typical load-slip curve shown in figure 7–3.

The loads obtained by the formula apply only for conditions where the side member and the member holding the nail point are of approximately the same density and where the nail penetrates into the member holding the point by not less than 10 times the nail

Table 7–4.—*Coefficients for computing loads for fasteners in seasoned wood* [1]

Specific gravity range [2]	Lateral load coefficient (K)		
	Nails [3]	Screws	Lag screws
HARDWOODS			
0.33–0.47	1,440	3,360	3,820
.48– .56	2,000	4,640	4,280
.57– .74	2,720	6,400	4,950
SOFTWOODS			
.29– .42	1,440	3,360	3,380
.43– .47	1,800	4,320	3,820
.48– .52	2,200	5,280	4,280

[1] Wood with a moisture content of 15 pct.
[2] Specific gravity based on ovendry weight and volume at 12 pct. moisture content.
[3] Coefficients based on load at joint slip of 0.015 in.

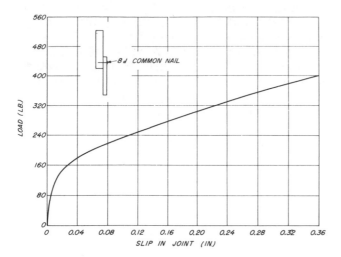

Figure 7–3.—Typical relation between lateral load and slip in the joint.

diameter for dense woods and 14 times the diameter for lightweight woods. The thickness of the side member should be about one-half the depth of penetration of the nail in the member holding the point. End distance should be no less than 15 times the nail diameter if the end is stressed or 12 times the nail diameter if unstressed. Edge distance should be no less than 10 times the nail diameter. When the side member is steel, an increase of about 25 percent can be applied to the lateral nail load because initiation of failure is forced to occur in the wood member holding the nail point.

Theoretical Analysis

A considerable amount of work has been done to evaluate the lateral resistance of nailed joints. For the interested designer a theoretical approach is available for determining the lateral load of a joint having a single nail or bolt.[1]

Results of the theoretical analysis, which considers the nail to be a beam supported on an elastic foundation (the wood), show that the lateral load up to the proportional limit for a two-member joint with one nail bearing parallel to the grain is given approximately by the formula:

$$P = 0.354 \ (k_0{}^{3/4}) \ (E^{1/4}) \ (I^{1/4}) \ (D^{5/4})S \qquad (7\text{–}3)$$

where P is the lateral load, k_0 is the elastic bearing constant of the wood, E is the modulus

of elasticity of the nail, I is the moment of inertia of the nail cross section, D is the nail size (cross-section dimension), and S is the joint slip. For a nail of circular cross section of diameter D, this formula reduces to

$$P = \frac{k_0{}^{3/4}E^{1/4}D^{7/4}S}{6} \qquad (7\text{–}4)$$

This equation has been shown, experimentally, to account for the different properties of the nail. The elastic bearing constant, k_0, is related to the average species specific gravity by the formula:

$$k_0 = 3{,}200{,}000G \qquad (7\text{–}5)$$

where G is the specific gravity based on volume at the moisture content at which the wood is used. This value of the elastic bearing constant is for smooth shank nails driven in prebored lead holes and loaded parallel to the grain.

Research Results

The lateral load for side-grain nailing given by the empirical formula $p = KD^{3/2}$ applies whether the load is in a direction parallel to the grain of the pieces joined or at right angles to it. When nails are driven into the end grain (parallel with the wood fibers), limited data on softwood species indicate that their maximum resistance to lateral displacement is about two-thirds that for nails driven into the side grain. Although the average proportional limit loads appear to be about the same for end- and side-grain nailing, the individual results are more erratic for end-grain nailing, and the minimum loads approach only 75 percent of corresponding values for side-grain nailing.

Nails driven into the side grain of unseasoned wood give maximum lateral resistance loads approximately equal to those obtained in seasoned wood, but the lateral resistance loads at 0.015-inch joint slip are somewhat less. To prevent excessive deformation, lateral loads obtained by the formula for seasoned wood should be reduced 25 percent for unseasoned wood that will remain wet or be loaded before seasoning takes place.

When nails are driven into green wood, their lateral proportional limit loads after the wood has seasoned are also less than when they are driven into seasoned wood and loaded. The erratic behavior of a nailed joint that has undergone one or more moisture content changes makes it difficult to establish a lateral load for a

---

[1] In Bibliography at the end of this chapter see reports by Kuenzi and by Wilkinson.

nailed joint under these conditions. Structural joints should be inspected at intervals, and if it is apparent that a loosening of the joint has ocurred during drying, the joint should be reinforced with additional nails.

Deformed-shank nails carry somewhat higher maximum lateral loads than common wire nails, but both perform similarly at small distortions in the joint.

Allowable Loads

The value of the lateral load at proportional limit obtained from tests must be reduced (to account for variability and duration of load effects) to arrive at allowable values (see eq. 7–2). A reduction factor of 1.6 has been used to arrive at a value for longtime loading. For normal loading, this value may be increased by 10 percent.

SPIKES

Common wire spikes are manufactured in the same manner as common wire nails. They have either a chisel point or a diamond point and are made in lengths of 3 to 12 inches. For corresponding lengths (3 to 6 in.), they have larger diameters (table 7–5) than the common wire nails, and beyond the sixtypenny size they are usually designated by inches of length.

The withdrawal and lateral resistance formulas and limitations given for common wire nails are also applicable to spikes, except that in calculating the withdrawal load for spikes, the depth of penetration should be reduced by two-thirds the length of the point.

STAPLES

Different types of staples have been developed with various modifications in points, shank treatment and coatings, gage, crown width, and length. These fasteners are available in clips or magazines to permit their use in pneumatically operated portable staplers. Most of the factors that affect the withdrawal and lateral loads of nails similarly affect the loads on staples. The withdrawal resistances, for example, vary almost directly with the circumference and depth of penetration when the type of point and shank are similar. Thus, equation (7–1) may be used to predict the withdrawal load for staples, and the same factors used for nails may be used to arrive at an allowable load.

Table 7-5.—*Sizes of common wire spikes*

Size	Length	Diameter	Size	Length	Diameter
	In.	*In.*		*In.*	*In.*
10d -----	3	0.192	40d -----	5	0.263
12d -----	3¼	.192	50d -----	5½	.283
16d -----	3½	.207	60d -----	6	.283
20d -----	4	.225	5/16 inch -	7	.312
30d -----	4½	.244	3/8 inch --	8½	.375

The load in lateral resistance varies about as the 3/2 power of the diameter when other factors, such as quality of metal, type of shank, and depth of penetration, are similar. The diameter of each leg of a two-legged staple must therefore be about two-thirds the diameter of a nail to provide a comparable load. Equation (7–2) may be used to predict the lateral resistance of staples and the same factors for nails may be used to arrive at allowable loads.

In addition to the immediate performance capability of staples and nails as determined by test, such factors as corrosion, sustained performance under service conditions, and durability in various uses should be considered in evaluating the relative usefulness of a connection.

DRIFT BOLTS

The ultimate withdrawal load of a round drift bolt or pin from the side grain of seasoned wood is given by the formula:

$$p = 6,600 G^2 DL \qquad (7\text{–}6)$$

where p is the ultimate withdrawal load in pounds, G is the specific gravity based on the ovendry weight and volume at 12 percent moisture content of the wood, D is the diameter of the drift bolt in inches, and L is the length of penetration of the bolt in inches.

This formula provides an average relationship for all species, and the withdrawal load of some species may be above or below the equation values. It also presumes that the bolts are driven into prebored holes having a diameter one-eighth inch less than the bolt diameter.

In lateral resistance, the load for a drift bolt driven into the side grain of wood should not exceed, and ordinarily should be taken as less than, that for a machine bolt of the same diameter. The drift bolt should normally be of greater length than the common bolt to compensate for the lack of washers and nut.

WOOD SCREWS

The common types of wood screws have flat, oval, or round heads. The flathead screw is most commonly used if a flush surface is desired. Ovalhead and roundhead screws are used for appearance, and roundhead screws are used when countersinking is objectionable. Besides the head, the principal parts of a screw are the shank, thread, and core (fig. 7–4). Wood screws are usually made of steel or brass or other metals, alloys, or with specific finishes such as nickel, blued, chromium, or cadmium. They are classified according to material, type, finish, shape of head, and diameter of the shank or gage.

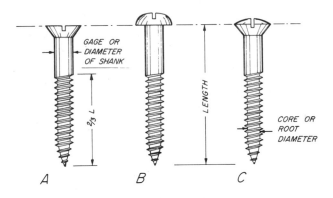

Figure 7–4.—Common types of wood screws: A, flathead; B, roundhead; and C, ovalhead.

Current trends in fastenings for wood also include tapping screws. Tapping screws have threads the full length of the shank and thus may have some advantage for certain specific uses.

Withdrawal Resistance

Experimental Loads

The resistance of wood screw shanks to withdrawal from the side grain of seasoned wood varies directly with the square of the specific gravity of the wood. Within limits, the withdrawal load varies directly with the depth of penetration of the threaded portion and the diameter of the screw, provided the screw does not fail in tension. The limiting length to cause screw failure decreases as the density of the wood increases. The longer lengths of standard screws are therefore superfluous in dense hardwoods.

The withdrawal resistance of type A tapping screws, commonly called sheet metal screws, is in general about 10 percent higher than for wood screws of comparable diameter and length of threaded portion. The ratio between the withdrawal resistance of tapping screws and wood screws is somewhat higher in the denser woods than in the lighter woods.

Ultimate test values for withdrawal loads of wood screws inserted into the side grain of seasoned wood may be expressed as:

$$p = 15,700G^2DL \qquad (7-7)$$

where p is the maximum withdrawal load in pounds, G is the specific gravity based on oven-dry weight and volume at 12 percent moisture content, D is the shank diameter of the screw in inches, and L is the length of penetration of the threaded part of the screw in inches.

This formula is applicable when screw lead holes in softwoods have a diameter of about 70 percent of the root diameter of the threads, and in hardwoods about 90 percent. The root diameter for most sizes of screws averages about two-thirds of the shank diameter.

The equation values are applicable to the following sizes of screws:

Screw length	Gage limits
Inches	
½	1 to 6
¾	2 to 11
1	3 to 12
1½	5 to 14
2	7 to 16
2½	9 to 18
3	12 to 20

For lengths and gages outside these limits, the actual values are likely to be less than the equation values. The withdrawal loads of screws inserted into the end grain of wood are somewhat erratic; but, when splitting is avoided, they should average 75 percent of the load sustained by screws inserted into the side grain.

Lubricating the surface of a screw is recommended to facilitate insertion, especially in the dense woods. It will have little effect on ultimate withdrawal resistance.

Allowable Loads

For allowable values, the practice has been to use one-sixth the ultimate load for longtime loading conditions. This also accounts for variability in test data. For normal duration of load, the allowable load may be increased 10 percent.

Lateral Resistance

Experimental Loads

The test proportional limit loads in lateral resistance for wood screws in the side grain of seasoned wood is given by the empirical formula:

$$p = KD^2 \qquad (7\text{-}8)$$

where p is the lateral load in pounds, D is the diameter of the screw shank in inches, and K is a coefficient depending on the inherent characteristics of the wood species. Values of screw shank diameters for various screw gages are:

Screw number or gage	Diameter
	Inches
4	0.112
5	.125
6	.138
7	.151
8	.164
9	.177
10	.190
11	.203
12	.216
14	.242
16	.268
18	.294
20	.320
24	.372

Values of K are based on ranges of specific gravity of hardwoods and softwoods and are given in table 7–4. They apply to wood at about 15 percent moisture content. Loads computed by substituting these constants in the equation are expected to allow a slip of 0.007 to 0.010 inch, depending somewhat on the species and density of the wood.

Formula (7–8) applies when the depth of penetration of the screw into the block receiving the point is not less than seven times the shank diameter and when the cleat (member not receiving point) and the block holding the point are approximately of the same density. The thickness of the side member should be about one-half the depth of penetration of the screw in the member holding the point. The end distance should be no less than the side member thickness, and the edge distances no less than one-half the side member thickness.

This depth of penetration (seven times shank diameter) gives an ultimate load of about four times the load obtained by the formula. For a depth of penetration of less than seven times the shank diameter, the ultimate load is reduced about in proportion to the reduction in penetration and the load at the proportional limit is reduced somewhat less rapidly. When the depth of penetration of the screw in the holding block is four times the shank diameter, the maximum load will be less than three times the load expressed by the formula, and the proportional limit load will be approximately equal to that given by the formula. When the screw holds metal to wood, the load can be increased by about 25 percent.

For these lateral loads, the part of the lead hole receiving the shank should be the same diameter or slightly smaller than the shank, and that receiving the threaded part the same diameter or slightly smaller than the root of the thread. The size of the lead hole may have to be varied with the density of the wood, the smaller lead holes being used with the lower density species.

Screws should always be turned in. They should never be started or driven with a hammer because this practice tears the wood fibers and injures the screw threads, seriously reducing the loadcarrying capacity of the screw.

Allowable Loads

For allowable values, the practice has been to reduce the proportional limit load by a factor of 1.6 to account for variability in test data and reduce the load to a longtime loading condition. For normal duration of load, the allowable load may be increased by 10 percent.

LAG SCREWS

Lag screws are commonly used because of their convenience, particularly where it would be difficult to fasten a bolt or where a nut on the surface would be objectionable. Commonly available lag screws range from about 0.2 to 1 inch in diameter and from 1 to 16 inches in length. The length of the threaded part varies with the length of the screw and ranges from three-fourths inch with the 1- and 1¼-inch screws to half the length for all lengths greater than 10 inches. The equations given here for withdrawal and lateral loads are based on lag screws having a base metal average tensile yield strength of about 45,000 pounds per square inch (p.s.i.) and an average ultimate tensile strength of 77,000 p.s.i. For metal lag screws having greater or lower yield and tensile strengths, the withdrawal loads should be adjusted in proportion to the tensile strength and the lateral loads in proportion to the square root of the yield-point stresses.

Withdrawal Resistance

Experimental Loads

The results of withdrawal tests have shown that the maximum load in direct withdrawal of lag screws from seasoned wood may be computed from the equation:

$$p = 8,100G^{3/2}D^{3/4}L \qquad (7\text{--}9)$$

where p is the maximum withdrawal load in pounds, D is the shank diameter in inches, G is the specific gravity of the wood based on ovendry weight and volume at 12 percent moisture content, and L is the length, in inches, of penetration of the threaded part.

Lag screws, like wood screws, require pre-bored holes of the proper size (fig. 7–5). The lead hole for the shank should be the same diameter as the shank. The diameter of the lead hole for the threaded part varies with the density of the wood: For the lightweight softwoods, such as the cedars and white pines, 40 to 70 percent of the shank diameter; for Douglas-fir and southern pine, 60 to 75 percent; and for dense hardwoods, such as the oaks, 65 to 85 percent. The smaller percentage in each range applies to lag screws of the smaller diameters, and the larger percentage to lag screws of larger diameters. Soap or similar lubricants should be used on the screws to facilitate turning, and lead holes slightly larger than those recommended for maximum efficiency should be used with lag screws of excessive length.

In determining the withdrawal resistance, the allowable tensile strength of the lag screw at the net (root) section should not be exceeded. Penetration of the threaded part to a distance about seven times the shank diameter in the denser species and 10 to 12 times the shank diameter in the less dense species will develop approximately the ultimate tensile strength of the lag screw.

The resistance to withdrawal of a lag screw from the end-grain surface of a piece of wood is about three-fourths as great as its resistance to withdrawal from the side-grain surface of the same piece.

Allowable Loads

For allowable values, the practice has been to use one-fifth the ultimate load to account for variability in test data and reduce the load to a longtime loading condition. For normal duration of load, the allowable load may be increased by 10 percent.

Lateral Resistance

Experimental Loads

The experimentally determined lateral loads for lag screws inserted in the side grain and loaded parallel to the grain of a piece of seasoned wood can be computed from the equation:

$$p = KD^2 \qquad (7\text{--}10)$$

where p is the proportional limit lateral load in pounds parallel to the grain, K is a coefficient depending on the species specific gravity, and D is the shank diameter of the lag screw in inches. Values for K for a number of specific gravity ranges can be found in table 7–4. These coefficients are based on average results for several ranges of specific gravity for hardwoods and softwoods. The loads given by this formula apply when the thickness of the attached member is 3.5 times the shank diameter of the lag screw, and the depth of penetration in the main member is seven times the diameter in the harder woods and 11 times the diameter in the softer woods. For other thicknesses, the computed loads should be multiplied by the following factors:

Figure 7–5.—A, clean-cut, deep penetration of thread made by lag screw turned into a lead hole of proper size, and B, rough, shallow penetration of thread made by lag screw turned into over-sized lead hole.

Ratio of thickness of attached member to shank diameter of lag screw	Factor
2	0.62
2½	.77
3	.93
3½	1.00
4	1.07
4½	1.13
5	1.18
5½	1.21
6	1.22
6½	1.22

The thickness of a solid wood side member should be about one-half the depth of penetration in the member holding the point.

When the lag screw is inserted into the side grain of wood and the load is applied perpendicular to the grain, the load given by the lateral resistance formula should be multiplied by the following factors:

Shank diameter of lag screw	Factor
Inches	
³⁄₁₆	1.00
¼	.97
⁵⁄₁₆	.85
⅜	.76
⁷⁄₁₆	.70
½	.65
⅝	.60
¾	.55
⅞	.52
1	.50

For other angles of loading, the loads may be computed from the parallel and perpendicular values by the use of the Scholten nomograph for determining the bearing strength of wood at various angles to the grain (fig. 7–6). The nomograph provides values comparable to those given by the Hankinson formula:

$$N = \frac{PQ}{P \sin^2 \theta + Q \cos^2 \theta} \qquad (7\text{–}11)$$

where P represents the load or stress parallel to the grain. Q, the load or stress perpendicular to the grain; and N, the load or stress at an inclination θ with the direction of the grain.

Example: P, the load parallel to grain is 6,000 pounds (lb.), and Q, the load perpendicular to the grain is 2,000 lb. N, the load at an angle of 40° to grain is found as follows: Connect with a straight line 6,000 lb. (a) on line OX of the nomograph with the intersection (b) on line OY of a vertical line through 2,000 lb. The point where this line (*ab*) intersects the line representing the given angle 40° is directly above the load, 3,285 lb.

Values for lateral resistance as computed by the preceding methods are based on complete penetration of the shank into the attached member but not into the foundation member. When the unthreaded portion of the shank penetrates the foundation member, the following increases in loads are permitted:

Ratio of penetration of shank into foundation member to shank diameter	Increase in load
	Percent
1	8
2	17
3	26
4	33
5	36
6	38
7	39

When lag screws are used with metal plates, the lateral loads parallel to the grain may be increased 25 percent, but no increase should be made in the loads when the applied load is perpendicular to the grain.

Lag screws should not be used in end grain, because splitting may develop under lateral load. If lag screws are so used, however, the loads should be taken as two-thirds those for lateral resistance when lag screws are inserted into side grain and the loads act perpendicular to the grain.

The spacings, end and edge distances, and net section for lag screw joints should be the same as those for joints with bolts of a diameter equal to the shank diameter of the lag screw.

Lag screws should always be inserted by turning with a wrench, not by driving with a hammer. Soap, beeswax, or other lubricants applied to the screw, particularly with the denser wood species, will facilitate insertion and prevent damage to the threads but will not affect the lag screw's performance.

Allowable Loads

For allowable loads, the accepted practice has been to reduce the proportional limit load by a factor of 2.25 to account for variability in test data and reduce the load to a longtime loading condition. For normal duration of load, the allowable load may be increased by 10 percent.

Table 7-5a Nail and Spike Sizes

Pennyweight	Size Desig-nation	Diameter, inches			
		Common Nail	Box Nail	Hardened Nail	Spike
6d	2	0.113	—	0.120	—
8d	2 1/2	0.131	0.113	0.120	—
10d	3	0.148	0.131	0.135	0.192
12d	3 1/4	0.148	0.148	0.135	0.192
16d	3 1/2	0.162	0.148	0.148	0.207
20d	4	1.192	0.162	0.177	0.225
30d	4 1/2	0.207	0.192	0.177	0.244
40d	5	0.225	0.207	0.177	0.263
50d	5 1/2	0.244	0.225	0.177	0.283
60d	6	0.263	0.244	0.177	0.283
70d	7	—	—	0.207	—
80d	8	—	—	0.207	—
90d	9	—	—	0.207	—
5/16"	7	—	—	0.207	—
3/8"	8 1/2	—	—	0.375	—

Table 7-5b   Lag Bolt Dimensions.

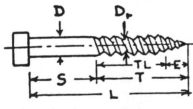

For Design use threaded length = TL = T − E

D. in.	Dr, in.	E, in.
3/16	0.120	5/32
1/4	0.173	3/16
5/16	0.227	1/4
3/8	0.265	1/4
7/16	0.328	9/32
1/2	0.371	5/16
9/16	0.435	3/8
5/8	0.471	3/8
3/4	0.579	7/16
7/8	0.683	1/2
1	0.78	9/16

L, in.	D, in.	S, in.	T, in.
1	3/16 - 1/2	1/4	3/4
1 1/2	3/16 - 1/2	3/8	1 1/8
2	3/16 - 5/8	1/2	1 1/2
2 1/2	3/16 - 1/4	1	1 1/2
2 1/2	5/16 - 3/8	7/8	1 5/8
2 1/2	7/16 - 5/8	3/4	1 3/4
3	3/16 - 1	1	2
4	3/16 - 1 1/4	1 1/2	2 1/2
5	3/16 - 1 1/4	2	3
6	3/16 - 1 1/4	2 1/2	3 1/2
7	3/16 - 1 1/4	3	4
8	3/16 - 1 1/4	3 1/2	4 1/2
9	3/16 - 1 1/4	4	5
10	3/16 - 1 1/4	4 3/4	5 1/4
11	3/16 - 1 1/4	5 1/2	5 1/2
12	3/16 - 1 1/4	6	6

Table 7-5c   Preboring Diameters for Wood Screws

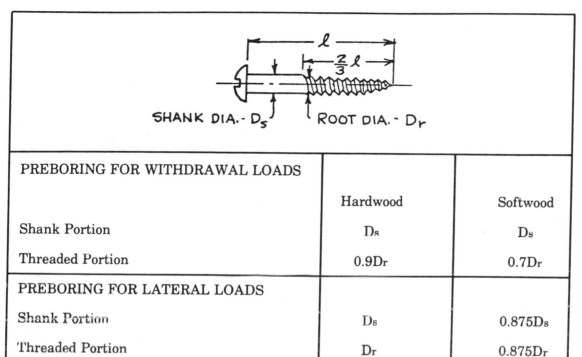

PREBORING FOR WITHDRAWAL LOADS	Hardwood	Softwood
Shank Portion	D_s	D_s
Threaded Portion	$0.9D_r$	$0.7D_r$
PREBORING FOR LATERAL LOADS		
Shank Portion	D_s	$0.875D_s$
Threaded Portion	D_r	$0.875D_r$

Table 7-5d   Screw Gages for Various Lengths

Gage Nos.	1-6	2-11	3-12	5-14	7-16	9-18	12-20
Lengths	1/2"	3/4"	1"	1 1/2"	2"	2 1/2"	3"

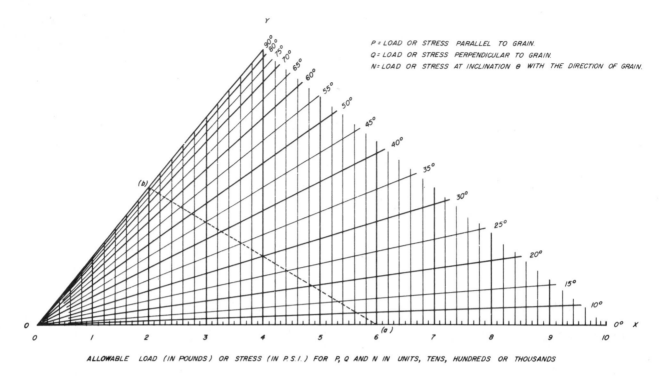

ALLOWABLE LOAD (IN POUNDS) OR STRESS (IN P.S.I.) FOR P, Q AND N IN UNITS, TENS, HUNDREDS OR THOUSANDS

Figure 7–6.—Scholten nomograph for determining the bearing stress of wood at various angles to the grain. The dotted line **ab** refers to the example given in the text.

BOLTS

Bearing Stress of Wood Under Bolts

The bearing stress under a bolt, computed by dividing the load by the product LD where L is length of bolt in main member and D is bolt diameter, is largest when the bolt does not bend, i.e. for joints with small L/D values.

The results of many tests of joints having a seasoned wood member and two steel splice plates show that bearing stress parallel to the grain at proportional limit loads approached 60 percent of the crushing strength of conifers and 80 percent of the crushing strength of hardwoods, where the crushing strengths are those obtained for clear, straight-grained specimens (table 4–2, ch. 4). The curve of figure 7–7 shows the reduction in bearing stress as L/D ratio increases. Thus the bolt-bearing stress parallel to the grain can be obtained for the crushing strength of small clear specimens, which is reduced by the above-mentioned factor for conifers or hardwoods, and an additional factor for L/D ratio. When wood splice plates were used, each splice plate half as thick as the center member, the bearing stresses were about 80 percent of those obtained with steel splice plates.

For bearing stresses perpendicular to the grain in seasoned wood, the stresses for short bolts (small L/D ratios) depend upon bolt diameter. Small bolts had higher proportional limit values than large bolts. The effect is presented in figure 7–8, wherein the ratio of proportional limit bearing stress to compression perpendicular to the grain pro-

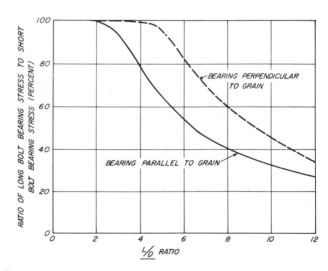

Figure 7–7.—Variation in the proportional limit bolt-bearing stress with L/D ratio.

portional limit stress is plotted as a function of bolt diameter. The variation with bolt length (L/D ratio) is given by the appropriate curve in figure 7–7. Thus, the bolt-bearing stress perpendicular to the grain can be obtained from the proportional limit stress perpendicular to grain of small clear specimens (table 4–2, ch. 4), this is increased by the factor given in figure 7–8 and then reduced by the factor given in figure 7–7. The same data were obtained with steel splice plates and wood splice plates, each half as thick as the main center member. Bearing was parallel to the grain in the splice plates.

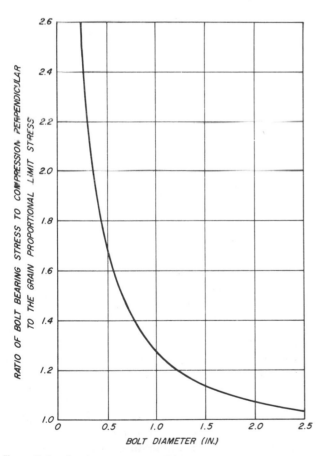

Figure 7–8.—Bearing stress perpendicular to the grain as affected by bolt diameter.

For loads applied at an angle intermediate between those parallel to the grain and perpendicular to the grain, the bolt bearing stress may be obtained from the nomograph in figure 7–6.

Effect of Bolt Quality on Joint Strength

Both the properties of the wood and the quality of the bolt are factors in determining the proportional limit strength of a bolted joint. The percentages of stresses given in figure 7–7 for calculating bearing stresses apply to steel machine bolts used in building construction. For high-strength bolts, such as aircraft bolts, the factors given in figure 7–7 would be conservative for the larger L/D ratios.

Design Details

The details of design required in the application of the loads for bolts may be summarized as follows:

(1) A load applied to only one end of a bolt, perpendicular to its axis, may be taken as one-half the symmetrical two-end load.

(2) The center-to-center distance along the grain between bolts acting parallel to the grain should be at least four times the bolt diameter. When a joint is in tension, the bolt nearest the end of a timber should be at a distance from the end of at least seven times the bolt diameter for softwoods and five times for hardwoods. When the joint is in compression, the end margin may be four times the bolt diameter for both softwoods and hardwoods. Any decrease in these spacings and margins will decrease the load in about the same ratio.

(3) For bolts bearing parallel to the grain, the distance from the edge of a timber to the center of a bolt should be at least 1.5 times the bolt diameter. This margin, however, will usually be controlled by (a) the common practice of having an edge margin equal to one-half the distance between bolt rows and (b) the area requirements at the critical section. (The critical section is that section of the member taken at right angles to the direction of load, which gives the maximum stress in the member based on the net area remaining after reductions are made for bolt holes at that section.) For parallel-to-grain loading in softwoods, the net area remaining at the critical section should be at least 80 percent of the total area in bearing under all the bolts in the particular joint under consideration; in hardwoods it should be 100 percent.

(4) For bolts bearing perpendicular to the grain, the margin between the edge toward which the bolt pressure is acting and the center of the bolt or bolts nearest this edge should be at least four times the bolt diameter. The

margin at the opposite edge is relatively unimportant. The minimum center-to-center spacing of bolts in the across-the-grain direction for loads acting through metal side plates need only be sufficient to permit the tightening of the nuts. For wood side plates, the spacing is controlled by the rules applying to loads acting parallel to grain if the design load approaches the bolt-bearing capacity of the side plates. When the design load is less than the bolt-bearing capacity of the side plates, the spacing may be reduced below that required to develop their maximum capacity.

Effect of Bolt Holes

The bearing strength of wood under bolts is affected considerably by the size and type of bolt hole into which the bolts are inserted. A bolt hole that is too large causes nonuniform bearing of the bolt; if the bolt hole is too small, the wood will split when the bolt is driven. Normally, bolts should fit neatly, so that they can be inserted by tapping lightly with a wood mallet. In general, the smoother the hole, the higher the bearing values will be (fig. 7–9). Deformations accompanying the load also increase with increase in the unevenness of the bolt-hole surface (fig. 7–10).

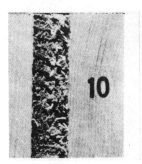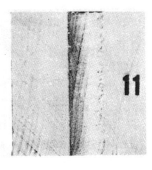

Figure 7–9.—Effect of rate of feed and drill speed on the surface condition of bolt holes drilled in Sitka spruce. The hole on the left was bored with a twist drill rotating at a peripheral speed of 300 inches per minute; the feed rate was 60 inches per minute. The hole on the right was bored with the same drill at a peripheral speed of 1,250 inches per minute; the feed rate was 2 inches per minute.

Rough holes are caused by using dull bits and improper rates of feed and drill speed. A twist drill operated at a peripheral speed of approximately 1,500 inches per minute produces uniform smooth holes at moderate feed rates. The rate of feed depends upon the diameter of the drill and the speed of rotation but should enable the drill to cut rather than tear the wood. The drill should produce shavings, not chips.

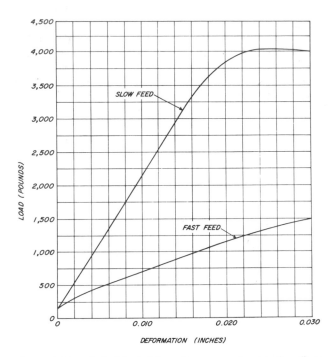

Figure 7–10.—Typical load-deformation curves showing the effects of surface condition of bolt holes, resulting from a slow feed rate and a fast feed rate, on the deformation in a joint when subjected to loading under bolts. The surface conditions of the bolt holes were similar to those illustrated in figure 7–9.

Multiple-Bolt Joints

Joints containing six or more bolts in a row have an uneven distribution of bolt loads. The two end bolts together usually carry over 50 percent of the load. More than six bolts in a row do not substantially increase the elastic strength of the joint, in that the additional bolts only tend to reduce the load on the less heavily loaded interior bolts.

A small misalinement of bolt holes may cause large shifts in bolt loads. Therefore, in field-fabricated joints, the exact distribution of bolt loads would be difficult to predict. The most even distribution of bolt loads occurs in a joint in which the extensional stiffness of the main member is equal to that of both splice plates.

A simplified method of analysis that closely predicts the load distribution among the bolts in a timber tension joint has been developed by Cramer (see Bibliography).

CONNECTOR JOINTS

Several types of connectors have been devised that increase joint bearing and shear areas by utilizing rings or plates around bolts holding joint members together. The primary load-carrying portions of these joints are the connectors; the bolts usually serve to prevent sideways separation of the members, but do contribute some load-carrying capacity.

The strength of the connector joint depends on the type and size of the connector, the species of wood, the thickness and width of the member, the distance of the connector from the end of the member, the spacing of the connectors, the direction of application of the load with respect to the direction of the grain of the wood, and other factors. Loads for wood joints with steel connectors—split-ring (fig. 7–11), toothed-ring (fig. 7–12), and shear-plate (fig. 7–13)—are discussed in this section. The split-ring and shear-plate connectors require closely fitting machined portions in the wood members. The toothed-ring connector is pressed into the wood.

Connector Joints Under Permanent Load

Allowable loads for the split-ring, shear-plate, and toothed-ring connectors provided in table 7–6 were derived from the results of tests of joints made from several wood species. The species were classified into four groups in accordance with their strength in connector joints. In establishing the loads, particular consideration was given to (1) the effect of long-continued loading as against the brief loading period involved in the test of joints and (2) allowance for variability in timber quality (ASTM 1761).

Since adequate data are not available on the effect of duration of stress on the strength of connector joints, insofar as the wood is con-

Figure 7–12.—Joint with toothed-ring connector.

sidered, the relation between the load at failure in a standard bending test of a few minutes' duration and the load that will cause failure under longtime loading was assumed to apply. Under constant load a beam will fail at a load only about nine-sixteenths as great as the breaking load found in the standard bending test (see ch. 6).

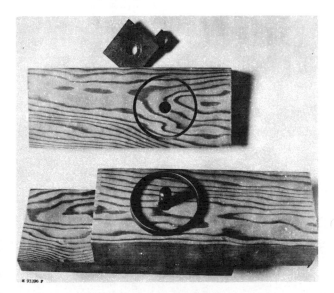

Figure 7–11.—Joint with split-ring connector showing connector, precut groove, bolt, washer, and nut.

Figure 7–13.—Joints with shear-plate connectors with A, wood side plates; and B, steel side plates.

Table 7–6.—*Allowable loads* [1] *for one connector in a joint*

Connector	Minimum thickness of wood member: With one connector only	Minimum thickness of wood member: With two connectors in opposite faces, one bolt [3]	Minimum width all members	Group 1 woods [2]: Load at 0° angle to grain	Group 1 woods [2]: Load at 90° angle to grain	Group 2 woods [2]: Load at 0° angle to grain	Group 2 woods [2]: Load at 90° angle to grain	Group 3 woods [2]: Load at 0° angle to grain	Group 3 woods [2]: Load at 90° angle to grain	Group 4 woods [2]: Load at 0° angle to grain	Group 4 woods [2]: Load at 90° angle to grain
	In.	*In.*	*In.*	*Lb.*	*Lb.*	*Lb.*	*Lb.*	*Lb.*	*Lb.*	*Lb.*	*Lb.*
Split Ring:											
2½-in. diameter, ¾-in. wide with ½-in. bolt	1	2	3½	1,785	1,055	2,085	1,230	2,480	1,475	2,875	1,725
4-in. diameter, 1-in. wide, with ¾-in. bolt	1½	3	5½	3,445	1,995	3,985	2,310	4,780	2,775	5,580	3,235
Toothed Ring:											
2-in. diameter, 0.94 in. wide with ½-in. bolt	1⅛	2	2½	860	[4] 570	990	[4] 660	1,100	[4] 735	1,210	[4] 805
2⅝-in. diameter, 0.94 in. wide with ⅝-in. bolt	1⅜	2½	3½	1,460	[4] 976	1,690	[4] 1,125	1,875	[4] 1,250	2,060	[4] 1,375
3⅜-in. diameter, 0.94 in. wide with ¾-in. bolt	1½	3	4⅝	2,055	[4] 1,370	2,370	[4] 1,580	2,630	[4] 1,755	2,895	[4] 1,930
4-in. diameter, 0.94 in. wide with ¾-in. bolt	1½	3	5½	2,385	[4] 1,590	2,750	[4] 1,835	3,055	[4] 2,035	3,360	[4] 2,240
Shear Plate:											
2⅝-in. diameter, 0.42 in. wide with ¾-in. bolt	1½	2⅝	3½	1,890	1,095	2,190	1,270	2,630	1,525	2,665	1,780
4-in. diameter, 0.64 in. wide with ¾- or ⅞-in. bolt	1¾	3⅝	5½	2,850	1,655	3,305	1,915	3,965	2,300	4,625	2,685

[1] The loads apply to seasoned timbers in dry, inside locations for a long-continued load. It is assumed also that the joints are properly designed with respect to such features as centering of connectors, adequate end distance, and suitable spacing.

[2] Group 1 woods provide the weakest connector joints, and group 4 woods the strongest. Groupings are given in table 7–7.

[3] A 3-member assembly with 2 connectors takes double the loads indicated in fifth to twelfth columns.

[4] These loads are for any angle from 45° to 90° to the grain.

Table 7–7.—*Species groupings for connector loads*

Connector load group	Species or species group		
Group 1	Aspen Western redcedar Eastern hemlock Sugar pine	Basswood Balsam fir Eastern white pine Western white pine	Cottonwood White fir Ponderosa pine Engelmann spruce
Group 2	Chestnut Alaska-cedar Red pine Sitka spruce	Yellow-poplar Port-Orford-cedar Redwood White spruce	Baldcypress Western hemlock Red spruce
Group 3	Elm, American Sweetgum Douglas-fir	Elm, slippery Sycamore Larch, western	Maple, soft Tupelo Southern pine
Group 4	Ash, white Elm, rock Oak	Beech Hickory	Birch Maple, hard

Tests have demonstrated that the density of the wood is a controlling factor in determining the strength of a joint. Consequently, the load carried by a connector in the laboratory test employing wood of average quality for a species was adjusted to allow for the lower than average material that might be in service.

The values listed in table 7–6 for connectors loaded parallel to the grain were derived by applying a reduction factor to the ultimate test load. For split-ring and shear-plate connectors, a reduction factor of 4 gave values that would not exceed five-eighths of the proportional limit test load. Because load-slip curves for joints with toothed connectors do not exhibit a well-defined proportional limit, a reduction factor of 4½ was applied to their ultimate loads.

Tests of connector joints under loads bearing perpendicular to grain, although less extensive than those for parallel bearing, have been sufficient to establish a generally applicable relationship between the two directions. This relationship was used in deriving loads for perpendicular bearing. Ultimate load was given less consideration for perpendicular than for parallel bearing, and greater dependence was placed on other factors, such as the load at proportional limit and at given slips of the joint.

The figures quoted as the ratios between tabulated loads and the loads found in tests are in no instance true factors of safety. For example, the reduction factor of 4 for split-ring and shear-plate connectors includes al-

lowances for duration of stress and for variability as well as a margin for safety. Thus, after the values from test are multiplied by a factor of $\frac{9}{16}$ as an allowance for a long-continued load and by $\frac{3}{4}$ to cover variability of the wood, the actual factor of safety for a connector joint is on the order of $1\frac{3}{4}$ ($4 \times \frac{9}{16} \times \frac{3}{4} = 1\frac{11}{16}$) if the load acts over a long period. The tests from which loads were derived were on specimens carefully made from seasoned material, under favorable conditions, and by experienced workmen.

For any joint assembly in which more than one connector is used in the contact faces with the same bolt axis, the total load is the sum of the loads of each connector. For example, in table 7–6 minimum actual thickness of the members is given for a joint assembly of three members employing two connectors in opposite faces with a common bolt; this assembly is equivalent to two connectors; therefore, the load will be twice the corresponding value shown for a one-connector assembly. The loads given apply only when the joints are properly designed with respect to such features as centering of connectors on the member axis, adequate end distances, and suitable spacing of connectors.

Modifications

Some of the factors that affect the loads of connectors were taken into account in deriving the tabular values. Other varied and extreme conditions require modification of the values.

Wind or Earthquake Loads

In designing for wind or earthquake forces acting alone, or acting in conjunction with dead and live loads, the loads for the various connectors may be increased by the following percentages, provided the number and size of connectors are not less than required for the combination of dead and live loads alone:

	Increase Percent[2]
Split-ring connector, any size, bearing in any direction	50
Shear-plate connector, any size, bearing parallel to grain	33½
Shear-plate connector, any size, bearing perpendicular to grain	50
Toothed-ring connector, 2-inch, bearing in any direction	50
Tooth-ring connector, 4-inch, bearing in any direction	25

[2] Percentages for shear-plate connectors bearing at intermediate angles and for toothed-ring connectors of other sizes can be obtained by interpolation.

Impact Forces

Impact may be disregarded up to the following percentage of the static effect of the live load producing the impact:

	Impact allowances Percent[2]
Split-ring connector, any size, bearing in any direction	100
Shear-plate connector, any size, bearing parallel to grain	66⅔
Shear-plate connector, any size, bearing perpendicular to grain	100
Toothed-ring connector, 2-inch, bearing in any direction	100
Toothed-ring connector, 4-inch, bearing in any direction	50

[2] One-half of any impact load that remains after disregarding the percentages indicated should be included with the other dead and live loads in obtaining the total force to be considered in designing the joint.

Factor of Safety Not Reduced

The procedures described for increasing the loads on connectors for forces suddenly applied and forces of short duration do not reduce the actual factor of safety of the joint but are realistic because of the favorable behavior of wood under suddenly applied forces. The differentiation among types and sizes of connector and directions of bearing is due to variations in the extent to which distortion of the metal, as well as the strength of the wood, affects the ultimate strength of the joint.

Exposure and Moisture Condition of Wood

The loads listed in table 7–6 apply to seasoned members used where they will remain dry. If the wood will be more or less continuously damp or wet in use, two-thirds of the tabulated values should be used. The amount by which the loads should be reduced to adapt them to other conditions of use depends upon the extent to which the exposure favors decay, the required life of the structure or part, the frequency and thoroughness of inspection, the original cost and the cost of replacements, the proportion of sapwood and the durability of the heartwood of the species, if untreated, and the character and efficiency of any treatment. These factors should be evaluated for each individual design. Industry recommendations for the use of connectors when the condition of the lumber is other than continuously wet or continuously dry are given in the National Design Specification for Stress-Grade Lumber and Its Fastenings.

Ordinarily, before fabrication of connector joints, members should be seasoned to a moisture content corresponding as nearly as practical to that which they will attain in service. This is particularly desirable for lumber for roof trusses and other structural units used in dry locations and in which shrinkage is an important factor. Urgent construction needs sometimes result in the erection of structures and structural units employing green or inadequately sesoned lumber with connectors. Since such lumber subsequently dries out in most buildings, causing shrinkage and opening the joints, it is essential that adequate maintenance measures be adopted. The maintenance for connector joints in green lumber should include inspection of the structural units and tightening of all bolts as needed during the time the units are coming to moisture equilibrium, which is normally during the first year.

Grade and Quality of Lumber

The lumber for which the loads for connectors are applicable should conform to the general requirements in regard to the quality of structural lumber given in the grading rule books of lumber manufacturers' associations for the various commercial species.

The loads for connectors were obtained for wood at the joints that were clear and free from checks, shakes, and splits. Cross grain at the joint should not be steeper than a slope of 1 in 10, and knots in the connector area should be accounted for as explained under "Net Section."

Loads at Angle With Grain

The loads for the split-ring and shear-plate connectors for angles of 0° to 90° between direction of load and grain may be obtained by Hankinson's formula (eq. 7–11) or by the nomograph in figure 7–6. With the toothed connectors, the load at an inclination to the grain of 0° to 45° may be obtained with the previously mentioned formula, but from 45° to 90° it is equal to the load perpendicular to the grain.

Thickness of Member

The relationship between the loads for the different thicknesses of lumber is based on test results for connector joints. The least thickness of member given in table 7–6 for the various sizes of connectors is the minimum to obtain optimum load. The loads listed for each type and size of connector are the maximum loads to be used for all thicker lumber. The loads for wood members of thicknesses less than those listed can be obtained by the percentage reductions indicated in figure 7–14. Thicknesses below those indicated by the curves should not be used.

Thicknesses of members containing one connector only are equal to half the thickness of a member containing one connector in each face plus one-eighth inch for split-ring and toothed-ring connectors. For split-ring connectors, the reduction in load for other thicknesses of members containing one connector only may be obtained by following the same rules as for shear plate and toothed-ring connectors.

Width of Member

The width of member listed for each type and size of connector is the minimum that should be used. When the connectors are bearing parallel to the grain, no increase in load occurs with an increase in width. When they are bearing perpendicular to the grain, the load increases about 10 percent for each 1-inch increase in width of member over the minimum widths required for each type and size of connector, up to twice the diameter of the connectors. When the connector is placed off center and the load is applied continuously in one direction only, the proper load can be determined by considering the width of member as equal to twice the edge distance (the distance between the center of the connector and the edge of the member toward which the load is acting). But the distance between the center of the connector and the opposite edge should not be less than one-half the permissible minimum width of the member.

Net Section

The net section is the area remaining at the critical section after subtracting the projected area of the connectors and bolt from the full cross-sectional area of the member. For sawed timbers, the stress in the net area (whether in tension or compression) should not exceed the stress for clear wood in compression parallel to the grain. In using this stress, it is assumed that knots do not occur within a length of one-half the diameter of the connector from the net section. If knots are present in the longitudinal projection of the net section within a length from the critical

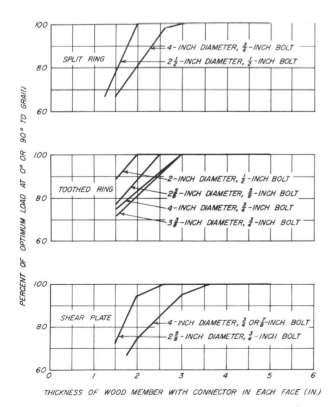

Figure 7–14.—Effect of thickness of wood member on the optimum load capacity of a timber connector.

section of one-half the diameter of the connector, the area of the knots should be subtracted from the area of the critical section.

In laminated timbers, knots may occur in the inner laminations at the connector location without being apparent from the outside of the member. It is impractical to assure that there are no knots at or near the connector. In laminated construction, therefore, the stress at the net section is limited to the compressive stress for the member, accounting for the effect of knots.

End Distance and Spacing

The load values in table 7–6 apply when the distance of the connector from the end of the member (end distance e) and the spacing (s) between connectors in multiple joints are

not factors affecting the strength of the joint (fig. 7–15, A). When the end distance or spacing for connectors bearing parallel to the grain is less than that required to develop the full load, the proper reduced load may be obtained by multiplying the loads in table 7–6 by the appropriate strength ratio given in table 7–8. For example, the load for a 4-inch split-ring connector bearing parallel to the grain, when placed 7 or more inches from the end of a Douglas-fir tension member that is 1½ inches thick, is 4,780 pounds. When the end distance is only 5¼ inches, the strength ratio obtained by direct interpolation from the values given in table 7–8 is 0.81, and the load equals 0.81 times 4,780 or 3,870 pounds.

Placement of Multiple Connectors

Preliminary investigations of the placement of connectors in a multiple joint, together with the observed behavior of single connector joints tested with variables that simulate those in a multiple joint, furnish a basis for some suggested design practices.

When two or more connectors in the same face of a member are in a line at right angles to the grain of the member and are bearing parallel to the grain (fig. 7–15, C), the clear distance (c) between the connectors should not be less than one-half inch.

When two or more connectors are acting perpendicular to the grain and are spaced on a line at right angles to the length of the member (fig. 7–15, B), the rules for the width of member and edge distances used with one connector are applicable to the edge distances for multiple connectors. The clear distance between the connectors (c) should be equal to the clear distance from the edge of the member toward which the load is acting to the connector nearest this edge (c).

In a joint with two or more connectors spaced on a line parallel to the grain and with the load acting perpendicular to the grain (fig. 7–15, D), the available data indicate that the load for multiple connectors is not equal to the sum of the loads for individual connectors. Somewhat more favorable results can be obtained if the connectors are staggered so that they do not act along the same line with respect to the grain of the transverse member. Industry recommendations for various angle-to-grain loadings and spacings are given in National Design Specifications.

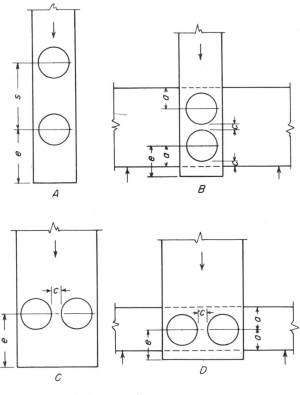

LEGEND:

e - END DISTANCE
S - SPACING PARALLEL TO GRAIN
a - EDGE DISTANCE
C - CLEAR DISTANCE

Figure 7–15.—Types of multiple-connector joints: A, joint strength dependent on end distance e and connector spacing s; B, joint strength dependent on end e, clear c, and edge a distances; C, joint strength dependent on end e and clear c distances; D, joint strength dependent on end e, clear c, and edge a distances.

Table 7–8.—*Strength ratio for connectors for various longitudinal spacings and end distances* [1]

Connector diameter	Spacing [2]	Spacing strength ratio	End distance [3] Tension member	End distance [3] Compression member	End distance strength ratio
In.	*In.*	*Pct.*	*In.*	*In.*	*Pct.*
SPLIT-RING					
2½	6¾ +	100	5½ +	4 +	100
2½	3⅜	50	2¾	2½	62
4	9 +	100	7 +	5½ +	100
4	4⅞	50	3½	3¼	62
SHEAR-PLATE					
2⅝	6¾ +	100	5½ +	4 +	100
2⅝	3⅜	50	2¾	2½	62
4	9 +	100	7 +	5½ +	100
4	4½	50	3½	3¼	62
TOOTHED-RING					
2	4 +	100	3½ +	2 +	100
2	2	50	2		67
2⅝	5¼ +	100	4⅝ +	2⅝ +	100
2⅝	2⅝	50	2⅝		67
3⅜	6¾ +	100	5⅞ +	3⅜ +	100
3⅜	3⅜	50	3⅜		67
4	8 +	100	7 +	4 +	100
4	4	50	4		67

[1] Strength ratio for spacings and end distances intermediate to those listed may be obtained by interpolation, and multiplied by the loads in table 7–6 to obtain design load. The strength ratio applies only to those connector units affected by the respective spacings or end distances. The spacings and end distances should not be less than the minimum shown.

[2] Spacing is distance from center to center of connectors (fig. 7 15, *A*).

[3] End distance is distance from center of connector to end of member (fig. 7–15, *A*).

Cross Bolts

Cross bolts or stitch bolts placed at or near the end of members joined with connectors or at points between connectors will provide additional safety. They may also be used to reinforce members that have, through change in moisture content in service, developed splits to an undesirable degree.

Examples of Connector-Joint Design

(1) Calculate the load of a tension joint of seasoned Douglas-fir in which two pieces 3½ inches thick and 5½ inches wide are joined end to end by side plates 1½ inches thick, 5½ inches wide, and 28 inches long, when four 4-inch split-ring connectors and two ¾-inch bolts are used. In this arrangement, two connectors and a concentric bolt are placed symmetrically on either side of the butt joint at a distance of 7 inches from the ends of the members and side plates. This end distance, as shown in table 7–8, is adequate to develop the full load.

The load given in table 7–6 for one 4-inch split-ring connector, when used in one face of a Douglas-fir member 1½ inches thick or as one of two connectors used in opposite faces of a member 3 inches thick, is 4,780 lb. The load of the joint for two connectors is twice 4,780 or 9,560 lb.

(2) Calculate the load of the joint in example (1) when the side plates are 16 inches instead of 28 inches long. By placing the connectors halfway between the ends of the side plates and the butt joint, the end distance is 4 inches. The strength ratio as interpolated from values given in table 7–8 for a 4-inch end distance is 0.68, and the load accordingly equals 0.68 times 9,560 or 6,500 lb.

(3) Calculate the load of a joint of seasoned southern pine in which two tension side members 1½ inches thick and 5½ inches wide are joined at right angles to opposite faces of a center timber 3½ inches thick and 5½ inches wide by means of two 4-inch split-ring connectors and a ¾-inch bolt.

The load for one of two 4-inch split-ring connectors used in opposite faces of a member 3 inches thick and 5½ inches wide and bearing perpendicular to the grain is 2,775 lb. (table 7–6). The load for one connector bearing parallel to the grain in one face of a side member 1½ inches thick and with an end distance of 7 inches is 4,780 lb. (table 7–6). The load of the joint, which is governed by the center member, is twice 2,775 or 5,550 lb.

(4) Calculate the load of the joint in example (3) when the distance from the end of the side plates overlapping the center member to the center of the bolt hole is 3½ instead of 7 inches.

The strength ratio for an end distance of 3½ inches is 0.62 (table 7–8). The load for one 4-inch split-ring connector in the side member, hence, equals 0.62 times 4,780 or 2,964 lb. This is larger than the load for one connector in the center member. The strength of the joint, therefore, is still governed by the center member and, as before, is 5,550 lb.

FASTENER HEAD EMBEDMENT

The bearing strength of wood under fastener heads is important in such applications as the

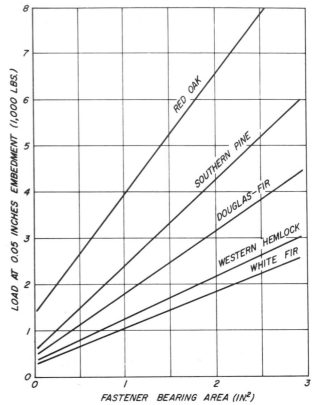

Figure 7–17.—Relation between load at 0.05 inch embedment and fastener bearing area for several species.

anchorage of building framework to foundation structures. When pressure tends to pull the framing member away from the foundation, the fastening loads could cause tensile failure of the fastenings, withdrawal of the fastenings from the framing member, or embedment of the fastener heads into the member. Possibly the fastener head could even be pulled completely through.

The maximum resistance to fastener head embedment is related to the fastener perimeter, while loads of low embedments (0.05 in.) are related to the fastener bearing area. These relations for several species at 10 percent moisture content are shown in figures 7–16 and 7–17.

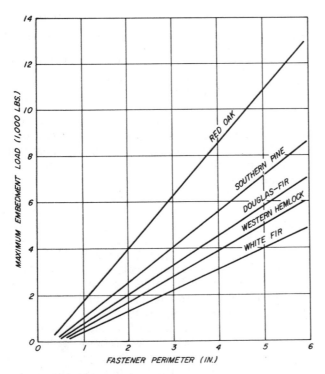

Figure 7–16.—Relation between maximum embedment load and fastener perimeter for several species of wood.

176

Table 7-9 — Grouping of Species for Determining Allowable
Loads for Lag Screws, Nails, Spikes, Wood
Screws, Drift Bolts.

Group	Species of Wood	Specific Gravity* (G)
I	Ash, Commercial White	.62
	Beech	.68
	Birch, Sweet & Yellow	.66
	Hickory & Pecan	.75
	Maple, Black & Sugar	.66
	Oak, Red & White	.67
II	Douglas Fir—Larch	.51
	Southern Pine	.55
	Sweetgum & Tupelo	.54
III	Aspen, Northern	.42
	California Redwood (Close grain)	.42
	Douglas Fir, South	.48
	Eastern Hemlock—Tamarack	.45
	Eastern Spruce	.43
	Hem—Fir	.44
	Idaho White Pine	.42
	Lodgepole Pine	.44
	Mountain Hemlock	.47
	Northern Pine	.46
	Ponderosa Pine—Sugar Pine	.42
	Red Pine	.42
	Sitka Spruce	.43
	Southern Cypress	.48
	Spruce—Pine—Fir	.42
	Western Hemlock	.48
	Yellow Poplar	.46
IV	Aspen	.40
	Balsam Fir	.38
	California Redwood (Open grain)	.37
	Coast Sitka Spruce and Coast Species	.39
	Cottonwood, Black	.33
	Cottonwood, Eastern	.41
	Eastern White Pine	.38
	Engelman Spruce	.37
	Northern Species	.36
	Northern White Cedar	.31
	Subalpine Fir	.34
	Western Cedars	.36
	Western White Pine	.40

*Based on weight and volume when oven-dry

Table 7-10   Lag Bolts or Lag Screws — Allowable Lateral Loads —
Normal Duration.

WOOD SIDE PIECES

(For Species in each Group, See Table 7-9)

Thickness of Side Member (inches)	Length of Lag Bolt (inches)	Diameter of Lag Bolt Shank (inches)	GROUP I Total Lateral Load per Lag Bolt in Single Shear (pounds)		GROUP II Total Lateral Load per Lag Bolt in Single Shear (pounds)		GROUP III Total Lateral Load per Lag Bolt in Single Shear (pounds)		GROUP IV Total Lateral Load per Lag Bolt in Single Shear (pounds)	
			Parallel to Grain	Perpendicular to Grain	Parallel to Grain	Perpendicular to Grain	Parallel to Grain	Perpendicular to Grain	Parallel to Grain	Perpendicular to Grain
1½"	4"	¼	200	190	170	170	130	120	100	100
		⁵⁄₁₆	280	230	210	180	150	130	120	100
		⅜	320	240	240	180	170	130	140	100
		⁷⁄₁₆	350	250	270	190	190	140	150	110
		½	390	250	290	190	210	140	170	110
		⅝	470	280	360	210	260	150	200	120
	5"	¼	230	220	200	190	180	170	160	150
		⁵⁄₁₆	330	280	290	250	230	200	190	160
		⅜	430	330	370	280	260	200	210	160
		⁷⁄₁₆	540	380	400	290	290	210	230	160
		½	580	380	440	280	310	200	250	160
		⅝	710	420	530	320	380	230	310	180
	6"	¼	270	260	230	220	200	200	180	180
		⁵⁄₁₆	380	320	330	280	290	250	260	220
		⅜	480	370	420	320	360	280	290	220
		⁷⁄₁₆	590	420	510	360	400	280	320	230
		½	700	460	600	390	430	280	340	220
		⅝	860	510	710	430	510	310	410	250
	7"	¼	280	270	240	230	210	210	190	180
		⁵⁄₁₆	400	340	350	300	310	270	280	230
		⅜	520	400	450	340	410	310	360	270
		⁷⁄₁₆	640	460	560	390	500	350	420	290
		½	760	500	660	430	560	360	450	290
		⅝	910	550	790	470	640	380	510	310
2½"	6"	⅜	450	340	370	280	270	200	210	160
		⁷⁄₁₆	570	400	430	310	310	220	250	180
		½	620	410	470	310	340	220	270	180
		⅝	730	440	550	330	390	240	320	190
		¾	820	450	620	340	440	240	360	200
		⅞	930	490	710	370	500	260	400	210
		1	1040	520	790	390	560	280	450	230
	7"	⅜	500	380	430	330	370	280	300	220
		⁷⁄₁₆	660	470	570	410	420	300	340	240
		½	830	540	650	420	460	300	370	240
		⅝	1000	600	750	450	540	320	430	260
		¾	1110	610	840	460	600	330	480	270
		⅞	1260	660	950	500	680	360	550	290
		1	1420	710	1070	540	770	380	620	310
	8"	⅜	550	420	480	360	430	320	380	290
		⁷⁄₁₆	720	510	630	440	550	390	440	310
		½	890	580	770	500	600	390	480	310
		⅝	1230	740	970	580	700	420	560	340
		¾	1430	790	1080	600	780	430	620	340
		⅞	1590	830	1200	630	860	450	690	360
		1	1800	900	1360	680	970	490	780	390
	9"	⅜	600	460	520	390	460	350	410	310
		⁷⁄₁₆	790	560	680	480	610	430	540	380
		½	970	630	840	540	750	490	600	390
		⅝	1310	790	1130	680	860	520	690	420
		¾	1670	920	1340	740	960	530	770	420
		⅞	1920	1000	1450	760	1040	540	830	430
		1	2170	1090	1640	820	1180	590	940	470

LAG-SCREW JOINTS

Table 7-11. — LAG BOLTS or LAG SCREWS — Allowable Lateral Loads — Normal Duration.
½″ METAL SIDE PIECES
(For Species in each Group, See Table 7-9)

Length of Lag Bolt (inches)	Diameter of Lag Bolt Shank (inches)	GROUP I Total Lateral Load per Lag Bolt in Single Shear (pounds)		GROUP II Total Lateral Load per Lag Bolt in Single Shear (pounds)		GROUP III Total Lateral Load per Lag Bolt in Single Shear (pounds)		GROUP IV Total Lateral Load per Lag Bolt in Single Shear (pounds)	
		Parallel to Grain	Perpendicular to Grain	Parallel to Grain	Perpendicular to Grain	Parallel to Grain	Perpendicular to Grain	Parallel to Grain	Perpendicular to Grain
3″	¼	240	185	210	160	155	120	125	100
	⁵⁄₁₆	355	240	265	180	190	130	155	105
	³⁄₈	420	255	320	195	230	140	180	110
	⁷⁄₁₆	485	275	370	210	265	150	210	120
	½	550	285	415	215	295	155	240	125
	⅝	645	310	490	235	350	170	280	135
4″	¼*	275	210	235	185	210	165	190	145
	⁵⁄₁₆	410	280	355	240	290	200	235	160
	³⁄₈	570	345	480	290	345	210	275	165
	⁷⁄₁₆	750	425	575	320	405	230	320	180
	½	830	430	625	325	450	235	360	185
	⅝	975	465	740	355	530	255	425	205
5″	⁵⁄₁₆	435	295	375	255	335	230	300	205
	³⁄₈	615	375	535	325	470	295	375	230
	⁷⁄₁₆	820	465	710	405	535	350	430	245
	½	1045	540	850	440	610	315	490	255
	⅝	1330	635	1005	480	720	345	580	280
	¾	1580	695	1190	525	855	375	690	305
6″	⁵⁄₁₆*	445	305	400	270	345	235	305	205
	³⁄₈	620	385	545	330	490	300	430	260
	⁷⁄₁₆	850	480	735	415	660	375	545	310
	½	1100	570	945	490	770	400	615	320
	⅝	1640	790	1250	600	900	430	720	345
	¾	1970	865	1480	650	1060	460	850	370
7″	³⁄₈*	645	390	555	340	500	305	440	270
	⁷⁄₁₆	865	490	750	425	670	380	590	335
	½	1120	580	970	505	865	450	745	385
	⅝	1700	820	1460	700	1020	490	900	430
	¾	2360	1040	2030	890	1290	570	1040	460
8″	⁷⁄₁₆*	875	500	760	430	680	385	600	340
	½	1140	590	985	510	880	455	775	400
	⅝	1750	840	1500	720	1325	635	1070	560
	¾	2475	1090	2130	935	1550	680	1250	555
	⅞	3280	1365	2720	1130	1950	810	1560	715
9″	½*	1150	600	990	515	885	460	780	405
	⅝	1770	850	1510	725	1360	650	1200	575
	¾	2520	1110	2160	950	1780	785	1435	630
	⅞	3350	1390	2880	1200	2060	855	1660	690
10″	⅝*	1800	865	1540	740	1380	660	1220	585
	¾	2540	1120	2190	965	1970	865	1625	715
	⅞	3420	1420	2960	1230	2340	970	1890	785
	1	4420	1770	3710	1485	2660	1065	2140	855
11″	¾*	2580	1130	2220	970	2000	880	1765	780
	⅞	3450	1430	2990	1240	2600	1080	2100	870
	1	4500	1800	3880	1550	2970	1190	2370	950
12″	⅞	3470	1440	3000	1250	2690	1120	2320	965
	1	4520	1810	3900	1560	3290	1320	2630	1050
	1⅛	5660	2260	4900	1960	3570	1430	2870	1150
13″	⅞*	3500	1455	3030	1260	2710	1130	2390	990
	1	4550	1820	3930	1570	3520	1410	2890	1155
	1⅛	5700	2280	4920	1970	3920	1570	3120	1250
14″	1	4570	1830	3950	1570	3530	1410	3120	1240
	1⅛	5740	2300	4950	1980	4380	1750	3500	1400
	1¼	7020	2800	6060	2420	4830	1930	3910	1560
15″	1	4580	1830	3960	1580	3550	1420	3130	1250
	1⅛	5770	2310	4980	1990	4460	1790	3820	1530
	1¼	7070	2830	6110	2450	5250	2100	4180	1670
16″	1*	4600	1840	3960	1580	3550	1420	3130	1250
	1⅛*	5800	2320	5000	2000	4470	1790	3950	1580
	1¼*	7120	2850	6150	2460	5500	2200	4520	1810

* Greater lengths do not provide higher loads.

Table 7-12. — LAG BOLTS OR LAG SCREWS — Allowable Withdrawal Loads — Normal Duration

Allowable load in withdrawal in pounds per inch of penetration of threaded part into side grain of member holding point.

D = the shank diameter in inches. G = specific gravity of the wood based on weight and volume when oven-dry.

Specific gravity G	Size (D)											
	¼	⁵⁄₁₆	⅜	⁷⁄₁₆	½	⁹⁄₁₆	⅝	¾	⅞	1	1⅛	1¼
	0.250	0.3125	0.375	0.4375	0.500	0.5625	0.625	0.750	0.875	1.000	1.125	1.250
.75	414	488	561	630	697	760	824	944	1060	1172	1282	1388
.68	357	422	484	543	601	655	710	814	914	1010	1103	1194
.67	349	412	472	531	587	640	693	795	893	986	1078	1166
.66	342	403	462	519	574	626	678	778	873	965	1054	1141
.62	311	367	421	473	523	570	618	708	795	878	960	1038
.55	264	312	356	402	443	484	524	601	675	745	815	881
.54	256	304	348	391	432	473	512	586	660	728	796	860
.51	232	274	313	352	389	425	460	528	593	655	716	774
.48	212	251	287	322	357	389	421	483	542	599	655	709
.47	205	242	278	312	345	376	407	467	525	580	633	685
.46	199	235	269	302	334	364	395	453	508	562	613	664
.45	192	227	260	292	323	353	382	438	492	544	594	643
.44	186	220	252	283	313	341	369	424	476	526	574	621
.43	180	213	244	274	303	331	358	411	461	509	556	602
.42	173	205	235	263	291	318	344	395	443	490	535	579
.41	167	197	226	254	281	306	332	380	427	472	515	558
.40	162	191	220	246	273	296	324	368	414	458	500	542
.39	156	184	211	237	262	284	311	354	397	439	480	520
.38	149	176	202	227	251	273	298	339	381	421	460	498
.37	143	169	194	218	241	263	285	326	367	405	442	479
.36	138	163	186	209	231	252	273	313	352	389	425	460
.34	126	149	171	192	212	231	251	287	323	356	389	421
.31	107	127	145	163	180	196	213	244	274	302	330	358

Table 7-13 — WOOD SCREWS — Allowable Lateral Loads — Normal Duration

Allowable lateral loads (shear) in pounds for screws embedded to approximately 7 times the shank diameter into the member holding the point. For less penetration, reduce loads in proportion. Penetration should not be less than 4 times the shank diameter.

SIZE OF SCREW

(For species in each group, see Table 7-9)		6	7	8	9	10	12	14	16	18	20	24
	g =											
	D =	0.138	0.151	0.164	0.177	0.190	0.216	0.242	0.268	0.294	0.320	0.372
	7D =	.966	1.057	1.148	1.239	1.330	1.512	1.694	1.876	2.058	2.240	2.604
	4D =	.552	.604	.656	.708	.760	.864	.968	1.072	1.176	1.280	1.488
Group I	=	91	109	129	150	173	224	281	345	415	492	664
Group II	=	75	90	106	124	143	185	232	284	342	406	548
Group III	=	62	74	87	101	117	151	190	233	280	332	448
Group IV	=	48	58	68	79	91	118	148	181	218	258	349

Table 7-14 — WOOD SCREWS — Allowable Withdrawal Loads — Normal Duration

Allowable load in withdrawal in pounds per inch of penetration of threaded part into side grain of member holding point.

g = gauge of screw. D = shank diameter in inches. G = specific gravity of the wood based on weight and volume when oven-dry

Specific gravity G		SIZE										
	g =	6	7	8	9	10	12	14	16	18	20	24
	D =	0.138	0.151	0.164	0.177	0.190	0.216	0.242	0.268	0.294	0.320	0.372
.75		222	242	263	284	306	347	388	430	471	514	597
.68		182	199	216	233	251	285	319	353	387	422	490
.67		177	193	210	227	243	276	310	343	376	410	476
.66		171	188	204	220	236	268	300	333	365	397	461
.62		151	166	180	194	208	237	265	294	322	351	407
.55		118	130	141	152	164	186	208	231	253	275	320
.54		114	126	136	147	158	180	201	224	245	266	310
.51		102	112	121	131	141	160	179	199	218	237	276
.48		91	99	108	116	125	142	159	176	193	210	244
.47		87	95	103	111	120	136	152	169	185	201	234
.46		83	91	99	107	115	130	146	162	177	193	224
.45		80	87	95	102	110	125	140	155	170	185	215
.44		76	83	91	97	105	119	133	148	162	177	205
.43		73	80	86	93	100	114	127	141	155	169	196
.42		69	76	82	89	95	109	121	135	148	161	187
.41		66	72	79	85	91	103	116	128	141	153	178
.40		65	71	77	83	89	102	113	126	138	151	175
.39		61	66	72	78	84	95	106	118	130	141	164
.38		57	62	67	73	78	89	99	110	121	132	153
.37		54	59	64	69	74	84	94	104	115	125	145
.36		51	56	60	65	70	80	89	99	109	118	137
.34		46	50	54	58	63	71	80	88	97	105	123
.31		38	42	45	48	53	59	67	73	81	88	103

Approximately two-thirds of the length of a standard wood screw is threaded.

BIBLIOGRAPHY

American Society for Testing and Materials
  Tentative methods of testing metal fasteners in wood. ASTM D 1761. (See current edition.) Philadelphia, Pa.

Anderson, L. O.
  1959. Nailing better wood boxes and crates. U.S. Dep. Agr., Handb. 160. 40 pp.

———
  1970. Wood-frame house construction. U.S. Dep. Agr., Agr. Handb. 73, rev. 223 pp.

Cramer, C. O.
  1968. Load distribution in multiple-bolt tension joints. Jour. Struct. Div., Amer. Soc. Civil Eng. 94(ST5): 1101–1117. (Proc. Pap. 5939.)

Doyle, D. V., and Scholten, J. A.
  1963. Performance of bolted joints in Douglas-fir. U.S. Forest Serv. Res. Pap. FPL 2. Forest Prod. Lab., Madison, Wis.

Fairchild, I. J.
  1926. Holding power of wood screws. U.S. Nat. Bur. Stand. Technol. Pap. 319.

Giese, H., and Henderson, S. M.
  1947. Effectiveness of roofing nails for application of metal building sheets. Iowa Agr. Exp. Sta. Res. Bull. 355: 525–592.

———, Body, L. L., and Dale, A. C.
  1950. Effect of moisture content of wood on withdrawal resistance of roofing nails. Agr. Eng. 31(4): 178–181, 183.

Goodell, H. R., and Philipps, R. S.
  1944. Bolt-bearing strength of wood and modified wood: Effect of different methods of drilling bolt holes in wood and plywood. Forest Prod. Lab. Rep. 1523.

Heebink, T. B.
  1962. Performance comparison of slender and standard spirally grooved pallet nails. Forest Prod. Lab. Rep. 2238.

Jordan, C. A.
  1963. Response of timber joints with metal fasteners to lateral impact loads. Forest Prod. Lab. Rep. 2263.

Kuenzi, E. W.
  1955. Theoretical design of a nailed or bolted joint under lateral load. Forest Prod. Lab. Rep. 1951.

Kurtenacker, R.S.
  1965. Performance of container fasteners subjected to static and dynamic withdrawal. U.S. Forest Serv. Res. Pap. FPL 29. Forest Prod. Lab., Madison, Wis.

Markwardt, L. J.
  1952. How surface condition of nails affects their holding power in wood. Forest Prod. Lab. Rep. D1927.

———, and Gahagan, J. M.
  1929. The grooved nail. Packing and Shipping 56(1): 12–14.

———, and Gahagan, J. M.
  1930. Effect of nail points on resistance to withdrawal. Forest Prod. Lab. Rep. 1226.

———, and Gahagan, J. M.
  1931. Mechanism of nail holding. Barrel and Box and Packages 36(8): 26–27.

———, and Gahagan, J. M.
  1952. Slant driving of nails. Does it pay? Packing and shipping 56(10): 7–9, 23, 25.

Martin, T. J., and Van Kleeck, A.
  1941. Fastening. U.S. Patent No. 2,268,323. U.S. Pat. Office, Offic. Gaz. 533: 1226.

National Forest Products Association
  National design specification for stress-grade lumber and its fastenings. (See current edition.) Washington, D.C.

Newlin, J. A., and Gahagan, J. M.
  1938. Lag screw joints: Their behavior and design. U.S. Dept. Agr. Tech. Bull. 597.

Perkins, N. S., Landsem, P., and Trayer, G. W.
  1933. Modern connectors for timber construction. U.S. Dep. Com., Nat. Com. Wood Util., and U.S. Dep. Agr, Forest Serv.

Scholten, J. A.
  1938. Modern connectors in wood construction. Agr. Eng. 19(5): 201–203.

———
  1940. Connector joints in wood construction. Railway Purchases and Stores 33(9): 431–435.

———
  1944. Timber-connector joints, their strength and designs. U.S. Dep. Agr. Tech. Bull. 865.

———
  1946. Strength of bolted timber joints. Forest Prod. Lab. Rep. R1202.

———
  1950. Nail-holding properties of southern hardwoods. Southern Lumberman 181(2273): 208–210.

———
  1965. Strength of wood joints made with nails, staples, and screws. U.S. Forest Serv. Res. Note FPL–0100. Forest Prod. Lab., Madison, Wis.

———, and Molander, E. G.
  1950. Strength of nailed joints in frame walls. Agr. Eng. 31(11): 551–555.

Stern, E. G.
  1940. A study of lumber and plywood joints with metal split-ring connectors. Pennsylvania Eng. Exp. Sta. Bull. 53.

———
  1950. Improved nails for building construction. Virginia Polytech. Inst. Eng. Exp. Sta. Bull. 76.

———
  1950. Nails in end-grain lumber. Timber News and Machine Woodworker 58(2138): 490–492.

Trayer, G. W.
  1932. Bearing strength of wood under bolts. U.S. Dep. Agr. Tech. Bull. 332.

U.S. Forest Products Laboratory.
  1962. General observations on the nailing of wood. Forest Prod. Lab. Tech. Note 243.

———
  1964. Nailing dense hardwoods. U.S. Forest Serv. Res. Note FPL–037. Madison, Wis.

———
  1965. Nail withdrawal resistance of American woods. U.S. Forest Serv. Res. Note FPL–093. Madison, Wis.

Wilkinson, T. L.
  1971. Theoretical lateral resistance of nailed joints. Jour. Struct. Div., Proc. Amer. Soc. Civil Eng., ST5(97): (Pap. 8121) 1381–1398.

———
  1971. Bearing strength of wood under embedment loading of fasteners. USDA Forest Serv. Res. Pap. FPL 163. Forest Prod. Lab., Madison, Wis.

———, and Laatsch, T. R.
  1970. Lateral and withdrawal resistance of tapping screws in three densities of wood. Forest Prod. J. 20(7): 34–41.

Chapter 8
WOOD STRUCTURAL MEMBERS

This chapter deals with fundamental considerations related to the simpler structural members, such as beams and columns. Members of several parts assembled with mechanical fastenings are included but details regarding glued structural members are discussed in chapter 10.

BEAMS

Wood beams are usually of rectangular cross section and of constant depth and width throughout their span. They should be large enough so that their deflection under load will not exceed the limit fixed by the intended use of the structure. Beams should also be large enough so that the following stresses do not exceed allowable values: Flexural stresses of compression on the concave portion and tension on the convex portion caused by bending moment, shear stresses caused by shear load, and compression across the grain at the end bearings. The actual, not nominal, sizes of the lumber must be used when computing the deflections and stresses.

Formulas for determining deflections and stresses in beams of sawn timber, of glued or mechanically fastened laminations at right angles to the neutral plane, or of glued laminations parallel to the neutral plane are given in this chapter. The property values, modulus of elasticity, and stresses for various wood species are given in chapter 4. Values of coefficients of variation are included in chapter 4 to indicate the variability and reliability of the data as an aid in selection of allowable property values.

Beam Deflections

The deflection of beams is often limited to $\frac{1}{360}$ of the span of framing over plastered ceilings and $\frac{1}{240}$ of the span over unplastered ceilings. Deflection of highway bridge stringers is often limited to $\frac{1}{200}$ of the span and stringers for railroad bridges and trestles to $\frac{1}{300}$ of the span.

Straight Beam Deflection

The deflection of straight beams (or long, slightly curved beams with the radius of curvature in the plane of bending), elastically stressed, and having a constant cross section throughout their length is given by the formula:

$$y = \frac{k_b W L^3}{EI} + \frac{k_s W L}{GA'} \qquad (8\text{--}1)$$

where y is deflection, W is total beam load acting perpendicular to beam neutral axis, L is beam span, k_b and k_s are constants dependent upon beam loading and location of point whose deflection is to be calculated, I is beam moment of inertia, A' is a modified beam area, E is beam modulus of elasticity (for beams having straight grain parallel to their axis $E = E_L$), and G is beam shear modulus (for beams with flat-grained vertical faces $G = G_{LT}$ and for beams with edge-grained vertical faces $G = G_{LR}$). Elastic property values are given in chapter 4.

The first term on the right side of formula (8–1) gives the bending deflection and the

Table 8–1.—*Values of k_b and k_s for several beam loadings*

Loading	Beam ends	Deflection at—	k_b	k_s
Uniformly distributed	Both simply supported	Midspan	$\frac{5}{384}$	$\frac{1}{8}$
	Both clamped	do	$\frac{1}{384}$	$\frac{1}{8}$
Concentrated at midspan	Both simply supported	do	$\frac{1}{48}$	$\frac{1}{4}$
	Both clamped	do	$\frac{1}{192}$	$\frac{1}{4}$
Concentrated at outer quarter span points	Both simply supported	do	$\frac{11}{768}$	$\frac{1}{8}$
	do	Load point	$\frac{1}{96}$	$\frac{1}{8}$
Uniformly distributed	Cantilever, 1 free, 1 clamped	Free end	$\frac{1}{8}$	$\frac{1}{2}$
Concentrated at free end	Cantilever, 1 free, 1 clamped	do	$\frac{1}{3}$	1

second term the shear deflection. Values of k_b and k_s for several beam loadings are given in table 8–1.

The moment of inertia, I, of the beams is given by the formulas:

$$I = \frac{bh^3}{12} \text{ for beam of rectangular cross section}$$

$$(8–2)$$

$$I = \frac{\pi d^4}{64} \text{ for beam of circular cross section}$$

where b is beam width, h is beam depth, and d is beam diameter. The modified area A' is given by the formulas:

$$A' = \frac{5}{6}bh \text{ for beam of rectangular cross section}$$

$$(8–3)$$

$$A' = \frac{9}{40}\pi d^2 \text{ for beam of circular cross section}$$

Tapered Beam Deflection

The bending deflection of beams tapering in depth but of constant width throughout their length can be determined from the nondimensional ordinates of the graphs given in figures 8–1 and 8–2. The graph axes are chosen so that the ordinate contains data usually known in a design such as elastic properties, deflection, span, and difference in beam height $(h_c - h_o)$, as required by roof slope or architectural effect. Then the value of the abscissa γ can be determined as shown by the example line on the graph to thus eventually compute the smallest beam depth, h_o. The graphs can also be used to determine maximum bending deflections if the abscissa is known. Tapered beams deflect due to shear distortion in addition to bending deflections and this shear deflection Δ_s can be closely approximated by the formulas:

$$\Delta_s = \frac{3WL}{20Gbh_o} \text{ for load uniformly distributed}$$

$$(8–4)$$

$$\Delta_s = \frac{3PL}{10Gbh_o} \text{ for midspan concentrated load}$$

The final beam design should include shear as well as bending deflection and it may be necessary to iterate to arrive at final beam dimensions.

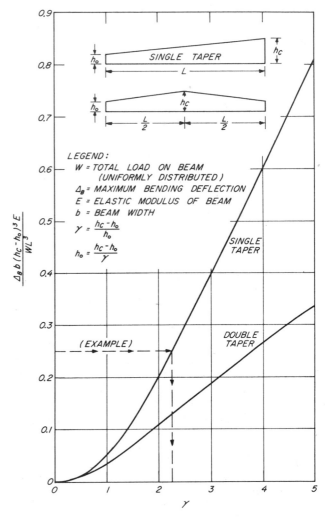

Figure 8–1.—Graph for determining tapered beam size based on deflection under uniformly distributed load.

Effects of Notches and Holes

The deflection of beams is increased if reductions in cross section dimensions occur such as are caused by holes or notches. The deflection of such beams can be determined by considering them of variable cross section along their length and appropriately solving the general differential equations of the elastic curves, $EI \dfrac{d^2y}{dx^2} = M$, to obtain deflection expressions or by the often simpler theorem of least work implicity in Castigliano's theorem. (These procedures are given in most texts on strength of materials.) If notches occur, their effective dimension along the beam axis for use in determining beam deflection is approximately equal to the length of the notch plus twice the depth of the notch.

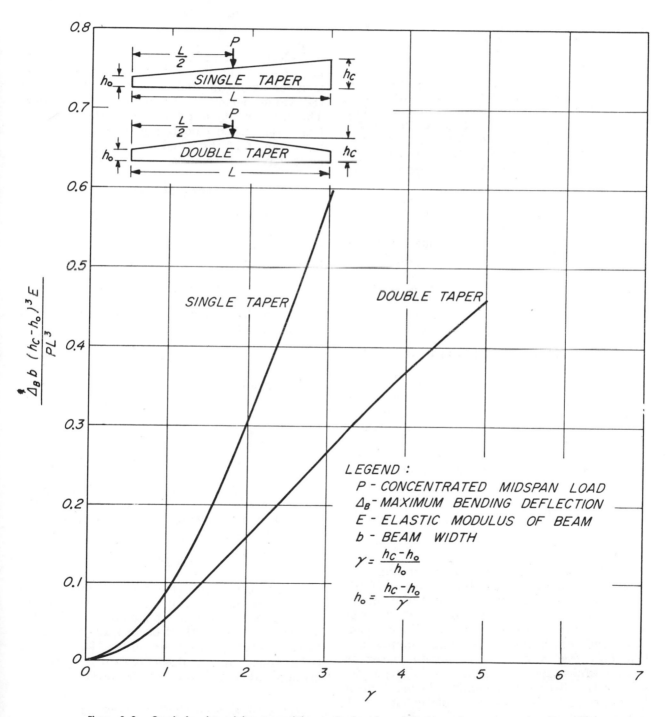

Figure 8–2.—Graph for determining tapered beam size based on deflection under concentrated midspan load.

Effect of Ponding Water

Ponding of water on roofs already deflected under normal loads can cause large increases in deflection. The deflection of a simply supported, uniformly loaded beam under ponding water can be estimated closely by multiplying deflection under design loading without ponding, by the following magnification factor:

$$\frac{1}{1 - \dfrac{W'L^3}{\pi^4 EI}} \qquad (8\text{-}5)$$

where W' is total load of 1 inch depth of water on the roof area supported by the beam; L is beam span; E is beam modulus of elasticity; and I is beam moment of inertia. As the second term in the denominator of formula (8–5) becomes unity, the magnification becomes infinite, thus denoting complete collapse of the beam.

The deflections of a beam with fixed ends under concentrated midspan load P plus ponding water can be estimated closely by multiplying deflection under design load without ponding, by the following magnification factor:

$$\frac{1}{1 - \dfrac{3W'L^3}{16\pi^4 EI}} \qquad (8\text{-}6)$$

Effect of End Loading

Addition of end loading (in a direction parallel to the beam length) to a beam under loads acting perpendicular to the beam neutral axis causes increase in deflection for added end compression and decrease in deflection for added end tension. The deflection under combined loading can be estimated closely by the formula:

$$\Delta = \frac{\Delta_0}{1 \pm \dfrac{P}{P_{cr}}} \qquad (8\text{-}7)$$

where the plus sign is chosen if the end load is tension and the minus sign is chosen if the end load is compression; Δ is deflection under combined loading; Δ_0 is beam deflection without end load; P is end load; and P_{cr} is buckling load of beam under end compressive load only (see section on Columns), based on beam stiffness about the neutral axis perpendicular to the direction of bending loads. P must be less than P_{cr} in order to avoid collapse if P is compression.

Deflection With Time

In addition to the elastic deflections previously discussed, wood beams usually sag in time; that is, the deflection increases beyond what it was immediately after the load was first applied. Green timbers, especially, will sag if allowed to dry under load, although partially dried material will also sag to some extent. In thoroughly dried beams, there are small changes in deflection with changes in moisture content but little permanent increase in deflection. If deflection under longtime load is to be limited, it has been customary to design for an initial deflection of about one-half the value permitted for longtime deflection. This can be done by doubling the longtime load value when calculating deflection, by using one-half of the usual value for modulus of elasticity or any equivalent method.

Beam Strength

The strength of beams is determined by flexural stresses caused by bending moment, shear stresses caused by shear load, and compression across the grain at the end bearings.

Bending Moment

The bending moment capacity of a beam is given by the formula:

$$M = fS \qquad (8\text{-}8)$$

where M is bending moment, S is beam section modulus (for a beam of rectangular cross section $S = \dfrac{bh^2}{6}$ and for a circular cross section $S = \dfrac{\pi d^3}{32}$), and f is stress. The significance of M in denoting allowable moment or maximum moment is dependent upon choice of the stress f at an allowable value or a modulus of rupture value.

Size Effect

It has been found that the modulus of rupture (maximum bending stress) of wood beams depends on beam size and method of loading and that the strength of clear, straight-grained beams decreases as size increases.[1] These effects were found to be describable by statistical strength theory involving "weakest link" hy-

---

[1] For further information see report by Bohannan in Bibliography.

potheses and can be summarized as: For two beams under two equal concentrated loads applied symmetrical to the midspan points, the ratio of the modulus of rupture of beam 1 to the modulus of rupture of beam 2 is given by the formula:

$$\frac{R_1}{R_2} = \left[\frac{h_2 L_2 \left(1 + \frac{ma_2}{L_2} \right)}{h_1 L_1 \left(1 + \frac{ma_1}{L_1} \right)} \right]^{1/m} \quad (8-9)$$

where subscripts 1 and 2 refer to beam 1 and beam 2; R is modulus of rupture; h is beam depth; L is beam span; a is distance between loads placed $\frac{a}{2}$ each side of midspan; and m is a constant. For clear, straight-grained Douglas-fir beams $m = 18$. If formula (8-9) is used for beam 2 of standard size (see ch. 4), loaded at midspan then $h_2 = 2$ inches, $L_2 = 28$ inches, and $a_2 = 0$ and formula (8-9) becomes

$$\frac{R_1}{R_2} = \left[\frac{50}{h_1 L_1 \left(1 + \frac{ma_1}{mL_1} \right)} \right]^{1/m} \quad (8-10)$$

Example: Determine modulus of rupture for a beam 10 inches deep, spanning 18 feet, and loaded at one-third span points compared with a beam 2 inches deep, spanning 28 inches, and loaded at midspan that had a modulus of rupture of 10,000 pounds per square inch. Assume $m = 18$. Substitution of the dimensions into formula (8-10) produces:

$$R_1 = 10,000 \left[\frac{56}{2,160(1 + 6)} \right]^{1/18}$$

$= 7,340$ pounds per square inch

Extrapolation of the theory of reference to beams under uniformly distributed load resulted in the following relationship between modulus of rupture of beams under uniformly distributed load and modulus of rupture of beams under concentrated loads:

$$\frac{R_u}{R_c} = \left[\frac{\left(1 + 18 \frac{a_c}{L_c} \right) h_c L_c}{3.876 h_u L_u} \right]^{1/18} \quad (8-11)$$

where subscripts u and c refer to beams under uniformly distributed and concentrated loads, respectively, and other symbols are as previously defined.

Effects of Notches and Holes

In beams having changes in cross sectional dimensions because of holes or notches, bending stresses can be calculated at the hole or notch by dividing the bending moment there by the section modulus of the material remaining. Values of this bending stress are not useful in design because the change in cross section also causes sheer stresses and stresses perpendicular to the beam neutral axis; the combination of these stresses with the bending stress can cause failure at low load. It is not known how to compute these stresses and therefore it would be wise to avoid notches in beams.

Effect of Ponding Water

Ponding of water on roofs can cause increases in bending stresses that can be computed by using the same magnification factors, formulas (8-5 and 8-6), as determined for deflection.

Compressive End Loading

Addition of compressive end loading to a beam under loads acting perpendicular to the beam's neutral axis increases compressive stress and decreases tensile stress; the stress due to combined loading can be estimated closely by the formulas:

Compressive stress, $f_c = \dfrac{f_o}{1 - \dfrac{P}{P_{cr}}} + \dfrac{P}{A}$ (8-12)

Tensile stress, $f_t = \dfrac{f_o}{1 - \dfrac{P}{P_{cr}}} - \dfrac{P}{A}$ (8-13)

where f_o is bending stress without end load, P is compressive end load, A is area of beam cross section; and P_{cr} is buckling load of beam under end compressive load only, based on beam stiffness about the neutral axis perpendicular to the direction of bending loads. P must be less than P_{cr} in order to avoid collapse.

Tensile End Loading

Addition of tensile end loading to a beam under loads acting perpendicular to the beam's neutral axis increases tensile stress and decreases compressive stress; the stress due to combined loading can be estimated closely by the formulas:

Compressive stress, $f_c = \dfrac{f_o}{1 + \dfrac{P'}{P_{cr}}} - \dfrac{P'}{A}$ (8–14)

Tensile stress, $f_t = \dfrac{f_o}{1 + \dfrac{P'}{P_{cr}}} + \dfrac{A}{P'}$ (8–15)

where P' is the tensile end load and the other symbols are as previously defined.

Shear Capacity

The shear capacity of a beam is given by the formula:

$$V = f_s A \qquad (8\text{–}16)$$

where V is shear load perpendicular to beam neutral axis, A is effective beam shear area (for a beam of rectangular cross section $A = \dfrac{2bh}{3}$ and for a beam of circular cross section $A = 3\pi\dfrac{d^2}{16}$), and f_s is shear stress at the neutral axis. The significance of V in denoting allowable shear or maximum shear depends on choice of the stress f_s at an allowable value or a strength value.

Increases in shear stresses caused by ponding of water on roofs can be computed by using the same magnification factors, formulas (8–5) and (8–6), as determined for deflection.

For beams that taper in depth but are of constant width, the shear stress can be a maximum at the tapered edge as well as at the neutral axis of the beam. The shear stress dis-

tribution for a beam of rectangular cross section, of constant width, tapering in depth linearly with spanwise distance, and loaded with concentrated loads to produce a reaction V, is shown in figure 8–3. For other loadings, the basic theory derived by Maki and Kuenzi can be used to determine shear stress distributions. The shear stress at the tapered edge can reach a maximum value as great as that at the neutral axis at a reaction. For the beam shown in figure 8–3, this maximum stress occurs at the cross section which is double the depth of the beam at the reaction. For other loadings, the location of the cross section with maximum shear stress at the tapered edge will be somewhat different.

For the beam shown in figure 8–3, the bending stress is also a maximum at the same cross section where the shear stress is maximum at the tapered edge. This stress situation also causes a stress in the direction perpendicular to the neutral axis that is maximum at the tapered edge. The effect of combined stresses at a point can be approximately accounted for by an interaction formula based on the Henky-von Mises theory of energy due to the change of shape. This theory applied to wood by Norris results in the formula:

$$\frac{f_x^2}{F_x^2} + \frac{f_{xy}^2}{F_{xy}^2} + \frac{f_y^2}{F_y^2} = 1 \qquad (8\text{–}17)$$

where f_x is the bending stress, f_y is the stress perpendicular to the neutral axis, and f_{xy} is the shear stress. Values of F_x, F_y, and F_{xy} are cor-

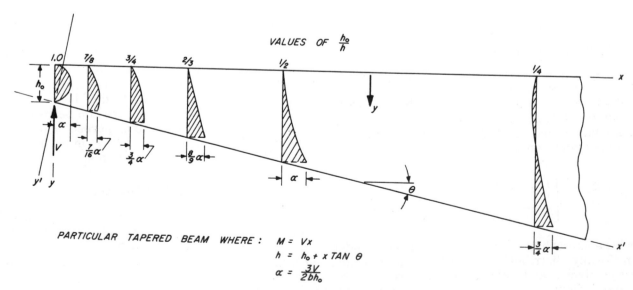

PARTICULAR TAPERED BEAM WHERE: $M = Vx$
$h = h_o + x\,TAN\,\theta$
$\alpha = \dfrac{3V}{2bh_o}$

Figure 8–3.—Shear stress distributions for a tapered beam.

responding stresses chosen at design allowable values or maximum values in accordance with allowable or maximum values being determined for the tapered beam. Maximum stresses in the beam shown in figure 8–3 are given by the formulas:

$$f_x = \frac{3M}{2bh_o{}^2} \qquad (8\text{--}18)$$

$$f_{xy} = f_x \tan \theta \qquad (8\text{--}19)$$

$$f_y = f_x \tan^2 \theta \qquad (8\text{--}20)$$

Substitution of these formulas into the interaction formula (8–17) will result in an expression for the moment capacity M of the beam. If the taper is on the beam tension edge, the values of f_x and f_y are tensile stresses.

Example: Determine the moment capacity of a tapered beam of width $b = 5$ inches, depth $h_o = 10$ inches, and taper $\tan \theta = \frac{1}{10}$. Substitution of these dimensions into formulas (8–18), (8–19), and (8–20) results in:

$$f_x = \frac{3M}{1,000}$$

$$f_{xy} = \frac{3M}{10,000}$$

$$f_y = \frac{3M}{100,000}$$

and substitution of these expressions into formula (8–17) and solving for M results finally in:

$$M = \frac{10^5}{3\left[\dfrac{10^4}{F_x{}^2} + \dfrac{10^2}{F_{xy}{}^2} + \dfrac{1}{F_y{}^2}\right]^{1/2}}$$

where appropriate allowable or maximum values of the F stresses are chosen.

End Bearing Area

The end bearing area of all beams as well as bearing areas of concentrated loads along the beam should be large enough to prevent stresses perpendicular to the grain in the beam from reaching chosen allowable or maximum values.

Lateral Buckling

The lateral buckling of beams can occur at low bending stresses and result in beam collapse if the flexural rigidity of the beam in the plane of bending is very large in comparison with its lateral rigidity and the beam is fairly long. Thus beam strength may be governed by lateral buckling rather than material strength per se. Lateral buckling depends on beam torsional rigidity as well as lateral rigidity, beam length, and manner of loading. The critical bending moment or load for beams of rectangular cross section have been derived for

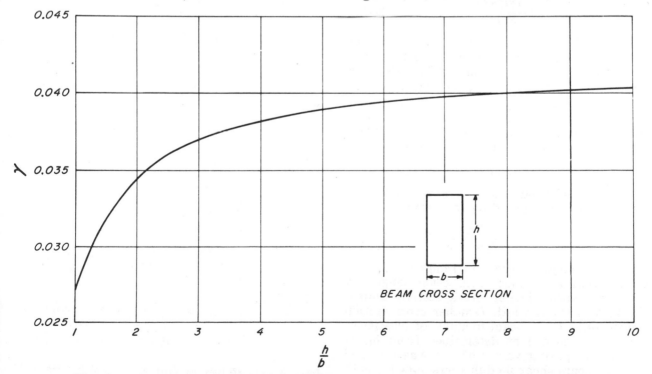

Figure 8–4.—Constant γ for determining lateral buckling and twist of wood beams of rectangular cross section.

189

several loadings by Trayer and March and Zahn. The results are summarized in the following formula:

$$P_{cr} \text{ or } W_{cr} \text{ or } \frac{M_{cr}}{L} = \beta\gamma E_L \frac{(hb^3)}{L^2} \quad (8\text{–}21)$$

where P_{cr} is total concentrated load; W_{cr} is total uniformly distributed load; M_{cr} is bending moment; E_L is beam modulus of elasticity; h is beam depth; b is beam width; L is beam span length; γ is beam torsion rigidity coefficient determined from the curve of figure 8–4 for a beam of rectangular cross section (assuming beam shear modulus is one-sixteenth of beam modulus of elasticity); and β is lateral buckling coefficient given in the following:

For a beam simply supported at the ends and under constant bending moment, $\beta = \pi$.

For a beam simply supported at the ends and under concentrated midspan load, $\beta = 16.9$.

For a beam simply supported at the ends and under uniformly distributed load, $\beta = 28.3$.

For a cantilever beam under concentrated end load, $\beta = 4$.

For a cantilever beam under uniformly distributed load, $\beta = 12.9$.

For beams laterally supported by sheathing, values of the coefficient β also depend on the parameter:

$$\frac{12CSL^2}{E_L hb^3} \quad (8\text{–}22)$$

where S is beam spacing and C is effective inplane shearing rigidity of the sheathing and its connections to the beams. This rigidity is determined as the ratio of a shearing force per unit length of sheathing edge to angular shearing distortion of the sheathing-beam diaphragm. Curves giving values of β for several beam loadings are given in figure 8–5.

Twist

The twist of wood beams of rectangular cross section can be computed by the formula:

$$\theta = \frac{TL}{12\gamma^2 hb^3 E_L} \quad (8\text{–}23)$$

where θ is angle of twist in radians; T is applied twisting torque; L is beam length; h is beam depth (larger cross section dimension); b is beam width (smaller cross section dimension); E_L is beam modulus of elasticity; and γ is a coefficient determined from figure 8–4. In computing values of γ it was assumed that the beam shear modulus was one-sixteenth of the beam modulus of elasticity.

COLUMNS

Columns should be designed so that stresses do not exceed allowable values and that buckling does not occur. This section contains formulas and graphs for determining buckling loads and stresses in wood columns of sawn timber, of glued or mechanically fastened laminations at right angles to the neutral plane having the smaller moment of inertia, or of glued laminations parallel to the neutral plane having the smaller moment of inertia. The property values for use in the formulas— modulus of elasticity and stresses for various wood species—are given in chapter 4. Values of coefficients of variation are included in chapter 4 to indicate the variability and reliability of the data as an aid in selecting allowable property values.

Concentrically Loaded Columns

A concentrically loaded column is stressed primarily in compression and, if this column is long, buckling can occur. The critical elas-

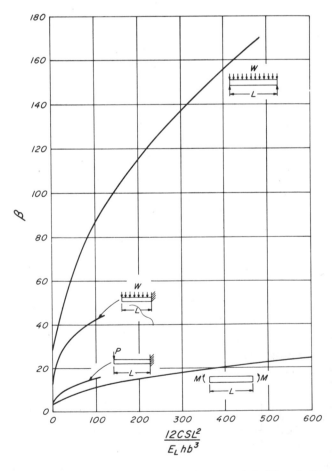

Figure 8–5.—Coefficients for determining lateral buckling loads of beams.

190

tic buckling stress is closely predictable by the Euler column formula for a hinge-ended column of uniform cross section throughout its length:

$$\left(f_{cr}\right)_A = \frac{\pi^2 E_L}{\left(\dfrac{L}{r}\right)^2} \qquad (8\text{--}24)$$

where E_L is the modulus of elasticity of the column, $\left(\dfrac{L}{r}\right)$ is the column slenderness ratio determined by the unsupported length L and the lesser radius of gyration (for a rectangular cross section with b as its smaller dimension $r = \dfrac{b}{2\sqrt{3}}$ and for a circular cross section $r = \dfrac{d}{4}$).

Short wood columns will not buckle elastically as predicted by formula (8–24) because they can be stressed beyond proportional limit values. Usually the short column range is explored empirically and appropriate formulas derived to extend for compressive strength values to the long column range. Material of this nature is presented in USDA Technical Bulletin 167. The final formula is a fourth-power parabolic function which can be written as:

$$\left(f_{cr}\right)_B = F_c \left\{ 1 - \frac{1}{3}\left[\frac{\left(\dfrac{L}{r}\right)}{\left(\dfrac{L}{r}\right)'} \right]^4 \right\} \qquad (8\text{--}25)$$

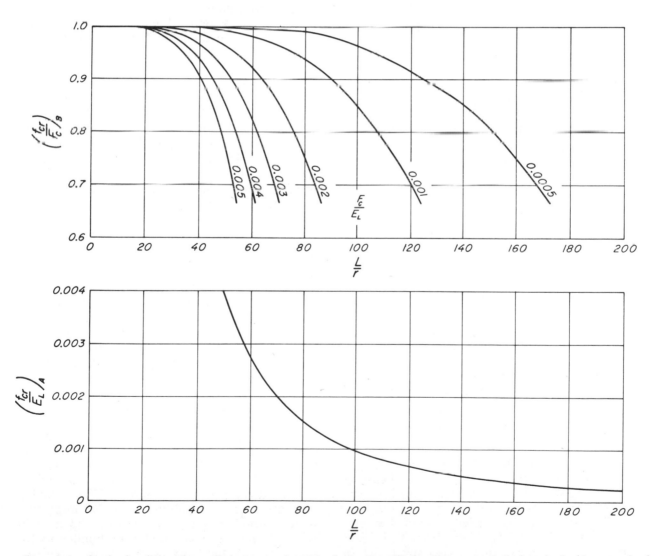

Figure 8–6.—Graphs for determining critical stresses of wood columns. Top, FPL fourth-power parabolic formula for short wood columns; bottom, hinged Euler column formula.

where F_c is compressive strength; $\left(\dfrac{L}{r}\right)$ is column slenderness ratio; and $\left(\dfrac{L}{r}\right)'$ is the slenderness ratio at which the Euler column buckling stress is two-thirds of F_c. A graphical presentation of formulas (8–24) and (8–25) is given in figure 8–6. These graphs may be used for any wood species by introducing the appropriate property values for modulus of elasticity and compressive stress. After entering the column slenderness ratio on the graphs of figure 8–6, the lesser values of f_{cr} as determined from the graphs is the critical stress. The curves in the top graph do not extend below $\left(\dfrac{f_{cr}}{F_c}\right)_B = 2/3$.

Example: Determine critical stresses for columns of a wood species for which $\dfrac{F_c}{E} = 0.002$.

For a column of slenderness ratio $\left(\dfrac{L}{r}\right) = 100$, the bottom graph gives a value of $\left(\dfrac{f_{cr}}{E}\right)_A = 0.001$ and the top graph has no solution at $\left(\dfrac{L}{r}\right) = 100$; therefore $f_{cr} = 0.001E$ for $\left(\dfrac{L}{r}\right) = 100$.

For a column of slenderness ratio $\dfrac{L}{r} = 60$, the bottom graph gives a value of $\left(\dfrac{f_{cr}}{E}\right)_A = 0.00276$ and top graph gives $\left(\dfrac{f_{cr}}{F_c}\right)_B = 0.92$ which, for $\dfrac{F_c}{E} = 0.002$, can be written as $\left(\dfrac{f_{cr}}{E}\right)_B = 0.00184$. Thus the lesser value of f_{cr} is given by the top graph as $f_{cr} = 0.00184E$.

Eccentrically Loaded Columns

An eccentrically loaded column is stressed in compression and bending. The eccentricity and end load determine the amount of bending stress induced and this must be added to the compressive stress caused by direct compression. The eccentrically loaded column does not buckle but continues to bend from the onset of loading. The maximum stress produced in a hinge-ended column is given by the formula:

$$f_{c\ max.} = \frac{P}{A}\left\{ 1 + \frac{ec}{r^2}\sec\left[\frac{P}{4EA}\left(\frac{L}{r}\right)^2\right]^{1/2} \right\}(8\text{–}26)$$

where P is end load; e is eccentricity (distance from neutral axis to point of application of load); c is distance from neutral axis to extreme fiber nearest the point of load application; r is radius of gyration of section about neutral axis; L is column length; A is area of cross section; and E is modulus of elasticity. For a column of circular cross section, $c = \dfrac{d}{2}$ and $r = \dfrac{d}{4}$ and for a rectangular cross section $c = \dfrac{b}{2}$ and $r = \dfrac{b}{2\sqrt{3}}$ or $c = \dfrac{h}{2}$ and $r = \dfrac{h}{2\sqrt{3}}$ for eccentricities about axes parallel to side h and b, respectively.

Stresses and deflections of columns with side loads are included in the previous section as beams with end loads.

Columns With Flanges

Columns with thin, outstanding flanges can fail by elastic instability of the outstanding flange, causing wrinkling of the flange and twisting of the column at stresses less than those for general column instability as given by formulas (8–24) and (8–25). For outstanding flanges of sections such as I, H, +, and L the flange instability stress can be estimated by the formula:

$$f_{cr} = 0.044E\,\frac{t^2}{b^2} \qquad (8\text{–}27)$$

where E is the column modulus of elasticity; t is the thickness of the outstanding flange; and b is the width of the outstanding flange. If the joints between the column members are glued and reinforced with glued fillets, the instability stress increases to as much as 1.6 times that given by formula (8–27).

Built-Up Columns

Build-up columns of nearly square cross section will not support as high loads if the lumber is nailed together as if it were glued together. The reason is that shear distortions can occur in the nailed joints. The shearing resistance of the column can be improved, so that previously presented formulas may be used, by nailing cover plates of lumber to the edges of the built-up layers. If the built-up column is of several spaced pieces, the spacer blocks should be placed close enough together, lengthwise in the column, so the unsupported portion of the spaced member will not buckle at the same or lower stress than that of the complete member. "Spaced columns" are designed with previously presented column formulas, considering each compression member as an unsupported simple column; the sum of column loads for all the members is taken as the column load for the spaced column. Sufficient net area should be provided in short columns so that compression failure does not occur. The net area is, of course, that area remaining after subtracting portions for connectors or bolts used to fasten the members together at the spacer blocks.

BIBLIOGRAPHY

Bohannan, Billy
    1966. Effect of size on bending strength of wood members. U.S. Forest Serv. Res. Pap. FPL 56. Forest Prod. Lab., Madison, Wis.
Kuenzi, E. W., and Bohannan, Billy
    1964. Increases in deflection and stresses caused by ponding of water on roofs. Forest Prod. J. 14(9): 421–424.

Maki, A. C., and Kuenzi, E. W.
    1965. Deflection and stresses of tapered wood beams. U.S. Forest Serv. Res. Pap. FPL 34. Forest Prod. Lab., Madison, Wis.

Newlin, J. A., and Gahagan, J. M.
    1930. Tests of large timber columns and presentation of the Forest Products Laboratory column formula. U.S. Dep. Agr. Tech. Bull. 167.

———, and Trayer, G. W.
    1924. Deflection of beams with special reference to shear deformations. U.S. Nat. Adv. Comm. Aeron. Rep. 180.

Norris, C. B.
    1950. Strength of orthotropic materials subjected to combined stresses. Forest Prod. Lab. Rep. 1816.

Trayer, G. W.
    1930. The torsion of members having sections common in aircraft construction. U.S. Nat. Adv. Comm. Aeron. Rep. 180.

———, and March, H. W.
    1931. Elastic instability of members having sections common in aircraft construction. U.S. Nat. Adv. Comm. Aeron. Rep. 382.

Zahn, J. J.
    1965. Lateral stability of deep beams with shear-beam support. U.S. Forest Serv. Res. Pap. FPL 43. Forest Prod. Lab., Madison, Wis.

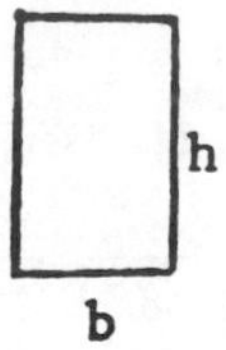

Table 8-2

PROPERTIES OF SECTIONS
BOARDS, DIMENSION, AND TIMBERS

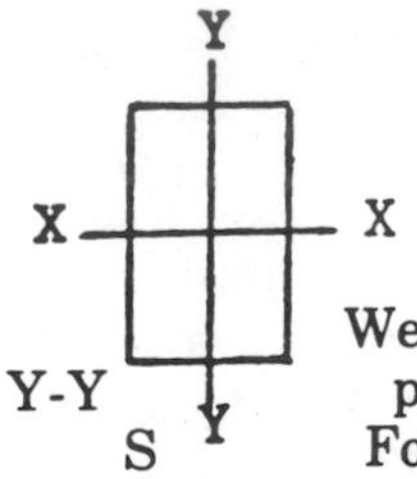

Nominal Size	Actual Size inches b x h	Area sq. in.	Axis X-X		Axis Y-Y		Weight per Foot* lbs.
			I in.^{4}	S in.^{3}	I in.^{4}	S in.^{3}	
1 x 3	¾ x 2½	1.88	0.98	0.78	0.09	0.23	0.47
1 x 4	¾ x 3½	2.63	2.68	1.53	0.12	0.33	0.64
1 x 6	¾ x 5½	4.13	10.40	3.78	0.19	0.52	1.00
1 x 8	¾ x 7¼	5.44	23.82	6.57	0.26	0.768	1.32
1 x 10	¾ x 9¼	6.94	49.47	10.70	0.33	0.87	1.69
1 x 12	¾ x 11¼	8.44	88.99	15.82	0.40	1.06	2.05
2 x 3	1½ x 2½	3.75	3.75	1.95	1.56	0.70	0.94
0.91							
2 x 4	1½ x 3½	5.25	5.36	3.06	0.98	1.31	1.28
2 x 6	1½ x 5½	8.25	20.80	7.56	1.55	2.06	2.00
2 x 8	1½ x 7¼	10.88	47.64	13.14	2.04	2.72	2.64
2 x 10	1½ x 9¼	13.88	98.93	21.39	2.60	3.47	3.37
2 x 12	1½ x 11¼	16.88	177.98	31.64	3.16	4.22	4.10
2 x 14	1½ x 13¼	19.88	290.78	43.89	3.73	4.97	4.83
3 x 3	2½ x 21	6.25	3.25	2.60	3.25	2.60	1.52
3 x 4	2½ x 3½	8.75	8.93	5.10	4.56	3.65	2.13
3 x 6	2½ x 5½	13.75	34.66	12.60	7.16	5.73	3.34
3 x 8	2½ x 7¼	18.13	79.39	21.90	9.44	7.55	4.41
3 x 10	2½ x 9¼	23.13	164.89	35.65	12.04	9.64	5.62
3 x 12	2½ x 11¼	28.13	296.63	52.73	14.65	11.72	6.84
3 x 14	2½ x 13¼	33.13	484.63	73.15	17.25	13.80	8.05
3 x 16	2½ x 15¼	38.13	738.87	96.90	20.18	16.15	9.27
4 x 4	3½ x 3½	12.25	12.51	7.15	12.51	7.15	2.98
4 x 6	3½ x 5½	19.25	48.53	17.65	19.65	11.23	4.68
4 x 8	3½ x 7¼	25.38	111.15	30.66	25.90	14.80	6.17
4 x 10	3½ x 9¼	32.38	230.84	49.91	33.05	18.89	7.87
4 x 12	3½ x 11¼	39.38	415.28	73.83	40.20	22.97	9.57
4 x 14	3½ x 13¼	46.38	678.46	102.41	47.44	27.11	11.28
4 x 16	3½ x 15¼	53.38	1034.40	135.66	54.60	31.20	12.98
6 x 6	5½ x 5½	30.25	76.26	27.73	76.26	27.73	7.35
6 x 8	5½ x 7½	41.25	193.36	51.56	103.98	37.81	10.03
6 x 10	5½ x 9½	52.25	392.96	82.73	131.71	47.90	12.70
6 x 12	5½ x 11½	63.25	697.07	121.23	159.44	57.98	15.37
6 x 14	5½ x 13½	74.25	1127.67	167.06	187.17	68.06	18.05
6 x 16	5½ x 15½	85.25	1706.78	220.23	214.90	78.15	20.72
6 x 18	5½ x 17½	96.25	2456.38	280.73	242.63	88.23	23.29
8 x 8	7½ x 7½	56.25	263.67	70.31	263.67	70.31	13.67
8 x 10	7½ x 9½	71.25	535.86	112.81	334.10	89.06	17.32
8 x 12	7½ x 11½	86.25	950.55	165.31	404.30	107.81	20.96
8 x 14	7½ x 13½	101.25	1537.73	227.81	474.61	126.56	24.61
8 x 16	7½ x 15½	116.25	2327.42	300.31	544.92	145.31	28.26
8 x 18	7½ x 17½	131.25	3349.61	382.81	615.23	164.06	31.90
10 x 10	9½ x 9½	90.25	678.76	142.90	678.76	142.90	21.94
10 x 12	9½ x 11½	109.25	1204.03	209.40	821.65	172.98	26.55

*Based on 35 lbs./cu. ft. actual volume and weight.

Table 8-2 (Continued)

Nominal Size	Actual Size inches b x h	Area sq. in.	Axis X-X I in.⁴	Axis X-X S in.³	Axis Y-Y I in.⁴	Axis Y-Y S in.³	Weight per Foot* lbs.
			Axis X-X		Axis Y-Y		
			I in.^{4}	S in.^{3}	I in.^{4}	S in.^{3}	
10 x 14	9½ x 13½	128.25	1947.80	288.56	964.55	203.06	31.17
10 x 16	9½ x 15½	147.25	2948.07	380.40	1107.44	233.15	35.79
10 x 18	9½ x 17½	166.25	4242.84	484.90	1250.34	263.23	40.41
12 x 12	11½ x 11½	132.25	1457.51	253.48	1457.51	253.48	32.14
12 x 14	11½ x 13½	155.25	2357.86	349.31	1710.98	297.56	37.73
12 x 16	11½ x 15½	178.25	3568.71	466.48	1964.46	341.65	43.33
12 x 18	11½ x 17½	201.25	5136.98	2217.94	385.73	48.92	

Note: Although the maximum size in this table is 12" x 18" larger timber sizes, while uncommon, can sometimes be obtained. Glued-laminated members deeper than 18" are quite common. Glued-laminated timber sizes always differ from those tabulated.

Chapter 9

GLUING OF WOOD

More and more gluing is done with wood in producing laminated wood and other built-up wood products, plywood, and sandwich materials. Modern adhesives, processes, and techniques vary as widely as the glued-wood products made with them, and many developments have been made in recent years. In general, however, the quality and serviceability of a glued-wood joint depends upon (1) kind of wood and its preparation for use, (2) kind and quality of the adhesive, (3) control exercised over the gluing process, (4) types of joints or construction, and (5) moisture-excluding effectiveness of the finish or protective treatment applied to the glued product. Besides these factors, conditions in use affect the performance of the joint.

GLUING PROPERTIES OF DIFFERENT WOODS

Table 9–1 broadly classifies the gluing properties of the woods most widely used for glued products. The classifications are based on the average quality of side-grain joints of lumber that is approximately average in density for the species, when glued with animal, casein, starch, urea-resin, and resorcinolresin adhesives at normal room temperatures.

A species is considered to be bonded satisfactorily when the strength of side-grain to side-grain joints is approximately equal to the strength of the wood. This criterion is considered reasonable for the conventional rigid wood adhesives. Whether it will be easy or difficult to obtain a satisfactory joint by this

Table 9–1.—*Classification of various hardwood and softwood species according to gluing properties*

Group 1 (Glue very easily with glues of wide range in properties and under wide range of gluing conditions)	Group 2 (Glue well with glues of fairly wide range in properties under a moderately wide range of gluing conditions)	Group 3 (Glue satisfactorily with good quality glue, under well-controlled gluing conditions)	Group 4 (Require very close control of glue and gluing conditions, or special treatment to obtain best results)
HARDWOODS			

Aspen
Chestnut, American
Cottonwood
Willow, black
Yellow-poplar
	Alder, red
	Basswood[1]
	Butternut[1,2]
	Elm:
	   American[2]
	   Rock[1,2]
	Hackberry
	Magnolia[1,2]
	Mahogany[2]
	Sweetgum[1]
		Ash, white[2]
		Cherry, black[1,2]
		Dogwood[2]
		Maple, soft[1,2]
		Oak:
		   Red[2]
		   White
		Pecan
		Sycamore[1,2]
		Tupelo:
		   Black[1]
		   Water[1,2]
		Walnut, black
			Beech, American
			Birch, sweet and yellow[2]
			Hickory[2]
			Maple, hard
			Osage-orange
			Persimmon

SOFTWOODS			

Baldcypress
Fir:
   White
   Grand
   Noble
   Pacific silver
   California red
Larch, western
Redcedar, western[3]
Redwood
Spruce, Sitka
	Douglas-fir
	Hemlock
	   Western[3]
	Pine:
	   Eastern white[3]
	   Southern[1]
	   Ponderosa
	Redcedar, eastern[2]
		Alaska-cedar[2]	

[1] Species is more subject to starved joints, particularly with animal glue, than the classification would otherwise indicate.

[2] Bonds more easily with resin adhesives than with nonresins.

[3] Bonds more easily with nonresin adhesives than with resin.

criterion depends on the density of the wood, its structure, moisture content during gluing, the presence of extractives or infiltrated materials, and the kind and quality of the adhesive.

In general, heavy woods require adhesives of superior quality and better control of bonding conditions than lightweight woods; hardwoods—particularly the denser ones—are generally more difficult to glue than softwoods, and heartwood is usually more difficult to glue than sapwood. Several species vary considerably in their gluing characteristics with different glues (table 9–1).

As a general rule, the gluability of tropical species can be estimated by comparison with a domestic species of similar density. Species high in resinous or oily extractives can be glued satisfactorily by freshly machining the surfaces just before gluing and carefully controlling the conditions used for bonding.

ADHESIVES

For Bonding Wood

The term "glue" was first applied to bonding materials of natural origin, while "adhesive" has been used to describe those of synthetic composition. Today the terms are often used interchangeably, but adhesive better covers all types of materials.

Table 9–2 describes briefly the characteristics, preparation and application, and uses of the adhesives most commonly used for bonding wood. The choice of the proper adhesive for a job will depend mainly on (1) species and type of joint (particularly on the kind and amount of stress likely to be encountered in service), (2) working properties of the adhesive (as dictated by the conditions under which gluing must be done), (3) degree of permanence required in service, and (4) cost.

Animal glues were long used extensively in wood bonding. Casein glue and vegetable protein glues, of which soybean is the most widely used, gained commercial importance during World War I for gluing lumber and veneer into products that required moderate water resistance. Glues of natural origin continue to hold an important place in bonding wood. Blood proteins are used by themselves, or in combination with soybean protein, or

with small amounts of phenol resins for making interior-type softwood plywood. Some ready-to-use liquid glues, based on either animal glue or fish glue, are still sold for use in home repair or small shop fabrication work. Casein glues, to be used after mixing with water, are also available for shop fabrication.

Synthetic resin adhesives were introduced before World War II, and now surpass most of the older glues in importance for wood bonding. Phenol-resin adhesives are widely used to produce softwood plywood for severe service conditions. Urea-resin adhesives are used extensively in producing hardwood plywood for furniture and interior paneling and for furniture assembly. Resorcinol and phenol-resorcinol resin adhesives are used mainly for gluing lumber into products that must withstand exposure to the weather. Polyvinyl-resin-emulsion adhesives are used in assembly joints of furniture. They are sometimes combined with urea resins to provide faster setting at normal shop temperatures. Thermosetting or modified polyvinyl resin-emulsion adhesives are used for assembly of interior woodwork, doors, furniture, and nonstructural finger joints, bonding decorative wall paneling to framing in mobile homes, and are being considered for applying cellulosic overlays to lumber and plywood.

Melamine resins are used primarily to improve the durability of urea-resin adhesives. Synthetic resin adhesives available for shop use include a moisture-resistant type based on urea resins and a waterproof type based on resorcinol resins.

Use of epoxy adhesives for bonding wood products has evolved slowly because of the high cost of materials when epoxies were first developed. Their potential properties, especially for gap filling, have led to specialty applications. Thus the epoxies find uses such as filling voids and cracks in timbers and panels, and for bonding other materials to wood.

Contact adhesives are essentially emulsions or solutions of natural or synthetic rubbers that are applied to both mating surfaces, partially dried either at room temperature or under forced-air heating, and then assembled carefully (because the joined pieces cannot be realined after they have come in contact) and pressed momentarily. These are used mainly for on-the-job bonding of plastic laminates or metal sheets to wood bases, or for continuous

Table 9-2.—Characteristics, preparation, and uses of the adhesives most commonly used for bonding wood

Class [1]	Form	Properties	Preparation and application	Typical uses for wood bonding
Animal	Many grades sold in dry form; liquid glues available.	High dry strength; low resistance to moisture and damp conditions.	Dry form mixed with water, soaked, and melted; solution kept warm during application; liquid forms applied as received; both pressed at room temperatures; adjustments in gluing procedures must be made for even minor changes in temperature.	Furniture assembly, use is declining.
Blood protein [2]	Primarily, dry soluble whole blood. Commonly now handled and used like soybean glues.	Moderate resistance to water and damp atmospheres. Moderate resistance to intermediate temperatures and to microorganisms.	Mixed with cold water, lime, caustic soda, and other chemicals; applied at room temperature; and pressed either at room temperature or in hot presses at 240° F. or higher.	Primarily for interior-type softwood plywood. Sometimes in combination with soybean protein.
Casein	Several brands sold in dry powder form; may also be prepared from raw materials by user.	Moderately high dry strength; moderate resistance to water, damp atmospheres, and intermediate temperatures; not suitable for exterior uses.	Mixed with water; applied and generally pressed at room temperature.	Laminated timbers for interior use.
Vegetable protein [3] (mainly soybean).	Protein sold in dry powder form (generally with small amounts of dry chemicals added) to be prepared for use by user.	Moderate to low dry strength; moderate to low resistance to water and damp atmospheres; moderate resistance to intermediate temperatures.	Mixed with cold water, lime, caustic soda, and other chemicals; applied and pressed at room temperatures, but more frequently hot-pressed.	Bonding softwood plywood for interior use.
Urea resin	Many brands sold as dry powders, others as liquids; may be blended with melamine or other resins.	High in both wet and dry strength; moderately durable under damp conditions; moderate to low resistance to temperatures in excess of 120° F.; white or tan.	Dry form mixed with water; hardeners, fillers, and extenders may be added by user to either dry or liquid form; applied at room temperatures, some formulas cure at room temperatures, others require hot pressing at about 250° F.	Hardwood plywood for interior use and furniture; interior particleboard; flush doors.
Melamine resin	Comparatively few brands available; usually marketed as a powder with or without catalyst.	High in both wet and dry strength; very resistant to moisture and damp conditions depending on type and amount of catalyst; white to tan.	Mixed with water and applied at room temperatures; heat required to cure (250° to 300° F.).	Primarily as fortifier for urea resins for hardwood plywood, end-jointing and edge-gluing of lumber, and scarf joining softwood plywood.

198

Type	Form and availability	Working properties	Preparation and application	Principal uses
Phenol resin	Many brands available, some dry powders, others as liquids, and at least one as dry film. Most commonly sold as aqueous, alkaline dispersions.	High in both wet and dry strength; very resistant to moisture and damp conditions; more resistant than wood to high temperatures; dark red; often combined with neoprene, polyvinyl butyral, nitrile rubber, or epoxy resins for bonding metals.	Film form used as received; powder form mixed with solvent, often alcohol and water, at room temperature; with liquid forms, modifiers and fillers are added by users; most common types require hot pressing at about 260° to 300° F.[4]	Exterior softwood plywood and particleboard.
Resorcinol resin and phenol-resorcinol resins.	Several brands available in liquid form; hardener supplied separately; some brands are combinations of phenol and resorcinol resins.	High in both wet and dry strength; very resistant to moisture and damp conditions; more resistant than wood to high temperatures; dark red.	Mixed with hardener and applied at room temperatures; resorcinol glues cure at room temperatures on most species; phenol-resorcinols cure at temperatures from 70° F. to 150° F., depending on curing period and species.	Primarily for laminated timbers and assembly joints that must withstand severe service conditions.
Polyvinyl acetate resin emulsions.	Several brands are available, varying to some extent in properties; marketed in liquid form ready to use.	Generally high in dry strength; low resistance to moisture and elevated temperatures; joints tend to yield under continued stress; white.	Marketed as a liquid ready to use; applied and pressed at room temperatures.[5]	Furniture assembly, flush doors, bonding plastic laminates. Assembly of panel systems (mobile homes).
Rubber-base adhesives A. Contact adhesives.	Typically a neoprene rubber base in organic solvents or water emulsion. Other elastomer systems are also available.	Initial joint strength develops immediately upon pressing, increases slowly over a period of weeks; dry strengths generally lower than those of conventional woodworking glues; water resistance and resistance to severe conditions variable.	Used as received; both surfaces spread and partially dried before pressing. Commonly used in roller presses for instantaneous bonding.	For some nonstructural bonds, as on-the-job bonding of decorative tops to kitchen counters. Useful for low-strength metal and some plastic bonding.
B. Mastics (elastomeric construction adhesives).	Puttylike consistency. Synthetic or natural rubber base usually in organic solvents; others solvent-free.	Gap filling. Develop strength slowly over several weeks. Water resistance and resistance to severe conditions variable.	Used as received. Extruded by calking guns in beads and ribbons, with and without supplemental nailing.	Lumber and plywood to floor joist and wall studs; laminating gypsum board, styrene and urethane foams, and other materials; assembly of panel systems.
Thermoplastic synthetic resins.	Solid chunks, pellets, ribbons, rods, or films; solvent-free.	Rapid bonding; gap filling; lower strength than conventional wood adhesives; minimal penetration; moisture resistant.	Melted for spreading; bond formation by cooling and solidification; requires special equipment for controlling bonding conditions.	Edge banding of panels; plastic lamination; patching; films and paper overlays.

Table 9-2.—*Characteristics, preparation, and uses of the adhesives most commonly used for bonding wood*—(cont.)

Class [1]	Form	Properties	Preparation and application	Typical uses for wood bonding
Epoxy resins	Several different chemical polymers of the general type available or possible; usually in 2 parts, both liquid, most common use in combination with other resins for bonding of metal and materials other than wood.	Completely reactive; no solvent or other volatiles in the liquid adhesives or evolved in curing; good adhesion to metals, glass, certain plastics, and wood products, permanence in wood joints not adequately established; most common use in combination with other resins for bonding metals, plastics, and materials other than wood, can be formulated for curing at either room or elevated temperatures.	Marketed in 2 parts, resin and curing agent, both liquid; mixed at the point of use; applied at room temperatures; cured at room or elevated temperatures, depending on formulation. Potlife and cure conditions vary widely with composition.	For bonding metals, certain plastics, and some masonry materials to themselves and to wood. Bonding wood-to-wood specialty items.

[1] Although starch (or vegetable) glues still are used in the United States, apparently little use is made of these glues in the wood industry. They have been replaced by urea-resin adhesives in gluing interior-type hardwood plywood and furniture.

[2] The older glues, referred to as "blood albumin glues" and dispersed in ammoniacal water solutions for use as hot-pressed plywood glues, are apparently no longer extensively used in the United States.

[3] Another principal type is a protein blend, primarily of blood and soybean proteins. These are mixed and used like the hot-press blood glues.

[4] Most types used in the United States are alkaline-catalyzed. The general statements refer to this type. Acid-catalyzed systems are also available, primarily for use at curing temperatures of 70° to 140° F., but are little used in the United States. Their principal limitation is the possible damage to wood by the acid catalyst.

[5] Modified vinyl-resin emulsions are available which involve addition of a curing agent at time of use, resulting in greatly improved resistance to heat and moisture.

bonding of such materials with a roller press. Their relatively low strength, and the creep and deformability of the glueline, make these contact adhesives suitable for nonstructural joints, but generally inadequate for highly stressed joints.

The elastomeric construction adhesives, extrudable from calking guns, are used in building construction to bond plywood to floor joists, a decorative wall paneling to studs, and in a variety of sealing and calking applications. They add strength, stiffness, and resiliency to the structure, and reduce the number of nails required. The adhesive fills the gaps between the members being joined. Construction adhesives are used in both factory and on-site building construction.

The thermoplastic synthetic resin adhesives, known as hot melts, are used for high-speed bonding in automatic production machines for edge-banding panels with lumber, veneer, plastic laminates, or film. Close control of melt temperature, amount spread, and time of assembly is essential, for bonding depends upon cooling and solidification of the melted adhesive after application and assembly of the joint. The adhesives are based upon various polymers such as polyolefins, vinyl acetate-olefin copolymers, polyamides, and polyurethanes.

Generally, the same adhesives that are suitable for wood-to-wood gluing may also be used for gluing wood-base materials, such as particleboard, hardboard, or fiberboard, to themselves or to wood. When such wood-base materials are produced in a hot press, as is much of the plywood, the heated surfaces are modified and adhesion is often impaired. This can usually be corrected by light sanding of such surfaces before bonding. Only a small amount of surface material need be removed.

For Bonding Wood to Metal

Adhesives capable of producing bonds of high strength and durability between wood and metal are comparatively new.

The contact-pressure adhesives mentioned previously may be used for moderate-strength joints between metal and wood, as in metal-faced plywood. Casein-latex adhesives are still used for this purpose, but to a lesser extent. When higher strength joints are required in structural applications, at least two types of adhesive systems are available—one-stage and two-stage. Neither type is yet highly developed for wood-to-metal bonding. They are essentially variations of systems originally developed for bonding metal to metal in aircraft.

Of the two types, one-stage systems are the most convenient. These involve applying a single adhesive to both metal and wood surfaces, force drying to remove solvent, and assembling and pressing the joint, usually at platen temperatures of 300° F. or higher. Typical adhesives used are combinations of phenol resins with thermoplastic resins or synthetic rubbers. More recently, adhesives curing at 200° to 250° F. have been offered. Some special epoxy formulations will cure and develop strength adequately at normal room temperatures, while others require hot-pressing at temperatures of 180° F. or more. A polyurethane adhesive system can be used to bond metal to wood in a one-stage process with room-temperature cure.

Two-stage systems are useful in bonding dissimilar materials where differences in material properties limit the bonding conditions or the adhesive formulations that may be used. For example, a two-stage system has advantages in bonding a metal-to-wood panel with metal on only one face. When this type of panel is bonded at high temperatures, the metal will contract far more than the wood as the panel cools; the result is an unbalanced panel. The two-stage system minimizes this effect. First, an adhesive primer, often an adhesive used for one-stage bonding, is applied to the metal surface. This primer is cured with heat and the primed material cooled to room temperature. Second, the primed metal surface is glued to the wood with a conventional room-temperature-setting wood glue, such as a resorcinol resin.

In bonding metal the surface must be specially prepared. Such surface preparation involves primarily the removal of rust and surface contamination to provide the necessary initial adhesion, but may also provide corrosion resistance or otherwise modify the surface so the resultant bond will be more durable in service. Instructions for cleaning metal surfaces are usually supplied by the adhesive manufacturer, and should be followed.

Because metal is impervious to moisture and solvent vapors, more care must be taken to remove volatile solvents from the adhesive layer before assembly and pressing of metal-to-wood joints than is necessary for wood-to-wood joints. Failure to remove solvents may result in excessive blisters or low-quality, frothy gluelines.

For Special Purposes

Various wood-base facing materials may be bonded to paper honeycomb and other types of cores for sandwich panels with conventional wood glues. Metal facings can usually be bonded to paper and other woodbase cores with some of the same adhesives described for metal-to-wood bonding.

A variety of plastic sheets and films may be bonded to wood or wood-base materials with specially selected adhesives. In some cases, such as with melamine- and phenol-resin-based paper laminates, conventional urea-resin, phenol-resin, polyvinyl-resin emulsion, or contact-setting adhesives may be used, depending upon the levels of strength and durability required. Other plastics, such as polyvinyl fluoride or polyvinyl chloride films, may require specially formulated adhesives to provide the necessary adhesion to the plastic surfaces, as well as to the wood.

PREPARATIONS FOR GLUING

Drying and Conditioning Wood for Gluing

The moisture content of wood at the time of gluing has much to do with the final strength of joints, the development of checks in the wood, and the dimensional stability of the glued members. Large changes in the moisture content of the wood after gluing cause shrinking or swelling stresses that may seriously weaken both the wood and the joints and cause warping, twisting, and other undesirable effects. It is generally impractical to glue green wood or wood at high moisture content, particularly the higher density hardwoods that have high coefficients of shrinkage due to changes in moisture content.

Essentially, the wood should be dry enough so that, even after some moisture is added in gluing, the moisture content is at about the level desired for service.

The choice of moisture content conditions for gluing according to this principle depends heavily on whether the gluing process involves heating, as in hot-pressing, or merely pressing at room temperature. In gluing 1-inch boards or thicker pieces at room temperature, the desired relationship can be attained by proper seasoning. In gluing veneer or other thin pieces pressed at room temperature, however, the moisture added by the glue frequently exceeds the moisture content of the wood in service. Under these conditions, the wood cannot be dried enough before gluing to achieve the desired moisture content and thus avoid redrying after gluing. The amount of moisture added to wood in room-temperature gluing varies from less than 1 percent in some lumber gluing to 45 percent or more in gluing plywood having thin veneers. Thickness of the wood, number of plies, density of the wood, glue mixture, quantity of glue spread, and gluing procedure (hot-pressing or cold-pressing) all affect the change in moisture content of the wood. In hot-pressing a significant proportion of water is volatilized, thus reducing the moisture content of the product when removed from the press.

In practice, adjustments cannot be made for all these widely varying factors, and it is seldom necessary to dry lumber to a moisture content below the 6 to 12 percent range. Lumber with a moisture content of 6 to 7 percent is generally satisfactory for cold-press gluing into furniture, interior millwork, and similar items. Lumber for outside use should generally contain 10 to 12 percent moisture before gluing. A moisture content of 3 to 5 percent in veneer at the time of gluing by hot pressing is satisfactory for thin plywood to be used in furniture, interior millwork, softwood plywood for construction and industrial uses, and similar products. For such uses as plywood for boxes, veneer at a moisture content of about 8 to 10 percent is acceptable for cold pressing.

Lumber that has been dried to the approximate average moisture content desired for gluing may still show moisture content differences between various boards and between the interior and the surfaces of individual pieces. Large differences in the moisture content of pieces that are glued together result eventually in considerable stress on glue joints and may result in delamination and warping of the product. Lumber that is to be glued should also be free from casehardening, warp, checks, and splits, to produce the highest quality bonded product.

Machining Lumber for Gluing

Wood surfaces that are to be glued should be smooth and true, free from machine marks, and have no chipped or loosened grain or other surface irregularities. Preferably, machining should be done just before gluing, so moisture changes cannot induce distortions in the surfaces before they are bonded. For uniform distribution of gluing pressure, each lamination or ply should be uniformly thick. A small variation in thickness in each lamination or ply may cause a considerable variation in the thickness of the assembly.

Surfaces made by saws are usually rougher than those made by planers, jointers, and other machines equipped with cutterheads. Some saws, however, produce a smoother cut than other types. Such saws save both labor and material by making it possible to glue sawed lumber for certain products. Joints approximately equal in strength to those between planed surfaces can be made between smoothly sawed surfaces. Joint quality in panels produced by edge gluing lumber direct from a straight-line ripsaw is considered quite satisfactory when properly controlled. Unless the saws are very well maintained, however, joints between sawed surfaces are generally weaker and less uniform in quality than those between well-planed or jointed surfaces.

Abrasive planing can also produce surfaces satisfactory for gluing when fine grit sizes are used, abrasive belts are kept clean and well maintained, and sander dust is thoroughly removed.

In the past, wood surfaces were intentionally roughened by some operators by tooth planing, scratching, or sanding with coarse sandpaper in the belief that rough surfaces were better for gluing. Tests of joints made using good gluing practices, however, generally show no benefit from roughening the surfaces.

Preparing Veneer for Gluing

Veneer is cut by rotary processes, slicing, or sawing. Sawed veneer is produced in long narrow strips, usually from flitches selected for figure and grain. The two sides of the sawn sheet are equally firm and strong, and either surface may be glued or exposed to view with the same results.

Sliced veneer is also cut in the form of long strips by moving a flitch or block against a heavy knife. Because the veneer is forced abruptly away from the flitch by the knife, it tends to have fine checks or breaks on the knife side. This checked surface is likely to show imperfections in finishing and therefore should be the glue side whenever possible. For matching face stock, where the checked side of part of the sheets must be the finish side, the veneer must be well cut. Fancy hardwood face veneers are generally sliced.

The rotary process produces continuous sheets of flat-grained veneer by revolving a log against a knife. When rotary-cut veneer is used for faces, the knife or checked side should be the glue side.

Because veneer usually is not resurfaced before it is glued, it must be carefully cut and dried. Well-cut veneer from any of the three processes will yield products with no appreciable difference in any property except appearance. Veneer selected to be glued should be (1) uniform in thickness, (2) smooth and flat, (3) free from large checks, decay, or other quality reducing features, and (4) have grain suitable for the intended product. For plywood of the lower grades, however, some of these requirements may be modified.

Veneers are normally dried rapidly after cutting, using continuous high-temperature dryers, heated either with steam or with hot air from oil- or gas-fired burners. Drying temperatures are usually from 340° to 400° F. for limited periods of time. Prolonged use of high drying temperatures is known to change the characteristics of the wood surfaces and interfere with subsequent gluing.

PROPER GLUING CONDITIONS

To produce a strong joint in wood with liquid glue, the glue must wet the wood surfaces completely and give a film of uniform thickness, free of foreign matter. Because different wood species vary in their absorptivity, a given glue mixture may penetrate into one wood more than into another under the same gluing conditions. A moderate amount of such penetration is not objectionable, and may even be desirable, if the wood surfaces tend to be somewhat torn and damaged. Excessive penetration, however, wastes glue and may result in starved gluelines.

Making strong joints with glues applied as liquids depends primarily upon a proper correlation between gluing pressure and glue consistency during pressing. Consistency of the glue mixture, once it is spread on wood, may vary appreciably. It depends upon such factors as the kind of glue; glue-water proportion of the mixture; age of the mixed glue; quantity of glue spread; moisture content and species of the wood; temperature of the glue, room, and wood; time elapsed after spreading; and the extent to which the glue-coated surfaces are exposed to the air. Room-temperature-setting glues usually thicken and harden steadily after spreading until they are fully cured. Hot-press glues often thin out during the initial heating period and then thicken and harden as curing progresses.

With dense species that are difficult to glue satisfactorily, a viscous glue generally gives the best results, as viscosity influences penetration of glue into the cell structure of the wood. Because glues are generally formulated for a variety of species, longer assembly periods (time between spreading and pressing) are usually required with dense than with light woods to allow the glue to thicken more before pressure is applied. Dense woods may also require longer assembly periods because they are less absorptive than light woods. Optimum penetration of glue for a specific species can also be influenced by proper selection of adhesive and its formulation.

Pressure is used to squeeze the glue into a thin continuous film between the wood layers, to force air from the joint, to bring the wood surfaces into intimate contact with the glue, and to hold them in this position during the setting or curing of the glue. Conversion of liquid to a strong solid film is achieved by physical action, by drying out of solvent, or by chemical action. During chemical action the individual molecules in the glue film become larger and more completely joined together. Chemical action is accelerated by increases in glueline temperature. This may be accomplished by applying external heat through use of hot air or hot platens or by high-frequency dielectric heating.

Light pressure should be used with a thin (low viscosity) glue or one that becomes thin during curing, heavy pressure with a thick glue, and corresponding variations in pressure with glues of intermediate consistency. The strongest joints usually result when the consistency of the glue permits the use of moderately high pressures (100 to 250 p.s.i.) to bring mating surfaces into close contact. Small areas with flat well-planed surfaces can be bonded satisfactorily at much lower pressures.

Lumber joints should be kept under pressure at least until they have enough strength to withstand the interior stresses that tend to separate the wood pieces. In cold-pressing operations, under favorable gluing conditions, this stage will be reached in 1 to 7 hours; time depends upon temperature of the glue room and the wood, upon curing characteristics of the glue, and upon the thickness, density, and absorptive characteristics of the wood. A longer pressing period is advisable, as a precautionary measure, when operating conditions permit. Cold-pressed softwood plywood, however, is usually kept under pressure for only 15 minutes.

In hot-pressing operations the time required varies with temperature of platens, thickness and kind of material being pressed, and glue formulation. In actual practice the variation is from about 2 minutes to as much as 30 minutes. The time under pressure may be reduced to a few seconds by heating the glue joint with high-frequency electrical energy, because it is possible to raise the glueline temperature very rapidly. High-frequency gluing is often used when bonding lumber joints but not in the production of softwood plywood.

Adhesives are sometimes supplied as dry films, which eliminates mixing and spreading operations. Because they add no moisture to the glueline and do not "bleed" through veneers, film glues are particularly suitable for gluing thin, figured, fragile veneers. Bonding these veneers requires special conditions that vary somewhat with the glue and the product, particularly as far as the moisture content of the wood is concerned.

GLUING TREATED WOOD

The advent of durable adhesives that will outlast wood itself under severe conditions has made it possible to glue wood treated with wood preservatives and fire retardants. Perhaps the most important applications have been in laminated beams in which the lumber was treated with preservatives before gluing, and in gluing fire-retardant-treated veneers and lumber for fire doors. Experience has shown that many types of preservative-treated lumber can be glued successfully with phenol-resorcinol-, resorcinol-, and melamine-resin adhesives under properly controlled gluing conditions. Generally, the preservative-treated wood surfaces should be resurfaced just before gluing to reduce interferences by oily solvents or other exudation of preservative material. Fire-retardant-treated material should also be resurfaced just before bonding to reduce interference by treating chemicals on the wood surfaces. It is usually necessary to control the gluing conditions more carefully in gluing treated wood than untreated wood of the same species; it may also be desirable to use a somewhat higher curing temperature, or a longer curing period, with the treated wood.

TYPES OF GLUED JOINTS

Side-Grain Surfaces

With most species of wood, straight, plain joints between side-grain surfaces (fig. 9–1, A) can be made substantially as strong as the wood itself in shear parallel to the grain, tension across the grain, and cleavage. The tongued-and-grooved joint (fig. 9–1, B) and other shaped joints have the theoretical ad-

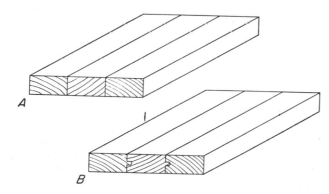

Figure 9–1.—Side-to-side-grain joints: A, plain; B, tongued-and-grooved.

vantage of larger gluing surfaces than the straight joints, but in practice they do not give higher strength with most woods. The theoretical advantage is often lost, wholly or partly, because the shaped joints are difficult to machine to obtain a perfect fit of the parts. Because of poor contact, the effective bonding area and strength may actually be less on a shaped joint than on a flat surface. The prin-

cipal advantage of the tongued-and-grooved and other shaped joints is that the parts can be more quickly alined in the clamps or press. A shallow tongue-and groove is usually as useful in this respect as a deeper cut and is less wasteful of wood.

End-Grain Surfaces

It is practically impossible with present water-based glues and techniques to make end-butt joints (fig. 9–2, A) sufficiently strong or permanent to meet the requirements of ordinary service. With the most careful gluing possible, not more than about 25 percent of the tensile strength of the wood parallel with the grain can be obtained in butt joints using conventional water-based adhesives. To approximate the tensile strength of various species, a scarf, finger, or other type of joint that approaches a side grain surface must be used. This side-grain area should be at least 10 times as large as the cross sectional area of the piece, because wood is approximately 10 times stronger in tension than in shear. In plywood scarfs and finger joints, a slope of 1 in 8 has been found adequate. For nonstructural, low-strength joints these requirements need not necessarily be met.

Finger joints may be cut with the profile showing either on the edge (horizontal joint)

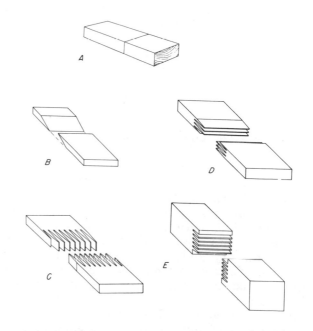

Figure 9–2.— End-to-end-grain joints: A, end butt; B, plain scarf; C, vertical structural finger joint; D, horizontal structural finger joint; E, nonstructural finger joint.

(fig. 9–2, *D*) or on the wide face (vertical joint) (fig. 9–2, *C*) of boards. There is greater leeway in design of a finger joint when a vertical joint is used, but a longer cutting head with more knives is needed. When the curing is done by high-frequency electrical energy, it can generally be done more rapidly with the vertical joint than with the horizontal joint.

The efficiencies of scarf joints of different slopes are discussed in chapter 10. Slopes 1:12 or flatter generally give the highest strength. This also holds true for finger joints, but in these the tip thickness also has to be as small as practical.

A well-manufactured glue joint (scarf, finger, or lap joint) exhibits fatigue behavior much like the wood that is joined. Curves showing stress versus cycles to failures are similar to those for unjointed wood when repeated stresses are expressed as a percent of static strength.

End-to-Side-Grain Surfaces

Plain end-to-side-grain glued joints (fig. 9–3, *A*) are difficult to design so they can carry an appreciable load. Also, joints in service face severe internal stresses in the members from unequal dimensional changes as moisture content changes; such stresses may be high enough to cause failure. It is therefore necessary to use joints of irregular shapes, dowels, tenons, rabbets, or other devices to reinforce a plain joint and bring side grain into contact with side grain or to secure larger gluing surfaces (fig. 9–3). All end-to-side-grain joints should be carefully protected from appreciable changes in moisture content in service.

CONDITIONING GLUED JOINTS

When boards are glued edge to edge, the wood at the joint absorbs moisture from the glue and swells. If the glued assembly is surfaced before this excess moisture is dried out or distributed, more wood is removed along the swollen joints than elsewhere. Later, when the joints dry and shrink, permanent depressions are formed that may be very conspicuous in a finished product. This is particularly important when using glues that contain large amounts of water, such as casein.

As pieces of lumber are glued edge to edge or face to face, the moisture added by the glue need not be dried out but simply allowed to

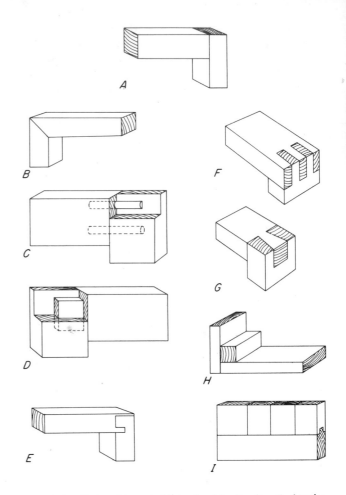

Figure 9–3.—End-to-side-grain joints; *A*, plain, *B*, miter; *C*, dowels; *D*, mortise and tenon; *E*, dado tongue and rabbet; *F*, slip or lock corner; *G*, dovetail; *H*, blocked; *I*, tongued-and-grooved.

distribute itself uniformly throughout the wood. Approximately uniform distribution of such moisture can usually be obtained by conditioning the stock after gluing for 24 hours at 160° F., 4 days at 120° F., or at least 7 days at room temperature; in each case the relative humidity must be adjusted to prevent significant drying.

In plywood, veneered panels, and other constructions made by gluing together thin layers of wood, it is advisable to condition the panels to the average moisture content they are likely to encounter in service. In room-temperature gluing operations, it is frequently necessary to dry out at least a part of the moisture added in gluing. The drying is most advantageously done under controlled conditions and time schedules. Drying room-temperature-glued products to excessively low moisture content materially increases warping, opening of joints, and checking. Panels will often be very dry after hot-press operations. In such instances, it may be desirable to replace moisture by spraying the panels lightly with water and tightly stacking them to allow moisture to distribute uniformly. However, this is not common practice in softwood plywood plants.

DURABILITY OF GLUED PRODUCTS

The durability of glued joints in wood members depends upon the type of glue, gluing technique, service conditions, finish or surface coating, and design and construction of the joints. Moisture conditions are particularly important, not only because of the effect of moisture on the glue itself, but because changes in moisture content affect the internal stresses developed on the glue joint. These internal stresses, and consequently the behavior of the joints in service, also depend upon the design of the joint, the thickness of the plies or laminations, and the density and shrinkage characteristics of the species used.

Available evidence indicates that joints well designed and well made with any of the commonly used woodworking glues will retain their strength indefinitely if the moisture content of the wood does not exceed approximately 15 percent and the temperature remains within the range of human comfort.

Low temperatures seem to have no significant effect on strength of glued joints, but some glues have shown evidence of deterioration when exposed either intermittently or continuously to temperatures much above 100° F. for long periods. Joints that were well made with phenol resin, resorcinol resin, or phenol-resorcinol resin adhesives have proved more durable than the normal unglued wood when exposed to water, to warmth and dampness, to alternate wetting and drying, and to temperatures sufficiently high to char the wood. These glues are entirely adequate for use in products that are exposed indefinitely to the weather.

High-temperature-setting, uncatalyzed melamine-resin adhesives have shown excellent durability for two decades in Douglas-fir lumber joints. Tests have shown that joints made with urea-resin adhesives are highly resistant to water and to wetting and drying at room temperatures. Significant decreases in joint quality have been noted with some urea-resin adhesives exposed at 140° and 158° F. even when the relative humidity of the atmosphere is as low as 20 to 25 percent. Joints with casein and soybeam glue will withstand short exposure to dampness or water without permanent loss in strength, but if the moisture content of the wood continuously or repeatedly exceeds about 18 percent, they will lose strength and eventually fail. The rate of strength loss may vary, depending on species and construction as well as the glue formulation; strength loss is generally more rapid with denser species, and more rapid with soybean than with casein glues.

Joints made with polyvinyl-resin-emulsion adhesives have moderate resistance to dampness, but low resistance to water; the tendency of the joints to yield under stress generally increases as the temperature and moisture content increase. Joints made with animal glue are not suited to damp service conditions. No general pattern has yet been established for the durability characteristics of epoxy resin and contact-pressure adhesives in wood joints, partly because of the wide variety of formulations available. At present it appears that at least some epoxy resin joints might be durable enough to use on lower density species even under exterior conditions. Joints made with contact adhesives tend to creep when subjected to dimensional movement of the wood, but at present seem promising for nonstructural service conditions.

Treatments that can be used to increase the durability of glued products include: (1) Coatings that reduce the moisture content changes in the wood, and (2) impregnation of the wood with preservatives. Moisture-excluding coatings reduce the shrinking and swelling stresses that occur during varying exposure conditions; the coatings do not protect wood effectively, however, during prolonged exposure to damp conditions. By impregnating glued members with preservatives, particularly those that will also prevent rapid moisture changes, the deteriorating effects of prolonged exposure to outdoor or damp conditions can be greatly reduced. Preservatives can reduce the rate of moisture exchange with the atmosphere and protect against attack by micro-organisms.

BIBLIOGRAPHY

Blomquist, R. F., and Olson, W. Z.
    1955. Durability of fortified urea-resin glues in plywood joints. Forest Prod. J. 5(1): 50–56.

————, and Olson, W. Z.
    1960. An evaluation of 21 rubber-base adhesives for wood. Forest Prod. J. 10(10): 494–502.

————, and Olson, W. Z.
    1964. Durability of fortified urea-resin glues exposed to exterior weathering. Forest Prod. J. 14(10): 461–466.

Eickner, H. W.
    1960. Adhesive-bonding properties of various metals as affected by chemical and anodizing treatments of the surfaces. Forest Prod. Lab. Rep. 1842.

Fleischer, H. O.
    1949. Experiments in rotary veneer cutting. Forest Prod. Res. Soc. Proc. 3: 20.

Gillespie, R. H., Olson, W. Z., and Blomquist, R. F.
    1964. Durability of urea-resin glues modified with polyvinyl acetate and blood. Forest Prod. J. 14(8): 343–349.

Olson, W. Z., and Blomquist, R. F.
    1953. Gluing techniques for beech. Northeastern Forest Exp. Sta. Beech Util. Ser. No. 5. Upper Darby, Pa.

————, and Blomquist, R. F.
    1955. Polyvinyl-resin emulsion woodworking glues. Forest Prod. J. 5(4): 219–226.

————, and Blomquist, R. F.
    1962. Epoxy-resin adhesives for gluing wood. Forest Prod. J. 12(2): 74–80.

Peck, E. C.
    1961. Moisture content of wood in use. Forest Prod. Lab. Rep. 1655.

Perry, T. D.
    1948. Modern plywood. Pitman Publ. Co., New York City.

Selbo, M. L.
    1952. Effectiveness of different conditioning schedules in reducing sunken joints in edge-glued lumber panels. Forest Prod. Res. Soc. 2(1): 8.

————
    1963. Effect of joint geometry on tensile strength of finger joints. Forest Prod. J. 13(9): 390–400.

————
    1964. Ten-year exposure of laminated beams treated with oilborne and waterborne preservatives. Forest Prod. J. 14(11):517–520.

————
    1965. Performance of melamine-resin adhesives in various exposures. Forest Prod. J. 15(12): 475–483.

————, Knauss, A. C., and Worth, H. E.
    1965. After two decades of service, glulam timbers show good performance. Forest Prod. J. 15(11): 466–472.

————, and Olson, W. Z.
    1953. Durability of woodworking glues in different types of assembly joints. Forest Prod. J. 3(5): 50–60.

U.S. Forest Products Laboratory
    1956. Durability of water-resistant woodworking glues. Forest Prod. Lab. Rep. 1530.

————
    1966. Synthetic-resin glues. USDA Forest Serv. Res. Note FPL–0141. Forest Prod. Lab., Madison, Wis.

————
    1967. Casein glues: Their manufacture, preparation, and application. U.S. Forest Serv. Res. Note FPL–0158. Forest Prod. Lab., Madison, Wis.

Wood, Andrew D., and Linn, T. C.
    1942. Plywoods, their development, manufacture, and application. Johnston Ltd., Edinburgh and London.

Chapter 10

GLUED STRUCTURAL MEMBERS

Glued structural members are of two types—glued-laminated timbers and glued wood-plywood members of built-up cross sections. Both types offer certain advantages.

GLUED-LAMINATED TIMBERS

Glued-laminated timbers in this chapter (fig. 10–1) refer to two or more layers of wood glued together with the grain of all layers or laminations approximately parallel. The laminations may vary as to species, number, size, shape, and thickness. Laminated wood was first used in the United States for furniture parts, cores of veneered panels, and sporting goods, but is now widely used for structural timbers in building.

The first use of glued-laminated timbers was in Europe, where as early as 1893 laminated arches (probably glued with casein glue) were erected for an auditorium in Basel, Switzerland. Improvements in casein glue during World War I aroused further interest in the manufacture of glued-laminated structural members, at first for aircraft and later as framing members of buildings. In the United States one of the early examples of glued-laminated arches designed according to engineering principles is in a building erected in 1934 at the Forest Products Laboratory. This installation was followed by many others in gymnasiums, churches, halls, factories, hangars, and barns. The development of very durable synthetic-resin glues during World War II permitted the

Figure 10–1.—Heavy laminated timber, glued from 44 nominal 2-inch laminations.

use of glued-laminated members in bridges, trucks, and marine construction where a high degree of resistance to severe service conditions is required. With growing public acceptance of glued-laminated construction, laminating has increased steadily until it now forms an important segment of the woodworking industry.

Glued structural timbers may be straight or curved. Curved arches have been used to span more than 300 feet in structures. Straight members spanning up to 100 feet are not uncommon, and some span up to 130 feet. Sections deeper than 7 feet have been used. Straight beams can be designed and manufactured with horizontal laminations (lamination parallel to neutral plane) or vertical laminations (laminations perpendicular to neutral plane). The horizontally laminated timbers are the most widely used. Curved members are horizontally laminated to permit bending of laminations during gluing.

Advantages of Glued-Laminated Timbers

The advantages of glued-laminated wood construction are many and significant. They include the following:

1. Ease of manufacturing large structural elements from standard commercial sizes of lumber.

2. Achievement of excellent architectural effects and the possibility of individualistic decorative styling in interiors, as nearly unlimited curved shapes are possible.

3. Minimization of checking or other seasoning defects associated with large one-piece wood members, in that the laminations are thin enough to be readily seasoned before manufacture of members.

4. The opportunity of designing on the basis of the strength of seasoned wood, for dry service conditions, inasmuch as the individual laminations can be dried to provide members thoroughly seasoned throughout.

5. The opportunity to design structural elements that vary in cross section along their length in accordance with strength requirements.

6. The possible use of lower grade material for less highly stressed laminations, without adversely affecting the structural integrity of the member.

7. The manufacture of large laminated structural members from smaller pieces is increasingly adaptable to future timber economy, as more lumber comes in smaller sizes and in lower grades.

Certain factors involved in the production of laminated timbers are not encountered, however, in producing solid timbers:

1. Preparation of lumber for gluing and the gluing operation usually raise the cost of the final laminated product above that of solid sawn timbers in sizes that are reasonably available.

2. As the strength of a laminated product depends upon the quality of the glue joints, the laminating process requires special equipment, plant facilities, and manufacturing skills not needed to produce solid sawn timbers.

3. Because several extra manufacturing operations are involved in manufacturing laminated members, as compared with solid members, greater care must be exercised in each operation to insure a product of high quality.

4. Large curved members are awkward to handle and ship by the usual carriers.

Avoiding Internal Stresses

For best results in manufacturing glued-laminated timbers, it is important to avoid the development of appreciable internal stresses when the member is exposed to conditions that change its moisture content. Differences in shrinking and swelling are the fundamental causes of internal stresses. Therefore laminations should be of such character that they shrink or swell similar amounts in the same direction. If laminations are of the same species or of species with similar shrinkage characteristics, if they are all flat-grained or all edge-grained material, and if they are of the same moisture content, the assembly will be reasonably free from internal stresses and have little tendency to change shape or to check. Laminations that have an abnormal tendency to shrink longitudinally because they have excessive cross grain or compression wood should not be included.

While observance of these principles is desirable, practical considerations may prevent exact conformance. In softwood structural timbers for interior use, for example, segregation of flat-grained from edge-grained material is generally unnecessary, and a range in moisture content of no greater than 5 percent among the laminations in the same assembly may be permitted without significant effect on serviceability. The average moisture content, however, should be the same or slightly lower than that which the timber will attain in service. A slight increase in moisture content generally causes no harm but rapid drying could result in checking.

Preservative Treatment

If the laminated timber is to be used under conditions that raise its moisture content to more than 20 percent, either the heartwood of durable species should be used or the wood should be treated with approved preservative chemicals and only waterproof glues should be used. Experience has shown that some oil-borne preservatives, besides providing protection from fungi and insects, also retard moisture changes at the surface of the wood, and thus inhibit checking. If the size and shape of the timbers permit, laminated timbers can be treated with preservatives after gluing, but penetration perpendicular to the planes of the glue joints will be distinctly retarded at the first glueline.

Laminated timbers treated with preservatives after gluing and fabrication have given excellent service in bridges and similar installations. Laminations also can be treated and then glued if suitable precautions are observed. The treated laminations should be conditioned and must be resurfaced just before gluing. Not all preservative-treated wood can be glued with all glues, but, if suitable glues and treatments are selected and the gluing is carefully done, laminated timbers can be produced that are entirely serviceable under moist, warm conditions that favor decay.

Species for Laminating

Softwoods, principally Douglas-fir and southern pine, are most commonly used for laminated timbers. Other softwoods used include western hemlock, larch, and redwood. Boat timbers, on the other hand, are often made of white oak because it is moderately durable under wet conditions. Red oak, treated with preservative, has also been laminated for ship and boat use. Other species can also be used, of course, when their mechanical and physical properties are suited for the purpose.

Quality of Glue Joints

The quality of glue joints in laminated timbers intended for service under dry conditions is usually evaluated by the block shear test. Acceptance criteria are often based on unit shear strength and percentage of wood failure considered satisfactory for the species being used. For laminated members that must withstand severe service conditions, however, the block shear test alone does not provide an adequate evaluation. To serve satisfactorily under severe conditions, the glue joints should be capable of withstanding, without significant delamination, high internal stresses that develop as a result of rapid wetting and drying. Standards covering the design and manufacture of glued structural laminated wood are available (see Bibliography) and at least two military specifications cover structural glued-laminated items for specific uses.

Strength and Stiffness

The bending strength of horizontally laminated timbers depends on the position of various grades of laminations. High-grade laminations may be placed in the outer portions of the member, where their high strength may be effectively used, and lower grade laminations in the inner portion, where their low strength will not greatly affect the overall strength of the member. By selective placement of the laminations, the knots can be scattered and improved strength can be obtained. Even with random assembly of laminations, studies have indicated that knots are unlikely to occur one above another in several adjacent laminations. by a careful selection of quality tension laminations.

Aside from the beneficial effect of dispersing imperfections, available test data do not indicate that laminating improves strength properties over those of a comparable solid piece. That is, gluing together pieces of wood does not, of itself, improve strength properties, unless the laminations are so thin that the glue bonds significantly affect the strength of the member. In laminating material of lumber thicknesses, that would not occur.

Most of the criteria relating strength to characteristics of the wood were developed in the 1940's from data on a large number of relatively small beams—12 inches in depth. Very large beams are now being manufactured. Current research is reevaluating the criteria for deriving design stresses for glued-laminated timbers. Specific research relates to the tensile strength of structural lumber. Significant improvement in beam strength can be obtained.

Criteria in the following sections are given only as general guides to factors that affect the strength of glued-laminated construction. The most current specific information on the strength of glued-laminated construction should be obtained from the Forest Products Laboratory or from the American Institute of Timber Construction.

The principal determinants of strength are knots, cross grain, and end-joint efficiency. The effects of these three factors are not cumulative; that is, the lowest of the three controls the strength. Where other effects, such as those of curvature or beam depth, are applicable, they should be applied in addition to those for knots, cross grain, and end joints.

The deflection characteristics of glued-laminated timbers can be computed with formulas given in chapter 8.

Effect of Knots on Bending Strength and Stiffness

The effect of knots on the bending strength and stiffness of laminated timbers depends upon the number, size, and position, with respect to the neutral axis of the member, of the knots close to the critical section. Specifically, the bending properties depend upon the sum of the moments of inertia, about the gravity axis of the full cross section, of the areas occupied by all knots within 6 inches of either side of the critical section. This sum may be represented by the symbol I_K. The moment of inertia of the full cross section of the member is represented by I_G. The relations between bending strength and stiffness and the ratio I_K/I_G are shown in figures 10–2 and 10–3. Procedures for calculating the ratio I_K/I_G are given in USDA Technical Bulletin 1069.

The curves shown in figures 10–2 and 10–3 were empirically derived from tests of laminated timbers containing knots in various concentrations.

The I_K/I_G strength relationship yields reliable results for relatively small members; however, for large members modifications are recommended. For large members, the effect of knots in tension laminations is not adequately defined by the I_K/I_G concept. A selected grade of tension laminations must be used for the higher strength large glued-laminated timbers. The current recommended grades of tension lam-

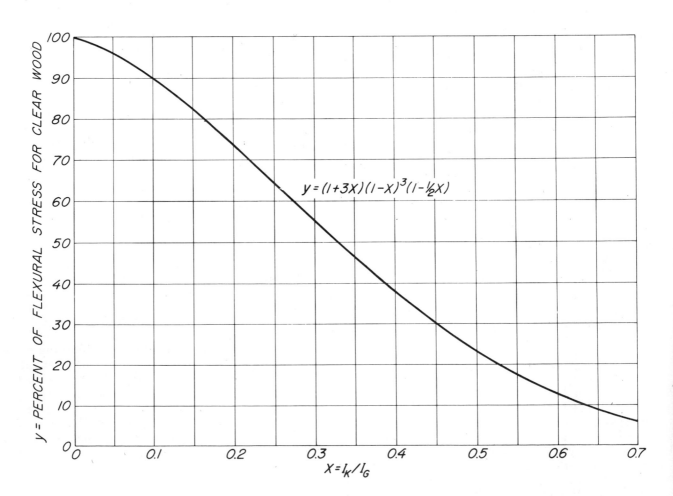

$$y = (1+3X)(1-X)^3(1-\tfrac{1}{2}X)$$

Figure 10–2.—Curve relating allowable flexural stress to moment of inertia of areas occupied by knots in laminations of laminated timbers.

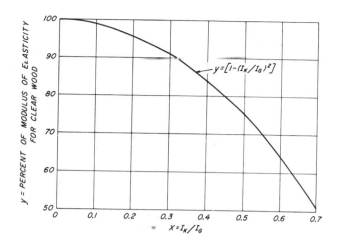

Figure 10–3.—Curve relating allowable modulus of elasticity to moment of inertia of areas occupied by knots in laminations of laminated timbers.

inations should be obtained from the American Institute of Timber Construction.

Effect of Knots on Compressive Strength

The compressive strength of laminated timbers depends upon the proportion of the cross-sectional area of each lamination occupied by the largest knot in the lamination. Figure 10–4 shows an empirically derived relationship between compressive strength and K/b, where b is the lamination width and K is the average of the largest knot sizes in each of the laminations.

Effect of Knots on Tensile Strength

Test data relating knot size to tensile strength of glued-laminated timbers are not

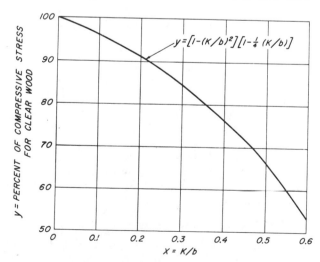

Figure 10–4.—Design curve relating allowable compressive stress to size of knots in laminations of laminated short columns.

available. Figure 10–5, however, represents a relation between tensile strength and K/b derived from figure 10–4.

Effect of Cross Grain on Strength

The effect of cross grain on strength is given in table 4–10, chapter 4. For laminated timbers, it is possible to vary the cross-grain limitations at different points in the depth of the beam in accordance with the stress requirements. That is, steeper cross grain may be permitted in laminations in the interior of the timber than in the laminations at and near the outside. The permitted variation should be

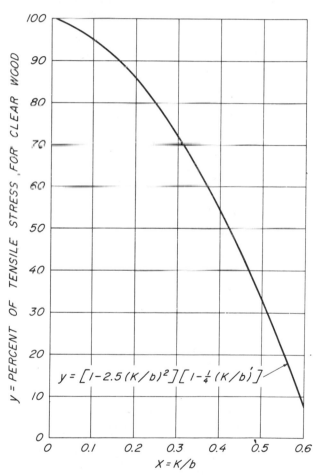

Figure 10–5.—Curve relating allowable tensile stress to size of knots in laminations of tension members.

based on the assumption of linear variation of stress across the depth.

Effect of End Joints on Strength

For a large proportion of laminated timbers, because of their size, pieces of wood must be joined end-to-end to provide laminations of sufficient length. In most cases, the strength of timbers is reduced by the presence of end joints.

The highest strength values are obtained with well-made plain scarf joints (ch. 9); the lowest values are with butt joints. This is because scarf joints with flat slopes have essentially side grain surfaces that can be well bonded and develop high strength while butt joints are end grain surfaces that cannot be bonded effectively. Finger joints are a compromise between scarf and butt joints and strength varies with joint design.

Structural end joints

Both plain scarf joints and finger joints can be manufactured with adequate strength for structural glued-laminated timbers. The adequacy is determined by joint efficiency principles in USDA Technical Bulletin 1069 or by physical testing procedures in Voluntary Product Standard PS 56. The physical testing procedures are more commonly used.

The joint efficiency or joint factor is the joint strength as a percentage of the strength of clear straight-grain material. The joint factor of scarf joints in the tension portion of bending members or in tension members is:

Scarf slope	Joint factor *Percent*
1 in 12 or flatter	90
1 in 10	85
1 in 8	80
1 in 5	65

These factors apply to interior laminations as well as to the lamination on the tension face. In a beam of 40 equal laminations, for example, the outer lamination should be stressed no higher than 90 percent of the clear wood stress value if a scarf joint sloping 1 in 12 is used. The stress at the outer face of the second lamination is 95 percent of the outer fiber stress, so a scarf joint of about 1 in 10 would be satisfactory. Usually, however, laminators use the same joint throughout the members for ease in manufacture.

Working stresses need not be reduced if laminations containing scarf joints sloping 1 in 5 or flatter are used in the compression portion of bending members or in compression members.

No general statement can be made regarding the joint factor of finger joints because finger joint strengths vary depending upon the type and configuration of joint and the manufacturing process. High-strength finger joints can be made when the design is such that the fingers have relatively flat slopes and sharp tips. Tips are essentially a series of butt joints and therefore reduce the effective sloping area

as well as being possible sources of stress concentration. As a result, the strength of even the best production finger joint developed to date has approached but not equaled the strength of a well-made plain scarf joint.

When the joint factor for a specific finger joint is adequately determined, allowable stresses may be calculated on the same basis as discussed for scarf joints.

The slope of scarf joints in compression members or in the compression portion of bending members should not be steeper than 1 in 5; a limitation of 1 in 10 is suggested for tension members and the tension portion of bending members. Because there is some question as to the durability of steep scarf joints for exterior use or other severe exposure, scarf joints with a slope steeper than 1 in 8 should not be used under such conditions.

Joints should be well scattered in portions of structural glued-laminated timbers highly stressed in tension. Where end joints in adjacent laminations are closely spaced, test results indicate that failure progresses more or less instantaneously from the joint in the outer lamination through the others. Adequate longitudinal separation of end joints in areas of high stress is, therefore, desirable.

No data are available by which to substantiate any proposed spacing requirements. Spacing requirements depend on joint quality and stress level. No spacing requirement should be necessary for well-made, high-strength joints or for joints stressed well below their strength.

Suggested spacings of end joints are given in Voluntary Product Standard PS 56.

Butt joints

Butt joints generally can transmit no tensile stress and can transmit compressive stress only after considerable deformation or if a metal bearing plate is tightly fitted between the abutting ends. In normal assembly operations, such fitting would not be done, and it is therefore necessary to assume that butt joints are ineffective in transmitting both tensile and compressive stresses. Because of this ineffectiveness, and because butt joints cause concentration of both shear stress and longitudinal stress, they are not recommended for use in horizontally laminated structural glued-laminated timbers.

Other types of end joints

In addition to scarf, finger, and butt joints, other types of end joints may be used in laminated members. In general, however, few data are available on their effect on strength. When data are not available for establishing spacing requirements, strength, and like factors for these joints, they may be treated as butt joints.

Effect of Edge Joints on Strength

It is sometimes necessary to join pieces edge-to-edge to provide laminations of sufficient width. For tension members, compression members, and horizontally laminated bending members, the strength of such joints is of little importance to the overall strength of the member. Therefore, from the standpoint of strength alone, it is unnecessary that edge joints be glued if they are not in the same location in adjacent laminations. Edge joints should be glued, however, where torsional loading is involved or where the load is applied parallel to the wide face of the lamination. Other considerations, such as the appearance of face laminations or the possibility that water will enter the unglued joints and promote decay, also may dictate that edge joints should be glued.

In vertically laminated beams (laminations at right angles to the neutral plane) sufficient laminations must be edge glued to provide adequate shear resistance in the beam. Not only is adequate initial strength required of such joints, but they must be durable enough to retain that strength under the conditions to which the beam is exposed in service.

Effect of Curvature on Bending Strength

Stress is induced when laminations are bent to curved forms, such as arches and curved rafters. While much of this stress is quickly relieved, some remains and tends to reduce the strength of a curved member. The ratio of the allowable design stress in curved members to that in straight members is

$$1.00 - \frac{2,000}{(R/t)^2} \qquad (10\text{--}1)$$

where t is the thickness of a lamination and R is the radius of the curve to which it is bent, both R and t being in the same units.

Effect of Lamination Thickness on Strength

Laminated timbers are typically made from nominal 1-inch and 2-inch lumber, and tests have indicated that this difference has no effect on the strength of straight timbers. Lamination thickness does, however, affect the strength of curved members, depending upon

the radius to which the lamination is bent (see preceding section).

If thin laminations of moderate or low grade are used, somewhat stronger straight members may be obtained than from 1-inch or 2-inch laminations of the same grade. This increase perhaps is due to the greater interaction between thinner laminations, but data do not yet permit design recommendations.

Effect of Shake, Checks, and Splits on Shear Strength

In general, shake, checks, and splits have little effect on the shear strength of laminated timbers. Shake generally occurs infrequently but should be excluded from material for laminations. Most laminated timbers are made from laminations that are thin enough to season readily without developing checks and splits.

Effect of Size on Bending Strength

The bending strength of wood beams has been shown to decrease as the size of beams increases. Laminated members can be of considerable size and the effect of size on strength should be considered in design. Design information is given in chapter 8.

Vertically Laminated Timbers

Data on vertically laminated timbers are limited because, until recently, there was relatively little commercial interest in their use. Evaluations were made of timbers laminated from dimension lumber of high strength ratios. The results indicated that the effect of knots may be taken into account by determining the strength ratio of each lamination (by the method described in the section on stress grades) and averaging these ratios to determine the strength ratio for the timber. Common practice is to increase the average stress 15 percent to account for the interaction of three or more members glued together. The effects of cross grain on strength and limitations on end joints are as described in preceding paragraphs.

Design of Glued-Laminated Timbers

Engineering formulas such as those given in chapter 8 for solid wood structures are applicable also to structures of laminated wood. The fact that laminated timbers can be made in curved form, however, introduces some circumstances not ordinarily met in the design of wood structures.

The application of ordinary engineering formulas to deep, sharply curved bending members may introduce appreciable error in the calculated stresses. For such cases, the methods applicable to curved timbers, as described in standard text books on mechanics, should be used.

When bending moments are applied to curved timbers, stresses are set up in a direction parallel to the radius of curvature (perpendicular to grain). Stresses so induced should be appropriately limited.

Procedures for establishing design stresses and factors affecting the design of glued-laminated structural members are discussed in detail by USDA Technical Bulletin 1069.

Voluntary Product Standard PS 56 for structural glued-laminated timber is widely used throughout the United States. It involves industry specifications that have been developed for several species and which recommend working stresses for members laminated from these species.

Many design details have been devised and presented by the American Institute of Timber Construction.

Allowable Unit Stresses For Clear Wood

Allowable unit stresses for clear wood for use in determining working stresses for the design of glued-laminated timbers are developed by principles given in ASTM standards and in other sections of this handbook. The average clear wood strength properties and variability of properties are given in ASTM Designation D 2555. These properties form the basis for allowable unit stresses for clear wood by considering appropriate adjustments for variability, duration of load, size, moisture content, density and rate of growth, and factor of safety. Discussions of these factors are given in ASTM Designation D 2555 and D 245 and in chapter 6 of this volume.

Bending Members

In designing bending members, a number of factors are involved in choosing an appropriate working stress—the effects of knots, cross grain, end joints, height or depth of the beam, and curvature of the laminations.

The basic allowable unit stress for clear wood is multiplied by the lowest of the first three factors to obtain a working stress for straight timbers up to 12 inches deep. If the depth is over 12 inches, this calculated stress is reduced further by the effect of size as given in chapter 8. For curved members, the reduced value is multiplied by the curvature factor.

Moment of inertia

The full cross section of the beam may be considered effective in design if there are no end joints or if end joints have been qualified by procedures given in PS–56. If butt joints are present, the moment of inertia should be reduced as described for them.

Deflections

Experience with figure 10–3 and statistically derived values of I_K/I_G indicates that, for timbers with relatively high grade laminations, the reduction of modulus of elasticity below the basic value will be less than 5 percent. However, there is a reduction in modulus of elasticity for lower grades of laminations (ASTM D 245). With known modulus of elasticity properties of each lamination, the modulus of elasticity of a completed glued-laminated beam can be calculated with transformed-section criteria.

The deflection calculated with modulus of elasticity values derived by any of the above methods will be only immediate values. Where defection is critical, consideration must be given to the added defection that occurs under long-time loading and that due to shear (see ch. 6 and 8).

Radial stresses

When curved members are subjected to bending moments, stresses are set up in a direction parallel to the radius of curvature (perpendicular to grain). If the moment increases the radius (makes the member straighter), the stress is tension; if it decreases the radius (makes the member more sharply curved), the stress is compression. For members of constant depth, the stress is a maximum at the neutral axis and is approximately

$$S_R = \frac{3}{2} \frac{M}{Rbh} \qquad (10\text{--}2)$$

where M is the bending moment, R is the radius of curvature of the centerline, and b and h are, respectively, the width and height of the cross section.

Values of S_R should be limited to those recommended by accepted industry standards.

Deep, Curved Members

It is known from the principles of mechanics that the stresses in sharply curved members subjected to bending are in error when computed by the ordinary formulas for straight timbers. The amount of the error depends upon the relation of the depth of the member to the centerline radius. Limitations on the sharpness to which laminations can be bent will limit the curvature of deep beams or arches. The analytical methods applicable to curved beams should be used, however, to determine whether or not the usual engineering formulas may be applied without significant error to deep, sharply curved members of glued-laminated wood.

Axially Stressed Members

Three factors must be considered in choosing working stresses for axially stressed members—the effects of knots, cross grain, and end joints. The factor giving the lowest strength ratio is the one that determines the working stress; the three factors are not to be combined.

Effective cross-sectional area

The full cross section may be considered effective in design if there are no end joints of if end joints are scarf joints of 1 in 5 slope or flatter. For finger joints that have been qualified by procedures given in PS–56 or with a tensile joint ratio of 50 percent or greater, it is probable that the full cross section could likewise be considered fully effective. If butt joints are present, the cross-sectional area should be reduced as was described for these joints.

Columns of different classes

The formulas for determining the load-carrying capacities of wood columns as given in chapter 8 should be used for laminated columns. For all classes of columns, the effect of joints should be considered in arriving at the proper value of the effective area.

Fastenings

Design loads or stresses for fastenings that are applicable to solid wood members (ch. 7) are also applicable to laminated timbers. Since greater depths are practical with laminated timbers than with solid timbers, however, design of fastenings for deep laminated timbers requires special consideration. It is desirable to have the moisture content of the timber as near as possible to that which it will attain in service. If moisture content is higer than the equilibrium moisture content in service, considerable shrinkage may occur between widely separated bolts; if the bolts are held in position by metal shoes or angles, large splitting forces may be set up. Slotting of the bolt holes in the metal fitting will tend to relieve the splotting stresses. Cross bolts will assist in preventing separation if splitting does occur; they will not, however, prevent splits.

WOOD-PLYWOOD GLUED STRUCTURAL MEMBERS

Highly efficient structural components or members can be produced by combining wood and plywood through gluing. The plywood is utilized quite fully in load-carrying capacity while also filling large opening spaces. These components, including box beams, I-beams, "stressed-skin" panels, and folded plate roofs, are discussed in detail and completed designs are also available in the many technical publications of the American Plywood Association. Details on structural design will be given in the following portion of this chapter for beams with plywood webs and for stressed-skin panels wherein the parts are glued together with a rigid, durable adhesive.

These highly efficient designs, while adequate structurally, may suffer from lack of resistance to fire and decay unless treatment or protection is provided. The rather thin portions of plywood, as compared with heavy, solid cross sections, are quite vulnerable to fire.

Beams With Plywood Webs

Box beams and I-beams with lumber or laminated flanges and glued plywood webs can be designed to provide desired stiffness, bending moment resistance, and shear resistance. The flanges resist bending moment, and the webs provide primary shear resistance. Either type of beam must not buckle laterally under design loads; thus, if lateral stability is a problem, the box beam design should be chosen because it is stiffer in lateral bending and in torsion than the I-beam. On the other hand, the I-beam should be chosen if buckling of the plywood web is of concern because its single web, double the thickness of that of a box beam, will offer greater buckling resistance.

Details of design are given for beams shown in cross section in figure 10–6. These beams have both flanges of the same thickness because a construction symmetrical about the neutral plane provides the greatest moment of inertia for the amount of material employed. The formulas given were derived by basic principles of engineering mechanics which can be extended to derive designs for unsymmetrical constructions if necessary.

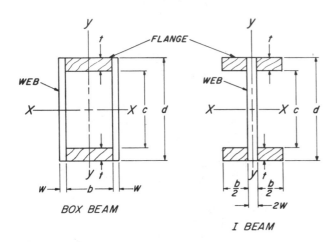

Figure 10–6.—Beams with plywood webs.

Beam Deflections

Beam deflections can be computed with formula (8–1) (ch. 8). The bending stiffness $(EI)_x$ and shear stiffness (GA') are given by the following formula for the box and I-beam shown in figure 10–6.

$$(EI)_x = \frac{1}{12}\left[E(d^3 - c^3)b + 2E_wWd^3 \right] \quad (10\text{–}3)$$

where E is flange modulus of elasticity and E_w is web modulus of elasticity. Values of E_w for the appropriate plywood construction and grain direction can be computed from formulas (11–2), (11–3), and (11–4) for edgewise compression modulus of elasticity of plywood (ch. 11).

Dimensions are as noted by symbols of figure 10–6.

$$(GA') = 2WcG \text{ (approx.)} \quad (10\text{–}4)$$

where G is plywood shear modulus for appropriate grain direction (see ch. 11). An improvement in shear stiffness can be made if the grain direction of the plywood webs is placed at 45° to the beam neutral plane rather than at 0° or 90° to the neutral plane. Data

in chapter 11 on shear modulus given by formulas (11–9) to (11–12) should be used.

Flange Compression and Tension Stresses

Flange compression and tension stresses at outer beam fibers are given by the formula

$$f_x = \frac{6M}{(d^3 - c^3)\dfrac{b}{d} + \dfrac{2E_wWd^2}{E}} \quad (10\text{–}5)$$

where M is bending moment and other symbols are as defined previously in this chapter.

Web Shear Stress

Web shear stress at the beam neutral plane is given by the formula

$$f_{xy} = \frac{3V}{4W}\left[\frac{E(d^2 - c^2)b + 2E_wWd^2}{E(d^3 - c^3)b + 2E_wWd^3} \right] \quad (10\text{–}6)$$

where V is shear load and other symbols are as defined previously in this chapter. The shear stress must not exceed values given by formulas in chapter 11 for plywood shear strength or the critical shear buckling stress, F_{scr}, given by formula (11–18) of chapter 11. To avoid web buckling, the web should be increased in thickness or the clear length of web should be broken by stiffeners glued to the webs.

Web edgewise bending stresses at the inside of the flanges can be computed by the formula

$$f_{xW} = \frac{6M}{\dfrac{E}{E_W}(d^3 - c^3)\dfrac{b}{c} + 2\dfrac{d^3}{c}W} \quad (10\text{–}7)$$

where the symbols are as previously defined in the chapter. It may be possible, but not very likely, for the web to buckle due to bending stresses. Thus the stresses given by formula (10–7) should not exceed those given by F_{bcr} of chapter 11, formula (11–19). Should buckling due to edgewise bending appear possible, the interaction of shear and edgewise bending buckling can be checked approximately by means of an interaction formula such as

$$\left(\frac{f_{xy}}{F_{scr}}\right)^2 + \left(\frac{f_{xW}}{F_{bcr}}\right)^2 = 1 \quad (10\text{–}8)$$

Web-to-flange glue shear stresses, f_{gl}, can be computed by the approximate formula

$$f_{gl} = \frac{3V}{2}\left[\frac{Eb(d + c)}{E(d^3 - c^3)b + 2E_wWd^3} \right] \quad (10\text{–}9)$$

218

where the symbols are as previously defined in this chapter. The stresses computed by formula (10–9) should be less than those given for glued joints. They should also be less than the rolling shear stresses for solid wood, because the thin plies of the plywood web allow the glue shear stresses to be transmitted to adjacent plies and could cause rolling shear failure in the wood.

Possible Lateral Buckling

Possible lateral buckling of the entire beam should be checked. The lateral buckling load is given by the formula

$$P_{cr} \text{ or } W_{cr} \text{ or } M_{cr}/L = \frac{\beta}{L^2}\sqrt{(EI)_y \, (JG)}$$
$$(10\text{–}10)$$

where P_{cr} is total concentrated load, W_{cr} is total uniformly distributed load; M_{cr} is bending moment; L is beam span length; β is a lateral buckling coefficient given in figure 8–5 (ch. 8) for various loadings; $(EI)_y$ is lateral stiffness of beam; and (JG) is beam torsional rigidity. Formulas for $(EI)_y$ and (JG) are as follows:

For box beams:

$$(EI)_y = \frac{1}{12}\Big\{ E(d - c)b^3 + E_w$$
$$\Big[(b + 2W)^3 - b^3 \Big] d \Big\} \quad (10\text{–}11)$$

$$(JG) = \Bigg[\frac{(d + c)\,(d^2 - c^2)\,(b + W)^2 W}{(d^2 - c^2) + 4(b + W)W} \Bigg] G$$
$$(10\text{–}12)$$

For I-beams:

$$(EI)_y = \frac{1}{12}\Big\{ E\Big[(b + 2W)^3 - (2W)^3 \Big]$$
$$(d - c) + E_{fw}(2W)^3 d \Big\} \quad (10\text{–}13)$$

where E_{fw} is flexural elastic modulus of the plywood web as computed by formulas (11–20) or (11–21) of chapter 11.

$$(JG) = \frac{1}{3}\Big[\frac{1}{4}(d - c)^3 b + d(2W)^3 \Big] G$$
$$(10\text{–}14)$$

In formulas (10–12) and (10–14) the shear modulus G can be assumed without great error to be about one-sixteenth of the flange modulus of elasticity, E_L. The resultant torsional stiffness (JG) will be slightly low if beam

webs have plywood grain at 45° to the neutral axis. The lateral buckling of I-beams will also be slightly conservative because bending rigidity of the flange has been neglected in writing the formulas given here. If buckling of the I-beam seems possible at design loads, the more accurate analysis of Forest Products Laboratory Report R1687 should be used before redesigning.

Stiffeners and Load Blocks

A determination of the number and sizes of stiffeners and load blocks needed in a particular construction does not lend itself to a rational procedure, but certain general rules can be given that will help the designer of a wood-plywood structure obtain a satisfactory structural member.

Stiffeners serve a dual purpose in a structural member of this type. One function is to limit the size of the unsupported panel in the plywood web, and the other is to restrain the flanges from moving toward each other as the beam is stressed.

Stiffeners should be glued to the webs and should be in contact with both flanges. No rational way of determining how thick the stiffener in contact with the web should be is available, but it appears, from tests of box beams made at the Forest Products Laboratory, that a thickness of at least six times the thickness of the plywood web is sufficient. Because stiffeners must also resist the tendency of the flanges to move toward each other, the stiffeners should be as wide as (extend to the edge of) the flanges.

The spacing of the stiffeners is relatively unimportant for the shear stresses that are allowed in plywood webs in which the grain of the wood in some plies is parallel and the grain of the wood in other plies is perpendicular to the axis of the member. Maximum allowable stresses are below those which will produce buckling. Stiffeners placed with a clear distance between stiffeners equal to or less than two times the clear distance between flanges are adequate.

Load blocks are special stiffeners placed along a wood-plywood structural member at points of concentrated load. Load blocks should be designed so that stresses caused by a load that bears against the side-grain material in the flanges do not exceed the design allowables for the flange material in compression perpendicular to grain.

Stressed-Skin Panels

Constructions consisting of plywood "skins" glued to wood stringers are often called stressed-skin panels. These panels offer efficient structural constructions for building floor, wall, and roof components. They can be designed to provide desired stiffness, bending moment resistance, and shear resistance. The plywood skins resist bending moment, and the wood stringers provide shear resistance.

Details of design are given for a panel cross section shown in figure 10–7. The formulas given were derived by basic principles of engineering mechanics.

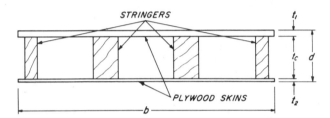

Figure 10–7.—Stressed-skin panel cross section.

Panel deflections can be computed with formula (8–1) chapter 8. The bending stiffness (EI) and shear stiffness (GA') are given by the following formulas for the stressed-skin panel shown in figure 10–7.

$$(EI) = \frac{b}{\left(\left(E_1t_1 + E_2t_2 + Et_c\dfrac{s}{b}\right)\right)}\Bigg\{ E_1t_1E_2t_2[(t_1 + t_c) + (t_2 + t_c)]^2 + E_1t_1Et_c$$

$$\frac{s}{b}(t_1 + t_c)^2 + E_2t_2Et_c\frac{s}{b}(t_2 + t_c)^2 \Bigg\}$$

$$+ \frac{b}{12}\left[E_{f1}t_1^3 + E_{f2}t_2^3 + Et_c^3\frac{s}{b}\right]$$

$$(10\text{--}15)$$

where E_1 and E_2 are modulus of elasticity values for skins 1 and 2 (compression values for plywood—see ch. 11); E_{f1} and E_{f2} are flexural modulus of elasticity values for skins 1 and 2 (also see ch. 11); E is stringer modulus of elasticity; and s is total width of all stringers in a panel. Other dimensions are as noted by symbols of figure 10–7.

$$(GA') = Gstc \text{ (approx.)} \quad (10\text{--}16)$$

where G is stringer shear modulus.

Skin Stresses

Skin tension and compression stresses are given by the formulas

$$f_{x1} = \frac{ME_1y_1}{(EI)}$$

$$f_{x2} = \frac{ME_2y_2}{(EI)} \quad (10\text{--}17)$$

where (EI) is given by formula (10–15); M is bending moment; other symbols are as defined previously in this chapter; and

$$y_1 = \frac{E_2t_2[(t_1 + t_c) + (t_2 + t_c)] + Et_c\dfrac{s}{b}(t_1 + t_c)}{2\left(E_1t_1 + E_2t_2 + Et_c\dfrac{s}{b}\right)}$$

$$y_2 = \frac{E_1t_1[(t_1 + t_c) + (t_2 + t_c)] + Et_c\dfrac{s}{b}(t_2 + t_c)}{2\left(E_1t_1 + E_2t_2 + Et_c\dfrac{s}{b}\right)}$$

The skins should be thick enough or the stringers spaced closely enough so that buckling does not occur in the compression skin. The buckling stress is given by F_{ccr} (formula (11–7), in ch. 11). The plywood tensile and compressive strength (see ch. 11) should not be exceeded.

Stringer Bending Stress

The stringer bending stress is the larger value given by the formulas

$$f_{sx1} = \frac{ME\left(y_1 - \dfrac{t_1}{2}\right)}{(EI)}$$

$$f_{sx2} = \frac{ME\left(y_2 - \dfrac{t_2}{2}\right)}{(EI)} \quad (10\text{--}18)$$

and these should not exceed appropriate values for the species.

The stringer shear stress is given by the formula

$$f_{sxy} = \frac{V(EQ)}{s(EI)} \quad (10\text{--}19)$$

where $(EQ) = \left(E_1 t_1 b + Es\dfrac{y_1}{2} \right) y_1$ and this should not exceed appropriate values for the species.

Glue Shear Stress

Glue shear stress in the joint between the skins and stringers is given by the formula

$$f_{gl} = \frac{V(EQ)}{s(EI)} \qquad (10\text{--}20)$$

where $(EQ) = E_1 t_1 b y_1$ and this should not exceed values for the glue and species. It should not exceed the wood stress, f_{TR} ("rolling" shear) for solid wood because the thin plywood plies allow the glue shear stresses to be transmitted to adjacent plies to cause rolling shear failure in the wood.

Buckling

Buckling of the stressed-skin panel of unsupported length, L, under end load applied in a direction parallel to the length of the stringers can be computed with the formula

$$P_{cr} = \frac{\pi^2 (EI)}{L^2} \qquad (10\text{--}21)$$

where L is unsupported panel length and (EI) is bending stiffness given by formula (10–15).

Compressive stress in the skins as given by the formula

$$f_{xc1} = \frac{PE_1}{(EA)}$$

$$(10\text{--}22)$$

$$f_{xc2} = \frac{PE_2}{(EA)}$$

and in the stringers as given by the formula

$$f_{sxc} = \frac{PE}{(EA)} \qquad (10\text{--}23)$$

where $(EA) = E_1 t_1 b + E_2 t_2 b + Et_c s$, should not exceed stress values for plywood or stringer material. The plywood stress should also be lower than the buckling stress given by F_{ccr} (formula (11–7), ch. 11).

Loadings

Normally directed loads, uniformly distributed or concentrated, on the panel skins produce deflections and stresses in the plywood

skins that can be computed by formulas for plywood panels given in chapter 11.

Racking loads on panels shear the skins, and buckling of the plywood should not occur due to the shear stress caused by racking. The buckling stress can be computed as F_{scr} from formula (11–18) of chapter 11.

Table 10-1 Spans for Conventional Nailed and Nailed-glued Floor Systems on 16-in. and 24-in. Support Spacings.

Species-Grade	Joist Size	16-in. Not Glued	5/8-in.* Plywood 16-in.	3/4-in.** Plywood 16-in.	24-in. Not Glued	3/4 in.*** Plywood 24-in.
Douglas Fir-Larch-No.2	2x6	9-11	10-6	10-6	8-6	8-7
	2x8	13-1	13-10	13-10	11-3	11-3
	2x10	16-9	17-7	17-7	14-4	14-5
	2x12	20-4	21-5	21-5	17-5	17-6
Douglas Fir-Larch-No.3	2x6	8-0	8-0	8-0	6-7	6-7
	2x8	10-7	10-7	10-7	8-6	8-8
	2x10	13-6	13-6	13-6	11-0	11-0
	2x12	16-5	16-5	16-5	12-7	13-5
Hem-Fir-No.1	2x6	9-6	10-3	10-3	8-4	8-5
	2x8	12-7	13-7	13-7	11-0	11-1
	2x10	16-0	17-2	17-4	14-0	14-2
	2x12	19-6	20-7	21-1	17-0	17-2
Hem-Fir-No.2	2x6	9-4	9-4	9-4	7-7	7-7
	2x8	12-3	12-4	12-4	10-0	10-0
	2x10	15-8	15-8	15-8	12-9	12-10
	2x12	19-1	19-1	19-1	15-7	15-7
Southern Pine No.2	2x6	9-4	9-6	9-6	7-9	7-9
	2x8	12-3	12-7	12-7	10-3	10-3
	2x10	15-8	16-0	16-0	13-0	13-1
	2x12	19-1	19-6	19-6	15-11	15-11
Southern Pine No.2 Medium Grain	2x6	9-9	10-6	10-6	8-5	8-7
	2x8	12-10	13-10	13-10	11-1	11-3
	2x10	16-5	17-5	17-7	14-1	14-5
	2x12	19-11	21-0	21-5	17-1	17-6

*UNDERLAYMENT INI-APA Group 1,2 or 3 may be used for combined subfloor-underlayment. UNDERLAYMENT grade may be 19/32". If separate underlayment or a structural finish floor is installed, C-D APA 5/8", 32/16 or 42/20 may be used.

**UNDERLAYMENT INT-APA Group 1,2,3 or 4 may be used for combined subfloor-underlayment. UNDERLAYMENT grade may be 25/32". If separate underlayment or a structural finish floor is applied, use 3/4" C-D APA.

***UNDERLAYMENT INT-APA Group 1 may be used for combined subfloor-underlayment. UNDERLAYMENT grade may be 25/32". If separate underlayment or structural finish floor is installed, C-D INT-APA 3/4" 48/24 may be used.

Table 10-2 Standard Widths of Laminated Structural Members

Nominal Width of Laminating Lumber	Net Finished Width of Laminated Member
3"	2 1/4"
4"	3 1/8"
6"	5 1/8"
8"	6 3/4"
10"	8 3/4"
12"	10 3/4"
14"	12 1/4"
16"	14 1/4"

BIBLIOGRAPHY

American Institute of Timber Construction
1966. Timber construction manual. John Wiley & Son, New York.

———— 1970. Standard specifications for structural glued-laminated timber of Douglas-fir, western larch, southern pine, and California redwood. AITC 203-70. Englewood, Colo.

American Plywood Association
1966-70. Plywood design specification
Supp. 1, Design of plywood curved panels
Supp. 2, Design of plywood beams
Supp. 3, Design of flat plywood stressed-skin panels
Supp. 4, Design of flat plywood sandwich panels
Tacoma, Wash.

———— 1971. Technical literature on plywood. Tacoma, Wash.

American Society for Testing and Materials
Method of test for integrity of glue joints in laminated wood products for exterior use. ASTM Designation D 1101. (See current edition.) Philadelphia, Pa.

———— Standard methods of testing veneer, plywood, and other glued veneer constructions. ASTM Designation D 805. (See current edition.) Philadelphia, Pa.

———— Standard methods for establishing structural grades and related allowable properties for visually graded lumber. ASTM Designation D 245. (See current edition.) Philadelphia, Pa.

———— Standard methods for establishing clear wood strength values. ASTM Designation D 2555. (See current edition.) Philadelphia, Pa.

Bohannan, Billy
1966. Effect of size on bending strength of wood members. U.S. Forest Serv. Res. Pap. FPL 56. Forest Prod. Lab., Madison, Wis.

———— 1966. Flexural behavior of large glued-laminated beams. U.S. Forest Serv. Res. Pap. FPL 72. Forest Prod. Lab., Madison, Wis.

———— and Moody, R. C.
1969. Large glued-laminated timber beams with two grades of tension laminations. USDA Forest Serv. Res. Pap. FPL 113. Forest Prod. Lab., Madison, Wis.

Douglas-fir Plywood Association
1948. Technical data on plywood. Sec. 7: 1-4, Sec. 9: 1-15. Tacoma, Wash.

Freas, A. D., and Selbo, M. L.
1954. Fabrication and design of glued laminated wood structural members. U.S. Dep. Agr. Tech. Bull. 1069. 220 pp.

Lewis, W. C., and Dawley, E. R.
1943. Stiffeners in box beams and details of design. Forest Prod. Lab. Rep. 1318-A.

————, Heebink, T. B., and Cottingham, W. S.
1944. Buckling and ultimate strengths of shear webs of box beams having plywood face grain direction parallel or perpendicular to the axis of the beams. Forest Prod. Lab. Rep. 1318-D.

————, Heebink, T. B., and Cottingham, W. S.
1944. Effects of certain defects and stress concentrating factors on the strength of tension flanges of box beams. Forest Prod. Lab. Rep. 1513.

————, Heebink, T. B., and Cottingham, W. S.
1945. Effect of increased moisture content on the shear strength at glue lines of box beams and on the glue-shear and glue-tension strengths of small specimens. Forest Prod. Lab. Rep. 1551.

————, Heebink, T. B., Cottingham, W. S., and Dawley, E. R.
1943. Buckling in shear webs of box and I-beams and the effect upon design criteria. Forest Prod. Lab. Rep. 1318-B.

————, Heebink, T. B., Cottingham, W. S., and Dawley, E. R.
1944. Additional tests of box and I-beams to substantiate further the design curves for plywood webs in box beams. Forest Prod. Lab. Rep. 1318-C.

Markwardt, L. J., and Freas, A. D.
1950. Approximate methods of calculating the strength of plywood. Forest Prod. Lab. Rep. 1630.

Moody, R. C., and Bohannan, Billy
1970. Large glued-laminated beams with AITC 301A-69 grade tension laminations. USDA Forest Serv. Res. Pap. FPL 146. Forest Prod. Lab., Madison, Wis.

Newlin, J. A., and Trayer, G. W.
1941. Deflection of beams with special reference to shear deformations. Forest Prod. Lab. Rep. 1309.

Selbo, M. L.
1963. Effect of joint geometry on tensile strength of finger joints. Forest Prod. J. 13(9): 390-400.

————, and Knauss, A. C.
1958. Glued laminated wood construction in Europe. Jour. Struct. Div., Amer. Soc. Civil Eng. Nov.

Timoshenko, S.
1936. Theory of elastic stability. McGraw-Hill, New York.

Trayer, G. W., and March, H. W.
1930. The torsion of members having sections common in aircraft construction. Nat. Adv. Comm. Aeron. Rep. 334.

————, and March, H. W.
1931. Elastic instability of members having sections common in aircraft construction. Nat. Adv. Comm. Aeron. Rep. 382, 42 pp.

U.S. Department of Commerce
Structural glued laminated timber. Voluntary Product Standard PS 56. (See current edition.)

U.S. Forest Products Laboratory
1943. Design of plywood webs for box beams. Forest Prod. Lab. Rep. 1318.

Wilson, T.R.C., and Cottingham, W. S.
1947. Tests of glued laminated wood beams and columns and development of principles of design. Forest Prod. Lab. Rep. R1687.

Chapter 11

PLYWOOD

Plywood is a glued wood panel made up of relatively thin layers, or plies, with the grain of adjacent layers at an angle, usually 90°. The usual constructions have an odd number of plies. If rather thick layers of wood are used as plies, these are often of two layers with the grain directions of each layer parallel. The plywood produced then is often called four-ply or six-ply. The outside plies are called faces or face and back plies, the inner plies are called cores or centers, and the plies immediately below the face and back are called crossbands. The core may be veneer, lumber, or particleboard, with the total panel thickness typically not less than $\frac{1}{16}$ inch or more than 3 inches. The plies may vary as to number, thickness, species, and grade of wood.

As compared with solid wood, the chief advantages of plywood are its approach to having properties along the length nearly equal to properties along the width of the panel, its greater resistance to splitting, and its form, which permits many useful applications where large sheets are desirable. Use of plywood may result in improved utilization of wood, because it covers large areas with a mimimum amount of wood fiber. This is because it is permissible to use plywood thinner than sawn lumber in some applications.

The properties of plywood depend on the quality of the different layers of veneer, the order of layer placement in the panel, the glue used, and the control of gluing conditions in the gluing process. The grade of the panel depends upon the quality of the veneers used, particularly of the face and back. The type of the panel depends upon the glue joint, particularly its water resistance. Generally, face veneers with figured grain that are used in panels where appearance is important have numerous short, or otherwise deformed, wood fibers. These may significantly reduce strength and stiffness of the panels. On the other hand, face veneers and other plies may contain certain sizes and distributions of knots, splits, or growth characteristics that have no undesirable effects on strength properties for specific uses. Such uses include structural applications such as sheathing for walls, roofs, or floors.

The plywood industry expands and develops new products with each passing year. Hence, the reader should always refer directly to current specifications on plywood and its use for specific details.

GRADES AND TYPES OF PLYWOOD

Broadly speaking, two classes of plywood are available—hardwood and softwood. In general, softwood plywood is intended for construction use and hardwood plywood for uses where appearance is important.

Originally, most softwood plywood was made of Douglas-fir, but western hemlock, larch, white fir, ponderosa pine, redwood, southern pine, and other species are now used.

Most softwood plywood used in the United States is produced domestically, and U.S. manufacturers export some material. Generally speaking, the bulk of softwood plywood is used where strength, stiffness, and construction convenience are more important than appearance. Some grades of softwood plywood are made with faces selected primarily for appearance and are used either with clear natural finishes or with pigmented finishes.

Hardwood plywood is made of many different species, both in the United States and overseas. Well over half of all hardwood panels used in the United States is imported. Hardwood plywood is normally used where appearance is more important than strength. Most of the production is intended for interior or protected uses, although a very small proportion is made with glues suitable for exterior service. A significant portion of all hardwood plywood is available completely finished.

The product standard for softwood plywood, construction and industrial, lists some hardwoods as qualifying for use in plywood made to the standard. Similarly, the product standard for hardwood and decorative plywood lists some softwoods.

The glues used in the hardwood and softwood plywood industries are quite different, but each type is selected to provide the necessary performance required by the appropriate specifications.

SPECIFICATIONS FOR PLYWOOD

The most commonly used specifications for plywood are the product standards established by the industry with the assistance of the U.S. Department of Commerce. The specifications are available from the National Bureau of Standards. Separate standards apply to softwood plywood and hardwood plywood. These specifications cover such items as the grading of veneer, the construction of panels, the glue-line performance requirements, and recommendations for use.

Imported plywood is generally not produced in conformance with United States product specifications. However, some countries have their own specifications for plywood manufactured for export to the United States, and these follow the requirements of the domestic product standards.

SPECIAL PLYWOOD

In addition to the all-wood panels, a variety of special plywood panels are made. Paper, plastic, and metal layers are combined with wood veneers, usually as the face layer, to provide special panel characteristics such as improved surface properties.

Special resin-treated papers called "overlays" can be bonded to plywood panels, either on one or both sides. These overlays are either of the "high density" or "medium density" types, and are intended to provide improved resistance to abrasion or wearing, better surfaces when appearance is important or for concrete-form use, or better paint-holding properties. Some panels are also overlaid, usually on the face only, with high-density paper-base decorative laminates, hardboards, or metal sheets. A backing sheet having the same properties (modulus of elasticity, dimensional stability, and vapor transmission rate) as the decorative face sheet must be used to provide a panel free from warp and twist as moisture changes occur. Overlays may be applied in the original layup or they may be applied to plywood after the panels have been surfaced. The two-step method permits a closer thickness tolerance. Requirements for certain types of overlaid softwood panels are included in the aforementioned product standards.

Other special products are embossed, grooved, and other textured panels. Texturing may be achieved by wire brushing or sand blasting. Such products are used primarily as interior paneling and exterior siding.

Prefinished plywood, particularly hardwood plywood, is available in wide variety. The finishes are normally applied in the plywood plants as clear or pigmented liquid finishes. Various printed film patterns are sometimes applied to plywood. Printed panels are also available using liquid finishing systems, and three-dimensional finishes can be achieved by passing the panels under an embossing roller. By these techniques, the appearance for such uses as furniture and wall paneling is improved. Some use is also made of clear, printed, and pigmented plastic films bonded to plywood for the same purpose.

Plywood may be purchased that has been specially treated for protection against fire or decay. It is technically feasible to treat the veneers with chemical solutions and then glue them into plywood. A more general practice is to treat the plywood after gluing, with either fire-retardant or wood-preservative solutions, and then redry the panels. Panels must be glued with durable glues of the exterior type to withstand such treatment.

Large-size panels for special purposes are made by end jointing standard-size panels with scarf joints or finger joints. This is done mainly with softwood panels for structural use as in boats or trailers. Requirements for joints are given in the product standards for conventional plywood.

Curved plywood is sometimes used, particularly in certain furniture items as a specialty product. Much curved plywood is made by gluing the individual veneers to the desired shape and curvature in special jigs or presses. Flat plywood can also be bent to simple curvatures after gluing using techniques similar to those for bending solid wood (see ch. 13).

FACTORS AFFECTING DIMENSIONAL STABILITY OF PLYWOOD

Arrangement of Plies

The tendency of crossbanded products to warp, as the result of uneven shrinking and swelling caused by moisture changes, is largely eliminated by balanced construction. Balanced construction involves plies arranged in similar pairs, a ply on each side of the core. Similar plies have the same thickness, kind of wood with particular reference to shrinkage and density, moisture content at the time of gluing, and grain direction. The importance of having the grain direction of similar plies parallel cannot be overemphasized. A face ply with its grain at a slight angle to the grain of the back ply may result in a panel that will twist excessively with moisture content changes.

An odd number of plies permits an arrangement that gives a substantially balanced effect; that is, when three plies are glued together with the grain of the outer two identical plies at right angles to the grain of the core, the panel is balanced and tends to remain flat through moisture content changes. With five, seven, or some other uneven number of plies, panels may be similarly balanced. Four- or six-ply panels made with the two center plies having grain parallel may also be balanced panels.

Balanced construction is highly important in panels that must remain flat. Conversely, in certain curved members, the natural cupping tendency of an even number of plies may be used to advantage.

Because the outer or face plies of a crossbanded construction are restrained on only one side, changes in moisture content induce relatively large stresses on the outer glue joints. The magnitude of stresses depends upon such factors as thickness of outer plies, density of the veneer involved, and the rate and amount of change in moisture content. In general, the thinner the face veneer the less problem with face checking.

Quality of Plies

In thin plywood where dimensional stability is important, all the plies affect the shape and permanence of form of the panel. All plies should be straight grained, smoothly cut, and of sound wood.

In thick five-ply lumber-core panels the crossbands, in particular, affect the quality and stability of the panel. In such panels thin and dissimilar face and back plies can be used without upsetting the stability of the panel. Imperfections in the crossbands, such as marked differences in the texture of the wood or irregularities in the surface, are easily seen through thin surface veneers. Cross grain that runs sharply through the crossband veneer from one face to the other causes the panels to cup. Cross grain that runs diagonally across the face of the crossband veneer causes the panel to twist unless the two crossbands are laid with their grain parallel. Failure to observe this simple precaution accounts for much warping in crossbanded construction.

In many hardwood plywood uses, both appearance and dimensional stability are important. The best woods for cores of high-grade hardwood panels are those of low density and low shrinkage, of slight contrast between earlywood and latewood, and of species that are easily glued. Edge-grained cores are better than flat-grained cores because they shrink less in width. In softwoods with pronounced latewood, moreover, edge-grained cores are better because the hard bands of latewood are less likely to show through thin veneer, and the panels show fewer irregularities in the surfaces. In most species, a core made entirely of either quartersawed or plainsawed material remains more uniform in thickness through moisture content changes than one in which the two types of material are combined.

Distinct distortion of surfaces has been noted, particularly in softwoods, when the core boards were neither distinctly flat-grained nor edge-grained.

For many uses of softwood plywood, as in sheathing, the appearance, moderate tendencies to warp, and small dimensional changes are of minor importance compared to the strength characteristics of the panel. Strength and stiffness in bending are particularly important. In such panels, veneers are selected mainly to provide strength properties. This selection often permits controlled amounts of knots, splits, and other irregularities that might be considered objectionable from an appearance standpoint.

Moisture Content

The tendency of plywood panels to warp is affected by changes in moisture content as a result of changes in atmospheric moisture conditions or is due to wetting of the surface by free water. Surface appearance may also be affected.

Most plywood is made in hot presses, but other panels are cold-pressed.

Hot-pressed panels come out of the press quite dry. The original moisture content of the veneer and the amount of water added by the glue must be kept low to avoid blister problems in hot-pressing the panels. In addition, water is lost from the glue and the wood during heating.

Cold-pressed panels are generally fairly high in moisture content when removed from the press, the actual values depending on the original moisture content of the veneer, the amount of water in the glue, and the amount of glue spread. Such panels may lose considerable moisture while reaching equilibrium in service.

Differences in the stability and appearance of plywood under service conditions may occur if a change is made from panels produced by one process to those made by the other. However, either type of panel may be used satisfactorily if it is properly designed for the service condition.

Expansion or Contraction

The dimensional stability of plywood, associated with moisture and thermal changes, involves not only cupping, twisting, and bowing but includes expansion or contraction. The usual swelling and shrinking of the wood is effectively reduced because grain directions of adjacent plies are placed at right angles. The low dimensional change parallel to the grain in one ply restrains the normal swelling and shrinking across the grain in the ply glued to it. An additional restraint results from a modulus of elasticity value parallel to the grain of about 20 times that across the grain (see ch. 4). The expansion or contraction of plywood can be closely approximated by the formula

$$\epsilon = \frac{\Sigma m_i E_i t_i}{\Sigma E_i t_i} \qquad (11\text{--}1)$$

where ϵ is the expansion or contraction of the plywood, m_i is the coefficient of expansion of the ith ply over the range of moisture or thermal increase desired, E_i is the modulus of elasticity of the ith ply, and t_i is the thickness of the ith ply. The units of ϵ correspond to the units of m. If m is percent, ϵ is percent; if m is total expansion, ϵ is total expansion. Values of m can be obtained from data given in chapter 3 in the sections on Shrinkage and Thermal Properties. Values of E are obtainable from data given in chapter 4. Plywood expansion or contraction as given by formula (11–1) will be about equal to the parallel-to-grain movement of the veneers.

For all practical purposes, the dimensional change of plywood in thickness does not differ from that of solid wood.

STRUCTURAL DESIGN OF PLYWOOD

The stiffness and strength of plywood can be computed by formulas relating the plywood properties to the construction of the plywood and properties of particular wood species in the component plies. Testing all of the many possible combinations of ply thickness, species, number of plies, and variety of structural components is impractical. The various formulas developed mathematically and presented here were checked by tests to verify their applicability.

Plywood may be used under loading conditions that require the addition of stiffeners to prevent it from buckling. It may also be used in the form of cylinders or curved plates. Such uses are beyond the scope of this handbook, but they are discussed in ANC–18 Bulletin.

It is obvious from its construction that a strip of plywood cannot be so strong in tension, compression, or bending as a strip of solid wood of the same size. Those plies having their grain direction oriented at 90° to the direction of stress can contribute only a fraction of the strength contributed by the corresponding areas of a solid strip, since they are stressed perpendicular to the grain. Strength properties in the directions parallel and perpendicular to the face grain tend to be equalized in plywood, since in some interior plies the grain direction is parallel to the face grain and in others it is perpendicular.

The formulas given in this handbook may be used, in general, for calculating the stiffness of plywood, stresses at proportional limit or ultimate, or for estimating working stresses, depending upon the veneer or wood species property that is substituted in the formulas. Values of the wood properties are given in tables in chapter 4. Chapter 4 also gives values of property coefficients of variation to indicate the variability and reliability of the data as an aid in selection of allowable property values. Modulus of elasticity values given in table 4–2 (ch. 4) should be increased by 10 percent before being used to predict plywood properties; values in table 4–2 represent test data wherein the measured deflection was attributed to bending stiffness and did not consider shear deflection, which effectively increases by about 10 percent the deflection of wood beams having a span-depth ratio of 14:1 as tested.

Properties in Edgewise Compression

Modulus of Elasticity

The modulus of elasticity of plywood in compression parallel to or perpendicular to the face-grain direction is equal to the weighted average of the moduli of elasticity of all plies parallel to the applied load. That is,

$$E_w \text{ or } E_x = \frac{1}{h} \sum_{i=1}^{i=n} E_i h_i \qquad (11\text{--}2)$$

where E_w is the modulus of elasticity of plywood in compression parallel to the face grain; E_x, the modulus of elasticity of plywood in compression perpendicular to the face grain; E_i, the modulus of elasticity parallel to the applied load of the veneer in ply i; h_i, the thickness of the veneer in ply i; h, the thickness of the plywood; and n, the number of plies.

When all plies are of the same thickness and wood species, the formula reduces to

$$E_w = \frac{1}{2n}\left[(E_L + E_T)n + (E_L - E_T)\right]$$
$$(11\text{--}3)$$
$$E_x = \frac{1}{2n}\left[(E_L + E_T)n - (E_L - E_T)\right]$$

where n is the number of plies (n is odd), E_L is the modulus of elasticity of the veneer parallel to the grain, and E_T is the modulus of elasticity of the veneer perpendicular to the grain (see ch. 4). If the veneer is rotary cut, the value of E_T is the modulus of elasticity in the tangential direction. For quarter-sliced veneer, the modulus of elasticity in the radial direction, E_R, should be substituted for E_T.

The modulus of elasticity in compression at angles to the facegrain direction other than 0° or 90° is given approximately by:

$$\frac{1}{E\theta} = \frac{1}{E_w}\cos^4\theta + \frac{1}{E_x}\sin^4\theta + \frac{1}{G_{wx}}\sin^2\theta\cos^2\theta \qquad (11\text{--}4)$$

where E_θ is the modulus of elasticity of a plywood strip in compression at an angle θ to the face grain; G_{wx} is the modulus of rigidity associated with plywood distortion under edgewise shearing forces along axes w (parallel to face grain) and x (perpendicular to face grain); and the other terms are as defined in the formula (11–3). Formulas for computing values of G_{wx} are given under "Properties in Edgewise Shear."

Strength

The compressive strength of plywood subjected to edgewise forces is given by:

$$F_{cw} = \frac{E_w}{E_{cL}} F_{cL}$$
$$(11\text{--}5)$$
$$F_{cx} = \frac{E_x}{E_{cL}} F_{cL}$$

where F_{cw} is the compressive strength of plywood parallel to the face grain; F_{cx}, the compressive strength of plywood perpendicular to the face grain; F_{cL}, the compressive strength of the veneer parallel to the grain; and E_{cL}, the modulus of elasticity of the veneer parallel to the grain. If more than one species is used in the longitudinal plies, values for the species having the lowest ratio of F_{cL}/E_{cL} should be used in the formulas given.

When plywood is loaded at an angle to the face grain, its compressive strength may be computed from:

$$F_{c\theta} \quad \frac{1}{\sqrt{\dfrac{\cos^4\theta}{F_{cw}^2} + \dfrac{\sin^4\theta}{F_{cx}^2} + \left(\dfrac{1}{F_{swx}^2} - \dfrac{1}{F_{cw}F_{cx}}\right)\sin^2\theta\cos^2\theta}} \qquad (11\text{--}6)$$

where $F_{c\theta}$ is the compressive strength of plywood at an angle θ to the face grain, and F_{swx} is the shear strength of plywood under edgewise shearing forces along axes w (parallel to face grain) and x (perpendicular to face grain) (eq. 11–13); and the other terms are as previously defined.

If the plywood is a thin panel, compressive edge loads can cause buckling and subsequent reduction of load-carrying capacity. Plywood panels in stressed-skin constructions must be designed so as to preclude buckling under edge compression loads or bending loads causing edgewise compression on one facing. The critical compressive buckling stress for a plywood panel with face grain parallel to edges and simply supported at four edges is given approximately by the formula:

$$F_{ccr} = \frac{\pi^2}{6}\left(\sqrt{E_{fa}E_{fb}} + 0.17E_L\right)\frac{h^2}{b^2} \quad (11\text{–}7)$$

$$\text{for } a \geq b\left(\frac{E_{fa}}{E_{fb}}\right)^{1/4}$$

where F_{ccr} is the critical compressive buckling stress; h is plywood thickness; b is width of plywood panel loaded edge; a is plywood panel length; E_{fa} and E_{fb} are flexural moduli of elasticity of the plywood in the a and b directions, respectively; and E_L is the modulus of elasticity of the plywood species. Formulas for computing E_f values are elsewhere in this section.

Properties in Edgewise Tension
Modulus of Elasticity

Values of modulus of elasticity in tension are the same as those in compression.

Strength

The strength of a plywood strip in tension parallel or perpendicular to the face grain may be taken as the sum of the strength values of the plies having their grain direction parallel to the applied load. For this purpose, the tensile strength may be taken as equal to the modulus of rupture.

The tensile strength parallel to the face grain will be designated as F_{tw} and the tensile strength perpendicular to the face grain as F_{tx}.

The tensile strength at an angle to the face grain may be computed from:

$$F_{t\theta} = \frac{1}{\sqrt{\dfrac{\cos^4\theta}{F_{tw}^2} + \dfrac{\sin^4\theta}{F_{tx}^2} + \left(\dfrac{1}{F_{swx}^2} - \dfrac{1}{F_{tw}F_{tx}}\right)\sin^2\theta\,\cos^2\theta}} \quad (11\text{–}8)$$

where $F_{t\theta}$ is the tensile strength of plywood at an angle θ to the face grain.

Properties in Edgewise Shear
Modulus of Rigidity

The modulus of rigidity of plywood may be calculated from:

$$G_{wx} = \frac{1}{h}\sum_{i=1}^{i=n}G_i h_i \quad (11\text{–}9)$$

where G_{wx} is the modulus of rigidity of plywood under edgewise shear; G_i is the modulus of rigidity of the ith ply; h_i is the thickness of the ith ply; and h is the plywood thickness.

When the plywood is made of a single species of wood:

$$G_{wx} = G_{LT} \text{ for rotary-cut veneers}$$

$$G_{wx} = G_{LR} \text{ for quarter-sliced veneer}$$

Values of G_{LT} and G_{LR} are given in terms of the modulus of elasticity parallel to grain (E_L) in table 4–2 of chapter 4.

The modulus of rigidity at an angle to the face grain may be computed from:

$$\frac{1}{G_\theta} = \frac{1}{G_{wx}} \cos^2 2\theta + \left[\frac{1}{E_w} + \frac{1}{E_x}\right] \sin^2 2\theta \quad (11\text{–}10)$$

This formula gives a maximum value for G_θ when $\theta = 45°$; thus shear deflections of constructions such as box- and I-beams, wherein plywood webs offer principal resistance to shear, can be reduced by orienting the plywood face grain at 45° to the beam axis. The formula for the special case of $\theta = 45°$ reduces to

$$G_{45°} = \frac{E_w E_x}{E_w + E_x} \quad (11\text{–}11)$$

This formula has a maximum value for plywood arranged to have the same area of parallel grain plies in the two principal directions to produce $E_w = E_x = \frac{1}{2}(E_L + E_T)$. This maximum 45° shear modulus is then

$$\max G_{45°} = \frac{1}{4}(E_L + E_T) \quad (11\text{–}12)$$

For quarter-sliced veneer, E_R is to be substituted for E_T.

Strength

The ultimate strength of plywood elements in shear, with the shearing forces parallel and perpendicular to the face-grain direction, is given by the empirical formula:

$$F_{swx} = 55\,\frac{n-1}{h} + \frac{9}{16h}\sum_{i=1}^{i=n} F_{swxi} h_i \quad (11\text{–}13)$$

where n is the number of plies and F_{swxi} is the shear strength of the ith ply.

In using this formula, the factor $(n-1)/h$ should not be assigned a value greater than 35.

In some commercial grades of plywood, gaps in the core or crossbands are permitted. These gaps reduce the shear strength of plywood, and the formula just given should be corrected to account for this effect. This may be done approximately by subtracting from the number of plies (n) in the first term twice the number of plies containing openings at any one section, and omitting from the summation in the second term all plies containing openings at any one section. Since the first term represents the contribution of the glue layers to shear, twice the number of plies containing openings at any one section is subtracted to account for the lack of glue on each side of the opening. The modification for the effect of core gaps just outlined represents a logically derived procedure not confirmed by test.

When the plywood is stressed in shear at an angle to the face grain, ultimate shear strength with face grain in tension or compression is given by the following formulas:

$$F_{s\theta t} = \frac{1}{\sqrt{\left(\dfrac{1}{F_{tw}^2} + \dfrac{1}{F_{tw}F_{cx}} + \dfrac{1}{F_{cx}^2}\right)\sin^2 2\theta + \dfrac{\cos^2 2\theta}{F_{swx}^2}}} \quad (11\text{–}14)$$

$$F_{s\theta c} = \frac{1}{\sqrt{\left(\dfrac{1}{F_{cw}^2} + \dfrac{1}{F_{cw}F_{tx}} + \dfrac{1}{F_{tx}^2}\right)\sin^2 2\theta + \dfrac{\cos^2 2\theta}{F_{swx}^2}}} \quad (11\text{–}15)$$

These formulas have maximum values for $\theta = 45°$ as did the modulus of rigidity formula. For $\theta = 45°$, the formulas reduce to:

$$F_{s45t} = \cfrac{1}{\sqrt{\cfrac{1}{F_{tw}{}^2} + \cfrac{1}{F_{tw}F_{cx}} + \cfrac{1}{F_{cx}{}^2}}} \quad (11\text{--}16)$$

$$F_{s45c} = \cfrac{1}{\sqrt{\cfrac{1}{F_{cw}{}^2} + \cfrac{1}{F_{cw}F_{tx}} + \cfrac{1}{F_{tx}{}^2}}} \quad (11\text{--}17)$$

If the plywood is a thin panel, edgewise shearing loads can cause buckling and subsequent reduction of load-carrying capacity. Plywood panels in structures, such as webs of I- or box-beams or walls subjected to racking, must be so designed as to preclude buckling due to shearing loads. The critical shear buckling stress for a plywood panel with face grain parallel to edges and simply supported at four edges is given approximately by the formula:

$$F_{scr} = \frac{K_s}{3} \; (E_{fa}E_{fb}{}^3)^{1/4} \; \frac{h^2}{b^2} \quad (11\text{--}18)$$

where F_{scr} is the critical shear buckling stress; h is plywood thickness; E_{fa} and E_{fb} are flexural moduli of elasticity of the plywood in the a and b directions, respectively; K_s is a buckling coefficient given by figure 11–1; and a and b panel dimensions are chosen so that the abscissa quantity in figure 11–1 is ≤ 1.0. If the shear buckling stress is too low for the intended use, the buckling stress can be increased considerably by placing the plywood in shear so that the face grain is in compression and at 45° to the panel edge. Details of design for grain directions other than parallel or perpendicular to panel edges are given in ANC–18 Bulletin.

Properties in Edgewise Bending

For the occasional use where plywood is subjected to edgewise bending, such as in plywood box- and I-beam webs, the values of modulus of elasticity, modulus of rigidity, and strength are the same as those for plywood in compression, tension, or shear, whichever loading is appropriate in the design. If the plywood is a thin panel, edgewise bending can cause buckling, which reduces load-carrying capacity. The

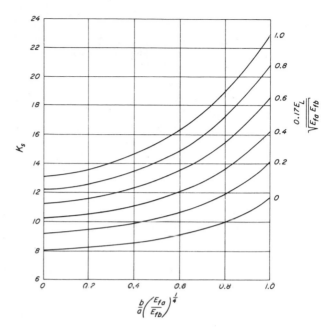

Figure 11–1.—Shear buckling coefficient for plywood panels having face grain parallel to panel edges.

critical buckling stress for a simply supported plywood panel under pure edgewise bending is approximately equal to six times the compression buckling stress; thus

$$F_{bcr} \approx 6F_{ccr} \quad (11\text{--}19)$$

$$\text{for a} \geq 0.7b \left(\frac{E_{fa}}{E_{fb}} \right)^{1/4}$$

Properties in Flexure

The following material pertains to flexure of plywood that causes curvature of the plane of the plywood sheet.

Modulus of Elasticity

The modulus of elasticity in flexure is equal to the average of the moduli of elasticity parallel to the span of the various plies weighted according to their moment of inertia about the neutral plane. That is,

$$E_{fw} \text{ or } E_{fx} = \frac{1}{I} \sum_{i=1}^{i=n} E_i I_i \quad (11\text{--}20)$$

where E_{fw} is the modulus of elasticity of plywood in bending when the face grain is parallel to the span; E_{fx}, the modulus of elasticity of plywood in bending when the face grain is perpendicular to the span; E_i, the modulus of elasticity of the ith ply in the span direction; I_i, the moment of inertia of the ith ply about the neutral plane of the plywood; and I, the moment of inertia of the total cross section about its centerline. When all plies are of the same thickness and wood species, the formula reduces to

$$E_{fw} = \frac{1}{2n^3}\left[(E_L + E_T)n^3 + (E_L - E_T)(3n^2 - 2)\right]$$

$$(11-21)$$

$$E_{fx} = \frac{1}{2n^3}\left[(E_L + E_T)n^3 - (E_L - E_T)(3n^2 - 2)\right]$$

where n is the number of plies (n is odd), E_L is the modulus of elasticity of the veneer parallel to the grain, and E_T is the modulus of elasticity of the veneer perpendicular to the grain (see ch. 4). If the veneer is rotary cut, the value of E_T is the modulus of elasticity in the tangential direction. For quarter-sliced veneer, the modulus of elasticity in the radial direction, E_R, should be substituted for E_T.

The effective moduli of elasticity E_{fw} and E_{fx} are useful in computing the deflections of plywood strips that are subjected primarily to bending on a long span. Deflections due to shear are low for strips on long spans but become important for short spans; they can be computed by analyses given in references by March in the Bibliography for this chapter.

The deflection of a plywood plate simply supported on all four edges also depends on plywood bending stiffness and plate aspect ratio. The center deflection of a plywood plate of width b and length a under a uniformly distributed load of intensity p is given by:

$$w = 0.155K\frac{pb^4}{E_{fb}h^3} \qquad (11-22)$$

where K is given in figure 11–2 and h is plywood thickness. The center deflection of a plywood plate of width b and length a under a center concentrated load P is given by:

$$w = 0.252K\left(\frac{E_{fb}}{E_{fa}}\right)^{1/4}\frac{Pb^2}{E_{fb}h^3} \quad (11-23)$$

where K is given in figure 11–2.

Strength

The resisting moment of plywood strips having face grain parallel to the span is given by:

$$M = 0.85\,\frac{E_{fw}}{E_L}\,\frac{F_b I}{c} \qquad (11-24)$$

For face grain perpendicular to the span,

$$M = 1.15\,\frac{E'_{fx}}{E_L}\,\frac{F_b I}{c} \text{ for three-ply plywood}$$

$$(11-25)$$

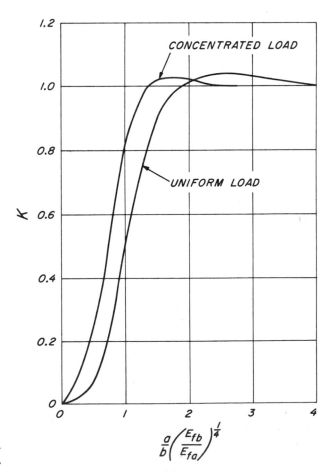

Figure 11–2.—Deflection coefficients for simply supported plywood plates under normal load.

232

$$M = \frac{E'_{fx} F_b I}{E_L \, c} \quad \text{for plywood having} \quad (11\text{–}26)$$

five or more plies

where M is the resisting moment of the plywood; F_b, the strength of the outermost longitudinal ply; c, the distance from the neutral plane to the outer fiber of the outermost longitudinal ply; and E'_{fx} is the same as E_{fx}, except that the outermost ply in tension is neglected, and the other terms are as defined previously.

For plywood having five or more plies, the use of E_{fx} in place of E'_{fx} in calculating the resisting moment will result in negligible error. It should be noted that E'_{fx} is used only in strength calculations and is not to be used in deflection calculations.

Other Design Considerations

Plywood of thin, crossbanded veneers is very resistant to splitting and therefore nails and screws can be placed close together and close to the edges of panels.

Highly efficient, rigid joints can, of course, be obtained by gluing plywood to itself or to heavier wood members such as needed in box-beams and stressed-skin panels. Glued joints should not be designed primarily to transmit load in tension normal to the plane of the plywood sheet because of the rather low tensile strength of wood in a direction perpendicular to the grain. Glued joints should be arranged to transmit loads through shear. It must be recognized that shear strength across the grain of wood (often called rolling shear strength because of the tendency to roll the wood fibers) is considerably less than parallel to the grain (see ch. 4). Thus sufficient area must be provided between plywood and flange members of box-beams and plywood and stringers of stressed-skin construction to avoid shearing failure perpendicular to the grain in the face veneer, in the crossband veneer next to the face veneer, or in the wood member. Various details of design are given in chapter 10.

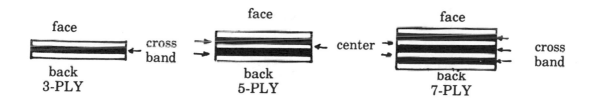

Fig. 11-3. Plywood Construction Terminology

Table 11-4. Panel Constructions

Panel Grades	Finished Panel Nominal Thickness Range (inch)	Minimum Number of Plys	Minimum Number of Layers
Exterior			
Marine	Through 3/8	3	3
Special Exterior (See para. 3,6,7-PSI-74)	Over 3/8, through 3/4	5	5
B-B concrete form	Over 3/4	7	7
High Density Overlay			
High Density concrete form overlay			
Interior			
N-N, N-A, N-B, N-D, A-A, A-B, A-D			
B-B, B-D			
Structural I (C-D, C-D Plugged and	Through 3/8	3	3
Underlayment)	Over 3/8, through 1/2	4	3
Structural II (C-D, C-D Plugged and	Over 1/2, through 7/8	5	5
Underlayment)	Over 7/8	6	5
Exterior			
A-A, A-B, A-C, B-B, B-C			
Structural I and Structural II			
(See para. 3.6.5-PSI-74)			
Medium Density and special overlays			
Interior	(including oracles withexterior glue)		
Underlayment	Through 1/2	3	3
	Over 1/2, through 3/4	4	3
Exterior	Over 3/4	5	5
C-C Plugged			
Interior	(including grades withexterior glue)		
C-D	Through 5/8	3	3
C-D Plugged	Over 5/8, through 3/4	4	3
	Over 3/4	5	5
Exterior			
C-C			

Note: The proportion of wood based on nominal finished panel thickness and dry veneer thickness before layup, as used, with grain running perpendicular to the panel face grain shall fall within the range of 33 percent. The combined thickness of all inner layers shall be not less than 1/2 of panel thickness based on nominal finished panel thickness and dry veneer thickness before layup, as used, for panels with 4 or more plys.

Plywood Grades

Two systems of grading and identification are used for construction and industrial plywood. These are the "Engineered Grades" and the "Appearance Grades." Neither of these grade systems is clearly and consistently stronger than the other, but the Engineered grades are generally more economically suitable for applications where appearance is not highly important.

Engineered grades are usually made from C and D grade veneer and are not sanded. Such panels are designated by thickness, face and back veneer grades, and an Identification Index. Veneer species group is not part of the grade mark for these grades.

Engineered grades are of prime interest to designers of wood structural systems. They are used for roof, floor and wall sheathing and for stress skin panels and built-up I and box beams, and other special structures.

Table 11-5   Veneer grades used in plywood
(Summary ... see PS 1 for complete specifications.)

Veneer Grade	Limiting Characteristics	
N Intended for Natural Finish	Presents smooth surface. Veneer shall be all heartwood or all sapwood free from knots, knotholes, open splits, pitch pockets, other open defects, and stain, but may contain pitch streaks averaging not more than 3/8″ wide blending with color of wood. If joined, not more than two pieces in 48″ width; not more than three pieces in wider panels. Joints parallel to panel edges and well-matched for color and grain. Repairs shall be neatly made, well-matched for color and grain, and limited to a total of six in number in any 4′ x 8′ sheet.	• Maximum of three "router" patches not exceeding 3/4″ x 3-1/2″ admitted. No overlapping. • Shims admitted not exceeding 12″ in length but may occur only at ends of panel. (Examples of permissible combinations: 3 router patches and 3 shims, 2 router patches and 4 shims, 1 router patch and 5 shims, or 6 shims). Suitable synthetic fillers may be used to fill 1/32″ wide checks, splits up to 1/16″ x 2″, and chipped areas or other openings not exceeding 1/8″ x 1/4″.

A	Presents smooth surface.
	 Admits—Pitch streaks blending with color of wood and averaging not more than ⅜″ in width.
	    —Sapwood.
	    —Discolorations.
	 Veneer shall be free from knots, knotholes, splits, pitch pockets and other open defects. If of more than one piece, veneer shall be well joined.
	 Repairs shall be neatly made, parallel to grain, and limited to 18 in number in any 4′ x 8′ sheet, excluding shims; proportionate limits on other sizes.
		Patches of "boat," "router," and "sled" type only, not exceeding 2-1/4″ in width, and may be die-cut if edges are cut clean and sharp. Radius of ends of boat patches shall not exceed 1/8″.
		 • Multiple patching limited to 2 patches, neither of which may exceed 7″ in length if either is wider than 1″.
		 • Shims admitted except over or around patches or as multiple repairs.
		 Suitable synthetic fillers may be used to fill 1/32″ wide checks, splits up to 1/16″ x 2″, and chipped areas or other openings not exceeding 1/8″ x 1/4″.

B	Presents solid surface.
	 Admits—Knots up to 1″ across the grain if both sound and tight.
	    —Pitch streaks averaging not more than 1″ in width.
	    —Discolorations.
	    —Slightly rough but not torn grain, minor sanding and patching defects, including sander skips not exceeding 5% of panel area.
	 Veneer shall be free from open defects except for splits not wider than 1/32″, vertical holes up to 1/16″ in diameter if not exceeding an average of one per square foot in number, and horizontal or surface tunnels up to 1/16″ in width and 1″ in length not exceeding 12 in num-
		ber in a 4′ x 8′ sheet (proportionately on other sizes). Repairs shall be neatly made and may consist of patches, plugs, synthetic plugs and shims.
		 • Patches may be "boat," "router," and "sled" type not exceeding 3″ in width individually when used in multiple repairs or 4″ in width when used as single repairs.
		 • Plugs may be "circular," "dog-bone," and "leaf-shaped," not exceeding 3″ in width when used in multiple repairs or 4″ in width when used as single repairs.
		 • Synthetic plugs shall present a solid, level, hard surface not exceeding above dimensions.
		 Suitable synthetic fillers may be used to fill small splits or openings up to 1/16″ x 2″, and chipped areas or other openings not exceeding 1/8″ x 1/4″.

C	Admits—Tight knots up to 1½″ across the grain.
	    —Knotholes not larger than 1″ across the grain. Also an occasional knothole not more than 1½″ measured across the grain, occurring in any section 12″ along the grain in which the aggregate width of all knots and knotholes occurring wholly within the section does not exceed 6″ in a 48″ width, and proportionately for other widths.
	    —Splits ½″ by one-half panel length; ⅜″ by any panel length if tapering to a point; ¼″ maximum where located within 1″ of parallel panel edge.
	    —Worm or borer holes up to ⅝″ x 1½″.
	    —Open pitch pockets not wider than 1″.
		Repairs shall be neatly made and may consist of patches, plugs, and synthetic plugs.
		 Patches ("boat," including die-cut) not exceeding 3″ in width individually when used in multiple repairs or 4″ in width when used as single repairs. Plugs may be circular, "dog-bone" and leaf-shaped. Synthetic plugs shall present a solid, level, hard surface not exceeding above dimensions.

C
 (plugged)
	Admits—Knotholes, worm or borer holes, and other open defects up to ¼″ x ½″.
	    —Sound tight knots up to 1½″ across the grain.
	    —Splits up to ⅛″ wide.
		—Ruptured and torn grain.
		 —Pitch pockets if solid and tight.
		 —Plugs, patches and shims.

D	D veneer used only in Interior type plywood and may contain plugs, patches, shims, worm or borer holes. Backs: Admits tight knots not larger than 2½″ measured across the grain and knotholes up to 2½″ in maximum dimension. An occasional tight knot larger than 2½″ but not larger than 3″ measured across the grain or knothole larger than 2½″ but not larger than 3″ maximum dimension, occurring in any section 12″ along the grain in which the aggregate width of all knots and knotholes occurring wholly within the section does not exceed 10″ in a 48″ width and proportionately for other widths. Inner Plys: Permits tight knots. Knotholes limited as for backs. —In sanded panels, knotholes not larger than 2½″ maximum dimension in veneer thicker than 1/8″.	—Knotholes not exceeding 3½″ maximum dimension in center ply of 5-ply STANDARD and C-D Plugged grades. All Plys: Pitch pockets not exceeding 2½″ measured across the grain. Splits up to 1″ except in backs only not more than one exceeding ½″; not exceeding ¼″ maximum width where located within 1″ of parallel panel edge; splits must taper to a point. White pocket in inner plys and backs, not exceeding three of the following characteristics in any combination in any area 24″ wide by 12″ long. (a) 6″ width heavy white pocket. (b) 12″ width light white pocket. (c) One knot or knothole or repair 1½″ to 2½″, or two knots or knotholes or repairs 1″ to 1½″.

Table 11-6. Classification of Species

Group 1	Group 2		Group 3	Group 4	Group 5 [a]
Apitong [b][c]	Cedar, Port Orford	Maple, Black	Alder, Red	Aspen	Basswood
Beech, American	Cypress	Mengkulang [b]	Birch, Paper	Bigtooth	Fir, Balsam
Birch	Douglas Fir 2 [d]	Meranti, Red [b][e]	Cedar, Alaska	Quaking	Poplar, Balsam
Sweet	Fir	Mersawa [b]	Fir, Subalpine	Cativo	
Yellow	California Red	Pine	Hemlock, Eastern	Cedar	
Douglas Fir 1 [d]	Grand	Pond	Maple, Bigleaf	Incense	
Kapur [b]	Noble	Red	Pine	Western Red	
Keruing [b][c]	Pacific Silver	Virginia	Jack	Cottonwood	
Larch, Western	White	Western White	Lodgepole	Eastern	
Maple, Sugar	Hemlock, Western	Spruce	Ponderosa	Black (Western	
Pine	Lauan	Red	Spruce	Poplar)	
Caribbean	Almon	Sitka	Redwood	Pine	
Ocote	Bagtikan	Sweetgum	Spruce	Eastern White	
Pine, Southern	Mayapis	Tamarack	Black	Sugar	
Loblolly	Red Lauan	Yellow-poplar	Engelmann		
Longleaf	Tangile		White		
Shortleaf	White Lauan				
Slash					
Tanoak					

(a) Design stresses for Group 5 not assigned.

(b) Each of these names represents a trade group of woods consisting of a number of closely related species.

(c) Species from the genus Dipterocarpus are marketed collectively Apitong if originating in the Philippines; Keruing if originating in Malaysia or Indonesia.

(d) Douglas fir from trees grown in the states of Washington, Oregon, California, Idaho, Montana, Wyoming, and the Canadian Provinces of Alberta and British Columbia shall be classed as Douglas fir No. 1. Douglas fir from trees grown in the states of Nevada, Utah, Colorado, Arizona and New Mexico shall be classed as Douglas fir No. 2.

(e) Red Meranti shall be limited to species having a specific gravity of 0.41 or more based on green volume and oven dry weight.

The Engineered grades are:

C-C Exterior
Structural I, C-C Exterior
Structural II, C-C Exterior
C-D Interior (with or without exterior glue)*
Structural I, C-D Interior, with exterior glue.
Structural II, C-D Interior, with exterior glue.

C-D Interior with exterior glue is the most commonly manufactured and universally available sheathing plywood. It may contain any species group veneer. The Identification Index placed on panels of this grade depends on the species group. The species group itself does not appear on the grade mark.

The Identification Index is in the form of a fraction. The numerator indicates the maximum panel span on roofs and the denominator shows the maximum panel span for floors.

The Structural I and Structural II versions of C-D Interior are both made with exterior adhesives and have some further restrictions on knot sizes and permissible repair or patching. Certain design properties are better than for C-D Interior. Structural I uses only Group 1 species. Structural II uses Group 1, 2 or 3. Structural II is not very common, and even Structural I is unfamiliar to distributors in some areas. Structural I has the highest shear strength (rolling shear) among the various grades, and is especially useful for built-up nailed glued beams, panels and gusset plates.

---

* C-D Interior has, until recently, been called standard and Structural I & II have been regarded as having C-D grade veneer unless otherwise specified.

Table 11-7. Effective Section Properties

All Plies From Same Species Group (Includes STRUCTURAL I and MARINE)

(1) NOMINAL THICKNESS (in.)	(2) APPROXIMATE WEIGHT (psf)	(3) EFFECTIVE THICKNESS FOR SHEAR (in.)	STRESS APPLIED PARALLEL TO FACE GRAIN				STRESS APPLIED PERPENDICULAR TO FACE GRAIN			
			(4) A AREA (in.²/ft)	(5) I MOMENT OF INERTIA (in.⁴/ft)	(6) KS EFF. SECTION MODULUS (in.³/ft)	(7) Ib/Q ROLLING SHEAR CONSTANT (in.²/ft)	(8) A AREA (in.²/ft)	(9) I MOMENT OF INERTIA (in.⁴/ft)	(10) KS EFF. SECTION MODULUS (in.³/ft)	(11) Ib/Q ROLLING SHEAR CONSTANT (in.²/ft)
UNSANDED PANELS										
5/16 - U	1.0	0.356	2.375	0.025	0.144	2.567	1.188	0.002	0.029	—
3/8 - U	1.1	0.371	2.226	0.041	0.195	3.107	1.438	0.003	0.043	—
1/2 - U	1.5	0.403	2.906	0.091	0.318	4.188	1.938	0.007	0.077	2.574
5/8 - U	1.8	0.434	3.464	0.155	0.433	5.268	2.438	0.015	0.122	3.238
3/4 - U	2.2	0.606	3.672	0.247	0.573	6.817	2.938	0.059	0.334	3.697
7/8 - U	2.6	0.776	4.388	0.346	0.690	6.948	3.510	0.192	0.584	5.086
1 - U	3.0	1.088	5.200	0.529	0.922	8.512	6.500	0.366	0.970	6.986
1-1/8 - U	3.3	1.119	6.654	0.751	1.164	9.061	5.542	0.503	1.131	8.675
SANDED PANELS										
1/4 -S	0.8	0.342	1.680	0.013	0.092	2.172	1.226	0.001	0.027	—
3/8 -S	1.1	0.373	1.680	0.038	0.177	3.382	2.126	0.007	0.078	—
1/2 -S	1.5	0.545	1.947	0.078	0.271	4.816	2.305	0.030	0.217	3.076
5/8 -S	1.8	0.576	2.280	0.131	0.361	6.261	2.929	0.077	0.343	3.887
3/4 -S	2.2	0.748	3.848	0.202	0.464	7.926	3.787	0.162	0.570	4.812
7/8 -S	2.6	0.778	3.952	0.288	0.569	7.539	5.759	0.275	0.798	5.671
1 -S	3.0	1.091	5.215	0.479	0.827	7.978	6.367	0.445	1.098	7.639
1-1/8 -S	3.3	1.121	5.593	0.623	0.955	8.840	6.611	0.634	1.356	9.031
TOUCH-SANDED PANELS										
1/2 -T	1.5	0.403	2.698	0.084	0.282	4.246	2.086	0.008	0.082	2.720
19/32 -T	1.7	0.567	3.127	0.124	0.349	5.390	2.899	0.030	0.212	3.183
5/8 -T	1.8	0.575	3.267	0.144	0.378	5.704	3.086	0.037	0.242	3.383
23/32 -T	2.1	0.598	3.337	0.201	0.469	6.582	3.625	0.057	0.322	3.596
3/4 -T	2.2	0.606	3.435	0.226	0.503	6.900	3.825	0.067	0.359	3.786

Table 11-8. Effective Section Properties for Plywood

Table 1. Face Plies of Different Species Group from Inner Plies (Includes all Product Standard Grades except those noted in Table 11-7)

(1) NOMINAL THICKNESS (in.)	(2) APPROXIMATE WEIGHT (psf)	(3) EFFECTIVE THICKNESS FOR SHEAR (in.)	STRESS APPLIED PARALLEL TO FACE GRAIN				STRESS APPLIED PERPENDICULAR TO FACE GRAIN			
			(4) A AREA (in.²/ft)	(5) I MOMENT OF INERTIA (in.⁴/ft)	(6) KS EFF. SECTION MODULUS (in.³/ft)	(7) Ib/Q ROLLING SHEAR CONSTANT (in.²/ft)	(8) A AREA (in.²/ft)	(9) I MOMENT OF INERTIA (in.⁴/ft)	(10) KS EFF. SECTION MODULUS (in.³/ft)	(11) Ib/Q ROLLING SHEAR CONSTANT (in.²/ft)
UNSANDED PANELS										
5/16-U	1.0	0.283	1.914	0.025	0.124	2.568	0.660	0.001	0.023	—
3/8 -U	1.1	0.293	1.866	0.041	0.162	3.108	0.799	0.002	0.033	—
1/2 -U	1.5	0.316	2.500	0.086	0.247	4.189	1.076	0.005	0.057	2.585
5/8 -U	1.8	0.336	2.951	0.154	0.379	5.270	1.354	0.011	0.095	3.252
3/4 -U	2.2	0.467	3.403	0.243	0.501	6.823	1.632	0.036	0.232	3.717
7/8 -U	2.6	0.757	4.109	0.344	0.681	7.174	2.925	0.162	0.542	5.097
1 -U	3.0	0.859	3.916	0.493	0.859	9.244	3.611	0.210	0.660	6.997
1-1/8 -U	3.3	0.877	4.621	0.676	1.047	10.008	3.464	0.307	0.821	8.483
SANDED PANELS										
1/4 -S	0.8	0.304	1.680	0.013	0.092	2.175	0.681	0.001	0.020	—
3/8 -S	1.1	0.313	1.680	0.038	0.176	3.389	1.181	0.004	0.056	—
1/2 -S	1.5	0.450	1.947	0.077	0.266	4.834	1.281	0.018	0.150	3.099
5/8 -S	1.8	0.472	2.280	0.129	0.356	6.293	1.627	0.045	0.234	3.922
3/4 -S	2.2	0.589	2.884	0.197	0.452	7.881	2.104	0.093	0.387	4.842
7/8 -S	2.6	0.608	2.942	0.278	0.547	8.225	3.199	0.157	0.542	5.698
1 -S	3.0	0.846	3.776	0.423	0.730	8.882	3.537	0.253	0.744	7.644
1-1/8 -S	3.3	0.865	3.854	0.548	0.840	9.883	3.673	0.360	0.918	9.032
TOUCH-SANDED PANELS										
1/2 -T	1.5	0.346	2.698	0.083	0.271	4.252	1.159	0.006	0.061	2.746
19/32 -T	1.7	0.491	2.618	0.123	0.337	5.403	1.610	0.019	0.150	3.220
5/8 -T	1.8	0.497	2.728	0.141	0.364	5.719	1.715	0.023	0.170	3.419
23/32 -T	2.1	0.503	3.181	0.196	0.447	6.600	2.014	0.035	0.226	3.659
3/4 -T	2.2	0.509	3.297	0.220	0.477	6.917	2.125	0.041	0.251	3.847
(2-4-1)1-1/8 -T	3.3	0.855	4.592	0.653	0.995	9.933	4.120	0.283	0.763	7.452

Table 11-9. Allowable Stresses for Plywood

Conforming to U.S. Product Standard PS-1-74 for Construction and Industrial Plywood. Normal Load Basis in PSI.

TYPE OF STRESS		SPECIES GROUP of FACE PLY	GRADE STRESS LEVEL *				
			S-1		S-2		S-3
			WET	DRY	WET	DRY	DRY ONLY
EXTREME FIBER STRESS IN BENDING (F_b) TENSION IN PLANE OF PLIES (F_t) FACE GRAIN PARALLEL OR PERPENDICULAR TO SPAN (AT 45° TO FACE GRAIN USE 1/6 F_t)	F_b & F_t	1	1430	2000	1190	1650	1650
		2, 3	980	1400	820	1200	1200
		4	940	1330	780	1110	1110
COMPRESSION IN PLANE OF PLIES. (F_c). PARALLEL OR PERPENDICULAR TO FACE GRAIN (AT 45° TO FACE GRAIN USE 1/3 F_c)	F_c	1	970	1640	900	1540	1540
		2	730	1200	680	1100	1100
		3	610	1060	580	990	990
		4	610	1000	580	950	950
SHEAR IN PLANE PERPENDICULAR TO PLIES PARALLEL OR PERPENDICULAR TO FACE GRAIN (AT 45° TO FACE GRAIN USE 2 F_v)	F_v	1	205	250	205	250	210
		2,3	160	185	160	185	160
		4	145	175	145	175	155
SHEAR, ROLLING, IN THE PLANE OF PLIES PARALLEL OR PERPENDICULAR TO FACE GRAIN (AT 45° TO FACE GRAIN USE 1 1/3 F_s)	F_s	MARINE and STRUCTURAL I	63	75	63	75	
		STRUCTURAL II and 2-4-1	49	56	49	56	55
		ALL OTHER	44	53	44	53	48
MODULUS OF RIGIDITY, OR SHEAR MODULUS IN PLANE OF PLIES	G	1	70,000	90,000	70,000	90,000	82,000
		2	60,000	75,000	60,000	75,000	68,000
		3	50,000	60,000	50,000	60,000	55,000
		4	45,000	50,000	45,000	50,000	45,000
BEARING (ON FACE) PERPENDICULAR TO PLANE OF PLIES	$F_{c\perp}$	1	210	340	210	340	340
		2,3	135	210	135	210	210
		4	105	160	105	160	160
MODULUS OF ELASTICITY IN BENDING IN PLANE OF PLIES. FACE GRAIN PARALLEL OR PERPENDICULAR TO SPAN	E	1	1,500,000	1,800,000	1,500,000	1,800,000	1,800,000
		2	1,300,000	1,500,000	1,300,000	1,500,000	1,500,000
		3	1,100,000	1,200,000	1,100,000	1,200,000	1,200,000
		4	900,000	1,000,000	900,000	1,000,000	1,000,000

To qualify for stress level S-1, gluelines must be exterior and only veneer grades N, A, and C are allowed in either face or back.

For stress level S-2, gluelines must be exterior and veneer grade B, C-plugged and D are allowed on the face or back.

Stress level S-3 includes all panels with interior or intermediate glue lines.

BIBLIOGRAPHY

American Society for Testing and Materials
  Definitions of terms relating to veneer and plywood. ASTM Designation D 1038. (See current edition.) Philadelphia, Pa.

Heebink, B. G.
  1959. Fluid-pressure molding of plywood. Forest Prod. Lab. Rep. 1624.

———
  1963. Importance of balanced construction in plastic-faced wood panels. U.S. Forest Serv. Res. Note FPL–021. Forest Prod. Lab., Madison, Wis.

———, Kuenzi, E. W., and Maki, A. C.
  1964. Linear movement of plywood and flakeboards as related to the longitudinal movement of wood. U.S. Forest Serv. Res. Note FPL–073. Forest Prod. Lab., Madison, Wis.

Liska, J. A.
  1955. Methods of calculating the strength and modulus of elasticity of plywood in compression. Forest Prod. Lab. Rep. 1315 rev.

March, H. W.
  1936. Bending of a centrally-loaded rectangular strip of plywood. Phys. 7(1): 32–41.

———, and Smith, C. B.
  1945. Buckling of flat sandwich panels in compression. Forest Prod. Lab. Rep. 1525.

Norris, C. B.
  1950. Strength of orthotropic materials subjected to combined stresses. Forest Prod. Lab. Rep. 1816.

———, Werren, F., and McKinnon, P. F.
  1948. The effect of veneer thickness and grain direction on the shear strength of plywood. Forest Prod. Lab. Rep. 1801.

Perry, T. D.
  1942. Modern plywood. 366 pp. Pitman Pub., New York.

U.S. Department of Commerce
  Softwood plywood—construction and industrial. U.S. Product Standard PS 1–66. (See current edition.)

———
  Hardwood and decorative plywood. Voluntary Product Standard PS 51–71.

U.S. Department of Defense
  1951. Design of wood aircraft structures. ANC–18 Bull. (Issued by Subcommittee on Air Force-Navy-Civil Aircraft Design Criteria, Aircraft Comm.) 2nd ed. Munitions Board. 234 pp.

U.S. Forest Products Laboratory
  1964. Manufacture and general characteristics of flat plywood. U.S. Forest Serv. Res. Note FPL–064. Madison, Wis.

———
  1964. Bending strength and stiffness of plywood. U.S. Forest Serv. Res. Note FPL–059. Madison, Wis.

———
  1966. Some causes of warping in plywood and veneered products. U.S. Forest Serv. Res. Note FPL–0136. Madison, Wis.

Zahn, J. J., and Romstad, K. M.
  1965. Buckling of simply supported plywood plates under combined edgewise bending and compression. U.S. Forest Serv. Res. Pap. FPL 50. Forest Prod. Lab., Madison, Wis.

11–11

Chapter 12

STRUCTURAL SANDWICH CONSTRUCTION

Structural sandwich construction is a layered construction formed by bonding two thin facings to a thick core (fig. 12–1). In this construction the facings resist nearly all the applied edgewise loads and flatwise bending moments. The thin, spaced facings provide nearly all the bending rigidity to the construction. The core spaces the facings and transmits shear between them so they are effective about a common neutral axis. The core also provides most of the shear rigidity of the sandwich construction. By proper choice of materials for facings and core, constructions with high ratios of stiffness to weight can be achieved. Sandwich construction is also economical, for only small amounts of the relatively expensive facing materials are used and the core materials are usually inexpensive. The materials are positioned so that each is used to its best advantage.

Specific nonstructural advantages can be incorporated in a sandwich construction by proper selection of facing and core materials. An impermeable facing can be employed to act as a moisture barrier for a wall or roof panel in a house; an abrasion-resistant facing can be used for the top facing of a floor panel; and decorative effects can be obtained by using panels with plywood or plastic facings for walls, doors, tables, and other furnishings. Core material can be chosen to provide thermal insulation, fire resistance, and decay resistance.

Care must be used in choice of sandwich components to avoid problems in sound transmission from room to room because of the light weight of the construction. Methods of joining panels to each other and to other structures must be planned so that the joints function properly and allow for possible dimension change due to temperature and moisture change.

The components of the sandwich construction should be compatible with service requirements. Moisture-resistant facings, cores, and adhesives should be employed if the construction is to be exposed to adverse moisture conditions. Similarly, heat-resistant or decay-resistant facings, cores, and adhesives should be used if exposure to elevated temperatures or decay organisms is expected.

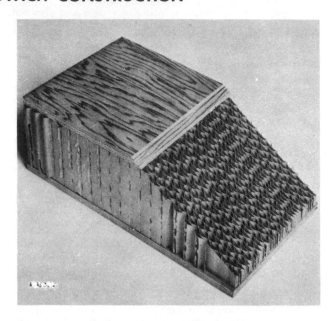

Figure 12–1.—A cutaway section of sandwich construction with plywood facings and a paper honeycomb core.

FABRICATION OF SANDWICH PANELS

Facing Materials

One of the advantages of sandwich construction is the great latitude it provides in choice of facings and the opportunity to use thin sheet materials because of the nearly continuous support by the core. The stiffness, stability, and, to a large extent, the strength of the sandwich are determined by the characteristics of the facings. Some of the different facing materials used include plywood, single veneers or plywood overlaid with a resin-treated paper, hardboard, asbestos cement board, particleboard, fiber-reinforced plastics or laminates, veneer bonded to metal, and such metals as aluminum, enameled steel, stainless steel, or magnesium.

Core Materials

Many lightweight materials, such as balsa wood, rubber or plastic foams, and formed sheets of cloth, metal, or paper, have been used as core for sandwich construction. Cores of formed sheet materials are often called honeycomb cores. By varying the sheet material, sheet thickness, cell size, and cell shape, cores of a wide range in density can be produced.

Various core configurations are shown in figures 12–2 and 12–3. The core cell configurations shown in figure 12–2 can be formed to moderate amounts of single curvature, but cores shown in figure 12–3 of configurations *A*, *B*, and *C* can be formed to severe single curvature and mild compound curvature (spherical).

Four types of readily formable cores are shown as configurations *D*, *E*, *F*, and *G* in figure 12–3. The type *D* and *F* cores form to cylindrical shape, the type *D* and *E* cores to spherical shape, and the type *D* and *G* cores to various compound curvatures.

If the sandwich panels are likely to be subjected to damp or wet conditions, a core of paper honeycomb should contain a synthetic resin. Paper with 15 percent phenolic resin provides good strength when wet, decay resistance, and desirable handling characteristics during fabrication. Resin amounts in excess of about 15 percent do not seem to produce a gain

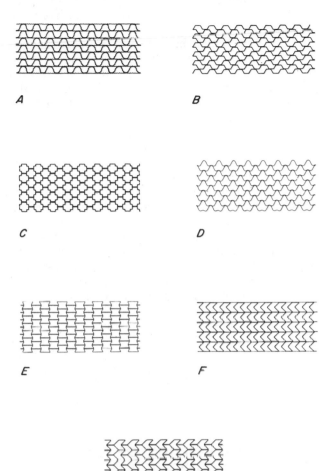

Figure 12–3.—Cell configurations for formable paper honeycomb cores.

in strength commensurate with the increased quantity of resin required. Smaller amounts of resin may be combined with fungicides to offer primary protection against decay.

Manufacturing Operations

The principal operation in the manufacture of sandwich panels is bonding the facings to the core. Special presses are needed for sandwich panel manufacture to avoid crushing lightweight cores; pressures required are usually lower than can be obtained in the range of good pressure control on presses ordinarily used for plywood or plastic manufacture. Because pressure requirements are low, however, simple and perhaps less costly presses could be used. Continuous roller presses or fluid pressure equipment may also be suitable. Certain

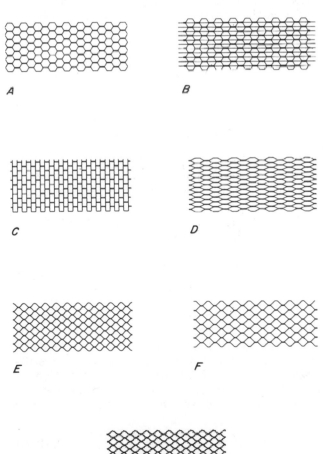

Figure 12–2.—Honeycomb core cell configurations.

special problems arise in the pressing of sandwich panels, but their manufacture is basically not complicated.

Adhesives must be selected to provide the necessary joint strength and permanence, as well as the working properties needed for fabrication of the panels. The facing materials, especially if metallic, may need special surface preparation before the adhesive is applied.

In certain sandwich panels, loading rails or edgings are placed between the facings at the time of assembly. Special fittings or equipment, such as heating coils, plumbing, or electrical wiring conduit, can be placed more easily in the panel during manufacture than after it is completed.

Some of the most persistent difficulties in the use of sandwich panels are caused by the necessity for edges, inserts, and connectors for panels. In some cases, the problem involves tying together thin facing materials without causing severe stress concentrations; in other cases, such as furniture manufacture, the problem is "showthrough" of core or inserts through decorative facings. These difficulties are minimized by a choice of materials in which the rate and degree of differential dimensional movement between core and insert are at a minimum.

STRUCTURAL DESIGN OF SANDWICH CONSTRUCTION

The structural design of sandwich construction may be compared to the design of an I-beam; the facings of the sandwich represent the flanges of the I-beam, and the sandwich core represents the I-beam web. The core of the sandwich, through the bonding adhesive, carries shearing loads and supports the thin facings against lateral wrinkling caused by compressive loads in the facings.

In general, the procedure is to provide facings thick enough to carry the compression and tension stresses, and then to space the facings with a core thick enough to impart stiffness and bending strength to the construction. The core should be strong enough to carry the shearing loads specified for the design. The construction should be checked for possible buckling, as for a column or panel in compression, and for possible wrinkling of the facings.

The core material itself is assumed to contribute nothing to the stiffness of the sandwich construction, because it usually has a low modulus of elasticity. The facing moduli of elasticity are usually at least 100 times as great as the core modulus of elasticity. The core material may also have a small shear modulus. This small shear modulus causes increased deflections of sandwich constructions subjected to bending and decreased buckling loads of columns and edge-loaded panels, compared to constructions in which the core shear modulus is large. The effect of this low shear modulus is greater for short beams and columns and small panels than it is for long beams and columns and large panels.

The bending stiffness of sandwich construction having facings of equal or unequal thickness is given by:

$$D = \frac{E_1 t_1 E_2 t_2 h^2}{E_1 t_1 + E_2 t_2} \qquad (12\text{--}1)$$

where D is the stiffness per unit width of sandwich construction (product of modulus of elasticity and moment of inertia of the cross section); E_1, E_2 are the moduli of elasticity of facings 1 and 2; t_1, t_2 are the facing thicknesses; and h is the distance between facing centroids.

The shear stiffness of sandwich, per unit width, is given by:

$$U = hG_c \qquad (12\text{--}2)$$

where G_c is the core shear modulus associated with distortion of the plane perpendicular to sandwich facings and parallel to the sandwich length.

The bending stiffness, D, and shear stiffness, U, of sandwich construction are used to compute deflections and buckling loads of sandwich panels.

The general expression for the deflection of flat sandwich beams is given by:

$$\frac{d^2 y}{dx^2} = -\frac{M_x}{D} + \frac{1}{U}\frac{dS_x}{dx} \qquad (12\text{--}3)$$

where y is deflection; x is distance along the beam, M_x is bending moment at point x (per unit beam width); S_x is shearing force at point x (per unit beam width); D is flexural stiffness; and U is shear stiffness. Integration of this formula leads to the following general expression for deflection of a sandwich beam:

$$y = \frac{k_b P a^3}{D} + \frac{k_s P a}{U} \qquad (12\text{--}4)$$

where y is deflection, P is total beam load per unit beam width, a is span, and k_b and k_s are constants dependent upon beam loading. The

Table 12–1.—*Values of k_b and k_s for several beam loadings*

Loading	Beam ends	Deflection at—	k_b	k_s
Uniformly distributed _____ { Both simply supported _____		Midspan _____	$\frac{5}{384}$	$\frac{1}{8}$
Both clamped _____		Midspan _____	$\frac{1}{384}$	$\frac{1}{8}$
Concentrated at midspan _____ { Both simply supported _____		Midspan _____	$\frac{1}{48}$	$\frac{1}{4}$
Both clamped _____		Midspan _____	$\frac{1}{192}$	$\frac{1}{4}$
Concentrated at outer quarter points __ { Both simply supported _____		Midspan _____	$\frac{11}{768}$	$\frac{1}{8}$
Both simply supported _____		Load point _____	$\frac{1}{96}$	$\frac{1}{8}$
Uniformly distributed _____	Cantilever, 1 free, 1 clamped _____	Free end _____	$\frac{1}{8}$	$\frac{1}{2}$
Concentrated at free end _____	Cantilever, 1 free, 1 clamped _____	Free end _____	$\frac{1}{3}$	1

first term in the right side of this formula gives the bending deflection and the second term the shear deflection. Values of k_b and k_s for several loadings are given in table 12–1.

If the sandwich panel is supported on all four edges instead of two ends, the deflections and stresses are decreased. For a complete treatment of sandwich plates under loads normal to the plane of the plate see the Bibliography at the end of this chapter.

The buckling load, per unit width, of a sandwich panel with no edge members and loaded as a simply supported column is given by the formula:

$$N = \frac{\pi^2 n^2 D}{a^2 \left(1 + \dfrac{\pi^2 n^2 D}{a^2 U}\right)} \quad (12\text{–}5)$$

where N is the buckling load per unit panel width, a is panel length, D is flexural stiffness, U is shear stiffness, and n is number of half waves into which the panel buckles. A minimum of N is obtained for $n = 1$ and this minimum is then

$$N_{cr} = \frac{\pi^2 D}{a^2 \left(1 + \dfrac{\pi^2 D}{a^2 U}\right)} \quad (n = 1) \quad (12\text{–}6)$$

This buckling form is often called "general buckling" and is illustrated in sketch A of figure 12–4. An upper limit is also obtainable from the formula for N and this is given for $n = \infty$. This limit is often called the "shear instability" limit because the formula for N becomes

$$N_s = U \quad (n = \infty) \quad (12\text{–}7)$$

The appearance of this buckling failure resembles a crimp as illustrated in sketch B of figure 12–4. "Shear instability" or "crimping" failure is always possible for edge-loaded sandwich and is a limit for general instability and not a localized failure.

If the sandwich panel under edge load has edge members, inserts perhaps, the edge members will carry a load proportional to their transformed area (area multiplied by ratio of

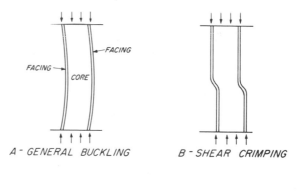

A - GENERAL BUCKLING   B - SHEAR CRIMPING

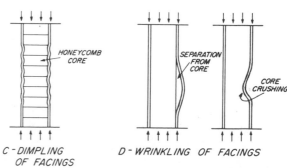

C - DIMPLING OF FACINGS   D - WRINKLING OF FACINGS

Figure 12–4.—Modes of failure of sandwich construction under edgewise loads. A, general buckling; B, shear crimping; C, dimpling of facings; D, wrinkling of facings either away from or into core.

edge member modulus of elasticity to the facing modulus of elasticity). Edge members will also raise the overall panel buckling load because of restraints at edges. Estimates of the effects of edge members can be obtained from Zahn and Cheng. If the edge members are rigid enough to provide simple support but do not clamp the panel edge, an increase in buckling load per unit panel width is obtained. For a sandwich of isotropic facings on an isotropic core, the buckling load for a panel with a length greater than its width (loaded edge) is given by the formula:

$$N_{cr} = \frac{4\pi^2 D}{b^2 \left(1 + \frac{\pi^2 D}{b^2 U} \right)^2} \quad \text{for} \quad \frac{\pi^2 D}{b^2 U} \leq 1$$

(12–8)

$$N_{cr} = U \quad \text{for} \quad \frac{\pi^2 D}{b^2 U} \geq 1$$

where N_{cr} is buckling load per unit panel width and b is panel width. More complete formulas including sandwich with both facings orthotropic, one facing isotropic and one orthotropic, and with orthotropic cores, are given by Kuenzi, Norris, and Jenkinson.

Buckling criteria for flat rectangular sandwich panels under loads other than compression have also been derived. Details for panels in edgewise shear are given by Kuenzi, Ericksen, and Zahn, and for panels in edgewise bending and under combined loading by Kimel.

Buckling of sandwich walls of cylinders has been derived for axial compression loading by Kimel and Stein and Mayers, for torsion loading by March and Kuenzi, and for external pressure by Kuenzi, Bohannan, and Stevens. All are covered in Military Handbook 23A.

Buckling of sandwich components has been emphasized because buckling causes complete failure, usually producing severe shear crimping at the edges of the buckles. Another important factor is the necessity that facing stresses be no more than allowable values at design loads. Facing stresses are obtained by dividing the load by the facing area under load. (Thus for sandwich in compression, the facing stress $f = \frac{N}{2t}$.)

In a strip of sandwich construction subjected to both bending moments and shear loads, the mean facing stresses are given by:

$$f_{cs} = \frac{M}{t_{1,2}h}$$

(12–9)

where $f_{1,2}$ is the mean compression or tension stress in facing 1 or 2; $t_{1,2}$ is the thickness of facing 1 or 2; and M is the bending moment per unit width of sandwich. The shear stress in the core is given by

$$f_{cs} = \frac{S}{h}$$

(12–10)

where f_{cs} is core shear stress and S is applied shear load per unit width of sandwich.

Localized failure of sandwich must be avoided. Such failure is shown as dimpling of the facings in sketch C and as wrinkling of the facings in sketch D of figure 12–4. The stress at which dimpling of the facings into a honeycomb core begins is given approximately by the formula:

$$f_d = 2E \left(\frac{t}{s} \right)^2$$

(12–11)

where f_d is facing stress at beginning of dimpling, E is facing modulus of elasticity, t is facing thickness, and s is cell size of honeycomb core (radius of inscribed circle). Increase in dimpling stress can be attained by decreasing the cell size. Wrinkling of the sandwich facings can occur because of instability of the thin facing supported by a lightweight core which acts as the facing elastic foundation. Wrinkling can occur because of poor facing-to-core bond, resulting in a separation of facing from the core (fig. 12–4, D). Increase in bond strength should produce wrinkling by core crushing. Thus a convenient rule of thumb is to require that the sandwich flatwise tensile strength (bond strength) is no less than flatwise compressive core strength. Approximate wrinkling stress for a fairly flat facing (precluding bond failure) is given by:

$$f_w = \frac{1}{2} (EE_cG_c)^{1/3}$$

(12–12)

where f_w is facing wrinkling stress, E_c is core modulus of elasticity in a direction perpendicular to facing, and G_c is core shear modulus. Localized failure is not accurately predictable and designs should be checked by ASTM tests of small specimens.

Because sandwich constructions are composed of several materials, it is often of interest to attempt to design a minimum weight of the construction for a particular component. Several components and a general analysis for stiffness are discussed by Kuenzi. For a sandwich with similar facings having a required bending stiffness, D, the dimensions to produce the minimum weight sandwich are given by:

$$h = 2 \left(\frac{Dw}{Ew_c} \right)^{1/3} \qquad (12\text{--}13)$$

$$t = \frac{w_c}{4w} h$$

where h is distance between facing centroids, t is facing thickness, E is facing modulus of elasticity, w is facing density, and w_c is core density. The resulting construction will have very thin facings on a very thick core and will be proportioned so that the total core weight is two-thirds the total sandwich weight minus bond weight. It may be a most impractical construction because the required exceedingly thin facings may not be available.

Many detailed design procedures necessary for rapid design of sandwich components for aircraft are summarized in Military Handbook 23A. Although the design of aircraft sandwich has many details not needed for other applications, the principles are broad and can be applied to components of structures other than aircraft.

DIMENSIONAL STABILITY, DURABILITY, AND BOWING OF SANDWICH PANELS

In a sandwich panel any dimensional movement of one facing with respect to the other due to changes in moisture content and temperature causes bowing of an unrestrained panel. Thus, although the use of dissimilar facings is often desirable from an economic or decorative standpoint, the dimensional instability of the facings during panel manufacture or exposure may rule out possible benefits. If dimensional change of both facings is equal, the length and width of the panel will increase or decrease but bowing will not result.

The problem of dimensional stability is chiefly related to the facings, because the core does not have enough stiffness to cause bowing of the panel or to cause it to remain flat. The magnitude of the bowing effect, however, depends on the thickness of the core.

It is possible to calculate mathematically the bowing of a sandwich construction if the per-cent expansion of each facing is known. The maximum deflection is given approximately by:

$$\Delta = \frac{ka^2}{800h} \qquad (12\text{--}14)$$

where k is the percent expansion of one facing as compared to the opposite facing; a, the length of the panel; and h, the total sandwich thickness.

In conventional construction, vapor barriers are often installed to block migration of vapor to the cold side of a wall. Various methods have been tried or suggested for reducing vapor movement through sandwich panels, which causes a moisture differential with resultant bowing of the panels. These include bonding metal foil within the sandwich construction; blending aluminum flakes with the resin bonding adhesives; and using plastic vapor barriers between veneers, overlay papers, special finishes, or metal or plastic facings. Because added cost is likely, some of these methods should not be resorted to unless need for them has been demonstrated.

A large test unit simulating use of sandwich panels in houses was constructed at the Forest Products Laboratory in 1947. The panels used, comprising various facings on paper honeycomb cores, were observed for bowing and general performance. The experimental assembly shown in figure 12–5 represents the type of construction used in the test unit. Details of observed behavior of various sandwich panel constructions after exposure in the test unit for 20 years are given by Sherwood. The major conclusions were that with proper combinations of facings, core, and adhesives, satisfactory sandwich panels can be assured by careful fabrication techniques. This was indicated by results of strength tests conducted on panels exposed for up to 20 years in the unit.

THERMAL INSULATION OF SANDWICH PANELS

Satisfactory thermal insulation can best be obtained with sandwich panels by using cores having low thermal conductivity, although the use of reflective layers on the facings is of some value. Paper honeycomb cores have thermal conductivity values (k values) ranging from 0.30 to 0.65 British thermal units per hour per 1° F. per square foot per inch of thickness, depending on the particular core construction. The k value does not vary linearly with core

thickness for a true honeycomb core because of direct radiation through the core cell opening from one facing to the other. Honeycomb with open cells can also have greater conductivity if cells are large enough (larger than about ⅜ inch) to allow convection currents to develop.

An improvement in the insulation value can be realized by filling the honeycomb core with fill insulation or a foamed-in-place resin.

FIRE RESISTANCE OF SÁNDWICH PANELS

In tests at the Forest Products Laboratory, the fire resistance of wood-faced sandwich panels was appreciably higher than that of hollow panels faced with the same thickness of plywood. Fire resistance was greatly increased when coatings that intumesce on exposure to heat were applied to the core material. The spread of fire through the honeycomb core depended to a large extent on the alinement of the flutes in the core. In panels having flutes perpendicular to the facings, only slight spread of flame occurred. In cores in which flutes were parallel to the length of the panel, the spread of flame occurred in the vertical direction along open channels. Resistance to flame spread could be improved by placing a barrier sheet at the top of the panel or at intervals in the panel height, or, if strength requirements permit, by simply turning the length of the core blocks at 90° to the vertical direction.

M 76939 F

Figure 12–5.—Experimental assembly used to investigate the performance of sandwich panels for house construction.

BIBLIOGRAPHY

American Society for Testing and Materials
  Methods of test for structural sandwich constructions. ASTM Designation. (See current editions.) Philadelphia, Pa.
Baird, P. K., Seidl, R. J., and Fahey, D. J.
  1949. Effect of phenolic resins on physical properties of kraft paper. Forest Prod. Lab. Rep. R1750.
Ericksen, W. S.
  1950. Effects of shear deformation in the core of a flat rectangular sandwich panel—deflection under uniform load of sandwich panels having facings of unequal thickness. Forest Prod. Lab. Rep. 1583-C.

———
  1951. Effects of shear deformation in the core of a flat rectangular sandwich panel—deflection under uniform load of sandwich panels having facings of moderate thickness. Forest Prod. Lab. Rep. 1583-D.
Fahey, D. J., Dunlap, M. E., and Seidl, R. J.
  1953. Thermal conductivity of paper honeycomb cores and sound absorption of sandwich panels. Forest Prod. Lab. Rep. 1952.
Jenkinson, P. M.
  1965. Effect of core thickness and moisture content on mechanical properties of two resin-treated paper honeycomb cores. U.S. Forest Serv. Res. Pap. FPL 35. Forest Prod. Lab., Madison, Wis.
Kimel, W. R.
  1956. Elastic buckling of a simply supported rectangular sandwich panel subjected to combined edgewise bending, compression, and shear. Forest Prod. Lab. Rep. 1857.

———
  1956. Elastic buckling of a simply supported rectangular sandwich panel subjected to combined edgewise bending and compression—results for panels with facings of either equal or unequal thickness and with orthotropic cores. Forest Prod. Lab. Rep. 1857-A.
Kuenzi, E. W.
  1970. Minimum weight structural sandwich. U.S. Forest Serv. Res. Note FPL-086, rev. Forest Prod. Lab., Madison, Wis.
———, Bohannan, B., and Stevens, G. H.
  1965. Buckling coefficients for sandwich cylinders of finite length under uniform external lateral pressure. U.S. Forest Serv. Res. Note FPL-0104. Forest Prod. Lab., Madison, Wis.
———, Ericksen, W. S., and Zahn, J. J.
  1962. Shear stability of flat panels of sandwich construction. Forest Prod. Lab. Rep. 1560, rev.

———, Norris, C. B., and Jenkinson, P. M.
  1964. Buckling coefficients for simply supported and clamped flat, rectangular sandwich panels under edgewise compression. U.S. Forest Serv. Res. Note FPL-070. Forest Prod. Lab., Madison, Wis.
Lewis, W. C.
  1968. Thermal insulation from wood for buildings: Effects of moisture control. USDA Forest Serv. Res. Pap. FPL 86. Forest Prod. Lab., Madison, Wis.
March, H. W., and Kuenzi, E. W.
  1958. Buckling of sandwich cylinders in torsion. Forest Prod. Lab. Rep. 1840, rev.
Norris, C. B.
  1964. Short-column compressive strength of sandwich constructions as affected by size of cells of honeycomb core materials. U.S. Forest Serv. Res. Note FPL-026. Forest Prod. Lab., Madison, Wis.
Raville, M. E.
  1955. Deflection and stresses in a uniformly loaded, simply supported, rectangular sandwich plate. Forest Prod. Lab. Rep. 1847.
Seidl, R. J.
  1952. Paper honeycomb cores for structural sandwich panels. Forest Prod. Lab. Rep. R1918.
———, Kuenzi, E. W., and Fahey, D. J.
  1951. Paper honeycomb cores for structural building panels: Effect of resins, adhesives, fungicide, and weight of paper on strength and resistance to decay. Forest Prod. Lab. Rep. R1796.
Sherwood, G. E.
  1970. Longtime performance of sandwich panels in Forest Products Laboratory experimental unit. USDA Forest Serv. Res. Pap. FPL 144. Forest Prod. Lab., Madison, Wis.
Stein, M., and Mayers, J.
  1952. Compressive buckling of simply supported curved plates and cylinders of sandwich construction. Nat. Adv. Comm. Aeron. Tech. Note 2601.
U.S. Department of Defense
  1968. Structural sandwich composites. Military Handbook 23A. Superintendent of Documents, Washington, D.C.
Zahn, J. J., and Cheng, S.
  1964. Edgewise compressive buckling of flat sandwich panels: Loaded ends simply supported and sides supported by beams. U.S. Forest Serv. Res. Note FPL-019. Forest Prod. Lab., Madison, Wis.
———, and Kuenzi, E. W.
  1963. Classical buckling of cylinders of sandwich construction in axial compression—orthotropic cores. U.S. Forest Serv. Res. Note FPL-018. Forest Prod. Lab., Madison, Wis.

Chapter 13

BENT WOOD MEMBERS

Bending can provide a variety of functional and esthetically pleasing wood members, ranging from large curved arches to small furniture components. Bent wood may be formed with or without softening or plasticizing treatments and with or without end pressure. The curvature of the bend, size of the member, and intended use of the product determine the production method.

LAMINATED MEMBERS

In the United States, curved pieces of wood were once laminated chiefly to produce such small items as parts for furniture and pianos. However, the principle was extended to the manufacture of arches for roof supports in farm, industrial, and public buildings (fig. 13–1) and other types of structural members.

Both softwoods and hardwoods are suitable for laminated bent structural members, and thin material of any species can be bent satisfactorily for such purposes. The choice of species and adhesive depends primarily on the cost, required strength, and demands of the application (see ch. 9 and 10).

Laminated curved members are produced from dry stock in a single bending and gluing operation. This process has several advantages over bending single-piece members:

(1) Bending thin laminations to the required radius involves only moderate stress and deformation of the wood fibers, eliminating the need for treatment with steam or hot water and associated drying and conditioning of the finished product.

(2) Because of the moderate stress induced in bending, stronger members are produced.

(3) The tendency of laminated members to change shape with changes in moisture content is less than that of single-piece bent members.

(4) Ratios of thickness of member to radius of curvature that are impossible to obtain by bending single pieces can be attained readily by laminating.

(5) Curved members of any desired length can be produced by staggering the joints in the laminations.

Design criteria for glued-laminated timbers are discussed in chapter 10.

Straight laminated members also can be steamed and bent after they are glued. However, this type of procedure requires an adhesive that will not be affected by the steaming or boiling treatment and complicates conditioning of the finished product.

CURVED PLYWOOD

Curved plywood is produced (1) by bending and gluing the plies in one operation, or (2) by bending previously glued flat plywood. Curved plywood made by method (1) is more stable in curvature than plywood curved by method (2).

Plywood Bent and Glued Simultaneously

In bending and gluing plywood in a single operation, glue-coated pieces of veneer are assembled and pressed over or between curved forms; pressure and sometimes heat are applied through steam or electrically heated forms until the glue sets and holds the assembly to the desired curvature. Some of the laminations are at an angle, usually 90°, to other laminations, as in the manufacture of flat plywood. The grain direction of the thicker laminations is normally parallel to the axis of the bend to facilitate bending.

A high degree of compound curvature can be obtained in an assembly comprising a considerable number of thin veneers. First, for both the face and back of the assembly, the two outer plies are bonded at 90° to each other in a flat press. The remaining veneers are then glue-coated and assembled at any desired angle to each other. The entire assembly is hot pressed to the desired curvature.

Bonding the two outer plies before molding allows a higher degree of compound curvature without cracking the face plies than could otherwise be obtained. Where a high degree of compound curvature is required, the veneer should be relatively thin, $1/32$ inch or less, with a moisture content of about 12 percent.

The advantages of bending and gluing plywood simultaneously to form a curved shape are similar to those for curved laminated members, and in addition, the cross plies give the curved members properties characteristic of cross-banded plywood. Curved plywood shells for furniture manufacture are examples of these bent veneer and glued products.

Molded Plywood

Although any piece of curved plywood may properly be considered to be molded, the term "molded" is usually reserved for plywood that is glued to the desired shape, either between curved forms or with fluid pressure. The molding of plywood with fluid pressure applied by flexible bags of some impermeable material produces plywood parts of various degrees of compound curvature. In "bag molding" fluid pressure is applied through a rubber bag by air, steam, or water. The veneer may be wrapped around a form and the whole assembly enclosed in a bag and subjected to pressure in an autoclave, the pressure in the bag being "bled." Or the veneer may be inserted inside a metal form and, after the ends have been attached and sealed, pressure applied by inflating a rubber bag. The form may be heated electrically or by steam.

Plywood Bent After Gluing

After the plies are glued together, flat plywood is often bent by methods that are somewhat similar to those used in bending solid wood. To bend plywood properly to shape, it must be plasticized by some means, usually moisture or heat, or a combination of both. The amount of curvature that can be introduced into a flat piece of plywood depends on numerous variables, such as moisture content, direction of grain, thickness and number of plies, species and quality of veneer, and the

Figure 13–1.—Curved laminated arch provides pleasing lines in this building. The laminations are bent without end pressure against a form and glued together.

technique applied in producing the bend. Plywood is normally bent over a form or a bending mandrel.

Flat plywood glued with a waterproof adhesive can be bent to compound curvatures after gluing. No simple criterion, however, is available for predetermining whether a specific compound curvature can be imparted to flat plywood. Soaking the plywood and the use of heat during forming are aids in manipulation. Normally the plywood to be postformed is first thoroughly soaked in hot water and then dried between heated male and female dies attached to a hydraulic press. If the use of postforming for bending flat plywood to compound curvatures is contemplated, exploratory trials to determine the practicability and the best procedure are recommended. It should be remembered that in postforming plywood to compound curvatures, all of the deformation must be by compression or shear, as plywood cannot be stretched. Hardwood species, such as birch, poplar, and gum, are normally used in plywood that is to be postformed.

VENEERED CURVED MEMBERS

Veneered curved members are usually produced by gluing veneer to one or both faces of a curved solid wood base. The bases are ordinarily bandsawed to the desired shape or bent from a piece grooved with saw kerfs on the concave side at right angles to the directions of bend. Pieces bent by making saw kerfs on the concave side are commonly reinforced and kept to the required curvature by gluing splines, veneer, or other pieces to the curved base.

Veneering over curved solid wood finds use mainly in furniture. The grain of the veneer is commonly laid in the same general direction as the grain of the curved wood base. The use of crossband veneers, that is, veneers laid with the grain at right angles to the grain of the base and face veneer, reduces the tendency of the member to split.

SOLID WOOD MEMBERS

With material thicker than veneer, some type of softening or plasticizing treatment normally is required to bend solid wood to sharp curvatures. Bent solid wood members are used primarily as furniture parts (fig. 13–2), boat frames, implement handles, and in manufacture of sporting goods.

Figure 13–2.—A chair with bent solid wood parts.

In general, hardwoods possess better bending quality than softwoods for this type of member. The species commonly used to produce solid bent members are white oak, red oak, elm, hickory, ash, beech, birch, maple, walnut, sweetgum, and mahogany. Some softwoods can be used, however, including yew and Alaska-cedar, and Douglas-fir, southern pine, northern and Atlantic white-cedar, and redwood are used for boat planking. Solid lumber for boat planks is often bent to moderate curvature after being steamed or soaked.

Bending Process

When a piece of wood is bent along its length, it is stretched in tension along the convex side of the bend and compressed along the concave side. If bending involves severe deformation, most of the deformation should be forced to take place as compression.

Wood softened with moisture and heat or by plasticizing with chemicals can be compressed considerably but stretched very little. In bending, therefore, the wood must be compressed lengthwise while restraining it from stretching along the convex side. Various devices have been developed to accomplish this, the most efficient being a tension strap complete with end blocks or clamps or a tension strap with a reversed lever and end blocks. In hot-plate pressing operations a metal pan is fitted with end bars to provide needed end pressure. Other devices have been developed for bending solid wood, and most involve forcing the wood member against a form.

Hand- and machine-bent members restrained by end pressure are cooled and dried while held in their curved shapes. When the bent member has dried to a moisture content suited for its application, the restraining devices are removed and the piece will hold its curved shape.

Bending Stock

Bending stock should be free from serious cross grain and distorted grain, such as may occur near knots. The slope of grain should not be greater than about 1 to 15. Knots, decay, surface checks, shake, pith, and exceptionally light or brashy wood should be avoided.

Although green wood can be bent to produce most curved members, difficulties are introduced in drying and fixing the bent piece and reducing the moisture content to a level suited for the end use. Bending stock that has been dried to a low moisture content (12 to 20 pct.) requires steaming or soaking to increase its moisture content to the point where it is sufficiently plastic for successful bending.

Wood of poor bending quality can be improved by gluing veneer of good bending quality to the surface that is to be concave. The veneer assumes the maximum amount of compressive deformation and supports the inner face of the wood. Wood treated with preservatives can be bent satisfactorily.

Plasticizing Bending Stock

Steaming at atmospheric or low gage pressure or soaking in boiling or nearly boiling water are satisfactory methods of plasticizing many wood species for bending. Heat and moisture are added to wood below 20 percent moisture content, and wood at moisture contents of 20 to 25 percent is heated while the moisture is retained. Steaming at high pressures causes wood to become plastic, but wood treated with high-pressure steam generally does does not bend as well as wood treated at low or atmospheric pressure.

Wood can be plasticized by a great variety of chemicals. Such chemicals behave like water, in that they are adsorbed and cause swelling. Common chemicals that plasticize wood include urea, dimethylol urea, low-molecular weight phenol-formaldehyde resin, dimethylol urea, low-molecular weight phenol-formaldehyde resin, dimethyl sulfoxide, and liquid ammonia. Urea and dimethylol urea have received limited commercial attention, and a free-bending process using liquid ammonia has been patented. Wood members after immersion in liquid ammonia or treatment under pressure with ammonia in the gas phase can be readily molded or shaped. As the ammonia evaporates, the wood stiffens and retains its new shape.

BIBLIOGRAPHY

Clark, W. M.
    1965. Veneering and wood bending in the furniture industry. Pergamon Press, New York. 120 pp.
Davidson, R. W.
    n.d. Plasticizing wood with anhydrous ammonia. New York State College of Forestry, Syracuse. 4 pp.
Dean, A. R.
    1967. Precompressed and flexible wood. Furniture Ind. Res. Assoc. (England). Bull. 20. Dec.
de Bat, A.
    1965. Bent wood best for curved components. Furniture Design and Manufacturing. Aug.
Fessel, F.
    1951. Problems in wood bending. Holz als Roh -und Werkstoff 9(4): 151–158.
Forest Products Research Laboratory
    1958. The steam bending properties of various timbers. FPRL Leafl. 45, Princes Risborough, England.

    ──────
    1959. The bending of solid timber. FPRL Leafl. 33, Princes Risborough, England.
Heebink, B. G.
    1959. Fluid-pressure molding of plywood. Forest Prod. Lab. Rep. 1624.
Hurst, K.
    1962. Plywood bending. Australian Timber J. June.

Jorgensen, R. N.
    1965. Furniture wood bending, Part I. Furniture Design and Manufacturing. Dec.

    ──────
    1966. Furniture wood bending, Part II. Furniture Design and Manufacturing. Jan.
McKean, H. B., Blumenstein, R. R., and Finnorm, W. F.
    1952. Laminating and steam bending of treated and untreated oak for ship timbers. Southern Lumberman 185: 2321.
Peck, E. C.
    1957. Bending solid wood to form. U.S. Dep. Agr. Handbook 125, 37 pp.
Perry, T. D.
    1951. Curves from flat plywood. Wood Prod. 56(4).
Schuerch, C.
    1964. Principles and potential of wood plasticization. Forest Prod. J. 14(9): 377–381.
Stevens, W. C., and Turner, N.
    1948. Solid and laminated wood bending. Her Majesty's Stationery Office, London. 67 pp.
──────, and Turner, N.
    1950. A method of improving the steam bending properties of certain timbers. Wood 15(3).
──────, and Turner, N.
    1964. Preservative-treated beech for bent work. Ship and Boat Builder Int. Oct.
Turner, N.
    1965. New and improved solid and laminated wood bending techniques developed at the FPRL. Timber and Plywood Annu. (England).

Chapter 14

CONTROL OF MOISTURE CONTENT AND DIMENSIONAL CHANGES

Correct seasoning, handling, and storage of wood will minimize moisture content changes that might occur in service. If moisture content is controlled within reasonable limits by such methods, major problems from dimensional changes will be avoided. Wood is subject naturally to dimensional changes. In the living tree, wood contains large quantities of water. As green wood dries, most of this water is removed. The moisture remaining in the wood tends to come to equilibrium with the relative humidity of the air. Also, when the moisture content is reduced below the fiber saturation point, shrinkage starts to occur.

This discussion is concerned with moisture content determination, recommended moisture content values, seasoning methods, methods of calculating dimensional changes, design factors affecting such changes in structures, and moisture content control during transit, storage, and construction. Data on green moisture content, fiber saturation point, shrinkage, and equilibrium moisture content are given with information on other physical properties in chapter 3.

Wood in service is virtually always undergoing at least slight changes in moisture content. The changes in response to daily humidity changes are small and usually of no consequence. Changes due to seasonal variation, although gradual, tend to be of more concern. Protective coatings retard changes but do not prevent them.

Generally, no significant dimensional changes will occur if wood is fabricated or installed at a moisture content corresponding to the average atmospheric conditions to which it will be exposed. When incompletely seasoned material is used in construction, some minor changes can be tolerated if the proper design is used.

DETERMINATION OF MOISTURE CONTENT

The amount of moisture in wood is ordinarily expressed as a percentage of the weight of the wood when ovendry. Four methods of determining moisture content are covered by Designation D 2016 of the American Society for Testing and Materials. Two of these, the ovendrying and the electrical method, are described here.

Ovendrying has been the most universally accepted method for determining moisture content, but it is slow and necessitates cutting the wood. In addition it gives values slightly higher than true moisture content with woods containing volatile extractives. The electrical method is rapid, does not require cutting the wood, and can be used on wood in place in a structure. However, considerable care must be taken to use and interpret the results correctly. Generally, use of the electrical method is limited to moisture content values below 30 percent.

Ovendrying Method

In the ovendrying method, specimens are taken from representative boards or pieces of a quantity of lumber or other wood units. With lumber, the specimens should be obtained at least 20 inches from the ends of the pieces. They should be free from knots and other irregularities, such as bark and pitch pockets. Specimens from lumber should be full cross sections 1 inch along the grain. Specimens from larger items may be representative sectors of such sections or subdivided increment borer or auger chip samples. Convenient amounts of chips and particles can be selected at random from larger batches, with care being taken to insure that the sample is representative of the batch. Samples of veneer should be selected from four or five locations in a sheet to insure that the sample average will accurately indicate the average of the sheet.

Each specimen should be weighed immediately, before any drying or reabsorption of moisture has taken place. If the specimen cannot be weighed immediately after it is taken, it should be placed in a plastic bag or tightly wrapped in metal foil to protect it from moisture change until it can be weighed. After weighing, the specimen is placed in an oven heated to 214° to 221° F. (101° to 105° C.) and kept there until constant weight is reached. A lumber section will reach a constant weight in 12 to 48 hours. Smaller specimens will take less time.

The constant or ovendry weight and the weight of the specimen when cut are used to determine the percentage moisture content with the following formula:

$$\text{Percent moisture content} =$$
$$\frac{\text{Weight when cut} - \text{Ovendry weight}}{\text{Ovendry weight}} \times 100$$
$$(14-1)$$

Electrical Method

The electrical method for determining moisture content makes use of such properties of wood as its resistance, dielectric constant, and power-loss factor. Accurate moisture meters for solid wood items are commercially available. The instruments determine the moisture content through its effect upon the direct-current electrical resistance of wood (resistance-type meters) or its effect on a capacitor in a high-frequency circuit in which the wood serves as the dielectric material (power-loss and capacitive admittance meters).

The principal advantages of the electrical method over the ovendrying method are its speed and convenience. Only a few seconds are required, and the piece of wood being tested is not cut or damaged, except for driving a few electrode needle points into the wood when using the resistance-type meters. Thus, the electrical method is adaptable to rapid sorting of lumber on the basis of moisture content, measuring the moisture content of wood installed in a building, or, when used in accordance with the ASTM Designation D 2016, establishing the moisture content of a quantity of lumber or other wood items.

For resistance meters, the $\frac{5}{16}$- to $\frac{7}{16}$-inch needle electrodes ordinarily supplied are appropriate for wood that has been in use for 6 months or longer, or for lumber up to $1\frac{1}{2}$ inches thick with a normal drying moisture gradient. For wood with normal moisture gradients, the pins should be driven to a depth of one-fifth to one-fourth of the wood thickness. If other than normal drying gradients are present, best accuracy can be obtained by exploring the gradient through readings made at various penetration depths.

Radiofrequency power loss meters are supplied with electrodes appropriate to the type and size of material to be tested. The field from the electrodes should penetrate roughly to the middle of the specimen.

Ordinarily, moisture meters should not be used on lumber with wet or damp surfaces, because the wet surface will cause inaccurate readings. A resistance meter with insulated-pin electrodes can be used, with caution, on such stock.

Although some meters have scales that go up to 120 percent, the range of moisture content that can be measured reliably is 0 to about 30 percent for radiofrequency meters and about 6 to 30 percent for resistance meters. The precision of the individual meter readings decreases near the limits of these ranges. Any readings above 30 percent must be considered only qualitative. When the meter is properly used on a quantity of lumber dried below fiber saturation, the average moisture content from the corrected meter readings should be within 1 percent of the true average.

To obtain accurate moisture content values, each instrument should be used in accordance with its manufacturer's instructions. The electrodes should be appropriate for the material being tested and properly oriented. The readings should be carefully taken as soon as possible after inserting the electrode. A species correction supplied with the instrument should be applied when appropriate. Temperature corrections then should be made for resistance-type meters if the temperature of the wood differs considerably from the temperature of calibration used by the manufacturer. Approximate corrections are to add or subtract about 0.5 percent for each 10° F. the wood differs from the calibration temperature; the correction factors are added to the readings for temperatures below the calibration temperature and subtracted from the readings for temperatures above this temperature.

RECOMMENDED MOISTURE CONTENT

Installation of wood at the moisture content percentages recommended here for different environments will reduce future changes in moisture content, thus minimizing dimensional changes after the wood is placed in service. The service condition to which the wood will be exposed—outdoors, in unheated buildings, or in heated and air-conditioned buildings—should be considered in determining moisture content requirements.

Timbers

Ideally, solid timbers should be seasoned to the average moisture content they will reach in service. While this optimum is possible with lumber less than 3 inches thick, it is seldom practical to obtain fully seasoned timbers, thick joists, and planks. When thick solid members are used, some shrinkage of the assembly should be expected. In the case of builtup assemblies such as roof trusses, it may be necessary to tighten the bolts or other fastenings from time to time as the members shrink.

Lumber

The moisture content requirements are more exacting for finish lumber and wood products used inside heated and air-conditioned buildings than those for lumber used outdoors or in unheated buildings. For general areas of the United States, the recommended moisture content values for wood used inside heated buildings are shown in figure 14–1. Values and tolerances both for interior and exterior use of wood in various forms are given in table 14–1. If the average moisture content value is within 1 percent of that recommended and all pieces fall within the individual limits, the entire lot is probably satisfactory.

General commercial practice is to kiln-dry wood for some products, such as flooring and furniture, to a slightly lower moisture content than service conditions demand, anticipating a moderate increase in moisture content during processing and construction. The practice is intended to assure uniform distribution of moisture among the individual pieces. Common grades of softwood lumber and softwood dimension are not normally seasoned to the moisture content values indicated in table 14–1. When they are not, shrinkage effects should be considered in the structural design and construction methods.

The American Softwood Lumber Standard requires that, to be classified as dry lumber, moisture content shall not exceed 19 percent. Much softwood dimension lumber meets this requirement. Some industry grading rules provide for even lower maximums. For example, to be grademarked KD (kiln dry) the maximum moisture content permitted is generally 15 percent.

Glued Wood Products

When veneers are bonded together with cold-setting glues to make plywood, they absorb comparatively large quantities of moisture. To keep the final moisture content low and to minimize redrying of the plywood, the initial moisture content of the veneer should be as low as practical. Very dry veneer, however, is difficult to handle without damage, so the minimum practical moisture content is about 4 percent. Freshly glued plywood intended for interior service should be dried to the moisture content values given in table 14–1.

Hot-pressed plywood and other board products, such as particleboard and hardboard, often do not arrive at the same equilibrium moisture content values given for lumber. The high temperatures used in hot presses cause these products to assume a lower moisture content for a given relative humidity. Since this lower equilibrium moisture content varies widely, depending on the specific type of hot-pressed product, it is recommended that such products be conditioned at 40 to 50 percent relative humidity for interior use and 65 percent for exterior use.

Lumber used in the manufacture of large laminated members should be dried to a moisture content slightly below the moisture content expected in service; thus, moisture absorbed from the glue will not cause the moisture content of the product to exceed the service value. The range of moisture content among laminations assembled into a single member should not exceed 5 percent. Although laminated members are often massive and respond rather slowly to changes in environmental conditions, it is desirable to follow the recommendations in table 14–1 for moisture content at time of installation.

SEASONING OF WOOD

Well-developed techniques have been established for removing the large amounts of moisture normally present in green wood (ch. 3). Seasoning is essentially a drying process but, for uses that require them, seasoning includes equalizing and conditioning treatments to improve moisture uniformity and relieve residual stresses and sets. Careful techniques are necessary, especially during the drying phase, to protect the wood from stain and decay and from excessive drying stresses that cause defects and degrade. The established seasoning methods are air drying, accelerated air drying, and kiln drying. Other methods, such as high-frequency dielectric heating, vapor drying, and solvent seasoning have been developed for special uses.

Drying reduces the weight of wood, with a resulting decrease in shipping costs; reduces or eliminates shrinkage, checking, and warping in service; increases strength and nail-holding power; decreases susceptibility to infection by blue stain and other fungi; reduces chance of attack by insects; and improves the

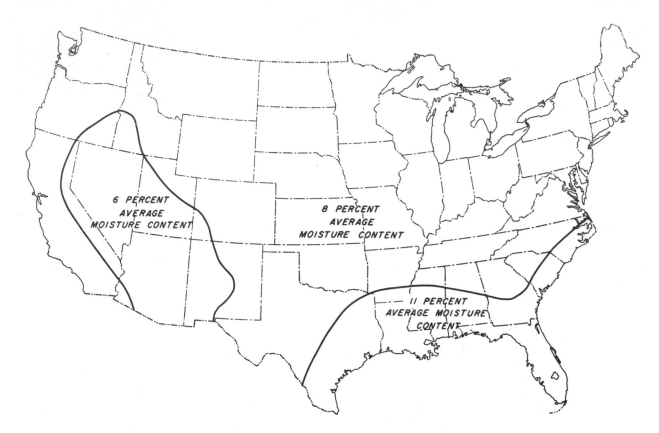

Figure 14–1.—Recommended average moisture content for interior use of wood products in various areas of the United States.

Table 14–1.—*Recommended moisture content values for various wood items at time of installation*

Use of wood	Moisture content for—					
	Most areas of United States		Dry southwestern area [1]		Damp, warm coastal areas [1]	
	Average [2]	Individual pieces	Average [2]	Individual pieces	Average [2]	Individual pieces
	Pct.	Pct.	Pct.	Pct.	Pct.	Pct.
Interior: Woodwork, flooring, furniture, wood trim, laminated timbers, cold-press plywood	8	6–10	6	4–9	11	8–13
Exterior: Siding, wood trim, framing, sheathing, laminated timbers	12	9–14	9	7–12	12	9–14

[1] Major areas are indicated in fig. 14–1.
[2] To obtain a realistic average, test at least 10 pct. of each item. If the amount of a given item is small, several tests should be made. For example, in an ordinary dwelling having about 60 floor joists, at least 10 tests should be made on joists selected at random.

capacity of wood to take preservative and fire-retardant treatment and to hold paint.

Sawmill Practice

It is common practice at most softwood sawmills to kiln dry all upper grade lumber intended for finish, flooring, and cut stock. Lower grade boards are often air dried. Dimension lumber is air dried or kiln dried, although some mills ship certain species without seasoning. Timbers are generally not held long enough to be considered seasoned, but some drying may take place between sawing and shipment or while they are held at a wholesale or distributing yard. Sawmills cutting hardwoods commonly classify the lumber for size and grade at the time of sawing. Some mills send all freshly sawed stock to the air-drying yard or an accelerated air-drying operation. Others kiln dry directly from the green condition. Air-dried stock is kiln dried at the sawmill, at a custom drying operation during transit, or at the remanufacturing plant before being made up into such finished products as furniture, cabinet work, interior finish, and flooring.

Air Drying

Air drying is not a complete drying process, except as preparation for uses for which the recommended moisture content is not more than 5 percent below that of the air-dry stock. Even when air-drying conditions are mild, air-dry stock used without kiln drying may have some residual stress and set that can cause distortions after nonuniform surfacing or machining. On the other hand, rapid air drying accomplished by low relative humidities produces a large amount of set that will assist in reducing warp during final kiln drying. Rapid surface drying also greatly decreases the incidence of chemical and sticker stain, blue strain, and decay.

Air drying is an economical seasoning method when carried out (1) in a well-designed yard or shed, (2) with proper piling practices, and (3) in favorable drying weather. In cold or humid weather, air drying is slow and cannot readily reduce wood moisture to levels suitable for rapid kiln drying or for use.

Accelerated air drying involves the use of fans to force the air through the lumber piles in a shed or under other protection from the weather. Sometimes small amounts of heat are used to reduce relative humidity and slightly increase temperature. Accelerated air drying to moisture content levels between 20 and 30 percent may take only one-half to one-fourth as long as ordinary air drying. Moisture content in the stock dried with such acceleration may vary somewhat more than that of stock air dried under natural conditions to the same average moisture level.

Kiln Drying

In kiln drying, higher temperatures and fast air circulation are used to increase the drying rate considerably. Average moisture content can be reduced to any desired value. Specific schedules are used to control the temperature and humidity in accordance with the moisture and stress situation within the wood, thus minimizing shrinkage-caused defects. For some purposes, equalizing and conditioning treatments are used to improve moisture content uniformity and relieve stresses and set at the end of drying, so the material will not warp when resawed or machined to smaller sizes or irregular shapes. Further advantages of kiln drying are the setting of pitch in resinous woods, the killing of staining or decay fungi or insects in the wood, and reductions in weight greater than those achieved by air-drying. At the end of kiln-drying, moisture-monitoring equipment is sometimes used to sort out moist stock for redrying and to insure that the material ready for shipment meets moisture content specifications.

Temperatures of ordinary kiln drying generally are between 110° and 180° F. Elevated-temperature (180° to 212° F.) and high-temperature (above 212° F.) kilns are becoming increasingly common, although some strength loss is possible with higher temperatures (see ch. 4).

Special Drying Methods

Wood has been dried for construction use by boiling in oily liquids, and vapor drying is being used for the partial drying of crossties prior to preservative treatment. Neither process can yet be considered a complete drying method, and each leaves some drying medium in the wood. Drying by high-frequency electrical energy is being used commercially for short blocks of permeable sapwood. Solvent seasoning, involving water-miscible chemicals that replace water as a first step in the process, has been technically developed for some western softwoods but has not been adopted commercially. High-frequency drying and solvent seasoning, properly carried out to dry the wood to low average moisture content values, may give as good results as kiln drying. Such other special drying methods as vacuum and infrared drying have been found unsuitable for commercial use.

Seasoning Degrade

The grading rules of the lumber associations specify the types and amounts of defects permitted in the various grades of dimension stock. Seasoning defects and other degrading factors are considered. The higher grades are practically free of defects, but high-quality material may not be needed for all uses. The defects permitted in the other approved grades have no detrimental effect on the wood's utility in many applications.

Seasoning defects that cause degrade may be classified into three main groups: (1) Those caused by unequal shrinkage (checks, honeycomb, warp, loosening of knots, and collapse); (2) those associated with the action of fungi (molds and stains); and (3) those associated with soluble wood constituents (brown stain and sticker stain). Collapse and warp affect appearance and ease of application. Checking and honeycombing may, in addition, reduce strength. Defects caused by fungi may affect both appearance and strength (see ch. 17).

Brown stain occurs in softwoods. It is a yellow to dark-brown discoloration apparently caused by concentration or chemical transformation of water-soluble materials in the wood. Sticker stain may occur in the air drying of both softwoods and hardwoods. It is caused by color changes of water-soluble materials in the wood under the stickers. Conversely, the chemicals through the rest of the board may undergo color changes while those under the stickers do not, causing light-colored "sticker marking." General blue-gray or gray-brown chemical stain of hardwoods may also be a problem. Although chemical stains cause considerable degrade because of appearance, none of them lower the strength.

Seasoning defects can be largely eliminated by good practice in either air drying or kiln drying. The period immediately after sawing is the most critical. Lumber subject to chemical or brown stain should not be solid piled. It should be piled on stickers under good drying conditions within 2 or 3 days after sawing.

Rapid surface drying decreases the incidence of mold, stain, and decay but sometimes additional measures are required (ch. 17). Too-rapid surface drying, however, may cause checking and splitting. Honeycombing and collapse are more likely to occur in hardwoods than in softwoods. These defects are more likely to occur under improper kiln-drying than under air drying, although very severe air-drying conditions can cause them to occur.

Moisture-resistant coatings are sometimes applied to the end-grain surfaces of green lumber to retard end drying and minimize the formation of end checks and end splits. To be effective, the coatings must be applied to the freshly trimmed green lumber before any checking has started. Sprayable wax emulsions are sometimes used on the ends of 1-inch lumber subject to considerable loss by end checking.

Moisture Content of Seasoned Lumber

The trade terms "shipping dry," "air dry," and "kiln dry," although widely used, may not have identical meanings as to moisture content in the different producing regions. Despite the wide variations in the use of these terms, they are sometimes used to describe seasoned lumber. The following statements, which are not exact definitions, outline the categories:

Shipping-dry Lumber.

Lumber that is partially dried to prevent stain or mold in brief periods of transit, preferably with the outer $\frac{1}{8}$ inch dried to 25 percent moisture content or below.

Air-dry lumber.

Lumber that has been dried by exposure to the air outdoors or in a shed, or by forced circulation of unhumidified air that has not been heated above 120° F. Commercial air-dry stock generally will have an average moisture content low enough for rapid kiln drying or for rough construction use. These values generally would be in the range of 20 to 25 percent for dense hardwoods and 15 to 20 percent for softwoods and low-density hardwoods. Extended exposure can bring 1- and 2-inch lumber within a percentage point or two of the average exterior equilibrium moisture content of the region. For much of the United States, the minimum moisture content of thoroughly air-dried lumber is 12 to 15 percent.

Kiln-dry lumber.

Lumber that has been dried in a kiln or by some special drying method to an average moisture content specified or understood to be suitable for a certain use. The average should have upper and lower tolerance limits, and all values should fall within the limits. Kiln-dry lumber generally has an average moisture content of 12 percent or below and can be specified to be free of drying stresses.

The importance of suitable moisture content values is recognized, and provisions covering them are now incorporated in some standards and grading rules. Moisture content values in the general grading rules may or may not be suitable for a specific use; if not, a special moisture content specification should be made.

MOISTURE CONTROL DURING TRANSIT AND STORAGE

Lumber and other wood items may change in moisture content and dimension while awaiting shipment, fabrication, or in transit, as well as when stored in a wholesale or retail yard.

When 1-inch dry softwood lumber is shipped in tightly closed boxcars or trucks, or in packages with complete and intact wrappers, average moisture content changes for a package can generally be held to 0.2 percent per month or less. In holds or between decks of ships, dry material absorbs usually about 1.5 percent moisture during normal shipping periods. If green material is included in the cargo, the moisture regain of the dry lumber may be doubled. On the top deck, the moisture regain may be as much as 7 percent.

When 1-inch softwood lumber, kiln-dried to 8 percent or less, is piled solid under a good pile roof in a yard in humid weather, average moisture content of a pile can increase at the rate of about 2 percent per month during the first 45 days. An absorption rate of about 1 percent per month then may be sustained throughout a humid season. Comparable initial and sustaining absorption rates are about 1 percent per month in open sheds and 0.3 percent per month in closed sheds. Stock piled in an open shed in a western location increased 2.7 percent on the inside of solid piles and 3.5 percent on the outside of the pile in a year.

All stock on which any manufacturing has been done should be protected from precipitation and spray, because water that gets into a solid pile tends to be absorbed by the wood instead of evaporating. The extent to which additional control of the storage environment is required depends upon the use to which the wood will be put and the corresponding moisture content recommendations. The moisture content and stress condition of all stock should be determined when it is received. If moisture content and stress condition are not as specified or required, stickered storage in an appropriate condition could ultimately bring the stock within the desired moisture content range. Such storage may also be helpful in relieving stresses in softwoods. If the degree of moisture change required is large, or the material is a hardwood with stress not appropriately relieved, the stock must be redried.

Sheathing and Structural Items

Green or only partially seasoned lumber and timbers should be open piled on stickers and protected from sunshine and precipitation by a tight roof. Framing lumber and plywood with 20 percent or less moisture content can be solid piled in a shed, which affords good protection against sunshine and direct or wind-driven precipitation. Preferably, stock above 12 percent moisture content should be sticker piled to bring its moisture content more in line with the moisture content in use. Dry lumber can be piled solid in the open for relatively short periods, but at least a minimum pile cover of waterproofed paper should be used whenever possible. Protective treatments containing a fungicide and water repellents reduce moisture absorption about 50 percent under exposure to intermittent short-term wetting, but do not protect against absorption when exposure to water is prolonged. Because it is difficult to keep rain out completely, long storage of solid-piled lumber in the open is not recommended. If framing lumber must be stored in the open for a long time, the lumber should be piled on stickers over good supports, and the piles should be roofed. Solid-piled material that has become wet again should be treated the same way.

Finish and Factory Items

Such kiln-dried items as exterior finish, siding, and exterior millwork can be stored in a closed but unheated shed. They should be placed on supports raised above the floor, 6 inches if the floor is paved, 12 inches if not paved.

Interior trim, flooring, cabinet work, and material for processing into furniture should be stored in a room or closed shed that is heated or dehumidified. Kiln-dried and machined hardwood dimension or softwood cut stock also should be stored under controlled humidity conditions. Under uncontrolled conditions, the ends of such stock may come to a higher moisture content than the balance of the length; then when the stock is straight-line ripped or jointed before edge gluing, subsequent shrinkage will cause splitting or open glue joints.

The simplest way to reduce relative humidity in storage areas of all sizes is to heat the space to a temperature slightly above that of the outside air. Dehumidifiers can be used in small, well-enclosed spaces. If the heating method is used, and there is no source of moisture except that contained in the air, the equilibrium moisture content can be maintained by using the data in the following tabulation:

Desired Equilibrium Moisture Content	Degrees Fahrenheit Above Outside Temperature
Percent	
6	25
7	19
8	15
9	12
10	8
11	5
12	3

Good control can be obtained by using data from the Weather Bureau on the average temperature to be expected for the next 15 or 30 days and setting an ordinary thermostat to control at the desired temperature. Adjustments should be made when actual weather conditions differ considerably from those anticipated. Precise control can be maintained by use of a wood-element hygrostat or relative humidity sensor.

When a dehumidifier is used, the average temperature in the storage space should be known or controlled and table 3–4 should be used to select the proper relative humidity to give the desired average moisture content.

Wood in a factory awaiting or following manufacture can become too dry if the area is heated to 70° F. or higher when there is a low outdoor temperature. Under such circumstances, exposed ends and surfaces of boards or cut pieces will tend to dry to the equilibrium moisture content condition, causing shrinkage and warping. Also an equilibrium moisture content of 4 percent or more below the moisture content of the core of freshly crosscut boards may cause end checking. Simple remedies are

to cover piles of partially manufactured items with plastic film and to use properly lowered shop temperatures during nonwork hours. More precise control can be obtained in critical shop and storage areas by humidification. In warm weather, cooling may increase relative humidity, and dehumidification may be necessary.

DIMENSIONAL CHANGES IN WOOD ITEMS

Dry wood undergoes small changes in dimension with normal changes in relative humidity. More humid air will cause slight swelling, and drier air will cause slight shrinkage. These changes are considerably smaller than those involved with shrinkage from the green condition. Approximate changes in dimension can be estimated by a simple formula involving a dimensional change coefficient when moisture content remains within the range of normal use.

Estimate Using Dimensional Change Coefficient

The change in dimension within the moisture content limits of 6 to 14 percent can be estimated satisfactorily by using a dimensional change coefficient based on the dimension at 10 percent moisture content:

$$\Delta D = D_I \left[C_T (M_F - M_I) \right] \quad (14\text{–}2)$$

where: ΔD = change in dimension,
D_I = dimension in inches or other units at start of change,
C_T = dimensional change coefficient, tangential direction (for radial direction, use C_R),
M_F = moisture content (percent) at end of change,
M_I = moisture content (percent) at start of change.

Values for C_T and C_R, derived from total shrinkage values, are given in table 14–2. When M_F is less than M_I, the quantity $(M_F - M_I)$ will be negative, indicating a decrease in dimension; when greater, it will be positive, showing an increase in dimension.

As an example, assuming the width of a flat-grained white fir board is 9.15 inches at 8 percent moisture content, its change in width at 11 percent moisture content is estimated as:

Table 14-2.—*Coefficients for dimensional change due to shrinkage or swelling within moisture content limits of 6 to 14 percent*

HARDWOODS

Species	Dimensional change coefficient [1]		Species	Dimensional change coefficient [1]	
	Radial C_R	Tangential C_T		Radial C_R	Tangential C_T
Alder, red	0.00151	0.00256	Locust, black	.00158	.00252
Apple	.00205	.00376	Madrone, Pacific	.00194	.00451
Ash:			Magnolia:		
Black	.00172	.00274	Cucumbertree	.00180	.00312
Oregon	.00141	.00285	Southern	.00187	.00230
Pumpkin	.00126	.00219	Sweetbay	.00162	.00293
White, green	.00169	.00274	Maple:		
Aspen, quaking	.00119	.00234	Bigleaf	.00126	.00248
Basswood, American	.00230	.00330	Red	.00137	.00289
Beech, American	.00190	.00431	Silver	.00102	.00252
Birch:			Sugar, black	.00165	.00353
Paper	.00219	.00304	Red oak:		
River	.00162	.00327	Commercial red	.00158	.00369
Yellow, sweet	.00256	.00338	California black	.00123	.00230
Buckeye, yellow	.00123	.00285	Water, laurel,		
Butternut	.00116	.00223	willow	.00151	.00350
Catalpa, northern	.00085	.00169	White oak:		
Cherry, black	.00126	.00248	Commercial white	.00180	.00365
Chestnut, American	.00116	.00234	Live	.00230	.00338
Cottonwood:			Oregon white	.00144	.00327
Black	.00123	.00304	Overcup	.00183	.00462
Eastern, southern	.00133	.00327	Persimmon, common	.00278	.00403
Elm:			Sassafras	.00137	.00216
American	.00144	.00338	Sweetgum	.00183	.00365
Rock	.00165	.00285	Sycamore, American	.00172	.00296
Slippery	.00169	.00315	Tanoak	.00169	.00423
Winged, cedar	.00183	.00419	Tupelo:		
Hackberry	.00165	.00315	Black	.00176	.00308
Hickory:			Water	.00144	.00267
Pecan	.00169	.00315	Walnut, black	.00190	.00274
True hickory	.00259	.00411	Willow:		
Holly, American	.00165	.00353	Black	.00112	.00308
Honeylocust	.00144	.00230	Pacific	.00099	.00319
			Yellow-poplar	.00158	.00289

SOFTWOODS

Species	Dimensional change coefficient [1]		Species	Dimensional change coefficient [1]	
	Radial C_R	Tangential C_T		Radial C_R	Tangential C_T
Baldcypress	.00130	.00216	Larch, western	.00155	.00323
Cedar:			Pine:		
Alaska-	.00095	.00208	Eastern white	.00071	.00212
Atlantic white-	.00099	.00187	Jack	.00126	.00230
Eastern redcedar	.00106	.00162	Loblolly, pond	.00165	.00259
Incense-	.00112	.00180	Lodgepole, Jeffrey	.00148	.00234
Northern white-[2]	.00101	.00229	Longleaf	.00176	.00263
Port-Orford-	.00158	.00241	Ponderosa, Coulter	.00133	.00216
Western redcedar[2]	.00111	.00234	Red	.00130	.00252
Douglas-fir:			Shortleaf	.00158	.00271
Coast-type	.00165	.00267	Slash	.00187	.00267
Interior north	.00130	.00241	Sugar	.00099	.00194
Interior west	.00165	.00263	Virginia, pitch	.00144	.00252
Fir:			Western white	.00141	.00259
Balsam	.00099	.00241	Redwood:		
California red	.00155	.00278	Old-growth[2]	.00120	.00205
Noble	.00148	.00293	Second-growth[2]	.00101	.00229
Pacific silver	.00151	.00327	Spruce:		
Subalpine, corkbark	.00088	.00259	Black	.00141	.00237
White, grand	.00112	.00245	Engelmann	.00130	.00248
Hemlock:			Red, white	.00130	.00274
Eastern	.00102	.00237	Sitka	.00148	.00263
Western	.00144	.00274	Tamarack	.00126	.00259

Species	Dimensional change coefficient [1]		Species	Dimensional change coefficient [1]	
	Radial C_R	Tangential C_T		Radial C_R	Tangential C_T
IMPORTED WOODS					
Andiroba, crabwood	.00137	.00274	Light red "Philippine mahogany"	.00126	.00241
Angelique	.00180	.00312	Limba	.00151	.00187
Apitong, keruing [2] (All *Dipterocarpus* spp.)	.00243	.00527	Lupuna	.00126	.00230
Avodire	.00126	.00226	Mahogany [2]	.00172	.00238
Balsa	.00102	.00267	Meranti	.00126	.00289
Banak	.00158	.00312	Nogal [2]	.00129	.00258
Cativo	.00078	.00183	Obeche	.00106	.00183
Emeri	.00106	.00169	Okoume	.00194	.00212
Greenheart [2]	.00390	.00430	Parana pine	.00137	.00278
Iroko [2]	.00153	.00205	Pau marfim	.00158	.00312
Ishpingo [2]	.00125	.00205	Primavera	.00106	.00180
Khaya	.00141	.00201	Ramin	.00133	.00308
Kokrodua [2]	.00148	.00297	Santa Maria	.00187	.00278
Lauans:			Spanish-cedar	.00141	.00219
Dark red "Philippine mahogany"	.00133	.00267	Teak [2]	.00101	.00186
			Virola	.00183	.00342
			Walnut, European	.00148	.00223

[1] Per 1 pct. change in moisture content, based on dimension at 10 pct. moisture content and a straightline relationship between the moisture content at which shrinkage starts and total shrinkage. (Shrinkage assumed to start at 30 pct. for all species except those indicated by footnote 2.)

[2] Shrinkage assumed to start at 22 pct. moisture content.

$$\Delta D = 9.15[0.00245(11-8)]$$
$$= 9.15[0.00735]$$
$$= 0.06725 \text{ or } 0.067 \text{ inch}$$
$$\text{Then dimension at end of change} = D_I + \Delta D$$
$$= 9.15 + 0.067$$
$$= 9.217 \text{ inches}$$

The thickness at 11 percent moisture content of the same board can be estimated by using the coefficient $C_R = 0.00112$.

The tangential coefficient, C_T, can be used for both width and thickness if a maximum estimate for dimensional change is desired. The dimension change for boards that are not truly flat- or quartersawn is most easily estimated by using the tangential coefficient, C_T.

Calculation Based on Green Dimensions

Approximate dimensional changes associated with moisture content changes larger than 6 to 14 percent, or when one moisture value is outside of those limits, can be calculated by:

$$\Delta D = \frac{D_I (M_F - M_I)}{\dfrac{30(100)}{S_T} - 30 + M_I} \quad (14\text{–}3)$$

where: ΔD = change in dimension, D_I = dimension in inches or other units at start of change, M_F = moisture content (percent) at end of change, M_I = moisture content (percent) at start of change, S_T = tangential shrinkage (percent) from green to ovendry (tables 3–5 and 3–6) (use radial shrinkage S_R when appropriate).

Neither M_I nor M_F should exceed 30, the assumed moisture content value when shrinkage starts for most species.

DESIGN FACTORS AFFECTING DIMENSIONAL CHANGE IN A STRUCTURE

Framing Lumber in House Construction

Ideally, house framing lumber should be seasoned to the moisture content it will reach in use, thus minimizing future dimensional changes due to frame shrinkage. This ideal condition is difficult to achieve, but some shrinkage of the frame may take place without being visible or causing serious defects after the house is completed. If, at the time the wall and ceiling finish is applied, the moisture content of the framing lumber is not more than about 5 percent above that which it will reach in service (table 14–1), there will be little or no evidence of defects caused by shrinkage of the frame. In heated houses in cold climates, joists over heated basements, studs, and ceiling joists may reach a moisture content as low as 6 to 7 percent. In mild climates the minimum moisture content will be higher.

The most common evidences of excessive shrinkage are cracks in plastered walls, open joints and nail pops in dry-wall construction, distortion of door openings, uneven floors, or loosening of joints and fastenings. The extent of vertical shrinkage after the house is completed is proportional to the depths of wood used as supports in a horizontal position, such as girders, floor joists, and plates. After all, shrinkage occurs primarily in the width of members, not the length.

Thorough consideration should be given to the type of framing best suited to the whole building structure. Methods should be selected that will minimize or balance the use of wood across the grain in vertical supports. These involve variations in floor, wall, and ceiling framing. The factors involved and details of construction are covered extensively in "Wood-Frame House Construction," USDA Agriculture Handbook 73.

Heavy Timber Construction

In heavy timber construction, a certain amount of shrinkage is to be expected. If not provided for in the design, it may cause weakening of the joints or uneven floors or both. One means of eliminating part of the shrinkage in mill buildings and similar structures is with metal post caps, separating the upper column from the lower column only by the metal in the post cap. This eliminates the shrinkage that would occur if the upper column bears directly on the wood girder. The same thing is accomplished by supporting the upper column on the lower column with wood corbels bolted to the side of the lower column to support the girders.

Where joist hangers are used, the top of the joist, when installed, should be above the top of the girder; otherwise, when the joist shrinks in the stirrup, the floor over the girder will be higher than that bearing upon the joist.

Heavy planking used for flooring should be near 12 percent in moisture content to minimize openings between boards as they approach moisture equilibrium. When 2- or 3-inch joists are nailed together to provide a laminated floor of greater depth for heavy design loads, the joist material should be somewhat below 12 percent moisture content if the building is to be heated.

Interior Finish

The normal seasonal changes in the moisture content of interior finish are not enough to cause serious dimensional change if the woodwork is carefully designed. Large members, such as ornamental beams, cornices, newel posts, stair stringers, and handrails, should be built up from comparatively small pieces. Wide door and window trim and base should be hollow-backed. Backband trim, if mitered at the corners, should be glued and splined before erection; otherwise butt joints should be used for the wide faces. Large, solid pieces, such as wood paneling, should be so designed and installed that the panels are free to move across the grain. Narrow widths are preferable.

Flooring

Flooring is usually dried to a suitable moisture content so that special design considerations are not necessary for installation in ordinary rooms. When used in basement, large hall, or gymnasium floors, however, enough space should be left around the edges to allow for some expansion.

WOOD CARE AND SCHEDULING DURING CONSTRUCTION

Lumber and Sheathing

Lumber and sheathing received at the building site should be protected from wetting and other damage. Construction lumber in place in a structure before it is enclosed may be wet during a storm, but the wetting is mostly on the exposed surface, and the lumber can dry out quickly. Dry lumber may be solid piled at the site, but the piles should be at least 6 inches off the ground and covered with canvas or waterproof paper laid to shed water from the top, sides, and ends of the pile.

Lumber that is green or nearly green, and lumber or plywood that has been used for concrete forms, should be piled on stickers under a roof for more thorough drying before it is built into the structure. The same procedure is required for preservative-treated lumber that has not been fully redried.

If framing lumber has higher moisture content when installed than that recommended in table 14–1, some shrinkage may be expected. Framing lumber, even thoroughly air-dried stock, will generally have a moisture content higher than that recommended when it is delivered to the building site. If carelessly handled in storage at the site, it may take up more moisture. Builders may schedule their work so an appreciable amount of seasoning can take place during the early stages of construction. This minimizes the effects of further drying and shrinkage after completion.

When the house has been framed, sheathed, and roofed, the framing is so exposed that in time it can dry to a lower moisture content than would ordinarily be expected in yard-dried lumber. The application of the wall and ceiling finish is delayed while wiring and plumbing are installed. If the delay is for about 30 days in warm, dry weather, framing lumber should lose enough moisture so that any further drying in place will be relatively unimportant. In cool, damp weather, or if unseasoned lumber is used, the period of exposure should be extended. Checking moisture content of door and window headers and floor and ceiling joists at this time with an electric moisture meter is good practice. When these members approach an average of 12 percent moisture content, interior finish and trim can normally be installed. Closing the house and using the heating system will hasten the rate of drying.

Before wall finish is applied, the frame should be examined and any defects that may have developed during drying, such as warped or distorted studs, shrinkage of lintels over openings, or loosened joints, should be corrected.

Exterior Trim and Millwork

Exterior trim such as cornice and rake moldings, fascia boards, and soffit material is normally installed before the shingles are laid. Trim, siding, and window and door frames should be protected on the site by storing in the house or garage if they are received some time before the contractor can use them. While items such as window frames and sash are usually treated with some type of water-repellent preservative to resist absorption of water, they should be stored in a protected area if they cannot be installed soon after delivery. Wood siding is often received in pack-

aged form and can ordinarily remain in the package until it is applied.

Finish Floor

Cracks develop in flooring if it absorbs moisture either before or after it is laid and then shrinks when the building is heated. Such cracks can be greatly reduced by observing the following practices: (1) Specify flooring manufactured according to association rules and sold by dealers that protect it properly during storage and delivery; (2) do not allow the flooring to be delivered before the masonry and plastering are completed and fully dry, unless a dry storage space is available; (3) have the heating plant installed before the flooring is delivered; (4) break open the flooring bundles and expose all sides of the flooring to the atmosphere inside the structure; (5) close up the house at night and raise the temperature about 15° F. above the outdoor temperature for about 3 days before laying the floor; (6) if the house is not occupied immediately after the floor is laid, keep the house closed at night or during damp weather and supply some heat if necessary.

Better and smoother sanding and finishing can be done when the house is warm and the wood has been kept dry.

Interior Finish

In a building under construction, the relative humidity will average higher than it will in an occupied house because of the moisture that evaporates from wet concrete, brickwork, and plaster, and even from the structural wood members. The average temperature will be lower, because workmen prefer a lower temperature than is common in an occupied house. Under such conditions the finish tends to have a higher moisture content during construction than it will have during occupancy.

Before any interior finish is delivered, the outside doors and windows should be hung in place so that they may be kept closed at night; in this way conditions of the interior can be held as close as possible to the higher temperature and lower humidity that ordinarily prevail during the day. Such protection may be sufficient during dry summer weather, but during damp or cool weather it is highly desirable that some heat be maintained in the house, particularly at night. Whenever possible, the heating plant should be placed in the house before the interior trim goes in, to be available for supplying the necessary heat. Portable heaters also may be used. The temperatures during the night should be maintained about 15° F. above outside temperatures and not be allowed to drop below about 70° F. during the summer or 62° F. when outside temperatures are below freezing.

After buildings have thoroughly dried, there is less need for heat, but unoccupied houses, new or old, should not be allowed to stand without some heat during the winter. A temperature of about 15° F. above outside temperatures and above freezing at all times will keep the woodwork, finish, and other parts of the house from being affected by dampness or frost.

Plastering

During a plastering operation in a moderate-sized six-room house approximately 1,000 pounds of water are used, all of which must be dissipated before the house is ready for the interior finish. Adequate ventilation to remove the evaporated moisture will avoid that moisture being adsorbed by the framework. In houses plastered in cold weather the excess moisture may also cause paint to blister on exterior finish and siding. During warm, dry, summer weather with the windows wide open, the moisture will be gone within a week after the final coat of plaster is applied. During damp, cold weather, the heating system or portable heaters are used to prevent freezing of plaster and to hasten its drying. Adequate ventilation should be provided at all times of the year, because a large volume of air is required to carry away the amount of water involved. Even in the coldest weather, the windows on the side of the house away from the prevailing winds should be opened 2 or 3 inches, preferably from the top.

BIBLIOGRAPHY

American Society for Testing and Materials
   Standard methods of test for moisture content of wood. ASTM Designation D 2016. (See current edition.) Philadelphia, Pa.
Anderson, L. O.
   1970. Wood-frame house construction. U.S. Dep. Agr., Agr. Handb. 73, rev. 223 pp.
Comstock, G. L.
   1965. Shrinkage of coast-type Douglas-fir and old-growth redwood boards. U.S. Forest Serv. Res. Pap. FPL 30. Forest Prod. Lab., Madison, Wis.
James, W. L.
   1963. Electric moisture meters for wood. U.S. Forest Serv. Res. Note FPL–08. Forest Prod. Lab., Madison, Wis.

———
   1968. Effect of temperature on readings of electric moisture meters. Forest Prod. J. 18(10): 23–31.

McMillen, J. M.
   1958. Stresses in wood during drying. Forest Prod. Lab. Rep. 1652.
Rasmussen, E. F.
   1961. Dry kiln operator's manual. U.S. Dep. Agr., Agr. Handb. 188, 197 pp.
Rietz, R. C., and Page, R. H.
   1971. Air drying of lumber: A guide to industry practices. U.S. Dep. Agr., Agr. Handb. 402, 110 pp.
U.S. Department of Commerce
   1970. American softwood lumber standard. NBS Voluntary Product Stand. PS 20–70. 26 pp.
U.S. Forest Products Laboratory
   1961. Wood floors for dwellings. U.S. Dep. Agr., Agr. Handb. 204, 44 pp.

———
   1972. Methods of controlling humidity in woodworking plants. USDA Forest Serv. Res. Note FPL–0218. Madison, Wis.

Chapter 15

FIRE RESISTANCE OF WOOD CONSTRUCTION

Wood construction for many years has been classified in building codes under three standard types—heavy timber, ordinary, and light-frame. Heavy timber and ordinary types have been used widely in educational, recreational, religious, industrial, commercial, and assembly buildings. However, light frame accounts for about 80 percent of the Nation's dwellings and some of the smaller commercial, educational, and industrial buildings. General principles of design of these types of construction, particularly as they affect fire prevention and control, are presented here.

The self-insulating qualities of wood, particularly in the large wood sections used in heavy timber construction, are an important factor in providing a good degree of fire resistance in wood construction. Good structural details, such as elimination of concealed spaces, also assure improved fire durability.

Light wood-frame construction can be protected to provide a high degree of fire performance through use of conventional gypsum board interior finish. Fire-resistance ratings of 1 hour or 2 hours are readily attained for such walls.

Treatment of wood with fire-retardant chemicals or fire-retardant coatings is also an effective means of preventing flame spread. Partitions and roof assemblies constructed of fire-retardant chemically treated wood framing are being accepted in "fire resistive" and "noncombustible" types of buildings.

HEAVY TIMBER CONSTRUCTION

Before the advent of glued-laminated construction, the sizes of solid sawn timbers that were available limited heavy timber construction. Even so, heavy timber construction was used extensively in multistory buildings. These have exterior walls of masonry, and interior columns, beams, and floors of wood in solid masses, with straight members of relatively short span and a minimum of surface or projections exposed to fire.

Glue-laminating techniques have since provided the means of manufacturing solid wood structural members with extremely long spans and of a variety of shapes. Thus, laminating has permitted the construction of heavy timber buildings with larger unobstructed areas

(figs. 15–1 and 15–2) and given the architect a wood product that he can use in ways never before possible.

Heavy timber construction is generally defined in building codes and standards by the following minimum sizes for the various members or portions of a building:

	Inches, nominal
Columns:	
Supporting floor loads	8 x 8
Supporting roof and ceiling loads only	6 x 8
Floor framing:	
Beams and girders	6 wide x 10 deep
Arches and trusses	8 in any dimension
Roof framing—not supporting floor loads:	
Arches springing from grade	6 x 8 lower half
	6 x 6 upper half
Arches, trusses, other framing springing from top of walls, etc.	4 x 6
Floor (covered with 1-inch nominal flooring, ½-inch plywood, or other approved surfacing):	
Splined or tongue-and-groove plank	3
Planks set on edge	4
Roof decks:	
Splined or tongue-and-groove plank	2
Plank set on edge	3
Tongue-and-groove plywood	1⅛

Roof arches and truss members may be spaced members—3 inches thick when blocked solidly throughout the intervening spaces, or when spaces are tightly closed by a continuous wood cover plate 2 inches thick secured to the underside of the members.

Although building code requirements vary somewhat for walls in heavy timber construction, they generally require that exterior and interior bearing walls be of 2-hour fire-resistive noncombustible construction. Also, use of fire-retardant-treated wood in exterior 2-hour fire-resistive bearing walls is permitted in some codes. However, 3-hour fire-resistive noncombustible construction is required for exterior walls when the distance from other buildings or the property line is 3 feet or less. This also applies to nonbearing walls, except that for a building or property line separation between 20 and 30 feet the fire resistance may be 1 hour and beyond 30 feet there is no fire resistance requirement. Consequently, today there is much heavy timber construction

Figure 15-1.—Use of glued-laminated beam construction provides large unobstructed floor area in school gymnasium.

in combination with glass or other nonrated exterior wall material when there is adequate separation from other buildings. Such construction is widely used for educational, religious, supermarket, and other buildings not built close to property lines.

Heavy timber construction is fire resistant because of the slow rate of burning of wood in massive form. The average rate of penetration of char under ASTM Designation E 119 time-temperature fire conditions is about 1½ inches per hour. When wood is first exposed to fire, there is some delay as it chars and eventually flames. Heating to ignition takes about 2 minutes under the ASTM standard fire test conditions, and then charring proceeds at a rate of approximately 1/30 inch per minute for the next 8 minutes. Thereafter, the char layer has an insulative effect, and the rate decreases to 1/40 inch per minute.

Considering the initial ignition delay, fast initial charring, and then slowing down to a constant rate, the average constant charring rate is about 1/40 inch per minute (or 1½ in. per hr.) for wood species of about 0.48 specific gravity at a moisture content of 7 percent. The rate of char penetration is inversely related to the wood's density and moisture content. The temperatures at the inner zone of char are approximately 550° F., and ¼ inch inward from that a maximum of 360° F. Therefore, when the surfaces of large wood members are directly exposed to fire for periods as long as 1 hour, the low thermal conductivity and slow penetration of fire by charring allow the members to maintain a high percentage of their original strength.

The overall fire resistance of heavy timber construction obviously varies depending on the sizes of timber used. Most building codes, however, recognize that heavy timber construction performs similarly to noncombustible construction with a 1-hour fire resistance, and permit its use in all fire districts for all types of occupancy. This acceptance is based on experience with the performance of heavy timber construction in actual fires, the lack of concealed spaces, and the high fire resistance of walls in this type of construction.

Fire-fighting operations at buildings of heavy timber construction are facilitated by the fact that the structural integrity of wood is well understood by firemen. Through long experience of observing wood under fire conditions, they can approximate the time wood will carry its load without the fear of sudden collapse and are familiar with the warning it gives before it loses its structural integrity. In addition, heavy timber construction simplifies fire-fighting operations because concealed spaces in which fire can begin and spread unnoticed are kept to a minimum.

The fire resistance of glued-laminated structural members, such as arches, beams, and columns, is approximately equal to the fire resistance of solid members of similar sizes. Available information indicates that laminated members glued with phenol, resorcinol, or melamine adhesives are at least equal in fire resistance to a one-piece member of the same size, and laminated members glued with casein have only slightly less fire resistance.

In tests at the Forest Products Laboratory, when the edges of the laminations in sections of laminated members bonded with casein glue were exposed to a gas fire, slightly deeper charring resulted at the glue joints than between the glue joints. When the broad face of a lamination was exposed to the fire, the outer lamination adhered to the rest of the member until the zone of char penetrated to the depth of the glueline. Available data indicate that a casein-glued member with ¾-inch-thick laminations will be penetrated by the char zone as much as 10 percent deeper than a solid beam of the same size after exposure to ASTM E 119 fire for 1 hour. The appearance of the casein-glued joints and the results of shear tests indi-

Figure 15–2.—Heavy timber construction in a warehouse.

cated, however, that little if any weakening of the glue joints occurred beyond the charred depth as a result of the fire exposure. Also, the performance of casein-glued-laminated members in actual fires is reported to demonstrate the integrity of casein-glued joints beyond the zone of char.

When the fire endurance required of a wood member is less than the time required for the zone of char to penetrate through the outer laminations, the type of adhesive is unimportant. At 1½ inches per hour penetration, laminated material with outer laminations not less than 1½ inches thick (such as nominal 2-in. lumber laminations) would be equivalent in fire resistance to solid members of the same actual size with a fire resistance up to 1 hour.

Thus, for use in the heavy-timber-construction classification, laminated members glued with phenol, resorcinol, or melamine adhesives or laminated members glued with casein and having nominal 2-inch outer laminations are considered equivalent to solid sawn members of the same actual size.

ORDINARY CONSTRUCTION

The term "ordinary construction" defines buildings with exterior walls of masonry and interior wood-joist frames with members not less than 2 inches (nominal) thick. This type of construction has been widely employed in commercial or public buildings up to five or six stories high. Ordinary construction differs from heavy timber construction in that exterior walls generally are not as heavy and interior framing is less massive. These differences are reflected in smaller heights and areas allowed. Ordinary construction differs from light-frame construction in its larger allowable heights and areas, its self-supporting masonry walls, and in a number of interior requirements appropriate to the occupancy. There are detailed code requirements for firestops (nominal 2-inch-thick wood or the equivalent) in concealed spaces in walls or ceilings through which fire might spread. Large attic spaces are divided by "draft stops," partitions made of ½-inch plywood or gypsum board, or the equivalent.

LIGHT-FRAME CONSTRUCTION

Most residential and some commercial, institutional, industrial, and assembly buildings of wood are of light-frame construction. Originally restricted to the conventional type of building with stud walls, joisted floors and ceilings, and raftered roofs, light-frame construction has been diversified by the introduction of prefabricated, panelized, or stressed-skin structural elements.

A type of wood construction known as "protected light frame," in which elements are designed to have a fire resistance of 1 hour, is commonly used. Based on the areas allowed, codes rate the fire performance of this type as intermediate between ordinary and heavy timber construction. There are many recognized assemblies involving wood framed walls, floors, and roofs that provide a 1-hour, and even a 2-hour, fire resistance.

Unprotected light-frame wood buildings do not have the natural fire resistance of the heavier wood frames. In these, as in all buildings, attention to good construction details is important to minimize fire hazards. Of particular importance are firestops, separation of wood from masonry around chimneys and fireplaces, and design of walls, ceilings, floors, roofs, stairways, and doors.

HEIGHT AND AREA LIMITATIONS

The model codes develop some fire safety in structures by limiting building areas and heights, dependent primarily upon the type of building construction. The occupancy, fire zone, sprinkler protection, fire-retardant treatment, distance to other structures, and availability to fronting on streets also are considered in establishing the height and area limitations. The National Building Code, which is fairly representative in this respect, establishes a maximum limit on height of 65 feet for heavy timber, 45 feet for ordinary construction, and 35 feet for wood frame. Maximum floor areas per story are 12,000 square feet (one story) and 8,000 square feet (multistory) for heavy timber; 9,000 and 6,000 square feet, respectively, for ordinary; and 6,000 and 4,000 square feet, respectively, for wood frame. This compares to 35 feet, 9,000 and 6,000 square feet, for unprotected noncombustible construction.

Some modifications include floor areas increased by 200 percent when the building is equipped with automatic sprinkler protection; an increase of 100 percent when all sides face toward public streets; and an increase of 50 percent for one-story heavy timber and ordinary construction located outside fire limits and with fire-retardant treatment. For wood-frame construction outside fire limits with fire-retardant treatment, the increase is 33⅓ percent. There are many variations of these limitations among the model codes, and careful consideration should be given to the specific code requirements.

IMPROVING FIRE RESISTANCE THROUGH DESIGN

The fire resistance of wood constructions, particularly that of light-frame construction, may be considerably improved by good design and construction details. Some of the more important details are covered in the following sections.

Firestops

Firestops are obstructions provided in concealed air spaces and are designed to interfere with the passage of flames up or across a building. Fire in buildings spreads by the movement of high-temperature air and gases through open channels. In addition to halls, stairways, and other large spaces, heated gases also follow the concealed spaces between floor joists, between studs in partitions and walls of frame construction, and between the plaster and the wall where the plaster is carried on furring strips. Obstruction of these hidden channels provides an effective means of restricting fire from spreading to other parts of the structure.

Wood of 2-inch nominal thickness or some noncombustible insulating material not less than 1 inch thick are effective firestops. Platform frame construction, which is commonly used in single family house construction, provides the firestopping. For balloon frame construction, good practice includes the use of: Firestops in exterior walls at each floor level, and at the level where the roof connects with the wall; firestops at each floor level in partitions that are continuous through two or more stories; headers at the top and bottom of the space between stair carriages; mineral wool asbestos, or an equivalent material, packed tightly around pipes or ducts that pass through a floor or a firestop; and self-closing doors on vertical shafts, such as clothes chutes. Figure 15–3 shows applications of firestops in an exterior wall in balloon frame construction.

Around Chimneys and Fireplaces

Good practice in the protection of wood from ignition by heat conducted through chimneys and fireplaces includes the following details:

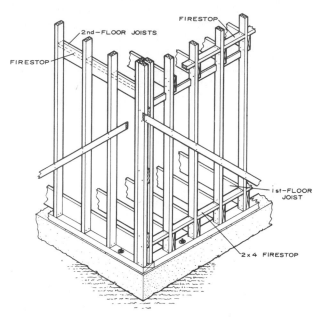

Figure 15–3.—Firestops in balloon frame construction.

1. If smoke pipes from furnaces pass through walls, they are protected by thimbles at least 8 inches larger in diameter than the pipe.
2. Smoke pipes do not pass through floors or ceilings, but join the chimney on the same floor where they originate.
3. Wood beams, joists, or rafters are separated from any chimney by a 2-inch space.
4. Wood furring strips placed around chimneys to support base or other trim are insulated from the masonry by asbestos paper at least ⅛ inch thick, and metal wall plugs or approved noncombustible nail-holding devices attached to the wall surface are used for nailing.
5. Wood construction is separated at least 4 inches from the back wall of any fireplace and at least 2 inches from the sides. The space between the walls of the fireplace and the wood construction should be filled with loose noncombustible material and suitably firestopped. Supporting wood header beams are placed at least 20 inches away from the face of the fireplace. A wood mantel or other woodwork is placed not less than 6 inches from either side nor less than 12 inches from the top of any fireplace opening. Fireplace hearths are of noncombustible material, not less than 18 inches wide measured from the face of the opening.
6. All spaces between the masonry of chimneys and wood joists, beams, headers, or trimmers are filled with noncombustible material.

Partitions

A fire starting in one room of a building will be confined to that room for a variable period of time, depending on the amount and distribution of combustible contents in the room, and the fire resistance of the walls, partitions, and doors, as well as the ceilings and floors. The fire resistance of wood frame walls and partitions depends to a considerable extent upon the materials used for faces, method of fastening facings to frame, the method of joining wall and partition units, the quality of workmanship, the type and quantity of any insulation that may be used, and the structural load that the element is supporting.

The following tabulation gives the fire resistance under ASTM Designation E 119 conditions of some typical bearing or nonbearing built-up wood partitions:

	Fire-resistance rating (Min.)
Hollow 2- by 4-inch wood stud wall panels, 16 inches on center, fire-stopped, with faces of:	
¼-inch plywood, exterior glue [1]	10
⅜-inch plywood, exterior glue [1]	20
½-inch plywood, exterior glue [1]	25
⅝-inch plywood, exterior glue [1]	35
¾-inch tongue-and-groove sheathing boards	20
¾-inch tongue-and-groove sheathing boards plus mineral wool filling	35
⅜-inch gypsum wallboard	25
⅜-inch gypsum wallboard (two layers)	60
½-inch gypsum wallboard	40
½-inch gypsum wallboard (two layers)	90
⅝-inch gypsum (type X) wallboard	60
¾-inch gypsum sand plaster 1:2, on metal lath	60
¾-inch gypsum sand plaster 1:2, on metal lath, plus mineral wool filling	90
½-inch gypsum sand plaster 1:2 on ⅜-inch perforated gypsum lath	60
Solid nonbearing partitions of 2- by 4-inch tongue-and-groove wood boards placed vertically	10
Solid nonbearing partitions of ¾-inch boards, 2½ to 6 inches wide, grooved, joined together with wood splines, nailed:	
Two board layers	15
Same, with 30-pound asbestos paper between layers	25
Three board layers	40
Solid nonbearing partitions of 3/16-inch plywood glued to 2-inch-thick wood core of glued tongue-and-groove construction	60

---

[1] Values obtained on walls with 1- by 3-inch wood studs.

Basement Ceilings

Since fires may start from heating plants located in basements, a fire-resistant separation of the furnace from the remainder of the building is desirable. Gypsum board, plaster on metal or gypsum lath placed on the basement joists affords an effective means of increasing the fire resistance of the basement ceiling and of retarding the rapid spread of flames. Particular attention should be given to the wood floor members directly above and near the furnace. With current improvements in furnace fuels, design, and control, the importance of ceiling protection has been reduced.

If, as is common, a basement stairway is directly under the stairway leading from the first to the second floor, it is good practice to protect the underside of the upper stairway with fire-resistant coverings, as suggested for basement ceilings, and to place firestops between the wood carriages at the top and bottom.

Floors

The conventional floor construction of joists, subfloor, and finish floor offers considerable resistance to the penetration of fire and will retain its load-carrying capacity in severe fire exposures up to 15 minutes. Prefabricated floor panels, in which the load-carrying capacity depends upon stressed covers, and floor systems supported by box girders with thin plywood webs may have more or less fire resistance depending on the dimensions of the elements, presence or absence of protective coverings, and other details.

Doors and Stairways

If a fire-resistant ceiling is placed on the basement joists, it is also desirable to have a self-closing door leading to the basement with fire resistance equal to the combined resistance of the ceiling and floor over the basement.

Enclosed stairways retard rapid spread of fire from floor to floor. If the interior design calls for an open stairway below, it can often be closed at the top with a solid-core wood flush door. Solid wood core or particleboard core wood flush doors provide up to 30 minutes endurance to fire penetration. Hollow-core flush doors offer less resistance to the penetration of fire unless the hollow spaces in the door are packed with an insulating material.

Wood Roof Coverings

The better grades of wood shingles are edge-grained and thick butted with five butts measuring at least 2 inches. Edge-grained shingles warp or curl less than flat-grained ones, thick-butted shingles less than thin ones, and narrow shingles less than wide ones. The use of good quality products and the accepted rules of good practice in laying shingles and shakes not only provides a long-lived economical roof but markedly reduces fire hazards. For improved fire performance, shingles or shakes treated with leach-resistant fire-retardants can be used. Installation of the shingles or shakes with asbestos paper interlay, underlay, or both, or gypsum between rafters and sheathing will further improve fire-resistance.

In modern building construction with efficient heating systems, better separation of structures, and improved fire protection, only 0.5 percent of all fires are attributed to sparks on roof covering of all types. Therefore, insurance penalties against the use of wood shingles and shakes have been eliminated in most states. For structures in frequently dry bushy areas, crowded areas, areas where it is difficult to supply fire protection, or within certain fire zones, restrictions are sometimes imposed requiring fire-retardant treatments of wood shingles and shakes. A limited number of leach-resistant fire-retardant treatments are available for this purpose.

Interior Finishes

The interior finish commonly referred to for building constructions includes the exposed interior surfaces where the surface is an integral part of the building or affixed thereto; examples are the materials for walls and ceilings, interior partitions, interior trim, paint, and wallpaper. Decorations and furnishings which are not affixed to the structure are not considered interior finish, and are not limited by building codes, even though they may furnish the primary source of fuel to an incipient fire.

The model building codes and the National Fire Protection Association Life Safety Code generally specify maximum flame-spread characteristics for interior finish, based on the building occupancy, location within building, and whether or not automatic sprinkler protection is available.

The flame-spread characteristics specified by these codes are generally based on results obtained in tests by the 25-foot tunnel furnace method (ASTM Designation E 84). This method involves the use of a 20-inch by 25-foot-long specimen exposed horizontally as the cover to a tunnel furnace operated under a forced-draft condition. A gas flame is introduced against the test surface at one end of the furnace. The time for the flames to reach the other end of the specimen or the distance traveled in 10 minutes of exposure is recorded. This flame spread is then compared to the flame travel over a red oak lumber specimen, which requires about 5½ minutes for travel over the entire length of the specimen. The red oak specimen is arbitrarily assigned a flame-spread index value of 100, and asbestos-cement board an index of 0. The values for other materials are then determined relative to the time or distances of the flame travel as compared to the red oak standard. For example, if flames reach the end of the specimen in one-half the time required on red oak, the flame-spread index is 200.

Materials are usually classified into groups based on their flame-spread index values—in Class A from 0 to 25, Class B from 26 to 75, Class C from 76 to 200, Class D from 201 to 500, and Class E over 500.

The requirements for surface flammability of interior finish generally prescribe Class A in the exitways of unsprinklered buildings intended for large assembly and institutional purposes, and Class B for school, small assembly, mercantile, and hotel buildings. In general, the next higher class (greater flammability) is permitted for the interior finish used in other areas of the building which are not considered exitways. Also, the next higher flame-spread classification is permitted for materials when they are protected by automatic sprinkler devices, except that Class C finish is usually the highest permitted in any area.

These requirements frequently exempt interior trim or permit up to 10 perecent of the total wall and ceiling surface areas in any use area or occupancy group to be materials with flame-spread classification as high as Class C. The exposed portions of structural members of heavy timber construction are also exempt from these flame-spread requirements in several types of occupancies. Furthermore, wallpaper, paint, and floor coverings may be exempt from these requirements, unless they are judged to be unusual fire hazards. Generally, the common paints and varnishes have only a slight effect on the flame-spread ratings of wood, usually lowering the values.

Most of the wood species have flame-spread index values of 90 to 160 by the ASTM E 84 method, and therefore are accepted for interior finish only for those applications requiring Class C interior finish. A few species have flame-spread index values of slightly less than 75, and these can be used for Class B applications. The Underwriters Laboratories, Inc., Card Data Service C60, U.L. 527, lists the flamespread index for various wood species.

Therefore, fire-retardant treatments are usually necessary for wood interior finish when Class A, and sometimes Class B, flame-spread performance is required.

FIRE-RETARDANT TREATMENTS

Two general methods are available for improving the fire performance of wood by the use of fire-retardant chemicals. One method consists of impregnating the wood with water-borne salts, using conventional vacuum-pressure methods, such as used in the wood-preserving industry. The second method involves the application of fire-retardant chemical paint coatings on the wood surface. The impregnation methods are usually the more effective and lasting and are intended for use on new wood construction. For wood in existing constructions, the surface application of the fire-retardant paints offers the principal means for increasing fire-retardant characteristics.

Chemical Impregnation

In the impregnation treatments, wood is pressure impregnated with water-soluble chemical solutions using full cell pressure processes similar to those used for chemical preservative treatments. Retentions of the fire-retardant salts must be fairly high ($2\frac{1}{2}$ to 5 lb. of dry salt per cubic foot of wood) to be effective.

The salts used in the current fire-retardant formulations are principally the same ones which have been known for their fire-retardant characteristics for over 50 years—monoammonium and diammonium phosphate, ammonium sulfate, zinc chloride, sodium tetraborate, and boric acid. These salts are combined in formulations to develop optimum fire performance characteristics and still have acceptable characteristics with regard to hygroscopicity, strength, corrosivity, machinability, surface appearance, gluability, paintability, and cost.

Some typical formulations as given in the American Wood-Preservers' Association Standard P10 are:

Type B	Percent
Zinc chloride	65.2
Ammonium sulfate	10.0
Boric acid	10.0
Sodium dichromate	14.8

Type C	
Diammonium phosphate	10.0
Ammonium sulfate	60.0
Sodium tetraborate (anhydrous)	10.0
Boric acid	20.0

Type D	
Zinc chloride	35.0
Ammonium sulfate	35.0
Boric acid	25.0
Sodium dichromate	5.0

Other commercial formulations of undisclosed composition are also available.

These formulations in water solutions at 10 to 18 percent concentration are impregnated into the wood under full-cell pressure treatment. American Wood-Preservers' Association Standards C20 and C27 prescribe recommended treating conditions for lumber and plywood. The wood is usually treated in the air- dried or kiln-dried condition, but certain species may be treated green if the wood is first given a steam treatment for periods up to 4 hours.

The treating characteristics of wood species vary considerably, as do sapwood and heart wood. Complete impregnation of the wood is necessary to obtain classification by some codes as being equivalent to "noncombustible." However, to reduce the surface flammability of thicker members, partial impregnation is the more common practice. For large wood members, impregnations of at least 0.5 inch are usually recommended. For some wood species, it is necessary to incise prior to treatment to consistently obtain this depth of treatment. Plywood sheets are treated without any need for incising as the knife checks and end-grain at panel edges improve the ease of impregnation. However, care should be taken that only exterior plywood is used so the plies will not delaminate as a result of the water penetration.

After the treated wood is removed from the treating solution, the wood must be carefully dried. Proper quality control may be obtained by following AWPA Standards C20 and C27, procuring under Military Specification MIL–L–19140C, or by obtaining a product evaluated, listed, and labeled by a rating laboratory such as the Underwriters Laboratories, Inc. The recent wider acceptance of fire-retardant-treated wood by code and insurance authorities has been largely atttributed to the fact that a quality product is insured, based on the inspection and labeling service of a recognized inspection agency.

The proper fire-retardant treatment of wood improves fire performance by greatly reducing the amount of flammable products released, thus reducing the rate at which flames spread over the surfaces. Treatment also reduces the amount of heat available or released in the volatiles during the initial stages of fire, and also results in the wood being self-extinguishing once the primary source of heat and fire is removed or exhausted.

The fire-retardant treatment of wood does not prevent the wood from decomposing and charring under fire exposure, and the rate of fire penetration through treated wood is approximately the same as for untreated wood. Slight improvement is obtained in the fire endurance of doors and walls where fire-retardant-treated wood is used. Most of this improvement is associated with the reduction in surface flammability, rather than any changes in charring rates. However, when walls or doors are improperly constructed or faulty, the use of fire-retardant wood can reduce the effect of the faults on fire endurance.

For most rating purposes, the surface flame-spread characteristics of interior finish materials are evaluated by ASTM Designation E 84 (25-ft. tunnel furnace method). Effective fire-tardant treatment can reduce the flame-spread index of lumber and most wood products to 25 or less by this method as compared to 100 for untreated red oak lumber.

Fire-retardant-treated wood and plywood is currently being used for interior finish and trim in rooms, auditoriums, and corridors where codes require materials with low surface flammability. In addition, many codes, including the model building codes, have accepted the use of fire-retardant-treated wood and plywood in fire-resistive and noncombustible constructions for the framing of nonload-bearing walls and roof assemblies, including decking. Fire-retardant-treated wood is also used for such special purposes as wood scaffolding, and for the framing and rails used in wooden fire doors. The use of fire-retardant treatment for all wood used in buildings over 150 feet high is also prescribed in New York City. Some building codes also permit increased floor area limits in heavy timber, ordinary, and wood-frame constructions when the structural wood members have been given fire-retardant treatment.

Durability

The chemicals used as fire retardants are inorganic salts, generally thermally stable at temperatures up to 330° F. Therefore, under normal interior conditions, the fire-retardant-treated wood remains durable and effective. This has been proven in fire tests of treated wood which has been in service for over 40 years.

The salts generally used as fire retardants are water soluble, and therefore, if exposed to exterior conditions or repeated washing, the effectiveness of the treatment will be diminished. Further, the treated wood is more hygroscopic than untreated wood; under prolonged exposure at relative humidities higher than 80 percent, the treated wood may actually exude moisture and chemical, thus slowly reducing the effectiveness. The use of a sealer topcoat can improve the resistance of leaching and the durability when treated wood is subjected to adverse moisture conditions.

New types of fire-retardant treatments have been developed for wood shingles and shakes and other exterior uses. These treatments have improved leach-resistance and do not add to the hygroscopic properties of the wood. Plywood sidings treated with such exterior fire-retardants have been recognized by code bodies for wall constructions where only non-combustible materials have been previously accepted.

Strength

Fire-retardant treatment results in some reduction of the strength properties of wood, but the reductions are not great. Current treatments have been observed, in tests at the Forest Products Laboratory, to decrease the modulus of elasticity values by 5 to 10 percent and modulus of rupture values by 10 to 20 percent as compared to untreated, matched controls. These values were obtained when both treated and untreated samples were conditioned at the same relative humidity conditions. As the hygroscopic characteristics of the treated wood are slightly greater, the wood density of equivalent cross sections of the treated sample was slightly less. This could account for an appreciable amount of the reduction in strength properties. Fire-retardant-treated wood is more brash than untreated wood. While this reduced resistance to impact is not usually considered in design, the work-to-maximum load, which measures brashness, may be decreased 30 percent or more.

There is no evidence that fire-retardant treatment will cause any further or progressive decrease in strength under temperature and humidity conditions in normal use.

As evidence indicates that there is some reduction in the strength properties of fire-retardant-treated wood, the national design specification for wood has reduced the allowable unit stress for design by 10 percent as compared to untreated wood.

Hygroscopicity

Wood treated with the inorganic fire-retardant salts is usually more hygroscopic than untreated wood, particularly at high relative humidities. For treated wood, increases in equilibrium moisture content will also depend upon the type of chemical, level of chemical retention, and size and species of wood involved. For wood treated with most fire-retardant formulations, the increase in equilibrium moisture content at 80° F. and 30 to 50 percent relative humidity is negligible. At 80° F. and 65 percent relative humidity, increases in moisture content for fire-retardant-treated wood are 2 to 8 percent. At 80° F. and 80 percent relative humidity, increases in moisture content range from 5 to 15 percent and may result in the exuding of chemical solution from the wood. Most current fire-retardant formulations are developed to be used at conditions up to 80 percent relative humidity without significant exuding of chemical solution. However, some of the new leach-resistant fire-retardant treatments are nonhygroscopic.

Corrosivity

Individually, some fire-retardant salts are quite corrosive to metals. However, combinations of these chemicals result in more neutral formulations. The addition of corrosion inhibitors, such as sodium dichromate, has generally reduced the corrosive action of the current types of fire-retardants for wood to an insignificant level.

Machinability

The presence of salt crystals in wood has an abrasive effect on cutting tools. Increased tool life can be obtained by using cutting and shaping tools tipped with tungsten carbide or similar abrasion-resistant alloys. When it is necessary to use regular high-speed steel tools, economy of cutting is practical only when a few hundred feet of the fire-retardant-treated wood is involved. The usual practice in preparing fire-retardant-treated wood for use in trim and moldings is to cut the material to approximate finish size before treatment so a minimum of machining after treatment is necessary.

Gluing Characteristics

Certain phases of the gluing of fire-retardant-treated woods still remain a problem. However, untreated veneer facings can be satisfactorily glued over treated plywood cores with the conventional hot-press phenolic adhesives. For assembly gluing of fire-retardant-treated wood for nonstructural purposes, adhesives such as casein, urea, and resorcinol types can be used. The major problem is in the structural bonding of fire-retardant-treated wood to provide bonds, in both interior and exterior performance tests, which are equivalent to those obtainable for the untreated wood. Special resorcinol-resin adhesives, which employ a high formaldehyde content hardener, have been developed for gluing fire-retardant-treated wood. Improved bonding can be obtained with this type of adhesive, when curing is done at temperatures of 150° F. or higher.

Paintability

The fire-retardant treatment of wood does not generally interfere with the adhesion of decorative paint coatings, unless the treated wood has extremely high moisture content because of its increased hygroscopicity. Moisture content of the treated wood should be at 12 percent or less at the time of the application of the paint coating. Natural finishes are not generally used for fire-retardant-treated wood as the treatment and subsequent drying often causes darkening and irregular staining. Decorative fire-retardant plywoods are usually prepared by treating the plywood core and then bonding a thin, untreated decorative veneer facing to these cores. This eliminates the stained surfaces, which may be difficult to finish properly to a natural wood finish. Crystals may appear on the surface of paint coatings applied over wood having high salt retentions, but only when the wood is exposed to high relative humidity for prolonged periods.

Fire-Retardant Coatings

Many commercial paint coating products are available to provide varying degrees of protection of wood against fire. These paint coatings generally have low surface flammability characteristics and "intumesce" to form an expanded low-density film upon exposure to fire, thus insulating the wood surface below from pyrolysis reactions. They have added ingredients to restrict the flaming of any released combustible vapors. Chemicals may also be present in these paints to promote the rapid decomposition of the wood surface to charcoal and water rather than forming intermediate volatile flammable products.

Fire-retardant paints include those based on water-soluble silicates, urea resins, carbohydrates and alginates, polyvinyl emulsions and oil-base alkyd, and pigmented types. In many of the water-soluble paints, ammonium phosphate or sodium borate is used in the formulation to obtain fire-retardant characteristics. The oil-base paints frequently make use of chlorinated paraffins and alkyds plus antimony trioxide to limit the flammability of any pyrolysis products produced. Inert materials, such as zinc borate, mica, kaolin, and inorganic pigments are also used in these formulations. Intumescence is obtained by the natural characteristics of some of the organic ingredients or special materials, such as isano oil, may be used. A limited number of clear fire-retardant finishes are available. Generally they do not have the effectiveness of the fire-retardant paints, as pigmentation and opaque chemical additives are usually necessary to gain greater effectiveness.

Many of the commercial formulations have been evaluated by ASTM Designation E 84 (25-ft. tunnel furnace) when applied over a substrate of Douglas-fir lumber. These coatings, when properly applied to lumber and wood products, can reduce the surface flame-spread index to 25 or less. To obtain this reduction in surface flammability, it is necessary to apply these coatings to much greater thicknesses (100 to 175 sq. ft. per gal.) than for conventional decorative coatings. Also, because of the added ingredients in these paints, many of them do not have as good brushing characteristics as the decorative paints.

Most of the fire-retardant coatings are intended for interior use, although some products on the market can be used on the exterior of a structure. The application of thin coatings of conventional paint products over the fire-retardant coatings has been one method to improve their durability. Most conventional decorative paint coating products will in themselves slightly reduce the flammability of wood products when applied in conventional film thicknesses.

More and more, fire-retardant coatings are being applied to panel products at the factory. Such application has been common for ceiling tile made of fiberboard.

BIBLIOGRAPHY

American Institute of Timber Construction
    1966. Timber construction manual. Wiley & Sons, New York.

    1962. What about fire? 12 pp., illus. Englewood, Colo.

American Insurance Association
    National Building Code. New York. (See current edition.)

American Plywood Association
    1965. Treated plywood roof system. Concepts No. 111. Tacoma, Wash.

    1965. Fire-resistive plywood floors and roofs. Concepts No. 112. Tacoma, Wash.

American Society for Testing and Materials
    Standard method of tests for surface burning characteristics of building materials. Designation E 84. (See current edition.) Philadelphia, Pa.

    Standard methods of fire tests of building construction and material. Designation E 119. (See current edition.) Philadelphia, Pa.

American Wood-Preservers' Association
    Standards for fire-retardant formulations. Stand. P10. (See current edition.) Washington, D.C.

    Structural lumber, fire-retardant treatment by pressure processes. Stand. C20. (See current edition.) Washington, D.C.

    Plywood, fire-retardant treatment by pressure processes. Stand. C27. (See current edition.) Washington, D.C.

Anderson, L. O.
    1970. Wood-frame house construction. U.S. Dep. Agr., Agr. Handb. 73, rev. 223 pp.

Browne, F. L.
    1958. Theories of the combustion of wood and its control. Forest Prod. Lab. Rep. 2136.

Bruce, H. D., and Fassnacht, D.
    1958. Wood houses can be fire-safe houses. Forests and People. Fourth Quart.

Building Officials and Code Administrators International, Inc.
    1970. The BOCA basic building code. 483 pp. Chicago.

———
1971. One and two family dwelling code. 228 pp. Chicago.
Degenkolb, J. G.
1965. Fire-retardant-treated wood framing 24-inch centers, nonbearing partition passes ASTM 1-hour fire test. Wood Preserving News, Dec., pp. 14–17.
Eickner, H. W.
1966. Fire-retardant-treated wood. ASTM Jour. Mater. 1(3): 625–644.
———, and Peters, C. C.
1963. Surface flammability of various decorative and fire-retardant coatings for wood as evaluated in FPL 8-foot tunnel furnace. Official Dig. Fed. and Soc. of Paint Tech. 35 (Aug.) pp. 800–813.
———, and Schaffer, E. L.
1967. Fire-retardant effects of individual chemicals on Douglas-fir plywood. Fire Tech. 3(2): 90–104.
Holmes, C. A.
1971. Evaluation of fire-retardant treatments for wood shingles. USDA Forest Serv. Res. Pap. FPL 158. Forest Prod. Lab., Madison, Wis.
International Conference of Building Officials
1970. Uniform building code. 651 pp.
Miniutti, V. P.
1958. Fire-resistance tests of solid wood flush doors. Forest Prod. J. 8(4): 141–144.
National Fire Protection Association
1969. Fire protection handbook, 13th ed., 2100 pp., illus. Boston.
———
1970. Code for safety to life from fire in buildings and structures. NFPA No. 101. 222 pp.
National Forest Products Association
National design specification for stress-grade lumber and its fastenings. (See current edition.) Washington, D.C.
National Safety Council
1960. Fire-retarding treatments for wood. Data Sheet 372, rev., Chicago, Ill.
Schaffer, E. L.
1968. A simple test for adhesive behavior in wood sections exposed to fire. USDA Forest

Serv. Res. Note FPL–0175. Forest Prod. Lab., Madison, Wis.
———
1967. Charring rate of selected wood—transverse to grain. USDA Forest Serv. Res. Pap. FPL 69. Forest Prod. Lab., Madison, Wis.
———
1966. Review of information related to the charring rate of wood. USDA Forest Serv. Res. Note FPL–0145. Forest Prod. Lab., Madison, Wis.
———, and Eickner, H. W.
1965. Effect of wall linings on fire performance within a partially ventilated corridor. USDA Forest Serv. Res. Pap. FPL 49. Forest Prod. Lab., Madison, Wis.
Southern Building Code Congress
1969. Southern standard building code. Birmingham, Ala.
Underwriters Laboratories, Inc.
Building Materials List. (Revised annually) Chicago, Ill.
———
1971. Card Data Service C60. Wood-fire hazard classification. Chicago, Ill.
U.S. Department of Defense
1964. Military specification, lumber and plywood, fire-retardant treated. MIL–L–19140C.
U.S. Forest Products Laboratory
1940. Fire resistance tests of plywood covered wall panels. Forest Prod. Lab. Rep. 1257.
———
1959. Fire-test methods used in research at the Forest Products Laboratory. Forest Prod. Lab. Rep. 1443.
———
1968. Surface flammability of various wood-base building materials. USDA Forest Serv. Res. Note FPL–0186.
U.S. National Bureau of Standards
1942. Fire resistance of building constructions. Build. Mater. and Struc. Rep. 92.
Yuill, C. H.
1963. An evaluation of performance of glued-laminated timber and steel structural members under equivalent fire exposure. Southwest Res. Inst. Rep. 1–923–3B.

Chapter 16

PAINTING AND FINISHING

Wood and wood products in a variety of species, grain patterns, textures, and colors can be finished effectively by several different methods. Painting, which totally obscures the wood grain with a coating, achieves a particular color decor. Penetrating preservatives and pigmented stains permit some or all of the wood grain and texture to show and provide a special color effect as well as natural or rustic appearance. Selection of a type of finish, painted or penetrating, depends on the appearance desired and substrate employed.

FACTORS AFFECTING FINISH PERFORMANCE

Satisfactory performance of finishes is achieved when full consideration is given to the many factors that affect finishes. These factors include the effect of the wood substrate, the properties of the finishing material, details of application, and severity of exposure to elements of the weather. This chapter reviews some of the more important considerations. Sources of more detailed information are given in the Bibliography at the end of this chapter.

Wood Properties

Wood surfaces that shrink and swell the least are best for painting. For this reason, vertical- or edge-grained surfaces are far better than flat-grained surfaces of any species, especially for exterior use where wide ranges in relative humidity and periodic wetting can produce wide ranges in swelling and shrinking.

Also, because the swelling of wood is directly proportional to density, low-density species are preferred over high-density species. However, even high-swelling and dense wood surfaces with flat grain can be stabilized with a resin-treated paper overlay (overlaid exterior plywood and lumber) to provide excellent surfaces for painting. Medium-density, stabilized fiberboard products with a uniform, low-density surface or paper overlay are also a good substrate for exterior use. Vertical-grained western redcedar and redwood, however, are probably the species most widely used as exterior siding to be painted. These species are classified in group I, those woods easiest to keep painted (table 16–1). Vertical-grain surfaces of all species actually are considered excellent for painting, but most species are generally available only as flat-grain lumber.

Species that are normally cut as flat-grained lumber, are high in density and swelling, or have defects such as knots or pitch, are classified in groups II through V, depending upon their general paint-holding characteristics. Many species in groups II through IV are commonly painted, particularly the pines, Douglas-fir, and spruce. These species generally require more care and attention than the group I species with vertical-grain surfaces. Exterior paint will be more durable on vertical-grain boards than on flat-grain boards for any species with marked differences in density between earlywood and latewood, even if the species is rated in group I. Flat-grain boards that are to be painted should be installed in areas protected from rain and sun.

Plywood for exterior use nearly always has a flat-grain surface. In addition, cycles of swelling and shrinking tend to check the face veneer of plywood much more than lumber. This checking extends through paint coatings to detract from their appearance and durability. Plywood with a resin-treated paper overlay, however, has excellent paintability and would be equal to or better than vertical-grain lumber surfaces of group I.

Before painting, resinous species should be thoroughly kiln dried at temperatures that will effectively set the pitch to reduce problems of resin exudation.

Such wood properties as high density, flat grain, and tight knots detract from paintability of boards but do not necessarily affect their finishing with penetrating preservatives and stains. These finishes penetrate into wood without forming a continuous film on the surface. Therefore, they will not blister or peel even if excessive moisture penetrates into wood.

---

[1] Mention of a chemical in this chapter does not constitute a recommendation; only those chemicals registered by the U.S. Environmental Protection Agency may be recommended, and then only for uses as prescribed in the registration and in the manner and at the concentration prescribed. The list of registered chemicals varies from time to time; prospective users, therefore, should get current information on registration status from the Environmental Protection Agency, Washington, D.C.

Table 16–1.—*Characteristics of woods for painting and finishing (omissions in the table indicate inadequate data for classification)*

Wood	Ease of keeping well painted; I—easiest, V—most exacting [1]	Weathering		Appearance	
		Resistance to cupping; 1—best, 4—worst	Conspicuousness of checking; 1—least, 2—most	Color of heartwood (sapwood is always light)	Degree of figure on flat-grained surface
SOFTWOODS					
Cedar:					
Alaska-	I	1	1	Yellow	Faint
California incense-	I	--	--	Brown	Do.
Port-Orford-	I	--	1	Cream	Do.
Western redcedar	I	1	1	Brown	Distinct
White-	I	1	--	Light brown	Do.
Cypress	I	1	1	----- do -----	Strong
Redwood	I	1	1	Dark brown	Distinct
Products [2] overlaid with resin-treated paper	I	--	1	--	--
Pine:					
Eastern white	II	2	2	Cream	Faint
Sugar	II	2	2	----- do --	Do.
Western white	II	2	2	----- do --	Do.
Ponderosa	III	2	2	----- do -----	Distinct
Fir, commercial white	III	2	2	White	Faint
Hemlock	III	2	2	Pale brown	Do.
Spruce	III	2	2	White	Do.
Douglas-fir (lumber and plywood)	IV	2	2	Pale red	Strong
Larch	IV	2	2	Brown	Do.
Lauan (plywood)	IV	2	2	----- do --	Faint
Pine:					
Norway	IV	2	2	Light brown	Distinct
Southern (lumber and plywood)	IV	2	2	----- do -----	Strong
Tamarack	IV	2	2	Brown	Do.
HARDWOODS					
Alder	III	--	--	Pale brown	Faint
Aspen	III	2	1	----- do -----	Do.
Basswood	III	2	2	Cream	Do.
Cottonwood	III	4	2	White	Do.
Magnolia	III	2		Pale brown	Do.
Yellow-poplar	III	2	1	----- do -----	Do.
Beech	IV	4	2	----- do -----	Do.
Birch	IV	4	2	Light brown	Do.
Gum	IV	4	2	Brown	Do.
Maple	IV	4	2	Light brown	Do.
Sycamore	IV	--	--	Pale brown	Do.
Ash	V or III	4	2	Light brown	Distinct
Butternut	V or III			----- do --	Faint
Cherry	V or III	--	--	Brown	Do.
Chestnut	V or III	3	2	Light brown	Distinct
Walnut	V or III	3	2	Dark brown	Do.
Elm	V or IV	4	2	Brown	Do.
Hickory	V or IV	4	2	Light brown	Do.
Oak, white	V or IV	4	2	Brown	Do.
Oak, red	V or IV	4	2	----- do -----	Do.

[1] Woods ranked in group V for *ease of keeping well painted* are hardwoods with large pores that need filling with wood filler for durable painting. When so filled before painting, the second classification recorded in the table applies.

[2] Plywood, lumber, and fiberboard with overlay or low-density surface.

Many wood products of lumber, plywood, shingles, and fiberboard are prepared with a roughsawn and absorptive surface that enhances the durability of stains by providing for better penetration.

Construction Details

House construction features that will minimize water damage of outside paint are: (a) Wide roof overhang, (b) wide flashing under shingles at roof edges, (c) effective vapor barriers, (d) adequate eave troughs and properly hung downspouts, (e) exhaust fans to remove excessive moisture, and (f) adequate insulation and ventilation of the attic. If these features are lacking in a new house, persistent paint blistering and peeling may occur and the structure then would best be finished with penetrating pigmented stains.

The proper application and nailing of wood siding does much to improve the appearance and durability of both wood and paint by reducing the tendency of the siding to split, crack, and cup with changes in moisture content. When possible, depending on the siding pattern, siding boards should be fastened so boards are free to shrink and swell, thereby reducing the tensile stresses that develop at fasteners.

Exterior siding and millwork should be installed with corrosion-resistant nails. Aluminum, hot-dipped galvanized, or stainless steel nails should be used for this purpose. Common iron nails or poor-quality galvanized nails corrode easily and will cause unsightly staining of the wood and paint. When the wood is to be left unfinished to weather or finished naturally with light-colored penetrating stains or water-repellent preservatives, only aluminum or stainless steel nails should be used.

Nails should be long enough to penetrate into studs and sheathing at least 1½ inches. Threaded nails should be used when nailing into plywood or lumber sheathing. When nails are near the end or edge of a board, nail holes should be predrilled or the tips of the nails blunted slightly to prevent splitting.

Recommended nailing practices for siding vary with the type of siding pattern and thickness and width of siding (fig. 16–1).

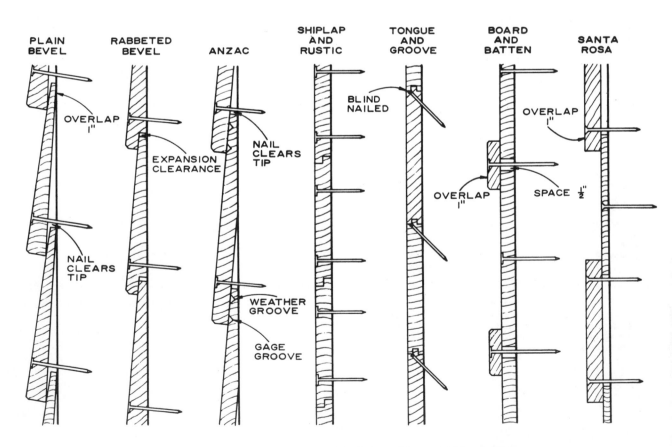

Figure 16–1.—Recommended nailing methods for various types of wood siding.

For plain bevel patterns, the siding should be face nailed, one nail per bearing, so that the nail clears the edge of the under course. Eightpenny or tenpenny nails are recommended for 1-inch-thick siding and sixpenny to eightpenny nails for thinner material.

Shiplap siding in 4- and 6-inch widths is face nailed with one nail per bearing a distance of 1 inch from the overlapping edge. Siding boards 8 inches or more in width should be nailed with two nails. Again, eightpenny nails should be used for siding 1 inch thick.

Tongued-and-grooved siding, 6 inches or less in width, is either face nailed with one eightpenny nail per bearing or blind nailed with one sixpenny finish nail through the tongue. Boards 6 inches or more in width are face nailed with two eightpenny nails.

In board-and-batten patterns, the under boards are spaced ½ inch apart and nailed with one eightpenny or ninepenny siding nail at the center of the board. The batten strip, 2½ inches wide, is nailed at the center with one tenpenny or twelvepenny nail. In board-on-board siding (fig. 16–1, Santa Rosa), the under board also is nailed with one nail at the center of the board. The outer boards, positioned to lap the under boards by 1 inch, are face nailed with two tenpenny or twelvepenny nails 1¼ inches from the edges.

Extractives and Impregnated Preservatives

Water-soluble color extractives occur naturally in western redcedar and redwood. It is to these substances that the heartwood of these species owes its attractive color, good stability, and natural decay resistance. Discoloration of paint occurs when the extractives are dissolved and leached from the wood by water. When the solution of extractives reaches the painted surface, the water evaporates, leaving the extractives as a reddish-brown stain. The water that gets behind the paint and causes moisture blisters also causes migration of extractives. The discoloration produced by water wetting siding from the back frequently forms a rundown or streaked pattern.

The latex paints and the so-called "breather" or low-luster oil paints are more porous than conventional oil paints. If these porous paint systems are used on new wood without a nonporous primer or a primer specifically designed to resist staining, or if any paint is applied too thinly on new wood (a skimpy two-coat paint job, for example), rain or even heavy dew can penetrate the coating and reach the wood.

When the water dries from the wood, the extractives are brought to the surface of the paint. Discoloration of paint by this process forms a diffused pattern which coincides with the dew and rain pattern of wetting.

On rough surfaces, such as shingles, machine-grooved shakes, and rough-sawn lumber sidings, it is difficult to obtain a uniformly thick coating on ridges and high points in the surface. Therefore, extractive staining is more likely to occur on such surfaces by water penetrating through defects in the coating.

Wood pressure-treated with waterborne chemicals, such as copper, chromium, and arsenic salts which react with the wood or form an insoluble residue, presents no major problems in painting if the wood is redried after treating. However, certain chemicals, which are not fixed in the wood but remain soluble, will diffuse to the surface under moist conditions to discolor paint.

Wood treated with solvent or oilborne preservative chemicals, such as pentachlorophenol, is not considered paintable until all the solvents have been removed. When heavy oil solvents with low volatility are used to treat wood under pressure, successful painting is usually impossible. Even special drying procedures for wood pressure-treated with the water-repellent preservative formulas that employ highly volatile solvents do not restore complete paintability.

Water-repellent preservatives introduced into wood by a vacuum-pressure process (NWMA Industry Standard 4–70) are paintable.

Coal-tar creosote or other dark oily preservatives tend to stain through paint, especially light-colored paint, unless the treated wood has weathered for many years before painting.

When it is necessary to finish pressure-treated wood, a dark-colored pigmented stain which penetrates the wood surface and forms no coating is recommended.

Moisture-Excluding Effectiveness of Finishes

The protection afforded by coatings in excluding moisture from wood depends on a great number of variables. Among them are film thickness, absence of defects and voids in the film, type of pigment, chemical composition of the vehicle, volume ratio of pigment to vehicle, vapor-pressure gradient across the film, and length of exposure period.

The relative effectiveness of several typical treating and finishing systems for wood in retarding adsorption of water vapor at 97 percent relative humidity is compared in table 16–2. Perfect protection, or no adsorption of water, would be represented by 100 percent effectiveness; complete lack of protection (as with unfinished wood) by 0 percent.

Values in table 16–2 are only representative and indicate the range in protection against moisture for some conventional finish systems. The degree of protection provided also depends on the kind of exposure. For example, the water-repellent preservative treatment, which may have 0 percent effectiveness after 2 hours at 80° F. and 97 percent relative humidity, would have an effectiveness of over 60 percent when tested after immersion in water for 30 minutes. The high degree of protection provided by a water-repellent preservative to short periods of wetting by water is the major reason they are recommended for exterior finishing.

Paints which are porous, such as the latex paints and low-luster or breather-type oil-base paints formulated at a pigment volume concentration usually above 40 percent, afford little protection against moisture. These paints permit rapid entry of water and so provide little protection against dew and rain unless applied over a nonporous primer.

FINISHING EXTERIOR WOOD

Weathering

The simplest of natural finishes for wood is weathering. Without paint or treatment of any kind, wood surfaces gradually change in color and texture and then may stay almost unaltered for a long time if the wood does not decay. Generally, the dark-colored woods become lighter and the light-colored woods become darker. As weathering continues, all woods become gray, accompanied by photodegradation of the wood cells at the surface. Exposed unfinished wood will wear away at the rate of about ¼ inch in 100 years.

The appearance of weathered wood exposed outdoors is usually affected by dark-colored spores and mycelia of fungi or mildew on the surface, which give the wood a dark gray, blotchy, and unsightly appearance. Highly colored wood extractives in such species as western redcedar and redwood also influence the color of weathered wood. The dark brown color may persist for a long time in areas not exposed to the sun and where the extractives are not removed by rain.

Table 16–2.—*Some typical values of moisture-excluding effectiveness of finishes. Wood was initially conditioned to 80° F. and 65 percent relative humidity and then exposed for 2 weeks to 80° F. and 97 percent relative humidity*

Coatings	Effectiveness
INTERIOR FINISHES	*Pct.*
Uncoated wood	0
3 coats of phenolic varnish	73
2 coats of phenolic varnish	49
1 coat of phenolic varnish (sealer)	5
3 coats of shellac	87
3 coats of cellulose lacquer	73
3 coats of lacquer enamel	76
3 coats of furniture wax	8
3 coats of linseed oil	21
2 coats of linseed oil	5
1 coat of linseed oil (sealer)	1
2 coats of latex paint	0
2 coats of semigloss enamel	52
2 coats of floor seal	0
2 coats of floor seal plus wax	10
EXTERIOR FINISHES	
1 coat water-repellent preservative [1]	0
1 coat of FPL natural finish (penetrating stain)	0
1 coat house paint primer	20
1 coat of house primer plus 2 coats of latex paint	22
1 coat of house primer plus 1 coat of TZ [2] linseed oil paint, 30 percent PVC [3]	60
1 coat of house primer plus 1 coat of TL [2] linseed oil paint, 30 percent PVC	65
1 coat of T-alkyd-oil, 30 percent PVC	45
1 coat of T-alkyd-oil, 40 percent PVC	3
1 coat of T-alkyd-oil, 50 percent PVC	0
2 coats of exterior latex paint	3
1 coat aluminum powder in long oil phenolic varnish	39
2 coats aluminum powder in long oil phenolic varnish	88
3 coats aluminum powder in long oil phenolic varnish	95

[1] The same product measured by immersing in water for 30 min. would have a water-repellency effectiveness of over 60 pct.
[2] The letters *T*, *L*, and *Z* denote paint's pigment with titanium dioxide, basic carbonated white lead, and zinc oxide, respectively.
[3] PVC denotes pigment volume concentration which is the volume percent of pigment in the nonvolatile portion of the paint.

Water-Repellent Preservatives

The natural weathering of wood may be modified by treatment with water-repellent finishes that contain a fungicide (usually pentachlorophenol), a small amount of resin, and a very small amount of water repellent which frequently is wax or waxlike in nature. The treatment, which penetrates the wood surface, retards the growth of fungi (mildew), reduces water staining of the ends of boards to a minimum, reduces warping, and protects species that have a low natural resistance to decay. A clear, golden tan color can be achieved on such sidings as smooth or rough-sawn western redcedar and redwood.

The preservative solution can be easily applied by dipping, brushing, or spraying. All lap and butt joints, edges, and ends of boards and panels should be liberally treated. Rough surfaces will absorb more solution than smoothly planed surfaces and the treatment will be more durable on them. Repeated brush and spray applications to the point of refusal will enhance durability and performance.

Because of the toxicity of pentachlorophenol, the commonly used fungicide in water-repellent preservative solutions, care should be exercised to avoid excessive contact with the solution or vapor, especially when spraying. Shrubs and plants should also be protected from contamination.

The initial application to smooth surfaces is usually short-lived. When a surface starts to show a blotchy discoloration due to extractives or mildew, it should be cleaned with detergent solution and retreated after drying. During the first few years, the finish may have to be applied every year or so. After sufficient time has elapsed so that the wood has weathered to a uniform color, however, the treatments are more durable and need refinishing only when the surface starts to become unevenly colored by fungi.

Inorganic pigments also can be added to the water-repellent preservative solutions to provide special color effects, and the mixture is then classified as a pigmented penetrating stain. Two to six fluid ounces of colors-in-oil or tinting colors can be added to each gallon of treating solution. Colors which match the natural color of the wood and extractives are usually preferred. The addition of pigment to the finish helps to stabilize the color and increases the durability of the finish.

Inorganic preservative chemicals which are soluble in water show promise as natural exterior finishes for wood. Such treatments retard the growth of fungi and also inhibit photodegradation of the surface fibers.

Pigmented Penetrating Stains

The pigmented penetrating stains are semi-transparent, permitting much of the grain pattern to show through, and penetrate into the wood without forming a continuous film on the surface. Therefore, they will not blister or peel even if excessive moisture enters the wood. Stains are made from both solvent resin and latex systems. Latex stains are a relatively new product.

Penetrating stains are suitable for both smooth and rough-textured surfaces; however, their performance is markedly improved if applied to rough-sawn, weathered, or rough-textured wood. They are especially effective on lumber and plywood that does not hold paint well, such as flat-grained surfaces of dense species. One coat of penetrating stain applied to smooth surfaces may last only 2 to 4 years; but the second application, after the surface has roughened by weathering, will last 8 to 10 years. A finish life of close to 10 years can be achieved initially on rough surfaces by applying two coats of stain. Two-coat staining also minimizes problems related to uneven stain application and lap marks. In two-coat staining, the second coat should always be applied before the first dries, usually within 30 to 60 minutes, so that both coats will penetrate.

Pigmented penetrating stains can be used effectively to finish such exterior surfaces as siding, trim, exposed decking, and fences.

One stain of this type is the Forest Products Laboratory natural finish. The finish has a linseed oil vehicle; a fungicide, pentachlorophenol, to protect the oil from mildew; and a water repellent, paraffin, wax, to protect the wood from excessive penetration of water. Durable red and brown iron oxide pigments simulate the natural colors of redwood and cedar. A variety of other colors also can be achieved with this type of finish; however, a pure white is not possible. Durability depends primarily on how much pigment penetrates into the wood surface.

Commercial finishes known as heavy-bodied or opaque stains are really considered to be more like a paint because of their film-forming characteristics. Such stains are finding wide success on textured surfaces and panel products. They should not be confused with the typical shake and shingle paints that are prone to check and peel.

Transparent Coatings

Clean coatings of conventional spar or marine varnishes, which are film-forming finishes, are not generally recommended for exterior use. Such coatings, embrittled by exposure to sunlight, develop severe cracking and peeling of the finish in less than 2 years. Areas that are protected from direct sunlight by overhang or are on the north side of the structure can be finished with exterior-grade varnishes. Even in protected areas, a minimum of three coats of varnish is recommended and the wood should be treated with water-repellent preservative before finishing. The use of pigmented stains and sealers as undercoats also will contribute to the life of the clear finish.

Expected major breakthroughs in polymer research will greatly enhance the longevity and use of exterior clear coatings. These developments will likely involve the use of new polymers which do not absorb ultraviolet light, such as certain vinyl fluoride, silicone, and acrylic polymers. Without the absorption of ultraviolet, the coating will undergo little photodegradation, and presumably could remain serviceable for many years. The use of coatings which are transparent to ultraviolet light will necessitate, however, the treating of wood surfaces with ultraviolet absorbers or the use of improved stable absorbers in the clear coating to protect the wood substrate from photodegradation. Treating the wood surface with salts of copper and chromium has been effective in protecting against photodegradation of the wood surface.

Painting Systems

Of all the finishes, paints provide the most protection for wood against surface erosion and offer the widest selection of colors. A nonporous paint film retards penetration of moisture and reduces the problem of discoloration by wood extractives, paint peeling, and checking and warping of the wood. Paint, however, is not a preservative; it will not prevent decay if conditions are favorable for fungal growth. Original and maintenance costs are usually higher for a paint finish than for a water-repellent preservative or penetrating stain finish.

The durability of paint coatings on exterior wood is affected both by variables in the wood surface and the type of paint.

Application of Paint

Wood to be painted should be handled like other building and finishing material to prevent contamination of surfaces from dirt, oil, and other foreign substances, and to avoid excessive wetting. Weathering of the wood surface before painting can detract from paintability and performance. It is therefore advisable to paint the surface promptly after installation.

Exterior wood surfaces can be very effectively painted by following a simple three-step procedure:

Step 1. Water-repellent-preservative treatment.—Wood siding and trim should be treated with water-repellent preservative to protect them against the entrance of rain and dew at joints. If treated exterior wood was not installed, the wood can be treated in place by brushing, dipping, or spraying. Care should be taken to brush well into lap and butt joints, especially treating ends of boards and panels. Two warm, sunny days should be allowed for adequate drying of the treatment before painting.

Care should be exercised in applying water-repellent preservatives, especially by spraying. The common fungicide used in these products is pentachlorophenol, which is toxic to both people and plants. The volatile solvents that are used are also toxic.

Step 2. Primer.—New wood should be given three coats of paint. The first, or prime, coat is the most important. It should be free of zinc-oxide pigment to reduce the tendency of the paint to blister, nonporous to prevent excessive entry of rain and dew, and flexible and thick enough to cover the wood grain so swelling stresses which develop on the wood surface are uniformly distributed. The primer should be applied soon after the wood is in place and treated; topcoats should be applied within 2 days to 2 weeks after the primer. Enough primer should be applied to obscure the wood grain. Many painters tend to spread primer too thinly. For best results, the spreading rates recommended by the manufacturer should be followed, or approximately 400 to 450 square feet should be covered per gallon with a paint that is about 85 percent solids by weight. A properly applied coat of a nonporous house paint primer will greatly reduce moisture blistering, peeling, and staining of paint by wood extractives.

For woods such as redwood and cedar, which contain water-soluble extractives, the best primers are good quality oil and alkyd-oil paints. For species free of extractives, such as the pines and Douglas-fir, high-quality acrylic latex paints (used for both primer and topcoat) hold great promise.

Primers effective on wood are not considered best for galvanized iron. Galvanized surfaces should be allowed to weather for several months and then primed with an appropriate primer, such as a linseed oil or resin-oil vehicle pigmented with metallic zinc dust (about 80 pct.) and zinc oxide (about 20 pct.)

Step 3. Finish coats.—Two coats of a good-quality latex, alkyd, or oil-base house paint should be applied over the nonporous primer. Two finish coats are particularly important for areas on east, south, and west sides of a structure that are exposed to maximum weather conditions, and not protected by overhang. One coat of a good house paint over a properly applied primer (a conventional two-coat paint system) will last only 4 or 5 years, but two coats should last 8 to 10 years.

High-quality acrylic latex paints and heavy-bodied stains are currently considered the best film-forming type finish for plywood panel products. These finishes are generally used without a primer.

To avoid future separation between coats of oil-base paint, or intercoat peeling, the first topcoat should be applied within 2 weeks after the primer and the second within 2 weeks of the first.

To avoid temperature blistering, oil-base paints should not be applied on a cool surface that will be heated by the sun within a few hours. Temperature blistering is most common with thickly applied paints of dark colors applied in cool weather. The blisters usually show up in the last coat of paint and occur within a few hours to 1 or 2 days after painting. They do not contain water.

To avoid wrinkling, fading, or loss of gloss of oil-base paints, and streaking of latex paints, paint should not be applied in the evenings of cool spring and fall days when heavy dews form before the surface of the paint has thoroughly dried.

Porches and Decks

Exposed flooring on porches and decks is commonly painted. The recommended procedure of treating with water-repellent preservative and primer is the same as for wood siding. After the primer, an undercoat and matching coat of porch and deck enamel should be applied.

Many fully exposed rustic-type decks are effectively finished with only water-repellent preservative or a penetrating-type pigmented stain. Because these finishes penetrate and form no film on the surface, they do not crack and peel. They may need more frequent refinishing than painted surfaces, but this is easily done because there is no need for laborious surface preparation as when painted surfaces start to peel.

RENEWING EXTERIOR PAINT SYSTEMS

Exterior wood surfaces need be repainted only when the old paint has worn thin and no longer protects the wood. In repainting with oil paint, one coat may be adequate if the old paint is in good condition. Dirty paint can often be freshened by washing with detergent. Paint in protected areas also should be washed well before repainting. Wood surfaces exposed by weathering, sanding, or scraping should be spot primed with a zinc-free oil-base primer before the finish coat is applied. Too-frequent repainting with oil-base systems produces an excessively thick film that is likely to crack abnormally across the grain of the wood. Complete paint removal is the only cure for cross-grain cracking. Latex films, based on either vinyl or acrylic polymers, have not been known to fail by cross-grain cracking.

When repainting with water-emulsion or latex paint, good adhesion of the latex to the badly chalked oil paint, is most important in preventing peeling. Vinyl and acrylic latexes are best for exterior exposure on wood.

This can be accomplished in several ways. The chalk layer can be removed by thorough washing and scrubbing with detergent solution. Or after mild steel wooling, the surface can be coated with a good oil-base primer. If the chalk layer is not excessively thick, certain latex paints can have adequate adhesion because of their modification with oil or alkyd-oil resins.

Porous latex paints, like porous low-luster oil-base paints, allow rain and dew to readily pass through the coating film. This can lead to problems of discoloration of the paint by wood extractives and peeling of old paint unless these paints are applied over a nonporous oil-base primer paint.

Avoiding Intercoat Peeling

To avoid intercoat peeling of oil-base paint, which indicates a weak bond between coats of paint, the old painted surface should be well cleaned and no more than 2 weeks allowed between coats in two-coat repainting. Sheltered areas, such as eaves and porch ceilings, need not be repainted every time the weathered body of the house is painted. Before repainting the sheltered areas where intercoat peeling frequently develops, the old paint surface should be washed with trisodium phosphate or detergent solution to remove surface contaminants that interfere with adhesion of the new coat of paint. After washing, the sheltered areas should be rinsed with large amounts of water and allowed to dry thoroughly before repainting. When intercoat peeling does occur, complete paint removal is the only satisfactory procedure to avoid future problems. Latex paints are less likely to fail by intercoat peeling than oil-base paints.

Blistering and Peeling

When too much water gets into paint or into the wood beneath the paint, the paint may either blister or peel.

Moisture blistering usually includes all of the paint down to the wood surface and indicates that the wood behind the paint is excessively wet. Blistering occurs in early spring and will occur first on only specific areas in heated buildings. These are areas that may enclose rooms with a high relative humidity in the winter or are areas wet because of ice dams and plugged gutters. If the blistering is severe, the paint may peel.

Moisture blistering is more likely in new, thin coatings of oil-base paint containing zinc-oxide pigment than in flat-alkyd or latex paints pigmented with titanium dioxide. Older and thicker coatings also are usually too rigid to swell enough to form blisters; instead, they are more prone to crack and peel after excessive wetting.

Peeling is a common type of water damage to paint and does not necessarily involve the formation of distinct blisters. Failure can occur both from interior and exterior source of moisture. Both heated and unheated buildings can peel where rain and dew wet the paint. Such failures are frequently associated with porous, flat, oil-alkyd, and latex paint systems which hold water on the surface, and so provide time for water to penetrate into the layers of paint. Peeling can occur at the wood interface or at some weak bond between layers of paint.

Cracking failures, followed by peeling at the ends of boards and the lower portion of horizontal siding, also indicate that rain and dew have been penetrating through paint or through cracks in paint. Such failures will occur on all sides of houses and also on unheated buildings. On the other hand, peeling failure or paint discoloration at localized areas, such as gable ends of heated buildings, indicates that the moisture is coming from within the building.

FINISHING INTERIOR WOOD

Interior finishing differs from exterior chiefly in that interior woodwork usually requires much less protection against moisture but more exacting standards of appearance and cleanability. Good finishes used indoors should last much longer than paint coatings on exterior surfaces. Veneered panels and plywood, however, present special finishing problems because of the tendency of these wood constructions to surface check.

Opaque Finishes

Interior surfaces may be easily painted by procedures similar to those for exterior surfaces. As a rule, however, smoother surfaces, better color, and a more lasting sheen are demanded for interior woodwork, especially wood trim; therefore, enamels or semigloss enamels rather than paints are used.

Before enameling, the wood surface should be sanded extremely smooth. Imperfections such as planer marks, hammer marks, and raised grain are accentuated by enamel finish. Raised grain is especially troublesome on flat-grained surfaces of the denser softwoods. It may occur because the hard bands of latewood have been crushed into the soft earlywood in planing, and later are pushed up again when the wood changes in moisture content. Before enameling, sponge softwoods with water, allow them to dry thoroughly, and then rub them lightly with new sandpaper. In new buildings, woodwork should be allowed adequate time to come to its equilibrium moisture content before finishing.

To effectively finish hardwoods with large pores, such as oak and ash, the pores must be filled with wood filler (see section on Fillers). After filling and sanding, successive applications of interior primer and sealer, undercoat, and enamel are used. Knots in the white pines, ponderosa pine, or southern pine should be sealed with shellac or a special knot sealer before priming. A coat of pigmented shellac or special knot sealer is also sometimes necessary over white pines and ponderosa pine to retard discoloration of light-colored enamels by colored matter present in the resin of the heartwood of these species.

One or two coats of enamel undercoat are applied; this should completely hide the wood and also present a surface that easily can be sandpapered smooth. For best results, the surface should be sandpapered before applying the finishing enamel; however, this step is sometimes omitted. After the finishing enamel has been applied, it may be left with its natural gloss, or rubbed to a dull finish. When wood trim and paneling are finished with a flat paint, the surface preparation is not nearly as exacting.

Transparent Finishes

Transparent finishes are used on most hardwood and some softwood trim and paneling, according to personal preference. Most finishing consists of some combination of the fundamental operations of staining, filling, sealing, surface coating, or waxing. Before finishing, planer marks and other blemishes on the wood surface that would be accentuated by the finish should be removed.

Stains

Both softwoods and hardwoods are often finished without staining, especially if the wood has a pleasing and characteristic color. When used, however, stain often provides much more than color alone because it is absorbed unequally by different parts of the wood; therefore, it accentuates the natural variations in grain. With hardwoods, such emphasis of the grain is usually desirable; the best stains for the purpose are dyes dissolved either in water or solvent. The water stains give the most pleasing results but raise the grain of the wood and require an extra sanding operation after the stain is dry.

The most commonly used stains are the "non-grain-raising" ones in solvents which dry quickly, and often approach the water stains in clearness and uniformity of color. Stains on softwoods color the earlywood more strongly than the latewood, reversing the natural gradation in color unless the wood has been sealed with a wash coat. Pigment-oil stains (essentially, thin paints) are less subject to this variation, and are therefore more suitable for softwoods. Alternatively, the softwood may be coated with clear sealer before applying the pigment-oil stain to give more nearly uniform coloring.

Fillers

If a smooth coating is desired in hardwoods with large pores, the pores must be filled (usually after staining) before varnish or lacquer is applied. The filler may be transparent and without effect on the color of the finish, or it may be colored to contrast with the surrounding wood.

For finishing purposes, the hardwoods may be classified as follows:

Hardwoods with large pores	Hardwoods with small pores
Ash	Alder, red
Butternut	Aspen
Chestnut	Basswood
Elm	Beech
Hackberry	Cherry
Hickory	Cottonwood
Khaya (African mahogany)	Gum
Lauans	Magnolia
Mahogany	Maple
Oak	Sycamore
Sugarberry	Yellow-poplar
Walnut	

Birch has pores large enough to take wood filler effectively when desired, but small enough as a rule to be finished satisfactorily without filling.

Hardwoods with small pores may be finished with paints, enamels, and varnishes in exactly the same manner as softwoods.

A filler may be paste or liquid, natural or colored. It is applied by brushing first across the grain and then by brushing with the grain. Surplus filler must be removed immediately after the glossy wet appearance disappears. Wipe first across the grain to pack the filler into the pores; then complete the wiping with a few light strokes with the grain. Filler should be allowed to dry thoroughly and sanded lightly before the finish coats are applied.

Sealers

Sealers are thinned varnish or lacquer and are used to prevent absorption of surface coatings and prevent the bleeding of some stains and fillers into surface coatings, especially lacquer coatings. Lacquer sealers have the advantage of being very fast drying.

Surface Coats

Transparent surface coatings over the sealer may be gloss varnish, semigloss varnish, nitrocellulose lacquer, or wax. Wax provides protection without forming a thick coating and without greatly enhancing the natural luster of the wood. Coatings of a more resinous nature, especially lacquer and varnish, accentuate the natural luster of some hardwoods and seem to permit the observer to look down into the wood. Shellac applied by the laborious process of French polishing probably achieves this impression of depth most fully, but the coating is expensive and easily marred by water. Rubbing varnishes made with resins of high refractive index for light (ability to bend light rays) are nearly as effective as shellac. Lacquers have the advantages of drying rapidly and forming a hard surface, but require more applications than varnish to build up a lustrous coating.

Varnish and lacquer usually dry with a highly glossy surface. To reduce the gloss, the surfaces may be rubbed with pumice stone and water or polishing oil. Waterproof sandpaper and water may be used instead of pumice stone. The final sheen varies with the fineness of the powdered pumice stone, coarse powders making a dull surface and fine powders a bright sheen. For very smooth surfaces with high polish, the final rubbing is done with rottenstone and oil. Varnish and lacquer made to dry to semigloss are also available.

Flat oil finishes commonly called Danish oils are currently very popular. This type of finish penetrates the wood and forms no noticeable film on the surface. Two coats of oil are usually applied, which may be followed with a paste wax. Such finishes are easily applied and maintained but are more subject to soiling than a film-forming type of finish.

Finishes for Floors

Wood possesses a variety of properties that make it a highly desirable flooring material for homes and industrial and public structures. A variety of wood flooring products permits a wide selection of attractive and serviceable wood floors. Selection is available not only from a variety of different wood species and grain characteristics, but also from a considerable number of distinctive flooring types and patterns.

The natural color and grain of wood floors make them inherently attractive and beautiful. Floor finishes enhance the natural beauty of wood, protect it from excessive wear and abrasion, and make the floors easier to clean. A complete finishing process may consist of four steps: Sanding the surface, applying a filler for certain woods, applying a stain to achieve a desired color effect, and applying a finish. Detailed procedures and specified materials depend largely on the species of wood used and individual preference in type of finish.

Careful sanding to provide a smooth surface is essential for a good finish because any irregularities or roughness in the wood surface will be magnified by the finish. Development of a top-quality surface requires sanding in several steps with progressively finer sandpaper, usually with a machine unless the area is small. The final sanding is usually done with a 2/0 grade paper. When sanding is complete, all dust must be removed with a vacuum cleaner or tack rag. Steel wool should not be used on floors unprotected by finish because minute steel particles left in the wood may later cause staining or discoloration.

A filler is required for wood with large pores, such as oak and walnut, if a smooth, glossy, varnish finish is desired.

Stains are sometimes used to obtain a more nearly uniform color when individual boards vary too much in their natural color. Stains may also be used to accent the grain pattern. If the natural color of the wood is acceptable, staining is omitted. The stain should be an oil-base or a non-grain-raising stain. Stains penetrate wood only slightly; therefore, the finish should be carefully maintained to prevent wearing through the stained layer. It is difficult to renew the stain at worn spots in a way that will match the color of the surrounding area.

Finishes commonly used for wood floors are classified either as sealers or varnishes. Sealers, which are usually thinned varnishes, are widely used in residential flooring. They penetrate the wood just enough to avoid formation of a surface coating of appreciable thickness. Wax is usually applied over the sealer; however, if greater gloss is desired, the sealed floor makes an excellent base for varnish. The thin surface coat of sealer and wax needs more frequent attention than varnished surfaces. However, rewaxing or resealing and waxing of high-traffic areas is a relatively simple maintenance procedure.

Varnish may be based on phenolic, alkyd, epoxy, or polyurethane resins. Varnish forms a distinct coating over the wood and gives a lustrous finish. The kind of service expected usually determines the type of varnish. Varnishes especially designed for homes, schools, gymnasiums, or other public buildings are available. Information on types of floor finishes can be obtained from the flooring associations or the individual flooring manufacturers.

Durability of floor finishes can be improved by keeping them waxed. Paste waxes generally give the best appearance and durability. Two coats are recommended and, if a liquid wax is used, additional coats may be necessary to get an adequate film for good performance.

BIBLIOGRAPHY

Allyn, Gerould
    1971. Acrylic latex paint systems give improved performance on southern pine. Amer. Paint J. 55(56): 54–75.

Anderson, L. O.
    1970. Wood-frame house construction. U.S. Dep. Agr., Agr. Handb. 73, rev.

————
    1972. Selection and use of wood products for home and farm building. U.S. Dep. Agr., Agr. Inf. Bull. 311, rev. 44 pp.

Browne, F. L.
    1959. Moisture content of wood as related to finishing of furniture. Forest Prod. Lab. Rep. 1722.

————, and Rietz, R. C.
    1959. Exudation of pitch and oils in wood. Forest Prod. Lab. Rep. 1735.

Cooper, G. A., and Barham, S. H.
    1971. The performance of a house paint on two overlays on cottonwood siding. Forest Prod. J. 21(3): 53–56.

Dost, W. A.
    1959. Attempts to modify the weathering of redwood. Forest Prod. J. 9(3): 18a–20a.

Kalnins, M. A.
    1966. Surface characteristics of wood as they affect durability of finishes. Part II. Photochemical degradation of wood. U.S. Forest Serv. Res. Pap. FPL 57. Forest Prod. Lab., Madison, Wis.

Keith, C. T.
    1964. Surface checking in veneered panels. Forest Prod. J. 14(10): 481.

Maass, W. B.
    1965. Coatings for galvanized steel. Amer. Paint J. 49(44): 37.

Marchessault, R. H., and Skaar, C.
    1967. Surfaces and coatings related to pulp and wood. Syracuse Univ. Press.

Michaels, A. C.
    1965. Water and the barrier film. Offic. Digest, J. Paint Technol. & Eng. 37(485): 638–653.

Miniutti, V. P.
    1963. Properties of softwoods that affect the performance of exterior paints. Official Digest, J. Paint Technol. & Eng. 35: 451–471.

————
    1964. Microscale changes in cell structure at softwood surfaces during weathering. Forest Prod. J. 15(12): 571–576.

National Woodwork Manufacturers Association
    1970. NWMA standard for water repellent preservatives, non-pressure treatment for millwork. Industry Standard 4–70. Chicago, Ill.

Newall, A. C., and Holtrop, W. F.
    1961. Coloring, finishing, and painting wood. Chas. A. Bennett Co., Peoria, Ill.

Panek, Edward
    1968. Study of paintability and cleanliness of wood pressure-treated with water-repellent preservative. Proc. Amer. Wood Preserv. Assoc. 64: 178–188.

Payne, H. F.
    1961. Organic coating technology. John Wiley & Sons. Vol. I and II.

Sinclair, R. M.
    1962. Comparison of the effect of wood preservative on paint durability. Jour. Oil & Color Chem. Assoc. 47(7): 491–528.

U.S. Forest Products Laboratory
    1961. Wood floors for dwellings. U.S. Dep. Agr., Agr. Handb. 204. 44 pp.

————
    1973. Wood siding: Installing, finishing, maintaining. U.S. Dep. Agr. Home and Garden Bull. 203. 13 pp.

Yan, M. M., and Lang, W. G.
    1960. What causes veneer checking and warping? Wood and Wood Prod. 65(8): 80.

Chapter 17

PROTECTION FROM ORGANISMS THAT DEGRADE WOOD [1]

Under proper conditions, wood will give centuries of service. Where conditions permit development of organisms that can degrade wood, however, protection must be provided in milling, merchandising, and building to insure maximum service life.

The principal organisms that can degrade wood are fungi, insects, bacteria, and marine borers.

Molds, most sapwood stains, and decay are caused by fungi, which are microscopic, threadlike plants that must have organic material to live. For some of them, wood offers the required food supply. The growth of fungi depends on suitably mild temperatures, dampness, and air (oxygen). Chemical stains, although they are not caused by organisms, are mentioned in this chapter because they resemble stains caused by fungi.

Insects also may damage wood, and in many situations must be considered in protective measures. Termites are the major insect enemy of wood, but, on a national scale, they are a less serious threat than fungi.

Bacteria in wood ordinarily are of little consequence, but some may make the wood excessively absorptive. Additionally, some may cause strength losses.

Marine borers are a fourth general type of wood-degrading organism. They can attack susceptible wood rapidly, and in salt-water harbors are the principal cause of damage to piles and other wood marine structures.

Wood degradation by organisms has been studied extensively and many preventive measures are well known and widely practiced. By taking ordinary precautions with the finished product, the user can contribute substantially to insuring a long service life.

FUNGUS DAMAGE AND CONTROL

Fungus damage to wood may be traced to three general causes: (1) Lack of suitable protective measures when storing logs or bolts;

(2) improper seasoning, storing, or handling of the raw material produced after storage; and (3) failure to take ordinary simple precautions in using the final product. The incidence and development of molds, decay, and stains caused by fungi depend heavily on temperature and moisture conditions.

Molds and Fungus Stains

Molds and fungus stains are confined largely to sapwood and are of various colors. The principal fungus stains are usually referred to as "sap stain" or "blue stain." The distinction between molding and staining is made largely on the basis of the depth of discoloration; with some molds and the lesser fungus stains there is no clear-cut differentiation. Typical sap stain or blue stain penetrates into the sapwood and cannot be removed by surfacing. Also, the discoloration as seen on a cross section of the wood often tends to exhibit some radial alinement corresponding to the direction of the wood rays. The discoloration may completely cover the sapwood or may occur as specks, spots, streaks, or patches of varying intensities of color. The so-called "blue" stains, which vary from bluish to bluish-black and gray to brown, are the most common, although various shades of yellow, orange, purple, and red are sometimes encountered. The exact color of the stain depends on the infecting organisms and the species and moisture condition of the wood. The brown stain mentioned here should not be confused with chemical brown stain.

Mold discolorations usually first become noticeable as largely fuzzy or powdery surface growths, with colors ranging from light shades to black. Among the brighter colors, green and yellowish hues are common. On softwoods—through the fungus may penetrate deeply—the discoloring surface growth often can easily be brushed or surfaced off. On hardwoods, however, the wood beneath the surface growth is commonly stained too deeply to be surfaced off. The staining tends to occur in spots of varying concentration and size, depending on the kind and pattern of the superficial growth.

Under favorable moisture and temperature conditions, staining and molding fungi may become established and develop rapidly in the sapwood of logs shortly after they are cut. In

---

[1] Mention of a chemical in this chapter does not constitute a recommendation; only those chemicals registered by the U.S. Environmental Protection Agency may be recommended, and then only for uses as prescribed in the registration and in the manner and at the concentration prescribed. The list of registered chemicals varies from time to time; prospective users, therefore, should get current information on registration status from the Environmental Protection Agency, Washington, D.C.

addition, lumber and such products as veneer, furniture stock, and millwork may become infected at any stage of manufacture or use if they become sufficiently moist. Freshly cut or unseasoned stock that is piled during warm, humid weather may be noticeably discolored within 5 or 6 days. Recommended moisture control measures are given in chapter 14.

Stains and molds should not be considered stages of decay as the causal fungi ordinarily do not attack the wood substance appreciably. Ordinarily, they affect the strength of the wood only slightly; their greatest effect is usually confined to strength properties that determine shock resistance or toughness (ch. 4).

Stain- and mold-infected stock is practically unimpaired for many uses in which appearance is not a limiting factor and a small amount of stain may be permitted by standard grading rules. Stock with stain and mold may not be entirely satisfactory for siding, trim, and other exterior millwork because the infected wood has greater water absorptiveness. Also, incipient decay may be present—though inconspicuous—in the discolored areas. Both of these factors increase the possibility of decay in wood that is rained on unless the wood has been treated with a suitable preservative.

Chemical Stains

One type of stain in unseasoned sapwood may resemble blue stain but is not caused by a fungus. This is called chemical stain or sometimes oxidation stain because it is brought about by a reaction between oxygen in the air and certain constituents of the exposed wood. Chemical brown stain is the only one of this type that is particularly serious in softwoods; however, many hardwoods are degraded by such stain. Chemical staining is largely a problem of seasoning; it can usually be prevented by rapid drying at low temperatures (ch. 14).

Decay

Decay-producing fungi may, under conditions that favor their growth, attack either heartwood or sapwood; the result is a condition variously designated as decay, rot, dote, or doze. Fresh surface growths of decay fungi may appear as fan-shaped patches, strands, or rootlike structures, usually white or brown. Sometimes fruiting bodies are produced that take the form of toadstools, brackets, or crusts.

The fungus, in the form of microscopic, threadlike strands, permeates the wood and uses parts of it as food. Some fungi live largely on the cellulose; others use the lignin as well as the cellulose.

Certain decay fungi attack the heartwood (causing "heartrot"), and rarely the sapwood of living trees, whereas others confine their activities to logs or manufactured products, such as sawed lumber, structural timbers, poles, and ties. Most of the tree-attacking groups cease their activities after the trees have been cut, as do the fungi causing brown pocket (peck) in baldcypress or white pocket in Douglas-fir. Relatively few continue their destruction after the trees have been cut and worked into products, and then only if conditions remain favorable for their growth.

Most decay can progress rapidly at temperatures that favor growth of plant life in general. For the most part decay is relatively slow at temperatures below 50° F. and much above 90° F. Decay essentially ceases when the temperature drops as low as 35° F. or rises as high as 100° F.

Serious decay occurs only when the moisture content of the wood is above the fiber saturation point (average 30 pct.). Only when previously dried wood is contacted by water, such as provided by rain, condensation, or contact with wet ground, will the fiber saturation point be reached. The water vapor in humid air alone will not wet wood sufficiently to support significant decay, but it will permit development of some mold. Fully air-dry wood usually will have a moisture content not exceeding 20 percent, and should provide a reasonable margin of safety against fungus damage. Thus wood will not decay if it is kept air dry—and decay already present from infection incurred earlier will not progress.

Wood can be too wet for decay as well as too dry. If it is water-soaked, there may be insufficient access of air to the interior of a piece to support development of typical decay fungi. For this reason, foundation piles buried beneath the water table and logs stored in a pond or under a suitable system of water sprays are not subject to decay by typical wood-decay fungi.

The early or incipient stages of decay are often accompanied by a discoloration of the wood, which is more evident on freshly exposed surfaces of unseasoned wood than on dry wood. Abnormal mottling of the wood color—with either unnatural brown or "bleached" areas—is often evidence of decay infection. Many fungi that cause heartrot in the standing tree produce incipient decay that differs only slightly from the normal color of the wood or gives a somewhat water-soaked appearance to the wood.

Typical or late stages of decay are easily recognized, because the wood has undergone definite changes in color and properties, the character of the changes depending on the organism and the substances it removes.

Two kinds of major decay are recognized—"brown rot" and "white rot." With brown rot, only the cellulose is extensively removed, the wood takes on a browner color, and it tends to crack across the grain and to shrink and collapse. With white rot, both lignin and cellulose usually are removed; the wood may lose color and appear "whiter" than normal, it does not crack across the grain, and until severely degraded it retains its outward dimensions and does not shrink or collapse.

Brown, crumbly rot, in the dry condition, is sometimes called "dry rot," but the term is incorrect because wood must be damp to decay, although it may become dry later. A few fungi, however, have water-conducting strands; such fungi are capable of carrying water (usually from the soil) into buildings or lumber piles, where they moisten and rot wood that would otherwise be dry. They are sometimes referred to technically as "dry rot fungi" or "water-conducting fungi." The latter term better describes the true situation as these fungi, like the others, must have water.

A third and generally less important kind of decay is known as soft rot. Soft rot is caused by fungi related to the molds rather than those responsible for brown and white rot. Soft rot typically is relatively shallow; the affected wood is greatly degraded and often soft when wet, but immediately beneath the zone of rot the wood may be firm. Because soft rot usually is rather shallow it is most likely to damage relatively thin pieces such as slats in cooling towers. It is favored by wet situations but is also prevalent on surfaces that have been alternately wet and dry over a substantial period. Heavily fissured surfaces—familiar to many as "weathered" wood—generally have been considerably degraded by soft rot fungi.

Decay Resistance of Wood

For a discussion of the natural resistance of wood to fungi and a grouping of species according to decay resistance, see chapter 3. Among decay-resistant domestic species, only the heartwood has significant resistance, because the natural preservative chemicals in wood that retard the growth of fungi are essentially restricted to the heartwood. Natural resistance of species to fungi is important only where conditions conducive to decay exist or may develop.

Effect of Decay on Strength of Wood

Incipient decay induced by some fungi is reflected immediately in pronounced weakening of the wood, whereas other fungi reduce strength much less. For example, white pocket (produced by *Fomes pini*) results in little or no loss in strength in its incipient stages. On the other hand, another cause of heartrot in standing softwoods, *Polyporus schweinitzii*, greatly reduces the strength of wood at a very early stage. In the later stages of decay any wood-damaging fungus will seriously reduce the strength of wood.

Control of Mold, Stain, and Decay

Logs, Poles, Piles, and Ties

The wood species, section of the country, and time of the year determine what precautions must be taken to avoid serious damage from fungi in poles, piles, ties, and similar thick products during seasoning or storage. In dry climates, rapid surface seasoning of poles and piles will retard development of mold, stains, and decay. First the bark is peeled from the pole and the peeled product is decked on high skids or piled on high, well-drained ground in the open to dry. In humid regions, such as the Gulf States, these products often do not air dry fast enough to avoid losses from fungi. Preseasoning treatments with approved preservative solutions can be helpful in these circumstances.

For logs, rapid conversion into lumber, or storage in water or under a water spray is the surest way to avoid fungus damage. Preservative sprays promptly applied in the woods will protect most timber species during storage for 2 to 3 months. For longer storage, an end coating is needed to prevent seasoning checks, through which infection can enter the log.

Lumber

Growth of decay fungi can be prevented in lumber and other wood products by rapidly drying them to a moisture content of 20 percent or less and keeping them dry. Standard air-drying practices will usually dry the wood fast enough to protect it, particularly if the protection afforded by drying is supplemented by surface treatment of the stock with an approved fungicidal solution. However, kiln drying is the most reliable method of rapidly reducing moisture content.

Dip or spray treatment of freshly cut lumber with suitable preservative solutions will prevent fungus infection during air drying. Successful control by this method depends not only upon immediate and adequate treatment but also upon the proper handling of the lumber after treatment.

Air drying yards should be kept as sanitary and as open as possible to air circulation. Recommended practice includes locating yards and sheds on well-drained ground; removing debris, which serves as a source of infection, and weeds, which reduce air circulation; and employing piling methods that permit rapid drying of the lumber and protect against wetting. Storage sheds should be constructed and maintained to prevent significant wetting of the stock; an ample roof overhang on open sheds is desirable. In areas where termites or water-conducting fungi may be troublesome, stock to be held for long periods should be set on foundations high enough so it can be inspected from beneath.

The user's best assurance of receiving lumber free from decay or other than light stain is to buy stock marked by a lumber association in a grade that eliminates or limits such quality-reducing features. Surface treatment for protection at the drying yard is only temporarily effective. Except for temporary structures, lumber to be used under conditions conductive to decay should be all heartwood of a naturally durable species or should be adequately treated with a wood preservative (ch. 18).

Buildings

The lasting qualities of properly constructed wood buildings are apparent in all parts of the country. Serious decay problems are almost always a sign of faulty design or construction or of lack of reasonable care in the handling of the wood.

Construction principles that assure long service and avoid decay in buildings include: (1) Build with dry lumber, free of incipient decay and not exceeding the amounts of mold and blue stain permitted by standard grading rules; (2) use designs that will keep the wood dry and accelerate rain runoff; (3) for parts exposed to above-ground decay hazards, use wood treated with a preservative or heartwood of a decay-resistant species; and (4) for the high-hazard situation associated with ground contact, use pressure-treated wood.

A building site that is dry or for which drainage is provided will reduce the possibility of decay. Stumps, wood debris, stakes, or wood concrete forms frequently lead to decay if left under or near a building.

Unseasoned or infected wood should not be enclosed until it is thoroughly dried. Unseasoned wood may be infected because of improper handling at the sawmill or retail yard, or after delivery on the job.

Untreated wood parts of substructures should not be permitted to contact the soil. A minimum of 8 inches clearance between soil and framing and 6 inches between soil and siding is recommended. An exception may be made for certain temporary constructions. If contact with soil is necessary, the wood should be pressure treated (ch. 18).

Sill plates and other wood resting on a concrete-slab foundation generally should be pressure treated, and additionally protected by installing beneath the slab a moisture-resistant membrane of heavy asphalt roll roofing or polyethylene. Girder and joist openings in masonry walls should be big enough to assure an air space around the ends of these wood members; if the members are below the outside soil level, moistureproofing of the outer face of the wall is essential.

In the crawl space of basementless buildings on damp ground, wetting of the wood by condensation during cold weather may result in serious decay damage. However, serious condensation leading to decay can be prevented by providing openings on opposite sides of the foundation walls for cross ventilation or by laying heavy roll roofing or polyethylene on the soil; both provisions may be helpful in very wet situations. To facilitate inspection and ventilation of the crawl space, at least an 18-inch clearance should be left under wood joists.

Porches, exterior steps, and platforms present a decay hazard that cannot be fully avoided by construction practices. Therefore, in the wetter climates the use of preservative-treated wood (ch. 18) or heartwood of a durable species usually is advisable for such items.

Protection from entrance or retention of rainwater or condensation in walls and roofs will prevent the development of decay in these areas. A fairly wide roof overhang (2 ft.) with gutters and downspouts that are never permitted to clog is very desirable. Sheathing papers under the siding should be of a "breathing" or vapor-permeable type (asphalt paper not exceeding 15-lb. weight). Vapor barriers should be near the warm face of walls and ceilings. Roofs must be kept tight, and cross ventilation in attics is desirable. The use of sound, dry lumber is important in all parts of buildings.

Where service conditions in a building are such that the wood cannot be kept dry, as in textile mills, pulp and paper mills, and cold-storage plants, lumber properly treated with an approved preservative or lumber containing all heartwood of a naturally decay-resistant species should be used.

In making repairs necessitated by decay, every effort should be made to correct the moisture condition leading to the damage. If the condition cannot be corrected, all infected parts should be replaced with treated wood or with all-heartwood lumber of a naturally decay-resistant wood species. If the sources of moisture that caused the decay are entirely eliminated, it is necessary only to replace the weakened wood with dry lumber.

Other Structures and Products

In general, the principles underlying the prevention of mold, stain, or decay damage to veneer, plywood, containers, boats, and other wood products and structures are similar to those described for buildings—dry the wood rapidly and keep it dry or treat it with approved protective and preservative solutions. Interior grades of plywood should not be used where the plywood will be exposed to moisture; the adhesives, as well as the wood, may be damaged by fungi and bacteria as well as being degraded by moisture. With either plywood or fiberboard of the exterior type, joint construction should be carefully designed to prevent the entrance of rainwater.

Wood boats present certain problems that are not encountered in other uses of wood. The parts especially subject to decay are the stem, knighthead, transom, and frameheads; these are reached by rainwater from above or condensation moisture from below. Faying surfaces are more liable to decay than exposed surfaces, and in salt-water service hull members just below the weather deck are more vulnerable than those below the waterline. Recommendations for avoiding decay include: (1) Use only heartwood of durable species, free of infection, and preferably below 20 percent in moisture content; (2) provide and maintain ventilation in the hull and all compartments; (3) keep water out as much as is practicable, especially fresh water; and (4) where it is necessary to use sapwood or nondurable heartwood, impregnate the wood with an approved preservative or treat the fully cut, shaped, and bored wood before installation by soaking it for a short time in preservative solution. Where such mild soaking treatment is used, the wood most subject to decay should also be flooded with an approved preservative at intervals of 2 or 3 years. When retreating, the wood should be dry so that joints are relatively loose.

BACTERIA

Most wood that has been wet for any considerable length of time probably will contain bacteria. The sour smell of logs that have been held under water for several months—or of lumber cut from them—manifests bacterial action. Usually bacteria have little effect on wood properties, except over long periods of time, but some may make the wood excessively absorptive. This effect has been a problem in the sapwood of millwork cut from pine logs that have been stored in ponds. There also is evidence that bacteria developing in pine veneer bolts held under water or water spray may cause noticeable changes in the physical character of the veneer—including some strength loss. Additionally, mixtures of different bacteria, and probably fungi also, were found capable of accelerating decay of treated cooling tower slats and mine timbers.

INSECT DAMAGE AND CONTROL

The more common types of degrade caused by wood-attacking insects are shown in table 17–1. Methods of controlling and preventing insect attack of wood are described in the following paragraphs.

Table 17–1.—*Common types of degrade caused by wood-attacking insects*

Type of degrade	Description	How and where made	Condition of degraded timber
Pinholes	Holes with dark streak in surrounding wood, $\frac{1}{100}$ to $\frac{1}{4}$ inch in diameter, usually circular and open (not grouped in given space): Hardwoods:		
	Stained area 1 inch or more long	By ambrosia beetles in living trees.	Wormholes, no living worms.
	Stained area less than 1 inch long	By ambrosia beetles in recently felled trees and green logs.	Do.
	Softwoods:		
	Stained area less than 1 inch long	By ambrosia beetles in sapwood of peeled trees, green logs, and green lumber.	Do.
	Holes usually without streak in surrounding wood of both softwoods and hardwoods (usually grouped in given space): Holes darkly stained, less than $\frac{1}{8}$ inch in diameter.	By ambrosia beetles in felled trees, green logs, and green lumber.	Do.
	Holes unstained, open, and variable in diameter:		
	Lined with a substance the color of wood: from $\frac{1}{100}$ to $\frac{1}{4}$ inch in diameter (hardwoods rarely with streaks in surrounding wood).	By timberworms in living and felled trees and green logs.	Do.
	Unlined, less than $\frac{1}{8}$ inch in diameter (not grouped in given space).	By ambrosia beetles in green logs and green lumber.	Do.
Grub holes	Holes $\frac{3}{8}$ to 1 inch in diameter, variable in shape, open or with boring dust: Holes stained, usually open	By wood-boring grubs in living trees.	Do.
	Holes unstained, usually with boring dust present:		
	Borings fine, granular, or fibrous ...	..... do	Do.
	Borings variable in character, sometimes absent.	By adults and larvae of beetles and other wood-boring insects in recently felled softwoods and hardwoods.	Do.
Powder-post	Holes mostly $\frac{1}{16}$ to $\frac{1}{4}$ inch in diameter, circular to broadly oval, filled with granular or powdery boring dust, and unstained: Hardwoods	By roundheaded borers and powder-post beetles in green or seasoned wood.	Wormholes.
	Softwoods	By flathead borers in living, recently felled, and dead trees.	Do.
		By roundheaded borers, weevils, and Anobium powder-post beetles in seasoned wood.	Do.
Pitch pocket		By various insects in living trees.	Wormholes, no living worms.
Black check		By the grubs of various insects in living trees.	Do.
Bluing	Stained area over 1 inch long	By fungus following insect wounds in living trees and recently felled sawlogs.	Do.
Pith fleck		By the maggots of flies or adult weevils in living trees.	Do.
Gum spot		By the grubs of various insects in living trees.	Do.
Ring distortions		By defoliating larvae or flathead cambium miners in living trees.	Do.

Beetles

Bark beetles may damage the components of log and other rustic structures on which the bark is left. They are reddish-brown to black and vary in length from about $\frac{1}{16}$ to $\frac{1}{4}$ inch. They bore through the outer bark to the soft inner part, where they make tunnels in which they lay their eggs. In making tunnels, bark beetles push out fine brownish-white sawdust-like particles. If many beetles are present, their extensive tunneling will loosen the bark and permit it to fall off in large patches, making the structure unsightly.

To avoid bark beetle damage, logs may be stored in water or under a water spray, or cut during the dormant season (October or November, for instance). If cut during this period, logs should immediately be piled off the ground where there will be good air movement, to promote rapid drying of the inner bark before the beetles begin to fly in the spring. Drying the bark will almost always prevent damage by insects that prefer freshly cut wood. Another protective measure is to thoroughly spray the logs with an approved insecticidal solution.

Ambrosia beetles, roundheaded and flatheaded borers, and some powder-post beetles that get into freshly cut timber can cause considerable damage to wood in rustic structures and some manufactured products. Certain beetles may complete development and emerge a year or more after the wood is dry, often raising a question as to the origin of the infestation. Proper cutting practices and spraying the material with an approved chemical solution, as recommended for bark beetles, will control these insects. Damage by ambrosia beetles can be prevented in freshly sawed lumber by dipping the product in a chemical solution. The addition of one of the sap-stain preventives approved for controlling molds, stains, and decay will keep the lumber bright.

Powder-post beetles attack both hardwoods and softwoods, and both freshly cut and seasoned lumber and timber. The powder-post beetles that cause most damage to dry hardwood lumber belong to the *Lyctus* species. They attack the sapwood of ash, hickory, oak, and other hardwoods as it begins to season. Eggs are laid in pores of the wood, and the larvae burrow through the wood, making tunnels from $\frac{1}{16}$ to $\frac{1}{12}$ inch in diameter, which they leave packed with a fine powder. Powder-post damage is indicated by holes left in the surface of the wood by the winged adults as they emerge and by the fine powder that may fall from the wood.

Susceptible hardwood lumber used for manufacturing purposes should be protected from powder-post beetle attack as soon as it is sawed and also when it arrives at the plant. An approved insecticide applied in water emulsion to the green lumber will provide protection. Such treatment may be effective even after the lumber is kiln dried—until it is surfaced.

Good plant sanitation is extremely important in alleviating the problem of infestations. Proper sanitation measures can often eliminate the necessity for other preventative steps. Damage to manufactured items frequently is traceable to infestations that occur before the products are placed on the market, particularly if a finish is not applied to the surface of the items until they are sold. Once wood is infested, the larvae will continue to work, even though the surface is subsequently painted, oiled, waxed, or varnished.

When selecting hardwood lumber for building or manufacturing purposes, any evidence of powder-post infestation should not be overlooked, for the beetles may continue to be active long after the wood is put to use. Sterilization of green wood with steam at 130° F. or sterilization of wood with a lower moisture content at 180° F. under controlled conditions of relative humidity for about 2 hours is effective for checking infestation or preventing attack of 1-inch lumber. Thicker material requires a longer time. A 3-minute soaking in a petroleum oil solution containing an insecticide is also effective for checking infestation or preventing attack of lumber up to 1 inch thick. Small dimension stock also can be protected by brushing or spraying with approved chemicals. For infested furniture or finished woodwork in a building, the same insecticides may be used but they should be dissolved in a refined petroleum oil, like mineral spirits.

As the *Lyctus* beetles lay their eggs in the open pores of wood, infestation can be prevented by covering the entire surface of each piece of wood with a suitable finish.

A roundheaded powder-post beetle, commonly known as the "old house borer," causes damage to seasoned pine floor joists. The larvae reduce the sapwood to a powdery or granular consistency and make a ticking sound while at work. When mature, the beetles make an oval hole about $\frac{1}{4}$ inch in diameter in the surface of the wood and emerge. Anobiid powder-post beetles, which make holes $\frac{1}{16}$ to $\frac{1}{8}$ inch in diameter, also cause damage to pine joists. Infested wood should be drenched with a solution of one of the currently recommended insecticides in a highly penetrating solvent. Beetles working in wood behind plastered or paneled walls can be eliminated by having a licensed operator fumigate the building.

Termites

Termites superficially resemble ants in size, general appearance, and habit of living in colonies. About 56 species are known in the United States. From the standpoint of their methods of attack on wood, they can be grouped into two main classes: (1) The ground-inhabiting or subterranean termites; and (2) the wood-inhabiting or nonsubterranean termites.

Subterranean Termites

Subterranean termites are responsible for most of the termite damage done to wood structures in the United States. This damage can be prevented. Subterranean termites are more prevalent in the southern states than in the northern states, where low temperatures do not favor their development (fig. 17–1). The hazard of infestation is greatest (1) beneath basementless buildings erected on a concrete-slab foundation or over a crawl space that is poorly drained and ventilated, and (2) in any substructure wood close to the ground or an earth fill (e.g. an earth-filled porch).

The subterranean termites develop their colonies and maintain their headquarters in

Figure 17–1.—A, The northern limit of recorded damage done by subterranean termites in the United States; B, the northern limit of damage done by dry-wood or nonsubterranean termites.

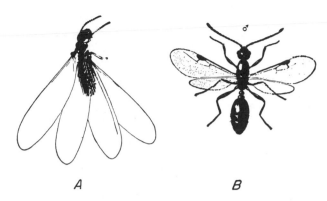

Figure 17–2.—A, Winged termite; B, winged ant (both greatly enlarged). The wasp waist of the ant and the long wings of the termite are distinguishing characteristics.

the ground. They build their tunnels through earth and around obstructions to get at the wood they need for food. They also must have a constant source of moisture. The worker members of the colony cause destruction of wood. At certain seasons of the year male and female winged forms swarm from the colony, fly a short time, lose their wings, mate, and, if successful in locating a suitable home, start new colonies. The appearance of "flying ants," or their shed wings, is an indication that a termite colony may be near and causing serious damage. Not all "flying ants" are termites; therefore suspicious insects should be identified before money is spent for their eradication (fig. 17–2).

Subterranean termites do not establish themselves in buildings by being carried there in lumber, but enter from ground nests after the building has been constructed. Telltale signs of their presence are the earthern tubes or runways built by these insects over the surfaces of foundation walls to reach the wood above. Another sign is the swarming of winged adults early in the spring or fall. In the wood itself, the termites make galleries that generally follow the grain, leaving a shell of sound wood to conceal their activities. As the galleries seldom show on the wood surfaces, probing with an ice pick or knife is advisable if the presence of termites is suspected.

The best protection where subterranean termites are prevalent is to prevent them from gaining access to the building. The foundations should be of concrete or other solid material through which the termites cannot penetrate. With brick, stone, or concrete blocks, cement mortar should be used, for termites can work through some other kinds of mortar.

Also, it is a good precaution to cap the foundation with about 4 inches of reinforced concrete. Posts supporting first-floor girders should, if they bear directly on the ground, be of concrete. If there is a basement, it should be floored with concrete. Untreated posts in such a basement should rest on concrete piers extending a few inches above the basement floor. However, pressure-treated posts can rest directly on the basement floor. With the crawl-space type of foundation, wood floor joists should be kept at least 18 inches and girders 12 inches from the earth and good ventilation provided beneath the floor.

Moisture condensation on the floor joists and subflooring, which may cause conditions favorable to decay and contribute to infestation by termites, can be avoided by covering the soil below with a moisture barrier. When concrete-slab floors are laid directly on the ground, all soil under the slab should be treated with an approved insecticide before the concrete is poured. Furthermore, insulation containing cellulose that is used as a filler in expansion joints should be impregnated with an approved chemical toxic to termites. Sealing the top ½ inch of the expansion joint with roofing-grade coal-tar pitch also provides effective protection from ground-nesting termites.

All concrete forms, stakes, stumps, and waste wood should be removed from the building site, for they are possible sources of infestation. In the main, the precautions effective against subterranean termites are also helpful against decay.

The principal method of protecting buildings in high termite hazard areas is to thoroughly treat the soil adjacent to the foundation walls and piers beneath the building with a soil insecticide. New modifications in soil treatment are currently under investigation and appear promising. Current references to termite control should be consulted.

To control termites already in a building, break any contact between the termite colony in the soil and the woodwork. This can be done by blocking the runways from soil to wood, by treating the soil, or both. Guard against possible reinfestations by frequent inspections for telltale signs that were listed previously.

Recently, the Formosan termite has become established in the United States. It is more active and voracious than the subterranean termites native to the United States. However, the conventional methods of protection seem to be effective.

Nonsubterranean termites have been found only in a narrow strip of territory extending from central California around the southern edge of continental United States to Virginia (fig. 17–1) and also in the West Indies and Hawaii. Their principal damage is confined to an area in southern California, to parts of southern Florida, notably Key West, and to the islands of Hawaii.

The nonsubterranean termites, especially the dry-wood type, do not multiply as rapidly as the subterranean termites, and have somewhat different colony life and habits. The total amount of destruction they cause in the United States is much less than that caused by the subterranean termites. The ability of dry-wood termites to live in dry wood without outside moisture or contact with the ground, however, makes them a definite menace in the regions where they occur. Their depredations are not rapid, but they can thoroughly riddle timbers with their tunnelings if allowed to work unmolested for a few years.

In constructing a building in localities where the dry-wood type of nonsubterranean termite is prevalent, it is good practice to inspect the lumber carefully to see that it was not infested before arrival at the building site. If the building is constructed during the swarming season, the lumber should be watched during the course of construction, because infestation by colonizing pairs can easily take place. Because paint is a good protection against the entrance of dry-wood termites, exposed wood (except that which is preservative treated) should be kept adequately painted. Fine screen should be placed over any openings through which access might be gained to the interior unpainted parts of the buildings. As in the case of ground-nesting termites, old stumps, posts, or wood debris of any kind that could serve as sources of infestation should be removed from the premises.

If a building is infested with dry-wood termites, badly damaged wood should be replaced. If the wood is only slightly damaged or difficult to replace, further termite activity can be arrested by forcing a teaspoonful of an approved pesticide dust into each nest. Also effective are approved liquid insecticides. Current recommendations for such formulations should be consulted.[2] Detached houses

---

[2] See footnote, p. 290.

heavily infested with nonsubterranean termites have been fumigated with success. This method is quicker and often cheaper than the use of poisonous liquids and dusts, but it does not prevent the termites from returning because no poisonous residue is left in the tunnels. Fumigation is very dangerous and should be conducted only by licensed professional fumigators.

In localities where dry-wood termites do serious damage to posts and poles, the best protection for these and similar forms of outdoor timbers is full-length pressure treatment with a preservative.

Naturally Termite-Resistant Woods

Only a limited number of woods grown in the United States offer any marked degree of natural resistance to termite attack. The close-grained heartwood of California redwood has some resistance, especially when used above ground. Very resinous heartwood of southern pine is practically immune, but wood of this type is not available in large quantities or suitable for many uses.

Carpenter Ants

Carpenter ants are black or brown. They occur usually in stumps, trees, or logs but sometimes damage poles, structural timbers, or buildings. One form is easily recognized by its giant size relative to other ants. Carpenter ants use wood for shelter rather than for food, usually preferring wood that is naturally soft or has been made soft by decay. They may enter a building directly, by crawling, or may be carried there in fuel wood. If left undisturbed they can, in a few years, enlarge their tunnels to the point where replacement or extensive repairs are necessary. The parts of dwellings they frequent most often are porch columns, porch roofs, window sills, and sometimes the wood plates in foundation walls. The logs of rustic cabins are also attacked.

Precautions that prevent attack by decay and termites are usually effective against carpenter ants. Decaying or infested wood, such as logs or stumps, should be removed from the premises, and crevices present in the foundation or woodwork of the building should be sealed. Particularly, leaks in porch roofs should be repaired, because the decay that may result makes the wood more desirable to the ants.

When carpenter ants are found in a structure, any badly damaged timbers should be replaced. Because the ant needs high humidity in its immature stages, alterations in the construction may also be required to eliminate moisture from rain or condensation. In wood not sufficently damaged to require replacement, the ants can be killed by dusting with an approved insecticide.

MARINE BORER DAMAGE AND CONTROL

Damage by marine-boring organisms to wood structures in salt or brackish waters is practically worldwide. Slight attack is sometimes found in rivers even above the region of brackishness. The rapidity of attack depends upon local conditions and the kinds of borers present. Along the Pacific, Gulf, and South Atlantic coasts of the United States attack is rapid, and untreated piling may be completely destroyed in a year or less. Along the coast of the New England States the rate of attack is slower but still sufficiently rapid, generally, to require protection of wood where long life is desired.

The principal marine borers from the standpoint of wood damage in the United States are described here. Control measures discussed in this section are those in use at the time this handbook was prepared. Regulations should be reviewed at the time control treatments are being considered so that approved practices will be followed.[3]

Shipworms

Shipworms are the most destructive of the marine borers. They are mollusks of various species that superficially are wormlike in form. The group includes several species of *Teredo* and several species of *Bankia*, which are especially damaging. These are readily distinguishable on close observation but are all very similar in several respects. In the early stages of their life they are minute, free-swimming organisms. Upon finding suitable lodgement on wood they quickly develop into a new form and bury themselves in the wood. A pair of boring shells on the head grows rapidly in size as the boring progresses, while the tail part or siphon remains at the original entrance. Thus, the animal grows in length and diameter within the wood but remains a prisoner in its burrow, which it lines with

---

[3] See footnote, p. 290

a shell-like deposit. It lives on the wood borings and the organic matter extracted from the sea water that is continuously being pumped through its system. The entrance holes never grow large, and the interior of a pile may be completely honeycombed and ruined while the surface shows only slight perforations. When present in great numbers, the borers grow only a few inches before the wood is so completely occupied that growth is stopped, but when not crowded they can grow to lengths of 1 to 4 feet according to species.

Pholads

Another group of wood-boring mollusks is *pholads*, which clearly resemble clams and therefore are not included with the shipworms. These are entirely encased in their double shells. The *Martesia* are the best-known species, but a second group is the *Xylophaga*. Like the shipworms, the *Martesia* enter the wood when very small, leaving a small entrance hole, and grow larger as they burrow into the wood. They generally do not exceed 2½ inches in length and 1 inch in diameter, but are capable of doing considerable damage. Their activities in the United States appear to be confined to the Gulf of Mexico.

Limnoria and Sphaeroma

Another distinct group of marine borers are crustaceans, which are related to lobsters and shrimp. The principal ones are species of *Limnoria* and *Sphaeroma*. Their attack differs from that of the shipworms and the *Martesia* in that it is quite shallow; the result is that the wood gradually is thinned through erosion by the combined action of the borers and water. Also, the *Limnoria* and *Sphaeroma* do not become imprisoned in the wood but may move freely from place to place.

Limnoria are small, about ⅛ to ⅙ inch long, and bore small burrows in the surface of piles. Although they can change their location, they usually continue to bore in one place. When great numbers are present, their burrows are separated by very thin walls of wood that are easily eroded by the motion of the water or damaged by objects floating upon it. This erosion causes the *Limnoria* to burrow continually deeper; otherwise the burrows would probably not become more than 2 inches long or more than ½ inch deep. As erosion is

greatest between tide levels, piles heavily attacked by *Limnoria* characteristically wear within such levels to an hourglass shape. Untreated piling can be destroyed by *Limnoria* within a year in heavily infested harbors.

Sphaeroma are somewhat larger, sometimes reaching a length of ½ inch and a width of ¼ inch. They resemble in general appearance and size the common sow bug or pill bug that inhabits damp places. *Sphaeroma* are widely distributed but not as plentiful as *Limnoria* and do much less damage. Nevertheless piles in some structures have been ruined by them. Occasionally they have been found working in fresh water. In types of damage, *Sphaeroma* action resembles that of *Limnoria*.

The average life of well-creosoted structures is many times the average life that could be obtained from untreated structures. However, even thorough creosote treatment will not always stop *Martesia*, *Sphaeroma*, and especially *Limnoria*.

Shallow or erratic creosote penetration affords but slight protection. The spots with poor protection are attacked, and from them the borers spread inward and destroy the untreated interior of the pile. Low retention fails to provide a reservoir of surplus preservative to compensate for depletion by evaporation and leaching.

When wood is to be used in salt water, avoidance of cutting or injuring the surface after treatment is even more important than when wood is to be used on land. No cutting or injury of any kind for any purpose should be permitted in the underwater part of the pile. Where piles are cut to grade above the waterline, the exposed surfaces should, of course, be protected from decay.

Resistance to Marine Borers

No wood is immune to marine-borer attack, and no commercially important wood of the United States has sufficient marine-borer resistance to justify its use untreated in any important structure in areas where borers are active. The heartwood of several foreign species, such as turpentine, greenheart, jarrah, azobe, totara, kasikasi, manbarklak, and several others, has shown resistance to marine-borer attack. Service records on these woods, however, do not always show uniform results and are affected by local conditions.

Protection of Permanent Structures

The best practical protection for piles in sea water where borer hazard is moderate is heavy treatment with coal-tar creosote or creosote-coal tar solution. Where severe borer hazard exists, dual treatment (copper arsenate[4] containing waterborne preservatives followed by coal-tar creosote) is recommended. The treatment must be thorough, the penetration as deep as possible, and the retention high to give satisfactory results in heavily infested waters. It is best to treat such piles by the full-cell process "to refusal;" that is, to force in all the preservative the piles can hold without using treatments that cause serious damage to the wood. The retentions recommended in chapter 18 are minimum values; when maximum protection against marine borers is desired, as much more preservative as is practicable should be injected. For highest retentions it is necessary to air dry the piling before treatment. Details of treatments are discussed in chapter 18.

The life of treated piles is influenced by the thoroughness of the treatment, the care and intelligence used in avoiding damage to the treated shell during handling and installation, and the severity of borer attack. Differences in exposure conditions, such as water temperature, salinity, dissolved oxygen, water depth, and currents, tend to cause wide variations in the severity of borer attack even within limited areas. The San Francisco Bay Marine Piling Committee has estimated a range of 15 to 30 years for the life of creosoted piles on the Pacific Coast. More recent service records show average-life figures of from 22 to 48 years on well-treated Douglas-fir piles in San Francisco Bay waters. In South Atlantic and Gulf of Mexico waters, creosoted piles are estimated to last 10 to 12 years, and frequently much longer. On the North Atlantic Coast, even longer life is to be expected.

Metal armor and concrete jacketing have been used with varying degrees of success for the protection of marine piles. The metal armor may be in the form of sheets, wire, or nails. Scupper-nailing with iron or steel furnishes some protection, particularly against *Limnoria*. Sheathing of piles with copper or muntz metal has been only partially successful, owing to difficulty in maintaining a continuous armor. Theft, damage in driving, damage by storm or driftwood, and corrosion have sooner

or later let in the borers, and in only a few cases has long life been reported. Attempts during World War II to electroplate wood piles with copper were not successful. Concrete casings are now in greater use than metal armor and appear to provide better protection when high-quality materials are used and are carefully applied. Unfortunately, they are readily damaged by ship impact. For this reason, concrete casings are less practical for fender piles than for foundation piles that are protected from mechanical damage.

Jacketing piles by wrapping them with heavy polyvinyl plastic is one of the most recent forms of supplementary protection. If properly applied, it will kill any borers that may have already become established, by rendering stagnant the water in contact with the piles. Like other materials, the plastic jacket is subject to mechanical damage.

Protection of Boats

Wood barges and lighters have been constructed with planking or sheathing pressure treated with creosote to provide hull protection from marine borers, and the results have been favorable. Although coal-tar creosote is an effective preservative for protecting wood against marine borers in areas of moderate borer hazard, it has disadvantages in many types of boats. Creosote adds considerably to the weight of the boat hull, and its odor is objectionable to boat crews. In addition, antifouling paints are difficult to apply over creosoted wood.

Some copper bottom paints protect boat hulls against marine-borer attack, but the protection continues only while the coating remains unbroken. As it is difficult to maintain an unbroken coating of antifouling paint, the U.S. Navy has found it desirable to impregnate the hull planking of some wood boats with certain copper-containing preservatives.[4] Such preservatives, when applied with high retentions (1.5 to 2.0 p.c.f.), have some effectiveness against marine borers and should help to protect the hull of a boat during intervals between renewals of the antifouling coating. These copper preservatives do not provide protection equivalent to that furnished by coal-tar creosote; their effectiveness in protecting boats is therefore best assured if the boats are dry docked at regular and frequent intervals and the antifouling coating

[4] See footnote, p. 290

maintained. However, the leach-resistant wood preservatives containing copper arsenates [5] have shown superior performance (at a retention of 2.5 p.c.f.) to creosote in tests conducted in areas of severe borer hazard.

Plywood as well as plank hulls can be protected against marine borers by preservative treatment. The plywood hull presents a surface that can be covered successfully with a protective membrane of reinforced plastic laminate. Such coverings should not be attempted on wood that has been treated with a preservative carried in oil, because the bond will be unsatisfactory.

[5] See footnote, p. 17–2.

BIBLIOGRAPHY

Anderson, L. O.
 1970. Wood-frame house construction. U.S. Dep. Agr., Agr. Handbook 73, rev. 223 pp.
Atwood, W. G., and Johnson, A. A.
 1924. Marine structures, their deterioration, and preservation. Nat. Res. Counc. Rep. Comm. on Marine Piling Invest., 534 pp., illus.
Baechler, R. H., and Roth, H. G.
 1961. Further data on the extraction of creosote from marine piles. Proc. Amer. Wood Preserv. Assoc. 57: 120–129.
Beal, J. A., and Massey, G. L.
 1945. Bark beetles and ambrosia beetles. Duke Univ. Forest. Bull. 10, 178 pp., illus.
Beal, R. H.
 1967. Formosan invader. Pest Control 35(2): 13–17.
Chellis, R. D.
 1948. Finding and fighting marine borers. Eng. News Rec. 140(12): 422–424, illus.; (14): 493–496, illus.
Federal Housing Administration
 1965. Minimum property standards for one and two living units. Pub. 300, 315 pp.
Furniss, R. L.
 1944. Carpenter ant control in Oregon. Oreg. Agr. Exp. Sta. Cir. 158, 12 pp.
Greaves, H.
 1969. Wood-inhabiting bacteria: General considerations. Commonwealth Sci. Ind. Res. Org. Forest Prod. Newsletter 359.
Hartley, C., and May, C.
 1943. Decay of wood in boats. U.S. Dep. Agr. Forest Path. Spec. Release 8, 12 pp.
Hill, C. L., and Kofoid, C. A.
 1927. Marine borers and their relation to marine construction on the Pacific Coast. San Francisco Bay Marine Piling Comm. Final Rep., 357 pp.
Hochman, H., Vind, H., Roe, T., Jr., Muraoka, J., and Casey, J.
 1956. The role of Limnoria tripunctata in promoting early failure of creosoted piling. Tech. Memorandum M–109, U.S. Naval Civil Eng. Res. and Evaluation Lab., 43 pp.
Johnston, H. R.
 1952. Control of insects attacking green logs and lumber. Southern Lumberman 184(2307): 37–39.

—— 1960. Soil treatments for subterranean termites. U.S. Dep. Agr. Occasional Pap. 152, rev. 6 pp.
——, Smith, R. H., and St. George, R. A.
 1955. Prevention and control of Lyctus powder-post beetles. Southern Lumberman. Mar.
Knuth, D. T., and McCoy, Elizabeth
 1962. Bacterial deterioration of pine logs in pond storage. Forest Prod. J. 12(9): 437–442.

Lewis, W. C.
 1968. Thermal insulation from wood for buildings: Effects of moisture and its control. USDA Forest Serv. Res. Pap. FPL 86. Forest Prod. Lab., Madison, Wis.
Light, S. F.
 1929. Termites and termite damage. Calif. Agr. Exp. Sta. Cir. 314, 28 pp.
MacLean, J. D.
 1950. Results of experiments on the effectiveness of various preservatives in protecting wood against marine-borer attack. Forest Prod. Lab. Rep. D1773.
Panek, Edward
 1963. Pretreatments for the protection of southern yellow pine poles during air seasoning. Proc. Amer. Wood Pres. Assoc. 59: 182–202.
Richards, A. P.
 1965. Marine biology handbook. U.S. Navy Department, NAVDOCKS MO–311.
——, and Clapp, W. F.
 1946. Control of marine borers in plywood. Amer. Soc. Mech. Eng. Wood Ind. Div. Pap. 46–A–43, 6 pp.
Scheffer, T. C., and Verrall, A. F.
 1973. Principles of protecting wood buildings from decay. USDA Forest Serv. Res. Pap. FPL 190. Forest Prod. Lab., Madison, Wis.
——, Wilson, T.R.C., Luxford, R. F., and Hartley, C.
 1941. Effect of certain heartrot fungi on the specific gravity and strength of Sitka spruce and Douglas-fir. U.S. Dep. Agr. Tech. Bull. 779, 24 pp.
St. George, R. A.
 1970. Protecting log cabins, rustic work, and unseasoned wood from injurious insects in the Eastern United States. U.S. Dep. Agr. Farmers' Bull. 2104, rev. 18 pp.
Snyder, T. E.
 1927. Defects in timber caused by insects. U.S. Dep. Agr. Bull. 1490, 46 pp.

—— 1966. Control of nonsubterranean termites. U.S. Dep. Agr. Farmers' Bull. 2018, rev. 16 pp.
U.S. Department of Agriculture
 1972. Subterranean termites, their prevention and control in buildings. U.S. Dep. Agr. Home & Garden Bull. 64, rev. 16 pp.

—— 1969. Wood decay in houses, how to prevent and control it. U.S. Dep. Agr. Home & Garden Bull. 73, rev. 17 pp.
U.S. Forest Products Laboratory
 1960. Making log cabins endure. Forest Prod. Lab. Rep. 982.
Verrall, A. F.
 1952. Control of wood decay in buildings. Agr. Eng. 33(4): 217–219.
——, and Scheffer, T. C.
 1949. Control of stain, mold, and decay in green lumber and other wood products. Forest Prod. Res. Soc. Proc. 3, 9 pp.

Chapter 18

WOOD PRESERVATION [1]

Wood can be protected from the attack of decay fungi, harmful insects, or marine borers by applying selected chemicals as wood preservatives. The degree of protection obtained depends on the kind of preservative used and on achieving proper penetration and retention of the chemicals. Some preservatives are more effective than others, and some are more adaptable to certain use requirements. The wood can be well protected only when the preservative substantially penetrates it, and some methods of treatment assure better penetration than others. There is also a difference in the treatability of various species of wood, particularly of their heartwood, which generally resists preservative treatment more than sapwood.

Good wood preservatives, applied with standard retentions and with the wood satisfactorily penetrated, substantially increase the life of wood structures, often by five or more times. On this basis the annual cost of treated wood in service is greatly reduced below that of similar wood without treatment. In considering preservative treatment processes and wood species, the combination must provide the required protection for the conditions of exposure and life of the structure.

WOOD PRESERVATIVES

Wood preservatives fall into two general classes: Oils, such as creosote and petroleum solutions of pentachlorophenol; and water-borne salts that are applied as water solutions.

Preservative Oils

The wood does not swell from the preservative oils, but it may shrink if it loses moisture during the treating process. Creosote and solutions with the heavier, less volatile petroleum oils often help protect the wood from weathering outdoors, but may adversely influence its

cleanliness, odor, color, paintability, and fire resistance in use. Preservative oils sometimes travel from treated studs or subflooring along nails and discolor adjacent plaster or finish flooring. Volatile oils or solvents with oil-borne preservatives, if removed after treatment, leave the wood somewhat cleaner than the heavier oils but may not provide as much protection. Wood treated with preservative oils can be glued satisfactorily, although special processing or cleaning may be required to remove surplus oils from surfaces before spreading the adhesive.

Coal-Tar Creosote

Coal-tar creosote, a black or brownish oil made by distilling coal tar, is one of the more important and useful wood preservatives. Its advantages are: (1) High toxicity to wood-destroying organisms; (2) relative insolubility in water and low volatility, which impart to it a great degree of permanence under the most varied use conditions; (3) ease of application; (4) ease with which its depth of penetration can be determined; (5) general availability and relative low cost (when purchased in wholesale quantities); and (6) long record of satisfactory use.

The character of the tar used, the method of distillation, and the temperature range in which the creosote fraction is collected all influence the composition of the creosote. The composition of the various coal-tar creosotes available, therefore, may vary to a considerable extent. Small differences in composition, however, do not prevent creosotes from giving good service; satisfactory results in preventing decay may generally be expected from any coal-tar creosote that complies with the requirements of standard specifications.

Although coal-tar creosote or creosote-coal tar solutions are well qualified for general outdoor service in structural timbers, they have properties that are disadvantageous for some purposes.

The color of creosote and the fact that creosote-treated wood usually cannot be painted satisfactorily make this preservative unsuitable for finish lumber or other lumber used where appearance and paintability are important.

The odor of creosoted wood is unpleasant to some persons. Also, creosote vapors are

[1] Mention of a chemical in this chapter does not constitute a recommendation; only those chemicals registered by the U.S. Environmental Protection Agency may be recommended, and then only for uses as prescribed in the registration and in the manner and at the concentration prescribed. The list of registered chemicals varies from time to time; prospective users, therefore, should get current information on registration status from the Environmental Protection Agency, Washington, D.C.

harmful to growing plants, and foodstuffs that are sensitive to odors should not be stored where creosote odors are present. Workmen sometimes object to creosoted wood because it soils their clothes and because it burns the skin of the face and hands of some individuals. With normal precautions to avoid direct dermal contact with creosote, there appears to be no danger to the health of workmen handling or working near the treated wood or on the health of the occupants of buildings in which the treated wood is used.

Freshly creosoted timber can be ignited easily and will burn readily, producing a dense smoke. However, after the timber has seasoned some months, the more volatile parts of the oil disappear from near the surface, and the creosoted wood usually is little, if any, easier to ignite than untreated wood. Until this volatile oil has evaporated, ordinary precautions should be taken to prevent fires. On the other hand, timber that has been kept sound by creosote treatment is harder to ignite than untreated wood that has started to decay. A preservative other than creosote should be used where fire hazard is highly important, unless the treated wood is also protected from fire.

A number of specifications prepared by different organizations are available for creosote oils of different kinds. Although the oil obtained under most of these specifications will probably be effective in preventing decay, the requirements of some organizations are more exacting than others. Federal Specification TT–C–645 for coal-tar creosote, adopted for use by the U.S. Government, will generally prove satisfactory; under normal conditions, this specification can be met without difficulty by most creosote producers. The requirements of this specification are similar to those of the American Wood-Preservers' Association (AWPA) Standard P1 for creosote, which is equally acceptable.

Federal Specification TT–C–645 provides for three classes of coal-tar creosote. Class I is for poles; class II is for ties, lumber, structural timbers, land or fresh water piles, and posts; and class III is for piles, lumber, and structural timbers for use in coastal waters.

Some pole users, to reduce bleeding in poles with high retentions of creosote, have specified lower retentions of coal-tar creosote fortified with 2 percent pentachlorophenol. Corrosion problems to treating plant equipment have accompanied the use of this preservative and are under investigation.

Coal-Tar Creosotes for Nonpressure Treatments

Special coal-tar creosotes are available for nonpressure treatments. They differ somewhat from regular commercial coal-tar creosote in (1) being crystal-free to flow freely at ordinary temperatures and (2) having low-boiling distillation fractions removed to reduce evaporation in thermal (hot-and-cold) treatments in open tanks. Federal Specification TT–C–655 covers coal-tar creosote for brush, spray, or open-tank treatments.

Other Creosotes

Creosotes distilled from tars other than coal tar are used to some extent for wood preservation, although they are not included in current Federal or AWPA specifications. These include wood-tar creosote, oil-tar creosote, and water-gas-tar creosote. These creosotes protect wood from decay and insect attack but are generally less effective than coal-tar creosote.

Tars

Coal tars are seldom used alone for preserving wood because good penetration is usually difficult to obtain and because they are less poisonous to wood-destroying fungi than the coal-tar creosotes. Service tests have demonstrated that surface coatings of tar are of little value. Coal tar has been used in the pressure treatment of crossties, but it has been difficult to get the highly viscous tar to penetrate wood satisfactorily. When good absorptions and deep penetrations are obtained, however, it is reasonable to expect a satisfactory degree of effectiveness from treatment with coal tar. The tar has been particularly effective in reducing checking in crossties in service.

Water-gas-tar is used less extensively than coal tar, but, in certain cases where the wood was thoroughly impregnated, the results were good.

Creosote Solution

For many years, either coal tar or petroleum oil has been mixed with coal-tar creosote, in various proportions, to lower preservative costs. These creosote solutions have a satisfactory record of performance, particularly for crossties where they have been most commonly used.

Federal Specification TT–C–650, "Creosote-Coal-Tar Solution," covers five classes of creosote-coal-tar solution. These classes contain not less than 80, 70, 60, 50, and 65 percent coal-tar distillate (creosote) (by volume), for classes I, II, III, IV, and V, respectively. Classes I and II are for land and fresh-water piles, posts, lumber, structural timber, and bridge ties. Classes III and IV are for crossties and switch ties. Class V, which has a 60 to 75 percent level of distillate, is for piles, lumber, and structural timber used in coastal waters.

AWPA Standard P2 includes four creosote-coal-tar solutions that must contain, respectively, not less than 80, 70, 60, or 50 percent by volume of coal-tar distillate and must also meet requirements as to physical and chemical properties. AWPA Standard P12 covers a creosote-coal-tar solution for the treatment of marine (coastal waters) piles and timbers. Federal Specification TT–W–568 and AWPA Standard P3 stipulate that creosote-petroleum oil solutions shall contain not less than 50 percent (by volume) of coal-tar creosote and the petroleum oil shall meet the requirements of AWPA's Standard P4.

Creosote-coal-tar solutions, compared to straight creosote, tend to reduce weathering and checking of the treated wood. The solutions may have a greater tendency to accumulate on the surface of the treated wood (bleed) and may penetrate the wood with greater difficulty, particularly because they generally are more viscous than straight creosote. Higher temperatures and pressures during treatment, when they can safely be used, will often improve penetration of solutions of high viscosity.

Even though petroleum oil and coal tar are less toxic to wood-destroying organisms than straight creosote, and their mixtures with creosote are also less toxic in laboratory tests, a reduction in toxicity does not imply less preservative protection. Creosote-petroleum solutions and creosote coal-tar solutions help to reduce checking and weathering of the treated wood. Frequently posts and ties treated with these solutions of standard formulation have shown better service than those similarly treated with straight coal-tar creosote.

Pentachlorophenol Solutions

Water-repellent solutions containing chlorinated phenols, principally pentachlorophenol, in solvents of the mineral spirits type, were first used in commercial treatments of wood by the millwork industry about 1931. Commercial pressure treatment with pentachlorophenol in heavy petroleum oils started on poles about 1941, and considerable quantities of various products were soon pressure treated. AWPA Standard P8 and Federal Specification TT–W–570 define the properties of pentachlorophenol and AWPA Standard P9 covers solvents for oil-borne preservatives. A commercial process using pentachlorophenol dissolved in liquid petroleum gas was introduced in 1961.

Pentachlorophenol solutions for wood preservation generally contain 5 percent (by weight) of this chemical although solutions with volatile solvents may contain lower or higher concentrations. The performance of pentachlorophenol and the properties of the treated wood are influenced by the properties of the solvent used. The heavy petroleum solvent included in AWPA Standard P9A is preferable for maximum protection, particularly where the wood treated with pentachlorophenol is used in contact with the ground, but further evaluation of newer volatile solvents is needed.

The heavy oils remain in the wood for a long time and do not usually provide a clean or paintable surface. The volatile solvents, such as liquefied petroleum gas and methylene chloride, are used with pentachlorophenol when the natural appearance of the wood must be retained and the treated wood requires a paint coating or other finish. Because of the toxicity of pentachlorophenol, care is necessary to avoid excessive personal contact with the solution or vapor in handling and using it.

A "bloom" preventive, such as ester gum or oil-soluble glycol, is generally required with volatile solvents to prevent crystals of pentachlorophenol from forming on the surface of the treated wood. Brushing or washing the surface with hot water or an alkaline solution has been used to remove the crystalline deposits.

The results of pole service and field tests on wood treated with 5 percent pentachlorophenol in a heavy petroleum oil are similar to those with coal-tar creosote. This similarity has been recognized in the preservative retention requirements of treatment specifications. Pentachlorophenol is ineffective against marine borers and is not recommended for the treatment of marine piling or timbers used in coastal waters.

Water-Repellent Preservatives

Preservative systems containing water-repellent components are sold under various trade names, principally for the dip or equivalent treatment of window sash and other millwork. Federal Specification TT–W–572 stipulates that such preservatives consist of volatile solvents, such as mineral spirits, that do not cause appreciable swelling of the wood, and that the treated wood be paintable and meet a performance test on water repellency. In pressure treatment with water-repellent preservative, however, considerable difficulty has been experienced in removing residual solvents and obtaining acceptable paintability.

The preservative chemicals in Federal Specification TT–W–572 may be not less than 5 percent of pentachlorophenol, not less than either 1 or 2 percent (for tropical conditions) of copper in the form of copper naphthenate, or not less than 0.045 percent copper in the form of copper-8-quinolinolate (for uses where foodstuffs will be in contact with the treated wood). Commercial Standard CS 262–63, covering the water-repellent preservative, non-pressure treatment for millwork, permits other preservatives provided their toxicity properties are as high as those of 5 percent (by weight) pentachlorophenol solution. Mixtures of other chlorinated phenols with pentachlorophenol meet this requirement according to tests by the National Woodwork Manufacturer's Association.

Water-repellent preservative containing copper-8-quinolinolate has been used in non-pressure treatment of wood containers, pallets, and other products for use in contact with foods. That preservative is also included in AWPA Standard P8. Here it is intended for use in volatile solvents to pressure-treat lumber for decking of trucks and cars or for related uses involving harvesting, storage, and transportation of foods.

Effective water-repellent preservatives will retard the ingress of water when wood is exposed above ground. They therefore help reduce dimensional changes in the wood due to moisture changes when the wood is exposed to rainwater or dampness for short periods. As with any wood preservative, their effectiveness in protecting wood against decay and insects depends upon the retention and penetration obtained in application.

Waterborne Preservatives

Standard wood preservatives used in water solution include acid copper chromate, ammoniacal copper arsenite, chromated copper arsenate (types I, II, and III), chromated zinc chloride, and fluor chrome arsenate phenol. These preservatives are often employed when cleanliness and paintability of the treated wood are required. The chromated zinc chloride and fluor chrome arsenate phenol formulations resist leaching less than preservative oils, and are seldom used where a high degree of protection is required for wood in ground contact or for other wet installations. Several formulations involving combinations of copper, chromium, and arsenic have shown high resistance to leaching and very good performance in service. The ammoniacal copper arsenite and chromated copper arsenate are now included in specifications for such items as building foundations, building poles, utility poles, marine piling, and piling for land and fresh water use.

Test results based on sea water exposure have shown that dual treatment (waterborne copper-containing salt preservatives followed by coal-tar-creosote) is possibly the most effective method of protecting wood against all types of marine borers. The AWPA standards have recognized this process as well as the treatment of marine piling with high retentions of ammoniacal copper arsenite or chromated copper arsenate. The recommended treatment and retention in pounds per cubic foot (p.c.f.) for round timber piles exposed to severe marine borer hazard are:

	Southern pine, red pine (P.c.f.)	Coastal Douglas-fir (P.c.f.)	AWPA standard
Severe borer hazard:			
Limnoria tripunctata only:			
Ammoniacal copper arsenite	2.5	2.5	C3 C18
Chromated copper arsenate	2.5	2.5	C3 C18
Limnoria tripunctata and Pholads (dual treatment):			
First treatment:			
Ammoniacal copper arsenite	1.0	1.0	C3 C18
Chromated copper arsenate	1.0	1.0	C3 C18
Second treatment:			
Creosote	20.0	20.0	C3 C18
Creosote-coal-tar	20.0	Not recommended	C3 C18

Waterborne preservatives leave the wood surface comparatively clean, paintable, and free from objectionable odor. With several exceptions, they must be used at low treating temperatures (100° to 150° F.) because they are unstable at the higher temperatures. This may involve some difficulty when higher temperatures are needed to obtain good treating results in such woods as Douglas-fir. Because water is added during treatment, the wood must be dried afterward to the moisture content required for use.

Waterborne preservatives, in the retentions normally specified for wood preservation, decrease the danger of ignition and rapid spread of flame, although formulations with copper and chromium stimulate and prolong glowing combustion in carbonized wood.

Acid Copper Chromate

Acid copper chromate (Celcure) contains, according to Federal Specification TT–W–546 and AWPA Standard P5, 31.8 percent copper oxide and 68.2 percent chromic acid. Equivalent amounts of copper sulfate, potassium dichromate, or sodium dichromate may be used in place of copper oxide. Tests on stakes and posts exposed to decay and termite attack indicate that wood well impregnated with Celcure gives good service. Tests by the Forest Products Laboratory and the U.S. Navy showed that wood thoroughly impregnated with at least 0.5 p.c.f. of Celcure has some resistance to marine borer attack. The protection against marine borers, however, is much less than that provided by a standard treatment with creosote.

Ammoniacal Copper Arsenite

According to Federal Specification TT–W–549 and AWPA Standard P5, ammoniacal copper arsenite (Chemonite) should contain approximately 49.8 percent copper oxide or an equivalent amount of copper hydroxide, 50.2 percent of arsenic pentoxide or an equivalent amount of arsenic trioxide, and 1.7 percent of acetic acid. The net retention of preservative is calculated as pounds of copper oxide plus arsenic pentoxide per cubic foot of wood treated within the proportions in the specification.

Service records on structures treated with ammoniacal copper arsenite show that this preservative provides very good protection against decay and termites. High retentions of preservative will provide extended service life to wood exposed to the marine environment, provided pholad-type borers are not present.

Chromated Copper Arsenate

Types I, II, and III of chromated copper arsenate are covered in Federal Specification TT–W–550 and AWPA Standard P5. The compositions of the three types according to that Federal specification are:

	Type I	Type II	Type III
		Parts by weight	
Chromium trioxide	61	35.3	47
Copper oxide	17	19.6	19
Arsenic pentoxide	22	45.1	34

The above types permit substitution of potassium or sodium dichromate for chromium trioxide; copper sulfate, basic copper carbonate, or copper hydroxide for copper oxide; and arsenic acid or sodium arsenate for arsenic pentoxide.

Type I (Erdalith, Greensalt, Tanalith, CCA)

Service data on treated poles, posts, and stakes installed in the United States since 1938 have shown excellent protection against decay fungi and termites. High retentions of copper chrome arsenate-treated wood have shown good resistance to marine borer attack when only Limnoria and teredo borers are present.

Type II (Boliden K–33)

This preservative has been used commercially in Sweden since 1950 and now throughout the world. It was included in stake tests in the United States in 1949 and commercial use in the United States started in 1964.

Type III (Wolman CCA)

Composition of this preservative was arrived at by AWPA technical committees in encouraging a single standard for chromated copper arsenate preservatives. Commercial preservatives of similar composition have been tested and used in England since 1954 and more recently in Australia, New Zealand, Malaysia, and in various countries of Africa and Central Europe and are performing very well.

Chromated Zinc Chloride

Chromated zinc chloride is covered in Federal Specification TT–W–551 and in AWPA Standard P5. Chromated zinc chloride (FR)[2]

---

[2] Designation for fire retardant.

is included, as a fire-retarding chemical, in AWPA Standard P10.

Chromated zinc chloride was developed about 1934. The specifications require that it contain 80 percent of zinc oxide and 20 percent of chromium trioxide. Zinc chloride may be substituted for the zinc oxide and sodium dichromate for the chromium trioxide. The preservative is only moderately effective in contact with the ground or in wet installations but has performed well under somewhat drier conditions. Its principal advantages are its low cost and ease of handling at treating plants.

Chromated zinc chloride (FR) contains 80 percent of chromated zinc chloride, 10 percent of boric acid, and 10 percent of ammonium sulfate. Retentions of from 1½ to 3 p.c.f. of wood provide combined protection from fire, decay, and insect attack.

Fluor Chrome Arsenate Phenol

The composition of fluor chrome arsenate phenol (FCAP) is included in Federal Specification TT–W–535 and the AWPA Standard P5. The active ingredients of this preservative are:

	Percent
Fluoride	22
Chromium trioxide	37
Arsenic pentoxide	25
Dinitrophenol	16

To avoid objectionable staining of building materials, sodium pentachlorophenate is sometimes substituted in equal amounts for the dinitrophenol.

Sodium or potassium fluoride may be used as a source of fluoride. Sodium chromate or dichromate may be used in place of chromium trioxide. Sodium arsenate may be used in place of arsenic pentoxide.

FCAP type I (Wolman salts) and FCAP type II (Osmosalts) have performed well in above-ground wood structures and given moderate protection when used in contact with the ground.

PRESERVATIVE EFFECTIVENESS

Preservative effectiveness is influenced not only by the protective value of the preservative chemical itself, but also by the method of application and extent of penetration and retention of the preservative in the treated wood. Even with an effective preservative, good protection cannot be expected with poor penetration and substandard retentions. The species of wood, proportion of heartwood and sapwood, heartwood penetrability, and moisture content are among the important variables influencing the results of treatment. For various wood products, the preservatives and retentions listed in Federal Specification TT–W–571 are given in table 18–1.

Results of service tests on various treated products that show the effectiveness of different wood preservatives are published periodically in the proceedings of the American Wood-Preservers' Association and elsewhere. Few service tests, however, include a variety of preservatives under comparable conditions of exposure. Furthermore, service tests may not show a good comparison between different preservatives due to the difficulty in controlling the above-mentioned variables. Such comparative data under similar exposure conditions, with various preservatives and retentions, are included in Forest Products Laboratory stake tests on southern pine sapwood. A summary of these FPL results is included in table 18–2.

PENETRABILITY OF DIFFERENT SPECIES

The effectiveness of preservative treatment is influenced by the penetration and distribution of the preservative in the wood. For maximum protection it is desirable to select species for which good penetration is best assured. The heartwood is commonly difficult to treat. With round members such as poles, posts, and piling, the penetrability of the sapwood is important in achieving a protective outer zone around the heartwood.

Table 18-1.—Preservatives and minimum retentions for various wood products [1]

Product and service condition	Coal tar creosote P.c.f.	Creosote coal tar solution P.c.f.	Creosote petroleum solution P.c.f.	Pentachloro-phenol in heavy AWPA P9 solvent P.c.f.	Acid copper chromate [2] P.c.f.	Ammoni-acal copper arsenite [2] P.c.f.	Chromated copper arsenate [2] Types I, II, or III P.c.f.	AWPA Standard
A. Ties (crossties, and switch ties)	8	8	8	0.4	---	---	---	C2 & C6
B. Lumber, plywood, and structural timbers (including glued laminated)	---	---	---	---	---	---	---	C2, C9, C14, C18 & C20
(1) For use in coastal waters: [3]								
Lumber (under 5 in. thick)	22	22	---	---	---	---	---	---
Timbers (5 in. or thicker):								
Southern pine	22	22	---	---	---	---	---	---
Coast Douglas-fir and western hemlock	22	6	---	---	---	---	---	---
Plywood	25	---	---	---	---	---	---	---
(2) For use in fresh water, in contact with ground, or for important structural members not in contact with ground or water	10	10	12	.5	---	0.60	0.60	---
Glued laminated timbers or laminates	12	12	12	.6	---	.60	.60	---
(3) For other uses not in contact with ground or water: [4]	8	8	8	.4	0.25	.25	.25	---
C. Piles	---	---	---	---	---	---	---	C3, C14, & C18
(1) For use in coastal waters: [3]								
Southern pine	---	25	---	---	---	2.5	2.5	---
Coast Douglas-fir	22	---	---	---	---	2.5	---	---

Species / Use							Reference
(2) For land or fresh water use:							
Southern and other pines	12	12	12	.6	.8	.8	C3 & C14
Douglas-fir and western larch	17	17	17	.85	1.0	1.0	
Oak	6	6	6	.3	—	—	C4
D. Poles (utility)							
Southern and ponderosa pine	7.5 & 9.0	—	—	.38 & .45	—	.60	.60
Red pine	10.5 & 13.5	—	—	.53 & .68	—	.60	.60
Jack and lodgepole pine	12.0 & 16.0	—	—	0.6 & 0.8	—	.60	.60
Coast Douglas-fir	12.0 & 15.0	—	—	0.6 & 0.75	—	.60	.60
Interior Douglas-fir and western larch	15	—	—	.75	—	.60	.60
Western redcedar	16	—	—	.8	—	—	—
Western redcedar, northern white-cedar, Alaska-cedar, lodgepole pine ("thermal" or hot and cold process)	20	—	—	1.0	—	—	—	C7, C8, & C10
E. Poles (Building, round)	15	—	—	.75	—	.70	.70	C23
F. Posts (round)							
Fence	6	6	6	.3	.50	.40	.40	C5 & C16
Building	15	15	15	.75	—	.70	.70	C23

[1] Retentions for lumber, timber, plywood, piles, poles, and fence posts are determined by assay of borings of a number and location as specified in Federal Specification TT-W-571 or in the Standards of the American Wood Preservers' Association referenced in last column.

[2] All waterborne preservatives retention are specified on an oxide basis.

[3] Dual treatments are recommended when marine borer activity is known to be high (see AWPA Standards C2, C3, C14, and C18 for details.)

[4] Additional preservatives recommended for this use, and their retention levels, include pentachlorophenol in AWPA P9 light or volatile solvent, 0.4 p.c.f.; plain chromated zinc chloride, 0.46 p.c.f.; and fluorchrome arsenate phenol, 0.22 p.c.f.

NOTE: Minimum retentions are those included in Federal Specification TT-W-571 and Standards of the American Wood Preservers' Association. The current issues of these specifications should be referred to for up-to-date recommendations and other details.

Table 18–2.—*Results of Forest Products Laboratory studies on stakes pressure-treated with commonly used wood preservatives—stakes 2 by 4 by 18 inches of southern pine sapwood, installed at Harrison Experimental Forest, Miss.*

Preservative	Average retention [1]	Average life or condition at last inspection
	P.c.f.	
Untreated stakes		1.8 to 3.6 years
Acid copper chromate	0.13	11.6 years
	.14	40 percent failed after 6 years
	.25	30 percent failed after 5 years
	.26	20 percent failed after 27 years
	.29	80 percent failed after 6 years
	.37	30 percent failed after 27 years
	.50	10 percent failed after 5 years
	.76	10 percent failed after 5 years
Ammoniacal copper arsenite	.24	30 percent failed after 28 years
	.51, 0.97, and 1.25	No failures after 28 years
Chromated copper arsenate		
Type I	.15	60 percent failed after 27 years
Type II	.29 and 0.44	No failures after 27 years
	.26, 0.37, 0.52, 0.79, and 1.04	No failures after 23 years
Chromated zinc chloride	.30	14.2 years
	.47	20.2 years
	.63	20.1 years
	.62	30 percent failed after 5 years
	.92	40 percent failed after 21 years
	1.78 and 3.67	No failures after 21 years
Copper naphthenate:		
0.11 percent copper in No. 2 fuel oil	10.3 solution	15.9 years
.29 percent copper in No. 2 fuel oil	10.2 solution	21.8 years
.57 percent copper in No. 2 fuel oil	10.6 solution	80 percent failed after 31 years
.86 percent copper in No. 2 fuel oil	9.6 solution	50 percent failed after 31 years
Copper 8 quinolinolate:		
0.1 percent in Stoddard solvent	.01	5.3 years
.2 percent in Stoddard solvent	.02	4.2 years
.6 percent in Stoddard solvent	.06	5.6 years
1.2 percent in Stoddard solvent	.12	7.8 years
.15 percent in AWPA P9 heavy oil	.01	No failures after 9 years
.3 percent in AWPA P9 heavy oil	.03	No failures after 9 years
.6 percent in AWPA P9 heavy oil	.59	No failures after 9 years
1.2 percent in AWPA P9 heavy oil	.12	No failures after 9 years
Creosote, coal-tar (regular type)	4.2	17.8 years
	8.0	40 percent failed after 32½ years
	8.3	No failures after 23 years
	11.8	10 percent failed after 32½ years
	16.5	No failures after 32½ years
	4.6	21.3 years
	10.0	60 percent failed after 32 years
	14.5	No failures after 32 years
	4.1	14.2 years
Creosote, coal-tar (special types):		
Low residue, straight run	8.0	17.8 years
Medium residue, straight run	8.0	18.8 years
High residue, straight run	7.8	20.3 years
Medium residue:		
Low in tar acids	8.1	19.4 years
Low in naphthalene	8.2	21.3 years
Low in tar acids and naphthalene	8.0	18.9 years
Low residue, low in tar acids and naphthalene	8.0	19.2 years
High residue, low in tar acids and naphthalene	8.2	20.0 years
English vertical retort	8.0	18.9 years
	5.3	60 percent failed after 22 years
	10.1	20 percent failed after 22 years
	15.0	No failures after 22 years
English coke oven	7.9	13.6 years
	4.7	16.3 years
	10.1	70 percent failed after 22 years
	14.8	70 percent failed after 22 years

Table 18–2.—*Results of Forest Products Laboratory studies on stakes pressure-treated with commonly used wood preservatives—stakes 2 by 4 by 18 inches of southern pine sapwood, installed at Harrison Experimental Forest, Miss.*—continued.

Preservative	Average retention [1]	Average life or condition at last inspection
	P.c.f.	
Fluor chrome arsenate phenol type I	.16	10.2 years
	.24	18.0 years
	.49	24.1 years
	.27, 0.38, and 0.57	No failures after 11 years
Pentachlorophenol (various solvents):		
Liquefied petroleum gas	.14	10 percent failed in 11½ years
	.19	30 percent failed in 11½ years
	.34	No failures in 11½ years
	.34	10 percent failed in 9 years
	.49	No failures in 9 years
	.58	No failures in 11½ years
	.65	No failures in 9 years
AWPA P9 (heavy petroleum)	.11	No failures in 11½ years
	.19	No failures in 11½ years
	.29	No failures in 11½ years
	.53	No failures in 9 years
	.67	No failures in 11½ years
Stoddard solvent (mineral spirits)	.14	30 percent failed in 11½ years
	.18	10 percent failed in 11½ years
	.38	No failures in 11½ years
	.67	No failures in 11½ years
	.2	13.7 years
	.2	9.5 years
	.4	15.5 years
Heavy gas oil (Mid-United States)	.2	22 percent failed in 22½ years
	.4	10 percent failed in 22½ years
	.6	10 percent failed in 22½ years
No. 4 aromatic oil (West Coast)	.2	60 percent failed in 21 years
	.4	20 percent failed in 21 years

[1] All waterborne salt preservative retentions are based on oxides.

Examples of species with sapwood that is easily penetrated when it is well dried and pressure treated are the pines, coast Douglas-fir, western larch, Sitka spruce, western hemlock, western redcedar, northern white-cedar, and white fir (*A. concolor*). Examples of species with sapwood and heartwood somewhat resistant to penetration are red and white spruces and Rocky Mountain Douglas-fir. Cedar poles are commonly incised to obtain satisfactory preservative penetration.

The sapwood and heartwood of several hardwood species, such as black jack oak, some of the lowland red oaks, and aspen often present a problem in getting uniform preservative penetration.

The heartwood of most species resists penetration of preservatives although well-dried white fir, western hemlock, northern red oak, the ashes, and tupelo are examples of species with heartwood reasonably easy to penetrate. The southern pines, ponderosa pine, redwood, Sitka spruce, coast Douglas-fir, beech, maples, and birches are examples of species with heartwood moderately resistant to penetration.

PREPARING TIMBER FOR TREATMENT

For satisfactory treatment and good performance thereafter, the timber must be sound and suitably prepared. Except in specialized treating methods involving unpeeled or green material, the wood should be well peeled and either seasoned or similarly conditioned before treatment. It is also highly desirable that all machining be completed before treatment. Machining may include incising to improve the preservative penetration in woods that are resistant to treatment, as well as the operations of cutting, framing, or boring of holes.

Peeling

Peeling round or slabbed products is necessary to enable the wood to dry quickly enough to avoid decay and insect damage and to permit the preservative to penetrate satisfactorily. (Processes in which a preservative is forced or permitted to diffuse through green wood lengthwise do not require peeling of the timber.) Even strips of the thin inner bark may prevent penetration. Patches of bark left on during treatment usually fall off in time and expose untreated wood, thus permitting decay to reach the interior of the member.

Careful peeling is especially important for wood that is to be treated by a superficial

Figure 18–1.—Machine peeling of poles. Here the outer bark had been removed by hand and the inner bark is being peeled by machine. Frequently the bark is completely removed by machine.

method. In the more thorough processes some penetration may take place both lengthwise and tangentially in the wood, and consequently small strips of bark are tolerated in some specifications. Machines of various types have been developed for peeling round timbers, such as poles, piling, and posts (fig. 18–1).

Drying

For treatment with waterborne preservatives by certain diffusion methods, high moisture content may be permitted. For treatment by other methods, however, drying before treatment is essential. Drying the material permits adequate penetration and distribution of the preservative and reduces the risk of checking that would expose unpenetrated wood after treatment. Good penetration of preservative is possible with wood at a moisture content as high as 40 to 60 percent, but serious checking after treatment can result when wood at that moisture level dries.

Air drying, despite the greater time, labor, and storage space required, is a widely used method of conditioning and is generally the cheapest and most effective, even for pressure treatment. Under wet, warm climatic conditions it is difficult to adequately air-dry wood without objectionable infection by stain, mold, and decay fungi. Such infected wood is often highly permeable; in rainy weather it can absorb a large quantity of water, which in turn prevents satisfactory treatment.

How long the lumber must be air dried before treatment depends on the climate, location, and condition of the seasoning yard, methods of piling, season of the year, size, and species of the timbers. The most satisfactory seasoning practice for any specific case will depend on the individual drying conditions and the preservative treatment to be used. Treating specifications therefore are not always specific as to moisture content requirements.

To prevent decay and other forms of fungus infection during air drying, the wood should be cut and dried when conditions are less favorable for fungus development (see ch. 17). If this is impossible, chances for infection can be minimized by prompt conditioning of the green material, careful piling and roofing during air drying, and pretreating the green wood with preservatives to protect during air drying.

Lumber, as well as southern pine poles, are often kiln dried before treatment, particularly in the southern United States where proper air seasoning is difficult. Kiln drying has the important added advantage of quickly reducing moisture content and thereby reducing transportation charges on poles.

Plants that treat wood by pressure processes can condition green material by means other than air drying and kiln drying. Thus they avoid a long delay and possible deterioration of the timber before treatment.

Conditioning Green Products for Pressure Treatment

When green wood is to be treated under pressure, one of several methods for conditioning may be selected. The steaming-and-vacuum process is employed mainly for southern pine, while the Boulton or boiling-under-vacuum process is used for Douglas-fir and sometimes for hardwoods.

In the steaming process the green wood is steamed in the treating cylinder for several hours, usually at a maximum temperature of 245° F. When the steaming is completed, a vacuum is immediately applied. During the steaming period the outer part of the wood is heated to a temperature approaching that of the steam; the subsequent vacuum lowers the boiling point so part of the water is evaporated or is forced out of the wood by the steam produced when the vacuum is applied. The steaming and vacuum periods employed depend upon the size, species, and moisture content of the wood. The steaming and vacuum usually reduce the moisture content of green wood slightly, and the heating assists greatly in getting the preservative to penetrate. A sufficient steaming period will also sterilize the wood. In the Boulton or boiling-under-vacuum method of partial seasoning, the wood is heated in the oil preservative under vacuum, usually at temperatures of about 180° to 220° F. This temperature range, lower than that of the steaming process, is a considerable advantage in treating woods that are especially susceptible to injury from high temperatures. The Boulton method removes much less moisture from heartwood than from the sapwood.

A third method of conditioning known as "vapor drying" has been patented and is used for seasoning railroad ties and other products. In the treating cylinder, the green wood is subjected to the vapors produced by boiling an organic chemical, such as xylene. The resulting mixed vapors of water and the chemical are then removed from the drying chamber. A small quantity of chemical remains in the wood, but the balance is recovered and reused. The wood is treated by standard pressure methods after the conditioning is completed.

Incising

Wood that is resistant to penetration by preservatives is often incised before treatment to permit deeper and more uniform penetration. To accomplish this, sawed or hewed timbers are passed through rollers equipped with teeth that sink into the wood to a predetermined depth, usually ½ to ¾ inch. The teeth are spaced to give the desired distribution of preservative with the minimum number of incisions. A machine of different design is required for deep incising the butts of poles (fig. 18–2).

The effectiveness of incising depends on the fact that preservatives usually penetrate into wood much farther in a longitudinal direction than in a direction perpendicular to the faces of the timber. The incisions expose end-grain surfaces and thus permit longitudinal penetration. It is especially effective in improving penetration in the heartwood areas of sawed or hewed surfaces.

Incising is practiced chiefly on Douglas-fir, western hemlock, and western larch ties and timbers for pressure treatment and on poles of cedar and Douglas-fir.

Cutting and Framing

All cutting, framing, and boring of holes should be done before treatment. Cutting into the wood in any way after treatment will fre-

Figure 18–2.—Deep incising permits better penetration.

quently expose the untreated interior of the timber and permit ready access to decay fungi or insects.

It is much more practical than is commonly supposed to design wood structures so all cutting and framing may be done before treatment. Railroads have followed the practice extensively and find it not only practical but economical. Many wood-preserving plants are equipped to carry on such operations as the adzing and boring of crossties; gaining, roofing, and boring of poles; and the framing of material for bridges and for specialized structures such as water tanks and barges.

Treatment of the wood with preservative oils involves little or no dimensional change. In the case of treatment with water-borne preservatives, however, some change in the size and shape may occur even though wood is redried to the moisture content it had before treatment. If precision fitting is necessary, the wood is cut and framed before treatment to its approximate final dimensions to allow for slight surfacing, trimming, and reaming of bolt holes. Grooves and bolt holes for timber connectors are cut before treatment and can be reamed out if necessary after treatment.

315

APPLYING PRESERVATIVES

Wood-preserving methods are of two general types: (1) Pressure processes, in which the wood is impregnated in closed vessels under pressures considerably above atmospheric, and (2) nonpressure processes, which vary widely as to procedures and equipment used. Pressure processes generally provide a closer control over preservative retentions and penetrations, and usually provide greater protection than nonpressure processes. Some nonpressure methods, however, are better than others and are occasionally as effective as pressure processes in providing good preservative retentions and penetrations.

Pressure Processes

In commercial practice, wood is most often treated by immersing it in preservative in high-pressure apparatus and applying pressure to drive the preservative into the wood. Pressure processes differ in details, but the general principle is the same. The wood, on cars, is run into a long steel cylinder (fig. 18–3), which is then closed and filled with preservative. Pressure forces preservative into the wood until the desired amount has been absorbed. Considerable preservative is absorbed, with relatively deep penetration. Two processes, the full-cell and empty-cell, are in common use.

Full-Cell

The full-cell (Bethel) process is used when the retention of a maximum quantity of preservative is desired. It is a standard procedure for timbers to be treated full-cell with creosote when protection against marine borers is required. Waterborne preservatives are generally applied by the full-cell process, and control over preservative retention is obtained by regulating the concentration of the treating solution.

Steps in the full-cell process are essentially:

1. The charge of wood is sealed in the treating cylinder, and a preliminary vacuum is applied for ½ hour or more to remove the air from the cylinder and as much as possible from the wood.

2. The preservative, previously heated to somewhat above the desired treating temperature, is admitted to the cylinder without admission of air.

3. After the cylinder is filled, pressure is applied until the required retention of preservative is obtained.

4. When the pressure period is completed, the preservative is withdrawn from the cylinder.

5. A short final vacuum may be applied to free the charge from dripping preservative.

When the wood is steamed before treatment, the preservative is admitted at the end of the vacuum period that follows steaming. When the timber has received preliminary conditioning by the Boulton or boiling-under-vacuum process, the cylinder can be filled and the pressure applied as soon as the conditioning period is completed.

A pressure treatment referred to commercially as the "Cellon" process usually employs the full-cell process. It uses a preservative such as pentachlorophenol in highly volatile liquefied petroleum gas, such as butane or propane, which are gases at atmospheric pressure and ordinary temperatures. A cosolvent is employed to obtain the required concentration of preservative in the treating liquid.

For closer control over preservative retention during the Cellon process, the empty-cell process may be used. If so, a noncombustible gas, such as nitrogen, is substituted for air during the initial air pressure in the conventional Rueping process.

Empty-Cell

The objective of empty-cell treatment is to obtain deep penetration with a relatively low net retention of preservative. For treatment with oil preservatives, the empty-cell process should always be used if it will provide the desired retention. Two empty-cell processes, the Rueping and the Lowry, are commonly employed; both use the expansive force of compressed air to drive out part of the preservative absorbed during the pressure period.

The Rueping empty-cell process has been widely used for many years, in both Europe and the United States. The following general procedure is employed:

1. Air under pressure is forced into the treating cylinder, which contains the charge of wood. The air penetrates some species easily, requiring but a few minutes' application of pressure. In the treatment of the more resistant species, common practice is to maintain air pressure from ½ to 1 hour before admitting the preservative, but the necessity for

long air-pressure periods does not seem fully established. The air pressures employed generally range between 25 and 100 p.s.i., depending on the net retention of preservative desired and the resistance of the wood.

2. After the period of preliminary air pressure, preservative is forced into the cylinder. As the preservative is pumped in, the air escapes from the treating cylinder into an equalizing or Rueping tank, at a rate that keeps the pressure constant within the cylinder. When the treating cylinder is filled with preservative, the treating pressure is raised above that of the initial air and is maintained until the wood will take no more preservative, or until enough has been absorbed to leave the required retention of preservative in the wood after the treatment.

Figure 18–3.—Interior view of treating cylinder at wood-preserving plant, with a load about to come in.

3. At the end of the pressure period the preservative is drained from the cylinder, and surplus preservative removed from the wood with a final vacuum. The amount recovered may be from 20 to 60 percent of the gross amount injected.

The Lowry is often called the empty-cell process without initial air pressure. Preservative is admitted to the cylinder without either an initial air pressure or a vacuum, and the air originally in the wood at atmospheric pressure is imprisoned during the filling period. After the cylinder is filled with the preservative, pressure is applied, and the remainder of the treatment is the same as described for the Rueping treatment.

The Lowry process has the advantage that equipment for the full-cell process can be used without other accessories; the Rueping process usually requires additional equipment, such as an air compressor and an extra cylinder or Rueping tank for the preservative, or a suitable pump to force the preservative into the cylinder against the air pressure. Both processes, however, have advantages, and both are widely and successfully used.

With poles and other products where bleeding of preservative oil is objectionable, the empty-cell process is followed by either heating in the preservative (expansion bath) at a maximum temperature of 220° F. or a final steaming, for a specified time limit at a maximum temperature of 240° F., prior to the final vacuum.

Treating Pressures and Preservative Temperatures

The pressures used in treatments vary from about 50 to 250 p.s.i., depending on the species and the ease with which the wood takes the treatment. Most commonly they are about 125 to 175 p.s.i. Many woods are sensitive to high treating pressures, especially when hot. AWPA standards, for example, permit a maximum pressure of 150 p.s.i. in the treatment of Douglas-fir, 125 p.s.i. for redwood, and 100 p.s.i. for western redcedar poles. In commercial practice even lower pressures are frequently used on such woods.

AWPA specifications commonly require that the temperature of creosote and creosote solutions during the pressure period shall not be more than 210° F. Pentachlorophenol solutions may be applied at somewhat lower temperatures. Since high temperatures are much more effective than low temperatures for treating resistant wood, it is common practice to use average temperatures between 190° and 200° F. with creosote and creosote solutions. With a number of waterborne preservatives, however, especially those containing chromium salts, maximum temperatures are limited to 120° to 150° F. to avoid premature precipitation of the preservative.

Preservative Penetration and Retention

Penetration and retention requirements are equally important in determining the quality of preservative treatment.

Penetrations vary widely, even in pressure-treated material. In most species, heartwood is more difficult to penetrate than sapwood. In addition, species differ greatly in the degree to which their heartwood may be penetrated. Incising tends to improve penetration of preservative in many refractory species, but those highly resistant to penetration will not have deep or uniform penetration even when incised. Penetrations in unincised heart faces of these species may occasionally be as deep as ¼ inch, but often are not more than ¹⁄₁₆ inch.

Long experience has shown that even slight penetrations have some value, although deeper penetrations are highly desirable to avoid exposing untreated wood when checks occur, particularly for important members of high replacement cost. The heartwood of coast-type Douglas-fir, southern pine, and various hardwoods, while resistant, will frequently show transverse penetrations of ¼ to ½ inch and sometimes considerably more.

Complete penetration of the sapwood should be the ideal in all pressure treatments. It can often be accomplished in small-size timbers of various commercial woods, and with skillful treatment it may often be obtained in piles, ties, and structural timbers. Practically, however, the operator cannot always insure complete penetration of sapwood in every piece when treating large pieces of round material with thick sapwood, for example poles and piles. Specifications therefore permit some tolerance; for instance, AWPA Standard C4 on southern pine poles requires that 2.5 inches, or 85 percent of the sapwood thickness, be penetrated in not less than 18 out of 20 poles sampled in a charge. This applies only to the smaller class of poles. The requirements vary somewhat depending on the species size class and specified retentions.

Preservative retentions, until recently, have been generally specified in terms of the weight of preservative per cubic foot of wood treated, based on total weight of preservative retained and the total volume of wood treated in a charge. Federal specifications for most products, however, stipulate a minimum retention of preservative as determined from chemical analysis of borings from specified zones of the treated wood.

The preservatives and minimum retentions listed in Federal Specification TT–W–571 are shown in table 18–1. Because the figures given in this table are minimums, it may often be desirable to use higher retentions. Higher preservative retentions are justified in products to be installed under severe climatic or exposure conditions. Heavy-duty transmission poles and items such as structural timbers and house foundations, with a high replacement cost, are required to be treated to higher retentions. Correspondingly deeper penetration is also necessary for the same reasons.

It may be necessary to increase retentions to assure satisfactory penetration, particularly when the sapwood is either unusually thick or is somewhat resistant to treatment. To reduce bleeding of the preservative, however, it may be desirable to use preservative-oil retentions lower than the stipulated minimum. Treatment to refusal is usually specified for woods that are resistant to treatment and will not absorb sufficient preservative to meet the minimum retention requirements. However, such a requirement does not assure adequate penetration of preservative and cannot be considered as a substitute for more thorough treatment.

Nonpressure Processes

The numerous nonpressure processes differ widely in the penetrations and retentions of preservative attained and consequently in the degree of protection they provide to the treated wood. When similar retentions and penetrations are achieved, wood treated by a nonpressure method should have a service life comparable to that of wood treated by pressure. Nevertheless, results of nonpressure treatments, particularly those involving superficial applications, are not generally as satisfactory as pressure treatment. The superficial processes do serve a useful purpose when more thorough treatments are either impractical or exposure conditions are such that little preservative protection is required.

Nonpressure methods, in general, consist of: (1) Superficial applications of preservative oils by spraying, brushing, or brief dipping; (2) soaking in preservative oils or steeping in solutions of waterborne preservatives; (3) diffusion processes with waterborne preservatives; (4) various adaptations of the thermal or hot-and-cold bath process; (5) vacuum treatment; and (6) a variety of miscellaneous processes.

Superficial Applications

The simplest treatment is to apply the preservative—creosote or other oils—to the wood with a brush or a spray nozzle. Oils that are throughly liquid when cold should be selected, unless it is possible to heat the preservative. The oil should be flooded over the wood, rather than merely painted upon it. Every check and depression in the wood should be thoroughly filled with the preservative, because any untreated wood left exposed provides ready access for fungi. Rough lumber may require as much as 10 gallons of oil per 1,000 square feet of surface, but surfaced lumber requires considerably less. The transverse penetrations obtained will usually be less than $\frac{1}{10}$ inch although, in easily penetrated species, end grain (longitudinal) penetration is considerably greater.

Brush and spray treatments should be used only when more effective treatments cannot be employed. The additional life obtained by such treatments over that of untreated wood will be affected greatly by the conditions of service; for wood in contact with the ground, it may be from 1 to 5 years.

Dipping for a few seconds to several minutes in a preservative oil gives greater assurance (than brushing or spraying) that all surfaces and checks are thoroughly coated with the oil; usually it results in slightly greater penetrations. It is a common practice to treat window sash, frames, and other millwork, either before or after assembly, by dipping for approximately 3 minutes in a water-repellent preservative. Such treatment is covered by Commercial Standard CS–262, which also provides for equivalent treatment by the vacuum process. The amount of preservative used may vary from about 6 to 17 gallons per thousand board feet (0.5 to 1.5 p.c.f.) of millwork treated.

The penetration of preservative into end surfaces of ponderosa pine sapwood is, in some cases, as much as 1 to 3 inches. End penetration in such woods as southern pine and Douglas-fir, however, is much less, particularly in the heartwood. Transverse penetration of the preservative applied by brief dipping is very shallow, usually only a few hundredths of an inch. Since the exposed end surfaces at joints are the most vulnerable to decay in millwork products, good end penetration is especially advantageous. Dip applications provide very limited protection to wood used in contact with the ground or under very moist conditions, and they provide very limited protection against attack by termites. They do have value, however, for exterior woodwork and millwork that is painted, that is not in contact with the ground, and that is exposed to moisture only for brief periods at a time.

Cold Soaking and Steeping

Cold soaking well-seasoned wood for several hours or days in low-viscosity preservative oils or steeping green or seasoned wood for several days in waterborne preservatives have provided varying success on fenceposts, lumber, and timbers.

Pine posts treated by cold soaking for 24 to 48 hours or longer, in a solution containing 5 percent of pentachlorophenol in No. 2 fuel oil, have shown an average life of 16 to 20 years or longer. The sapwood in these posts was well penetrated and preservative solution retentions ranged from 2 to 6 p.c.f. Most species do not treat as satisfactorily as the pines by cold soaking, and test posts of such woods as birch, aspen, and sweetgum treated by this method have failed in much shorter times.

Preservative penetrations and retentions obtained by cold soaking lumber for several hours are considerably better than those obtained by brief dipping of similar species. Preservative retentions, however, seldom equal those obtained in pressure treatment except in cases such as sapwood of pines that has become highly absorptive through mold and stain infection.

Steeping with waterborne preservatives has very limited use in the United States but has been employed for many years in Europe. In treating seasoned wood both the water and the preservative salt in the solution soak into the wood. With green wood, the preservative enters the water-saturated wood by diffusion. Preservative retentions and penetrations vary over a wide range, and the process is not generally recommended when more reliable treatments are practical.

Diffusion Processes

In addition to the steeping process, diffusion processes are used with green or wet wood. These processes employ waterborne preservatives that will diffuse out of the water of the treating solution or paste into the water of the wood.

The double-diffusion process developed by the Forest Products Laboratory has shown very good results in post tests, particularly on full-length immersion treatments. It consists of steeping green or partially seasoned wood first in one chemical and then in another. The two chemicals diffuse into the wood and then react to precipitate an effective preservative with high resistance to leaching. The process has had commercial application in cooling towers where preservative protection is needed to avoid early replacement.

Other diffusion processes involve applying preservatives to the butts or around the groundline of posts or poles. In standing-pole treatments the preservative may be injected into the pole at groundline with a special tool, applied on the pole surface as a paste or bandage, poured into holes bored in the pole at the groundline, or poured on the surface of the pole and into an excavation several inches deep around the groundline of the pole. These treatments have recognized value for application to untreated standing poles and to treated poles where preservative retentions are determined to be inadequate.

Adaptations of Thermal Process

The hot-and cold bath, referred to commercially as thermal treatment, with coal-tar creosote or pentachlorophenol in heavy petroleum oil is also an effective nonpressure process; the thoroughness of treatment obtainable in some cases approaches that of the pressure processes. The wood is heated in the preservative in an open tank for several hours, then quickly submerged in cold preservative and allowed to remain for several hours.

During the hot bath, the air in the wood expands and some is forced out. Heating the wood also improves the penetration of the preservative. In the cooling bath, the air in the wood contracts and a partial vacuum is created, so liquid is forced into the wood by atmospheric pressure. Some preservative is absorbed by the wood during the hot bath, but more is taken up during the cooling bath.

The chief use of the hot-and-cold process is for treating poles of some thin sapwood species, such as incised western redcedar and lodgepole pine, for utility poles (fig. 18–4). The process is also useful for fenceposts and for lumber or timbers for other purposes when circumstances do not permit the more effective pressure treatments. Coal-tar creosote and pentachlorophenol solutions are the preservatives ordinarily chosen for posts and poles. For the preservatives that cannot safely be heated, the process must be modified.

With coal-tar creosote, hot-bath temperatures up to 235° F. may be employed, but usually a temperature of 210° to 220° F. is sufficient. In the commercial treatment of cedar poles, temperatures of from 190° to 235° F., for not less than 6 hours, are specified with creosote and pentachlorophenol solutions. In the cold bath or cooling bath the specified temperature is not less than 90° F. nor more than 150° F. for not less than 2 hours.

The immersion time in both baths must be governed by the ease with which the timber takes treatment. With well-seasoned timber that is moderately easy to treat, a hot bath of 2 or 3 hours and a cold bath of like duration is probably sufficient. Much longer periods are required with resistant woods. With preservative oils, the objective is to obtain as deep penetration as possible, but with a minimum amount of oil.

Preservative retentions are often very high in the hot-and-cold bath treatments of posts of woods such as southern yellow pine, particularly if those posts contain molds, blue stain, and incipient decay. One method of reducing preservative retentions is to employ a final heating or "expansion" bath with the creosote at 200° to 220° F. for an hour or two, and to remove the wood while the oil is hot. This second heating expands the oil and air in the wood, and some of the oil is thus recovered. The expansion bath also leaves the wood cleaner than when it is removed directly from cold oil.

Vacuum Process

The vacuum process has been used to treat millwork with water-repellent preservatives and construction lumber with waterborne and water-repellent preservatives.

In treating millwork, the objective is to use a limited quantity of water-repellent preserva-

Figure 18–4.—A commercial plant for the hot-and-cold-bath (thermal) treatment of utility and building poles.

tive and obtain retentions and penetrations similar to those obtained by dipping for 3 minutes. The treatment is included in Commercial Standard CS–262 for "Water-Repellent Preservative Nonpressure Treatment of Millwork." Here a quick, low initial vacuum is followed by brief immersion in the preservative, and then a high final or recovery vacuum. The treatment is advantageous over the 3-minute-dip treatment because the surface of wood is quickly dried—thus expediting the glazing, priming, and painting operations. The vacuum treatment is also reported to be less likely than dip treatment to leave objectionably high retentions in bacteria-infected wood referred to as "sinker stock."

For buildings, lumber has been treated by the vacuum process, either with a waterborne preservative or a water-repellent pentachlorophenol solution, with preservative retentions usually lower than those required for pressure treatment. The process differs from that used in treating millwork in employing a higher initial vacuum and a long immersion or soaking period.

A study of the process by the Forest Products Laboratory employed an initial vacuum of 27.5 inches for 30 minutes, a soaking period of 8 hours, and a final or recovery vacuum of 27.5 inches for 2 hours. The study showed good penetration of preservative in the sapwood of dry lumber of easily penetrated species such as the pines; however, in heartwood and unseasoned sapwood of pine and heartwood of seasoned and unseasoned coast Douglas-fir, penetration was much less than that obtained in pressure treatment. Preservative retention was less controllable in vacuum than in empty-cell pressure treatment. Good control over retentions is possible, in vacuum treatment with a waterborne preservative, by adjusting concentration of the treating solution.

Miscellaneous Nonpressure Processes

A number of other nonpressure methods of various types have been used to a limited extent. Several of these involve the application of waterborne preservatives to living trees. The Boucherie process for the treatment of green, unpeeled poles has been used for many years in Europe. The process involves attaching liquid-tight caps to the butt ends of the poles. Then, through a pipeline or hose leading to the cap, a waterborne preservative is forced into the pole under hydrostatic pressure.

A tire-tube process is a simple adaptation of the Boucherie process used for treating green, unpeeled fenceposts. In this treatment a section of used inner tube is fastened tightly around the butt end of the post to make a bag that holds a solution of waterborne preservative.

Effect of Treatment on Strength

Coal-tar creosote, creosote-coal-tar mixtures, creosote-petroleum oil mixtures, and pentachlorophenol dissolved in petroleum oils are practically inert to wood and have no chemical influence that would affect its strength. Likewise, solutions containing standard waterborne preservatives, in the concentrations commonly used in preservative treatment, have limited or no important effect on the strength of wood.

Although wood preservatives are not harmful in themselves, injecting them into the wood may result in considerable loss in wood strength if the treatment is unusually severe or not properly carried out. Factors that influence the effect of the treating process on strength include (1) species of wood, (2) size and moisture content of the timbers treated, (3) heating medium used and its temperature, (4) length of the heating period in conditioning the wood for treatment and time the wood is in the hot preservative, and (5) amount of pressure used. Most important of these factors are the severity and duration of the heating conditions used. The effect of temperature on the strength of wood is covered in chapter 4.

HANDLING AND SEASONING TIMBER AFTER TREATMENT

Treated timber should be handled with sufficient care to avoid breaking through the treated areas. The use of pikes, cant hooks, picks, tongs, or other pointed tools that dig deeply into the wood should be prohibited. Handling heavy loads of lumber or sawed timber in rope or cable slings may crush the corners or edges of the outside pieces. Breakage or deep abrasions may also result from throwing the lumber or dropping it. If damage results, the exposed places should be retreated as thoroughly as conditions permit. Long storage of treated wood before installation should be avoided because such storage encourages deep and detrimental checking and may also result in significant loss of some preservatives. Treated wood that must be stored before use should be covered for protection from the sun and weather.

Although cutting wood after treatment is highly undesirable, it cannot always be avoided. When cutting is necessary, the damage may be partly overcome in timber for land or fresh-water use by a thorough application of a grease containing 10 percent pentachlorophenol. This provides a protective reservoir of preservative on the surface, some of which may slowly migrate into the end grain of the wood. Thoroughly brushing the cut surfaces with two coats of hot creosote is also helpful, although brush coating cut surfaces gives little protection against marine borers. A special device is available for pressure treating bolt holes bored after treatment. For wood treated with waterborne preservatives, where the use of creosote or pentachlorophenol solution on the cut surfaces is not practicable, a 5 percent solution of the waterborne preservative in use should be substituted.

For treating the end surfaces of piles where they are cut off after driving, at least two generous coats of creosote should be applied. A coat of asphalt or similar material may well applied over the creosote, followed by some protective sheet material, such as metal, roofing felt, or saturated fabric, fitted over the pile head and brought down the sides far enough to protect against damage to the top treatment and against the entrance of storm water. AWPA standard M4 contains instructions for the care of pressure-treated wood after treatment.

Wood treated with preservative oils should generally be installed as soon as practicable after treatment but some times cleanliness of the surface can be improved by exposure to the weather for a limited time before use. Waterborne preservatives or pentachlorophenol in a volatile solvent, however, are best suited to uses where cleanliness or paintability are of great importance.

With waterborne preservatives, seasoning after treatment is important for wood to be used in buildings or other places where shrinkage after placement in the structure would be undesirable. Injecting waterborne preservatives puts large amounts of water into the wood, and considerable shrinkage is to be expected as subsequent seasoning takes place. For best results, the wood should be dried to approximately the moisture content it will ultimately reach in service. During drying, the wood should be carefully piled, and whenever possible, restrained by sufficient weight on the top of the pile to avoid warping.

With some waterborne preservatives, seasoning after treatment is recommended for all treated wood. During this seasoning period, volatile chemicals escape and the chemical reactions are completed within the wood; thus, the resistance of the preservative to leaching by water is increased.

QUALITY ASSURANCE FOR TREATED WOOD

Treating Conditions and Specifications

Specifications on the treatment of various wood products by pressure processes and on the hot-and-cold bath (thermal) treatment of cedar poles have been developed by AWPA. These specifications limit pressures, temperatures, and time during conditioning and treatment to avoid conditions that will cause serious injury to the wood. They also contain minimum requirements as to preservative penetrations and retentions and recommendations for handling wood after treatment, to provide a quality product.

The specifications are rather broad in some respects, allowing the purchaser some latitude in specifying the details of his individual requirements. The purchaser should exercise great care, however, not to limit the operator of the treating plant so he cannot do a good treating job, and not to require treating conditions so severe that they will damage the wood. Federal Specification TT–W–571 lists treatment practices for use on U.S. Govern-ment orders for pressure-treated wood products; other purchasers have specifications similar to those of AWPA.

Inspection

Inspection of timber for quality and grade before treatment is desirable. Grademarked lumber, plywood, and timber graded at the producing mill can be obtained in many instances. When inspection prior to treatment is impractical, the purchaser can usually inspect for quality and grade after treatment; if this is to be done, however, it should be made clear in the purchase order.

Currently, the inspection of treatment of complete charges is generally specified at the time of treatment at the treating plant; however, the option is generally available whereby the purchaser could determine the quality of treatment from selected samples of the treated product at destination or within a specified time after treatment. The purchaser should recognize, however, that a sample selected from a charge at the treating plant is likely to be different than a sample taken at destination from a few items from a much larger charge. Furthermore, the nature and quantity of the preservative in the wood change as the period of service increases, so samples of treated wood taken at treatment may not be the same as those taken later. Destination inspection requires consideration of these questions and the details have not yet been worked out for all treated products.

The treating industry, with the assistance of the Federal Housing Administration and the Forest Products Laboratory, has developed a quality-control and grademarking program for treated products, such as lumber, timbers, plywood, and marine piling. This quality control program, administered through the American Wood Preservers' Bureau, promises to assist the user in securing well-treated material; otherwise the purchaser must either accept the statements or certificate of the treating-plant operator or have an inspector at the treating plant to inspect the treated products and insure compliance with the specifications. Railroad companies and other corporations that purchase large quantities of treated timber usually maintain their own inspection services. Commercial inspection and consulting service is available for purchasers willing to pay an inspection fee but not using enough treated timber to justify employing inspectors of their own. Experienced, competent, and reliable inspectors can assure compliance with material and treating standards and thus reduce risk of premature failure of the material.

Penetration measurements should be made at the treating plant if inspection service is provided, but can be made by the purchaser at any time after the timber has been treated. They give about the best single measure of the thoroughness of the treatment.

The depth of penetration of creosote and other dark-colored preservatives can be determined directly by observing a core removed by an increment borer. The core should usually be taken at about midlength of the piece, or at least several feet from the end of the piece, to avoid the unrepresentative end portion that is sometimes completely treated by end penetration. Since preservative oils tend to creep over cut surfaces, the observation should be made promptly after the borer core is taken. Holes made for penetration measurements should be tightly filled with thoroughly treated wood plugs.

The penetration of preservatives that are practically colorless must be determined by chemical dips or sprays that show the penetration by color reactions.

How to Purchase Treated Wood

To receive optimum service from wood when it is exposed to biological deterioration, it should be treated with an effective preservative. The use of treated wood will reduce the maintenance and replacement costs of wood components in structures.

For the purchaser to obtain a treated wood product of high quality, he must avail himself of the appropriate specifications. Specifications and standards of importance here are: Federal Specification TT–W–571, "Wood Preservation—Treating Practices;" F.S. TT–W–572, "Wood Preservation—Water Repellent;" Official Quality Control Standards of the American Wood Preservers' Bureau; and the American Wood-Preservers' Association Book of Standards. The inspection of material for conformity to the minimum requirements listed in the above specifications should be in accordance with the American Wood Preservers' Standard M2, "Standard for Inspection of Treated Timber Products."

BIBLIOGRAPHY

American Wood-Preservers' Association
Annual proceedings. (Reports of Preservations and Treatment Committees contain information on new wood preservatives considered in the development of standards.) (See current issue.)

——— Book of Standards. (Includes standards on preservatives, treatments, methods of analysis, and inspection.) (American Wood Preservers' Bureau Official quality control standards.) (See current issue.)

Baechler, R. H., Gjovik, L. R., and Roth, H. G.
1969. Assay zones for specifying preservative-treated Douglas-fir and southern pine timbers. Proc. AWPA: 114–123.

———, Gjovik, L. R., and Roth, H. G.
1970. Marine tests on combination-treated round and sawed specimens. Proc. AWPA: 249–257.

———, and Roth, H. G.
1964. The double-diffusion method of treating wood: A review of studies. Forest Prod. J. 14(4): 171–178.

Best, C. W., and Martin, G. E.
1969. Deep treatment in Douglas-fir poles. Proc. AWPA: 223–228.

Blew, J. O.
1956. Study on the preservative treatment of lumber. Proc. AWPA 52: 78–117.

———, and Davidson, H. L.
1971. Preservative retentions and penetration in the treatment of white fir. Proc. AWPA: 204–221.

———, and Kulp, J. W.
1964. Service records on treated and untreated posts. USDA Forest Serv. Res. Note FPL–068. Forest Prod. Lab., Madison, Wis.

———, and Panek, E.
1964. Problems in the production of clean treated wood. Proc. AWPA 60: 89–97.

———, Panek, E., and Roth, H. G.
1970. Vacuum treatment of lumber. Forest Prod. J. 20(2): 40–47.

Fahlstrom, G. B., Gunning, P. E., and Carlson, J. A.
1967. A study of the influence of composition on leachability. Forest Prod. J. 17(7): 17–22.

Gjovik, L. R., and Baechler, R. H.
1970. Treated wood foundations for buildings. Forest Prod. J. 20(5): 45–48.

———, and Davidson, H. L.
1973. Comparison of wood preservatives in Mississippi post study. USDA Forest Serv. Res. Note FPL–01. Forest Prod. Lab., Madison, Wis.

———
1973. Comparison of wood preservatives in stake tests. USDA Forest Serv. Res. Note FPL–02. Forest Prod. Lab., Madison, Wis.

———, Roth, H. G., and Davidson, H. L.
1972. Treatment of Alaskan species by double-diffusion and modified double-diffusion methods. USDA Forest Serv. Res. Pap. FPL 182. Forest Prod. Lab., Madison, Wis.

Henry, W. T., and Jeroski, E. B.
1967. Relationship of arsenic concentration to the leachability of chromated copper arsenate formulations. Proc. AWPA: 187–196.

Hochman, Harry
1967. Creosoted wood in the marine environment—a summary report. Proc. AWPA: 138–150.

Hunt, George M., and Garratt, George A.
1967. Wood preservation. 3d ed. McGraw Hill, New York.

MacLean, J. D.
1960. Preservative treatment of wood by pressure methods. USDA Agr. Handb. No. 40.

National Forest Products Association
The all-weather wood foundation. NFPA Tech. Rep. No. 7. (See current issue).

Panek, E.
1968. Study of paintability and cleanliness of wood pressure treated with water-repellent preservative. Proc. AWPA: 178–188.

Verrall, A. F.
1965. Preserving wood by brush, dip, and short-soak methods. USDA Tech. Bull. No. 1334.

Weir, T. P.
1958. Lumber treatment by the vacuum process. Forest Prod. J. 8(3): 91–95.

U.S. Department of Commerce
Water-repellent preservative nonpressure treatment for millwork. Commercial Standard CS–262 63. (See current revision.)

U.S. Federal Supply Service
Wood preservation treating practices. Fed. Spec. TT–W–571. (See current revision.)

———
Wood preservatives: Water-repellent. Fed. Spec. TT–W–572. (See current revision).

Chapter 19
POLES, PILES, AND TIES

Industrial wood products such as railroad ties, poles for transmission and distribution lines and buildings, and piles for bridge, wharf, and building construction continue to be important in the United States. Prime factors in the selection of particular wood species for this type of product are availability in quantity, strength and weight, the natural shape of the tree, and the ability of the wood to receive and retain commercial preservative treatments.

POLES

Principal Species Used for Poles

The principal species used for poles are southern pines, Douglas-fir, western redcedar, and lodgepole pine. Miscellaneous species used for poles are ponderosa pine, red pine, jack pine, northern white-cedar, other cedars, and western larch. Most poles are pressure treated with preservatives although cedar poles are treated principally by the hot-and-cold (thermal) process or are used occasionally without treatment.

Hardwood species can be used for poles when the trees are of suitable size and form; their use is limited, however, by their weight, excessive checking, and by lack of experience in preservative treatment of hardwoods.

Southern Pines

The southern pines (principally loblolly, longleaf, shortleaf, and slash) account for the highest percentage of poles treated in the United States. The thick and easily treated sapwood of these species, their favorable strength properties and form, and their availability in popular pole sizes over a wide area account for their extensive use. In longer lengths, southern pine poles are in limited supply so Douglas-fir, and to some extent western redcedar, are used to meet requirements for 50-foot and longer transmission poles.

Southern pine poles are pressure treated full length generally with preservatives recommended in Federal Specification TT–W–571. Well-treated southern pine poles can be expected to have an average service life of 35 years or longer.

Douglas-Fir

Douglas-fir constitutes approximately 6 percent of total poles treated, for the most part by pressure, with preservatives listed in Federal Specification TT–W–571. This species is widely used in the United States for transmission poles and specifically on the Pacific Coast for distribution and building poles. The sapwood of this species averages about 1.3 inches in thickness in the Interior North Region and from 1.6 to 2.0 inches along the Pacific Coast. Since the heartwood has limited decay and termite resistance, it is important that the sapwood be well treated and poles adequately seasoned or conditioned before treatment to minimize checking after treatment. With these precautions the poles should compare favorably with treated southern pine poles in serviceability.

Western Redcedar

About 3 percent of the poles treated in the United States are of western redcedar, produced mostly in British Columbia. A small number of poles of this species are used without treatment. The poles have comparatively thin sapwood and the heartwood is naturally decay resistant, although without treatment an average pole life somewhat less than 20 years can be expected.

Except when used in dry areas not conductive to "shellrot" in the tops, western redcedar poles are treated full length. Treatment is generally with the oil type preservatives listed in the Federal Specification TT–W–571 by the thermal (hot-and-cold) process, although some poles are pressure treated. The poles are mostly for utility line use, although well-treated western redcedar poles could be used effectively in pole-type buildings. In the northern and western United States, where they are used most, western redcedar poles when well treated full length compare favorably in service life with poles of other species.

Lodgepole Pine

Approximately 2 percent of poles treated are of lodgepole pine. The majority are full length pressure treated or full length thermal (hot-and-cold) treated with approved preservatives. The poles are used both for utility lines and for pole-type buildings. Good service can be expected from well-treated lodgepole pine poles. Special attention is necessary, however, to obtain poles with sufficient sapwood thickness to insure adequate penetration of preservative, because the heartwood is not usually pentrated and is not decay resistant. The poles must also be well seasoned before treatment to avoid checking and exposure of the unpenetrated heartwood to attack by decay fungi.

Other Species

Western larch, ponderosa pine, Atlantic white-cedar, northern white-cedar, jack pine, red pine, eastern redcedar, redwood, spruce, and hemlock are occasionally used for poles.

Western larch poles produced in Montana and Idaho came into use following World War II because of their favorable size, shape, and strength properties. Western larch requires preservative treatment full length for use in most areas, must be selected for adequate sapwood thickness, and must be well seasoned prior to treatment.

Ponderosa pine has been used to some extent because of its availability, favorable shape, and thick sapwood that is easily penetrated with preservatives.

Redwood, Atlantic white-cedar, jack pine, red pine, and eastern redcedar are used to a slight extent, for the most part locally in the areas where they are produced. All of these species generally require preservative treatment.

Other species having local use include tamarack, baldcypress, black locust, ash, elm, and cottonwood. With the exception of black locust, none of these last long without preservative treatment.

Weight and Volume of Poles and Piles

The weight of a pole depends on the species, size, moisture content, and preservative treatment. Weights per cubic foot of the various species of wood may be calculated from the data described in chapters 3 and 4. Poles may be green when first produced, but in service moisture content above ground falls below 30 percent in most areas.

Volumes of poles and piles may be computed by two methods given in American Wood-Preservers' Association Standard F3.

Using method 1, the volume can be calculated by the formula

$$V = 3L\left(\frac{C_m}{\pi}\right)^2 0.001818 \qquad (19\text{--}1)$$

where V is volume in cubic feet, L is length in feet, and C_m is midlength circumference in inches. By method 2, the volume can be calculated by the formula

$$V = 0.001818L(D^2 + d^2 + Dd) \qquad (19\text{--}2)$$

where D is the top diameter (in inches) and d is the butt diameter. V and L definitions are the same as for formula (19–1). If method 2 is used, a correction factor as indicated below must be used for certain species:

Oak piles	0.82
Southern pine piles	.98
Southern pine and red pine poles	.95

Method 1 is the AWPA official method except for Douglas-fir, for which either method can be used. Volume tables for both methods are given in AWPA Standard F3.

The volume of a pole shows little difference whether green or dry. Drying of poles causes checks to open, but there is little reduction of the gross diameter of the pole.

Engineering Properties of Wood Poles

Round poles used for transmission and distribution lines and buildings are specified by species, class, and length. Specifications for wood poles for transmission and distribution lines, adopted by the National Electric Safety Code, are given in the American National Standards Institute (ANSI) Standard 05.1, "Specifications and Dimensions of Wood Poles." Specifications for wood poles for farm buildings are given in the American Society of Agricultural Engineers Tentative Recommendation R299T, "Construction Poles—Preservative-Treated Wood."

Species of timber commonly used for poles and their fiber stress in bending are given in ANSI Standard 05.1. These values are the near ultimate fiber stress in the outer fibers of the pole at failure in flexure as a cantilever beam. It is customary to reduce the stresses for use in design to provide a factor of safety in accordance with the type of construction in which the poles are used. Recommended reductions are given in the National Electric Safety Code.

Life of Poles

The life of poles can vary within wide limits depending on their growth and use conditions, kind and quality of the preservative, method of treatment, penetration and distribution of preservative, and mechanical damage. Service life, due to line changes and obsolescence, is often somewhat less than the physical life of poles.

It is common to report the "average" life of untreated or treated poles based on observations over a period of years. These average life values are useful as a rough guide to what physical life may be expected from a group of poles. However, it should be kept in mind that, within a given group, 60 percent of the poles will have failed before reaching an age equal to the average life.

Early or premature failure of treated poles can generally be attributed to one or more of three factors: (1) Poor penetration and distribution of preservative; (2) an inadequate retention of preservatives; or (3) a substandard or untried preservative.

Preservative Treatment of Poles

Federal Specification TT–W–571 covers the preservative treatment of utility and building poles and includes the principal requirements of AWPA Standards C1 and C4 for pressure treatment, C8 for full-length thermal (hot-and-cold) treatment of western redcedar, C10 for full-length thermal (hot-and-cold) treatment for lodgepole pine poles, and C23 for pressure treatment in pole building construction.

Seasoning or conditioning requirements are included in these specifications. For western redcedar poles to be treated by the thermal process, incising at the groundline is required to meet the preservative penetration requirement. Penetration and retention requirements for pressure treatment vary for different pole species and service conditions also for group A poles (less than 37.5 in. in circumference 6 ft. from the butt) and group B (larger poles). The zones from which borings are to be taken for retention by assay also differ for different species treated by pressure. Table 18–1 (ch. 18) includes minimum preservative retention figures for poles of different types.

Some treated poles exude preservative sufficiently to make the surface oily in spots. This "bleeding" is not likely to reduce the life of the poles, but it may prove objectionable to men who work on the poles or to anyone who may come in contact with a bleeding pole. Methods of completely preventing bleeding have not been established. The use of lower retentions, low-residue creosotes, or selected oil to act as preservative carriers, and a final heating or expansion bath following treatment, however, will help to produce clean poles. There is an increasing use of the waterborne preservatives as covered in TT–W–571 where cleanliness and paintability of poles are required.

In pole-type structures where siding or other exterior covering is applied, the poles are generally set with the taper to the interior side of the structures to provide a vertical exterior surface. Another common practice is to modify the round poles by slabbing to provide a continuous flat face. The slabbed face permits more secure attachment of sheathing and framing members and facilitates the alinement and setting of intermediate wall and corner poles. The slabbing consists of a minimum cut to provide a single continuous flat face from groundlines to top of intermediate wall poles, and two continuous flat faces at right angles to one another from groundline to top of corner poles.

It should be recognized that preservative penetration is generally limited to the sapwood of most species. Thus slabbing, particularly in the groundline area of poles, with thin sapwood may result in somewhat less protection than that of an unslabbed pole. All cutting and sawing should be confined to that portion of the pole above groundline and should be performed before treatment.

Treatment to Retard Decay in Standing Poles

Preservative applications have been made to the groundline zone of untreated poles to retard decay. A study in Canada on six different preservative applications to untreated cedar showed that from 6 to 13 years of additional service was provided. Studies by the Forest Products Laboratory indicated that groundline treatments had questionable value for well-treated poles with a good reserve supply of preservative in the outer zone or for treated poles with heartwood decay. For treated poles with light surface decay or other evidence of an inadequate supply or quality of preservative in the outer part of the pole, some groundline treatments showed promise of providing additional pole service.

The untreated above-ground sections of butt-treated poles that have started to show sapwood decay or "shellrot" are frequently sprayed with solutions of preservative containing 5 to 10 percent active chemical.

PILES

Choice of Species for Piles

The properties desirable in piles include sufficient strength and straightness to withstand driving and to carry the weight of structures built on them, and in some instances to resist bending stresses. Decay resistance or ease of penetration by preservatives is also important except in piles for temporary use or piles that will be in fresh water and entirely below the permanent water level.

Southern pine, Douglas-fir, and oak are among the principal species used for piles, but western redcedar and numerous other species also are used.

Specifications for timber piles covering kinds of wood, general quality, resistance to decay, dimensions, tolerance, manufacture, inspection, delivery, and shipment have been published in the American Railway Engineering Association (AREA) Manual. Specifications for timber piles have also been prepared by the American Association of State Highway Officials, the American Society for Testing and Materials, and the Federal Supply Service.

Bearing Loads for Piles

Bearing loads on piles are sustained by earth friction along the sides of the pile, by bearing of the tip on a solid stratum, or by a combination of the two. Wood piles, because of their tapered form, are particularly efficient in supporting loads by side friction. Bearing values that depend upon side friction are related to the stability of the soil and generally do not approach the ultimate strength of the pile. Where wood poles sustain foundation loads by bearing of the tip on a solid stratum, loads may be limited by the compressive strength of the wood parallel to the grain. If a large proportion of the length of a pile extends above ground, its bearing value may be limited by its strength as a long column. Side loads may also be applied to piles extending above ground. In such instances, however, bracing is often used to reduce the unsupported column length or to resist the side loads.

There are several ways of determining bearing capacity of piles. Engineering formulas can be used for estimating bearing values from the penetration under blows of known energy from the driving hammer. Some engineers prefer to estimate bearing capacity from experience or observation of the behavior of pile foundations under similar conditions or from the results of static-load tests.

Working stresses for piles are governed by building code requirements by recommendations of the American Society for Testing and Materials.

Eccentric Loading and Crooked Columns

The reduction in strength of a wood column resulting from crooks, eccentric loading, or any other condition that will result in combined bending and compression is not as great as might be expected. Tests have shown that a timber, when subjected to combined bending and compression, develops a higher stress at both the proportional limit and maximum load than when subjected to compression only. This does not imply that crooks and eccentricity should be without restriction, but it should relieve anxiety as to the influence of crooks, such as those found in piles.

Design procedures for eccentrically loaded columns are given in chapter 8.

Seasoning Effect on Pile Driving

Under usual conditions of service, wood piles will be wet, but they may be driven in either the green or the seasoned condition. Because of the increased strength resulting from drying, seasoned piles either treated or untreated are likely to stand driving better than are green or unseasoned ones. This is particularly true of treated piles; tests have demonstrated that, while the strength of green wood may be considerably reduced by pretreatment conditioning, thoroughly seasoned treated wood may be nearly as strong as seasoned untreated wood. Under the same drying conditions, however, untreated wood loses moisture more rapidly than does treated wood.

Decay Resistance and Preservative Treatment of Piles

Species most commonly used for piles generally have rather thick sapwood and consequently low decay resistance. High natural decay resistance will be found only when the piles have thin sapwood and are of species that have decay-resistant heartwood.

Because wood that remains completely submerged in water does not decay, decay resistance is not necessary in piles so used; resistance to decay is necessary in any part of the piles that may extend above the permanent water level. When piles that support the foundations of bridges or buildings are to be cut off above the permanent water level, they should be treated to conform to recognized specifications (such as Federal Specification TT–W–571 and AWPA Standards C1 and C3). The untreated surfaces exposed at the cutoffs should also be protected by thoroughly brushing the cut surface with coal-tar creosote. A coat of pitch, asphalt, or similar material may then be applied over the creosote and a protective sheet material, such as metal, roofing felt, or saturated fabric, fitted over the pile head.

Piles driven into earth that is not constantly wet are subject to about the same service conditions as apply to poles, but are generally expected to last longer and therefore require higher preservative retentions than poles (table 18–1).

Piles used in salt water are, of course, subject to destruction by marine borers even though they do not decay below the waterline. Up to this time the best practical protection against marine borers has been a treatment to refusal with coal-tar creosote or creosote-coal-tar solution. Recent experiments with dual treatments (pressure treatment, first with a waterborne preservative followed after seasoning with a creosote treatment) show promise of providing greater protection. Federal Specification TT–W–571 and AWPA Standard C3 cover the preservative treatment of marine piles (table 18–1).

TIES

Strength and Other Requirements for Ties

Many species of wood are used for ties. The more common are oaks, gums (tupelo and sweetgum), Douglas-fir, mixed hardwoods, hemlock, southern pine, and mixed softwoods. Their relative suitability depends largely upon their strength, wearing qualities, treatability with wood preservatives, and to some extent their natural resistance to decay and tendency to check, although availability and cost must also be considered.

The chief strength properties considered in a wood for crossties are (1) bending strength, (2) end hardness and strength in compression parallel to grain (which indicate resistance to spike pulling and the lateral thrust of spikes), and (3) side hardness and compression perpendicular to the grain (which indi-

cate resistance to wear under the rail or the tieplate).

Sizes of crossties range from 6 by 7 to 7 by 9 inches; lengths are usually 8, 8½, or 9 feet. With heavier traffic and higher speeds of trains, the present tendency is toward increasing use of the larger sizes.

Specifications for crossties covering general quality, resistance to wear, resistance to decay, design, manufacture, inspection, delivery, and shipment have been published in the AREA Manual and in Federal Specification MM–T–371.

Life of Ties

The service conditions under which ties are exposed are severe. The life of ties in service therefore depends on their ability to resist decay and the extent to which they are protected from mechanical destruction by breakage, loosening of spikes, and rail or plate wear. Under sufficiently light traffic, heartwood ties of naturally durable wood, even if of low strength, may give 10 or 15 years average service without preservative treatment; under heavy traffic without adequate mechanical protection, the same ties might fail through mechanical wear in 2 or 3 years. The life of treated ties is affected also by the preservative used and the thoroughness of treatment. As a result, the life of individual groups of ties may vary widely from the general average depending on the local circumstances.

With these limitations, the following rough estimates are given: Ties well treated according to the specifications cited in this chapter should last from 25 to 40 years on an average when protected against mechanical destruction. Untreated white oak ties have lasted 10 to 12 years on an average in the northern United States.

Records on the life of treated and untreated ties are published from time to time in the annual proceedings of AREA and AWPA.

Decay Resistance and Preservative Treatment of Ties

Although the majority of ties used are given preservative treatment before installation, a few are used untreated and, for these, natural decay resistance is important. In ties given preservative treatment, variations in natural decay resistance are less important than ability to accept treatment.

The majority of ties treated are pressure treated with coal-tar creosote, creosote-coal-tar solutions, or creosote-petroleum mixtures. Federal Specification TT–W–571 includes listings for treatment of crossties, switch ties, and bridge ties (table 18–1). AWPA Standards C2 and C6 and specifications of AREA also cover the preservative treatment of crossties and switch ties.

BIBLIOGRAPHY

American Association of State Highway Officials
  Standard specifications for highway bridges. (See current edition.) Washington, D.C.
American National Standards Institute
  ANSI standard specifications and dimensions for wood poles. ANSI Standard 05.1 (See current edition.)
American Railway Engineering Association
  Manual of the American Railway Engineering Association, Chicago, Ill. (Looseleaf manual, revised annually.)
American Society of Agricultural Engineers
  Construction poles—Preservative-treated wood. ASAE tentative recommendations. ASAE R299T. (See current edition.)
American Society for Testing and Materials
  Standard specification for round timber piles. ASTM Designation D 25. (See current edition.) Philadelphia, Pa.

————
  Establishing design stresses for round timber piles. ASTM Designation D 2899. (See current edition.) Philadelphia, Pa.
American Wood-Preservers' Association.
  All timber products—Preservative treatments by pressure processes. AWPA Stand. C1. (See current edition.)

————
  Lumber, timbers, bridge ties and mine ties—Preservative treatment by pressure processes. AWPA Stand. C2. (See current edition.)

————
  Piles—Preservative treatment by pressure processes. AWPA Stand. C3. (See current edition.)

————
  Poles—Preservative treatment by pressure processes. AWPA Stand. C4. (See current edition.)

————
  Crossties and switch ties—Preservative treatment by pressure processes. AWPA Stand. C6. (See current edition.)

————
  Standard for the full-length thermal process treatment of western redcedar poles. AWPA Stand. C8. (See current edition.)

————
  Lodgepole pine poles—Preservative treatment by the full-length thermal process. AWPA Stand. C10. (See current edition.)

————
  Pole building construction—Preservative treatment by pressure processes. AWPA Stand. C23. (See current edition.)

————
  Standard volumes of round forest products. AWPA Stand. F3. (See current edition.)

————
  1970. Wood preservation statistics 1969. AWPA Proc. 66: 271–303.

Blew, J. O.
  1970. Pole groundline preservative treatments evaluated. Transmission and Distribution, Aug.
Chellis, R. D.
  1961. Pile foundations. McGraw-Hill Book Co., Inc., New York.
Chrisholm, T. H., and Suggitt, N. A.
  1959. An evaluation of six preservative systems on cedar test stubs after 16 years' field exposure. Ontario Hydro Res. News 11(2): 25–28.
Markwardt, L. J.
  1930. Comparative strength properties of woods grown in the United States. U.S. Dep. Agr. Tech. Bull. 158. 39 pp.
Newlin, J. A., and Trayer, G. W.
  1924. Stresses in wood members subjected to combined column and beam action. Nat. Adv. Comm. Aeron. Rep. 188. 13 pp.
Panek, Edward
  1960. Results of groundline treatments one year after application to western redcedar posts. Amer. Wood-Preserv. Assoc. Proc. 56: 225–235.
————, Blew, J. O., and Baechler, R. H.
  1961. Study of groundline treatments applied to five pole species. Forest Prod. Lab. Rep. 2227.
Patterson, Donald
  1969. Pole building design. American Wood Preserv. Instit.
U.S. Department of Commerce
  National electric safety code. Nat. Bur. Stand. Handb., Washington, D.C.
U.S. Federal Supply Service
  Piles: Wood. Fed. Specif. MM–P–371. (See current edition.)

————
  Ties, railroad, wood (cross and switch). Fed. Specif. MM–T–371C. (See current edition.)

————
  Wood preservation: Treating practices. Fed. Specif. TT–W–571. (See current edition.)
Wilkinson, Thomas Lee
  1968. Strength evaluation of round timber piles. USDA Forest Serv. Res. Pap. FPL 101. Forest Prod. Lab., Madison, Wis.
Wood, Lyman W., Erickson, E.C.O., and Dohr, A. W.
  1960. Strength and related properties of wood poles. Amer. Soc. Testing and Mater.
———— and Markwardt, L. J.
  1965. Derivation of fiber stress from strength values of wood poles. U.S. Forest Serv. Res. Pap. FPL 39. Forest Prod. Lab., Madison, Wis.

Chapter 20

MOISTURE RELATIONSHIPS SIGNIFICANT IN DECAY CONTROL

An understanding of wood-moisture relationships is essential to the application of decay preventive and control measures, because decay-producing fungi must have water.

Wood swells and shrinks with changes in moisture content caused by changes in humidity and by absorption and loss of water. Small dimensional changes of wood in use are unavoidable; excessive changes commonly are associated with poor design, careless workmanship, or inadequate maintenance. These conditions can lead to warping, checking, or nail pulling that opens joints or creates new avenues for the entry of rainwater and thus increases the chance of decay.

Wetting and drying of wood and the relative ease with which it can be treated with preservatives are influenced by absorptivity (the capacity of wood to pick up moisture) and permeability (the relative ease with which moisture can move inward under a pressure gradient). Oftentimes no distinction is made between these two properties.

Moisture Vapor

Wood will give off or take on moisture from the surrounding atmosphere until the amount it contains balances that in the atmosphere. The moisture content of the wood at the point of balance is called the equilibrium moisture content and is expressed as a percentage of the ovendry weight of the wood. In the usual temperature range, the equilibrium moisture content depends chiefly on the relative humidity of the atmosphere.

Water vapor is adsorbed only in the walls of the wood cells. The moisture content when the cell walls are saturated but there is no water in the cell cavities is known as the fiber saturation point. At room temperature this point amounts to approximately 30 percent of the ovendry weight of the wood. The amount of moisture at fiber saturation is the maximum that can be adsorbed from moisture vapor. The fiber saturation point is the approximate lower moisture limit for attack of wood by decay fungi.

As the cell walls take up or lose water, they swell or shrink accordingly. Correspondingly, the wood swells or shrinks; if the dimensional changes are uneven, warping, twisting, or cupping occurs. For each 1 percent loss in moisture content below the fiber saturation point,

wood shrinks about one-thirtieth of its possible shrinkage.

Liquids

Water, such as rain or droplets of condensate, may be drawn into the wood by capillary action. It moves inward most rapidly through the tubelike cell cavities. With a continuing source of supply, the water will continue inward at a significant rate until the capillary forces are offset by built-up air pressure and by the drag imposed by friction. While the water travels through the cell cavities, it will also enter the cell walls and saturate them.

Oils are similarly absorbed into wood but with a very important exception: Oils move into the cell cavities but do not enter the cell wall. Therefore, absorption of oil does not cause wood to swell.

Water generally moves into wood and wood structures much faster than it escapes through subsequent evaporation. Consequently, a relatively brief period of wetting may create a decay-promoting moisture level that will remain for a considerable time.

Penetrability of Wood

The movement of a liquid along the pathway of the cell cavities is influenced by the cellular structure. Most important are the pits in the cell walls, through which the bulk of the liquid must pass to get from one cell to another. Some pits are readily traversed; others are relatively impenetrable. Therefore, some woods are easily impregnated with preservatives, whereas others are practically impossible to impregnate.

In most species, heartwood is more difficult to penetrate than sapwood because in the process of heartwood formation certain changes occur in the pits that tend to plug them. In some hardwoods, the vessels become plugged with obstructions known as tyloses, which reduce permeability. Tyloses, for example, are responsible for the marked impermeability of white oak wood, which has made this a wood favored for barrel construction. Back pressure is also significant in liquid absorption. Air in wood becomes compressed while a liquid is being absorbed until finally back pressure may become the dominant factor determining the rate of absorption.

A liquid can move into wood more easily along the grain than across it. Wood cells are much longer than they are wide, thus the liquid must move through many more cells when going across than when going in the direction of the grain. Therefore, water in contact with the side of a board is absorbed in far smaller amount in the same length of time than is water in contact with the end of the same board. This great difference in absorptivity between side grain and end grain is an important factor in determining the decay susceptibility of different kinds of wood construction.

Irrespective of differences in penetrability, the total amount of a given liquid that wood can hold varies with specific gravity. The denser the wood, the larger is the amount of space taken up by the cell walls and the smaller is the volume of cell cavities available to receive the liquid. Douglas-fir with an average specific gravity of about 0.45 can absorb about 160 percent water; that is, 160 pounds per 100 pounds of ovendry wood. A denser wood such as oak with a specific gravity of about 0.60 can absorb only about 100 percent, but a very light wood such as balsa with a specific gravity of about 0.25 can absorb as much as 300 percent.

Effects of Fungus Infection on Absorptivity

Fungus infection tends to increase absorptivity of wood. The increased absorptivity may be pronounced even if the wood is not noticeably infected. The earliest changes in wood structure brought about by infection occur in the wood elements that affect absorptivity most. As the fungi move into wood, they pass through the pits and, in doing so, remove portions of the pit membrane and occluding substances; thus they enlarge the pit openings. In addition, they commonly remove portions of the cells of the wood rays, through which liquids flow into wood radially (fig. 2). Although degradation of the pits and the wood rays by fungus infection does not appreciably reduce wood strength, the increased absorptivity can be disadvantageous to wood exposed to the weather because it promotes pickup of rainwater and thereby increases vulnerability to decay.

Natural Decay Resistance

The structural elements of wood consist of lignins, celluloses, hemicelluloses, and ash-forming minerals. These have no effect on decay resistance, but the celluloses and lignins, along with stored starches and sugars, are the main food source for wood-destroying fungi.

Wood also contains extractives which are not part of the wood structure but impart color, odor, and taste. Some woods contain extractives that are toxic to fungi and act as natural preservatives; the type and amount of these extractives determine the degree of susceptibility or resistance to decay.

The wood extractives effective against decay are chiefly phenolics. They vary in specific composition and in potency as preservatives. They are present in effective amounts only in heartwood. In the heartwood of many durable species, decay-retarding extractives tend to decrease in quantity from the outer heartwood to the center of the tree. At a given radial position in the tree, they usually decrease progressively from the base to the top of the trunk. These within-tree differences generally are greater as the tree increases in size and age; the greatest decay resistance in an individual tree is most likely to occur in the outer basal heartwood and the least resistance in the inner basal heartwood.

The radial decline in resistance from outer to inner heartwood results largely from chemical changes in the protective extractives over long periods. The potent preservatives deposited when the heartwood is formed gradually alter with age and become less effective. Extractives that are effective against fungi are not necessarily the same as those effective against insects or marine organisms.

If long life is desired from untreated wood under conditions suitable for decay organisms, only heartwood should be depended on regardless of species. Under conditions that are suitable for decay organisms, the sapwood usually will be rotted quickly. In a few species, such as the spruces and the true firs (not Douglas-fir) the color of the sapwood and the heartwood is so similar that frequently the two cannot be distinguished easily. Generally, both the heartwood and the sapwood of these species have low resistance to decay.

Relatively young, second-growth trees usually contain a higher proportion of sapwood than does

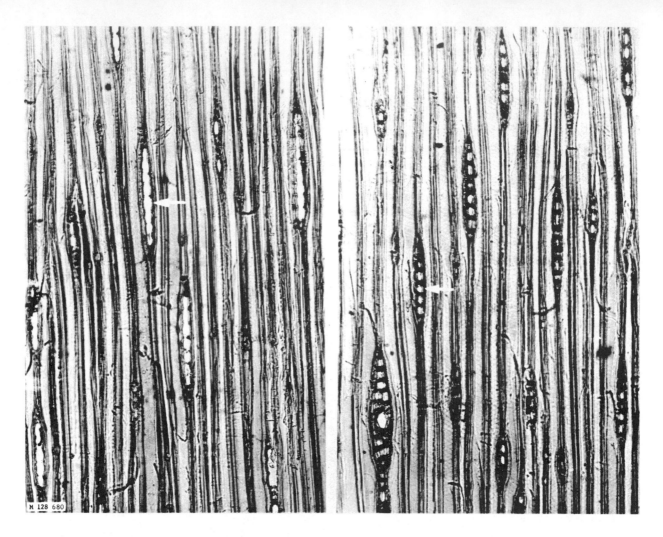

Figure 2.--Photomicrographs of tangential surface of pine showing wood rays opened by mold infection (arrow at left) and normal rays with cells intact (arrow at right); wood-invading bacteria affect ray cells similarly.

virgin timber. Because most lumber coming from areas east of the Rocky Mountains is from second-growth timber, it contains large percentages of sapwood. For example, there are but few remaining all-heartwood supplies of baldcypress lumber.

Heartwood of virgin redwood and western red-cedar is the chief source of domestic softwood lumber with high natural decay resistance. Douglas-fir heartwood is only moderately decay resistant in contact with the ground but in most exterior service above ground it will considerably outperform other structural woods such as the spruces, the true firs, and the hemlocks. White oak heartwood has been extensively used in ship-building because of its combination of strength and decay resistance, but it is not available in sufficient amount for general use in land structures.

Black locust, with its unusually durable heartwood, is suitable primarily for posts.

Comparisons of wood species for decay resistance by ascribing to each a single class of resistance lack precision because wood of the same species may vary considerably in resistance. This simple basis of comparison can be useful, however, if it is recognized that in specific cases the resistance may vary considerably from the average.

In table 1, common species native to the United States are grouped according to the average decay resistance of the heartwood. Quantitative significance may be assigned to the groupings by noting that fence posts of substantially resistant or very resistant heartwood should last 10 to 20 or more years and those with nonresistant heartwood about

Table 1. –Heartwood decay resistance of domestic woods

Resistant or very resistant	Moderately resistant	Slightly or nonresistant
Baldcypress (old growth)[1]	Baldcypress (young growth)[1]	Alder
Catalpa	Douglas-fir	Ashes
Cedars	Honeylocust[2]	Aspens
Cherry, black	Larch, western	Basswood
Chestnut	Oak, swamp chestnut	Beech
		Birches
Cypress, Arizona	Pine, eastern white[1]	Buckeye[2]
Junipers	Pine, longleaf[1]	Butternut
Locust, black[3]	Pine, slash[1]	Cottonwood
Mesquite	Tamarack	Elms
Mulberry, red[3]		Hackberry
Oak, bur		Hemlocks
Oak, chestnut		Hickories
Oak, Gambel		Magnolia
Oak, Oregon white		Maples
Oak, post		Oak (red and black species)[2]
Oak, white		
Osage-orange[3]		Pines (most other species)[2]
Redwood		
Sassafras		Poplar
Walnut, black		Spruces
Yew, Pacific[3]		Sweetgum[2]
		Sycamore
		Willows
		Yellow-poplar

[1] Southern and eastern pines and baldcypress are now largely second growth with a large proportion of sapwood. Consequently, it is no longer practicable to obtain substantial quantities of heartwood lumber in these species for general building purposes.

[2] These species or certain species within the groups have higher decay resistance than most woods in this grouping.

[3] Exceptionally high decay resistance.

5 years or less. A wood that lasts longer than another in the ground can also be expected to last longer when used above ground. The actual years of service--and correspondingly, the differences in service--will, of course, normally be greater above ground.

A number of foreign hardwoods have very high decay resistance. The best known is teak. Among the mahoganies, those from Central and South America (species of Swietenia) are generally classified as decay resistant. African mahogany (Khaya species) seems to have moderate decay resistance. The Philippine mahoganies (species of Shorea, Parashorea, and Dipterocarpus) have exhibited no more than moderate or slight resistance. An exception is Shorea guiso, which has showed high resistance. The woods classed as red lauans usually are somewhat more resistant than the white lauans.

The season of the year when wood is cut is not known to have any effect on decay resistance. Wood cut or peeled in late fall or winter is usually safer from immediate damage than that cut in warm weather. In cold weather, freshly cut timber is exposed at a time when fungi are not active; thus, it may be moved to a safe place or it may become dry enough to avoid attack by fungi before warm weather begins.

335

Drying wood to be used untreated in contact with the ground does not increase natural resistance to decay. Drying may, however, have a very important influence on the life-expectancy of wood used in certain confined parts of buildings or in other unexposed places. Wood that is not dried before being built into spaces in which it cannot dry rapidly may retain moisture so long that decay organisms may harm it before it becomes dry.

It should be observed that no wood species classed as very decay resistant is uniformly resistant to all decay fungi. The water-conducting fungi (see "Protecting Buildings After Construction") are notorious for their ability to decay all-heart cypress, redwood, and other decay resistant woods. Therefore, when decay-resistant woods are used to repair a structure with decay, one should make certain that a water-conducting fungus is not present.

BIOLOGICAL DETERIORATION OF WOOD

Most measures to protect wood are directed against three types of destruction: Biological (destructive utilization of wood by various organisms), fire, and physical damage by breakage or by deformation. By observing simple and inexpensive precautions, the effects of these destructive forces can be held to a low level.

This report is concerned primarily with biological deterioration caused by decay fungi. Mold and stain fungi and bacteria are considered briefly, particularly as they influence decay. Such physical damage as warping and paint peeling are also mentioned as affecting decay or as signs of a possible decay hazard.[4]

Kinds of Damage

Four primary types of damage by wood-attacking fungi are commonly recognized: Sap stain (chiefly the dark type, commonly known as blue stain), mold, decay, and soft rot. The distinctions among types are generally useful although not always sharp. Also, bacteria degrade wood under special conditions.

Fungi destroy more wood than do any other organisms. Those causing decay, or rot, are by far the most destructive. Fungi in their simple growing stage are threadlike, and the individual strands, called hyphae, are invisible to the naked eye except in mass. These hyphae penetrate and ramify within wood.

Botanically, the fungi are a low form of plant life. They have no chlorophyll; therefore, they cannot manufacture their own food but must depend on food, such as wood, already elaborated by green plants. Fungi convert wood they are invading into simple digestible products, and in the process the wood loses weight and strength.

Sap Stain

Sap stain is a discoloration that occurs mainly in logs and pulpwood during storage and in lumber during air drying. As its name implies, it is a discoloration of the sapwood. The "blue-stain" type, which dominates and is the only one of the sap stains of much importance commercially, is caused by the dark color of the invading fungus. Sap stain can go deep into the wood causing a permanent blemish that cannot be surfaced off. The color of sap stain usually ranges from a brownish or a steel gray to almost black, depending on the fungus and the wood species.

Sap stain alone ordinarily does not seriously affect the strength of wood, but heavily stained wood is objectionable where strength is of prime importance. Its presence signifies that moisture and temperature have been suitable for the development of decay fungi; hence, early decay often is present though masked by the stain. Stained wood is more permeable to rainwater; thus wood in exterior service is more subject to decay infection.

Mold

Molds cause discoloration that is largely superficial and can be removed by brushing or shallow planing. In coniferous woods the discoloration imparted by mold fungi typically is caused by the color of surface spore masses (green, black, orange); in hardwoods the wood itself often is superficially discolored by dark spots of various sizes.

Mold hyphae, however, penetrate wood deeply and increase permeability, sometimes very markedly. Also, heavy molding often is accompanied by hidden incipient decay.

Decay

Under favorable conditions, decay can rapidly destroy wood substance and seriously reduce the strength of wood. This may happen before there is any pronounced change in external appearance of the wood; thus even early decay can dangerously weaken a wood structural component. Advanced decay, of course, can render it entirely useless.

Two major kinds of decay are recognized-- brown rot and white rot. Different types of fungi are responsible for each. With brown rot, only the cellulose is extensively removed; the wood takes on a browner color and tends to crack across the grain, shrink, and collapse (fig. 3). With white rot, both lignin and cellulose usually are removed, the wood may lose color and appear whiter than normal, it does not crack across the grain, and until it has been severely degraded, it does not shrink or collapse (fig. 4).

Soft Rot

Soft rot, although not identified and classified until recent years, is another form of severe wood degradation. Soft rot often can be identified by these characteristics: It tends to be shallower than typical decay; the transition between the rotted wood and the sound wood beneath is often abrupt; and if the damaged wood is scraped off with a knife, the blade may suddenly strike wood that has almost normal hardness. Additional aids in identifying are a comparatively slow, inward development of the rot, and wood surfaces exposed to the elements tend to be profusely cracked and fissured both in and across the grain direction like weatherbeaten driftwood and old unpainted buildings.

The fungi that cause soft rot are in a completely different group than the more familiar and destructive decay fungi; physiologically, they tolerate both wetter and drier conditions. The wettest parts of cooling towers, for example, are subject to soft rot. Some of the molds and the sap stain fungi can progress far enough to cause soft rot if the wood is wet for a long enough time-- which usually is substantially greater than the time involved in producing air-dry lumber.

Because soft rot ordinarily is relatively shallow and progresses inward slowly, it ordinarily is not of great concern to the wood user. But where the wood member is comparatively thin, as in the fill members of a cooling tower, even rather shallow rot can weaken the wood members severely.

Bacterial Damage

Bacteria, as well as fungi, can invade wood. Bacteria generally have been troublesome in logs stored in ponds or under continuous water sprays. They remove certain sapwood constituents, which increases the permeability of the wood. The greater permeability adds, like that caused by fungi, to susceptibility to serious wetting by rain. Moreover, millwork of bacteria-invaded wood absorbs excessive amounts of preservative solution when dip-treated. Generally bacteria do not appreciably weaken wood. Serious bacterial weakening has been observed, however, in thin items like cooling-tower slats and in piling wetted by fresh water for several decades.

Rot in the Living Tree

All types of wood-inhabiting fungi and bacteria occur in living trees. The decay fungi cause heartrot. One familiar heartrot occurs in pecky cypress, used for decorative paneling; another is the so-called white pocket common in low-grade

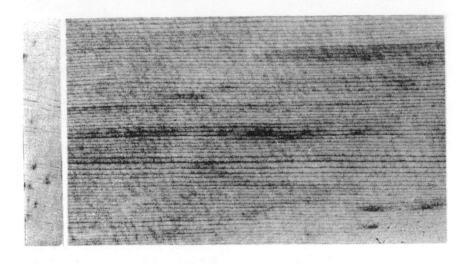

Figure 3.--Typical brown rot: Top, Early stage, evidenced by discoloration in flat-grain board and, at left, in end grain. Bottom, Late stage, with cracked and collapsed wood. Brown rot, as the name suggests, ultimately imparts an abnormal brown color to the wood.

lumber from over-mature stands of Douglas-fir. Heartrot fungi typically do not invade healthy sapwood because the wood is too wet and, possibly, because many of the sapwood cells are still living. Heartrot can be damaging to wood cut from the tree depending on the stage of development of the rot.

Heartrot fungi generally do not attack wood products. Therefore, further damage by them after wood is placed in service in unlikely. Conversely, most fungi that attack wood in service do not attack wood in the living tree to any significant extent.

Figure 4.--Intermediate stage of white rot: Top, Decay on end grain. White rot is indicated by abnormally light discoloration--often with mottling of the wood--and often by dark-colored "zone lines" bordering the discoloration (see arrows). It does not cause the wood to crack as does the brown rot. Bottom, Decay on side grain.

Initiation and Spread of Decay

Decay fungi penetrate and proliferate inside wood from one wood cell to another through the natural openings, the pits, or through the small openings they make called bore holes. They destroy the cell walls they contact.

The fungi may spread from one piece of wood to another by two methods:

1. Direct contact of sound wood with decayed wood.

2. Spores or unspecialized fungus fragments.

The tiny spores, which are analogous to seeds, are transported in large numbers--often for great distances--by wind and insects; those that light on susceptible wood may grow and create new infections.

The cycle of decay infection is illustrated in figure 5. The highly specialized, sexual spores of decay fungi are produced in so-called fruit bodies; the familiar field mushrooms are an example. Others are the shelf- or bracket-like bodies that appear on tree trunks, logs, dead branches, and rotting structures. Millions of spores can be produced by a single fruit body.

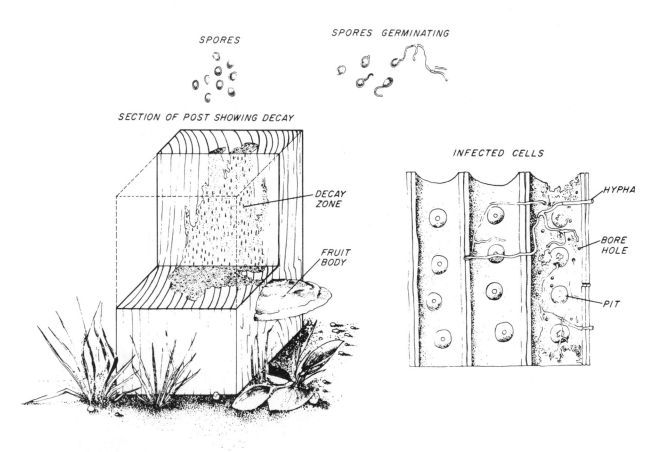

Figure 5.--Cycle of decay: Thousands of spores (top left) developed in fruit body (bottom left) of the decay fungus are distributed by wind or insects. On contracting susceptible wood they germinate and the fungus penetrates the wood infecting it cell by cell (right). After causing a certain amount of decay, a fruit body is developed and the cycle is repeated.

Basic Conditions for Decay

Decay fungi have the same four primary needs as all of the higher forms of life, namely: Food, air (oxygen), favorable temperature, and water. By excluding or limiting any one of these, decay can be prevented or restricted.

Fungus-Susceptible Wood

Decay fungi depend on the wood cellulose or on both the cellulose and the lignin for their nourishment. Wood can be made unavailable as food by keeping it dry or by poisoning it with a preservative chemical. The heartwood of some timber species contains naturally occurring preservative chemicals that variously resist fungus infection.

Stain and mold fungi do not depend on the major wood constituents for food; rather, much of their development is supported by the nonstructural elements--mainly sugars and starch. Heartwood generally is not suitable for stain and mold fungi because it does not contain these nutritional elements in sufficient amounts.

Air in the Wood

All wood-attacking fungi require air as a source of oxygen. However, they need comparatively little oxygen and can maintain essentially a normal rate of development at levels considerably below the amount of oxygen in ordinary air. If wood is held under water, it cannot receive sufficient oxygen to support heavy fungus attack. This explains why many foundation piles have served for decades without preservative treatment when driven to a depth completely below the water table. Conversely, portions exposed to air by a lowering of the water table will rot. Foundation piles should be treated to ensure protection.

One of the earliest and simplest means of protecting stored logs is to submerge them in freshwater ponds. Similarly effective restriction of air can be obtained by keeping logs wet with a water spray. Certain specialized fungi and bacteria can invade wood under water, but their effect generally is one of increased permeability rather than of typical decay.

Moderate Temperatures

Decay fungi require moderate temperatures for rapid development. They can do little harm to wood at near freezing or above 100° F. Usually, the rate of decay is slow at temperatures below 50° F. and falls off markedly with temperatures above 90° F. In the main, decay fungi attack most rapidly in the range of 75° to 90° F.

Naturally occurring subzero temperatures merely inactivate fungi, but high temperatures kill them. The lethal effect of a high temperature depends on the specific temperature and the length of time it is applied. A temperature below about 150° F. probably would be impractical as an eradication measure because it would have to be applied for an excessive length of time. Temperatures generally reached in commercial kiln drying and pressure treatment sterilize wood.

Adequate Wood Moisture

Serious decay will occur only when the moisture content of wood is above the fiber saturation point (approximately 30 percent, ovendry basis). This amount of moisture cannot be acquired from humid air but requires wetting of the wood by water. Initially dry wood kept dry under shelter and protected against condensation will not rot.

An excellent rule for drying lumber is to reduce the moisture content as soon as practicable to 20 percent or less. This is substantially below the approximate 30 percent minimum needed to support decay organisms, but the lower figure is advisable because it provides a margin of safety. Although an average moisture content of 20 percent can be easily attained in a load of lumber by air-drying, a number of boards may have a moisture content considerably higher than the average. At construction time, lumber should have a moisture content substantially below 20 percent to minimize dimensional changes as well as to prevent decay.

The moisture content at which wood is most susceptible to decay lies in a broad range from not far above fiber saturation to somewhere between 60 and 100 percent. The upper level for rapid decay is determined by the specific gravity and cross-sectional size of the wood. Both of these factors determine the rate of air exchange between the inside and the outside of the piece.

The commonly used term "dry rot" is an unfortunate misnomer because it implies that wood can decay without being wet, which it cannot. The notion of rot in dry wood may have originated from the appearance of dried wood after it has been decayed by a brown-rot fungus. Wood with brown rot often looks unusually dry, being brown and cracked as though it has been severely heated. It is better to avoid the term "dry rot" and simply use "rot" or "decay."

The term "dry rot" also is misapplied to decay caused by the water-conducting fungi. The necessary moisture in this case is imparted to initially dry wood by the attacking fungus itself. These specialized fungi conduct water through vinelike growths to considerable distances from an external source (such as the soil) to the wood they decay.

To prevent attack of a water-conducting fungus, measures against wetting must be aimed directly at the water-conducting fungus rather than at one of the usual sources of moisture (see Protecting Building After Construction). Fortunately, decay by the water-conducting fungi, although often rapid and severe, is a comparatively minor part of the overall problem. There are only two species of these specialized fungi; one occurs predominantly in the United States and the other mainly in Europe.

Effect of Climate on Decay

Temperature and amount of rainfall and their distribution throughout the year are climatic factors that affect the amount of decay in exterior structures exposed to the weather. Warm weather during many months of the year promotes decay more than hot weather for a few months and cold weather during the remainder of the year. Similarly, prolonged rains are more conducive to decay than the same amounts delivered in heavy but relatively brief showers.

To relate the climate of geographical areas to its decay-contributing potential and to establish a measure to determine the kind and amount of protection needed for various building components influenced by climate, a "climate index" was developed. The climate index is derived from a formula based on mean monthly temperatures and frequencies of rainfall (as published by the U.S. Weather Bureau in Local Climatological Data), relation of decay rate to temperature, and measured actual rates of decay in different areas. It is designed to keep the indexes in the United States largely within the range of 0 to 100. The formula is:

$$\text{Climate Index} = \frac{\sum_{\text{Jan.}}^{\text{Dec.}} (T - 35)(D - 3)}{30}$$

where T is the mean monthly temperature (°F), D is the mean number of days in the month with 0.01 inches or more of precipitation, and $\sum_{\text{Jan.}}^{\text{Dec.}}$ indicates the summation of the products for each of the 12 months, January through December. The climate index for Madison, Wis., for example, is derived as follows: Monthly products $(T - 35)$ x $(D - 3)$: ^{Jan} 0:0:0:72: 168 : 248 : 252 :210:156: 75: 0:0 ^{Dec}. The products are then added, giving 1,181, which when divided by 30 gives an index of 39.

The map in figure 6 is derived from the formula. It is useful for estimating the climate index of localities where differences in elevation are insufficient to cause abrupt differences in climate over comparatively short distances. Three climate-index levels are shown on the map: (1) Index less than 35, where the least preservative protection is needed; (2) an index of 35 to 65, where moderate protection is needed; and (3) an index greater than 65, where the most protection is needed. In regions where mountains cause marked variations in climate, the formula can be used more reliably than the map for indexing a particular locality.

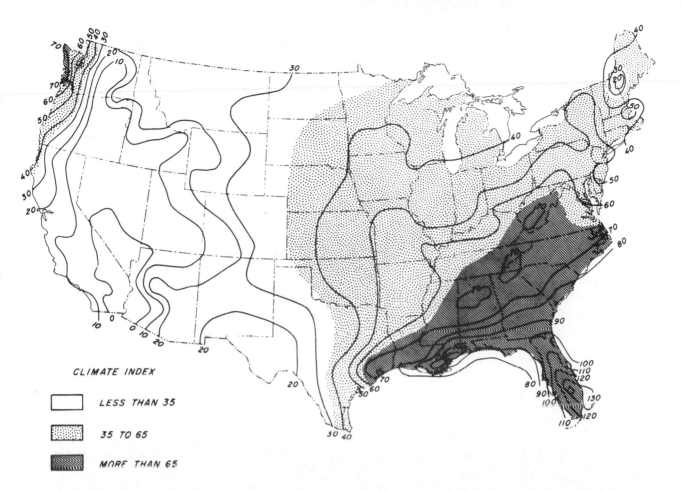

Figure 6.--Levels of decay potential for wood exposed to the weather in aboveground service based on a climate index derived from standard temperature and rainfall data: Darkest areas, wettest climates, most suitable for decay; index greater than 65. Lightest areas, driest climates, least suitable for decay; index less than 35. Gray areas, moderately wet climates, moderately suitable for decay organisms; index 35 to 65.

HOW WOOD IN BUILDINGS BECOMES WET

To protect wood in buildings from decay, the first concern is for practicable measures to ensure use of dry wood and to keep it dry after it is in the structure. Therefore, it is important to understand how wood in buildings becomes sufficiently wet to decay.

Water to support decay can arise from five main sources: Original moisture in unseasoned wood, ground moisture, rainwater, condensate, and piped water. In addition, water is created as a byproduct of the fungal breakdown of wood, but this happens only where decay already has started and so is not of primary concern.

Original Moisture and Wet Lumber

Freshly sawed green lumber has a moisture content that will support attack by decay fungi. As observed earlier, one should aim to dry lumber down to a moisture content of at least 20 percent, never leaving more than 30 percent, if it is to be safe from decay. When wet lumber is placed in a building it usually dries before there is much decay. Preferably, however, drying should be accomplished before installation in a building, in order to minimize later dimensional changes from

343

shrinkage. Also before wet lumber dries, potentially harmful stain, mold, and incipient decay may develop if the wood is not fungus resistant. If lumber that is both infected and wet is enclosed within a relatively vapor-tight wall or in a damp crawl space, it may become damaged by further progress of the decay (fig. 7).

Installation of wet or incompletely dried lumber can cause trouble in another way. In drying, there is danger of splitting and warping with accompanying loosening of joints that, in exposed places, promote rain seepage and decay infection.

Fully air-dried or kiln-dried lumber can, of course, be rewetted sufficiently to decay if not properly protected from rain wetting at the lumber yard, during transportation, or at the building site. This is discussed later under "General Protective Measures."

Wetting From Ground Moisture

Damp soil can cause serious wetting in a frame building. Also, damp soil usually harbors a variety of decay fungi that can directly infect wood in contact with it. Ground moisture can get to wood parts of buildings in four ways:

(1) By direct movement into wood in contact with the soil. The more common direct contacts are substructure members next to dirt-filled porches and terraces, forms left on concrete foundations, and basal exterior woodwork where grade levels have been raised.

(2) By condensation of vapor in crawl spaces as described in the section on condensation.

(3) By being transported through the conducting strands of the rare water-conducting fungi

Figure 7.--White surface growth of a decay fungus in an exterior wall. The decay resulted from using infected and incompletely dried studs and closing them in shortly after they were installed.

344

(see "Protecting Buildings After Construction").

(4) By indirect transfer from soil to wood through concrete or masonry. Some wetting of plates and other wood resting on groundline slabs can occur by capillary transfer through the concrete, but most serious wetting occurs in basements on wet sites where hydrostatic pressure adds to the capillary movement.

Wetting from Rain

Siding and other exterior woodwork may get wet from rain driven directly against it, from roof runoff, or from water splashed from the ground. Rainwater enters largely by capillary movement and is trapped in the interfacial space between joint members. In the exposed part of a building, the joints are especially vulnerable to wetting and decay. Some rainwater may be carried into wall joints by wind pressure and gravity flow. Winds less than 40 miles per hour, however, have surprisingly little effect on the total rain seepage into siding. The chief effect of strong wind is to reduce the protective influence of roof overhang and gutters by driving rain onto the siding, windows, and doors.

Joints of end grain where siding abuts trim or where a porch rail meets a post are the most absorptive. Though a joint may be painted, the paint seal rarely remains tight; moreover, water can enter even through hairline checks in the paint. Longitudinal splits in siding lumber occasionally may result in troublesome amounts of water entering a wall.

The amount of rainwater accumulating in exterior woodwork depends on the extent of the wetting and the rapidity of drying. The extent of the wetting is a function of the amount and the frequency of rainfall, the prevalence and velocity of wind with the rains, the water-shedding protection afforded by the building design (such as roof overhang), and the degree to which water-trapping joints and crevices are avoided by the type of construction.

The rate of drying is governed mainly by the lengths of periods of dry weather between rains and by construction details determining the amount of air exchange in the wetted zone and the corresponding rate of loss of water vapor. In exterior walls, for example, evaporation of any water that gets into the wall is governed largely by the permeability of the sheathing paper used.

Damaging amounts of rain seepage may occur in any wood exposed on the surface of the building. The greatest danger is to roof edges (fig. 8), appendages (porches and exterior stairs), and exposed structural members. Except in dry climates (fig. 6), significant amounts of decay may also occur in siding, trim, sash, and decorative structures such as shutters. Siding of the comparatively durable cedar and redwood heartwood is rarely subject to decay. Sapwood siding, generally of pine species, is subject to molds and stains. These molds and stains generally can be more troublesome than decay.

Wall decay can be caused by runoff from the roof splashing from the ground, a lower roof, or a canopy (fig. 8). This trouble is most likely in regions where rainfall is frequent. Ground splash is greatest when the runoff strikes a flat, hard surface, such as a concrete walk. The amount of wetting also varies with foundation height and the height and width of eave. Splash on masonry walls with permeable joints occasionally leads to decay of wood joists or beams embedded in the walls.

Even where wetting does not afford conditions for decay, it can cause paint failure. Paint failure in siding may develop some distance from the leakage, because after the water enters a joint at the end of a siding board it can flow along the top edge of the board to a different location. Local accumulations of water near joints may raise the humidity of the air between siding and sheathing, then cooling at night results in condensation at various points on the back of the siding. A thin film of water on the back of wood siding can put heavy stresses on the paint coating.

Wetting from Condensation

Condensation results from cooling of air on contact with a cold surface. The amount of water that air will hold varies chiefly with temperature. The warmer the air, the more water vapor it will hold. When air is cooled, it eventually reaches the saturation point, and further cooling causes some of the vapor to condense. The temperature at which condensation begins is the dewpoint temperature. Any surface at or below the dewpoint temperature of an atmosphere will become wet with condensate.

Under a given set of vapor and temperature conditions, the amount of condensate collecting on a given wood surface will be governed by two factors: (1) How rapidly the water vapor can permeate

Figure 8.--Wall and roof edge discoloration from decay promoted by water runoff from roof: Top left, Decay in lower sheathing and plate induced by splash from sidewalk. Top right, Wall decay caused by leak in gravel stop. Bottom left, Runoff from porch roof caused dangerous wetting and decay behind siding adjacent to the window. Bottom right, Sheathing decayed at roof edge because it lacked metal flashing.

materials between the condensing surface and the atmosphere or materials beyond and (2) how rapidly the condensate can escape from the surface during any periods when the temperature is above the dewpoint.

Water vapor moves through a structure from an area of high vapor pressure toward one of lower vapor pressure. In buildings, vapor pressure gradients are created by temperature differences. During the heating season, the gradient is outward from the warm interior; when air conditioning is employed, the gradient is reversed. Most materials used in constructing buildings, such as wood, plywood, asbestos-cement, cork, plaster, and concrete, are permeable to water vapor to varying degrees. Where a dewpoint temperature exists and unless entry of vapor into walls, ceiling, and floors is restricted, damaging condensation will occur within these structures.

When conditions promoting condensation occur within a structure, the use of thermal insulation without a vapor barrier will not prevent condensation but merely change the location of the point where the dewpoint temperature occurs.

Critical wetting by condensation may occur in four areas: (1) Near the perimeter of crawl spaces in cold weather; (2) in floors, walls, and ceilings of cold-storage rooms; (3) in areas where sizable amounts of steam are released or unintentionally escape, such as from leaking steam pipes or radiators; and (4) in the floor below air-conditioned rooms over a damp crawl space. Also, certain climatic and occupancy conditions would favor condensation on slab foundations.

Condensation in Crawl Spaces

Wetting of the perimeter substructure wood by condensate can be serious in crawl spaces under a heated building during cold weather. Two conditions are responsible for this type of condensation: (1) Warm, damp air in the crawl space associated with moist or wet ground; and (2) prevailing outside temperatures of about 50° F. or below. Under these conditions, the relatively cold air outside cools the sills, plates, and joists on or near the foundation creating a dewpoint temperature on the exposed, inside surfaces of these members (fig. 9). The wettest places ordinarily are the corners where air movement in the crawl space is slowest and the wood correspondingly gets coldest. Enough condensate can ultimately be absorbed by the wood to support decay.

Figure 9.--Cold-weather condensation in corner of crawl space; the lumber had recently been installed to replace decayed lumber. The condensation, which continued and threatened further decay, was easily eliminated by laying an inexpensive vapor-resistant cover on the damp ground

There is substantial evidence that crawl-space condensation is most likely to promote decay if the wetted wood is already infected when installed. The lumber may have dried out after reaching the building site, but the dormant decay infection can again become active when the wood is rewetted.

Condensation sufficient to support decay in floors immediately above the crawl space can be caused by overcooling with summer air conditioning. Because the entire floor is cooled, the central as well as perimeter floor area may be affected.

Similarly, although rarely, hazardous condensation may occur during cold weather in floors above a wet crawl space in an unheated building; the wetting again is not restricted to perimeter locations. Where this occurs, the air in the crawl space is slightly warmed by the residual heat in the ground. This condition, coupled with a lower temperature of the unheated floor--cooled by the ambient air--brings about a dewpoint temperature on the underside of the floor.

Condensation in Walls

Except in cold-storage rooms, condensation in walls seldom leads to decay, but--like rain seepage--it can cause paint problems. Troublesome condensation sometimes occurs in exterior walls during the winter, mainly in the North, although it occasionally occurs as far south as the Gulf Coast.

During cold weather the air in living quarters moves outward at a rate determined by the vapor-pressure differential and the permeability of the wall components; then, if the heated air on reaching the sheathing or siding is cooled to the dewpoint temperature, condensation will take place on the cold wood.

Where average temperatures for January are 35° F. or lower (fig. 10) vapor barriers should be installed in exterior walls of all new wood-frame buildings at the time of construction.

Wetting is most likely to occur if vapor-resistant rather than vapor-permeable paper is used under the siding or if the siding itself is vapor impervious. In walls containing no material with high resistance to passage of vapor from the wall space to the outside, condensate accumulation is rare. In northerly climates the possiblity of wall condensation is increased by incorporating thermal insulation within the wall. The presence of the insulation is not likely to be troublesome, however, if a vapor barrier is installed on the warm side of the wall and if winter humidifying is properly regulated.

Condensation in an exterior wall or in an attic is sometimes aggravated by damp air from the crawl space entering the wall void. Warming the air in the wall space, sometimes in part by the sun, tends to create a stack effect causing the air to rise in the wall and new damp air to move in.

Condensation sometimes occurs on cold water pipes in walls, but seldom to a bothersome degree. Occasionally, a restricted amount of decay is found where pipes penetrate a sill plate. Similarly, steam exhaust pipes, if of insufficient length, may lead to steam condensing on a wall. With central air conditioning, condensation occurs on imperfectly insulated ducts and pipes. Most trouble is water drip resulting in mold, but important decay has been found in wood supporting the heating-cooling unit and in wall plates where refrigerant pipes enter the building.

Winter condensation often occurs on the room side of exterior walls and results in unsightly molding of the surface. This is chiefly a problem of small, tightly constructed homes in which moisture released in the living quarters has limited opportunity to escape.

Figure 10.--Moisture problems from winter condensation in exterior walls occur most often in the North where the average temperature for January is lower than 35° F.

Condensation Associated With Heat Radiation

Exposed surfaces cool at night by heat radiation. On a clear, still night the surface temperature of a building may drop 10° F. or more below that of the surrounding air. In the Southern States and the tropics, this leads to condensation on exterior surfaces. The condensate is absorbed by unpainted wood but collects as a fine film on painted surfaces; this commonly leads to surface molding and nonadhesion of oil paints on repainted surfaces.

Although not completely verified, condensation induced by heat radiation probably explains mold on the underside of thin roofs, as in carports, stoops, or exposed unboxed eaves. In some localities, condensation is heavy on screens, streaking

the wire and wood below or soaking into the frames as it runs down.

Condensation on Ground-line Slab Foundations

In humid climates, condensation often occurs on slab foundations. Cooled by the supporting ground, the slab acquires a dewpoint temperature. The same condition can occur in a building artificially humidified. Because condensation may be conspicuous at times, it has been suspected of dangerously wetting wood resting on the slab. Evidence does not support this.

In northern climates the periphery of the slab may cool below the dewpoint temperature during the winter. This presumably can induce condensation wetting of basal plates of the outer walls, but here, also, evidence is lacking.

Wetting By Piped Water

To a minor extent certain components of a building get wet enough to decay because plumbing leaks are neglected or tapwater is used carelessly. Fortunately, most plumbing leaks are found and corrected before serious damage results. An exception is the economically constructed shower stall that receives heavy usage-- particularly the type in many military barracks. Observations of decay in wood framing and sheathing near the showers--particularly in the wood under the stalls--suggest unfamiliarity with the construction needed to maintain a watertight lining. Related to this type of hazard has been minor decay commonly occurring around tubs, kitchen sinks, toilets, and washtubs from water seepage into adjoining wood. Occasionally, a small water pipe leak will not be found until appreciable decay has occurred.

In lawn sprinkling, frequent, heavy wetting of siding can lead to a moisture problem. This is a possibility mainly in arid and semiarid areas where sprinkling is the main source of lawn moisture. Wetting sufficient to support decay of sheathing and framing can occur through a stucco facing.

Frequent and excessive washing of wood floors by hosing or mopping can lead to sufficient moisture accumulation to support decay; this has been observed in kitchens, school gymnasiums, and military drill halls.

Miscellaneous Wetting

Overflow from cooling towers for air conditioning sometimes results in excessive wetting of a wall. On a sloping roof, water may flow continuously over the eave, or the mist from the condenser may be blown against the wall.

An additional factor promoting decay is the metabolic water produced by the fungus itself as it breaks down the wood. This metabolic water weighs about half as much as the destroyed wood; therefore, it may be sufficient to support decay in poorly ventilated space despite slow losses by evaporation. In some crawl spaces, cold-weather condensation further retards drying of the wood.

PROTECTING FOUNDATION AND SUBSTRUCTURE WOOD

The foundation can do much to protect a wood building from decay organisms because it separates the wood from the ground. If properly constructed, the foundation bars moisture movement from the ground into the wood substructure. The kind of foundation--crawl space, slab, or basement--determines the type of protection needed.

Foundations With Crawl Space

A building set over a crawl space usually rests on a perimeter foundation wall; in addition, interior supports or piers may be used or, for large buildings, supplementary walls.

Foundation Materials

The perimeter foundation usually is of concrete or of brick or stone masonry. The poured concrete wall is less likely than the more loosely constructed masonry wall to have internal openings for the water-conducting decay fungus to pass through unnoticed. Masonry foundations and piers of hollow block or brick can be greatly improved by capping them with a minimum of 4 inches of reinforced poured concrete. A waterproof membrane such as polyethylene on top of a masonry foundation resting on damp ground has been reported to prevent capillary movement of water into wood resting on the wall.

Post supports are constructed of concrete, masonry, or wood. If of untreated wood, they should stand on well-elevated concrete footings. For this purpose precast footings are widely available. Pressure-treated wood pile supports below grade are, of course, used under large buildings in many areas.

There is a trend to use wood as the major foundation component, with the wood resting directly on the ground. Basic types of wood foundation are pole, post-and-beam, joists-and-header, and plywood-frame (box beam). Wood foundations--particularly if expensive to replace--should be protected by the best preservative treatment that can practicably be used. The expense and effort that is justified to provide superior treatment will depend on the cost of the structure and the service life expected of the foundation.

Wood or wood-product forms should not be left on concrete foundation walls or posts. Forms make a bridge between the ground and the wood above it (fig. 11), creating easy access to the wood by water-conducting fungi.

Figure 11.--Wood forms left on a concrete pillar create a bridge between ground and substructure wood, making the wood vulnerable to attack by a water-conducting fungus.

Height of Foundation

The top of the foundation should be at least 8 inches above the finished grade and have at least 6 inches of foundation exposed below the siding in areas subject to some decay (fig. 12). A foundation height above grade of 12 to 18 inches is advocated for areas where frequent hard rains can cause significant splash wetting of wood siding, sheathing ends, or sills. In the crawl space at least 18 inches between joists and ground and 12 inches between ground and girders should be allowed. The greater interior spacing between ground and wood, as compared with the exterior spacing, allows for crawl-space inspection. The need for ample crawl space cannot be too strongly emphasized--a substructure that is difficult to reach may not receive adequate inspection.

Dry, Uninfected Lumber

The importance of using dry, uninfected lumber in the crawl space has been emphasized. Partially dry lumber in the substructure may not only be wet enough to support fungus growth, but may already be infected by a decay fungus. If moist, infected lumber is used and enclosed so it cannot dry readily, it is likely to decay rapidly--assuming, of course, temperatures also are favorable. Decay incurred in this manner is not prevalent, but when it does occur it can be costly because the affected wood usually is not readily accessible.

Precautions Against Condensation in the Crawl Space

Sufficient condensation to promote decay of sills, headers, joists, and subflooring may result from winter condensation or that associated with air conditioning by refrigeration. Winter condensation and most of that associated with air conditioning can be prevented by keeping the crawl space dry, with soil drainage, with ventilation, or with a soil cover. The method needed varies with the type of condensation. Where drying is impractical, the use of treated wood is necessary.

Soil drainage.--If the surface soil in the crawl space can be kept dusty dry by good drainage, the danger of condensation is small even if ventilation is substandard.

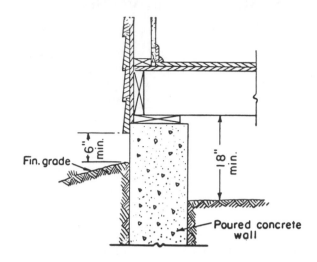

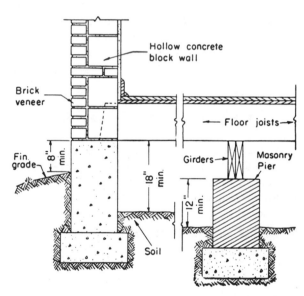

Figure 12.--Generally the distances shown are adequate spacing between ground and wood in substructure and siding; greater outside clearance is recommended for high-rainfall areas. Similar spacing is recommended if the foundation is a concrete slab.

Ventilation.--Good ventilation of the crawl space is a safeguard against damage by decay fungi. Vent openings in the perimeter foundation should provide cross movement of air if they are placed uniformly around the perimeter and as near corners as possible without reducing the strength of the wall. Near-corner positioning is desirable because dead air is most common there. The total effective opening of the vents should be proportional to the size of the space; openings totaling 1/160 of the ground area are adequate. Accordingly, a crawl space of 900 square feet would require a total vent area of about 5.6 square feet.

Screening reduces passage of air through vents by 25 percent or more. To compensate for this the screened vent area should be about a third larger than the 1/160. Also, if vents are below grade level, a somewhat larger size opening is needed. Shrubbery placed in front of a vent can materially reduce its effectiveness.

Vent areas meeting the above standards are effective against winter condensation but not necessarily against condensation under air-conditioned buildings in high-humidity coastal areas (see "Protecting Special Rooms"). Also, since vents often are closed in cold weather, ventilation alone seldom can be relied on as a sole means of condensation control in cold climates. For this reason ground covers were developed.

Ground covers.--A good ground cover will effectively limit condensation if vents are closed during winter. A cover should be considered particularly where winter temperatures commonly are 50° F or below and ground surfaces are prevalently damp, or where there is air conditioning over a damp crawl space.

A good ground cover (fig. 13) will generally keep the substructure wood dry despite damp ground. If a cover is used, the vent opening area can be substantially less than has been recommended. A cover should not only have adequate vapor resistance but also sufficient strength to permit some traffic on it, because it may be necessary to crawl on the cover while making inspections or repairing plumbing, heating ducts, and electrical wiring. The crawl space also may be used to store certain materials, which place an additional burden on the cover.

Materials that have both vapor resistance and strength, yet are not costly, are 45-pound or heavier roll roofing and 6-mil polyethylene sheeting. A 4-mil polyethylene film often is used with a layer of sand or gravel on top to reduce the chance of physical damage.

The strips of ground cover should be overlapped slightly and the outer edges carried to the foundation wall. Special measures to limit escape of water vapor at the seams or edges are not necessary. The ground does not have to be perfectly flat, for in time the cover will conform to moderate surface irregularities.

Figure 13.--Polyethylene sheeting, preferably 6 mil thick, or roll roofing, 45 pound or heavier, and certain durable duplex vapor-barrier materials provide excellent ground cover on damp soil in crawl spaces. A cover reduces the hazard of winter condensation (fig. 9) and of summer condensation in flooring of air-conditioned rooms above.

Concrete Slab Foundations

Slab Construction

In some areas the concrete slab-on-ground foundation has wide acceptance. Because it cannot be inspected beneath, the slab should be made as crackproof as practicable, and necessary openings through it should ultimately be tightly closed. Cracking can be minimized by good reinforcement of the slab. Joints can be avoided by making the slab a monolithic type in which the floor and footing are poured in a single operation. Joints and openings made for plumbing and conduits should be filled with high-quality coal-tar pitch or coal-tar plastic cement.

Elevation of Slab

The top of the slab foundation should be at least 8 inches above the finish grade, and wood or wood-product siding should come no closer to the grade than 6 inches, as recommended for the perimeter wall of the crawl-space foundation. Where considerable rain splashing is likely, as in the Gulf States, a greater clearance is desirable; clearances of 18 inches are not uncommon.

Vapor Barrier Under the Slab

A membranous vapor barrier should be placed under the slab if conventional construction practice is followed. A membrane with suitably high vapor resistance will ensure that ground moisture does not migrate through the slab and warp wood floors resting on the slab, degrade the flooring finish, or damage the adhesive where floors are bonded to the slab. There is still some doubt whether, in the absence of a vapor barrier, ground moisture will penetrate a conventional slab at a sufficient rate to support decay.

Membranes may be heavy asphalts or polyethylene films. Effective barriers can also be installed above the slab, but this limits the finish flooring to a type that can be installed on flush sleepers.

Slab Insulation

In the colder climates, it is desirable to install thermal insulation around the perimeter of the slab and beneath it. This not only keeps the floors warm, but also limits condensation on the slab and wetting of wood in contact with the slab. The exact location of the insulation will depend on how the slab is constructed. The insulation for floor slabs should be nonabsorbent; have high resistance to heat transmission; have resistance to breakdown by moisture, micro-organisms, or insects; and have mechanical strength to withstand superimposed loads or expansion forces. Cellular glass, bonded glass fibers, and insulating concrete are available for the purpose. A thick gravel fill under the slab helps to reduce heat loss.

Because condensation may be conspicuous at times, it has been suspected of dangerously wetting wood resting on slabs. Evidence does not support this; nevertheless, because of the load-bearing importance and inaccessibility of substructure components resting on the slab, it is considered a wise precaution to pressure treat them. As just noted, the tendency for condensation can be substantially reduced by insulating the perimeter and the ground side of the slab.

One should not attempt to prevent condensation on a slab by placing a vapor barrier on top of the wood. No ordinary barrier will entirely prevent condensation on the slab and what does occur will be prevented from evaporating by the overlying barrier. In addition, a moisture barrier so placed can be doubly dangerous if any significant amount of moisture comes through the slab from below.

Basement Foundations

Wood in the substructure is easiest to protect from decay organisms if the building is set on a basement foundation. The basement permits frequent inspection beneath the building. Also, the ordinary basement has a concrete floor and walls; thus there is no damp soil to cause condensation on substructure wood as in crawl spaces. However, damaging amounts of water can enter through cracks or joints by hydrostatic pressure. Therefore, adequate waterproofing of basement walls is essential, particularly on wet sites.

Any wood bearing posts in the basement should be elevated above the floor on footings to keep the lower end dry (fig. 14). The basement wall should extend above the exterior grade level at least 8 inches with at least 6 inches of the exterior surface exposed, and precautions of construction should be as noted for crawl-space walls.

Porch Foundations

The same precautions should be observed for a porch foundation as for the main foundation. If the porch has a wood floor, provision should be made for ventilation and inspection beneath it, and the height of the foundation above grade should be at least 8 inches. The bottom of platform joists should be high enough above the ground to permit inspection from beneath. Whether the porch floor is of wood or concrete, it should slope away from the building to allow rainwater to drain.

Porches constructed by pouring a slab platform on a dirt fill can create a special kind of decay hazard. The proximity of the dirt fill to the building substructure violates the guiding principle of keeping wood and soil well separated. Unless an appropriate barrier is placed between the porch and the building, fungi, especially the water-conducting type, can move into the building sub-structure along the line of juncture.

POOR PRACTICE

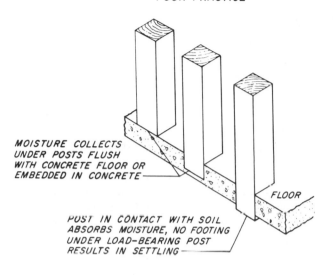

MOISTURE COLLECTS
UNDER POSTS FLUSH
WITH CONCRETE FLOOR OR
EMBEDDED IN CONCRETE —

FLOOR

POST IN CONTACT WITH SOIL
ABSORBS MOISTURE, NO FOOTING
UNDER LOAD-BEARING POST
RESULTS IN SETTLING —

GOOD PRACTICE

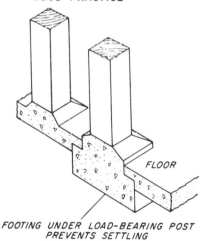

FLOOR

FOOTING UNDER LOAD-BEARING POST
PREVENTS SETTLING

Figure 14.--Poor (top) and good (bottom) methods for installing wood posts on a concrete floor at ground level.

The dirt-filled porch can be made reasonably safe by building it independent of the house at all points and providing a 2- or 3-inch airspace that can be covered at the top. This necessitates adding a porch foundation at the end toward the house; it is well worth the extra cost. An alternative is to pour the porch slab and building foundation as a monolithic concrete structure without joints. Also, a metal shield can be used to protect the sill area but care is needed to ensure adequate termite protection (see Additional Reading at the end of this report). With a slab-on-ground building no danger exists if the porch slab is at a lower level than the building slab.

The above precautions also apply to terraces, steps, and other horizontal surfaces adjacent to the building foundation.

For the safest type of porch, step, or terrace, the dirt fill is avoided and a self-supporting slab is used for the platform. Provision must be made for removal of any wood forms under the slab.

Preservative Treatment of Wood in Foundation and Substructure

Foundation wood, or substructure wood that is in direct contact with the ground or laid on a moisture-resistant ground cover, should be given the best practicable preservative treatment, especially if it is intended to serve for the life of a permanent building. The preservative type and amount should meet the highest standards prescribed for land service by Federal Specification TT-W-571[5], or by standard LP-22[6] except that the retention shall be 0.60 pounds per cubic foot.

A lesser quality of treatment would, of course, be appropriate for a temporary building, and it might also be justified in some situations where the item could be replaced without high labor costs.

Generally, wood items in or on a slab foundation or on other concrete laid on the ground should be pressure treated unless experience indicates that less complete protection will adequately serve. Items in contact with an elevated foundation wall ordinarily do not require preservative treatment, but if they are closer than 8 inches to the ground, and in a warm damp climate, pressure treatment is suggested.

[5] Federal Supply Service, General Services Administration. Federal Specification TT-W-571, Wood Preservation, Treating Practices.

[6] American Wood Preservers Institute. LP-22, Standard for Softwood Lumber, Timber, and Plywood with Water-Borne Preservatives for Ground Contact. McLean, Va. 22101.

If part of the crawl space is used as a plenum to conduct air for heating or cooling, wood used in the substructure should be treated with one of the waterborne chemicals and not with an oil. Oil fumes could contaminate the air within the building.

PROTECTING EXTERIOR WALLS, BUILDING APPENDAGES, AND ASSOCIATED WOODWORK

Although in individual cases the most damaging and costly decay can occur in load-bearing components of the substructure, in total the most prevalent and costly decay occurs in exterior items such as windows, doors, porches, steps, railings, walls, and associated woodwork. The greater prevalence of decay in these exterior parts is chiefly caused by exposure to wetting from rain. Less common, but significant, sources of wetting are condensation, use of wet lumber, careless sprinkling, ground contact, and ground splash.

General Protective Measures

Use Dry, Uninfected Lumber

Proper moisture content and freedom from infection can be obtained most reliably by using lumber that has been kiln-dried and kept under cover.

For best performance, siding and trim should have a moisture content when installed of 7 to 12 percent in the dry Southwest and of 9 to 14 percent in the remainder of the United States (table 2).

Table 2. –Moisture contents recommended for wood at time of installation

Use	Moisture content (based on ovendry weight)					
	West[1]		Coastal areas[2]		Other areas	
	Average	Range[3]	Average	Range[3]	Average	Range[3]
	Pct.	Pct.	Pct.	Pct.	Pct.	Pct.
Interior finish	6	4-9	11	8-13	8	5-10
Softwood flooring	6	4-9	11	8-13	8	5-10
Hardwood flooring	6	5-8	10	9-12	7	6-9
Siding, exterior trim, sheathing, framing	9	7-12	12	9-14	12	9-14

[1] Most of eastern California, Nevada, Utah, and western Arizona and contiguous parts of Oregon and Idaho.

[2] Mainly the warmer coastal regions: California south of the Bay area, and the South Atlantic south of Chesapeake Bay and across the Gulf States to Texas.

[3] Range is generally more important than average. If pieces fall within the range, condition will be satisfactory regardless of average.

Protect Stored Lumber

The importance of proper lumber storage cannot be overemphasized. Much of the rapid deterioration of wood in use can be traced to incipient fungus infections originating during air-drying or storage. Lumber and other wood items should be protected against wetting from the ground and from rain; this is necessary to prevent incipient fungus infections and to ensure that the lumber is maintained at the moisture content suitable for a particular use (table 2). Even for temporary storage, as at a building site, lumber should be stored off the ground.

At permanent storage areas, lumber is usually placed under a roof for protection; in transit or at a building site, it may be kept under tarpaulins or plastic covers. Finish items should not be delivered to a building site until the structure has a tight roof and walls that provide indoor storage.

Commonly available on the market is lumber, mainly framing and sheathing, that has been dip-treated with emulsions of water-repellents dyed red, yellow, purple, etc. The water repellents used do not meet Federal Specification TT-W-572[7] and contain no fungicide. These treatments are intended solely to reduce rainwetting during transit, storage at the building site, and the early stages of construction. However, they do not eliminate the need for protecting lumber from rain wetting and soil contact during storage. Where moderate protection of the completed structure is required, they should not be used in lieu of water-repellent treatments meeting Federal Specification TT-W-572.

Lumber at permanent storage sites should be placed on foundations of concrete or treated wood or in a dry shed with a dry concrete floor where it may safely be kept on untreated skids or pallets. If the soil has not been treated with an approved insecticide in areas with appreciable termite hazard, the foundations should be high enough to permit inspection below the piles.

Lumber with no more than 20 percent moisture can be bulk-piled under cover with safety. Wet lumber should be put in a ventilated space on dry stickers as in a regular air-drying pile.

When receiving lumber, do not assume that lumber supplied as "dry" is as dry as it should be. It is highly desirable to make sure with an electrical moisture meter; with its needlelike penetrating electrodes the instrument rapidly indicates the interior moisture content (fig. 15). Moisture meters that do not require penetrating electrodes also are available for use with finish items and plaster.

Figure 15.--Resistance-type moisture meters showing needlelike electrodes that are driven into wood to measure moisture content. Electrodes are commonly 3/8 inch long (left) but longer (right) are available for deeper penetration to determine moisture contents when searching for hazardous amounts of moisture in building members.

Provide Roof Overhang

The extent of roof overhang is the single most important design feature determining the amount of wetting of walls, windows, and doors in any climate. The addition of gutters reduces wetting from splash and wind-driven roof runoff. The protective value of roof overhang varies with the width of projections, amount of rainfall, duration of rains, and the amount of wind accompanying the rain. An overhang of at least 2 feet on a one-story building is recommended in areas of substantial rainfall.

Avoid Wetting By Splash

Splash damage occurs where roof runoff strikes

---

[7] Federal Supply Service, General Services Administration. Federal Specification TT-W-572, Wood Preservative; Water Repellent.

the ground or a lower roof and wets the wall (fig. 8).

The logical control of splash wetting is a wide roof overhang or an efficient eave gutter. Where splash is limited to a small area, such as a stoop, a roof-edge baffle can be used, although this is less satisfactory than a gutter. Also, splash can be reduced by making the foundation high enough to provide good separation between the ground and the wood.

Minimize Construction that Traps Water

The serious rain wetting of wood in buildings occurs primarily at joints. Therefore, in exterior construction, joints and other forms of contacting surfaces should be avoided where possible. Exterior bracing, for example, can be a water-trapping hazard, whereas interior bracing is safe (fig. 16). Similarly, an exterior step newel constructed of two 2 by 4's is more hazardous than one 4 by 4, or the use of subflooring for stoops is more hazardous than a single floor (fig. 16). Of course, when double construction is accomplished by laminating with a waterproof glue, a piece is essentially solid and without joints so long as the glueline is intact.

Tight joinery, paint, and calking are aids rather than primary means to prevent rain seepage. On structures exposed to heavy rainwash, they should not be exclusively relied on to prevent excessive seepage.

Where there is heavy rainwash or some other form of intermittent, heavy wetting of the surface, a good water repellent properly applied can aid considerably to keep joints from getting wet. Although it will not prevent gravity flow of water into loose joints or prevent penetration of rainwater when driven by unusual wind pressures, the water repellent nevertheless effectively interferes with capillary movement of water into reasonably tight joints. (See "Degree of Preservative Protection Needed.")

Design to Shed Water

Insofar as possible, building components should be designed to: (1) Divert rainwater from joints and other water-trapping places exposed to the weather, and (2) allow whatever water reaches critical surfaces to drain rather than stand long enough to soak into the wood.

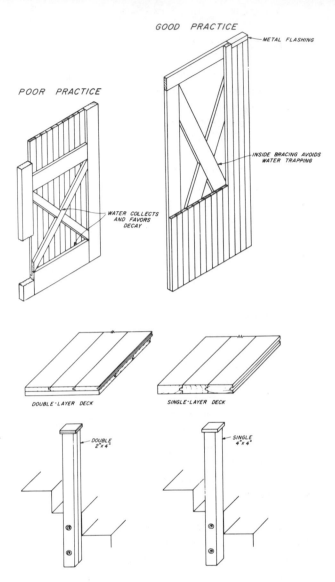

Figure 16.--Poor (left) and good (right) practices in construction to avoid trapping rainwater. Double thicknesses of lumber and exposed contacting surfaces-- especially those required by architectural frills--should be avoided as much as possible

Good examples of how the absence or presence of water-shedding construction can favor or deter accumulation of rainwater are found in the building of exterior boxed beams. When a roof beam is boxed, the soffit is sometimes extended beyond the exposed facia and rainwater collects on the ledge thus formed and seeps into the boxing. If the facia is extended down beyond the soffit, this is prevented (fig. 17).

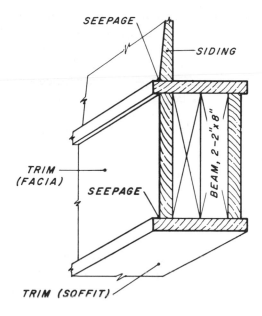

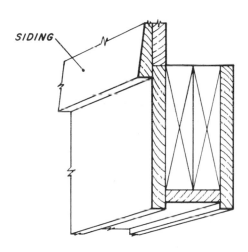

Figure 17.--Boxed beam (top) has horizontal ledge that traps rainwater. Boxed beam (bottom) illustrates good construction; boxed similarly to the standard eave, it is less subject to rain seepage because it does not have the rain-trapping ledge.

For minimum maintenance, roofs should be sloped sufficiently to allow rainwater to run off rapidly. The number, location, and size of drains and the needed roof and valley pitch vary with climatic areas and particularly with the amount of rainfall. Long, low-pitched valleys on large buildings are particularly hazardous. Interior drains, strategically located, often are desirable. Roof design should prevent runoff from striking an exterior wall or another roof below. This is accomplished by providing for adequate overhang, gutters, or baffles--according to circumstances.

Flat roofs must be especially well protected against leakage to offset their poor drainage features; hence they need regular, careful inspections coupled with timely maintenance.

Components, such as step treads, porch or stoop flooring, and window sills, all with relatively broad, flat surfaces exposed to weather, will wet easily and be susceptible to decay if the upper surface is not sloped to facilitate drainage. Flat-sawed lumber tends to shrink and swell more than quartersawed (edge grained) and this disadvantage is particularly pronounced in wide boards. Consequently, flatsawed step treads and flooring are more apt to cup and hold water; flatsawed siding and trim are more apt to split, warp, or twist after attachment and break paint films or crack calking. Most of this trouble is in proportion to the severity of exposure. Much lumber is flat-sawed, but quartersawed material is available if specified.

Unnecessary horizontal projections from the building wall, such as water tables and lookouts, frequently become decayed and should be avoided in the wetter climates unless treated with preservative or made of a naturally decay-resistant wood.

Install Flashing Where Special Safeguarding is Needed

Noncorrosive metal or durable waterproof felts are used as flashing to prevent rainwater from entering critical junctures of exterior components of buildings. These include joints where siding joins the roof (dormers, porches, and canopies), the top of window and door trim where unprotected by sufficient roof overhang, the juncture of siding and a concrete porch slab, horizontal joints between panel siding, roof valleys, and at the roof edge--between roof cover (shingles) and sheathing. Generally, flashing should be used wherever a horizontal projection from the wall occurs in an exposed location (fig. 18).

In applying flashing, nails should not be exposed, except in unusual cases where the nailing is through a surface that will not be wetted by rain. Nailing through the exposed vertical part of roof-edge flashing can lead to serious leakage. Roof-edge flashing should be attached below the underlay paper. This is particularly necessary in a tile

357

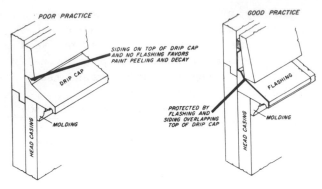

Figure 18.--Joining of siding to a drip cap; note how the "poor practice" (left) allows for entry of rainwater into several decay-vulnerable joints, whereas "good practice" (right) provides protection by flashing and by siding overlapping top of drip cap.

roof in a tropical area because during heavy rainfall water flows over the tile cup edge and drains over the underlay paper to the roof edge.

Flashing is sometimes used to protect exterior load-bearing items such as projecting roof beams and laminated wood arches. If this style of construction is wanted, it is much safer to use adequately treated wood rather than to depend on the more temporary protection offered by flashing.

Treat With Water-Repellent Preservative

Exterior woodwork is often made additionally resistant to rain wetting and decay by dip treating the finished or knocked-down item in an oil solution of a water-repellent preservative. Some items commonly protected by this treatment are screens, sash, shutters, doors, siding, and trim. High-hazard items, such as porch rails and outdoor steps, may need pressure treatment in areas most conducive to decay (fig. 6). Limited protection to siding, trim, and other items under moderate-to-light wetting exposures can be attained in all areas by flooding with brush or spray after the wood is on the building, with particular attention given to joints.

Water-repellent treatments are discussed in "Preservative Treatments for Building Lumber."

Install Vapor Barrier for Walls in Cold Climate

In areas with long, cold winters (fig. 10), con-

densation that results chiefly in paint failure may occur in walls. This happens particularly where artificial humidification is used without appropriate precautions and in small homes in which humidity is high because of crowded occupancy.

This type of condensation can be prevented by using a vapor barrier (table 3) near the inner surface (warm side) of the wall; by avoiding excessive artificial humidification of the living quarters; and by improving--when needed--ventilation of small living quarters.

Only two situations warrant placing the moisture barrier elsewhere than on the inside of the wall. In warm humid climates when severe air conditioning by artificial cooling is used for long periods or in cold-storage rooms with refrigerant cooling, the vapor gradient is reversed, and the barrier should be near the outer face of the wall.

Protecting Siding

Wetting of siding tends to be troublesome mainly by causing paint problems, but a large amount may nevertheless promote decay of the more susceptible woods. Similarly, wetting can induce mold discoloration of paint and of unpainted surfaces. Protection of siding is not difficult.

Good Construction for Siding

Lumber siding.--If lumber siding is vertically alined (boards and battens or boards with interlapping joints), no special construction details are necessary to protect it.

If lumber siding is horizontally alined, as is most common, certain helpful precautions can be taken. Added protection can be given to drop siding, where appearance permits, by applying trim over the ends of the siding. Standard drop siding is less subject to rain seepage than such patterns as tongue-and-groove with a beveled lower edge. With bevel siding, the conventional metal cover applied over the corners of each siding course gives good protection against rain seepage. Bevel siding with smooth back surfaces wets much less from rain seepage than does rough-backed siding. This is particularly true of redwood and cedar siding. These additional protective measures are of particular value in high-rainfall areas.

If narrow plywood or hardboard is used to simulate horizontal lumber siding, the protective needs are essentially those of the lumber siding. The material must be strictly of exterior grade.

Where siding abuts a roof, as on a dormer, it is advisable to leave a 2-inch clearance between siding and shingles in high-rainfall areas and 1 inch elsewhere, thus exposing 1 to 2 inches of flashing. This prevents roof runoff from contacting the end grain of the siding. Also, it is a good practice to leave a minimum of 6 to 12 inches of clearance between siding and gradeline (depending on climate) and 1 inch between siding and slabs of porches. These suggestions also apply to shingle and panel siding.

Except for cold storage rooms and occasionally air-conditioned quarters, only the breathing type of sheathing paper (table 3) should be used under wood siding.

Shingle siding.--Most shingles are of naturally decay-resistant wood. Consequently, decay problems are few if good building practices are followed. With a single coursing, the overlap must be sufficient to ensure a continuous double layer to prevent water from seeping behind the shingles. Any nailing strips should be of decay-resistant or preservative-treated wood.

In applying asbestos-cement shingles, special attention should be given to proper application of felt backer strips at all vertical joints, to proper flashing at corners and trim, and to calking at juncture of siding and trim.

Panel siding.--For siding of plywood or of hardboard panels, only exterior-grade material is acceptable. The glue bond between veneers or between wood particles in interior-grade material will not withstand wetting. The following procedures are recommended for installing wood panel siding:

1. After the panels are in place, flood the joints with double-strength preservative (10 percent pentachlorophenol plus water repellents). A squirt-type oil can or small pressure-type sprayer can be used to advantage.

2. Prime the joints with a good quality paint primer of white lead in linseed oil, with no zinc pigment (Federal Specification TT-P-25[8]).

3. Protect the horizontal joints with flashing commercially available for the purpose. Do not use horizontal batten strips.

4. Protect the vertical joints by filling with the best grade of weather-resisting mastic or calking compound (allowing space in the joint) or by covering the joints with batten strips of decay-

Table 3. - Principal uses of building papers and specifications

Vapor barrier	Use	Federal specification	FHA Conformance (according to "FHA Minimum Property Standards one and two living units," May 1963)	Commerical products not covered in specifications
Heavy duty[1]	Refrigerators, ground covers, vapor barriers under concrete slabs, swimming pools, shower rooms, garbage rooms	UU-B-790a[2], Type I, Grade A 45-lb. or heavier smooth surface roll roofing (polyethylene membrane[3])	Passes FHA "Test Procedure for Vapor Barrier Materials under Concrete Slabs and for Ground Cover in Crawl Spaces," (Sept. 20, 1957)	Various existing and new products being developed to meet FHA requirements
Ordinary service	Side walls and under attics or roofs	UU-B-790a[2], Type I, Grade A	Transmission rate not exceeding 1 perm	Foils mounted on paper or on back of sheet rock. Mylar 2-mil polyethylene
Sheathing paper	Under wood, shingle, asbestos stucco siding[4]	HH-R-595[5], Type I or II	Saturated felt. Complies with ASTM D-226 or F.S. HH-R-595[5]	Tarred or asphalt felt, 15 lb. per square
Floor liner	Wood floors frequently scrubbed	UU-B-790a[2], Type I, Grade D		
	Wood floors infrequently scrubbed	UU-B-790a[2], any class, any type; HH-R-595[5], any type	No. 15 asphalt-saturated felt	Building paper, deadening effect
Roll roofing	Roofing to temporary buildings	....		Roll roofing available in weights of 45, 55, 65 lb. or more per square (108 per sq. ft.)
Roof underlayment	Waterproof membranes used under slats, tile, or asphalt shingles	UU-B-790a[2], Type I, Grade C HH-R-595[5], Type I or II	Saturated felt. Complies with ASTM D-226 or F.S. HH-R-595[5]	....

[1] Represents material having a perm value of 0.25 or less and sufficiently rugged to withstand some abuse in service.

[2] Building paper, vegetable fiber (Kraft, waterproofed, water repellent, and fire resistant).

[3] Also Commercial Standard CS 238--6 mil polyethylene membrane (construction, industrial, and agricultural applications).

[4] 30-lb. felt is sometimes used under some kinds of siding. Because this material is much more resistant to movement of water vapor than 15-lb. felt, vapor barriers used in combination with sheathing paper consisting of 30-lb. felt should have a perm value of 0.25 or less.

[5] Roofing felt (coal-tar and asphalt-saturated felt rolls; for use in roofing and waterproofing).

[8] Federal Supply Service, General Services Administration. Federal Specification TT-P-25, Primer Paint, Exterior (Undercoat for Wood, Ready-Mixed, White and Tints). 1965.

resistant wood. Both calking and battens may be needed in the wettest climates.

5. Carry the bottom edge of the lowest panel down over the supporting framing to form a drip edge.

Treating Siding With Water-Repellent Preservative

Treatment of board siding with a water-repellent preservative can be beneficial in the wetter climates. The treatment minimizes entry of rainwater behind the siding, thereby limiting damage from decay and sap stain and reducing the tendency toward paint blistering (fig. 19). This is especially pertinent for two-story houses and one-story houses with a roof overhang of less than 2 feet, particularly if there are no gutters.

Treating ordinarily cannot be justified solely for protection against decay and stain if the siding is of naturally resistant woods, such as the heart-wood of western redcedar or redwood. However, a good water-repellent preservative protects siding of all species against paint blistering. Because blistering is a major problem of paint mainten-ance, the use of a water-repellent preservative is advisable wherever the rainfall and type of roof construction subject the siding to considerable wetting.

Treating relatively unsheltered siding will pro-vide worthwhile protection in all tropical regions and in areas within the United States with long warm, wet periods. The value of this protection will be greater where the rainfall is usually in pro-longed periods rather than in short, heavy showers, and where winds often are sufficient to blow the rain against the siding.

In treating new pine siding before it is installed, a 3-minute dip or an equivalent vacuum treatment in a water-repellent preservative is recom-mended. The ends of the boards cut on the job should be retreated by dipping or by liberal brush-ing with the preservative solution. The siding may also be spray treated before it is installed or after it is in place, but spraying is less effective than dipping before installation, especially for protec-tion against decay. Treating in place involves ap-plying the preservative solution generously around board ends at windows, doors, and corners, and to the lap joints between the boards.

Figure 19.--Condition of siding after various treatments, 5 years after repainting: Top, Unsatisfactory. Spray treated with a water-repellent preservative over the old paint. Center, Unsatisfactory. Old paint removed but not followed by spray treatment. Bottom, Satisfactory. Spray treated after first removing old paint.

Cedar and redwood siding seem to be adequately treated by applying the preservative solution to the outer face or even just to the edges of the siding. Treating of siding of these woods is largely by brush application on the construction site.

Plywood panel siding ideally should be treated in accordance with Commercial Standard 262.[9] This treatment presumably would be given by the manufacturer. If plywood siding is destined for the tropics, it should be pressure treated.

[9] U.S. Department of Commerce. Water Repellent Non-Pressure Treatment for Millwork. Commercial Standard 262-62.

360

Protecting Roof Edges

The roof edge is especially subject to decay because it is in the open and exposed both to the direct wetting from rain and to wetting from roof runoff (fig. 8). Because it has one of the highest potentials for decay of any building part, it requires careful attention both in designing and during maintenance.

Roof runoff, particularly with asphalt shingles, tends to travel around the shingle butts and wet the various components of the roof edge. Corroded, undersized, occluded, or sagging eave gutters can lead to overflow at the back edge of the gutters and to wood wetting. Heavy rains sometimes cause water to overflow the joint channel in roof tiles and flow over the underlayment. When this occurs, the roof edge is subject to more wetting than it is from surface flow over the protruding tile edge.

Serious decay frequently occurs in (1) facia, especially at joints between two facia boards; (2) molding at the roof edge; and (3) rafter ends and sheathing edge, particularly when no facia is used. In inexpensive or essentially temporary buildings, roll roofing is brought down over the roof edge and nailed to facia, either with or without a nailing strip. The nailing strip is subject to direct wetting by water running off the roof; eventually water penetrates the roofing at nail punctures, leading to decay of the underlying wood.

Unless specially protected, exposed rafter and beam ends are vulnerable to decay in wet climates. If decay of facia or sheathing edges progresses unchecked, it eventually will involve rafter ends, and greatly increase the cost of repairs.

Construction to Protect Roof Edges

The following five construction features are recommended to protect roof edges against rain wetting:

1. Extend shingles, tile, slate, or metal roof coverings at least 1 inch beyond any wood at the eave and rake edges.

The gable end, or rake, can be additionally protected by laying under the shingles a cant strip of bevel siding 1/2 by 6 or 8 inches, thin edge inward, along the gable edge, to guide the water away from the edge.

2. Use corrosion-resistant metal flashing on the eave and rake edges. With roofs covered with gravel or other aggregate, a gravel stop serves as the flashing. With other types of roofs, an L-shaped flashing is used (fig. 20). Also at the eave, place the underlay over the top of the flashing. Do not nail through the exposed vertical part of the flashing. Let the flashing extend away from the building sufficiently to provide a free drip edge. Bending the bottom of the flashing out from the facia is not sufficient for this purpose; insertion of a 1-inch strip between the flashing and the facia provides better protection (fig. 20). Occasionally the flashing is brought down to cover the facia completely.

3. In northerly areas where ice dams cause trouble, install flashing to protect against water backed up by the dams. Lay smooth-surface 55-pound roll roofing on the roof sheathing from the eave upward 6 inches beyond the inside face of the stud line (fig. 21).

4. Particularly if the roof is flat, have the soffit area appropriately ventilated (fig. 22).

5. Avoid nailing roll roofing to the facia (either with or without an exposed nailing strip) unless the facia and nailing strip are of naturally decay-resistant or preservative-impregnated wood. The exposed edges of roll roofing should be secured, if possible, with a suitable adhesive rather than with nails.

Preservative Treatment for Roof Edges

Facia, nailing strips on roll roofing, and exposed molding can be given maximum protection by deep preservative treatment. Two treatments that will leave the wood paintable are: (1) Impregnation with pentachlorophenol (penta) in a volatile petroleum solvent and (2) impregnation with a waterborne salt. In areas of moderate rainfall, naturally decay-resistant wood or a 3-minute dip in a water-repellent preservative usually protects these items.

For exposed rafter ends in the wet climates, a minimum treatment would consist of dipping the ends or liberally brushing or spraying them with a water-repellent preservative (if pentachlorophenol, a penta concentration of 10 percent, or a double application of the conventional 5 percent solution, is desirable).

361

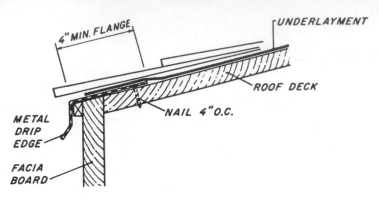

METAL DRIP EDGE – EAVES SECTION

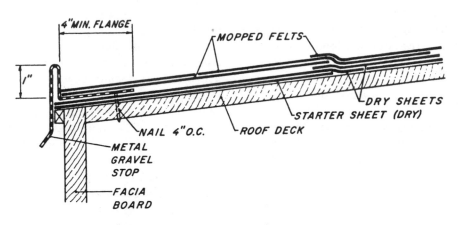

GRAVEL STOP AT EAVES

Figure 20.--Appropriately designed metal flashing (top) and similar gravel stop (bottom) provide good protection to edges of built-up roofs. The turned-out lip of the flashing diverts much of the rainwater from the fascia and wall.

Protecting Building Appendages

Of all exterior components of buildings, appendages are most subject to maximum wetting from rain and to decay. Most decay in outside steps and stairs, porches, platforms, and similar appendages results from rain wetting. Soil contact also can lead to even more hazardous wetting of carriages and supports of steps and outside stairs, and except in the driest climates should be strictly avoided unless treated with preservative.

Construction for Protection of Appendages

In the wetter climates (fig. 6) design alone can do little to prevent decay in appendages of wood buildings. In other climates, however, proper design can materially lengthen service life. Basic features consist of providing cover by a very wide eave, sloping the porch or similar deck to drain off rainwater, and keeping the construction simple to minimize joints and contacting surfaces in and between which water can be trapped (fig. 16). Depending on the wetness of the climate, these precautions can be supplemented by using treated or naturally decay-resistant woods.

Several procedures can minimize trouble from rain wetting of porches, stoops, platforms, and other flat decks exposed to the weather. Perhaps most important is to avoid double flooring (fig. 16). Others are to: Calk the joints in the flooring with white lead in oil if naturally resistant or treated wood is not used; promote drainage away from

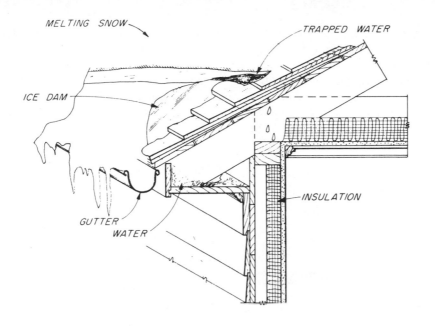

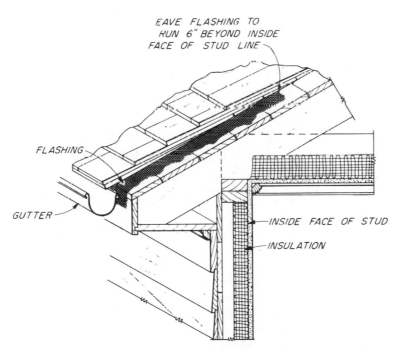

Figure 21.--Top, How exterior walls can be wetted by melted snow trapped behind an ice dam. Bottom, How flashing prevents wetting of exterior walls from snow water.

the building; and provide drain channels in the bottom rail of screens to keep the rail and window ledge dry.

As little molding or other trim as possible should be used. Railings should have a simple design with few joints into which water can seep, and components should be so joined that they can be easily replaced if decay does occur. In boxing the roof beam across the front of a porch, the facia board should extend below the soffit (fig. 17); thus there will be no horizontal ledge to hold water. Step rails should be placed over the top of the newel--not abutted to its side. An eave gutter should be provided to prevent roof runoff from striking porches or other decks.

For porches a self-supporting concrete slab is much preferable to one on a dirt fill. If a dirt fill is used, the porch should be separated from the building or it should be constructed as described in "Protecting Foundation and Substructure Wood." All forms used in pouring concrete should be removed.

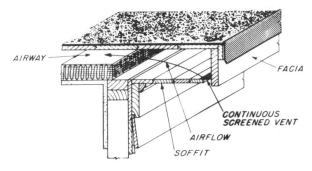

Figure 22.--Good roof-edge construction for a flat roof with a soffit at the eave. The soffit is continuously vented, and airflow is increased by an airway beneath the roof.

Preservative Treatment of Appendages

Because wood steps, stairs, porches, and other appendages cannot be adequately protected in a wet climate by design alone, they will have a longer service life if they are of naturally decay-resistant wood or are treated with a preservative. Framing or other items to be left unpainted can be pressure treated with any preservative covered by Federal Specification TT-W-571[5] that will be suitably free of discoloration or odor; for items that must be essentially free of oil for painting, only the waterborne salts or pentachlorophenol in volatile petroleum solvent should be used. These preservatives also are safest where there is possibility of discoloring flooring or other wood by oil creep along nails.

Water-repellent as well as fungicidal protection is often desirable. Suitable water repellents cannot be incorporated in a water solution to be applied by pressure, and if paintability is desired they should not be in an oil solution to be applied by pressure. But they can be added later by dipping or by spraying.

A short-soak (dip) treatment of 3 to 15 minutes in a mixture of pentachlorophenol and water re-pellent in light oil, such as mineral spirits (Commercial Standard CS 262[9]), can add a great deal to the service life of many appendages (fig. 23). Woods with moderate natural decay resistance, like Douglas-fir, will give especially good service after dip treatment in a water-repellent preservative.

Protecting Exposed Load-Bearing Members

For architectural effects or for ease of construction, load-bearing members are sometimes exposed to wetting from rain (fig. 24). Laminated or solid arches in churches and gymnasiums are extended beyond the roof edge to a concrete abutment; purlins or roof beams are extended beyond walls; rafters are extended beyond roof decking; and sometimes the sides of heavy edge rafters or beams, particulary in flat beam-and-plank roofs, serve as facia and are exposed to the weather and subject to a high seepage and decay hazard. Protruding parts of arches have been destroyed by decay within 4 years in warm, moist coastal areas.

Construction to Protect Load-Bearing Members

If load-bearing members, such as arches, rafters, and roof beams, must be exposed to appreciable wetting, they should be pressure treated. Metal caps or flashing will give considerable protection but should not be depended on to give long-time protection in high-rainfall areas. Metal sockets in which wood arches or columns rest should have a drain hole at the lowest point to prevent their serving as a reservoir for water.

Preservative Treatment of Exposed Load-Bearing Members

Only pressure-treated wood should be used for load-bearing members exposed to considerable wetting. Because of the relatively high cost and the load-supporting requirements, exposed arches should <u>always</u> be given a high-quality treatment. The preservative should be chosen for appearance, for paintability, and, in glued laminates, for its effect on gluing. For laminated wood, the fabricating company should be consulted for chemicals acceptable for treatment.

Exposed load-bearing items other than arches may be benefited in areas of low-to-moderate rainfall--and in wet areas if considerable protection is also afforded by roof overhang--by brushing heavily or by soaking with a water-repellent preservative. This treatment would precede any installing of a metal cap. Penta-grease treatment (see "Preservative Treatments for Building Lumber") should be especially effective on the end grain of exposed members.

Figure 23.--Contrast between dip-treated steps (top) and untreated (bottom) after 5 years of exposure near Gulfport, Miss. On-site dip treatment consisted of immersing the precut lumber 3 minutes in a light oil solution of 5 percent pentachlorophenol and water repellent.

Figure 24.--Load-bearing wood members exposed to weather: Top, Laminated arches. Bottom, Exposed rafters. Exposed load-bearing wood in a permanent building should be pressure treated.

SPECIAL ROOMS PROTECTING

Shower, cold-storage, laundry, and air-conditioned rooms, kitchens, and enclosed swimming pool areas can have special moisture and decay problems. Decay in these areas, though small overall, is occasionally costly; the affected wood ordinarily is not visible, hence damage may not be apparent until a wood member fails. Fortunately, there are usually indications of wetting before any significant decay occurs, and most difficulties can be easily avoided by good design.

Shower Rooms

Sources of moisture that promote decay in shower rooms are: (1) An occasional plumbing leak; (2) condensation in walls, ceiling, and floor; and (3) leaks through walls and floors. Even minor leaks can lead to costly decay. Wetting and decay are most acute in barracks and dormitories where shower rooms and stalls are heavily used. Condensation may be especially serious where rooms next to showers are air conditioned. Moisture may be reduced at its source by forced mechanical ventilation.

Decay-Resistant Construction for Shower Rooms

Shower rooms can be kept free of serious decay by the following conditions: (1) Watertight plumbing; (2) watertight lining of the walls and floor; (3) an effective vapor barrier as near as possible to the warm side of the walls, floors, and ceiling (also applicable to adjacent dressing rooms); and (4) decay-resistant wood framing and sheathing in the walls and floor, and resistant window sash and frames (if present).

Although the vapor barrier and the decay-resistant wood may not be needed for decay control in the average home; the vapor barrier nevertheless frequently helps to prevent early paint failures on adjacent siding. Regulated forced-draft ventilation is a good supplementary precaution to limit the buildup of vapor in a shower area.

For the home, individual shower stalls are often prefabricated of metal or fiberglass. Therefore they are usually watertight. In other construction the shower pan becomes a waterproof layer on the floor (fig. 25) except for slab-on-ground construction on the first floor.

Vapor-barrier material to be installed on the warm side of the walls, ceilings, and floors of the heavily used shower area will be of adequate quality if it meets Federal Specification UU-B-790[10], Type I, Grade A. The shower pan itself serves as a vapor barrier under the area covered by it. If the barrier membrane is attached to wood treated with an oil preservative, it should be polyethylene. Oils may damage asphaltic materials.

Walls should be waterproof, and the upturned pan edge overlapped at least 2 inches. A satisfactory wall material is waterproof plaster on galvanized wire lath, particularly when covered with tile or sheet metal with waterproof crimped or soldered joints. The effectiveness of panel liners will depend on how well holes made by the attaching nails are sealed. Special grades of smooth-surface asbestos-cement panels are satisfactory if the joints are properly sealed. Without precautions, leakage through joints between the panels can occur in the shower room. Joint seals can be made by using well-designed metal connectors or--if the appearance is acceptable--by gluing on batten strips of asbestos cement.

Holes for the entry of water pipes also are weak points in shower-room walls--particularly in walls with panel liners. These are not easily sealed. Locating entry holes as high as possible helps. In buildings where appearance is not a consideration, entry holes can, of course, be avoided by using surface piping. Shower and dressing rooms in particular can benefit from forced-draft ventilation, especially in air-conditioned buildings.

In cold weather, window ledges and frames in showers and adjacent rooms may rot from condensation running off the panes. This can be reduced by using double panes or storm windows. In general, windows should be avoided in shower rooms or over tubs fitted with shower heads.

Preservative Treatment for Shower Rooms

Wood window sash and frames in shower rooms and framing or sheathing in floors and walls should be pressure treated with a wood preservative. Trim above tubs with showers should also be pressure treated. A waterborne chemical or pentachlorophenol in volatile petroleum solvent is preferable (see Federal Specification TT-W-571[5]). Vapor barriers should be polyethylene if attached to studs or joists that are treated with an oil preservative.

---

[10] Federal Supply Service, General Services Administration. Federal Specification UU-B-790, Building Paper, Vegetable Fiber (Kraft, Waterproofed with Water Repellent and Fire Resistant). 1968.

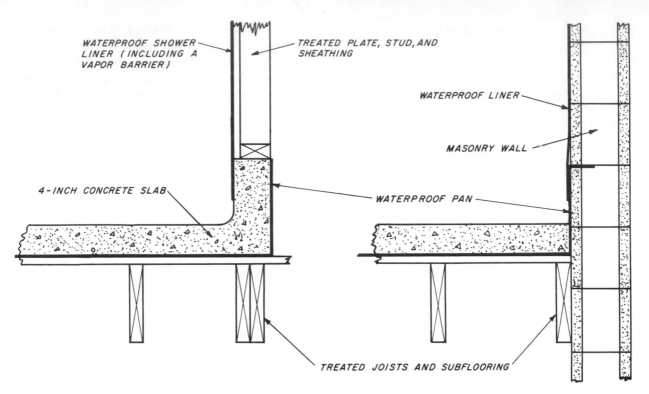

WATERPROOF SHOWER LINER (INCLUDING A VAPOR BARRIER)

TREATED PLATE, STUD, AND SHEATHING

WATERPROOF LINER

MASONRY WALL

4-INCH CONCRETE SLAB

WATERPROOF PAN

TREATED JOISTS AND SUBFLOORING

Figure 25.--Basic designs for heavily used shower rooms: Left, With wood-frame wall. A continuous waterproof pan encloses the slab and 8-inch wall curb. A waterproof liner overhangs the curb by several inches. Where an adjoining room also has slab floor, the slab can be broken to allow installations of the pan as shown. Right, With masonry wall. For masonry or concrete walls, the pan must be tied into the walls to prevent water that runs down the wall from getting back of the pan and seeping below the shower slab. The pan can be bent into one of the masonry joints or it can be secured to side of masonry wall and overlapped from above by a waterproof shower-wall liner. All adjacent wood is pressure treated.

Cold-Storage Rooms

Wood in walls, floor, or ceiling of a cold-storage room is subject to decay associated with condensation unless certain simple elements of design are incorporated. Also, fibrous or other hygroscopic thermal insulation may become waterlogged, and metal parts rust. Surface molding is common. The construction designs discussed here can be applied to any refrigerated space held at temperatures below 65° F.

In cold-storage rooms there are two distinct zones of condensation: (1) On inner wall surfaces and on stored products, and (2) within the walls, floor, and ceiling.

Vapor moves into a cold-storage room when a door is opened. Worn or inadequate gaskets on door jambs and on pipe openings through walls are also points of entry for water vapor. The walls and stored products near the door, being colder than the incoming vapor, act as condensing surfaces and become wet. If doors are used frequently, some surfaces may be wet for long periods and become heavily molded. This is objectionable even though not a decay problem. Wet wall surfaces and room contents will gradually dry, however, if the door is left closed sufficiently long. The drying occurs because the temperature of the cooling coils is several degrees below that of the room and the coils act as condensers on which moisture in the room sooner or later collects.

Most materials used in constructing cold-storage rooms, such as wood, plywood, asbestos-cement, cork, plaster, and concrete, are, to varying degrees, permeable to water vapor; unless entry of vapor into the wall, ceiling, and floor is restricted, damaging condensation within these areas can be expected.

Construction to Minimize Condensation in Cold-Storage Rooms

A cold-storage room of typical frame construction, built next to an outside wall, is shown in figure 26; the vapor barrier and the thermal insulation are particularly essential.

Vapor barrier.--A vapor barrier must be installed in floors, ceilings, and all walls except those between rooms of the same temperature. All joints should be lapped at least 2 inches on walls, ceilings, floors, and at the junctions of walls with ceilings and floors.

The vapor barrier should be the heavy-duty type

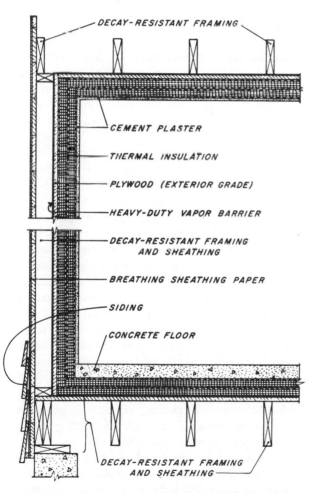

DECAY-RESISTANT FRAMING

CEMENT PLASTER

THERMAL INSULATION

PLYWOOD (EXTERIOR GRADE)

HEAVY-DUTY VAPOR BARRIER

DECAY-RESISTANT FRAMING AND SHEATHING

BREATHING SHEATHING PAPER

SIDING

CONCRETE FLOOR

DECAY-RESISTANT FRAMING AND SHEATHING

Figure 26.--Design for a cold-storage room to avoid moisture accumulation and decay in walls, ceiling, and floor. The vapor barrier is outside the insulation, conforming to the basic rule that a barrier used for controlling condensation must be placed on the warm side of a partition.

with a permeability not to exceed 0.25 perm and rugged enough to withstand considerable abuse (see table 3). Typical barriers of this kind are smooth-surface roll roofing weighing at least 45 pounds per square, duplex material, and 6-mil polyethylene film. If the barrier is to be placed against wood that has been creosoted or treated with a preservative containing a petroleum solvent, polyethylene film should be used.

The exact location of the vapor barrier will vary with the type of construction, but it must be on the warm side of thermal insulation. With frame construction, the vapor barrier can be installed on the inner faces of the studs and ceiling joists before the plywood sheathing is attached. In the floor, the barrier usually is laid over the supporting subfloor before the thermal insulation is installed.

A vapor barrier is also needed on the warm side of doors of wood or other vapor-permeable material. Steel bumper plates mounted flush on the inner sides of doors are not recommended because they act as vapor barriers on the wrong side of the door. It is safer to have an air-space between bumper and door. Flush steel bumpers might be safe if liberally perforated to permit vapor passage.

Insulation and room liner.--Thermal insulation is not needed in partitions between rooms of the same temperature but is needed in all other walls, floors, ceilings, and doors. It must be located on the cold side of the vapor barrier. To ensure that the dewpoint temperature will always be on the cold side of the vapor barrier, the amount of insulation needed should be calculated on the basis of the lowest temperatures expected.

If the vapor barrier is effective and the thermal insulation adequate, the type of the inner wall surface is immaterial. However, as an additional precaution it is well to have lining materials that are as vapor permeable as possible. If vapor leaks past the barrier, and can escape to the inside, it will then not build up within the wall. Unpainted cement plaster is a good liner.

An entrance area, or vestibule, to the refrigerated space will minimize the inflow of warm humid air because the first door can be closed before the second is opened. Vestibules such as these need tight-fitting entrance doors.

Preservative treatment for cold-storage rooms.--All framing and sheathing used in constructing a cold-storage room and the wood trim and doors should be pressure treated with an approved wood preservative. The safest preservative is a waterborne salt because it presents no problem of odor, oil creepage, or injury to barrier material. If the preservative is a type that leaves an oil residue in the wood, the vapor barrier in contact with the treated wood should be polyethylene.

Air-Conditioned Rooms

Air conditioning by refrigeration can create conditions in which condensation and even decay can result. Although these conditions are not widely prevalent, they are discussed here because of the rapidly expanding use of air conditioning and because little has been written about its potential for decay-favoring condensation.

The more moderate degree of cooling with air conditioning generates less condensation in spaces surrounding living quarters than is found in spaces around cold-storage rooms. Some commercial storage rooms may carry a moderately cool temperature of 65° F. or somewhat lower. Such rooms, although not much cooler than air-conditioned living quarters, should be treated in humid climates as cold-storage space.

Damage from Condensation

Damage from condensation caused by air conditioning occurs in the following forms:

1. Dimensional changes in flooring. Cupping of individual floor boards is the usual deformation; the boards remain attached, but turn up at the edges because the lower surfaces swell more than the upper, making slight ridges at the longitudinal seams. Cups will show through most sheet or tile coverings. Greater and more general deformation may occur if the flooring is laid at low moisture contents without appropriate spacing. The subsequent swelling may buckle the floor and create ridges several inches high in a number of boards.

Cupping or buckling should be taken as a warning to beware of condensation, not as evidence of its presence; these flooring defects can be produced without condensation by elevated wood moistures in equilibrium with the damp crawl-space atmosphere. Crawl-space air next to the floor is cooled and its relative humidity thereby increased. Thus, if the increase in relative humidity adjacent to the floor is great enough, the moisture vapor

entering the wood may cause objectionable swelling without condensation.

2. Paint peeling, paint blistering, and molding on walls or ceilings.

3. Plaster failure and rusting of lath. Only occasionally does wall paneling cup or buckle.

4. Loosening of tile or linoleum. The loosening occurs if condensate accumulates at the adhesive line.

5. Decay. Usually is restricted to the top of the subfloor but sometimes includes all of the subfloor, the joists, and the lower face of the finish floor.

Construction and Temperature Regulation to Minimize Floor Condensation

To minimize condensation in floors cooled by air conditioning, dewpoint temperatures in the floors should be prevented. Three general recommendations can be offered: (1) Have a moderate temperature in the air-cooled space, and maintain it no longer than necessary, (2) keep the humidity of the crawl-space atmosphere as low as practicable, and (3) install a vapor barrier on the warm, under side of the floor to limit vapor migration to the cooled surfaces.

The temperature of air-cooled rooms should be no lower than is necessary for comfort, and it should be maintained only when needed. Assuming a dry crawl space, indoor temperatures of 75° F. or above are safe even with continuous air conditioning. Some minor floor deformation may occur, particularly if floors are laid tight at a low moisture content. In rooms where many people congregate, the temperature may be safely lowered during periods of heavy use to maintain reasonable comfort. Also when conditioners are operated only during daytime, lower temperatures usually are safe. Intermittent cooling effectively precludes condensate accumulation by permitting drying when the conditioners are not in use.

A crawl space can be made dry by site drainage, ventilation, and the use of a vapor-resistant soil cover (as described in "Protecting Foundation and Substructure Wood"). A soil cover is most effective.

The condensate from air-conditioning units should be drained to the outside and provisions made to lead it away from the building, to prevent crawl-space wetting.

It is especially important to keep the crawl

space dry if the crawl space contains hot pipes. Hot pipes can increase temperatures in the space to such a level that, even with comparatively low relative humidities, the dewpoint temperature may be considerably above 70° F. In this case, troublesome condensation in the floor can occur despite moderate air conditioning. The condensation hazard is, of course, increased if the pipes carry steam and are leaking.

Vapor barriers properly placed and correctly installed prevent condensation caused by air conditioning when dewpoint temperatures cannot be avoided. They are especially appropriate for use in walls between an air-conditioned and an adjacent room; they also can protect floors against wetting from the crawl-space atmosphere.

When a vapor barrier is needed below the floor in a crawl space, practical difficulties arise. Installation is complicated by bridging, beams, pipes, and electrical conduits. Pliable films are extremely difficult to maintain if applied to the lower edges of joists; consequently, this type of application is not recommended. Rigid insulation board with a vapor barrier attached is much more dependable. It should be installed with the vapor barrier on the underside. Even better is blanket insulation material with a vapor barrier on one side; this is attached flush against the underface of the subfloor between joists with the vapor barrier exposed on the lower surface.

Construction to Avoid Condensation in Walls

Outside walls of an air-conditioned room do not normally require special construction to avoid condensation. Significant condensation in the outside walls is evidence that the air conditioning is too continuous, the temperature is set too low, or that both conditions exist. There should be little trouble in walls between kitchen or laundry and an air-conditioned room if the warmer room is reasonably ventilated and its walls are painted with one of the more vapor-impermeable paints, and if materials comprising the cold side of the wall are as vapor permeable as possible (to prevent such vapor as gets past the warm face from being trapped within the wall). Plaster on studs with the cold face painted with a latex-emulsion type paint having relatively low resistance to vapor would, for example, be good vapor-

permeable construction. Refrigerant pipes within walls must be adequately insulated or condensate can run down them and seriously wet wall plates.

If large volumes of steam or vapor are released in an adjoining room, forced-draft ventilation of that room is recommended; it may be additionally helpful to install a vapor barrier on the warm side of the separating wall. Requirements are the same as described for the barrier to prevent winter condensation in outside walls.

Indoor Swimming Pool Areas

Frame walls and the roof enclosing an indoor swimming pool are especially subject to condensation, decay, and paint problems unless preventive measures are taken (fig. 27). The basic difficulty with the swimming pool area arises, of course, because the pool has a large surface of heated water that gives off a great deal of warm moisture. In cool or cold weather, the situation is essentially the reverse of cold-storage rooms or air-conditioned rooms: The warm moist atmosphere is in the swimming pool area, and the relatively cold temperatures are on the outside.

Wall and Roof Construction for Swimming Pools

To avoid serious condensation in the swimming pool area five measures are suggested (1) Ventilate the area as much as practicable to decrease accumulation of moisture in the air. Forced-draft ventilation is especially helpful and should generally be provided in temperate climates; (2) install an appropriate thickness of thermal insulation within the walls and roof to prevent a dewpoint temperature being reached on the inner surfaces; (3) install a vapor-barrier membrane on the warm side of the wall studding and as near the warm surface of the roof structure as possible to reduce the amount of moisture vapor leaking into the walls and roof; (4) avoid highly vapor-retardant material in the outer part of the wall; and (5) provide for ventilation just below the roof cover. Precautionary measures (4) and (5) allow water vapor to escape to the outside if the vapor-barrier membrane on the inside does not wholly exclude vapor in amounts that might produce condensation.

Preservative Treatment for Wood Enclosing Swimming Pools

If the foregoing precautions are followed, it should not be necessary to use preservative-treated wood for sheltering the swimming pool. If treated wood is available at reasonable cost, it might be justified to ensure satisfactory performance, especially for the roof members.

Figure 27.--Condensation promoted decay in inner roof boards (dark color) of a swimming pool. Light-colored boards are replacements for decayed boards. Watermarking on the beams evidences extreme wetness.

PRESERVATIVE TREATMENTS FOR BUILDING LUMBER

If a part of a building is vulnerable to decay and cannot be protected by keeping it dry, the alternative is to use wood that is naturally decay-resistant or has been treated with preservative.

Cost is often the decisive factor in determining whether to treat wood, but the cost of not treating may be greater. An example of savings to be realized from use of treated wood as compared with untreated wood is given in figure 28.

In the preceding sections treated wood has been recommended for particular items under certain climatic and building conditions. Many factors will determine the type of preservative and treating method best suited for a particular part of a building. Details on chemicals and treating methods can be found in the Additional Reading at the end of the text. The following summarizes the information particularly applicable to buildings.

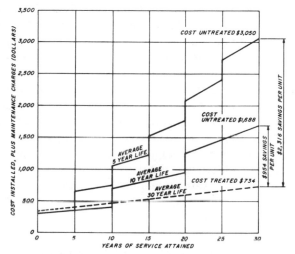

Figure 28.--Example of savings obtainable by using wood treated with a preservative, compared to cost of untreated material.

Types of Preservatives

A wide variety of preservatives are available to meet all needs for protecting wood in buildings. Since the acceptable usage of wood preservatives undergoes continuing review, the most recent registered usages must be followed. State or Federal pesticide regulation officials should be consulted about acceptable preservative usages.

Creosote, Creosote-Petroleum Mixtures, and Pentachlorophenol in a Heavy Oil

These have a high degree of permanence but have a persistent odor which may be objectionable. Wood treated with them usually cannot be painted or glued satisfactorily, and the oil may creep along nails and stain untreated wood attached to it. These preservatives are used mainly for commercial pressure treating.

Pentachlorophenol in Liquefied Petroleum Gas

Pentachlorophenol also is dissolved for pressure treating in a liquefied (under pressure) petroleum gas. When pressure is released, the gas volatilizes rapidly at ordinary temperature. Consequently, items treated with pentachlorophenol in liquefied petroleum gas do not present paintability or other problems commonly associated with material pressure-treated with the preservative in an oil carrier.

Pentachlorophenol in Light Petroleum Solvents

Light oil solvents, such as mineral spirits, evaporate and leave a clean, paintable surface if applied by such simple methods as brushing, dipping, and spraying. An additional advantage--resistance to rain wetting and the resultant improved dimensional stability and decay resistance--can be secured if a water-repellent ingredient is incorporated in the treating solution.

Ready-to-use solutions of 5 percent pentachlorophenol in light oil with a water repellent meeting Federal Specification TT-W-572 are available through a variety of local retailers as "water-repellent penta." Also "1 to 5" and "1 to 10" concentrates of pentachlorophenol are available which can be diluted with mineral spirits when a solution stronger than 5 percent is needed to compensate for particularly shallow penetration.

Light-oil solutions are used mainly for on-site treating, but millwork treated with such solutions and including a water-repellent is available on the market.

Pentachlorophenol in a Grease-Like Base

"Penta grease" is used mainly for on-site treating. Surfaces close to the point where penta grease has been applied may not be paintable for several weeks. However, penta-grease permits substantially deeper penetration than can be obtained by other simple treating methods.

Waterborne Salts

Wood treated with a waterborne salt and then dried is clean, paintable, and free of objectionable odor. The treated wood must be redried and can develop additional checking in service from wetting and drying. Waterborne salts are used mainly for commercial pressure treating; such treated wood has many uses in buildings. Some of the waterborne salts are among the most effective and permanent preservatives available. They are especially recommended for wood foundations in contact with the ground.

Degree of Preservative Protection Needed

Types of treatment suitable for building items in various usages and exposures are given in tables 4, 5, and 6. The items are placed in the respective tables according to whether their exposure to wetting is influenced by their distance from the ground, by the amount of roof overhang, or primarily by climate alone. Climate is a factor in all cases. Treatment needs for items other than those listed can be judged from the similarities of service and exposure. The thoroughness of treatment suggested, types A (pressure treatment), B (nonpressure treatment), and C (no treatment), is based in part on the climate index. A formula for climate index appears in "Effect of Climate on Decay" and a contour map of climate indexes in the continental United States appears in figure 6.

For appropriate treatment of items that require pressure treatment with the principal oilborne or waterborne preservatives, details can be found in Federal Specification TT-W-571[5] and in the standards of the American Wood-Preservers' Association. These guides give the minimum retentions considered necessary for various conditions of service. Where treating may differ with wood species, this also is evaluated.

On-Site Treating

On-site treating refers to application of the preservative at the building site. If treating is done after construction, it is called in-place treating. Because on-site treating can be timed and tailored to fit a variety of needs, it can be a significant and valuable adjunct to the total measures for protecting buildings. Also it often can be a convenient and adequate means of providing precautionary protection for lumber that is used as a replacement for decayed lumber in situations where the contributing moisture situation presumably has been corrected. It has a place in the maintenance of structures that show signs of needing protection.

Superficial applications of preservatives should be used only when other methods are either impractical or not required (see tables 4, 5, and 6). All surfaces of the lumber must be flooded with the preservative solution. For treating wood in place a spray is most convenient. A paint spray gun can be used, adjusted to deliver a fine solid stream which can be directed into joints between boards. A water-repellent solution should be used.

Dip and Short-Period Soaking

Dipping and short soaking are the most appropriate methods to treat lumber on the building site before it is placed in the structure. The preservative ordinarily would be water-repellent pentachlorophenol, Federal Specification TT-W-572,[7] which is rigidly limited to items in aboveground service and preferably to material that has had all the necessary cutting and boring.

Where pre-cutting or pre-boring is not practicable, untreated surfaces exposed at the building site should be protected (the same as with pressure-treated wood) by follow-up treating-- dipping, brushing, spraying, or spreading with penta grease--whichever is the most practical.

As the name implies, short soaking consists of relatively brief immersion of the item in a preservative solution. For building items, the conventional preservative for this treatment consists of a 5 percent solution of pentachlorophenol with water repellents carried in mineral spirits or in an oil of comparable volatility. The soaking period ideally should not be less than 3 minutes, a standard time widely used in commercial treating of window and exterior door components (Commercial Standard CS 262[9]). Short soaking is commonly referred to as "dipping" rather than "soaking." Longer treating times up to about 15 minutes are desirable where practicable for many items of exterior woodwork.

The equipment for short soaking may be no more than a trough large enough to accommodate

Table 4. -Suggested Protective measures[1,2] for interior building items relative to climate and to distance to ground

Item	Climate index[3] greater than 65 and wood-to-ground distance of--		Climate index[3] 33 to 65 and wood-to-ground distance of	Climate index[3] less than 35 and wood-to-ground distance of
	Less than 8 inches	8 inches or more		
Sleepers in or on concrete laid on ground	A[4]	A[4]	A[4]	A[4]
Furring strips below grade level	A[4]	A[4]	B[5]	A[4]
Sills or plates on concrete laid on ground	A[4]	A[4]	A[4]	A[4]
Sills or plates on concrete or masonry wall foundation	A[4]	C	C	C
Joists, girders, and beams in contact with concrete or masonry wall foundation	A[4]	C	C	C
Permanent wood foundation or foundation components in ground contact or separated from ground by only a water-resistant membrane	A	A	A	A
Wood piers in crawl space (on concrete footing)	A[4]	C	C	C
Wood windows, framing and sheathing of shower room walls and floors (where showers are subject to heavy usage)	A[6]	A[6]	A[6]	A[6]
Framing and sheathing of cold-storage rooms	A[6]	A[6]	A[6]	A[6]

[1] A, Pressure treat according to Federal Specification TT-W-471 or published standards of American Wood Preservers Association. For items to be painted or that for other reasons must be free of residual oil, preservative should be the waterborne type or pentachlorophenol applied in volatile petroleum solvent.
B, Nonpressure treat according to standards that equal or exceed those of Commercial Standard CS 263-63 or use specified naturally durable wood.
C, No treatment.

[2] Untreated surfaces exposed by cutting, boring, or shaping after an item has been treated should be brush treated at the building site, preferably with double-strength preservative. A protection may be used instead of B. The preservative in some cases obviously must be noncreeping and of a type that will leave the wood clean and paintable.

[3] See formula under "Effect of Climate on Decay;" also fig. 6.

[4] Protection is desirable until additional information is obtained to substantiate or disprove need.

[5] If foundation is brick or concrete block, use A treatment.

[6] Climate or ground-distance does not ordinarily apply.

the lumber, preferably with a drainboard attached to recover excess solution carried out on the treated stock. Heating facilities are unnecessary. Ordinary construction lumber that is air dry-- therefore dry enough to be soak treated--will float. A simple means can be used to keep it submerged for the desired time.

Penta-Grease Treating

Preconstruction treating with penta grease will be most appropriate for items needing protection chiefly on cross sections where the end grain is exposed. The base of a post or column that is to rest on concrete exposed to the weather is a good example of the part of an item that might be particularly benefited by penta-grease treating. The grease should be spread on the cross section in a 1/4- to 1/2-inch layer, the thickness depending on the decay vulnerability and on the time that can be allowed for the preservative to be absorbed. Essentially complete absorption is indicated when the layer of penta grease ceases to shrink in thickness; this will typically require several days, although the absorption obtained over a period of several

Table 5. -Suggested protective measures[1,2] for exterior building items not much affected by roof overhang, in various climates

Item	Climate index greater than 65	Climate index 35 to 65	Climate index less than 35
Posts set in ground	A	A	A
Columns (porch, carport)	A	B[3]	C[3]
Board panels (fence, carport louvers)	A	B	C
Porch flooring and joists	A	B	C
Rails (porch, step, fence)	A	B	C
Treads and stringers (carriages)	A	B	C
Roof edges (exposed sheathing, molding, fascia)	A	B	C
Exposed rafters	A	B[3]	C
Exposed arches	A	A	A[4]
Access panels	B	B	C
Access frames	A	B	C

[1] A, Pressure treat according to Federal Specification TT-W-571 or published standards of American Wood Preservers Association. For items to be painted or that for other reasons must be free of residual oil, preservative should be the waterborne type or pentachlorophenol applied in volatile petroleum solvent.
B, Nonpressure treat according to standards that equal or exceed those of Commercial Standard CS 263-63 or use specified naturally durable wood.
C, No treatment.

[2] Untreated surfaces exposed by cutting, boring, or shaping after an item has been treated should be brush treated at the building site, preferably with double-strength preservative. A protection may be used instead of B. The preservative in some cases obviously must be noncreeping and of a type that will leave the wood clean and paintable.

[3] Penta-grease treatment of ends (end grain) would be a good precaution.

[4] Protection is desirable until additional information is obtained to substantiate or disprove need.

hours will be worthwhile.

Experimentation indicates that penta grease can be useful in treating joints and contact zones after construction by laying it in the exterior angles of the joint. The preservative can be applied rapidly from a calking gun with a nozzle of about 3/4 inch. Commercial applicators variously designed for different forms of spreading are available. The greater the number of external angles receiving penta grease, the more likely the preservative will reach the deepest part of the joint. With external treating, however, one should not expect the preservative to penetrate further into the joint than about 1 inch.

Precautions For On-Site Treating

To treat lumber on the building site, some planning is necessary to schedule and to complete the treating without delaying construction. The following procedures should be observed if the treating is in conjunction with construction:

1. The item should be cut to size and shaped and bored prior to treating if feasible.

2. For treating with a solution, the items should

Table 6. - Suggested protective measures [1,2] for exterior bulding items relative to climate and to amount of roof overhang

Item	Climate index greater than 65 and roof overhang of--		Climate index 35 to 65 and roof overhang of less than 2 feet	Climate index 35 to 65 and roof overhang of any amount
	Less than 2 feet	2 feet or more		
Exposed, load-bearing structures	A	B[3]	B	B[4]
Siding and trim	B[5]	B	C	C
Window sash	B[5]	B	B[6]	C
Frames and trim (windows, screens, doors)	B[5]	B	B[6]	C
Shutters	B[5]	B	B[6]	C

[1] A, Pressure treat according to Federal Specification TT-W-571 or published standards of American Wood Preservers Association. For items to be painted or that for other reasons must be free of residual oil, preservative should be the waterborne type or pentachlorophenol applied in volatile petroleum solvent.
B, Nonpressure treat according to standards that equal or exceed those of Commercial Standard CS 263-63 or use specified naturally durable wood.
C, No treatment.

[2] Untreated surfaces exposed by cutting, boring, or shaping after an item has been treated should be brush treated at the building site, preferably with double-strength preservative. A protection may be used instead of B. The preservative in some cases obviously must be noncreeping and of a type that will leave the wood clean and paintable.

[3] Protection valid only for 1-story house; if house is higher, value of 2-ft. overhang is diminished, and situation should be considered same as for overhang less than 2 ft.

[4] Protection is desirable until additional information is obtained to substantiate or disprove need.

[5] Decay hazard is greatest at joints of item where most serious wetting is by end-grain absorption. Therefore adequate penetration of preservative can be obtained by protection B. Exception: Items going into tropical service may need A protection.

[6] Where overhang is 2 ft. or more and the house has but 1 story, no preservative treatment is needed.

be submerged not less than 3 minutes; 15 minutes is preferable.

3. The item should be set aside briefly after removal from the treating solution to allow the surface to dry before it is handled by carpenters.

4. If any cutting must be done after treating, the cut surface should be retreated. The best procedure is to brush-coat or spray the newly exposed surface, using at least double strength or double application of the preservative. Most siding and trim items are cut to length just before they are attached, which usually necessitates delaying the retreating--by brushing or spraying the end grain joints--until after the items are attached to a building.

Unless enough time is available to get the preservative solution deep into joints by repeated flooding, in-place treating of exterior parts of older buildings does not appear worthwhile. Items like porches, steps, and railings have numerous joints that are particularly vulnerable to decay, but flooding treatment--as with a coarse spray-- usually does not get the preservative deep enough into the joints to reach all existing infections. Another complicating factor is paint in the joints, which acts as a barrier to the preservative.

Flooding treatment and external application of penta grease have some potential for treating new aboveground structures in place. Success with these types of in-place treating apparently varies with types of construction and forms of joints. Generally, however, results cannot be expected to be as good as those obtained by treating before construction. It is reasonably well established that the end cuts of pretreated board siding can be effectively protected by flooding the vertical joints after the siding has been attached. The butt joints of panel siding can be similarly protected.

If a natural-finish or a painted or stained surface is desired, a "clean" treatment is necessary. Five percent pentachlorophenol with water repellents in mineral spirits does not discolor wood and if applied without pressure it leaves the surface paintable. If a stain is desired, a pigment can be added to the solution.

Good paintability generally can be assured if the water-repellent preservative is applied by short soaking or by short vacuum treatment. Common commercial treatment of this kind for exterior woodwork such as window sash and frames or doors is covered by Commercial Standard CS 262.[9]

Occasionally, a single board will be abnormally permeable because of bacterial or inconspicuous mold infection and will absorb enough oil by a dip treatment to be unpaintable. These boards frequently can be detected after treating by a dark oil-soaked appearance; they can be set apart and used in places where paintability is not important. This overtreating induced by fungal and bacterial action is largely a problem of pine sapwood from logs that have been stored under water for considerable time.

A simple test can show whether treated wood is sufficiently oil-free to be successfully painted. A cross section of the treated lumber is taken at least 3 feet from the end of the piece, and placed end grain down on a sheet of asphalt laminated paper (30-30-30 weight). If the paper under the wood discolors in 15 minutes at 100° F., a paint-harming amount of residual oil remains in the wood. The disadvantage of this test is that a freshly cut cross section is required; therefore, testing cannot be done on assembled structures.

Safety Precautions.--The following safety precautions should be followed:

1. Prevent workmen from breathing dusts or sprays or from allowing any significant amount of them to come in contact with their faces.

2. Workmen should wear protective gauntlets and aprons when treating or when handling lumber that is still wet with treating solution.

3. Gauntlets should be washed on the inside with soap and water frequently.

4. Hands and other skin areas wetted by preservative should be washed with soap and water immediately.

5. Workmen who treat manually for the first time should be watched for special sensitivity to the preservative. If they exhibit undue skin sensitivity they should be removed from the treating.

Regular Inspections

The cost to protect a frame home or other building from serious decay can be minimal if buildings are inspected regularly and trouble is corrected early.

At least once a year, places and items most vulnerable to wetting should be inspected. Particular attention should be paid to roofs, roof edges (facia, soffits, rafter ends), joints in and adjacent to window and door frames, and appendages such as porches, steps, and rails. Any signs of repeated wetting and traces of decay should be investigated.

The crawl space, though it can be difficult to inspect, should not be neglected. In milder climates it is vulnerable to decay by the water-conducting type of fungus. In colder climates it is subject to wetting by winter condensation--especially in the corners. During the summer in warm humid climates the floor may get wet from condensation created by summer air conditioning. The framing in the crawl space merits particular attention because it makes up the basic load-bearing members of the building. Any evidence of plumbing leaks should be looked for and any leaks promptly corrected.

Inspection for winter condensation in the crawl space should be made in the late fall and winter, and for condensation from air conditioning in the late summer. Wetting caused by air conditioning should be looked for especially on the subfloor. Cupping or buckling of the finish floor is a sign that condensation may be occurring.

The presence of a water-conducting fungus may or may not be revealed by strandlike surface growths of the fungus.

Inside a building, places to watch most carefully are shower rooms--for water leakage into walls or floor; kitchen areas--for plumbing leaks; and cold-storage rooms--for condensation within the walls, floor, or ceiling. To find the source of a leak is sometimes baffling, because water may travel some distance from the point of leakage within the walls or beneath flooring before it is noticeable. Watermarking of walls or floors near outside walls and damp-appearing surfaces indicate leaking water. Often a damp surface will be accompanied by molding. Steam radiators should be routinely checked for leaks.

A moisture meter (fig. 15) is helpful in locating wet zones within walls or in similar places. A type of meter employing nonpenetrating electrodes can be used to measure moisture content of plaster and finish items.

Recognition of Decay and Serious Wetting

Persistently wet places should be noted, and the cause corrected promptly. Various indications of wetting and decay are shown in figure 29. Advanced decay is easily recognized; early decay may not be. Yet even a small amount of decay is cause for concern because serious strength losses usually accompany it; hence those who build or maintain wood structures should be acquainted with the ordinary signs of decay infection, particularly in load-bearing items.

Discoloration of the wood.--As decay progresses it usually imparts an abnormal color to wood. This change in color can be a useful diagnostic of decay if the inspector is reasonably familiar with the color or color shades of the sound wood. On surfaced wood the discoloration commonly shows as some shade of brown deeper than that of the sound wood. Some decays, however, produce a lighter than normal shade of brown, and this change may progress to a point where the surface might be called white or bleached. If this bleaching is accompanied by fine black lines, "zone lines," decay is virtually certain (fig. 4). Often, an abnormal variation in color creating a mottled appearance is more helpful in detecting early decay than actual hue or shade of discoloration. Highly indicative of decay, and especially conspicuous, is variable bleaching on a dark background of blue stain or mold.

Accompanying the color change, there may be an absence of normal sheen on the surface of infected wood. Here also, familiarity with the normal appearance of the wood can be of great help in recognizing the loss of sheen. Occasionally, in relatively damp situations, the presence of decay infection will be denoted by surface growth of the attacking fungus; in these cases the wood beneath usually is weakened--at least superficially.

Stain showing through paint films, particularly on exterior woodwork, is evidence of serious

Figure 29.--Various indications in building components of moisture accumulation and decay: A, Rust around nail. B and C, Surface mold on wood, and failure of paint on areas near absorptive end grain. D, Sunken surface or visibly deteriorated wood, often with paint failure. E, Fruit bodies of fungi (evidence of advanced decay). F, Nail pulling (evidence of alternate shrinking and swelling).

wetting and probable decay beneath the film. Rust around nail heads suggests that wetting has been sufficient for decay to occur.

Loss of wood toughness and hardness.--Wood can also be examined for decay by simple tests for toughness of the fibers and for hardness. Toughness is the strength property most severely reduced by early decay. The pick test is a helpful and widely used simple means of detecting diminished toughness. It is made most reliably on wet wood. An ice pick, small chisel, sharpened screwdriver, or similar sharp-pointed or edged instrument of tough steel is jabbed a short distance into the wood and a sliver pried out of the surface. The resistance offered by the wet wood to prying and the character of the sliver when it finally breaks are indicative of toughness.

In the pick test, sound wood tends to break out as one or two relatively long slivers and the breaks are of a splintering type (fig. 30): Where loss of toughness has been appreciable, the wood

tends to lift out with less than usual resistance and usually as two relatively short pieces. Moreover, these short pieces break brashly at points of fracture; that is, abruptly across the grain with virtually no small splinters protruding into the fracture zone.

On planed lumber, the reduced toughness of wood with early decay is sometimes indicated by abnormally rough or fibrous surfaces. Similarly the end grain of a board or timber may be rougher than usual after sawing.

Toughness may also be reduced by certain other factors such as compression wood, tension wood, or compression failures. There usually is little doubt of decay infection if the weakening is accompanied by a decay-induced type of discoloration.

In many cases the reduced hardness of infected wood can be detected by prodding the wood with a sharp tool. Softening, however, usually is not so obvious or so easily detectable with early decay

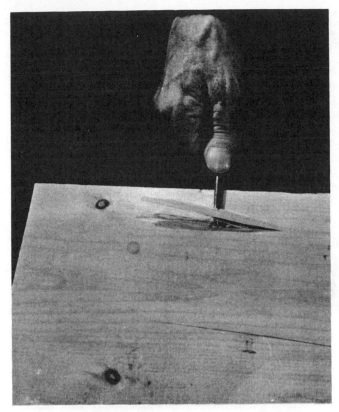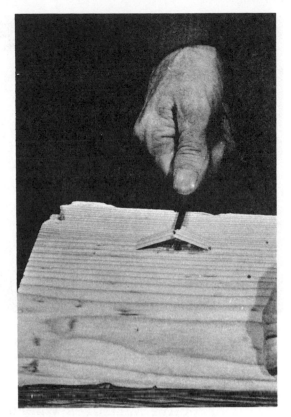

Figure 30.--The "pick test" for early decay. Wetted wood if sound (left) lifts as a long sliver or breaks by splintering; if infected (right) it tends to lift in short lengths and to break abruptly across the grain without splintering.

as is a decrease in toughness.

Shrinkage and collapse.--Decay in the more advanced stages frequently causes wood to shrink and collapse. Under paint, this may first be manifested by a depression in the surface. Often the paint will acquire a brown-to-black discoloration from soluble materials migrating from the interior zone of decay to the outside; also, fruit bodies of the attacking fungus sometimes appear on the surface.

In exterior woodwork, wetting is often evidenced by brown-to-black discoloration or loosening of paint, particularly at joints. If there also has been substantial decay, the discoloration may be associated with interior collapse and surface depression.

Surface growths.--Decay in crawl spaces invaded by a water-conducting fungus may be evidenced by fanlike growths, vinelike strands, or by a sunken surface of wood resting on foundation walls or piers. Such decay is usually most advanced near the foundation, because the fungus usually starts there. The fanlike growths are papery, of a dirty white with a yellow tinge. They may spread over the surface of moist wood, or--more commonly--between sub- and finish-flooring or between joists and subfloor (fig. 31). These growths may further appear under carpets, in cupboards, or in other protected places. Water-conducting, vinelike strands grow over the foundation, framing, the underside of flooring, inside hollow concrete blocks, or in wall voids.

The fungus carries water through these strands from the damp ground or other source to the normally dry wood being attacked. Usually the main water conductors are 1/4 to 1/2 inch wide although they sometimes reach 2 inches. They are similar in color to the fanlike growths although they sometimes turn brown to black. During dry weather, shrinkage cracks in floors often outline the extent of an attack. Rotted joists and subflooring in relatively dry crawl spaces usually have a sound shell even when the interior wood is essentially destroyed.

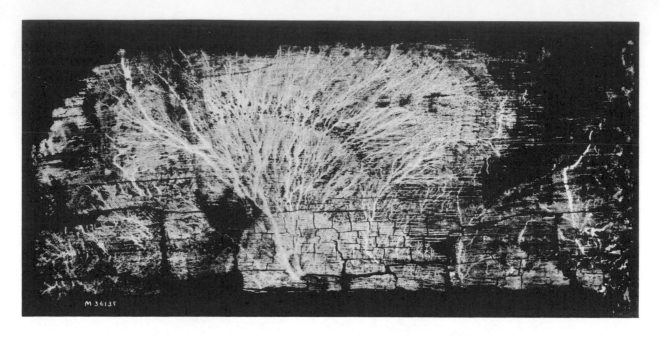

Figure 31.--Advanced stage of decay by water-conducting fungus <u>Poria incrassata</u> showing fanlike surface growth and typical decay pattern of wood.

Corrective Measures Where Wetting or Decay Occurs

Eliminating Conditions Responsible for Wetting

The cause of excessive wetting of any wood members in a building should be investigated, and measures to correct the situation should be taken as soon as practicable. Ordinarily, alterations or repairs to stop the wetting and to keep the damaged item dry are sufficient. If there is any question about not eliminating the wetting, the replacement wood should--as a precautionary measure--be treated with preservative. For preventing decay-producing situations, it is important to remember that wood will not decay if its moisture content is no more than about 20 percent. To decay it must be contacted by water; moisture imparted by damp air alone can cause objectionable swelling, but cannot support decay.

In making repairs required because of decay, it usually is necessary to replace only wood so weakened that it is no longer serviceable. Infected wood will not endanger adjacently placed sound wood so long as both are kept dry.

Because of the high decay hazard of the roof edge, it should be watched for signs of wetting and decay. With flat roofs, gravel stops cannot be kept watertight for appreciable lengths of time unless regular attention is given to the joint seals. These should be resealed at the first sign of leakage. If eave gutters show corrosion, they should be covered with a corrosion-inhibiting paint. This is particularly necessary with recessed gutters because considerable hidden decay can occur before leakage is evident. Where no edge flashing has been used, it should be installed if there is evidence of sufficient wetting to cause staining of the last sheathing board, rafter ends, or facia.

Maintenance by Preservative Treatment of Items in Place

New wood for replacing areas of decay in buildings should receive the same type of preservative treatment that would be recommended for new construction in those areas. Flooding treatment with brush or spray of items that have been in service for a considerable time is not likely to prevent or arrest decay if the job is so big that it must be done on a large scale. It can be effective, however, if it is a home operation, where the owner can give the time needed to get the preservative solution sufficiently deep into joints and cracks.

In flood treating the object is to get the preservative deep into joints and crevices where rainwater is likely to be trapped and to have the penetration equal that reached by the water. Treating every few years will add to the margin of safety. Examples of items that can be protected by flood treating in place are the bases of porch pillars and carport posts resting on concrete, plank porch floors, shutters, window boxes, and lookouts. Where rain seepage into a joint is indicated by failure of nearby paint, trouble from this source can be minimized if the joint is flooded by spraying or brushing with a water-repellent preservative prior to repainting. In treating, the preservative should be kept off plants and grass and not permitted to accumulate on the skin. A suitable preservative for most maintenance treating is 5 percent pentachlorophenol with water repellents in mineral spirits. It is widely obtainable and is identifiable by the label.

Eradicating the Water-Conducting Fungus

Decay caused by the water-conducting fungus is easily prevented by incorporating the building procedures that have been discussed in previous sections. However, if attack has already occured, special control measures are required. If the fungus is well established and conditions supporting it are not removed, large areas of flooring or walls may have to be repeatedly replaced. Cases are reported in which replacements were necessary at 1- and 2-year intervals.

The water-conducting fungus is susceptible to drying; therefore, it should be permanently separated from its source of water. When this is accomplished, the affected wood soon dries, and the fungus dies within a few weeks. Then only wood that has been too weakened to safely support its load needs replacement. However, if there is any doubt that the fungus has not been eliminated, it is safest to replace all infected wood with wood that has been pressure treated with a preservative.

Usually the fungus gets its water from the ground or--less frequently--from wet wood or masonry in the general area where the decay occurs. The following are the most common measures to control the water-conducting fungus; some will be recognized as measures that should have been taken at the time of construction.

1. Irrespective of the location of the decay, seek out and remove any wood forms left from pouring concrete steps or foundations and any other wood, building paper, insulation board, or similar cellu-losic material that may be making a direct bridge from the soil to the wood of the building. Also, eliminate stumps and all building debris. If necessary, regrade the crawl space or soil outside the building to provide wood-soil clearance.

2. Provide for drainage of surface water away from the outside foundation and the crawl space. If the ground in the crawl space is not dusty dry, include better ventilation or a soil cover, with the aim of making the air in the space dry enough to restrain the fungus in its development of water-conducting strands. A polyethylene film sheet is better than roll roofing for a soil cover when the water-conducting fungus is present, because this fungus will attack asphaltic papers.

3. Open the foundation of the porch and remove enough soil from the fill to expose the entire sill under the slab. The opening should be sufficient to permit inspection of the sill and provide ventilation to it. If the sill needs replacement, use pressure-treated wood. Termite-control operators are familiar with the technique of excavating fills.

4. If water-conducting strands or other growths of the fungus are observed on concrete or brick foundations, scrape off the larger strands and brush off the remainder with a steel brush. Finally, the cleaned surfaces should be thoroughly flooded with a preservative. This treatment also is applicable to situations in which the fungus is found on concrete exposed by excavating fills or by removal of forms. Always examine the treated areas and re-treat if any evidence of new growth appears. Where it is apparent that water-conducting strands are hidden inside concrete blocks or in loose mortar in brickwork, insert a metal shield between the foundation and substructure wood or, if the construction is brick, reset a few upper courses using cement mortar.

5. If the source of the decay is traced to a plumbing leak, repair the leak. Where the trouble is associated with shower stalls, a completely new watertight lining may be needed. If the framing and sheathing for the floor and walls of the shower are exposed during repairs, replace them with pressure-treated wood. The most dangerous leaks are the small ones that are difficult to detect.

6. If attack occurs in a slab-on-ground supported building that does not meet waterproofing and ground-clearance standards, replace all basal plates with pressure-treated wood and use nonwood flooring. Provide as good outside clearance as possible and chemically treat the slab edge and adjacent soil. Attack seldom occurs in slab-supported houses with adequate waterproofing of

the slab and adequate ground clearance, except when unusual wetting occurs, such as that caused by excessive lawn sprinkling or by elevated flower beds.

7. If attack occurs in a basement, replace wood in contact with a wall or the floor with pressure-treated wood. Do not have any enclosed stairs, partitions finished on both sides, cupboards, or paneling on the outside walls in moist basements. This type of construction creates "dead air" spaces that promote growth of the fungus.

8. If preservative-treated wood is specified, use creosote, pentachlorophenol, or a noncopper waterborne compound, applied under pressure. Although all heartwood of decay-resistant species is acceptable for some items and exposures in well-designed new buildings, do not use such wood to replace wood that has been attacked by the water-conducting fungus. This fungus will attack the heartwood of most decay-resistant woods including redwood and cedar.

Chapter 21

THERMAL INSULATION

The contribution of thermal insulation to comfort of occupants and economies of heating and cooling have been recognized and the ecological implications are being assessed. In the future the amount of insulation required in buildings may well be governed by factors other than what is economical and will provide a desirable level of comfort for occupants. These other considerations include air pollution from burning fossil fuels, be it in the building proper or at a power station some distance away, thermal pollution from wasted heat, and the need to conserve energy and fuels.

The inflow of heat through outside walls and roofs in hot weather and its outflow during cold weather have important effects upon the comfort of the occupants of a building. During cold weather, heat flow also governs fuel consumption to a great extent. Wood itself is a good insulator but commercial insulating materials are usually incorporated into exposed walls, ceilings, and floors to increase resistance to heat passage. The use of insulation in warmer climates is usually justified with air conditioning not only to reduce cooling costs but also to permit use of smaller capacity units.

Commercial insulating materials are manufactured in a number of forms and types. Each has advantages for specific uses; some are more resistive to heat flow than others, and no one type is best for all applications.

INSULATING MATERIALS

For purposes of description, materials commonly used for insulation may be grouped in five general classes: (1) Rigid insulation, (a) structural or (b) nonstructural; (2) flexible insulation, (a) blanket or (b) batt; (3) loose-fill insulation; (4) reflective insulation; and (5) miscellaneous types.

Rigid Insulation

Structural Insulating Board

Structural insulating board is made by reducing wood, cane, or other lignocellulosic material to a coarse pulp and then felting it into large panels. When these panels are pressed and dried, they have densities ranging from about 10 to 31 pounds per cubic foot (p.c.f.). They provide thermal insulation while also providing some strength or other physical property required for a particular use.

Structural insulating board is fabricated in the following forms, which are described in more detail, as related to the use, in chapter 22:

Building boards; insulating roof deck; roof insulation; wallboard; ceiling tile and lay-in panels; plank; sheathing—regular density, intermediate, nail-base; shingle backer; insulating formboard; and sound-deadening board.

Roof insulation primarily provides thermal resistance to heat flow in roof constructions. It is applied on top of the structural deck and roofing is applied over it. Although products like roof deck, ceiling tile, plank, sheathing, shingle backer, insulating formboard, and sound-deadening board (when used in exterior constructions) are fabricated primarily to serve other functions, they also help to resist heat flow. Building board and wallboard are products for remanufacture and are used in diverse ways.

It is common practice for the insulating board industry to supply sheathing products, where the thermal insulation is of importance, with rated insulating values.

Nonstructural (Block) Insulation

Nonstructural rigid insulation is often called "block insulation." The slabs or blocks are small rigid units, sometimes 1 inch thick but generally thicker, and vary in size usually up to 24 by 28 inches. The types made from wood-base materials are cork blocks, wood fiber blocks, and fiberboard slabs. Cork blocks are made by bonding small pieces of cork together in blocks or slabs 1 to 6 inches thick. They are used for cold-storage insulation and for insulating flat roofs of industrial and commercial buildings.

Wood-fiber blocks are made by bonding wood fibers with some inorganic bonding agent, such as portland cement. They are made in thicknesses of 1 to 3 inches and in various widths and lengths. Principal uses are for roof-deck in industrial buildings, structural floor and ceiling slabs, and nonbearing partitions.

Fiberboard slabs are made by laminating insulating board products to produce rigid blocks. Mineral-wool slabs or blocks are made both of rock wool and glass wool with suitable binders for low-temperature insulation and specialty uses. Other types of blocks and slabs include cellular glass, plastic, cellular-rubber products, and vermiculite or expanded mica with asphalt binder.

Flexible Insulation

Flexible insulation is manufactured as blanket and batt. Blanket insulation is furnished in rolls of convenient length and in various widths suited to standard stud and joist spacing. The usual thicknesses are 1½, 2, and 3 inches. Each roll covers from 70 to 140 square feet. The body of the blanket is usually made of loosely felted mats of mineral or vegetable fibers, such as rock, slag, or glass wool, wood fiber, and cotton. Organic fiber mats are chemically treated to make them resistant to fire, decay, insects, and vermin. Most blanket insulation is provided with a covering sheet of paper on one or both sides and with tabs on the edges for fastening the blanket in place. The covering sheet on one side may be of a type which serves as a vapor barrier. In some cases the covering sheet is surfaced with aluminum foil or other reflective material.

Batt insulation also is made of loosely felted fibers, generally of mineral-wool products. It is also made in widths suitable for fitting between standard framing spaces. Thicknesses are usually 4 and 6 inches. Some batts have no covering; others are covered on one side with a vapor barrier similar to that used for blanket insulation. In walls batt insulation is installed in the same manner as blanket insulation. In ceilings it is placed between joists with the vapor barrier facing downward toward the warm side.

Fill Insulation

Loose-fill insulation is usually composed of materials used in bulk form, supplied in bags or bales, and intended to be poured or blown into place or packed by hand. It is used to fill stud spaces or to build up any desired thicknesses on horizontal surfaces. Loose-fill insulation includes rock, glass, and slag wool, wood fibers, granulated cork, ground or macerated woodpulp products, vermiculite, perlite, sawdust, and wood shavings.

Reflective Insulation

Most materials reflect radiant heat, and certain of them have this property to a high degree. Some emit less heat than others. For reflective insulation, high reflectivity and low emissivity are required, as provided by aluminum foil, sheet metal coated with an alloy of lead and tin, and paper products coated with a reflective oxide composition. Aluminum foil is available in sheets mounted on paper, in corrugated form supported on paper, or mounted on the back of gypsum lath or paper-backed wire lath. Reflective insulation is installed with the reflective surface facing or exposed to an air space. It is generally considered to be effective only when the air space is between ¾ and 4 inches deep. Reflective surfaces in contact with other surfaces lose their reflective properties.

Miscellaneous Insulation

Some insulation does not fit in the classifications used here, such as insulation blankets made up of multiple layers of creped or reflective paper. Other types, such as lightweight vermiculite and perlite aggregates, are sometimes used in plaster as a means of reducing heat transmission. Lightweight aggregates made from blast-furnace slag, burned-clay products, and cinders are used in concrete and concrete blocks. The thermal conductivity of concrete products made of such lightweight aggregates is substantially lower than that of concrete products made of gravel and stone aggregates.

Other materials are foamed-in-place insulations, which include sprayed plastic foam types. Other sprayed insulation is usually inorganic fibrous material blown against a clean surface to which a coat of adhesive has been applied. Sprayed insulation is usually applied up to a thickness of 2 inches and the surface lightly tamped to obtain uniformity; density is usually 1 to 3½ p.c.f. It is often left exposed to serve as an acoustical treatment as well as insulation.

Polystyrene and urethane plastic foams may be molded or foamed-in-place. Urethane insulation may also be applied by spraying. These materials can be used in the field and are applied as roof, wall, and floor insulation. Expanded polystyrene is most commonly used in board form in thicknesses of ½ to 2 inches.

The methods of applying these materials are undergoing rapid changes because of improving technology, and manufacturers' recommendations should be consulted before using them as building insulations.

METHODS OF HEAT TRANSFER

Heat seeks to attain a balance with surrounding conditions, just as water will flow from a higher to a lower level. When occupied buildings are heated to maintain inside temperature in the comfort range, there is a difference in temperature between inside and outside. Heat will therefore be transferred through walls, floors, ceilings, windows, and doors at a rate that bears some relation to the temperature difference and to the resistance to heat flow of intervening materials. The transfer of heat takes place by one or more of three methods—conduction, convection, and radiation (fig. 21-1).

Conduction is defined as the transmission of heat through solid materials; for example, the conduction of heat along a metal rod when one end is heated in a fire. Convection involves transfer of heat by air currents; for example, air moving across a hot radiator carries heat to other parts of the room or space. Heat also may be transmitted from a warm body to a cold body by wave motion through space, and this process is called radiation because it rep-resents radiant energy. Heat obtained from the sun is radiant heat.

Heat transfer through a structural unit composed of a variety of materials may include one or more of the three methods described. Consider a frame house with an exterior wall composed of gypsum lath and plaster, 2- by 4-inch studs, sheathing, sheathing paper, and bevel siding. In such a house, heat is transferred from the room atmosphere to the plaster by radiation, conduction, and convection, and through the lath and plaster by conduction. Heat transfer across the stud space is by radiation and convection. By radiation, it moves from the back of the gypsum lath to the colder sheathing; by convection, the air warmed by the lath moves upward on the warm side of the stud space, and that cooled by the sheathing moves downward on the cold side. Heat transfer through sheathing, sheathing paper, and siding is by conduction. Some small air spaces will be found back of the siding, and the heat transfer across these spaces is principally by radiation. Through the studs from gypsum lath to sheathing, heat is transferred by conduction and from the outer surface of the wall to the atmosphere, it is transferred by convection and radiation.

The thermal conductivity of a material is an inverse measure of the insulating value of that material. The customary measure of heat conductivity is the amount of heat in British thermal units that will flow in 1 hour through 1 square foot of a layer 1 inch thick of a homogeneous material, per 1° F. temperature difference between surfaces of the layer. This is usually expressed by the symbol k.

Where a material is not homogenous in structure, such as one containing air spaces like hollow tile, the term conductance is used instead of conductivity. The conductance, usually designated by the symbol C, is the amount of heat in British thermal units that will flow in 1 hour through 1 square foot of the material or combination of materials per 1° F. temperature difference between surfaces of the material, or the equivalent of a surface layer, or a dead air space with or without a reflective surface.

Resistivity and resistance (direct measures of the insulating value) are the reciprocals of transmission (conductivity or conductance) and are represented by the symbol R. Resistivity, which is unit resistance, is the reciprocal of k and is given the same symbol R as resistance in the technical literature because resistances

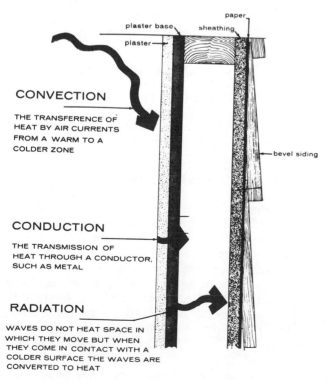

CONVECTION

THE TRANSFERENCE OF HEAT BY AIR CURRENTS FROM A WARM TO A COLDER ZONE

CONDUCTION

THE TRANSMISSION OF HEAT THROUGH A CONDUCTOR, SUCH AS METAL

RADIATION

WAVES DO NOT HEAT SPACE IN WHICH THEY MOVE BUT WHEN THEY COME IN CONTACT WITH A COLDER SURFACE THE WAVES ARE CONVERTED TO HEAT

plaster base
plaster
paper
sheathing
bevel siding

Figure 21-1.—Methods of heat transfer.

are added together to calculate the total for any construction. The overall coefficient of heat transmission through a wall or similar unit air to air, including surface resistances, is represented by the symbol U. U defines the movement in British thermal units per hour, per square foot, per 1° F. The total resistance of a construction would be $R = \dfrac{1}{U}$.

HEAT LOSS

Through Walls

The heat loss through walls and roofs made of different materials can be found by computing the overall coefficients of heat transmission, or U values, of the construction assemblies. To determine the U value by test would be impractical in most cases, but it is a simple matter to calculate this value for most combinations of materials commonly used in building construction whose thermal properties are known.

Table 21-1 gives conductivity and conductance values with corresponding resistivity and resistance values used in calculating the thermal properties of construction units. No values are included for reflective materials because they depend on the reflectivity and permanence of the surface brightness, the direction of heat flow, and to a lesser extent the depth of the adjoining air space. For information on calculating the amount to be credited to reflective insulations consult the American Society of Heating, Refrigerating, and Air-Conditioning Engineers' ASHRAE Handbook of Fundamentals.

To compute the U value: Add the resistance of each material, exposed surface, and air space in the given section, using values given in table 21-1. The sum of these resistances divided into 1 (reciprocal of the sum) gives the coefficient U.

Example: Calculate the U factor for winter (heating) conditions through the stud space of a conventional wood frame wall consisting of ½-inch gypsum board, air space, ½-inch regular density insulating board sheathing, and beveled wood siding, ¾ by 10 inches lapped:

SUMMARY OF RESISTANCE

Interior surface (still air)	0.68
Gypsum board	.45
Air space	1.16
Insulating board sheathing	1.32
Siding	1.05
Outside surface (15 m.p.h. air)	.17
Overall resistance	4.83

$$U = \frac{1}{R} = \frac{1}{4.83} = 0.21 \text{ Btu per hour per}$$

square foot per °F. difference in temperature (through the stud space).

For the U value through the stud, substitute the resistivity of wood based on the depth of the stud for the resistance value of the air space. For example, a species of wood that has a k value of 0.88 has a resistivity of 1.14. The resistance of a nominal 2- by 4-inch stud is:

$$3\tfrac{1}{2} \times 1.14 = 3.99$$

Substituting the value of 3.99 for the air space value of 1.16 gives an overall resistance value of 7.66 or a U value of 0.13. Assuming the area of the studs, plates, and headers represents 15 percent of the wall area, the corrected transmission value becomes:

$$U = \frac{0.21 \times 85 + 0.13 \times 15}{100} = 0.20$$

If 2-inch mineral wool blanket insulation with one 1½-inch air space is used in the stud space, the resistance contributed by the stud space becomes 8.56 instead of the single air space value of 1.16, and the overall resistance of the wall through the stud space is 12.23, compared with 4.83 without insulation, and the U value becomes 0.08 instead of 0.21.

Through Doors and Windows

In determining heat loss for houses, the loss through doors and windows should be included in the computations. Table 21-2 gives heat transmission values for doors and windows.

WHERE TO INSULATE

Insulation is used to retard the flow of heat through ceilings, walls, and floors if wide temperature differences occur on opposite sides of these structural elements (see fig. 21-2). In dwellings, for example, insulation should be used in the ceiling of those rooms just below an unheated attic. If the attic is heated, the insulation should be placed in the attic ceiling and in the dwarf walls extending from the roof to the floor. All exterior walls should be insulated. Floors over unheated basements, crawl spaces, porches, or garages should also be insulated.

Table 21-1.—*Conductivities (k), conductances (C), and resisitivities or resistances (R) of building and insulating materials (design values)*[1 2]

Material	Description	Thickness	Density	Conductivity or conductance (k)	Conductivity or conductance (C)	Resistivity or resistance (R) $\frac{1}{k}$	Resistivity or resistance (R) $\frac{1}{C}$
		In.	*P.c.f.*				
Air spaces[3]	Horizontal position:						
	Heat flow up:						
	Winter	¾ to 4	----	----	1.02	----	0.98
	Summer	¾ to 4	----	----	1.28	----	.78
	Heat flow down:						
	Winter	¾ to 4	----	----	.68	----	1.47
	Summer	¾ to 4	----	----	1.09	----	.92
	Sloping, 45° position						
	Heat flow up, winter	¾ to 4	----	----	.96	----	1.04
	Heat flow down, summer	¾ to 4	----	----	1.15	----	.87
	Vertical position; heat flow horizontal						
	Winter	¾ to 4	----	----	.86	----	1.16
	Summer	¾ to 4	----	----	1.18	----	.85
Air surfaces:[4] Still air	Horizontal position:						
	Heat flow up	----	----	----	1.63	----	.61
	Heat flow down	----	----	----	1.08	----	.92
	Sloping, 45° position:						
	Heat flow up	----	----	----	1.60	----	.62
	Heat flow down	----	----	----	1.32	----	.76
	Vertical position; heat flow horizontal	----	----	----	1.46	----	.68
Moving air	Any position; heat flow any direction:						
15-m.p.h. wind	Winter	----	----	----	6.00	----	.17
7½-m.p.h. wind	Summer	----	----	----	4.00	----	.25
	Asbestos-cement board	----	120	4.0	----	.25	----
		⅛	----	----	33.00	----	.03
	Gypsum or plaster board	⅜	50	----	3.10	----	.32
		½	50	----	2.25	----	.45
		----	34	.80	----	1.25	----
	Plywood	¼	34	----	3.20	----	.31
		⅜	34	----	2.13	----	.47
		½	34	----	1.60	----	.62
	Plywood or wood panels	¾	----	----	1.07	----	.93
	Insulating board	----	18	.40	----	2.50	----
	Regular-density sheathing	½	18	----	.76	----	1.32
		25/32	18	----	.49	----	2.06
	Intermediate-density sheathing	½	25	----	.82	----	1.22
	Nail-base sheathing	½	27	----	.88	----	1.14
	Shingle backer	5/16	18	----	1.28	----	.78
		⅜	18	----	1.06	----	.94
	Sound-deadening board	½	15	----	.74	----	1.35

(Continued)

Building boards, panels, sheathing, etc.[5]	Tile, plain or acoustic	----	18	.40	----	2.50	----
		½	18	----	.80	----	1.25
		¾	18	----	.53	----	1.89
	Insulating roof deck (nominal)	1½	----	----	.24	----	4.17
		2	----	----	.18	----	5.56
		3	----	----	.12	----	8.33
	Laminated paperboard, homogeneous repulped paperboard	----	30	.50	----	2.00	----
		⅜	30	----	1.33	----	.75
	Medium-density hardboard siding	7/16	40	----	1.49	----	.67
	Other medium-density hardboard	----	50	.73	----	1.37	----
	High-density hardboard:						
	Service, tempered service	----	55	.82	----	1.22	----
	Standard, tempered	----	63	1.00	----	1.00	----
	Particleboard:						
	Low-density	----	37	.54	----	1.85	----
	Medium-density	----	50	.94	----	1.06	----
	High-density	----	62.5	1.18	----	.85	----
	Wood:						
	Fir or pine sheathing	¾	----	----	1.06	----	.94
	Fir or pine	1½	----	----	.53	----	1.89
Building paper	Vapor-permeable felt	----	----	----	16.70	----	.06
	Vapor seal:						
	2 layers of mopped 15-pound felt	----	----	----	8.35	----	.12
	Plastic film	----	----	----	----	----	Negligible
Flooring materials	Carpet and fibrous pad	----	----	----	.48	----	2.08
	Carpet and rubber pad	----	----	----	.81	----	1.23
	Cork tile	----	25	.45	----	2.22	----
		⅛	----	----	3.60	----	.28
	Felt, flooring	----	----	----	16.70	----	.06
	Floor tile or roll material, average value for asphalt, linoleum, rubber, vinyl	⅛	----	----	20.00	----	.05
	Plywood subfloor	⅝	----	----	1.28	----	.78
	Wood subfloor	¾	----	----	1.04	----	.96
	Wood, hardwood finish	¾	----	----	1.47	----	.68
	Hardwood underlayment	.22	55	----	3.73	----	.26
	Particleboard underlayment	⅝	40	----	1.22	----	.82
	Terrazzo	1	----	----	12.50	----	.08
Insulating materials Blanket and batt	Cotton fiber[6]	----	0.8 to 2.0	.26	----	3.85	----
	Mineral wool, fibrous form, processed from rock, slag, or glass[6]	----	1.5 to 4.0	.27	----	3.70	----
	Wood fiber[6]	----	2.0 to 3.5	.30	----	3.33	----
		----	9.5	.25	----	4.00	----
	Roof insulation board (wood and mineral fiber)	½	15	----	.72	----	1.39
		1	----	----	.36	----	2.78
		1½	----	----	.24	----	4.17
		2	----	----	.19	----	5.26
		2½	----	----	.15	----	6.67
		3	----	----	.12	----	8.33
Glass fiber Block and panel	Wood wool (excelsior) cement	----	22	.60	----	1.67	----
	Expanded polyurethane	----	1.5	.16	----	6.25	----
	Polystyrene:						
	Extruded (R 12 blown)	----	3.5	.19	----	5.26	----
	Beads, molded	----	1.0	.28	----	3.57	----

Table 21-1.—*Conductivities (k), conductances (C), and resistivities or resistances (R) of building and insulating materials (design values)*[1][2]—continued

Material	Description	Thickness (In.)	Density (P.c.f.)	Conductivity or conductance (k)	Conductivity or conductance (C)	Resistivity or resistance (R) 1/k	Resistivity or resistance (R) 1/C
Glass fiber Block and panel—continued	Cellular glass	---	9.0	.40	---	2.50	---
	Cork board (without added binder)	---	6.5 to 8.0	.27	---	3.70	---
	Hog hair (with asphalt binder)	---	8.5	.33	---	3.03	---
Loose fill	Macerated paper or pulp products	---	2.5 to 3.5	.27	---	3.70	---
	Mineral wool (glass, slag, or rock)	---	2.0 to 5.0	.30	---	3.33	---
	Sawdust or shavings	---	8.0 to 15.0	.45	---	2.22	---
	Vermiculite (expanded)	---	7.0 to 8.0	.47	---	2.13	---
	Wood fiber: Redwood, hemlock, or fir	---	2.0 to 3.5	.30	---	3.33	---
	Perlite (expanded)	---	5.0 to 8.0	.37	---	2.70	---
Masonry materials	Cement mortar		116	5.0	---	.20	---
	Gypsum-fiber concrete 87½ percent gypsum, 12½ percent wood chips		51	1.66	---	.60	---
	Lightweight aggregates including expanded shale, clay, or slate; expanded slags, cinders, pumice, perlite, vermiculite; also cellular concretes		120	5.2	---	.19	---
			100	3.6	---	.28	---
			80	2.5	---	.40	---
			60	1.7	---	.59	---
			40	1.15	---	.86	---
			30	.90	---	1.11	---
			20	.70	---	1.43	---
	Sand and gravel or stone aggregate: Ovendried		140	9.0	---	.11	---
	Not dried		140	12.0	---	.08	---
	Stucco		116	5.0	---	.20	---
Masonry units	Brick: Common		120	5.0	---	.20	---
	Face		130	9.0	---	.11	---
	Clay tile, hollow: 1 cell deep	3	---	---	1.25	---	.80
		4	---	---	.90	---	1.11
	2 cells deep	6	---	---	.66	---	1.52
		8	---	---	.54	---	1.85
	3 cells deep	10	---	---	.45	---	2.22
		12	---	---	.40	---	2.50
	Concrete blocks, three oval core: Sand and gravel aggregate	4	---	---	1.40	---	.71
		8	---	---	.90	---	1.11
		12	---	---	.78	---	1.28
	Cinder aggregate	3	---	---	1.16	---	.86
		4	---	---	.90	---	1.11
		8	---	---	.58	---	1.72
		12	---	---	.53	---	1.89
	Lightweight aggregate (expanded shale, slag, etc.) 3-core	3	75	---	.79	---	1.27
		4	---	---	.67	---	1.50
	2-core	8	75	---	.50	---	2.00
	3-core	12	---	---	.44	---	2.27

(Continued)

Category	Material	Thickness (in.)	Density (lb per cu ft)	Conductivity (k)	Conductance (C)	Resistance per inch (1/k)	Resistance for thickness listed (1/C)
Metals	Gypsum partition tile:						
	3 by 12 by 30 inches, solid				.79		1.26
	3 by 12 by 30 inches, 4-cell				.74		1.35
	4 by 12 by 30 inches, 3-cell				.60		1.67
	Stone masonry, limestone, or sandstone			12.50		.08	
	See chapter 30, table 3, of 1972 ASHRAE Handbook of Fundamentals.						
Plastering materials	Cement plaster:						
	Sand aggregate		116	5.0		.20	
	3/8	3/8			13.3		.08
	3/4	3/4			6.66		.15
	Gypsum plaster:						
	Lightweight aggregate	1/2	45		3.12		.32
		5/8	45		2.67		.39
		3/4			2.13		.47
	On metal lath						
	Perlite aggregate		45	1.5		.67	
	Vermiculite aggregate		45	1.7		.59	
	Sand aggregate	3/8	105	5.6	14.9	.18	.07
		1/2	105		11.10		.09
		5/8	105		9.10		.11
	On metal lath	3/4	105		7.70		.13
	On wood lath	3/4			2.50		.40
Roofing	Asbestos-cement shingles		120		4.76		.21
	Asphalt roll roofing		70		6.50		.15
	Asphalt shingles		70		2.27		.44
	Built-up roofing	3/8	70		3.00		.33
	Slate	1/2			20.00		.05
	Sheet metal			400+			Negligible
	Wood shingles				1.06		.94
Siding materials (on flat surface)	Shingles:						
	Wood:						
	16-inch, 7½-inch exposure				1.15		.87
	Double, 16-inch, 12-inch exposure				.84		1.19
	Plus 5/16-inch insulation backer board				.71		1.41
	Siding:						
	Asbestos-cement, ¼-inch, lapped				4.76		.21
	Asphalt roll siding				6.50		.15
	Asphalt insulating siding (½-inch board)				.69		1.45
	Wood:						
	Drop, 1 by 8 inches				1.27		.79
	Bevel: ½ by 8 inches, lapped				1.23		.81
	¾ by 10 inches, lapped				.95		1.05
	Plywood, 3/8 inch, lapped				1.59		.63
	Architectural glass				10.00		.10
	Medium-density hardboard (panel or lapped)	7/16	40		1.49		.67
Wood[7]	Maple and oak		43	1.11		.90	
	Douglas-fir		32	.88		1.14	

[1] Table based on values obtained at Forest Products Laboratory for wood and wood-base materials, and from table 3A, chapter 20 of ASHRAE Handbook of Fundamentals for materials of other base.

[2] Representative values for dry materials at 75° F., mean temperature, intended for design and comparison, not for specifications.

[3] Values are dependent on thickness of air space and temperature and temperature differences across air space; ones presented are mean values for typical conditions. For more accurate values consult chapter 20 of ASHRAE Handbook of Fundamentals. Air space resistance values are for spaces faced both sides with ordinary nonreflective building materials.

[4] Surface resistance values are for ordinary nonreflective materials.

[5] See also insulating materials (block and panel) and siding materials.

[6] Includes paper backing and faces, if any, having nonreflective surfaces.

[7] Thermal properties of other woods are discussed in chapter 3.

CONDENSATION AND VAPOR BARRIERS

Two types of condensation create a problem in buildings during cold weather; that which collects on the inner surfaces of windows, ceilings, and walls, and that which collects within walls or roof spaces. Surface condensation is quite common in industrial buildings where relative humidity is high. In a factory or warehouse, water dripping from a ceiling may seriously damage manufactured materials and machinery. "Sweating" walls and windows also are a serious nuisance.

Condensation may collect on the indoor surface of exterior walls of houses, particularly behind furniture, or in outside closets, causing damage to finish, furniture, and flooring. It may also collect on windows, particularly those unprotected by storm sash. Water running off the windows may create conditions favorable to decay in wood sash, cause rust in steel sash, and damage window finish and

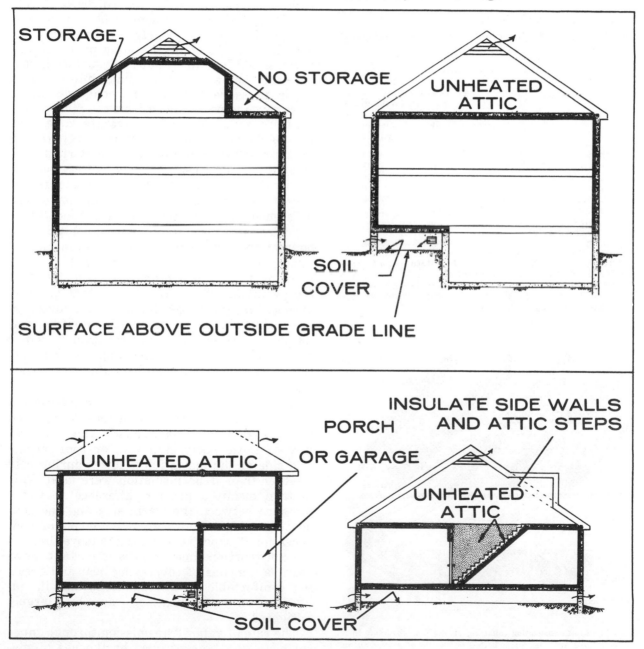

Figure 21-2. —Insulation should be installed in side walls between heated rooms and outdoors, in walls and floors between unheated garages and porches and heated rooms, in floors in basementless houses, in ceilings below unheated attics, in roofs over heated rooms, and in side walls and below stairs leading to unheated attics.

walls and floors below the windows.

To prevent surface condensation, the relative humidity in the building must be reduced or the surface temperature must be raised above the dew point of the atmosphere. Adding insulation to a wall or roof reduces heat transfer through the unit, and the inside surface temperature is increased accordingly. The amount of insulation required for given conditions can be calculated. Storm sash or double glazing reduces condensation on windows.

Table 21-2.—*Coefficients of transmission (U) commonly used for doors, windows, and glass blocks* [1]

	U value	
Item	Exposed unit alone	With glass storm door or storm window
1⅛-inch solid wood, door [2] ____	0.55	0.34
1⅜-inch solid wood door [2] ____	.48	.31
1⅝-inch solid wood door [2] ____	.43	.28
Single glass wood window ____	1.02	.50
Single glass metal window ___	1.13	.56
Glass block [3] _____	.60	____
Single glass _____	1.13	____
Double glass (insulated ¼-in. space) _____	.65	.38

[1] Table based in part on the data from 1972 ASHRAE Handbook of Fundamentals.
[2] For doors containing thin wood panels or glass, use the same U value as shown for single windows.
[3] Blocks are 6 by 6 by 4 in.

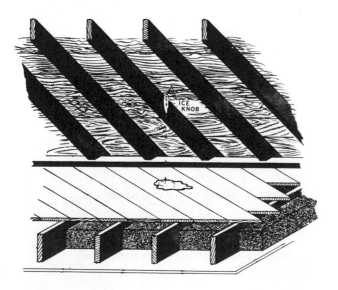

Figure 21-3.—**Condensation collecting on the roof sheathing in an unventilated attic. Ice has collected on the tip of protruding nail (arrow). When outdoor temperatures rise after a cold spell or when the sun strikes the roof the ice melts and drips down to the ceiling below.**

Moisture sometimes condenses within the walls or roof spaces of buildings if relative humidity is comparatively high during cold weather (fig. 21-3). In walls, this type of condensation may result in decay of wood, rusting of steel, and damage to exterior paint coatings. In roofs and ceilings it may cause stained finish and loosened plaster and increase the chances of decay in structural members.

When outdoor temperatures are low, water vapor will sometimes pass through permeable inner surface materials and condense within a wall or roof space on some cold surface. Key to this situation is a surface temperature below the dew point of the atmosphere on the warm side. When the condensing surface is considerably below the dew point, differences in vapor pressure between the cold and warm sides cause vapor to move from the high-vapor-pressure zone to the low-pressure zone. The rate of movement is more or less proportional to the resistance of interposed materials. The amount of condensation that collects on the condensing surface depends upon the resistance to permeance of intervening materials, differences in vapor pressure, and time. There will also be some difference in vapor pressure between the condensing surface and the outdoor atmosphere. Some part of the water vapor reaching the condensing surface will therefore escape outside through materials that are permeable. Materials used for side wall coverings preferably should be permeable. Roofing materials are generally highly resistant to vapor transmission. Wood shingles applied over narrow roof boards are not resistant to vapor transmission.

Insulation can cause increased condensation under certain conditions. Heat flow is reduced by insulation, and consequently the temperature of those parts of a wall or roof on the cold side of the insulation is lower during cold weather than if no insulation were used. This in turn means a greater difference in vapor pressure between the warm side and the condensing surface and a greater amount of condensation if vapor is permitted to move through the construction. Insulation is important, however, as a means of conserving heat and creating comfortable living conditions and its influence on condensation can be largely mitigated.

The rate of vapor transmission through inner surfaces may be controlled by the use of materials having high resistance to vapor movement. Such vapor barriers should be located on the warm surface of the wall (fig. 21-4) so that the temperature of the barrier will always be above the dew point of the heated space. Broken barriers of poor installation around outlet boxes in exterior walls will cause escape of water vapor and often condensation problems. The use of a full wall covering with a vapor barrier (enveloping) and calking around the perimeter of the outlet boxes will minimize these problems. Vapor barrier requirements differ with geographic climates, so local conditions determine what is required. More detailed information than is possible to provide in this handbook is presented in references listed in the bibliography.

For new construction, the barrier may be any one of several materials, such as asphalt-impregnated and surface-coated paper applied over the face of the studs, plastic films, gypsum lath with aluminum-foil backing, blanket insulation with vapor-resistive cover, and reflective insulation. For existing construction, certain types of paint coatings add materially to the resistance to vapor transfer. One coat of aluminum primer followed by two decorative coats of flat paint or lead and oil seems to offer satisfactory resistance.

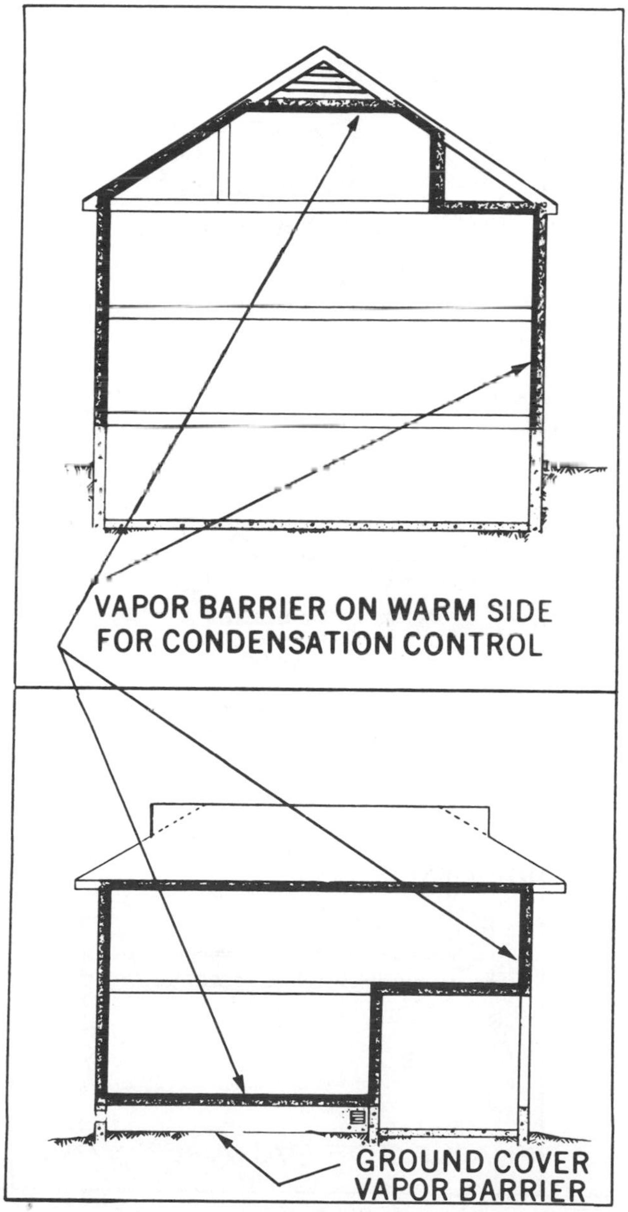

VENTILATION

Attics and roof spaces are generally provided with suitable openings for ventilation, partly for summer cooling and partly to prevent winter condensation (fig. 21-5). For gable roofs, louvered openings are provided in the gable ends, allowing at least 1 square foot of louver opening (minimum net area) for each 300 square feet of projected ceiling area. For hip roofs, inlet openings are usually provided under the over-hanging eaves with a globe or ridge ventilator at or near the peak for an outlet. The inlets should equal 1 square foot to each 900 square foot of projected ceiling area and the outlets 1 square foot (net area) to each 1,600 square feet. Ventilation for flat roofs should be developed to suit the method of construction.

VAPOR BARRIER ON WARM SIDE FOR CONDENSATION CONTROL

GROUND COVER VAPOR BARRIER

Figure 21-4.—To control cold weather condensation, a vapor barrier should always be installed on the warm side of the construction.

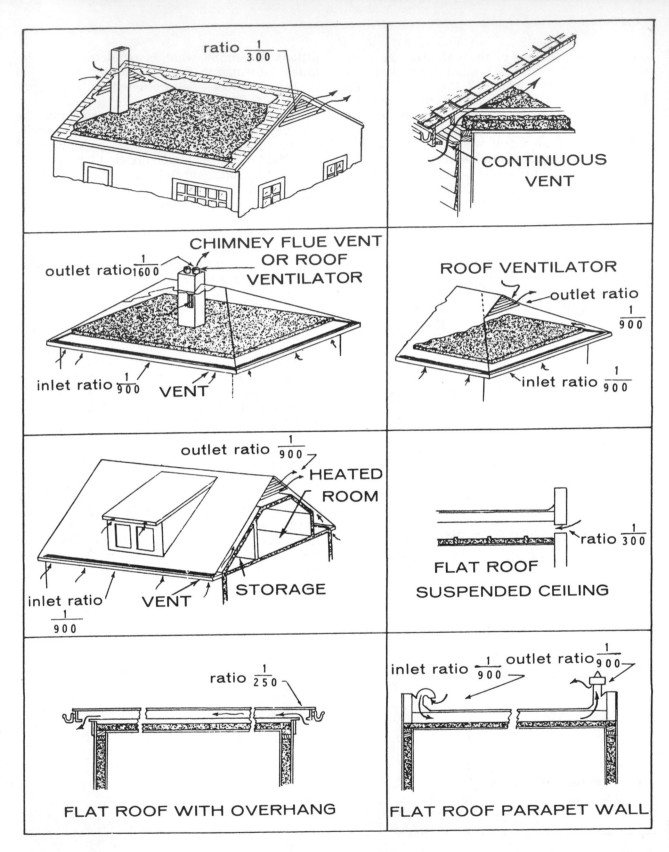

Figure 21-5.—Methods used to ventilate attics and roof spaces. Air inlet openings under the eaves of pitched roofs in addition to outlet openings near peak provide air movement independent of the effect of wind. For flat roofs where joists are used to support ceiling and roof, continuous vents, open to each joist space, are needed for both inlets and outlets. The dormer has inlets at eave and a roof space opens into attic. All dormers should be carefully framed to assure means of ventilation in the roof space. The sketches show the ratio of free opening in louvers and vents to the area of the ceiling in the rooms below.

Insulation Thickness

Given the material and form of insulation most widely used in the typical wood-frame house, how shall we determine how much to use? Most previous methods of design emphasize comfort criteria, i.e., the essential design

requirement being that occupants of a building are entitled to a thermally comfortable environment. It is presumed to follow that this would also be a healthful environment. Among widely used rules of thumb to achieve average comfort conditions are:

1. "All Weather Comfort Standard," as developed by manufacturers, equipment suppliers, and power companies. This standard provides a range of choice of insulation thickness based on three arbitrary weather zones (6), using the degree-day[1]/ heating requirement to define these zones.

2. Comfort criteria based on the average difference between desired room air temperature and that of enclosing surfaces. An average balance of heat gained by convection and heat lost by radiation is thus obtained, and criteria are applied from experience as to what this difference should be (4).

These standards address the fundamental requirements of comfort and health of the occupants but provide little guidance to the designer or builder on the economic use of insulation or on energy conservation. Illustrations are sometimes provided to show the savings in heating cost which can be made by adding another inch of insulation, but no attempt is made to arrive at an economic optimum.

A publication of the National Association of Home Builders (5) recognizes the growing use of air conditioning in American homes and provides an excellent guide for the designer or builder in analyzing heat losses and energy costs. It provides heating and cooling worksheets whereby the builder may enter trial combinations of insulation and openings, determining total heat transfer and equipment sizes. Simple cost calculations are provided, establishing heating and cooling costs for a given heat transfer in any part of the United States. It does not, however, identify the optimum economic thickness of insulation which results in the least total cost (insulation plus operating cost) to the owner. Rather, it leaves it to the builder to choose constructions in a manner to result in least first-cost, and to use what he feels is a reasonable level of insulation thickness acceptable to a buyer.

The first-cost approach has never been conducive to the best interests of the typical home buyer, since the few hundred dollars saved by eliminating or minimizing insulation can cost the owner a few thousand dollars over the life of the building in increased energy costs. We can no longer afford this kind of energy waste.

---

[1]/ A degree-day is a unit used to predict seasonal fuel consumption for heating. For 1 day, the number of degree-days is equal to the number of degrees that the mean temperature for that day is below 65° F. For the heating season, the number of degree-days is the sum of degrees for all days that the mean temperature falls below 65° F. The average seasonal total of degree-days over a number of years is useful to estimate average annual heating costs for a given locality.

Optimum Thickness of Insulation

Minimum total costs result when first-cost of insulation plus the corresponding cost in energy over the useful life of the building are a minimum. A relatively simple engineering and cost analysis can be used to identify optimum thicknesses of any kind of insulation for local energy costs in any climate. This is no more than the same cost-effectiveness approach applied to many materials and types of equipment in industry and government. The method is commonly applied in heavy construction to many aspects of design, yet it has not been widely used in insulation design.

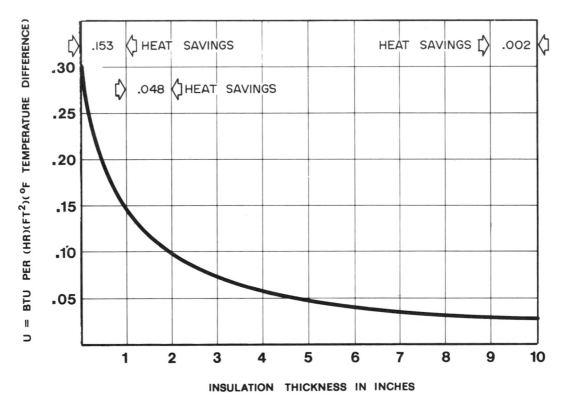

Fig. 21-6. — *Effect of increasing insulation thickness on heat losses in a typical roof-ceiling construction.*

The design of insulation for optimum total cost economy, as applied in this section, does not at first seem to satisfy the objective of maximum conservation. Optimum conservation of energy would seem to imply maximum energy savings, perhaps regardless of costs. In fact, though, cost does enter the picture. As energy becomes less available, its cost will rise, and the development of new energy sources will very probably increase costs. As the cost of energy increases, so will the thickness of insulation needed to effect minimum total cost, and conservation automatically increases.

The economic method of insulation design presented here results in greater amounts of insulation and greater energy savings than previous methods. It is a practical method of obtaining maximum conservation which is economically realistic, as the fuel conserved per inch of insulation rapidly diminishes beyond the point of optimum economy. This can be readily seen in fig. 21-6, where heat loss in a ceiling is plotted against insulation thickness. The heat saved by the 1st inch of insulation is 3 times that saved by the 2d inch and 75 times that saved by the 10th inch. Excessive quantities of insulation would quickly reach a point of no return, where the energy consumed in insulation manufacture, transportation, and installation, plus that which would be incurred by additional framing, exceeds the energy saved.

A Comparison of Design Methods

The foregoing method of selecting insulation thickness in a wood-framed home, based on optimum economy, generally results in a greater insulation requirement than methods previously in use. This can be seen in table 21-3, which compares with the minimum thickness of ceiling insulation as determined by several criteria. The method using the difference between air temperature and inside surface temperature gives very inadequate thicknesses for the various locations, although a temperature difference of only 4° F was used here for a relatively high degree of comfort. The Industry All-Weather Comfort Standard results in a range of choices, obviously leaving this judgment to the local designer; but the high end of the range does not result in good economy in most areas. Only the new FHA Minimum Property Standards give reasonable economic thicknesses, though the accuracy of these standards is variable in different locations.

Table 21-3. Minimum insulation thickness for a wood-framed ceiling construction under various criteria
(inches)

Location	San Diego	Seattle	St. Louis	Duluth
Minimum comfort criteria, based on allowable ceiling temperature	1	1	2	3
Industry All-Weather Comfort Standard, National Forest Products Association	2-3	2-3	2-3	3-5
Federal Housing Administration Minimum Property Standards, 1974, new construction	3	5	5	6
Optimum economy (to next higher full inch)	3	6	7	9

Heat loss and heat gain in a typical, uninsulated wood-framed house are not limited to that passing through wall and ceiling constructions, though the latter may amount to some 50 to 60 percent of total loss or gain. Other sources of loss include floor or ground, windows and doors, ventilators or cracks, and miscellaneous, such as may occur from ducts, chimneys, and water pipes. The relative percentage of loss occurring from each source, of course, will vary with areas involved and their insulation or sealing.

GROUND LOSSES AND CRAWL SPACES

Several types of construction are used below the first floor of a home, either because of local practice or specific design advantages. With a full or partial basement, minimal heat losses to the basement occur in most climates, and the basement is warmed sufficiently for occasional use for laundry, workshop or storage, and to protect water pipes in extreme weather. Such a nominally unheated basement does not require insulation in the first floor. However, when the basement is designed for normal occupancy, floor and wall insulation may be required in the basement itself for economical heating. Heat transfer through basement floors and walls depends on their construction, ground temperatures at their outside surfaces, and the heat conductivity of the ground. The latter two factors are variable and may be unknown, and engineering assumptions based on local practice are usually necessary. Lower average wall temperatures in winter will normally require more insulation for walls than for floors. Because the temperature differences between outside and inside surfaces may not be accurately known over the heating season and because insulation costs are generally higher for masonry constructions, a determination of optimum amount is not practical. The objective for masonry walls should be to limit inside surface temperatures to comfort levels.

For concrete floor slabs in contact with the ground, experiments indicate that heat losses occur principally at the edges of the slabs and are negligible in interior areas. Therefore, insulation of the perimeter of the slab is sufficient to reduce heat loss to an economical level. In colder climates, as much as 2 inches of rigid insulation may be used around the perimeter, extending from the top of the slab, down the foundation, and under the slab about 2 feet. An alternate method used in milder climates is to place the insulation vertically on the outside of the foundation, extending from the top of the foundation to a point 6 inches below grade, using perhaps 1-inch thickness. Insulation placed in or on the ground in this manner must be waterproof, such as foamed plastic or glass board which does not absorb water. Slabs on the ground may usually be made comfortable enough for children's play by using 4 to 6 inches of gravel under the floor--also desired for moisture control--or by such floor surfaces as wood or thick pads and carpets. None of these measures may be needed, however, in milder southern climates where these slabs are most common.

Crawl space construction typically uses floor joists, supported by foundations and interior beams and posts, over a shallow excavation sufficient for access to heating ducts and pipes. A crawl space may be vented to the exterior the year around for moisture control, if ground moisture conditions are severe. However, control of ground moisture is usually possible with a ground cover of polyethylene sheet or roofing felts, so that the ventilating apertures in the outside wall may be closed off during the heating season.

When a closed crawl space can be used, insulation is not used in the floor, and ducts and hot-water pipes need not be insulated. Small amounts of heat from these sources keep the crawl space reasonably warm in mild winter climates and protect other piping. In severe winter climates, insulation may be needed; this is usually placed vertically on the inside of the foundation walls to limit heat loss. Again, plotting an optimum thickness for typical conditions is not possible, as average ground temperature, the proportion of the wall above grade, and kind of insulation will vary. The University of Illinois Small Homes Council (10) recommends 1 inch of foamed plastic or glass fiberboard for climates with an average outside air minimum temperature of -10° F or higher, with greater thickness desired for lower minimum temperature. A greater thickness may well be used to pack the header blocks at the ends of the joists to a depth equal to the optimum wall insulation thickness for the region.

When ground moisture conditions are severe, a crawl space may have to be vented continuously to control humidity and protect the structure. A ground cover can become ineffective if liquid water runs under it and through the laps; it then becomes a liability by retaining pools of water on the top as ground water recedes. Under these conditions, maximum ventilation is desired, and temperatures within the crawl space may approach that of the outside air.

In cold areas, water pipes should be insulated to prevent freezing, and heating ducts to limit heat loss. For comfortable floors and limited loss of heat from the area above, insulation in the floor structure is required. This is usually placed between the floor joists in the form of batts or blankets. In a new house, batts may be placed from above before the subfloor is laid, stapling flanges to the joists much as is done in walls. In existing construction, blankets may be placed from the crawl space, taking care that the moisture-barrier is up, and retained in place by spring clips, chicken wire, or other moisture resistant material nailed to the bottom of the joists.

Heat losses through a floor are not as great as losses through a ceiling, because the average inside-outside temperature difference is less. Average crawl space temperatures are moderated by ground temperature when there is little wind for ventilation and there is no solar load to provide for in summer cooling. Hence, insulation thickness in fully ventilated floors may be somewhat less than that of ceilings. For new construction where insulation cost would be about the same in each case, it is recommended to use optimum ceiling thickness for the climate, less 2 inches. This will result in a thickness close to the optimum for average open crawl space conditions.

STORM WINDOWS AND DOORS

Windows and doors account for a very high percentage of heat loss in a typical residence, which may be as much as 30 percent of the total heating and cooling load. The single-pane window is a particular problem, with very low resistance to the flow of heat through the glass and high transparency to radiant heat from the sun.

Windows, however, offer important amenities which outweigh problems of thermal control, furnishing daylight, ventilation, and an outward view. Windowless buildings have been designed and built but are not successful when continuous human occupancy is involved. Most building codes require that glass areas be at least 10 percent of floor areas, and the University of Illinois Small Homes Council recommends glass areas of at least 20 percent of floor areas (9).

Thermal problems related to windows may be greatly reduced by good design. Orientation to reduce solar loads is an important factor. If principal window areas face south, greater control of solar heat can be achieved by overhangs or solar shades which permit the entry of low winter rays but shade the window from the high summer sun. East and west windows are more difficult to shade in summer because of the sun's lower inclination, and loads are higher. Although drapes or venetian blinds may be used, these limit outward view and natural lighting. North windows take no advantage of winter sunlight, though they may be preferred in very hot climates to limit summer heating by radiation.

Horizontal windows high in the wall are often easier to shade than vertical windows extending lower. Moreover, they give better light from the sky with greater privacy and allow greater freedom in the placement of furniture.

Trees may be an important means of shading openings. Deciduous trees, in particular, have the advantage of shading in summer while allowing warming by the sun's rays in winter.

FHA Minimum Property Standards require that windows and doors in climates with more than 4,500 degree-days in the heating season be protected by double glazed units or storm windows and doors. This standard is reasonable in limiting energy use and in improving comfort as it is affected by radiation losses. Double glazed windows and doors or double doors of other material may reduce heat losses through openings some 40 percent. They do not, however, give optimum economy as these units do not fully recover their amortized cost in the 40-year useful life of a home in most areas of the United States.

It is important to note here, however, that the investment is by no means necessarily a bad one. It will effect a more comfortable interior environment, both in reduced radiant heat losses from the body during extremely cold outside weather and in increased acoustical privacy. It does this at a very small price, since most of the investment is returned in energy cost savings in 40 years, and it represents good energy conservation at the same small price. Also, inflation in energy costs may well greatly exceed inflation in other costs, a contingency that would improve the economy of the units.

Chicago has, of course, a rather severe climate with average heating degree days of 6,600. The example indicates an economic break-even point of some 7,500 degree-days for storm windows, but some north-central locations have as high as 10,000 degree-days, where the storm windows would pay for themselves in less than 40 years at current prices.

Similar cost comparisons can be shown for storm doors. Costs for sealed double glass windows and doors are generally higher than for single units plus storm sash, and the energy savings are about the same. However, the double glass units do offer greater convenience, as there is no interference with the unit's ventilating function and only two surfaces require cleaning. In new construction, particularly, they merit consideration because of these conveniences.

Reflective glass has been used for some years in air-conditioned commercial buildings to reduce solar heat gain. It can be obtained in single glass or sealed double glass units and deserves consideration in areas of high solar load.

AIR CHANGE AND INFILTRATION

Air leakage may account for very significant energy losses in a home. This usually occurs as infiltration around window and door units or leakage between the frames and wall surfaces and, of course, as doors are opened for entry or egress or windows are opened for ventilation. Other ventilating units, such as air conditioning systems, provide for variable amounts of air change, as do kitchen or bathroom fan ventilators when in use. Cumulatively, these and other miscellaneous sources of leakage provide the air change necessary for health and for combustion of stoves or furnaces, but they usually do so at an unnecessary and excessive rate, wasting energy in the process.

Excessive ventilation by careless use of doors or windows is an obvious source of waste. Cracks around poorly fitted window sash or doors are major causes of unwanted heat loss or gain, particularly in older homes inadequately weatherstripped, or perhaps poorly adjusted or warped.

Permanent, metal or plastic channel weatherstripping of windows and doors is an economical device for reduction of the considerable energy lost through these cracks. It may often be installed by the homeowner and pays for itself in energy cost savings in a few years in most climates. Storm windows and doors reduce crack losses even when weatherstripping is not used, but additional reductions can be made by weatherstripping the inside unit. The storm unit should be less tightly fitted, since some ventilation to the outside is desired to limit condensation between the units. Crack losses can be reduced further if inside windows are kept locked, for the locking devices usually operate to pull sliding sash together or to tighten swinging sash to the frames.

MISCELLANEOUS LOSSES

Energy losses from ducts, pipes, and chimneys can be significant when they are inadequately insulated or sealed.

Hot or cold air ducts and returns which pass through unheated attics or open crawl spaces should be insulated with the equivalent of at least 1 inch of air-cell asbestos. Greater thickness is justified in extreme climates where the inside-outside temperature difference is large, and a 2-inch wrapping of mineral wool is common in northern locations. Steam and hot water pipes should be similarly covered when they pass through unheated spaces. In basements, ducts and pipes are frequently left uninsulated so that they contribute some heat to the basement area, but insulating cold water pipes to prevent dripping from condensation in the summer is desirable.

When heating ducts or radiant heating pipes or wires are incorporated in concrete slabs on the ground, the entire slab should be insulated from the outside walls and the ground, with moisture-proof insulation. When warm air perimeter ducts are used, it is satisfactory to limit the insulated area to the perimeter of the slab under the ducts, using at least 2 inches of insulation extending from the top of the slab, down the outside wall, and 24 inches under the ducts and slab.

Modern open fireplaces, commonly regarded as heating units, are insidious heat wasters. They are very inefficient, delivering little more than 10 percent of the energy generated to a room while pouring the larger portion of the heat up the chimney. Used in a home heated by other means, they draw much of the heated air from other parts of the house, reducing the effectiveness of the principal heater. They have a necessary function in places where other heating units may not be available and ample supplies of firewood exist, and they can be designed and located to yield a larger portion of their heat to a living area. In mild climates, they may also be a handy, occasional heat source when operation of a central heat source is not needed.

The fireplace today is a sentimental segment of tradition, emanating a cheerful, pleasant warmth to the family circle, while providing visual interest, lively sounds, and a pleasing aroma. These values cannot be measured in economic terms, and reasonable use of the fireplace is an individual choice. Moreover, it can be used as an indoor barbecue and is an excellent ventilator when large numbers of people are present.

Occasional use of the fireplace is least wasteful of heat when the room can be closed off from the rest of the dwelling to avoid drain on the central heater. A window may be opened a crack to provide draft within the room. If the fire is started sometime during the day or late afternoon and can be allowed to burn down early in the evening, it may be completely extinguished so the damper may be closed for the night. The damper should not be closed while any coals remain and very often is allowed to remain open all night to draw heat from the entire house. The fireplace damper should fit tightly, must be properly balanced so that the wind cannot open it, and should be kept closed when the fireplace is not in use. If a persistent draft is evident when the damper is closed, the fireplace opening may be closed off with a piece of plywood or other building board.

BIBLIOGRAPHY

Acoustical and Insulating Materials Association
    Fundamentals of building insulation. A.I.A. File 37.
        (Revised annually). Park Ridge, Ill.
American Society of Heating, Refrigerating, and Air-
    Conditioning Engineers
    1972. ASHRAE Handbook of Fundamentals—
        Heating, Ventilating, and Air-Condition-
        ing. 688 pp. New York.
Anderson, L. O.
    1970. Wood-frame house construction. U.S. Dep.
        Agr., Agr. Handbook 73, rev. 223 pp.

———

    1972. Condensation problems: their prevention and
        solution. USDA Forest Serv. Res. Pap.
        FPL 132. Forest Prod. Lab., Madison, Wis.
Building Research Advisory Board
    1952. Condensation control in buildings. Nat. Res.
        Counc. 118 pp. Washington, D.C.
Close, P. D.
    1951. Sound Control and Thermal Insulation.
        Rheinhold Publishing Corp., New York.
Lewis, W. C.
    1967. Thermal conductivity of wood-base fiber and
        particle panel materials. USDA Forest
        Serv. Res. Pap. FPL 77. Forest Prod.
        Lab., Madison, Wis.

———

    1968. Thermal insulation from wood for buildings:
        effects of moisture and its control. USDA
        Forest Serv. Res. Pap. FPL 86. Forest
        Prod. Lab., Madison, Wis.
MacLean, J. D.
    1941. Thermal conductivity of wood. Trans. Amer.
        Soc. Heating and Ventil. Eng. 47: 323.
Rogers, T. S.
    1964. Thermal Design of Buildings. John Wiley &
        Sons, New York.
U.S. Forest Products Laboratory
    1973. Wood siding: installing, finishing, maintain-
        ing. U.S. Dep. Agr. Home and Garden
        Bull. 203. 13 pp.

Chapter 22

WOOD-BASE FIBER AND PARTICLE PANEL MATERIALS

The group of materials generally classified as "wood-base fiber and particle panel materials" includes such familiar products as insulation boards, hardboards, particleboards, and laminated paperboards. In some instances they are known by such proprietary names as "Masonite," "Celotex," "Insulite," and "Beaver board" or, in the instance of particleboards, by the kind of particle used such as flakeboard, chipboard, chipcore, or shavings board.

These panel materials are all reconstituted wood (or some other lignocellulose like bagasse) in that the wood is first reduced to small fractions and then put back together by special forms of manufacture into panels of relatively large size and moderate thickness. These board or panel materials in final form retain some of the properties of the original wood but, because of the manufacturing methods, gain new and different properties from those of the wood. Because they are manufactured, they can be and are "tailored" to satisfy a use-need, or a group of needs.

Generally speaking, the wood-base panel materials are manufactured either (1) by converting wood substance essentially to fibers and then interfelting them together again into the panel material classed as building fiberboard, or (2) by strictly mechanical means of cutting or breaking wood into small discrete particles and then, with a synthetic resin adhesive or other suitable binder, bonding them together again in the presence of heat and pressure. These latter products are appropriately called particleboards.

Building fiberboards, then, are made essentially of fiberlike components of wood that are interfelted together in the reconstitution and are characterized by a bond produced by that interfelting. They are frequently classified as fibrous-felted board products. At certain densities under controlled conditions of hot-pressing, rebonding of the lignin effects a further bond in the panel product produced. Binding agents and other materials may be added during manufacture to increase strength, resistance to fire, moisture, or decay, or to improve some other property. Among the materials added are rosin, alum, asphalt, paraffin, synthetic and natural resins, preservative and fire-resistant chemicals, and drying oils.

References for greater detail on methods of manufacture of building fiberboards are included in the bibliography at the end of this chapter. Evaluation procedures for basic properties are reasonably well standardized.

Particleboards are manufactured from small components of wood that are glued together with a thermosetting synthetic resin or equivalent binder. Wax sizing is added to all commercially produced particleboard to improve water resistance. Other additives may be introduced during manufacture to improve some property or provide added resistance to fire, insects such as termites, or decay. Particleboard is among the newest of the wood-base panel materials. It has become a successful and economical panel product because of the availability and economy of thermosetting synthetic resins, which permit blends of wood particles and the synthetic resin to be consolidated and the resin set (cured) in a press that is heated.

Thermosetting resins used are primarily urea formaldehyde and phenol-formaldehyde. Urea-formaldehyde is lowest in cost and is the binder used in greatest quantity for particleboard intended for interior or other nonsevere exposures. Where moderate water or heat resistance is required, melamine-urea-formaldehyde resin blends are being used. For severe exposures like exteriors or where some heat resistance is required, phenolics are generally used.

The kinds of wood particles used in the manufacture of particleboard range from specially cut flakes an inch or more in length (parallel to the grain of the wood) and only a few hundredths of an inch thick to fine particles approaching fibers or flour in size. The synthetic resin solids are usually between 5 and 10 percent by weight of the dry wood furnish. These resins are set by heat as the wood particle-resin blend is compressed either in flat-platen presses, similar to those used for hot-pressing hardboard and plywood, or in extrusion presses where the wood-resin mixture is squeezed through a long, wide, and thin die that is heated to provide the energy to set the resin.

Particleboards produced by flat-platen presses are called "mat-formed" or "platen-pressed." Those produced in an extrusion press are called "extruded" particleboards.

Building fiberboards and particleboards are produced from small components of wood, hence the raw material need not be in log form. Many processes for manufacture of board materials start with wood in the form of pulp chips. Coarse residues from other primary forest products manufacture therefore are an important source of raw materials for both kinds of wood-based panel products. Particleboards, and to a lesser extent building fiberboards, use fine residues as raw material. For instance, planer shavings are significant in manufacture of particleboard. Overall, about 70 percent of the raw material requirements for wood-base fiber and particle panel materials are satisfied by residues. Bagasse, the fiber residue from sugar cane, and wastepaper are used also as raw material for board products.

In total, the wood-base fiber and particle panel materials form an important part of the forest products industry in the United States. Not only are they valuable from the standpoint of integrated utilization, but the total production of more than 7 billion square feet is important in terms of forest products consumed. Along with softwood plywood they are among the fastest growing components of the industry and production has been doubling about each 10 years.

FIBER PANEL MATERIALS

Broadly, the wood-base fiber panel materials (building fiberboards) are divided into two groups—insulation board (the lower density products) and hardboard, which requires consolidation under heat and pressure as a separate step in manufacture. The dividing point between an insulation board and a hardboard, on a density basis, is a specific gravity of 0.5 (about 31 p.c.f.). Practically, because of the range of uses and specially developed products within the broad classification, further breakdowns are necessary to classify the various products adequately. The following breakdown by density places the building fiberboards in their various groups:

	DENSITY	
	G. per cc.	P.c.f.
Insulation board	0.02 to 0.50	1.2 to 31
Semirigid insulation board	.02 to .16	1.2 to 10
Rigid insulation board	.16 to .5	10 to 31
Hardboard	.5 to 1.45	31 to 90
Medium-density hardboard	.5 to .8	31 to 50
High-density hardboard	.8 to 1.20	50 to 75
Special densified hardboard	1.35 to 1.45	84 to 90
Laminated paperboard	.50 to .59	31 to 37

Laminated paperboards require a special classification because the density of these products is slightly greater than the maximum for non-hot-pressed, fibrous-felted, wood-base panel materials. Also, because these products are made by laminating together plies of paper about $\frac{1}{16}$ inch thick, they have different properties along the direction of the plies than across the machine direction. The other fibrous-felted products have nearly equal properties along and across the panel.

PARTICLE PANEL MATERIALS

Mat-formed particleboards, because of differences in properties and uses, are generally classified by density into low, medium, and high categories:

	DENSITY	
	G. per cc.	P.c.f.
Low density	Less than 0.59	Less than 37
Medium density	0.59 to 0.80	37 to 50
High density	More than 0.80	More than 50

All mat-formed particleboards are hot-pressed to cure the resins used as binders.

These mat-formed particleboards are further described as being homogeneous (the same kind, size, and quality of particle throughout the thickness), graduated (a gradation of particle size from coarsest in the center of the thickness to finest at each surface), or "three layer" (the material on and near each surface is different than that in the core). These boards may be also described by the predominant kind of particle, as shavings, flakes, slivers, or the combination in the instance of layered construction as "flake-faced" or "fines-surfaced" boards.

Extruded boards account for less than 5 percent of total production of particleboard, and standards have not been developed for them to any appreciable extent. Most extruded particleboard can be classed as being of medium density because the compression applied to the wood particles during extruding does not increase the density beyond 50 p.c.f. Particleboards thicker than about $\frac{5}{8}$ inch can be extruded with hollow core sections similar to those molded in concrete blocks. Because of these hollow core sections the equivalent density may fall below that of medium-density, mat-formed particleboard. Extruded particleboards of that type, usually called "fluted" particleboards, are commonly classified on the basis of weight per square foot for a specified thickness.

PROPERTIES AND USES

Properties of the various wood-base panel materials to a considerable extent either suggest or limit the uses. In the following sections the fiber- and particle-based panel materials are divided into the various categories suggested by kind of manufacture, properties, and use.

Semirigid Insulation Board

Semirigid insulation board is the term applied to fiberboard products manufactured primarily for use as insulation and cushioning. These very low-density fiberboards have about the same heat-flow characteristics as conventional blanket or batt insulation but have sufficient stiffness and strength to maintain their position and form without being attached to the structure. They may be bent around curves or corners and, when cemented, mechanically fastened, or placed between framing members, will hold their shape and position even though subjected to considerable vibration.

The semirigid insulation boards are manufactured in sheets from ½ to 1½ inches thick. When greater thicknesses are required, two or more sheets are cemented together. Sheet sizes vary from 1 by 2 feet to 4 by 4 feet. The thermal conductivity factors (k) range from 0.24 to 0.27 Btu · in./h · ft² · deg F (British thermal units per inch of thickness per hour per square foot of surface per degree Fahrenheit) difference in temperature.

Semirigid insulation boards are used for heat insulation in truck and bus bodies, automobiles, refrigerators, railway cars, on the outside of ductwork, and wherever vibrations are so severe that loosefill or batt insulation may pack or shift. They are also used as sound insulation for the walls of telephone booths and around speakers in radios, public address systems, and phonographs, and as cushioning in furniture, mattresses, and special packaging applications.

Rigid Insulation Board

Rigid insulation board is the oldest of the wood-base fiber and particle panel materials. It has been produced in the United States for more than a half century and, when produced to the various standards developed by the insulation board industry, is classed as structural insulating board. Structural insulating board is manufactured mainly for specific uses in construction although some is fabricated for special padding and blocking in packaging and a wide variety of other industrial uses. Insulation board is produced in two general types, interior and sheathing. Interior-quality boards are for uses where high water resistance is not required, but a light-colored product is desired. Sheathing-quality boards are used where water resistance is required and are manufactured with added water-resistant materials (usually asphalt). Density is somewhat greater for sheathing-quality boards than for interior boards. Some sheathing-quality boards are coated with asphalt as well as impregnated.

Important strength and related properties for insulating board are included in table 22-1 with those for other building fiberboards. The two basic insulating board products, with only minor modification in manufacture as relates to composition, are fabricated into a group of products designed to satisfy specific use requirements in construction. These requirements may call for structural strength and either high thermal insulation or good acoustical properties, or both. Since individual structural insulating board products are use-oriented, the name of the product describes the use. Principal ones are described as follows:

General Purpose Board

There are two general purpose structural insulating boards—"Building Board" and "Wallboard" (sometimes called "Thin Board" in the trade because it is either ⁵⁄₁₆ or ⅜ inch thick while most other insulating board is ½ inch or thicker). Both of these general purpose boards may be converted for a multiplicity of uses not specifically covered in the other products. Some material could be classed as industrial board because it is used by others in some other article of commerce. The general use boards are usually furnished with a factory-applied, flame-resistant finish. "Building Board" is ½ inch thick and may be obtained in panels 4 feet by 8, 9, 10, or 12 feet with square edges. "Wallboard" is furnished regularly 4 feet wide in either 8- or 10-foot lengths. Quality limits are set for these and other regular products in the standards.

Property	Value for structural insulating board	Value for medium-density hardboard	Value for high-density hardboard	Value for tempered hardboard	Value for special densified hardboard	Unit
Density	10–30	33–50	50–80	60–80	85–90	P.c.f.
Specific gravity	0.16–0.42	0.53–0.80	0.80–1.28	0.93–1.28	1.36–1.44	----
Modulus of elasticity (bending)	25–125	325–700	400–800	650–1,100	1,250	1,000 p.s.i.
Modulus of rupture	200–800	1,900–6,000	3,000–7,000	5,600–10,000	10,000–12,500	P.s.i.
Tensile strength parallel to surface	200–500	1,000–4,000	3,000–6,000	3,600–7,800	7,800	Do.
Tensile strength perpendicular to surface	10–25	40–200	75–400	160–450	500	Do.
Compressive strength parallel to surface	----	1,000–3,500	1,800–6,000	3,700–6,000	26,500	Do.
Shear strength (in plane of board)	----	100–475	300–600	430–850	----	Do.
Shear strength (across plane of board)	----	600–2,500	2,000–3,000	2,800–3,400	----	Do.
24-hour water absorption	1–10	----	----	----	----	Pct. by volume
24-hour water absorption	----	5–20	3–30	3–20	0.3–1.2	Pct. by weight
Thickness swelling, 24-hour soaking	----	2–10	10–25	8–15	----	Pct.
Linear expansion from 50 to 90 percent relative humidity [2]	0.2–0.5	0.2–0.4	0.15–0.45	0.15–0.45	----	Do.
Thermal conductivity at mean temperature of 75° F.	0.27–0.45	0.54–0.75	0.75–1.40	0.75–1.50	1.85	Btu · in./h · ft² · deg F

[1] The data presented are general round-figure values, accumulated from numerous sources; for more exact figures on a specific product, individual manufacturers should be consulted or actual tests made. Values are for general laboratory conditions of temperature and humidity.

[2] Measurements made on material at equilibrium at each condition at room temperature.

Insulating Roof Deck

Insulating roof deck is a laminated structural insulating board product maufactured of several layers of sheathing-grade board and one layer of factory-finished interior board (either perforated or plain). It is used in exposed-beam ceiling constructions where the factory-finished interior board is applied face down. Insulating roof deck is regularly made in 1½-, 2-, and 3-inch nominal thickness in 2- by 8-foot panels. The 1½-inch-thick panel is made to span 24 inches, the 2-inch-thick panel to span 32 inches, and the 3-inch-thick material to span 48 inches. Panel ends are square and sides are tongued and grooved.

In climates where condensation due to winter cold can be a problem, insulating roof deck is furnished with a vapor-barrier membrane installed in the glueline between the layer of ½-inch-thick interior-finish board and the first sheathing-quality layer. In this construction the roof decking furnishes the structural rigidity to support snow and water or wind loads, besides providing the interior ceiling finish and thermal insulation. Thermal conductance factors for the various thicknesses are specified at 0.24, 0.18, and 0.12 Btu per hour per square foot of area per °F. difference in temperature for the thickness, respectively, of 1½, 2, and 3 inches. For flat roofs a builtup roof is applied directly to the top surface of the deck. When pitch of the roof is sufficient, asphalt shingles may be attached to 2- and 3-inch-thick decking with special annular grooved nails.

Roof Insulation

Above-deck thermal insulation made of structural insulating board has been a major product for many years. It competes well from the cost and service standpoint with other products made for that use and is included with them in the specifications for such materials. Roof insulation is manufactured in blocks 23 by 47 or 24 by 48 inches in ½-inch multiples of thickness between ½ and 3 inches. The blocks are usually multiple ½-inch thicknesses of insulation board and may be laminated or stapled together in the greater thicknesses. Thicknesses are nominal and may be less than or greater than nominal by any amount necessary to give certified conductance values for the thicknesses:

Nominal thickness In.	"C" value Btu/h·ft²·deg F
½	0.72
1	.36
1½	.24
2	.19
2½	.15
3	.12

Insulation board roof insulation is applied above decks, where the final roofing is of the builtup variety. It is secured in place by hot asphalt or roofing pitch or by mechanical fasteners, and has enough internal bond strength to resist uplift forces on the roof structure.

Ceiling Tile and Lay-In Panels

Ceiling tile, either plain or perforated, is an important use for structural insulating board. Such board has a paint finish applied in the factory to provide resistance to flame spread. Interior-finish insulating board, when perforated or provided with special fissures or other sound traps, will also provide a substantial reduction in noise reflectance. The fissures and special sound traps are designed to provide improved appearance over that of the conventional perforations while satisfying the requirements for sound absorption. The manufacturers of insulation board long have recognized the appeal of esthetically pleasing ceiling finishes. Each of them offers finishes in designs that blend with either traditional or contemporary architecture and furnishings.

Generally ceiling tiles are 12 by 12 and 12 by 24 inches in size, ½ inch thick, and have tongue and groove or butt and chamfered edges. They are applied to nailing strips with nails, staples, or special mechanical fastenings, or directly to a surface with adhesives.

A panel product similar to tile, but nominally 24 by 24 or 24 by 48 inches, is gaining popularity. These panels, commonly called "lay-in ceiling panels," are installed in metal tees and angles in suspended ceiling systems. These lay-in panels are usually ½ inch thick and are supported in place along all four edges. They are frequently used in combination with translucent plastic panels that conceal light fixtures. Finishes and perforation treatments for sound absorption are the same as for regular ceiling tile. Producers of insulating board are extending their manufacture to specially embossed ceiling panels that can be applied with butt joint edges and ends that present an essentially unbroken surface. Plastic films are being used increasingly for surfacing ceiling tile for applications like kitchens and bathrooms where repeated washability and resistance to moisture is desired. These products are especially adaptable for remodeling.

Plank

Structural insulating board plank is installed on side walls, often in remodeling, where it is used in conjunction with ceiling tile installations. Plank is manufactured 12 inches wide with matching long edges and is finished with a flame-resistant paint applied at the factory. Because of its low density, it is subject to abrasion when used for the lower part of walls where chairs or other furniture can bump it. Frequently it is used in conjunction with wainscoting of wood paneling or one of the other wood-base panel materials like hardboard or particleboard.

Sheathing

Insulation board is used more than any other material for sheathing houses in the United States (fig. 22-1). Sheathing is regularly manufactured in three grades: Regular-density, intermediate, and nail-base. Regular-density sheathing is manufactured in both ½- and 25/32-in thicknesses. Intermediate and nail base are made only ½ inch thick. Regular-density sheathing is furnished in two sizes, 2 by 8 feet with long edges matched, or 4 by 8, 9, 10, or 12 feet with edges square; the other two grades are furnished only in 4-foot widths and 8- or 9-foot lengths with square edges.

Regular-density sheathing is usually about 18 p.c.f. in density and is sold with a thermal resistance (R) rating of 2.06 (deg F·h·ft² ·Btu) for 25/32-inch material and 1.32 for the ½-inch thickness. When the 2- by 8-foot material is used as sheathing, it is applied with long edges horizontal. The 4-foot-wide material is recommended for application with long edges vertical. When 25/32-inch-thick regular-density sheathing is applied with the long edges vertical and adequate fastening (either nails or staples) around the perimeter and along intermediate framing, requirements for racking resistance of the wall construction are usually satisfied. Horizontal applications with the 25/32-inch material require additional bracing in the wall system to meet code requirements for rigidity, as do some applications of the ½-inch-thick regular-density sheathing applied with long edges vertical. Regular-density sheathing,

Figure 22-1.—Proper nailing provides racking resistance with insulation board sheathing.

because of its resistance to heat flow and air infiltration, provides sufficient resistance so that added thermal insulation is not required to meet minimum standards in many parts of the country.

Costs of heating and requirements for air-conditioning from summer heat may justify added thermal insulation over that required by minimum standards. When such added thermal insulation is used in construction of walls, intermediate and nail-base sheathing (with lower thermal resistance) are used. They are applied with long edges vertical. With recommended fastening, such sheathing provides the racking rigidity and strength for the wall without added bracing.

Intermediate sheathing is usually about 22 p.c.f. in density; nail-base is about 25. The insulation board industry provides intermediate density and nail-base sheathing with rated thermal resistance (R) values also. The R values for those materials are 1.22 and 1.14, respectively. Nail-base sheathing has adequate nail-holding strength so that asbestos and wood shingles for weather course (siding) can be attached directly to the nail-base sheathing with special annular grooved nails. With the other grades of sheathing, siding materials must be nailed directly to framing members or to nailing strips attached through the sheathing to framing. Because the method and amount of fastening is critical to racking resistance, local building codes should be consulted for requirements in different areas.

Shingle Backer

Shingle backer is a specially manufactured sheathing-grade structural insulating board $\frac{5}{16}$ or $\frac{3}{8}$ inch thick, 11¾, 13½, or 15 inches wide, and 4 feet long. It is used with coursed wood shingles on sidewalls. The shingle backer is installed beneath each course of shingles to provide the deep shadow line so desirable for

411

that type of sidewall finish; by nailing through the added thickness of insulation board, as well as sheathing, adequate nail holding is provided to keep the shingles in position. Special long deformed nails (usually annularly grooved) are used for the fastening.

Insulating Formboard

Insulating formboard is a special insulating board for permanent in-place forms; lightweight aggregate cement or gypsum roof decks are poured on it. The formboard has sufficient stiffness, when installed as recommended, to support the wet deck material until it sets. After cure it remains in place to provide thermal insulation; it also shares in carrying roof loads because the poured deck and the formboard act as a composite beam. Standard thicknesses of insulating formboard are 1 and 1½ inches; widths are 24, 32, and 48 inches; and lengths are between 4 and 12 feet, as required. When formboard is to be exposed in the final construction, the exposed surface may be furnished with a factory-applied paint finish.

Sound-Deadening Board

Sound-deadening board is specially manufactured to provide a meaningful reduction in sound transmission through walls. Standard sizes are ½ inch thick, 4 feet wide, and 8 or 9 feet long. In light-frame construction, sound-deadening board is usually applied to the wall framing; the final wall finish, such as gypsum board, is then applied to the outside faces of the sound-deadening board. Acoustic efficiency of walls constructed with sound-deadening board depends on tight construction with no air leaks around the edges of panels, and close adherence to prescribed methods of installation.

Medium-Density Hardboard

Medium-density hardboard, formerly classified as medium-density building fiberboard, is the newest of the wood-based panel products. Nearly all of the material being manufactured by the conventional methods used for other hardboard is being tailored for use as house siding (fig. 22-2). Medium-density hardboard for house siding use is mostly ⅜ and ⁷⁄₁₆ inch thick and is fabricated for application as either panel or lap siding.

Medium-density hardboard is also manufactured by a newer process that involves radio-frequency energy for curing thicker panels (usually about ¾ in. although it is possible to make panels as thick as 3 in.) for use mainly in furniture and cabinets as corestock or panel stock. Properties of medium-density hardboard are summarized in table 22-1.

Panel siding is 4 feet wide and commonly furnished in 8-, 9-, or 10-foot lengths. Surfaces may be grooved 2 inches or more on center parallel to the long dimension to simulate reversed board and batten or may be pressed with ridges simulating a raised batten.

Lap siding is frequently 12 inches wide with lengths to 16 feet and is applied in the same way as conventional wood lap siding. Some manufacturers offer their lap siding products with special attachment systems that provide either concealed fastening or a wider shadow line at the bottom of the lap.

Most siding is furnished with some kind of a factory-applied finish. At least the surface and edges are given a prime coat of paint. Finishing is completed later by application of at least one coat of paint. Two coats of additional paint, one of a second primer and one of topcoat, provide for a longer interval before

Figure 22-2.—A 16-foot hardboard panel is applied quickly to a sidewall of a two-story building.

repainting. There is a trend for complete pre-finishing of medium-density hardboard siding. The complete prefinishing ranges from several coats of liquid finishes to cementing various films to the surfaces and edges of boards. Surfaces of medium-density hardboard for house siding range from very smooth to textured; one of the newest ones simulates weathered wood with the latewood grain raised as though earlywood has been eroded away.

Small amounts of medium-density hardboard are being prefinished for interior paneling along with the high-density hardboards. Siding is the most important use and others will not become extensive until that market is fully developed and exploited. The experience with medium-density hardboard has been good. Dimensional movement with moisture change has not produced major problems in service. When hardboard is properly painted with high-quality paints, it has required little paint maintenance. Medium-density hardboard siding has been accepted largely on an individual manufacturer basis. Code authorities and others have recognized the evidence submitted by manufacturers on the performance of medium-density hardboard siding.

The properties of medium-density hardboard make it a desirable material for industrial use. Since possible uses are many, the American hardboard industry offers this product for industrial use under the name of "Industrialite." A summary of the range in properties for this material is included in table 22-1.

High-Density Hardboard

Manufacture of high-density hardboard has grown rapidly since World War II. Numerous older uses are well established and new ones are being developed continually. Properties are summarized in table 22-1, but in the trade the various qualities are subdivided in smaller groups beyond those shown in the table. There is an overlapping of properties as shown by the limiting values in the various standards for hardboard.

Originally there were two basic qualities, standard and tempered, for high-density hardboard. These are still the two qualities used in greatest quantity. Standard hardboard is a panel product with a density of about 60 to 65 p.c.f., usually unaltered except for humidification as it is produced by hot pressing. Tempered hardboard is a standard-quality hardboard that is treated with a blend of

siccative resins (drying oil blends or synthetics) after hot-pressing. The resins are stabilized by baking after the board has been impregnated. Usually about 5 percent solids are required to produce a hardboard of tempered quality. Tempering improves water resistance, hardness, and strength appreciably, but embrittles the board and makes it less shock resistant.

A third hardboard, service quality, has become important. This is a product of lower density than standard, usually 50 to 55 p.c.f., made to satisfy needs where the higher strength of standard quality is not required. Because of its lower density, service-quality hardboard has better dimensional stability than the denser products.

When service hardboard is given the tempering treatment it is classed as tempered service, and property limits have been set for specifications. It is used where water resistance is required but the higher strength of regular treatment is not. Underlayment is service-quality hardboard, nominally ¼ inch thick, that is sanded or planed on the back surface to provide a thickness of 0.215 ± 0.005 inch.

These are the regular qualities of high-density hardboard; because a substantial amount of this hardboard is manufactured for industrial use, special qualities are made with different properties dictated by the specific use. For example, hardboard manufactured for concrete forms is frequently given a double tempering treatment. For some uses where high impact resistance is required, like backs of television cabinets, boards are formulated from specially prepared fiber and additives. Where special machining properties like die punching or post-forming requirements must be satisfied, the methods of manufacturing and additives used are modified to produce the desired properties.

High-density and medium-density conventional hardboard are manufactured in several ways and the result is reflected in the appearance of the final product. Hardboard is described as being screen-backed or S–2–S (smooth two sides). When the mat from which the board is made is formed from a water slurry (wet-felted) and the wet mat is hot-pressed, a screen is required to permit steam to escape. In the final board the reverse impression of the screen is apparent on the back of the board, hence the screen-back designation. A screen is similarly required with mats formed from an air suspension (air-felted) when moisture contents are sufficiently high going into the hot press so that venting is required.

In some variations of hardboard manufacture, a wet-felted mat is dried before being hot-pressed. With this variation it is possible to hot-press without using the screen, and an S–2–S board is produced. In air-felting hardboard manufacture, it is possible also to press without the screen, if moisture content of mats entering the hot press is low. In a new adaptation of pressing hardboard mats, a caul with slots or small circular holes is used to vent steam; the board produced has a series of small ridges or circular nubbins which, when planed or sanded off, yield an S–2–S board.

Medium-density hardboard produced by the newer process using radio-frequency curing is produced from relatively dry fiber-resin blends. The mats are pressed between heated platens where the high-frequency heat provides additional heat energy to cure the resin binder (usually urea-formaldehyde as compared to phenolics when used with the more conventional hardboards). This kind of hardboard is S–2–S.

Commercial thicknesses of high-density hardboard generally range from $\frac{1}{2}$ to $\frac{3}{8}$ inch. Not all thicknesses are produced in all grades. The thicknesses of $\frac{1}{10}$ and $\frac{1}{12}$ inch are regularly produced only in the standard grade. Tempered hardboards are produced regularly in thicknesses between $\frac{1}{8}$ and $\frac{5}{16}$ inch. Service and tempered service are regularly produced in fewer thicknesses, none less than $\frac{1}{8}$ inch and not by all manufacturers or in screen-back and S–2–S types. The appropriate standard specification or source of material should be consulted for specific thicknesses of each kind.

High-density hardboards are produced in 4- and 5-foot widths with the more common width being 4 feet. Standard commercial lengths are 4, 6, 8, 12, and 16 feet with an 18-foot length being available in the 4-foot width. Most manufacturers maintain cut-to-size departments for special orders. Retail lumberyards and warehouses commonly stock 8-foot lengths, except for underlayment, which is usually 4 feet square.

About 15 percent of the hardboard used in the United States is imported. Foreign-made board may or may not be manufactured to the same standards as domestically produced products. Before substituting a foreign-made product in a use where specific properties are required, it should be determined that it has properties required for the use. Canadian products are usually produced to the same standards as United States products.

In addition to the standard smooth-surface hardboards, special products are made using patterned cauls so the surface is striated or produced with a relief to simulate ceramic tile, leather, basket weave, etched wood, or other texture. Hardboards are punched to provide holes for anchoring fittings for shelves and fixtures (perforated board) or with holes comprising 15 percent or more of the area for installation in ceilings with sound-absorbent material behind it for acoustical treatments or as air diffusers above plenums. Cutout designs are fabricated in hardboard to produce "filigree" effects when the hardboard is used in screens.

More and more effort is being put forth by industry to modify and finish hardboard so it can be used in more ways with less "on-the-job" cost of installation and finishing and to permit industrial users a saving in final product. Most important is prefinishing, particularly wood graining, where the surface of the board is finished with lithographic patterns of popular cabinet woods printed in two or more colors.

The uses for hardboard are diverse. It has been claimed that "hardboard is the grainless wood of 1,000 uses, and can be used wherever a dense, hard panel material in the thicknesses as manufactured will satisfy a need better or more economically than any other material." Because of its density it is harder than most natural wood, and because of its grainless character it has nearly equal properties in all directions in the plane of the board. It is not so stiff nor as strong as natural wood along the grain, but is substantially stronger and stiffer than wood across the grain. Specific properties in table 22–1 can be compared with similar properties for wood, wood-base, and other materials. Hardboard retains some of the properties of wood; it is hygroscopic and shrinks and swells with changes in moisture content.

Changes in moisture content due to service exposures may be a limiting factor in satisfactory performance. Correct application and attachment as well as prior conditioning to a proper moisture content may permit satisfactory service when improper application or conditioning precludes it. Proper moisture conditioning prior to assembly is of particular importance in glued assemblies.

Product development in hardboard has held generally to the line of class and type of board product, in contrast with structural insulating board which deals with specific items for particular uses. During the past few years much of the success of hardboard resulted because the industry developed certain products for a specific use and had treatments, fabrication, and finishes required by the use. Typical are prefinished paneling, house siding, underlayment, and concrete form hardboard.

Many uses for hardboard have been listed but generally they can be subdivided according to uses developed for construction, furniture and furnishings, cabinet and store fixture work, appliances, and automotive and rolling stock.

In construction, hardboard is used as floor underlayment to provide a smooth undercourse under plastic or linoleum flooring, as a facing for concrete forms for architectural concrete, facings for flush doors, as insert panels and facings for garage doors, and material punched with holes for wall linings in storage walls and in built-ins where ventilation is desired. Prefinished hardboard, either with baked finishes or the regular ones like those used generally in wood-grain printing, is used for wall lining in kitchens, bathrooms, family rooms, and recreation rooms.

In furniture, furnishings, and cabinetwork, conventional hardboard is used extensively for drawer bottoms, dust dividers, case goods and mirror backs, insert panels, television, radio, and stereo cabinet sides, backs (die-cut openings for ventilation), and as crossbands and balancing sheets in laminated or overlay panels. Hardboard also has use as a core material for relatively thin panels overlaid with films and thin veneers, and as backup material for metal panels. In appliances other than television, radio, and stereo cabinets, it is used wherever the properties of the dense, hard sheet satisfy a need economically. Because it can be post-formed to single curvature (and in some instances to mild double curvature) by the application of heat and moisture, it is used in components of appliances requiring that kind of forming.

In automobiles, trucks, buses, and railway cars, hardboard is commonly used in interior linings. Door and interior sidewall panels of automobiles are frequently hardboard, post-formed, and covered with cloth or plastic. The base for sun visors is often hardboard, as are the platforms between seats and rear windows. Molded hardboard also has been used for three-dimensional-shaped components like door panels and armrests. Ceilings of station wagons and truck cabs are often enameled or vinyl-covered thin hardboard.

Medium-density hardboard corestock about ¾ inch thick is used with veneer or other overlays in the same way as particleboard, but it may not be necessary to edge band because of the smooth edge. Places in furniture and cabinets where it finds most use are in tops, doors, drawer fronts, and shelves.

Special Densified Hardboard

This special building fiberboard product is manufactured mainly as diestock and electrical panel material. It has a density of 85 to 90 p.c.f. and is produced in thicknesses between ⅛ and 2 inches in panel sizes of 3 by 4, 4 by 6, and 4 by 12 feet.

Special densified hardboard is machined easily with machine tools and its low weight as compared with metals (aluminum alloys about 170 p.c.f.) makes it a useful material for templates and jigs for manufacturing. It is relatively stable dimensionally from moisture change because of low rates of moisture absorption. It is more stable for changes in temperature than the metals generally used for those purposes. The ⅛-inch-thick board is specially manufactured for use as lofting board.

As diestock, it finds use for stretch- and press-forming and spinning of metal parts, particularly when few of the manufactured items are required and where the cost of making the die itself is important in the choice of material.

The electrical properties of the special densified hardboard meet many of the requirements set forth by the National Electrical Manufacturers Association for insulation resistance and dielectric capacity in electrical components so it is used extensively in electronic and communication equipment.

Other uses where its combination of hardness, abrasion resistance, machinability, stability, and other properties are important include cams, gears, wear plates, laboratory work surfaces, and welding fixtures.

Laminated Paperboards

Laminated paperboards are made in two general qualities, an interior and a weather-resistant quality. The main differences be-

Table 22-2.—*Strength and mechanical properties of laminated paperboard* [1]

Property	Value	Unit
Density	32–33	P.c.f.
Specific gravity	0.52–0.53	- - - -
Modulus of elasticity (compression):		
Along the length of the panel [2]	300–390	1,000 p.s.i.
Across the length of the panel [2]	100–140	Do.
Modulus of rupture:		
Span parallel to length of panel [2]	1,400–1,900	P.s.i.
Span perpendicular to length of panel [2]	900–1,100	Do.
Tensile strength parallel to surface:		
Along the length of the panel [2]	1,700–2,100	Do.
Across the length of the panel [2]	600–800	Do.
Compressive strength parallel to surface:		
Along the length of the panel [2]	700–900	Do.
Across the length of the panel [2]	500–800	Do.
24-hour water absorption	10–170	Pct. by weight
Linear expansion from 50 to 90 percent relative humidity: [3]		
Along the length of the panel [2]	0.2–0.3	Pct.
Across the length of the panel [2]	1.1–1.3	Do.
Thermal conductivity at mean temperature of 75° F.	0.51	Btu in./h · ft² · deg F

[1] The data presented are general round-figure values, accumulated from numerous sources; for more exact figures on a specific product, individual manufacturers should be consulted or actual tests made. Values are for general laboratory conditions of temperature and humidity.

[2] Because of directional properties, values are presented for two principal directions, along the usual length of the panel (machine direction) and across it.

[3] Measurements made on material at equilibrium at each condition at room temperature.

tween the two qualities are in the kind of bond used to laminate the layers together and in the amount of sizing used in the pulp stock from which the individual layers are made. For interior-quality boards, the laminating adhesives are commonly of starch origin while for the weather-resistant board synthetic resin adhesives are used. Laminated paperboard is regularly manufactured in thicknesses of 3/16, 1/4, and 3/8 inch for construction uses although for such industrial uses as dust dividers in case goods, furniture, and automotives liners, 1/8-inch thickness is common. Important properties are presented in table 22-2.

For building use, considerable amounts go into the prefabricated housing and mobile home construction industry as interior wall and ceiling finish. In the more conventional building construction market, interior-quality boards are also used for wall and ceiling finish, often in remodeling to cover cracked plaster.

Water-resistant grades are manufactured for use as sheathing, soffit linings, and other "exterior protected" applications like porch and carport ceilings. Soffit linings and lap siding are specially fabricated in widths commonly used and are prime coated with paint at the factory.

The common width of laminated paperboard is 4 feet, although 8-foot widths are available in 12-, 14-, 16-foot, and longer lengths for such building applications as sheathing entire walls. Laminated paperboards for use where a surface is exposed have the surface ply coated with a high-quality pulp to improve surface appearance and performance. Surface finish may be smooth or textured.

Particleboard

Important properties for mat-formed particleboards are presented in table 22-3. Similar values are not presented for extruded particleboards since they are never used without facings of some kind glued to them, and the facings influence the physical and strength properties. Extruded particleboards have a distinct zone of weakness across the length of the panel as extruded. They also have a strong tendency to swell in the lengthwise (extruded) direction because of the compression and orientation of particles from the extrusion pressures (ram or screw that forces the particle mass through the extruder). Consequently, extruded particleboards are always used as corestock; mat-formed boards are used both as corestock and as panel stock where the only thing added to the surface is finish.

From the selling standpoint, mat-formed particleboards are classified as to class and type, as are hardboards. For certain uses where special requirements must be satisfied, additional specifications outline the requirements for the particleboard. Particleboard is manufactured in both 4- and 5-foot widths and wider, although for industrial sales much is cut to size for the purchaser. In construction, the common size panel is 4 by 8 feet, as for other panel materials.

The uses for particleboard are still developing, but the main ones parallel those for lumber core in veneered or overlaid construction and plywood. The two properties of particleboard that have the greatest positive influence on its selection for a use are (1) uniform surface and (2) that the panel stays flat as manufactured, particularly in applications where edges are not fastened to a rigid framework.

For the majority of uses where exposures are interior or equivalent (furniture, cabinetwork, interior doors, and most floor underlayment), urea-formaldehyde resins are used. Boards with that kind of bond are classed as type 1 in specifications. Where greater resistance to heat, moisture, or a combination of heat and moisture is required, type 2 boards generally bonded with phenolic resin are required.

In general, particleboards are manufactured in about the same thicknesses as softwood plywood; most manufacture is in thicknesses between ½ and 1 inch, although there are notable use developments for particleboard thinner than ½ inch and thicker than 1 inch. Much extruded particleboard is of the "fluted" kind in thicknesses that satisfy the need for cores for flush doors. Similarly low-density, mat-formed particleboard is manufactured for solid core doors in thicknesses so that, when facings are applied, final door thicknesses are the standard 1⅜ and 1¾ inches.

While particleboard manufacturers formerly concentrated production in thick boards, there has been a decided trend toward thinner products. Thicknesses of ¼ and ⅜ inch are becoming more common, both for special and general use in the United States. In other countries where alternate thin materials are not available, thicknesses of less than ⅛ inch are produced.

The two uses for particleboard that are major in terms of volume are furniture and cabinet core and floor underlayment (fig. 22-3). As corestock, particleboard has moved into a market formerly held by lumber core and, to a limited extent, plywood. For example, certain grades of hardwood plywood now permit the use of particleboard as the core ply where formerly lumber core was specified.

Table 22-3.—*Strength and mechanical properties of mat-formed (platen-pressed) wood particleboard* [1]

Property	Value for low-density particleboard	Value for medium-density particleboard	Value for high-density particleboard	Unit
Density	[2] 25–37	37–50	50–70	P.c.f.
Specific gravity	[2] 0.40–0.59	0.59–0.80	0.80–1.12	----
Modulus of elasticity (bending)	[3] 150–250	250–700	350–1,000	1,000 p.s.i.
Modulus of rupture	[3] 800–1,400	1,600–8,000	2,400–7,500	P.s.i.
Tensile strength parallel to surface	----	500–4,000	1,000–5,000	Do.
Tensile strength perpendicular to surface	[3] 20–30	40–200	125–450	Do.
Compression strength parallel to surface	----	1,400–3,000	3,500–5,200	Do.
Shear strength (in the plane of board)	----	100–450	200–800	Do.
Shear strength (across the plane of the board)	----	200–1,800	----	Do.
24-hour water absorption	----	10–50	15–40	Pct. by weight
Thickness swelling from 24-hour soaking	----	5–50	15–40	Pct.
Linear expansion [4] from 50 to 90 percent relative humidity	[5] 0.30	0.2–0.6	0.2–0.85	Do.
Thermal conductivity at a mean temperature of 75° F.	0.55–0.75	0.75–1.00	1.00–1.25	Btu · in./h · ft² · deg F

[1] The data presented are general round-figure values, accumulated from numerous sources; for more exact figures on a specific product, individual manufacturers should be consulted or actual tests made. Values are for general laboratory conditions of temperature and humidity.

[2] Lower limit is for boards as generally manufactured; lower density products with lower properties may be made.

[3] Only limited production of low-density particleboard so values presented are specification limits.

[4] Measurements made on material at equilibrium at each condition at room temperature.

[5] Maximum permitted by specification.

In built-up constructions where particleboard is used as the core, both three and five plies are employed. Extruded particleboard nearly always requires five-ply construction because of the board's instability and low strength in the one direction. A relatively thick crossband with the grain direction parallel to the extruded direction stiffens, strengthens, and stabilizes the core. Thinner face plies are laid with the grain at right angles to the crossband to provide the final finish.

With mat-formed particleboard corestock, the use of three- or five-ply construction depends on the class and type of particleboard core (stiffness and strength), kind of facings being applied (plastic or veneer), and the requirements of the final construction. Balanced construction in lay-ups using particleboard is important; otherwise facings or crossbands with different properties can cause objectionable warping, cupping, or twisting in service. Edge bonding of wood or the facing material is frequently employed in panelized units using particleboard as corestock.

As floor underlayment, particleboard provides (1) the leveling, (2) the thickness of construction required to bring the final floor to elevation, and (3) the indentation-resistant smooth surface necessary as the base for resilient finish floors of linoleum, rubber, vinyl, and vinyl asbestos tile and sheet material. Particleboard for this use is produced in 4- by 8-foot panels commonly $\frac{1}{4}$, $\frac{3}{8}$, or $\frac{5}{8}$ inch thick. Separate use specifications cover particleboard floor underlayment. In addition, all manufacturers of particleboard floor underlayment provide individual application instructions and guarantees because of the importance of proper application and the interaction effects of joists, subfloor, underlayment, adhesives, and finish flooring. Particleboard underlayment is sold under a certified quality program where established grade marks clearly identify the use, quality, grade, and originating mill.

Other uses for particleboard have special requirements, as for phenol-formaldehyde, a more durable adhesive, in the board. Particleboard for siding, combined siding-sheathing, and use as soffit linings and ceilings for carports, porches, and the like requires this more durable adhesive. For these uses, type 2 medium-density board is required. In addition, such agencies as Federal Housing Administration have established requirements for particleboard for such use. The satisfactory perform-

Figure 22-3.—Joints in particleboard floor underlayment are filled and sanded smooth before application of resilient finish floor.

ance of particleboard in exterior exposure depends not only on the manufacture and kind of adhesive used, but on the protection afforded by finish. Manufacturers recognize the importance by providing both paint-primed panels and those completely finished with liquid paint systems or factory-applied plastic films.

Mobile home manufacture and factory-building of conventional housing are increasing. These industries are important and increasing users of forest products. Since particleboard is actually manufactured in hot presses as large as 8 by 40 feet, larger sized panels are available than those generally used for conventional construction. With mechanical handling available in factories, large-sized panels can be attached effectively and economically. Two particleboard products have been developed to satisfy these uses. Mobile home decking is used for combined subfloor and underlayment. It is a board with a type 1 bond, but is protected from moisture in use by a "bottom board," so is generally regarded as giving satisfactory service for the limited life of a mobile home. Particleboard decking for factory-built housing is similar to mobile home decking but it has a type 2 bond because of the greater durability requirements for regular housing. The United States industry has established separate standards for these products. They are marketed under a certified product quality program with each product adequately identified.

Another use that commonly requires the more durable bond of a type 2 quality is a special corestock where laminated plastic sheets are formed on the face and back of a particleboard at the same time as they are bonded to it. Usually a high-density particleboard is used for that purpose. The temperatures used in curing the resin-impregnated plastic sheets may reduce the strength of the bonds in boards made with the urea-formaldehyde resins.

A special particleboard is fabricated for finish flooring. Here a board with a void-free, tight surface is required to prevent dirt build-up in service. To simulate service conditions of such flooring, special test procedures have been developed. Specifications with limiting values based on those procedures have been established for finish flooring of particleboard.

The properties of particleboard depend on the shape and quality of particle used, as well as the kinds and amount of resin binder. While most particleboard is produced using particles that yield a board of intermediate strength and stiffness, a substantial production uses flakes or other "engineered" particles for a board of higher strength and stiffness. In the specifications, the boards of intermediate stiffness and bending strength are called the class 1 and those of the higher stiffness and bending strength are called class 2. Class 2 particleboards are naturally more expensive than those in class 1, but they are usually justified for uses where the greater stiffness and strength are required. The same applies to those applications where a special surface like "fines surfaced" will provide either a better base for finishing or less "showthrough" of an overlaid construction.

Particleboard in molded, three-dimensional shapes is being found in commerce. At present, relatively high die costs and limited flow of the particle-resin blend during molding restrict the kinds of items that can be molded profitably. Such items as bowls, trays, and counter sections with limited relief are presently being molded from conventional size particles. Toilet seats, croquet balls, hamper tops, and similar high-density items are being compression-molded from small particles approaching wood flour in size and synthetic resin.

BIBLIOGRAPHY

Acoustical and Insulating Materials Association
1972. Recommended product and application specification, ½-inch fiberboard, nail-base sheathing. AIMA–IB Specif. No. 2, 3 pp. Park Ridge, Ill.

———— 1969. Why settle for less—only insulation board sheathing offers this 3-in-1 package. A.I.A. File No. 19–D3, 7 pp. Park Ridge, Ill.

———— 1972. Recommended product and application specification; ½-inch intermediate fiberboard sheathing. AIMA–IB Specif. No. 3, 3 pp. Park Ridge, Ill.

———— 1970. Recommended product and application specification; structural insulating roof deck. AIMA–IB Specif. No. 1, 12 pp. Park Ridge, Ill.

———— 1969. Product specification for sound deadening board in wall assemblies. AIMA–IB Specif. No. 4, 2 pp. Park Ridge, Ill.

———— Fundamentals of building insulation. A.I.A. File No. 37, (Revised annually.) Park Ridge, Ill.

———— 1967. Noise control with insulation board for homes, apartments, motels, and offices. Ed. 3, A.I.A. File No. 39–B, 20 pp. Park Ridge, Ill.

American Hardboard Association
n.d. The wonderful world of hardboard, the engineered wood. 24 pp.

———— 1967. What builders should know about hardboard. Nat. Assoc. of Homebuilders, Jour. Homebuilding, pp. 50–63. June.

———— 1970. Hardboard partitions for sound control, 8 pp. Jan.

———— 1971. Hardboard industry standard. AHA IS–71, 18 pp.

American Society for Testing and Materials
Standard methods of testing structural insulating board made from vegetable fibers. Designation C 209. (See current issue.) Philadelphia, Pa.

———— Standard specifications for structural insulating board made from vegetable fibers. Designation C 208. (See current issue.) Philadelphia, Pa.

———— Standard specification for fiberboard nail-base sheathing. Designation D 2277. (See current issue.) Philadelphia, Pa.

———— Standard specification for structural insulating formboard made from vegetable fibers. Designation C 532. (See current issue.) Philadelphia, Pa.

———— Standard definitions of terms relating to wood-base fiber and particle panel materials. Designation D 1554 (See current issue.) Philadelphia, Pa.

———— Standard methods of test for evaluating the properties of wood-base fiber and particle panel materials. Designation D 1037. (See current issue.) Philadelphia, Pa.

———— Standard method of test for structural insulating roof deck. Designation D 2164. (See current issue.) Philadelphia, Pa.

———— Standard methods of test for simulated service testing of wood and wood-base finish flooring. Designation D 2394. (See current issue.) Philadelphia, Pa.

Gatchell, C. J., Heebink, B. G., and Hefty, F. V.
1966. Influence of component variables on properties of particleboard for exterior use. Forest Prod. J. 16(4): 46–59.

Godshall, W. D., and Davis, J. H.
1969. Acoustical absorption of wood-base panel materials. USDA Forest Serv. Res. Pap. FPL 104. Forest Prod. Lab., Madison, Wis.

Food and Agriculture Organization of United Nations
1958. Fiberboard and particle board. 180 pp.

Hann, R. A., Black, J. M., and Blomquist, R. F.
1962. How durable is particle board. Part I. Forest Prod. J. 12(12): 577–584.

————, Black, J. M., and Blomquist, R. F.
1963. How durable is particle board. Part II. Forest Prod. J. 13(5): 169–174.

Heebink, B. G.
1960. Exploring molded particle boards. Wood and Wood Prod. 65(5): 36.

———— 1963. Importance of balanced construction in plastic-faced wood panels. U.S. Forest Serv. Res. Note FPL–021. Forest Prod. Lab., Madison, Wis.

———— 1967. A look at degradation in particleboards for exterior use. Forest Prod. J. 17(1): 59–66.

————, Hann, R. A., and Haskell, H. H.
1965. Particleboard quality as affected by planer shaving geometry. Forest Prod. J. 14(10): 486–494.

————, and Hefty, F. V.
1968. Steam post-treatments to reduce thickness swelling of particleboard. U.S. Forest Serv. Res. Note FPL–0187. Forest Prod. Lab., Madison, Wis.

————, Kuenzi, E. W., and Maki, A. C.
1964. Linear movement of plywood and flakeboards as related to the longitudinal movement of wood. U.S. Forest Serv. Res. Note FPL–073. Forest Prod. Lab., Madison, Wis.

————, and Lewis, W. C.
1967. Thick particleboards with pulp chip cores— possibilities as roof decking. U.S. Forest Serv. Res. Note FPL–0174. Forest Prod. Lab., Madison, Wis.

Lewis, W. C.
1964. Board materials from wood residues. U.S. Forest Serv. Res. Note FPL–045. Forest Prod. Lab., Madison, Wis.

———— 1967. Thermal conductivity of wood-base fiber and particle panel materials. U.S. Forest Serv. Res. Pap. FPL 77. Forest Prod. Lab., Madison, Wis.

————, and Schwartz, S. L.
1965. Insulating board, hardboard, and other structural fiberboards. U.S. Forest Serv. Res. Note FPL–077. Forest Prod. Lab., Madison, Wis.

Masonite Corporation
Benelex 70. Structural and die-stock grade. 4 pp. Chicago, Ill.

McNatt, J. D.
1969. Rail shear test for evaluating edgewise shear properties of wood-base panel products. USDA Forest Serv. Res. Pap. FPL 117. Forest Prod. Lab., Madison, Wis.

———— 1970. Design stresses for hardboard—effect of rate, duration, and repeated loading. Forest Prod. J. 20(1): 53–60.

National Particleboard Association
1963. Working with particleboard. A.I.A. File No. 19–F–1, 2 pp. June.

———— 1965. Particleboard design and use manual. A.I.A. File No. 23–L, 25 pp. July.

———— 1970. Standard for particleboard for mobile home decking. NPA 1–70, 2 pp. Oct.

———— 1971. Standard for particleboard for decking for factory-built housing. NPA 2–70, 2 pp.

U.S. Air Force
1956. Fiberboard, solid, dunnage, multipurpose cushioning and blocking applications. Mil. Specif. MIL–F–26862 (USAF), 10 pp.

U.S. Department of Commerce
1966. Mat-formed wood particle board. Commercial Stand. CS 236–66, 8 pp. Apr. 15.

U.S. Forest Products Laboratory
1964. Particle board. U.S. Forest Serv. Res. Note FPL–072. Forest Prod. Lab., Madison, Wis.

Chapter 23

MODIFIED WOODS AND PAPER-BASE LAMINATES

Materials with properties subst. itially different than the base material are obtained by chemically treating a wood or wood-base material, compressing it under specially controlled conditions, or by combining the processes of chemical treatment and compression. Sheets of paper treated with chemicals (plastics) are laminated and hot-pressed into thicker panels that have the appearance of plastic rather than paper, and they are used in special applications because of their structural properties and on items requiring hard, impervious, and decorative surfaces.

Modified woods, modified wood-base materials, and paper-base laminates are normally more expensive than wood because of the cost of the chemicals and the special processing required to produce them. Thus, their use is generally limited to special applications where the increased cost is justified by the special properties needed. Volume manufacture of the paper-base plastic laminates has made them the most commonly used surfacing for kitchen and bathroom countertops and similar applications.

MODIFIED WOODS

Wood is treated with chemicals to increase its decay, fire, and moisture resistance. Application of water-resistant chemicals to the surface of wood, or impregnation of the wood with such chemicals dissolved in volatile solvents, reduces the rate of swelling and shrinking of the wood when in contact with water. Such treatments may also reduce the rate at which wood changes dimension because of humidity changes, even though they do not affect the final dimension changes caused by long-duration exposures. Paints, varnishes, lacquers, wood-penetrating water repellents, and plastic and metallic films retard the rate of moisture absorption, but have little effect on total dimension change if exposures are long enough.

Resin-Treated Wood (Impreg)

Permanent stabilization of the dimensions of wood is needed for certain specialty uses. This can be accomplished by depositing a bulking agent within the swollen structure of the wood fibers. The most successful bulking agents that have been commercially applied are highly water-soluble, thermosetting, phenol-formaldehyde resin-forming systems, with initially low molecular weights. No thermoplastic resins have been found that effectively stabilize the dimensions of wood.

Wood treated with a thermosetting fiber-penetrating resin and cured without compressing is known as impreg. The wood (preferably green veneer to facilitate resin pickup) is soaked in the aqueous resin solution or, if air-dry, is impregnated with the solution under pressure until the resin content equals 25 to 35 percent of the weight of dry wood. The treated wood is allowed to stand under non-drying conditions for a day or two to permit uniform distribution of the solution throughout the wood. The resin-containing wood is dried at moderate temperatures to remove the water and then heated to higher temperatures to set the resin.

Uniform distribution of the resin has been effectively accomplished with solid wood [1] only in sapwood of readily penetrated species in lengths up to 6 feet. Drying resin-treated solid wood may result in checking and honeycombing. For these reasons, treatments should be confined to veneer and the treated and cured veneer be used to build up the desired products. Any species can be used for the veneer except the resinous pines. The stronger the original wood, the stronger will be the product.

Impreg has a number of properties differing from those of normal wood and ordinary plywood. These are given in table 23-1, together with similar generalized findings for other modified woods. Table 23-2 gives data for the strength properties of birch impreg. Table 23-3 presents information on thermal expansion properties of "bone-dry" impreg.

Dimensional stability as a basic property of impreg has been the basis of one use where its cost was no deterrent to its acceptability. Wood dies of all parts of automobile bodies serve as the master from which the metal forming dies for actual manufacture of parts are made. Small changes in moisture content, even with the most dimensionally stable wood, produce

---

[1] Veneers with thickness up to 1/3 inch have been successfully treated, although treating times increases rapidly with increases in thickness. Boards of greater thickness and lengths of more than a few inches are classified here as "solid" wood.

changes in dimension and curvature of an un-modified wood die; such changes create major problems in making the metal forming dies where close final tolerances are required. The substitution of impreg, with its high antishrink efficiency, almost entirely eliminated the problem of dimensional change during the entire period that the wood master dies were needed. Despite the tendency of the resins to dull cutting tools, patternmakers accepted the impreg readily because it machined more easily than the unmodified wood.

Patterns made from impreg also were superior to unmodified wood in resisting heat when used with shell molding techniques where temperatures as high as 400° F. were required to cure the resin in the molding sand.

Resin-Treated Compressed Wood (Compreg)

Compreg is similar to impreg except that it is compressed before the resin is cured within the wood. The resin-forming chemicals (usually phenol-formaldehyde) act as plasticizers for the wood so that it can be compressed under modest pressures (1,000 p.s.i.) to a specific gravity of 1.35. Some of its properties are similar to those of impreg, and others vary considerably (tables 23-1 and 23-2). Its advantages over impreg are its natural lustrous finish that can be developed on any cut surface by sanding with fine-grit paper and buffing, its greater strength properties, and the fact that it can be molded (tables 23-1 and 23-2). Thermal expansion coefficients of bone-dry compreg are increased also (table 23-3).

Compreg can be molded by: (1) Gluing blocks of resin-treated (but still uncured) wood with a phenolic glue so that the gluelines and resin within the plies are only partially set; (2) cutting to the desired length and width but two to three times the desired thickness; and (3) compressing in a split mold at about 300° F. Only a small flash squeezeout at the parting line between the two halves of the mold need be machined off. This technique was used for molding motor-test "club" propellers and airplane antenna masts during World War II.

A more generally satisfactory molding technique, known as expansion molding, has been developed. The method consists of rapidly precompressing dry but uncured single sheets of resin-treated veneer in a cold press after preheating them to 200° to 240° F. The heat-plasticized wood responds to compression before cooling. The heat is insufficient to cure

the resin, but the subsequent cooling sets the resin temporarily. These compressed sheets are cut to the desired size, and the assembly of plies is placed in a split mold of the final desired dimensions. Because the wood was precompressed, the filled mold can be closed and locked. When the mold is heated, the wood is again plasticized and tends to recover its uncompressed dimensions. This exerts an internal pressure in all directions against the mold equal to about half of the original compressing pressure. On continued heating, the resin is set. After cooling, the object may be removed from the mold in finished form. Metal inserts or metal surfaces can be molded to compreg or compreg handles molded onto tools by this means. Compreg bands have been molded to the outside of turned wood cylinders without compressing the core. Compreg tubes and small airplane propellers have been molded in this way.

Past uses for compreg once related largely to aircraft—such items as adjustable pitch propellers for training airplanes, antenna masts, and spar and connector plates. Compreg is a suitable material where bolt-bearing strength is required, as in connector plates, because of its good specific strength (strength per unit of weight). Layers of veneer making up the compreg for such uses are often cross-laminated (alternate plies at right angles to each other, as in plywood) to give nearly equal properties in all directions.

Compreg is useful also as supporting blocks for refrigerators, where the combination of load-bearing strength and relatively low thermal conductivity is advantageous. Compreg is extremely useful for aluminum drawing and forming dies, drilling jigs, and jigs for holding parts in place while welding, because of its excellent strength properties, dimensional stability, low thermal conductivity, and ease of fabrication.

Compreg also can be used in silent gears, pulleys, water-lubricated bearings, fan blades, shuttles, bobbins and picker sticks for looms, instrument bases and cases, clarinets, electrical insulators, tool handles, and various novelties. Compreg at present finds considerable use in handles for knives and other cutlery. Both the expansion molding techniques of forming and curing the compreg around the metal parts of the handle and attaching previously made compreg with rivets are used.

Experimental designs with compreg for bowling alley approaches have proved satisfactory. In this instance, the surface sliding

Table 23-1.—*Properties of modified woods*

Property	Impreg	Compreg	Staypak
Specific gravity	15 to 20 percent greater than normal wood.	Usually 1.0 to 1.4	1.25 to 1.40.
Equilibrium swelling and shrinking.	$\frac{1}{4}$ to $\frac{1}{3}$ that of normal wood.	$\frac{1}{4}$ to $\frac{1}{3}$ that of normal wood at right angles to direction of compression, greater in direction of compression but very slow to attain.	Same as normal wood at right angles to compression, greater in direction of compression but very slow to attain.
Springback	None	Very small when properly made	Moderate when properly made.
Face checking	Practically eliminated	Practically eliminated for specific gravities below 1.3.	About the same as in normal wood.
Grain raising	Greatly reduced	Greatly reduced for uniform-texture woods, considerable for contrasting-grain woods.	About the same as in normal wood.
Surface finish	Similar to normal wood	Varnished-like appearance for specific gravities above about 1.0. Cut surfaces can be given this surface by sanding and buffing.	Varnished-like appearance. Cut surfaces can be given this surface by sanding and buffing.
Permeability to water vapor.	About $\frac{1}{10}$ that of normal wood	No data, but much lower than impreg	No data but presumably lower than impreg.
Decay and termite resistance.	Considerably better than normal wood	Considerably better than normal wood	Normal, but decay occurs somewhat more slowly.
Acid resistance	Considerably better than normal wood	Better than impreg because of impermeability.	Better than normal wood because of impremeability but not as good as compreg.
Alkali resistance	Same as normal wood	Somewhat better than normal wood because of impermeability.	Somewhat better than normal wood because of impermeability.
Fire resistance	Same as normal wood	Same as normal wood for long exposure, somewhat better for short exposure.	Same as normal wood for long exposures, somewhat better for short exposures.
Heat resistance	Greatly increased	Greatly increased	No data.
Electrical conductivity	$\frac{1}{10}$ that of normal wood at 30 percent relative humidity; $\frac{1}{1000}$ that of normal wood at 90 percent relative humidity.	Slightly more than impreg at low relative humidity values due to entrapped water.	No data.

Table 23–1.—*Properties of modified woods*—Continued

Property	Impreg	Compreg	Staypak
Heat conductivity	Slightly increased	Increased about in proportion to specific gravity increase.	No data, but should increase about in proportion to specific gravity increase.
Compressive strength	Increased more than proportional to specific gravity increase.	Increased considerably more than proportional to specific gravity increase.	Increased about in proportion to specific gravity increase parallel to grain, increased more perpendicular to grain.
Tensile strength	Decreased significantly	Increased less than proportional to specific gravity increase.	Increased about in proportion to specific gravity increase.
Flexural strength	Increased less than proportional to specific gravity increase.	Increased less than proportional to specific gravity increase parallel to grain, increased more perpendicular to grain.	Increased proportional to specific gravity increase parallel to grain, increased more perpendicular to grain.
Hardness	Increased considerably more than proportional to specific gravity increase.	10 to 20 times that of normal wood	10 to 18 times that of normal wood.
Impact strength: Toughness	About ½ of value for normal wood but very susceptible to the variables of manufacture.	½ to ¾ of value for normal wood but very susceptible to the variables of manufacture.	Same to somewhat greater than normal wood.
Izod	About ⅕ of value for normal wood	⅓ to ¾ of value for normal wood	Same to somewhat greater than normal wood.
Abrasion resistance (tangential).	About ½ of value for normal wood	Increased about in proportion to specific gravity increase.	Increased about in proportion to specific gravity increase.
Machinability	Cuts cleaner than normal wood, but dulls tools more.	Requires metalworking tools and metal-working-tool speeds.	Requires metalworking tools and metal-working-tool speeds.
Moldability	Cannot be molded, but can be formed to single curvatures at time of assembly.	Can be molded by compression and expansion molding methods.	Cannot be molded.
Gluability	Same as normal wood	Same as normal wood after light sanding, or, in the case of thick stock, machining surfaces plane.	Same as normal wood after light sanding, or, in the case of thick stock, machining surfaces plane.

Table 23-2.—*Strength properties of normal and modified laminates* [1] *of yellow birch and a laminated paper plastic*

Property	Normal [2] laminated wood	Impreg [3] (impregnated, uncompressed)	Compreg [3] (impregnated, highly compressed)	Staypak [2] (unimpregnated, highly compressed)	Papreg [4] (impregnated, highly compressed)
Thickness (t) of laminate _____ in.	0.94	1.03	0.63	0.48	0.126 0.512
Moisture content at time of test _____ pct.	9.2	5.0	5.0	4.0	----
Specific gravity (based on weight and volume at test) _	0.7	0.8	1.3	1.4	1.4

PARALLEL LAMINATES

Property	Normal laminated wood	Impreg	Compreg	Staypak	Papreg
Flexure—grain parallel to span (flatwise): [5]					
Proportional limit stress _____ p.s.i.	11,500	15,900	26,700	20,100	15,900
Modulus of rupture _____ p.s.i.	20,400	18,800	36,300	39,400	36,600
Modulus of elasticity _____ 1,000 p.s.i.	2,320	2,380	3,690	4,450	3,010
Flexure—grain perpendicular to span (flatwise): [5]					
Proportional limit stress _____ p.s.i.	1,000	1,300	4,200	3,200	10,500
Modulus of rupture _____ p.s.i.	1,900	1,700	4,600	5,000	24,300
Modulus of elasticity _____ 1,000 p.s.i.	153	220	626	602	1,480
Compression parallel to grain (edgewise): [6]					
Proportional limit stress _____ p.s.i.	6,400	10,200	16,400	9,700	7,200
Ultimate strength _____ p.s.i.	9,500	15,400	26,100	19,100	20,900
Modulus of elasticity _____ 1,000 p.s.i.	2,300	2,470	3,790	4,670	3,120
Compression perpendicular to grain (edgewise): [6]					
Proportional limit stress _____ p.s.i.	670	1,000	4,800	2,600	4,200
Ultimate strength _____ p.s.i.	2,100	3,600	14,000	9,400	18,200
Modulus of elasticity _____ 1,000 p.si.	162	243	571	583	1,600
Compression perpendicular to grain (flatwise) [5]					
maximum crushing strength _____ p.s.i.	----	4,280	16,700	13,200	42,200
Tension parallel to grain (lengthwise):					
Ultimate strength _____ p.s.i.	22,200	15,800	37,000	45,000	35,600
Modulus of elasticity _____ 1,000 p.s.i.	2,300	2,510	3,950	4,610	3,640
Tension perpendicular to grain (crosswise):					
Ultimate strength _____ p.s.i.	1,400	1,400	3,200	3,300	20,000
Modulus of elasticity _____ 1,000 p.s.i.	166	227	622	575	1,710
Shear strength parallel to grain (edgewise): [6]					
Johnson, double shear across laminations __ p.s.i.	2,980	3,460	7,370	6,370	17,800
Cylindrical, double shear parallel to laminates _____ p.s.i.	3,030	3,560	5,690	3,080	3,000
Shear modulus:					
Torsion method _____ 1,000 p.s.i.	182	255	454	----	----
Plate shear method (FPL test) _____ 1,000 p.s.i.	----	----	----	385	909
Toughness (FPL test, edgewise) [6] _____ in.-lb.	235	125	145	250	----
Do _____ in.-lb. per in. of width	250	120	230	515	----
Impact strength (Izod)—grain lengthwise:					
Flatwise (notch in face) __ ft.-lb. per in. of notch	14.0	2.3	4.3	12.7	4.7
Edgewise (notch in face) __ ft.-lb. per in. of notch	11.3	1.9	[7] 3.2	----	0.67
Hardness (Rockwell, flatwise) [5] _____ M-numbers	----	−22	84	----	110
Load to imbed 0.444-inch steel ball to ½ its diameter _____ lb.	1,600	2,400	----	----	----
Hardness modulus (H_M) [8] _____ p.s.i.	5,400	9,200	41,300	43,800	35,600
Abrasion-Navy wear-test machine (flatwise), [5] wear per 1,000 revolutions _____ in.	0.030	0.057	0.018	0.015	0.018
Water absorption (24-hr. immersion), increase in weight _____ pct.	43.6	13.7	2.7	4.3	2.2
Dimensional stability in thickness direction:					
Equilibrium swelling _____ pct.	9.9	2.8	8.0	29	----
Recovery from compression _____ pct.	----	0	0	4	----

Property	Normal [2] laminated wood	Impreg [3] (impregnated, uncompressed)	Compreg [3] (impregnated, highly compressed)	Staypak [2] (unimpregnated, highly compressed)	Papreg [4] (impregnated, highly compressed)
CROSSBAND LAMINATES					
Flexure—face grain parallel to span (flatwise): [5]					
Proportional limit stress _____ p.s.i.	6,900	8,100	14,400	11,400	12,600
Modulus of rupture _____ p.s.i.	13,100	11,400	22,800	25,100	31,300
Modulus of elasticity _____ 1,000 p.s.i.	1,310	1,670	2,480	2,900	2,240
Compression parallel to face grain (edgewise): [6]					
Proportional limit stress _____ p.s.i.	3,300	5,200	8,700	5,200	5,000
Ultimate strength _____ p.s.i.	5,800	11,400	23,900	14,000	18,900
Modulus of elasticity _____ 1,000 p.s.i	1,360	1,500	2,300	2,700	2,370
Tension parallel to face grain (lengthwise):					
Ultimate strength _____ p.s.i.	12,300	7,900	16,500	24,500	27,200
Modulus of elasticity _____ 1,000 p.s.i.	1,290	1,460	2,190	2,570	2,700
Toughness (FPL test edgewise) [6] _____ in.-lb. per in. of width	105	40	115	320	-----

[1] Laminates made from 17 plies of 1/16-in. rotary-cut yellow birch veneer.

[2] Veneer conditioned at 80° F. and 65 pct. relative humidity before assembly with phenol resin film glue.

[3] Impregnation, 25 to 30 pct. of water-soluble phenol-formaldehyde resin based on the dry weight of untreated veneer.

[4] High-strength paper (0.003-in. thickness) made from commercial unbleached black spruce pulp (Mitscherlich sulfite), phenol resin content 36.3 pct., based on weight of treated paper. Izod impact, abrasion, flatwise compression, and shear specimens, all on 1/2-in.-thick papreg.

[5] Load applied to the surface of the original material (parallel to laminating pressure direction).

[6] Load applied to the edge of the laminations (perpendicular to laminating pressure direction).

[7] Values as high as 10.0 ft.-lb per in. of notch have been reported for compreg made with alcohol-soluble resins and 7.0 ft.-lb. with water-soluble resins.

[8] Values based on the average slope of load-penetration plots, where H_M is an expression for load per unit of spherical area of penetration of the 0.444-in. steel ball expressed in pounds per square inch:

$$H_M = \frac{P}{2\pi rh} \text{ or } 0.717\,\frac{P}{h}.$$

Table 23-3.—*Coefficients of linear thermal expansion per degree Celsius of wood, hydrolyzed wood, and paper products* [1]

Material [2]	Specific gravity of product	Glue plus resin content [3]	Linear expansion per °C. by 10^6			Cubical expansion per °C. by 10^6
			Fiber or machine direction	Perpendicular to fiber or machine direction in plane of laminations	Pressing direction	
		Pct.				
Yellow birch laminate	0.72	3.1	3.254	40.29	36.64	80.18
Yellow birch staypak laminate	1.30	4.7	3.406	37.88	65.34	106.63
Yellow birch impreg laminate	.86	33.2	4.648	35.11	37.05	76.81
Yellow birch compreg laminate	1.30	24.8	4.251	39.47	59.14	102.86
Do	1.31	34.3	4.931	39.32	54.83	99.08
Sitka spruce laminate	.53	[4] 6.0	3.837	37.14	27.67	68.65
Parallel-laminated papreg	1.40	36.5	5.73	15.14	65.10	85.97
Crossbanded papreg	1.40	36.5	10.89	[5] 11.00	62.20	84.09
Molded hydrolyzed-wood plastic	1.33	25	42.69	42.69	42.69	128.07
Hydrolyzed-wood sheet laminate	1.39	18	13.49	24.68	77.41	115.58

[1] These coefficients refer to bone-dry material. Generally, air-dry material has a negative thermal coefficient of expansion, because the shrinkage resulting from the loss in moisture is greater than the normal thermal expansion.

[2] All wood laminates made from rotary-cut veneer, annual rings in plane of sheet.
[3] On basis of dry weight of product.
[4] Approximate.
[5] Calculated value.

coefficient so necessary for proper delivery of a bowling ball remained more consistent than for conventional approaches, with practically no maintenance. It was found that surfaces could be maintained with only occasional renewing by light sanding. No top-dressing was required, and the surface was not sensitive to humidity changes. It was never "sticky." Other flooring uses show promise where low maintenance can offset higher initial cost as compared with more conventional flooring materials. Compreg can replace fabric-reinforced plastics in a number of uses because of its better strength properties and lower cost. It should be significantly less expensive because veneer costs less than fabric on a weight basis and about 50 percent less resin is used per unit weight of compreg than for fabric laminates.

Veneer of any nonresinous species can be used for making compreg. Most properties depend upon the specific gravity to which the wood is compressed rather than the species used. Up to the present, however, compreg has been made almost exclusively from yellow birch or sugar maple.

Untreated Compressed Wood (Staypak)

Resin-treated wood in both the uncompressed (impreg) and compressed (compreg) forms is more brittle than the original wood. To meet the demand for a tougher compressed product than compreg, a compressed wood containing no resin (staypak) was developed. It will not lose its compression under swelling conditions as will ordinary compressed untreated wood. In making staypak, the compressing conditions are modified so the lignin cementing material between the cellulose fibers flows sufficiently to eliminate internal stresses.

Staypak is not as water resistant as compreg, but it is about twice as tough and has higher tensile and flexural strength properties, as shown in tables 23-1 and 23-2. The natural finish of staypak is almost equal to that of compreg. Under weathering conditions, however, it is definitely inferior to compreg. For outdoor use a good synthetic resin varnish or paint finish should be applied to staypak.

Staypak can be used in the same way as compreg where extremely high water resistance is not needed. It shows promise for use in tool handles, forming dies, connector plates, propellors, and picker sticks and shuttles for weaving, where high impact strength is needed. As staypak is not impregnated, it can be made from solid wood as well as from veneer. It should cost less than compreg.

Staypak is not being manufactured at the present time. Several companies, however, are prepared to make it if the demand becomes appreciable. Two commercial applications (both patented) of the staypak principle are densification of the corners of desk legs (places subject to wear and slivering) and of ball-line areas of bowling pins where major impacts reduce service life.

Polyethylene-Glycol (PEG) Treated Wood

The dimensional stabilization of wood with polyethylene glycol-1000 (PEG) is accomplished by bulking the fibers to keep the wood in a partially swollen condition. PEG acts in the same manner as does the previously described phenolic resin. It cannot be further cured. The only reason for heating the wood after treatment is to drive off water. PEG remains water soluble. Above 60 percent relative humidity it is a strong humectant; and unless used with care and properly protected, PEG-treated wood can become sticky at these high relative humidities.

Treatment with PEG is facilitated by using green wood. Here pressure is without effect. Treating times are such that uniform uptakes of 25 to 30 percent of chemical are achieved (based on dry weight). The time necessary for this uptake depends on the thickness of the wood and may require weeks. This treatment is being effectively used for walnut gunstocks for high-quality rifles. The dimensional stability of such gunstocks greatly enhances the continued accuracy of the guns. Tabletops of high-quality furniture stay remarkably flat and dimensionally stable when made from PEG-treated wood.

Another application of this chemical is to reduce the checking of green wood during drying. For this application a high degree of polyethylene glycol penetration is not required. This method of treatment has been used to reduce checking during drying of small wood blanks or turnings.

Cracking and distortion that old, waterlogged wood undergoes when it is dried can be substantially reduced by treating the wood with polyethylene glycol. The process was used to dry 200-year-old waterlogged wood boats raised from Lake George, N.Y. The "Vasa," a Swedish ship that sunk on its initial trial voyage in 1628, has also been treated after it was raised.

Wood-Plastic Combination (WPC)

In the modified wood products previously discussed, most of the chemical resides in cell walls; the lumens are essentially empty. If wood is vacuum impregnated with certain liquid vinyl monomers that do not swell wood, and which are later polymerized by gamma radiation or catalyst heat systems, the resulting polymer resides almost exclusively in the lumens. Methyl methacrylate is a common monomer used for a wood-plastic combination. It is converted to polymethyl methacrylate. Such wood-plastic combinations (WPC) with resin contents of 75 to 100 percent (based on the dry weight of wood) resist moisture movement through them. Moisture movement is extremely slow so that normal equilibrium swelling is reached very slowly.

The main commercial use of this modified wood at present is as parquet flooring where it is produced in squares about $5\frac{1}{2}$ inches on a side from strips about $\frac{7}{8}$ inch wide and $\frac{5}{16}$ inch thick. It has a specific gravity of 1.0. Comparative tests with conventional wood flooring indicate WPC material resisted indentation from rolling, concentrated, and impact loads better than white oak. This is largely attributed to improved hardness, which was increased 40 percent in regular wood-plastic combination and 20 percent in the same material treated with a fire retardant. Abrasion resistance was no better than white oak; but because the finish was built-in (buffing is all that is required for finishing), finish is easily maintained even under severe traffic conditions.

Wood-plastic combinations are also being used in sporting goods. One producer of archery equipment is using WPC in bows.

PAPER-BASE PLASTIC LAMINATES

Papreg

Commercially, papreg has become the most important of the resin-impregnated wood-base materials. In thicknesses of about $\frac{1}{16}$ inch, quantities of papreg approaching $\frac{1}{2}$ billion square feet are used as facings for doors, walls, and tops for counters, tables, desks, and other furniture. In other instances, thinner laminates are formed directly on a core material. Here resin-impregnated sheets of paper are hot-pressed, cured, and bonded to the core, which is usually of hardboard or particleboard. Because balanced construction is required if panels are to remain flat in service, a backing is formed at the same time.

Papreg is more generally known as decorative laminate or, technically, as laminated thermosetting decorative sheets. In the trade, it is often referred to by the brand name of the manufacturer. Such names as Consoweld, Formica, Micarta, Textolite, and Panelyte are typical. These decorative laminates are usually composed of a combination of phenolic- and melamine-impregnated sheets of paper. The phenolic-impregnated sheets are brown because of the impregnating resins and comprise most of the built-up thickness of the laminate. The phenolic sheets are overlaid with ones impregnated with melamine resin, which is water-white and transparent. One sheet of the overlay is usually a relatively thick one of high opacity and has the color or design printed on it. Then one or more tissue-thin sheets, which become transparent after the resin is cured, are overlaid on the printed sheet to protect it in service.

Paper-base plastic laminates inherit their final properties from the paper from which they are made. High-strength papers yield higher strength plastics than do low-strength papers. Papers with definite directional properties result in plastics with definite directional properties unless they are cross-laminate (alternate sheets oriented with the machine direction at 90° to each other).

Strength and some other properties of commerical laminates show directional effects introduced from the paper. Tables 23-2 and 23-3 show the properties of parallel- and cross-laminated papreg manufactured for the structural aspect alone (not decorative) or for such purposes as electrical insulation, the oldest use.

Improving the paper used has helped develop paper-base laminates suitable for structural use. Pulping under milder conditions and operating the paper machines to give optimum orientation of the fibers in one direction, together with the desired absorbency, contribute markedly to improvements in strength.

Phenolic resins are the most suitable resins for impregnating the paper from the standpoint of high water resistance, low swelling and shrinking, and high-strength properties (except for impact). Phenolics are also lower in cost than other resins that give comparable properties. Water-soluble resins of the type used for impreg impart the highest water resistance and compressive strength properties to the product, but they unfortunately make the product brittle (low impact strength). Advanced phenolic resins produce a considerably tougher product, but the resins fail to penetrate the fibers as well as water-soluble resins and thus impart less water resistance and dimensional stability to the product. In practice, compromise alcohol-soluble phenolic resins are generally used.

Table 23-2 gives the strength properties of high-strength paper-base laminates. The strength properties of paper-base laminates compare favorably with those for the wood laminates, compreg and staypak, and are superior to those for fabric laminates, except in the edgewise Izod impact test. Fabric laminates have one advantage—they can be molded to greater double curvatures. As paper is considerably less expensive than fabric and can be molded at considerably lower pressures, the paper-base laminates should have an appreciable price advantage over fabric laminates.

Physical properties of the paper are imparted to the paper plastic. Of course, paper will absorb or give off moisture, depending upon conditions of exposure. This moisture change causes paper to shrink and swell, usually more across the machine direction than along it. Likewise, the laminated paper plastics shrink and swell, although at a much slower rate. Cross-laminating minimizes the amount of this shrinking and swelling. In many uses in furniture where laminates are bonded to cores, these changes in dimension due to moisture changes in service with the change of seasons are different than those of the core material. To balance the construction, a paper plastic with similar properties may be glued to the opposite face of the core to prevent bowing or cupping from the moisture changes. Plastics made from different papers will have different shrinking and swelling properties, as shown in table 23-4 for six typical commercial products.

Papreg was used during World War II for molding nonstructural and semistructural airplane parts, such as gunner's seats and turrets, ammunition boxes, wing tabs, and the surfaces of cargo aircraft flooring and "catwalks." It was tried to a limited extent for the skin surface of airplane structural parts, such as wing tips. One major objection to its use for such parts is that it is more brittle than aluminum and requires special fittings.

Papreg has been used to some extent for heavy-duty truck floors and industrial processing trays for nonedible materials, and with melamine-treated papers for decorative table-

Table 23-4.—*Rate of length change of papreg specimens exposed at 80° F., 90 percent relative humidity* [1]

Type of material	Position of specimen with respect to length of sheet	Time (days)						
		1	2	4	7	14	29	56
		Pct.	Pct.	Pct.	Pct.	Pct.	Pct.	Pct.
Decorative laminate	Parallel	0.055	0.075	0.105	0.115	0.100	0.085	0.075
Do	Perpendicular	.110	.170	.260	.325	.345	.365	.350
Do	Parallel	.030	.035	.035	.045	.025	.015	−.005
Do	Perpendicular	.080	.110	.155	.230	.255	.270	.245
Do	Parallel	.080	.115	.140	.145	.135	.135	.130
Do	Perpendicular	.170	.265	.360	.410	.410	.435	.420
Do	Parallel	.065	.100	.145	.175	.180	.190	.180
Do	Perpendicular	.135	.210	.310	.400	.420	.515	.500
Average parallel		.058	.081	.106	.120	.110	.106	.095
Average perpendicular		.124	.189	.271	.341	.358	.496	.379
Backing sheet	Parallel	.130	.115	.105	.105	.090	.085	.075
Do	Perpendicular	.370	.425	.430	.445	.430	.425	.405
Do	Parallel	.110	.090	.080	.075	.065	.045	.030
Do	Perpendicular	.435	.420	.430	.440	.445	.420	.385
Average parallel		.120	.103	.093	.090	.078	.065	.053
Average perpendicular		.403	.423	.430	.443	.438	.423	.395

[1] Material at equilibrium at 80° F., 30 pct. relative humidity before exposure began.

tops. Papreg also appears suitable for pulleys, gears, bobbins, and many other objects for which fabric laminates are used. Because it can be molded at low pressures and is made from thin papers, it is advantageous for use where very large single sheets or accurate control of panel thickness are required.

Lignin-Filled Laminates

The cost of phenolic resins resulted in considerable effort to find impregnating and bonding agents that were more inexpensive and yet readily available. Lignin-filled laminates made with lignin recovered from the spent liquor of the soda pulping process have been produced as a result of this search. Lignin is precipitated from solution within the pulp or added in a pre-precipitated form before the paper is made. The lignin-filled sheets of paper can be laminated without the addition of other resins, but their water resistance is considerably enhanced when some phenolic resin is used. The water resistance can also be improved by merely impregnating the surface sheet with phenolic resin. It is also possible to introduce lignin, together with phenolic resin, into untreated paper sheets with an impregnating machine.

The lignin-filled laminates are always dark brown or black. Their strength properties except for toughness are, in general, lower than those of papreg. The Izod impact values are usually twice those for papreg. In spite of the fact that lignin is somewhat thermoplastic, the loss in strength on heating to 200° F. is proportionately no more than for papreg.

Reduction in costs of phenolic resins has virtually eliminated the lignin-filled laminates from American commerce. They have a number of potential applications, however, where a cheaper laminate with less critical properties than papreg can be used.

Paper-Face Overlays

Paper has found considerable use as an overlay material for veneer or plywood. Overlays can be classified into three different types according to their use—masking, decorative, and structural. Masking overlays are used to cover minor defects in plywood, such as face checks and patches, minimize grain raising, and provide a more uniform paintable surface, thus making possible the use of lower grade veneer. Paper for this purpose need not be of high strength, as the overlays need not add strength to the product. For adequate masking a single surface sheet with a thickness of 0.015 to 0.030 inch is desirable. Paper impregnated with phenolic resins to about 25 percent of the weight of the paper gives the best all-around product. Higher resin contents make the product too costly and tend to make the overlay more transparent. Appreciably lower resin contents give a product with low scratch and abrasion resistance, especially when the panels are wet or exposed to high relative humidities.

The paper faces can be applied at the same time that the veneer is assembled into plywood in a hot press. Undue thermal stresses that might result in checking are not set up if the machine direction of the paper overlays is at right angles to the grain direction of the face plies of the plywood.

Commercially, most of the paper-face overlays are applied to softwood plywood; however, for masking purposes overlays of paper impregnated with resin but only partially cured are available to apply to lumber or other wood-base material like particleboard or hardboard. The curing of the resin is completed at the same time the paper is bonded to the wood or wood-base material in a hot press.

Specific plywood grades with paper-face overlays are available commercially. These are of three types—high density, medium density, and special overlay. Although they are designed for either exterior or interior service, all commercial overlaid plywood conforming to the Product Standards is made in the exterior type.

By specification, the high-density type is one in which the surface on the finished product is hard, smooth, and of such character that further finishing by paint or varnish is not required. It consists of a cellulose fiber sheet or sheets, in which not less than 45 percent of the weight of the laminate is a thermosetting resin of the phenolic or melamine type. The resin-impregnated material cannot be less than 0.012 inch thick before pressing and must weigh not less than 60 pounds per 1,000 square feet of single face before hot-pressing (including both resin and fiber).

By specification also, the medium-density type must present a smooth, uniform surface suitable for high-quality paint finishes. It consists of a cellulose-fiber sheet in which not less than 17 percent resin solids by weight of the laminate for a beater-loaded sheet (or 22 pct. by weight if impregnated) is a thermosetting resin of either the phenolic or melamine type. The resin-impregnated material is at least 0.012 inch thick after application and weighs not less than 58 pounds per 1,000 square feet of single face before hot-pressing (including both resin and fiber). An integral phenolic resin is applied to one surface of the facing material to bond it to the plywood.

The main difference between the two kinds of paper-face overlays for plywood is that the medium-density overlay face is opaque (of solid color) and not translucent like the high-density one. Some evidence of the underlying grain may appear, but compared to the high-density surface, there is no consistent show-through.

Special overlays are those surfacing materials with special characteristics that do not fit the exact description of high- or medium-density overlay types but otherwise meet the test requirements for overlaid plywood.

Thin, transparent (when cured) papers impregnated with melamine resins are used for covering and providing permanent finishes for decorative veneers in furniture and similar articles of wood. In this use the impregnated sheet is bonded to the wood surface in hot presses at the same time the resin is cured. The heat and stain resistance and the strength of this kind of film make it a superior finish.

Masking paper-base overlays unimpregnated with resin are used for such applications as wood house siding that is painted. Vulcanized fiber is the most important commercially. These overlays mask defects in the wood, prevent bleedthrough of resins and extractives in the wood, provide a better substrate for paint, and improve the across-the-board stability from changes in dimension due to changes in moisture content.

BIBLIOGRAPHY

Anonymous
1962. Resin treated, radiation cured wood. Nucleonics 20(3): 94.

Erickson, E.C.O.
1947. Mechanical properties of laminated modified woods. Forest Prod. Lab. Rep. 1639.

Heebink, B. G.
1963. Importance of balanced construction in plastic-faced wood panels. U.S. Forest Serv. Res. Note FPL–021. Forest Prod. Lab., Madison, Wis.

———, and Haskell, H. H.
1962. Effect of heat and humidity on the properties of high-pressure laminates. Forest Prod. J. 12(11): 542–548.

Meyer, J. A.
1965. Treatment of wood-polymer systems using catalyst-heat techniques. Forest Prod. J. 15(9): 362–364.

———, and Loos, W. E.
1969. Treating southern pine wood for modification of properties. Forest Prod. J. 19(12): 32–38.

Mitchell, H. L.
1972. How PEG helps the hobbyist who works with wood. Forest Prod. Lab. unnumbered pub.

———, and Iverson, E. S.
1961. Seasoning green wood carvings with polyethylene glycol–1000. Forest Prod. J. 11(1): 6–7.

———, and Wahlgren, H. E.
1959. New chemical treatment curbs shrink and swell of walnut gunstocks. Forest Prod. J. 9(12): 437–441.

Seborg, R. M., and Inverarity, R. B.
1962. Preservation of old, waterlogged wood by treatment with polyethylene glycol. Science 136(3516): 649–650.

———, Millett, M. A., and Stamm, A. J.
1945. Heat-stabilized compressed wood (staypak). Mech. Eng. 67(1): 25–31.

———, Tarkow, Harold, and Stamm, A. J.
1962. Modified woods. Forest Prod. Lab. Rep. 2192.

———, and Vallier, A. E.
1954. Application of impreg for patterns and die models. Forest Prod. J. 4(5): 305–312.

Seidl, R. J.
1947. Paper and plastic overlays for veneer and plywood. Forest Prod. Res. Soc. Proc. 1: 23–32.

Siau, J. F., Meyer, J. A., and Skaar, C.
1965. A review of developments in dimensional stabilization of wood using radiation techniques. Forest Prod. J. 15(4): 162–166.

Stamm, A. J.
1948. Modified woods. Mod. Plast. Encycl., 40 pp. New York.

———
1964. Wood and cellulose science. Ronald Press Co., New York.

———
1959. Effect of polyethylene glycol on dimensional stability of wood. Forest Prod. J. 9(10): 375–381.

———, and Seborg, R. M.
1962. Forest Products Laboratory resin-treated wood (impreg), rev. Forest Prod. Lab. Rep. 1380.

———, and Seborg, R. M.
1951. Forest Products Laboratory resin-treated, laminated, compressed wood (compreg). Forest Prod. Lab. Rep. 1381.

U.S. Forest Products Laboratory
1962. Physical and mechanical properties of lignin-filled laminated paper plastic, rev. Forest Prod. Lab. Rep. 1579.

———
1943. Preparation of lignin-filled paper for laminated plastics. Forest Prod. Lab. Rep. 1577.

———
1966. Basic properties of yellow birch laminates modified with phenol and urea resins. U.S. Forest Serv. Res. Note FPL–0140. Forest Prod. Lab., Madison, Wis.

Weatherwax, R. C., and Stamm, A. J.
1945. Electrical resistivity of resin-treated wood (impreg and compreg), hydrolyzed-wood sheet (hydroxylin), and laminated resin-treated paper (papreg). Forest Prod. Lab. Rep. 1385.

———, and Stamm, A. J.
1946. The coefficients of thermal expansion of wood and wood products. Trans. Amer. Soc. Mech. Eng. 69(44): 421–432.

GLOSSARY

ADHESIVE. A substance capable of holding materials together by surface attachment. It is a general term and includes cements, mucilage, and paste, as well as glue.

AIR-DRIED. (*See* SEASONING.)

ALLOWABLE PROPERTY.—The value of a property normally published for design use. Allowable properties are identified with grade descriptions and standards, reflect the orthotropic structure of wood, and anticipate certain end uses.

ALLOWABLE STRESS. (*See* ALLOWABLE PROPERTY.)

AMERICAN LUMBER STANDARDS. American lumber standards embody provisions for softwood lumber dealing with recognized classifications, nomenclature, basic grades, sizes, description, measurements, tally, shipping provisions, grademarking, and inspection of lumber. The primary purpose of these standards is to serve as a guide in the preparation or revision of the grading rules of the various lumber manufacturers' associations. A purchaser must, however, make use of association rules as the basic standards are not in themselves commercial rules.

ANISOTROPIC. Not isotropic; that is, not having the same properties in all directions. In general, fibrous materials such as wood are anisotropic.

ANNUAL GROWTH RING. The layer of wood growth put on a tree during a single growing season. In the temperature zone the annual growth rings of many species (e.g., oaks and pines) are readily distinguished because of differences in the cells formed during the early and late parts of the season. In some temperate zone species (black gum and sweetgum) and many tropical species, annual growth rings are not easily recognized.

BALANCED CONSTRUCTION. A construction such that the forces induced by uniformly distributed changes in moisture content will not cause warping. Symmetrical construction of plywood in which the grain direction of each ply is perpendicular to that of adjacent plies is balanced construction.

BARK POCKET. An opening between annual growth rings that contains bark. Bark pockets appear as dark streaks on radial surfaces and as rounded areas on tangential surfaces.

BASTARD SAWN. Lumber (primarily hardwoods) in which the annual rings make angles of 30° to 60° with the surface of the piece.

BEAM. A structural member supporting a load applied transversely to it.

BENDING, STEAM. The process of forming curved wood members by steaming or boiling the wood and bending it to a form.

BENT WOOD. (*See* BENDING, STEAM.)

BIRD PECK. A small hole or patch of distorted grain resulting from birds pecking through the growing cells in the tree. In shape, bird peck usually resembles a carpet tack with the point towards the bark; bird peck is usually accompanied by discoloration extending for considerable distance along the grain and to a much lesser extent across the grain.

BIRDSEYE. Small localized areas in wood with the fibers indented and otherwise contorted to form few to many small circular or elliptical figures remotely resembling birds' eyes on the tangential surface. Sometimes found in sugar maple and used for decorative purposes; rare in other hardwood species.

BLOOM. Crystals formed on the surface of treated wood by exudation and evaporation of the solvent in preservative solutions.

BLUE STAIN. (*See* STAIN.)

BOARD. (*See* LUMBER.)

BOARD FOOT. A unit of measurement of lumber represented by a board 1 foot long, 12 inches wide, and 1 inch thick or its cubic equivalent. In practice, the board foot calculation for lumber 1 inch or more in thickness is based on its nominal thickness and width and the length. Lumber with a nominal thickness of less than 1 inch is calculated as 1 inch.

BOLE. The main stem of a tree of substantial diameter—roughly, capable of yielding sawtimber, veneer logs, or large poles. Seedlings, saplings, and small-diameter trees have stems, not boles.

BOLT. (1) A short section of a tree trunk; (2) in veneer production, a short log of a length suitable for peeling in a lathe.

BOW. The distortion of lumber in which there is a deviation, in a direction perpendicular to the flat face, from a straight line from end-to-end of the piece.

BOX BEAM. A built-up beam with solid wood flanges and plywood or wood-base panel product webs.

BOXED HEART. The term used when the pith falls entirely within the four faces of a piece of wood anywhere in its length. Also called boxed pith.

BRASHNESS. A condition that causes some pieces of wood to be relatively low in shock resistance for the species and, when broken in bending, to fail abruptly without splintering at comparatively small deflections.

BREAKING RADIUS. The limiting radius of curvature to which wood or plywood can be bent without breaking.

BRIGHT. Free from discoloration.

BROAD-LEAVED TREES. (*See* HARDWOODS.)

BROWN ROT. In wood, any decay in which the attack concentrates on the cellulose and associated carbohydrates rather than on the lignin, producing a light to dark brown friable residue—hence loosely termed "dry rot." An advanced stage where the wood splits along rectangular planes, in shrinking, is termed "cubical rot."

BROWN STAIN. (*See* STAIN.)

BUILT-UP TIMBERS. An assembly made by joining layers of lumber together with mechanical fastenings so that the grain of all laminations is essentially parallel.

BURL. (1) A hard, woody outgrowth on a tree, more or less rounded in form, usually resulting from the entwined growth of a cluster of adventitious buds. Such burls are the source of the highly figured burl veneers used for purely ornamental purposes. (2) In lumber or veneer, a localized severe distortion of the grain generally rounded in outline, usually resulting from overgrowth of dead branch stubs, varying from ½ inch to several inches in diameter; frequently includes one or more clusters of several small contiguous conical protuberances, each usually having a core or pith but no appreciable amount of end grain (in tangential view) surrounding it.

BUTT JOINT. (*See* JOINT.)

BUTTRESS. A ridge of wood developed in the angle between a lateral root and the butt of a tree, which may extend up the stem to a considerable height.

CAMBIUM. A thin layer of tissue between the bark and wood that repeatedly subdivides to form new wood and bark cells.

CANT. A log that has been slabbed on one or more sides. Ordinarily, cants are intended for resawing at right angles to their widest sawn face. The term is loosely used. (*See* FLITCH.)

CASEHARDENING. A condition of stress and set in dry lumber characterized by compressive stress in the outer layers and tensile stress in the center or core.

CELL. A general term for the structural units of plant tissue, including wood fibers, vessel members, and other elements of diverse structure and function.

CELLULOSE. The carbohydrate that is the principal constituent of wood and forms the framework of the wood cells.

CHECK. A lengthwise separation of the wood that usually extends across the rings of annual growth and commonly results from stresses set up in wood during seasoning.

CHEMICAL BROWN STAIN. (*See* STAIN.)

CHIPBOARD. A paperboard used for many purposes that may or may not have specifications for strength, color, or other characteristics. It is normally made from paper stock with a relatively low density in the thickness of 0.006 inch and up.

CLOSE GRAINED. (*See* GRAIN.)

COARSE GRAIN. (*See* GRAIN.)

COLD-PRESS PLYWOOD. (*See* PLYWOOD.)

COLLAPSE. The flattening of single cells or rows of cells in heartwood during the drying or pressure treatment of wood. Often characterized by a caved-in or corrugated appearance of the wood surface.

COMPARTMENT KILN. (*See* KILN.)

COMPOUND CURVATURE. Wood bent to a compound curvature has curved surfaces, no element of which is a straight line.

COMPREG. Wood in which the cell walls have been impregnated with synthetic resin and compressed to give it reduced swelling and shrinking characteristics and increased density and strength properties.

COMPRESSION FAILURE. Deformation of the wood fibers resulting from excessive compression along the grain either in direct end compression or in bending. It may develop in standing trees due to bending by wind or snow or to internal longitudinal stresses developed in growth, or it may result from stresses imposed after the tree is cut. In surfaced lumber compression failures may appear as fine wrinkles across the face of the piece.

COMPRESSION WOOD. Wood formed on the lower side of branches and inclined trunks of softwood trees. Compression wood is identified by its relatively wide annual rings, usually eccentric, relatively large amount of summerwood, sometimes more than 50 percent of the width of the annual rings in which it occurs, and its lack of demarcation between springwood and summerwood in the same annual rings. Compression wood shrinks excessively lengthwise, as compared with normal wood.

CONIFER. (*See* SOFTWOODS.)

CONNECTOR, TIMBER. Metal rings, plates, or grids which are embedded in the wood of adjacent members, as at the bolted points of a truss, to increase the strength of the joint.

COOPERAGE. Containers consisting of two round heads and a body composed of staves held together with hoops, such as barrels and kegs.

    Slack cooperage. Cooperage used as containers for dry, semidry or solid products. The staves are usually not closely fitted and are held together with beaded steel, wire, or wood hoops.

    Tight cooperage. Cooperage used as containers for liquids, semi-solids, and heavy solids. Staves are well fitted and held tightly with cooperage grade steel hoops.

CORBEL. A projection from the face of a wall or column supporting a weight.

CORE STOCK. A solid or discontinuous center ply used in panel-type glued structures (such as furniture panels and solid or hollowcore doors).

CROOK. The distortion of lumber in which there is a deviation, in a direction perpendicular to the edge, from a straight line from end-to-end of the piece.

CROSSBAND. To place the grain of layers of wood at right angles in order to minimize shrinking and swelling; also, in plywood of three or more plies, a layer of veneer whose grain direction is at right angles to that of the face plies.

CROSS BREAK. A separation of the wood cells across the grain. Such breaks may be due to internal stress resulting from unequal longitudinal shrinkage or to external forces.

CROSS GRAIN. (*See* GRAIN.)

CUP. A distortion of a board in which there is a deviation flatwise from a straight line across the width of the board.

CURE. To change the properties of an adhesive by chemical reaction (which may be condensation, polymerization, or vulcanization) and thereby develop maximum strength. Generally accomplished by the action of heat or a catalyst, with or without pressure.

CURLY GRAIN. (*See* GRAIN.)

CUT STOCK. A term for softwood stock comparable to dimension stock in hardwoods. (*See* DIMENSION STOCK.)

CUTTINGS. In hardwoods, a portion of a board or plank having the quality required by a specific grade or for a particular use. Obtained from a board by crosscutting or ripping.

DECAY. The decomposition of wood substance by fungi.

Advanced (or typical) decay. The older stage of decay in which the destruction is readily recognized because the wood has become punky, soft and spongy, stringy, ringshaked, pitted, or crumbly. Decided discoloration or bleaching of the rotted wood is often apparent.

Incipient decay. The early stage of decay that has not proceeded far enough to soften or otherwise perceptibly impair the hardness of the wood. It is usually accompanied by a slight discoloration or bleaching of the wood.

DELAMINATION. The separation of layers in a laminate through failure within the adhesive or at the bond between the adhesive and the laminae.

DELIGNIFICATION. Removal of part or all of the lignin from wood by chemical treatment.

DENSITY. As usually applied to wood of normal cellular form, density is the mass of wood substance enclosed within the boundary surfaces of a wood-plus-voids complex having unit volume. It is variously expressed as pounds per cubic foot, kilograms per cubic meter, or grams per cubic centimeter at a specified moisture content.

DENSITY RULES. A procedure for segregating wood according to density, based on percentage of latewood and number of growth rings per inch of radius.

DEW POINT. The temperature at which a vapor begins to deposit as a liquid. Applies especially to water in the atmosphere.

DIAGONAL GRAIN. (*See* GRAIN.)

DIFFUSE-POROUS WOOD. Certain hardwoods in which the pores tend to be uniform in size and distribution throughout each annual ring or to decrease in size slightly and gradually toward the outer border of the ring.

DIMENSION. (*See* LUMBER.)

DIMENSION STOCK. A term largely superseded by the term "hardwood dimension lumber." It is hardwood stock processed to a point where the maximum waste is left at the mill, and the maximum utility is delivered to the user. It is stock of specified thickness, width, and length, or multiples thereof. According to specification it may be solid or glued up, rough or surfaced, semifabricated or completely fabricated.

DIMENSIONAL STABILIZATION. Special treatment of wood to reduce the swelling and shrinking that is caused by changes in its moisture content with changes in relative humidity.

DOTE. "Dote," "doze," and "rot" are synonymous with "decay" and are any form of decay that may be evident as either a discoloration or a softening of the wood.

DRESSED LUMBER. (*See* LUMBER.)

DRY-BULB TEMPERATURE. The temperature of air as indicated by a standard thermometer. (*See* PSYCHROMETER.)

DRY KILN. (*See* KILN.)

DRY ROT. A term loosely applied to any dry, crumbly rot but especially to that which, when in an advanced stage, permits the wood to be crushed easily to a dry powder. The term is actually a misnomer for any decay, since all fungi require considerable moisture for growth.

DRY WALL. Interior covering material, such as gypsum board, hardboard, or plywood, which is applied in large sheets or panels.

DURABILITY. A general term for permanence or resistance to deterioration. Frequently used to refer to the degree of resistance of a species of wood to attack by wood-destroying fungi under conditions that favor such attack. In this connection the term "decay resistance" is more specific.

EARLYWOOD. The portion of the annual growth ring that is formed during the early part of the growing season. It is usually less dense and weaker mechanically than latewood.

EDGE GRAIN. (*See* GRAIN.)

EDGE JOINT. (*See* JOINT.)

EMPTY-CELL PROCESS. Any process for impregnating wood with preservatives or chemicals in which air, imprisoned in the wood under pressure, expands when pressure is released to drive out part of the injected preservative or chemical. The distinguishing characteristic of the empty-cell process is that no vacuum is drawn before applying the preservative. The aim is to obtain good preservative distribution in the wood and leave the cell cavities only partially filled.

ENCASED KNOT. (*See* KNOT.)

END GRAIN. (*See* GRAIN.)

END JOINT. (*See* JOINT.)

EQUILIBRIUM MOISTURE CONTENT. The moisture content at which wood neither gains nor loses moisture when surrounded by air at a given relative humidity and temperature.

EXTERIOR PLYWOOD. (*See* PLYWOOD.)

EXTRACTIVE. Substances in wood, not an integral part of the cellular structure, that can be removed by solution in hot or cold water, ether, benzene, or other solvents that do not react chemically with wood components.

FACTORY AND SHOP LUMBER. (*See* LUMBER.)

FEED RATE. The distance that the stock being processed moves during a given interval of time or operational cycle.

FIBER, WOOD. A wood cell comparatively long ($\frac{1}{25}$ or less to $\frac{1}{3}$ in.), narrow, tapering, and closed at both ends.

FIBERBOARD. A broad generic term inclusive of sheet materials of widely varying densities manufactured of refined or partially refined wood (or other vegetable) fibers. Bonding agents and other materials may be added to increase strength, resistance to moisture, fire, or decay, or to improve some other property.

FIBER SATURATION POINT. The stage in the drying or wetting of wood at which the cell walls are saturated and the cell cavities free from water. It applies to an individual cell or group of cells, not to whole boards. It is usually taken as approximately 30 percent moisture content, based on ovendry weight.

FIBRIL. A threadlike component of cell walls, visible under a light microscope.

FIDDLEBACK. (*See* GRAIN.)

FIGURE. The pattern produced in a wood surface by annual growth rings, rays, knots, deviations from regular grain such as interlocked and wavy grain, and irregular coloration.

FILLER. In woodworking, any substance used to fill the holes and irregularities in planed or sanded surfaces to decrease the porosity of the surface before applying finish coatings.

FINE GRAIN. (*See* GRAIN.)

FINGER JOINT. (*See* JOINT.)

FINISH (FINISHING). Wood products such as doors, stairs, and other fine work required to complete a building, especially the interior. Also, coatings of paint, varnish, lacquer, wax, etc., applied to wood surfaces to protect and enhance their durability or appearance.

FIRE ENDURANCE. A measure of the time during which a material or assembly continues to exhibit fire resistance under specified conditions of test and performance.

FIRE RESISTANCE. The property of a material or assembly to withstand fire or to give protection from it.

FIRE RETARDANT. A chemical or preparation of chemicals used to reduce flammability or to retard spread of a fire over the surface.

FLAKE. A small flat wood particle of predetermined dimensions, uniform thickness, with fiber direction essentially in the plane of the flake; in overall character resembling a small piece of veneer. Produced by special equipment for use in the manufacture of flakeboard.

FLAKEBOARD. A particleboard composed of flakes.

FLAT GRAIN. (*See* GRAIN.)

FLAT-SAWN. (*See* GRAIN, FLAT.)

FLECKS. (*See* RAYS, WOOD.)

FLITCH. A portion of a log sawn on two or more faces—commonly on opposite faces, leaving two waney edges. When intended for resawing into lumber, it is resawn parallel to its original wide faces. Or, it may be sliced or sawn into veneer, in which case the resulting sheets of veneer laid together in the sequence of cutting are called a flitch. The term is loosely used. (See also Cant.)

FRAMING. Lumber used for the structural member of a building, such as studs and joists.

FULL-CELL PROCESS. Any process for impregnating wood with preservatives or chemicals in which a vacuum is drawn to remove air from the wood before admitting the preservative. This favors heavy adsorption and retention of preservative in the treated portions.

GELATINOUS FIBERS. Modified fibers that are associated with tension wood in hardwoods.

GIRDER. A large or principal beam of wood or steel used to support concentrated loads at isolated points along its length.

GRADE. The designation of the quality of a manufactured piece of wood or of logs.

GRAIN. The direction, size, arrangement, appearance, or quality of the fibers in wood or lumber. To have a specific meaning the term must be qualified.

  Close-grained wood. Wood with narrow, inconspicuous annual rings. The term is sometimes used to designate wood having small and closely spaced pores, but in this sense the term "fine textured" is more often used.

  Coarse-grained wood. Wood with wide conspicuous annual rings in which there is considerable difference between springwood and summerwood. The term is sometimes used to designate wood with large pores, such as oak, ash, chestnut, and walnut, but in this sense the term "coarse textured" is more often used.

  Cross-grained wood. Wood in which the fibers deviate from a line parallel to the sides of the piece. Cross grain may be either diagonal or spiral grain or a combination of the two.

Grain (con't.)

  Curly-grained wood. Wood in which the fibers are distorted so that they have a curled appearance, as in "birdseye" wood. The areas showing curly grain may vary up to several inches in diameter.

  Diagonal-grained wood. Wood in which the annual rings are at an angle with the axis of a piece as a result of sawing at an angle with the bark of the tree or log. A form of cross-grain.

  Edge-grained lumber. Lumber that has been sawed so that the wide surfaces extend approximately at right angles to the annual growth rings. Lumber is considered edge grained when the rings form an angle of 45° to 90° with the wide surface of the piece.

  End-grained wood. The grain as seen on a cut made at a right angle to the direction of the fibers (e.g., on a cross section of a tree).

  Fiddleback-grained wood. Figure produced by a type of fine wavy grain found, for example, in species of maple, such wood being traditionally used for the backs of violins.

  Fine-grained wood. (*See* **Close-grained wood.**)

  Flat-grained wood. Lumber that has been sawed parallel to the pith and approximately tangent to the growth rings. Lumber is considered flat grained when the annual growth rings make an angle of less than 45° with the surface of the piece.

  Interlocked-grained wood. Grain in which the fibers put on for several years may slope in a right-handed direction, and then for a number of years the slope reverses to a left-handed direction, and later changes back to a right-handed pitch, and so on. Such wood is exceedingly difficult to split radially, though tangentially it may split fairly easily.

  Open-grained wood. Common classification for woods with large pores, such as oak, ash, chestnut, and walnut. Also known as "coarse textured."

  Plainsawed lumber. Another term for flat-grained lumber.

  Quartersawed lumber. Another term for edge-grained lumber.

  Side-grained wood. Another term for flat-grained lumber.

  Slash-grained wood. Another term for flat-grained lumber.

  Spiral-grained wood. Wood in which the fibers take a spiral course about the trunk of a tree instead of the normal vertical course. The spiral may extend in a right-handed or left-handed direction around the tree trunk. Spiral grain is a form of cross grain.

  Straight-grained wood. Wood in which the fibers run parallel to the axis of a piece.

  Vertical-grained lumber. Another term for edge-grained lumber.

  Wavy-grained wood. Wood in which the fibers collectively take the form of waves or undulations.

GREEN. Freshly sawed or undried wood. Wood that has become completely wet after immersion in water would not be considered green, but may be said to be in the "green condition."

GROWTH RING. (*See* ANNUAL GROWTH RING.)

GUM. A comprehensive term for nonvolatile viscous plant exudates, which either dissolve or swell up in contact with water. Many substances referred to as gums such as pine and spruce gum are actually oleoresins.

HARDBOARD. A generic term for a panel manufactured primarily from interfelted ligno-cellulosic fibers (usually wood), consolidated under heat and pressure in a hot press to a density of 31 pounds per cubic foot or greater, and to which other materials may have been added during manufacture to improve certain properties.

HARDNESS. A property of wood that enables it to resist indentation.

HARDWOODS. Generally one of the botanical groups of trees that have broad leaves in contrast to the conifers or softwoods. The term has no reference to the actual hardness of the wood.

HEART ROT. Any rot characteristically confined to the heartwood. It generally originates in the living tree.

HEARTWOOD. The wood extending from the pith to the sapwood, the cells of which no longer participate in the life processes of the tree. Heartwood may contain phenolic compounds, gums, resins, and other materials that usually make it darker and more decay resistant than sapwood.

HEMICELLULOSE. A celluloselike material (in wood) that is easily decomposable as by dilute acid, yielding several different simple sugars.

HOLLOW-CORE CONSTRUCTION. A panel construction with faces of plywood, hardboard, or similar material bonded to a framed-core assembly of wood lattice, paperboard rings, or the like, which support the facing at spaced intervals.

HONEYCOMB CORE. A sandwich core material constructed of thin sheet materials or ribbons formed to honeycomblike configurations.

HONEYCOMBING. Checks, often not visible at the surface, that occur in the interior of a piece of wood, usually along the wood rays.

HORIZONTALLY LAMINATED. (*See* **LAMINATED WOOD.**)

IMPREG. Wood in which the cell walls have been impregnated with synthetic resin so as to reduce materially its swelling and shrinking. Impreg is not compressed.

INCREMENT BORER. An augerlike instrument with a hollow bit and an extractor, used to extract thin radial cylinders of wood from trees to determine age and growth rate. Also used in wood preservation to determine the depth of penetraiton of a preservative.

INTERGROWN KNOT. (*See* **KNOT.**)

INTERLOCKED-GRAINED WOOD. (*See* **GRAIN.**)

INTUMESCE. To expand with heat to provide a low-density film; used in reference to certain fire-retardant coatings.

JOINT. The junction of two pieces of wood or veneer.
  Butt joint. An end joint formed by abutting the squared ends of two pieces.
  Edge joint. The place where two pieces of wood are joined together edge to edge, commonly by gluing. The joints may be made by gluing two squared edges as in a plain edge joint or by using machined joints of various kinds, such as tongued-and-grooved joints.
  End joint. The place where two pieces of wood are joined together end to end, commonly by scarf or finger jointing.

Joint (con't.)
  Finger joint. An end joint made up of several meshing wedges or fingers or wood bonded together with an adhesive. Fingers are sloped and may be cut parallel to either the wide or edge faces of the piece.
  Lap joint. A joint made by placing one member partly over another and bonding the overlapped portions.
  Scarf joint. An end joint formed by joining with glue the ends of two pieces that have been tapered or beveled to form sloping plane surfaces, usually to a feather edge, and with the same slope of the plane with respect to the length in both pieces. In some cases, a step or hook may be machined into the scarf to facilitate alinement of the two ends, in which case the plane is discontinuous and the joint is known as a stepped or hooked scarf joint.
  Starved joint. A glue joint that is poorly bonded because an insufficient quantity of glue remained in the joint.

JOINT EFFICIENCY OR FACTOR. The strength of a joint expressed as a percentage of the strength of clear straight-grained material.

JOIST. One of a series of parallel beams used to support floor and ceiling loads and supported in turn by larger beams, girders, or bearing walls.

KILN. A chamber having controlled air-flow, temperature, and relative humidity, for drying lumber, veneer, and other wood products.
  Compartment kiln. A kiln in which the total charge of lumber is dried as a single unit. It is designed so that, at any given time, the temperature and relative humidity are essentially uniform throughout the kiln. The temperature is increased as drying progresses, and the relative humidity is adjusted to the needs of the lumber.
  Progressive kiln. A kiln in which the total charge of lumber is not dried as a single unit but as several units, such as kiln truckloads, that move progressively through the kiln. The kiln is designed so that the temperature is lower and the relative humidity higher at the end where the lumber enters than at the discharge end.

KILN DRIED. (*See* **SEASONING.**)

KNOT. That portion of a branch or limb which has been surrounded by subsequent growth of the stem. The shape of the knot as it appears on a cut surface depends on the angle of the cut relative to the long axis of the knot.
  Encased knot. A knot whose rings of annual growth are not intergrown with those of the surrounding wood.
  Intergrown knot. A knot whose rings of annual growth are completely intergrown with those of the surrounding wood.
  Loose knot. A knot that is not held firmly in place by growth or position and that cannot be relied upon to remain in place.
  Pin knot. A knot that is not more than ½ inch in diameter.
  Sound knot. A knot that is solid across its face, at least as hard as the surrounding wood, and shows no indication of decay.
  Spike knot. A knot cut approximately parallel to its long axis so that the exposed section is definitely elongated.

LAMINATE. A product made by bonding together two or more layers (laminations) of material or materials.

LAMINATE, PAPER-BASE. A multilayered panel made by compressing sheets of resin-impregnated paper together into a coherent solid mass.

LAMINATED WOOD. An assembly made by bonding layers of veneer or lumber with an adhesive so that the grain of all laminations is essentially parallel. (*See* **Built-up timbers.**)

   Horizontally laminated wood. Laminated wood in which the laminations are so arranged that the wider dimension of each lamination is approximately perpendicular to the direction of load.

   Vertically laminated wood. Laminated wood in which the laminations are so arranged that the wider dimension of each lamination is approximately parallel to the direction of load.

LAP JOINT. (*See* **JOINT.**)

LATEWOOD. The portion of the annual growth ring that is formed after the earlywood formation has ceased. It is usually denser and stronger mechanically than earlywood.

LIGNIN. The second most abundant constituent of wood, located principally in the secondary wall and the middle lamella, which is the thin cementing layer between wood cells. Chemically it is an irregular polymer of substituted propylphenol groups, and thus no simple chemical formula can be written for it.

LONGITUDINAL. Generally, parallel to the direction of the wood fibers.

LOOSE KNOT. (*See* **KNOT.**)

LUMBER. The product of the saw and planing mill not further manufactured than by sawing, resawing, passing lengthwise through a standard planing machine, crosscutting to length, and matching.

   Boards. Lumber that is nominally less than 2 inches thick and 2 or more inches wide. Boards less than 6 inches wide are sometimes called strips.

   Dimension. Lumber with a nominal thickness of from 2 up to but not including 5 inches and a nominal width of 2 inches or more.

   Dressed size. The dimensions of lumber after being surfaced with a planing machine. The dressed size is usually ½ to ¾ inch less than the nominal or rough size. A 2- by 4-inch stud, for example, actually measures about 1½ by 3½ inches.

   Factory and shop lumber. Lumber intended to be cut up for use in further manufacture. It is graded on the basis of the percentage of the area that will produce a limited number of cuttings of a specified minimum size and quality.

   Matched lumber. Lumber that is edge dressed and shaped to make a close tongued-and-grooved joint at the edges or ends when laid edge to edge or end to end.

   Nominal size. As applied to timber or lumber, the size by which it is known and sold in the market; often differs from the actual size. (See also, Dressed size.)

   Patterned lumber. Lumber that is shaped to a pattern or to a molded form in addition to being dressed, matched, or shiplapped, or any combination of these workings.

   Rough lumber. Lumber which has not been dressed (surfaced) but which has been sawed, edged, and trimmed.

Lumber (con't.)

   Shiplapped lumber. Lumber that is edge dressed to make a lapped joint.

   Shipping-dry lumber. Lumber that is partially dried to prevent stain and mold in transit.

   Side lumber. A board from the outer portion of the log—ordinarily one produced when squaring off a log for a tie or timber.

   Structural lumber. Lumber that is intended for use where allowable properties are required. The grading of structural lumber is based on the strength of the piece as related to anticipated uses.

   Surface lumber. Lumber that is dressed by running it through a planer.

   Timbers. Lumber that is nominally 5 or more inches in least dimension. Timbers may be used as beams, stringers, posts, caps, sills, griders, purlins, etc.

   Yard lumber. A little-used term for lumber of all sizes and patterns that is intended for general building purposes having no design property requirements.

LUMEN. In wood anatomy, the cell cavity.

MANUFACTURING DEFECTS. Includes all defects or blemishes that are produced in manufacturing, such as chipped grain, loosened grain, raised grain, torn grain, skips in dressing, hit and miss (series of surfaced areas with skips between them), variation in sawing, miscut lumber, machine burn, machine gouge, mismatching, and insufficient tongue or groove.

MATCHED LUMBER. (*See* **LUMBER.**)

MILLWORK. Planed and patterned lumber for finish work in buildings, including items such as sash, doors, cornices, panelwork, and other items of interior or exterior trim. Does not include flooring, ceiling, or siding.

MINERAL STREAK. An olive to greenish-black or brown discoloration of undermined cause in hardwoods.

MODIFIED WOOD. Wood processed by chemical treatment, compression, or other means (with or without heat) to impart properties quite different from those of the original wood.

MOISTURE CONTENT. The amount of water contained in the wood, usually expressed as a percentage of the weight of the ovendry wood.

MOLDED PLYWOOD. (*See* **PLYWOOD.**)

MOLDING. A wood strip having a curved or projecting surface, used for decorative purposes.

MORTISE. A slot cut into a board, plank, or timber, usually edgewise, to receive the tenon of another board, plank, or timber to form a joint.

NAVAL STORES. A term applied to the oils, resins, tars, and pitches derived from oleoresin contained in, exuded by, or extracted from trees, chiefly species of pines (genus *Pinus*). Historically, these were important items in the stores of wood sailing vessels.

NOMINAL-SIZE LUMBER. (*See* **LUMBER.**)

OLD GROWTH. Timber in or from a mature, naturally established forest. When the trees have grown during most if not all of their individual lives in active competition with their companions for sunlight and moisture, this timber is usually straight and relatively free of knots.

OLEORESIN. A solution of resin in an essential oil that occurs in or exudes from many plants, especially softwoods. The oleoresin from pine is a solution of pine resin (rosin) in turpentine.

OPEN GRAIN. (*See* GRAIN.)

ORTHOTROPIC. Having unique and independent properties in three mutually orthogonal (perpendicular) planes of symmetry. A special case of anisotropy.

OVENDRY WOOD. Wood dried to a relatively constant weight in a ventilated oven at 101° to 105°C.

OVERLAY. A thin layer of paper, plastic, film, metal foil, or other material bonded to one or both faces of panel products or to lumber to provide a protective or decorative face or a base for painting.

PALLET. A low wood or metal platform on which material can be stacked to facilitate mechanical handling, moving, and storage.

PAPERBOARD. The distinction between paper and paperboard is not sharp, but broadly speaking, the thicker (over 0.012 in.), heavier, and more rigid grades of paper are called paperboard.

PAPREG. Any of various paper products made by impregnating sheets of specially manufactured high-strength paper with synthetic resin and laminating the sheets to form a dense, moisture-resistant product.

PARENCHYMA. Short cells having simple pits and functioning primarily in the metabolism and storage of plant food materials. They remain alive longer than the tracheids, fibers, and vessel segments, sometimes for many years. Two kinds of parenchyma cells are recognized—those in vertical strands, known more specifically as axial parenchyma, and those in horizontal series in the rays, known as ray parenchyma.

PARTICLEBOARD. A generic term for a panel manufactured from lignocellulosic materials—commonly wood—essentially in the form of particles (as distinct from fibers). These materials are bonded together with synthetic resin or other suitable binder, under heat and pressure, by a process wherein the interparticle bonds are created wholly by the added binder.

PATTERNED LUMBER. (*See* LUMBER.)

PECK. Pockets or areas of disintegrated wood caused by advanced stages of localized decay in the living tree. It is usually associated with cypress and incense-cedar. There is no further development of peck once the lumber is seasoned.

PEEL. To convert a log into veneer by rotary cutting.

PHLOEM. The tissues of the inner bark, characterized by the presence of sieve tubes and serving for the transport of elaborate foodstuffs.

PILE. A long, heavy timber, round or square cut, that is driven deep into the ground to provide a secure foundation for structures built on soft, wet, or submerged sites; e.g., landing stages, bridge abutments.

PIN-KNOT. (*See* KNOT.)

PITCH POCKET. An opening extending parallel to the annual growth rings and containing, or that has contained, pitch, either solid or liquid.

PITCH STREAKS. A well-defined accumulation of pitch in a more or less regular streak in the wood of certain conifers.

PITH. The small, soft core occurring near the center of a tree trunk, branch, twig, or log.

PITH FLECK. A narrow streak, resembling pith on the surface of a piece; usually brownish, up to several inches in length; resulting from burrowing of larvae in the growing tissues of the tree.

PLAINSAWED. (*See* GRAIN.)

PLANING MILL PRODUCTS. Products worked to pattern, such as flooring, ceiling, and siding.

PLANK. A broad board, usually more than 1 inch thick, laid with its wide dimension horizontal and used as a bearing surface.

PLASTICIZING WOOD. Softening wood by hot water, steam, or chemical treatment to increase its moldability.

PLYWOOD. A composite panel or board made up of cross-banded layers of veneer only or veneer in combination with a core of lumber or of particleboard bonded with an adhesive. Generally the grain of one or more plies is roughly at right angles to the other plies, and almost always an odd number of plies are used.

> **Cold-pressed plywood.** Refers to interior-type plywood manufactured in a press without external applications of heat.
>
> **Exterior plywood.** A general term for plywood bonded with a type of adhesive that by systematic tests and service records has proved highly resistant to weather; micro-organisms; cold, hot, and boiling water; steam; and dry heat.
>
> **Molded plywood.** Plywood that is glued to the desired shape either between curved forms or more commonly by fluid pressure applied with flexible bags or blankets (bag molding) or other means.
>
> **Postformed plywood.** The product formed when flat plywood is reshaped into a curve configuration by steaming or plasticizing agents.

POCKET ROT. Advanced decay that appears in the form of a hole or pocket, usually surrounded by apparently sound wood.

PORE. (*See* VESSELS.)

POROUS WOODS. Hardwoods having vessels or pores large enough to be seen readily without magnification.

POSTFORMED PLYWOOD. (*See* PLYWOOD.)

PRESERVATIVE. Any substance that, for a reasonable length of time, is effective in preventing the development and action of wood-rotting fungi, borers of various kinds, and harmful insects that deteriorate wood.

PRESSURE PROCESS. Any process of treating wood in a closed container whereby the preservative or fire retardant is forced into the wood under pressures greater than 1 atmosphere. Pressure is generally preceded or followed by vacuum, as in the vacuum-pressure and empty cell processes respectively, or they may alternate, as in the full cell and alternating-pressure processes.

PROGRESSIVE KILN. (*See* KILN.)

PSYCHROMETER. An instrument for measuring the amount of water vapor in the atmosphere. It has both a dry-bulb and wet-bulb thermometer. The bulb of the wet-bulb thermometer is kept moistened and is, therefore, cooled by evaporation to a temperature lower than that shown by the dry-bulb thermometer. Because evaporation is greater in dry air, the difference between the two thermometer readings will be greater when the air is dry than when it is moist.

QUARTERSAWED. (*See* GRAIN.)

RADIAL. Coincident with a radius from the axis of the tree or log to the circumference. A radial section is a lengthwise section in a plane that passes through the centerline of the tree trunk.

RAFTER. One of a series of structural members of a roof designed to support roof loads. The rafters of a flat roof are sometimes called roof joists.

RAISED GRAIN. A roughened condition of the surface of dressed lumber in which the hard summerwood is raised above the softer springwood but not torn loose from it.

RAYS, WOOD. Strips of cells extending radially within a tree and varying in height from a few cells in some species to 4 or more inches in oak. The rays serve primarily to store food and transport it horizontally in the tree. On quartersawed oak, the rays form a conspicuous figure, sometimes referred to as flecks.

REACTION WOOD. Wood with more or less distinctive anatomical characters, formed typically in parts of leaning or crooked stems and in branches. In hardwoods this consists of tension wood and in softwoods of compression wood.

RELATIVE HUMIDITY. Ratio of the amount of water vapor present in the air to that which the air would hold at saturation at the same temperature. It is usually considered on the basis of the weight of the vapor but, for accuracy, should be considered on the basis of vapor pressures.

RESILIENCE. The property whereby a strained body gives up its stored energy on the removal of the deforming force.

RESIN. Inflammable, water-soluble, vegetable substances secreted by certain plants or trees, and characterizing the wood of many coniferous species. The term is also applied to synthetic organic products related to the natural resins.

RESIN DUCTS. Intercellular passages that contain and transmit resinous materials. On a cut surface, they are usually inconspicuous. They may extend vertically parallel to the axis of the tree or at right angles to the axis and parallel to the rays.

RETENTION BY ASSAY. The determination of preservative retention in a specific zone of treated wood by extraction or analysis of specified samples.

RING FAILURE. A separation of the wood during seasoning, occurring along the grain and parallel to the growth rings. (See also, Shake.)

RING-POROUS WOODS. A group of hardwoods in which the pores are comparatively large at the beginning of each annual ring and decrease in size more or less abruptly toward the outer portion of the ring, thus forming a distinct inner zone of pores, known as the earlywood, and an outer zone with smaller pores, known as the latewood.

RING SHAKE. (*See* SHAKE.)

RIP. To cut lengthwise, parallel to the grain.

ROT. (*See* DECAY.)

ROTARY-CUT VENEER. (*See* VENEER.)

ROUGH LUMBER. (*See* LUMBER.)

SANDWICH CONSTRUCTION. (*See* STRUCTURAL SANDWICH CONSTRUCTION.)

SAP STAIN. (*See* STAIN.)

SAPWOOD. The wood of pale color near the outside of the log. Under most conditions the sapwood is more susceptible to decay than heartwood.

SASH. A frame structure, normally glazed (e.g., a window), that is hung or fixed in a frame set in an opening.

SAWED VENEER. (*See* VENEER.)

SAW KERF. (1) Grooves or notches made in cutting with a saw; (2) that portion of a log, timber, or other piece of wood removed by the saw in parting the material into two pieces.

SCARF JOINT. (*See* JOINT.)

SCHEDULE, KILN DRYING. A prescribed series of dry- and wet-bulb temperatures and air velocities used in drying a kiln charge of lumber or other wood products.

SEASONING. Removing moisture from green wood to improve its serviceability.

Air-dried. Dried by exposure to air in a yard or shed, without artificial heat.

Kiln-dried. Dried in a kiln with the use of artificial heat.

SECOND GROWTH. Timber that has grown after the removal, whether by cutting, fire, wind, or other agency, of all or a large part of the previous stand.

SET. A permanent or semipermanent deformation.

SHAKE. A separation along the grain, the greater part of which occurs between the rings of annual growth. Usually considered to have occurred in the standing tree or during felling.

SHAKES. In construction, shakes are a type of shingle usually hand cleft from a bolt and used for roofing or weatherboarding.

SHAVING. A small wood particle of indefinite dimensions developed incidental to certain woodworking operations involving rotary cutterheads usually turning in the direction of the grain. This cutting action produces a thin chip of varying thickness, usually feathered along at least one edge and thick at another and generally curled.

SHEAR. A condition of stress or strain where parallel planes slide relative to one another.

SHEATHING. The structural covering, usually of boards, building fiberboards, or plywood, placed over exterior studding or rafters of a structure.

SHIPLAPPED LUMBER. (*See* LUMBER.)

SHIPPING-DRY LUMBER. (*See* LUMBER.)

SHOP LUMBER. (*See* LUMBER.)

SIDE-GRAIN. (*See* GRAIN.)

SIDE LUMBER. (*See* LUMBER.)

SIDING. The finish covering of the outside wall of a frame building, whether made of horizontal weatherboards, vertical boards with battens, shingles, or other material.

SLASH GRAINED. (*See* GRAIN.)

SLICED VENEER. (*See* VENEER.)

SOFT ROT. A special type of decay developing under very wet conditions (as in cooling towers and boat timbers) in the outer wood layers, caused by cellulose-destroying microfungi that attack the secondary cell walls and not the intercellular layer.

SOFTWOODS. Generally, one of the botanical groups of trees that in most cases have needlelike or scalelike leaves; the conifers, also the wood produced by such trees. The term has no reference to the actual hardness of the wood.

SOUND KNOT. (*See* KNOT.)

SPECIFIC GRAVITY. As applied to wood, the ratio of the ovendry weight of a sample to the weight of a volume of water equal to the volume of the sample at a specified moisture content (green, air-dry, or ovendry).

SPIKE KNOT. (*See* KNOT.)

SPIRAL GRAIN. (*See* GRAIN.)

SPRINGWOOD. (*See* EARLYWOOD.)

STAIN. A discoloration in wood that may be caused by such diverse agencies as micro-organisms, metal, or chemicals. The term also applies to materials used to impart color to wood.

Blue stain. A bluish or grayish discoloration of the sapwood caused by the growth of certain dark-colored fungi on the surface and in the interior of the wood; made possible by the same conditions that favor the growth of other fungi.

Stain (con't.)

    Brown stain. A rich brown to deep chocolate-brown discoloration of the sapwood of some pines caused by a fungus that acts much like the blue-stain fungi.

    Chemical brown stain. A chemical discoloration of wood, which sometimes occurs during the air-drying or kiln-drying of several species, apparently caused by the concentration and modification of extractives.

    Sap stain. (*See* **Blue stain.**)

    Sticker stain. A brown or blue stain that develops in seasoning lumber where it has been in contact with the stickers.

STARVED JOINT. (*See* **JOINT.**)

STATIC BENDING. Bending under a constant or slowly applied load; flexure.

STAYPAK. Wood that is compressed in its natural state (that is, without resin or other chemical treatment) under controlled conditions of moisture, temperature, and pressure that practically eliminate springback or recovery from compression. The product has increased density and strength characteristics.

STICKERS. Strips or boards used to separate the layers of lumber in a pile and thus improve air circulation.

STICKER STAIN. (*See* **STAIN.**)

STRAIGHT GRAINED. (*See* **GRAIN.**)

STRENGTH. (1) The ability of a member to sustain stress without failure. (2) In a specific mode of test, the maximum stress sustained by a member loaded to failure.

STRENGTH RATIO. The hypothetical ratio of the strength of a structural member to that which it would have if it contained no strength-reducing characteristics (knots, cross-grain, shake, etc.).

STRESSED-SKIN CONSTRUCTION. A construction in which panels are separated from one another by a central partition of spaced strips with the whole assembly bonded so that it acts as a unit when loaded.

STRINGER. A timber or other support for cross members in floors or ceilings. In stairs, the support on which the stair treads rests.

STRUCTURAL LUMBER. (*See* **LUMBER.**)

STRUCTURAL SANDWICH CONSTRUCTION. A layered construction comprising a combination of relatively high-strength facing materials intimately bonded to and acting integrally with a low-density core material.

STRUCTURAL TIMBERS. Pieces of wood of relatively large size, the strength of which is the controlling element in their selection and use. Trestle timbers (stringers, caps, posts, sills, bracing, bridge ties, guardrails); car timbers (car framing, including upper framing, car sills); framing for building (posts, sills, girders); ship timber (ship timbers, ship decking); and cross-arms for poles are examples of structural timbers.

STUD. One of a series of slender wood structural members used as supporting elements in walls and partitions.

SUMMERWOOD. (*See* **LATEWOOD.**)

SURFACED LUMBER. (*See* **LUMBER.**)

SYMMETRICAL CONSTRUCTION. Plywood panels in which the plies on one side of a center ply or core are essentially equal in thickness, grain direction, properties, and arrangement to those on the other side of the core.

TANGENTIAL. Strictly, coincident with a tangent at the circumference of a tree or log, or parallel to such a tangent. In practice, however, it often means roughly coincident with a growth ring. A tangential section is a longitudinal section through a tree or limb perpendicular to a radius. Flat-grained lumber is sawed tangentially.

TENON. A projecting member left by cutting away the wood around it for insertion into a mortise to make a joint.

TENSION WOOD. A form of wood found in leaning trees of some hardwood species and characterized by the presence of gelatinous fibers and excessive longitudinal shrinkage. Tension wood fibers hold together tenaciously, so that sawed surfaces usually have projecting fibers, and planed surfaces often are torn or have raised grain. Tension wood may cause warping.

TEXTURE. A term often used interchangeably with grain. Sometimes used to combine the concepts of density and degree of contrast between springwood and summerwood. In this handbook texture refers to the finer structure of the wood (*see* Grain) rather than the annual rings.

THERMOPLASTIC GLUES AND RESINS. Glues and resins that are capable of being repeatedly softened by heat and hardened by cooling.

THERMOSETTING GLUES AND RESINS. Glues and resins that are cured with heat but do not soften when subsequently subjected to high temperatures.

TIMBERS, ROUND. Timbers used in the original round form, such as poles, piling, posts, and mine timbers.

TIMBER, STANDING. Timber still on the stump.

TIMBERS. (*See* **LUMBER.**)

TOUGHNESS. A quality of wood which permits the material to absorb a relatively large amount of energy, to withstand repeated shocks, and to undergo considerable deformation before breaking.

TRACHEID. The elongated cells that constitute the greater part of the structure of the softwoods (frequently referred to as fibers). Also present in some hardwoods.

TRANSVERSE. Directions in wood at right angles to the wood fibers. Includes radial and tangential directions. A transverse section is a section through a tree or timber at right angles to the pith.

TREENAIL. A wooden pin, peg, or spike used chiefly for fastening planking and ceiling to a framework.

TRIM. The finish materials in a building, such as moldings, applied around openings (window trim, door trim) or at the floor and ceiling of rooms (baseboard, cornice, and other moldings).

TRUSS. An assembly of members, such as beams, bars, rods, and the like, so combined as to form a rigid framework. All members are interconnected to form triangles.

TWIST. A distortion caused by the turning or winding of the edges of a board so that the four corners of any face are no longer in the same plane.

TYLOSES. Masses of parenchyma cells appearing somewhat like froth in the pores of some hardwoods, notably the white oaks and black locust. Tyloses are formed by the extension of the cell wall of the living cells surrounding vessels of hardwood.

VAPOR BARRIER. A material with a high resistance to vapor movement, such as foil, plastic film, or specially coated paper, that is used in combination with insulation to control condensation.

VENEER. A thin layer or sheet of wood.

    Rotary-cut veneer. Veneer cut in a lathe which rotates a log or bolt, chucked in the center, against a knife.

    Sawed veneer. Veneer produced by sawing.

    Sliced veneer. Veneer that is sliced off a log, bolt, or flitch with a knife.

VERTICAL GRAIN. (*See* **Grain.**)

VERTICALLY LAMINATED WOOD. (*See* **Laminated wood.**)

VESSELS. Wood cells of comparatively large diameter that have open ends and are set one above the other to form continuous tubes. The openings of the vessels on the surface of a piece of wood are usually referred to as pores.

VIRGIN GROWTH. The original growth of mature trees.

WANE. Bark or lack of wood from any cause on edge or corner of a piece.

WARP. Any variation from a true or plane surface. Warp includes bow, crook, cup, and twist, or any combination thereof.

WAVY-GRAIN. (*See* **Grain.**)

WEATHERING. The mechanical or chemical disintegration and discoloration of the surface of wood caused by exposure to light, the action of dust and sand carried by winds, and the alternate shrinking and swelling of the surface fibers with the continual variation in moisture content brought by changes in the weather. Weathering does not include decay.

WET-BULB TEMPERATURE. The temperature indicated by the wet-bulb thermometer of a psychrometer.

WHITE-ROT. In wood, any decay or rot attacking both the cellulose and the lignin, producing a generally whitish residue that may be spongy or stringy rot, or occur as pocket rot.

WOOD FLOUR. Wood reduced to finely divided particles approximately those of cereal flours in size, appearance, and texture, and passing a 40–100 mesh screen.

WOOD SUBSTANCE. The solid material of which wood is composed. It usually refers to the extractive-free solid substance of which the cell walls are composed, but this is not always true. There is no wide variation in chemical composition or specific gravity between the wood substance of various species, the characteristic differences of species being largely due to differences in extractives and variations in relative amounts of cell walls and cell cavities.

WORKABILITY. The degree of ease and smoothness of cut obtainable with hand or machine tools.

XYLEM. The portion of the tree trunk, branches, and roots that lies between the pith and the cambium.

YARD LUMBER. (*See* **LUMBER.**)

Part II

WOOD–FRAME HOUSE CONSTRUCTION

CHAPTER 24

LOCATION AND EXCAVATION

Condition at Site

Before excavating for the new home, determine the subsoil conditions by test borings or by checking existing houses constructed near the site. A rock ledge may be encountered, necessitating costly removal; or variation from standard construction practices will increase the cost of the *foundation* and footings. Thus a high water table may require design changes from a full basement to crawl space or concrete slab construction. If the area has been filled, the *footings* [2] should always extend through to undisturbed soil. Any it is good practice to examine the type of foundations used in neighboring houses—this might influence the design of the new house.

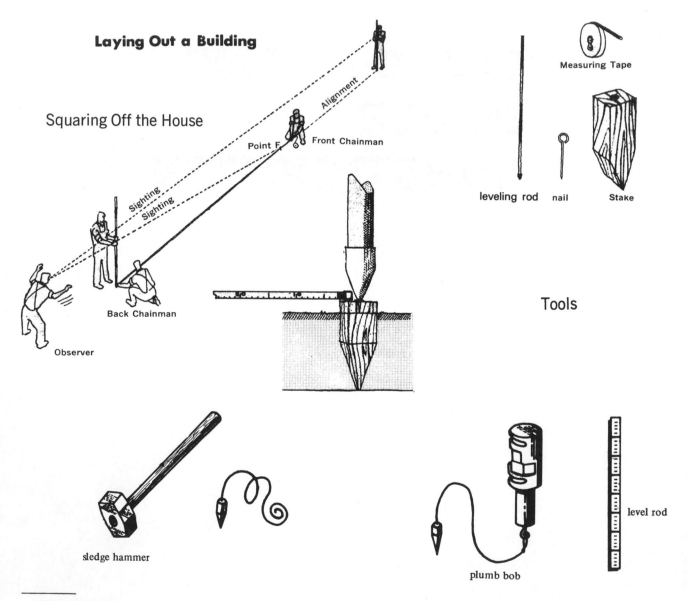

Laying Out a Building

Squaring Off the House

Alignment

Point F

Front Chainman

Sighting

Sighting

Back Chainman

Observer

Measuring Tape

leveling rod   nail   Stake

Tools

sledge hammer

plumb bob

level rod

[1] Maintained at Madison, Wis., in cooperation with the University of Wisconsin.

[2] Key words in italics appear in the glossary

Squaring the Layout

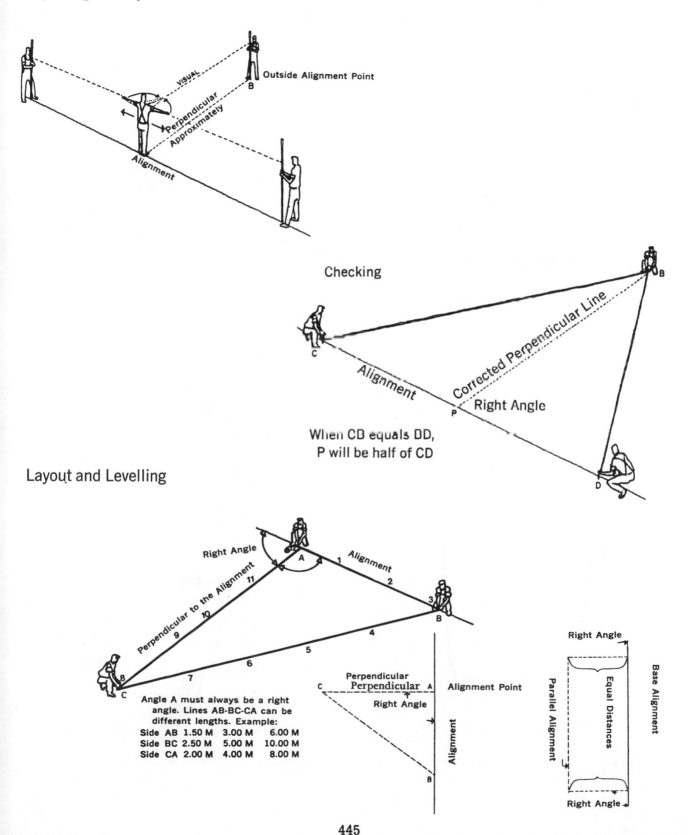

VISUAL

Outside Alignment Point

B

Perpendicular Approximately

Alignment

Checking

B

Corrected Perpendicular Line

C

Alignment

Right Angle

P

When CB equals DD,
P will be half of CD

D

Layout and Levelling

Right Angle

Perpendicular to the Alignment

Alignment

A

B

C

Angle A must always be a right
angle. Lines AB-BC-CA can be
different lengths. Example:

Side AB 1.50 M   3.00 M   6.00 M
Side BC 2.50 M   5.00 M   10.00 M
Side CA 2.00 M   4.00 M   8.00 M

Perpendicular
Perpendicular A   Alignment Point

C

Right Angle

Alignment

B

Parallel Alignment

Right Angle

Equal Distances

Base Alignment

Parallel Alignment

Right Angle

445

Establishing a level between two points

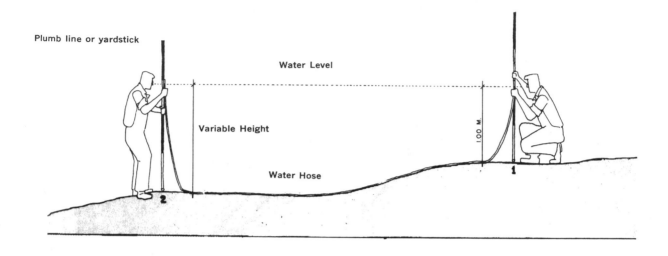

Plumb line or yardstick

Water Level

Variable Height

100 M.

Water Hose

2   1

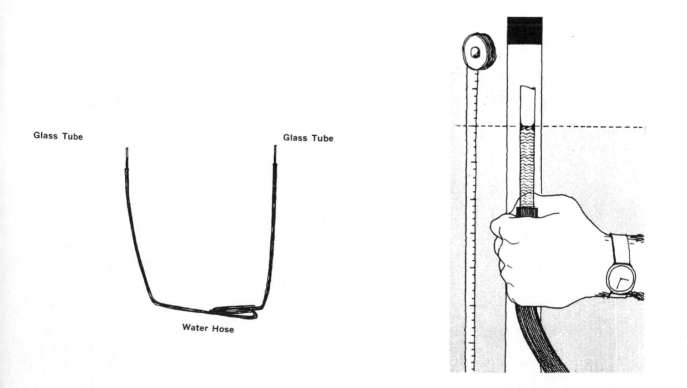

Glass Tube   Glass Tube

Water Hose

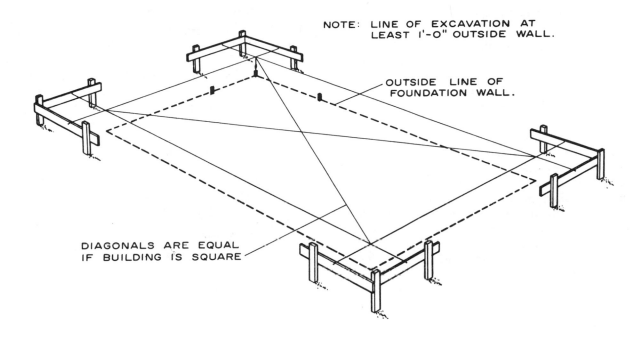

NOTE: LINE OF EXCAVATION AT LEAST 1'-0" OUTSIDE WALL.

OUTSIDE LINE OF FOUNDATION WALL.

DIAGONALS ARE EQUAL IF BUILDING IS SQUARE

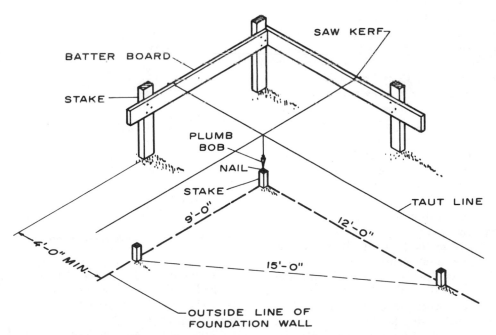

SAW KERF

BATTER BOARD

STAKE

PLUMB BOB

NAIL

STAKE

TAUT LINE

9'-0"

12'-0"

4'-0" MIN.

15'-0"

OUTSIDE LINE OF FOUNDATION WALL

Figure 24.1.—Staking and laying out the house.

Placement of the House

After the site is cleared, the location of the outer walls of the house is marked out. In general, the surveyor will mark the corners of the lot after making a survey of the plot of land. The corners of the proposed house also should be roughly marked by the surveyor.

Before the exact location of the house is determined, check local codes for minimum setback and side-yard requirements; the location of the house is usually determined by such codes. In some cases, the setback may be established by existing houses on adjacent property. Most city building regulations require that a plot plan be a part of the house plans so its location is determined beforehand.

The next step, after the corners of the house have been established, is to determine lines and grades as aids in keeping the work level and true. The *batter board* (fig. 24. 1) is one of the methods used to locate and retain the outline of the house. The height of the boards is sometimes established to conform to the height of the foundation wall.

Small stakes are first located accurately at each corner of the house with nails driven in their tops to indicate the outside line of the foundation walls. To assure square corners, measure the diagonals to see if they are the same length. The corners can also be squared by measuring along one side a distance in 3-foot units such as 6, 9, and 12 and along the adjoining side the same number of 4-foot units as 8, 12, and 16. The diagonals will then measure the equal of 5-foot units such as 10, 15, and 20 when the unit is square. Thus, a 9-foot distance on one side and a 12-foot distance on the other should result in a 15-foot diagonal measurement for a true 90° corner.

After the corners have been located, three 2- by 4-inch or larger stakes of suitable length are driven at each location 4 feet (minimum) beyond the lines of the foundation; then 1- by 6- or 1- by 8-inch boards are nailed horizontally so the tops are all level at the same grade. Twine or stout string (carpenter chalkline) is next held across the top of opposite boards at two corners and adjusted so that it will be exactly over the nails in the corner stakes at either end; a *plumb* bob is handy for setting the lines. Saw kerfs at the outside edge are cut where the lines touch the boards so that they may be replaced if broken or disturbed. After similar cuts are located in all eight batter boards, the lines of the house will be established. Check the diagonals again to make sure the corners are square. An "L" shaped plan, for example, can be divided into rectangles, treating each separately or as an extension of one or more sides.

Height of Foundation Walls

It is common practice to establish the depth of the excavation, and consequently the height of the foundation, on ungraded or graded sites, by using the highest elevation of the excavation's perimeter as the control point (fig. 24.2).This method will insure good drainage if sufficient foundation height is allowed for the sloping of the final grade (fig. 3). Foundation walls at least 7 feet 4 inches high are desirable for full basements, but 8-foot walls are commonly used.

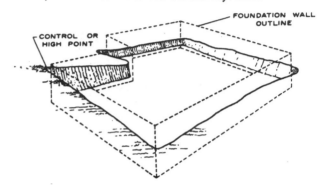

Figure 24.2.—**Establishing depth of excavation.**

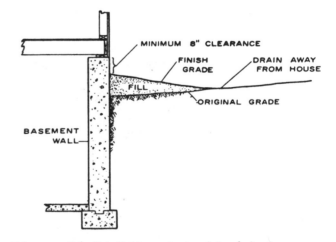

Figure 24.3.—**Finish grade sloped for drainage.**

Foundation walls should be extended above the finished grade around the outside of the house so that the wood finish and framing members will be adequately protected from soil moisture and be well above the grass line. Thus, in termite-infested areas, there will be an opportunity to observe any termite tubes between the soil and the wood and take protective measures before damage develops. Enough height should be provided in crawl spaces to permit periodic inspection for termites and for installation of soil covers to minimize the effects of ground moisture on framing members.

The top of the foundation wall should usually be at least 8 inches above the finish grade at the wall line. The finish grade at the building line might be 4 to 12 inches or more above the original ground level. In lots sloping upward from front to rear (fig. 24. 3), this distance may amount to 12 inches or more. In very steeply sloped lots, a retaining wall to the rear of the wall line is often necessary.

For houses having crawl space, the distance between the ground level and underside of the *joist* should be at least 18 inches above the highest point within the area enclosed by the foundation wall. Where the interior ground level is excavated or otherwise below the outside finish grade, adequate precautionary measures should be made to assure positive drainage at all times.

Excavation

Excavation for basements may be accomplished with one of several types of earth-removing equipment. Top soil is often stockpiled by bulldozer or front-end loader for future use. Excavation of the basement area may be done with a front-end loader, power shovel, or similar equipment.

Power trenchers are often used in excavating for the walls of houses built on a slab or with a crawl space, if soil is stable enough to prevent caving. This eliminates the need for forming below grade when footings are not required.

Excavation is preferably carried only to the top of the footings or the bottom of the basement floor, because some soil becomes soft upon exposure to air or water. Thus it is advisable *not* to make the final excavation for footings until nearly time to pour the concrete unless formboards are to be used.

Excavation must be wide enough to provide space to work when constructing and waterproofing the wall and laying drain tile, if it is necessary in poor drainage areas (fig.24.4).The steepness of the back slope of the excavation is determined by the subsoil encountered. With clay or other stable soil, the back slope can be nearly vertical. When sand is encountered, an inclined slope is required to prevent caving.

Some contractors, in excavating for basements, only roughstake the perimeter of the building for the removal of the earth. When the proper floor elevation has been reached, the footing layout is made and the earth removed. After the concrete is poured and set, the building wall outline is then established on the footings and marked for the formwork or concrete block wall.

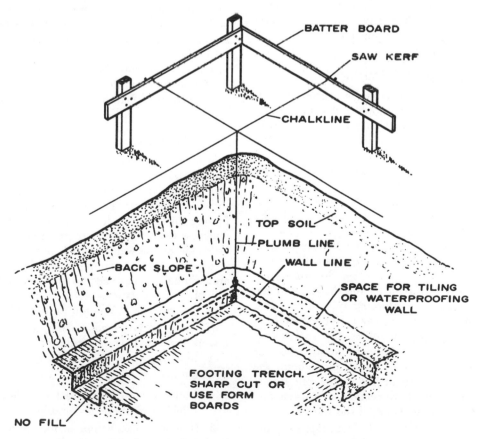

Figure 24.4.—**Establishing corners for excavation and footings.**

CHAPTER 25

CONCRETE AND MASONRY

Concrete and *masonry* units such as concrete block serve various purposes in most house designs, including concrete-slab and crawl-space houses which have poured concrete or concrete block foundation walls of some type. However, developments in treated wood foundation systems will permit all-weather construction and provide reliable foundations for crawl-space houses.

A great amount of concrete is supplied by ready-mix plants, even in rural areas. Concrete in this form is normally ordered by the number of bags per cubic yard, in addition to aggregate size and water-content requirements. Five-bag mix is considered minimum for most work, and where high strength or reinforcing is used, six-bag mix is commonly specified.

The size of gravel or crushed rock which can be obtained varies in different locations and it may be necessary to change the cement ratio normally recommended. Generally speaking, when gravel size is smaller than the normal $1\frac{1}{2}$- to $\frac{1}{4}$-inch size, it is good practice to use a higher cement ratio. When gravel size is a maximum of 1 inch, add one-quarter sack of cement to the 5-bag mix; when gravel size is a maximum of $\frac{3}{4}$-inch, add one-half bag; and for $\frac{3}{8}$-inch size add one bag.

Mixing and Pouring

Proportions of fine and coarse aggregate, amount of cement, and water content should follow the recommendations of the American Concrete Institute. Mixing plants are normally governed by these quantities. It is common practice to limit the amount of water to not more than $7\frac{1}{2}$ gallons for each sack of cement, including that contained in the sand and gravel. Tables of quantities for field mixing on small jobs are available. For example, one combination utilizing a 1-inch maximum size of coarse aggregate uses: 5.8 sacks of cement per cubic yard, 5 gallons of water per sack of cement, and a cement to fine aggregate to coarse aggregate ratio of 1 to $2\frac{1}{2}$ to $3\frac{1}{2}$. Size of coarse aggregate is usually governed by the thickness of the wall and the spacing of reinforcing rods, when used. The use of 2-inch coarse aggregate, for example, is not recommended for slabs or other thin sections.

Concrete should be poured continuously wherever possible and kept practically level throughout the area being poured. All vertical joints should be keyed. *Rod* or *vibrate* the concrete to remove air pockets and force the concrete into all parts of the forms.

In hot weather, protect concrete from rapid drying. It should be kept moist for several days after pouring.

Rapid drying lowers its strength and may injure the exposed surfaces of sidewalks and drives.

In very cold weather, keep the temperature of the concrete above freezing until it has set. The rate at which concrete sets is affected by temperature, being much slower at 40° F. and below than at higher temperatures. In cold weather, the use of heated water and aggregate during mixing is good practice. In severe weather, insulation or heat is used until the concrete has set.

Footings

The *footings* act as the base of the foundation and transmit the superimposed load to the soil. The type and size of footings should be suitable for the soil condition, and in cold climates the footings should be far enough below ground level to be protected from frost action. Local codes usually establish this depth, which is often 4 feet or more in northern sections of the United States.

Poured concrete footings are more dependable than those of other materials and are recommended for use in house foundations. Where fill has been used, the foundations should extend below the fill to undisturbed earth. In areas having adobe soil or where soil moisture may cause soil shrinkage, irregular settlement of the foundation and the building it supports may occur. Local practices that have been successful should be followed in such cases.

Wall Footings

Well-designed wall footings are important in preventing settling or cracks in the wall. One method of determining the size, often used with most normal soils, is based on the proposed wall thickness. The footing thickness or depth should be equal to the wall thickness (fig. 25. 1a) Footings should project beyond each side of the wall one-half the wall thickness. This is a general rule, of course, as the footing bearing area should be designed to the load capacity of the soil. Local regulations often relate to these needs. This also applies to column and fireplace footings.

If soil is of low load-bearing capacity, wider reinforced footings may be required.

A few rules that apply to footing design and construction are:

1. Footings must be at least 6 inches thick, with 8 inches or more preferable.
2. If footing excavation is too deep, fill with concrete—never replace dirt.

3. Use formboards for footings where soil conditions prevent sharply cut trenches.
4. Place footings *below* the frostline.
5. Reinforce footings with steel rods where they cross pipe trenches.
6. Use key slot for better resistance to water entry at wall location.
7. In freezing weather, cover with straw or supply heat.

Pier, Post, and Column Footings

Footings for *piers, posts,* or *columns* (fig. 25.1.b) should be square and include a *pedestal* on which the member will bear. A protruding steel pin is ordinarily set in the pedestal to anchor a wood post. Bolts for the bottom plate of steel posts are usually set when the pedestal is poured. At other times, steel posts are set directly on the footing and the concrete floor poured around them.

Footings vary in size depending on the allowable soil pressure and the spacing of the piers, posts, or columns. Common sizes are 24 by 24 by 12 inches and 30 by 30 by 12 inches. The pedestal is sometimes poured after the footing. The minimum height should be about 3 inches above the finish basement floor and 12 inches above finish grade in crawl-space areas.

Footings for fireplaces, furnaces, and chimneys should ordinarily be poured at the same time as other footings.

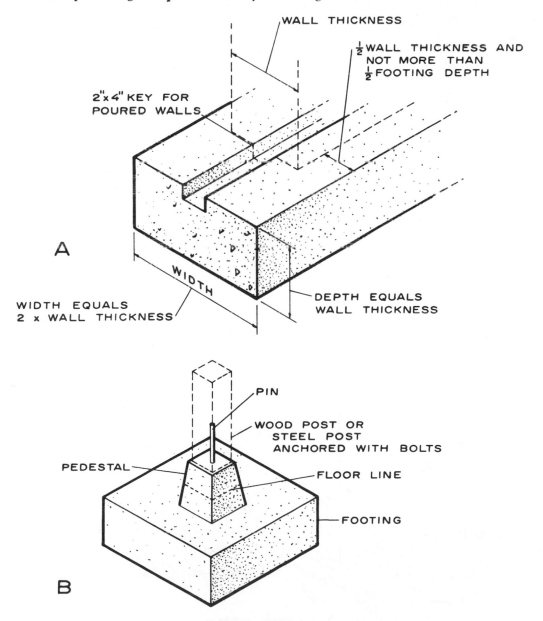

Figure 25.1.—Concrete footing: A, Wall footing; B, post footing.

Stepped Footings

Stepped footings are often used where the lot slopes to the front or rear and the garage or living areas are at basement level. The vertical part of the step should be poured at the same time as the footing. The bottom of the footing is always placed on undisturbed soil and located below the frostline. Each run of the footing should be level.

The vertical step between footings should be at least 6 inches thick and the same width as the footings (fig. 25.2) The height of the step should not be more than three-fourths of the adjacent horizontal footing. On steep slopes, more than one step may be required. It is good practices, when possible, to limit the vertical step to 2 feet. In very steep slopes, special footings may be required.

Draintile

Foundation or footing drains must often be used around foundations enclosing basements, or habitable spaces below the outside finish grade (fig. 25.3) This may be in sloping or low areas or any location where it is

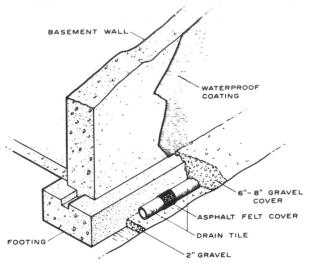

Figure 25.3.

Draintile for soil drainage at outer wall.

necessary to drain away subsurface water. This precaution will prevent damp basements and wet floors. Draintile is often necessary where habitable rooms are developed in the basement or where houses are located

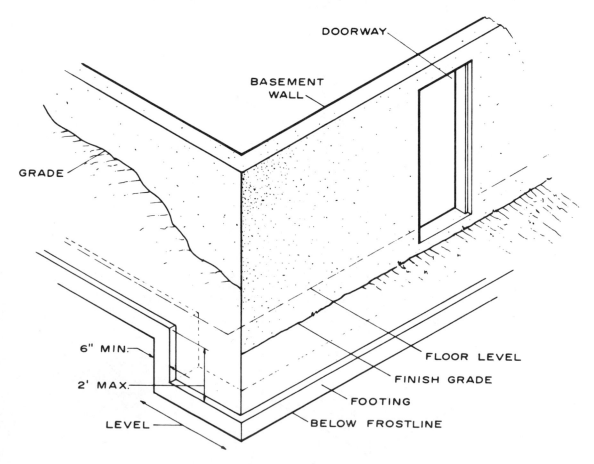

Figure 25.2.—Stepped footings.

near the bottom of a long slope subjected to heavy runoff.

Drains are installed at or below the area to be protected, and drain toward a ditch or into a sump where the water can be pumped to a storm sewer. Clay or concrete draintile, 4 inches in diameter and 12 inches long, is ordinarily placed at the bottom of the footing level on top of a 2-inch gravel bed (fig. 25.3). Tile are placed end to end and spaced about $\frac{1}{8}$ inch apart. The top of the joint between the tile is covered with a strip of asphalt felt or similar paper; 6 to 8 inches of gravel is used over the tile. Drainage is toward the outfall or ditch. Dry wells for drainage water are used only when the soil conditions are favorable for this method of disposal. Local building regulations vary somewhat and should be consulted before construction of drainage system is started.

CHAPTER 26

FOUNDATION WALLS AND PIERS

Foundation walls form an enclosure for basements or crawl spaces and carry wall, floor, roof, and other building loads. The two types of walls most commonly used are *poured concrete* and *concrete block*. Treated wood foundations might also be used when accepted by local codes.

Preservative-treated posts and poles offer many possibilities for low-cost foundation systems and can also serve as a structural framework for the walls and roof (6, 7,).[3]

Wall thicknesses and types of construction are ordinarily controlled by local building regulations. Thicknesses of poured concrete basement walls may vary from 8 to 10 inches and concrete block walls from 8 to 12 inches, depending on story heights and length of unsupported walls.

Clear wall height should be no less than 7 feet from the top of the finish basement floor to the bottom of the joists; greater clearance is usually desirable to provide adequate headroom under girders, pipes, and ducts. Many contractors pour 8-foot-high walls above the footings, which provide a clearance of 7 feet 8 inches from the top of the finish concrete floor to the bottom of the joists. Concrete block walls, 11 courses above the footings with 4-inch solid cap-block, will produce about a 7-foot 4-inch height to the joists from the basement floor.

Poured Concrete Walls

Poured concrete walls (fig. 26.1) require forming that must be tight and also braced and tied to withstand the forces of the pouring operation and the fluid concrete.

Poured concrete walls should be double-formed (formwork constructed for each wall face). Reusable forms are used in the majority of poured walls. Panels may consist of wood framing with plywood facings and are fastened together with clips or other ties (fig. 26.1) Wood sheathing boards and studs with horizontal members and braces are sometimes used in the construction of forms in small communities. As in reusable forms, formwork should be plumb, straight, and braced sufficiently to withstand the pouring operations.

Frames for cellar windows, doors, and other openings are set in place as the forming is erected, along with forms for the beam pockets which are located to support the ends of the floor beam.

Reusable forms usually require little bracing other than horizontal members and sufficient blocking and bracing to keep them in place during pouring operations. Forms constructed with vertical studs and waterproof plywood or lumber sheathing require horizontal whalers and bracing.

Level marks of some type, such as nails along the form, should be used to assure a level foundation top. This will provide a good level sill plate and floor framing.

Concrete should be poured continuously without interruption and constantly puddled to remove air pockets and work the material under window frames and other blocking. If wood spacer blocks are used, they should be removed and not permitted to become buried in the concrete. Anchor bolts for the sill plate should be placed while the concrete is still plastic. Concrete should always be protected when temperatures are below freezing.

Forms should not be removed until the concrete has hardened and acquired sufficient strength to support loads imposed during early construction. At least 2 days (and preferably longer) are required when temperatures are well above freezing, and perhaps a week when outside temperatures are below freezing.

Poured concrete walls can be dampproofed with one heavy cold or hot coat of tar or asphalt. It should

---

[3] Numbers in parentheses refer to Literature Cited at the end of the Handbook.

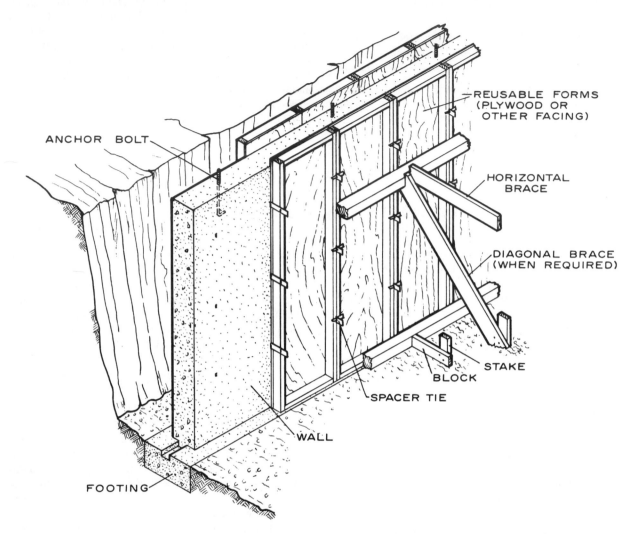

ANCHOR BOLT

REUSABLE FORMS
(PLYWOOD OR
OTHER FACING)

HORIZONTAL
BRACE

DIAGONAL BRACE
(WHEN REQUIRED)

STAKE

BLOCK

SPACER TIE

WALL

FOOTING

Figure 26.1.—Forming for poured concrete walls.

be applied to the outside from the footings to the finish gradeline. Such coatings are usually sufficient to make a wall watertight against ordinary seepage (such as may occur after a rainstorm), but should not be applied until the surface of the concrete has dried enough to assure good adhesion. In poorly drained soils, a membrane (such as described for concrete block walls) may be necessary.

Concrete Block Walls

Concrete blocks are available in various sizes and forms, but those generally used are 8, 10, and 12 inches wide. Modular blocks allow for the thickness and width of the mortar joint so are usually about $7\frac{5}{8}$ inches high by $15\frac{5}{8}$ inches long. This results in blocks which measure 8 inches high and 16 inches long from centerline to centerline of the mortar joints.

Concrete block walls require *no* formwork. Block courses start at the footing and are laid up with about $\frac{3}{8}$-inch mortar joints, usually in a common bond (fig. 26.2) Joints should be tooled smooth to resist water seepage. Full bedding of mortar should be used on all contact surfaces of the block. When *pilasters* (column-like projections) are required by building codes or to strengthen a wall, they are placed on the interior side of the wall and terminated at the bottom of the beam or girder supported.

Basement door and window frames should be set with keys for rigidity and to prevent air leakage (fig. 26. 2).

Block walls should be capped with 4 inches of solid masonry or concrete reinforced with wire mesh. *Anchor bolts* for sills are usually placed through the top two rows of blocks and the top cap. They should be anchored with a large plate washer at the bottom and the block openings filled solidly with mortar or concrete. (fig. 26. 2).

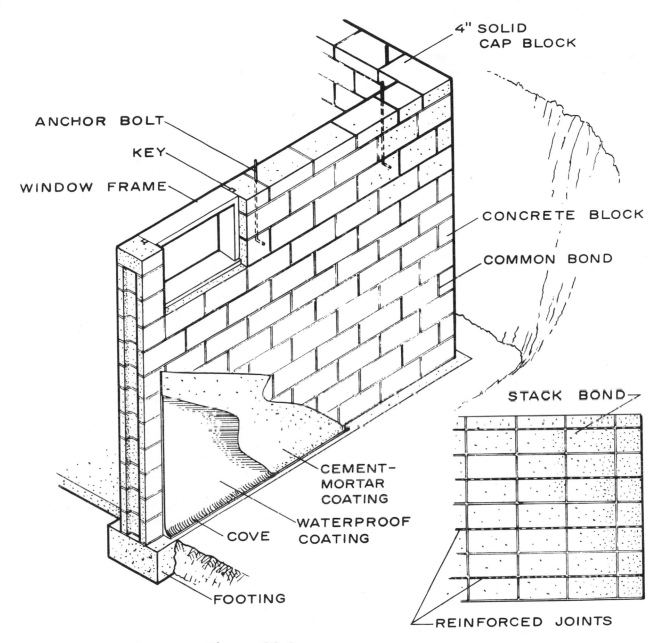

ANCHOR BOLT

KEY

WINDOW FRAME

4" SOLID CAP BLOCK

CONCRETE BLOCK

COMMON BOND

STACK BOND

CEMENT-MORTAR COATING

WATERPROOF COATING

COVE

FOOTING

REINFORCED JOINTS

Figure 26.2.—Concrete block walls.

When an exposed block foundation is used as a finished wall for basement rooms, the *stack bond pattern* may be employed for a pleasing effect. This consists of placing blocks one above the other, resulting in continuous vertical mortar joints. However, when this system is used, it is necessary to incorporate some type of joint reinforcing every second course. This usually consists of small diameter steel longitudinal and cross rods arranged in a grid pattern. The common bond does not normally require this reinforcing, but when additional strength is desired, it is good practice to incorporate this bonding system into the wall.

Freshly laid block walls should be protected in temperatures below freezing. Freezing of the mortar before it has set will often result in low adhesion, low strength, and joint failure.

To provide a tight, waterproof joint between the footing and wall, an elastic calking compound is often used. The wall is waterproofed by applying a coating of cement-mortar over the block with a cove formed at the juncture with the footing (fig. 26.2). When the mortar is dry, a coating of asphalt or other waterproofing will normally assure a dry basement.

For added protection when wet soil conditions may be encountered, a waterproof membrane of roofing

455

felt or other material can be mopped on, with shingle-style laps of 4 to 6 inches, over the cement-mortar coating. Hot tar or hot asphalt is commonly used over the membrane. This covering will prevent leaks if minor cracks develop in the blocks or joints between the blocks.

Masonry Construction for Crawl Spaces

In some areas of the country, the crawl-space house is often used in preference to those constructed over a basement or on a concrete slab. It is possible to construct a satisfactory house of this type by using (a) a good soil cover, (b) a small amount of ventilation, and (c) sufficient insulation to reduce heat loss. These details will be covered in later chapters.

One of the primary advantages of the crawl-space house over the full basement house is, of course, the reduced cost. Little or no excavation or grading is required except for the footings and walls. In mild climates, the footings are located only slightly below the finish grade. However, in the northern States where frost penerates deeply, the footing is often located 4 or more feet below the finish grade. This, of course, requires more masonry work and increases the cost. The footings should always be poured over undisturbed soil and never over fill unless special piers and grade beams are used.

The construction of a masonry wall for a crawl space is much the same as those required for a full basement (figs. 26.1 and.2) except that no excavation is required within the walls. Waterproofing and draintile are normally *not* required for this type of construction. The masonry pier replaces the wood or steel posts of the basement house used to support the center beam. Footing size and wall thicknesses vary somewhat by location and soil conditions. A common minimum thickness for walls in single-story frame houses is 8 inches for hollow concrete block and 6 inches for poured concrete. The minimum footing thickness is 6 inches and the width is 12 inches for concrete block and 10 inches for the poured foundation wall for crawl-space houses. However, in well constructed houses, it is common practice to use 8-inch walls and 16- by 8-inch footings.

Poured concrete or concrete block piers are often used to support floor beams in crawl-space houses. They should extend at least 12 inches above the groundline. The minimum size for a concrete block pier should be 8 by 16 inches with a 16- by 24- by 8-inch footing. A solid cap block is used as a top course. Poured concrete piers should be at least 10 by 10 inches in size with a 20- by 20- by 8-inch footing. Unreinforced concrete piers should be no greater in height than 10 times their least dimension. Concrete block piers should be no higher than four times the least dimension. The spacing of piers should not exceed 8 feet on center under exterior wall beams and interior girders set at right angles to the floor joists, and 12 feet on center under exterior wall beams set parallel to the floor joists. Exterior wall piers should not extend above grade more than four times their least dimension unless supported laterally by masonry or concrete walls. As for wall footing sizes, the size of the pier footings should be based on the load and the capacity of the soil.

Sill Plate Anchors

In wood-frame construction, the *sill plate* should be anchored to the foundation wall with ½-inch bolts hooked and spaced about 8 feet apart (fig. 26.3a). In some areas, sill plates are fastened with masonry nails, but such nails do not have the uplift resistance of bolts. In high-wind and storm areas, well-anchored plates are very important. A *sill sealer* is often used under the sill plate on poured walls to take care of any irregularities which might have occured during curing of the concrete. Anchor bolts should be embedded 8 inches or more in poured concrete walls and 16 inches or more in block walls with the core filled with concrete. A large plate washer should be used at the head end of the bolt for the block wall. If termite shields are used, they should be installed under the plate and sill sealer.

Although not the best practice, some contractors construct wood-frame houses without the use of a sill plate. Anchorage of the floor system must then be provided by the use of steel strapping, which is placed during the pour or between the block joints. Strap is bent over the joist or the header joist and fastened by nailing (fig. 26.3b). The use of a concrete or mortar beam fill provides resistance to air and insect entry.

Reinforcing in Poured Walls

Poured concrete walls normally do not require steel *reinforcing* except over window or door openings located below the top of the wall. This type of construction requires that a properly designed steel or reinforced-concrete *lintel* be built over the frame (fig. 26.4a). In poured walls, the rods are laid in place while the concrete is being poured so that they are about 1½ inches above the opening. Frames should be prime painted or treated before installation. For concrete block walls, a similar reinforced poured concrete or a precast lintel is commonly used.

Where concrete work includes a connecting porch or garage wall not poured with the main basement wall, it is necessary to provide reinforcing-rod ties (fig. 26.4b). These rods are placed during pouring of the main wall. Depending on the size and depth, at least three ½-inch deformed rods should be used at the intersection of each wall. Keyways may be used in addition to resist lateral movement. Such connecting walls should extend below normal frostline and be

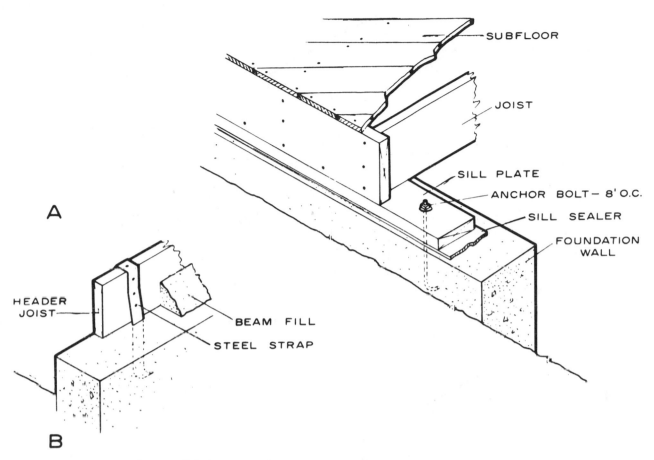

Figure 26.3. Anchoring floor system to concrete or masonry walls: A, With sill plate; B, without sill plate.

supported by undisturbed ground. Wall extensions in concrete block walls are also of block and are constructed at the same time as the main walls over a footing placed below frostline.

Masonry Veneer Over Frame Walls

If *masonry veneer* is used for the outside finish over wood-frame walls, the foundation must include a supporting ledge or offset about 5 inches wide (fig. 26. 5). This results in a space of about 1 inch between the masonry and the sheathing for ease in laying the brick. A base flashing is used at the brick course below the bottom of the sheathing and framing, and should be lapped with sheathing paper. Weep holes, to provide drainage, are also located at this course and are formed by eliminating the mortar in a vertical joint. Corrosion-resistant metal ties—spaced about 32 inches apart horizontally and 16 inches vertically—should be used to bond the brick veneer to the framework. Where other than wood sheathing is used, secure the ties to the studs.

Brick and stone should be laid in a full bed of mortar; avoid dropping mortar into the space between the veneer and sheathing. Outside joints should be tooled to a smooth finish to get the maximum resistance to water penetration.

Masonry laid during the cold weather should be protected from freezing until after the mortar has set.

Notch for Wood Beams

When basement beams or girders are wood, the wall notch or pocket for such members should be large enough to allow at least ½ inch of clearance at sides and ends of the beam for ventilation (fig. 26.6). Unless the wood is treated there is a decay hazard where beams and girders are so tightly set in wall notches that moisture cannot readily escape. A waterproof membrane, such as roll roofing, is commonly used under the end of the beam to minimize moisture absorption (fig. 26. 6).

Protection Against Termites

Certain areas of the country, particularly the Atlantic Coast, Gulf States, Mississippi and Ohio Valleys, and southern California, are infested with wood-destroying termites. In such areas, wood construction

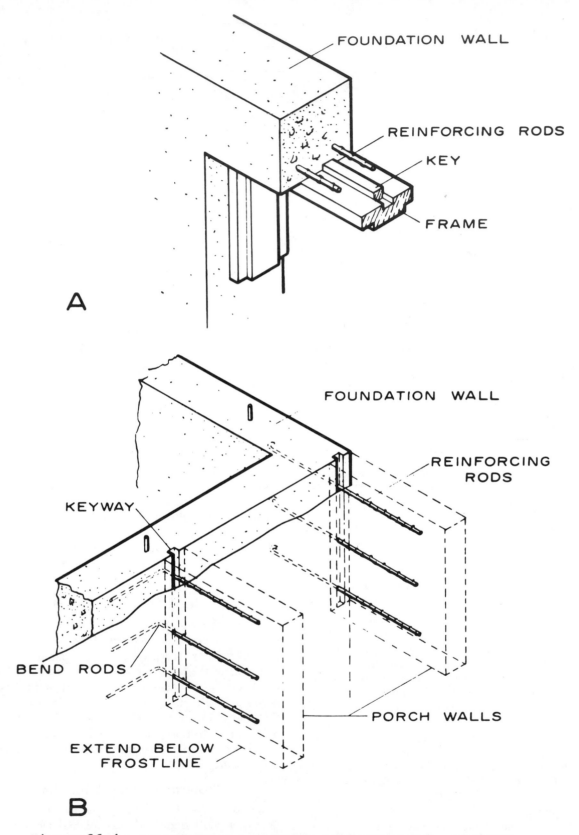

A

B

Figure 26.4.—Steel reinforcing rods in concrete walls: *A*, Rods used over window or doorframes; *B*, rod ties used for porch or garage walls.

458

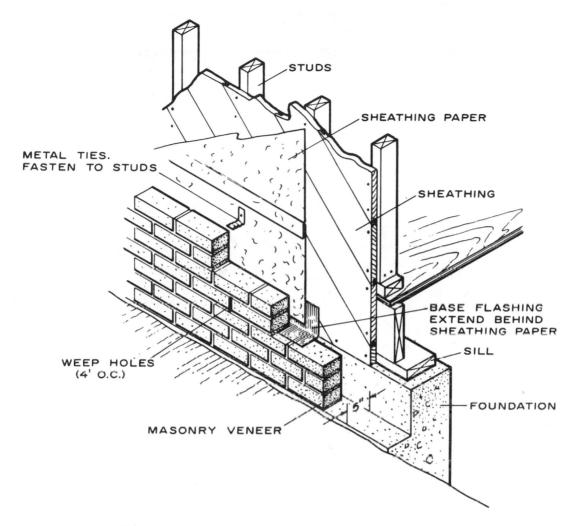

Figure 26.5.—Wood-frame wall with masonry veneer.

over a masonry foundation should be protected by one or more of the following methods:

1. Poured concrete foundation walls.
2. Masonry unit foundation walls capped with reinforced concrete.
3. Metal shields made of rust-resistant material. (Metal shields are effective only if they extend beyond the masonry walls and are continuous, with no gaps or loose joints. This shield is of primary importance under most conditions.)
4. Wood-preservative treatment. (This method protects only the members treated.)
5. Treatment of soil with soil poison. (This is perhaps one of the most common and effective means used presently.)

See Chapter 55 for further details on protection against termites.

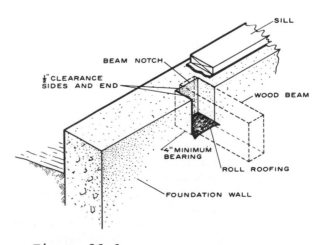

Figure 26.6.—Notch for wood beam.

CHAPTER 27

CONCRETE FLOOR SLABS ON GROUND

The number of new one-story houses with full basements has declined in recent years, particularly in the warmer areas of the United States. This is due in part to lower construction costs of houses without basements and an apparent decrease in need for the basement space.

The primary function of a basement in the past has been to provide space for a central heating plant and for the storage and handling of bulk fuel and ashes. It also houses laundry and utilities. With the wide use of liquid and gas fuels, however, the need for fuel and ash storage space has greatly diminished. Because space can be compactly provided on the ground-floor level for the heating plant, laundry, and utilities, the need for a basement often disappears.

Types of Floor Construction

One common type of floor construction for basementless houses is a concrete slab over a suitable foundation. Sloping ground or low areas are usually not ideal for slab-on-ground construction because structural and drainage problems would add to costs. Split-level houses often have a portion of the foundation designed for a grade slab. In such use, the slope of the lot is taken into account and the objectionable features of a sloping ground become an advantage.

The finish flooring for concrete floor slabs on the ground was initially asphalt tile laid in *mastic* directly on the slab. These concrete floors did not prove satisfactory in a number of instances, and considerable prejudice has been built up against this method of construction. The common complaints have been that the floors are cold and uncomfortable and that condensation sometimes collects on the floor, near the walls in cold weather, and elsewhere during warm, humid weather. Some of these undesirable features of concrete floors on the ground apply to both warm and cold climates, and others only to cold climates.

Improvements in methods of construction based on past experience and research have materially reduced the common faults of the slab floor but consequently increased their cost.

Floors are cold principally because of loss of heat through the floor and the foundation walls, with most loss occurring around the exterior walls. Suitable insulation around the perimeter of the house will help to reduce the heat loss. *Radiant floor heating* systems are effective in preventing cold floors and floor condensation problems. Peripheral warm-air heating ducts are also effective in this respect. Vapor barriers over a gravel fill under the floor slab prevent soil moisture from rising through the slab.

Basic Requirements

Certain basic requirements should be met in the construction of concrete floor slabs to provide a satisfactory floor. They are:

1. Establish finish floor level high enough above the natural ground level so that finish grade around the house can be sloped away for good drainage. Top of slab should be no less than 8 inches above the ground and the siding no less than 6 inches.

2. Top soil should be removed and sewer and water lines installed, then covered with 4 to 6 inches of gravel or crushed rock well-tamped in place.

3. A vapor barrier consisting of a heavy plastic film, such as 6-mil polyethylene, asphalt laminated duplex sheet, or 45-pound or heavier roofing, with minimum of ½-perm rating should be used under the concrete slab. Joints should be lapped at least 4 inches and sealed. The barrier should be strong enough to resist puncturing during placing of the concrete.

4. A permanent, waterproof, nonabsorptive type of rigid insulation should be installed around the perimeter of the wall. Insulation may extend down on the inside of the wall vertically or under the slab edge horizontally.

5. The slab should be reinforced with 6- by 6-inch No. 10 wire mesh or other effective reinforcing. The concrete slab should be at least 4 inches thick and should conform to information in Chapter 25, "Concrete and Masonry." A monolithic slab (fig. 27. 1) is preferred in termite areas.

6. After leveling and screeding, the surface should be floated with wood or metal floats while concrete is still plastic. If a smooth dense surface is needed for the installation of wood or resilient tile with adhesives, the surface should be steel troweled.

Combined Slab and Foundation

The combined slab and foundation, sometimes referred to as the *thickened-edge slab*, is useful in warm climates where frost penetration is not a problem and where soil conditions are especially favorable. It consists of a shallow perimeter reinforced footing poured integrally with the slab over a vapor barrier (fig. 27. 1) The bottom of the footing should be at least 1 foot below the natural gradeline and supported on solid, unfilled, and well-drained ground.

Independent Concrete Slab and Foundation Walls

When ground freezes to any appreciable depth during winter, the walls of the house must be supported

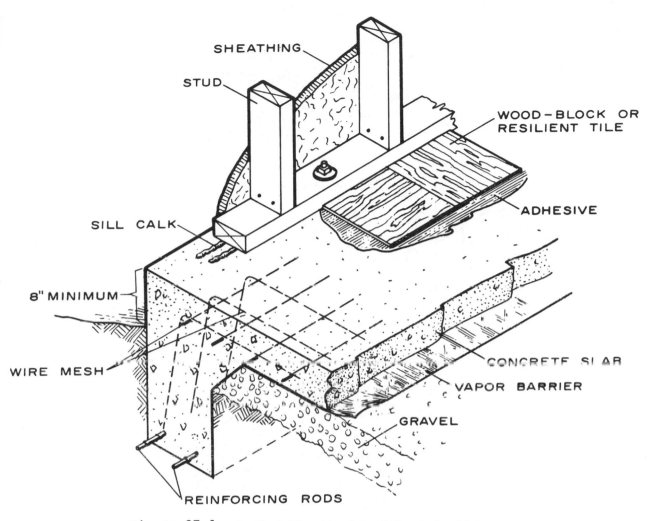

SHEATHING

STUD

WOOD-BLOCK OR
RESILIENT TILE

ADHESIVE

SILL CALK

8" MINIMUM

WIRE MESH

CONCRETE SLAB

VAPOR BARRIER

GRAVEL

REINFORCING RODS

*Figure 27.1.—*Combined slab and foundation (thickened edge slab.)

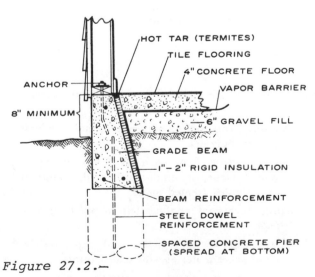

HOT TAR (TERMITES)

TILE FLOORING

4" CONCRETE FLOOR

VAPOR BARRIER

ANCHOR

8" MINIMUM

6" GRAVEL FILL

GRADE BEAM

1"-2" RIGID INSULATION

BEAM REINFORCEMENT

STEEL DOWEL
REINFORCEMENT

SPACED CONCRETE PIER
(SPREAD AT BOTTOM)

Figure 27.2.—

Reinforced grade beam for concrete slab. Beam spans
between concrete piers located below frostline.

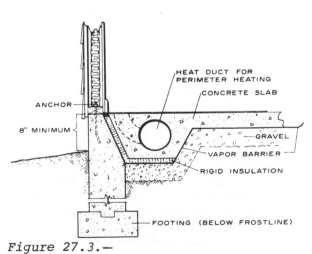

HEAT DUCT FOR
PERIMETER HEATING

CONCRETE SLAB

ANCHOR

8" MINIMUM

GRAVEL

VAPOR BARRIER

RIGID INSULATION

FOOTING (BELOW FROSTLINE)

Figure 27.3.—

Full foundation wall for cold climates. Perimeter heat
duct insulated to reduce heat loss.

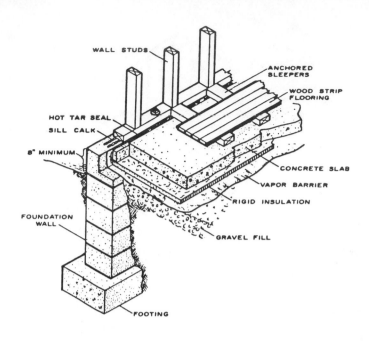

WALL STUDS
ANCHORED SLEEPERS
WOOD STRIP FLOORING
HOT TAR SEAL
SILL CALK
8" MINIMUM
CONCRETE SLAB
VAPOR BARRIER
RIGID INSULATION
FOUNDATION WALL
GRAVEL FILL
FOOTING

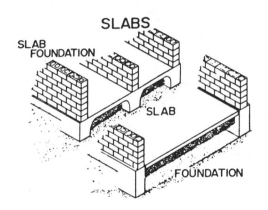

SLABS
SLAB FOUNDATION
SLAB
FOUNDATION

Figure 27.4.—

Independent concrete floor slab and wall. Concrete block is used over poured footing which is below frostline. Rigid insulation may also be located along the inside of the block wall.

by foundations or piers which extend below the frost-line to solid bearing on unfilled soil. In such construction, the concrete slab and foundation wall are usually separate. Three typical systems are suitable for such conditions (figs. 27.2, .3, and .4).

Vapor Barrier Under Concrete Slab

The most desirable properties in a vapor barrier to be used under a concrete slab are: (a) Good vapor-transmission rating (less than 0.5 perm); (b) resistance to damage by moisture and rot; and (c) ability to withstand normal usage during pouring operations.

Such properties are included in the following types of materials:

1. 55-pound roll roofing or heavy asphalt laminated duplex barriers.

2. Heavy plastic film, such as 6-mil or heavier polyethylene, or similar plastic film laminated to a duplex treated paper.

3. Three layers of roofing felt mopped with hot asphalt.

4. Heavy asphalt impregnated and vapor-resistant rigid sheet material with sealed joints.

Insulation Requirements for Concrete Floor Slabs on Ground

The use of perimeter insulation for slabs is necessary to prevent heat loss and cold floors during the heating season, except in warm climates. The proper locations for this insulation under several conditions are shown in figures 27. 2 to 27. 4 .

The thickness of the insulation will depend upon requirements of the climate and upon the materials used. Some insulations have more than twice the insulating value of others (see Chapter 38). The resistance (R) per inch of thickness, as well as the heating design temperature, should govern the amount required. Perhaps two good general rules to follow are:

1. For average winter low temperatures of 0° F. and higher (moderate climates), the total R should be about 2.0 and the depth of the insulation or the width under the slab not less than 1 foot.

2. For average winter low temperatures of −20° F. and lower (cold climates), the total R should be about 3.0 without floor heating and the depth or width of insulation not less than 2 feet.

Table 1 shows these factors in more detail. The values shown are minimum and any increase in insulation will result in lower heat losses.

TABLE 1.—*Resistance values used in determining minimum amount of edge insulation for concrete floors slabs on ground for various design temperatures.*

Low temperatures	Depth insulation extends below grade	Resistance (R) factor	
		No floor heating	Floor heating
°F.	Ft.		
−20	2	3.0	4.0
−10	1½	2.5	3.5
0	1	2.0	3.0
+10	1	2.0	3.0
+20	1	2.0	3.0

Insulation Types

The properties desired in insulation for floor slabs are: 1) High resistance to heat transmission, 2) permanent durability when exposed to dampness and frost, and 3) high resistance to crushing due to floor loads, weight of slab, or expansion forces. The slab should also be immune to fungus and insect attack, and should not absorb or retain moisture. Examples of materials considered to have these properties are:

1. *Cellular-glass insulation board,* available in slabs 2, 3, 4, and 5 inches thick. R factor, or resistivity, 1.8 to 2.2 per inch of thickness. Crushing strength, approximately 150 pounds per square inch. Easily cut and worked. The surface may spall (chip or crumble) away if subjected to moisture and freezing. It should be dipped in roofing pitch or asphalt for protection. Insulation should be located above or inside the vapor barrier for protection from moisture (figs.27.2 to27. 4) This type of insulation has been replaced to a large extent by the newer foamed plastics such as polystyrene and polyurethane.

2. *Glass fibers with plastic binder,* coated or uncoated, available in thicknesses of ¾, 1, 1½, and 2 inches. R factor, 3.3 to 3.9 per inch of thickness. Crushing strength, about 12 pounds per square inch. Water penetration into coated board is slow and inconsequential unless the board is exposed to a constant head of water, in which case this water may disintegrate the binder. Use a coated board or apply coal-tar pitch or asphalt to uncoated board. Coat all edges. Follow manufacturer's instructions for cutting. Placement of the insulation inside the vapor barrier will afford some protection.

3. *Foamed plastic* (polystyrene, polyurethane, and others) insulation in sheet form, usually available in thicknesses of ½, 1, 1½, and 2 inches. At normal temperatures the R factor varies from 3.7 for polystyrenes to over 6.0 for polyurethane for a 1-inch thickness. These materials generally have low water-vapor transmission rates. Some are low in crushing strength and perhaps are best used in a vertical position (fig. 15) and not under the slab where crushing could occur.

4. *Insulating concrete.* Expanded mica aggregate, 1 part cement to 6 parts aggregate, thickness used as required. R factor, about 1.1 per inch of thickness. Crushing strength, adequate. It may take up moisture when subject to dampness, and consequently its use should be limited to locations where there will be no contact with moisture from any source.

5. *Concrete made with lightweight aggregate,* such as expanded slag, burned clay, or pumice, using 1 part cement to 4 parts aggregate; thickness used as required. R factor, about 0.40 per inch of thickness. Crushing strength, high. This lightweight aggregate may also be used for foundation walls in place of stone or gravel aggregate.

Under service conditions there are two sources of moisture that might affect insulating materials: (1) Vapor from inside the house and (2) moisture from soil. Vapor barriers and coatings may retard but not entirely prevent the penetration of moisture into the insulation. Dampness may reduce the crushing strength of insulation, which in turn may permit the edge of the slab to settle. Compression of the insulation, moreover, reduces its efficiency. Insulating materials should perform satisfactorily in any position if they do not change dimensions and if they are kept dry.

Protection Against Termites

In areas where termites are a problem, certain precautions are necessary for concrete slab floors on the ground. Leave a countersink-type opening 1-inch wide and 1-inch deep around plumbing pipes where they pass through the slab, and fill the opening with hot tar when the pipe is in place. Where insulation is used between the slab and the foundation wall, the insulation should be kept 1 inch below the top of the slab and the space should also be filled with hot tar (fig. 27.2) Further discussion of protection against termites, such as soil poisoning, is given in Chapter 55.

Finish Floors Over Concrete Slabs on the Ground

A natural concrete surface is sometimes used for the finish floor, but generally is not considered wholly satisfactory. Special dressings are required to prevent dusting. Moreover, such floors tend to feel cold. Asphalt or vinyl-asbestos tile laid in mastic in accordance with the manufacturer's recommendations is comparatively economical and easy to clean, but it also feels cold. Wood tile in various forms and wood parquet flooring may be used, also laid in mastic (fig. 27.1) in accordance with the manufacturer's recommendations. Tongued-and-grooved wood strip flooring 25/32 inch thick may be used but should be used over pressure-treated wood sleepers anchored to the slab (fig. 27. 4). For existing concrete floors, the use of a vaporproof coating before installation of the treated sleepers is good practice.

CHAPTER 28

FLOOR FRAMING

The *floor framing* in a wood-frame house consists specifically of the posts, beams, sill plates, joists, and subfloor. When these are assembled properly on a foundation, they form a level anchored platform for the rest of the house. The posts and center beams of wood or steel, which support the inside ends of the joists, are sometimes replaced with a woodframe or masonry wall when the basement area is divided into rooms. Wood-frame houses may also be constructed upon a concrete floor slab or over a crawl-space area with floor framing similar to that used for a full basement.

Factors in Design

One of the important factors in the design of a wood floor system is to equalize shrinkage and expansion of the wood framing at the outside walls and at the center beam. This is usually accomplished by using approximately the same total depth of wood at the center beam as the outside framing. Thus, as beams and joists approach moisture equilibrium or the moisture content they reach in service, there are only small differences in the amount of shrinkage. This will minimize plaster cracks and prevent sticking doors and other inconveniences caused by uneven shrinkage. If there is a total of 12 inches of wood at the foundation wall (including joists and sill plate), this should be balanced with about 12 inches of wood at the center beam.

Moisture content of beams and joists used in floor framing should not exceed 19 percent. However, a moisture content of about 15 percent is much more desirable. Dimension material can be obtained at these moisture contents when so specified. When moisture contents are in the higher ranges, it is good practice to allow joists and beams to approach their moisture equilibrium before applying inside finish and trim, such as baseboard, base shoe, door jambs, and casings.

Grades of dimension lumber vary considerably by species. For specific uses in this publication, a sequence of first, second, third, fourth, and sometimes fifth grade material is used. In general, the first grade is for a high or special use, the second for better than average, the third for average, and the fourth and fifth for more economical construction. Joists and girders are usually second grade material of a species, while sills and posts are usually of third or fourth grade. Specific recommendations for each species are available (5).

Recommended Nailing Practices

Of primary consideration in the construction of a house is the method used to fasten the various wood members together. These connections are most commonly made with nails, but on occasions metal straps, lag screws, bolts, and adhesives may be used.

Proper fastening of frame members and covering materials provides the rigidity and strength to resist severe windstorms and other hazards. Good nailing is also important from the standpoint of normal performance of wood parts. For example, proper fastening of intersecting walls usually reduces plaster cracking at the inside corners.

The schedule in table 2 outlines good nailing practices for the framing and sheathing of a well-constructed wood-frame house. Sizes of common wire nails are shown in figure 28. 1.

When houses are located in hurricane areas, they should be provided with supplemental fasteners. Details of these systems are outlined in "Houses Can Resist Hurricanes" (7).

Posts and Girders

Wood or steel posts are generally used in the basement to support wood girders or steel beams. Masonry piers might also be used for this purpose and are commonly employed in crawl-space houses.

The round steel post can be used to support both wood girders and steel beams and is normally supplied with a steel bearing plate at each end. Secure anchoring to the girder or beam is important (fig. 28. 2).

Wood posts should be solid and not less than 6 by 6 inches in size for freestanding use in a basement. When combined with a framed wall, they may be 4 by 6

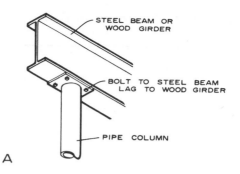

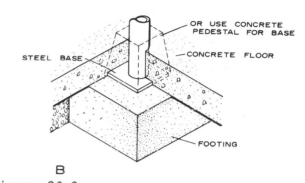

Figure 28.2.
Steel post for wood or steel girder: **A**, Connection to beam; **B**, base plate also may be mounted on and anchored to a concrete pedestal.

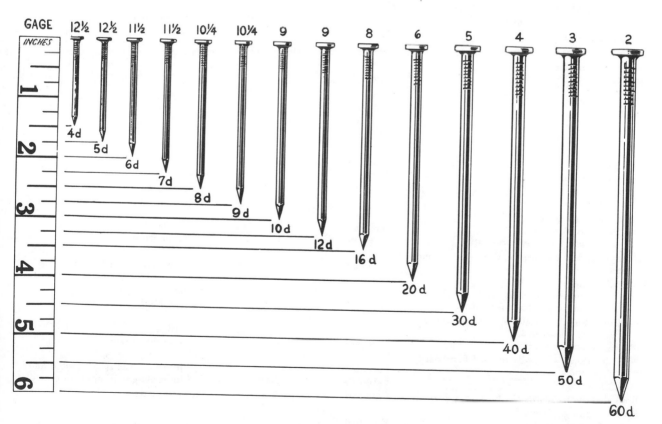

Figure 28.1.—Sizes of common wire nails.

TABLE 2.—*Recommended schedule for nailing the framing and sheathing of a well-constructed wood-frame house*

Joining	Nailing method	Number	Size	Placement
			Nails	
Header to joist	End-nail	3	16d	
Joist to sill or girder	Toenail	2	10d or	
		3	8d	
Header and stringer joist to sill	Toenail		10d	16 in. on center
Bridging to joist	Toenail each end	2	8d	
Ledger strip to beam, 2 in. thick		3	16d	At each joist
Subfloor, boards:				
1 by 6 in. and smaller		2	8d	To each joist
1 by 8 in.		3	8d	To each joist
Subfloor, plywood:				
At edges			8d	6 in. on center
At intermediate joists			8d	8 in. on center
Subfloor (2 by 6 in., T&G) to joist or girder	Blind-nail (casing) and face-nail	2	16d	
Soleplate to stud, horizontal assembly	End-nail	2	16d	At each stud
Top plate to stud	End-nail	2	16d	
Stud to soleplate	Toenail	4	8d	
Soleplate to joist or blocking	Face-nail		16d	16 in. on center
Doubled studs	Face-nail, stagger		10d	16 in. on center
End stud of intersecting wall to exterior wall stud	Face-nail		16d	16 in. on center
Upper top plate to lower top plate	Face-nail		16d	16 in. on center
Upper top plate, laps and intersections	Face-nail	2	16d	
Continuous header, two pieces, each edge			12d	12 in. on center
Ceiling joist to top wall plates	Toenail	3	8d	
Ceiling joist laps at partition	Face-nail	4	16d	
Rafter to top plate	Toenail	2	8d	
Rafter to ceiling joist	Face-nail	5	10d	
Rafter to valley or hip rafter	Toenail	3	10d	
Ridge board to rafter	End-nail	3	10d	
Rafter to rafter through ridge board	Toenail	4	8d	
	Edge-nail	1	10d	
Collar beam to rafter:				
2 in. member	Face-nail	2	12d	
1 in. member	Face-nail	3	8d	
1-in. diagonal let-in brace to each stud and plate (4 nails at top)		2	8d	
Built-up corner studs:				
Studs to blocking	Face-nail	2	10d	Each side
Intersecting stud to corner studs	Face-nail		16d	12 in. on center
Built-up girders and beams, three or more members	Face-nail		20d	32 in. on center, each side
Wall sheathing:				
1 by 8 in. or less, horizontal	Face-nail	2	8d	At each stud
1 by 6 in. or greater, diagonal	Face-nail	3	8d	At each stud
Wall sheathing, vertically applied plywood:				
3/8 in. and less thick	Face-nail		6d	6 in. edge
1/2 in. and over thick	Face-nail		8d	12 in. intermediate
Wall sheathing, vertically applied fiberboard:				
1/2 in. thick	Face-nail			1½ in. roofing nail} 3 in. edge and
25/32 in. thick	Face-nail			1¾ in. roofing nail} 6 in. intermediate
Roof sheathing, boards, 4-, 6-, 8-in. width	Face-nail	2	8d	At each rafter
Roof sheathing, plywood:				
3/8 in. and less thick	Face-nail		6d }	6 in. edge and 12 in. intermediate
1/2 in. and over thick	Face-nail		8d }	

466

inches to conform to the depth of the studs. Wood posts should be squared at both ends and securely fastened to the girder (fig. 28.3) The bottom of the post should rest on and be pinned to a masonry pedestal 2 to 3 inches above the finish floor. In moist or wet conditions it is good practice to treat the bottom end of the post or use a moisture-proof covering over the pedestal.

Both wood girders and steel beams are used in present-day house construction. The standard *I-beam* and wide *flange beam* are the most commonly used steel beam shapes. Wood girders are of two types—*solid* and *built up*. The built-up beam is preferred because it can be made up from drier dimension material and is more stable. Commercially available glue-laminated beams may be desirable where exposed in finished basement rooms.

The built-up girder (fig. 28. 4) is usually made up of two or more pieces of 2 inch dimension lumber spiked together, the ends of the pieces joining over a supporting post. A two-piece girder may be nailed from one side with tenpenny nails, two at the end of each piece and others driven stagger fashion 16 inches apart. A three-piece girder is nailed from each side with twentypenny nails, two near each end of each piece and others driven stagger fashion 32 inches apart.

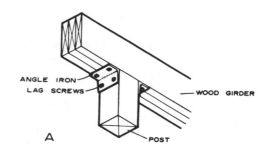

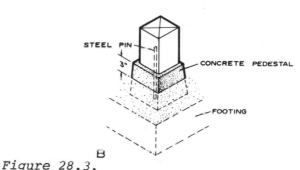

Figure 28.3.

Wood post for wood girder: A, Connection to girder; B, base

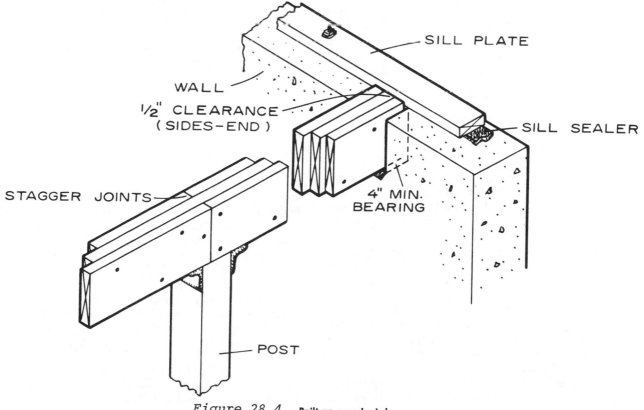

Figure 28.4.—**Built-up wood girder.**

467

Ends of wood girders should bear at least 4 inches on the masonry walls or pilasters. When wood is untreated, a ½-inch air space should be provided at each end and at each side of wood girders framing into masonry (fig. 28.4). In termite-infested areas, these pockets should be lined with metal. The top of the girder should be level with the top of the sill plates on the foundation walls, unless *ledger strips* are used. If steel plates are used under ends of girders, they should be of full bearing size.

Girder-joist Installation

Perhaps the simplest method of floor-joist framing is one where the joists bear directly on the wood girder or steel beam, in which case the top of the beam coincides with the top of the anchored sill (fig. 28.4). This method is used when basement heights provide adequate headroom below the girder. However, when wood girders are used in this manner, the main disadvantage is that shrinkage is usually greater at the girder than at the foundation.

For more uniform shrinkage at the inner beam and the outer wall and to provide greater headroom, joist hangers or a supporting ledger strip are commonly used. Depending on sizes of joists and wood girders, joists may be supported on the ledger strip in several ways (fig. 28.5). Each provides about the same depth of wood subject to shrinkage at the outer wall and at the center wood girder. A continuous horizontal tie between exterior walls is obtained by nailing notched joists together (fig. 28.5a). Joists must always bear on the ledgers. In figure 28.5b, the connecting scab at each pair of joists provides this tie and also a nailing area for the subfloor. A steel strap is used to tie the joists together when the tops of the beam and the joists are level (fig. 28.5c). It is important that a small space be allowed above the beam to provide for shrinkage of the joists.

When a space is required for heat ducts in a partition supported on the girder, a *spaced wood girder* is sometimes necessary (fig. 28.6). Solid blocking is used at intervals between the two members. A single post support for a spaced girder usually requires a bolster, preferably metal, with sufficient span to support the two members.

Joists may be arranged with a steel beam generally the same way as illustrated for a wood beam. Perhaps the most common methods, depending on joist sizes, are:

1. The joists rest directly on the top of the beam.
2. Joists rest on a wood ledger or steel angle iron, which is bolted to the web (fig. 28.7a).
3. Joists bear directly on the flange of the beam (fig. 28.7b).

In the third method, wood blocking is required between the joists near the beam flange to prevent overturning.

Wood Sill Construction

The two general types of wood sill construction used over the foundation wall conform either to platform or balloon framing. The *box sill* is commonly used in platform construction. It consists of a 2-inch or thicker plate anchored to the foundation wall over a sill sealer which provides support and fastening for the joists and header at the ends of the joists (fig. 28.8). Some houses are constructed without benefit of an anchored sill plate although this is not entirely desirable. The floor framing should then be anchored with metal strapping installed during pouring operations (fig. 26.3b).

Balloon-frame construction uses a nominal 2-inch or thicker wood sill upon which the joists rest. The studs also bear on this member and are nailed both to the floor joists and the sill. The subfloor is laid diagonally or at right angles to the joists and a fire-stop added between the studs at the floorline (fig. 28.9). When diagonal subfloor is used, a nailing member is normally required between joists and studs at the wall lines.

Because there is less potential shrinkage in exterior walls with balloon framing than in the platform type, balloon framing is usually preferred over the platform type in full two-story brick or stone veneer houses.

Floor Joists

Floor joists are selected primarily to meet strength and stiffness requirements. Strength requirements depend upon the loads to be carried. Stiffness requirements place an arbitrary control on deflection under load. Stiffness is also important in limiting vibrations from moving loads—often a cause of annoyance to occupants. Other desirable qualities for floor joists are good nail holding ability and freedom from warp.

Wood floor joists are generally of 2-inch (nominal) thickness and of 8-, 10-, or 12-inch (nominal) depth. The size depends upon the loading, length of span, spacing between joists, and the species and grade of lumber used. As previously mentioned, grades in species vary a great deal. For example, the grades generally used for joists are "Standard" for Douglas-fir, "No. 2 or No. 2KD" for southern pine, and comparable grades for other species.

Span tables for floor joists, which are published by the Federal Housing Administration (*11*) or local building codes can be used as guidelines. These sizes are of course often minimum, and it is sometimes the practice in medium- and higher priced houses to use the next larger size than those listed in the tables.

Joist Installation

After the sill plates have been anchored to the foundation walls or piers, the joists are located according to the house design. (Sixteen-inch center-to-center spacing is most commonly used.)

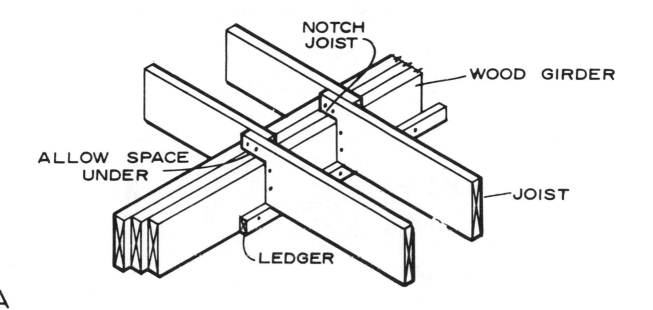

NOTCH JOIST

WOOD GIRDER

ALLOW SPACE UNDER

JOIST

LEDGER

A

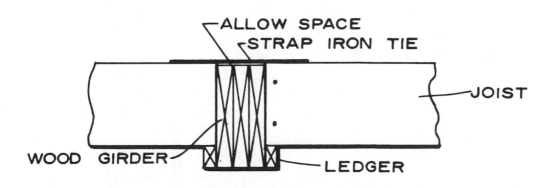

SPACE

SCAB

JOIST

WOOD GIRDER

LEDGER

B

ALLOW SPACE

STRAP IRON TIE

JOIST

WOOD GIRDER

LEDGER

C

Figure 28.5.—Ledger on center wood girder: A, Notched joist; B, scab tie between joist; C, flush joist.

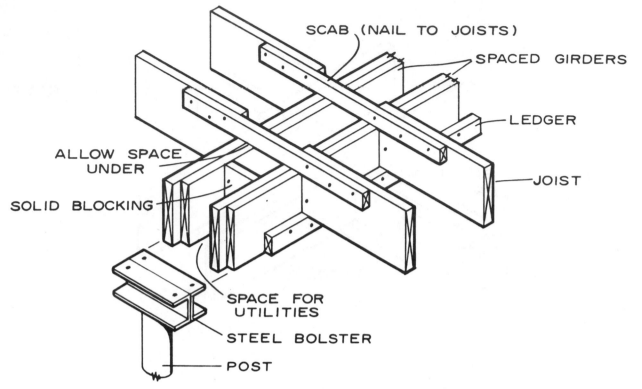

Figure 28.6.—Spaced wood girder.

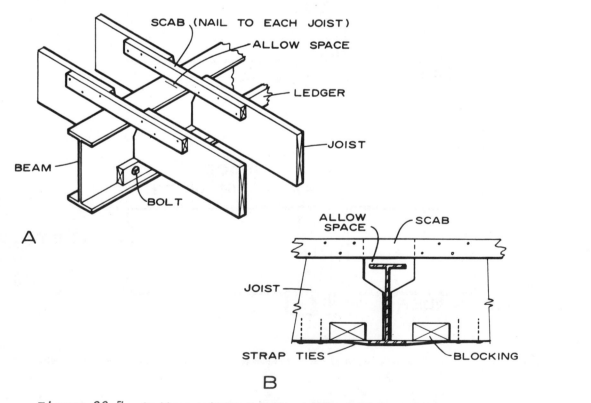

Figure 28.7.—Steel beam and joists: A, Bearing on ledger; B, bearing on flange.

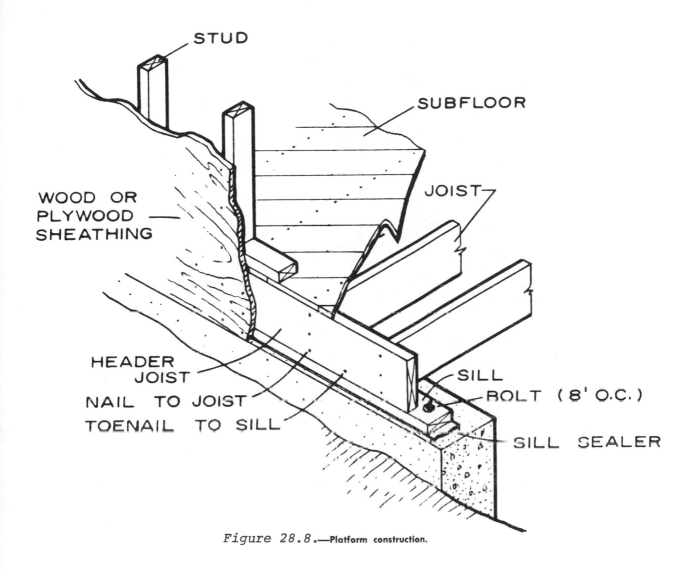

STUD

SUBFLOOR

JOIST

WOOD OR
PLYWOOD
SHEATHING

HEADER
JOIST
NAIL TO JOIST
TOENAIL TO SILL

SILL

BOLT (8' O.C.)

SILL SEALER

Figure 28.8.—**Platform construction.**

Any joists having a slight bow edgewise should be so placed that the crown is on top. A crowned joist will tend to straighten out when subfloor and normal floor loads are applied. The largest edge knots should be placed on top, since knots on the upper side of a joist are on the compression side of the member and will have less effect on strength.

The header joist is fastened by nailing into the end of each joist with three sixteenpenny nails. In addition, the header joist and the stringer joists parallel to the exterior walls in platform construction (fig. 28.10) are toenailed to the sill with tenpenny nails spaced 16 inches on center. Each joist should be toenailed to the sill and center beam with two tenpenny or three eightpenny nails; then nailed to each other with three or four sixteenpenny nails when they lap over the center beam. If a nominal 2-inch scab is used across butt-ended joists, it should be nailed to each joist with at least three sixteenpenny nails at each side of the joint. These and other nailing patterns and practices are outlined in table 2.

The "in-line" joist splice is sometimes used in framing for floor and ceiling joists. This system normally allows the use of one smaller joist size when center supports are present. Briefly, it consists of uneven length joists, the long overhanging joist is cantilevered over the center support, then spliced to the supported joist (fig. 28.11). Overhang joists are alternated. Depending on the span, species, and joist size, the overhang varies between about 1 foot 10 inches and 2 feet 10 inches. Plywood splice plates are used on each side of the end joints.[4]

It is good practice to double joists under all parallel bearing partition walls; if spacing is required for heat ducts, solid blocking is used between the joists (fig. 28.10).

[4] Details of this type of construction can be obtained from builders, lumber dealers, or architects, or by contacting the American Plywood Association, Tacoma, Wash. 98401.

471

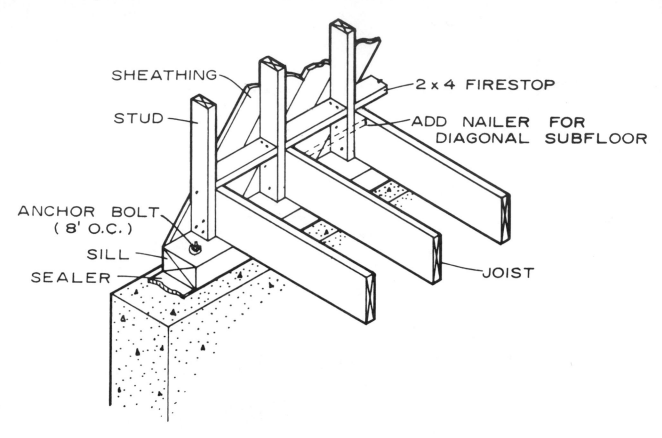

Figure 28.9.—Sill for balloon framing.

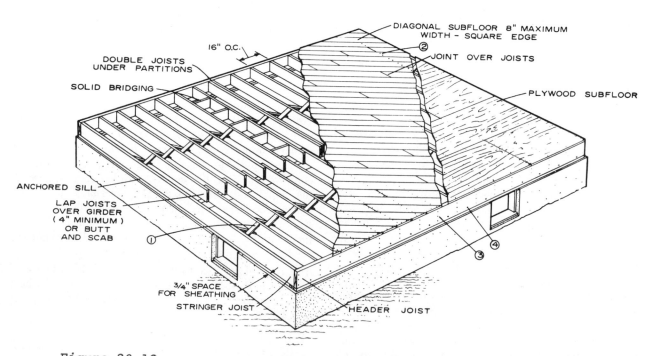

Figure 28.10—Floor framing: (1) Nailing bridging to joists; (2) nailing board subfloor to joists; (3) nailing header to joists; (4) toenailing header to sill.

472

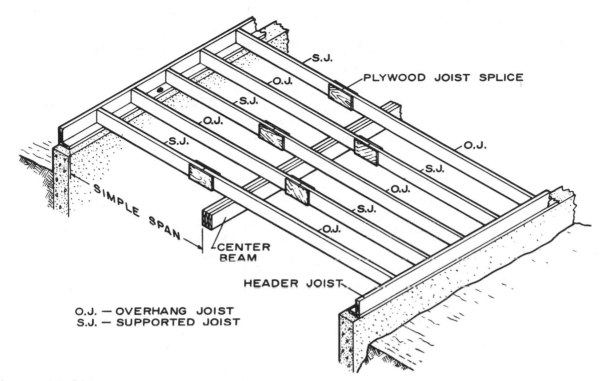

Figure 28.11.

"In-Line" joist system. Alternate extension of joists over the center support with plywood gusset joint allows the use of a smaller joist size.

Details At Floor Openings

When framing for large openings such as stairwells, fireplaces, and chimneys, the joists and headers around the opening should be doubled. A recommended method of framing and nailing is shown in (fig. 28.12).

Joist hangers and short sections of angle iron are often used to support headers and tail beams for large openings. For further details on stairwells, see Chapter 49—"Stairs."

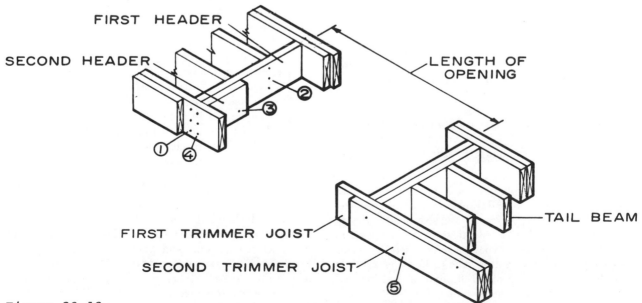

Figure 28.12.

Framing for floor openings: (1) Nailing trimmer to first header; (2) nailing header to tail beams; (3) nailing header together; (4) nailing trimmer to second header; (5) nailing trimmers together.

Bridging

Cross-bridging between wood joists has often been used in house construction, but research by several laboratories has questioned the benefits of bridging in relation to its cost, especially in normal house construction. Even with tight-fitting, well-installed bridging, there is no significant ability to transfer loads after subfloor and finish floor are installed. However, some building codes require the use of cross-bridging or *solid bridging* (table 2).

Solid bridging is often used between joists to provide a more rigid base for partitions located above joist spaces. Well-fitted solid bridging securely nailed to the joists will aid in supporting partitions above them (fig. 28.10). Load-bearing partitions should be supported by doubled joists.

Subfloor

Subflooring is used over the floor joists to form a working platform and base for finish flooring. It usually consists of (a) square-edge or tongued-and-grooved boards no wider than 8 inches and not less than ¾ inch thick or (b) plywood ½ to ¾ inch thick, depending on species, type of finish floor, and spacing of joists (fig. 28.10).

Boards

Subflooring may be applied either *diagonally* (most common) or at *right angles* to the joists. When subflooring is placed at right angles to the joists, the finish floor should be laid at right angles to the subflooring. Diagonal subflooring permits finish flooring to be laid either parallel or at right angles (most common) to the joists. End joints of the boards should always be made directly over the joists. Subfloor is nailed to each joist with two eightpenny nails for widths under 8 inches and three eightpenny nails for 8-inch widths.

The joist spacing should not exceed 16 inches on center when finish flooring is laid parallel to the joists, or where parquet finish flooring is used; nor exceed 24 inches on center when finish flooring at least $25/32$ inch thick is at right angles to the joists.

Where balloon framing is used, blocking should be installed between ends of joists at the wall for nailing the ends of diagonal subfloor boards (fig. 28.9).

Plywood

Plywood can be obtained in a number of grades designed to meet a broad range of end-use requirements. All Interior-type grades are also available with fully waterproof adhesive identical with those used in Exterior plywood. This type is useful where a hazard of prolonged moisture exists, such as in underlayments or subfloors adjacent to plumbing fixtures and for roof sheathing which may be exposed for long periods during construction. Under normal conditions and for sheathing used on walls, Standard sheathing grades are satisfactory.

Plywood suitable for subfloor, such as Standard sheathing, Structural I and II, and C-C Exterior grades, has a panel identification index marking on each sheet. These markings indicate the allowable spacing of rafters and floor joists for the various thicknesses when the plywood is used as roof sheathing or subfloor. For example, an index mark of 32/16 indicates that the plywood panel is suitable for a maximum spacing of 32 inches for rafters and 16 inches for floor joists. Thus, no problem of strength differences between species is involved as the correct identification is shown for each panel.

Normally, when some type of underlayment is used over the plywood subfloor, the minimum thickness of the subfloor for species such as Douglas-fir and southern pine is ½ inch when joists are spaced 16 inches on center, and ⅝-inch thick for such plywood as western hemlock, western white pine, ponderosa pine, and similar species. These thicknesses of plywood might be used for 24-inch spacing of joists when a finish $25/32$-inch strip flooring is installed at right angles to the joists. However, it is important to have a solid and safe platform for workmen during construction of the remainder of the house. For this reason, some builders prefer a slightly thicker plywood subfloor especially when joist spacing is greater than 16 inches on center.

Plywood can also serve as combined plywood subfloor and underlayment, eliminating separate underlayment because the plywood functions as both structural subfloor and a good substrate. This applies to thin resilient floorings, carpeting, and other nonstructural finish flooring. The plywood used in this manner must be tongued and grooved or blocked with 2-inch lumber along the unsupported edges. Following are recommendations for its use:

Grade: Underlayment, underlayment with exterior glue, C-C plugged

Spacing and thickness: (a) For species such as Douglas-fir (coast type), and southern pine— ½ inch minimum thickness for 16-inch joist spacing, ⅝ inch for 20-inch joist spacing, and ¾ inch for 24-inch joist spacing.

(b) For species such as western hemlock, western white pine, and ponderosa pine—⅝ inch minimum thickness for 16-inch joist spacing, ¾ inch for 20-inch joist spacing, and ⅞ inch for 24-inch joist spacing.

Plywood should be installed with the grain direction of the outer plies at right angles to the joists and be staggered so that end joints in adjacent panels break over different joists. Plywood should be nailed to the joist at each bearing with eightpenny common or sevenpenny threaded nails for plywood ½ inch to ¾ inch thick. Space nails 6 inches apart along all edges and 10 inches along intermediate members. When plywood serves as both subfloor and underlay-

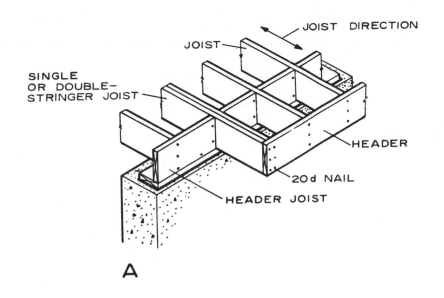

A

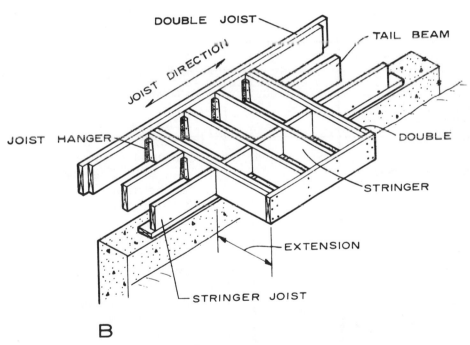

B

Figure 28.13.—Floor framing at wall projections: A, Projection of joists for bay window extensions; B, projection at right angles to joists.

ment, nails may be spaced 6 to 7 inches apart at all joists and blocking. Use eight- or ninepenny common nails or seven- or eightpenny threaded nails.

For the best performance, plywood should *not* be laid up with tight joints whether used on the interior or exterior. The following spacings are recommendations by the American Plywood Association on the basis of field experience:

Plywood location and use	Spacing	
	Edges (In.)	Ends (In.)
Underlayment or interior wall lining	1/32	1/32
Panel sidings and combination sub-floor underlayment	1/16	1/16
Roof sheathing, subflooring, and wall sheathing (Under wet or humid conditions, spacing should be doubled.)	1/8	1/16

475

Minimum Uniformly Distributed Floor Live Loads

Occupancy or Use	Live Load, psf	Occupancy or Use	Live Load, psf
Apartments (see Residential)		Residential	
Armories and drill rooms	150	Multifamily houses	
Assembly halls and other places of assembly:		Private apartments	40
		Public rooms	100
Fixed seats	60	Corridors	60
Movable seats	100	Dwellings	
Balconies (exterior)	100	First floors	40
Bowling alleys, poolrooms, and similar recreational areas	75	Second floors and habitable attics	30
		Uninhabitable attics	20
Corridors		Hotels	
First floor	100	Guest rooms	40
Other floors, same as occupancy served except as indicated		Public rooms	100
		Corridors serving public rooms	100
Dance halls	100	Public corridors	60
Dining rooms and restaurants	100	Private corridors	40
Dwellings (see Residential)		Reviewing stands and bleachers	100
Garages (passenger cars)	50	Schools	
Floors should be designed to carry 150% of the maximum wheel load anywhere on the floor		Classrooms	40
		Corridors	100
Grandstands (see Reviewing stands)		Sidewalks, vehicular driveways, and yards subject to trucking	250
Gymnasiums, main floors and balconies	100	Skating rinks	100
Hospitals		Stairs, fire escapes, and exitways	100
Operating rooms	60	Storage warehouses, light	125
Private rooms	40	Storage warehouses, heavy	250
wards	40	Stores	
Hotels (see Residential)		Retail	
Libraries		First floors, rooms	100
Reading rooms	60	Upper floors	75
Stack rooms	150	Wholesale	125
Manufacturing	125	Theaters	
Marquees	75	Aisles, corridors and lobbies	100
Office Buildings		Orchestra floors	60
Offices	50	Balconies	60
Lobbies	100	Stage floors	150
Penal institutions			
Cell blocks	40		
Corridors	100		

Concentrated Live Loads

Location	Load, Lbs.[1]
Office floors (on a 2.5 sq. ft. area)	2000
Garages, passenger car (on a 2.5 sq. ft. area)	1.5 x max. wheel load
Structures where loaded trucks will be used	maximum wheel load
Stair treads (on a 2.5 sq. ft. area)	300
Finish light floor construction (on a 1 sq. in. area)	200
Elevator machine room grating (on a 4 sq. in. area)	300

[1]Located to produce maximum stress.

The duration of load for roof live loads is quite variable, and usually of less cumulative duration than the normal load condition assumed for floor live loads.

Snow Loads

Snow loads are a form of roof live load that requires the designer's consideration. Snow loads may be full or unbalanced depending on the geometry of the roof. Portions with different pitches will retain different snow loads. Curved roofs may shed snow in some areas and not in others. The roof surface smoothness will affect snow shedding, with metal roofs shedding at different pitches than roofs with rougher surfaces.

Insulation, as it affects heat loss through roofs,,has a great influence on snow retention. Overhangs tend to accumulate snow and ice because of the unheated lower surface. Sometimes electric resistance heaters are placed at overhangs to prevent ice dams and water accumulation.

Valleys formed by dormers and angularly arranged gable roofs will produce pockets for the localized accumulation of snow.

The Uniform Building Code does not attempt to recommend snow loads, but assigns the task to the local building code authorities, which is wise considering the variable nature of terrain and climate in many regions. It is not unusual for design snow loads to vary as much as threefold in a given county in mountainous regions. The local building official must establish snow loads on the basis of climatological information from the weather bureau, his own observations, and the information gleaned from long-time residents of the area in question.

UBC does recommend a snow load reduction practice related to pitch of the roof. Snow loads in excess of 20 PSF may be reduced by S/40 minus ½ for each degree of pitch exceeding 20 degrees. S is the total snow load PSF.

Example: Snow load for design is 45 PSF. For a roof pitch of 30 degrees this may be reduced by

$$\left(\frac{45}{40} - \frac{1}{2} \right)\left(30 - 20 \right) = 6.25 \text{ PSF}$$

The design snow load becomes 38.75 PSF.

Floor Framing at Wall Projections

The framing for wall projections such as a *bay window* or first or second floor extensions beyond the lower wall should generally consist of projection of the floor joists (fig. 28.13). This extension normally should not exceed 24 inches unless designed specifically for greater projections, which may require special anchorage at the opposite ends of the joists. The joists forming each side of the bay should be doubled. Nailing, in general, should conform to that for stair

openings. The subflooring is carried to and sawed flush with the outer framing member. Rafters are often carried by a header constructed in the main wall over the bay area, which supports the roofload. Thus the wall of the bay has less load to support.

Projections at right angles to the length of the floor joists should generally be limited to small areas and extensions of not more than 24 inches. In this construction, the stringer should be carried by doubled joists (fig. 28.13b) Joist hangers or a ledger will provide good connections for the ends of members.

CHAPTER 29

WALL FRAMING

The floor framing with its subfloor covering has now been completed and provides a convenient working platform for construction of the wall framing. The term "wall framing" includes primarily the vertical studs and horizontal members (soleplates, top plates, and window and door headers) of exterior and interior walls that support ceilings, upper floors, and the roof. The wall framing also serves as a nailing base for wall covering materials.

The wall framing members used in conventional construction are generally nominal 2- by 4-inch studs spaced 16 inches on center (*11*). Depending on thickness of covering material, 24-inch spacing might be considered. Top plates and soleplates are also nominal 2 by 4 inches in size. Headers over doors or windows in load-bearing walls consist of doubled 2- by 6-inch and deeper members, depending on span of the opening.

Requirements

The requirements for wall-framing lumber are good stiffness, good nail-holding ability, freedom from warp, and ease of working (*5*). Species used may include Douglas-fir, the hemlocks, southern pine, the spruces, pines, and white fir. As outlined under "Floor Framing," the grades vary by species, but it is common practice to use the third grade for studs and plates and the second grade for headers over doors and windows.

All framing lumber for walls should be reasonably dry. Material at about 15 percent moisture content is desirable, with the maximum allowable considered to be 19 percent. When the higher moisture content material is used (as studs, plates, and headers), it is advisable to allow the moisture content to reach in-service conditions before applying interior trim.

Ceiling height for the first floor is 8 feet under most conditions. It is common practice to rough-frame the wall (subfloor to top of upper plate) to a height of 8 feet $1\frac{1}{2}$ inches. In platform construction, precut studs are often supplied to a length of 7 feet $8\frac{5}{8}$ inches for plate thickness of $1\frac{5}{8}$ inches. When dimension material is $1\frac{1}{2}$ inches thick, precut studs would be 7 feet 9 inches long. This height allows the use of 8-foot-high dry-wall sheets, or six courses of rock lath, and still provides clearance for floor and ceiling finish or for plaster grounds at the floor line.

Second-floor ceiling heights should not be less than 7 feet 6 inches in the clear, except that portion under sloping ceilings. One-half of the floor area, however, should have at least a 7-foot 6-inch clearance.

As with floor construction, two general types of wall framing are commonly used—platform construction and balloon-frame construction. The platform method is more often used because of its simplicity. Balloon framing is generally used where stucco or masonry is the exterior covering material in two-story houses, as outlined in the chapter "Floor Framing."

Platform Construction

The wall framing in platform construction is erected above the subfloor which extends to all edges of the building (fig. 29.1). A combination of platform construction for the first floor sidewalls and full-length studs for end walls extending to end rafters of the gable ends is commonly used in single-story houses.

One common method of framing is the horizontal assembly (on the subfloor) or "tilt-up" of wall sections. When a sufficient work crew is available, full-length wall sections are erected. Otherwise, shorter length sections easily handled by a smaller crew can be used. This system involves laying out precut studs, window and door headers, cripple studs (short-length studs), and windowsills. Top and soleplates are then nailed to all vertical members and adjoining studs to headers and sills with sixteenpenny nails. Let-in corner bracing should be provided when required. The entire section is then erected, plumbed, and braced (fig. 29.1).

A variation of this system includes fastening the studs only at the top plate and, when the wall is erected, toenailing studs to the soleplates which have been previously nailed to the floor. Corner studs and headers are usually nailed together beforehand to form a single unit. Many contractors will also install sheathing before the wall is raised in place. Complete finished walls with windows and door units in place and most of the siding installed can also be fabricated in this manner.

When all exterior walls have been erected, plumbed, and braced, the remaining nailing is completed. Soleplates are nailed to the floor joists and headers or stringers (through the subfloor), corner braces (when used) are nailed to studs and plates, door and window headers are fastened to adjoining studs, and corner studs are nailed together. These and other recommended nailing practices are shown in table 2 and figure 29.1.

In hurricane areas or areas with high winds, it is often advisable to fasten wall and floor framing to the anchored foundation sill when sheathing does not provide this tie. Fig. 29.2 illustrates one system of anchoring the studs to the floor framing with steel straps (7).

Several arrangements of studs at outside corners can be used in framing the walls of a house. Fig. 29.1 shows one method commonly used. Blocking between two corner studs is used to provide a nailing edge for

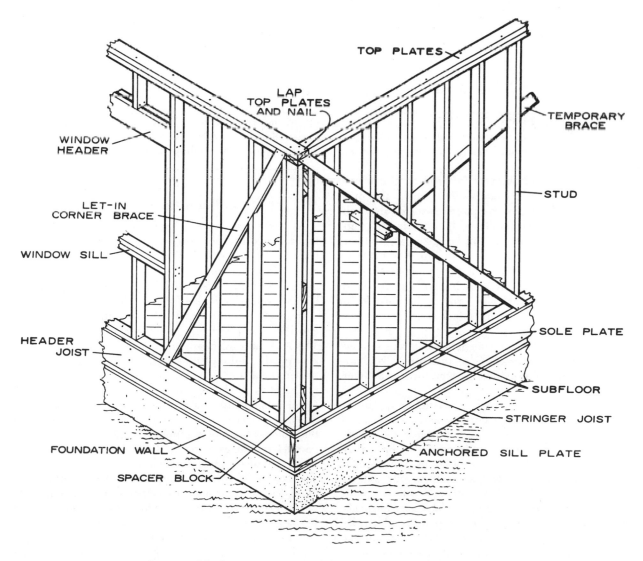

Figure 29.1. —**Wall framing used with platform construction.**

479

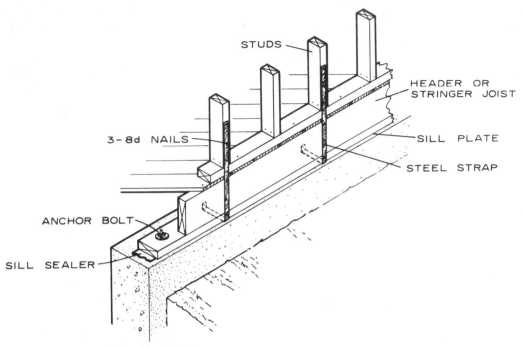

Figure 29.2.—Anchoring wall to floor framing.

interior finish (fig. 29.3a). Fig. 29.3b and c show other methods of stud arrangement to provide the needed interior nailing surfaces as well as good corner support.

Interior walls should be well fastened to all exterior walls they intersect. This intersection should also provide nailing surfaces for the plaster base or dry-wall finish. This may be accomplished by doubling the outside studs at the interior wall line (fig. 29.4a). Another method used when the interior wall joins the exterior wall between studs is shown in figure 29.4b.

Short sections of 2- by 4-inch blocking are used between studs to support and provide backing for a 1- by 6-inch nailer. A 2- by 6-inch vertical member might also be used.

The same general arrangement of members is used at the intersection or crossing of interior walls. Nailing surfaces must be provided in some form or another at all interior corners.

After all walls are erected, a second top plate is added that laps the first at corners and wall intersections (fig. 29.1). This gives an additional tie to the framed walls. These top plates can also be partly fastened in place when the wall is in a horizontal position. Top plates are nailed together with sixteenpenny nails spaced 16 inches apart and with two nails at each wall interesection (table 2). Walls are normally plumbed and alined before the top plate is added. By using 1- by 6- or 1- by 8-inch temporary braces on the studs between intersecting partitions, a straight wall is assured. These braces are nailed to the studs at the

top of the wall and to a 2- by 4-inch block fastened to the subfloor or joists. The temporary bracing is left in place until the ceiling and the roof framing are completed and sheathing is applied to the outside walls.

Balloon Construction

As described in the chapter on "Floor Framing," the main difference between platform and balloon framing is at the floor-lines. The balloon wall studs extend from the sill of the first floor to the top plate or end rafter of the second floor, whereas the platform-framed wall is complete for each floor.

In balloon-frame construction, both the wall studs and the floor joists rest on the anchored sill (fig. 29.5). The studs and joists are toenailed to the sill with eightpenny nails and nailed to each other with at least three tenpenny nails.

The ends of the second-floor joists bear on a 1- by 4-inch ribbon that has been let into the studs. In addition, the joists are nailed with four tenpenny nails to the studs at these connections (fig. 29.5). The end joists parallel to the exterior on both the first and second floors are also nailed to each stud.

Other nailing details should conform in general to those described for platform construction and in table 2.

In most areas, building codes require that *firestops* be used in balloon framing to prevent the spread of fire through the open wall passages. These firestops are ordinarily of 2- by 4-inch blocking placed between the studs (fig. 29.5) or as required by local regulations.

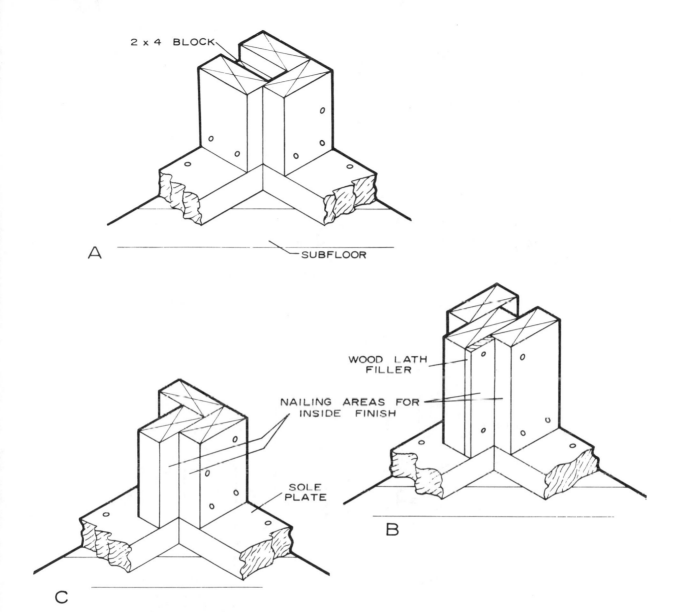

Figure 29.3.—Examples of corner stud assembly: A, Standard outside corner; B, special corner with lath filler; C, special corner without lath filler.

Window and Door Framing

The members used to span over window and door openings are called *headers* or *lintels* (fig. 29.6).As the span of the opening increases, it is necessary to increase the depth of these members to support the ceiling and roofloads. A header is made up of two 2-inch members, usually spaced with ⅜-inch lath or wood strips, all of which are nailed together. They are supported at the ends by the inner studs of the double-stud joint at exterior walls and interior bearing walls. Two headers of species normally used for floor joists are usually appropriate for these openings in normal light-frame construction. The following sizes might be used as a guide for headers:

Maximum span (Ft.)	Header size (In.)
3½	2 by 6
5	2 by 8
6½	2 by 10
8	2 by 12

For other than normal light-frame construction, independent design may be necessary. Wider openings often require trussed headers, which may also need special design.

Location of the studs, headers, and sills around window openings should conform to the rough open-

481

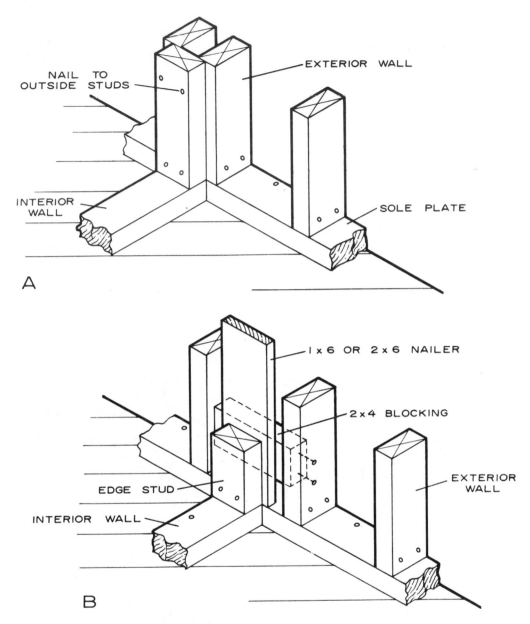

NAIL TO OUTSIDE STUDS

EXTERIOR WALL

INTERIOR WALL

SOLE PLATE

A

1 x 6 OR 2 x 6 NAILER

2 x 4 BLOCKING

EDGE STUD

INTERIOR WALL

EXTERIOR WALL

B

Figure 29.4.—Intersection of interior wall with exterior wall: A, With doubled studs on outside wall; B, Partition between outside studs.

ing sizes recommended by the manufacturers of the millwork. The framing height to the bottom of the window and door headers should be based on the door heights, normally 6 feet 8 inches for the main floor. Thus to allow for the thickness and clearance of the head jambs of window and door frames and the finish floor, the bottoms of the headers are usually located 6 feet 10 inches to 6 feet 11 inches above the subfloor, depending on the type of finish floor used.

Rough opening sizes for exterior door and window frames might vary slightly between manufacturers, but the following allowances should be made for the stiles and rails, thickness of jambs, and thickness and slope of the sill:

Double-Hung Window (Single Unit)
    Rough opening width = glass width plus 6 inches
    Rough opening height = total glass height plus 10 inches

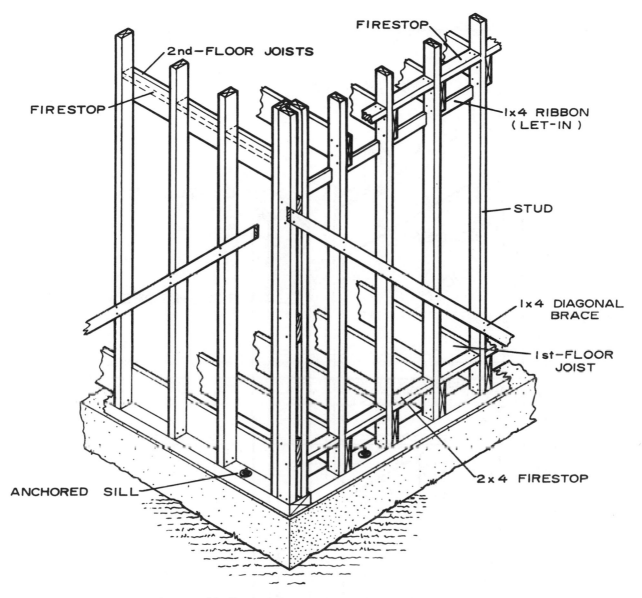

Figure 29.5.—Wall framing used in balloon construction.

For example, the following tabulation illustrates several glass and rough opening sizes for double-hung windows:

Window glass size (each sash)			*Rough frame opening*		
Width	*Height*		*Width*	*Height*	
(In.)	*(In.)*		*(In.)*	*(In.)*	
24	by	16	30	by	42
28	by	20	34	by	50
32	by	24	38	by	58
36	by	24	42	by	58

Casement Window (One Pair—Two Sash)

Rough opening width = total glass width plus 11¼ inches

Rough opening height = total glass height plus 6⅜ inches

Doors

Rough opening width = door width plus 2½ inches

Rough opening height = door height plus 3 inches

End-wall Framing

The framing for the end walls in platform and balloon construction varies somewhat. Fig. 29.7 shows a commonly used method of wall and ceiling framing for platform construction in 1½- or 2-story houses with finished rooms above the first floor. The edge floor joist is toenailed to the top wall plate with eight-penny nails spaced 16 inches on center. The subfloor,

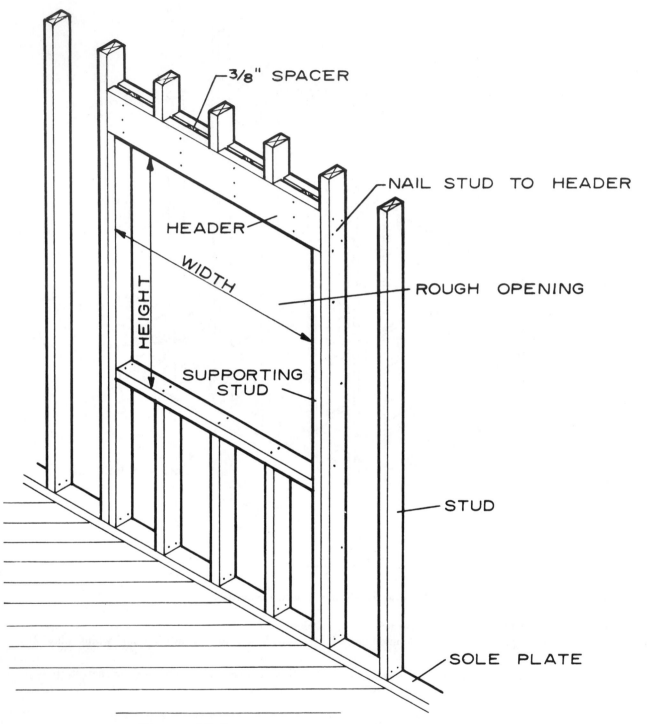

3/8" SPACER

NAIL STUD TO HEADER

HEADER

ROUGH OPENING

WIDTH

HEIGHT

SUPPORTING STUD

STUD

SOLE PLATE

Figure 29.6.—Headers for windows and door openings.

484

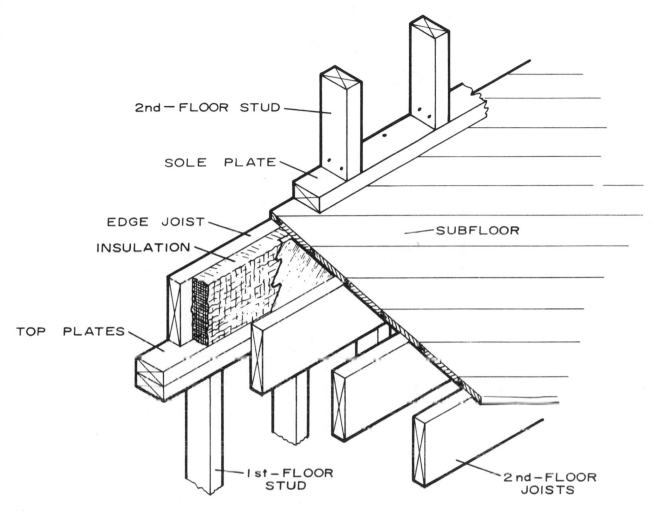

2nd—FLOOR STUD

SOLE PLATE

EDGE JOIST

INSULATION

TOP PLATES

SUBFLOOR

1st—FLOOR STUD

2nd—FLOOR JOISTS

Figure 29.7.—End-wall framing for platform construction (junction of first-floor ceiling with upper-story floor framing).

soleplate, and wall framing are then installed in the same manner used for the first floor.

In balloon framing, the studs continue through the first and second floors (fig. 29.8) . The edge joist can be nailed to each stud with two or three tenpenny nails. As for the first floor, 2- by 4-inch firestops are cut between each stud. Subfloor is applied in a normal manner. Details of the sidewall supporting the ends of the joists are shown in figure 29.5 .

Interior Walls

The interior walls in a house with conventional joist and rafter roof construction are normally located to serve as bearing walls for the ceiling joists as well as room dividers. Walls located parallel to the direction of the joists are commonly nonload bearing. Studs are nominal 2 by 4 inches in size for load-bearing walls but can be 2 by 3 inches in size for nonload-bearing

walls. However, most contractors use 2 by 4's through-out. Spacing of the studs is usually controlled by the thickness of the covering material. For example, 24-inch stud spacing will require ½-inch gypsum board for dry wall interior covering.

The interior walls are assembled and erected in the same manner as exterior walls, with a single bottom (sole) plate and double top plates. The upper top plate is used to tie intersecting and crossing walls to each other. A single framing stud can be used at each side of a door opening in nonload-bearing partitions. They must be doubled for load-bearing walls, however, as shown in fig. 29. 6. When trussed rafters (roof trusses) are used, no load-bearing interior partitions are required. Thus, location of the walls and size and spacing of the studs are determined by the room size desired and type of interior covering selected. The bottom chords of the trusses are used to fasten and

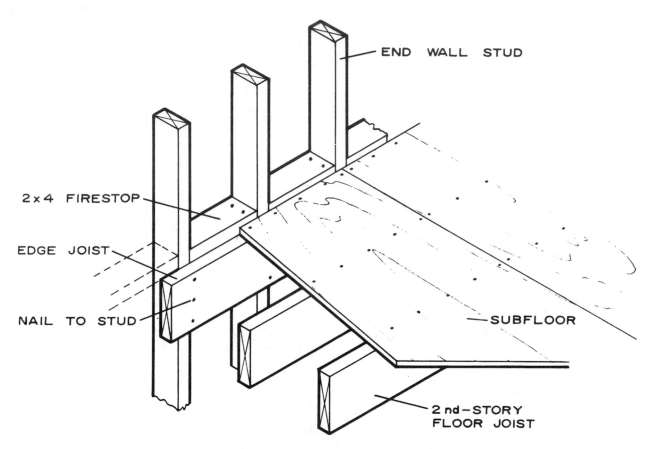

END WALL STUD

2 x 4 FIRESTOP

EDGE JOIST

NAIL TO STUD

SUBFLOOR

2 nd—STORY FLOOR JOIST

Figure 29.8.—End-wall framing for balloon construction (junction of first-floor ceiling and upper-story floor framing).

anchor crossing partitions. When partition walls are parallel to and located between trusses, they are fastened to 2- by 4-inch blocks which are nailed between the lower chords.

Lath Nailers

During the framing of walls and ceilings, it is necessary to provide for both vertical and horizontal fastening of plaster-base lath or dry wall at all inside corners. Figs. 29.3 and .4, which illustrate corner and intersecting wall construction, also show methods of providing lath nailers at these areas.

Horizontal lath nailers at the junction of wall and ceiling framing may be provided in several ways. Fig. 29.9a shows doubled ceiling joists above the wall, spaced so that a nailing surface is provided by each joist. In fig. 29.9b the parallel wall is located between two ceiling joists. A 1- by 6-inch lath nailer is placed and nailed to the top plates with backing blocks spaced on 3- to 4-foot centers. A 2- by 6-inch member might also be used here in place of the 1 by 6.

When the partition wall is at a right angle to the ceiling joists, one method of providing lath nailers is to let in 2- by 6-inch blocks between the joists (fig. 29.9c). They are nailed directly to the top plate and toenailed to the ceiling joists.

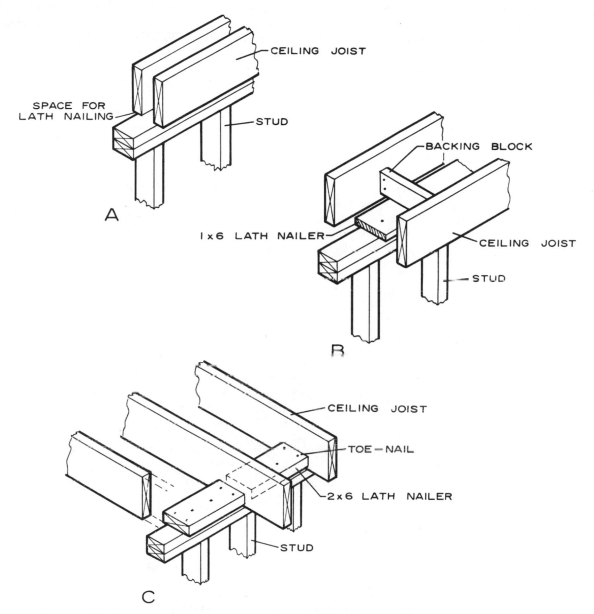

Figure 29.9 —Horizontal lath catchers at ceiling: A, Using ceiling joists over stud wall;
B, lath nailer between ceiling joists; C, stud wall at right angle to joist.

CHAPTER 30

CEILING AND ROOF FRAMING

Ceiling Joists

After exterior and interior walls are plumbed, braced, and top plates added, ceiling joists can be positioned and nailed in place. They are normally placed across the width of the house, as are the rafters. The partitions of the house are usually located so that ceiling joists of even lengths (10, 12, 14, and 16 ft. or longer) can be used without waste to span from exterior walls to load-bearing interior walls. The sizes of the joists depend on the span, wood species, spacing between joists, and the load on the second floor or attic. The correct sizes for various conditions can be found in joist tables or designated by local building requirements (7). When preassembled trussed rafters (roof trusses) are used, the lower chord acts as the ceiling joist. The truss also eliminates the need for load-bearing partitions.

Second grades of the various species are commonly used for ceiling joists and rafters (5). This has been more fully described in Chapter 28,"Floor Framing." It is also desirable, particularly in two-story houses and when material is available, to limit the moisture content of the second-floor joists to no more than 15 percent. This applies as well to other lumber used throughout the house. Maximum moisture content for dimension material should be 19 percent.

Ceiling joists are used to support ceiling finishes. They often act as floor joists for second and attic floors and as ties between exterior walls and interior partitions. Since ceiling joists also serve as tension members to resist the thrust of the rafters of pitched roofs, they must be securely nailed to the plate at outer and inner walls. They are also nailed together, directly or with wood or metal cleats, where they cross or join at the load-bearing partition (fig. 30.1a) and to the rafter at the exterior walls (fig. 30.1b). Toenail at each wall.

In areas of severe windstorms, the use of metal strapping or other systems of anchoring ceiling and roof framing to the wall is good practice. When ceiling joists are perpendicular to rafters, collar beams and cross ties should be used to resist thrust. Recommended sizes and spacing of nails for the framing are listed in table 2. The in-line joist system as shown in Fig. 28.11 and described in the section on joist installation can also be adapted to ceiling or second floor joists.

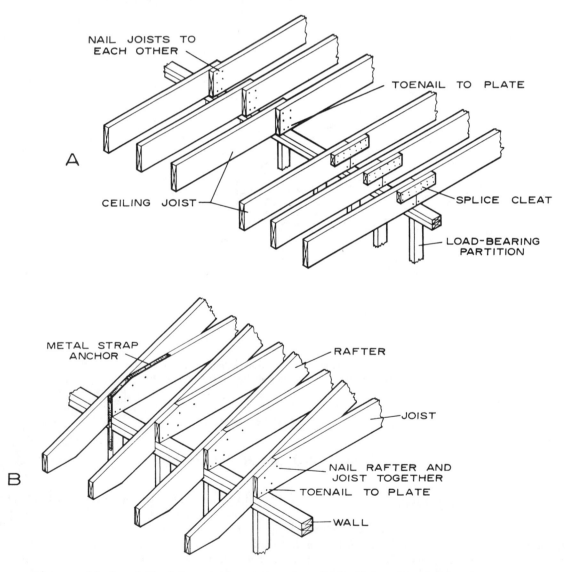

Figure 30.1.—Ceiling joist connections: A, At center partition with joists lapped or butted; B, at outside wall.

Flush Ceiling Framing

In many house designs, the living room and the dining or family room form an open "L." A wide, continuous ceiling area between the two rooms is often desirable. This can be created with a flush beam, which replaces the load-bearing partitions used in the remainder of the house. A nail-laminated beam, designed to carry the ceiling load, supports the ends of the joists. Joists are toenailed into the beam and supported by metal joist hangers (fig. 30.2a) or wood hangers (fig. 30.2b). To resist the thrust of the rafters for longer spans, it is often desirable to provide added resistance by using metal strapping. Strapping should be nailed to each opposite joist with three or four eightpenny nails.

Post and Beam Framing

In contemporary houses, *exposed beams* are often a part of the interior design and may also replace interior and exterior load-bearing walls. With post and beam construction, exterior walls can become fully glazed panels between posts, requiring no other support. Areas below interior beams within the house can remain open or can be closed in with wardrobes, cabinets, or light curtain walls.

This type of construction, while not adaptable to many styles of architecture, is simple and straightforward. However, design of the house should take into account the need for shear or racking resistance of the exterior walls. This is usually accomplished by solid masonry walls or fully sheathed frame walls

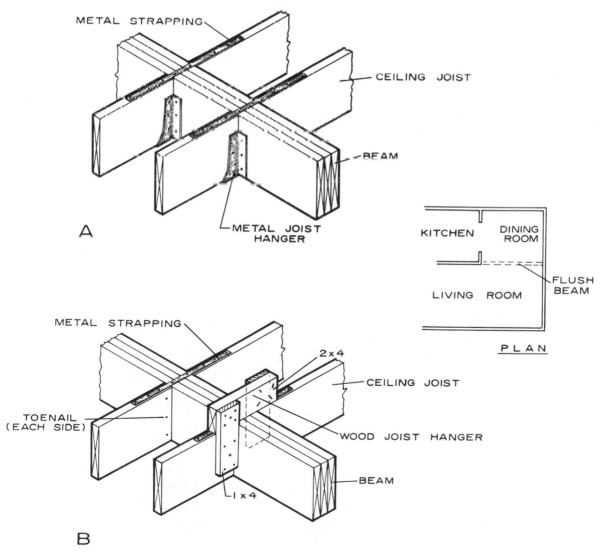

Figure 30.2.—Flush ceiling framing: A, Metal joist hanger; B, wood hanger.

489

between open glazed areas.

Roofs of such houses are often either flat or low-pitched, and may have a conventional rafter-joist combination or consist of thick wood decking spanning between beams. The need for a well-insulated roof often dictates the type of construction that might be used.

The connection of the supporting posts at the floor plate and beam is important to provide uplift resistance. Fig. 30.3 shows connections at the soleplate and at the beam for solid or spaced members. The solid post and beam are fastened together with metal angles nailed to the top plate and to the soleplate as well as the roof beam (fig. 30.3a). The spaced beam and post are fastened together with a 3/8-inch or thicker ply-

wood cleat extending between and nailed to the spaced members (fig. 30.3b). A wall header member between beams can be fastened with joist hangers.

Continuous headers are often used with spaced posts in the construction of framed walls or porches requiring large glazed openings. The beams should be well fastened and reinforced at the corners with lag screws or metal straps. Fig. 30.4a illustrates one connection method using metal strapping.

In low-pitch or flat roof construction for a post and beam system, wood or fiberboard decking is often used. Wood decking, depending on thickness, is frequently used for beam spacings up to 10 or more feet. However, for the longer spans, special application instructions are required (1). Depending on the type,

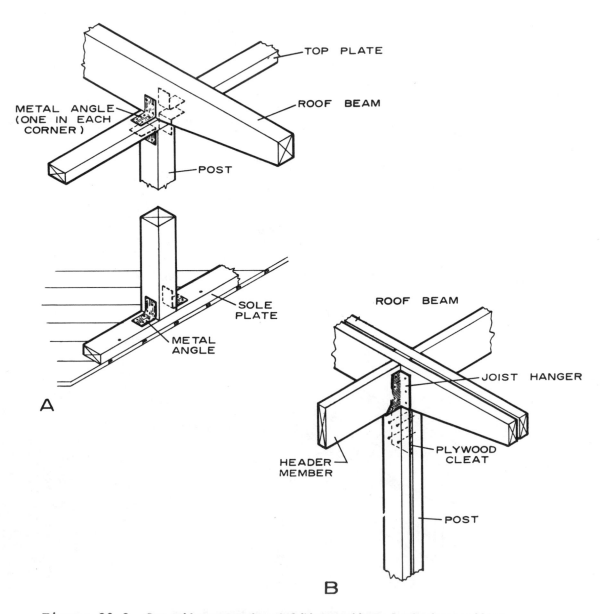

Figure 30.3.—Post and beam connections: A, Solid post and beam; B, spaced post and beam.

490

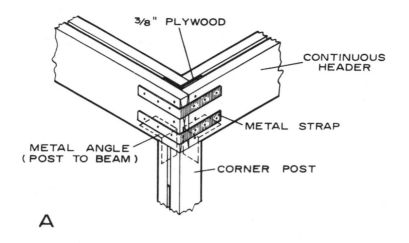

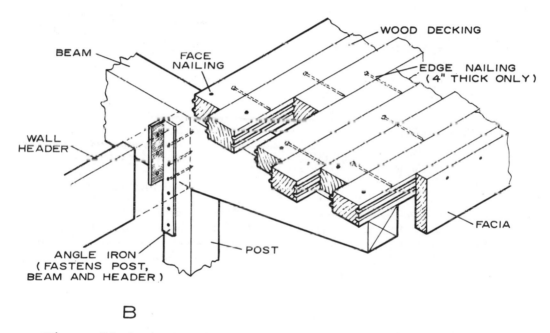

Figure 30.4.—Post and beam details: A, Corner connection with continuous header; B, with roof decking.

2- to 3-inch thick fiberboard decking normally is limited to a beam or purlin spacing of 4 feet.

Tongued-and-grooved solid wood decking, 3 by 6 and 4 by 6 inches in size, should be toe-nailed and face-nailed directly to the beams and edge-nailed to each other with long nails used in predrilled holes (fig. 30.4b). Thinner decking is usually only face-nailed to the beams. Decking is usually square end-trimmed to provide a good fit. If additional insulation is required for the roof, fiberboard or an expanded foamed plastic in sheet form is fastened to the decking before the built-up or similar type of roof is installed. The moisture content of the decking should be near its in-service condition to prevent joints opening later as the wood dries.

Roof Slopes

The architectural style of a house often determines the type of roof and roof slope which are best suited. A contemporary design may have a flat or slightly pitched roof, a rambler or ranch type an intermediate slope, and a Cape Cod cottage a steep slope. Generally, however, the two basic types may be called *flat* or *pitched*, defined as (a) flat or slightly pitched roofs in which roof and ceiling supports are furnished by one type of member, and (b) pitched roofs where both ceiling joists and rafters or trusses are required.

The slope of the roof is generally expressed as the number of inches of vertical rise in 12 inches of horizontal run. The rise is given first, for example, 4 in 12.

A further consideration in choosing a roof slope is

the type of roofing to be used. However, modern methods and roofing materials provide a great deal of leeway in this. For example, a built-up roof is usually specified for flat or very low-pitched roofs, but with different types of asphalt or coal-tar pitch and aggregate surfacing materials, slopes of up to 2 in 12 are sometimes used. Also, in sloped roofs where wood or asphalt shingles might be selected, doubling the underlay and decreasing the exposure distance of the shingles will allow slopes of 4 in 12 and less.

Second grades of the various wood species are normally used for rafters. Most species of softwood framing lumber are acceptable for roof framing, subject to maximum allowable spans for the particular species, grade, and use. Because all species are not equal in strength properties, larger sizes, as determined from the design, must be used for weaker species for a given span (*11*).

All framing lumber should be well seasoned. Lumber 2 inches thick and less should have a moisture content not over 19 percent, but when obtainable, lumber at about 15 percent is more desirable because less shrinkage will occur when moisture equilibrium is reached.

Flat Roofs

Flat or low-pitched roofs, sometimes known as *shed roofs*, can take a number of forms, two of which are shown in fig. 30.5. Roof joists for flat roofs are commonly laid level or with a slight pitch, with roof sheathing and roofing on top and with the underside utilized to support the ceiling. Sometimes a slight roof slope may be provided for roof drainage by tapering the joist or adding a cant strip to the top.

The house design usually includes an overhang of the roof beyond the wall. Insulation is sometimes used in a manner to provide for an airways just under the roof sheathing to minimize condensation problems in winter. Flat or low-pitched roofs of this type require larger sized members than steeper pitched roofs because they carry both roof and ceiling loads.

The use of solid wood decking often eliminates the need for joists. Roof decking used between beams serves as: (a) Supporting members, (b) interior finish, and (c) roof sheathing. It also provides a moderate amount of insulation. In cold climates, rigid insulating materials are used over the decking to further reduce heat loss.

When overhang is involved on all sides of the flat roof, lookout rafters are ordinarily used (fig. 30.6). Lookout rafters are nailed to a doubled header and toenailed to the wallplate. The distance from the doubled header to the wall line is usually twice the overhang. Rafter ends may be finished with a nailing header which serves for fastening soffit and facia boards. Care should be taken to provide some type of ventilation at such areas.

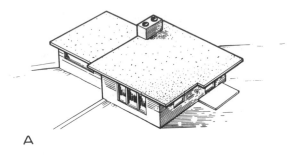

A

B

Figure 30.5.

Roofs using single roof construction: A, Flat roof; B, low-pitched roof.

Pitched Roofs

Gable Roof

Perhaps the simplest form of the pitched roof, where both rafters and ceiling joists are required because of the attic space formed, is the gable roof (fig. 30.7a). All rafters are cut to the same length and pattern and erection is relatively simple, each pair being fastened at the top to a *ridge board*. The ridge board is usually a 1- by 8-inch member for 2- by 6-inch rafters and provides support and a nailing area for the rafter ends.

A variation of the gable roof, used for Cape Cod or similar styles, includes the use of shed and gable *dormers* (fig. 30.7b). Basically, this is a one-story house because the majority of the rafters rest on the first-floor plate. Space and light are provided on the second floor by the shed and gable dormers for bedrooms and bath. Roof slopes for this style may vary from 9 in 12 to 12 in 12 to provide the needed headroom.

A third style in roof designs is the *hip roof* (fig. 30.7c). Center rafters are tied to the ridge board, while hip rafters supply the support for the shorter jack rafters. Cornice lines are carried around the perimeter of the building.

While these roof types are the most common, others may include such forms as the *mansard* and the *A-frame* (where wall and roof members are the same members).

492

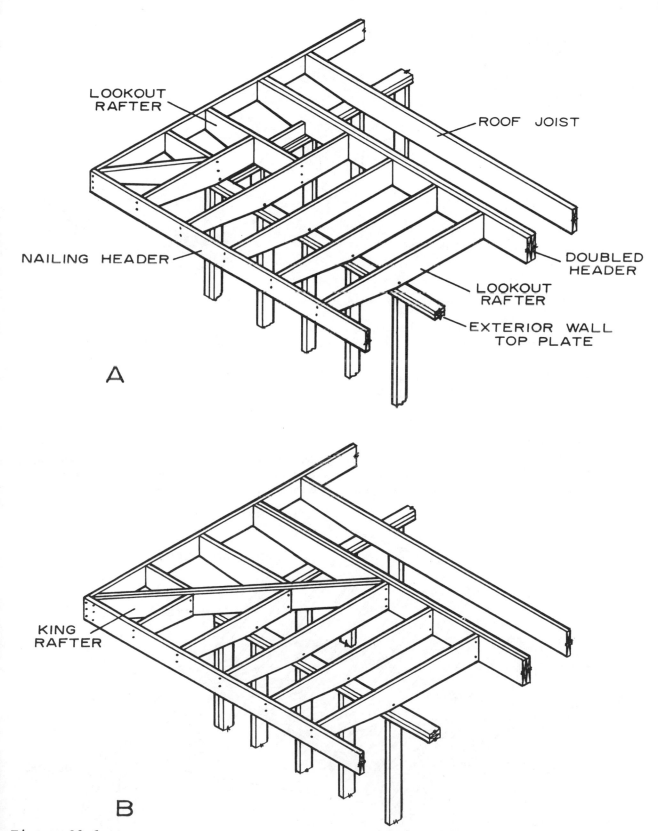

LOOKOUT RAFTER

ROOF JOIST

NAILING HEADER

DOUBLED HEADER

LOOKOUT RAFTER

EXTERIOR WALL TOP PLATE

A

KING RAFTER

B

Figure 30.6.

Typical construction of flat or low-pitched roof with side and end overhang of: A, Less than 3 feet; B, more than 3 feet.

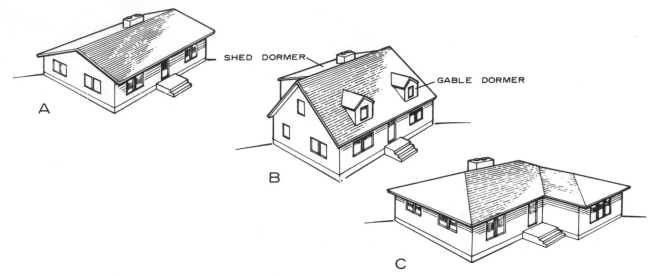

Figure 30.7.—Types of pitched roofs: A, Gable; B, gable with dormers; C, hip.

In normal pitched-roof construction, the ceiling joists are nailed in place after the interior and the exterior wall framing are complete. Rafters should not be erected until ceiling joists are fastened in place, as the thrust of the rafters will otherwise tend to push out the exterior walls.

Rafters are usually precut to length with proper angle cut at the ridge and eave, and with notches provided for the top plates (fig. 30.8a). Rafters are erected in pairs. Studs for gable end walls are cut to fit and nailed to the end rafter and the topplate of the end wall soleplate (fig. 30.8b). With a gable (rake) overhang, a fly rafter is used beyond the end rafter and is fastened with blocking and by the sheathing. Additional construction details applicable to roof framing are given in Chapter 33, "Exterior Trim and Millwork."

Hip Roof

Hip roofs are framed the same as a gable roof at the center section of a rectangular house. The ends are framed with hip rafters which extend from each outside corner of the wall to the ridge board at a 45°

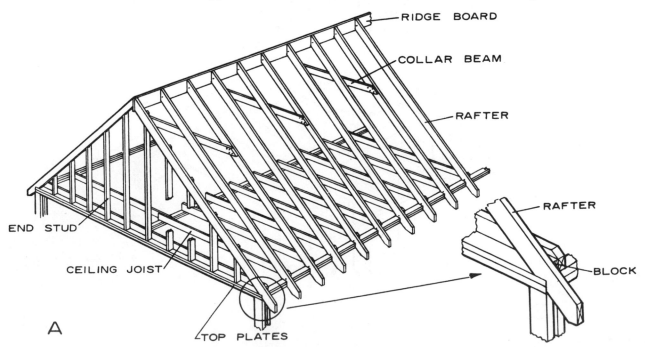

Figure 30.8a.—Ceiling and roof framing: A, Overall view of gable roof framing.

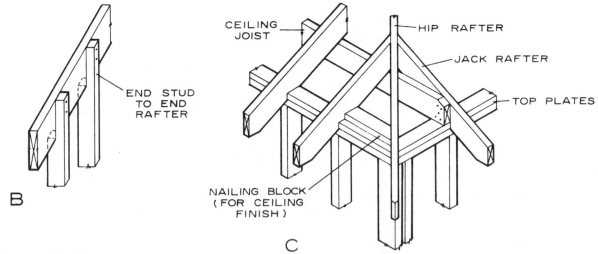

Figure 30.8b.

Ceiling and roof framing: B, Connection of gable end studs to end rafter; C, detail of corner of hip roof.

angle. Jack rafters extend from the top plates to the hip rafters (fig. 30.8c).

When roofs spans are long and slopes are flat, it is common practice to use collar beams between opposing rafters. Steeper slopes and shorter spans may also require collar beams but only on every third rafter. Collar beams may be 1- by 6-inch material. In 1½-story houses, 2- by 4-inch members or larger are used at each pair of rafters which also serve as ceiling joists for the finished rooms.

Good practices to be followed in the nailing of rafters, ceiling joists, and end studs are shown in figure 30.8 and table 2.

Valleys

The *valley* is the internal angle formed by the junction of two sloping sides of a roof. The key member of valley construction is the *valley rafter*. In the intersection of two equal-size roof sections, the valley rafter is doubled (fig. 30.9) to carry the roofload, and is 2 inches deeper than the common rafter to provide full contact with jack rafters. Jack rafters are nailed to the ridge and toenailed to the valley rafter with three tenpenny nails.

Dormers

In construction of small gable dormers, the rafters at each side are doubled and the side studs and the short valley rafter rest on these members (fig. 30.10). Side studs may also be carried past the rafter and bear on a soleplate nailed to the floor framing and subfloor. This same type of framing may be used for the sidewalls of shed dormers. The valley rafter is also tied to the roof framing at the roof by a header. Methods of fastening at top plates conform to those

previously described. Where future expansion is contemplated or additional rooms may be built in an attic, consideration should be given to framing and enclosing such dormers when the house is built.

Overhangs

In two-story houses, the design often involves a projection or overhang of the second floor for the purpose of architectural effect, to accommodate brick veneer on the first floor, or for other reasons. This overhang may vary from 2 to 15 inches or more. The overhang should ordinarily extend on that side of the house where joist extensions can support the wall framing (fig. 30.11). This extension should be provided with insulation and a vapor barrier.

When the overhang parallels the second-floor joists, a doubled joist should be located back from the wall at a distance about twice the overhang. These details

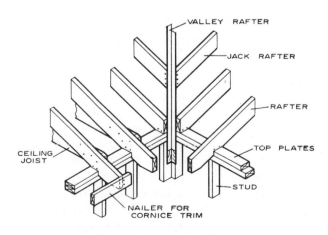

*Figure 30.9.—***Framing at a valley.**

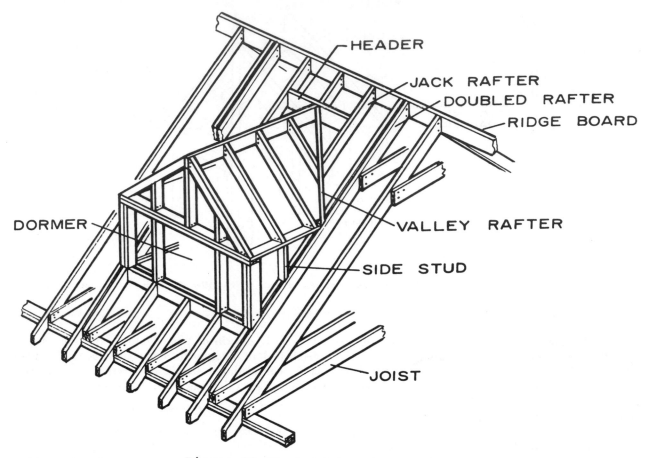

HEADER

JACK RAFTER

DOUBLED RAFTER

RIDGE BOARD

DORMER

VALLEY RAFTER

SIDE STUD

JOIST

Figure 30.10.—**Typical dormer framing.**

are similar to those shown in fig. 28.13 under "Floor Framing."

Ridge Beam Roof Details

In low-slope roof designs, the style of architecture often dictates the use of a *ridge beam*. These solid, glue-laminated, or nail-laminated beams span the open area and are usually supported by an exterior wall at one end and an interior partition wall or a post at the other. The beam must be designed to support the roof load for the span selected. Wood decking can serve both as supporting and sheathing. Spaced rafters placed over the ridge beam or hung on metal joist hangers serve as alternate framing methods. When a ridge beam and wood decking are used (fig. 30.12a), good anchoring methods are needed at the ridge and outer wall. Long ringshank nails and supplemental metal strapping or angle iron can be used at both bearing areas.

A combination of large spaced rafters (purlin rafters) which serve as beams for longitudinal wood or structural fiberboard decking is another system which might be used with a ridge beam. Rafters can be supported by metal hangers at the ridge beam (fig. 30.12b) and extend beyond the outer walls to form an overhang. Fastenings should be supplemented by strapping or metal angles.

Lightweight Wood Roof Trusses

The *simple truss* or *trussed rafter* is an assembly of members forming a rigid framework of triangular shapes capable of supporting loads over long spans without intermediate support. It has been greatly refined during its development over the years, and the *gusset* and other preassembled types of wood trusses are being used extensively in the housing field. They save material, can be erected quickly, and the house can be enclosed in a short time.

Trusses are usually designed to span from one exterior wall to the other with lengths from 20 to 32 feet or more. Because no interior bearing walls are required. the entire house becomes one large workroom. This allows increased flexibility for interior planning, as partitions can be placed without regard to structural requirements.

Wood trusses most commonly used for houses include the *W-type* truss, the *King-post,* and the *scissors*

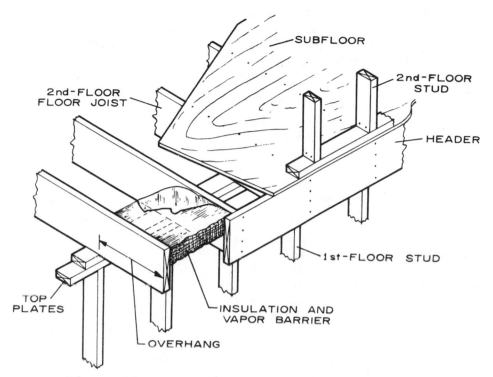

SUBFLOOR

2nd-FLOOR STUD

2nd-FLOOR FLOOR JOIST

HEADER

1st-FLOOR STUD

TOP PLATES

INSULATION AND VAPOR BARRIER

OVERHANG

Figure 30.11.—Construction of overhang at second floor.

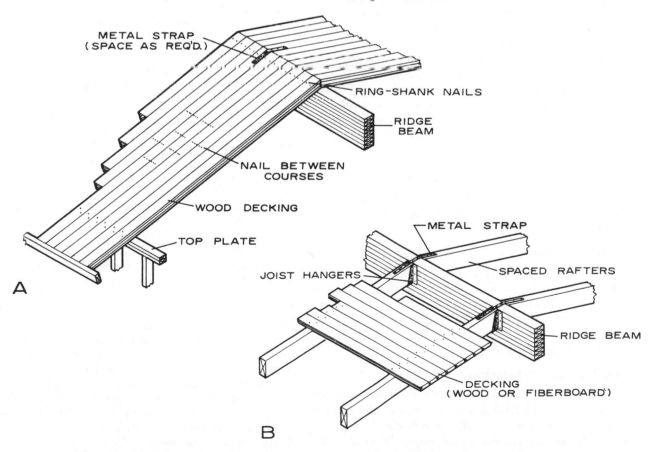

METAL STRAP (SPACE AS REQ'D.)

RING-SHANK NAILS

RIDGE BEAM

NAIL BETWEEN COURSES

WOOD DECKING

TOP PLATE

A

METAL STRAP

SPACED RAFTERS

JOIST HANGERS

RIDGE BEAM

DECKING (WOOD OR FIBERBOARD)

B

Figure 30.12.—Ridge beam for roof: A, With wood decking; B, with rafters and decking.

(fig. 30.13). These and similar trusses are most adaptable to houses with rectangular plans so that the constant width requires only one type of truss. However, trusses can also be used for L plans and for hip roofs as special hip trusses can be provided for each end and valley area.

Trusses are commonly designed for 2-foot spacing, which requires somewhat thicker interior and exterior sheathing or finish material than is needed for conventional joist and rafter construction using 16-inch spacing. Truss designs, lumber grades, and construction details are available from several sources including the American Plywood Association (2).

W-Type Truss

The W-type truss (fig. 30.13a) is perhaps the most popular and extensively used of the light wood trusses. Its design includes the use of three more members than the King-post truss, but distances between connections are less. This usually allows the use of lower grade lumber and somewhat greater spans for the same member size.

King-post truss.—The King-post is the simplest form of truss used for houses, as it is composed only of upper and lower chords and a center vertical post (fig. 30.13b). Allowable spans are somewhat less than for the W-truss when the same size members are used, because of the unsupported length of the upper chord. For short and medium spans, it is probably more economical than other types because it has fewer pieces and can be fabricated faster. For example, under the same conditions, a plywood gusset King-post truss with 4 in 12 pitch and 2-foot spacing is limited to about a 26-foot span for 2- by 4-inch members, while the W-type truss with the same size members and spacing could be used for a 32-foot span. Furthermore, the grades of lumber used for the two types might also vary.

Local prices and design load requirements (for snow, wind, etc.) as well as the span should likely govern the type of truss to be used.

Scissors Truss

The scissors truss (fig. 30.13c) is a special type used for houses in which a sloping living room ceiling is desired. Somewhat more complicated than the W-type truss, it provides good roof construction for a "cathedral" ceiling with a saving in materials over conventional framing methods.

Design and Fabrication

The design of a truss not only includes snow and windload considerations but the weight of the roof itself. Design also takes into account the slope of the roof. Generally, the flatter the slope, the greater the stresses. This results not only in the need for larger

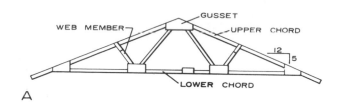

A

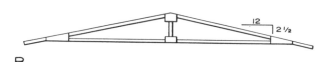

B

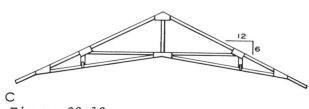

C

Figure 30.13.

Light wood trusses: A, W-type; B, King-post; C, scissors.

members but also in stronger connections. Consequently, all conditions must be considered before the type of truss is selected and designed.

A great majority of the trusses used are fabricated with gussets of plywood (nailed, glued, or bolted in place) or with metal gusset plates. Others are assembled with split-ring connectors. Designs for standard W-type and King-post trusses with plywood gussets are usually available through a local lumber dealer (2). Information on metal plate connectors for wood trusses is also available. Many lumber dealers are able to provide the builder or homeowner with completed trusses ready for erection.

To illustrate the design and construction of a typical wood W-truss more clearly, the following example is given:

The span for the nail-glued gusset truss (fig. 30.14) is 26 feet, the slope 4 in 12, and the spacing 24 inches. Total roof load is 40 pounds per square foot, which is usually sufficient for moderate to heavy snow belt areas. Examination of tables and charts (2) shows that the upper and lower chords can be 2 by 4 inches in size, the upper chord requiring a slightly higher grade of material. It is often desirable to use dimension material with a moisture content of about 15 percent with a maximum of 19 percent.

Plywood gussets can be made from ⅜- or ½-inch

standard plywood with exterior glueline or exterior sheathing grade plywood (18). The cutout size of the gussets and the general nailing pattern for nail-gluing are shown in fig. 30.14. More specifically, fourpenny nails should be used for plywood gussets up to 3/8 inch thick and sixpenny for plywood 1/2 to 7/8 inch thick. Three-inch spacing should be used when plywood is no more than 3/8 inch thick and 4 inches for thicker plywood. When wood truss members are nominal 4 inches wide, use two rows of nails with a 3/4-inch edge distance. Use three rows of nails when truss members are 6 inches wide. Gussets are used on both sides of the truss.

For normal conditions and where relative humidities in the attic area are inclined to be high, such as might occur in the southern and southeastern States, a *resorcinol glue* should be used for the gussets. In dry and arid areas where conditions are more favorable, a casein or similar glue might be considered.

Glue should be spread on the clean surfaces of the gusset and truss members. Either nails or staples might be used to supply pressure until the glue has set, although only nails are recommended for plywood 1/2 inch and thicker. Use the nail spacing previously outlined. Closer or intermediate spacing may be used to insure "squeezeout" at all visible edges. Gluing should be done under closely controlled temperature conditions. This is especially true if using the resorcinol adhesives. Follow the assembly temperatures recommended by the manufacturer.

Handling

In handling and storage of completed trusses, avoid placing unusual stresses on them. They were designed to carry roofloads in a vertical position, and it is important that they be lifted and stored in an upright position. If they must be handled in a flat position, enough men or supports should be used along their length to minimize bending deflections. Never support them only at the center or only at each end when in a flat position.

Erection

One of the important details in erecting trusses is the method of *anchoring*. Because of their single member thickness and the presence of plywood gussets at the wallplates, it is usually desirable to use some type of metal connector to supplement the toenailings. Plate anchors are available commercially or can be formed from sheet metal. Resistance to uplift stresses as well as thrust must be considered. Many dealers supply trusses with a 2- by 4-inch soffit return at the end of each upper chord to provide nailing areas for the soffit.

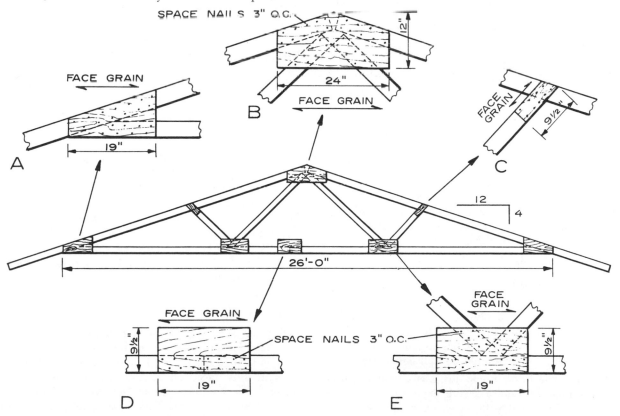

Figure 30.14—Construction of a 26-foot W truss: A, Bevel-heel gusset; B, peak gusset; C, upper chord intermediate gusset; D, splice of lower chord; E, lower chord intermediate gusset.

CHAPTER 31

WALL SHEATHING

Wall sheathing (7) is the outside covering used over the wall framework of studs, plates, and window and door headers. It forms a flat base upon which the exterior finish can be applied. Certain types of sheathing and methods of application can provide great rigidity to the house, eliminating the need for corner bracing. Sheathing serves to minimize air infiltration and, in certain forms, provides some insulation.

Some sheet materials serve both as sheathing and siding. Sheathing is sometimes eliminated from houses in the mild climates of the South and West. It is a versatile material and manufacturers produce it in many forms. Perhaps the most common types used in construction are: boards, plywood, structural insulating board, and gypsum sheathing.

Types of Sheathing

Wood Sheathing

Wood sheathing is usually of nominal 1-inch boards in a shiplap, a tongued-and-grooved, or a square-edge pattern. Resawn $1\frac{1}{16}$-inch boards are also allowed under certain conditions. The requirements for wood sheathing are easy working, easy nailing, and moderate shrinkage. Widths commonly used are 6, 8, and 10 inches. It may be applied horizontally or diagonally (fig. 31.1a). Sheathing is sometimes carried only to the subfloor (fig. 31.1b), but when diagonal sheathing or sheet materials are placed as shown in figure 34.1c greater strength and rigidity result. It is desirable to limit the wood moisture content to 15 percent to minimize openings between matched boards when shrinkage occurs.

Some manufacturers produce random-length side- and end-matched boards for sheathing. Most softwood species, such as the spruces, Douglas-fir, southern pine, hemlock, the soft pines, and others, are suitable for sheathing. Grades vary between species, but sheathing is commonly used in the third grade (5).

Refer to the chapter on "Floor Framing."

Plywood Sheathing

Plywood is used extensively for sheathing of walls, applied vertically, normally in 4- by 8-foot and longer sheets (fig. 31.2). This method of sheathing eliminates the need for diagonal corner bracing; but, as with all sheathing materials, it should be well nailed (table 2).

Standard sheathing grade (18) is commonly used for sheathing. For more severe exposures, this same plywood is furnished with an exterior glueline. While the minimum plywood thickness for 16-inch stud spacing is $\frac{5}{16}$ inch, it is often desirable to use $\frac{3}{8}$ inch and thicker, especially when the exterior finish must be nailed to the sheathing. The selection of plywood thickness is also influenced somewhat by standard jamb widths in window and exterior door frames. This may occasionally require sheathing of $\frac{1}{2}$-inch or greater thicknesses. Some modification of jambs is required and readily accomplished when other plywood thicknesses are used.

Structural Insulating Board Sheathing

The three common types of insulating board (structural fiberboards) used for sheathing include *regular density*, *intermediate density*, and *nail-base*. Insulating board sheathings are coated or impregnated with asphalt or given other treatment to provide a water-resistant product. Occasional wetting and drying that occur during construction will not damage the sheathing materially.

Regular-density sheathing is manufactured in $\frac{1}{2}$- and $\frac{25}{32}$-inch thicknesses and in 2- by 8-, 4- by 8-, and 4- by 9-foot sizes. Intermediate-density and nail-base sheathing are denser products than regular-density. They are regularly manufactured only in $\frac{1}{2}$-inch thickness and in 4- by 8- and 4- by 9-foot sizes. While 2- by 8-foot sheets with matched edges are used horizontally, 4- by 8-foot and longer sheets are usually installed with the long dimension vertical.

Corner bracing is required on horizontally applied sheets and usually on applications of $\frac{1}{2}$-inch regular-density sheathing applied vertically. Additional corner bracing is usually not required for regular-density insulating board sheathing $\frac{25}{32}$ inch thick or for intermediate-density and nail-base sheathing when properly applied (fig. 31.2) with long edges vertical. Naturally fastenings must be adequate around the perimeter and at intermediate studs, and adequately fastened (nails, staples, or other fastening system). Nail-base sheathing also permits shingles to be applied directly to it as siding if fastened with special annular-grooved nails. Galvanized or other corrosion-resistant fasteners are recommended for installation of insulating-board sheathing.

Gypsum Sheathing

Gypsum sheathing is $\frac{1}{2}$ inch thick, 2 by 8 feet in size, and is applied horizontally for stud spacing of 24 inches or less (fig. 31.3). It is composed of treated gypsum filler faced on two sides with water-resistant paper, often having one edge grooved, and the other with a matched *V* edge. This makes application easier, adds a small amount of tie between sheets, and provides some resistance to air and moisture penetration.

Corner Bracing

The purpose of corner bracing is to provide rigidity to the structure and to resist the racking forces of wind. Corner bracing should be used at all external corners of houses where the type of sheathing used does not provide the bracing required (figs. 29.1 and 31.1a). Types of sheathing that provide adequate bracing are: (a) Wood sheathing, when applied diagonally; (b) plywood, when applied vertically in sheets 4 feet wide by 8 or more feet high and where attached with nails or staples spaced not more than 6 inches apart on all edges and not more than 12 inches at intermediate supports; and (c) structural insulating board sheathing 4 feet wide and 8 feet or longer ($25\frac{}{}/_{32}$-inch-thick regular grade and $1/2$-inch-thick intermediate-density or nail-base grade) applied with long edges vertical with nails or staples spaced 3 inches along all edges and 6 inches at intermediate studs.

Another method of providing the required rigidity and strength for wall framing consists of a $1/2$-inch plywood panel at each side of each outside corner and $1/2$-inch regular-density fiberboard at intermediate areas. The plywood must be in 4-foot-wide sheets and applied vertically with full perimeter and intermediate stud nailing.

Where corner bracing is required, use 1- by 4-inch or wider members let into the outside face of the studs, and set at an angle of 45° from the bottom of the soleplate to the top of the wallplate or corner stud. Where window openings near the corner interfere with 45° braces, the angle should be increased but the full-length brace should cover at least three stud spaces. Tests conducted at the Forest Products Laboratory showed a full-length brace to be much more effective than a K-brace, even though the angle was greater than that of a 45° K-brace.

Installation of Sheathing

Wood Sheathing

The minimum thickness of wood sheathing is generally $3/4$ inch. However, for particular uses, depending on exterior coverings, resawn boards of $11\frac{}{}/_{16}$-inch thickness may be used as sheathing. Widths commonly used are 6, 8, and 10 inches. The 6- and 8-inch widths will have less shrinkage than greater widths, so that smaller openings will occur between boards.

The boards should be nailed at each stud crossing with two nails for the 6- and 8-inch widths and three nails for the 10- and 12-inch widths. When diagonal sheathing is used, one more nail can be used at each stud; for example, three nails for 8-inch sheathing. Joints should be placed over the center of studs (fig. 31.1a) unless end-matched (tongued-and-grooved) boards are used. End-matched tongued-and-grooved boards are applied continuously, either horizontally or diagonally, allowing end joints to fall where they may, even if between studs (fig. 31.1a). However, when end-matched boards are used, no two adjoining boards should have end joints over the same stud space and each board should bear on at least two studs.

Two arrangements of floor framing and soleplate location may be used which affect wall sheathing application. The first method has the soleplate set in from the outside wall line so that the sheathing is flush with the floor framing (fig. 31.1b). This does not provide a positive tie between wall and floor framing and in high wind areas should be supplemented with metal strapping (fig. 29.2) placed over the sheathing. The second method has the sill plate located the thickness of the sheathing in from the edge of the foundation wall (fig. 31.1c). When vertically applied plywood or diagonal wood sheathing is used, a good connection between the wall and floor framing is obtained. This method is usually preferred where good wall-to-floor-to-foundation connections are desirable.

Wood sheathing (fig. 31.1a) is commonly applied horizontally because it is easy to apply and there is less lumber waste than with diagonal sheathing. Horizontal sheathing, however, requires diagonal corner bracing for wall framework.

Diagonal sheathing (fig. 31.1a) should be applied at a 45° angle. This method of sheathing adds greatly to the rigidity of the wall and eliminates the need for corner bracing. There is more lumber waste than with horizontal sheathing because of angle cuts, and application is somewhat more difficult. End joints should be made over studs. This method is often specified in hurricane areas along the Atlantic Coast and in Florida.

Structural Insulating Board, Plywood, and Other Sheathing in 4-foot and Longer Sheets

Vertical application of structural insulating board (fig. 31.2) in 4- by 8-foot sheets is usually recommended by the manufacturer because perimeter nailing is possible. Depending on local building regulations, spacing nails 3 inches on edges and 6 inches at intermediate framing members usually eliminates the need for corner bracing when $25\frac{}{}/_{32}$-inch structural insulating board sheathing or $1/2$-inch medium-density structural insulating board sheathing is used. Use $1\frac{3}{4}$-inch galvanized roofing nails for the $25\frac{}{}/_{32}$-inch sheathing and $1\frac{1}{2}$-inch nails for the $1/2$-inch sheathing (table 2). The manufacturers usually recommend $1/8$-inch spacing between sheets. Joints are centered on framing members.

Plywood used for sheathing should be 4 by 8 feet or longer and applied vertically with perimeter nailing to eliminate the need for corner bracing (fig. 31.2). Sixpenny nails are used for plywood $3/8$ inch or less in thickness. Use eightpenny nails for plywood $1/2$ inch and more in thickness. Spacing should be a minimum of 6 inches at all edges and 12 inches at intermediate framing members (table 2).

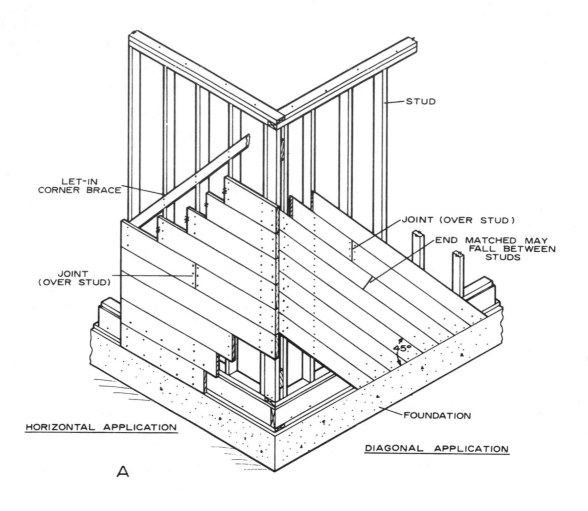

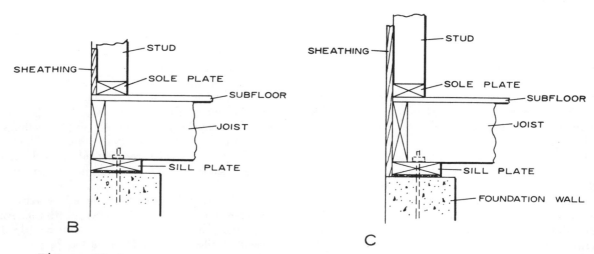

Figure 31.1.—Application of wood sheathing: A, Horizontal and diagonal; B, started at subfloor;
C, started at foundation wall.

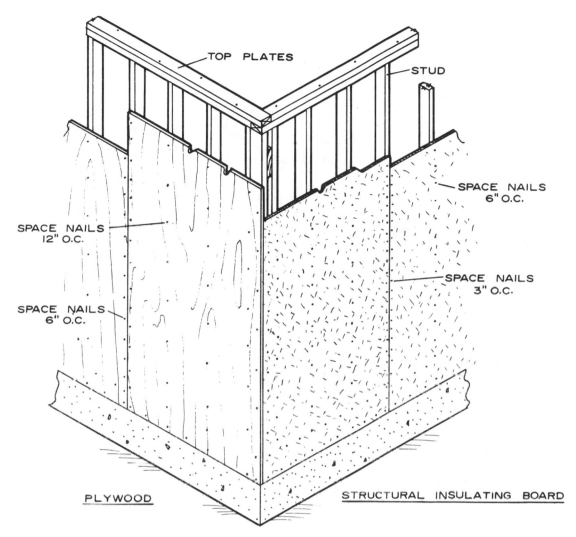

TOP PLATES

STUD

SPACE NAILS
6" O.C.

SPACE NAILS
3" O.C.

SPACE NAILS
12" O.C.

SPACE NAILS
6" O.C.

PLYWOOD

STRUCTURAL INSULATING BOARD

Figure 31.2.—Vertical application of plywood or structural insulating board sheathing.

Plywood may also be applied horizontally, but not being as efficient from the standpoint of rigidity and strength, it normally requires diagonal bracing. However, blocking between studs to provide for horizontal edge nailing will improve the rigidity and usually eliminate the need for bracing. When shingles or similar exterior finishes are employed, it is necessary to use threaded nails for fastening when plywood is only $\frac{5}{16}$ or $\frac{3}{8}$-inch thick. Allow $\frac{1}{8}$-inch edge spacing and $\frac{1}{16}$-inch end spacing between plywood sheets when installing.

Particleboard, hardboard, and other sheet materials may also be used as a sheathing. However, their use is somewhat restricted because cost is usually substantially higher than the sheet materials previously mentioned.

Insulating Board and Gypsum Sheathing in 2-by 8-foot Sheets

Gypsum and insulating board sheathing in 2- by 8-foot sheets applied horizontally require corner bracing (fig. 31.3). Vertical joints should be staggered. The $\frac{25}{32}$-inch board should be nailed to each crossing stud with $1\frac{3}{4}$-inch galvanized roofing nails spaced about $4\frac{1}{2}$ inches apart (six nails in the 2-foot height).

The $\frac{1}{2}$-inch gypsum and insulating board sheathing should be nailed to the framing members with $1\frac{1}{2}$-inch galvanized roofing nails spaced about $3\frac{1}{2}$ inches apart (seven nails in the 2-foot height).

When wood bevel or similar sidings are used over plywood sheathing less than $\frac{5}{8}$ inch thick, and over insulating board and gypsum board, nails must usually be located so as to contact the stud. When wood

503

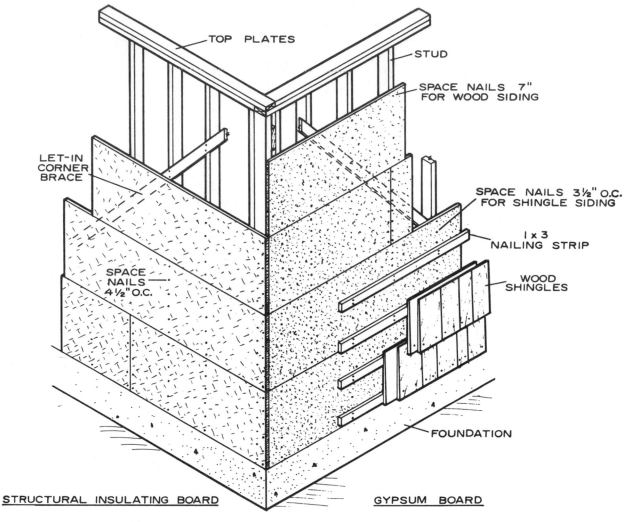

LET-IN CORNER BRACE

TOP PLATES

STUD

SPACE NAILS 7"
FOR WOOD SIDING

SPACE NAILS 3½" O.C.
FOR SHINGLE SIDING

1 x 3
NAILING STRIP

WOOD
SHINGLES

SPACE
NAILS
4½" O.C.

FOUNDATION

STRUCTURAL INSULATING BOARD   GYPSUM BOARD

Figure 31.3.—Horizontal application of 2-by 8-foot structural insulating board or gypsum sheathing.

shingles and similar finishes are used over gypsum and regular density insulating board sheathing, the walls are stripped with 1- by 3-inch horizontal strips spaced to conform to the shingle exposure. The wood strips are nailed to each stud crossing with two eightpenny or tenpenny threaded nails, depending on the sheathing thickness (fig. 31.3). Nail-base sheathing board usually does *not* require stripping when threaded nails are use.

Sheathing Paper

Sheathing paper should be water-resistant but not vapor-resistant. It is often called "breathing" paper as it allows the movement of water vapor but resists

entry of direct moisture. Materials such as 15-pound asphalt felt, rosin, and similar papers are considered satisfactory. Sheathing paper should have a "perm" value of 6.0 or more (4),(17). It also serves to resist air infiltration.

Sheathing paper should be used behind a stucco or masonry veneer finish and over wood sheathing. It should be installed horizontally starting at the bottom of the wall. Succeeding layers should lap about 4 inches. Ordinarily, it is not used over plywood, fiberboard, or other sheet materials that are water-resistant. However, 8-inch or wider strips of sheathing paper should be used around window and door openings to minimize air infiltration.

CHAPTER 32

ROOF SHEATHING

Roof sheathing is the covering over the rafters or trusses and usually consists of nominal 1-inch lumber or plywood. In some types of flat or low-pitched roofs with post and beam construction, wood roof planking or fiberboard roof decking might be used. Diagonal wood sheathing on flat or low-pitched roofs provides racking resistance where areas with high winds demand added rigidity. Plywood sheathing provides the same desired rigidity and bracing effect. Sheathing should be thick enough to span between supports and provide a solid base for fastening the roofing material.

Roof sheathing boards are generally the third grades of species such as the pines, redwood, the hemlocks, western larch, the firs, and the spruces (5). It is important that thoroughly seasoned material be used with asphalt shingles. Unseasoned wood will dry out and shrink in width causing buckling or lifting of the shingles, along the length of the board. Twelve percent is a desirable maximum moisture content for wood sheathing in most parts of the country. Plywood for roofs is commonly standard sheathing grade.

Lumber Sheathing

Closed Sheathing

Board sheathing to be used under such roofing as asphalt shingles, metal-sheet roofing, or other materials that require continuous support should be laid closed (without spacing) (fig. 32.1) . Wood shingles can also be used over such sheathing. Boards should be matched, shiplapped, or square-edged with joints made over the center of rafters. Not more than two adjacent

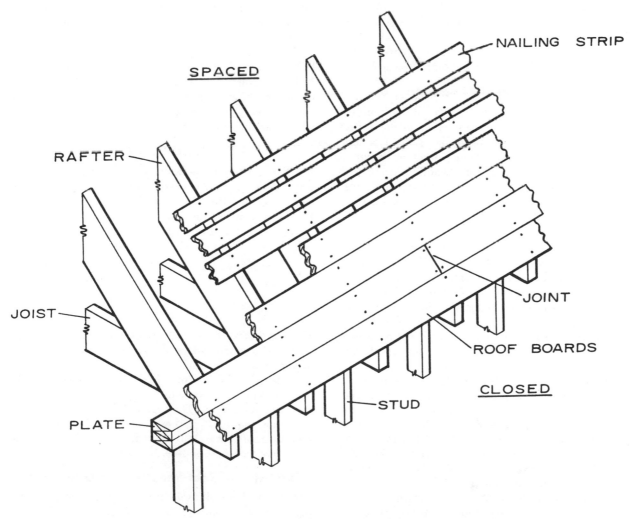

Figure 32.1.—Installation of board roof sheathing, showing both closed and spaced types.

505

boards should have joints over the same support. It is preferable to use boards no wider than 6 or 8 inches to minimize problems which can be caused by shrinkage. Boards should have a minimum thickness of ¾ inch for rafter spacing of 16 to 24 inches, and be nailed with two eightpenny common or sevenpenny threaded nails for each board at each bearing. End-matched tongued-and-grooved boards can also be used and joints made between rafters. However, in no case should the joints of adjoining boards be made over the same rafter space. Each board should be supported by at least two rafters.

Use of long sheathing boards at roof ends is desirable to obtain good framing anchorage, especially in gable roofs where there is a substantial rake overhang.

Spaced Sheathing

When wood shingles or shakes are used in damp climates, it is common to have spaced roof boards (fig. 32.1).Wood nailing strips in nominal 1- by 3- or 1- by 4-inch size are spaced the same distance on centers as the shingles are to be laid to the weather. For example, if shingles are laid 5 inches to the weather and nominal (surfaced) 1- by 4-inch strips are used, there would be spaces of 1⅜ to 1½ inches between each board to provide the needed ventilation spaces.

Plywood Roof Sheathing

When plywood roof sheathing is used, it should be laid with the face grain perpendicular to the rafters (fig. 32.2). Standard sheathing grade plywood is commonly specified but, where damp conditions occur, it is desirable to use a standard sheathing grade with exterior glueline. End joints are made over the center of the rafters and should be staggered by at least one rafter 16 or 24 inches, or more.

For wood shingles or shakes and for asphalt shingles, ⁵⁄₁₆-inch-thick plywood is considered to be a minimum thickness for 16-inch spacing of rafters. When edges are blocked to provide perimeter nailing, ⅜-inch-thick plywood can be used for 24-inch rafter spacing. A system which reduces costs by eliminating the blocking is acceptable in most areas for ⅜-inch plywood when rafters are spaced 24 inches on center. This is with the use of plyclips or similar H clips between rafters instead of blocking.

To provide better penetration for nails used for the shingles, better racking resistance, and a smoother roof appearance, it is often desirable to increase the minimum thicknesses to ⅜ and ½ inch. U.S. Product Standard PS 1–66 (18) provides that standard grades be marked for allowable spacing of rafters. For slate and similar heavy roofing materials, ½-inch plywood is considered minimum for 16-inch rafter spacing.

Plywood should be nailed at each bearing, 6 inches on center along all edges and 12 inches on center along intermediate members. A sixpenny common

nail or fivepenny threaded nail should be used for ⁵⁄₁₆- and ⅜-inch plywood, and eightpenny common or sevenpenny threaded nail for greater thicknesses. Unless plywood has an exterior glueline, raw edges should not be exposed to the weather at the gable end or at the cornice, but should be protected by the trim. Allow a ⅛-inch edge spacing and ¹⁄₁₆-inch end spacing between sheets when installing.

Plank Roof Decking

Plank roof decking, consisting of 2-inch and thicker tongued-and-grooved wood planking, is commonly used in flat or low-pitched roofs in post and beam construction. Common sizes are nominal 2- by 6-, 3- by 6-, and 4- by 6-inch V-grooved members, the thicker planking being suitable for spans up to 10 or 12 feet. Maximum span for 2-inch planking is 8 feet when continuous over two supports, and 6 feet over single spans in grades and species commonly used for this purpose. Special load requirements may reduce these allowable spans. Roof decking can serve both as an interior ceiling finish and as a base for roofing. Heat loss is greatly reduced by adding fiberboard or other rigid insulation over the wood decking.

The decking is blind-nailed through the tongue and also face-nailed at each support. In 4- by 6-inch size, it is predrilled for edge nailing Fig. 30.4b . For thinner decking, a vapor barrier is ordinarily installed between the top of the plank and the roof insulation when planking does not provide sufficient insulation.

Fiberboard Roof Decking

Fiberboard roof decking is used the same way as wood decking, except that supports are spaced much closer together. Planking is usually supplied in 2- by 8-foot sheets with tongued-and-grooved edges. Thicknesses of the plank and spacing of supports ordinarily comply with the following tabulation:

Minimum thickness (In.)	Maximum joist spacing (In.)
1½	24
2	32
3	48

Manufacturers of some types of roof decking recommend the use of 1⅞-inch thickness for 48-inch spacing of supports.

Nails used to fasten the fiberboard to the wood members are corrosion-resistant and spaced not more than 5 inches on center. They should be long enough to penetrate the joist or beam at least 1½ inches. A built-up roof is normally used for flat and low-pitched roofs having wood or fiberboard decking.

Extension of Roof Sheathing at Gable Ends

Method of installing board or plywood roof sheathing at the gable ends of the roof is shown in fig. 32.3 Where the gable ends of the house have little or no

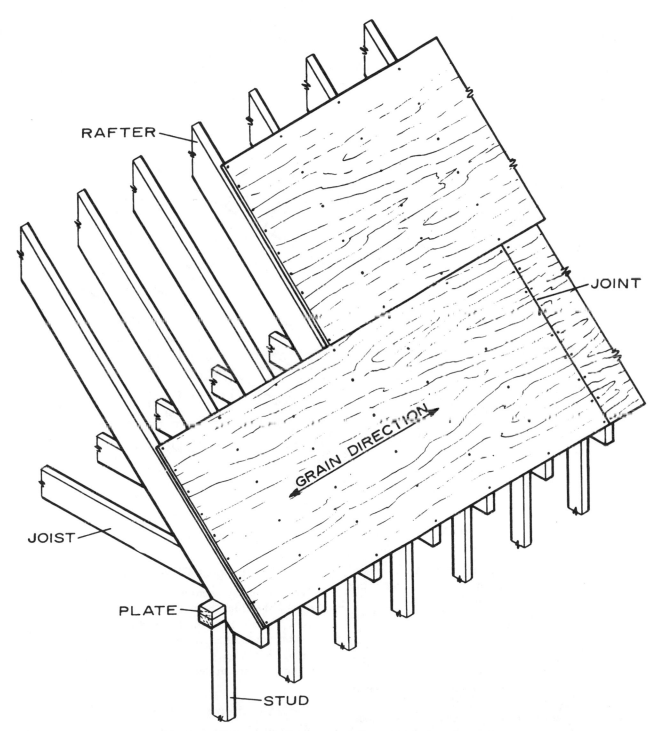

RAFTER

JOINT

GRAIN DIRECTION

JOIST

PLATE

STUD

Figure 32.2.—Application of plywood roof sheathing.

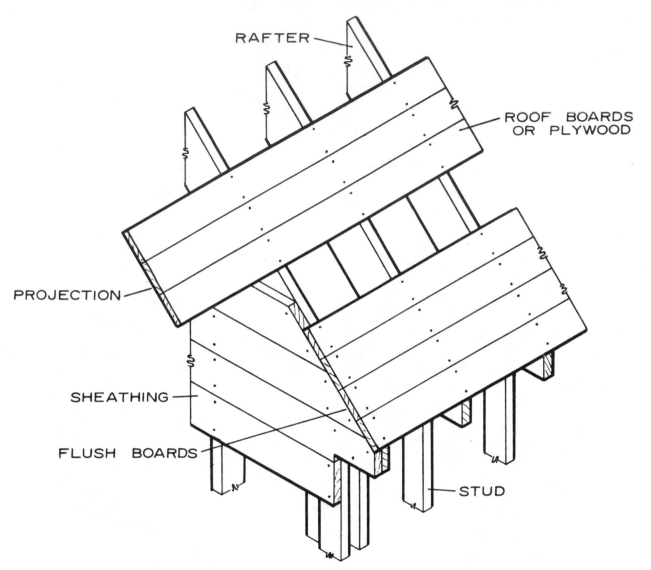

Figure 32.3.—Board roof sheathing at ends of gable.

extension (rake projection), roof sheathing is placed flush with the outside of the wall sheathing. (See Chapter 33, "Exterior Trim and Millwork.")

Roof sheathing that extends beyond end walls for a projected roof at the gables should span not less than three rafter spaces to insure anchorage to the rafters and to prevent sagging (fig. 32.3.) .When the projection is greater than 16 to 20 inches, special ladder framing is used to support the sheathing, as described in Chapter 33, "Exterior Trim and Millwork."

Plywood extension beyond the end wall is usually governed by the rafter spacing to minimize waste. Thus, a 16-inch rake projection is commonly used when rafters are spaced 16 inches on center. Butt joints of the plywood sheets should be alternated so they do not occur on the same rafter.

Sheathing at Chimney Openings

Where chimney openings occur within the roof area, the roof sheathing and subfloor should have a clearance of ¾ inch from the finished masonry on all sides (fig. 32.4, sec. A–A). Rafters and headers around the opening should have a clearance of 2 inches from the masonry for fire protection.

Sheathing at Valleys and Hips

Wood or plywood sheathing at the valleys and hips should be installed to provide a tight joint and should be securely nailed to hip and valley rafters (fig. 32.4). This will provide a solid and smooth base for metal flashing.

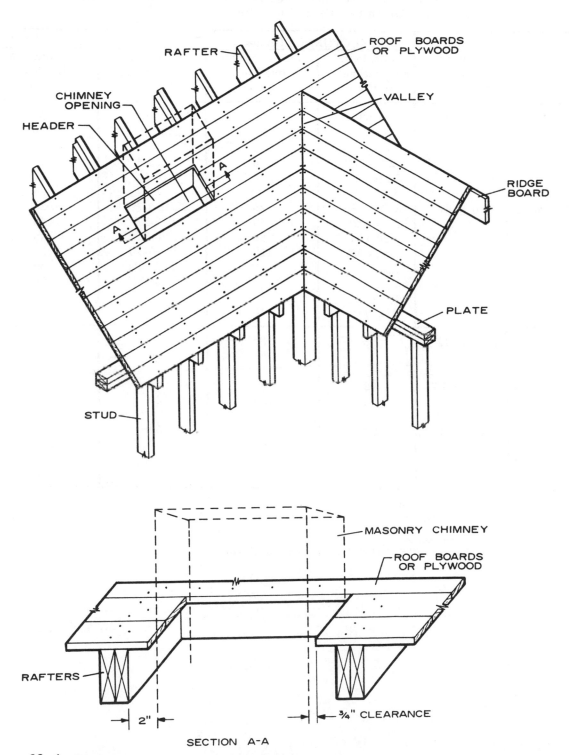

RAFTER

ROOF BOARDS
OR PLYWOOD

CHIMNEY
OPENING

HEADER

VALLEY

A

A

RIDGE
BOARD

PLATE

STUD

MASONRY CHIMNEY

ROOF BOARDS
OR PLYWOOD

RAFTERS

2"

¾" CLEARANCE

SECTION A-A

*Figure 32.4.—*Board roof sheathing detail at valley and chimney openings. Section A-A shows clearance from masonry.

Allowable Live Loads for Plywood Roof Sheathing with Face Grain Parallel to Supports.

Surface	Number & length of spans	STRUCTURAL I Grades		Grades other than STRUCT I	
		Ident. Index and thickness (inches)	Max. allowable live load (psf)	Ident. Index and thickness (inches)	Max. allowable live load (psf)
Unsanded	Four @ 12"	24/0-3/8	35	24/0-1/2* 24/0-1/2**	50 160
	Three @ 16"	32/16-1/2** 32/16-1/2 +	115 75	24/0-1/2** 32/16-1/2**	65 70
	Two @ 24"	32/16-1/2 + 32/16-1/2** 42/20-5/8 48/24-3/4	25 40 80 110	32/16-1/2** ‡ 32/16-5/8** 42/20-5/8** 42/20-3/4 48/24-3/4 48/24-7/8	25 45 45 50 50 120
	One @ 48"	1 1/8	45	2-4-1	25
Sanded#	Four @ 12"	3/8	80	3/8 1/2	50 195
	Three @ 16"	3/8 1/2	30 135	1/2 5/8	80 190
	Two @ 24"	1/2 5/8	45 105	5/8 3/4	65 115
	One @ 48"	1 1/8	55	1 1/8	30

Three-ply and four-ply construction.
**Five-ply construction, only
+Four-ply construction
‡Solid blocking recommended.
#Sanded panels other than STRUCTURAL I assumed to be Group 1.

CHAPTER 33

EXTERIOR TRIM AND MILLWORK

Exterior trim is usually considered as being that part of the exterior finish other than the wall covering. It includes such items as window and door trim, cornice moldings, facia boards and soffits, rake or gable-end trim, and porch trim and moldings. Contemporary house designs with simple cornices and moldings will contain little of this material, while traditionally designed houses will have considerably more. Much of the exterior trim, in the form of finish lumber and moldings, is cut and fitted on the job. Other materials or assemblies such as shutters, louvres, railings, and posts are shop-fabricated and arrive on the job ready to be fastened in place.

Material Used for Trim

The properties (5) desired in materials used for trim are good painting and weathering characteristics, easy working qualities, and maximum freedom from warp. Decay resistance is also desirable where materials may absorb moisture in such areas as the caps and the bases of porch columns, rails, and shutters. Heartwood of the cedars, cypress, and redwood has high decay resistance. Less durable species may be treated to make them decay resistant.

Many manufacturers predip at the factory such materials as siding, window sash, window and door frames, and trim using a water-repellent preservative. On-the-job dipping of end joints or miters cut at the building site is recommended when resistance to water entry and increased protection are desired.

Fastenings used for trim, whether nails or screws, should preferably be rust-resistant, i.e., galvanized, stainless steel, aluminum, or cadmium-plated. When a natural finish is used, nails should be stainless steel or aluminum to prevent staining and discoloration. Cement-coated nails are not rust-resistant.

Siding and trim are normally fastened in place with a standard siding nail, which has a small flat head. However, finish or casing nails might also be used for some purposes. If not rust-resistant, they should be set below the surface and puttied after the prime coat of paint has been applied. Most of the trim along the shingle line, such as at gable ends and cornices, is installed before the roof shingles are applied.

Material used for exterior trim should be of the better grade (5). Moisture content should be approximately 12 percent, except in the dry Southwestern States, where it should average about 9 percent.

Cornice Construction

The *cornice* of a building is the projection of the roof at the eave line that forms a connection between the roof and sidewalls. In gable roofs it is formed on each side of the house, and in hip roofs it is continuous around the perimeter. In flat or low-pitched roof designs, it is usually formed by the extension of the ceiling joists which also serve as rafters.

The three general cornice types might be considered to be the *box*, the *close* (no projection), and the *open*. The box cornice is perhaps the most commonly used in house design and not only presents a finished appearance, but also aids in protecting the sidewalls from rain. The close cornice with little overhang does not provide as much protection. The open cornice may be used in conjunction with exposed laminated or solid beams with wood roof decking and wide overhangs in contemporary or rustic designs or to provide protection to side walls at a reasonable cost in low-cost houses.

Narrow Box Cornice

The narrow box cornice is one in which the projection of the rafter serves as a nailing surface for the soffit board as well as the facia trim (fig. 33.1) Depending on the roof slope and the size of the rafters, this extension may vary between 6 and 12 or more inches. The soffit provides a desirable area for inlet ventilators. (See Chapter 42, "Ventilation.")

A frieze board or a simple molding is often used to terminate the siding at the top of the wall. Some builders slope the soffit slightly outward, leaving a $\frac{1}{4}$-inch open space behind the facia for drainage of water that might enter because of snow and ice dams on the overhang. However, good attic ventilation and proper cornice ventilators, in addition to good insulation, will minimize ice dams under normal conditions.

Wide Box Cornice (With Lookouts)

A wide box cornice normally requires additional members for fastening the soffit. This is often supplied by lookout members which can be toenailed to the wall and face-nailed to the ends of the rafter extensions (fig. 33.2). Soffit material is often lumber, plywood, paper-overlaid plywood, hardboard, medium-density fiberboard, or other sheet materials. Thicknesses should be based on the distance between supports, but $\frac{3}{8}$-inch plywood and $\frac{1}{2}$-inch fiberboard are often used for 16-inch rafter spacing. A nailing header at the ends of the joists will provide a nailing area for soffit and facia trim. The nailing header is sometimes eliminated in moderate cornice extensions when a rabbeted facia is used. Inlet ventilators, often narrow

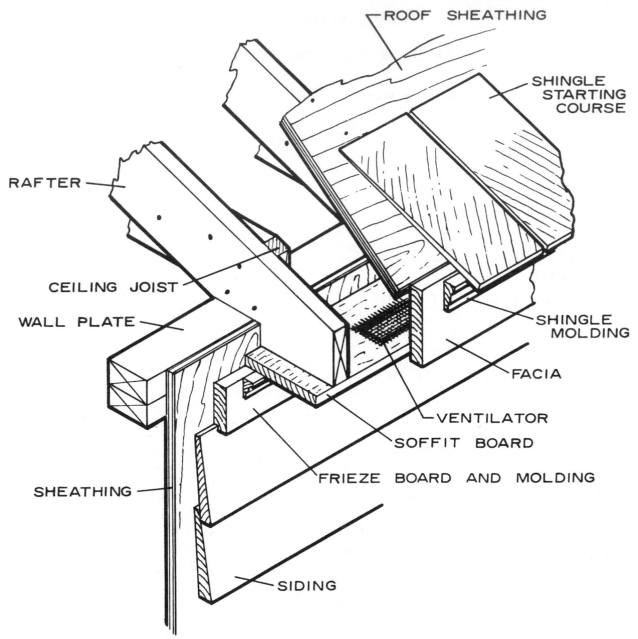

ROOF SHEATHING

SHINGLE STARTING COURSE

RAFTER

CEILING JOIST

WALL PLATE

SHINGLE MOLDING

FACIA

VENTILATOR

SOFFIT BOARD

SHEATHING

FRIEZE BOARD AND MOLDING

SIDING

Figure 33.1.—Narrow box cornice.

continuous slots, can be installed in the soffit areas. This type cornice is often used in a hip-roofed house.

The projection of the cornice beyond the wall should not be so great as to prevent the use of a narrow frieze or a frieze molding above the top casing of the windows. A combination of a steeper slope and wide projection will bring the soffit in this type of cornice too low, and a box cornice, without the lookouts, should be used.

Boxed Cornice Without Lookouts

A wide boxed cornice without lookouts provides a sloped soffit and is sometimes used for houses with wide overhangs (fig. 33.3) The soffit material is nailed directly to the underside of the rafter extensions. In gable houses, this sloping soffit extends around the roof extension at each gable end. Except for elimination of the lookout members, this type of cornice is much the same as the wide box cornice previously described. Inlet ventilators, singly or a continuous screened slot, are installed in the soffit area.

Open Cornice

The open cornice is much the same structurally as the wide box cornice without lookouts (fig.33.3), except that the soffit is eliminated. It might be used on post

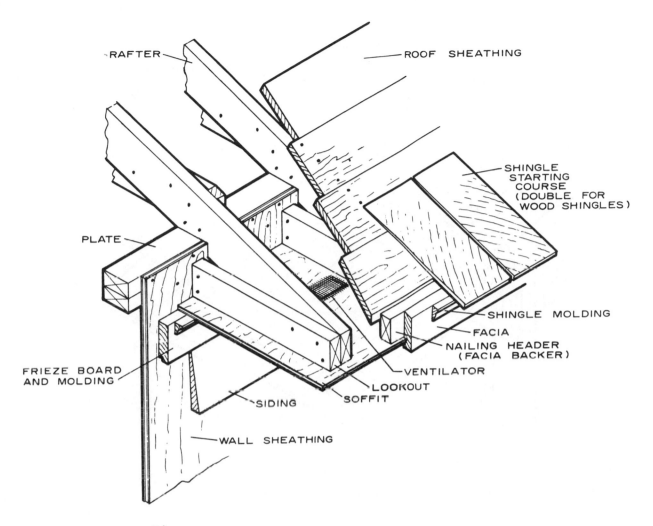

RAFTER

ROOF SHEATHING

SHINGLE STARTING COURSE (DOUBLE FOR WOOD SHINGLES)

PLATE

SHINGLE MOLDING

FACIA

NAILING HEADER (FACIA BACKER)

VENTILATOR

FRIEZE BOARD AND MOLDING

LOOKOUT

SOFFIT

SIDING

WALL SHEATHING

Figure 33.2.—Wide box cornice (with horizontal lookouts)

and beam construction in which spaced rafters extend beyond the wall line. In widely spaced rafters, the roof sheathing may consist of wood decking, the underside of which would be visible. When rafters are more closely spaced, paper-overlaid plywood or V-grooved boards might be used for roof sheathing at the overhanging section. This type of cornice might also be used for conventionally framed low-cost houses, utility buildings, or cottages, with or without a facia board.

Close Cornice

A close cornice is one in which there is no rafter projection beyond the wall (fig. 33.4). Sheathing is often carried to the ends of the rafters and ceiling joists. The roof is terminated only by a frieze board and shingle molding. While this cornice is simple to build, it is not too pleasing in appearance and does not provide much weather protection to the sidewalls or a

convenient area for inlet ventilators. Appearance can be improved somewhat by the use of a formed wood gutter.

Rake or Gable-end Finish

The *rake section* is the extension of a gable roof beyond the end wall of the house. This detail might be classed as being (a) a close rake with little projection or (b) a boxed or open extension of the gable roof, varying from 6 inches to 2 feet or more. Sufficient projection of the roof at the gable is desirable to provide some protection to the sidewalls. This usually results in longer paint life.

When the rake extension is only 6 to 8 inches, the facia and soffit can be nailed to a series of short lookout blocks (fig. 33.5a). In addition, the facia is further secured by nailing through the projecting roof sheath-

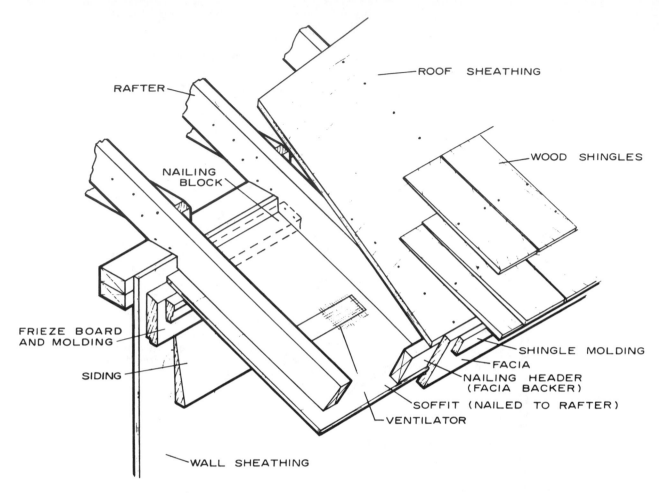

Figure 33.3.—**Wide box cornice (without lookouts)**

ing. A frieze board and appropriate moldings complete the construction.

In a moderate overhang of up to 20 inches, both the extending sheathing and a *fly rafter* aid in supporting the rake section (fig. 33.5b). The fly rafter extends from the ridge board to the nailing header which connects the ends of the rafters. The roof sheathing boards or the plywood should extend from inner rafters to the end of the gable projection to provide rigidity and strength.

The roof sheathing is nailed to the fly rafter and to the lookout blocks which aid in supporting the rake section and also serve as a nailing area for the soffit. Additional nailing blocks against the sheathing are sometimes required for thinner soffit materials.

Wide gable extensions (2 feet or more) require rigid framing to resist roof loads and prevent deflection of the rake section. This is usually accomplished by a series of *purlins* or lookout members nailed to a fly rafter at the outside edge and supported by the end wall and a doubled interior rafter (fig. 33.6 a and b).

This framing is often called a "ladder" and may be constructed in place or on the ground or other convenient area and hoisted in place.

When ladder framing is preassembled, it is usually made up with a header rafter on the inside and a fly rafter on the outside. Each is nailed to the ends of the lookouts which bear on the gable end wall. When the header is the same size as the rafter, be sure to provide a notch for the wall plates the same as for the regular rafters. In moderate width overhangs, nailing the header and fly rafter to the lookouts with supplemental toenailing is usually sufficiently strong to eliminate the need for the metal hangers shown in fig. 33. 6b. The header rafters can be face-nailed directly to the end rafters with twelvepenny nails spaced 16 to 20 inches apart.

Other details of soffit, facia, frieze board, and moldings can be similar to those used for a wide gable overhang. Lookouts should be spaced 16 to 24 inches apart, depending on the thickness of the soffit material.

A close rake has no extension beyond the end wall

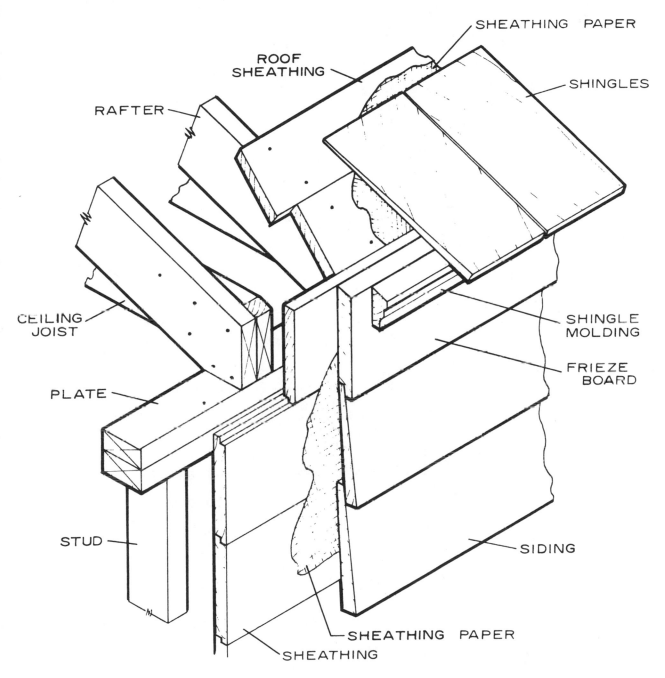

Figure 33.4.—Close cornice.

other than the frieze board and moldings. Some additional protection and overhang can be provided by using a 2- by 3- or 2- by 4-inch facia block over the sheathing (fig. 33.6c). This member acts as a frieze board, as the siding can be butted against it. The facia, often 1 by 6 inches, serves as a trim member. Metal roof edging is often used along the rake section as flashing.

Cornice Return

The *cornice return* is the end finish of the cornice on a gable roof. In hip roofs and flat roofs, the cornice is usually continuous around the entire house. In a gable house, however, it must be terminated or joined with the gable ends. The type of detail selected depends to a great extent on the type of cornice and the projection of the gable roof beyond the end wall.

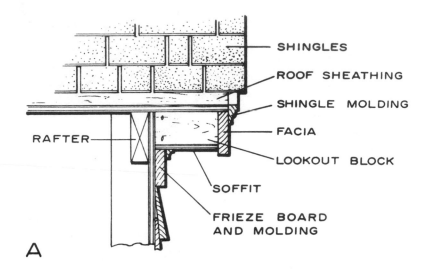

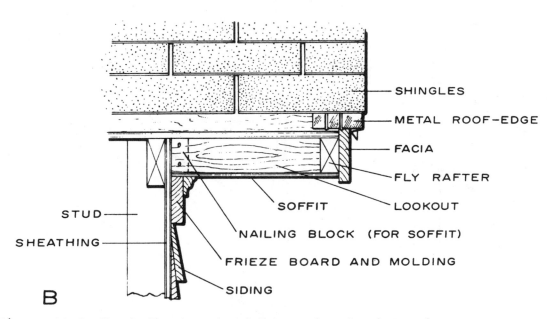

Figure 33.5.—Normal gable-end extensions: *A,* Narrow overhang; *B,* moderate overhang.

A narrow box cornice often used in houses with Cape Cod or colonial details has a boxed return when the rake section has some projection (fig. 33.7a). The facia board and shingle molding of the cornice are carried around the corner of the rake projection.

When a wide box cornice has no horizontal lookout members (fig. 33.3), the soffit of the gable-end overhang is at the same slope and coincides with the cornice soffit (fig. 33.7b). This is a simple system and is often used when there are wide overhangs at both sides and ends of the house.

A close rake (a gable end with little projection) may be used with a narrow box cornice or a close cornice. In this type, the frieze board of the gable end, into which the siding butts, joints the frieze board or facia of the cornice (fig. 33.7c).

While close rakes and cornices with little overhang are lower in cost, the extra material and labor required for good gable and cornice overhangs are usually justified. Better sidewall protection and lower paint maintenance costs are only two of the benefits derived from good roof extensions.

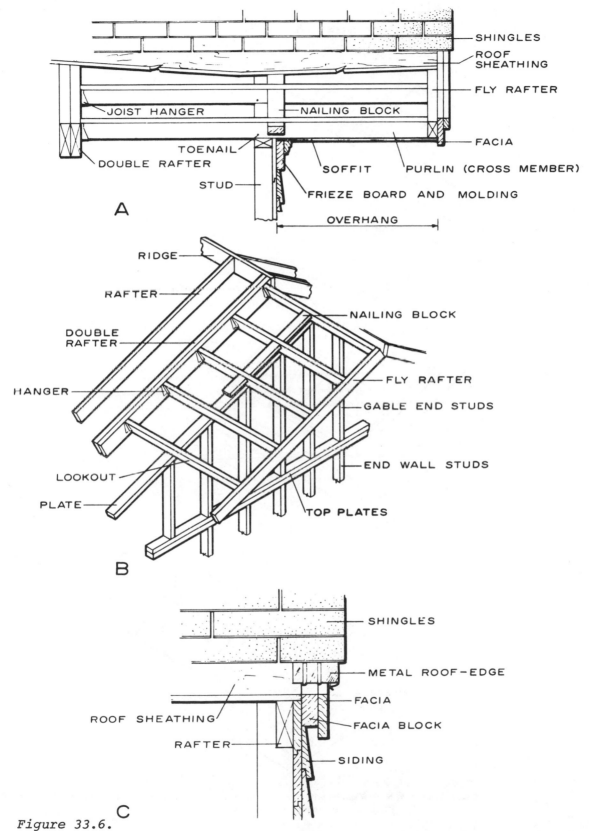

SHINGLES
ROOF SHEATHING
FLY RAFTER
JOIST HANGER
NAILING BLOCK
FACIA
TOENAIL
DOUBLE RAFTER
SOFFIT
PURLIN (CROSS MEMBER)
STUD
FRIEZE BOARD AND MOLDING
OVERHANG

A

RIDGE
RAFTER
NAILING BLOCK
DOUBLE RAFTER
HANGER
FLY RAFTER
GABLE END STUDS
END WALL STUDS
LOOKOUT
PLATE
TOP PLATES

B

SHINGLES
METAL ROOF-EDGE
FACIA
ROOF SHEATHING
FACIA BLOCK
RAFTER
SIDING

C

Figure 33.6.

Special gable-end extensions: *A*, Extra wide overhang; *B*, ladder framing for wide overhang; *C*, close rake.

517

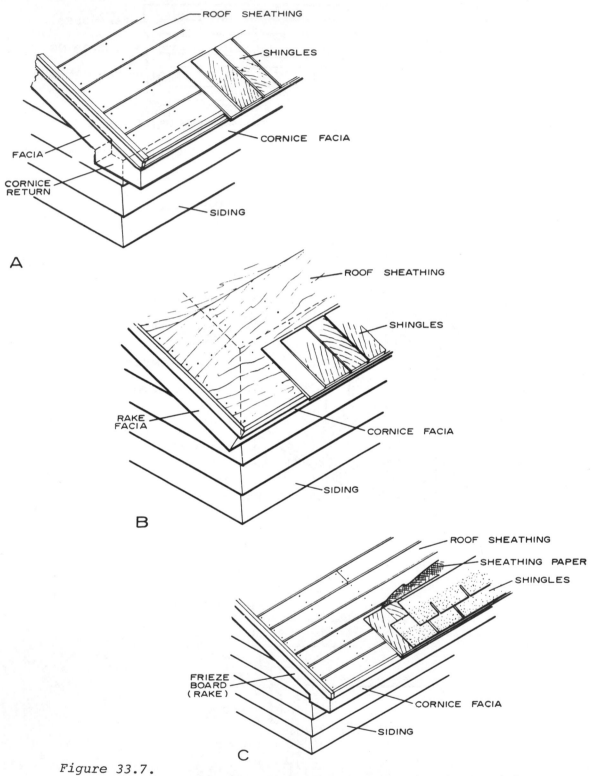

Figure 33.7.

Cornice returns: *A*, Narrow cornice with boxed return; *B*, wide overhang at cornice and rake; *C*, narrow box cornice and close rake.

518

CHAPTER 34

ROOF COVERINGS

Roof coverings should provide a long-lived water-proof finish that will protect the building and its contents from rain, snow, and wind. Many materials have withstood the test of time and have proved satisfactory under given service conditions.

Materials

Materials used for pitched roofs are wood, asphalt, and asbestos shingles, and also tile and slate. Sheet materials such as roll roofing, galvanized iron, aluminum, copper, and tin are also used. Perhaps the most common covering for flat or low-pitched roofs is the built-up roof with a gravel topping or cap sheet. Plastic films, often backed with an asbestos sheet, are also being applied on low-slope roofs. While these materials are relatively new, it is likely that their use will increase, especially for roofs with unusual shapes. However, the choice of roofing materials is usually influenced by first cost, local code requirements, house design, or preferences based on past experience.

In shingle application, the exposure distance is important and the amount of exposure generally depends on the roof slope and the type of material used. This may vary from a 5-inch exposure for standard size asphalt and wood shingles on a moderately steep slope to about 3½ inches for flatter slopes. However, even flatter slopes can be used for asphalt shingles with double underlay and triple shingle coverage. Built-up construction is used mainly for flat or low-pitched roofs but can be adapted to steeper slopes by the use of special materials and methods.

Roof underlay material usually consists of 15- or 30-pound asphalt-saturated felt and should be used in moderate and lower slope roofs covered with asphalt, asbestos, or slate shingles, or tile roofing. It is not commonly used for wood shingles or shakes. In areas where moderate to severe snowfalls occur, cornices without proper protection will often be plagued with ice dams (fig. 34.1a). These are formed when snow melts, runs down the roof, and freezes at the colder cornice area. Gradually, the ice forms a dam that backs up water under the shingles. Under these conditions, it is good practice to use an undercourse (36-in. width) of 45-pound or heavier smooth-surface roll roofing along the eave line as a flashing (fig. 34.1b). This will minimize the chance of water backing up and entering the wall. However, good attic ventilation and sufficient ceiling insulation are of primary importance in eliminating this harmful nuisance. These details are described in Chapter 16, "Ventilation."

Metal roofs (tin, copper, galvanized iron, or aluminum) are sometimes used on flat decks of dormers,

porches, or entryways. Joints should be watertight and the deck properly flashed at the juncture with the house. Nails should be of the same metal as that used on the roof, except that with tin roofs, steel nails may be used. All exposed nailheads in tin roofs should be soldered with a rosin-core solder.

Wood Shingles

Wood shingles of the types commonly used for house roofs are No. 1 grade. Such shingles (5) are all-heartwood, all-edgegrain, and tapered. Second grade shingles make good roofs for secondary buildings as well as excellent sidewalls for primary buildings. Western redcedar and redwood are the principal commercial shingle woods, as their heartwood has high decay resistance and low shrinkage.

Four bundles of 16-inch shingles laid 5 inches "to the weather" will cover 100 square feet. Shingles are of random widths, the narrower shingles being in the lower grades. Recommended exposures for the standard shingle sizes are shown in table 3.

TABLE 3.—*Recommended exposure for wood shingles*[1]

Shingle length	Shingle thickness (Green)	Maximum exposure	
		Slope less[2] than 4 in 12	Slope 5 in 12 and over
In.		*In.*	*In.*
16	5 butts in 2 in.	3¾	5
18	5 butts in 2¼ in.	4¼	5½
24	4 butts in 2 in.	5¾	7½

[1] As recommended by the Red Cedar Shingle and Handsplit Shake Bureau.
[2] Minimum slope for main roofs—4 in 12.
Minimum slope for porch roofs—3 in 12.

Fig. 34.2 illustrates the proper method of applying a wood-shingle roof. Underlay or roofing felt is not required for wood shingles except for protection in ice-dam areas. Spaced roof boards under wood shingles are most common, although spaced or solid sheathing is optional.

The following general rules should be followed in the application of wood shingles:

1. Shingles should extend about 1½ inches beyond the eave line and about ¾ inch beyond the rake (gable) edge.

2. Use two rust-resistant nails in each shingle; space them about ¾ inch from the edge and 1½ inches above the butt line of the next course. Use threepenny

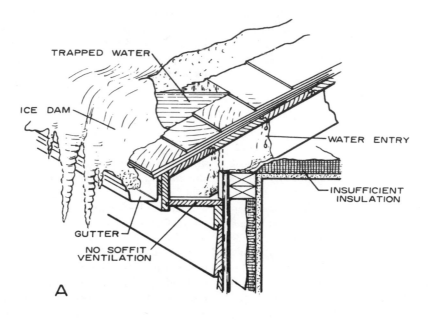

TRAPPED WATER

ICE DAM

WATER ENTRY

INSUFFICIENT
INSULATION

GUTTER

NO SOFFIT
VENTILATION

A

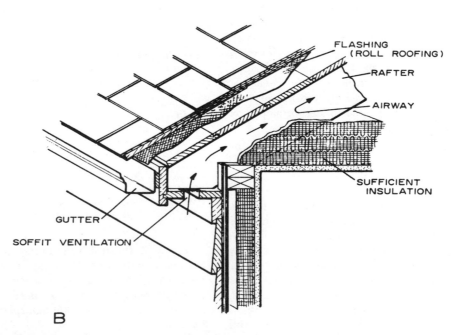

FLASHING
(ROLL ROOFING)

RAFTER

AIRWAY

SUFFICIENT
INSULATION

GUTTER

SOFFIT VENTILATION

B

Figure 34.1.—Snow and ice dams: *A,* Ice dams often build up on the overhang of roofs and
in gutters, causing melting snow water to back up under shingles and under the facia
board of closed cornices. Damage to ceilings inside and to paint outside results. *B,* Eave
protection for snow and ice dams. Lay smooth-surface 45-pound roll roofing on roof sheathing over the eaves extending upward well above the inside line of the wall.

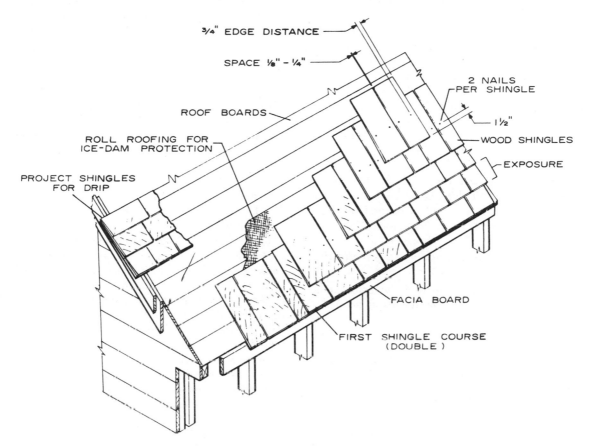

Figure 34.2.—Installation of wood shingles.

nails for 16- and 18-inch shingles and fourpenny for 24-inch shingles in new construction. A *ring-shank nail* (threaded) is often recommended for plywood roof sheathing less than ½ inch thick.

3. The first course of shingles should be doubled. In all courses, allow ⅛- to ¼-inch space between each shingle for expansion when wet. The joints between shingles should be offset at least 1½ inches from the joints between shingles in the course below. Further, the joints in succeeding courses should be spaced so that they do not directly line up with joints in the second course below.

4. When valleys are present, shingle away from the valleys, selecting and precutting wide valley shingles.

5. A metal edging along the gable end will aid in guiding the water away from the sidewalls.

6. In laying No. 1 all-heartwood edge-grain shingles no splitting of wide shingles is necessary.

Wood shakes are applied much the same as wood shingles. Because shakes are much thicker (longer shakes have the thicker butts), long galvanized nails are used. To create a rustic appearance, the butts are often laid unevenly. Because shakes are longer than shingles, they have a greater exposure. Exposure dis-

tance is usually 7½ inches for 18-inch shakes, 10 inches for 24-inch shakes, and 13 inches for 32-inch shakes. Shakes are not smooth on both faces, and because wind-driven snow might enter, it is essential to use an underlay between each course. An 18-inch-wide layer of 30-pound asphalt felt should be used between each course with the bottom edge positioned above the butt edge of the shakes a distance equal to double the weather exposure. A 36-inch wide starting strip of the asphalt felt is used at the eave line. Solid sheathing should be used when wood shakes are used for roofs in areas where wind-driven snow is experienced.

Asphalt Shingles

The usual minimum recommended weight for asphalt shingles is 235 pounds for square-butt strip shingles. This may change in later years, as 210 pounds (weight per square) was considered a minimum several years ago. Strip shingles with a 300-pound weight per square are available, as are lock-type and other shingles weighing 250 pounds and more. Asphalt shingles are also available with seal-type tabs for wind resistance. Many contractors apply a small spot of asphalt roof cement under each tab after installation of regular asphalt shingles to provide similar protection.

The square-butt strip shingle is 12 by 36 inches, has three tabs, and is usually laid with 5 inches exposed to the weather. There are 27 strips in a bundle, and three bundles will cover 100 square feet. Bundles should be piled flat for storage so that strips will not curl when the bundles are opened for use. The method of laying an asphalt-shingle roof is shown in figure 34.3a. A metal edging is often used at the gable end to provide additional protection (fig. 34.3b).

Data such as that in table 4 are often used in determining the need for and the method of applying *underlayment* for asphalt shingles on roofs of various slopes. Underlayment is commonly 15-pound saturated felt.

TABLE 4.—*Underlayment requirements for asphalt shingles*

(Headlap for single coverage of underlayment should be 2 inches and for double coverage 19 inches.)

Underlayment	Minimum roof slope	
	Double coverage [1] shingles	Triple coverage [1] shingles
Not required	7 in 12	[2] 4 in 12
Single	[2] 4 in 12	[3] 3 in 12
Double	2 in 12	2 in 12

[1] Double coverage for a 12- by 36-in. shingle is usually an exposure of about 5 in. and about 4 in. for triple coverage.
[2] May be 3 in 12 for porch roofs.
[3] May be 2 in 12 for porch roofs.

An asphalt-shingle roof can also be protected from ice dams by adding an initial layer of 45-pound or heavier roll roofing, 36 inches wide, and insuring good ventilation and insulation within the attic space (fig. 34.1b)

A course of wood shingles or a metal edging should be used along the eave line before application of the asphalt shingles. The first course of asphalt shingles is doubled; or, if desired, a starter course may be used under the first asphalt-shingle course. This first course should extend downward beyond the wood shingles (or edging) about ½ inch to prevent the water from backing up under the shingles. A ½-inch projection should also be used at the rake.

Several chalklines on the underlay will help aline the shingles so that tab notches will be in a straight line for good appearance. Each shingle strip should be fastened securely according to the manufacturer's directions. The use of six 1-inch galvanized roofing nails for each 12- by 36-inch strip is considered good practice in areas of high winds. A sealed tab or the use of asphalt sealer will also aid in preventing wind damage during storms. Some contractors use four

nails for each strip when tabs are sealed. When a nail penetrates a crack or knothole, it should be removed, the hole sealed, and the nail replaced in sound wood; otherwise, it will gradually work out and cause a hump in the shingle above it.

Built-up Roofs

Built-up roof coverings are installed by roofing companies that specialize in this work. Roofs of this type may have 3, 4, or 5 layers of roofer's felt, each mopped down with tar or asphalt, with the final surface coated with asphalt and covered with gravel embedded in asphalt or tar, or covered with a cap sheet. For convenience, it is customary to refer to built-up roofs as 10-, 15-, or 20-year roofs, depending upon the method of application.

For example, a 15-year roof over a wood deck (fig. 34.4a) may have a base layer of 30-pound saturated roofer's felt laid dry, with edges lapped and held down with roofing nails. All nailing should be done with either (a) roofing nails having ⅜-inch heads driven through 1-inch-diameter tin caps or (b) special roofing nails having 1-inch-diameter heads. The dry sheet is intended to prevent tar or asphalt from entering the rafter spaces. Three layers of 15-pound saturated felt follow, each of which is mopped on with hot tar rather than being nailed. The final coat of tar or asphalt may be covered with roofing gravel or a cap sheet of roll roofing.

The cornice or eave line of projecting roofs is usually finished with metal edging or flashing, which acts as a drip. A metal gravel strip is used in conjunction with the flashing at the eaves when the roof is covered with gravel (fig. 34.4b). Where built-up roofing is finished against another wall, the roofing is turned up on the wall sheathing over a cant strip and is often also flashed with metal (fig. 34.4c). This flashing is generally extended up about 4 inches above the bottom of the siding.

Other Roof Coverings

Other roof coverings, including asbestos, slate, tile, metal and others, many of which require specialized applicators, are perhaps less commonly used than wood or asphalt shingles and built-up roofs. Several new materials, such as plastic films and coatings, are showing promise for future moderate-cost roof coverings. However, most of them are more expensive than the materials now commonly being used for houses. These newer materials, however, as well as other new products, are likely to come into more general use during the next decade.

Finish at the Ridge and Hip

The most common type of ridge and hip finish for wood and asphalt shingles is known as the *Boston ridge*. Asphalt-shingle squares (one-third of a 12- by

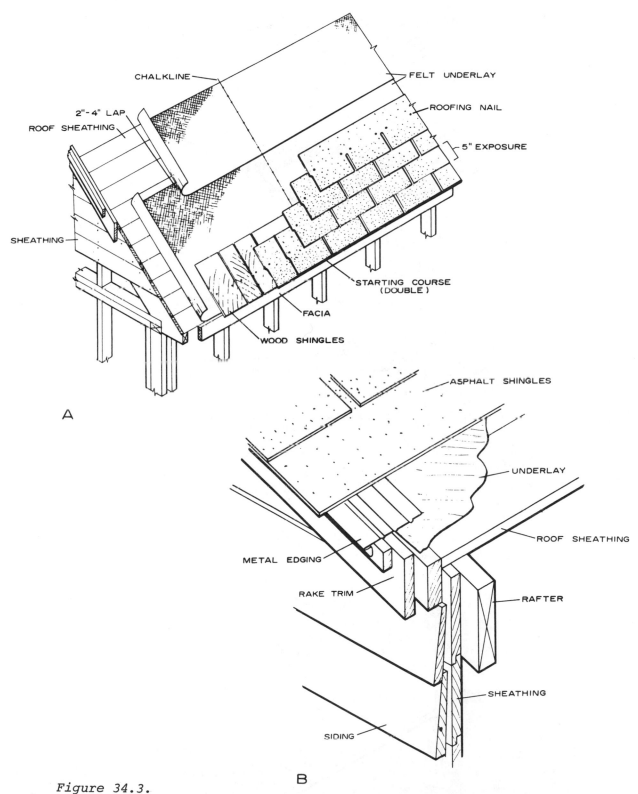

CHALKLINE

FELT UNDERLAY

ROOFING NAIL

2"-4" LAP

ROOF SHEATHING

5" EXPOSURE

SHEATHING

STARTING COURSE
(DOUBLE)

FACIA

WOOD SHINGLES

A

ASPHALT SHINGLES

UNDERLAY

ROOF SHEATHING

METAL EDGING

RAKE TRIM

RAFTER

SHEATHING

SIDING

B

Figure 34.3.

Application of asphalt shingles: *A*, Normal method with strip shingles; *B*, metal edging at gable end.

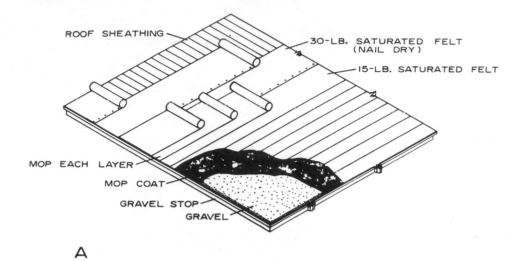

ROOF SHEATHING

30-LB. SATURATED FELT (NAIL DRY)

15-LB. SATURATED FELT

MOP EACH LAYER

MOP COAT

GRAVEL STOP

GRAVEL

A

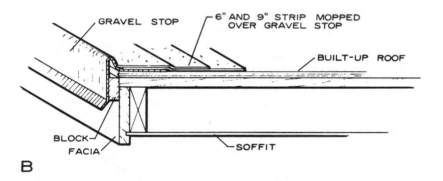

GRAVEL STOP

6" AND 9" STRIP MOPPED OVER GRAVEL STOP

BUILT-UP ROOF

BLOCK

FACIA

SOFFIT

B

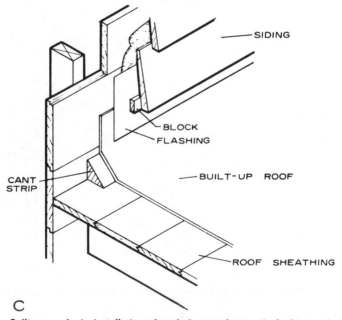

SIDING

BLOCK

FLASHING

BUILT-UP ROOF

CANT STRIP

ROOF SHEATHING

C

Figure 34.4.—Built-up roof: A, Installation of roof; B, gravel stop; C, flashing at building line.

36-inch strip) are used over the ridge and blind-nailed (fig. 34.5a). Each shingle is lapped 5 to 6 inches to give double coverage. In areas where driving rains occur, it is well to use metal flashing under the shingle ridge. The use of a ribbon of asphalt roofing cement under each lap will also greatly reduce the chance of water penetration.

A wood-shingle roof (fig. 34.5b) also should be finished in a Boston ridge. Shingles 6 inches wide are alternately lapped, fitted, and blind-nailed. As shown, the shingles are nailed in place so that exposed trimmed edges are alternately lapped. Pre-assembled hip and ridge units are available and save both time and money.

A metal ridge roll can also be used on asphalt-shingle or wood-shingle roofs (fig. 34.5c). This ridge is formed to the roof slope and should be copper, galvanized iron, or aluminum. Some metal ridges are formed so that they provide an outlet ventilating area. However, the design should be such that it prevents rain or snow blowing in.

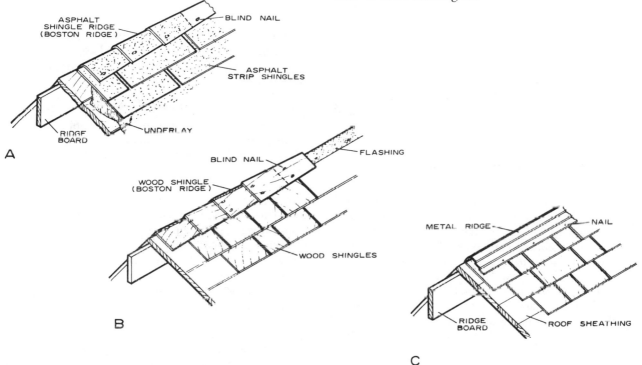

*Figure 34.5.—*Finish at ridge: A, Boston ridge with asphalt shingles; B, Boston ridge with wood shingles; C, metal ridge.

CHAPTER 35

EXTERIOR FRAMES, WINDOWS, AND DOORS

Windows, doors, and their frames are millwork items that are usually fully assembled at the factory. Window units, for example, often have the sash fitted and weatherstripped, frame assembled, and exterior casing in place. Standard combination storms and screens or separate units can also be included. Door frames are normally assembled ready for use in the building. All such wood components are treated with a water-repellent preservative at the factory to provide protection before and after they are placed in the walls.

Windows are mainly to allow entry of light and air, but may also be an important part of the architectural design. Some variation may occur, but normally in habitable rooms the glass area should be not less than 10 percent of the floor area. Natural ventilation should be not less than 4 percent of the floor area in a habitable room unless a complete air-conditioning system is used.

Types of Windows

Windows are available in many types, each having advantages. The principal types are double-hung, casement, stationary, awning, and horizontal sliding.

They may be made of wood or metal. Heat loss through metal frames and sash is much greater than through similar wood units. Glass blocks are sometimes used for admitting light in places where transparency or ventilation is not required.

Insulated glass, used both for stationary and moveable sash, consists of two or more sheets of spaced glass with hermetically-sealed edges. This type has more resistance to heat loss than a single thickness and is often used without a storm sash.

Wood sash and door and window frames should be made from a clear grade of all-heartwood stock of a decay-resistant wood species or from wood which is given a preservative treatment. Species commonly used include ponderosa and other pines, the cedars, cypress, redwood, and the spruces.

Tables showing glass size, sash size, and rough opening size are available at lumber dealers, so that the wall openings can be framed accordingly. Typical openings for double-hung windows are shown in the chapter "Wall Framing."

Double-hung Windows

The double-hung window is perhaps the most familiar window type. It consists of an upper and lower sash that slide vertically in separate grooves in the side jambs or in full-width metal weatherstripping (fig. 35.1). This type of window provides a maximum face opening for ventilation of one-half the total window area. Each sash is provided with springs, balances, or *compression weatherstripping* to hold it in place in any location. Compression weatherstripping, for example, prevents air infiltration, provides tension, and acts as a counterbalance; several types allow the sash to be removed for easy painting or repair.

The *jambs* (sides and top of the frames) are made of nominal 1-inch lumber; the width provides for use with dry-wall or plastered interior finish. Sills are made from nominal 2-inch lumber and sloped at about 3 in 12 for good drainage (fig. 35.1d). Sash are normally 1⅜ inches thick and wood combination storm and screen windows are usually 1⅛ inches thick.

Sash may be divided into a number of lights by small wood members called *muntins*. A ranch-type house may provide the best appearance with top and bottom sash divided into two horizontal lights. A colonial or Cape Code house usually has each sash divided into six or eight lights. Some manufacturers provided preassembled dividers which snap in place over a single light, dividing it into six or eight lights. This simplifies painting and other maintenance.

Assembled frames are placed in the rough opening over strips of building paper put around the perimeter to minimize air infiltration. The frame is plumbed and nailed to side studs and header through the cas-

ings or the blind stops at the sides. Where nails are exposed, such as on the casing, use the corrosion-resistant type.

Hardware for double-hung windows includes the sash lifts that are fastened to the bottom rail, although they are sometimes eliminated by providing a finger groove in the rail. Other hardware consists of sash locks or fasteners located at the meeting rail. They not only lock the window, but draw the sash together to provide a "windtight" fit.

Double-hung windows can be arranged in a number of ways—as a single unit, doubled (or mullion) type, or in groups of three or more. One or two double-hung windows on each side of a large stationary insulated window are often used to effect a window wall. Such large openings must be framed with headers large enough to carry roofloads.

Casement Windows

Casement windows consist of side-hinged sash, usually designed to swing outward (fig. 35.2) because this type can be made more weathertight than the inswinging style. Screens are located inside these outswinging windows and winter protection is obtained with a storm sash or by using insulated glass in the sash. One advantage of the casement window over the double-hung type is that the entire window area can be opened for ventilation.

Weatherstripping is also provided for this type of window, and units are usually received from the factory entirely assembled with hardware in place. Closing hardware consists of a rotary operator and sash lock. As in the double-hung units, casement sash can be used in a number of ways—as a pair or in combinations of two or more pairs. Style variations are achieved by divided lights. Snap-in muntins provided a small, multiple-pane appearance for traditional styling.

Metal sash are sometimes used but, because of low insulating value, should be installed carefully to prevent condensation and frosting on the interior surfaces during cold weather. A full storm-window unit is sometimes necessary to eliminate this problem in cold climates.

Stationary Windows

Stationary windows used alone or in combination with double-hung or casement windows usually consist of a wood sash with a large single light of insulated glass. They are designed to provide light, as well as for attractive appearance, and are fastened permanently into the frame (fig. 35.3). Because of their size, (sometimes 6 to 8 feet wide) 1¾-inch-thick sash is used to provide strength. The thickness is usually required because of the thickness of the insulating glass.

Other types of stationary windows may be used

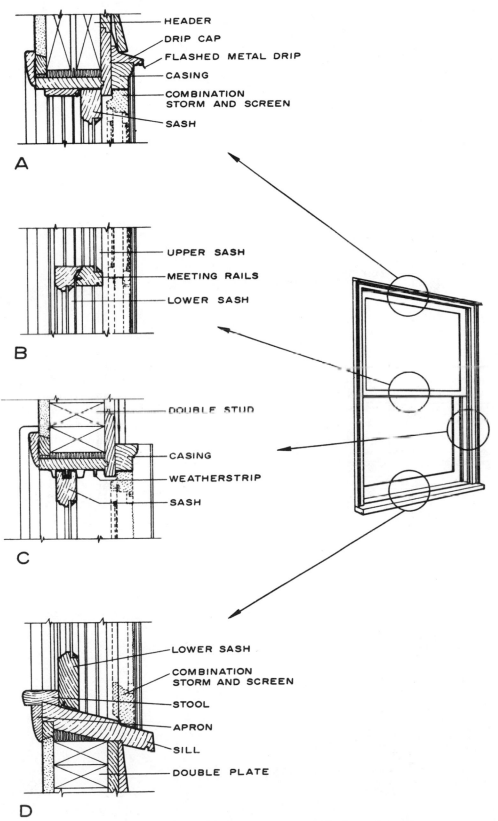

HEADER
DRIP CAP
FLASHED METAL DRIP
CASING
COMBINATION
STORM AND SCREEN
SASH

A

UPPER SASH
MEETING RAILS
LOWER SASH

B

DOUBLE STUD
CASING
WEATHERSTRIP
SASH

C

LOWER SASH
COMBINATION
STORM AND SCREEN
STOOL
APRON
SILL
DOUBLE PLATE

D

Figure 35.1.—Double-hung windows. Cross sections: A, Head jamb; B, meeting rails;
C, side jambs; D, sill.

527

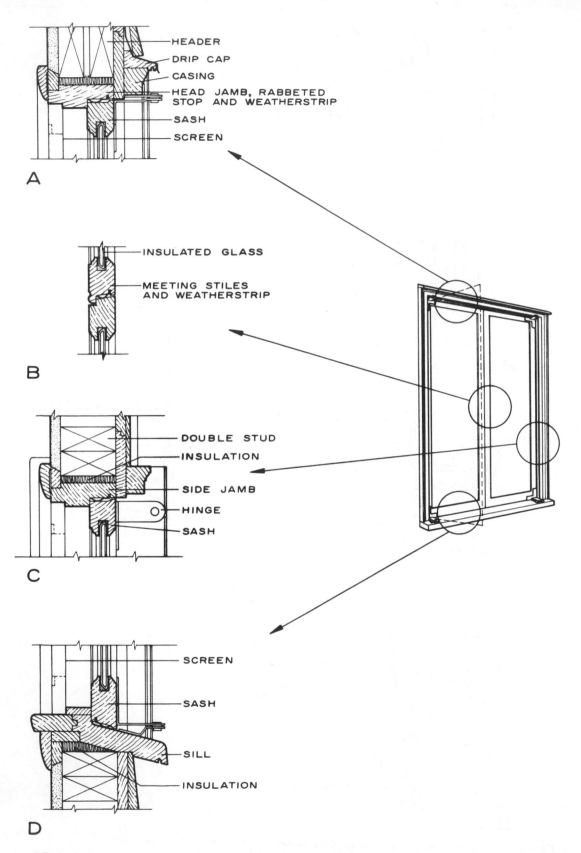

A

- HEADER
- DRIP CAP
- CASING
- HEAD JAMB, RABBETED STOP AND WEATHERSTRIP
- SASH
- SCREEN

B

- INSULATED GLASS
- MEETING STILES AND WEATHERSTRIP

C

- DOUBLE STUD
- INSULATION
- SIDE JAMB
- HINGE
- SASH

D

- SCREEN
- SASH
- SILL
- INSULATION

Figure 35.2.—Outswinging casement sash. Cross sections: *A*, Head jamb; *B*, meeting stiles; *C*, side jambs; *D*, sill.

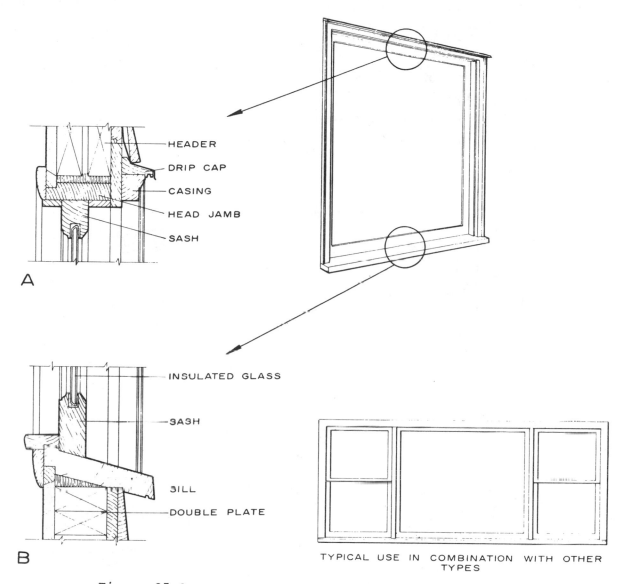

Figure 35.3.—Stationary window. Cross sections: A, Head jamb; B, sill.

without a sash. The glass is set directly into rabbeted frame members and held in place with stops.
As with all window-sash units, back puttying and
face puttying of the glass (with or without a stop)
will assure moisture-resistance.

Awning Windows

An awning window unit consists of a frame in
which one or more operative sash are installed (fig.
35.4) They often are made up for a large window wall
and consist of three or more units in width and
height.

Sash of the awning type are made to swing outward at the bottom. A similar unit, called the hopper
type, is one in which the top of the sash swings inward. Both types provide protection from rain when
open.

Jambs are usually $1\frac{1}{16}$ inches or more thick because they are rabbeted, while the sill is at least
$1\frac{5}{16}$ inches thick when two or more sash are used in
a complete frame. Each sash may also be provided
with an individual frame, so that any combination
in width and height can be used. Awning or hopper
window units may consist of a combination of one
or more fixed sash with the remainder being the
operable type. Operable sash are provided with hinges,
pivots, and sash supporting arms.

Weatherstripping and storm sash and screens are
usually provided. The storm sash is eliminated when
the windows are glazed with insulated glass.

Horizontal-sliding Window Units

Horizontal-sliding windows appear similar to casement sash. However, the sash (in pairs) slide hori-

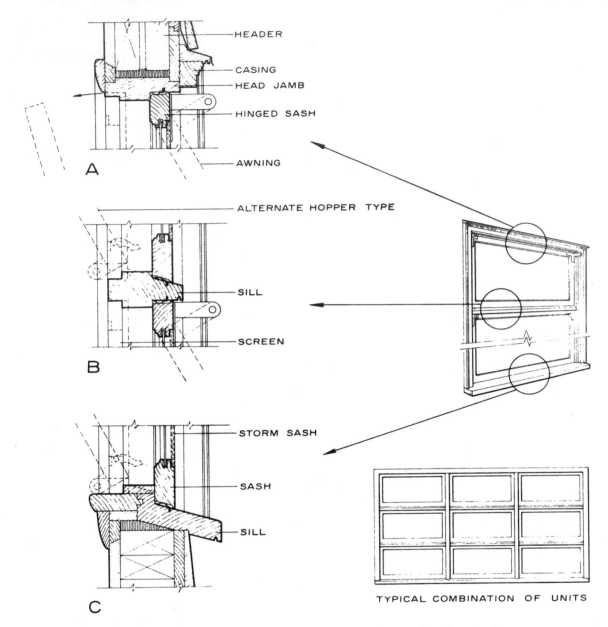

Figure 35.4.—Awning window. Cross sections: A, Head jamb; B, horizontal mullion; C, sill.

zontally in separate tracks or guides located on the sill and head jamb. Multiple window openings consist of two or more single units and may be used when a window-wall effect is desired. As in most modern window units of all types, weatherstripping, water-repellent preservative treatments, and sometimes hardware are included in these fully factory-assembled units.

Exterior Doors and Frames

Exterior doors are 1¾ inches thick and not less than 6 feet 8 inches high. The main entrance door is 3 feet wide and the side or rear service door 2 feet 8 inches wide.

The frames for these doors are made of 1⅛-inch or thicker material, so that rabbeting of side and head jambs provides stops for the main door (fig. 35.5) The wood sill is often oak for wear resistance, but when softer species are used, a metal nosing and wear strips are included. As in many of the window units, the outside casings provide space for the 1⅛-inch combination or screen door.

The frame is nailed to studs and headers of the rough opening through the outside casing. The sill

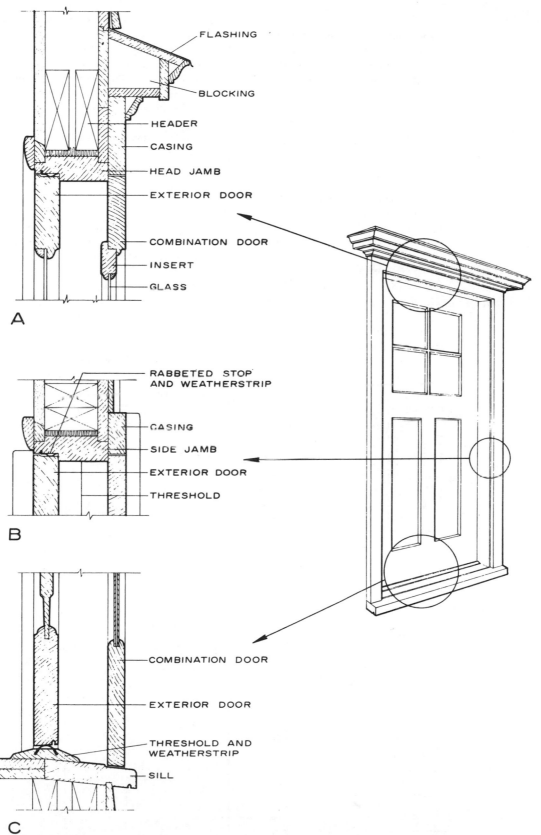

FLASHING

BLOCKING

HEADER

CASING

HEAD JAMB

EXTERIOR DOOR

COMBINATION DOOR

INSERT

GLASS

A

RABBETED STOP
AND WEATHERSTRIP

CASING

SIDE JAMB

EXTERIOR DOOR

THRESHOLD

B

COMBINATION DOOR

EXTERIOR DOOR

THRESHOLD AND
WEATHERSTRIP

SILL

C

Figure 35.5.—Exterior door and frame. Exterior-door and combination-door (screen and storm) cross sections: *A*, Head jamb; *B*, side jamb; *C*, sill.

must rest firmly on the header or stringer joist of the floor framing, which commonly must be trimmed with a saw and hand ax or other means. After finish flooring is in place, a hardwood or metal threshold with a plastic weatherstop covers the joints between the floor and sill.

The exterior trim around the main entrance door can vary from a simple casing to a molded or plain pilaster with a decorative head casing. Decorative designs should always be in keeping with the architecture of the house. Many combinations of door and entry designs are used with contemporary houses, and manufacturers have millwork which is adaptable to this and similar styles. If there is an entry hall, it is usually desirable to have glass included in the main door if no other light is provided.

Types of Exterior Doors

Exterior doors and outside combination and storm doors can be obtained in a number of designs to fit the style of almost any house. Doors in the traditional pattern are usually the *panel type* (fig. 35. 6a) They consist of *stiles* (solid vertical members), *rails* (solid cross members), and *filler panels* in a number of designs. Glazed upper panels are combined with raised wood or plywood lower panels. For methods of hanging doors and installing hardware, see Chapter 47, "Interior Doors, Frames, and Trim."

Exterior flush doors should be of the solid-core type rather than hollow-core to minimize warping during the heating season. (Warping is caused by a difference in moisture content on the exposed and unexposed faces.)

Flush doors consist of thin plywood faces over a framework of wood with a woodblock or particle board core. Many combinations of designs can be obtained, ranging from plain flush doors to others with a variety of panels and glazed openings (fig. 35. 6b).

Wood combination doors (storm and screen) are available in several styles (fig. 35. 6c). Panels which include screen and storm inserts are normally located in the upper portion of the door. Some types can be obtained with self-storing features, similar to window combination units. Heat loss through metal combination doors is greater than through similar type wood doors.

Weatherstripping of the 1¾-inch-thick exterior door will reduce both air infiltration and frosting of the glass on the storm door during cold weather.

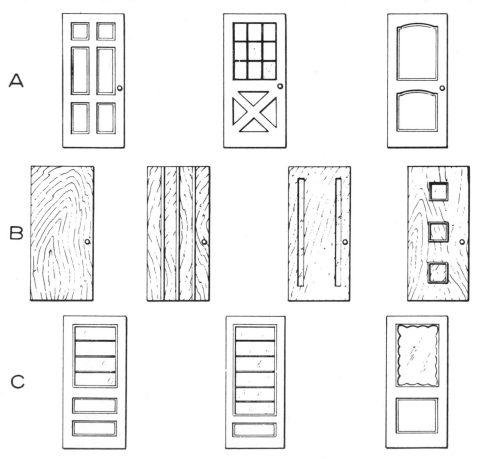

Figure 35.6.—Exterior doors: A, Traditional panel; B, flush; C, combination.

532

CHAPTER 36

EXTERIOR COVERINGS

Because siding and other types of coverings used for exterior walls have an important influence on the appearance as well as on the maintenance of the house, a careful selection of the pattern should be made. The homeowner now has a choice of many wood and wood-base materials which may be used to cover exterior walls. Masonry, veneeers, metal or plastic siding, and other nonwood materials are additional choices. Wood siding can be obtained in many different patterns and can be finished naturally, stained, or painted. Wood shingles, plywood, wood siding or paneling, fiberboard, and hardboard are some of the types and as exterior coverings. Many prefinished sidings are available, and the coatings and films applied to several types of base materials presumably eliminate the need of refinishing for many years.

Wood Siding

One of the materials most characteristic of the exteriors of American houses is *wood siding*. The essential properties required for siding are good painting characteristics, easy working qualities, and freedom from warp. Such properties are present to a *high* degree in the cedars, eastern white pine, sugar pine, western white pine, cypress, and redwood; to a *good* degree in western hemlock, ponderosa pine, the spruces, and yellow-poplar; and to a *fair* degree in Douglas-fir, western larch, and southern pine (5).

Material used for exterior siding which is to be painted, should preferably be of a high grade and free from knots, pitch pockets, and waney edges. Vertical grain and mixed grain (both vertical and flat) are available in some species such as redwood and western redcedar.

The moisture content at the time of application should be that which it would attain in service. This would be approximately 10 to 12 percent except in the dry Southwestern States where the moisture content should average about 8 to 9 percent. To minimize seasonal movement due to changes in moisture content, vertical-grain (edge-grain) siding is preferred. While this is not as important for a stained finish, the use of edge-grain siding for a paint finish will result in longer paint life. A 3-minute dip in a water-repellent preservative (Federal Specification TT-W-572) before siding is installed will not only result in longer paint life, but also will resist moisture entry and decay. Some manufacturers supply siding with this treatment. Freshly cut ends should be brush-treated on the job.

Horizontal Sidings

Some wood siding patterns are used only horizontally and others only vertically. Some may be used

in either manner if adequate nailing areas are provided. Following are descriptions of each of the general types.

Bevel Siding

Plain bevel siding can be obtained in sizes from $\frac{1}{2}$ by 4 inches to $\frac{1}{2}$ by 8 inches, and also in sizes of $\frac{3}{4}$ by 8 inches and $\frac{3}{4}$ by 10 inches (fig. 36.1). "Anzac" siding (fig. 36.1) is $\frac{3}{4}$ by 12 inches in size. Usually the finished width of bevel siding is about $\frac{1}{2}$ inch less than the size listed. One side of bevel siding has a smooth planed surface, while the other has a rough resawn surface. For a stained finish, the rough or sawn side is exposed because wood stain is most successful and longer lasting on rough wood surfaces.

Dolly Varden Siding

Dolly Varden siding is similar to true bevel siding except that shiplap edges are used, resulting in a constant exposure distance (fig. 36.1) Because is lies flat against the studs, it is sometimes used for garages and similar buildings without sheathing. Diagonal bracing is then needed to provide racking resistance to the wall.

Other Horizontal Sidings

Regular *drop sidings* can be obtained in several patterns, two of which are shown in fig. 36.1 This siding, with matched or shiplap edges, can be obtained in 1- by 6- and 1- by 8-inch sizes. This type is commonly used for lower cost dwellings and for garages, usually without benefit of sheathing. Tests conducted at the Forest Products Laboratory have shown that the tongued-and-grooved (matched) patterns have greater resistance to the penetration of wind-driven rain than the shiplap patterns, when both are treated with a water-repellent preservative.

Fiberboard and *hardboard sidings* are also available in various forms. Some have a backing to provide rigidity and strength while others are used directly over sheathing. Plywood horizontal lap siding, with medium density overlaid surface, is also avaiable as an exterior covering material. It is usually $\frac{3}{8}$ inch thick and 12 and 16 inches wide. It is applied in much the same manner as wood siding, except that a shingle wedge is used behind each vertical joint.

Sidings for Horizontal or Vertical Applications

A number of siding or paneling patterns can be used horizontally or vertically (fig. 36.1). These are manufactured in nominal 1-inch thicknesses and in

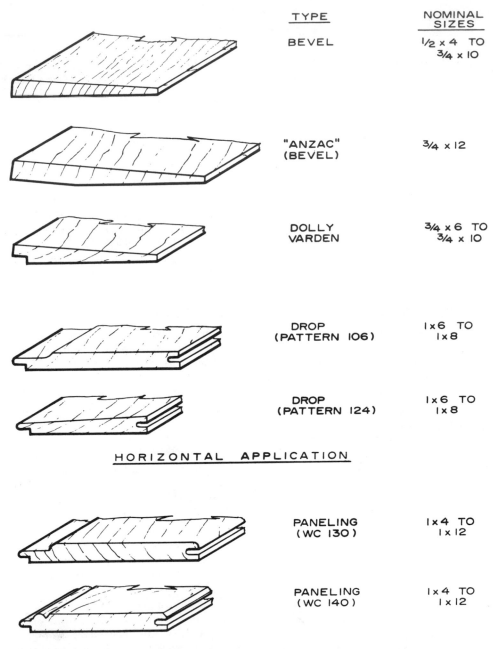

TYPE	NOMINAL SIZES
BEVEL	1/2 x 4 TO 3/4 x 10
"ANZAC" (BEVEL)	3/4 x 12
DOLLY VARDEN	3/4 x 6 TO 3/4 x 10
DROP (PATTERN 106)	1 x 6 TO 1 x 8
DROP (PATTERN 124)	1 x 6 TO 1 x 8

HORIZONTAL APPLICATION

TYPE	NOMINAL SIZES
PANELING (WC 130)	1 x 4 TO 1 x 12
PANELING (WC 140)	1 x 4 TO 1 x 12

HORIZONTAL OR VERTICAL APPLICATION

Figure 36.1.—Wood siding types.

widths from 4 to 12 inches. Both dressed and matched and shiplapped edges are available. The narrow and medium-width patterns will likely be more satisfactory when there are moderate moisture content changes. Wide patterns are more successful if they are vertical grain to keep shrinkage to a minimum. The correct moisture content is also important when tongued and grooved siding is wide, to prevent shrinkage to a point where the tongue is exposed.

Treating the edges of both drop and the matched and shiplapped sidings with water-repellent preservative usually prevents wind-driven rain from penetrating the joints if exposed to weather. In areas under wide overhangs, or in porches or other protected sections, this treatment is not as important. Some manufacturers provide siding with this treatment applied at the factory.

534

Sidings for Vertical Application

A method of siding application, popular for some architectural styles, utilizes rough-sawn boards and battens applied vertically. These boards can be arranged in several ways: (a) Board and batten, (b) batten and board, and (c) board and board (fig. 36.2) As in the vertical application of most siding materials, nominal 1-inch sheathing boards or plywood sheathing ⅝ or ¾ inch thick should be used for nailing surfaces. When other types of sheathing materials or thinner plywoods are used, nailing blocks between studs commonly provide the nailing areas. Nailers of 1 by 4 inches, laid horizontally and spaced from 16 to 24 inches apart vertically, can be used over nonwood sheathing. However, special or thicker casing is some-

times required around doors and window frames when this system is used. It is good practice to use a building paper over the sheathing before applying the vertical siding.

Sidings with Sheet Materials

A number of sheet materials are now available for use as siding. These include plywood in a variety of face treatments and species, paper-overlaid plywood, and hardboard. Plywood or paper-overlaid plywood is sometimes used without sheathing and is known as panel siding with ⅜-inch often considered the minimum thickness for such use for 16-inch stud spacing. However, from the standpoint of stiffness and strength, better performance is usually obtained by using ½- or ⅝-inch thickness.

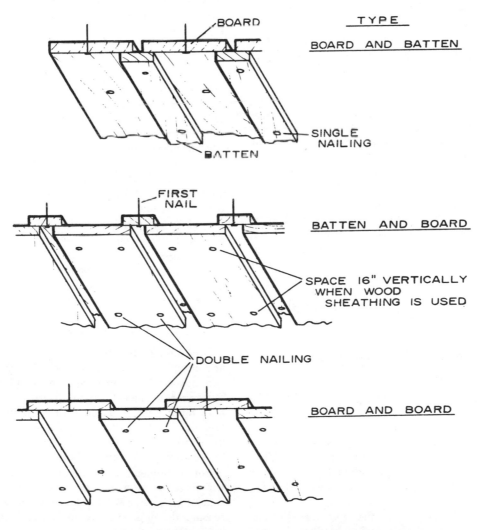

NOTE : NAIL FOR FIRST BOARD - 8d OR 9d
NAIL FOR SECOND BOARD - 12d

*Figure 36.2.—*Vertical board siding.

These 4- by 8-foot and longer sheets must be applied vertically with intermediate and perimeter nailing to provide the desired rigidity. Most other methods of applying sheet materials require some type of sheathing beneath. When horizontal joints are necessary, they should be protected by a simple flashing.

An exterior-grade plywood should always be used for siding, and can be obtained in such surfaces as grooved, brushed, and saw-textured. These surfaces are usually finished with some type of stain. If shiplap or matched edges are not provided, some method of providing a waterproof joint should be used. This often consists of calking and a batten at each joint and a batten at each stud if closer spacing is desired for appearance. An edge treatment of water-repellent preservative will also aid in reducing moisture penetration. Allow $1/16$-inch edge and end spacing when installing plywood in sheet form.

Exterior grade particleboard might also be considered for panel siding. FHA Material Use Bulletin No. 32 (9) lists the requirements when this material is used. Normally $5/8$-inch thickness is required for 16-inch stud spacing and $3/4$ inch for 24-inch stud spacing. The finish must be with an approved paint, and the stud wall behind must have corner bracing.

Paper-overlaid plywood has many of the advantages of plywood with the addition of providing a very satisfactory base for paint. A medium-density, overlaid plywood is most commonly used.

Hardboard sheets used for siding are applied the same way as plywood, that is, by using battens at vertical points and at intermediate studs. Medium-density fiberboards might also be used in some areas as exterior coverings over certain types of sheathing.

Many of these sheet materials resist the passage of water vapor. Hence, when they are used, it is important that a good vapor barrier, well installed, be employed on the warm side of the insulated walls. These factors are described in Chapter 41, "Thermal Insulation and Vapor Barriers."

Wood Shingles and Shakes

Grades and Species

Wood shingles and shakes are desirable for sidewalls in many styles of houses. In Cape Cod or Colonial houses, shingles may be painted or stained. For ranch or contemporary designs, wide exposures of shingles or shakes often add a desired effect. They are easily stained and thus provide a finish which is long-lasting on those species commonly used for shingles.

Western redcedar is perhaps the most available species, although northern white-cedar, baldcypress, and redwood are also satisfactory. The heartwood of these species has a natural decay resistance which is desirable if shingles are to remain unpainted or unstained.

Western redcedar shingles can be obtained in three grades. The first-grade (No. 1) is all heartwood, edge grain, and knot free; it is primarily intended for roofs but is desirable in double-course sidewall application where much of the face is exposed.

Second-grade shingles (No. 2) are most often used in single-course application for sidewalls, since only three-fourths of the shingle length is blemish-free. A 1-inch width of sapwood and mixed vertical and flat grain are permissible.

The third-grade shingle (No. 3) is clear for 6 inches from the butt. Flat grain is acceptable, as are greater widths of sapwood. Third-grade shingles are likely to be somewhat thinner than the first and second grades; they are used for secondary buildings and sometimes as the undercourse in double-course application.

A lower grade than the third grade, known as under-coursing shingle, is used only as the under and completely covered course in double-course sidewall application.

Shingle Sizes

Wood shingles are available in three standard lengths—16, 18, and 24 inches. The 16-inch length is perhaps the most popular, having five butt thicknesses per 2 inches when green (designated a 5/2). These shingles are packed in bundles with 20 courses on each side. Four bundles will cover 100 square feet of wall or roof with an exposure of 5 inches. The 18- and 24-inch-length shingles have thicker butts, five in $2\frac{1}{4}$ inches for the 18-inch shingles and four in 2 inches for the 24-inch lengths.

Shakes are usually available in several types, the most popular being the split-and-resawn. The sawed face is used as the back face. The butt thickness of each shake ranges between $3/4$ and $1\frac{1}{2}$ inches. They are usually packed in bundles (20 sq. ft.), five bundles to the square.

Other Exterior Finish

Nonwood materials, such as asbestos-cement siding and shingles, metal sidings, and the like are available and are used in some types of architectural design. Stucco or a cement plaster finish, preferably over a wire mesh base, is most often seen in the Southwest and the West Coast areas. Masonry veneers may be used effectively with wood siding in various finishes to enhance the beauty of both materials.

Some homebuilders favor an exterior covering which requires a minimum of maintenance. While some of the nonwood materials are chosen for this reason, developments by the paint industry are providing comparable long-life coatings for wood-base materials. Plastic films on wood siding or plywood are also promising, so that little or no refinishing is indicated for the life of the building.

Installation of Siding

One of the important factors in successful performance of various siding materials is the type of fasteners used. Nails are the most common of these, and it is poor economy indeed to use them sparingly. Corrosion-resistant nails, galvanized or made of aluminum, stainless steel, or similar metals, may cost more, but their use will insure spot-free siding under adverse conditions.

Two types of nails are commonly used with siding, the finishing nail having a small head and the siding nail having a moderate-size flat head. The small-head finishing nail is set (driven with a nail set) about $\frac{1}{16}$ inch below the face of the siding, and the hole is filled with putty after the prime coat of paint is applied. The flathead siding nail, most commonly used, is driven flush with the face of the siding and the head later covered with paint.

Ordinary steel-wire nails tend to rust in a short time and cause a disfiguring stain on the face of the siding. In some cases, the small-head nails will show rust spots through the putty and paint. Noncorrosive nails that will not cause rust are readily available.

Siding to be "natural finished" with a water-repellent preservative or stain should be fastened with stainless steel or aluminum nails. In some types of prefinished sidings, nails with color-matched heads are supplied.

In recent years, nails with modified shanks have become quite popular. These include the *annularly* threaded shank nail and the *helically* threaded shank nail. Both have greater withdrawal resistance than the smooth shank nail and, for this reason, a shorter nail is often used.

Exposed nails in siding should be driven just flush with the surface of the wood. Overdriving may not only show the hammer mark, but may also cause objectionable splitting and crushing of the wood. In sidings with prefinished surfaces or overlays, the nails should be driven so as not to damage the finished surface.

Bevel Siding

The minimum lap for bevel siding should not be less than 1 inch. The average exposure distance is usually determined by the distance from the underside of the window sill to the top of the drip cap (fig. 36.3) From the standpoint of weather resistance and appearance, the butt edge of the first course of siding above the window should coincide with the top of the window drip cap. In many one-story houses with an overhang, this course of siding is often replaced with a frieze board (fig. 33.2). It is also desirable that the bottom of a siding course be flush with the underside of the window sill. However, this may not always be possible because of varying window heights and types that might be used in a house.

One system used to determine the siding exposure width so that it is about equal both above and below the window sill is described below:

Divide the overall height of the window frame by the approximate recommended exposure distance for the siding used (4 for 6-inch-wide siding, 6 for 8-inch siding, 8 for 10-inch siding, and 10 for 12-inch siding). This will result in the number of courses between the top and bottom of the window. For example, the overall height of our sample window from top of the drip cap to the bottom of the sill is 61 inches. If 12-inch siding is used, the number of courses would be $61/10 = 6.1$ or six courses. To obtain the exact exposure distance, divide 61 by 6 and the result would be $10\frac{1}{6}$ inches. The next step is to determine the exposure distance from the bottom of the sill to just below the top of the foundation wall. If this is 31 inches, three courses at $10\frac{1}{3}$ inches each would be used. Thus, the exposure distance above and below the window would be almost the same (fig. 36.3).

When this system is not satisfactory because of big differences in the two areas, it is preferable to use an equal exposure distance for the entire wall height and notch the siding at the window sill. The fit should be tight to prevent moisture entry.

Siding may be installed starting with the bottom course. It is normally blocked out with a starting strip the same thickness as the top of the siding board (fig. 36.3). Each succeeding course overlaps the upper edge of the lower course. Siding should be nailed to each stud or on 16-inch centers. When plywood or wood sheathing or spaced wood nailing strips are used over nonwood sheathing, sevenpenny or eightpenny nails ($2\frac{1}{4}$ and $2\frac{1}{2}$ in. long) may be used for $\frac{3}{4}$-inch-thick siding. However, if gypsum or fiberboard sheathing is used, the tenpenny nail is recommended to penetrate into the stud. For $\frac{1}{2}$-inch-thick siding, nails may be $\frac{1}{4}$ inch shorter than those used for $\frac{3}{4}$-inch siding.

The nails should be located far enough up from the butt to miss the top of the lower siding course (fig. 36.4). This clearance distance is usually $\frac{1}{8}$ inch. This allow for slight movement of the siding due to moisture changes without causing splitting. Such an allowance is especially required for the wider sidings of 8 to 12 inches wide.

It is good practice to avoid butt joints whenever possible. Use the longer sections of siding under windows and other long stretches and utilize the shorter lengths for areas between windows and doors. If unavoidable, butt joints should be made over a stud and staggered between courses as much as practical (fig. 36.3).

Siding should be *square-cut* to provide a good joint at window and door casings and at butt joints. Open joints permit moisture to enter, often leading to paint deterioration. It is good practice to brush

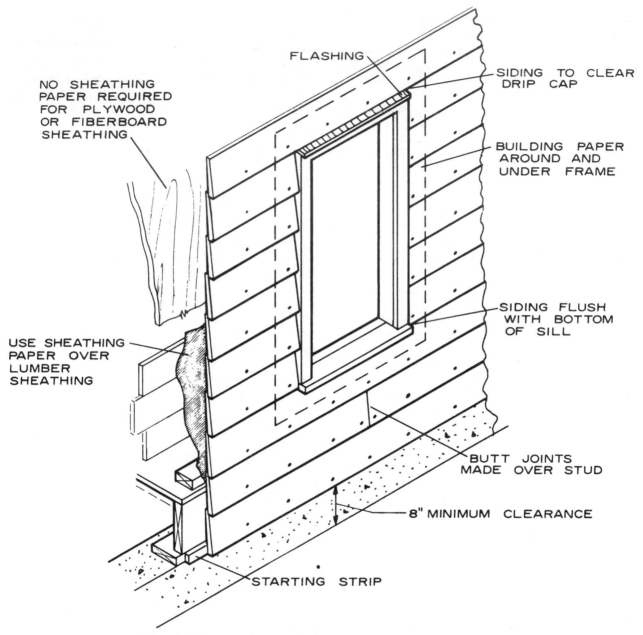

NO SHEATHING PAPER REQUIRED FOR PLYWOOD OR FIBERBOARD SHEATHING

FLASHING

SIDING TO CLEAR DRIP CAP

BUILDING PAPER AROUND AND UNDER FRAME

USE SHEATHING PAPER OVER LUMBER SHEATHING

SIDING FLUSH WITH BOTTOM OF SILL

BUTT JOINTS MADE OVER STUD

8" MINIMUM CLEARANCE

STARTING STRIP

Figure 36.3. —Installation of bevel siding.

or dip the fresh-cut ends of the siding in a water-repellent preservative before boards are nailed in place. Using a small finger-actuated oil can to apply the water-repellent preservative at end and butt joints after siding is in place is also helpful.

Drop and Similar Sidings

Drop siding is installed much the same as lap siding except for spacing and nailing. Drop, Dolly Varden, and similar sidings have a constant exposure distance. This face width is normally $5\frac{1}{4}$ inches for

1- by 6-inch siding and $7\frac{1}{4}$ inches for 1- by 8-inch siding. Normally, one or two eightpenny or ninepenny nails should be used at each stud crossing, depending on the width (fig. 36.4). Length of nail depends on type of sheathing used, but penetration into the stud or through the wood backing should be at least $1\frac{1}{2}$ inches.

Horizontally applied matched paneling in narrow widths should be blind-nailed at the tongue with a corrosion-resistant finishing nail (fig. 36.4). For widths greater than 6 inches. an additional nail should be used as shown.

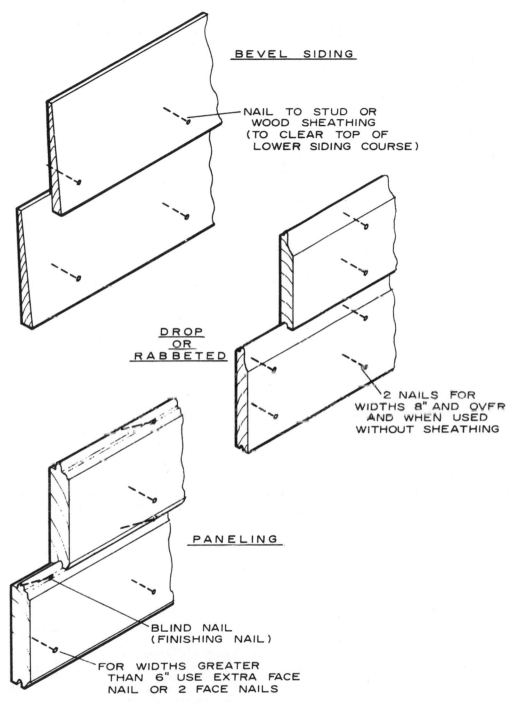

BEVEL SIDING

NAIL TO STUD OR
WOOD SHEATHING
(TO CLEAR TOP OF
LOWER SIDING COURSE)

DROP
OR
RABBETED

2 NAILS FOR
WIDTHS 8" AND OVER
AND WHEN USED
WITHOUT SHEATHING

PANELING

BLIND NAIL
(FINISHING NAIL)

FOR WIDTHS GREATER
THAN 6" USE EXTRA FACE
NAIL OR 2 FACE NAILS

Figure 36.4.—Nailing of siding.

Other materials such as plywood, hardboard, or medium-density fiberboard, which are used horizontally in widths up to 12 inches, should be applied in the same manner as lap or drop siding, depending on the pattern. Prepackaged siding should be applied according to manufacturers' directions.

Vertical Sidings

Vertically applied matched and similar sidings having interlapping joints are nailed in the same manner as when applied horizontally. However, they should be nailed to blocking used between studs or to wood or plywood sheathing. Blocking is spaced from 16 to

24 inches apart. With plywood or nominal 1-inch board sheathing, nails should be spaced on 16-inch centers.

When the various combinations of boards and battens are used, they should also be nailed to blocking spaced from 16 to 24 inches apart between studs, or closer for wood sheathing. The first boards or battens should be fastened with one eightpenny or ninepenny nail at each blocking, to provide at least 1½-inch penetration. For wide under-boards, two nails spaced about 2 inches apart may be used rather than the single row along the center (fig. 36.2). The second or top boards or battens should be nailed with twelvepenny nails. Nails of the top board or batten should always miss the under-boards and not be nailed through them (fig. 36.2). In such applications, double nails should be spaced closely to prevent splitting if the board shrinks. It is also good practice to use a sheathing paper, such as 15 pound asphalt felt, under vertical siding.

Plywood and Other Sheet Siding

Exterior grade plywood, paper-overlaid plywood, and similar sheet materials used for siding are usually applied vertically. When used over sheathing, plywood should be at least ¼ inch thick, although ⁵⁄₁₆ and ⅜ inch will normally provide a more even surface. Hardboard should be ¼ inch thick and materials such as medium-density fiberboard should be ½ inch.

All nailing should be over studs and total effective penetration into wood should be at least 1½ inches. For example, ⅜-inch plywood siding over ¾-inch wood sheathing would require about a sevenpenny nail, which is 2¼ inches long. This would result in a 1⅛-inch-penetration into the stud, but a total effective penetration of 1⅞ inches into wood.

Plywood should be nailed at 6-inch intervals around the perimeter and 12 inches at intermediate members. Hardboard siding should be nailed at 4- and 8-inch intervals. All types of sheet material should have a joint calked with mastic unless the joints are of the interlapping or matched type or battens are installed. A strip of 15-pound asphalt felt under uncalked joints is good practice.

Corner Treatment

The method of finishing wood siding or other materials at exterior corners is often influenced by the overall design of the house. A mitered corner effect on horizontal siding or the use of corner boards are perhaps the most common methods of treatment.

Mitering corners (fig. 36.5a) of bevel and similar sidings, unless carefully done to prevent openings, is not always satisfactory. To maintain a good joint, it is necessary that the joint fit tightly the full depth of the miter. It is also good practice to treat the ends

with a water-repellent preservative prior to nailing.

Metal corners (fig. 36.5b) are perhaps more commonly used than the mitered corner and give a mitered effect. They are easily placed over each corner as the siding is installed. The metal corners should fit tightly without openings and be nailed on each side to the sheathing or corner stud beneath. If made of galavanized iron, they should be cleaned with a mild acid wash and primed with a metal primer before the house is painted to prevent early peeling of the paint. Weathering of the metal will also prepare it for the prime paint coat.

Corner boards of various types and sizes may be used for horizontal sidings of all types (fig. 36.5c) They also provide a satisfactory termination for plywood and similar sheet materials. Vertical applications of matched paneling or of boards and battens are terminated by lapping one side and nailing into the edge of this member, as well as to the nailing members beneath. Corner boards are usually 1⅛- or 1⅜-inch material and for a distinctive appearance might be quite narrow. Plain outside casing commonly used for window and door frames can be adapted for corner boards.

Prefinished shingle or shake exteriors sometimes are used with color-matched metal corners. They can also be lapped over the adjacent corner shingle, alternating each course. This is called "lacing." This type of corner treatment usually requires that some kind of flashing be used beneath.

When siding returns against a roof surface, such as at a dormer, there should be a clearance of about 2 inches (fig. 36.5d). Siding cut tight against the shingles retains moisture after rains and usually results in paint peeling. Shingle flashing extending well up on the dormer wall will provide the necessary resistance to entry of wind-driven rain. Here again, a water-repellent preservative should be used on the ends of the siding at the roofline.

Interior corners (fig. 36.5e) are butted against a square corner board of nominal 1¼- or 1⅜-inch size, depending on the thickness of the siding.

Material Transition

At times, the materials used in the gable ends and in the walls below differ in form and application. The details of construction used at the juncture of the two materials should be such that good drainage is assured. For example, if vertical boards and battens are used at the gable end and horizontal siding below, a drip cap or similar molding might be used (fig. 36.6). Flashing should be used over and above the drip cap so that moisture will clear the gable material.

Another method of material transition might also be used. By extending the plate and studs of the gable end out from the wall a short distance, or by

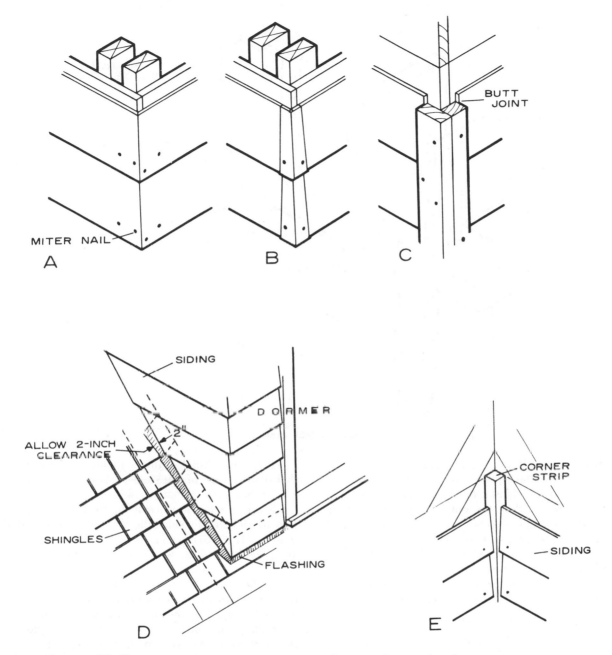

Figure 36.5 —Siding details: *A*, Miter corner; *B*, metal corners; *C*, corner boards; *D*, siding return at roof; *E*, interior corner.

the use of furring strips, the gable siding will project beyond the wall siding and provide good drainage (fig. 36. 7).

Installation of Wood Shingles and Shakes

Wood shingles and shakes are applied in a single- or double-course pattern. They may be used over wood or plywood sheathing. If sheathing is ⅜-inch plywood, use threaded nails. For nonwood sheathing, 1- by 3- or 1- by 4-inch wood nailing strips are used as a base. In the single-course method, one course is simply laid over the other as lap siding is applied. The shingles can be second grade because only one-half or less of the butt portion is exposed (fig. 36. 8) Shingles should not be soaked before application but should usually be laid up with about ⅛ to ¼ inch space between adjacent shingles to allow for expansion during rainy weather. When a "siding effect" is desired, shingles should be laid up so that they are only lightly in contact. Prestained or treated shingles provide the best results for this system.

In a double-course system, the undercourse is ap-

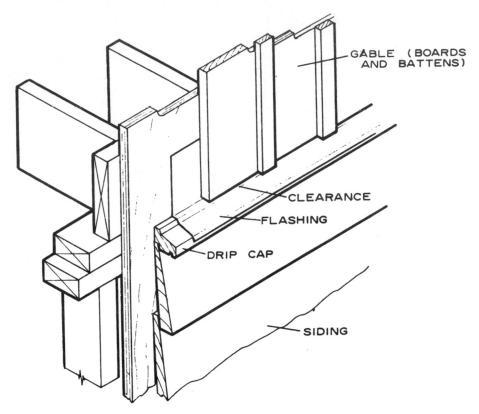

Figure 36.6.—Gable-end finish (material transition).

plied over the wall and the top course nailed directly over a ¼- to ½-inch projection of the butt (fig. 36. 9) The first course should be nailed only enough to hold it in place while the outer course is being applied. The first shingles can be a lower quality, such as third grade or the undercourse grade. The top course, because much of the shingle length is exposed, should be first-grade shingles.

Exposure distance for various length shingles and shakes can be guided by the recommendations in table 5.

TABLE 5.—*Exposure distances for wood shingles and shakes on sidewalls*

Material	Length	Single coursing	Double coursing	
			No. 1 grade	No. 2 grade
	In.	*In.*	*In.*	*In.*
Shingles	16	7½	12	10
	18	8½	14	11
	24	11½	16	14
Shakes (hand split and resawn)	18	8½	14	--------
	24	11½	20	--------
	32	15		--------------

As in roof shingles, joints should be "broken" so that butt joints of the upper shingles are at least 1½ inches from the under-shingle joints.

Closed or open joints may be used in the application of shingles to sidewalls at the discretion of the builder (fig. 36.8). Spacing of ¼ to ⅜ inch produces an individual effect, while close spacing produces a shadow line similar to bevel siding.

Shingles and shakes should be applied with rust-resistant nails long enough to penetrate into the wood backing strips or sheathing. In single coursing, a threepenny or fourpenny zinc-coated "shingle" nail is commonly used. In double coursing, where nails are exposed, a fivepenny zinc-coated nail with a small flat head is used for the top course and threepenny or fourpenny size for the undercourse. Use building paper over lumber sheathing.

Nails should be placed in from the edge of the shingle a distance of ¾ inch (fig. 36. 8). Use two nails for each shingle up to 8 inches wide and three nails for shingles over 8 inches. In single-course applications, nails should be placed 1 inch above the butt line of the next higher course. In double coursing, the use of a piece of shiplap sheathing as a guide allows the outer course to extend ½ inch below the undercourse, producing a shadow line (fig. 36.9). Nails should be placed 2 inches above the bottom of the single or shake. Rived or fluted processed shakes,

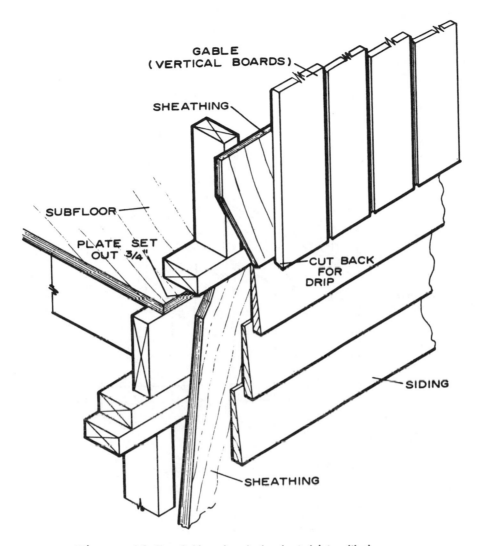

Figure 36.7.—Gable-end projection (material transition).

usually factory-stained, are available and have a distinct effect when laid with closely fitted edges in a double-course pattern.

Nonwood Coverings

Asbestos-Cement Shingles

Asbestos-cement shingles and similar nonwood exterior coverings should be applied in accordance with the manufacturer's directions. They are used over wood or plywood sheathing or over spaced nailing strips. Nails are of the noncorrosive type and usually are available to match the color of the shingles. Manufacturers also supply matching color corners.

Cement-Plaster Finish

Stucco and similar cement-mortar finishes, most commonly used in the Southwest, are applied over a coated expanded-metal lath and, usually, over some type of sheathing. However, in some areas where local building regulations permit, such a finish is applied to metal lath fastened directly to the braced wall framework. Waterproof paper is used over the studs before the metal lath is applied.

When a plastered exterior is applied to two-story houses, balloon framing is recommended (fig. 29. 5) If platform framing is used for one-story houses (fig. 29.1), shrinkage of joists and sills may cause an unsightly bulge or break in the cement-plaster at those points unless joists have reached moisture equilibrium. This stresses the need for proper moisture content of the framing members when this type of finish is used.

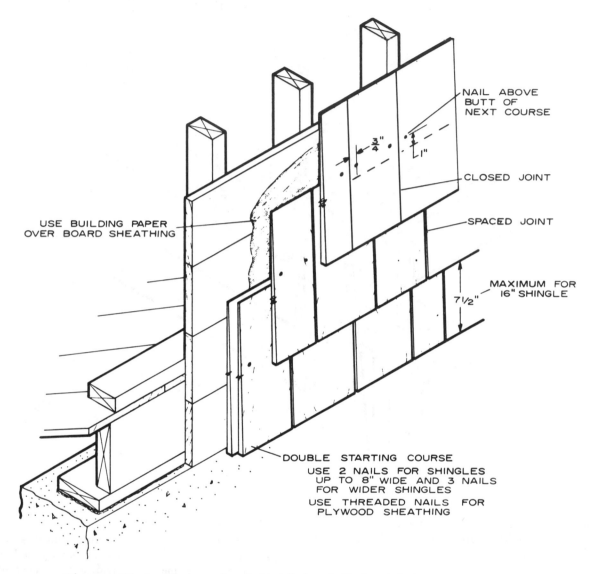

NAIL ABOVE
BUTT OF
NEXT COURSE

$\frac{3}{4}"$ 1"

CLOSED JOINT

SPACED JOINT

MAXIMUM FOR
16" SHINGLE

7½"

USE BUILDING PAPER
OVER BOARD SHEATHING

DOUBLE STARTING COURSE
USE 2 NAILS FOR SHINGLES
UP TO 8" WIDE AND 3 NAILS
FOR WIDER SHINGLES
USE THREADED NAILS FOR
PLYWOOD SHEATHING

Figure 36.8.—Single-coursing of sidewalls (wood shingles - shakes)

Masonry Veneer

Brick or stone veneer is used for all or part of the exterior wall finish for some styles of architecture. The use of balloon framing for brick-veneered two-story houses will prevent cracks due to shrinkage of floor joists. It is good practice, when possible, to delay applying the masonry finish over platform framing until the joists and other members reach moisture equilibrium. The use of a waterproof paper backing and sufficient wall ties is important. Details on the installation of masonry veneer are shown in figure 26.5.

544

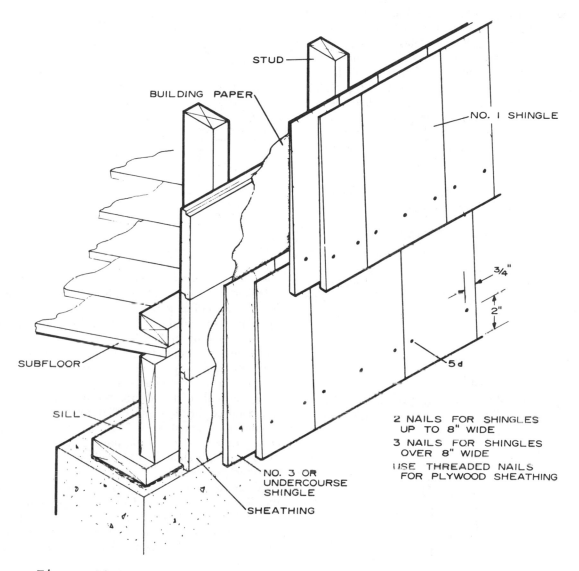

STUD

BUILDING PAPER

NO. I SHINGLE

3/4"

2"

5 d

SUBFLOOR

SILL

2 NAILS FOR SHINGLES UP TO 8" WIDE
3 NAILS FOR SHINGLES OVER 8" WIDE
USE THREADED NAILS FOR PLYWOOD SHEATHING

NO. 3 OR UNDERCOURSE SHINGLE

SHEATHING

Figure 36.9.—**Double-coursing of sidewalls (wood shingles-shakes).**

CHAPTER 37

FRAMING DETAILS FOR PLUMBING, HEATING, AND OTHER UTILITIES

It is desirable, when framing a house, to limit cutting of framing members for installation of plumbing lines and other utilities. A little planning before framing is started will reduce the need for cutting joists and other members. This is more easily accomplished in one-story houses, however, than in two-story houses. In a single-story house, many of the connections are made in the basement area; in two-story houses they must be made between the first-floor ceiling joists. Thus, it is sometimes necessary to cut or notch joists, but this should be done in a manner least detrimental to their strength (12).

Plumbing Stack Vents

One wall of the bath, kitchen, or utility room is normally used to carry the water, vent, and drainage lines. This is usually the wall behind the water closet where connections can be easily made to the tub or

545

shower and to the lavatory. When 4-inch cast-iron bell pipe is used in the soil and vent stack, it is necessary to use 2- by 6- or 2- by 8-inch plates to provide space for the pipe and the connections. Some contractors use a double row of studs placed flatwise so that no drilling is required for the horizontal runs (fig. 37.1a).

Building regulations in some areas allow the use of 3-inch pipe for venting purposes in one-story houses. When this size is used, 2- by 4-inch plates and studs may be employed. However, it is then necessary to reinforce the top plates, which have been cut, by using a *double scab* (fig. 37.1b). Scabs are well nailed on each side of the stack and should extend over two studs. Small angle irons can also be used.

Bathtub Framing

A bathtub full of water is heavy; so floor joists must be arranged to carry the load without excessive deflection. Too great a deflection will sometimes cause an opening above the edge of the tub. Joists should be doubled at the outer edge (fig. 37.2) The intermediate joist should be spaced to clear the drain. Metal hangers or wood blocking support the inner edge of the tub at the wall line.

Cutting Floor Joists

Floor joists should be cut, notched, or drilled only where the effect on strength is minor. While it is always desirable to prevent cutting joists whenever

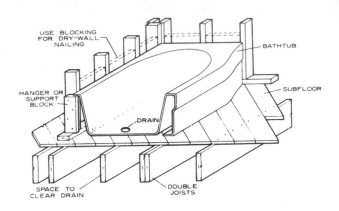

Figure 37.2.—Framing for bathtub.

possible, sometimes such alterations are required. Joists or other structural members should then be reinforced by nailing a reinforcing scab to each side or by adding an additional member. Well-nailed plywood scabs on one or both sides of altered joists also provide a good method of reinforcing these members.

Notching the top or bottom of the joist should only be done in the end quarter of the span and not more than one-sixth of the depth. When greater alterations are required, headers and tail beams should be added around the altered area. This may occur where the closet bend must cross the normal joist locations. In other words, it should be framed out similar to a stair opening (fig. 28.12).

When necessary, holes may be bored in joists if the diameter is no greater than 2 inches and the edge of the hole is not less than $2\frac{1}{2}$ to 3 inches from the top or bottom edge of the joists (fig. 37.3). This usually limits the joist size to a 2 by 8 or larger member.

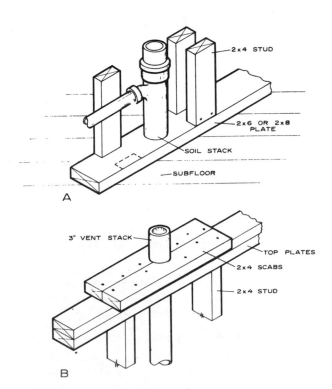

Figure 37.1.—Plumbing stacks: A, 4-inch cast-iron stack; B, 3-inch pipe for vent.

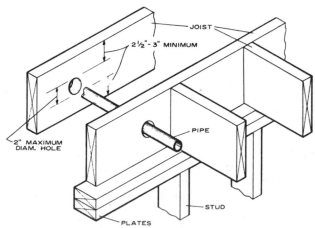

Figure 37.3.—Drilled holes in joists.

Such a method of installation is suitable where joist direction changes and the pipe can be inserted from the long direction, such as from the plumbing wall to a tub on the second floor. Connections for first-floor plumbing can normally be made without cutting or drilling of joists.

Alterations for Heating Ducts

A number of systems are used to heat a house, from a multi-controlled hot-water system to a simple floor or wall furnace. Central air conditioning combined with the heating system is becoming a normal part of house construction. Ducts and piping should be laid out so that framing or other structural parts can be adjusted to accommodate them. However, the system which requires heat or cooling ducts and return lines is perhaps the most important from the standpoint of framing changes required.

Supply and Cold Air Return Ducts

The installation of ducts for a forced-warm-air or air-conditioning system usually requires the removal of the soleplate and the subfloor at the duct location. Supply ducts are made to dimensions that permit them to be placed between studs. When the same duct system is used for heating and cooling, the duct sizes are generally larger than when they are designed for heating alone. Such systems often have two sets of registers; one near the floor for heat and one near the ceiling for more efficient cooling. Both are furnished with dampers for control.

Walls and joists are normally located so that they do not have to be cut when heating ducts are installed. This is especially true when partitions are at right angles to the floor joists.

When a load-bearing partition requires a doubled parallel floor joist as well as a warm-air duct, the joists can be spaced apart to allow room for the duct (fig. 37.4) This will eliminate the need for excessive cutting of framing members or the use of intricate pipe angles.

Cold-air returns are generally located in the floor between joists or in the walls at floor level (fig. 37.5) They are sometimes located in outside walls, in which case they should be lined with metal. *Unlined ducts* in exterior walls have been known to be responsible for exterior-wall paint failures, especially those from a second-floor room.

The elbow from the return duct below the floor is usually placed between floor joists. The space between floor joists, when enclosed with sheet metal, serves as a cold-air return. Other cold-air returns may connect with the same joist-space return duct.

Framing for Convectors

Convectors and hot-water or steam radiators are sometimes recessed partly into the wall to· provide

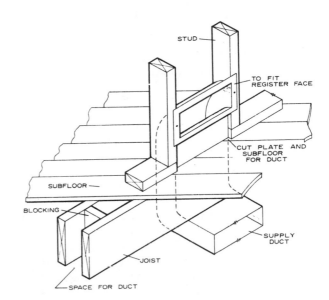

Figure 37.4.—Spaced joists for supply ducts.

more usable space in the room and improve appearance by the installation of a decorative grill. Such framing usually requires the addition of a doubled header to carry the wall load from the studs above (fig. 37.6) Size of the headers depends on the span and should be designed the same as those for window or door openings. The sizes in the tabulation listed under window and door framing in Chapter 29, "Wall Framing", should be used to determine the correct

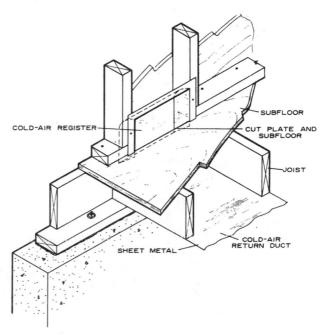

Figure 37.5.—Cold-air return.

sizes. Because only 1⅝ inches of space in the wall is available for insulation, a highly efficient insulation (one with a low "k" value) is sometimes used.

Wiring

House wiring for electrical services is usually started some time after the house has been closed in. The initial phase, of it, termed "roughing in," includes the installation of conduit or cable and the location of switch, light, and outlet boxes with wires ready to connect. This roughing-in work is done before the plaster base or dry-wall finish is applied, and before the insulation is placed in the walls or ceilings. The placement of the fixtures, the switches, and switch plates is done after plastering.

Framing changes for wiring are usually of a minor nature and, for the most part, consist of holes drilled in the studs for the flexible conduit. Although these holes are small in diameter, they should comply with locations shown in fig. 37.3. Perhaps the only area which requires some planning to prevent excessive cutting or drilling is the location of wall switches at entrance door frames. By spacing the doubled framing studs to allow for location of multiple switch boxes, little cutting will be required.

Switches or convenience outlet boxes on exterior walls must be sealed to prevent water vapor move-

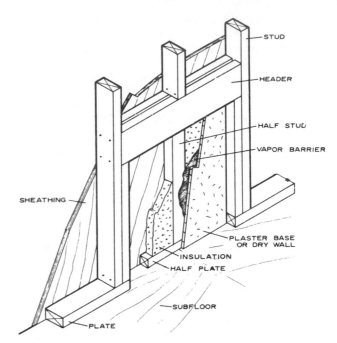

Figure 37.6.—Framing for a convector recess.

ment. Sealing of the vapor barrier around the box is important and will be discussed further in Chapter 41, "Thermal Insulation and Vapor Barriers."

CHAPTER 38

WATER SUPPLY SYSTEMS

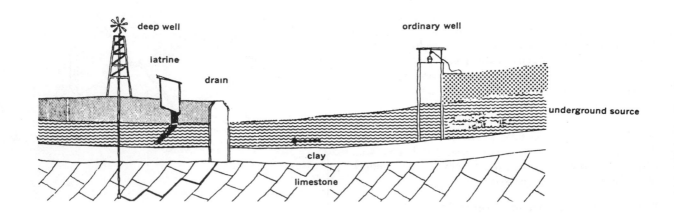

Well Designs

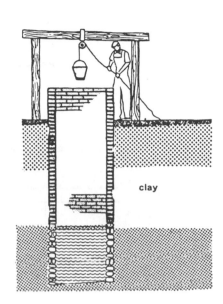

soft earth

clay

sand aquifer

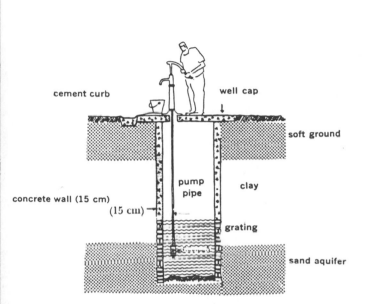

cement curb

well cap

soft ground

concrete wall (15 cm)

(15 cm)

pump pipe

clay

grating

sand aquifer

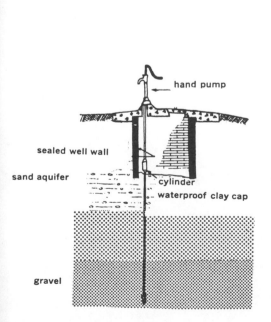

hand pump

sealed well wall

sand aquifer

cylinder

waterproof clay cap

gravel

sealed well

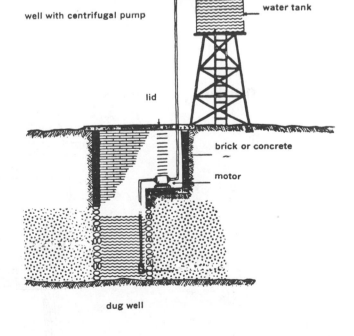

well with centrifugal pump

water tank

lid

brick or concrete

motor

dug well

549

House Plumbing System

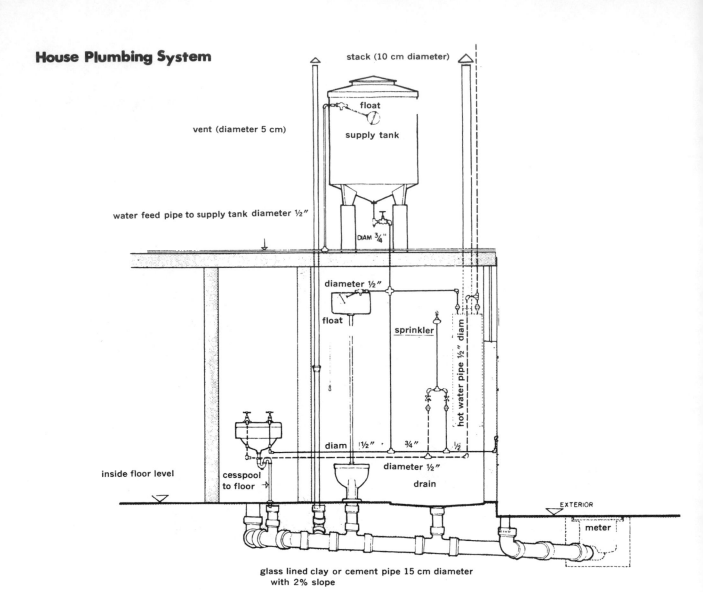

stack (10 cm diameter)

float

supply tank

vent (diameter 5 cm)

water feed pipe to supply tank diameter ½"

DIAM ¾"

diameter ½"

float

sprinkler

hot water pipe ½" diam

diam 1½"   ¾"   ½"

inside floor level

cesspool
to floor

diameter ½"

drain

EXTERIOR

meter

glass lined clay or cement pipe 15 cm diameter
with 2% slope

Water Purification

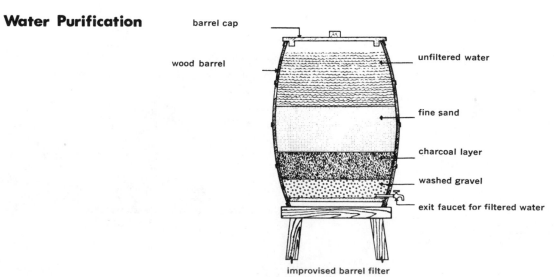

barrel cap

wood barrel

unfiltered water

fine sand

charcoal layer

washed gravel

exit faucet for filtered water

improvised barrel filter

Kitchen

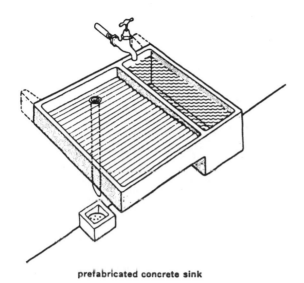

prefabricated concrete sink

faucet

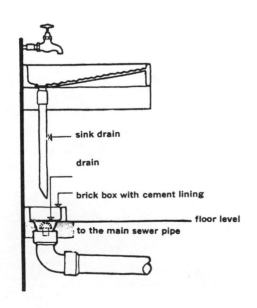

sink drain

drain

brick box with cement lining

floor level

to the main sewer pipe

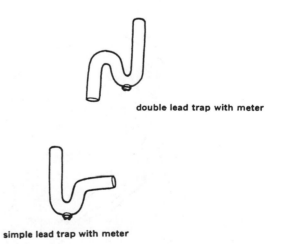

double lead trap with meter

simple lead trap with meter

Bathroom

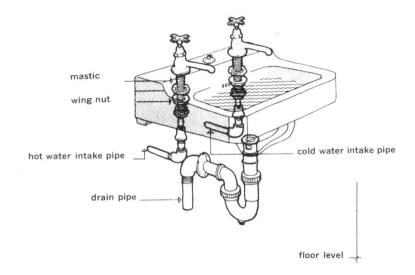

mastic

wing nut

hot water intake pipe

cold water intake pipe

drain pipe

floor level

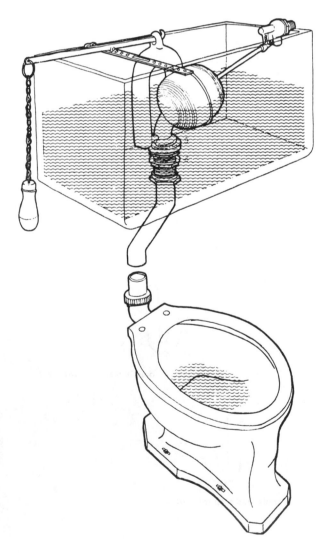

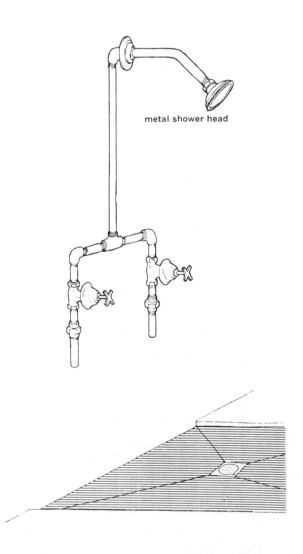

metal shower head

552

CHAPTER 39

SEWERAGE DISPOSAL SYSTEMS

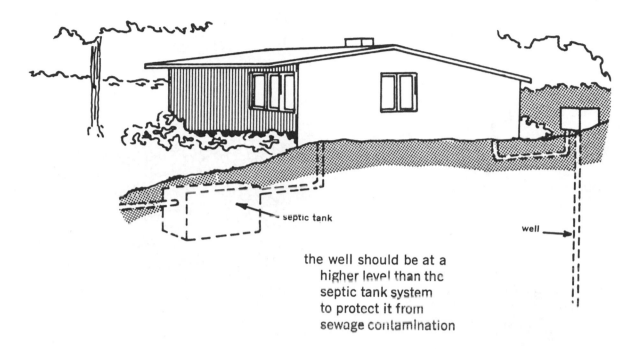

the well should be at a
higher level than the
septic tank system
to protect it from
sewage contamination

Household Septic Tank

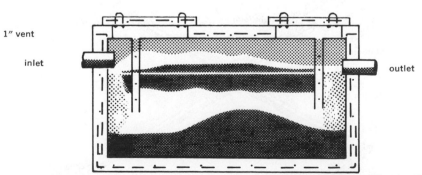

Maximum for manhole cover as shown. For greater
distances below ground, provide extension collars.

manhole cover

1" vent

inlet

outlet

Penetration of outlet baffle generally 40% of
liquid depth (for rectangular tanks)

NB: make inlet at least 1-3" above outlet

Septic Tank Disposal System

Absorption Trench and Lateral

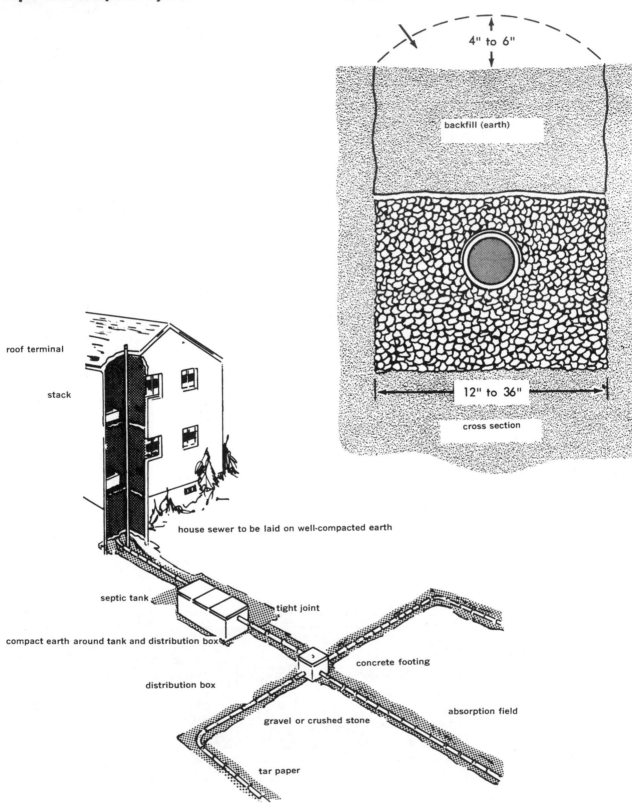

4" to 6"

backfill (earth)

12" to 36"

cross section

roof terminal

stack

house sewer to be laid on well-compacted earth

septic tank

tight joint

compact earth around tank and distribution box

concrete footing

distribution box

absorption field

gravel or crushed stone

tar paper

Sewer

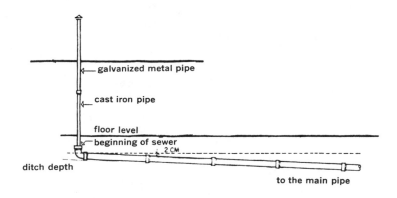

galvanized metal pipe

cast iron pipe

floor level
beginning of sewer
2 CM
ditch depth

to the main pipe

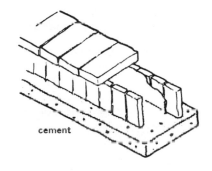

cement

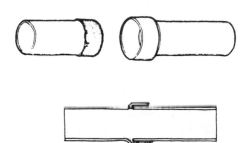

hole to protect the sewer from possible collapse

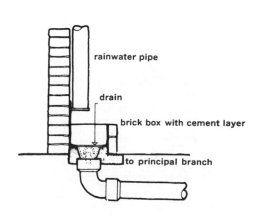

rainwater pipe

drain

brick box with cement layer

to principal branch

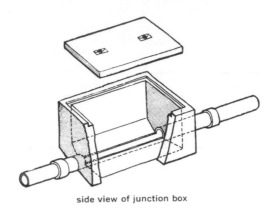

side view of junction box

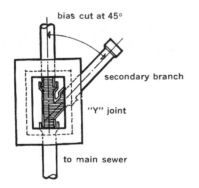

bias cut at 45°

secondary branch

"Y" joint

to main sewer

principal branch entry into junction box

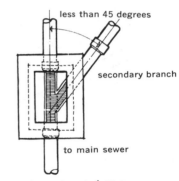

less than 45 degrees

secondary branch

to main sewer

connection at less than 45 degrees

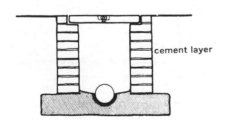

cement layer

removable cap

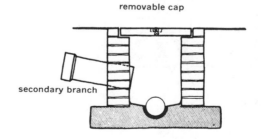

secondary branch

depth of medium sized junction box

Outhouses or Latrines

Bucket Latrine

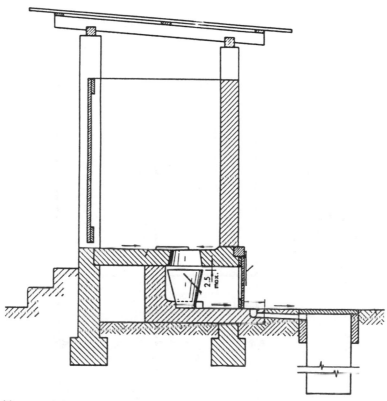

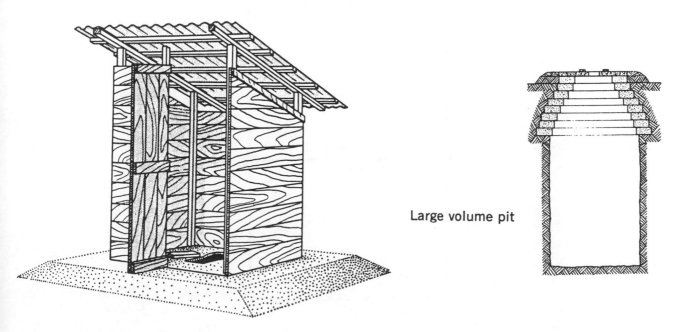

latrine of cut lumber with corrugated metal or
asbestos cement roof

Large volume pit

CHAPTER 40

ELECTRICAL SYSTEMS

A three wire system is used in homes to provide both 120 and 240 volts. 120 volts is used for large appliances such as electric stoves, water heaters, etc.

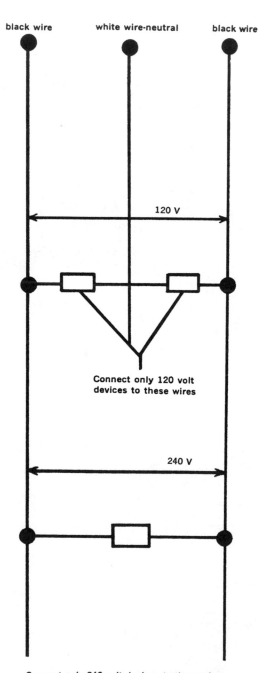

black wire · white wire-neutral · black wire

120 V

Connect only 120 volt devices to these wires

No voltage if wires are connected to same terminal. Device will not operate.

240 V

Connect only 240 volt devices to these wires

Wiring

Figures shown indicate the number
and types of wires used
in this example:
$2 \times 12 + 2 \times 18$ means that two
#12 wires and 2 #18 wires were
used.

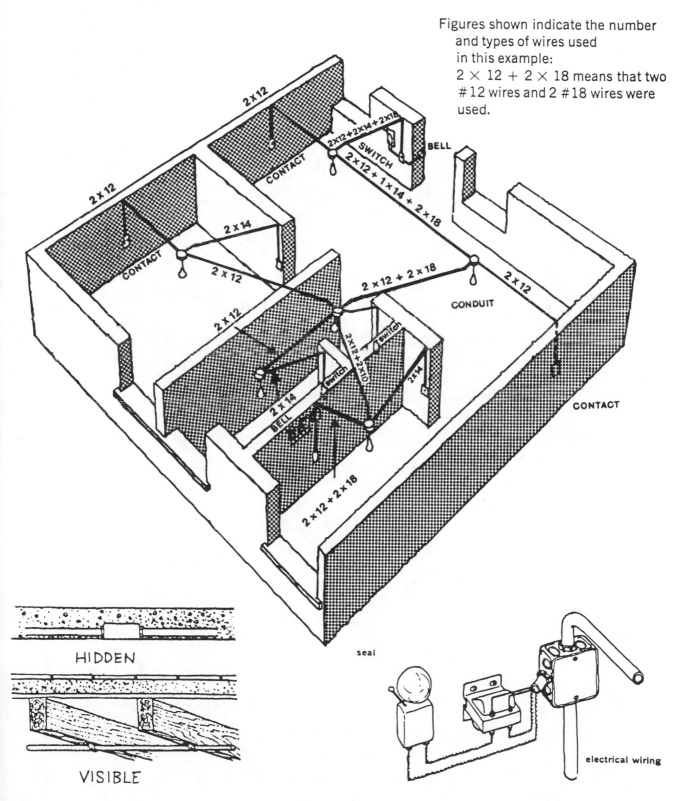

2×12

2×12

2×12

2×14

2×12

CONTACT

CONTACT

2×12+2×14+2×18   SWITCH   BELL

2×12 + 1×14 + 2×18

2×12 + 2×18   CONDUIT

2×12

CONTACT

2×12+2×10   switch

switch

2×14

2×14

BELL

BELL

2×12 + 2×18

seal

HIDDEN

VISIBLE

electrical wiring

559

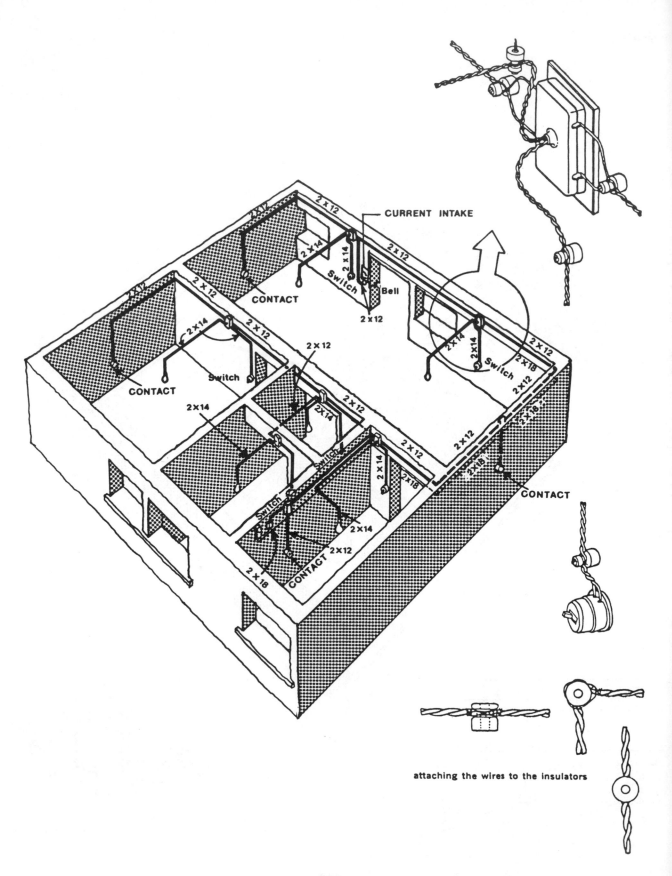

CURRENT INTAKE

2 x 12

2 x 12

2 x 14

2 x 14

Switch

Bell

2 x 12

CONTACT

2 x 12

2 x 14

2 x 12

CONTACT

Switch

2 x 14

2 x 14

Switch

2 x 12

2 x 18

2 x 12

CONTACT

2 x 12

2 x 14

2 x 12

2 x 18

2 x 12

2 x 14

2 x 18

Switch

2 x 14

2 x 14

Switch

2 x 18

Switch

2 x 14

2 x 12

CONTACT

2 x 18

attaching the wires to the insulators

560

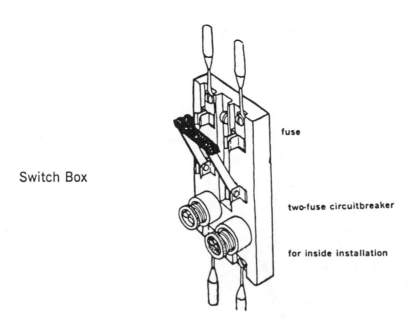

Switch Box

fuse

two-fuse circuitbreaker

for inside installation

Switch Installation

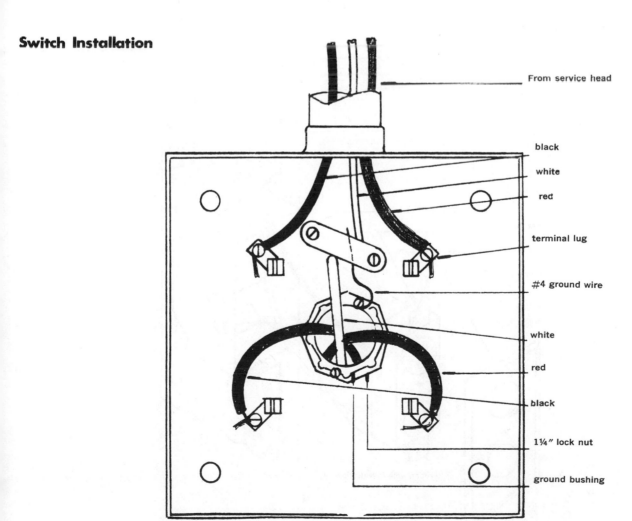

From service head

black

white

red

terminal lug

#4 ground wire

white

red

black

1¼" lock nut

ground bushing

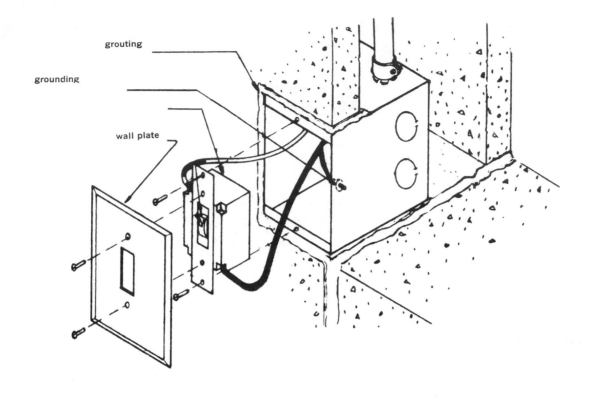

grouting

grounding

wall plate

switch box

receptacle

green screw

wall plate

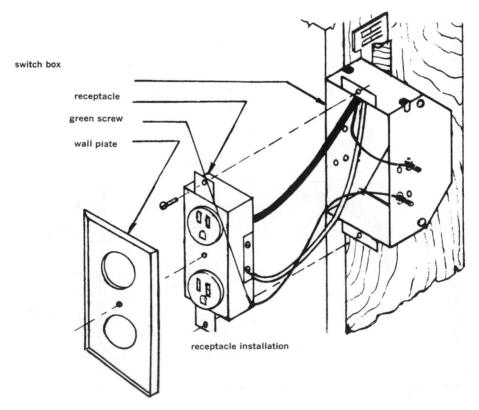

receptacle installation

CHAPTER 41

THERMAL INSULATION AND VAPOR BARRIERS

Most materials used in houses have some insulating value. Even air spaces between studs resist the passage of heat. However, when these stud spaces are filled or partially filled with a material high in resistance to heat transmission, namely thermal insulation, the stud space has many times the insulating value of the air alone.

The inflow of heat through outside walls and roofs in hot weather or its outflow during cold weather have important effects upon (a) the comfort of the occupants of a building and (b) the cost of providing either heating or cooling to maintain temperatures at acceptable limits for occupancy. During cold weather, high resistance to heat flow also means a saving in fuel. While the wood in the walls provides good insulation, commercial insulating materials are usually incorporated into exposed walls, ceilings, and floors to increase the resistance to heat passage. The use of insulation in warmer climates is justified with air conditioning, not only because of reduced operating costs but also because units of smaller capacity are required. Thus, whether from the standpoint of thermal insulation alone in cold climates or whether for the benefit of reducing cooling costs, the use of 2 inches or more of insulation in the walls can certainly be justified.

Average winter low-temperature zones of the United States are shown in fig. 41.1. These data are used in determining the size of heating plant required after calculating heat loss. This information is also useful in selecting the amount of insulation for walls, ceilings, and floors.

Insulating Materials

Commercial insulation is manufactured in a variety of forms and types, each with advantages for specific uses. Materials commonly used for insulation may be grouped in the following general classes: (1) Flexible insulation (blanket and batt); (2) loose-fill insulation; (3) reflective insulation; (4) rigid insulation (structural and nonstructural); and (5) miscellaneous types.

The thermal properties of most building materials are known, and the rate of heat flow or coefficient of transmission for most combinations of construction can be calculated (4). This coefficient, or *U-value,* is a measure of heat transmission between air on the warm side and air on the cold side of the construction unit. The insulating value of the wall will vary with different types of construction, with materials used in construction, and with different types and thickness of insulation. Comparisons of U-values may be made and used to evaluate different combinations of materials and insulation based on overall heat loss, potential fuel savings, influence on comfort, and installation costs.

Air spaces add to the total resistance of a wall section to heat transmission, but an air space is not as effective as it would be if filled with an insulating material. Great importance is frequently given to dead-air spaces in speaking of a wall section. Actually, the air in never dead in cells where there are differences in temperature on opposite sides of the space, because the difference causes convection currents.

Information regarding the calculated U-values for typical constructions with various combinations of insulation may be found in "Thermal Insulation from Wood for Buildings: Effects of Moisture and Its Control" (*15*).

Flexible Insulation

Flexible insulation is manufactured in two types, *blanket* and *batt.* Blanket insulation (fig. 41.2a) is furnished in rolls or packages in widths suited to 16- and 24-inch stud and joist spacing. Usual thicknesses are 1½, 2, and 3 inches. The body of the blanket is made of felted mats of mineral or vegetable fibers, such as rock or glass wool, wood fiber, and cotton. Organic insulations are treated to make them resistant to fire, decay, insects, and vermin. Most blanket insulation is covered with paper or other sheet material with tabs on the sides for fastening to studs or joists. One covering sheet serves as a vapor barrier to resist movement of water vapor and should always face the warm side of the wall. Aluminum foil or asphalt or plastic laminated paper are commonly used as barrier materials.

Batt insulation (fig. 41.2b) is also made of fibrous material preformed to thicknesses of 4 and 6 inches for 16- and 24-inch joist spacing. It is supplied with or without a vapor barrier. One friction type of fibrous

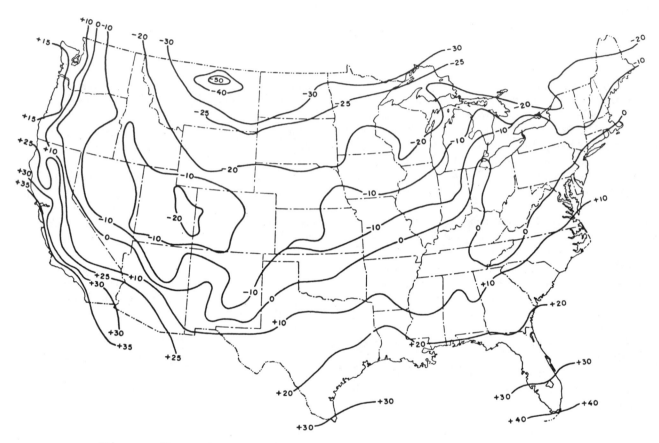

Figure 41.1.—Average outside design temperature zones of the United States.

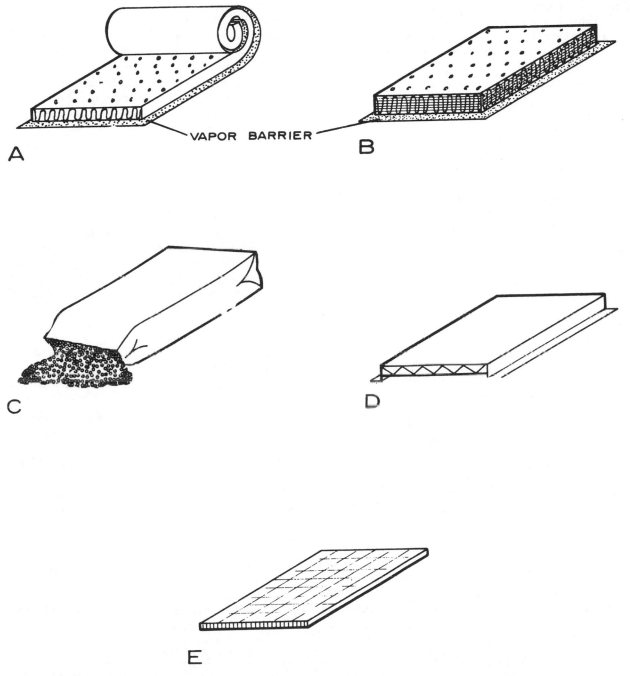

VAPOR BARRIER

A

B

C

D

E

Figure 41.2.—Types of Insulation: A, Blanket; B, batt; C, fill; D, reflective (one type), E, rigid.

glass batt is supplied without a covering and is designed to remain in place without the normal fastening methods.

Loose Fill Insulation

Loose fill insulation (fig. 41. 2c) is usually composed of materials used in bulk form, supplied in bags or bales, and placed by pouring, blowing, or packing by hand. This includes rock or glass wool, wood fibers, shredded redwood bark, cork, wood pulp products, vermiculite, sawdust, and shavings.

Fill insulation is suited for use between first-floor ceiling joists in unheated attics. It is also used in sidewalls of existing houses that were not insulated during construction. Where no vapor barrier was installed during construction, suitable paint coatings, as described later in this chapter, should be used for vapor barriers when blown insulation is added to an existing house.

Reflective Insulation

Most materials reflect some radiant heat, and some materials have this property to a very high degree (4). Materials high in reflective properties include aluminum foil, sheet metal with tin coating, and paper products coated with a reflective oxide composition. Such materials can be used in enclosed stud spaces, in attics, and in similar locations to retard heat transfer by radiation. These reflective insulations are effective only when used where the reflective surface faces an air space at least ¾ inch or more deep. Where a reflective surface contacts another material, the reflective properties are lost and the material has little or no insulating value.

Reflective insulations are equally effective regardless of whether the reflective surface faces the warm or cold side. However, there is a decided difference in the equivalent conductance and the resistance to heat flow. The difference depends on (a) the orientation of the reflecting material and the dead air space, (b) the direction of heat flow (horizontal, up, or down), and (c) the mean summer or winter temperatures. Each possibility requires separate consideration. However, reflective insulation is perhaps more effective in preventing summer heat flow through ceilings and walls. It should likely be considered more for use in the southern portion of the United States than in the northern portion.

Reflective insulation of the foil type is sometimes applied to blankets and to the stud-surface side of gypsum lath. Metal foil suitably mounted on some supporting base makes an excellent vapor barrier. The type of reflective insulation shown in (fig. 41.2d) includes reflective surfaces and air spaces between the outer sheets.

Rigid Insulation

Rigid insulation is usually a fiberboard material manufactured in sheet and other forms (fig. 41.2e). However, rigid insulations are also made from such materials as inorganic fiber and glass fiber, though not commonly used in a house in this form. The most common types are made from processed wood, sugarcane, or other vegetable products. Structural insulating boards, in densities ranging from 15 to 31 pounds per cubic foot, are fabricated in such forms as building boards, roof decking, sheathing, and wallboard. While they have moderately good insulating properties, their primary purpose is structural.

Roof insulation is nonstructural and serves mainly to provide thermal resistance to heat flow in roofs. It is called "slab" or "block" insulation and is manufactured in rigid units ½ to 3 inches thick and usually 2 by 4 feet in size.

In house construction, perhaps the most common forms of rigid insulation are sheathing and decorative coverings in sheets or in tile squares. Sheathing board is made in thicknesses of ½ and $^{25}/_{32}$ inch. It is coated or impregnated with an asphalt compound to provide water resistance. Sheets are made in 2- by 8-foot size for horizontal application and 4- by 8-feet or longer for vertical application.

Miscellaneous Insulation

Some insulations do not fit in the classifications previously described, such as insulation blankets made up of multiple layers of corrugated paper. Other types, such as lightweight vermiculite and perlite aggregates, are sometimes used in plaster as a means of reducing heat transmission.

Other materials are foamed-in-place insulations, which include sprayed and plastic foam types. Sprayed insulation is usually inorganic fibrous material blown against a clean surface which has been primed with an adhesive coating. It is often left exposed for acoustical as well as insulating properties.

Expanded *polystyrene* and *urethane* plastic foams may be molded or foamed-in-place. Urethane insulation may also be applied by spraying. Polystyrene and urethane in board form can be obtained in thicknesses from ½ to 2 inches.

Values in table 6 will provide some comparison of the insulating value of the various materials. These are expressed as "k" values or heat conductivity and are defined as the amount of heat, in British thermal units, that will pass in 1 hour through 1 square foot of material 1 inch thick per 1° F. temperature differ-

TABLE 6.—*Thermal conductivity values of some insulating materials*

Insulation group		"k" range (conductivity)
General	Specific type	
Flexible		0.25 — 0.27
Fill	Standard materials	.28 — .30
	Vermiculite	.45 — .48
Reflective (2 sides)		(¹)
Rigid	Insulating fiberboard	.35 — .36
	Sheathing fiberboard	.42 — .55
Foam	Polystyrene	.25 — .29
	Urethane	.15 — .17
Wood	Low density	.60 — .65

¹ Insulating value is equal to slightly more than 1 inch of flexible insulation. (Resistance, "R" = 4.3)

ence between faces of the material. Simply expressed, "k" represents heat loss; the lower this numerical value, the better the insulating qualities.

Insulation is also rated on its resistance or "R" value, which is merely another expression of its insulating value. The "R" value is usually expressed as the total resistance of the wall or of a thick insulating blanket or batt, whereas "k" is the rating per inch of thickness. For example, a "k" value of 1 inch of insulation is 0.25. Then the resistance, "R" is $\frac{1}{0.25}$ or 4.0. If there is three inches of this insulation, the total "R" is three times 4.0, or 12.0.

The "U" value is the overall heat-loss value of all materials in the wall. The lower this value, the better the insulating value. Specific insulating values for various materials are also available (*4, 15*). For comparison with table 6, the "U" value of window glass is:

Glass	U value
Single	1.13
Double	
Insulated, with ¼-inch air space	.61
Storm sash over single glazed window	.53

Where to Insulate

To reduce heat loss from the house during cold weather in most climates, all walls, ceilings, roofs, and floors that separate heated from unheated spaces should be insulated.

Insulation should be placed on all outside walls and in the ceiling (fig. 41.3a). In houses involving unheated crawl spaces, it should be placed between the floor joists or around the wall perimeter. If a flexible type of insulation (blanket or batt) is used, it should be well-supported between joists by slats and a galvanized wire mesh, or by a rigid board with the vapor barrier installed toward the subflooring. Press-fit or friction insulations fit tightly between joists and require only a small amount of support to hold them in place. Reflective insulation is often used for crawl spaces, but only one dead-air space should be assumed in calculating heat loss when the crawl space is ventilated. A ground cover of roll roofing or plastic film such as polyethylene should be placed on the soil of crawl spaces to decrease the moisture content of the space as well as of the wood members.

In 1½-story houses, insulation should be placed along all walls, floors, and ceilings that are adjacent to unheated areas (fig. 41. 3b). These include stairways, dwarf (knee) walls, and dormers. Provisions should be made for ventilation of the unheated areas.

Where attic space is unheated and a stairway is included, insulation should be used around the stairway as well as in the first-floor ceiling (fig. 41.3c). The door leading to the attic should be weather-

stripped to prevent heat loss. Walls adjoining an unheated garage or porch should also be insulated.

In houses with flat or low-pitched roofs (fig. 41.3d), insulation should be used in the ceiling area with sufficient space allowed above for clear unobstructed ventilation between the joists. Insulation should be used along the perimeter of houses built on slabs. A vapor barrier should be included under the slab.

In the summer, outside surfaces exposed to the direct rays of the sun may attain temperatures of 50° F. or more above shade temperatures and, of course, tend to transfer this heat toward the inside of the house. Insulation in the walls and in attic areas retards the flow of heat and, consequently, less heat is transferred through such areas, resulting in improved summer comfort conditions.

Where air-conditioning systems are used, insulation should be placed in all exposed ceilings and walls in the same manner as insulating against cold-weather heat loss. Shading of glass against direct rays of the sun and the use of insulated glass will aid in reducing the air-conditioning load.

Ventilation of attic and roof spaces is an important adjunct to insulation. Without ventilation, an attic space may become very hot and hold the heat for many hours. (See Chapter 42, "Ventilation.") Obviously, more heat will be transmitted through the ceiling when the attic temperature is 150° F. than if it is 100° to 120° F. Ventilation methods suggested for protection against cold-weather condensation apply equally well to protection against excessive hot-weather roof temperatures.

The use of storm windows or insulated glass will greatly reduce heat loss. Almost twice as much heat loss occurs through a single glass as through a window glazed with insulated glass or protected by a storm sash. Furthermore, double glass will normally prevent surface condensation and frost forming on inner glass surfaces in winter. When excessive condensation persists, paint failures or even decay of the sash rail or other parts can occur.

How to Install Insulation

Blanket insulation or batt insulation with a vapor barrier should be placed between framing members so that the tabs of the barrier lap the edge of the studs as well as the top and bottom plates. This method is not often popular with the contractor because it is more difficult to apply the dry wall or rock lath (plaster base). However, it assures a minimum amount of vapor loss compared to the loss when tabs are stapled to the sides of the studs. To protect the head and soleplate as well as the headers over openings, it is good practice to use narrow strips of vapor barrier material along the top and bottom of the wall (fig. 41.4a). Ordinarily, these areas are not covered too well by the barrier on the blanket or

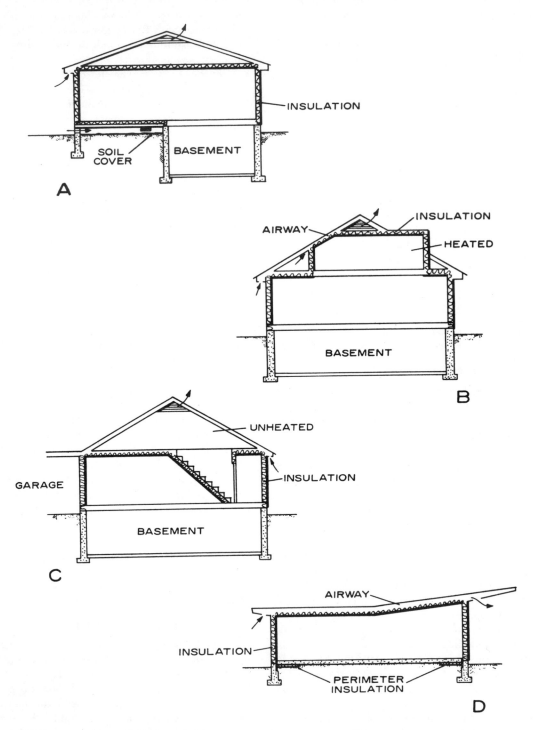

Figure 41.3.—Placement of insulation: *A,* In walls, floor, and ceiling; *B,* in 1½-story house; *C,* at attic door; *D,* in flat roof.

batt. A hand stapler is commonly used to fasten the insulation and the barriers in place.

For insulation without a barrier (press-fit or friction type), a plastic film vapor barrier such as 4-mil polyethylene is commonly used to envelop the entire exposed wall and ceiling (fig. 41.4b). It covers the openings as well as window and door headers and edge studs. This system is one of the best from the standpoint of resistance to vapor movement. Furthermore, it does not have the installation inconveniences encountered when tabs of the insulation are stapled over the edges of the studs. After the dry wall is installed or plastering is completed, the film is trimmed around the window and door openings.

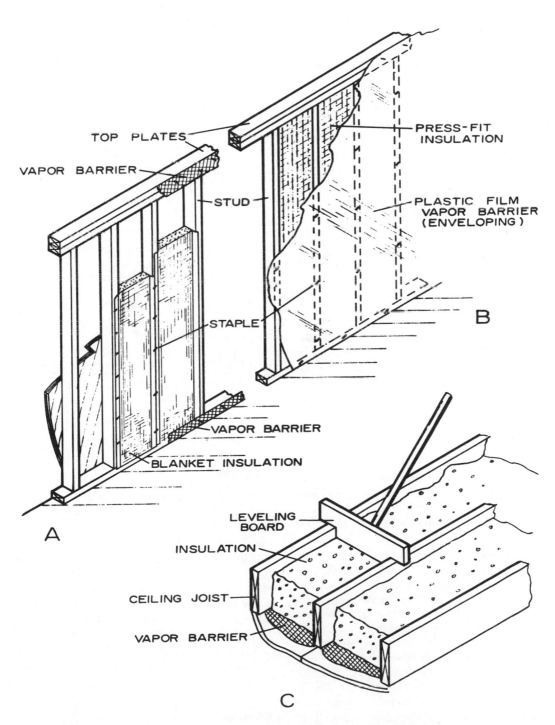

Figure 41.4.—Application of insulation: A, Wall section with blanket type; B, wall section with "press-fit" insulation; C, ceiling with full insulation.

Reflective insulation, in a single-sheet form with two reflective surfaces, should be placed to divide the space formed by the framing members into two approximately equal spaces. Some reflective insulations include air spaces and are furnished with nailing tabs. This type is fastened to the studs to provide at least a ¾-inch space on each side of the reflective surfaces.

Fill insulation is commonly used in ceiling areas and is poured or blown into place (fig. 41. 4c). A vapor barrier should be used on the warm side (the bottom, in case of ceiling joists) before insulation is placed. A leveling board (as shown) will give a constant insulation thickness. Thick batt insulation is also used in ceiling areas. Batt and fill insulation might also be combined to obtain the desired thickness with the vapor barrier against the back face of the ceiling finish. Ceiling insulation 6 or more inches thick greatly reduces heat loss in the winter and also provides summertime protection.

Precautions in Insulating

Areas over door and window frames and along side and head jambs also require insulation. Because these areas are filled with small sections of insulation, a vapor barrier must be used around the opening as well as over the header above the openings (fig. 41.5a). Enveloping the entire wall eliminates the need for this type of vapor barrier installation.

In 1½- and 2-story houses and in basements, the area at the joist header at outside walls should be insulated and protected with a vapor barrier (fig. 41.5b).

Insulation should be placed behind electrical outlet boxes and other utility connections in exposed walls to minimize condensation on cold surfaces.

Vapor Barriers

Some discussion of vapor barriers has been included in the previous sections because vapor barriers are usually a part of flexible insulation. However, further information is included in the following paragraphs.

Most building materials are permeable to water vapor. This presents problems because considerable water vapor is generated in a house from cooking, dishwashing, laundering, bathing, humidifiers, and other sources. In cold climates during cold weather, this vapor may pass through wall and ceiling materials and condense in the wall or attic space; subsequently, in severe cases, it may damage the exterior paint and interior finish, or even result in decay in structural members. For protection, a material highly resistive to vapor transmission, called a *vapor barrier*, should be used on the warm side of a wall or below the insulation in an attic space.

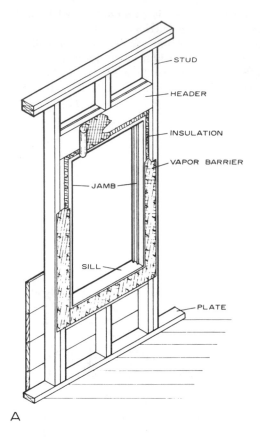

A

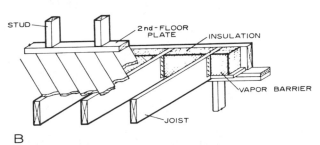

B

Figure 41.5.
—Precautions in insulating: A, Around openings; B, joist space in outside walls.

Among the effective vapor-barrier materials are asphalt laminated papers, aluminum foil, and plastic films. Most blanket and batt insulations are provided with a vapor barrier on one side, some of them with paper-backed aluminum foil. Foil-backed gypsum lath or gypsum boards are also available and serve as excellent vapor barriers.

The perm values of vapor barriers vary (17), but ordinarily it is good practice to use those which have values less than ¼ (0.25) perm. Although a value of ½ perm is considered adequate, aging reduces the effectiveness of some materials.

Some types of flexible blanket and batt insulations

have a barrier material on one side. Such flexible insulations should be attached with the tabs at their sides fastened on the inside (narrow) edges of the studs, and the blanket should be cut long enough so that the cover sheet can lap over the face of the sole-plate at the bottom and over the plate at the top of the stud space. However, such a method of attachment is not the common practice of most installers. When a positive seal is desired, wall-height rolls of plastic-film vapor barriers should be applied over studs, plates, and window and door headers. This system, called "enveloping," is used over insulation having no vapor barrier or to insure excellent protection when used over any type of insulation. The barrier should be fitted tightly around outlet boxes and sealed if necessary. A ribbon of sealing compound around an outlet or switch box will minimize vapor loss at this area. Cold-air returns in outside walls should consist of metal ducts to prevent vapor loss and subsequent paint problems.

Paint coatings on plaster may be very effective as vapor barriers if materials are properly chosen and applied. They do not however, offer protection during the period of construction, and moisture may cause paint blisters on exterior paint before the interior paint can be applied. This is most likely to happen in buildings that are constructed during periods when outdoor temperatures are 25° F. or more below inside temperatures. Paint coatings cannot be considered a substitute for the membrane types of vapor barriers, but they do provide some protection for houses where other types of vapor barriers were not installed during construction.

Of the various types of paint, one coat of *aluminum primer* followed by two decorative coats of *flat wall* or *lead and oil* paint is quite effective. For rough plaster or for buildings in very cold climates, two coats of the aluminum primer may be necessary. A primer and sealer of the pigmented type, followed by decorative finish coats or two coats of rubber-base paint, are also effective in retarding vapor transmission.

Because no type of vapor barrier can be considered 100 percent resistive, and some vapor leakage into the wall may be expected, the flow of vapor to the outside should not be impeded by materials of relatively high vapor resistance on the cold side of the vapor barrier. For example, sheathing paper should be of a type that is waterproof but not highly vapor resistant. This also applies to "permanent" outer coverings or siding. In such cases, the vapor barrier should have an equally low perm value. This will reduce the danger of condensation on cold surfaces within the wall.

CHAPTER 42

VENTILATION

Condensation of moisture vapor may occur in attic spaces and under flat roofs during cold weather. Even where vapor barriers are used, some vapor will probably work into these spaces around pipes and other inadequately protected areas and some through the vapor barrier itself. Although the amount might be unimportant if equally distributed, it may be sufficiently concentrated in some cold spot to cause damage. While wood shingle and wood shake roofs do not resist vapor movement, such roofings as asphalt shingles and built-up roofs are highly resistant. The most practical method of removing the moisture is by adequately ventilating the roof spaces.

A warm attic that is inadequately ventilated and insulated may cause formation of *ice dams* at the cornice. During cold weather after a heavy snowfall, heat causes the snow next to the roof to melt (fig. 34.1). Water running down the roof freezes on the colder surface of the cornice, often forming an ice dam at the gutter which may cause water to back up at the eaves and into the wall and ceiling. Similar dams often form in roof valleys. Ventilation thus provides part of the answer to the problems. With a well-insulated ceiling and adequate ventilation, attic temperatures are low and melting of snow over the attic space will be greatly reduced (15).

In hot weather, ventilation of attic and roof spaces offers an effective means of removing hot air and thereby materially lowering the temperature in these spaces. Insulation should be used between ceiling joists below the attic or roof space to further retard heat flow into the rooms below and materially improve comfort conditions.

It is common practice to install louvered openings in the end walls of gable roofs for ventilation. Air movement through such openings depends primarily on wind direction and velocity, and no appreciable movement can be expected when there is no wind or unless one or more openings face the wind. More positive air movement can be obtained by providing openings in the soffit areas of the roof overhang in addition to openings at the gable ends or ridge. Hip-roof houses are best ventilated by inlet ventilators in the soffit area and by outlet ventilators along the ridge.

The differences in temperature between the attic and the outside will then create an air movement independent of the wind, and also a more positive movement when there is wind.

Where there is a crawl space under house or porch, ventilation is necessary to remove moisture vapor rising from the soil. Such vapor may otherwise condense on the wood below the floor and facilitate decay. A permanent vapor barrier on the soil of the crawl space greatly reduces the amount of ventilating area required.

Tight construction (including storm window and storm doors) and the use of humidifiers have created potential moisture problems which must be resolved through planning of adequate ventilation as well as the proper use of vapor barriers. Blocking of ventilating areas, for example, must be avoided as such practices will prevent ventilation of attic spaces. Inadequate ventilation will often lead to moisture problems which can result in unnecessary costs to correct.

Area of Ventilators

Types of ventilators and minimum recommended sizes have been generally established for various types of roofs. The minimum net area for attic or roof-space ventilators is based on the projected ceiling area of the rooms below (fig. 42.1). The ratio of ventilator openings as shown are net areas, and the actual area must be increased to allow for any restrictions such as louvers and wire cloth or screen. The screen area should be double the specified net area shown in figures 42.1 to 42.3.

To obtain extra area of screen without adding to the area of the vent, use a frame of required size to hold the screen away from the ventilator opening. Use as coarse a screen as conditions permit, not smaller than No. 16, for lint and dirt tend to clog fine-mesh screens. Screens should be installed in such a way that paint brushes will not easily contact the screen and close the mesh with paint.

Gable Roofs

Louvered openings are generally provided in the end walls of gable roofs and should be as close to the ridge as possible (fig. 42.1a). The net area for the openings should be 1/300 of the ceiling area (fig. 42.1a). For example, where the ceiling area equals 1,200 square feet, the minimum total net area of the ventilators should be 4 square feet.

As previously explained, more positive air move-

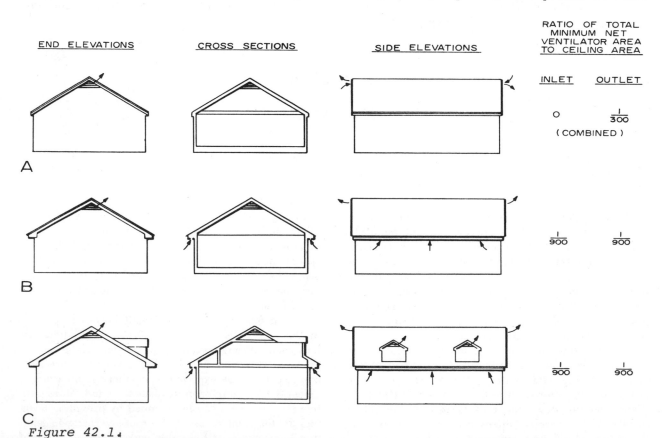

Figure 42.1.

Ventilating areas of gable roofs: A, Louvers in end walls; B, louvers in end walls with additional openings in soffit area; C, louvers at end walls with additional openings at eaves and dormers. Cross section of C shows free opening for air movement between roof boards and ceiling insulation of attic room.

ment can be obtained if additional openings are provided in the soffit area. The minimum ventilation areas for this method are shown in fig. 42.1b

Where there are rooms in the attic with sloping ceilings under the roof, the insulation should follow the roof slope and be so placed that there is a free opening of at least 1½ inches between the roof boards and insulation for air movement (fig. 42. 1c).

Hip Roofs

Hip roofs should have air-inlet openings in the soffit area of the eaves and outlet openings at or near the peak. For minimum net areas of openings see fig. 42. 2a . The most efficient type of inlet opening is the continuous slot, which should provide a free opening of not less than ¾ inch. The air-outlet opening near the peak can be a globe-type metal ventilator or several smaller roof ventilators located near the ridge. They can be located below the peak on the rear slope of the roof so that they will not be visible from the front of the house. Gabled extensions of a hip-roof house are sometimes used to provide efficient outlet ventilators (fig. 42.2b).

Flat Roofs

A greater ratio of ventilating area is required in some types of flat roofs than in pitched roofs because the air movement is less positive and is dependent upon wind. It is important that there be a clear open space above the ceiling insulation and below the roof sheathing for free air movement from inlet to outlet openings. Solid blocking should *not* be used for bridging

or for bracing over bearing partitions if its use prevents the air circulation.

Perhaps the most common type of flat or low-pitched roof is one in which the rafters extend beyond the wall, forming an overhang (fig. 42.3a). When soffits are used, this area can contain the combined inlet-outlet ventilators, preferably a continuous slot. When single ventilators are used, they should be distributed evenly along the overhang.

A parapet-type wall and flat roof combination may be constructed with the ceiling joists separate from the roof joists or combined. When members are separate, the space between can be used for an airway (fig. 42.3b). Inlet and outlet vents are then located as shown, or a series of outlet stack vents can be used along the centerline of the roof in combination with the inlet vents. When ceiling joists and flat rafters are served by one member in parapet construction, vents may be located as shown in fig. 42.3c . Wall inlet ventilators combined with center stack outlet vents is another variable in this type of roof.

Types and Location of Outlet Ventilators

Various styles of gable-end ventilators are available ready for installation. Many are made with metal louvers and frames, while others may be made of wood to fit the house design more closely. However, the most important factors are to have sufficient net ventilating area and to locate ventilators as close to the ridge as possible without affecting house appearance.

One of the types commonly used fits the slope of the roof and is located near the ridge (fig. 42.4a). It can

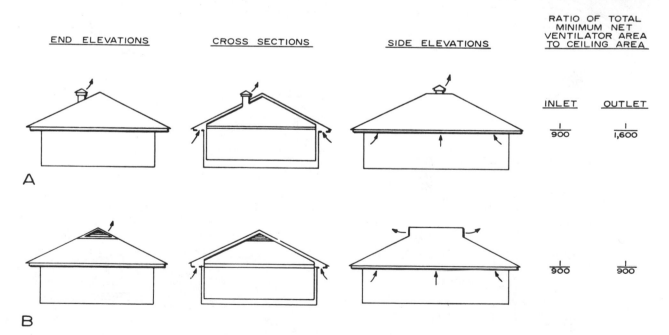

Figure 42.2.—Ventilating areas of hip roofs: A, Inlet openings beneath eaves and outlet vent near peak; B, inlet openings beneath eaves and ridge outlets.

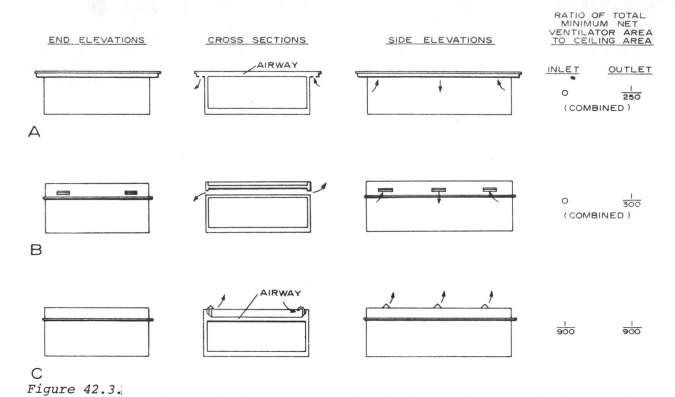

END ELEVATIONS   CROSS SECTIONS   SIDE ELEVATIONS   RATIO OF TOTAL MINIMUM NET VENTILATOR AREA TO CEILING AREA

INLET   OUTLET

A   O $\frac{1}{250}$
(COMBINED)

B   O $\frac{1}{300}$
(COMBINED)

C $\frac{1}{900}$ $\frac{1}{900}$

AIRWAY

Figure 42.3.

Ventilating area of flat roofs: *A*, Ventilator openings under overhanging eaves where ceiling and roof joists are combined; *B*, for roof with a parapet where roof and ceiling joists are separate; *C*, for roof with a parapet where roof and ceiling joists are combined.

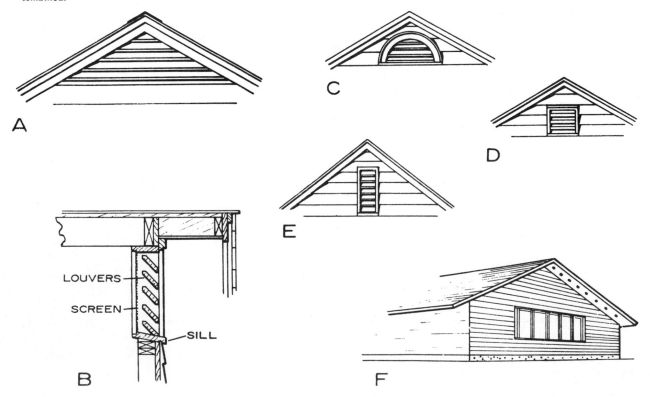

LOUVERS
SCREEN
SILL

Figure 42.4.—Outlet ventilators: *A*, Triangular; *B*, typical cross section; *C*, half-circle; *D*, square; *E*, vertical; *F*, soffit.

be made of wood or metal; in metal it is often adjustable to conform to the roof slope. A wood ventilator of this type is enclosed in a frame and placed in the rough opening much as a window frame (fig. 42.4b). Other forms of gable-end ventilators which might be used are shown in figs. 42.4 c, d, and e.

A system of attic ventilation which can be used on houses with a wide roof overhang at the gable end consists of a series of small vents or a continuous slot located on the underside of the soffit areas (fig. 42. 4f). Several large openings located near the ridge might also be used. This system is especially desirable on low-pitched roofs where standard wall ventilators may not be suitable.

It is important that the roof framing at the wall line does not block off ventilation areas to the attic area. This might be accomplished by the use of a "ladder" frame extension. A flat nailing block used at the wall line will provide airways into the attic (fig. 33.6b). This can also be adapted to narrower rake sections by providing ventilating areas to the attic.

Types and Location of Inlet Ventilators

Small, well-distributed ventilators or a continuous slot in the soffit provide inlet ventilation. These small

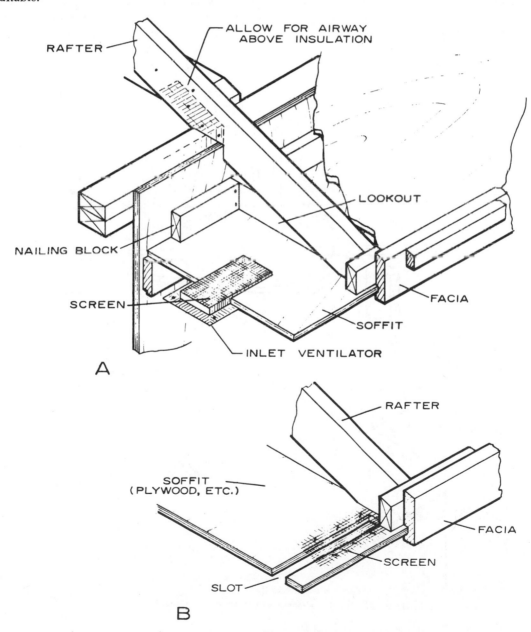

Figure 42.5.—Inlet ventilators: A, Small insert ventilator; B, slot ventilator.

louvered and screened vents can be obtained in most lumberyards or hardware stores and are simple to install.

Only small sections need to be cut out of the soffit; these can be sawed out before the soffit is applied. It is more desirable to use a number of smaller well-distributed ventilators than several large ones (fig. 42.5a). Any blocking which might be required between rafters at the wall line should be installed so as to provide an airway into the attic area.

A continuous screened slot, which is often desirable, should be located near the outer edge of the soffit near the facia (fig. 42.5b). Locating the slot in this area will minimize the chance of snow entering. This type may also be used on the extension of flat roofs.

Crawl-space Ventilation and Soil Cover

The crawl space below the floor of a basementless house and under porches should be ventilated and protected from ground moisture by the use of a *soil cover* (fig. 42.6). The soil cover should be a vapor barrier with a perm value of less than 1.0. This includes such barrier materials as plastic films, roll roofing, and asphalt laminated paper. Such protection will minimize the effect of ground moisture on the wood framing members. High moisture content and humidity encourage staining and decay of untreated members.

Where there is a partial basement open to a crawl-space area, no wall vents are required if there is some type of operable window. The use of a soil cover in the crawl space is still important, however. For crawl

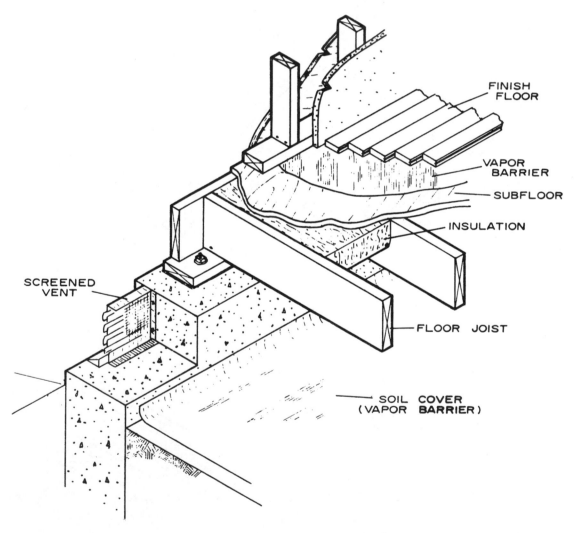

*Figure 42.6.—*Crawl-space ventilator and soil cover.

spaces with no basement area, provide at least four foundation-wall vents near corners of the building. The total free (net) area of the ventilators should be equal to 1/160 of the ground area when no soil cover is used. Thus, for a ground area of 1,200 square feet, a total net ventilating area of about 8 square feet is required, or 2 square feet for each of four ventilators. More smaller ventilators having the same net ratio is satisfactory.

When a vapor barrier ground cover is used, the required ventilating area is greatly reduced. The net ventilating area required with a ground cover is 1/1600 of the ground area, or for the 1,200-square-foot house, an area of 0.75 square foot. This should be divided between two small ventilators located on opposite sides of the crawl space. Vents should be covered (fig. 42.6) with a corrosion-resistant screen of No. 8 mesh.

The use of a ground cover is normally recommended under all conditions. It not only protects wood framing members from ground moisture but also allows the use of small, inconspicuous ventilators.

CHAPTER 43

SOUND INSULATION

Development of the "quiet" home or the need for incorporating sound insulation in a new house is becoming more and more important. In the past, the reduction of sound transfer between rooms was more important in apartments, motels, and hotels than in private homes. However, house designs now often incorporate a family room or "active" living room as well as "quiet" living room. It is usually desirable in such designs to isolate these rooms from the remainder of the house. Sound insulation between the bedroom area and the living area is usually desirable, as is isolation of the bathrooms and lavatories. Isolation from outdoor sounds is also often advisable. Thus, sound control has become a vital part of house design and construction, and will be even more important in the coming years.

How Sound Travels

How does sound travel, and how is it transferred through a wall or floor? Airborne noises inside a house, such as loud conversation or a barking dog, create sound waves which radiate outward from the source through the air until they strike a wall, floor, or ceiling. These surfaces are set in vibration by the fluctuating pressure of the sound wave in the air. Because the wall vibrates, it conducts sound to the other side in varying degrees, depending on the wall construction.

The resistance of a building element, such as a wall, to the passage of airborne sound is rated by its *Sound Transmission Class* (STC). Thus, the higher the number, the better the sound barrier. The approximate effectiveness of walls with varying STC numbers is shown in the following tabulation:

STC No.	Effectiveness
25	Normal speech can be understood quite easily
35	Loud speech audible but not intelligible
45	Must strain to hear loud speech
48	Some loud speech barely audible
50	Loud speech not audible

Sound travels readily through the air and also through some materials. When airborne sound strikes a conventional wall, the studs act as sound conductors unless they are separated in some way from the covering material. Electrical switches or convenience outlets placed back-to-back in a wall readily pass sound. Faulty construction, such as poorly fitted doors, often allows sound to travel through. Thus, good construction practices are important in providing sound-resistant walls, as well as those measures commonly used to stop ordinary sounds.

Thick walls of dense materials such as masonry can stop sound. But in the wood-frame house, an interior masonry wall results in increased costs and structural problems created by heavy walls. To provide a satisfactory sound-resistant wall economically has been a problem. At one time, sound-resistant frame construction for the home involved significant additional costs because it usually meant double walls or suspended ceilings. However, a relatively simple system has been developed using sound-deadening insulating board in conjunction with a gypsum board outer covering. This provides good sound-transmission resistance suitable for use in the home with only slight additional cost. A number of combinations are possible with this system, providing different STC ratings.

Wall Construction

As the preceding STC tabulation shows, a wall providing sufficient resistance to airborne sound transfer likely has an STC rating of 45 or greater. Thus, in construction of such a wall between the rooms of a house, its cost as related to the STC rating should be considered. As shown in   fig. 43.1, details *A*, with gypsum wallboard, and *B*, with plastered wall, are those commonly used for partition walls. However, the hypothetical rating of 45 cannot be obtained in this construction. An 8-inch concrete block wall (fig. 43.1c) has the minimum rating, but this construction is not always practical in a wood-frame house.

Good STC ratings can be obtained in a wood-frame wall by using the combination of materials shown in figure 43.1 d and e. One-half-inch sound-deadening board nailed to the studs, followed by a lamination

WALL DETAIL	DESCRIPTION	STC RATING
A	½" GYPSUM WALLBOARD	32
	⅝" GYPSUM WALLBOARD	37
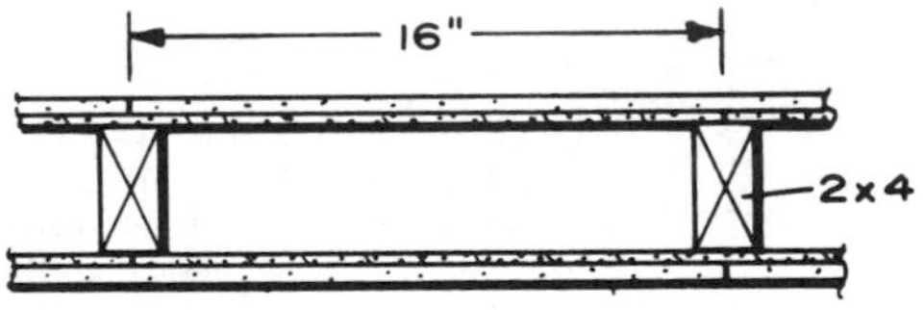 B	⅜" GYPSUM LATH (NAILED) PLUS ½" GYPSUM PLASTER WITH WHITECOAT FINISH (EACH SIDE)	39
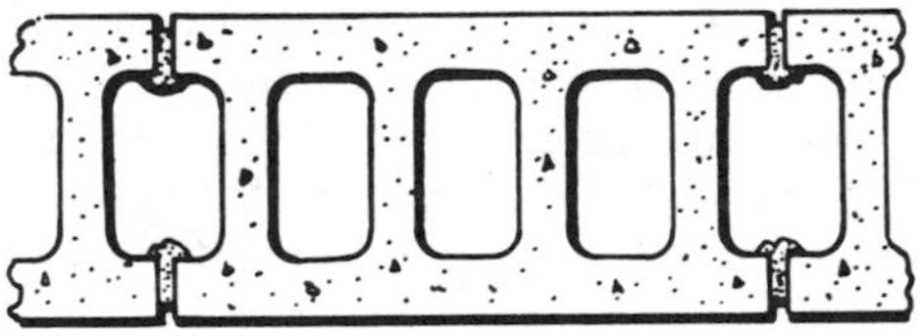 C	8" CONCRETE BLOCK	45
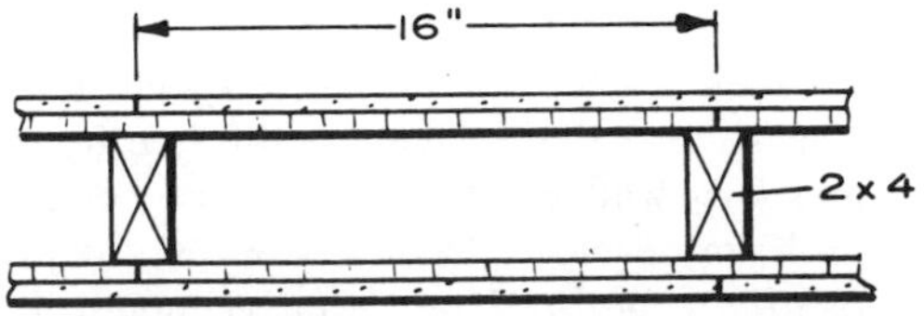 D	½" SOUND DEADENING BOARD (NAILED) ½" GYPSUM WALLBOARD (LAMINATED) (EACH SIDE)	46
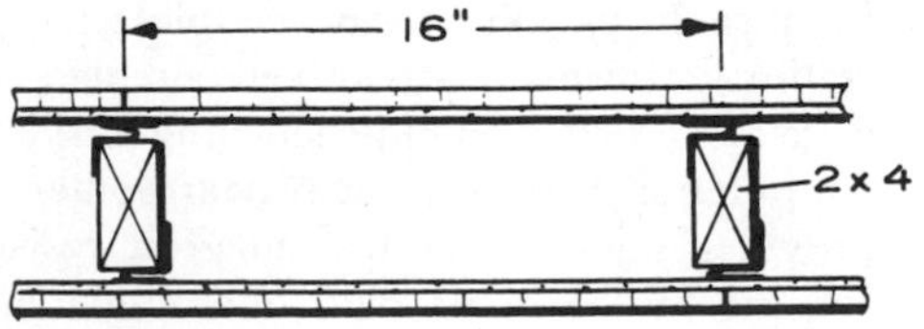 E	RESILIENT CLIPS TO ⅜" GYPSUM BACKER BOARD ½" FIBERBOARD (LAMINATED) (EACH SIDE)	52

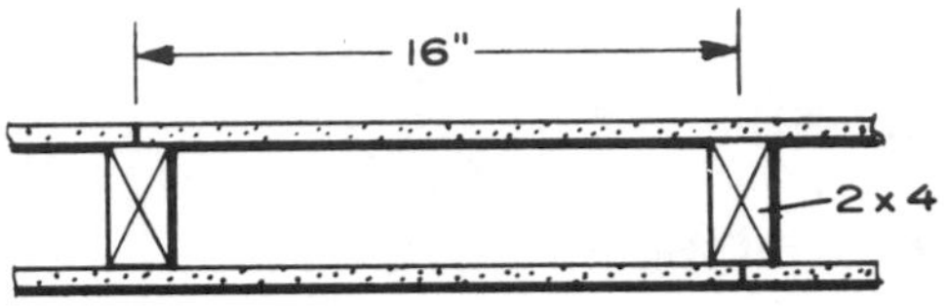

Figure 43.1.—Sound insulation of single walls.

of ½-inch gypsum wallboard, will provide an STC value of 46 at a relatively low cost. A slightly better rating can be obtained by using ⅝-inch gypsum wallboard rather than ½-inch. A very satisfactory STC rating of 52 can be obtained by using resilient clips to fasten gypsum backer boards to the studs, followed by adhesive-laminated ½-inch fiberboard (fig. 43.1e). This method further isolates the wall covering from the framing.

A similar isolation system consists of resilient channels nailed horizontally to 2- by 4-inch studs spaced 16 inches on center. Channels are spaced 24 inches apart vertically and ⅝-inch gypsum wallboard is screwed to the channels. An STC rating of 47 is thus obtained at a moderately low cost.

The use of a double wall, which may consist of a 2 by 6 or wider plate and staggered 2- by 4-inch studs, is sometimes desirable. One-half-inch gypsum wallboard on each side of this wall (fig. 43.2a). results in an STC value of 45. However, two layers of ⅝-inch gypsum wallboard add little, if any, additional sound-transfer resistance (fig. 43.2b).. When 1½-inch blanket insulation is added to this construction (fig. 43.2c), the STC rating increases to 49. This insulation may be installed as shown or placed between studs on one wall. A single wall with 3½ inches of insulation will show a marked improvement over an open stud space and is low in cost.

The use of ½-inch sound-deadening board and a lamination of gypsum wallboard in the double wall will result in an STC rating of 50 (fig. 43.2d). The addition of blanket insulation to this combination will likely provide an even higher value, perhaps 53 or 54.

Floor-Ceiling Construction

Sound insulation between an upper floor and the

WALL DETAIL	DESCRIPTION	STC RATING
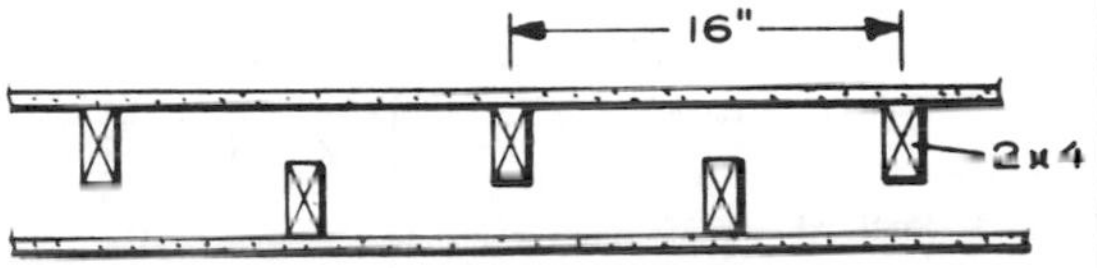 A	½" GYPSUM WALLBOARD	45
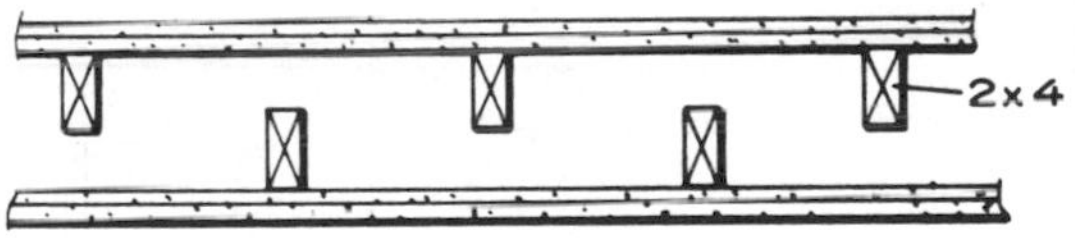 B	⅝" GYPSUM WALLBOARD (DOUBLE LAYER EACH SIDE)	45
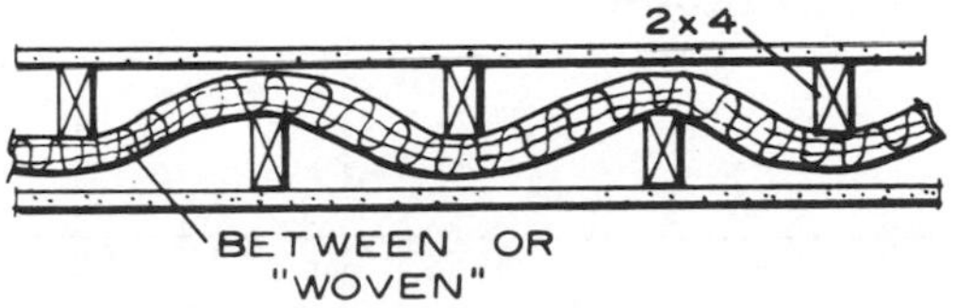 C	½" GYPSUM WALLBOARD 1½" FIBROUS INSULATION	49
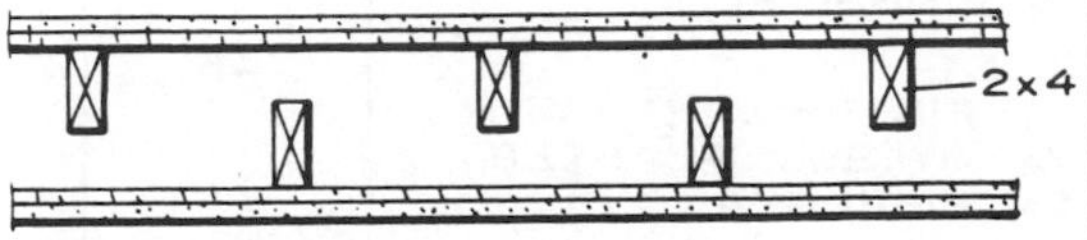 D	½" SOUND DEADENING BOARD (NAILED) ½" GYPSUM WALLBOARD (LAMINATED)	50

Figure 43.2.—Sound insulation of double walls.

ceiling of a lower floor not only involves resistance of airborne sounds but also that of impact noises. Thus, impact noise control must be considered as well as the STC value. Impact noise is caused by an object striking or sliding along a wall or floor surface, such as by dropped objects, footsteps, or moving furniture. It may also be caused by the vibration of a dish-washer, bathtub, food-disposal apparatus, or other equipment. In all instances, the floor is set into vibra-tion by the impact or contact and sound is radiated from both sides of the floor.

A method of measuring impact noise has been developed and is commonly expressed as the *Impact Noise Ratings (INR)*.[6] The greater the positive value of the INR, the more resistant is the floor to impact noise transfer. For example, an INR of -2 is better than one of -17, and one of $+5$ INR is a further improvement in resistance to impact noise transfer.

Fig. 43.3 shows STC and approximate INR(db)

[6] INR ratings in some publications are being abandoned in favor of IIC (Impact Insulation Class) ratings. See Glossary.

DETAIL	DESCRIPTION	ESTIMATED VALUES	
		STC RATING	APPROX. INR
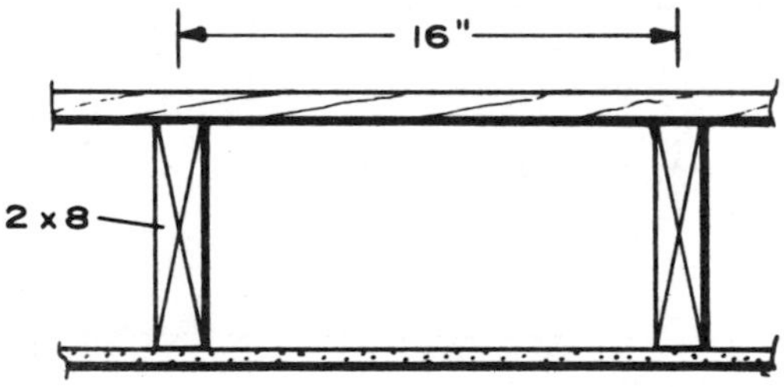 A	FLOOR ⅞" T. & G. FLOORING CEILING ⅜" GYPSUM BOARD	30	-18
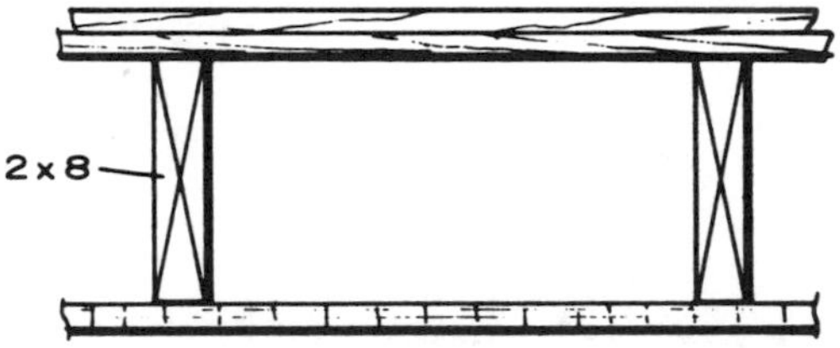 B	FLOOR ¾" SUBFLOOR ¾" FINISH FLOOR CEILING ¾" FIBERBOARD	42	-12
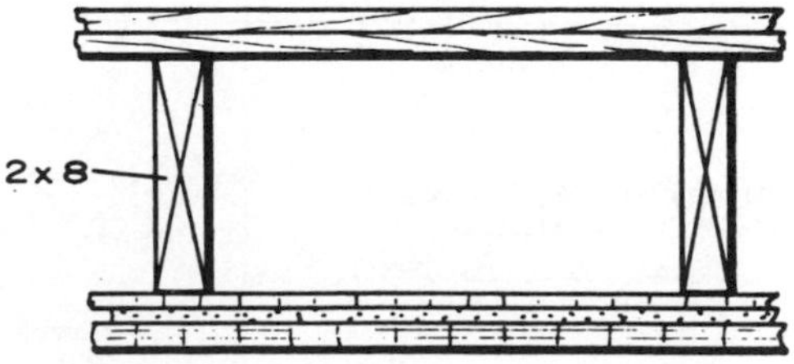 C	FLOOR ¾" SUBFLOOR ¾" FINISH FLOOR CEILING ½" FIBERBOARD LATH ½" GYPSUM PLASTER ¾" FIBERBOARD	45	-4

Figure 43.3.—Relative impact and sound transfer in floor-ceiling combinations (2- by 8-in. joists).

values for several types of floor constructions. Figure **43.3a** perhaps a minimum floor assembly with tongued-and-grooved floor and ⅜-inch gypsum board ceiling, has an STC value of 30 and an approximate INR value of −18. This is improved somewhat by the construction shown in figure 43.3b, and still further by the combination of materials in figure **43.3c.**

The value of isolating the ceiling joists from a gypsum lath and plaster ceiling by means of spring clips is illustrated in figure **43.4a.** An STC value of 52 and an approximate INR value of −2 result.

DETAIL	DESCRIPTION	ESTIMATED VALUES	
		STC RATING	APPROX. INR

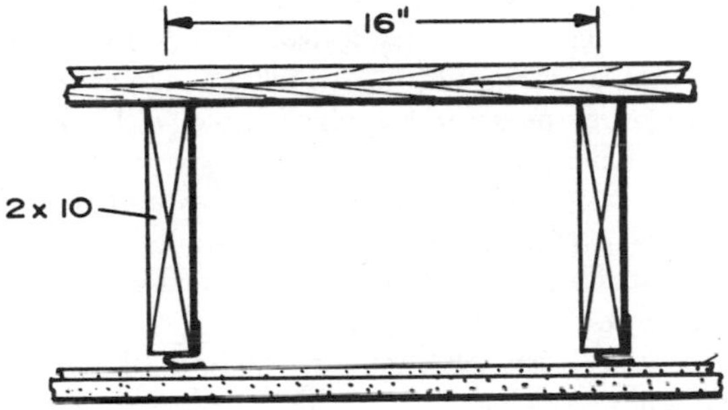 A
2 x 10
16"
	FLOOR
	  ¾" SUBFLOOR
	    (BUILDING PAPER)
	  ¾" FINISH FLOOR
	CEILING
	  GYPSUM LATH AND
	    SPRING CLIPS
	  ½" GYPSUM PLASTER
		52	− 2

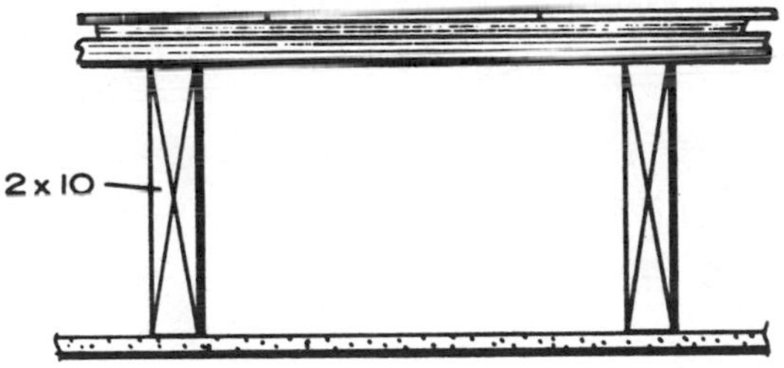 B
2 x 10
	FLOOR
	  ⅝" PLYWOOD SUBFLOOR
	  ½" PLYWOOD
	    UNDERLAYMENT
	  ⅛" VINYL-ASBESTOS
	    TILE
	CEILING
	  ½" GYPSUM WALLBOARD
		31	−17

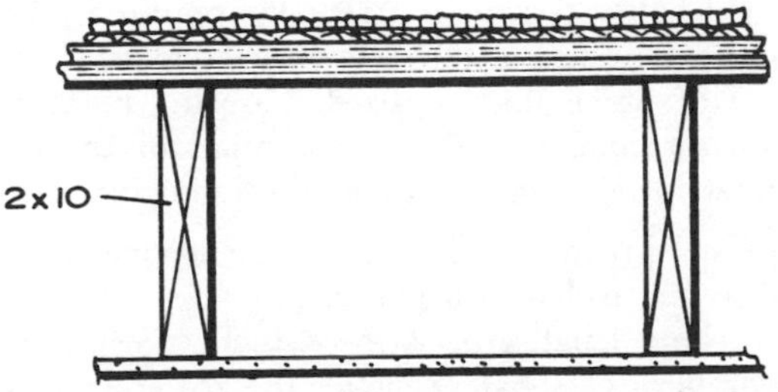 C
2 x 10
	FLOOR
	  ⅝" PLYWOOD SUBFLOOR
	  ½" PLYWOOD
	    UNDERLAYMENT
	  FOAM RUBBER PAD
	  ⅜" NYLON CARPET
	CEILING
	  ½" GYPSUM WALLBOARD
		45	+ 5

Figure 43.4.—Relative impact and sound transfer in floor-ceiling combinations
(2- by 10-in. joists).

Foam-rubber padding and carpeting improve both the STC and the INR values. The STC value increases from 31 to 45 and the approximate INR from −17 to +5 (fig. 43.4 b and c). This can likely be further improved by using an isolated ceiling finish with spring clips. The use of sound-deadening board and a lamination of gypsum board for the ceiling would also improve resistance to sound transfer.

An economical construction similar to (but an improvement over) figure 43.4c, with a STC value of 48 and an approximate INR of +18, consists of the following: (a) A pad and carpet over 5/8-inch tongued-and-grooved plywood underlayment, (b) 3-inch fiber-glass insulating batts between joists, (c) resilient channels spaced 24 inches apart, across the bottom of the joists, and (d) 5/8-inch gypsum board screwed to the bottom of the channels and finished with taped joints.

The use of separate floor joists with staggered ceiling joists below provides reasonable values but adds a good deal to construction costs. Separate joists with insulation between and a soundboard between sub-floor and finish provide an STC rating of 53 and an approximate INR value of −3.

Sound Absorption

Design of the "quiet" house can incorporate another system of sound insulation, namely, sound absorption. Sound-absorbing materials can minimize the amount of noise by stopping the reflection of sound back into a room. Sound-absorbing materials do not necessarily have resistance to airborne sounds. Perhaps the most commonly used sound-absorbing material is acoustic tile. Wood fiber or similar materials are used in the manufacture of the tile, which is usually processed to provide some fire resistance and designed with numerous tiny sound traps on the tile surfaces. These may consist of tiny drilled or punched holes, fissured surfaces, or a combination of both.

Acoustic tile is most often used in the ceiling and areas where it is not subjected to excessive mechanical damage, such as above a wall wainscoting. It is normally manufactured in sizes from 12 by 12 to 12 by 48 inches. Thicknesses vary from 1/2 to 3/4 inch, and the tile is usually factory finished ready for application. Paint or other finishes which fill or cover the tiny holes or fissures for trapping sound will greatly reduce its efficiency.

Acoustic tile may be applied by a number of methods—to existing ceilings or any smooth surface with a mastic adhesive designed specifically for this purpose, or to furring strips nailed to the underside of the ceiling joists. Nailing or stapling tile is the normal application method in this system. It is also used with a mechanical suspension system involving small "H," "Z," or "T" members. Manufacturers' recommendations should be followed in application and finishing.

CHAPTER 44

BASEMENT ROOMS

Many houses are now designed so that one or more of the rooms in lower floors are constructed on a concrete slab. In multilevel houses, this area may include a family room, a spare bedroom, or a study. Furthermore, it is sometimes necessary to provide a room in the basement of an existing house. Thus, in a new house or in remodeling the basement of an existing one, several factors should be considered, including insulation, waterproofing, and vapor resistance.

Floors

In the construction of a new building having basement rooms, provision should be made for reduction of heat loss and for prevention of ground moisture movement. As previously described in Chapter 27, "Concrete Floor Slabs on Ground," perimeter insulation reduces heat loss and a vapor barrier under the slab will prevent problems caused by a concrete floor damp from ground moisture (fig. 44.1). Providing these essential details, however, is somewhat more difficult in existing construction than in new construction.

The installation of a vapor barrier over an existing unprotected concrete slab is normally required when the floor is at or below the outside ground level and some type of finish floor is used. Flooring manufacturers often recommend that preparation of the slab for wood strip flooring consist of the following steps:

1. Mop or spread a coating of tar or asphalt mastic followed by an asphalt felt paper.

2. Lay short lengths of 2- by 4-inch screeds in a coating of tar or asphalt, spacing the rows about 12 inches apart, starting at one wall and ending at the opposite wall.

3. Place insulation around the perimeter, between screeds, where the outside ground level is near the basement floor elevation.

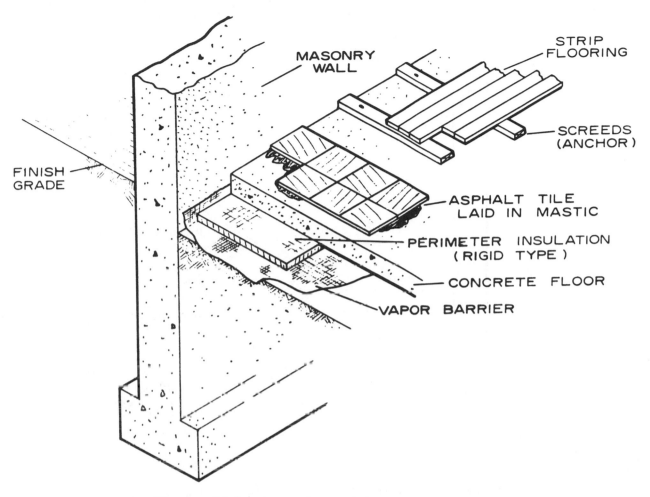

Figure 44.1.—Basement floor details for new construction.

4. Install wood strip flooring across the wood screeds.

This system can be varied somewhat by placing a conventional vapor barrier of good quality directly over the slab. Two- by four-inch furring strips spaced 12 to 16 inches apart are then anchored to the slab with concrete nails or with other types of commercial anchors. Some leveling of the 2 by 4's might be required. Strip flooring is then nailed to the furring strips after perimeter insulation is placed (fig. 44.2). If a wood block flooring is desired under these conditions, a plywood subfloor may be used over the furring strips. Plywood, 1/2 or 5/8 inch thick, is normally used if the edges are unblocked and furring strips are spaced 16 inches or more apart.

When insulation is not required around the perimeter because of the height of the outside grade above the basement floor, a much simpler method can be used for wood block or other type of tile finish. An asphalt mastic coating, followed by a good vapor

barrier, serves as a base for the tile. An adhesive recommended by the flooring manufacturer is then used over the vapor barrier, after which the wood tile is applied. It is important that a smooth vapor-tight base be provided for the tile.

It is likely that such floor construction should be used only under favorable conditions where draintile is placed at the outside footings and soil conditions are favorable. When the slab or walls of an existing house are inclined to be damp, it is often difficult to insure a dry basement. Under such conditions, it is often advisable to use resilient tile or similar finish over some type of stable base such as plywood. This construction is to be preceded by installation of vapor barriers and protective coatings.

Walls

The use of an interior finish over masonry basement walls is usually desirable for habitable rooms. Furthermore, if the outside wall is partially exposed, it is advisable to use insulation between the wall and the

583

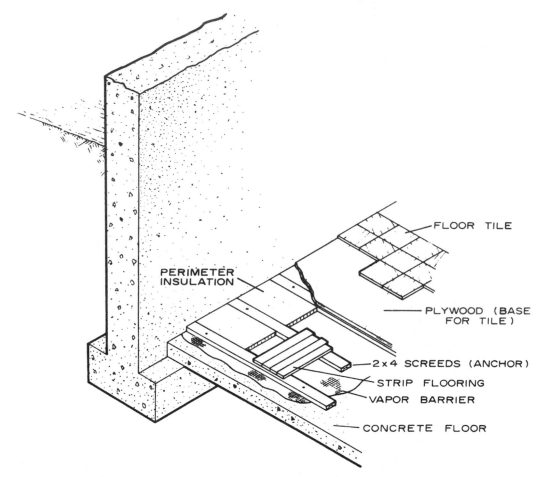

FLOOR TILE

PERIMETER
INSULATION

PLYWOOD (BASE
FOR TILE)

2 x 4 SCREEDS (ANCHOR)

STRIP FLOORING

VAPOR BARRIER

CONCRETE FLOOR

*Figure 44.2.—*Basement floor details for existing construction.

inside finish. Waterproofing the wall is important if there is any possibility of moisture entry. It can be done by applying one of the many waterproof coatings available to the inner surface of the masonry.

After the wall has been waterproofed, furring strips are commonly used to prepare the wall for interior finish. A 2- by 2-inch bottom plate is anchored to the floor at the junction of the wall and floor. A 2- by 2-inch or larger top plate is fastened to the bottom of the joists, to joist blocks, or anchored to the wall (fig. 44.3). Studs or furring strips, 2 by 2 inches or larger in size are then placed between top and bottom plates, anchoring them at the center when necessary with concrete nails or similar fasteners (fig. 44.3). Electrical outlets and conduit should be installed and insulation with vapor barrier placed between the furring strips. The interior finish of gypsum board, fiberboard, plywood, or other material is then installed. Furring strips are commonly spaced 16 inches on center, but this, of course, depends on the type and thickness of the interior finish.

Foamed plastic insulation is sometimes used on masonry walls without furring. It is important that the inner face of the wall be smooth and level without protrusions when this method is used. After the wall has been waterproofed, ribbons of adhesive are applied to the wall and sheets of foam insulation installed (fig. 44.4). Dry-wall adhesive is then applied and the gypsum board, plywood, or other finish pressed in place. Manufacturers' recommendations on adhesives and methods of installation should be followed. Most foam-plastic insulations have some vapor resistance in themselves, so the need for a separate vapor barrier is not as great as when blanket type insulation is used.

Ceilings

Some type of finish is usually desirable for the ceiling of the basement room. Gypsum board, plywood, or fiberboard sheets may be used and nailed directly to the joists. Acoustic ceiling tile and similar materials normally require additional nailing areas. This may be supplied by 1- by 2-inch or 1- by 3-inch strips nailed across the joists, and spaced to conform to

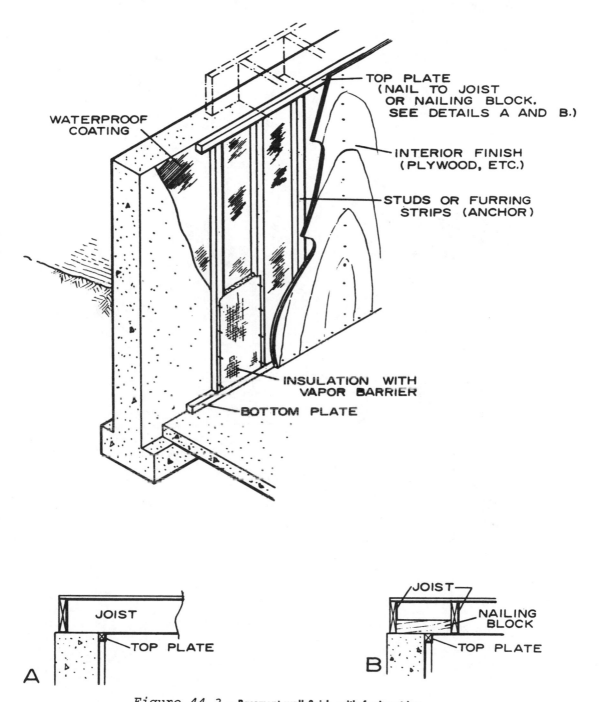

WATERPROOF COATING

TOP PLATE
(NAIL TO JOIST
OR NAILING BLOCK.
SEE DETAILS A AND B.)

INTERIOR FINISH
(PLYWOOD, ETC.)

STUDS OR FURRING
STRIPS (ANCHOR)

INSULATION WITH
VAPOR BARRIER

BOTTOM PLATE

JOIST

TOP PLATE

A

JOIST

NAILING
BLOCK

TOP PLATE

B

Figure 44.3.—Basement wall finish, with furring strips.

the size of the ceiling tile (fig. 44.5).

A *suspended ceiling* may also be desirable. This can consist of a system of light metal angles hung from the ceiling joists. Tiles are then dropped in place. This will also aid in decreasing sound transfer from the rooms above. Remember to install ceiling lights, heat supply and return ducts, or other utilities before finish is applied.

585

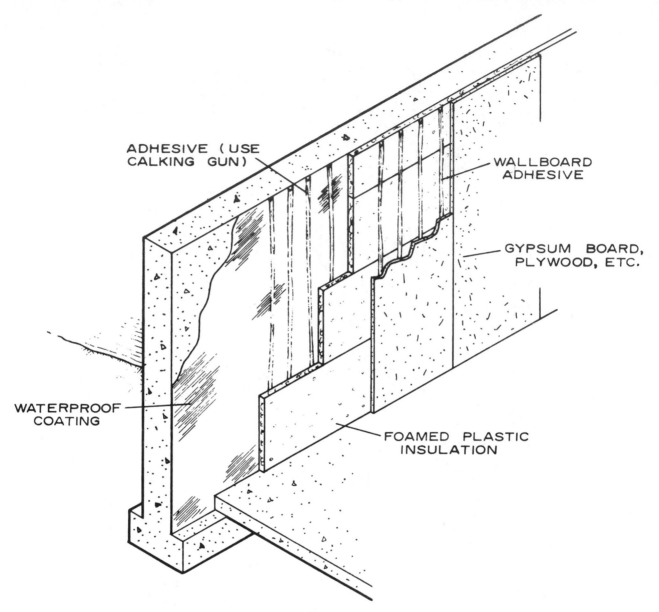

ADHESIVE (USE
CALKING GUN)

WALLBOARD
ADHESIVE

GYPSUM BOARD,
PLYWOOD, ETC.

WATERPROOF
COATING

FOAMED PLASTIC
INSULATION

Figure 44.4.—Basement wall finish, without furring strips.

CHAPTER 45

INTERIOR WALL AND CEILING FINISH

Interior finish is the material used to cover the interior framed areas or structures of walls and ceilings. It should be prefinished or serve as a base for paint or other finishes including wallpaper. Depending on whether it is wood, gypsum wallboard, or plaster, size and thickness should generally comply with recommendations in this handbook. Finishes in bath and kitchen areas should have more rigid requirements because of moisture conditions. Several types of interior finishes are used in the modern home, mainly:

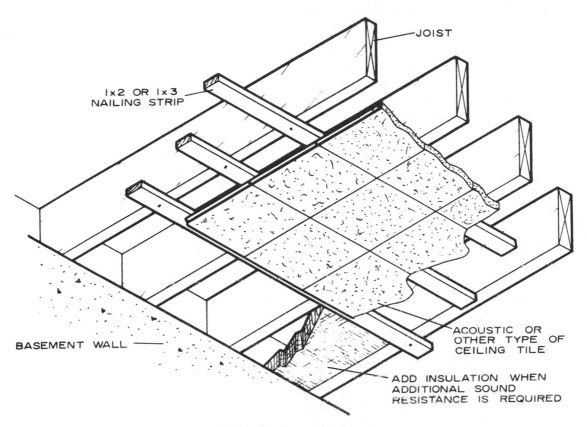

JOIST

1x2 OR 1x3 NAILING STRIP

ACOUSTIC OR OTHER TYPE OF CEILING TILE

BASEMENT WALL

ADD INSULATION WHEN ADDITIONAL SOUND RESISTANCE IS REQUIRED

*Figure 45.1.—*Installation of ceiling tile.

(a) Lath and plaster, (b) wood paneling, fiberboard, or plywood, and (c) gypsum wallboard.

Types of Finishes

Though lath and plaster finish is widely used in home construction, use of dry-wall materials has been increasing. Dry wall is often selected by builders because there is usually a time saving in construction. A plaster finish, being a wet material, requires drying time before other interior work can be started—dry-wall finish does not. However, a gypsum dry wall demands a moderately low moisture content of the framing members to prevent "nail-pops." These result when frame members dry out to moisture equilibrium, causing the nailhead to form small "humps" on the surface of the board. Furthermore, stud alinement is more important for single-layer gypsum finish to prevent a wavy, uneven appearance. Thus, there are advantages to both plaster and gypsum dry-wall finishes and each should be considered along with the initial cost and future maintenance involved.

A plaster finish requires some type of base upon which to apply the plaster. *Rock lath* is perhaps the most common, *Fiberboard lath* is also used, and wood lath, quite common many years ago, is permitted in some areas. *Metal lath* or similar mesh forms are normally used only in bathrooms and as reinforcement, but provide a rigid base for plaster finish. They usually cost more, however, than other materials. Some of the rigid foam insulations cemented to masonry walls also serve as plaster bases.

There are many types of dry-wall finishes, but one of the most widely used is gypsum board in 4- by 8-foot sheets and in lengths up to 16 feet which are used for horizontal application. Plywood, hardboard, fiberboard, particleboard, wood paneling, and similar types, many in prefinished form, are also used.

Lath and Plaster

Plaster Base

A plaster finish requires some type of base upon which the plaster is applied. The base must have bonding qualities so that plaster adheres, or is keyed to the base which has been fastened to the framing members.

One of the most common types of plaster base, that may be used on sidewalls or ceilings, is *gypsum lath*, which is 16 by 48 inches and is applied horizontally across the framing members. It has paper faces with a gypsum filler. For stud or joist spacing of 16 inches on center, ⅜-inch thickness is used. For 24-inch on-

587

center spacing, ½-inch thickness is required. This material can be obtained with a foil back that serves as a vapor barrier. If the foil faces an air space, it also has reflective insulating value. Gypsum lath may be obtained with perforations, which, by improving the bond, would lengthen the time the plaster would remain intact when exposed to fire. The building codes in some cities require such perforation.

Insulating fiberboard lath in ½-inch thickness and 16 by 48 inches in size is also used as a plaster base. It has greater insulating value than the gypsum lath, but horizontal joints must usually be reinforced with metal clips.

Metal lath in various forms such as diamond mesh, flat rib, and wire lath is another type of plaster base. It is usually 27 by 96 inches in size and is galvanized or painted to resist rusting.

Installation of Plaster Base

Gypsum lath should be applied horizontally with joints broken (fig. 45.2). Vertical joints should be made over the center of studs or joists and nailed with 12- or 13-gage gypsum-lathing nails 1½ inches long and with a ⅜-inch flat head. Nails should be spaced 5 inches on center, or four nails for the 16-inch height, and used at each stud or joist crossing. Some manufacturers specify the ring-shank nails with a slightly greater spacing. Lath joints over heads of openings should not occur at the jamb lines (fig. 45.2).

Insulating lath should be installed much the same as gypsum lath, except that slightly longer blued nails should be used. A special waterproof facing is provided on one type of gypsum board for use as a ceramic tile base when the tile is applied with an adhesive.

Metal lath is often used as a plaster base around tub recesses and other bath and kitchen areas (fig. 45.3). It is also used when a ceramic tile is applied over a plastic base. It must be backed with water-resistant sheathing paper over the framing. The metal lath is applied horizontally over the waterproof backing with side and end joints lapped. It is nailed with No. 11 and No. 12 roofing nails long enough to provide about 1½-inch penetration into the framing member or blocking.

Plaster Reinforcing

Because some drying usually takes place in wood framing members after a house is completed, some shrinkage can be expected; in turn, this may cause plaster cracks to develop around openings and in corners. To minimize, if not eliminate, this cracking, expanded metal lath is used in certain key positions over the plaster-base material as reinforcement. Strips of expanded metal lath may be used over window and

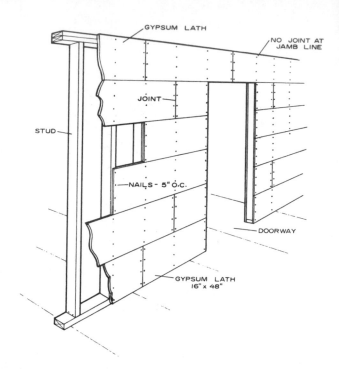

Figure 45.2.—Application of gypsum lath.

door openings (fig. 45.4a). A strip about 10 by 20 inches is placed diagonally across each upper corner of the opening and tacked in place.

Metal lath should also be used under flush ceiling beams to prevent plaster cracks (fig. 45.4b). On wood drop beams extending below the ceiling line, the metal

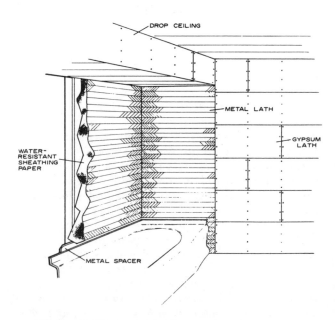

Figure 45.3.—Application of metal lath.

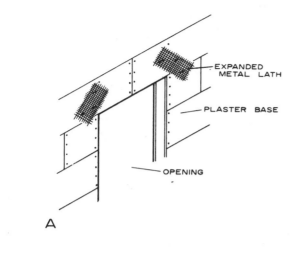

A

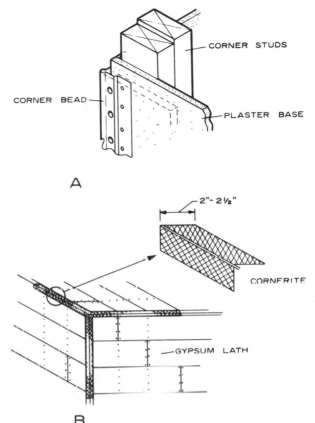

A

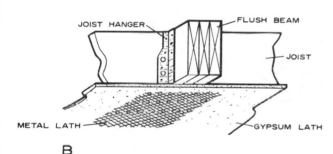

B

Figure 45.4.
Metal lath used to minimize cracking: A, At door and window openings; B, under flush beams.

Figure 45.5.
Reinforcing of plaster at corners:
A, Outside; B, inside.

lath is applied with self-furring nails to provide space for keying of the plaster.

Corner beads of expanded metal lath or of perforated metal should be installed on all exterior corners (fig. 45.5).. They should be applied plumb and level. The bead acts as a leveling edge when walls are plastered and reinforces the corner against mechanical damage. To minimize plaster cracks, inside corners at the juncture of walls and of ceilings should also be reinforced. Metal lath or wire fabric (*cornerites*) are tacked lightly in place in these areas. Cornerites provide a key width of 2 to 2½ inches at each side for plaster.

Plaster Grounds

Plaster grounds are strips of wood used as guides or strike-off edges when plastering and are located around window and door openings and at the base of the walls. Grounds around interior door openings are often full-width pieces nailed to the sides over the studs and to the underside of the header (fig. 45.6a). They are 5¼ inches in width, which coincides with

standard jamb widths for interior walls with a plaster finish. They are removed after plaster has dried. Narrow strip grounds might also be used around these interior openings (fig. 45. 6b).

In window and exterior door openings, the frames are normally in place before plaster is applied. Thus, the inside edges of the side and head jamb can, and often do, serve as grounds. The edge of the window sill might also be used as a ground, or a narrow ⅞-inch-thick ground strip is nailed to the edge of the 2- by 4-inch sill. Narrow ⅞- by 1-inch grounds might also be used around window and door openings (fig. 45.6c). These are normally left in place and are covered by the casing.

A similiar narrow ground or screed is used at the bottom of the wall in controlling thickness of the gypsum plaster and providing an even surface for the baseboard and molding (fig. 45.6a). These strips are also left in place after plaster has been applied.

Plaster Materials and Method of Application

Plaster for interior finishing is made from combinations of sand, lime, or prepared plaster and water. Waterproof-finish wall materials (*Keene's cement*)

589

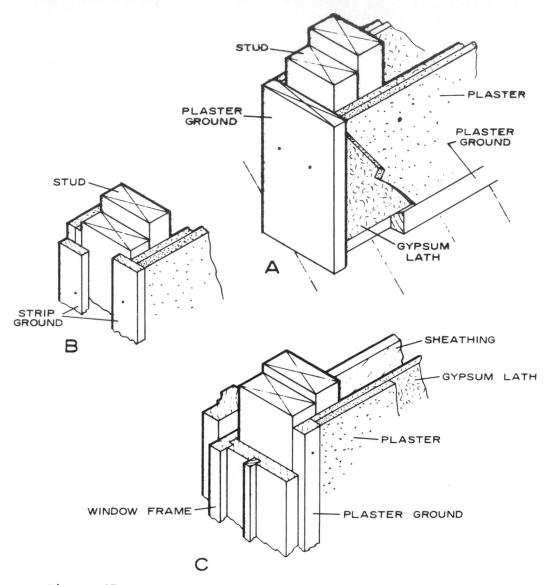

STUD

PLASTER
GROUND

PLASTER

PLASTER
GROUND

GYPSUM
LATH

A

STUD

STRIP
GROUND

B

SHEATHING

GYPSUM LATH

PLASTER

WINDOW FRAME

PLASTER GROUND

C

Figure 45.6.—Plaster grounds: A, At doorway and floor; B, strip ground at doorway; C, ground at window.

are available and should be used in bathrooms, especially in showers or tub recesses when tile is not used, and sometimes in the kitchen wainscot.

Plaster should be applied in three-coat or two-coat double-up work. The minimum thickness over ⅜-inch gypsum lath should be about ½ inch. The first plaster coat over metal lath is called the scratch coat and is scratched, after a slight set has occurred, to insure a good bond for the second coat. The second coat is called the brown or leveling coat, and leveling is done during the application of this coat.

The double-up work, combining the scratch and brown coat, is used on gypsum or insulating lath, and leveling and plumbing of walls and ceilings are done during application.

The final or finish coat consists of two general types —the *sand-float* and the *putty* finish. In the sand-float finish, lime is mixed with sand and results in a textured finish, the texture depending on the coarseness of the sand used. Putty finish is used without sand and has a smooth finish. This is common in kitchens and bathrooms where a gloss paint or enamel finish is used, and in other rooms where a smooth finish is desired. Keene's cement is often used as a finish plaster in bathrooms because of its durability.

The plastering operation should not be done in freezing weather without constant heat for protection from freezing. In normal construction, the heating unit is in place before plastering is started.

Insulating plaster, consisting of a vermiculite, per-

590

lite, or other aggregate with the plaster mix, may also be used for wall and ceiling finishes.

Dry-wall Finish

Dry-wall finish is a material that requires little, if any, water for application. More specifically, dry-wall finish includes gypsum board, plywood, fiberboard, or similar sheet material, as well as wood paneling in various thicknesses and forms.

The use of thin sheet materials such as gypsum board or plywood requires that studs and ceiling joists have good alinement to provide a smooth, even surface. Wood sheathing will often correct misalined studs on exterior walls. A "strong back" provides for alining of ceiling joists of unfinished attics (fig. 45.7a) and can be used at the center of the span when ceiling joists are uneven.

Table 7 lists thicknesses of wood materials commonly used for interior covering.

TABLE 7.—*Minimum thicknesses for plywood, fiberboard, and wood paneling.*

Framing spaced (inches)	Thickness		
	Plywood	Fiberboard	Paneling
	In.	In.	In.
16	1/4	1/2	3/8
20	3/8	3/4	1/2
24	3/8	3/4	5/8

Gypsum Board

Gypsum board is a sheet material composed of a gypsum filler faced with paper. Sheets are normally 4 feet wide and 8 feet in length, but can be obtained in lengths up to 16 feet. The edges along the length are usually tapered, although some types are tapered on all edges. This allows for a filled and taped joint. This material may also be obtained with a foil back which serves as a vapor barrier on exterior walls. It is also available with vinyl or other prefinished surfaces. In new construction, 1/2-inch thickness is recommended for single-layer application. In laminated two-ply applications, two 3/8-inch-thick sheets are used. The 3/8-inch thickness, while considered minimum for 16-inch stud spacing in single-layer applications, is normally specified for repair and remodeling work.

Table 8 lists maximum member spacing for the various thicknesses of gypsum board.

When the single-layer system is used, the 4-foot-wide gypsum sheets are applied vertically or horizontally on the walls after the ceiling has been covered. Vertical application covers three stud spaces when studs are spaced 16 inches on center, and two when spacing is 24 inches. Edges should be centered on studs, and

TABLE 8.—*Gypsum board thickness (single layer)*

Installed long direction of sheet	Minimum thickness	Maximum spacing of supports (on center)	
		Walls	Ceilings
	In.	In.	In.
Parallel to framing members	3/8	16	
	1/2	24	16
	5/8	24	16
Right angles to framing members	3/8	16	16
	1/2	24	24
	5/8	24	24

only moderate contact should be made between edges of the sheet.

Fivepenny cooler-type nails (1 5/8 in. long) should be used with 1/2-inch gypsum, and fourpenny (1 3/8 in. long) with the 3/8-inch-thick material. Ring-shank nails, about 1/8 inch shorter, can also be used. Some manufacturers often recommend the use of special screws to reduce "bulging" of the surface ("nail-pops" caused by drying out of the frame members). If moisture content of the framing members is less than 15 percent when gypsum board is applied, "nail-pops" will be greatly reduced. It is good practice, when framing members have a high moisture content, to allow them to approach moisture equilibrium before application of the gypsum board. Nails should be spaced 6 to 8 inches for sidewalls and 5 to 7 inches for ceiling application (fig. 45.7b). Minimum edge distance is 3/8 inch.

The horizontal method of application is best adapted to rooms in which full-length sheets can be used, as it minimizes the number of vertical joints. Where joints are necessary, they should be made at windows or doors. Nail spacing is the same as that used in vertical application. When studs are spaced 16 inches on center, horizontal nailing blocks between studs are normally not required when stud spacing is not greater than 16 inches on center and gypsum board is 3/8 inch or thicker. However, when spacing is greater, or an impact-resistant joint is required, nailing blocks may be used (fig. 45.7c).

Another method of gypsum-board application (laminated two-ply) includes an undercourse of 3/8-inch material applied vertically and nailed in place. The finish 3/8-inch sheet is applied horizontally, usually in room-size lengths, with an adhesive. This adhesive is either applied in ribbons, or is spread with a notched trowel. The manufacturer's recommendations should be followed in all respects.

Nails in the finish gypsum wallboard should be driven with the heads slightly below the surface. The crowned head of the hammer will form a small dimple in the wallboard (fig. 45.8a). A nail set should *not*

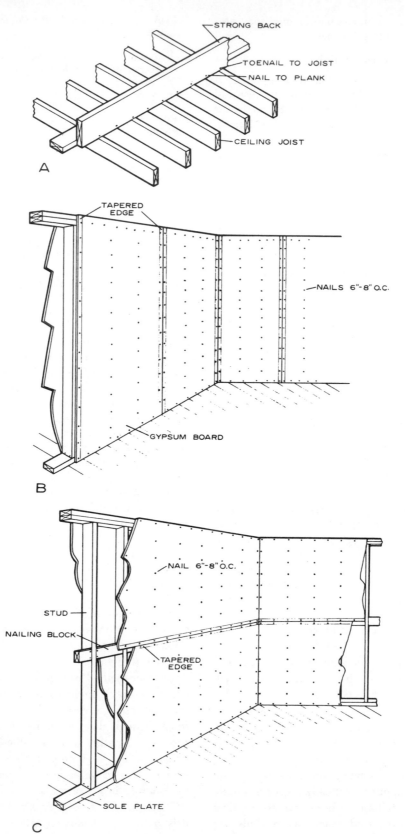

Figure 45.7.
Application of gypsum board finish: *A,* Strong back; *B,* vertical application; *C,* horizontal application.

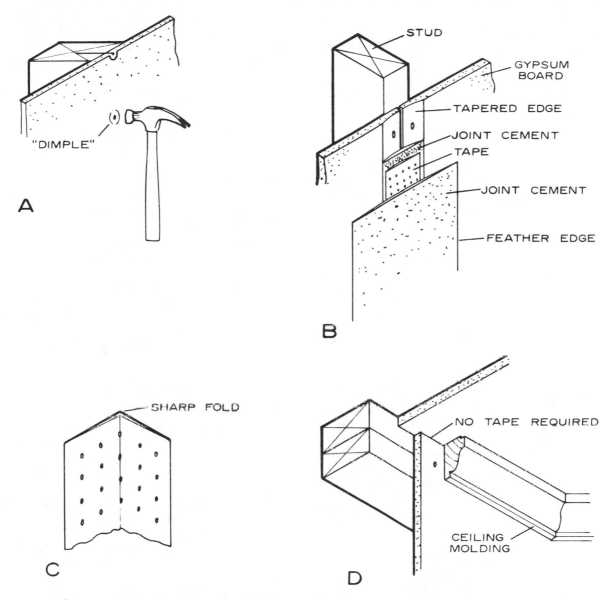

Figure 45.8.—Finishing gypsum dry wall: *A*, Nail set with crowned hammer; *B*, cementing and taping joint; *C*, taping at inside corners; *D*, alternate finish at ceiling.

be used, and care should be taken to avoid breaking the paper face.

Joint cement, "spackle," is used to apply the tape over the tapered edge joints and to smooth and level the surface. It comes in powder form and is mixed with water to a soft putty consistency so that it can be easily spread with a trowel or putty knife. It can also be obtained in premixed form. The general procedure for taping (fig. 45.8b) is as follows:

1. Use a wide spackling knife (5 in.) and spread the cement in the tapered edges, starting at the top of the wall.

2. Press the tape into the recess with the putty knife until the joint cement is forced through the perforations.

3. Cover the tape with additional cement, feathering the outer edges.

4. Allow to dry, sand the joint lightly, and then apply the second coat, feathering the edges. A steel trowel is sometimes used in applying the second coat. For best results, a third coat may be applied, feathering beyond the second coat.

5. After the joint cement is dry, sand smooth (an electric hand vibrating sander works well).

6. For hiding hammer indentations, fill with joint cement and sand smooth when dry. Repeat with the second coat when necessary.

Interior corners may be treated with tape. Fold the tape down the center to a right angle (fig. 45. 8c). and (1) apply cement at the corner, (2) press the tape in place, and (3) finish the corner with joint cement. Sand smooth when dry and apply a second coat.

The interior corners between walls and ceilings may also be concealed with some type of molding (fig. 45.'8 d). When moldings are used, taping this joint is not necessary. Wallboard corner beads at exterior corners will prevent damage to the gypsum board. They are fastened in place and covered with the joint cement.

Plywood

Prefinished plywood is available in a number of species, and its use should not be overlooked for accent walls or to cover entire room wall areas. Plywood for interior covering may be used in 4- by 8-foot and longer sheets. They may be applied vertically or·horizontally, but with solid backing at all edges. For 16-inch frame-member spacing, $\frac{1}{4}$-inch thickness is considered minimum. For 20- or 24-inch spacing, $\frac{3}{8}$-inch plywood is the minimum thickness. Casing or finishing nails $1\frac{1}{4}$ to $1\frac{1}{2}$ inches long are used. Space them 8 inches apart on the walls and 6 inches apart on ceilings. Edge nailing distance should be not less than $\frac{3}{8}$ inch. Allow $\frac{1}{32}$-inch end and edge distance between sheets when installing. Most wood or wood-base panel materials should be exposed to the conditions of the room before installation. Place them around the heated room for at least 24 hours.

Adhesives may also be used to fasten prefinished plywood and other sheet materials to wall studs. These panel adhesives usually eliminate the need for more than two guide nails for each sheet. Application usually conforms to the following procedure: (a) Position the sheet and fasten it with two nails for guides at the top or side, (b) remove plywood and spread contact or similar adhesive on the framing members, (c) press the plywood in place for full contact using the nails for positioning, (d) pull the plywood away from the studs and allow adhesive to set, and (e) press plywood against the framing members and tap lightly with a rubber mallet for full contact. Manufacturers of adhesives supply full instructions for application of sheet materials.

Hardboard and Fiberboard

Hardboard and fiberboard are applied the same way as plywood. Hardboard must be at least $\frac{1}{4}$ inch when used over open framing spaced 16 inches on center. Rigid backing of some type is required for $\frac{1}{8}$-inch hardboard.

Fiberboard in tongued-and-grooved plank or sheet form must be $\frac{1}{2}$ inch thick when frame members are spaced 16 inches on center and $\frac{3}{4}$ inch when 24-inch spacing is used, as previously outlined. The casing or finishing nails must be slightly longer than those used for plywood or hardboard; spacing is about the same. Fiberboard is also used in the ceiling as acoustic tile and may be nailed to strips fastened to ceiling joists. It is also installed in 12- by 12-inch or larger tile forms on wood or metal hangers which are hung from the ceiling joists. This system is called a "suspended ceiling."

Wood Paneling

Various types and patterns of woods are available for application on walls to obtain desired decorative effects. For informal treatment, knotty pine, white-pocket Douglas-fir, sound wormy chestnut, and pecky cypress, finished natural or stained and varnished, may be used to cover one or more sides of a room. Wood paneling should be thoroughly seasoned to a moisture content near the average it reaches in service (fig. 45.9), in most areas about 8 percent. Allow the material to reach this condition by placing it around the wall of the heated room. Boards may be applied horizontally or vertically, but the same general methods of application should pertain to each. The following may be used as a guide in the application of matched wood paneling:

1. Apply over a vapor barrier and insulation when application is on the exterior wall framing or blocking (fig. 45.10).
2. Boards should not be wider than 8 inches except when a long tongue or matched edges are used.
3. Thickness should be at least $\frac{3}{8}$ inch for 16-inch spacing of frame members, $\frac{1}{2}$ inch for 20-inch spacing, and $\frac{5}{8}$ inch for 24-inch spacing.
4. Maximum spacing of supports for nailing should be 24 inches on center (blocking for vertical applications).
5. Nails should be fivepenny or sixpenny casing or finishing nails.

Use two nails for boards 6 inches or less wide and three nails for 8-inch and wider boards. One nail can be blind-nailed in matched paneling.

Wood paneling in the form of small plywood squares can also be used for an interior wall covering (fig. 45.11).When used over framing and a vapor barrier, blocking should be so located that each edge has full bearing. Each edge should be fastened with casing or finish nails. When two sides are tongued and grooved, one edge (tongued side) may be blind-nailed. When paneling (16 by 48 in. or larger) crosses studs, it should also be nailed at each intermediate bearing. Matched (tongued-and-grooved) sides should be used when no horizontal blocking is provided or paneling is not used over a solid backing.

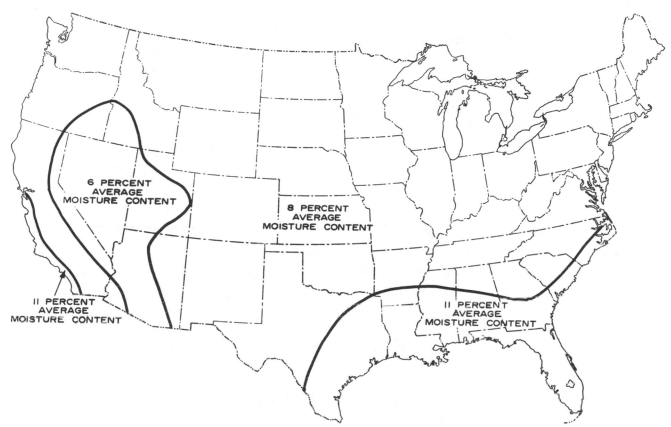

Figure 45.9.—Recommended average moisture content for interior finish woodwork in different parts of the United States.

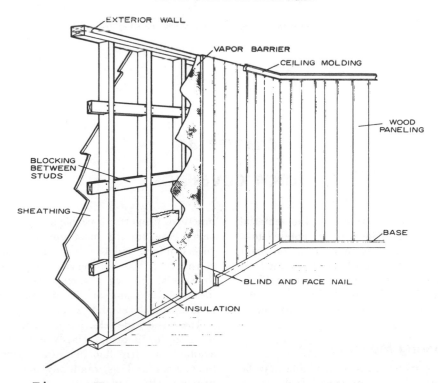

Figure 45.10.—Blocking between studs for vertical wood paneling.

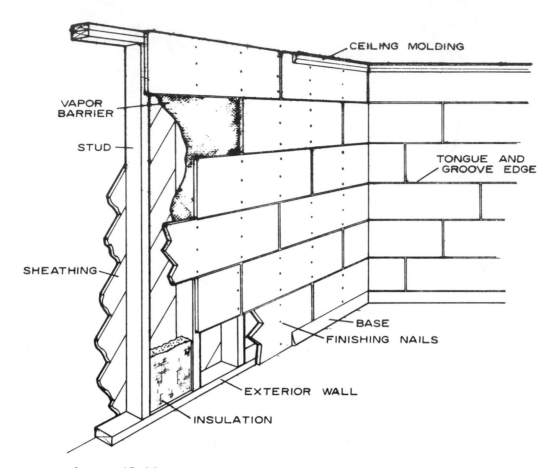

Figure 45.11.—**Application of tongued-and-grooved paneling over studs.**

Labels in figure: CEILING MOLDING, VAPOR BARRIER, STUD, TONGUE AND GROOVE EDGE, SHEATHING, BASE, FINISHING NAILS, EXTERIOR WALL, INSULATION

CHAPTER 46

FLOOR COVERINGS

The term "finish flooring" refers to the material used as the final wearing surface that is applied to a floor. Perhaps in its simplest form it might be paint over a concrete floor slab. One of the many resilient tile floorings applied directly to the slab would likely be an improvement from the standpoint of maintenance, but not necessarily from the comfort standpoint.

Flooring Materials

Numerous flooring materials now available may be used over a variety of floor systems. Each has a property that adapts it to a particular usage. Of the practical properties, perhaps durability and maintenance ease are the most important. However, initial cost, comfort, and beauty or appearance must also be considered. Specific service requirements may call for special properties, such as resistance to hard wear in warehouses and on loading platforms, or comfort to users in offices and shops.

There is a wide selection of wood materials that may be used for flooring. Hardwoods and softwoods are available as strip flooring in a variety of widths and thicknesses and as random-width planks and block flooring. Other materials include linoleum, asphalt, rubber, cork, vinyl, and other materials in tile or sheet forms. Tile flooring is also available in a particleboard which is manufactured with small wood particles combined with resin and fabricated under high pressure. Ceramic tile and carpeting are used in many areas in ways not thought practical a few years ago. Plastic floor coverings used over concrete or stable wood subfloor are another variation in the types of finishes available.

Wood-strip Flooring

Softwood finish flooring costs less than most hardwood species and is often used to good advantage in bedroom and closet areas where traffic is light. It might also be selected to fit the interior decor. It is less dense than the hardwoods, less wear-resistant, and shows surface abrasions more readily. Softwoods most commonly used for flooring are southern pine, Douglas-fir, redwood, and western hemlock.

Table 9 lists the grades and description of softwood strip flooring. Softwood flooring has tongued-and-grooved edges and may be hollow-backed or grooved. Some types are also end-matched. Vertical-grain flooring generally has better wearing qualities than flat-grain flooring under hard usage.

Hardwoods most commonly used for flooring are red and white oak, beech, birch, maple, and pecan. Table 9 lists grades, types, and sizes. Manufacturers supply both prefinished and unfinished flooring.

Perhaps the most widely used pattern is a $25/32$- by $2\frac{1}{4}$-inch *strip flooring*. These strips are laid lengthwise in a room and normally at right angles to the floor joists. Some type of a subfloor of diagonal boards or plywood is normally used under the finish floor. Strip flooring of this type is tongued-and-grooved and end-matched (fig. 46.1). Strips are random length and may vary from 2 to 16 feet or more. End-matched strip flooring in $25/32$-inch thickness is generally hollow backed (fig. 46.1a). The face is slightly wider than the bottom so that tight joints result when flooring is laid. The tongue fits tightly into the groove to prevent movement and floor "squeaks," All of these details are designed to provide beautiful finished floors that require a minimum of maintenance.

Another matched pattern may be obtained in $3/8$- by 2-inch size ((fig. 46.1b) . This is commonly used for remodeling work or when subfloor is edge-blocked or thick enough to provide very little deflection under loads.

Square-edged strip flooring (fig. 46.1c) might also be used occasionally. It is usually $3/8$ by 2 inches in size and is laid up over a substantial subfloor. Face-nailing is required for this type.

Wood block flooring (fig. 46.2) is made in a number of patterns. Blocks may vary in size from 4 by 4 inches to 9 by 9 inches and larger. Thickness varies by type from $25/32$ inch for laminated blocking or plywood

TABLE 9.—*Grade and description of strip flooring of several species and grain orientation*

Species	Grain orientation	Size		First grade	Second grade	Third grade
		Thickness	Width			
		In.	*In.*			
SOFTWOODS						
Douglas-fir and hemlock	Edge grain	$25/32$	$2\frac{3}{8}$–$5\frac{3}{16}$	B and Better	C	D
	Flat grain	$25/32$	$2\frac{3}{8}$–$5\frac{3}{16}$	C and Better	D	----------------
Southern pine	Edge grain and Flat grain	$5/16$–$1\frac{5}{16}$	$1\frac{3}{4}$–$5\frac{7}{16}$	B and Better	C and Better	D (and No. 2)
HARDWOODS						
Oak	Edge grain Flat grain	$25/32$	$1\frac{1}{2}$–$3\frac{1}{4}$	Clear	Select	----------------
		$3/8$	$1\frac{1}{2}$, 2	Clear	Select	No. 1 Common
		$1/2$	$1\frac{1}{2}$, 2			
Beech, birch, maple, and pecan[1]		$25/32$	$1\frac{1}{2}$–$3\frac{1}{4}$	First grade	Second grade	----------------
		$3/8$	$1\frac{1}{2}$, 2			
		$1/2$	$1\frac{1}{2}$, 2			

[1] Special grades are available in which uniformity of color is a requirement.

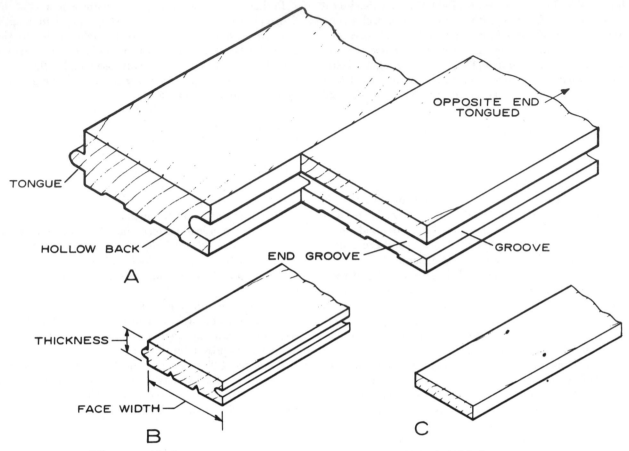

Figure 46.1.—Types of strip flooring: A, Side- and end-matched—25/32-inch;
B, thin flooring strips—matched; C, thin flooring strips—square-edged.

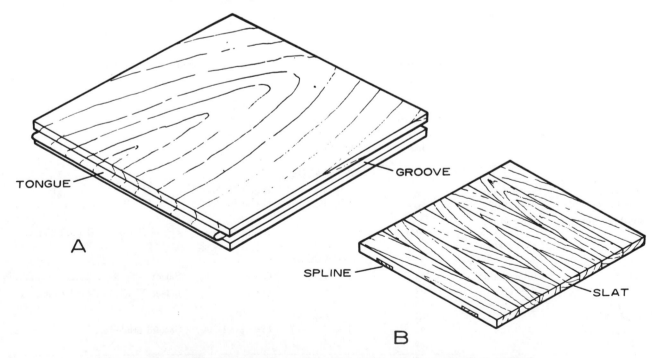

Figure 46.2.—Wood block flooring: A, Tongued-and-grooved; B, square-edged—splined.

block tile (fig. 46.2a) to ⅛-inch stabilized veneer. Solid wood tile is often made up of narrow strips of wood splined or keyed together in a number of ways. Edges of the thicker tile are tongued and grooved, but thinner sections of wood are usually square-edged (fig. 46.2b). Plywood blocks may be ⅜ inch and thicker and are usually tongued-and-grooved. Many block floors are factory-finished and require only waxing after installation. While stabilized veneer squares are still in the development stage, it is likely that research will produce a low-cost wood tile which can even compete with some of the cheaper nonwood resilient tile now available.

Installation of Wood Strip Flooring

Flooring should be laid after plastering or other interior wall and ceiling finish is completed and dried out, windows and exterior doors are in place, and most of the interior trim, except base, casing, and jambs, are applied, so that it may not be damaged by wetting or by construction activity.

Board subfloors should be clean and level and covered with a deadening felt or heavy building paper. This felt or paper will stop a certain amount of dust, will somewhat deaden sound, and, where a crawl space is used, will increase the warmth of the floor by preventing air infiltration. To provide nailing into the joists wherever possible, location of the joists should be chalklined on the paper as a guide. Plywood subfloor does not normally require building paper.

Strip flooring should normally be laid crosswise to the floor joists (fig. 46.3a).). In conventionally designed houses, the floor joists span the width of the building over a center supporting beam or wall. Thus, the finish flooring of the entire floor area of a rectangular house will be laid in the same direction. Flooring with "L" or "T" shaped plans will usually have a direction change at the wings, depending on joist direction. As joists usually span the short way in a living room, the flooring will be laid lengthwise to the room. This is desirable appearance-wise and also will reduce shrinkage and swelling effects on the flooring during seasonal changes.

Flooring should be delivered only during dry weather and stored in the warmest and driest place available in the house. The recommended average moisture content for flooring at time of installation varies somewhat in different sections of the United States. The moisture content map (fig. 45.9) outlines these recommendations. Moisture absorbed after delivery to the house site is one of the most common causes of open joints between flooring strips that appear after several months of the heating season.

Floor squeaks are usually caused by movement of one board against another. Such movement may occur because: (a) Floor joists are too light, causing excessive deflection, (b) sleepers over concrete slabs are not held down tightly, (c) tongues are loose fitting, or (d) nailing is poor. Adequate nailing is an important means of minimizing squeaks, and another is to apply the finish floors only after the joists have dried to 12 percent moisture content or less. A much better job results when it is possible to nail the finish floor through the subfloor into the joists than if the finish floor is nailed *only* to the subfloor.

Various types of nails are used in nailing different thicknesses of flooring. For $^{25}/_{32}$-inch flooring, it is best to use eightpenny flooring nails; for ½-inch, sixpenny; and for ⅜-inch fourpenny casing nails. (All the foregoing are blind-nailed.) For thinner square-edge flooring, it is best to use a 1½-inch flooring brad and face-nail every 7 inches with two nails, one near each edge of the strip, into the subfloor.

Other types of nails, such as the ring-shank and screw-shank type, have been developed in recent years for nailing of flooring. In using them, it is well to check with the floor manufacturer's recommendations as to size and diameter for specific uses. Flooring brads are also available with blunted points to prevent splitting of the tongue.

Fig. 46. 3b shows the method of nailing the first strip of flooring placed ½ to ⅝ inch away from the wall. The space is to allow for expansion of the flooring when moisture content increases. The nail is driven straight down through the board at the groove edge. The nails should be driven into the joist and near enough to the edge so that they will be covered by the base or shoe molding. The first strip of flooring can also be nailed through the tongue. Fig. 46. 4a shows in detail how nails should be driven into the tongue of the flooring at an angle of 45° to 50°. The nail should not be driven quite flush so as to prevent damaging the edge by the hammerhead (fig. 46.4b) The nail can be set with the end of a large-size nail set or by laying the nail set flatwise against the flooring (fig. 46.4b). Nailing devices using standard flooring or special nails are often used by flooring contractors. One blow of the hammer on the plunger drives and sets the nail.

To prevent splitting the flooring, it is sometimes desirable to predrill through the tongue, especially at the ends of the strip. For the second course of flooring from the wall, select pieces so that the butt joints will be well separated from those in the first course. Under normal conditions, each board should be driven up tightly. Crooked pieces may require wedging to force them into alinement or may be cut and used at the ends of the course or in closets. In completing the flooring, a ½- to ⅝-inch space is provided between the wall and the last flooring strip. Because of the closeness of the wall, this strip is usually face-nailed so that the base or shoe covers the set nailheads.

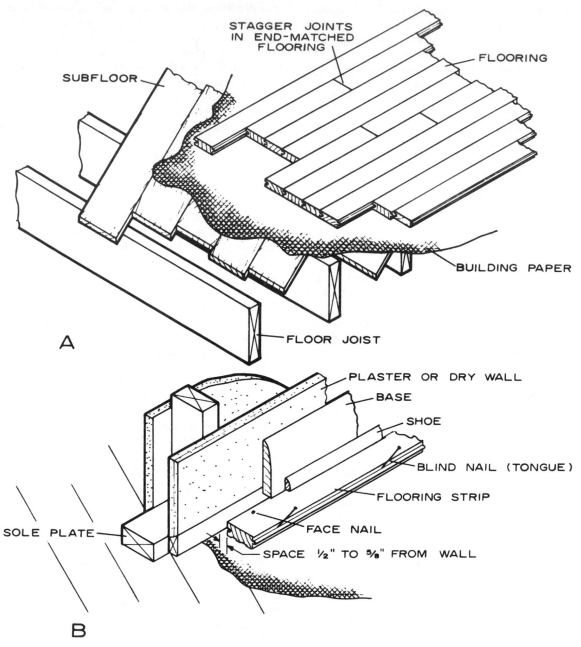

STAGGER JOINTS
IN END-MATCHED
FLOORING

FLOORING

SUBFLOOR

BUILDING PAPER

FLOOR JOIST

A

PLASTER OR DRY WALL

BASE

SHOE

BLIND NAIL (TONGUE)

FLOORING STRIP

FACE NAIL

SOLE PLATE

SPACE ½" TO ⅝" FROM WALL

B

Figure 46.3.—**Application of strip flooring: A, General application; B, starting strip.**

Installation of Wood Flooring Over Concrete Slabs

Installation of wood floor over concrete slabs was described briefly in the chapter "Concrete Floor Slabs on Ground" and illustrated in fig. 27.2 . As outlined, one of the important factors in satisfactory performance is the use of a good vapor barrier under the slab to resist the movement of ground moisture and vapor. The vapor barrier is placed under the slab during construction. However, an alternate method must be used when the concrete is already in place (fig. 44.2)

Another system of preparing a base for wood flooring when there is no vapor barrier under the slab is shown in fig. 46.5. . To resist decay, treated 1- by 4-inch furring strips are anchored to the existing slab, shimming when necessary to provide a level base. Strips should be spaced no more than 16 inches on center. A good waterproof or water-vapor resistant coating on the concrete before the treated strips are applied is usually recommended to aid in reducing moisture movement. A vapor barrier, such as a 4-mil polyethylene or similar membrane, is then laid over the anchored 1-by 4-inch wood strips and a second set of 1 by 4's nailed to the first. Use 1½-inch-long nails spaced 12 to 16 inches apart in a staggered pattern. The moisture content of these second members should be about the same as that of the strip flooring to be applied (6 to 11 pct., fig.45.9). Strip flooring can then be installed as previously described.

When other types of finish floor, such as a resilient tile, are used, plywood is placed over the 1 by 4's as a base.

Wood and Particleboard Tile Flooring

Wood and particleboard tile are, for the most part, applied with adhesive on a plywood or similar base. The exception is $^{25}/_{32}$-inch wood block floor, which has tongues on two edges and grooves on the other two edges. If the base is wood, these tiles are commonly nailed through the tongue into the subfloor. However, wood block may be applied on concrete slabs with an adhesive. Wood block flooring is installed by changing the grain direction of alternate blocks. This minimizes the effects of shrinking and swelling of the wood.

One type of wood floor tile is made up of a number of narrow slats to form 4- by 4-inch and larger squares. Four or more of these squares, with alternating grain direction, form a block. Slats, squares, and blocks are held together with an easily removed membrane. Adhesive is spread on the concrete slab or underlayment with a notched trowel and the blocks installed immediately. The membrane is then removed and the blocks tamped in place for full adhesive contact. Manufac-

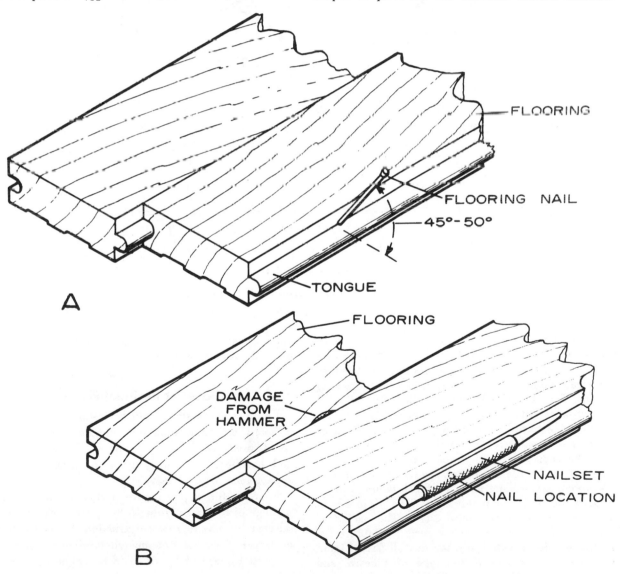

Figure 46.4.—Nailing of flooring: A, Nail angle; B, setting of nail.

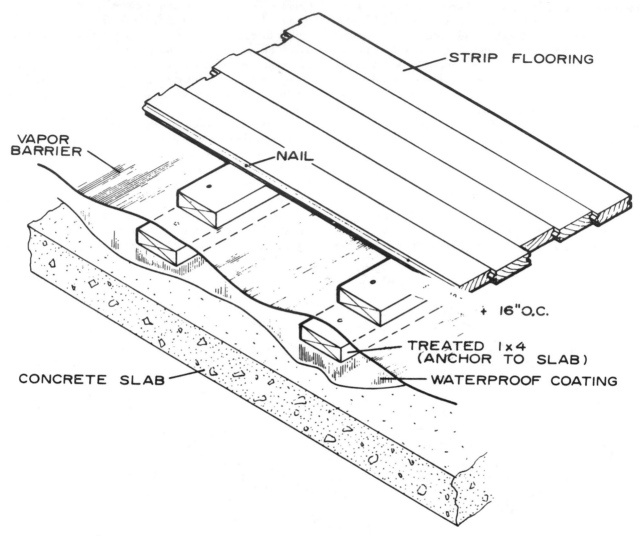

STRIP FLOORING

VAPOR BARRIER

NAIL

16"O.C.

TREATED 1x4
(ANCHOR TO SLAB)

WATERPROOF COATING

CONCRETE SLAB

Figure 46.5.—Base for wood flooring on concrete slab (without an underlying vapor barrier).

turer's recommendations for adhesive and method of application should always be followed. Similar tile made up of narrow strips of wood are fastened together with small rabbeted cleats, tape or similar fastening methods. They too are normally applied with adhesive in accordance with manufacturer's directions.

Plywood squares with tongued-and-grooved edges are another popular form of wood tile. Installation is much the same as for the wood tile previously described. Usually, tile of this type is factory-finished.

A wood-base product used for finish floors is particleboard tile. It is commonly 9 by 9 by 3/8 inches in size with tongued-and-grooved edges. The back face is often marked with small saw kerfs to stabilize the tile and provide a better key for the adhesive. Manufacturer's directions as to the type of adhesive and method of installation are usually very complete; some even include instructions on preparation of the base

upon which the tile is to be laid. This tile should not be used over concrete.

Base for Resilient Floors

Resilient floors should *not* be installed directly over a board or plank subfloor. Underlayment grade of wood-based panels such as plywood, particleboard, and hardboard is widely used for suspended floor applications (fig. 46.6a).

Four- by 8-foot plywood or particleboard panels, in a range of thickness from 3/8 to 3/4 inch, are generally selected for use in new construction. Four- by 4-foot or larger sheets of untempered hardboard, plywood, or particleboard of 1/4- or 3/8-inch thickness is used in remodeling work because of the floor thicknesses involved. The underlayment grade of particleboard is

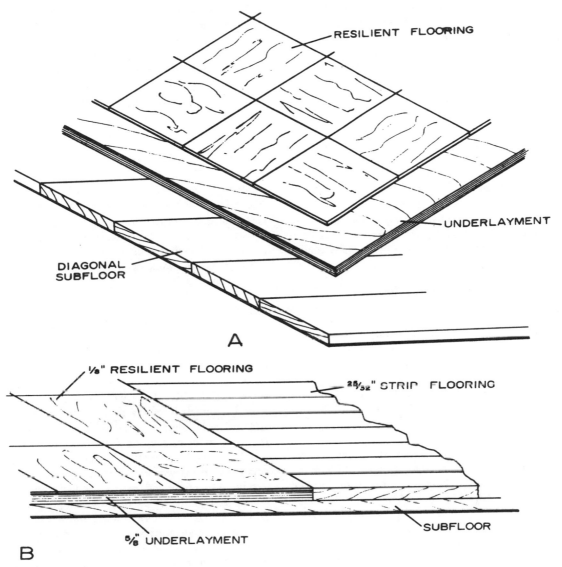

RESILIENT FLOORING

UNDERLAYMENT

DIAGONAL SUBFLOOR

A

⅛" RESILIENT FLOORING

²⁵/₃₂" STRIP FLOORING

⅝" UNDERLAYMENT

SUBFLOOR

B

*Figure 46.6.—*Base for wood flooring and concrete slab (without an underlying vapor barrier).

a standard product and is available from many producers. Manufacturer's instructions should be followed in the care and use of the product. Plywood underlayment is also a standard product and is available in interior types, exterior types, and interior types with an exterior glueline. The underlayment grade provides for a sanded panel with a C-plugged or better face ply and a C-ply or better immediately under the face. This construction resists damage to the floor surface from concentrated loads such as chair legs, etc.

Generally, underlayment panels are separate and installed over structurally adequate subfloors. Combination subfloor-underlayment panels of plywood construction find increasing usage. Panels for this dual purpose use generally have tongued-and-grooved or blocked edges and C-plugged or better faces to provide a smooth, even surface for the resilient floor covering.

The method of installing plywood combination subfloor and underlayment has been covered in the section on Plywood Subfloor. Underlayment should be laid up as outlined in that section with ⅟₃₂-inch edge and end spacing. Sand smooth to provide a level base for the resilient flooring. To prevent nails from showing on the surface of the tile, joists and subfloor should have a moisture content near the average value they reach in service.

The thickness of the underlayment will vary somewhat, depending on the floors in adjoining rooms. The installation of tile in a kitchen area, for example, is usually made over a ⅝-inch underlayment when finish floors in the adjoining living or dining areas are $^{25}/_{32}$-inch strip flooring (fig. 46.6b). When thinner wood floors are used in adjoining rooms, adjustments are made in the thickness of the underlayment.

Concrete for resilient floors should be prepared as shown in figures 27.1, 2, or 3, with a good vapor barrier installed somewhere between the soil and the finish floor, preferably just under the slab. Concrete should be leveled carefully when a resilient floor is to be used directly on the slab to minimize dips and waves.

Tile should not be laid on a concrete slab until it has completely dried. One method which may be used to determine this is to place a small square of polyethylene or other low-perm material on the slab overnight. If the underside is dry in the morning, the slab is usually considered dry enough for the installation of the tile.

Types of Resilient Floors

Linoleum

Linoleum may be obtained in various thicknesses and grades, usually in 6-foot-wide rolls. It should not be laid on concrete slabs on the ground. Manufacturer's directions should be followed. After the linoleum is laid, it is usually rolled to insure complete adhesion to the floor.

Asphalt Tile

Asphalt tile is one of the lower cost resilient coverings and may be laid on a concrete slab which is in contact with the ground. However, the vapor barrier under the slab is still necessary. Asphalt tile is about $^{1}/_{8}$ inch thick and usually 9 by 9 or 12 by 12 inches in size. Because most types are damaged by grease and oil, it is not used in kitchens.

Asphalt tile is ordinarily installed with an adhesive spread with a notched trowel. Both the type of adhesive and size of notches are usually recommended by the manufacturer.

Other Tile Forms

Vinyl, vinyl asbestos, rubber, cork, and similar coverings are manufactured in tile form, and several types are available for installation in 6-foot-wide rolls. These materials are usually laid over some type of underlayment and not directly on a concrete slab. Standard tile size is 9 by 9 inches but it may also be obtained in 12- by 12-inch size and larger. Decorative strips may be used to outline or to accent the room's perimeter.

In installing all types of square or rectangular tile, it is important that the joints do not coincide with the joints of the underlayment. For this reason, it is recommended that a layout be made before tile is laid. Normally, the manufacturer's directions

include laying out a base line at or near the center of the room and parallel to its length. The center or near center, depending on how the tile will finish at the edges, is used as a starting point. This might also be used as a point in quartering the room with a second guideline at exact right angles to the first. The tile is then laid in quarter-room sections after the adhesive is spread.

Seamless

A liquid-applied seamless flooring, consisting of resin chips combined with a urethane binder, is a relatively new development in floor coverings. It is applied in a 2-day cycle and can be used over a concrete base or a plywood subfloor. Plywood in new construction should be at least a C-C plugged exterior grade in $^{5}/_{8}$-inch thickness, or $^{3}/_{8}$-inch plywood over existing floors. This type of floor covering can be easily renewed.

Carpeting

Carpeting many areas of a home from living room to kitchen and bath is becoming more popular as new carpeting materials are developed. The cost, however, may be considerably higher than a finished wood floor, and the life of the carpeting before replacement would be much less than that of the wood floor. Many wise home builders will specify oak floors even though they expect to carpet some areas. The resale value of the home is then retained even if the carpeting is removed. However, the advantage of carpeting in sound absorption and resistance to impact should be considered. This is particularly important in multifloor apartments where impact noise reduction is an extremely important phase of construction. If carpeting is to be used, subfloor can consist of $^{5}/_{8}$-inch (minimum) tongued and grooved plywood (over 16-inch joist spacing). Top face of the plywood should be C plugged grade or better. Mastic adhesives are also being used to advantage in applying plywood to floor joists. Plywood, particleboard, or other underlayments are also used for a carpet base when installed over a subfloor.

Ceramic Tile

Ceramic tile and similar floor coverings in many sizes and patterns for bath, lavatory, and entry areas may be installed by the cement-plaster method or by the use of adhesives. The cement-plaster method requires a concrete-cement setting bed of $1\frac{1}{4}$ inches minimum thickness (fig. 46.7). Joists are chamfered (beveled) and cleats used to support waterproof plywood subfloor or forms cut between the joists. The cement base is reinforced with woven wire fabric or expanded metal lath.

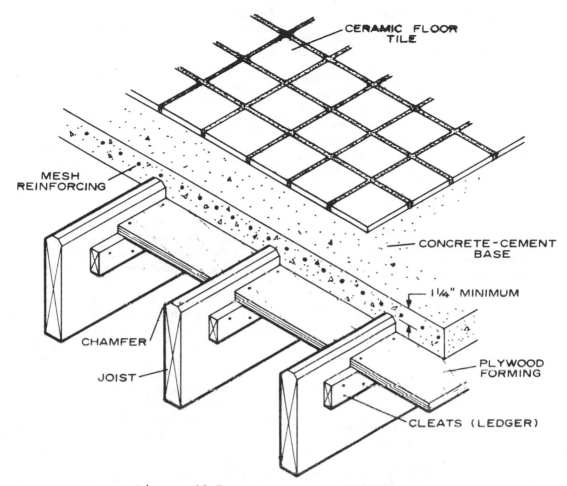

*Figure 46.7.—*Cement base for ceramic floor tile.

Tile should be soaked before it is installed. It is pressed firmly in place in the still plastic setting bed, mortar is compressed in the joints, and the joints tooled the same day tile is laid. Laying tile in this manner normally requires a workman skilled in this system. It should then be covered with waterproof paper for damp curing.

Adhesive used for ceramic floor tile should be the type recommended by the manufacturer. When installed over wood joists, a waterproof plywood ¾-inch thick with perimeter and intermediate nailing provides a good base. Before installing tile, a waterproof sealer or a thin coat of tile adhesive is applied to the plywood. Tile should be set over a full covering of adhesive using the "floating method" with a slight twisting movement for full embedment. "Buttering" or using small pats of adhesive on each tile is not acceptable. Tile should not be grouted until volatiles from the adhesive have evaporated. After grouting, joints should be fully tooled.

CHAPTER 47

INTERIOR DOORS, FRAMES, AND TRIM

Interior trim, doorframes, and doors are normally installed after the finish floor is in place. Cabinets, built-in bookcases and fireplace mantels, and other millwork units are also placed and secured at this time. Some contractors may install the interior door-frames *before* the finish floor is in place, allowing for the flooring at the bottom of the jambs. This is usually done when the jambs act as plaster grounds. However, because excessive moisture is present and edges of the jambs are often marred, this practice is usually undesirable.

Decorative Treatment

The decorative treatment for interior doors, trim, and other millwork may be paint or a natural finish with stain, varnish, or other non-pigmented material. The paint or natural finish desired for the woodwork in various rooms often determines the type of species of wood to be used. Interior finish that is to be painted should be smooth, close-grained, and free from pitch streaks. Some species having these requirements in a high degree include ponderosa pine, northern white pine, redwood, and spruce. When hardness and resistance to hard usage are additional requirements, species such as birch, gum, and yellow-poplar are desirable.

For natural finish treatment, a pleasing figure, hardness, and uniform color are usually desirable. Species with these requirements include ash, birch, cherry, maple, oak, and walnut. Some require staining· for best appearance.

The recommended moisture content for interior finish varies from 6 to 11 percent, depending on the climatic conditions. The areas of varying moisture content in the United States are shown in fig. 45.9.

Trim Parts for Doors and Frames

Doorframes

Rough openings in the stud walls for interior doors are usually framed out to be 3 inches more than the door height and $2\frac{1}{2}$ inches more than the door width. This provides for the frame and its plumbing and leveling in the opening. Interior doorframes are made up of two side *jambs* and a head jamb and include stop moldings upon which the door closes. The most common of these jambs is the one-piece type (fig. 47.1a). Jambs may be obtained in standard $5\frac{1}{4}$-inch widths for plaster walls and $4\frac{5}{8}$-inch widths for walls with $\frac{1}{2}$-inch dry-wall finish. The two-and three-piece adjustable jambs are also standard types (fig. 47.1b and c). Their principal advantage is in being adaptable to a variety of wall thicknesses.

Some manufacturers produce interior doorframes with the door fitted and prehung, ready for installing. Application of the casing completes the job. When used with two- or three-piece jambs, casings can even be installed at the factory.

Common minimum widths for single interior doors are: (a) Bedroom and other habitable rooms, 2 feet 6 inches; (b) bathrooms, 2 feet 4 inches; (c) small closet and linen closets, 2 feet. These sizes vary a great deal, and sliding doors, folding door units, and similar types are often used for wardrobes and may be 6 feet or more in width. However, in most cases, the jamb, stop, and casing parts are used in some manner to frame and finish the opening.

Standard interior and exterior door heights are 6 feet 8 inches for first floors, but 6-foot 6-inch doors are sometimes used on the upper floors.

Casing

Casing is the edge trim around interior door openings and is also used to finish the room side of windows and exterior door frames. Casing usually varies in width from $2\frac{1}{4}$ to $3\frac{1}{2}$ inches, depending on the style. Casing may be obtained in thicknesses from $\frac{1}{2}$ to $\frac{3}{4}$ inch, although $\frac{11}{16}$ inch is standard in many of the narrow-line patterns. Two common patterns are shown in figure 47. 1d and e.

Interior Doors

As in exterior door styles, the two general interior types are the flush and the panel door. Novelty doors, such as the folding door unit, might be flush or louvered. Most standard interior doors are $1\frac{3}{8}$ inches thick.

The flush interior door is usually made up with a hollow core of light framework of some type with thin plywood or hardboard (fig. 47. 2a). Plywood-faced flush doors may be obtained in gum, birch, oak, mahogany, and woods of other species, most of which are suitable for natural finish. Nonselected grades are usually painted as are hardboard-faced doors.

The panel door consists of solid *stiles* (vertical side members), *rails* (cross pieces), and *panel filters* of various types. The five-cross panel and the Colonial-type panel doors are perhaps the most common of this style (fig. 47.2 b and c). The louvered door (fig. 47. 2 d) is also popular and is commonly used for closets because it provides some ventilation. Large openings for wardrobes are finished with sliding or folding doors, or with flush or louvered doors (fig. 47. 2 e). Such doors are usually $1\frac{1}{8}$ inches thick.

Hinged doors should open or swing in the direction of natural entry, against a blank wall whenever possible, and should not be obstructed by other swinging doors. Doors should *never* be hinged to swing into a hallway.

Doorframe and Trim Installation

When the frame and doors are not assembled and prefitted, the side jambs should be fabricated by nailing through the notch into the head jamb with three sevenpenny or eightpenny coated nails (fig. 47.1 a). The assembled frames are then fastened in the rough openings by shingle wedges used between the side jamb and the stud (fig. 47. 3 a). One jamb is plumbed and leveled using four or five sets of shingle wedges for the height of the frame. Two eightpenny finishing nails are used at each wedged area, one driven so that the doorstop will cover it (fig. 47.3 a). The opposite side jamb is now fastened in place with shingle wedges and

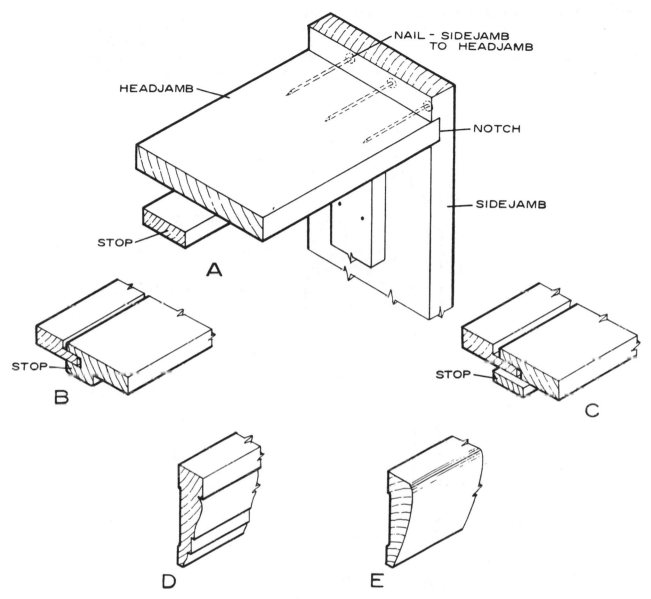

Figure 47.1.—Interior door parts: *A*, Door jambs and stops; *B*, two-piece jamb;
C, three-piece jamb; *D*, Colonial casing; *E*, ranch casing.

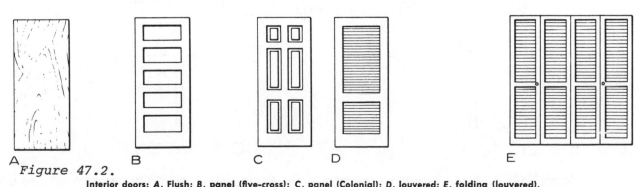

Figure 47.2.
Interior doors: *A*, Flush; *B*, panel (five-cross); *C*, panel (Colonial); *D*, louvered; *E*, folding (louvered).

finishing nails, using the first jamb as a guide in keeping a uniform width.

Casings are nailed to both the jamb and the framing studs or header, allowing about a $\frac{3}{16}$-inch edge distance from the face of the jamb (fig. 47.3 a). Finish or casing nails in sixpenny or sevenpenny sizes, depending on the thickness of the casing, are used to nail into the stud. Fourpenny or fivepenny finishing nails or 1½-inch brads are used to fasten the thinner edge of the casing to the jamb. In hardwood, it is usually advisable to predrill to prevent splitting. Nails in the casing are located in pairs (fig. 47.3 a) and

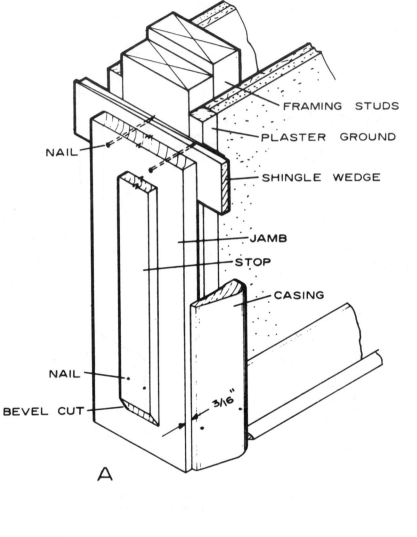

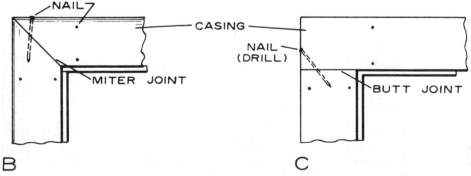

Figure 47.3.—Doorframe and trim: A, Installation; B, miter joint for casing; C, butt joint for casing.

608

spaced about 16 inches apart along the full height of the opening and at the head jamb.

Casing with any form of molded shape must have a mitered joint at the corners (fig. 47.3 b) When casing is square-edged, a butt joint may be made at the junction of the side and head casing (fig. 47.3 c). If the moisture content of the casing is well above that recommended in fig. 45.9, a mitered joint may open slightly at the outer edge as the material dries. This can be minimized by using a small glued spline at the corner of the mitered joint. Actually, use of a spline joint under any moisture condition is considered good practice, and some prefitted jamb, door, and casing units are provided with splined joints. Nailing into the joint after drilling will aid in retaining a close fit (fig. 47.3 b and c).

The door opening is now complete except for fitting and securing the hardware and nailing the stops in proper position. Interior doors are normally hung with two 3½- by 3½-inch loose-pin butt hinges. The door is fitted into the opening with the clearances shown in figure 134. The clearance and location of hinges, lock set, and doorknob may vary somewhat, but they are generally accepted by craftsmen and conform to most millwork standards. The edge of the lock stile should be beveled slightly to permit the door to clear the jamb when swung open. If the door is to swing across heavy carpeting, the bottom clearance may be slightly more.

Thresholds are used under exterior doors to close the space allowed for clearance. Weather strips around exterior door openings are very effective in reducing air infiltration.

In fitting doors, the stops are usually temporarily nailed in place until the door has been hung. Stops for doors in single-piece jambs are generally 7/16 inch thick and may be ¾ to 2¼ inches wide. They are installed with a mitered joint at the junction of the side and head jambs. A 45° bevel cut at the bottom of the stop, about 1 to 1½ inches above the finish floor, will eliminate a dirt pocket and make cleaning or refinishing of the floor easier (fig. 47. 3 a)

Some manufacturers supply prefitted door jambs and doors with the hinge slots routed and ready for installation. A similar door buck of sheet metal with formed stops and casing is also available.

Installation of Door Hardware

Hardware for doors may be obtained in a number of finishes, with brass, bronze, and nickel perhaps the most common. Door sets are usually classed as: (a) Entry lock for exterior doors, (b) bathroom set (inside lock control with safety slot for opening from the outside), (c) bedroom lock (keyed lock), and (d) passage set (without lock).

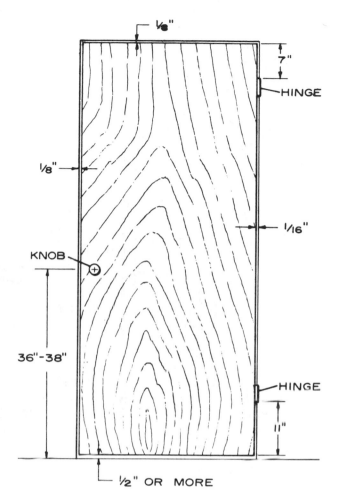

Figure 47.4.—Door clearances.

Hinges

Using three hinges for hanging 1¾-inch exterior doors and two hinges for the lighter interior doors is common practice. There is some tendency for exterior doors to warp during the winter because of the difference in exposure on the opposite sides. The three hinges reduce this tendency. Three hinges are also useful on doors that lead to unheated attics and for wider and heavier doors that may be used within the house.

Loose-pin butt hinges should be used and must be of the proper size for the door they support. For 1¾-inch-thick doors, use 4- by 4-inch butts; for 1⅜-inch doors, 3½- by 3½-inch butts. After the door is fitted to the framed opening, with the proper clearances, hinge halves are fitted to the door. They are routed into the door edge with about a 3/16-inch back distance (fig. 47.5 a). One hinge half should be set flush with the surface and must be fastened square with the edge of the door. Screws are included with each pair of hinges.

609

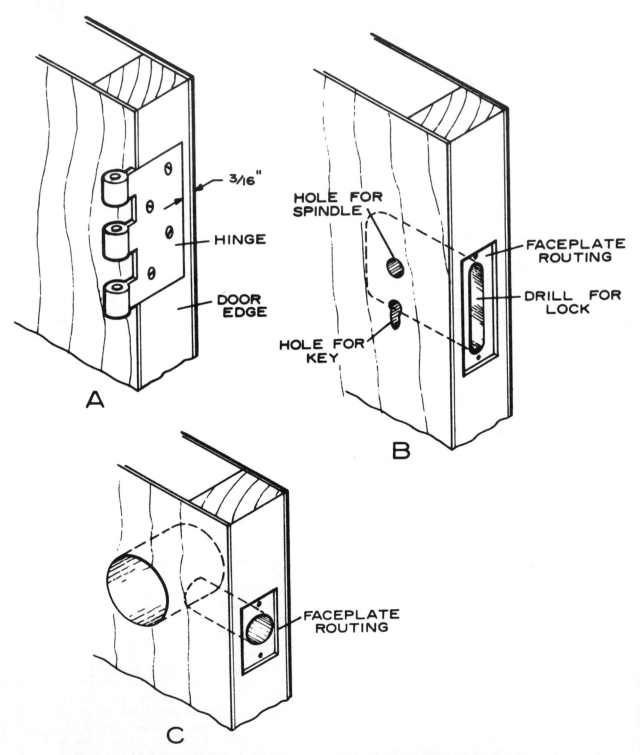

Figure 47.5.—Installation of door hardware: A, Hinge; B, mortise lock; C, bored lock set.

610

The door is now placed in the opening and blocked up at the bottom for proper clearance. The jamb is marked at the hinge locations, and the remaining hinge half is routed and fastened in place. The door is then positioned in the opening and the pins slipped in place. If hinges have been installed correctly and the jambs are plumb, the door will swing freely.

Locks

Types of door locks differ with regard to installation, first cost, and the amount of labor required to set them. Lock sets are supplied with instructions that should be followed for installation. Some types require drilling of the edge and face of the door and routing of the edge to accommodate the lock set and faceplate (fig. 47.5 b). A more common bored

type (fig. 47.5 c) is much easier to install as it requires only one hole drilled in the edge and one in the face of the door. Boring jigs and faceplate markers are available to provide accurate installation. The lock should be installed so that the doorknob is 36 to 38 inches above the floorline. Most sets come with paper templates marking the location of the lock and size of the holes to be drilled.

Strike Plate

The strike plate, which is routed into the door jamb, holds the door in place by contact with the latch. To install, mark the location of the latch on the door jamb and locate the strike plate in this way. Rout out the marked outline with a chisel and also rout for the latch (fig. 47.6 a). The strike plate

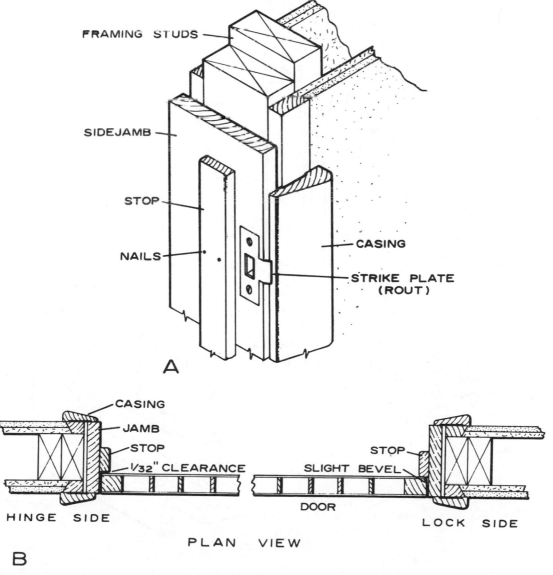

Figure 47.6.—Door details: A, Installation of strike plate; B, location of stops.

should be flush with or slightly below the face of the door jamb. When the door is latched, its face should be flush with the edge of the jamb.

Doorstops

The stops which have been set temporarily during fitting of the door and installation of the hardware may now be nailed in place permanently. Finish nails or brads, 1½ inches long, should be used. The stop at the lock side should be nailed first, setting it tight against the door face when the door is latched. Space the nails 16 inches apart in pairs **(fig. 47.6 a)**.

The stop behind the hinge side is nailed next, and a 1/32-inch clearance from the door face should be allowed (fig. 47.3 b) to prevent scraping as the door is opened. The head-jamb stop is then nailed in place. Remember that when door and trim are painted, some of the clearances will be taken up.

Wood-trim Installation

The casing around the window frames on the interior of the house should be the same pattern as that used around the interior door frames. Other trim which is used for a double-hung window frame includes the sash stops, stool, and apron (fig. 47.7 a). Another method of using trim around windows has the entire opening enclosed with casing (fig. 47.7 b). The stool is then a filler member between the bottom sash rail and the bottom casing.

The *stool* is the horizontal trim member that laps the window sill and extends beyond the casing at the sides, with each end notched against the plastered

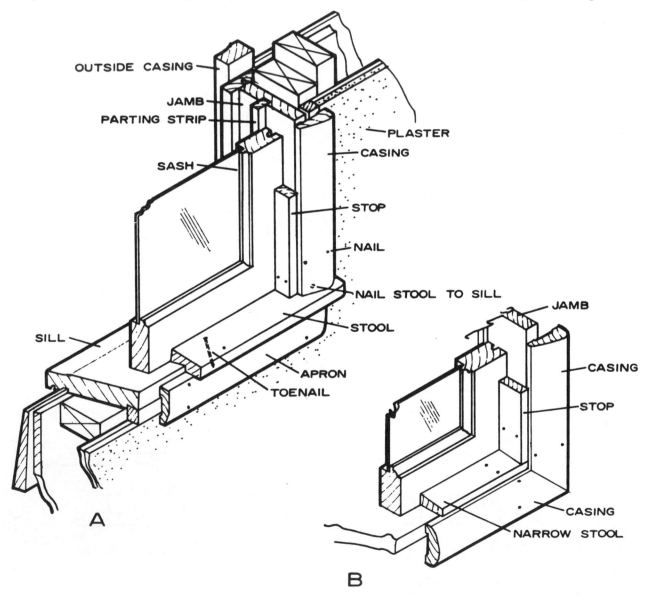

Figure 47.7.—Installation of window trim: A, With stool and apron; B, enclosed with casing.

wall. The *apron* serves as a finish member below the stool. The window stool is the first piece of window trim to be installed and is notched and fitted against the edge of the jamb and the plaster line, with the outside edge being flush against the bottom rail of the window sash (fig. 47.7 a) . The stool is blind-nailed at the ends so that the casing and the stop will cover the nailheads. Predrilling is usually necessary to prevent splitting. The stool should also be nailed at midpoint to the sill and to the apron with finishing nails. Face-nailing to the sill is sometimes substituted or supplemented with toenailing of the outer edge to the sill (fig. 47. 7 a) .

The casing is applied and nailed as described for doorframes (fig. 47.3a) , except that the inner edge is flush with the inner face of the jambs so that the stop will cover the joint between the jamb and casing. The window stops are then nailed to the jambs so that the window sash slides smoothly. Channel-type weather stripping often includes full-width metal

subjambs into which the upper and lower sash slide, replacing the parting strip. Stops are located against these instead of the sash to provide a small amount of pressure. The apron is cut to a length equal to the outer width of the casing line (fig. 47.7 a). It is nailed to the window sill and to the 2- by 4-inch framing sill below.

When casing is used to finish the bottom of the window frame as well as the sides and top, the narrow stool butts against the side window jamb. Casing is then mitered at the bottom corners (fig. 47.7 b) . and nailed as previously described.

Base and Ceiling Moldings

Base Moldings

Base molding serves as a finish between the finished wall and floor. It is available in several widths and forms. Two-piece base consists of a baseboard topped with a small base cap (fig. 47.8 a) . When plaster is not straight and true, the small base molding will

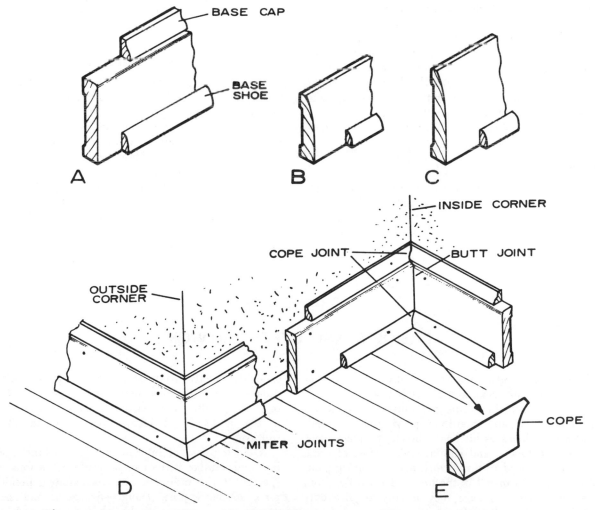

Figure 47.8.—Base molding: A, Square-edge base; B, narrow ranch base; C, wide ranch base; D, installation; E, cope.

613

conform more closely to the variations than will the wider base alone. A common size for this type of baseboard is ⅝ by 3¼ inches or wider. One-piece base varies in size from ⁷⁄₁₆ by 2¼ inches to ½ by 3¼ inches and wider (fig. 47.8 b and c). Although a wood member is desirable at the junction of the wall and carpeting to serve as a protective "bumper", wood trim is sometimes eliminated entirely.

Most baseboards are finished with a base shoe, ½ by ¾ inch in size (fig. 47.8 a, b, and c). A single-base molding without the shoe is sometimes placed at the wall-floor junction, especially where carpeting might be used.

Installation of Base Molding

Square-edged baseboard should be installed with a butt joint at inside corners and a mitered joint at outside corners (fig. 47.8 d). It should be nailed to each stud with two eightpenny finishing nails. Molded single-piece base, base moldings, and base shoe should have a coped joint at inside corners and a mitered joint at outside corners. A coped joint is one in which the first piece is square-cut against the plaster or base and the second molding coped. This is accomplished by sawing a 45° miter cut and with a coping saw trimming the molding along the inner line of the miter (fig. 47.8 e). The base shoe should be nailed into the subfloor with long slender nails and not into the baseboard itself. Thus, if there is a small amount of shrinkage of the joists, no opening will occur under the shoe.

Ceiling Moldings

Ceiling moldings are sometimes used at the junction of wall and ceiling for an architectural effect or to terminate dry-wall paneling of gypsum board or wood (fig. 47.9 a). As in the base moldings, inside

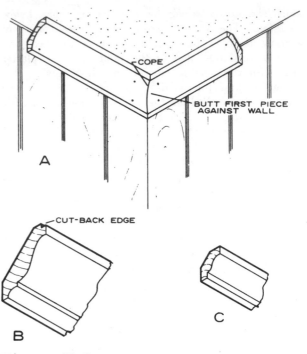

Figure 47.9.

Ceiling moldings; A, Installation (inside corner); B, crown molding; C, small crown molding.

corners should also be cope-jointed. This insures a tight joint and retains a good fit if there are minor moisture changes.

A cutback edge at the outside of the molding will partially conceal any unevenness of the plaster and make painting easier where there are color changes (fig. 47.9 b). For gypsum dry-wall construction, a small simple molding might be desirable (fig. 47.9 c). Finish nails should be driven into the upper wall-plates and also into the ceiling joists for large moldings when possible.

CHAPTER 48

CABINETS AND OTHER MILLWORK

Millwork, as a general term, usually includes most of those wood materials and house components which require manufacturing. This not only covers the interior trim, doors, and other items previously described, but also such items as kitchen cabinets, fireplace mantels, china cabinets, and similar units. Most of these units are produced in a millwork manufacturing plant and are ready to install in the house. They differ from some other items because they usually require only fastening to the wall or floor.

While many units are custom made, others can be ordered directly from stock. For example, kitchen cabinets are often stock items which may be obtained in 3-inch-width increments, usually beginning at widths of 12 or 15 inches and on up to 48 inch widths.

As in the case of interior trim, the cabinets, shelving, and similar items can be made of various wood species. If the millwork is to be painted, ponderosa pine, southern pine, Douglas-fir, gum, and similar species may be used. Birch, oak, redwood, and knotty pine, or other species with attractive surface variations, are some of the woods that are finished with varnish or sealers.

Recommended moisture content for book cases and other interior millwork may vary from 6 to 11 percent in different parts of the country. These areas, together with the moisture contents, are shown on the moisture-content map (fig. 45.9).

Kitchen Cabinets

The kitchen usually contains more millwork than the rest of the rooms combined. This is in the form of wall and base cabinets, broom closets, and other items. An efficient plan with properly arranged cabinets will not only reduce work and save steps for the housewife, but will often reduce costs because of the need for a smaller area. Location of the refrigerator, sink, dishwasher, and range, together with the cabinets, is also important from the standpoint of plumbing and electrical connections. Good lighting, both natural and artificial, is also important in designing a pleasant kitchen.

Kitchen cabinets, both base and wall units, should be constructed to a standard of height and depth. Figure 48.1 shows common base cabinet counter heights and depths as well as clearances for wall cabinets. While the counter height limits range from 30 to 38 inches, the standard height is usually 36 inches.

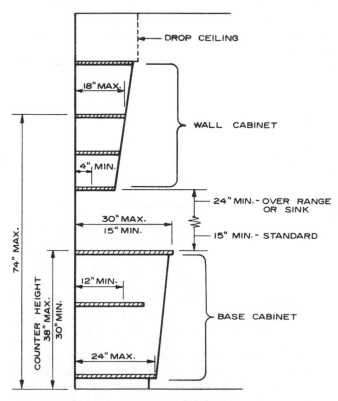

Figure 48.1.—Kitchen cabinet dimensions.

Wall cabinets vary in height depending on the type of installation at the counter. The tops of wall cabinets are located at the same height, either free or under a 12- to 14-inch drop ceiling or storage cabinet. Wall cabinets are normally 30 inches high, but not more than 21 inches when a range or sink is located under them. Wall cabinets can also be obtained in 12-, 15-, 18-, and 24-inch heights. The shorter wall cabinets are usually placed over refrigerators.

Narrow wall cabinets are furnished with single doors and the wider ones with double doors (fig. 48.2 a). Base cabinets may be obtained in full-door or full-drawer units or with both drawers and doors (fig. 48.2 b). Sink fronts or sink-base cabinets, corner cabinets, broom closets, and desks are some of the special units which may be used in planning the ideal kitchen. Cabinets are fastened to the wall through cleats located at the back of each cabinet. It is good practice to use long screws to penetrate into each wall stud.

Four basic layouts are commonly used in the design of a kitchen. The *U-type* with the sink at the bottom of the U and the range and refrigerator on opposite sides is very efficient (fig. 48.3 a).

The *L-type* (fig. 48.3 b), with the sink and range on one leg and the refrigerator on the other, is sometimes used with a dining space in the opposite corner.

The "parallel wall" or *pullman kitchen plan* (fig. 48.3 c) is often used in narrow kitchens and can be quite efficient with proper arrangement of the sink, range, and refrigerator.

The *sidewall type* (fig. 48.3 d) usually is preferred for small apartments. All cabinets, the sink, range, and refrigerator are located along one wall. Counter space is usually somewhat limited in this design when kitchens are small.

Closets and Wardrobes

The simple clothes closet is normally furnished with a shelf and a rod for hanging clothes. Others may have small low cabinets for the storage of shoes and similar items. Larger wardrobes with sliding or folding doors may be combined with space for hanging clothes as well as containing a dresser complete with drawers and mirror. Many built-in combinations are possible, all of which reduce the amount of bedroom furniture needed.

Linen closets may be simply a series of shelves behind a flush or panel door. Others may consist of an open cabinet with doors and drawers built directly into a notch or corner of the wall located near the bedrooms and bath.

Mantels

The type of *mantel* used for a fireplace depends on the style and design of the house and its interior finish. The contemporary fireplace may have no mantel

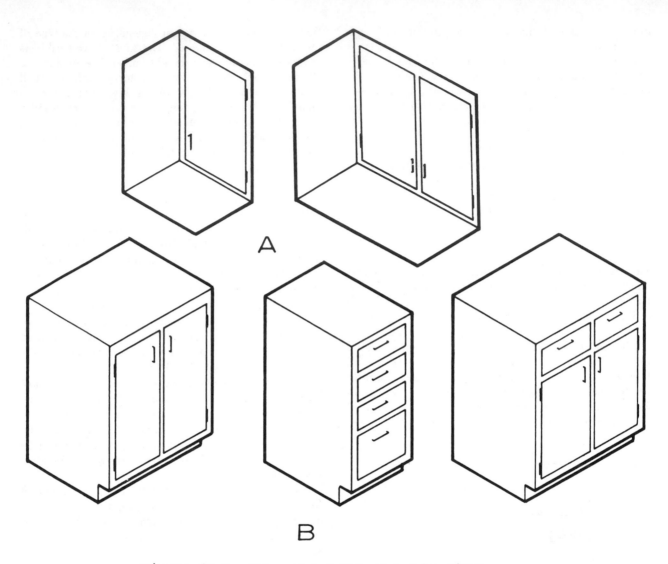

Figure 48.2.—Kitchen cabinets: A, Wall cabinets; B, base cabinets.

at all, or at best a simple wood molding used as a transition between the masonry and the wall finish. However, the colonial or formal interior usually has a well-designed mantel enclosing the fireplace opening. This may vary from a simple mantel (fig. 48.4). to a more elaborate unit combining paneling and built-in cabinets along the entire wall. In each design, however, it is important that no wood or other combustible material be placed within 3½ inches of the edges of the fireplace opening. Furthermore, any projection more than 1½ inches in front of the fireplace, such as the mantel shelf, should be at least 12½ inches above the opening. Mantels are fastened to the header and framing studs above and on each side of the fireplace.

China Cases

Another millwork item often incorporated in the dining room of a formal or traditional design is the china case. It is usually designed to fit into one or two corners of the room. This corner cabinet often has glazed doors above and single- or double-panel doors below (fig. 48.5). It may be 7 feet or more high with a drop ceiling above with a face width of about 3 feet. Shelves are supplied in both the upper and lower cabinets.

China cases or storage shelves in dining rooms of contemporary houses may be built in place by the contractor. A row of cabinets or shelves may act as a separator between dining room and kitchen and serve as a storage area for both rooms.

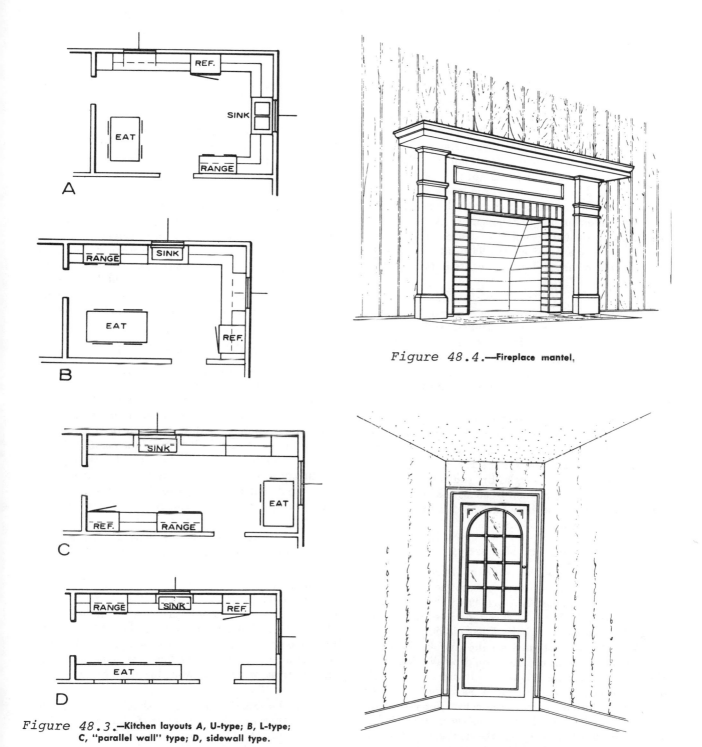

A

B

C

D

Figure 48.3.—Kitchen layouts A, U-type; B, L-type;
C, "parallel wall" type; D, sidewall type.

Figure 48.4.—Fireplace mantel,

Figure 48.5.—Corner china case.

CHAPTER 49

STAIRS

Stairways in houses should be designed and constructed to afford safety and adequate headroom for the occupants as well as space for the passage of furniture (*11*). The two types of stairs commonly used in houses are (a) the finished main stairs leading to the second floor or split-level floors and (b) the basement or service stairs leading to the basement or garage area. The main stairs are designed to provide easy ascent and descent and may be made a feature of the interior design. The service stairs to basement areas are usually somewhat steeper and are constructed of less expensive materials, although safety and convenience are still prime factors in their design.

Construction

Most finish and service stairs are constructed in place. The main stairs are assembled with prefabricated parts, which include housed stringers, treads, and risers. Basement stairs may be made simply of 2-by 12-inch carriages and plank treads. In split-level design or a midfloor outside entry, stairways are often completely finished with plastered walls, handrails, and appropriate moldings.

Wood species appropriate for main stairway components include oak, birch, maple, and similar hardwoods. Treads and risers for the basement or service stairways may be of Douglas-fir, southern pine, and similar species. A hardwood tread with a softwood or lower grade hardwood riser may be combined to provide greater resistance to wear.

Types of Stairways

Three general types of stairway runs most commonly used in house construction are the straight run (fig. 49.1 a), the long "L" (fig. 49.1 b) and the narrow "U" (fig. 49.2 b). Another type is similar to the Long "L" except that "winders" or "pie-shaped" treads (fig. 49.2 b) are substituted for the landing. This type of stairs is not desirable and should be avoided whenever possible because it is obviously not as convenient or as safe as the long "L." It is used where the stair run is not sufficient for the more conventional stairway containing a landing. In such instances, the winders should be adjusted to replace the landings so that the width of the tread, 18 inches from the narrow end, will not be less than the tread width on the straight run (fig. 49.3 a). Thus if the standard tread is 10 inches wide, the winder tread should be at least 10 inches wide at the 18-inch line.

Another basic rule in stair layout concerns the landing at the top of a stairs when the door opens into the stairway, such as on a stair to the basement. This landing, as well as middle landings, should not be less than 2 feet 6 inches long (fig. 49.3 b).

Sufficient headroom in a stairway is a primary requisite. For main stairways, clear vertical distance should not be less than 6 feet 8 inches (fig. 49.4 a). Basement or service stairs should provide not less than a 6-foot 4-inch clearance.

The minimum tread width and riser height must also be considered. For closed stairs, a 9-inch tread width and an $8\frac{1}{4}$-inch riser height should be considered a minimum even for basement stairways (fig. 49.4 b). Risers with less height are always more desirable. The nosing projection should be at least $1\frac{1}{8}$ inches; however, if the projection is too much greater, the stairs will be awkward and difficult to climb.

Ratio of Riser to Tread

There is a definite relation between the height of a riser and the width of a tread, and all stairs should be laid out to conform to well-established rules governing these relations. If the combination of run and rise is too great, there is undue strain on the leg muscles and on the heart of the climber; if the combination is too small, his foot may kick the riser at each step and an attempt to shorten stride may be tiring. Experience has proved that a riser $7\frac{1}{2}$ to $7\frac{3}{4}$ inches high with appropriate tread width combines both safety and comfort.

A rule of thumb which sets forth a good relation between the height of the riser and the width of the tread is:

The tread width multiplied by the riser height in inches should equal to 72 to 75. The stairs shown in figure 49.4 b would conform to this rule—9 times $8\frac{1}{4} = 74\frac{1}{4}$. If the tread is 10 inches, however, the riser should be $7\frac{1}{2}$ inches, which is more desirable for common stairways. Another rule sometimes used is: The tread width plus twice the riser height should equal about 25.

These desirable riser heights should, therefore, be used to determine the number of steps between floors. For example, 14 risers are commonly used for main stairs between the first and second floors. The 8-foot ceiling height of the first floor plus the upper-story floor joists, subfloor, and finish floor result in a floor-to-floor height of about 105 inches. Thus, 14 divided into 105 is exactly $7\frac{1}{2}$ inches, the height of each riser. Fifteen risers used for this height would result in a 7-inch riser height.

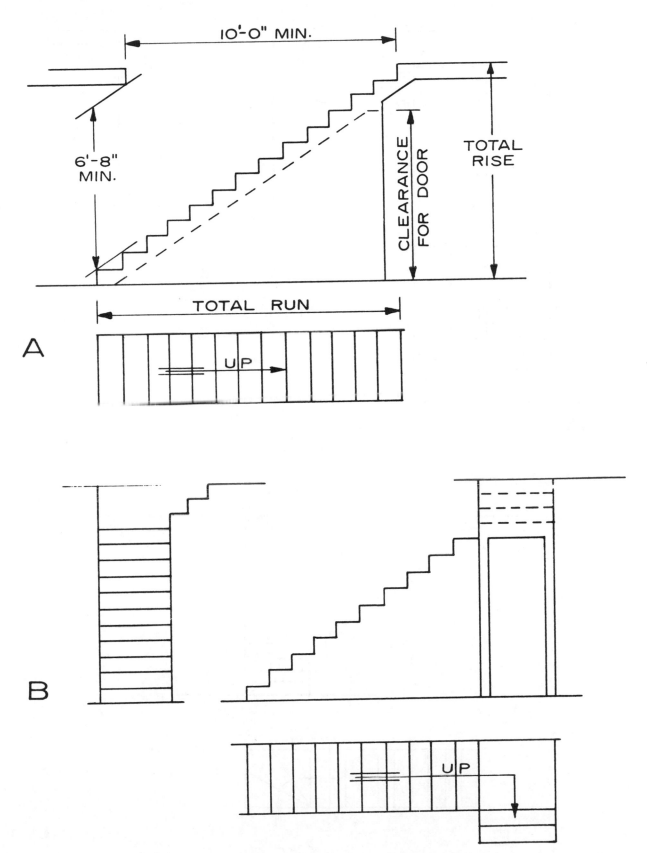

Figure 49.1.—Common types of stair runs: A, Straight; B, long "L."

619

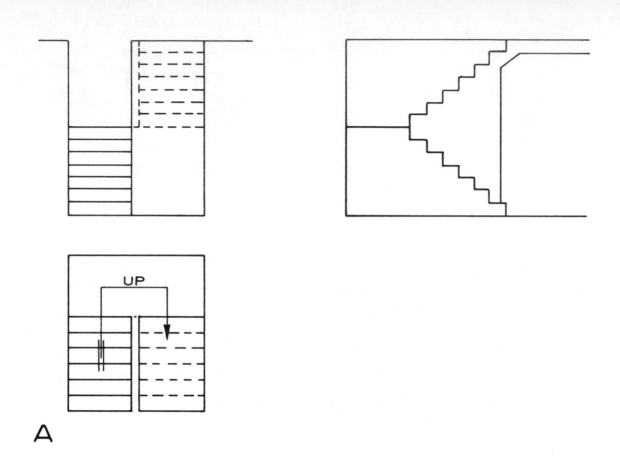

A

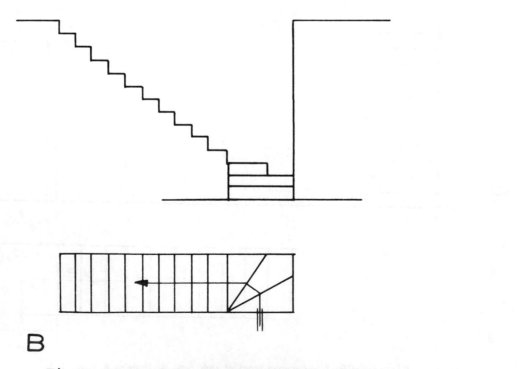

B

Figure 49.2.—Space-saving stairs: A, Narrow "U"; B, winder.

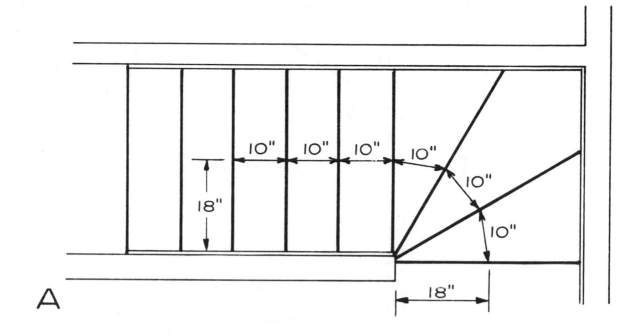

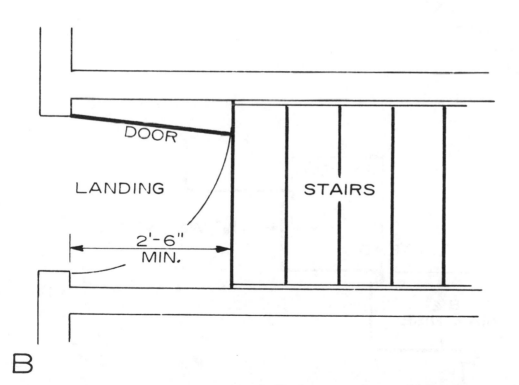

Figure 49.3.—Stair layout: A, Winder treads; B, landings.

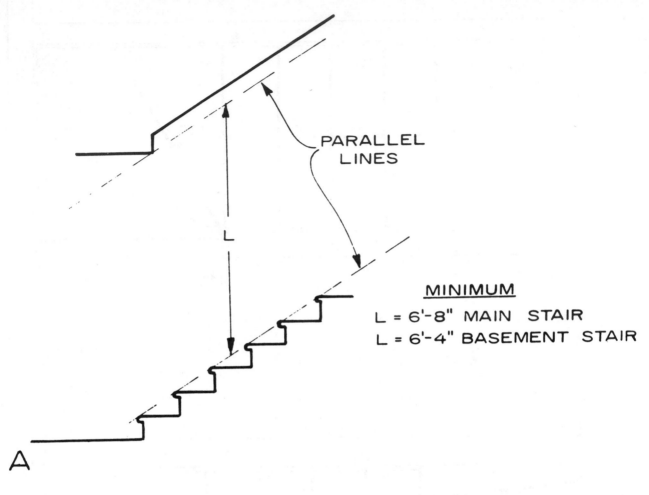

PARALLEL LINES

L

MINIMUM

L = 6'-8" MAIN STAIR

L = 6'-4" BASEMENT STAIR

A

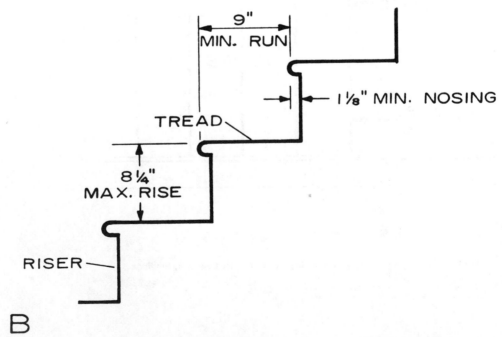

9"
MIN. RUN

1⅛" MIN. NOSING

TREAD

8¼"
MAX. RISE

RISER

B

Figure 49.4.—Stairway dimensions: A, Minimum headroom; B, closed stair dimensions.

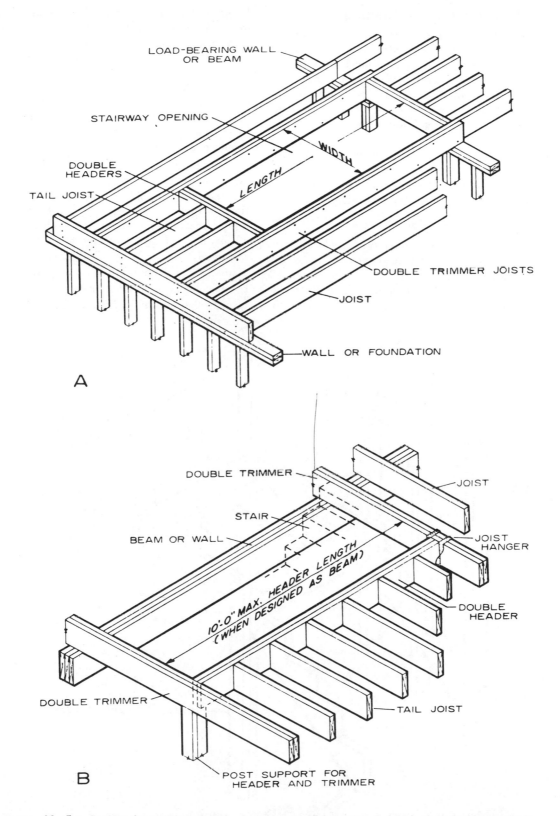

LOAD-BEARING WALL
OR BEAM

STAIRWAY OPENING

WIDTH

LENGTH

DOUBLE
HEADERS

TAIL JOIST

DOUBLE TRIMMER JOISTS

JOIST

WALL OR FOUNDATION

A

DOUBLE TRIMMER

JOIST

STAIR

BEAM OR WALL

JOIST
HANGER

10'-0" MAX. HEADER LENGTH
(WHEN DESIGNED AS BEAM)

DOUBLE
HEADER

DOUBLE TRIMMER

TAIL JOIST

POST SUPPORT FOR
HEADER AND TRIMMER

B

Figure 49.5.—Framing for stairs: *A,* Length of opening parallel to joists; *B,* length of opening perpendicular to joists.

623

Stair Widths and Handrails

The width of the main stairs should be not less than 2 feet 8 inches clear of the handrail. However, many main stairs are designed with a distance of 3 feet 6 inches between the centerline of the enclosing sidewalls. This will result in a stairway with a width of about 3 feet. Split-level entrance stairs are even wider. For basement stairs, the minimum clear width is 2 feet 6 inches.

A continuous handrail should be used on at least one side of the stairway when there are more than three risers. When stairs are open on two sides, there should be protective railings on each side.

Framing for Stairs

Openings in the floor for stairways, fireplaces, and chimneys are framed out during construction of the floor system (figs. 28.10 and 12). The long dimension of stairway openings may be either parallel or at right angles to the joists. However, it is much easier to frame a stairway opening when its length is parallel to the joists. For basement stairways, the rough openings may be about 9 feet 6 inches long by 32 inches wide (two joist spaces). Openings in the second floor for the main stair are usually a minimum of 10 feet long. Widths may be 3 feet or more. Depending on the short header required for one or both ends, the opening is usually framed as shown in figure 49.5 a when joists parallel the length of the opening. Nailing should conform to that shown in figs. 28.10 and 28.12.

When the length of the stair opening is perpendicular to the length of the joists, a long doubled header is required (fig. 49.5 b). A header under these conditions without a supporting wall beneath is usually limited to a 10-foot length. A load-bearing wall under all or part of this opening simplifies the framing immensely, as the joists will then bear on the top plate of the wall rather than be supported at the header by joist hangers or other means. Nailing should conform to that shown in figs. 28.10 and 28.12.

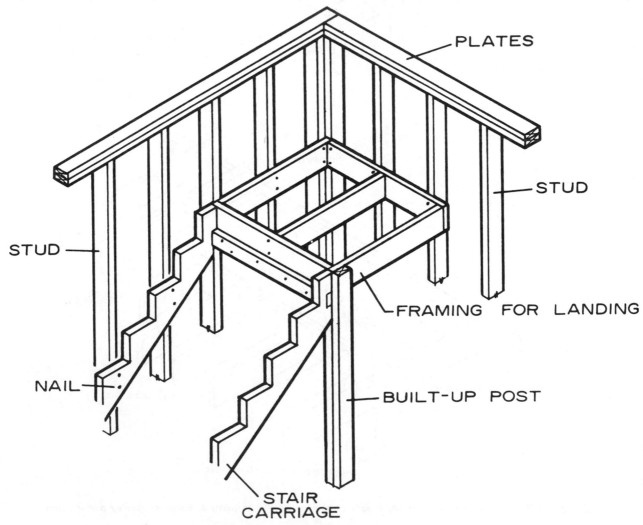

PLATES

STUD

STUD

FRAMING FOR LANDING

NAIL

BUILT-UP POST

STAIR CARRIAGE

Figure 49.6.—Framing for stair landing.

The framing for an L-shaped stairway is usually supported in the basement by a post at the corner of the opening or by a load-bearing wall beneath. When a similar stair leads from the first to the second floor, the landing can be framed-out (fig. 49.6). The platform frame is nailed into the enclosing stud walls and provides a nailing area for the subfloor as well as a support for the stair carriages.

Stairway Details

Basement Stairs

Stair carriages which carry the treads and support the loads on the stair are made in two ways. Rough stair carriages commonly used for basement stairs are made from 2- by 12-inch planks. The effective depth below the tread and riser notches must be at least $3\frac{1}{2}$ inches (fig. 49.7a). Such carriages are usually placed only at each side of the stairs; however, an intermediate carriage is required at the center of the stairs when the treads are $1\frac{1}{16}$ inches thick and the stairs wider than 2 feet 6 inches. Three carriages are also required when treads are $1\frac{5}{8}$ inches thick and stairs are wider than 3 feet. The carriages are fastened to the joist header at the top of the stairway or rest on a supporting ledger nailed to the header (fig. 49.7b).

Firestops should be used at the top and bottom of all stairs, as shown (fig. 49.7a).

Perhaps the simplest system is one in which the carriages are not cut out for the treads and risers. Rather, cleats are nailed to the side of the unnotched carriage and the treads nailed to them. This design, however, is likely not as desirable as the notched carriage system when walls are present. Carriages can also be supported by walls located below them.

The bottom of the stair carriages may rest and be anchored to the basement floor. Perhaps a better method is to use an anchored 2- by 4- or 2- by 6-inch treated kicker plate (fig. 49.7c).

Basement stair treads can consist of simple $1\frac{1}{2}$-inch-thick plank treads without risers. However, from the standpoint of appearance and maintenance, the use of $1\frac{1}{8}$-inch finished tread material and nominal 1-inch boards for risers is usually justified. Finishing nails fasten them to the plank carriages.

A somewhat more finished staircase for a fully enclosed stairway might be used from the main floor to the attic. It combines the rough notched carriage with a finish stringer along each side (fig. 49.8a). The finish stringer is fastened to the wall, before carriages are fastened. Treads and risers are cut to fit snugly between the stringers and fastened to the rough carriage with finishing nails (fig. 49.8a). This may be varied somewhat by nailing the rough carriage directly to the wall and notching the finished stringer to fit (fig. 49.8b). The treads and risers are installed as previously described.

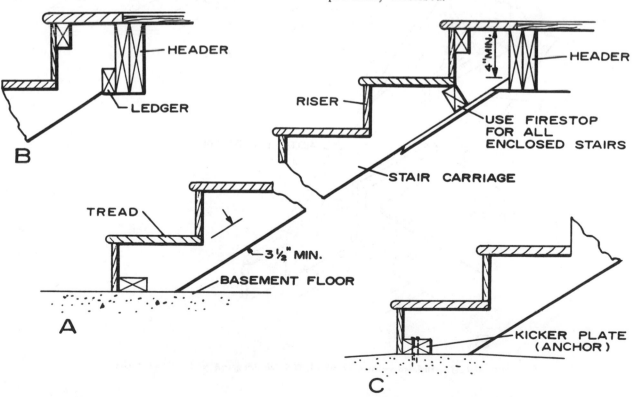

Figure 49.7.—Basement stairs: A, Carriage details; B, ledger for carriage; C, kicker plate.

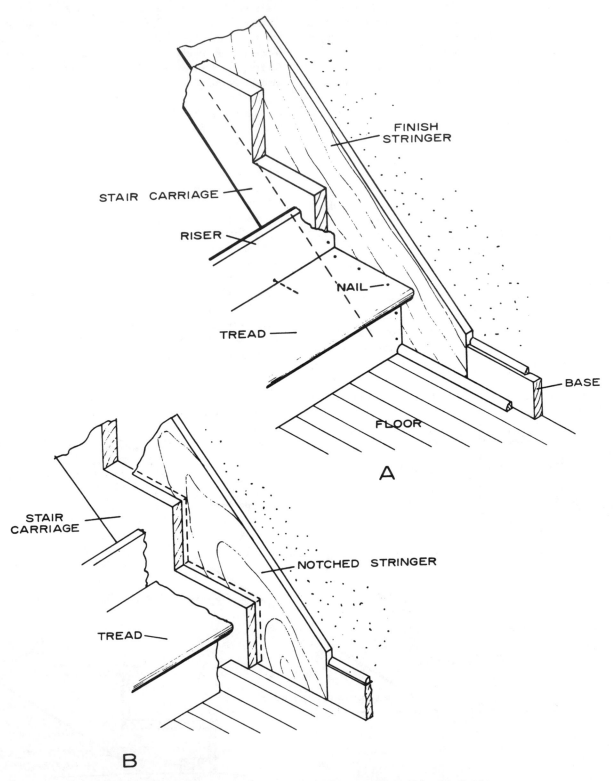

FINISH STRINGER

STAIR CARRIAGE

RISER

TREAD

NAIL

BASE

FLOOR

A

STAIR CARRIAGE

NOTCHED STRINGER

TREAD

B

Figure 49.8.—Enclosed stairway details: A, With full stringer; B, with notched stringer.

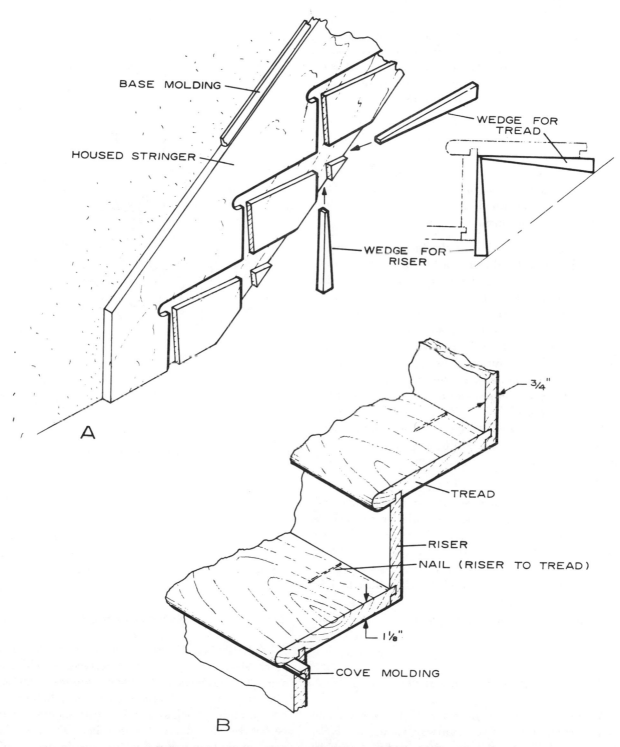

BASE MOLDING

HOUSED STRINGER

WEDGE FOR TREAD

WEDGE FOR RISER

A

3/4"

TREAD

RISER

NAIL (RISER TO TREAD)

1⅛"

COVE MOLDING

B

Figure 49.9.—Main stair detail with: A, Housed stringer; B, combination of treads and risers.

627

Main Stairway

An open main stairway with its railing and balusters ending in a *newel* post can be very decorative and pleasing in the traditional house interior. It can also be translated to a contemporary stairway design and again result in a pleasing feature.

The main stairway differs from the other types previously described because of: (a) The housed stringers which replace the rough plank carriage; (b) the routed and grooved treads and risers; (c) the decorative railing and balusters in open stairways; and (d) the wood species, most of which can be given a natural finish.

The supporting member of the finished main stairway is the housed stringer (fig. 49.9a). One is used on each side of the stairway and fastened to the plastered or finished walls. They are routed to fit both the tread and riser. The stair is assembled by means of hardwood wedges which are spread with glue and driven under the ends of the treads and in back of the risers. Assembly is usually done from under and the rear side of the stairway. In addition, nails are used to fasten the riser to the tread between the ends of the step (fig. 49.9b). When treads and risers are wedged and glued into housed stringers, the maximum allowable width is usually 3 feet 6 inches. For wider stairs, a notched carriage is used between the housed stringers.

When stairs are open on one side, a railing and balusters are commonly used. Balusters may be fastened to the end of the treads which have a finished return (fig. 49.10). The balusters are also fastened to a railing which is terminated at a newel post. Balusters may be turned to form doweled ends, which fit into drilled holes in the treads and the railing. A stringer and appropriate moldings are used to complete the stairway trim.

Attic Folding Stairs

Where attics are used primarily for storage and where space for a fixed stairway is not available, hinged or folding stairs are often used and may be purchased ready to install. They operate through an opening in the ceiling of a hall and swing up into the attic space, out of the way when not in use. Where such stairs are to be installed, the attic floor joists should be designed for limited floor loading. One com-

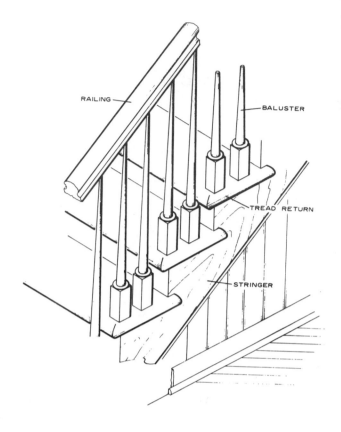

Figure 49.10.—Details of open main stairway.

mon size of folding stairs requires only a 26- by 54-inch rough opening. These openings should be framed out as described for normal stair openings.

Exterior Stairs

Proportioning of risers and treads in laying out porch steps or approaches to terraces should be as carefully considered as the design of interior stairways. Similar riser-to-tread ratios can be use; however, the riser used in principal exterior steps should normally be between 6 and 7 inches in height. The need for a good support or foundation for outside steps is often overlooked. Where wood steps are used, the bottom step should be concrete or supported by treated wood members. Where the steps are located over backfill or disturbed ground, the foundation should be carried down to undisturbed ground.

CHAPTER 50

FLASHING AND OTHER SHEET METAL WORK

In house construction, the *sheet-metal work* normally consists of flashing, gutters, and downspouts, and sometimes attic ventilators. Flashing (*11*) is often provided to prevent wicking action by joints between moisture-absorbent materials. It might also be used to provide protection from wind-driven rain or from action of melting snows. For instance, damage from ice dams is often the result of inadequate flashing. Thus, proper installation of these materials is important, as well as their selection and location.

Gutters are installed at the cornice line of a pitched-roof house to carry the rain or melted snow to the downspouts and away from the foundation area. They are especially needed for houses with narrow roof overhangs. Where positive rain disposal cannot be assured, downspouts should be connected with storm sewers or other drains. Poor drainage away from the wall is often the cause of wet basements and other moisture problems.

Materials

Materials most commonly used for sheet-metal work are galvanized metal, terneplate, aluminum, copper, and stainless steel. Near the seacoast, where the salt in the air may corrode galvanized sheet metal, copper or stainless steel is preferred for gutters, downspouts, and flashings. Molded wood gutters, cut from solid pieces of Douglas-fir or redwood, are also used in coastal areas because they are not affected by the corrosive atmosphere. Wood gutters can be attractive in appearance and are preferred by some builders.

Galvanized (zinc-coated) sheet metal is used in two weights of zinc coatings: 1.25 and 1.50 ounces per square foot (total weight of coating on both sides). When the lightly coated 1.25-ounce sheet is used for exposed flashing and for gutters and downspouts, 26-gage metal is required. With the heavier 1.50-ounce coating, a 28-gage metal is satisfactory for most metal work, except that gutters should be 26-gage.

Aluminum flashing should have a minimum thickness of 0.019 inch, the same as for roof valleys. Gutters should be made from 0.027-inch-thick metal and downspouts from 0.020-inch thickness. Copper for flashing and similar uses should have a minimum thickness of 0.020 inch (16 oz.). Aluminum is not normally used when it comes in contact with concrete or stucco unless it is protected with a coat of asphaltum or other protection against reaction with the alkali in the cement.

The types of metal fastenings, such as nails and screws, and the hangers and clips used with the various metals, are important to prevent corrosion or deterioration when unlike metals are used together. For aluminum, only aluminum or stainless steel fasteners should be used. For copper flashing, use copper nails and fittings. Galvanized sheet metal or terneplate should be fastened with galvanized or stainless-steel fasteners.

Flashing

Flashing should be used at the junction of a roof and a wood or masonry wall, at chimneys, over exposed doors and windows, at siding material changes, in roof valleys, and other areas where rain or melted snow may penetrate into the house.

Material Changes

One wall area which requires flashing is at the intersection of two types of siding materials. For example, a stucco-finish gable end and a wood-siding lower wall should be flashed (fig. 50.1a). A wood molding such as a drip cap separates the two materials and is covered by the flashing which extends behind the stucco. The flashing should extend at least 4 inches above the intersection. When sheathing paper is used, it should lap the flashing (fig. 50.1a).

When a wood-siding pattern change occurs on the same wall, the intersection should also be flashed. A vertical board-sided upper wall with horizontal siding below usually requires some type of flashing (fig. 50.1b). A small space above the molding provides a drip for rain. This will prevent paint peeling which could occur if the boards were in tight contact with the molding. A drip cap is sometimes used as a terminating molding (fig. 36.6) When the upper wall, such as a gable end, projects slightly beyond the lower wall (fig. 36.7), flashing is usually *not* required.

Doors and Windows

The same type of flashing shown in figure 50.1a should be used over door and window openings exposed to driving rain. However, window and doorheads protected by wide overhangs in a single-story house with a hip roof do not ordinarily require such flashing. When building paper is used on the sidewalls, it should lap the top edge of the flashing. To protect the walls behind the window sill in a brick veneer exterior, flashing should extend under the masonry sill up to the underside of the wood sill.

Flat Roof

Flashing is also required at the junctions of an exterior wall and a flat or low-pitched built-up roof

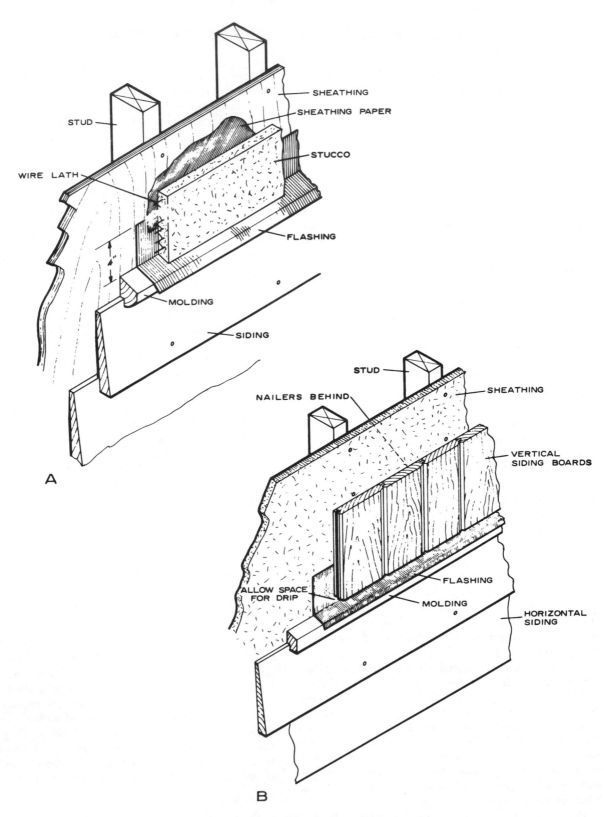

STUD

SHEATHING

SHEATHING PAPER

STUCCO

WIRE LATH

FLASHING

MOLDING

SIDING

A

STUD

NAILERS BEHIND

SHEATHING

VERTICAL
SIDING BOARDS

FLASHING

ALLOW SPACE
FOR DRIP

MOLDING

HORIZONTAL
SIDING

B

Figure 50.1.—Flashing at material changes: *A,* Stucco above, siding below;
B, vertical siding above, horizontal below.

(fig. 34.4 c). When a metal roof is used, the metal is turned up on the wall and covered by the siding. A clearance of 2 inches should be allowed at the bottom of the siding for protection from melted snow and water.

Ridge and Roof

Ridge flashing should be used under a Boston ridge in wood shingle or shake roofs to prevent water entry (fig. 34.5b). The flashing should extend about 3 inches on each side of the ridge and be nailed in place only at the outer edges. The ridge shingles or shakes, which are 6 to 8 inches wide, cover the flashing.

Stack vents and roof ventilators are provided with flashing collars which are lapped by the shingles on the upper side. The lower edge of the collar laps the shingles. Sides are nailed to the shingles and calked with a roofing mastic.

Valley

The valley formed by two intersecting rooflines is usually covered with metal flashing. Some building regulations allow the use of two thicknesses of mineral-surfaced roll roofing in place of the metal flashing. As an alternate, one 36-inch-wide strip of roll roofing with closed or woven asphalt shingles is also allowed. This type of valley is normally used only on roofs with a slope of 10 in 12 or steeper.

Widths of sheet-metal flashing for valleys should not be less than:

(a) 12 inches wide for roof slopes of 7 in 12 and over.

(b) 18 inches wide for 4 in 12 to 7 in 12 roof slopes.

(c) 24 inches wide for slopes less than 4 in 12.

The width of the valley between shingles should increase from the top to the bottom (fig. 50.2 a) The minimum open width at the top is 4 inches and should be increased at the rate of about $\frac{1}{8}$ inch per foot. These widths can be chalklined on the flashing before shingles are applied.

When adjacent roof slopes vary, such as a low-slope porch roof intersecting a steeper main roof, a 1-inch crimped standing seam should be used (fig. 50.2b). This will keep heavy rains on the steeper slopes from overrunning the valley and being forced under the shingles on the adjoining slope. Nails for the shingles should be kept back as far as possible to eliminate holes in the flashing. A ribbon of asphalt-roofing mastic is often used under the edge of the shingles. It is wise to use the wider valley flashings supplemented by a width of 15- or 30-pound asphalt felt where snow and ice dams may cause melting snow water to back under shingles.

Roof-Wall Intersections

When shingles on a roof intersect a vertical wall, shingle flashing is used at the junction. These tin or galvanized-metal shingles are bent at a 90° angle and extend up the side of the wall over the sheathing a minimum of 4 inches (fig. 50.3 a). When roofing felt is used under the shingle, it is turned up on the wall and covered by the flashing. One piece of flashing is used at each shingle course. The siding is then applied over the flashing, allowing about a 2-inch space between the bevel edge of the siding and the roof.

If the roof intersects a brick wall or chimney, the same type of metal shingle flashing is used at the end of each shingle course as described for the wood-sided wall. In addition, counterflashing or brick flashing is used to cover the shingle flashing (fig. 50.3 b). This counterflashing is often preformed in sections and is inserted in open mortar joints. Unless soldered together, each section should overlap the next a minimum of 3 inches with the joint calked. In laying up the chimney or the brick wall, the mortar is usually raked out for a depth of about 1 inch at flashing locations. Lead wedges driven into the joint above the flashing hold it in place. The joint is then calked to provide a watertight connection. In chimneys, this counterflashing is often preformed to cover one entire side.

Around small chimneys, chimney flashing often consists of simple counterflashing on each side. For single-flue chimneys, the shingle flashing on the high side should be carried up under the shingles. The vertical distance at top of the flashing and the upturned edge should be about 4 inches above the roof boards (fig. 50.4 a).

A wood *saddle* usually constructed on the high side of wide chimneys for better drainage, is made of a ridgeboard and post and sheathed with plywood or boards (fig. 50.4 b). It is then covered with metal, which extends up on the brick and under the shingles. Counterflashing at the chimney is then used (as previously described) by lead plugging and calking. A very wide chimney may contain a partial gable on the high side and be shingled in the same manner as the main roof.

Roof Edge

The cornice and the rake section of the roof are sometimes protected by a metal edging. This edging forms a desirable drip edge at the rake and prevents rain from entering behind the shingles (fig. 34.3 b).

At the eave line, a similar metal edging may be used to advantage (fig. 50.5 a). This edging, with the addition of a roll roof flashing (fig. 34.1 b) will aid in resisting water entry from ice dams. Variations of it are shown in figure 50.5 b and c. They form a good drip edge and prevent or minimize the chance of rain

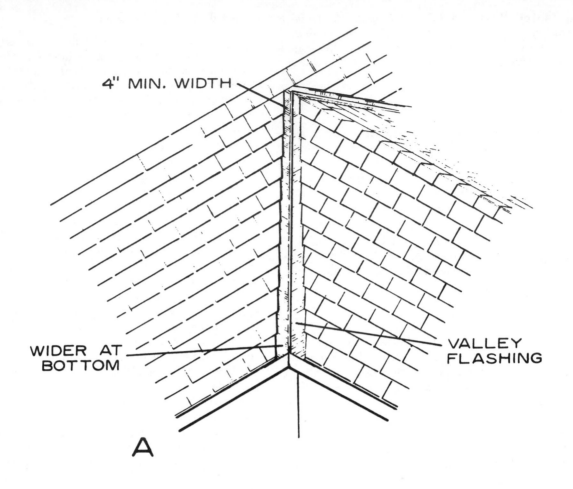

4" MIN. WIDTH

WIDER AT
BOTTOM

VALLEY
FLASHING

A

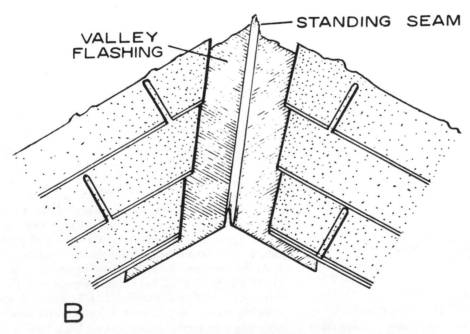

VALLEY
FLASHING

STANDING SEAM

B

Figure 50.2.—Valley flashing: A, Valley; B, standing seam.

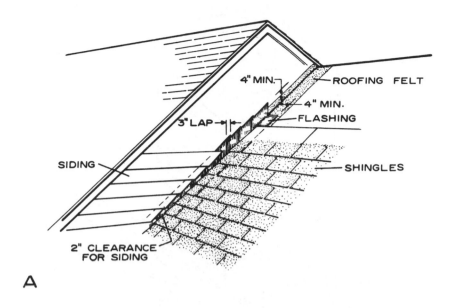

A

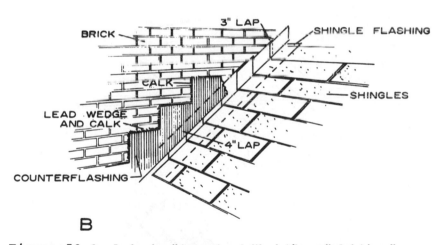

B

Figure 50.3.—Roof and wall intersection: A, Wood siding wall; B, brick wall.

being blown back under the shingles. This type of drip edge is desirable whether or not a gutter is used.

Gutters and Downspouts

Types

Several types of gutters are available to guide the rainwater to the downspouts and away from the foundation. Some houses have built-in gutters in the cornice. These are lined with sheet metal and connected to the downspouts. On flat roofs, water is often drained from one or more locations and carried through an inside wall to an underground drain. All downspouts connected to an underground drain should contain basket strainers at the junction of the gutter.

Perhaps the most commonly used gutter is the type hung from the edge of the roof or fastened to the edge of the cornice facia. Metal gutters may be the half-round (fig. 50.6 a) or the formed type (fig. 50.6 b) and may be galvanized metal, copper, or aluminum. Some have a factory-applied enamel finish.

Downspouts are round or rectangular (fig. 50.6 c and d), the round type being used for the half-round gutters. They are usually corrugated to provide extra stiffness and strength. Corrugated patterns are less likely to burst when plugged with ice.

Wood gutters have a pleasing appearance and are fastened to the facia board rather than being carried by hangers as are most metal gutters. The wood should

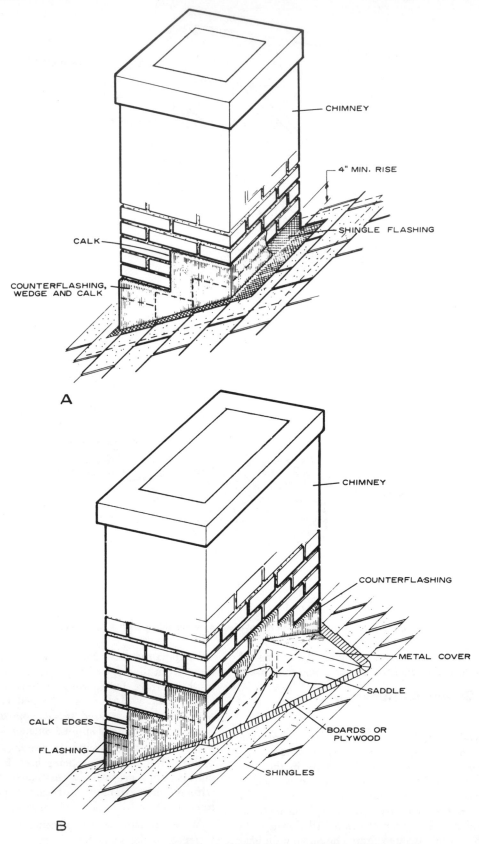

Figure 50.4.—Chimney flashing: *A*, Flashing without saddle; *B*, chimney saddle.

634

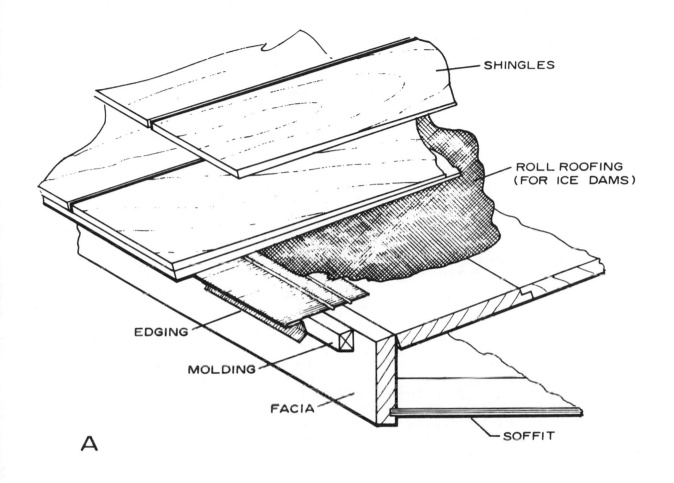

SHINGLES

ROLL ROOFING
(FOR ICE DAMS)

EDGING

MOLDING

FACIA

SOFFIT

A

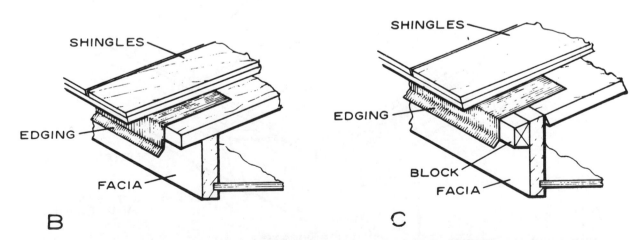

SHINGLES

EDGING

FACIA

B

SHINGLES

EDGING

BLOCK

FACIA

C

Figure 50.5.—Cornice flashing: A, Formed flashing; B, flashing without wood blocking; C, flashing with wood blocking.

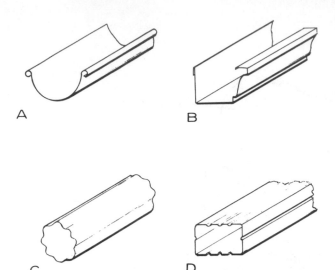

Figure 50.6.

Gutters and downspouts: A, Half-round gutter; B, formed gutter; C, round downspout; D, rectangular downspout.

Size

The size of gutters should be determined by the size and spacing of the downspouts used. One square inch of downspout is required for each 100 square feet of roof. When downspouts are spaced up to 40 feet apart, the gutter should have the same area as the downspout. For greater spacing, the width of the gutter should be increased.

Installation

On long runs of gutters, such as required around a hip-roof house, at least four downspouts are desirable. Gutters should be installed with a slight pitch toward the downspouts. Metal gutters are often suspended from the edge of the roof with hangers (fig. 50.7a). Hangers should be spaced 48 inches apart when made of galvanized steel and 30 inches apart when made of copper or aluminum. Formed gutters may be mounted on furring strips, but the gutter should be reinforced with wrap-around hangers at 48-inch intervals. Gutter splices, downspout connections, and corner joints should be soldered or provided with watertight joints.

Wood gutters are mounted on the facia using furring blocks spaced 24 inches apart (fig. 50.7b). Rust-proof screws are commonly used to fasten the gutters to the blocks and facia backing. The edge shingle should be located so that the drip is near the center of the gutter.

Downspouts are fastened to the wall by straps or hooks (fig. 51.1a). Several patterns of these fasteners allow a space between the wall and downspout. One common type consists of a galvanized metal strap with

be clear and free of knots and preferably treated, unless made of all heartwood from such species as redwood, western redcedar, and cypress. Continuous sections should be used wherever possible. When splices are necessary, they should be square-cut butt joints fastened with dowels or a spline. Joints should be set in white lead or similar material. When untreated wood gutters are used, it is good practice to brush several generous coats of water-repellent preservative on the interior.

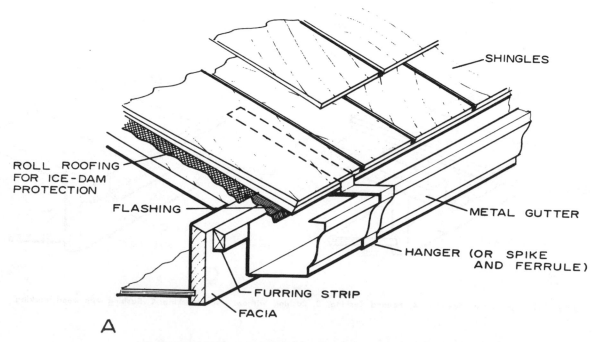

SHINGLES

ROLL ROOFING
FOR ICE-DAM
PROTECTION

FLASHING

METAL GUTTER

HANGER (OR SPIKE
AND FERRULE)

FURRING STRIP

FACIA

A

Figure 50.7.—**Gutter installation: A, Formed metal gutter.**

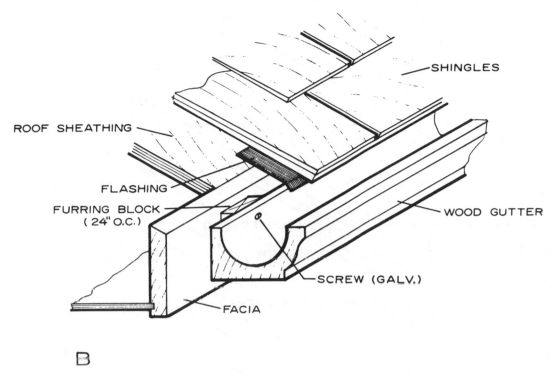

ROOF SHEATHING

SHINGLES

FLASHING

FURRING BLOCK
(24" O.C.)

WOOD GUTTER

SCREW (GALV.)

FACIA

B

Figure 50.8.—Gutter installation: B, Wood gutter.

a spike and spacer collar. After the spike is driven through the collar and into the siding and backing stud, the strap is fastened around the pipe. Downspouts should be fastened at the top and bottom. In addition, for long downspouts a strap or hook should be used for every 6 feet of length.

An elbow should be used at the bottom of the downspout, as well as a splash block, to carry the water away from the wall. However, a vitrified tile line is sometimes used to carry the water to a storm sewer. In such installations, the splash block is not required (fig. 50.1b).

CHAPTER 51

PORCHES AND GARAGES

An attached porch or garage which is in keeping with the house design usually adds to overall pleasing appearance. Thus, any similar attachments to the house after it has been built should also be in keeping structurally and architecturally with the basic design. In such additions, the connections of the porch or garage to the main house should be by means of the framing members and roof sheathing. Rafters, ceiling joists, and studs should be securely attached by nailing to the house framing.

When additions are made to an existing house, the siding or other finish is removed so that framing members can be easily and correctly fastened to the house. In many instances, the siding can be cut with a skill saw to the outline of the addition and removed only where necessary. When concrete

foundations, piers, or slabs are added, they should also be structurally correct. Footings should be of sufficient size, the bottoms located below the frostline, and the foundation wall anchored to the house foundation when possible.

Porches

There are many types and designs of porches, some with roof slopes continuous with the roof of the house itself. Other porch roofs may have just enough pitch to provide drainage. The fundamental construction principles, however, are somewhat alike no matter what type is built. Thus, a general description, together with several construction details, can apply to several types.

Figure 51.2 shows the construction details of a

637

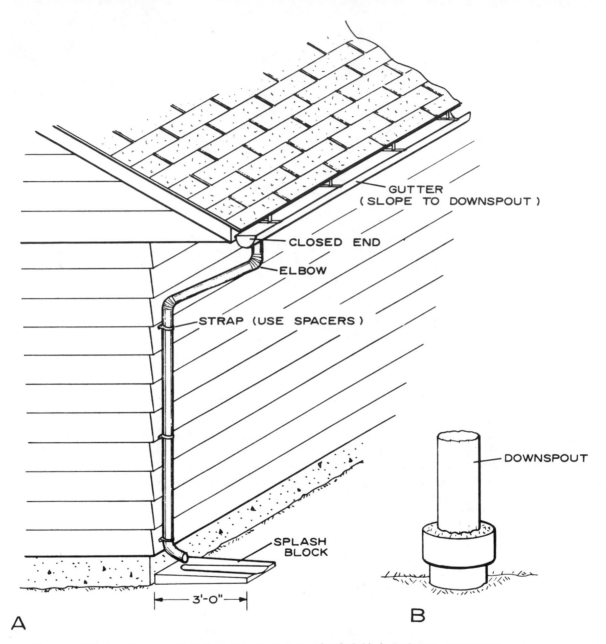

GUTTER
(SLOPE TO DOWNSPOUT)

CLOSED END

ELBOW

STRAP (USE SPACERS)

DOWNSPOUT

SPLASH
BLOCK

3'-0"

A

B

Figure 51.1.—Downspout installation: *A*, Downspout with splash block; *B*, drain to storm sewer.

typical flat-roofed porch with a concrete slab floor. An attached porch can be open or fully enclosed; or it can be constructed with a concrete slab floor, insulated or uninsulated. A porch can also be constructed using wood floor framing over a crawl space (fig. 51.3). Most details of such a unit should comply with those previously outlined for various parts of the house itself.

Porch Framing and Floors

Structural framing for the floors and walls should comply with the details given in Chapter 28, "Floor Framing," and Chapter 29, "Wall Framing." General details of the ceiling and roof framing are covered in Chapter 30, "Ceiling and Roof Framing."

Porch floors, whether wood or concrete, should have sufficient slope away from the house to provide good drainage. Weep holes or drains should be provided in any solid or fully sheathed perimeter wall. Open wood balusters with top and bottom railings should be constructed so that the bottom rail is free of the floor surface.

Floor framing for wood floor construction should be at least 18 inches above the soil. The use of a soil cover of polyethylene or similar material under a partially open or a closed porch is good practice.

638

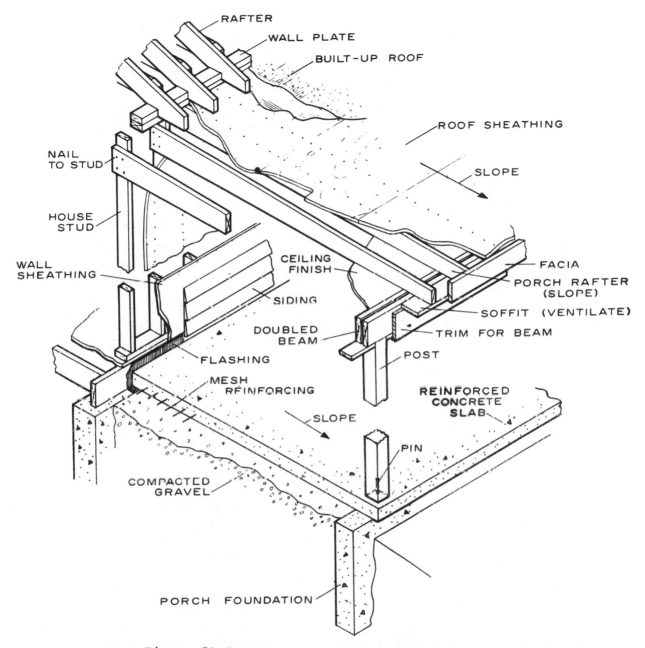

Figure 51.2.—Details of porch construction for concrete slab.

Slats or grillwork used around an open crawl space should be made with a removable section for entry in areas where termites may be present. (See Chapter 55, "Protection Against Decay and Termites.") A fully enclosed crawl-space foundation should be vented or have an opening to the basement.

Wood species used for finish porch floor should have good decay and wear resistance, be nonsplintering, and be free from warping. Species commonly used are cypress, Douglas-fir, western larch, southern pine, and redwood. Only treated material should be used where moisture conditions are severe.

Porch Columns

Supports for enclosed porches usually consist of fully framed stud walls. The studs are doubled at openings and at corners. Because both interior and exterior finish coverings are used, the walls are constructed much like the walls of the house. In open or partially open porches, however, solid or built-up posts or columns are used. A more finished or cased column is often made up of doubled 2 by 4's which are covered with 1- by 4-inch casing on two opposite sides and 1- by 6-inch finish casing on the other sides (fig. 51.4 a). Solid posts, normally 4 by 4 inches in

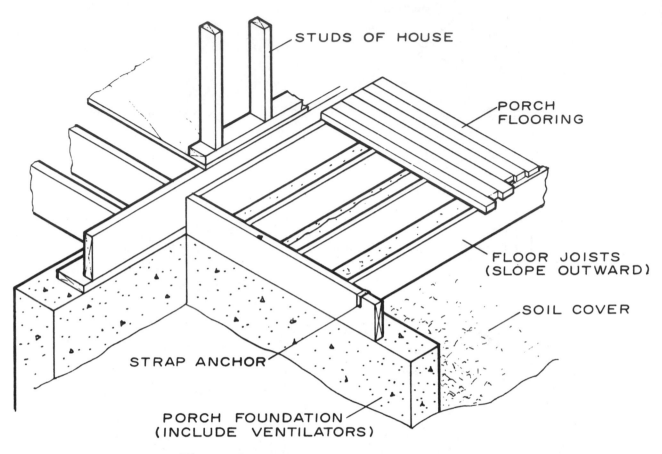

*Figure 51.3.—*Porch floor with wood framing.

size, are used mainly for open porches. An open railing may be used between posts.

A formal design of a large house entrance often includes the use of round built-up columns topped by *Doric* or *Ionic* capitals. These columns are factory-made and ready for installation at the building site.

The base of posts or columns in open porches should be designed so that no pockets are formed to retain moisture and encourage decay. In single posts, a steel pin may be used to locate the post and a large galvanized washer or similar spacer used to keep the bottom of the post above the concrete or wood floor (fig. 51.4b). The bottom of the post should be treated to minimize moisture penetration. Often single posts of this type are made from a decay-resistant wood species. A cased post can be flashed under the base molding (fig. 51.4c). Post anchors which provide connections to the floor and to the post are available commercially, as are post caps.

Balustrade

A porch *balustrade* usually consists of one or two railings with *balusters* between them. They are de-

signed for an open porch to provide protection and to improve the appearance. There are innumerable combinations and arrangements of them. A closed balustrade may be used with screens or combination windows above (fig. 51.5a). A balustrade with decorative railings may be used for an open porch (fig. 51.5b). This type can also be used with full-height removable screens.

All balustrade members that are exposed to water and snow should be designed to shed water. The top of the railing should be tapered and connections with balusters protected as much as possible (fig. 51.6a). Railings should not contact a concrete floor but should be blocked to provide a small space beneath. When wood must be in contact with the concrete, it should be treated to resist decay.

Connection of the railing with a post should be made in a way that prevents moisture from being trapped. One method provides a small space between the post and the end of the railing (fig. 51.6b). When the railing is treated with paint or water-repellent preservative, this type connection should provide good service. Exposed members, such as posts, balusters, and railings, should be all-heart-

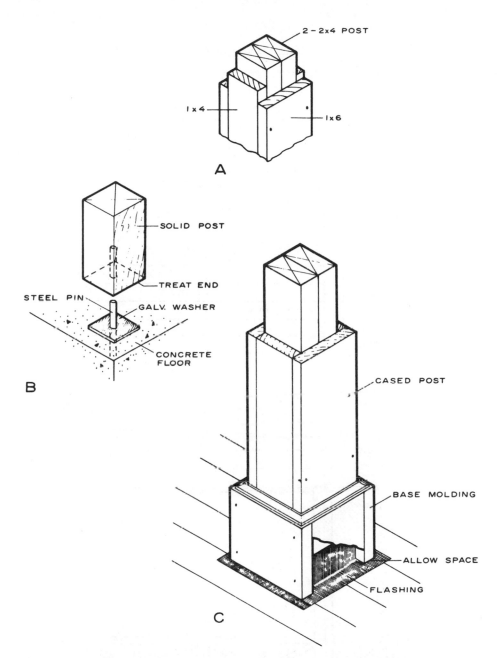

Figure 51.4.—Post details: A, Cased post; B, pin anchor and spacer; C, flashing at base.

wood stock of decay-resistant or treated wood to minimize decay.

Garages

Garages can be classified as attached, detached, basement, or carport. The selection of a garage type is often determined by limitations of the site and the size of the lot. Where space is not a limitation, the attached garage has much in its favor. It may give better architectural lines to the house, it is warmer during cold weather, and it provides covered protection to passengers, convenient space for storage, and a short, direct entrance to the house.

Building regulations often require that detached garages be located away from the house toward the rear of the lot. Where there is considerable slope to a lot, basement garages may be desirable, and generally such garages will cost less than those above grade.

Carports are car-storage spaces, generally attached to the house, that have roofs and often no sidewalls. To improve the appearance and utility of this type of structure, storage cabinets are often used on a side and at the end of the carport.

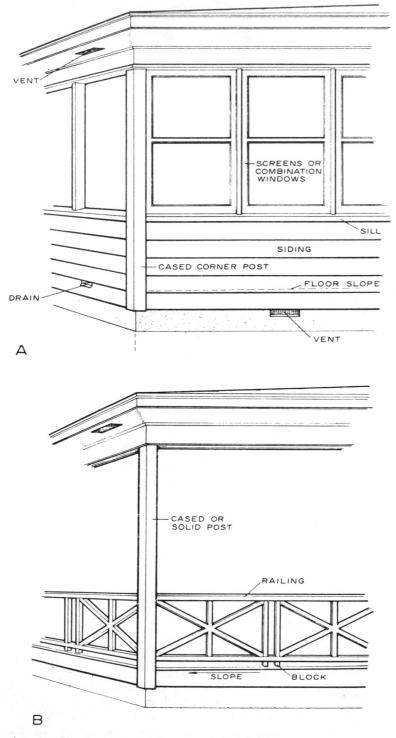

VENT

SCREENS OR
COMBINATION
WINDOWS

SILL

SIDING

CASED CORNER POST

FLOOR SLOPE

DRAIN

VENT

A

CASED OR
SOLID POST

RAILING

SLOPE   BLOCK

B

Figure 51.5.—Types of balustrades: A, Closed; B, open.

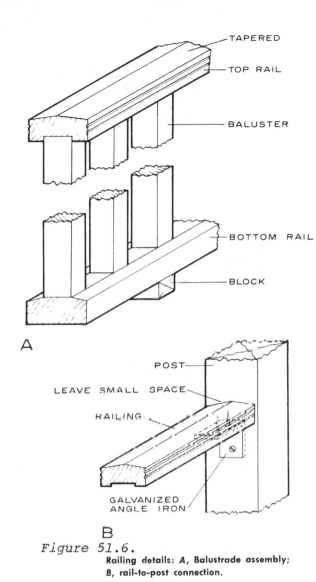

Figure 51.6.
Railing details: *A*, Balustrade assembly;
B, rail-to-post connection.

Size

It is a mistake to design the garage too small for convenient use. Cars vary in size from the small import models to the large foreign and domestic sedans. Many popular models are now up to 215 inches long, and the larger and more expensive models are usually over 230 inches—almost 20 feet in length. Thus, while the garage need not necessarily be designed to take all sizes with adequate room around the car, it is wise to provide a minimum distance of 21 to 22 feet between the inside face of the front and rear walls. If additional storage or work space is required at the back, a greater depth is required.

The inside width of a single garage should never be less than 11 feet with 13 feet much more satisfactory.

The minimum outside size for a single garage,

therefore, would be 14 by 22 feet. A double garage should be not less than 22 by 22 feet in outside dimensions to provide reasonable clearance and use. The addition of a shop or storage area would increase these minimum sizes.

For an attached garage, the foundation wall should extend below the frostline and about 8 inches above the finish-floor level. It should be not less than 6 inches thick, but is usually more because of the difficulty of trenching this width. The sill plate should be anchored to the foundation wall with anchor bolts spaced about 8 feet apart, at least two bolts in each sill piece. Extra anchors may be required at the sides of the main door. The framing of the sidewalls and roof and the application of the exterior covering material of an attached garage should be similar to that of the house.

The interior finish of the garage is often a matter of choice. The studs may be left exposed or covered with some type of sheet material or they may be plastered. Some building codes require that the wall between the house and the attached garage be made of fire-resistant material. Local building regulations and fire codes should be consulted before construction is begun.

If fill is required below the floor, it should preferably be sand or gravel well-compacted and tamped. If other types of soil fill are used, it should be wet down so that it will be well compacted and can then be well-tamped and time allowed before pouring. Unless these precautions are taken, the concrete floor will likely settle and crack.

The floor should be of concrete not less than 4 inches thick and laid with a pitch of about 2 inches from the back to the front of the garage. The use of wire reinforcing mesh is often advisable. The garage floor should be set about 1 inch above the drive or apron level. It is desirable at this point to have an expansion joint between the garage floor and the driveway or apron.

Garage Doors

The two overhead garage doors most commonly used are the sectional and the single-section swing types. The swing door **(fig. 51.7 a)** is hung with side and overhead brackets and an overhead track, and must be moved outward slightly at the bottom as it is opened. The sectional type **(fig. 51.7 b)**, in four or five horizontal hinged sections, has a similar track extending along the sides and under the ceiling framing, with a roller for the side of each section. It is opened by lifting and is adaptable to automatic electric opening with remote control devices. The standard desirable size for a single door is 9 feet in width by 6½ or 7 feet in height. Double doors are usually 16 by 6½ or 7 feet in size.

bricks at each course (fig. 52.1b), and a 12- by 12-inch flue (13 by 13 in. in outside dimension) by eight bricks (fig. 52.1c), and so on. Each fireplace should have a separate flue and, for best performance, flues should be separated by a 4-inch-wide brick spacer (withe) between them (fig. 52.2a).

The greater the difference in temperature between chimney gases and outside atmosphere, the better the draft. Thus, an interior chimney will have better draft because the masonry will retain heat longer. The height of the chimney as well as the size of the flue are important factors in providing sufficient draft.

The height of a chimney above the roofline usually depends upon its location in relation to the ridge. The top of the extending flue liners should not be less than 2 feet above the ridge or a wall that is within 10 feet (fig. 52.2b). For flat or low-pitched roofs, the chimney should extend at least 3 feet above the highest point of the roof. To prevent moisture from entering between the brick and flue lining, a concrete cap is usually poured over the top course of brick (fig. 52.2c). Precast or stone caps with a cement wash are also used.

Flashing for chimneys is illustrated in figure 50.4. Masonry chimneys should be separated from wood framing, subfloor, and other combustible materials. Framing members should have at least a 2-inch clearance and should be firestopped at each floor with asbestos or other types of noncombustible material (fig. 52.3). Subfloor, roof sheathing, and wall sheathing should have a ¾-inch clearance. A cleanout door is included in the bottom of the chimney where there are fireplaces or other solid fuel-burning equipment as

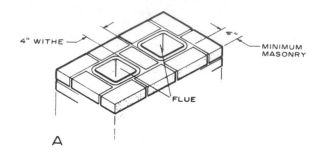

A

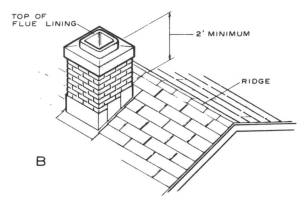

B

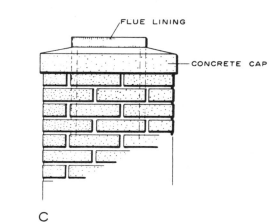

C

Figure 52.2.
Chimney details: A, Spacer between flues; B, height of chimneys; C, chimney cap.

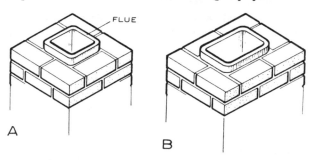

A   B

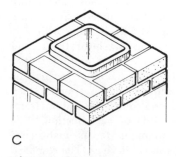

C

Figure 52.1.
Brick and flue combinations: A, 8- by 8-inch flue lining; B, 8- by 12-inch flue lining; C, 12- by 12-inch flue lining.

well as at the bottom of other flues. The cleanout door for the furnace flue is usually located just below the smokepipe thimble with enough room for a soot pocket.

Flue Linings

Rectangular fire-clay flue linings (previously described) or round vitrified tile are normally used in all chimneys. Vitrified (glazed) tile or a stainless-steel lining is usually required for gas-burning equipment. Local codes outline these specific requirements. A fireplace chimney with at least an 8-inch-thick masonry wall ordinarily does not require a flue lining. However, the cost of the extra brick or masonry and the labor involved are most likely greater than the

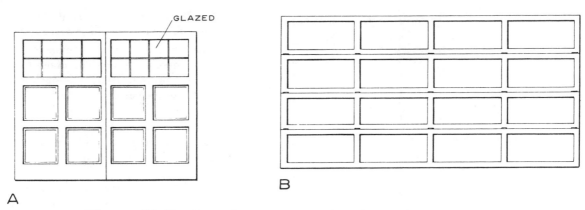

Figure 51.7.—Garage doors: A, One-section swing; B, sectional.

Doors vary in design, but those most often used are the panel type with solid stiles and rails and panel fillers. A glazed panel section is often included. Clearance above the top of the door required for overhead doors is usually about 12 inches. However, low-head-room brackets are available when such clearance is not possible.

The header beam over garage doors should be designed for the snow load which might be imposed on the roof above. In wide openings, this may be a steel I-beam or a built-up wood section. For spans of 8 or 9 feet, two doubled 2 by 10's of high-grade Douglas-fir or similar species are commonly used when only snow loads must be considered. If floor loads are also imposed on the header, a steel I-beam or wide-flange beam is usually selected.

CHAPTER 52

CHIMNEYS AND FIREPLACES

Chimneys are generally constructed of masonry units supported on a suitable foundation. A chimney must be structurally safe, capable of producing sufficient draft for the fireplace, and capable of carrying away harmful gases from the fuel-burning equipment and other utilities. Lightweight, prefabricated chimneys that do not require masonry protection or concrete foundations are now accepted for certain uses by fire underwriters. Make certain, however, they are approved and listed by Underwriters' Laboratories, Inc.

Fireplaces should not only be safe and durable but should be so constructed that they provide sufficient draft and are suitable for their intended use. From the standpoint of heat-production efficiency, which is estimated to be only 10 percent, they might be considered a luxury. However, they add a decorative note to a room and a cheerful atmosphere. Improved heating efficiency and the assurance of a correctly proportioned fireplace can usually be obtained by the installation of a factory-made circulating fireplace. This metal unit, enclosed by the masonry, allows air to be heated and circulated throughout the room in a system separate from the direct heat of the fire.

Chimneys

The chimney should be built on a concrete footing of sufficient area, depth, and strength for the imposed load. The footing should be below the frostline. For houses with a basement, the footings for the walls and fireplace are usually poured together and at the same elevation.

The size of the chimney depends on the number of flues, the presence of a fireplace, and the design of the house. The house design may include a room-wide brick or stone fireplace wall which extends through the roof. While only two or three flues may be required for heating units and fireplaces, several "false" flues may be added at the top for appearance. The flue sizes are made to conform to the width and length of a brick so that full-length bricks can be used to enclose the flue lining. Thus an 8- by 8-inch flue lining (about $8\frac{1}{2}$ by $8\frac{1}{2}$ in. in outside dimensions) with the minimum 4-inch thickness of surrounding masonry will use six standard bricks for each course (fig. **52.1a**). An 8- by 12-inch flue lining ($8\frac{1}{2}$ by 13 in. in outside dimensions) will be enclosed by seven

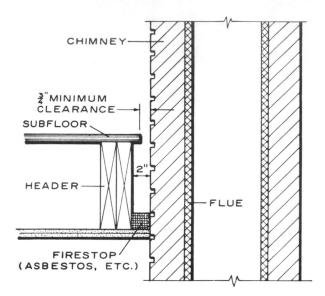

Figure 52.3.—Clearances for wood construction.

cost of flue lining. Furthermore, a well-installed flue lining will result in a safer chimney.

Flue liners should be installed enough ahead of the brick or masonry work, as it is carried up, so that careful bedding of the mortar will result in a tight and smooth joint. When diagonal offsets are necessary, the flue liners should be beveled at the direction change in order to have a tight joint. It is also good practice to stagger the joints in adjacent tile.

Flue lining is supported by masonry and begins at least 8 inches below the thimble for a connecting smoke or vent pipe from the furnace. In fireplaces, the flue liner should start at the top of the throat and extend to the top of the chimney.

Rectangular flue lining is made in 2-foot lengths and in sizes of 8 by 8, 8 by 12, 12 by 12, 12 by 16, and up to 20 by 20 inches. Wall thicknesses of the flue lining vary with the size of the flue. The smaller sizes have a ⅝-inch-thick wall, and the larger sizes vary from ¾ to 1⅜ inches in thickness. Vitrified tiles, 8 inches in diameter, are most commonly used for the flues of the heating unit, although larger sizes are also available. This tile has a bell joint.

Fireplaces

A fireplace adds to the attractiveness of the house interior, but one that does not "draw" properly is a detriment, not an asset. By following several rules on the relation of the fireplace opening size to flue area, depth of the opening, and other measurements, satisfactory performance can be assured. Metal circulating fireplaces, which form the main outline of the opening and are enclosed with brick, are designed for proper functioning when flues are the correct size.

One rule which is often recommended is that the depth of the fireplace should be about two-thirds the height of the opening. Thus, a 30-inch-high fireplace would be 20 inches deep from the face to the rear of the opening.

The flue area should be at least one-tenth of the open area of the fireplace (width times height) when the chimney is 15 feet or more in height. When less than 15 feet, the flue area in square inches should be one-eighth of the opening of the fireplace. This height is measured from the throat to the top of the chimney. Thus, a fireplace with a 30-inch width and 24-inch height (720 sq. in.) would require an 8- by 12-inch flue, which has an inside area of about 80 square inches. A 12- by 12-inch flue liner has an area of about 125 square inches, and this would be large enough for a 36- by 30-inch opening when the chimney height is 15 feet or over.

The back width of the fireplace is usually 6 to 8 inches narrower than the front. This helps to guide the smoke and fumes toward the rear. A vertical back-wall of about a 14-inch height then tapers toward the upper section or "throat" of the fireplace (fig. 172). The area of the throat should be about 1¼ to 1⅓ times the area of the flue to promote better draft. An adjustable damper is used at this area for easy control of the opening.

The *smoke shelf* (top of the throat) is necessary to prevent back drafts. The height of the smoke shelf should be 8 inches above the top of the fireplace opening (fig. 52.4). The smoke shelf is concave to retain any slight amount of rain that may enter.

Steel angle iron is used to support the brick or masonry over the fireplace opening. The bottom of the inner hearth, the sides, and the back are built of a heat-resistant material such as firebrick. The outer hearth should extend at least 16 inches out from the face of the fireplace and be supported by a reinforced concrete slab (fig. 52.4). This outer hearth is a precaution against flying sparks and is made of noncombustible materials such as glazed tile. Other fireplace details of clearance, framing of the wall, and cleanout opening and ash dump are also shown. Hangers and brackets for fireplace screens are often built into the face of the fireplace.

Fireplaces with two or more openings (fig. 52.5) require much larger flues than the conventional fireplace. For example, a fireplace with two open adjacent faces (fig. 52.5 a) would require a 12- by 16-inch flue for a 34- by 20- by 30-inch (width, depth, and height, respectively) opening. Local building regulations usually cover the proper sizes for these types of fireplaces.

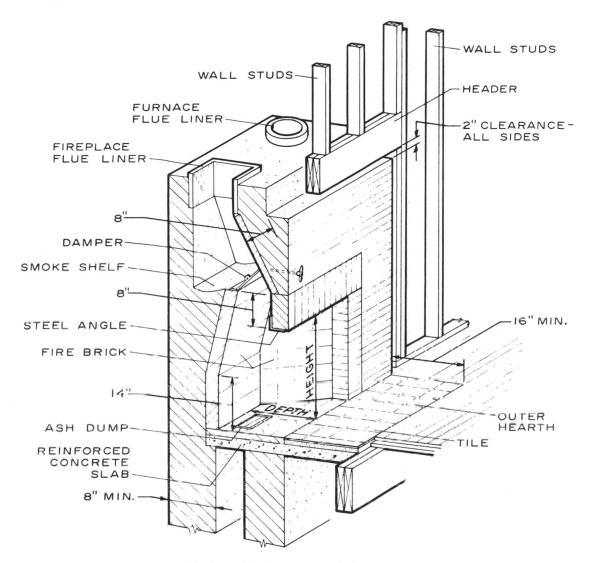

WALL STUDS

FURNACE
FLUE LINER

FIREPLACE
FLUE LINER

8"

DAMPER

SMOKE SHELF

8"

STEEL ANGLE

FIRE BRICK

14"

ASH DUMP

REINFORCED
CONCRETE
SLAB

8" MIN.

WALL STUDS

HEADER

2" CLEARANCE-
ALL SIDES

16" MIN.

HEIGHT

DEPTH

OUTER
HEARTH

TILE

Figure 52.4.—Masonry fireplace.

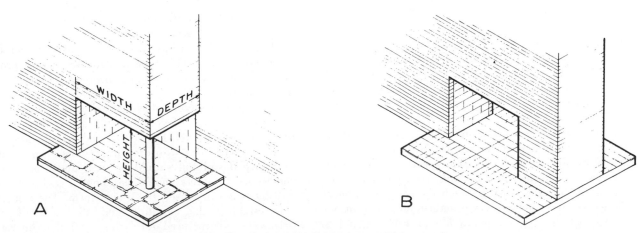

WIDTH

DEPTH

HEIGHT

A

B

Figure 52.5.—Dual-opening fireplace: *A*, Adjacent opening;
B, through fireplace.

CHAPTER 53

DRIVEWAYS, WALKS, AND BASEMENT FLOORS

A new home is not complete until driveways and walks have been installed so that landscaping can be started. Landscaping includes final grading, planting of shrubs and trees, and seeding or sodding of lawn areas. Because the automobile is an important element in American life, the garage is usually a prominent part of house design. This in turn establishes the location of driveways and walks.

Concrete and bituminous pavement are most commonly used in the construction of walks and drives, especially in areas where snow removal is important. In some areas of the country, a gravel driveway and a flagstone walk may be satisfactory and would reduce the cost of improvements.

Basements are normally finished with a concrete floor of some type, whether or not the area is to contain habitable rooms. These floors are poured after all improvements such as sewer and waterlines have been connected. Concrete slabs should *not* be poured on recently filled areas.

Driveways

The grade, width, and radius of curves in a driveway are important factors in establishing a safe entry to the garage. Driveways for attached garages, which are located near the street on relatively level property, need only be sufficiently wide to be adequate. Driveways that have a grade more than 7 percent (7-ft. rise in 100 ft.) should have some type of pavement to prevent wash. Driveways that are long and require an area for a turnaround should be designed carefully. **Figure 53.1** shows a drive way and turnaround which allow the driver to back out of a single or double garage into the turn and proceed to the street or highway in a forward direction. This, in areas of heavy traffic, is much safer than having to back into the street or roadway. A double garage should be serviced by a wider entry and turnaround.

Driveways that are of necessity quite steep should have a near-level area in front of the garage for safety, from 12 to 16 feet long.

Two types of paved driveways may be used, (a) the more common *slab* or full-width type and (b) the *ribbon* type. When driveways are fairly long or steep, the full-width type is the most practical. The ribbon driveway is cheaper and perhaps less conspicuous. because of the grass strip between the concrete runners. However, it is not always practical for all locations.

The width of the single-slab type drive should be 9 feet for the modern car, although 8 feet is often considered minimum **(fig. 53. 2 a)**. When the driveway is

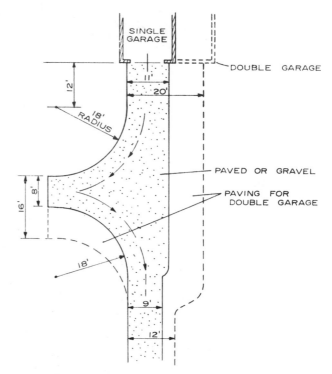

Figure 53.1.—Driveway turnaround.

also used as a walk, it should be at least 10 feet wide to allow for a parked car as well as a walkway. The width should be increased by at least 1 foot at curves. The radius of the drive at the curb should be at least 5 feet **(fig. 53.2 a)**. Relatively short double driveways should be at least 18 feet wide, and 2 feet wider when they are also to be used as a walk from the street.

The concrete strips in a ribbon driveway should be at least 2 feet in width and located so that they are 5 feet on center **(fig. 53.2 b)**. When the ribbon is also used as a walk, the width of strips should be increased to at least 3 feet. This type of driveway is not practical if there is a curve or turn involved or the driveway is long.

Pouring a concrete driveway over an area that has been recently filled is poor practice unless the fill, preferably gravel, has settled and is well tamped. A gravel base is not ordinarily required on sandy undisturbed soil but should be used under all other conditions. Concrete should be about 5 inches thick. A 2 by 6 is often used for a side form. These members establish the elevation and alinement of the driveway and are used for striking off the concrete. Under most conditions, the use of steel reinforcing is good practice. Steel mesh, 6 by 6 inches in size,

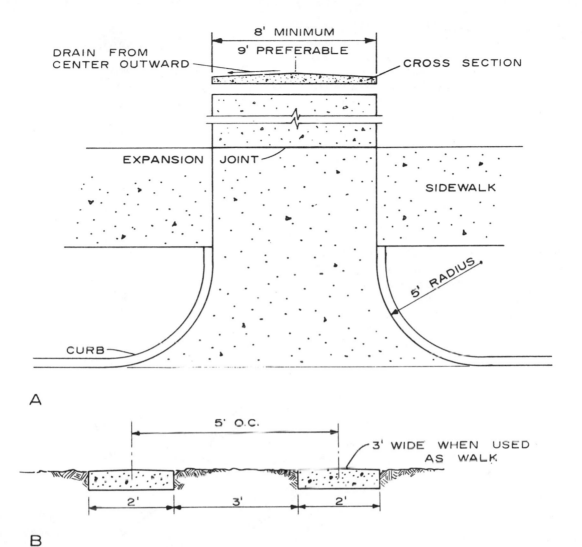

Figure 53.2.—Driveway details: *A*, Single-slab driveway; *B*, ribbon-type driveway.

will normally prevent or minimize cracking of the concrete. Expansion joints should be used (a) at the junction of the driveway with the public walk or curb, (b) at the garage slab, and (c) about every 40 feet on long driveways. A 5- or 5½-bag commercial mix is ordinarily used for driveways. However, a 5½- to 6-bag mix containing an air-entraining mixture should be used in areas having severe winter climates.

Contraction joints should be provided at 10- to 12-foot intervals. These crosswise grooves, cut into the partially set concrete, will allow the concrete to open along these lines during the cold weather rather than in irregular cracks in other areas.

Blacktop driveways, normally constructed by paving contractors, should also have a well-tamped gravel or crushed rock base. Top should be slightly crowned for drainage.

Sidewalks

Main sidewalks should extend from the front entry to the street or front walk or to a driveway leading to the street. A 5 percent grade is considered maximum for sidewalks; any greater slope usually requires steps. Walks should be at least 3 feet wide.

Concrete sidewalks should be constructed in the same general manner as outlined for concrete driveways. They should not be poured over filled areas unless they have settled and are very well tamped. This is especially true of the areas near the house after basement excavation backfill has been completed.

The minimum thickness of the concrete over normal undisturbed soil is usually 4 inches. As described for concrete driveways, contraction joints should be used and spaced on 4-foot centers.

When slopes to the house are greater than a 5 percent grade, stairs or steps should be used. This may be accomplished with a ramp sidewalk, a flight of stairs at a terrace, or a continuing sidewalk (fig. 53.3 a). These stairs have 11-inch treads and 7-inch risers when the stair is 30 inches or less in height. When the rise is more than 30 inches, the tread is 12 inches and the riser 6 inches. For a moderately uniform slope, a stepped ramp may be satisfactory (fig. 53. 3 b). Generally, the rise should be about 6 to 6½ inches and the length between risers sufficient for two or three normal paces.

Walks can also be made of brick, flagstone, or other types of stone. Brick and stone are often placed directly over a well-tamped sand base. However, this system is not completely satisfactory where freezing of the soil is possible. For a more durable walk in cold climates, the brick or stone topping is embedded in a freshly laid reinforced concrete base (fig. 53.4).

As in all concrete sidewalks and curbed or uncurbed driveways, a slight crown should be included in the walk for drainage. Joints between brick or stone may be filled with a cement mortar mix or with sand.

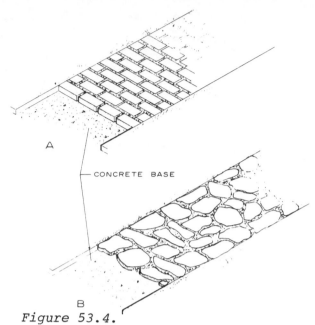

CONCRETE BASE

Figure 53.4.

Masonry paved walks: A, Brick; B, flagstone.

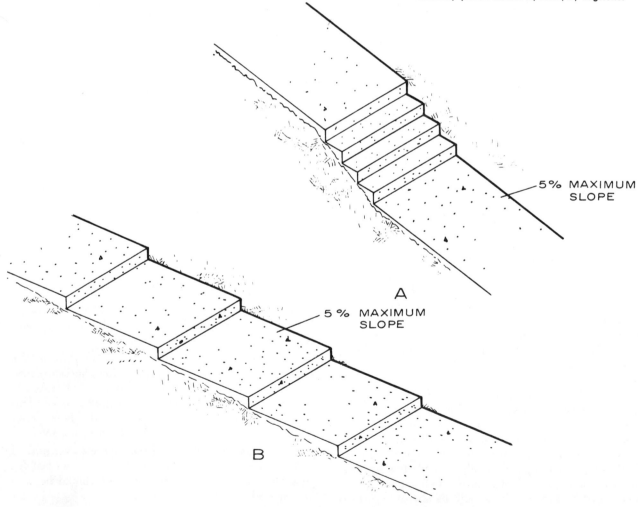

5% MAXIMUM SLOPE

A

5% MAXIMUM SLOPE

B

Figure 53.3.—Sidewalks on slopes: A, Stairs; B, stepped ramp.

650

Basement Floors

Basement floor slabs should be no less than 3½ inches thick and sloped toward the floor drains. A 2 by 4 (3½ in. wide) is often used on edge for form work. At least one floor drain should be used, and for large floor areas, two are more satisfactory. One should be located near the laundry areas.

For a dry basement floor, the use of a polyethylene film or similar vapor barrier under the concrete slab is usually justified. However, basement areas or multi-level floors used only for utility or storage do not require a vapor barrier unless soil conditions are adverse. When finished rooms have concrete floors, the use of a vapor barrier is normally required. Details of basement floor or concrete-slab construction are outlined in Chapter 27, "Concrete Floor Slabs on Ground," and in Chapter 44, "Basement Rooms."

CHAPTER 54

PAINTING AND FINISHING

Wood and wood products in a variety of species, grain patterns, texture, and colors are available for use as exterior and interior surfaces. These wood surfaces can be finished quite effectively by several different methods. Painting, which totally obscures the wood grain, is used to achieve a particular color decor. Penetrating-type preservatives and pigmented stains permit some or all of the wood grain and texture to show and provide a special color effect as well as a natural or rustic appearance. The type of finish, painted or natural, often depends on the wood to be finished.

Effect of Wood Properties

Wood surfaces that shrink and swell the least are best for painting. For this reason, vertical- or edge-grained surfaces are far better than flat-grained surfaces of any species. Also, because the swelling of wood is directly proportional to density, low-density species are preferred over high-density species. However, even high-swelling and dense wood surfaces with flat grain have been stabilized with a resin-treated paper overlay, such as overlaid exterior plywood and lumber, to make them excellent for painting.

Medium-density fiberboard products fabricated with a uniform, low-density surface for exterior use are often painted, but little is known of their long-time performance. The most widely used species for exterior siding to be painted are vertical-grained western redcedar and redwood. These species are classified in Group I, those woods easiest to keep painted (table 10). Other species in Group I are excellent for painting but are not generally available in all parts of the country.

Species that are not normally cut as vertical-grained lumber, are high in density (swelling), or have defects such as knots or pitch are classified in Groups II through V, depending upon their general paint-holding characteristics. Many species in Groups II through IV are commonly painted, particularly the pines, Douglas-fir, and spruce; but these species generally require more care and attention than the species in Group I. Resinous species should be thoroughly kiln dried at temperatures that will effectively set the pitch.

The properties of wood that detract from its paintability do not necessarily affect the finishing of such boards naturally with penetrating preservatives and stains. These finishes penetrate into wood without forming a continuous film on the surface. Therefore, they will not blister, crack, or peel even if excessive moisture penetrates into wood. One way to further improve the performance of penetrating finishes is to leave the wood surface rough sawn. Allowing the high-density, flat grained wood surfaces of lumber and plywood to weather several months also roughens the surface and improves it for staining. Rough-textured surfaces absorb more of the preservative and stain, insuring a more durable finish.

Natural Finishes for Exterior Wood

Weathered Wood

The simplest of natural finishes for wood is natural weathering. Without paint or treatment of any kind, wood surfaces change in color and texture in a few months or years, and then may stay almost unaltered for a long time if the wood does not decay. Generally, the dark-colored woods become lighter and the light-colored woods become darker. As weathering continues, all woods become gray, accompanied by degradation of the wood cells at the surface. Unfinished wood will wear away at the rate of about ¼ inch in 100 years.

The appearance of weathered wood is affected by dark-colored spores and mycelia of fungi or mildew on the surface, which give the wood a dark gray, blotchy, and unsightly appearance. Highly-colored wood extractives in such species as western redcedar and redwood also influence the color of weathered wood. The dark brown color may persist for a long time in areas not exposed to the sun and where the

TABLE 10.—*Characteristics of woods for painting and finishing (omission in the table indicate inadequate data for classification)*

| Wood | Ease of keeping well-painted I—easiest V—most exacting [1] | Weathering | | Appearance | |
		Resistance to cupping 1—best 4—worst	Conspicuousness of checking 1—least 2—most	Color of heartwood (sapwood is always light)	Degree of figure on flat-grained surface
SOFTWOODS					
Cedar:					
Alaska	I	1	1	Yellow	Faint
California incense	I			Brown	Do.
Port-Orford	I		1	Cream	Do.
Western redcedar	I	1	1	Brown	Distinct
White	I	1		Light brown	Do.
Cypress	I	1	1	____do	Strong
Redwood	I	1	1	Dark brown	Distinct
Pine:					
Eastern white	II	2	2	Cream	Faint
Sugar	II	2	2	____do	Do.
Western white	II	2	2	____do	Do.
Ponderosa	III	2	2	____do	Distinct
Fir, commercial white	III	2	2	White	Faint
Hemlock	III	2	2	Pale brown	Do.
Spruce	III	2	2	White	Do.
Douglas-fir (lumber and plywood)	IV	2	2	Pale red	Strong
Larch	IV	2	2	Brown	Do.
Pine:					
Norway	IV	2	2	Light brown	Distinct
Southern (lumber and plywood)	IV	2	2	____do	Strong
Tamarack	IV	2	2	Brown	Do.
HARDWOODS					
Alder	III			Pale brown	Faint
Aspen	III	2	1	____do	Do.
Basswood	III	2	2	Cream	Do.
Cottonwood	III	4	2	White	Do.
Magnolia	III	2		Pale brown	Do.
Poplar	III	2	1	____do	Do.
Beech	IV	4	2	____do	Do.
Birch	IV	4	2	Light brown	Do.
Gum	IV	4	2	Brown	Do.
Maple	IV	4	2	Light brown	Do.
Sycamore	IV			Pale brown	Do.
Ash	V or III	4	2	Light brown	Distinct
Butternut	V or III			____do	Faint
Cherry	V or III			Brown	Do.
Chestnut	V or III	3	2	Light brown	Distinct
Walnut	V or III	3	2	Dark brown	Do.
Elm	V or IV	4	2	Brown	Do.
Hickory	V or IV	4	2	Light brown	Do.
Oak, white	V or IV	4	2	Brown	Do.
Oak, red	V or IV	4	2	____do	Do.

[1] Woods ranked in group V for *ease of keeping well-painted* are hardwoods with large pores that need filling with wood filler for durable painting. When so filled before painting, the second classification recorded in the table applies.

extractives are not removed by rain.

With naturally weathered wood, it is important to avoid the unsightly effect of rusting nails. Iron nails rust rapidly and produce a severe brown or black discoloration. Because of this, only aluminum or stainless steel nails should be used for natural finishes.

Water-Repellent Preservatives

The natural weathering of wood may be modified by treatment with water-repellent finishes that contain a preservative (usually pentachlorophenol), a small amount of resin, and a very small amount of a water repellent which frequently is wax or waxlike in nature. The treatment, which penetrates the wood surface, retards the growth of mildew, prevents water staining of the ends of boards, reduces warping, and protects species that have a low natural resistance to decay. A clear, golden tan color can be achieved on such popular sidings as smooth or rough-sawn western redcedar and redwood.

The preservative solution can be easily applied by dipping, brushing, or spraying. All lap and butt joints, edges, and ends of boards should be liberally treated. Rough surfaces will absorb more solution than smoothly planed surfaces and be more durable.

The initial application to smooth surfaces is usually short-lived. When the surfaces start to show a blotchy discoloration due to extractives or mildew, clean them with detergent solution and re-treat following thorough drying. During the first 2 to 3 years, the finish may have to be applied every year or so. After weathering to uniform color, the treatments are more durable and need refinishing only when the surface becomes unevenly colored.

Pigmented colors can also be added to the water-repellent preservative solutions to provide special color effects. Two to six fluid ounces of colors in oil or tinting colors can be added to each gallon of treating solution. Light-brown colors which match the natural color of the wood and extractives are preferred. The addition of pigment to the finish helps to stabilize the color and increases the durability of the finish. In applying pigmented systems, a complete course of siding should be finished at one time to prevent lapping.

Pigmented Penetrating Stains

The pigmented penetrating stains are semitransparent, permitting much of the grain pattern to show through, and penetrate into the wood without forming a continuous film on the surface. Therefore, they will not blister, crack, or peel even if excessive moisture enters the wood.

Penetrating stains are suitable for both smooth and rough-textured surfaces; however, their performance is markedly improved if applied to rough-sawn, weathered, or rough-textured wood. They are especially effective on lumber and plywood that does not hold paint well, such as flat-grained surfaces of dense species. One coat of penetrating stains applied to smooth surfaces may last only 2 to 4 years, but the second application, after the surface has roughened by weathering, will last 8 to 10 years. A finish life of close to 10 years can be achieved initially by applying two coats of stain to rough-sawn surfaces. Two-coat staining is usually best for the highly adsorptive rough-sawn or weathered surfaces to reduce lapping or uneven stain application. The second coat should always

be applied the same day as the first and before the first dries.

An effective stain of this type is the Forest Products Laboratory natural finish (13). This finish has a linseed oil vehicle; a fungicide, pentachlorophenol, that protects the oil from mildew; and a water repellent, paraffin wax, that protects the wood from excessive penetration of water. Durable red and brown iron oxide pigments simulate the natural colors of redwood and cedar. A variety of colors can be achieved with this finish, but the more durable ones are considered to be the red and brown iron oxide stains.

Paints for Exterior Wood

Of all the finishes, paints provide the most protection for wood against weathering and offer the widest selection of colors. A nonporous paint film retards penetration of moisture and reduces discoloration by wood extractives, paint peeling, and checking and warping of the wood. Paint, however, is *not* a preservative; it will not prevent decay if conditions are favorable for fungal growth. Original and maintenance costs are usually higher for a paint finish than for a water-repellent preservative or penetrating stain finish.

The durability of paint coatings on exterior wood is affected both by variables in the wood surface and type of paint.

Application

Exterior wood surfaces can be very effectively painted by following a simple 3-step procedure:

Step 1. Water-repellent preservative treatment.— make sure wood siding and trim have been treated with water-repellent preservative to protect them against the entrance of rain and heavy dew at joints. If treated exterior woodwork was not installed, treat it by brushing or spraying in place. Care should be taken to brush well into lap and butt joints, especially retreating cut ends. Allow 2 warm, sunny days for adequate drying of the treatment before painting.

Step 2. Primer.—New wood should be given three coats of paint. The first, or prime, coat is the most important and should be applied soon after the woodwork is erected; topcoats should be applied within 2 days to 2 weeks. Use a nonporous linseed oil primer free of zinc pigments (Federal Specification TT-P-25). Apply enough primer to obscure the wood grain. Many painters tend to spread primer too thinly. For best results, follow the spreading rates recommended by the manufacturer, or approximately 400 to 450 square feet per gallon for a paint that is about 85 percent solids by weight. A properly applied coat of a nonporous house paint primer will greatly reduce moisture blistering, peeling, and staining of paint by wood extractives.

The wood primer is *not* suitable for galvanized iron.

Allow such surfaces to weather for several months and then prime with an appropriate primer, such as a linseed oil or resin vehicle pigmented with metallic zinc dust (about 80 pct.) and zinc oxide (about 20 pct.).

Step 3. Finish Coats.—Keep the following points in mind when applying topcoats over the primer on new wood and galvanized iron:

(1) Use two coats of a wood-quality latex, alkyd, or oil-base house paint over the nonporous primer. This is particularly important for areas that are fully exposed to the weather, such as the south side of a house.

(2) To avoid future separation between coats of paint, or intercoat peeling, apply the first topcoat within 2 weeks after the primer and the second within 2 weeks of the first.

(3) To avoid temperature blistering, *do not* apply oil-base paints on a cool surface that will be heated by the sun within a few hours. Follow the sun around the house. Temperature blistering is most common with thickly applied paints of dark colors. The blisters usually show up in the last coat of paint and occur within a few hours to 1 or 2 days after painting. They do not contain water.

(4) To avoid the wrinkling, fading, or loss of gloss of oil-base paint, and streaking of latex paints, do not paint in the evenings of cool spring and fall days when heavy dews are frequent before the surface of the paint has dried.

Repainting

(1) Repaint only when the old paint has worn thin and no longer protects the wood. Faded or dirty paint can often be freshened by washing. Where wood surfaces are exposed, spot prime with a zinc-free linseed oil primer before applying the finish coat. Too-frequent repainting produces an excessively thick film that is more sensitive to the weather and also likely to crack abnormally across the grain of the paint. The grain of the paint is in the direction of the last brush strokes. Complete paint removal is the only cure for cross-grain cracking.

(2) Use the same brand and type of paint originally applied for the topcoat. A change is advisable only if a paint has given trouble. When repainting with latex paint, apply a nonporous, oil-base primer overall before applying the latex paint.

(3) To avoid intercoat peeling, which indicates a weak bond between coats of paint, clean the old painted surface well and allow no more than 2 weeks between coats in two-coat repainting. Do not repaint sheltered areas, such as eaves and porch ceilings, every time the weathered body of the house is painted. In repainting sheltered areas, wash the old paint surface with trisodium phosphate or detergent solution to remove surface contaminants that will interfere with adhesion of the new coat of paint. Following washing, rinse sheltered areas with large amounts of water and let dry thoroughly before repainting. When intercoat peeling does occur, complete paint removal is the only satisfactory procedure.

Blistering and Peeling

When too much water gets into paint or wood, the paint may blister and peel. The moisture blisters normally appear first and peeling follows. But sometimes the paint peels without blistering. At other times the blisters go unnoticed. Moisture blisters usually contain water when they form, or soon afterward, and eventually dry out. Small blisters may disappear completely on drying; however, fairly large ones may leave a rough spot on the surface. If the blistering is severe, the paint may peel.

New, thin coatings are more likely to blister because of too much moisture under them than old, thick coatings. The older and thicker coatings are too rigid to stretch, as they must do to blister, and so are more prone to cracking and peeling.

House construction features that will *minimize* water damage of outside paint are: (a) Wide roof overhang, (b) wide flashing under shingles at roof edges, (c) effective vapor barriers, (d) adequate eave troughs and properly hung downspouts, and (e) adequate ventilation of the house. If these features are lacking in a new house, persistent blistering and peeling may occur.

Discoloration by Extractives

Water-soluble color extractives occur naturally in western redcedar and redwood. It is to these substances that the heartwood of these two species owes its attractive color, good stability, and natural decay resistance. Discoloration occurs when the extractives are dissolved and leached from the wood by water. When the solution of extractives reaches the painted surface, the water evaporates, leaving the extractives as a reddish-brown stain. The water that gets behind the paint and causes moisture blisters also causes migration of extractives. The discoloration produced by water wetting the siding from the back side frequently forms a rundown or streaked pattern.

The emulsion paints and the so-called "breather" or low-luster oil paints are more porous than conventional oil paints. If these are used on new wood without a good oil primer, or if any paint is applied too thinly on new wood (a skimpy two-coat paint job, for example), rain or even heavy dew can penetrate the coating and reach the wood. When the water dries from the wood, the extractives are brought to the surface of the paint. Discoloration of paint by this process forms a diffused pattern.

On rough surfaces, such as shingles, machine-

654

grooved shakes, and rough-sawn lumber sidings, it is difficult to obtain an adequately thick coating on the high points. Therefore, extractive staining is more likely to occur on such surfaces by water penetrating through the coating. But the reddish-brown extractives will be less conspicuous if dark-colored paints are used.

Effect of Impregnated Preservatives on Painting

Wood treated with the water-soluble preservatives in common use can be painted satisfactorily after it is redried. The coating may not last quite as long as it would have on untreated wood, but there is no vast difference. Certainly, a slight loss in durability is not enough to offer any practical objection to using treated wood where preservation against decay is necessary, protection against weathering desired, and appearance of painted wood important. When such treated wood is used indoors in textile or pulpmills, or other places where the relative humidity may be above 90 percent for long periods, paint may discolor or preservative solution exude. Coal-tar creosote or other dark oily preservatives tend to stain through paint unless the treated wood has been exposed to the weather for many months before it is painted.

Wood treated with oilborne, chlorinated phenols can be painted only when the solvent oils have evaporated completely from the treated wood. If volatile solvents that evaporate rapidly are used for the treating solution, such as in water-repellent preservatives, painting can be done only after the treated wood has dried.

Finishes for Interior Woodwork

Interior finishing differs from exterior chiefly in that interior woodwork usually requires much less protection against moisture, more exacting standards of appearance, and a greater variety of effects. Good interior finishes used indoors should last much longer than paint coatings on exterior surfaces. Veneered panels and plywood, however, present special finishing problems because of the tendency of these wood constructions to surface check.

Opaque Finishes

Interior surfaces may be painted with the materials and by the procedures recommended for exterior surfaces. As a rule, however, smoother surfaces, better color, and a more lasting sheen are demanded for interior woodwork, especially the wood trim; therefore, enamels or semigloss enamels rather than paints are used.

Before enameling, the wood surface should be extremely smooth. Imperfections, such as planer marks, hammer marks, and raised grain, are accentuated by enamel finish. Raised grain is especially troublesome on flat-grained surfaces of the heavier softwoods because the hard bands of summerwood are sometimes crushed into the soft springwood in planing, and later are pushed up again when the wood changes in moisture content. It is helpful to sponge softwoods with water, allow them to dry thoroughly, and then sandpaper them lightly with sharp sandpaper before enameling. In new buildings, woodwork should be allowed adequate time to come to its equilibrium moisture content before finishing.

For hardwoods with large pores, such as oak and ash, the pores must be filled with wood filler before the priming coat is applied. The priming coat for all woods may be the same as for exterior woodwork, or special priming paints may be used. Knots in the white pines, ponderosa pine, or southern yellow pine should be shellacked or sealed with a special knot sealer after the priming coat is dry. A coat of knot sealer is also sometimes necessary over wood of white pines and ponderosa pine to prevent pitch exudation and discoloration of light-colored enamels by colored matter apparently present in the resin of the heartwood of these species.

One or two coats of enamel undercoat are next applied; this should completely hide the wood and also present a surface that can easily be sandpapered smooth. For best results, the surface should be sandpapered before applying the finishing enamel; however, this operation is sometimes omitted. After the finishing enamel has been applied, it may be left with its natural gloss, or rubbed to a dull finish. When wood trim and paneling are finished with a flat paint, the surface preparation is not nearly as exacting.

Transparent Finishes

Transparent finishes are used on most hardwood and some softwood trim and paneling, according to personal preference. Most finishing consists of some combination of the fundamental operations of staining, filling, sealing, surface coating, or waxing. Before finishing, planer marks and other blemishes of the wood surface that would be accentuated by the finish must be removed.

Both softwoods and hardwoods are often finished without staining, especially if the wood is one with a pleasing and characteristic color. When used, however, stain often provides much more than color alone because it is absorbed unequally by different parts of the wood; therefore, it accentuates the natural variations in grain. With hardwoods, such emphasis of the grain is usually desirable; the best stains for the purpose are dyes dissolved either in water or in oil. The water stains give the most pleasing results, but raise the grain of the wood and require an extra sanding operation after the stain is dry.

The most commonly used stains are the "non-grain-raising" ones which dry quickly, and often approach the water stains in clearness and uniformity of color. Stains on softwoods color the springwood more strongly than the summerwood, reversing the natural gradation in color in a manner that is often garish. Pigment-oil stains, which are essentially thin paints, are less subject to this objection, and are therefore more suitable for softwoods. Alternatively, the softwood may be coated with clear sealer before applying the pigment-oil stain to give more nearly uniform coloring.

In hardwoods with large pores, the pores must be filled before varnish or lacquer is applied if a smooth coating is desired. The filler may be transparent and without effect on the color of the finish, or it may be colored to contrast with the surrounding wood.

Sealer (thinned out varnish or lacquer) is used to prevent absorption of subsequent surface coatings and prevent the bleeding of some stains and fillers into surface coatings, especially lacquer coatings. Lacquer sealers have the advantage of being very fast drying.

Transparent surface coatings over the sealer may be of gloss varnish, semigloss varnish, nitrocellulose lacquer, or wax. Wax provides a characteristic sheen without forming a thick coating and without greatly enhancing the natural luster of the wood. Coatings of a more resinous nature, especially lacquer and varnish, accentuate the natural luster of some hardwoods and seem to permit the observer to look down in the wood. Shellac applied by the laborious process of *French polishing* probably achieves this impression of depth most fully, but the coating is expensive and easily marred by water. Rubbing varnishes made with resins of high refractive index for light are nearly as effective as shellac. Lacquers have the advantages of drying rapidly and forming a hard surface, but require more applications than varnish to build up a lustrous coating.

Varnish and lacquer usually dry with a highly glossy surface. To reduce the gloss, the surfaces may be rubbed with pumice stone and water or polishing oil. Waterproof sandpaper and water may be used instead of pumice stone. The final sheen varies with the fineness of the powdered pumice stone, coarse powders making a dull surface and fine powders a bright sheen. For very smooth surfaces with high polish, the final rubbing is done with rotten-stone and oil. Varnish and lacquer made to dry to semigloss are also available.

Flat oil finishes are currently very popular. This type of finish penetrates the wood and forms no noticeable film on the surface. Two coats of oil are usually applied, which may be followed with a paste wax. Such finishes are easily applied and maintained but are more subject to soiling than a film-forming type of finish.

Filling Porous Hardwoods Before Painting

For finishing purposes, the hardwoods may be classified as follows:

Hardwoods with large pores	*Hardwoods with small pores*
Ash	Alder, red
Butternut	Aspen
Chestnut	Basswood
Elm	Beech
Hackberry	Cherry
Hickory	Cottonwood
Khaya (African mahogany)	Gum
Mahogany	Magnolia
Oak	Maple
Sugarberry	Poplar
Walnut	Sycamore

Birch has pores large enough to take wood filler effectively when desired, but small enough as a rule to be finished satisfactorily without filling.

Hardwoods with small pores may be finished with paints, enamels, and varnishes in exactly the same manner as softwoods. Hardwoods with large pores require wood filler before they can be covered smoothly with a film-forming finish. Without filler, the pores not only appear as depressions in the coating, but also become centers of surface imperfections and early failure.

Finishes for Floors

Interior Floors

Wood possesses a variety of properties that make it a highly desirable flooring material for home, industrial, and public structures. A variety of wood flooring products permit a wide selection of attractive and serviceable wood floors. Selection is available not only from a variety of different wood species and grain characteristics, but also from a considerable number of distinctive flooring types and patterns.

The natural color and grain of wood floors make them inherently attractive and beautiful. It is the function of floor finishes to enhance the natural beauty of wood, protect it from excessive wear and abrasion, and make the floors easier to clean. A complete finishing process may consist of four steps: Sanding the surface, applying a filler for certain woods, applying a stain to achieve a desired color effect, and applying a finish. Detailed procedures and specified materials depend largely on the species of wood used and individual preference in type of finish.

Careful sanding to provide a smooth surface is essential for a good finish because any irregularities or roughness in the base surface will be magnified by the finish. The production of a satisfactory surface requires sanding in several steps with progressively finer sandpaper, usually with a machine, unless the area is small. The final sanding is usually done with a 2/0 grade paper. When sanding is complete, all dust must be removed by vacuum cleaner or tack rag. Steel wool should not be used on floors unprotected by finish because minute steel particles left in the wood may later cause staining or discoloration.

A filler is required for wood with large pores, such as oak and walnut, if a smooth, glossy, varnish finish is desired. A filler may be paste or liquid, natural or colored. It is applied by brushing first across the grain and then by brushing with the grain. Surplus filler must be removed immediately after the glossy wet appearance disappears. Wipe first across the grain to pack the filler into the pores; then complete the wiping with a few light strokes with the grain. Filler should be allowed to dry thoroughly before the finish coats are applied.

Stains are sometimes used to obtain a more nearly uniform color when individual boards vary too much in their natural color. Stains may also be used to accent the grain pattern. If the natural color of the wood is acceptable, staining is omitted. The stain should be an oil-base or a non-grain-raising stain. Stains penetrate wood only slightly; therefore, the finish should be carefully maintained to prevent wearing through the stained layer. It is difficult to renew the stain at worn spots in a way that will match the color of the surrounding area.

Finishes commonly used for wood floors are classified either as sealers or varnishes. Sealers, which are usually thinned out varnishes, are widely used in residential flooring. They penetrate the wood just enough to avoid formation of a surface coating of appreciable thickness. Wax is usually applied over the sealer; however, if greater gloss is desired, the sealed floor makes an excellent base for varnish. The thin surface coat of sealer and wax needs more frequent attention than varnished surfaces. However, rewaxing or resealing and waxing of high traffic areas is a relatively simple maintenance procedure.

Varnish may be based on *phenolic, alkyd, epoxy,* or *polyurethane* resins. They form a distinct coating over the wood and give a lustrous finish. The kind of service expected usually determines the type of varnish. Varnishes especially designed for homes, schools, gymnasiums, and other public buildings are available. Information on types of floor finishes can be obtained from the flooring associations or the individual flooring manufacturers.

Durability of floor finishes can be improved by keeping them waxed. Paste waxes generally give the best appearance and durability. Two coats are recommended and, if a liquid wax is used, additional coats may be necessary to get an adequate film for good performance.

Porches and Decks

Exposed flooring on porches and decks is commonly painted. The recommended procedure of treating with water-repellent preservative and primer is the same as for wood siding. After the primer, an undercoat and matching coat of porch and deck enamel should be applied.

Many fully exposed rustic-type decks are effectively finished with only water-repellent preservative or a penetrating-type pigmented stain (13). Because these finishes penetrate and form no film on the surface, they do not crack and peel. They may need more frequent refinishing than painted surfaces, but this is easily done because there is no need for laborious surface preparation as when painted surfaces start to peel.

Moisture-excluding Effectiveness of Coatings

The protection afforded by coatings in excluding moisture vapor from wood depends on a number of variables. Among them are film thickness, absence of defects and voids in the film, type of pigment, type of vehicle, volume ratio of pigment to vehicle, vapor pressure gradient across the film, and length of exposure period.

The relative effectiveness of several typical treating and finishing systems for wood in retarding adsorption of water vapor at 97 percent relative humidity is compared in table 11. Perfect protection, or no adsorption of water, would be represented by 100 percent effectiveness; complete lack of protection (as with unfinished wood) by 0 percent.

Paints which are porous, such as the latex paints and low-luster or breather-type oil-base paints formulated at a pigment volume concentration above 40 percent, afford little protection against moisture vapor. These porous paints also permit rapid entry of water and so provide little protection against dew and rain unless applied over a nonporous primer.

CHAPTER 55

PROTECTION AGAINST DECAY AND TERMITES

Wood used under conditions where it will always be dry, or even where it is wetted briefly and rapidly redried, will not decay. However, all wood and wood products in construction use are susceptible to decay if kept wet for long periods under temperature conditions favorable to the growth of decay organisms. Most

TABLE 11.—*Some typical values of moisture-excluding effectiveness of coatings after 2 weeks' exposure of wood initially conditioned from 80° F. and 65 percent relative humidity to 80° F. and 97 percent relative humidity*

Coatings			Coatings		
Type	Number of coats	Effectiveness	Type	Number of coats	Effectiveness
		Pct.			*Pct.*
INTERIOR COATINGS			EXTERIOR COATINGS		
Uncoated wood		0	Water-repellent preservative	1	0
Latex paint	2	0	FPL natural finish (stain)	1	0
Floor seal	2	0	Exterior latex paint	2	3
Floor seal plus wax	2	10	House paint primer:	1	20
Linseed oil	1	1	Plus latex paint	2	22
Do.	2	5	Plus titanium-zinc linseed oil paint (low-luster oil base) (30 pct. PVC) [1]	1	65
Do.	3	21	Titanium-alkyd oil:		
Furniture wax	3	8	30 pct. PVC [1]	1	45
Phenolic varnish	1	5	40 pct. PVC [1]	1	3
Do.	2	49	50 pct. PVC [1]	1	0
Do.	3	73	Aluminum powder in long oil phenolic varnish	1	39
Semigloss enamel	2	52	Do.	2	88
Cellulose lacquer	3	73	Do.	3	95
Lacquer enamel	3	76			
Shellac	3	87			

[1] PVC (pigment volume concentration) is the volume of pigment, in percent, in the nonvolatile portion of the paint.

of the wood as used in a house is not subjected to such conditions. There are places where water can work into the structure, but such places can be protected. Protection is accomplished by methods of design and construction, by use of suitable materials, and in some cases by using treated material.

Wood is also subject to attack by termites and some other insects. Termites can be grouped into two main classes—*subterranean* and *dry-wood*. Subterranean termites are important in the northernmost States where serious damage is confined to scattered, localized areas of infestation (fig. 55.1). Buildings may be fully protected against subterranean termites by incorporating comparatively inexpensive protection measures during construction. The Formosan subterranean termite has recently (1966) been discovered in several locations in the South. It is a serious pest because its colonies contain large numbers of the worker caste and cause damage rapidly. Though presently in localized areas, it could spread to other areas. Controls are similar to those for other subterranean species. Dry-wood termites are found principally in Florida, southern California, and the Gulf Coast States. They are more difficult to control, but the damage is less serious than that caused by subterranean termites.

Wood has proved itself through the years to be desirable and satisfactory as a building material. Damage from decay and termites has been small in proportion to the total value of wood in residential structures, but it has been a troublesome problem to many homeowners. With changes in building-design features and use of new building materials, it becomes pertinent to restate the basic safeguards to protect buildings against both decay and termites.

Decay

Wood decay is caused by certain fungi that can utilize wood for food. These fungi, like the higher plants, require air, warmth, food, and moisture for growth. Early stages of decay caused by these fungi may be accompanied by a discoloration of the wood. Paint also may become discolored where the underlying wood is rotting. Advanced decay is easily recognized because the wood has then undergone definite changes in properties and appearance. In advanced stages of building decay, the affected wood generally is brown and crumbly, and sometimes may be comparatively white and spongy. These changes may not be apparent on the surface, but the loss of sound wood inside often is reflected by sunken areas on the surface or by a "hollow" sound when the wood is tapped with a hammer. Where the surrounding atmosphere is very damp, the decay fungus may grow out on the surface—appearing as white or brownish growths in patches or strands or in special cases as vine-like structures.

Fungi grow most rapidly at temperatures of about 70° to 85° F. Elevated temperatures such as those used in kiln-drying of lumber kill fungi, but low temperatures, even far below zero, merely cause them to remain dormant.

Moisture requirements of fungi are within definite

Figure 55.1.—The northern limit of damage in the United States by subterranean termites, Line A;
by dry-wood or nonsubterranean termites, line B.

limitations. Wood-destroying fungi will not become established in dry wood. A moisture content of 20 percent (which can be determined with an electrical moisture meter) is safe. Moisture contents greater than this are practically never reached in wood sheltered against rain and protected, if necessary, against wetting by condensation or fog. Decay can be permanently arrested by simply taking measures to dry out the infected wood and to keep it dry. Brown crumbly decay, in the dry condition, is sometimes called "dry rot," but this is a misnomer. Such wood must necessarily be damp if rotting is to occur.

The presence of mold or stain fungi should serve as a warning that conditions are or have been suitable for decay fungi. Heavily molded or stained lumber, therefore, should be examined for evidence of decay. Furthermore, such discolored wood is not entirely satisfactory for exterior millwork because it has greater water absorptiveness than bright wood.

The natural decay resistance of all common native species of wood lies in the heartwood. When untreated, the sapwood of all species has low resistance to decay and usually has a short life under decay-producing conditions. Of the species of wood commonly used in house construction, the heartwood of baldcypress, red-

wood, and the cedars is classified as being highest in decay resistance. All-heartwood, quality lumber is becoming more and more difficult to obtain, however, as increasing amounts of timber are cut from the smaller trees of second-growth stands. In general, when substantial decay resistance is needed in load-bearing members that are difficult and expensive to replace, appropriate preservative-treated wood is recommended.

Subterranean Termites

Subterranean termites are the most destructive of the insects that infest wood in houses. The chance of infestation is great enough to justify preventive measures in the design and construction of buildings in areas where termites are common.

Subterranean termites are common throughout the southern two-thirds of the United States except in mountainous and extremely dry areas.

One of the requirements for subterranean-termite life is the moisture available in the soil. These termites become most numerous in moist, warm soil containing an abundant supply of food in the form of wood or other cellulosic material. In their search for additional food (wood), they build earthlike shelter tubes over foundation walls or in cracks in the walls, or on pipes or supports leading from the soil to the house. These tubes are from $\frac{1}{4}$ to $\frac{1}{2}$ inch or more in width and flattened, and serve to protect the termites in their travels between food and shelter.

Since subterranean termites eat the interior of the wood, they may cause much damage before they are discovered. They honeycomb the wood with definite tunnels that are separated by thin layers of sound wood. Decay fungi, on the other hand, soften the wood and eventually cause it to shrink, crack, and crumble without producing anything like these continuous tunnels. When both decay fungi and subterranean termites are present in the same wood, even the layers between the termite tunnels will be softened.

Dry-wood Termites

Dry-wood termites fly directly to and bore into the wood instead of building tunnels from the ground as do the subterranean termites. Dry-wood termites are common in the tropics, and damage has been recorded in the United States in a narrow strip along the Atlantic Coast from Cape Henry, Va., to the Florida Keys, and westward along the coast of the Gulf of Mexico to the Pacific Coast as far as northern California (fig. 55.1) Serious damage has been noted in southern California and in localities around Tampa, Miami, and Key West, Fla. Infestations may be found in structural timber and other woodwork in buildings, and also in furniture, particularly where the surface is not adequately protected by paint or other finishes.

Dry-wood termites cut across the grain of the wood and excavate broad pockets, or chambers, connected by tunnels about the diameter of the termite's body. They destroy both springwood and the usually harder summerwood, whereas subterranean termites principally attack springwood. Dry-wood termites remain hidden in the wood and are seldom seen, except when they make dispersal flights.

Safeguards Against Decay

Except for special cases of wetting by condensation or fog, a dry piece of wood, when placed off the ground under a tight roof with wide overhang, will stay dry and never decay. This principle of "umbrella protection," when applied to houses of proper design and construction, is a good precaution. The use of dry lumber in designs that will keep the wood dry is the simplest way to avoid decay in buildings.

Most of the details regarding wood decay have been included in earlier chapters, but they are given here as a reminder of their relationship to protection from decay and termites.

Untreated wood should *not* come in contact with the soil. It is desirable that the foundation walls have a clearance of at least 8 inches above the exterior finish grade, and that the floor construction have a clearance of 18 inches or more from the bottom of the joists to the ground in basementless spaces. The foundation should be accessible at all points for inspection. Porches that prevent access should be isolated from the soil by concrete or from the building proper by metal barriers or aprons (fig. 55.2).

Steps and stair carriages, posts, wallplates, and sills should be insulated from the ground with concrete or masonry. Sill plates and other wood in contact with concrete near the ground should be protected by a moistureproof membrane, such as heavy roll roofing or 6-mil polyethylene. Girder and joists openings in masonry walls should be big enough to assure an air space around the ends of these members.

Design Details

Surfaces like steps, porches, door and window frames, roofs, and other protections should be sloped to promote runoff of water (Ch. 51, "Porches and Garages"). Noncorroding flashing should be used around chimneys, windows, doors, or other places where water might seep in (Ch. 50, "Flashing and Sheet Metal"). Roofs with considerable overhang give added protection to the siding and other parts of the house. Gutters and downspouts should be placed and maintained to divert water away from the buildings. Porch columns and screen rails should be shimmed above the floor to allow quick drying; or posts should slightly overhang raised concrete bases (Ch. 51, "Porches and Garages").

Exterior steps, rails, and porch floors exposed to rain need protection from decay, particularly in warm, damp parts of the country. Pressure treatment of the wood in accordance with the recommendation of Federal Specification TT–W–571 provides a high degree of protection against decay and termite attack (12). Where the likelihood of decay is relatively small or where pressure-treated wood is not readily obtainable, on-the-job application of water-repellent preservatives by dipping or soaking has been found to be worthwhile. The wood should by dry, cut to final dimensions, and then dipped or soaked in the preservative solution (19). Soaking is the best of these nonpressure methods, and the ends of the boards should be soaked for a minimum of 3 minutes. It is important to protect the end grain of wood at joints, for this area absorbs water easily and is the most common infection point. The edges of porch flooring should be coated with thick white lead or other durable coating as it is laid.

Leaking pipes should be remedied immediately to prevent damage to the house, as well as to guard against possible decay.

Green or Partially Seasoned Lumber

Construction lumber that is green or partially seasoned may be infected with one or more of the stain-

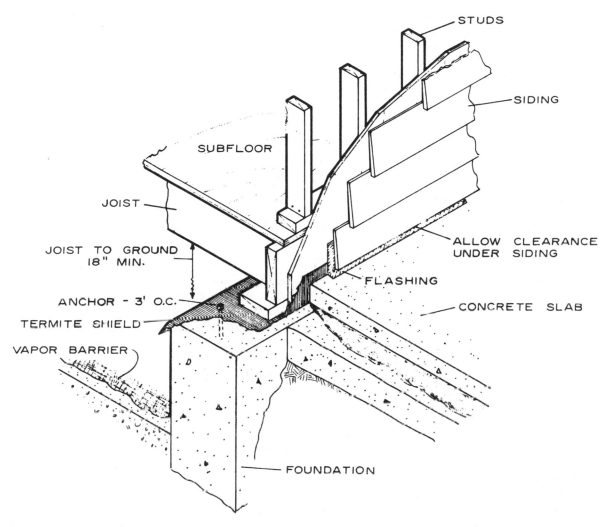

Figure 55.2.—Metal shield used to protect wood at porch slab.

ing, molding, or decay fungi and should be avoided. Such wood may contribute to serious decay in both the substructure and exterior parts of buildings. If wet lumber must be used, or if wetting occurs during construction, the wood should not be fully enclosed or painted until thoroughly dried.

Water Vapor from the Soil

Crawl spaces of houses built on poorly-drained sites may be subjected to high humidity. During the winter when the sills and outer joists are cold, moisture condenses on them and, in time, the wood absorbs so much moisture that it is susceptible to attack by fungi. Unless this moisture dries out before temperatures favorable for fungus growth are reached, considerable decay may result. However, this decay may progress so slowly that no weakening of the wood becomes apparent for a few years. Placing a layer of 45 pound or heavier roll roofing or a 6-mil sheet of poly-

ethylene over the soil to keep the vapor from getting into the crawl space would prevent such decay. This might be recommended for all sites where, during the cold months, the soil is wet enough to be compressed in the hand.

If the floor is uninsulated, there is an advantage in closing the foundation vents during the coldest months from the standpoint of fuel savings. However, unless the crawl space is used as a heat plenum chamber, insulation is usually located between floor joists. The vents could then remain open. Crawl-space vents can be very small when soil covers are used; only 10 percent of the area required without covers (Ch. 42, "Ventilation").

Water Vapor from Household Activities

Water vapor is also given off during cooking, washing, and other household activities. This vapor can pass through walls and ceilings during very cold weather and condense on sheathing, studs, and rafters, causing condensation problems. A vapor barrier of an approved type is needed on the warm side of walls. (See section on "Vapor Barriers," Ch. 41.) It is also important that the attic space be ventilated (17), as previously discussed in **Chapter 42,** "Ventilation."

Water Supplied by the Fungus Itself

In the warmer coastal areas principally, some substructure decay is caused by a fungus that provides its own needed moisture by conducting it through a vinelike structure from moist ground to the wood. The total damage caused by this water-conducting fungus is not large, but in individual instances it tends to be unusually severe. Preventive and remedial measures depend on getting the soil dry and avoiding untreated wood "bridges" such as posts between ground and sills or beams.

Safeguards Against Termites

The best time to provide protection against termites is during the planning and construction of the building. The first requirement is to remove all woody debris like stumps and discarded form boards from the soil at the building site before and after construction. Steps should also be taken to keep the soil under the house as dry as possible.

Next, the foundation should be made impervious to subterranean termites to prevent them from crawling up through hidden cracks to the wood in the building above. Properly reinforced concrete makes the best foundation, but unit-construction walls or piers capped with at least 4 inches of reinforced concrete are also satisfactory. No wood member of the structural part of the house should be in contact with the soil.

The best protection against subterranean termites is by treating the soil near the foundation or under an entire slab foundation. The effective soil treatments are water emulsions of aldrin (0.5 pct.), chlordane (1.0 pct.), dieldrin (0.5 pct.), and heptachlor (0.5 pct.). The rate of application is 4 gallons per 10 linear feet at the edge and along expansion joints of slabs or along a foundation. For brick or hollow-block foundations, the rate is 4 gallons per 10 linear feet for each foot of depth to the footing. One to 1½ gallons of emulsion per 10 square feet of surface area is recommended for overall treatment before pouring concrete slab foundations. Any wood used in secondary appendages, such as wall extensions, decorative fences, and gates, should be pressure-treated with a good preservative.

In regions where dry-wood termites occur, the following measures should be taken to prevent damage:

1. All lumber, particularly secondhand material, should be carefully inspected before use. If infected, discard the piece.

2. All doors, windows (especially attic windows), and other ventilation openings should be screened with metal wire with not less than 20 meshes to the inch.

3. Preservative treatment in accordance with Federal Specification TT–W–571 ("Wood Preservatives: Treating Practices," available through GSA Regional Offices) can be used to prevent attack in construction timber and lumber.

4. Several coats of house paint will provide considerable protection to exterior woodwork in buildings. All cracks, crevices, and joints between exterior wood members should be filled with a mastic calking or plastic wood before painting.

5. The heartwood of foundation-grade redwood, particularly when painted, is more resistant to attack than most other native commercial species.

Pesticides used improperly can be injurious to man, animals, and plants. Follow the directions and heed all precautions on the labels.

Store pesticides in original containers—out of reach of children and pets—and away from foodstuff.

Apply pesticides selectively and carefully. Do not apply a pesticide when there is danger of drift to other areas. Avoid prolonged inhalation of a pesticide spray or dust. When applying a pesticide it is advisable that you be fully clothed.

After handling a pesticide, do not eat, drink or smoke until you have washed. In case a pesticide is swallowed or gets in the eyes, follow the first aid treatment given on the label, and get prompt medical attention. If the pesticide is spilled on your skin or clothing, remove clothing immediately and wash skin thoroughly.

Dispose of empty pesticide containers by wrapping them in several layers of newspaper and placing them in your trash can.

It is difficult to remove all traces of a herbicide (weed killer) from equipment. Therefore, to prevent injury to desirable plants do not use the same equipment for insecticides and fungicides that you use for a herbicide.

NOTE: Registrations of pesticides are under constant review by the U. S. Department of Agriculture. Use only pesticides that bear the USDA registration number and carry directions for home and garden use.

FOUR OBVIOUS SIGNS OF A TERMITE INFESTATION

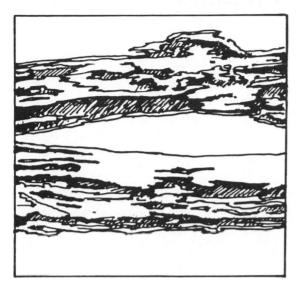

Hollow wood underneath a finished surface. Wood underneath will have earthfilled galleries, but termites are not always present.

Earth tubes between the soil and some wooden structure under the house is a sure sign of subterranean termites. These tubes can go up the foundation wall, along plumbing, or directly from the soil to the wood.

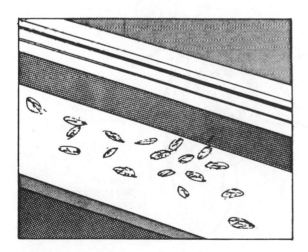

Termite wings on the window sill or the appearance of winged termites in and around the home. Soon after flying, termites lose their wings, so the only evidence they leave is scattered wings.

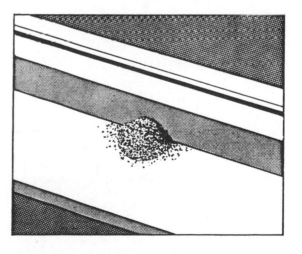

Piles of fecal pellets on the outside of the finished wood are a sure sign of dry wood termite attack. Pellets look like sawdust; but upon close examination they appear very symmetrical with six grooves along the side.

DISTINCTION BETWEEN FLYING TERMITES & FLYING ANTS

Termite

1. Antenna straight and beadlike

2. Thorax and abdomen broadly joined

3. Wings similar in shape, size, and pattern — many small veins

Ant

1. Antenna "elbowed"

2. Thorax and abdomen joined by a narrow waist.

3. Wings not alike in shape, size, or pattern — few veins

Termites, like ants, live in colonies, which are usually headed by a king and queen. These you will seldom see except in the flying stage. The workers however, are responsible for the damage done by termites. They are the individuals that seek wood out and remove it from your home in very small quantities every day. Soldiers are responsible for the protection of the colony from ants and other insect enemies.

THE LIFE CYCLE OF A TERMITE COLONY

EGGS — The tiny termite egg is almost transparent. During the incubation period the egg is groomed and tended by workers. The larva hatches from the egg and is about the same size.

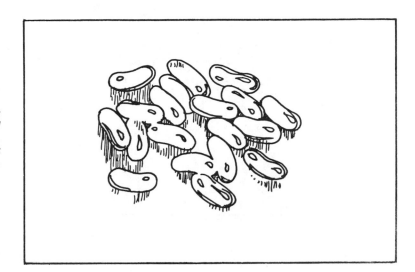

LARVA — During early larval stages, all young termites "look alike," and all are fed by attendant workers. The larvae can develop into one of four castes.

WORKER — This termite is the one which forages from the nest to the wood supply and returns with food for the colony.

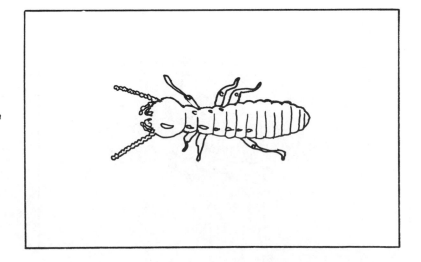

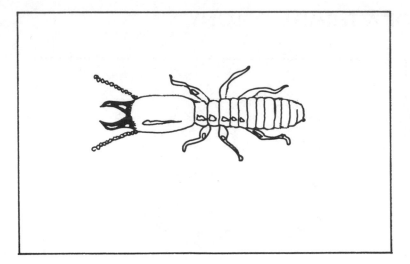

SOLDIER — *Defender of the colony, the soldier termite develops a long, armored head and mandibles capable of cutting an enemy ant in half. The soldier also sounds the alarm by banging his head against the side of a tunnel.*

SECONDARY REPRODUCTIVE — *This caste is capable of reproducing and perpetuating the life of the colony, should the king or queen die. In very large colonies these reproductives produce offspring in addition to that of the king and queen.*

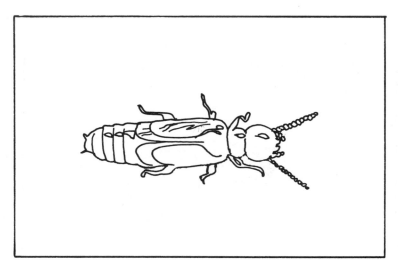

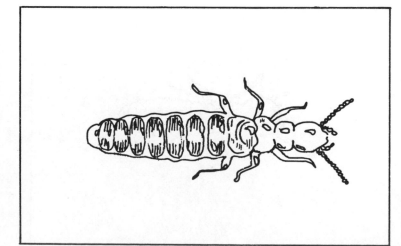

NYMPH — *The nymph is the most vital individual to the perpetuation of the species since it may become a king or queen of a new colony. Physically, it is longer than the other termites. Before it leaves the nest to breed, its body turns black and all four wings extend to about twice its body length.*

When nymphs reach maturity and become swarmers (kings and queens), they all leave the colony at about the same time, usually in spring or fall. The swarmers fly very poorly and most of them flutter for only a few yards before falling to the ground.

When the short flight is finished, the swarmers drop their wings and the males begin a frenzied search for compatible mates. Because the swarmers are exposed and are prey to predator birds and insects, very few ever survive to establish a new colony.

KING AND QUEEN — In a young colony, the king and queen are the actively reproducing termites in the colony. Their only function is the production of eggs.

15 MOST FREQUENT DANGER AREAS

3/ Concrete porches with earthfill underneath pose a special hazard. Often, wood framing members are in contact with the fill.

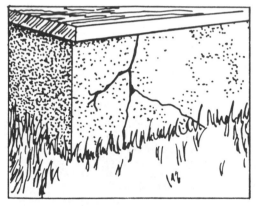

1/ Cracks in concrete foundations allow termites hidden access from soil to sill.

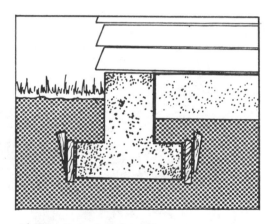

4/ Form boards left in place contribute to termite food supply.

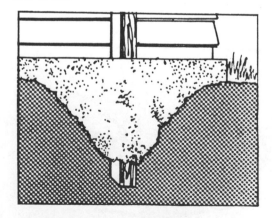

2/ Posts set in concrete may give a false sense of security. What if they are in contact with the soil underneath?

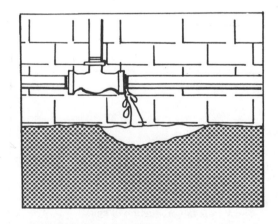

5/ Leaking pipes and dripping faucets keep the soil moist. Excess irrigation has the same effect. Downspouts should carry water away from the building.

6/ *Shrubbery which blocks the air flow through vents causes the air underneath house to remain warm and moist — an ideal climate for termites!*

9/ *Stucco or brick veneer carried down over concrete foundation permits hidden entrance between exterior and foundation, if bond fails.*

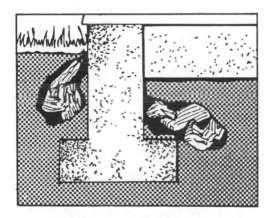

7/ *Debris under house supports termite colony until population becomes large enough to attack the house itself.*

10/ *Planters built against the foundation allow direct access to unprotected veneer, siding, or cracked stucco.*

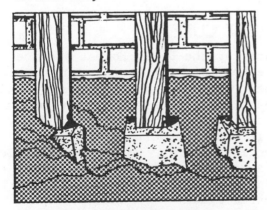

8/ *Foundation walls or footings which are too low, permit wood to contact the soil.*

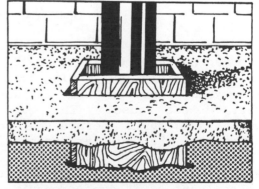

11/ *Forms left in hole of slab where bathtub drain enters building provide a direct route to inner walls.*

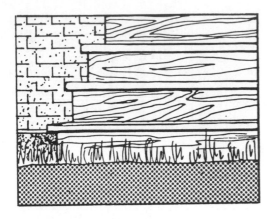

12/ *Porch steps in contact with untreated soil literally offer termites a stairway to your home.*

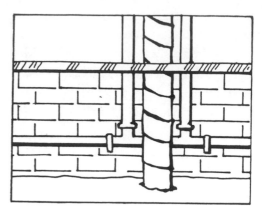

14/ *Paper is made of wood. Don't leave paper collars around pipes.*

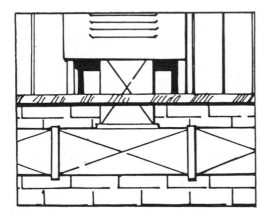

13/ *Heating unit accelerates termite development by maintaining warm soil for colony on a year-round basis.*

15/ *In a house that was once termite-free, do-it-yourselfers can provide access to termites by building trellises and other adornments which provide a direct link from soil to wood.*

PREVENTION IS THE BEST CURE

As already seen, most of the danger areas can be rendered harmless by any homeowner simply by moving soil as far away as possible from any wooden member of the house. But careful! Chances are that, before your house was built, the builder treated the soil underneath and adjacent to the slab or pier foundations with a termite toxicant. If done properly, the treatment will prevent termites from entering through or around the foundation for many years. In fact, research by the U. S. Forest Service has revealed four chemicals—aldrin, chlordane, dieldrin and heptachlor—which create a termite-proof barrier for at least 25 years. None have failed yet! So if you transport much soil from the immediate vicinty of foundation piers or a concrete slab foundation, the remaining soil should be chemically treated again.

If you find or suspect termites call a responsible pest control operator. As said before, don't panic. Take your time and get two or three cost estimates from established firms. Check references of the operator and beware of firms that:

• quote a price based on gallonage of material used(get estimate of total price for the job).

• profess to have a secret formula or ingredient for termite control. Chemicals tested by the U. S. Forest Service are the best known to man and are not expensive.

• have no listed phone number.

• show up without an invitation and use evidence of termites in trees as an excuse to inspect the house.

• also want to trim trees and do general repair work as part of the contract.

WHAT HOMEOWNERS ASK ABOUT TERMITES

Scientists at the Wood Products Insect Laboratory in Gulfport, Mississippi are among the nation's leading authorities on the biological habits of termites and methods of termite prevention and control. Here are their answers to some of the questions they receive from around the world:

How can I tell if the damage to my door jamb is caused by termites or decay?

Strip away the outer surface of the board. If termites have been there, you'll find that the softer springwood hasn't been eaten away while the harder summerwood is intact, giving a honeycomb effect. Termites also leave light brown specks of excrement and earth in their path. When decayed, the wood is soft to the touch and, of course, there's no earth or tunnels inside

I prefer not to use chemicals. Is there any biological control for termites?

Not yet. We and other scientists are searching for biological controls but none have been found to be very effective so far. However, all four recommended chemicals are registered for use against termites at this time. The U.S. Forest Service has shown that these chemicals have moved only a few inches laterally or downward through sandy loam soil after two decades of heavy rainfall. Thus, the risk of contaminating water resources is minimal. Moreover, most of the insecticide is under the building and not exposed to the environment. In the event that any or all of these insecticides become unavailable to pest control operators, the Forest Service is continuing its research effort to find effective substitutes.

I plan to build my house on a concrete slab. Isn't this the safest from termite attack?

Actually, it's one of the most susceptible types of construction used today. The homeowner has a false sense of security. With time most slabs will crack. Termites can enter through tiny cracks in the concrete, over the edge of the slab, or easier yet, through openings around plumbing. Be sure the soil under your house is treated with the right type of chemical before pouring the slab and that the pest control operator uses the proper amount of the chemical.

CHAPTER 56

PROTECTION AGAINST FIRE

Fire hazards exist to some extent in nearly all houses. Even though the dwelling is of the best fire-resistant construction, hazards can result from occupancy and the presence of combustible furnishings and other contents.

The following tabulation showing the main causes of fires in one- and two-family dwellings is based on an analysis (16) made of 500 fires by the National Fire Protection Association.

Cause of fire	Percent of Total
Heating equipment	23.8
Smoking materials	17.7
Electrical	13.8
Children and matches	9.7
Mishandling of flammable liquids	9.2
Cooking equipment	4.9
Natural gas leaks	4.4
Clothing ignition	4.2
Combustibles near heater	3.6
Other miscellaneous	8.7
	100.0

Fire-protection engineers generally recognize that a majority of fires begin in the contents, rather than in the building structure itself. Proper housekeeping and care with smoking, matches, and heating devices can reduce the possibility of fires. Other precautions to reduce the hazards of fires in dwellings—such as fire stops, spacing around heating units and fireplaces, and protection over furnaces—are recommended elsewhere in this handbook.

Fire Stops

Fire stops are intended to prevent drafts that foster movement of hot combustible gases from one area of the building to another during a fire. Exterior walls of wood-frame construction should be fire-stopped at each floor level (figs. 28.9, 29.5, and 29.8), at the top-story ceiling level, and at the foot of the rafters.

Fire stops should be of noncombustible materials or wood not less than 2 inches in nominal thickness. The fire stops should be placed horizontally and be well fitted to completely fill the width and depth of the spacing. This applies primarily to balloon-type frame construction. Platform walls are constructed with top and bottom plates for each story (fig. 29.1). Similar fire stops should be used at the floor and ceiling of interior stud partitions, and headers should be used at the top and bottom of stair carriages (fig. 49.7).

Noncombustible fillings should also be placed in any spacings around vertical ducts and pipes passing through floors and ceilings, and self-closing doors should be used on shafts, such as clothes chutes.

When cold-air return ducts are installed between studs or floor joists, the portions used for this purpose should be cut off from all unused portions by tight-fitting stops of sheet metal or wood not less than 2 inches in nominal thickness. These ducts should be constructed of sheet metal or other materials no more flammable than 1-inch (nominal) boards.

Fire stops should also be placed vertically and horizontally behind any wainscoting or paneling applied over furring, to limit the formed areas to less than 10 feet in either dimension.

With suspended ceilings, vertical panels of noncombustible materials from lumber of 2-inch nominal thickness or the equivalent, should be used to subdivide the enclosed space into areas of less than 1,000 square feet. Attic spaces should be similarly divided into areas of less than 3,000 square feet.

Chimney and Fireplace Construction

The fire hazards within home construction can be reduced by insuring that chimney and fireplace constructions are placed in proper foundations and properly framed and enclosed. (See Ch. 52, "Chimneys and Fireplaces.") In addition, care should be taken that combustibles are not placed too close to the areas of high temperature. Combustible framing should be no closer than 2 inches to chimney construction; however, when required, this distance can be reduced to $\frac{1}{2}$ inch, provided the wood is faced with a $\frac{1}{4}$-inch-thick asbestos sheeting.

For fireplace construction, wood should not be placed closer than 4 inches from the backwall nor within 8 inches of either side or top of the fireplace opening. When used, wood mantels should be located at least 12 inches from the fireplace opening.

Heating Systems

Almost 25 percent of fires are attributed to faulty construction or to improper use of heating equipment, and the greater proportion of these fires originate in the basement. Combustible products should not generally be located nearer than 24 inches from a hot air, hot water, or steam heating furnace; however, this distance can be reduced in the case of properly insulated furnaces or when the combustible materials are protected by gypsum board, plaster, or other materials with low flame spread. Most fire-protection agencies limit to 170° F. the temperature to which combustible wood products should be exposed for long periods of time, although experimentally, ignition does not occur until much higher temperatures have been reached.

In confining a fire to the basement of a home, added protection can be obtained with gypsum board, asbestos board, or plaster construction on the basement ceiling, either as the exterior surface or as backings for decorative materials. These ceiling surfaces are frequently omitted to reduce costs, but particular attention should be given to protection of the wood members directly above and near the furnace.

Flame Spread and Interior Finish

In some areas of a building, flame-spread ratings are assigned to limit spread of fire on wall and ceiling surfaces. Usually, these requirements do *not* apply to private dwellings because of their highly combustible content, particularly in furnishings and drapes usually found in this type of structure.

To determine the effect of the flammability of wall linings on the fire hazards within small rooms, burnout tests have been made at the Forest Products Laboratory in Madison, Wis. For this purpose, an 8- by 12- by 8-foot high room was furnished with an average amount of furniture and combustible contents. This room was lined with various wall panel products, plywood, fiber insulation board, plaster on fiberboard lath, and gypsum wallboard. When a fire was started in the center of this room, the time to reach the *critical* temperature (when temperature rise became very rapid) or the flashover temperature (when everything combustible burst into flames) was not signifi-

cantly influenced by either combustible or noncombustible wall linings. In the time necessary to reach these critical temperatures (usually less than 10 minutes) the room would already be unsafe for human occupancy.

Similar recent tests in a long corridor, partially ventilated, showed that the "flashover" condition would develop for 60 to 70 feet along a corridor ceiling within 5 to 7 minutes from the burning of a small amount of combustible contents. This "flashover" condition developed in approximately the same time, whether combustible or noncombustible wall linings were used, and before any appreciable flame spread along wall surfaces.

Wood paneling, treated with fire-retardant chemicals or fire-resistant coatings as listed by the Underwriters' Laboratory, Inc. or other recognized testing laboratories, can also be used in areas where flame-spread resistance is especially critical. Such treatments, however, are not considered necessary in dwellings, nor can the extra cost of treatment be justified.

Fire-resistant Walls

Whenever it is desirable to construct fire resistant walls and partitions in attached garages and heating rooms, information on fire resistance ratings using wood and other materials is readily available through local code authorities. Wood construction assemblies can provide ½ hour to 2 hour fire resistance under recognized testing methods, depending on the covering material.

CHAPTER 57

METHODS OF REDUCING BUILDING COSTS

The average homebuilder is interested in reducing the overall cost of his house but not at the expense of its livability or resale value. This is often somewhat difficult to do for a single custom-built house.

Operators of large housing developments often build hundreds of houses each year. Because of their need for huge volumes of materials, they buy direct from the manufacturer. They also develop the building sites from large sections of land. Much of the work, such as installation of the roofing, application of gypsum board interiors, and painting, is done by subcontractors. Their own crews are specialists, each crew becoming proficient in its own phase of the work. Central shops are established where all material is cut to length and often preassembled before being trucked to the site. These methods reduce the cost of the individual house in a large building project, but few of them can be applied to an individual house built by the owner.

If a home builder pays attention to various construction details and to the choice of materials, however, this information will usually aid in reducing costs. The following suggestions are intended as possible ways for the owner to lower the cost of his house.

Design

The first area where costs of the house may be reduced somewhat is during the design stages. However, such details should not affect the architectural lines or appearance of the house, but rather the room arrangement and other factors. The following design elements might be considered before final plans are chosen:

1. The size of the house, width and length, should be such that standard-length joists and rafters and standard spacings can be used without wasting material. The architect or contractor will have this informa-

tion available. Also reflected in the house size is the use of standard-width sheets of sheathing materials on the exterior as well as in the interior. Any waste or ripping required adds to both the labor and material costs.

The rooms should be arranged so that the plumbing, water, and heating lines are short and risers can serve more than one room. An "expandable" house may mean the use of a steeper pitched roof to provide space for future rooms in the attic area. It might also be desirable to include second-floor dormers in the original design. Additional rooms can thus be provided at a much lower cost than by adding to the side or rear of the house at a future date. Roughing in plumbing and heating lines to the second floor will also reduce future costs when the second floor is completed, yet not add appreciably to the original construction costs.

While a rectangular plan is the most economical from many standpoints, it should not always govern final design. A rectangular plan of the house proper, with a full basement, can be made more desirable by a garage or porch wing of a different size or alinement. Such attachments require only shallow footings, without the excavation necessary for basement areas.

2. The type of foundation to be used, such as slab, crawl space, or basement, is an important consideration. Base this selection on climatic conditions and needs of the family for storage, hobby, or recreation space. While space in the basement is not so desirable as in areas above grade, its cubic-foot cost is a great deal lower. The design of a slab-type house usually includes some additional space for heating, laundry, and storage. This extra area may often cost as much as a full basement. Many multilevel houses include habitable rooms over concrete slabs as well as a full-basement. Consult local architects or contractors for their opinions on the most desirable type of home in your area from the cost standpoint.

3. Many contemporary house designs include a flat or low-pitched roof which allows one type of member to serve as both ceiling joists and rafters. This generally reduces the cost compared to that of a pitched roof, both in materials and labor. However, all styles of houses are not adaptable to such a roof. Many contractors incur savings by using preassembled roof trusses for pitched roofs. Dealers who handle large quantities of lumber are usually equipped to furnish trusses of this type.

Pitched roofs are of *gable* or *hip* design, with the *gambrel* roof a variation of each. While the hip roof is somewhat more difficult to frame than the gable roof, it usually requires less trim and siding. Furthermore, painting is much simpler in the hip roof because of less wall area by elimination of the gable and because of accessibility. In the gambrel roof, which is

adapted to two-story houses, roof shingles serve also as siding over the steep-pitched portions. Furthermore, a roof of this type provides a greater amount of headroom (perhaps the original purpose of this design) than the common gable type.

Choice of Materials

The type and grade of materials used in a house can vary greatly and savings can be effected in their choice. It is poor practice to use a low grade or an inferior material which could later result in excessive maintenance costs. On the other hand, it is not economical to use a material of too high a grade when not needed for strength or appearance (5).

Several points might be considered as a means of reducing costs. (Your contractor or lumber dealer who is familiar with these costs will aid you in your final selection.)

1. Consider the use of concrete blocks for foundation walls as opposed to the use of poured concrete. It is less costly to provide a good water-resistant surface on a poured wall than on a block wall. On the other hand, a common hollow concrete block has better insulating properties than a poured concrete wall of equal thickness. Costs often vary by areas.

2. If *precast* blocks are available, consider them for chimneys. These blocks are made to take flue linings of varied sizes and are laid up more rapidly than brick. Concrete block units are also used in laying up the base for a first-floor fireplace, rather than bricks. Prefabricated, lightweight chimneys that require no masonry may also save money.

3. Dimension material varies somewhat in cost by species and grades. Use the better grades for joists and rafters and the lower cost grades for studs. Do not use better grades of lumber than are actually needed. Conversely, grades that involve excessive cutting and selection would dissipate the saving by increased labor costs. Proper moisture content is an important factor.

4. Conventional items such as cabinets, moldings, windows, and other millwork, which are carried as stock or can be easily ordered, also reduce costs. Any special, nonstandard materials which require extra machine setups will be much more expensive. This need not restrict the homebuilder in his design, however, as there are numerous choices of millwork components from many manufacturers.

5. The use of a single material for wall and floor covering will provide a substantial saving. A combination subfloor underlayment of 5/8- or 3/4-inch tongued and grooved plywood will serve both as subfloor and as a base for resilient tile or similar material, as well as for carpeting. Panel siding consisting of 4-foot-wide full-height sheets of plywood or similar material serves both as sheathing and a finish siding. For example,

Table 12. Weights of Construction Materials.

Material	Weight, psf	Material	Weight, psf		
		ROOFS (Continued)	2 in nom.	3 in nom.	4 in nom.
CEILINGS		Pine, Southern	4.9	8.2	11.4
Acoustical fiber tile	1.0	ponderosa	3.7	6.2	8.7
Channel suspended system	1.0	lodgepole	3.7	6.2	8.7
Plaster and lath (see		white	3.6	5.9	8.3
WALLS AND PARTITIONS)		Spruce, Sitka	3.6	6.1	8.5
		white	3.6	6.1	8.5
FLOORS		Engelmann	3.1	5.1	7.2
Hardwood, 1 in. nominal	4.0				
Plywood, per inch of		Aluminum	Flat	Corrug. (1½ & 2½ in)	
thickness	3.0	(includes laps)			
Asphalt mastic, per inch					
of thickness	12.0	12 American	1.2		
Cement finish, per inch		or B&S ga			
of thickness	12.0	14 or B&S ga	0.9	1.1	
Ceramic or quarry tile,		16 or B&S ga	0.7	0.0.9	
¾ in.	10.0	18 or B&S ga	0.6	0.7	
Concrete, per inch of		20 or B&S ga	0.5	0.6	
thickness		22 or B&S ga		0.4	
Lightweight	6.0 to 10.0	20 or B&S ga			
Reinforced	12.5				
Stone	12.5				
Cork tile, 1/16 in.	0.5				
Flexicore, 6 in. slab	46.0	Galvanized steel	Flat	Corrug. 2½ & 3 in.)	
Linoleum, ¼ in.	1.0	(includes laps)			
Terrazo finish, 1½ in.	19.0				
Vinyl tile, ⅛ in.	1.4	12 U.S. Std. ga	4.5	4.9	
Timber decking		14 U.S. Std. ga	3.3	3.6	
		16 U.S. Std. ga	2.7	2.9	
		18 U.S. Std. ga	2.2	2.4	
ROOFS		20 U.S. Std. ga	1.7	1.8	
Lumber sheathing, 1 in.		22 U.S. Std. ga	1.4	1.5	
nominal	2.5	24 U.S. Std. ga	1.2	1.3	
Plywood sheathing, per		26 U.S. Std. ga	0.9	1.0	
inch of thickness	3.0				

Material	2 in nom.	3 in nom.	4 in nom.	Material	Weight, psf
Timber decking				Concrete plank,	
15 % MC				per in. thickness	6.5
Cedar, Alaska	4.0	6.6	9.3	Insulrock	2.7
western red	3.1	5.1	7.2	Petrical	2.7
Hemlock, western	4.0	6.7	9.4	Porex	2.7
eastern	3.7	6.2	8.6	Poured gypsum	6.5
Fir, true	3.6	5.9	8.3	Tectum	2.0
Douglas fir	4.3	7.1	10.0	Vermiculite concrete	2.6
Larch, western	4.6	7.7	10.8		

(continued on following page)

Table 12. (Continued)

Material	Weight, psf	Material	Weight, psf
Asbestos, corrugated		WALLS AND PARTITIONS	
¼ in.	3.0	Masonry, per 4 inches	
Felt, 3 ply	1.5	of thickness	
Felt, 3 ply with gravel	5.5	Brick	38.0
Felt, 5 ply	2.5	Concrete block	30.0
Felt, 5 ply with gravel	6.5	Cinder concrete block	20.0
Insulation, per inch of		Hollow clay tile,	
Rigid fiberboard,		load bearing	23.0
wood base	1.5	Hollow clay tile,	
Rigid fiberboard,		nonbearing	18.0
mineral base	2.1	Hollow gypsum block	13.0
Expanded polystyrens	0.2	Limestone	55.0
Fiberglass, rigid	1.5		
Terra-cotta tile 25.0			
Fiberglass, loose	0.5	Stone	55.0
Roll roofing	1.0	Plaster, 1 in.	8.0
Shingles		Plaster, 1 in., on wood	
Asphalt, approx. ¼"	2.0	lath	10.0
Book tile, 2"	12.0	'Plaster, 1 in., on	
Book tile, 3"	20.0	metal lath	8.5
Cement asbestos,		Porcelain-enameled	
approx. ⅜ in.	4.0	steel	3.0
Clay tile (for mortar		Stucco, ⅞ in.	10.0
add 10 lb)	9.0 to 14.0	Windows, glass, frame,	
Ludowici	10.0	and sash	8.0
Roman	12.0		
Slate, ¼"	10.0		
Spanish	19.0		
Wood, 1"	3.0		
WALLS AND PARTITIONS			
Wood paneling, 1 in.	2.5		
Wood studs, 2 x 4			
12 in. o.c.	2.1		
16 in. o.c.	1.7		
24 in. o.c.	1.3		
Glass block, 4 in.	18.0		
Glass, plate, ¼ in.	3.3		
Glazed tile	18.0		
Gypsum board, per in.	4.0		
Marble or marble			
waincoting	15.0		

From "Timber Construction Mannual" by permission of AITC. (Weights of timber decking changed to conform to current lumber size standards.)

exterior particleboard with a painted finish and corner bracing on the stud wall may also qualify as a panel siding. Plywood may be obtained with a paper overlay, as well as rough sawn, striated, reverse board and batten, brushed, and other finishes.

6. In planning a truly low-cost house where each dollar is important, a crawl space design with the use of a treated wood post foundation is worth investigating. This construction utilizes treated wood foundation posts bearing on concrete footings. The post support floor beams upon which the floor joists rest. A variation of this design includes spacing of the beams on 48-inch centers and the use of $1\frac{1}{8}$-inch-thick tongue and groove plywood eliminating the need for joists as such.

7. Costs of exterior siding or other finish materials often vary a great deal. Many factory-primed sidings are available which require only finish coats after they are applied. A rough-sawn, low-grade cedar or similar species in board and batten pattern with a stained finish will often reduce the overall cost of exterior coverings. Many species and textures of plywood are available for the exterior. When these sheet materials are of the proper thickness and application, they might also serve as sheathing. Paintability of species is also important (5). Edge-grained boards or paper-overlaid plywood provide good bases for paint.

In applying all exterior siding and trim, galvanized or other rust-resistant nails reduce the need for frequent treatment or refinishing. Stainless steel or aluminum nails on siding having a natural finish are a must. Corrosion-resistant nails will add slightly to the cost but will save many dollars in reduced maintenance costs.

8. Interior coverage also deserves consideration. While gypsum board dry-wall construction may be lower in cost per square foot, it requires decorating before it can be considered complete; plaster walls do not require immediate decorating. These costs vary by areas, depending largely on the availability of the various trades. However, prefinished or plastic-faced gypsum board (available in a number of patterns) with a simple "V" joint or with a joint flap of the same covering, and the use of adhesive for application, will result in an economical wall and ceiling finish.

9. There are many cost-related considerations in the choice of flooring, trim, and other interior finish. Areas which will be fully carpeted do not require a finish floor. However, there is a trend to provide a finish floor under the carpeting. The replacement cost of the carpeting may be substantially greater than the cost of the original finish floor.

Species of trim, jambs, and other interior moldings vary from a relatively low-cost softwood to the higher cost hardwoods such as oak or birch. Softwoods are ordinarily painted, while the hardwoods have a natural finish or are lightly stained. The softwoods, though lower in cost, are less resistant to blows and impacts.

Another consideration is the selection of panel and flush doors. Flush doors can be obtained in a number of species and grades. Unselected gum, for example, might have a paint finish while the more costly woods are best finished with a varnish or sealer. Hollow-core flush doors are lower in cost and are satisfactory for interior use, but exterior flush doors should be solid core to better resist warping. The standard exterior panel door can be selected for many styles of architecture.

Construction

Methods of reducing construction costs are primarily based on reducing on-site labor time. The progressive contractor often accomplishes this in several ways, but the size of the operation generally governs the method of construction. A contractor might use two carpenter crews—one for framing and one for interior finishing. Close cooperation with the subcontractors—such as plumbers, plasterers, and electricians—avoids wasting time. Delivery of items when needed so that storage is not a problem also reduces on-site costs. Larger operators may preassemble components at a central shop to permit rapid on-site erection. While the small contractor building individual houses cannot always use the same cost-saving methods, he follows certain practices:

1. Power equipment, such as a radial-arm saw, skill saw, or an automatic nailer, aids in reducing the time required for framing and is used by most progressive contractors. Such equipment not only reduces assembly time for floor, wall, and roof framing and sheathing, but is helpful in applying siding and exterior and interior trim. For example, with a radial-arm saw on the job, studs can be cut to length, headers and framing members prepared, and entire wall sections assembled on the floor and raised in place. Square cuts, equal lengths, and accurate layouts result in better nailing and more rigid joints.

2. Where a gypsum-board dry-wall finish is used, many contractors employ the horizontal method of application. This brings the taped joint below eye level, and large room-size sheets may be used. Vertical joints may be made at window or door openings. This reduces the number of joints to be treated and results in a better-looking wall.

3. Staining and painting of the exterior and interior surfaces and trim are important. For example, one cost study of interior painting indicated that prestaining of jambs, stops, casing, and other trim before application would result in a substantial saving. These are normally stained or sealed after they have been fitted and nailed.

4. During construction, the advantages of a simple

plan and the selection of an uncomplicated roof will be obvious. There will be less waste by cutting joists and rafters, and erection will be more rapid than on a house where intricate construction is involved.

CHAPTER 58

PROTECTION AND CARE OF MATERIALS AT THE BUILDING SITE

Many building contractors arrange for the materials needed for a house to be delivered just before construction begins. Perhaps the first load, after the foundation has been completed, would include all materials required for the wood-floor system. A second load, several days later, would provide the materials for framing and sheathing the walls, and a third load for roof and ceiling framing and roof sheathing. In this manner, storage of framing and sheathing materials on the site would not be as critical as when all materials were delivered at once. On the other hand, materials for factory-built or preassembled houses may be delivered in one large truckload, because a crew erects the house in a matter of hours. This practically eliminates the need for protection of materials on the site.

Protection Requirements

Unfortunately, the builder of a single house may not be able to have delivery coincide with construction needs. Thus, some type of protection may be required at the building site. This is especially true for such millwork items as window and door frames, doors, and moldings. Finished cabinets, floor underlayment, flooring, and other more critical items should be delivered only after the house is enclosed and can provide complete protection from the weather.

During fall, winter, and spring months, the interior of the house should be heated so that finished wood materials will not be affected. Exposure to damp and cold conditions will change the dimensions of such materials as flooring and cause problems if they are installed at too high a moisture content. Thus, care of the materials after they arrive at the site and the conditions to which they are exposed are important to most materials used in house construction.

Protection of Framing Materials

In normal construction procedures, after excavation is complete, some dimension lumber and sheathing materials are delivered on the job. After delivery, it is the builder's responsibility to protect these materials against wetting and other damage. Structural and framing materials in place on a house before it is enclosed may become wet during a storm, but exposed surfaces can dry out quickly in subsequent dry weather without causing damage.

Lumber should *not* be stored in tight piles without some type of protection. Rather, if lumber is not to be used for several days or a week, it should be unloaded on skids with a 6-inch clearance above the soil. The pile should then be covered with waterproof paper, canvas, or polyethylene so that it sheds water. However, the cover should allow air to circulate and not enclose the pile to the groundline. In a tight enclosure, moisture from the ground may affect the moisture content of lumber. The use of a polyethylene cover over the ground before lumber is piled will reduce moisture rise. The same type of protection should be given to sheathing grade plywood.

After the framing and the wall and roof sheathing have been completed, the exterior roof trim, such as the cornice and rake finish, is installed. During this period, the shingles may have been delivered. Asphalt shingles should be stored so that bundles lie flat without bending; curved or buckled shingles often result in a poor looking roof. Wood shingles can be stored with only moderate protection from rain.

Window and Door Frames

Window and exterior door frames should not be delivered until they can be installed. In normal construction procedures, these frames are installed after the roof is completed and roofing installed. Generally, window units are ready for installation with sash and weatherstrip in place, and all wood protected by a dip treatment with a water-repellent preservative. Such units are premium items and, even though so treated, should be protected against moisture or mechanical damage. If it not possible to install frames when they arrive, place them on a dry base in an upright position and cover them.

Siding and Lath

Siding materials can be protected by storing temporarily in the house or garage. Place them so they will not be stepped on and split. Wood bevel siding is usually bundled with the pieces face to face to protect the surfaces fom mechanical damage and soiling. Some manufacturers treat their siding with a water-repellent material and pack in bundles with an outer protective wrap. All siding materials that cannot be installed immediately should be protected against exposure to conditions that could appreciably change their moisture content.

Insulation and rock lath should be stored inside the house. These materials are generally not installed until the electrical, heating, and plumbing trades have completed the roughing-in phases of their work.

Plastering in Cold Weather

During winter months, and in colder spring and fall weather in northern areas of the country, the heating unit of a house should be in operation before plastering is started. In fact, if the wood-framing members are much above 15 percent moisture content, it is good practice to let them dry out somewhat before rock lath is applied. This normally presents no problem, as the plumber and electrician do the rough-in work shortly after the house is closed in. Heat will allow the plaster to dry more readily, but because much moisture is driven off during this period, windows should be opened slightly.

Interior Finish

Millwork, floor underlayment, flooring, and interior trim manufactured by reputable companies are normally shipped at a moisture content satisfactory for immediate use. However, if storage conditions at the lumber company or in an unheated house during the inclement seasons are not satisfactory, wood parts will pick up moisture. Results may not be apparent immediately. If material is installed at too high a moisture content, the following heating season openings will show up between flooring strips and poorly matched joints in the trim because members have dried out and shrunk.

In flooring, for instance, the recommended moisture content at installation varies from 10 percent in the damp southern States to 6 and 7 percent for other localities. In examining wood floors with objectionable cracks between the boards, it has been found that in most cases the material had picked up moisture *after* manufacture and *before* it was installed. As such material redries during the heating season, it shrinks and the boards separate. Some of the moisture pickup may occur before the flooring is delivered to the building, but often such pickup occurs after delivery and before laying.

In an unheated building under construction, the relative humidity will average much higher than that in an occupied house. Thus, the flooring and finish tend to absorb moisture. To prevent moisture pickup at the building and to dry out any excess moisture picked up between time of manufacture and delivery, the humidity must be reduced below that considered normal in an unheated house. This may be accomplished by maintaining a temperature above the outdoor temperature even during the warmer seasons.

Before any floor underlayment, flooring or interior finish is delivered, the outside doors and windows should be hung and the heating plant installed to supply heat. For warm-weather control, when the workmen leave at night, the thermostat should be set to maintain a temperature of 15°F. above the average outdoor temperature. In the morning when the workmen return, the thermostat can be set back so that the burner will not operate. During the winter, fall, and spring, the temperature should be kept at about 60° F.

Several days before flooring is to be laid, bundles should be opened and the boards spread about so that their surfaces can dry out evenly. This will permit the drying of moisture picked up before delivery. Wood wall paneling and floor underlayment should also be exposed to the heated conditions of the house so the material will approach the moisture content it reaches in service. Actually, exposure of all interior finish to this period of moisture adjustment is good practice. Supplying some heat to the house in damp weather, even during the summer months, will be justified by improved appearance and owner satisfaction.

CHAPTER 59

MAINTENANCE AND REPAIR

A well constructed house will require comparatively little maintenance if adequate attention was paid to details and to choice of materials, as presented in previous chapters. A house may have an outstanding appearance, but if construction details have not been correct, the additional maintenance that might be required would certainly be discouraging to the owner. This may mean only a little attention to some apparently unimportant details. For example, an extra $10 spent on corrosion-resistant nails for siding and trim may save $100 or more annually because of the need for less frequent painting. The use of edge-grained rather than flat-grained siding will provide a longer paint life, and the additional cost of the edge-grained boards then seems justified.

The following sections will outline some factors

relating to maintenance of the house and how to reduce or eliminate conditions that may be harmful as well as costly. These suggestions can apply to both new and old houses.

Basement

The basement of a poured block wall may be damp for some time after a new house has been completed. However, after the heating season begins, most of this dampness from walls and floors will gradually disappear if construction has been correct. If dampness or wet walls and floors persist, the owner should check various areas to eliminate any possibilities for water entry.

Possible sources of trouble:

1. Drainage at the downspouts. The final grade around the house should be away from the building and a splash block or other means provided to drain water away from the foundation wall.

2. Soil settling at the foundation wall and forming pockets in which water may collect. These areas should be filled and tamped so that surface water can drain away.

3. Leaking in a poured concrete wall at the form tie rods. These leaks usually seal themselves, but larger holes should be filled with a cement mortar or other sealer. Clean and slightly dampen the area first for good adhesion.

4. Concrete-block or other masonry walls exposed above grade often show dampness on the interior after a prolonged rainy spell. A number of waterproofing materials on the market will provide good resistance to moisture penetration when applied to the inner face of the basement wall. If the outside of below-grade basement walls is treated correctly during construction, waterproofing the interior walls is normally not required. (See Ch. 26, "Foundation Walls and Piers.")

5. There should be at least a 6-inch clearance between the bottom of the siding and the grass. This means that at least 8 inches should be allowed above the finish grade before sod is laid or foundation plantings made. This will minimize the chance of moisture absorption by siding, sill plates, or other adjacent wood parts. Shrubs and foundation plantings should also be kept away from the wall to improve air circulation and drying. In lawn sprinkling, it is poor practice to allow water to spray against the walls of the house.

6. Check areas between the foundation wall and the sill plate. Any openings should be filled with a cement mixture or a calking compound. This filling will decrease heat loss and also prevent entry of insects into the basement, as well as reduce air infiltration.

7. Dampness in the basement in the early summer months is often augmented by opening the windows for ventilation during the day to allow warm, moisture-laden outside air to enter. The lower temperature of the basement will cool the incoming air and frequently cause condensation to collect and drip from cold-water pipes and also collect on colder parts of the masonry walls and floors. To air out the basement, open the windows during the night.

Perhaps the most convenient method of reducing humidity in basement areas is with *dehumidifiers*. A mechanical dehumidifier is moderate in price and does a satisfactory job of removing moisture from the air during periods of high humidity. Basements containing living quarters and without air conditioners may require more than one dehumidifier unit. When they are in operation, all basement windows should be closed.

Crawl-space Area

Crawl-space areas should be checked as follows:

1. Inspect the crawl-space area annually, for termite activity. Termite tubes on the walls or piers are an indication of this. In termite areas, soil in the crawl space or under the concrete slab is normally treated with some type of chemical to prevent termite damage. A house should contain a termite shield under the wood sill with a 2-inch extension on the interior. It must be well installed to be effective. Examine the shield for proper projection, and also any cracks in the foundation walls, as such cracks form good channels for termites to enter (Ch. 55, "Protection Against Decay and Termites").

2. While in the crawl space, check exposed wood joists and beams for indications of excessive moisture. In older houses where soil covers had not been used in the past, signs of staining or decay may be present. Use a penknife to test questionable areas.

3. Soil covers should be used to protect wood members from ground moisture (Ch. 42, "Ventilation"). These may consist of plastic films, roll roofing, or other suitable materials. A small amount of ventilation is desirable to provide some air movement. If the crawl space is not presently covered, install a barrier for greater protection.

Roof and Attic

The roof and the attic area of both new and older houses might be inspected with attention to the following:

1. Humps which occur on an asphalt-shingle roof are often caused by movement of roofing nails which have been driven into knots, splits, or unsound wood. Remove such nails, seal the holes, and replace the nails with others driven into sound wood. Blind-nail such replacements so that the upper shingle tab covers the nailhead.

A line of buckled shingles across the roof of a relatively new house is often caused by shrinkage of wide roof boards. It is important to use sheathing boards not over 8 inches wide and at a moisture content not exceeding 12 to 15 percent. When moisture content is greater, boards should be allowed to dry out for several days before shingles are applied. Time and hot weather tend to reduce buckling. Plywood sheathing would eliminate this problem altogether.

2. A dirt streak down the gable end of a house with a close rake section can often be attributed to rain entering and running under the edge of the shingles. This results from insufficient shingle overhang or the lack of a metal roof edge, such as shown in figure 34.3b. The addition of a flashing strip to form a drip edge will usually minimize this problem.

3. In winters with heavy snows, ice dams may form at the eaves, often resulting in water entering the cornice and walls of the house. The immediate remedy is to remove the snow on the roof for a short distance above the gutters and, if necessary, in the valleys. Additional insulation between heated rooms and roof space, and increased ventilation in the overhanging eaves to lower the general attic temperature, will help to decrease the melting of snow on the roof and thus minimize ice formation. Deep snow in valleys also sometimes forms ice dams that cause water to back up under shingles and valley flashing (fig. 34.1a and b).

4. Roof leaks are often caused by improper flashing at the valley, ridge, or around the chimney. Observe these areas during a rainy spell to discover the source. Water may travel many feet from the point of entry before it drips off the roof members.

5. The attic ventilators are valuable year round; in summer, to lower the attic temperature and improve comfort conditions in the rooms below; in winter, to remove water vapor that may work through the ceiling and condense in the attic space and to minimize ice dam problems. The ventilators should be open both in winter and summer.

To check for sufficient ventilation during cold weather, examine the attic after a prolonged cold period. If nails protruding from the roof into the attic space are heavily coated with frost, ventilation is usually insufficient. Frost may also collect on the roof sheathing, first appearing near the eaves on the north side of the roof. Increase the size of the ventilators or place additional ones in the soffit area of the cornice. This will improve air movement and circulation. (See Ch. 42, "Ventilation," and figs. 42.1, 42.2, and 42.3 for proper size and location.)

Exterior Walls

One of the maintenance problems which sometimes occurs with a wood-sided house involves the exterior paint finish. Several reasons are known for peeling and poor adherence of paint. One of the major ones perhaps can be traced to moisture in its various forms. Paint quality and methods of application are other reasons. Another factor involves the species of wood and the direction of grain. Some species retain paint better than others, and edge grain provides a better surface for paint than flat grain. Chapter 54, "Painting and Finishing," covers correct methods of application, types of paint, and other recommendations for a good finish. Other phases of the exterior maintenance that the owner may encounter with his house are as follows:

1. In applying the siding, if bright steel nails have been used rather than galvanized, aluminum, stainless steel, or other noncorrosive nails, rust spots may occur at the nailhead. These spots are quite common where nails are driven flush with the heads exposed. The spotting may be remedied somewhat, in the case of flush nailing, by setting the nailhead below the surface and puttying. The puttying should be preceded by a priming coat.

2. Brick and other types of masonry are not always waterproof, and continued rains may result in moisture penetration. Masonry veneer walls over a sheathed wood frame are normally backed with a waterproof sheathing paper to prevent moisture entry. When walls do not have such protection and the moisture problem persists, use a waterproof coating over the exposed masonry surfaces. Transparent waterproof materials can be obtained for this purpose.

3. Calking is usually required where a change in materials occurs on a vertical line, such as that of wood siding abutting against brick chimneys or walls. The wood should normally have a prime coating of paint for proper adhesion of the calking compound. Calking guns with cartridges are the best means of waterproofing these joints. Many permanent-type calking materials with a neoprene, elastomer or other type base are available.

4. Rainwater may work behind wood siding through butt and end joints and sometimes up under the butt edge by capillarity when joints are not tight. Setting the butt and end joints in white lead is an old-time custom that is very effective in preventing water from entering. Painting under the butt edges at the lap adds mechanical resistance to water ingress. However, moisture changes in the siding cause some swelling and shrinking that may break the paint film. Treating the siding with a water repellent before it is applied is effective in reducing capillary action. For houses already built, the water repellent could be applied under the butt edges of bevel siding or along the joints of drop siding and at all vertical joints. Such water repellents are often combined with a preservative and can be purchased at your local paint dealers as a water-repellent preservative. In-place application is often done with a plunger-type oil can. Excess repellent on the face of painted surfaces should be wiped off.

Interior

Plaster

The maintenance of plastered interior surfaces normally is no problem in a properly constructed house. However, the following points are worthy of attention:

1. Because of the curing (aging) period ordinarily required for plastered walls, it is not advisable to apply oil-base paints until at least 60 days after plastering is completed. Water-mix or resin-base paints may be applied without the necessity of an aging period.

2. In a newly constructed house, small plaster cracks may develop during or after the first heating season. Such cracks are ordinarily caused by the drying of framing members that had too high a moisture content when the plaster was applied. The cracks usually occur at interior corners and also above windows and doors because of shrinkage of the headers. For this reason, it is usually advisable to wait for a part of the heating season before painting plaster so that such cracks can be filled first.

3. Large plaster cracks in houses, new or old, often indicate a structural weakness in the framing or column footings. This may be due to excessive deflection or to settling of beam supports. Common areas for such defects might be along the center beam or around the basement stairway. In such cases, the use of an additional post and pedestal may be required. (See **Ch. 28,** "Floor Framing," for recommended methods of framing.)

Moisture on Windows

Moisture on inside surfaces of windows may often occur during the colder periods of the heating season. The following precautions and corrections should be observed during this time:

1. During cold weather, condensation and, in cold climates, frost will collect on the inner face of single-glazed windows. Water from the condensation or melting frost runs down the glass and soaks into the wood sash to cause stain, decay, and paint failure. The water may rust steel sash. To prevent such condensation, the window should be provided with a storm sash. Double glazing will also minimize this condensation. If it still presists on double-glazed windows, it usually indicates that the humidity is too high. If a humidifier is used, it should be turned off for a while or the setting lowered. Other moisture sources should also be reduced enough to remedy the problem. Increasing the inside temperature will also reduce surface condensation.

2. Occasionally, in very cold weather, frost may form on the inner surfaces of the storm windows. This may be caused by (a) loose-fitting window sash that allows moisture vapor from the house to enter the space between the window and storm sash, (b) high relative humidity in the living quarters, or (c) a combination of both. Generally, the condensation on storm sash does not create a maintenance problem, but it may be a nuisance. Weather-stripping the inner sash offers resistance to moisture flow and may prevent this condensation. Lower relative humidities in the house are also helpful.

Problems with Exterior Doors

Condensation may occur on the glass or even on the interior surface of exterior doors during periods of severe cold. Furthermore, warping may result. The addition of a tight-fitting storm or combination door will usually remedy both problems. A solid-core flush door or a panel door with solid stiles and rails is preferred over a hollow-core door to prevent or minimize this warping problem.

Openings in Flooring

Laying finish-strip flooring at too high a moisture content or laying individual boards with varying moisture contents may be a source of trouble to the homeowner. As the flooring dries out and reaches moisture equilibrium, spaces will form between the boards. These openings are often very difficult to correct. If the floor has a few large cracks, one expedient is to fit matching strips of wood between the flooring strips and glue them in place. In severe cases, it may be necessary to replace sections of the floor or to refloor the entire house.

Another method would be to cover the existing flooring with a thin flooring $5/16$ or $3/8$ inch thick. This would require removal of the base shoe, fitting the thin flooring around door jambs, and perhaps sawing off the door bottoms. (For proper methods of laying floors to prevent open joints in new houses, see Ch. **46,** "Floor Coverings.")

Unheated Rooms

To lower fuel consumption and for personal reasons, some homeowners close off unused rooms and leave them unheated during the winter months. These factors of low temperatures and lack of heat, unfortunately, are conducive to trouble from condensation. Certain corrective or protective measures can be taken to prevent damage and subsequent maintenance expense, as follows:

1. Do not operate humidifiers or otherwise intentionally increase humidity in heated parts of the house.

2. Open the windows of unheated rooms during bright sunny days for several hours for ventilation. Ventilation will help draw moisture out of the rooms.

3. Install storm sash on all windows, including those in unheated rooms. This will materially reduce heat loss from both heated and unheated rooms and will minimize the condensation on the inner glass surfaces.

LITERATURE CITED

(1) American Institute of Timber Construction
   1965. Standard for heavy timber roof decking. AITC 112. Washington, D.C.

(2) American Plywood Association
   1964. Plywood truss designs, APA 64–650 11 pp., illus.

(3) ———
   1967. Plywood in apartments. APA 67–310.

(4) American Society of Heating, Refrigerating, and Air-Conditioning Engineers
   1967. ASHRAE handbook of fundamentals. 544 pp., illus.

(5) Anderson, L. O.
   1967. Selection and use of wood products for home and farm building. U.S. Dep. Agr., Agr. Inform. Bull. 311, 41 pp., illus.

(6) ———
   1969. Low-cost wood homes for rural America—construction manual. Agr. Handb. 364, 112 pp.

(7) ———, and Smith, W,
   1965. Houses can resist hurricanes. U.S. Forest Serv. Res. Pap. FPL 33. Forest Prod. Lab., Madison, Wis., 44 pp. illus.

(8) Berendt, R.D., and Winzer, G.E.
   1954. Sound insulation of wall, floor, and door constructions. Nat. Bur. of Stand. Monogr. No. 77, 49 pp.

(9) Federal Housing Administration
   1961. Mat-formed particleboard for exterior use. FHA Use of Materials Bull. No. UM-32. June 19.

(10) ———
   1963. A guide to impact noise control in multifamily dwellings. FHA No. 750, 86 pp.

(11) ———
   1964. Minimum property standards for one and two family living units. FHA No. 300, 315 pp.

(12) Forest Products Laboratory, Forest Service, U.S. Dept. of Agriculture.
   1955. Wood handbook. U.S. Dep. Agr., Agr. Handb. 72, 528 pp., illus.

(13) ———
   1964. FPL natural finish. U.S. Forest Serv. Res. Note FPL–046, Madison, Wis., 5 pp.

(14) Insulation Board Institute
   1963. Noise control with insulation board. Chicago, Ill., 15 pp., illus.

(15) Lewis, Wayne C.
   1968. Thermal insulation from wood for buildings: Effects of moisture and its control. U.S.D.A. Forest Serv. Res. Pap. FPL 86. Forest Prod. Lab., Madison, Wis.

(16) National Fire Protection Association
   1962. Occupancy fire record: One-and two-family dwellings. Fire Rec. Bull. FR56–2A, Boston, Mass., 20 pp., illus.

(17) Teesdale, L.V.
   1955. Remedial measures for building condensation difficulties. U.S. Forest Prod. Lab. Rep. 1710, 15 pp., illus.

(18) U.S. Department of Commerce
   1966. Softwood plywood—construction and industrial. U.S. Prod. Stand. PS 1–66, 28 pp., illus.

(19) Verrall, A.F.
   1961. Brush, dip, and soak treatment with water-repellent preservatives. Forest Prod. J. 11(1): 23–26.

HOUSE PLAN no. 1

This house was developed to provide a good livable home for a cost much lower than most houses now being constructed. It is 24 by 32 feet in size and contains 768 square feet of living area. In spite of its relatively small size, this home has three bedrooms, affording desirable privacy for a family with 3 to 5 children. There is little waste space and the bath and kitchen, as well as the living room and dining area, are conveniently arranged.

AREA - 768 sq. ft.

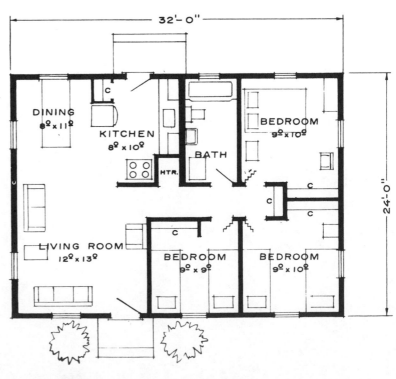

This crawl-space house has a treated wood post foundation which reduces the cost without reducing its performance. It can be constructed on sloping sites without costly grading and masonry work. Most of the materials used can be obtained at local lumber yards or small local mills.

The panel siding exterior and trim are finished with long-lasting stains which can be obtained in many contrasting colors. The wide overhangs at the cornice and gable ends provide a good appearance as well as excellent protection for the side walls. Insulation in the walls, floors, and ceiling reduce heating costs as well as providing a cool house during the hot summer months. The open living-dining-kitchen area gives a feeling of spaciousness not possible when walls separate these rooms.

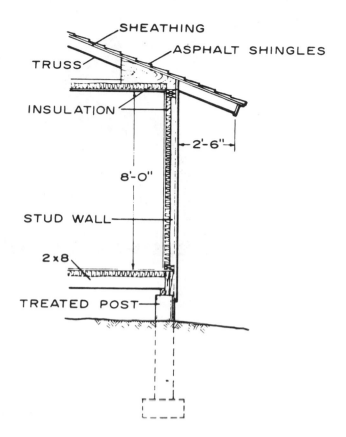

SHEATHING
ASPHALT SHINGLES
TRUSS
INSULATION
2'-6"
8'-0"
STUD WALL
2x8
TREATED POST

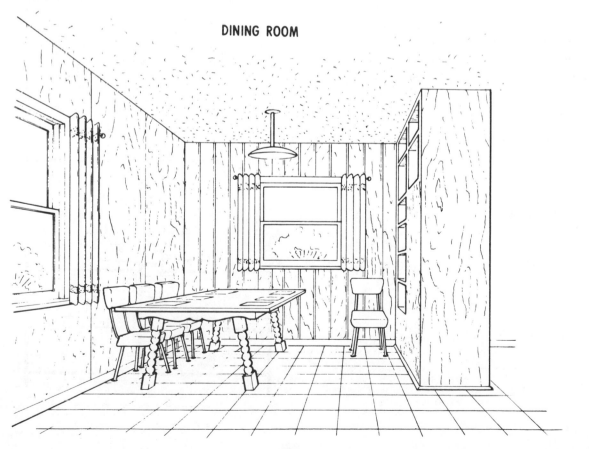

DINING ROOM

HOUSE PLAN no. 2

This home (Plan 2) was developed for a large family of up to 12 children at a reasonable cost. It is 24 by 36 feet in size and is one and one-half stories. The first floor has 864 square feet, consisting of three bedrooms, a bath, and a living-dining-kitchen area. The second floor contains about 540 square feet and consists of two large dormitory-type bedrooms. Each is divided by a wardrobe-type closet which, in effect, contains space for two single beds on each side.

AREA = 1404 sq. ft.

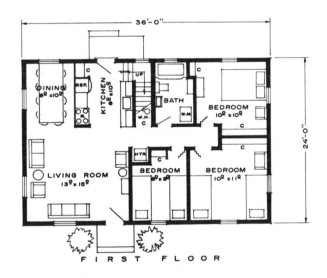

FIRST FLOOR

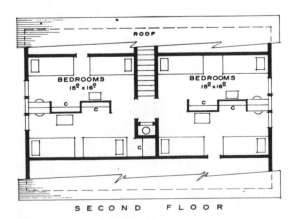

SECOND FLOOR

The design features which aid in reducing the cost and in providing maximum space for the overall size are (a) the treated post foundation with a crawl space and (b) the steeply sloped roof. The long-lived treated foundation posts can be installed with little or no costly grading and leveling. They serve as a rugged base for the beams and joists of the floor system. The one-half pitch (12 in 12 slope) roof with a 4-foot-high knee wall encloses two large 15- by 16-foot dormitory bedrooms. Windows are installed at the gable ends of the second floor.

The panel siding on exterior walls and the trim and shutters are finished with a pigmented stain which can be obtained in a variety of colors for contrast. The wide overhangs at both the gable ends and the cornice provide desirable protection to the side and end walls. Floors, walls, and ceiling are insulated to reduce heat loss. The thickness of this insulation can be varied, and the amount usually depends on whether the house is located in the northern part of the United States or in a milder climate.

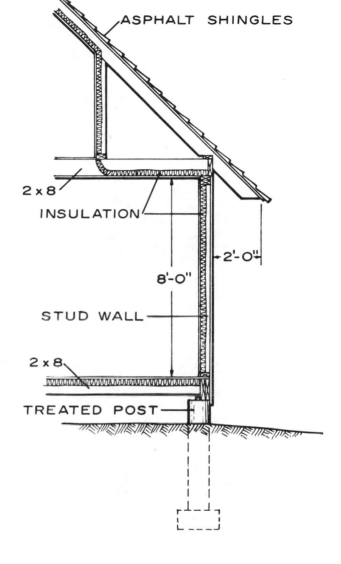

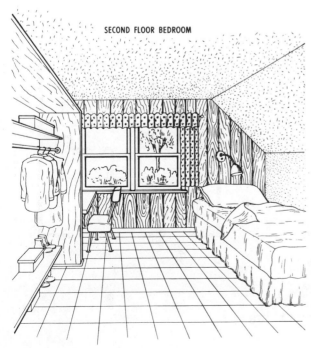

SECOND FLOOR BEDROOM

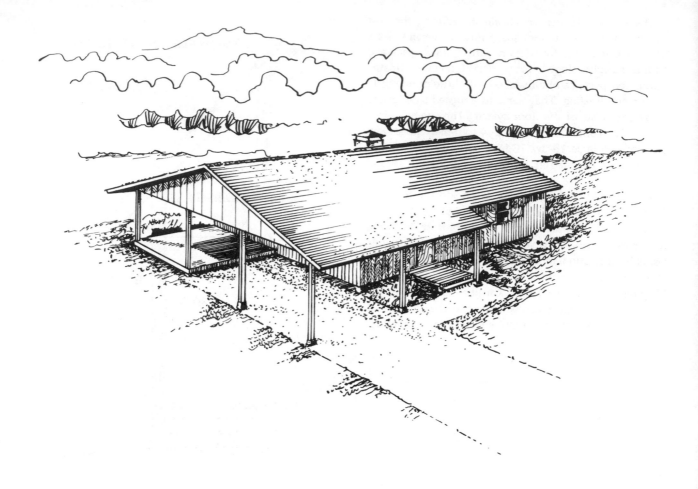

HOUSE PLAN no. 3

This house plan (3) was developed to serve a small family or senior citizens who require only one main bedroom and a spare room which can serve as a sewing room, workroom, or also as a second bedroom. The kitchen and living room are open with a drop beam between them. The main house is 24 by 24 feet in size and has 576 square feet of living area. As shown in the plan, an 8- by 12-foot porch is included as well as a carport. When a minimum house is desired, the porch and the carport can be eliminated.

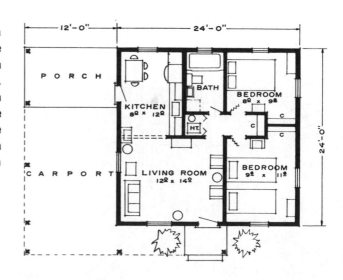

AREA - 576 sq. ft.

This crawl space house has a treated wood post foundation which reduces the cost of the house without reducing its performance. This type of house can be constructed on a sloping lot, eliminating costly grading which is normally necessary for houses with other designs.

The floors, walls, and ceilings are well insulated, insuring comfort both winter and summer. The small forced-air heating unit can be either the oil or the gas-fired type. The bathroom contains space for a small washing machine. Closet space is sufficient for a small family. More storage area can be provided by the installation of small plywood wardrobes in the bedrooms.

The panel siding and other exterior wood can be finished with a pigmented stain. These materials are not only easily applied, but they are long lasting. Many colors are available. The interior has a gypsum board wall and ceiling finish which are painted. In addition, the bath and part of the living room walls have prefinished plywood for a desirable contrast.

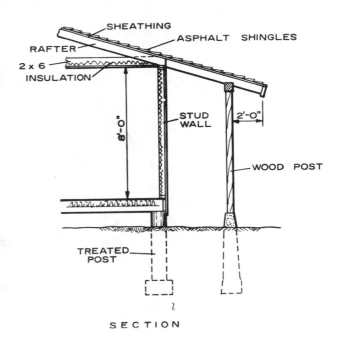

SECTION

KITCHEN

HOUSE PLAN no. 4

AREA = 768 sq. ft. expandable to 1228 sq. ft.

This home (Plan 4) is an expandable type. With its steeply pitched roof, there is more than adequate space on the second floor for two dormitory-type bedrooms, which can accommodate up to eight children. The working drawings also provide for an additional bath on the second floor if it is desired. The house is 24 by 32 feet in size with an area of 768 square feet on the first floor and about 460 square feet of usable space on the second. The first floor contains a moderate-size living room, a compact kitchen with a large adjoining dining area, two bedrooms, and a bath. Storage space is adequate with four closets on the first floor and five on the second. The first-floor bath is arranged to accommodate a washing machine.

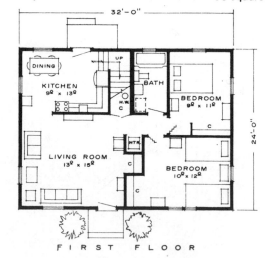

FIRST FLOOR

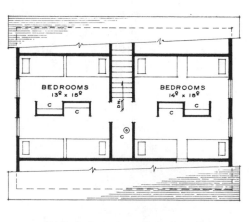

SECOND FLOOR

There are several important factors which aid in reducing the cost of this home. One is the fact that it is a crawl-space house, which eliminates the need for extensive excavation and grading. In addition, the floor framing is supported by long-lived treated wood foundation posts resting on concrete footings. A more costly masonry foundation is included in the working drawings as an alternate. The use of a single covering material for the subfloor and the exterior walls also leads to reduced costs. The subfloor consists of tongued-and-grooved plywood or square-edge plywood with edge blocking and serves as a base for a resilient floor covering. The panel siding, with perimeter nailing, eliminates the need for corner bracing as well as the need for sheathing. Such coverings are usually rough-textured exterior-grade plywood, which can be finished with a pigmented stain. Suitable stains are available in many colors, and contrasts can be obtained by treating the trim and shutters with a different color or shade.

Further cost reductions are obtained by eliminating much of the exterior trim as well as some of the less important interior millwork. However, these refinements can be added in the future. An adequate forced-air heating unit with relatively short heat runs is also a part of the design. Insulation is included in the floor, wall, and ceiling areas; the thickness selected depending on the location of the home. In the colder climates, the ceiling and floor insulation might be 4 inches or thicker with 2-inch-thick blanket insulation in the walls.

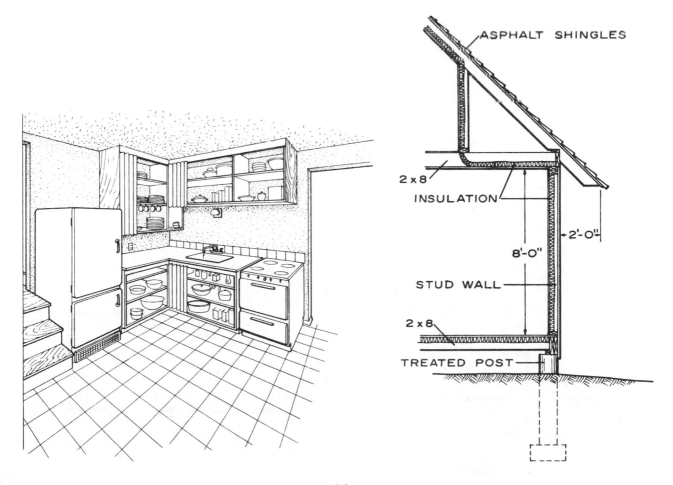

HOUSE PLAN no. 5

AREA = 672 sq. ft. expandable to 1042 sq. ft.

Plan 5 provides for the construction of either a single-story two-bedroom home or an expandable type with two additional bedrooms on the second floor. The basic house is 24 by 28 feet in size with an area of 672 square feet. The expandable plan provides an additional area of about 370 square feet on the second floor. The second-floor bedrooms can be completed with the rest of the house or left unfinished until later. Both plans include a kitchen, bath, living room, and two bedrooms on the first floor.

To accommodate the stairway to the second-floor bedrooms in the expandable plan, the first-floor bed-

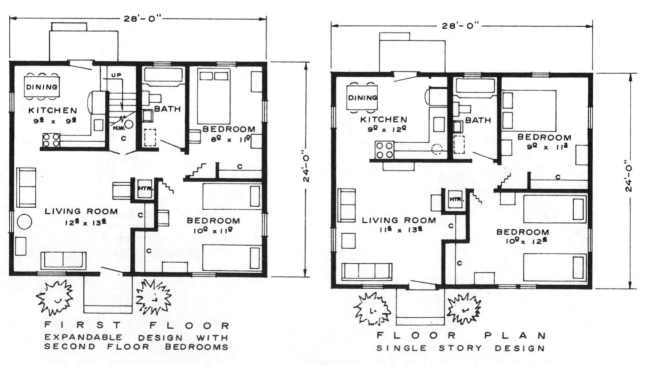

FIRST FLOOR
EXPANDABLE DESIGN WITH
SECOND FLOOR BEDROOMS

FLOOR PLAN
SINGLE STORY DESIGN

rooms and the kitchen are slightly smaller than those in the one-story plan. One second-floor bedroom is 9-1/2 by 14 feet and the other is 13 by 14 feet. The larger bedroom may be divided by the addition of a wardrobe wall, which provides two closets and also serves as a room divider.

Both plans have a front entrance closet and a closet for each bedroom. The expandable house also has a storage area under the stairway in which the hot water heater is located. The heating unit is located in a small closet adjacent to the bath-bedroom hallway. Walls, floors, and ceiling areas are insulated. Dining space is provided for in each kitchen.

There are a number of factors which aid in reducing the cost of these homes. They are designed as crawl-space houses with post or pier foundations, which eliminate the need for extensive grading on sloping building sites. The single floor covering serves as a base for resilient tile or a low-cost linoleum rug. It can be painted if further cost reductions are required. Panel siding is used for exterior wall finish which eliminates sheathing and the need for a braced wall. Exteriors are finished with long-lasting pigmented stains. Many contrasting colors are available in this type of finish. Exterior and interior trim and millwork have been reduced to a minimum. However, many of these refinements can be made at anytime after the house has been completed.

AREA = 1024 sq. ft.

HOUSE PLAN no. 6

This design, intended for a flat site, encloses 1,024 square feet. The square plan provides much more usable space, within the same exterior walls, than a typical rectangular plan.

While essentially conventional, the design features a novel floating-wood-floor system. The space under the floor serves as a return-air plenum from a centrally located forced-air furnace. Air will flow into each room through the opening above the door, then enter a space behind the baseboard for return under the floor to the furnace. Alternate designs are available for a conventional wood floor over a crawl space and for a concrete slab floor.

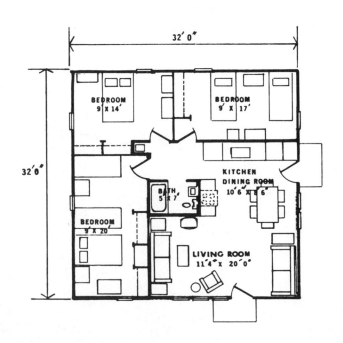

Large bedrooms, closets, built-in desks, and shelving in the bedrooms are bonus features. These bedrooms are really multi-purpose rooms for play, study and sleeping.

Economical, nonbearing partition walls of particle board have replaced conventional hollow-core interior walls. Being nonbearing, they can be located in various positions without affecting the structure.

Plywood combination siding and sheathing on the exterior walls can be finished with modern natural finishes and stains for durability and easy maintenance.

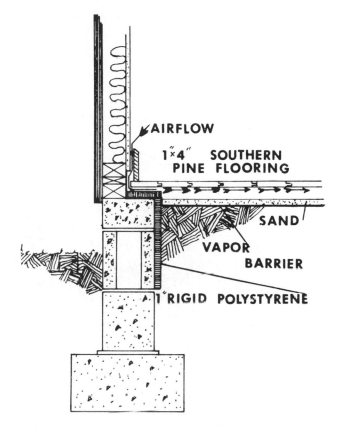

AIRFLOW

1"×4" SOUTHERN PINE FLOORING

SAND

VAPOR BARRIER

1" RIGID POLYSTYRENE

BUNK-BEDROOM

SHOWING CLOSET/DESK AREA

HOUSE PLAN no. 7

AREA = 1008 sq. ft.

Wood piers are used for the foundation to make this plan particularly suited to sloping sites. The piers are pressure treated with a clean, non-leaching preservative. A carport and storage area under the house might be considered for steeply sloping sites.

The play area is a bonus feature in a three-bedroom house of this size. Elimination of a central hall has provided more usable living space in the 1,008 square feet of floor area. The large bedrooms will also provide extra, multi-purpose space for large families.

Conventional framing, stock-sized windows and doors, and the simple rectangular plan were selected with the small contractor in mind. Rafters and joists, instead of roof trusses, help to open up the attic space for convenient storage. Plywood combination sheathing and siding on exterior walls is finished with a natural finish or stain to provide atrractive, economical, easy-to-

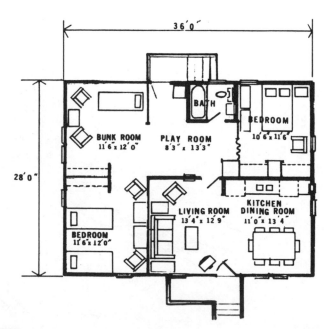

maintain exterior walls. The skirting could be placed around the foundation piers in cold climates to lower the heating bill. If this is done, the floor insulation and insulation board under the floor joists could be eliminated.

Heat is provided by a centrally located forced-air wall furnace, or by electric heaters.

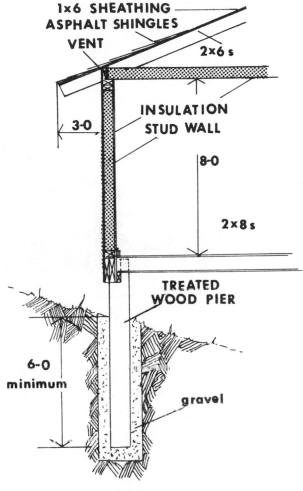

PLAYROOM and BATH — UTILITY AREA

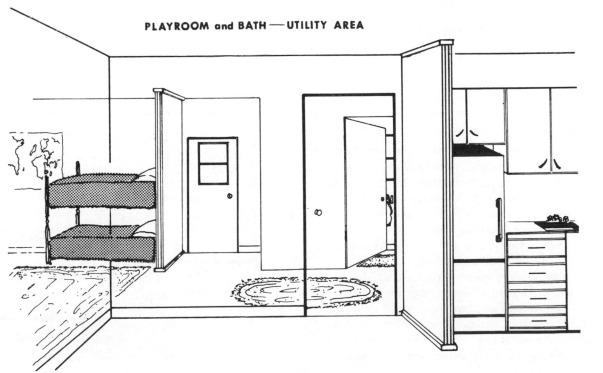

AREA = 1008 sq. ft.

HOUSE PLAN no. 8

Strong, durable, pressure-treated wood poles support this house, which is particularly suitable for sloping sites.

This attractive house was designed expecially for large families wanting a low-cost three-bedroom home. The bedrooms and dining room will accommodate a large family without crowding. A utility room has been provided next to the bathroom, and the laundry tub will be useful as a second lavatory. The sizable open space of the kitchen-dining-living area will make this house seem much larger than its 1008 square feet.

When this house is constructed on a sloping site, a carport and storage room under the house are bonus features. The furnace and hot water heater will usually be located in the storage room under the house; however, if the site is level, the furnace could be in the linen closet location, and the under-counter type of hot

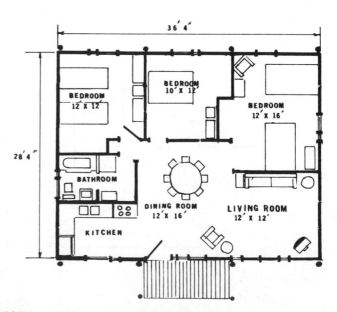

water heater could be installed in the kitchen.

This pole-truss structural system has several advantages over conventional foundation and framing systems. It is economical, very little site grading will be needed, and construction need not be delayed by weather or frozen ground. In addition, all walls are non-load bearing. This makes all walls easy to prefabricate and erect, and it provides design flexibility in that the interior walls can be easily moved to alternate locations.

Since all of the walls, except those around the plumbing core, are prefabricated in 4 x 8 foot panels, window and door panels have also been designed for use with the wall panels. This prefabricated panel system of construction can be used very effectively with pole-frame construction.

Southern pine flooring is used throughout the house. A single layer of tongue and groove, 1 x 4 inch strip flooring is fastened to the joists with nails and adhesive. This provides a strong, stiff, and durable floor. Inlaid linoleum protects the floor in the bath and utility area.

Plywood combination siding and sheathing, finished with a natural finish or stain, provides attractive exterior walls, that are economical and easy to maintain.

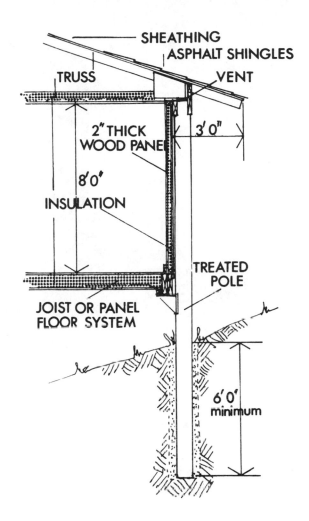

KITCHEN - DINING AREA
CUT OUT TO SHOW UTILITY AND BATH AREA

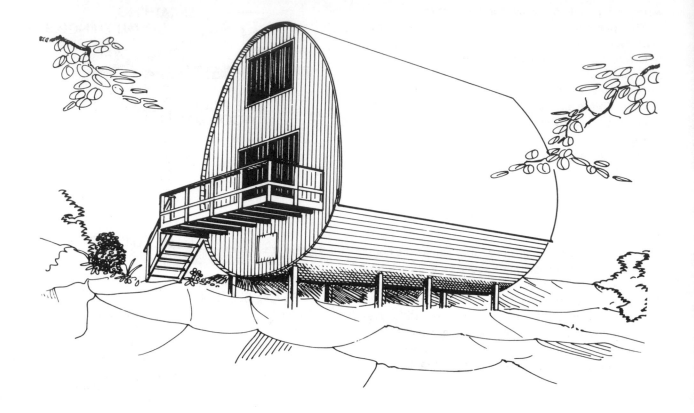

HOUSE PLAN no. 9

A TUBULAR HOME OF WOOD

This unusual home offers attractive living space within its curved walls. It is primarily intended for sloping sites in rural areas. It also has particular advantages for second homes on lakeside or ocean front sites. Very little site disturbance is necessary for construction. This home provides a total of 1,000 square feet of floor area, with either two or three bedrooms on the second floor. The estimated cost of this home is about one-half the cost of equivalent conventional construction. Lumber and plywood requirements are also about half of such conventional construction.

Foundation System

Eight preservative-treated 6x6 square posts support this home. Treated round poles could be used instead. The posts extend up to the second floor and are bolted to the laminated ribs and the floor beams. Embedment and footing requirements will depend on local soil conditions and the height of the house above the ground.

Exterior Curved Walls

Eight egg-shaped, glue-laminated wood ribs form the curved exterior walls. The ribs are manufactured in half sections. They should be produced by an experienced laminator. A complete rib is located at the front and back faces of each pair of foundation posts. The ribs are bolted to the posts and to the ends of each floor beam.

Tongue-and-grooved 2x6 wood decking is nailed to the ribs around the entire circumference of the house. The decking is covered on the outside with foamed-in-place rigid polyurethane down to the first floor level. The foam provides excellent thermal insulation and a water-tight roof covering. It is painted on the outside with one or two coats of aluminum-filled asphalt emulsion, or a roof paint of similar quality. Below the first floor, the foam is placed on the inside of the decking to insulate and seal the crawl-space. If a forced-air furnace is used, the sealed crawl-space can serve as an air plenum. The crawl-space also provides a large storage area, accessible from the outside.

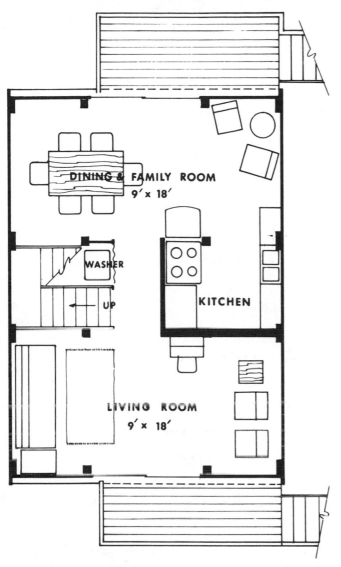

FIRST FLOOR

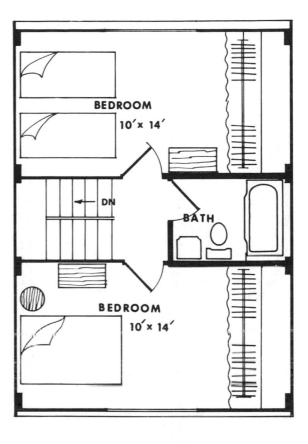

SECOND FLOOR

Exterior End Walls

Exterior 3/8-inch plywood is placed over 2x4 framing to enclose the end walls. These walls will resist racking loads and stiffen the entire structure. Interior finish is 3/8-inch gypsum board, or 1x6 wood paneling placed horizontally. Conventional insulation is placed in the walls. The frame members are installed flatwise so that they can be face nailed to the ribs and floor joists. A cantilevered deck is provided at each end of the home by extending the first-floor joists through the end walls. Joists on the second floor can also be extended for an upper deck if desired.

Floor System

Single layer wood-strip flooring is placed directly on the floor joists over a 1/4-inch bead of construction adhesive and nailed at each joist location. Construction adhesive will insure a stiffer floor and help to eliminate squeaking. The joists are exposed, and the lower side of the flooring on the second floor is the ceiling of the first floor.

Interior Partitions

Single-layer, self-supporting, panels of particleboard or plywood are used to partition interior spaces. Tongue-and-grooved solid wood paneling can also be used. Partitions are installed without framing except where a hollow wall is needed to enclose wiring, pipes, or ducts.

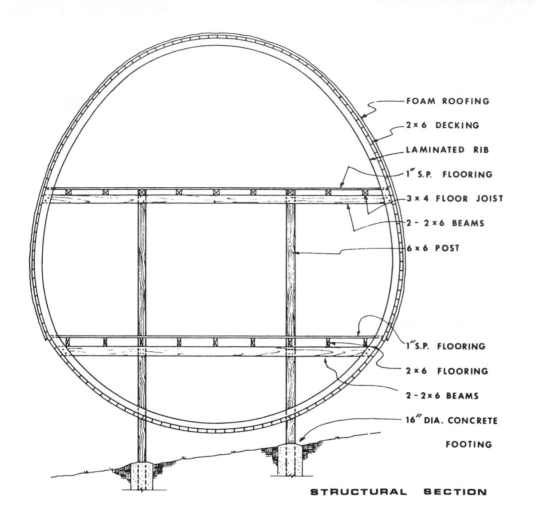

FOAM ROOFING

2 × 6 DECKING

LAMINATED RIB

1" S.P. FLOORING

3 × 4 FLOOR JOIST

2 – 2 × 6 BEAMS

6 × 6 POST

1" S.P. FLOORING

2 × 6 FLOORING

2 – 2 × 6 BEAMS

16" DIA. CONCRETE

FOOTING

STRUCTURAL SECTION

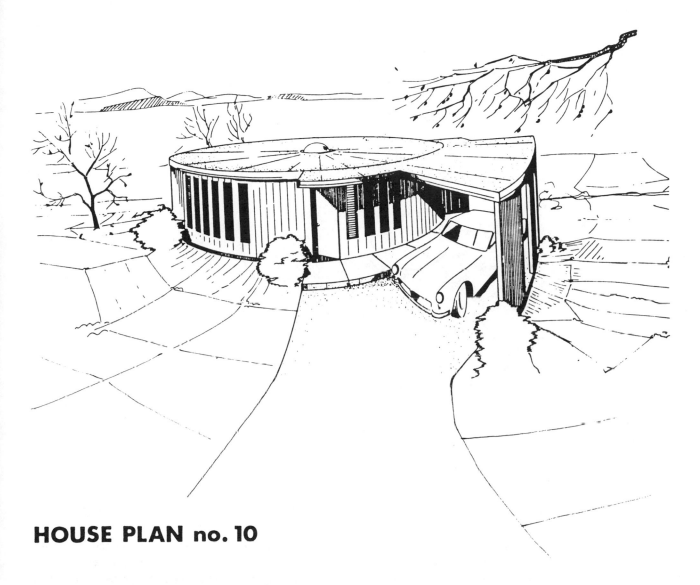

HOUSE PLAN no. 10

A ROUND HOUSE OF WOOD

This unique design provides a three-bedroom home with 1,134 square feet of living area. It is designed for a flat site. A smaller version provides three bedrooms and a total area of 804 square feet.

General Background

Round homes are an efficient means of providing housing space. In this design, interior walls are spaced radially from a central atrium hall. The design permits good arrangements of rooms and furniture.

AREA = 1134 sq. ft.

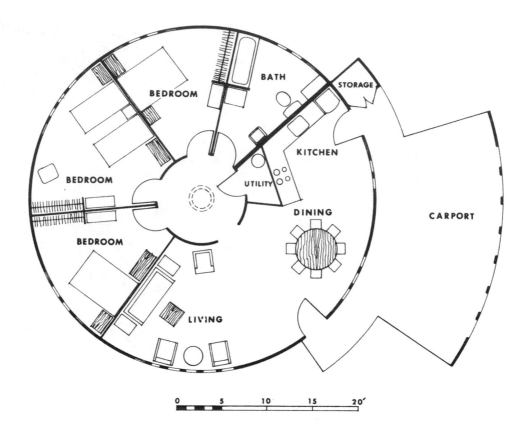

BEDROOM · BATH · STORAGE · BEDROOM · KITCHEN · UTILITY · DINING · CARPORT · BEDROOM · LIVING

0   5   10   15   20'

It is estimated the round house will cost about half as much as a conventional house with an equal amount of floor space. The house should be easy to build and suitable for self-help programs.

Floor System

A circular concrete slab is placed within a low brick foundation wall. The perimeter insulation and vapor barrier are conventional. Where needed, the soil is treated or poisoned to prevent attack by termites and other insects. A preservative-treated wood member is fixed around the edge of the slab. The ends of the exterior plank walls are nailed to this member.

Partition Walls

The interior partition walls are particleboard panels located under the roof beams and fitted into slots in 2x2

members at the vertical joints. The panels can be moved to alternate locations to provide two, three, or four bedrooms. Closet shelving and sidewalls serve to stiffen the particleboard walls. The walls may be finished with conventional wall paints or with natural finishes, either pigmented or clear.

Roof System

The roof system, which is essentially flat, consists of radially placed 4x6 beams, rough sawn or finished, inserted into slots cut in the tops of the perimeter planks and those around the atrium. The beams are covered with 1x6 tongue-and-grooved lumber decking in a herringbone pattern. The lower surface of the decking and the exposed beams are finished with natural stains. The roof surface of the decking is covered with a 1-inch layer of foamed-in-place rigid polyurethane insulation (2 pounds per cubic foot). The polyurethane is then

Additional Information

Various room arrangements, door and window placement, and interior details can be varied, because the interior walls are all nonbearing. Other construction details must be followed closely, however, because they affect structural strength and performance. Experienced builders can generally modify the plans and specifications satisfactorily. Although the design was intended primarily for warmer southern climates, some modification and insulation of the exterior walls should make the house suitable for northern climates.

Certain experimental features may not meet all requirements of some building codes. Prospective builders should confer with local code officials to determine the applicability of the design for the particular area in which the house is to be built.

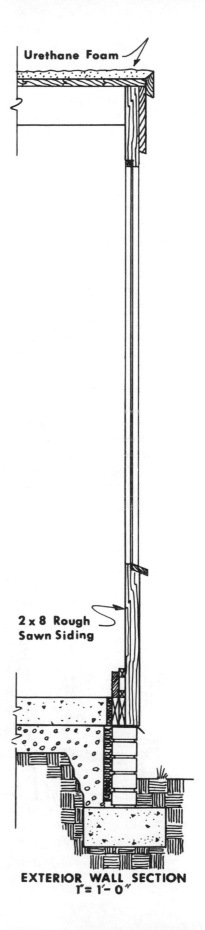

Urethane Foam

2 x 8 Rough Sawn Siding

EXTERIOR WALL SECTION
1" = 1'- 0"

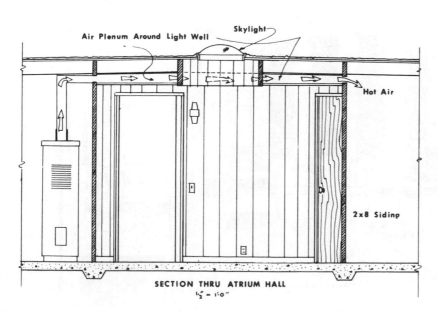

Air Plenum Around Light Well

Skylight

Hot Air

2x8 Siding

SECTION THRU ATRIUM HALL
½" = 1'-0"

overlaid with one or two coats of aluminum-filled asphalt emulsion, or a roof paint of similar quality. When thus protected, the foam provides thermal insulation, a good seal against moisture, and roofing in one operation. A clear plastic dome may be placed over the atrium hall, if natural illumination is desired.

Mechanical Systems

Heating is provided either with electric baseboard units, or with a furnace located in the utility room. If a furnace is used, a heat duct carries warm air to a plenum chamber created by dropping the ceiling in the atrium hall. Openings through the plank wall carry warm air to each room. A minimum of sheet metal work is required. Electrical outlets around the exterior walls and along the interior partitions are provided in a wood baseboard-raceway system. A water heater is installed in the utility room. Some of these features are illustrated in the sketch of the atrium wall section.

Exterior Wall System

Exterior walls consist of rough-sawn 2x8 softwood planks, placed on end around the slab and joined with hardboard splines inserted in slots in the plank edges. To minimize dimensional changes in service, moisture content of the planks should be no more than 12 to 16 percent when they are installed. A fascia board is fitted

around the edge of the roof. The wall of the atrium hall is of similar plank construction. Exterior wall surfaces are finished with a pigmented natural finish containing a preservative and water repellent. This attractive finish is easy to apply and maintain and it has good durability. Interior wall surfaces are finished with pigmented or natural stains.

Window System

Sections of two adjacent planks are omitted at appropriate locations in the exterior wall, and fixed glass panels are fitted into the openings, framed with a simple wood frame, and properly sealed. Ventilation is provided by a system of vertically sliding hardboard panels. As an alternate method, single-hung aluminum windows are fitted into the plank openings.

Carport

The extended roof over the carport is supported by another plank wall, with appropriate openings. The roofline is also extended over exterior doors by canti-levering interior roof beams out beyond the exterior walls.

HOUSE PLAN no. 11

AREA = 900 sq. ft. each unit

A HILLSIDE DUPLEX OF WOOD

This interesting design for a two-family home is intended particularly for sloping sites. It provides a total of 900 square feet in each of the two units, approximately half on each of two floors. The design is based on a pole-frame structure combined with wood arches that can be built in a simple shop. The compact design gives surprisingly open space, with complete privacy for each family. A pleasing wood deck is provided. One unit can be built separately for single-family homes. The simple design makes it particularly attractive for second homes to be built in isolated areas. Estimated costs are about

one half that of conventional construction of the same size. Lumber and plywood requirements are about half of such conventional construction.

Foundation System

Preservative-treated poles or posts are set into the ground with a minimum of soil disturbance. Posts extend to the second-floor level. Wood girders are bolted to the posts for each floor, and a conventional wood-joist system is installed. One layer of softwood tongue-and-grooved flooring is nailed to the joists, with a bead of construction mastic adhesive applied to the top of joists for increased stiffness and to reduce floor squeaking. No subfloor is required. Joist spaces of the first floor are insulated, and the lower joist surfaces are covered with plywood or other suitable sheet material.

Roof System

The unique feature of the design is a simple Gothic-type arch, made in two sections, as illustrated on pages 3 and 4. An arch consists of 1x2 lumber flanges nailed to short length of 2x4's with spacing between them. Arch sections can be assembled on a simple jig. Each section weighs only about 30 pounds. The sections are assembled on the ground, and the joist for the second floor is nailed to the arch. The sections are joined at the peak with a plywood gusset. The assembly is then raised into place and fastened to the ends of the first-floor joist.

Several roof sheathings may be used. In one prototype, erected in Athens, Georgia, 1/2-inch softwood plywood was used, with arches 24 inches on center. The plywood was covered with building paper and asphalt shingles with a stapled lower tab so that shingles were tight on the reverse curvature of the lower roof sections. Alter-nate sheathing might be bevelled siding nailed to the arches and finished with a water-repellent pigmented stain. The latter system is easier to install without scaffolding, and the problem of bending plywood in large sheets to the slight curvature of the arches is avoided. Insulation is installed between arches, and the inner surface is covered with thin gypsum board, plywood, or lap siding to provide various interior treatments.

End Walls and Partitions

All walls are essentially nonbearing, although end walls do provide stiffening of the arch frame. End walls are 2-inch thick lumber-framed panels, erected in place, covered with plywood on the outside, and finished with natural or pigmented stains. The inner surface may be covered with plywood or other panel materials. Insulation is provided in the spaces between framing. Aluminum or wood windows are installed as desired. Sliding patio doors in each unit open to the cantilevered deck.

Interior partitions may be of a single thickness of particleboard with panels fitted into slots in vertical 2x2 member at joints. The walls can also be of conventional construction.

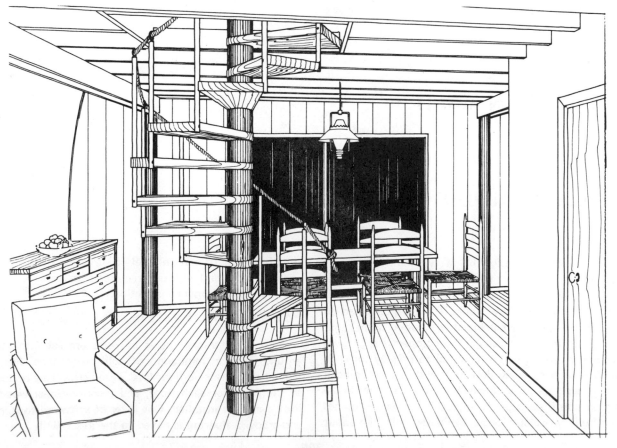

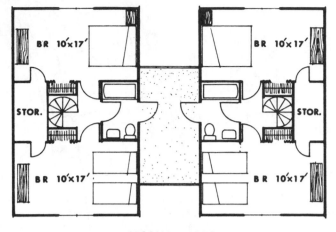

SECOND FLOOR

BR 10'×17'   BR 10'×17'

STOR.   STOR.

BR 10'×17'   BR 10'×17'

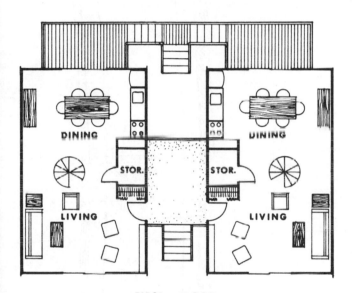

FIRST FLOOR

DINING   DINING

STOR.   STOR.

LIVING   LIVING

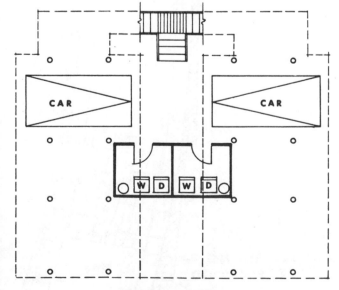

GROUND LEVEL

CAR   CAR

W   D   W   D

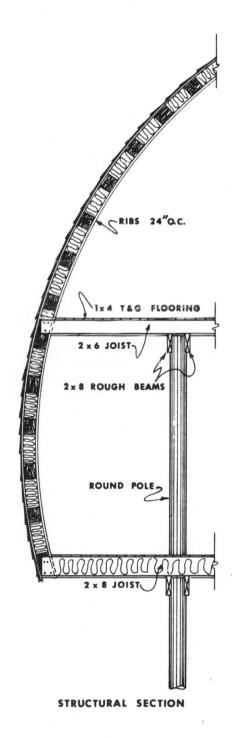

STRUCTURAL SECTION

RIBS 24" O.C.

1×4 T&G FLOORING

2×6 JOIST

2×8 ROUGH BEAMS

ROUND POLE

2×8 JOIST

Spiral-Wood Stairs

Although the units can be provided with conventional wood stairs, a spiral-stair unit was designed for easy fabrication on the site. It is estimated to cost about one-third as much as conventional metal spiral-stair units. The stairway is shown in the sketch of the interior. It is constructed of spacer blocks cut from wood poles, and drilled to receive a metal rod threaded at the ends for bolts and washers. Treads are cut from 2x12 timbers, and sandwiched between the spacers with mastic adhesive. Metal pipe balusters are fastened to edges of adjacent treads, front to back, and capped with a wood plug into which a large screw eye is inserted to take a rope banister or rail. Once in place, the unit is tightened with the bolts at top and bottom. The wood can then be stained.

Utility Space and Utilities

An enclosure for utility space is provided under both sections below the first floor. This is an on-grade concrete slab, over which either a concrete block wall, or a conventional wood-frame wall of treated lumber and plywood is erected. This provides space, entered from outside, for the water heater and heating system and for washing and ironing.

Electrical wiring is installed conventionally. Horizontal runs are easy to install through openings between the block web members in the arches. Plumbing and heating is conventional, as required by local codes.

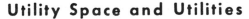

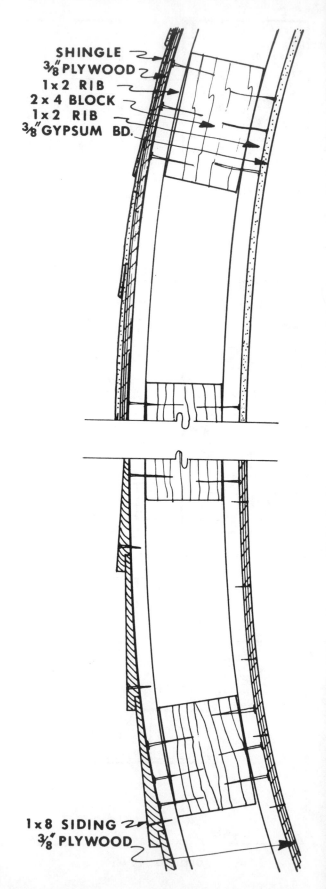

SHINGLE
3/8" PLYWOOD
1x2 RIB
2x4 BLOCK
1x2 RIB
3/8" GYPSUM BD.

1x8 SIDING
3/8" PLYWOOD

TYPES OF SHELTERS

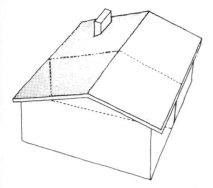 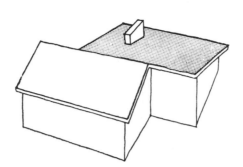 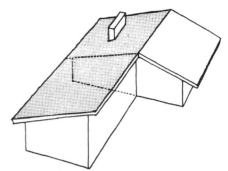

Units for Rainy Climates

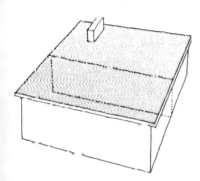 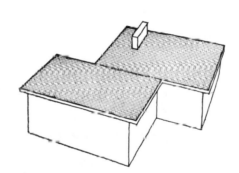 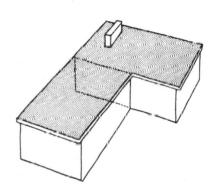

Units for Hot Climates

 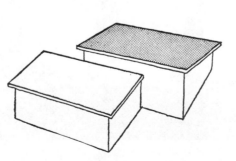 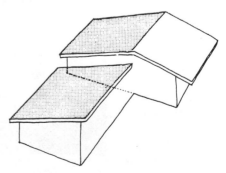

Units on Sloped Terrain

EFFICIENT and ECONOMICAL FEATURES

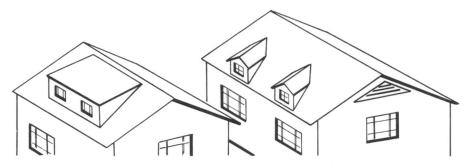

Shed dormers provide more usable space than the gable type.

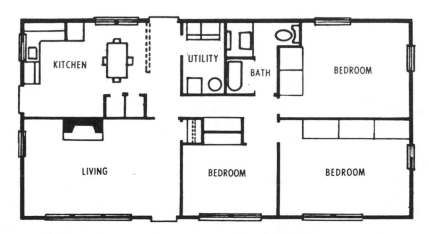

Rectangular floor plans are the most economical. Other economical features of this arrangement are the combination kitchen and dining area and the compact utility room.

Well-drained, gently sloping sites are ideal for basements. Economical features of this house include the one-story rectangular design and the plain gable roof.

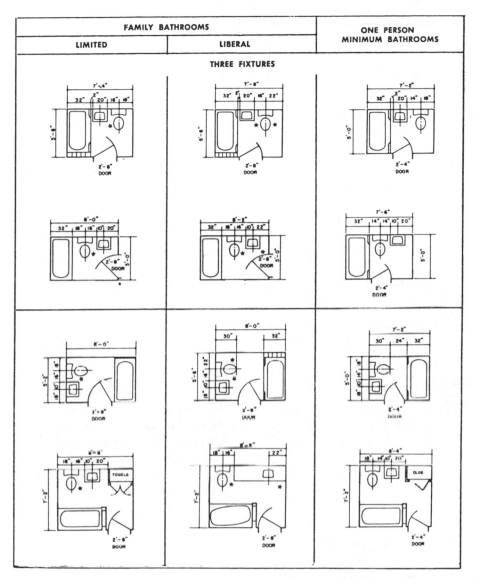

Many different bathroom arrangements are possible. In the top group, all fixtures and plumbing are along one wall. This is the most economical arrangement. Fixtures and plumbing are along two walls in the bottom group.

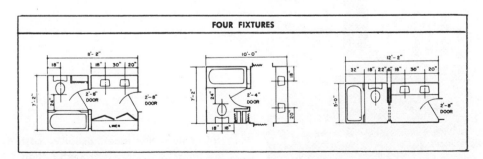

Compartmented bathrooms are more economical than two separate bathrooms.

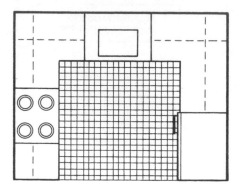

U-shaped kitchens are very efficient arrangements.

Wide, shallow closets are convenient for storing frequently used items. Note the additional storage provided on the wide-opening, double-hung doors.

Prefabricated fireplaces usually cost less and heat more efficiently than comparable size masonry fireplaces.

Part III

SPECIAL LOW COST METHODS IN DESIGN AND CONSTRUCTION

CHAPTER 61

SPECIAL LOW COST METHODS IN DESIGN AND CONSTRUCTION

A MATTER OF COSTS

In the building of a low-cost home, the matter of cost must be considered at every step—in design, selection of materials, and in construction. Each square foot of area added to a plan increases the cost substantially. It is often wise to omit some features, even though desirable, and to add these at some opportune future time. Selection of satisfactory alternate materials can also account for substantial savings.

The unit cost of framing and enclosing a basic wood house does not vary a great deal and is generally determined by square footage. However, the type of foundation, materials used on the inside and outside of the basic wood frame, the type of windows and doors, number of kitchen cabinets and amount of other millwork materials, floor covering, and the caliber of the utilities included can vary a great deal and generally govern the overall cost of the house. For example, in an average single-story house, kitchen cabinets,

interior doors and trim, and hardwood floors account for about 15 percent of the total cost, and substitute materials or deletions in these areas result in substantial savings. The plumbing, electrical, and heating installations also account for about 15 percent of the total house cost, and savings can also be made by eliminating or delaying some of these phases of construction until later.

The cost of a basement in an average house can also amount to 15 percent of the total cost. It seems justified, therefore, in instances where cost is important, to eliminate a basement and have a crawl space consisting of a foundation of treated wood posts or masonry piers. In a small house, this could result in a saving of up to 50 percent of the cost of a conventional foundation.

With a good plan and adequate construction details, a small house can be constructed at a reasonable cost and yet provide for good family living and be as pleasing in appearance as higher cost houses.

MAJOR HOUSE PARTS

Figure 1 is an exploded view of a single-*story*, wood-frame house showing the major parts. The floor system, interior and exterior walls, and the roof are the major components of such a house. Houses with flat or low-sloped roofs are usually variations of these systems.

Floor System

Figure 1 shows a floor system constructed over a *crawl space* area. Supporting *beams* are fastened to treated posts embedded in the soil or to masonry *piers*. In the South, Central, and Coastal areas, provisions must be made for protection from termites. Construction of this type of support for the floor joists has a great advantage because grading is not required and thus it can be used on relatively steep or uneven slopes. Floor *joists* are fastened to these beams and the *subfloor* nailed to the joists. This results in a level, sturdy platform upon which the rest of the house is constructed.

Exterior Walls

Exterior walls, often assembled flat on the subfloor and raised in "tilt-up" fashion, are fastened to the perimeter of the floor platform. Exterior coverings and window and door units are included after walls are plumbed and braced.

Interior Walls

Interior walls are usually the next components to be erected unless *trussed* rafters (roof trusses) are used. Trussed rafters are designed to span from one exterior sidewall to the other and do not require support from interior *partitions*. This allows partitions to be placed as required for room dividers. When ceiling joists and rafters are used, a *bearing partition* near the center of the width is necessary.

Roof Trusses or Roof Framing

Several systems can be used to provide a roof over the house. One consists of normal ceiling joists and rafters which require some type of load-bearing wall between the sidewalls (fig. 2A). Another is the trussed rafter system (figs. 1 and 2B) (commonly called *trusses*). This design requires no load-bearing walls between the sidewalls. A third design consists of thick wood roof decking (fig. 2C). A fourth is open beams and decking which span between the exterior walls and a center wall or ridge beam (fig. 2D). The truss and the conventional joist-and-rafter construction require some type of finish for the ceiling. The decking (fig. 2C) or the beam and decking (fig. 2D) combinations can serve both as interior finish and as a surface to apply the roofing material.

717

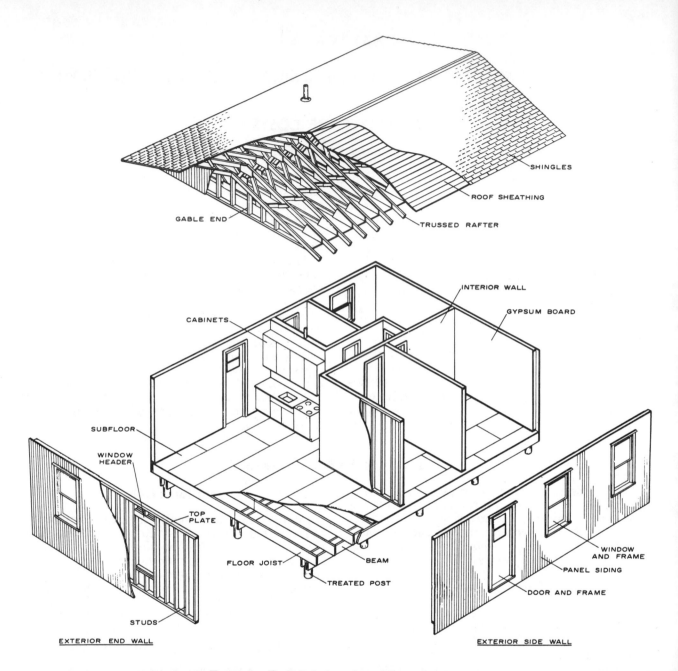

FIGURE 1.—Exploded view of wood-frame house.

MATERIAL SELECTION

There are hundreds of materials on the market which can be used somewhere in the construction of a house. Many are costly and are meant primarily for use in the most expensive homes. Others may not be suitable for all of the intended uses. However, among these building materials are many that are reasonable in cost and perform efficiently. Most manufacturers recommend particular uses and application methods for each of their products, and few problems will occur if such recommendations are followed.

Wood

Wood in its various forms is perhaps the most common and well-known material used in house construction. It is used for framing of floors, walls, and roofs. It is sometimes used in board form as a covering material, but more often the covering materials take the form of plywood or other *panel* wood products. Wood is also used as *siding* or exterior covering, as interior covering, as interior and exterior trim, as flooring, in the many forms

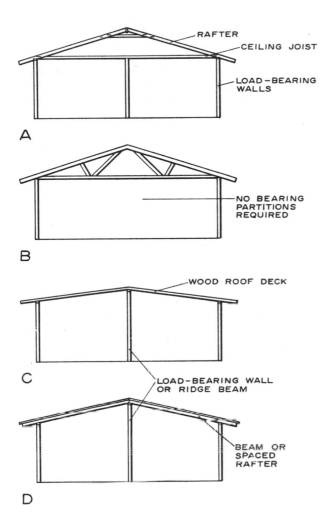

FIGURE 2.—Types of roof construction. *A*, Rafter and ceiling joists—sloped roofs; *B*, trussed rafter—sloped roofs; *C*, wood roof decking—low-sloped roofs; *D*, beam with wood or fiberboard decking.

and types of millwork, and also as shingles to cover roofs and sidewalls.

Wood is easy to form, saw, nail, and fit; even with simple handtools, and with proper use and protection will give excellent service. The *moisture content of wood* used in various parts of a house is important, and recommended moisture contents will be outlined in later sections of this handbook.

There are a number of basic standard wood and wood products used in the *construction* of wood-frame houses. The selection in the proper type and grade for each use is important.[4] The materials can be divided into groups by their use in the construction of a house: some require good strength, others workability, and still others necessitate good appearance.

4 Anderson, L. O. Selection and Use of Wood Products for Home and Farm Building. U.S. Dept. Agr. Agr. Inform. Bul. 311. 1967.

Treated Posts

Wood posts or poles which are embedded in the soil and used for support of the house should be pressure treated. A number of species are used for these round members. The pressure treatment should conform to Federal Specification TT–W–571. Pressure treatments normally utilize the oil type of *preservatives* (empty-cell process) or leach-resistant waterborne salt preservatives (full-cell process.)

Dimension Material

Surfaced *dimension* material wood members 2 to 4 inches thick and other wood parts are not full size as they are received from the *lumber yard*. For example, a nominal 2 by 4 may have a finished thickness of $1\frac{1}{2}$ to $1\frac{9}{16}$ inches and a width of $3\frac{1}{2}$ to $3\frac{9}{16}$ inches, depending on the moisture content. These materials are sawn from green logs and must be surfaced as well as dried to a usable moisture content. These processes account for the difference in size between a finished dry member and a rough green member.

The following tabulation of sizes is being recommended to the American Lumber Standards Committee by the Southern Pine Inspection Bureau, West Coast Lumber Inspection Bureau, and Western Wood Products Association.

Nominal (inches)	Dry (inches)	Green (inches)
1	$\frac{3}{4}$	$\frac{25}{32}$
2	$1\frac{1}{2}$	$1\frac{9}{16}$
4	$3\frac{1}{2}$	$3\frac{9}{16}$
6	$5\frac{1}{2}$	$5\frac{5}{8}$
8	$7\frac{1}{4}$	$7\frac{1}{2}$
10	$9\frac{1}{4}$	$9\frac{1}{2}$
12	$11\frac{1}{4}$	$11\frac{1}{2}$
14	$13\frac{1}{4}$	$13\frac{1}{2}$
16	$15\frac{1}{4}$	$15\frac{1}{2}$

The green-dry size relationship should hold for dimension lumber up to 4 inches in nominal thickness.

The first materials to be used, after the foundation is in place, are the floor *joists* and *beams* upon which the joists rest. These require adequate strength in bending and moderate stiffness. The sizes used depend on a number of factors; the *span*, spacing, species, and grade. Recommended sizes are listed in most working plans. The second grade of a species, such as southern pine, western hemlock, or Douglas-fir, is commonly selected for these uses. In lower cost houses, the third grade is usually acceptable.

For best performance, the *moisture content* of most dimension materials should not exceed 19 percent.

Wall *studs* (the structural members making up the wall framing) are usually nominal 2 by 4 inches in size and spaced 16 or 24 inches apart. Their strength and stiffness are not as important as for the floor joists, and in low-cost houses the third grade of a species such as Douglas-fir or southern pine is satisfactory. Slightly higher grades of other species, such as white fir, eastern white pine, spruce, the western white pines, and others, are normally used.

Members used for trusses, rafters, beams, and ceiling joists in the roof framing have about the same requirements as those listed for floor joists. For low-cost houses, the second grade can be used for trusses if the additional strength reduces the amount of material required.

Covering Materials

Floor *sheathing* (subfloor) consists of board *lumber* or *plywood*. Here, too, the spacing of the joists, the species of boards or plywood, and the intended use determine thickness of the subfloor. A single layer may serve both as subfloor and top surface material; for example, 5/8-inch or thicker tongued-and-groved plywood of Douglas-fir, southern pine, or other species in a slightly greater thickness can be used when joists are spaced no more than 20 inches on center. While Douglas-fir and southern pine plywoods are perhaps the most common, other species are equally adaptable for floor, wall, and roof coverings. The "Identification Index" system of marking each sheet of plywood provides the allowable rafter or roof truss and floor joist spacing for each thickness of a standard grade suitable for this purpose. A nominal 1-inch board subfloor normally requires a top covering of some type.

Roof sheathing, like the subfloor, most commonly consists of plywood or *board lumber*. Where exposed wood beams spaced 2 to 4 feet apart are used for low-pitched roofs, for example, wood decking, fiberboard roof deck, or composition materials in 1- to 3-inch thicknesses might be used. The thickness varies with the spacing of the supporting rafters or beams. These sheathing materials often serve as an interior finish as well as a base for roofing.

Wall sheathing, if used with a siding or secondary covering material, can consist of plywood, lumber, structural insulating board, or gypsum board. The type and method of sheathing application normally determine whether *corner bracing* is required in the wall. When 4- by 8-foot sheets of 25/32-inch regular or 1/2-inch-thick, medium-density, insulating fiberboard or 5/16-inch or thicker plywood are used vertically with proper nailing all around the edge, no bracing is required for the rigidity and strength needed to resist windstorms. There are plywood materials available with grooved or roughened surfaces which serve both as sheathing and finishing materials. Horizontal application of plywood, insulating fiberboard, lumber, and other materials usually requires some type of diagonal *brace* for rigidity and strength.

Exterior Trim

Some exterior trim, such as *facia* boards at cornices or gable-end overhangs, is placed before the roofing is applied. Using only those materials necessary to provide good utility and satisfactory appearance results in a cost saving. These trim materials are usually wood and, if relatively clear of *knots*, can be painted without problems. Lower grade boards with a rough-sawn surface can be stained.

Roofing

One of the lowest cost roofing materials which provides satisfactory service for sloped roofs is mineral-surfaced *asphalt roll roofing*. Asphalt shingles also give good service. Both are available in a number of colors. The material cost of asphalt *shingles* is about twice that of surfaced roll roofing. In a small, 24- by 32-foot house, use of surfaced roll roofing may mean a saving of $40 for material and about $20 for labor, which is less than 1 percent of the total cost of the house. However, an asphalt shingle would normally last longer and have a better apperance than the roll roofing.

Wood shingles have a pleasing appearance for sloped roofs. Although they are usually more costly than composition roofing, they could be used where availability, cost, and application conditions were favorable.

Window and Door Frames

Double-hung, casement, or awning wood windows normally consist of prefitted sash in assembled frames ready for installation. A double-hung window is one in which the upper and lower sash slide vertically past each other. A *casement sash* is hinged at the side and swings in or out. An awning window is hinged at the top and swings out. Separate sash in 1 1/8- or 1 3/8-inch thickness can be used, but some type of frame must be made that includes *jambs*, stops, *sill*, casing, and the necessary hardware. A low-cost, factory-built unit which requires only fastening in place may be the most economical. A fixed sash or a large window glass can be fastened by stops to a prepared frame and generally costs less than a movable-type window. It is normally more economical to use one larger window unit than two smaller ones. Screens should ordinarily be supplied for all operable windows and for doors. In the colder climates, storm windows and storm or *combination doors* are also desirable. Combination units, with screen and storm inserts, are commonly used.

Exterior Coverings

Exterior coverings such as horizontal wood *siding*, vertical boards, boards and *battens*, and similar forms of siding usually require some type of backing in the form of *sheathing* or nailers between studs. In mild climates, nominal 1-inch and thicker sidings are often used over a waterproof paper applied directly to the braced stud wall. There are many sidings of this type on the market in both wood and nonwood materials.

Combination sheathing-siding materials (panel siding) usually consist of 4-foot-wide sheets of plywood, exterior particleboard, or hardboard. Applied vertically before installation of window and door frames, such materials serve very well for exteriors. Plywood may be stained or painted, and the other materials should be painted. Paper-overlaid plywood also serves as a dual-purpose exterior covering material and takes paint well. Wood shingles and *shakes* and similar materials normally require a solid backing or spaced boards of some type.

Insulation

Most houses, even those of lowest cost, should have some type of *insulation* to resist the cold and to increase comfort during hot weather. There are various types of insulation, from insulating fiberboard to fill types, which can be used in the construction of a house. Perhaps the most common *thermal insulations* are the flexible (blanket and batt) and the fill types

A blanket insulation might be used between the floor joists or studs. Batt insulation of various types might be used between floor joists or in the ceiling areas. Most flexible insulations are supplied with a vapor barrier which resists movement of water vapor through the wall and minimizes *condensation* problems. A friction-type batt insulation is also available for use in floors, walls, or ceilings. Fill-type insulation is most commonly used in attic-ceiling areas.

The structural insulating board often serves as sheathing in the wall or as a fair insulating material under a plywood floor. Each material has its place, and selection should be based on climate as well as on cost and utility.

Interior Coverings

Many *dry-wall* (unplastered) interior coverings are available, from gypsum board to prefinished plywood. Perhaps the most economical are the gypsum board products. They are normally applied vertically in 4- by 8-foot sheets or horizontally in room-length sheets with the joint at midwall heights. They are also used for ceilings. Thicknesses range from 3/8 to 5/8 inch. *Butt joints* and corners require the use of tape and joint compound, or a *corner bead*, and add somewhat to labor costs over prefinished materials. Plastic-covered gypsum board is also available at additional cost, but must usually be installed with an adhesive. Hardboards, insulation board, plywood, and other sheet materials are available, as are wood and fiberboard paneling. The choice must be based on overall cost of material and labor as well as on ease of mainte-

nance. Prefinished ceiling tile in 12- by 12-inch sizes and larger is also available.

Interior Finish and Millwork

Interior finish and *millwork* consist of doors and door frames, *base moldings*, window and door *trim*, kitchen and other *cabinets*, flooring, and similar items. The type and grade selected determine the cost to a great extent. Selection of simple *moldings*, lower cost species for jambs and other wood members, simple kitchen shelving, low-cost floor coverings, and the elimination of doors where practical will often make a difference of hundreds of dollars in the total cost of the house.

The most commonly used doors are the flush-type and the panel-type. The flush-type consists of thin plywood or similar facings with a solid or hollow core. The panel door consists of solid side stiles and cross*rails* with plywood or other panel fillers. For exterior types, both may be supplied with openings for glass. Exterior doors are usually 1¾ inches thick and interior doors 1⅜ inches thick.

Door jambs, casings, moldings, and similar millwork of a number of wood species can be obtained. Select the lower cost materials, yet those that will still give good service. Some species in this class are the pines, the spruces, and Douglas-fir.

Factory-built kitchen cabinets are expensive and can cost several hundred dollars in a moderate-size house. The use of open shelving which can be curtained and a good counter is almost a must in a low-cost house. Doors can be added at a later date.

Wood strip flooring or wood tile of hardwood species might be too costly to consider in the original construction, but could be installed at a future time. Softwood floorings or the lower cost hardwood floorings might be within the original budget. The use of lower cost asphalt tile or even a painted finish may be the best initial choice. However, when a woodboard subfloor is used, some type of underlayment is required under the tile. Particleboard, hardboard, and plywood are the most common materials for this use.

Nails and Nailing

In a wood-frame house, nailing is the most common method of fastening the various parts together. Nailing should be done correctly because even the highest grade member often does not serve its purpose without proper nailing. Thus, it is well to follow established rules in nailing the various wood members together. While most of the nailing will be described in future sections, table 1 lists recommended practices used for framing and application of covering materials. Figure 3 shows the sizes of common nails. Most finish and siding nails have the same equivalent lengths. For example, an eight*penny* common nail is the same length as an eightpenny galvanized siding nail, but not necessarily the same diameter.

Joining	Nailing method	Nails		
		Number	Size	Placement
Header to joist	End-nail	3	16d	
Joist to sill or girder	Toenail	2–3	10d or 8d	
Header and stringer joist to sill	Toenail		10d	16 inches on center.
Bridging to joist	Toenail each end	2	8d	
Ledger strip to beam, 2 inches thick		3	16d	At each joist.
Subfloor, boards:				
1 by 6 inches and smaller		2	8d	To each joist.
1 by 8 inches		3	8d	To each joist.
Subfloor, plywood:				
At edges			8d	6 inches on center.
At intermediate joists			8d	8 inches on center.
Subfloor (2 by 6 inches, T&G) to joist or girder	Blind-nail (casing) and face-nail.	2	16d	
Soleplate to stud, horizontal assembly	End-nail	2	16d	At each stud.
Top plate to stud	End-nail	2	16d	
Stud to soleplate	Toenail	4	8d	
Soleplate to joist or blocking	*Face-nail*		16d	16 inches on center.
Doubled studs	Face-nail, stagger		10d	16 inches on center.
End stud of intersecting wall to exterior wall stud	Face-nail		16d	16 inches on center.
Upper top plate to lower top plate	Face-nail		16d	16 inches on center.
Upper top plate, laps and intersections	Face-nail	2	16d	
Continous header, 2 pieces, each edge			12d	12 inches on center.
Ceiling joist to top wall plates	Toenail	3	8d	
Ceiling joist laps at partition	Face-nail	4	16d	
Rafter to top plate	Toenail	2	8d	
Rafter to ceiling joist	Face-nail	5	10d	
Rafter to valley or hip rafter	Toenail	3	10d	
Ridge board to rafter	End-nail	3	10d	
Rafter to rafter through ridge board	{Toenail	4	8d	
	{Edge-nail	1	10d	
Collar beam to rafter:				
2-inch member	Face-nail	2	12d	
1-inch member	Face-nail	3	8d	
1-inch diagonal let-in brace to each stud and plate (4 nails at top).		2	8d	
Built-up corner studs:				
Studs to blocking	Face-nail	2	10d	Each side.
Intersecting stud to corner studs	Face-nail		16d	12 inches on center.
Built-up girders and beams, 3 or more members	Face-nail		20d	32 inches on center, each side.
Wall sheathing:				
1 by 8 inches or less, horizontal	Face-nail	2	8d	At each stud.
1 by 6 inches or greater, diagonal	Face-nail	3	8d	At each stud.
Wall sheathing, vertically applied plywood:				
⅜ inch and less thick	Face-nail		6d	}6-inch edge.
½ inch and over thick	Face-nail		8d	}12-inch intermediate.
Wall sheathing, vertically applied fiberboard:				
½ inch thick	Face-nail			1½-inch roofing nail.[1]
²⁵⁄₃₂ inch thick	Face-nail			1¾-inch roofing nail.[1]
Roof sheathing, boards, 4-, 6-, 8-inch width	Face-nail	2	8d	At each rafter.
Roof sheathing plywood:				
⅜ inch and less thick	Face-nail		6d	}6-inch edge and 12-inch intermediate.
½ inch and over thick	Face-nail		8d	}

[1] 3-inch edge and 6-inch intermediate.

Painting and Finishing

There are many satisfactory *paints* and finishes for exterior use. *Pigmented stain* is one of the easiest types to apply and is also long lasting. It is available in many colors from light to dark and is generally one of the best finishes for rough or sawn wood surfaces. Exterior paints used on smooth-surfaced siding and trim or on the trim as an accent for stained walls should be applied in several coats for best service. The first may consist of a nonporous linseed oil *primer*. Following coats can consist of latex, alkyd, or oil-base exterior paints. A *water-repellent preservative* provides a natural clear finish for wood surfaces.

Many interior paints are suitable for walls and ceilings. Latex and alkyd types are perhaps the most common, but the oil types are also suitable.

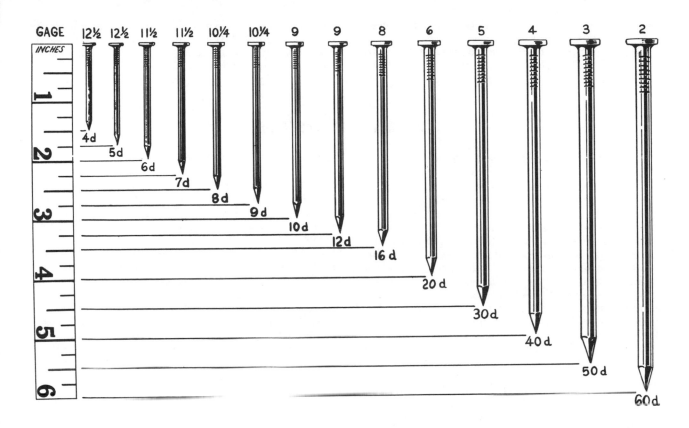

FIGURE 3.—Nail size table.

Chimney

Some type of chimney will be required for the heating unit. A masonry chimney requires a rigid concrete base, bricks or other masonry, *flue lining*, plus labor to erect it. A manufactured chimney which is supported by the ceiling joists or rafters may be the best choice from the standpoint of overall cost.

Utilities

Cost of the heating, plumbing, and wiring phases of house construction is usually a high percentage of total cost. These costs can be reduced to a great extent by careful selection and planning. Electricity is readily available in many areas and should be included in most low-cost houses. A minimum number of circuits and few switched outlets will aid in reducing the overall installation cost.

Heating units might consist of a low-cost space heater for wood, oil, or gas, or a small, central forced-air system with a minimum amount of duct work. The difference between the two may amount to several hundred dollars. The space heater may be more than sufficient for houses constructed in the milder climates.

Water supply and sewage disposal systems are often the most costly and difficult to install of all the utilities. When municipal or other systems are available for water and sewer service, there are no problems except the cost of installation. When good water can be obtained from a shallow well, a pump and pressure tank will provide a water supply at a low cost. However, in areas where a deep well is needed, costs may be too great. In such cases, one well might be used to supply several houses when they are ideally grouped. Wells should ordinarily be located a minimum of 50 feet from a septic tank and 100 feet from an absorption field.

Disposal of sewage in areas where public systems are available presents no problem. Where public systems are not available, the use of a septic tank and absorption system is required. The satisfactory performance of such a system depends on drainage, soil types, and other factors. It is sometimes necessary, when costs are critical, to provide for future installation of a disposal system. Roughed-in connections for plumbing facilities can and should be made in the house during its construction when sewer connections are not immediately available or costs are too great. Long-term

Floor and *deck paints* provide long wearing surfaces. One of the most common finishes for wood floors is the floor *sealer*, which provides a natural transparent surface.

costs of sewer connection will be more reasonable if provisions are made during construction.

Water supply and sewage disposal systems are specialized phases of house construction when these facilities are not available from a municipal or central source. Advice and guidance of a local health officer and engineer from your county office should be requested.

FOUNDATION SYSTEMS

Site Selection and Layout

One of the first essentials in house construction is to select the most desirable property site for its location. A lot in a smaller city or community presents few problems. The front set-back of the house and side-yard distances are either controlled by local regulations or should be governed by other houses in the neighborhood. However, if the site is in a rural or outlying area, care should be taken in staking out the house location.

Good drainage is essential. Be certain that natural drainage is away from the house or that such drainage can easily be assured by modification of ground slope. Low areas should be avoided. Soil conditions should be favorable for excavation for the treated posts or masonry piers of the foundation. Large rocks or other obstructions may require changes in the type of footings or *foundation*.

After the site for the house has been selected, all plant growth and sod should be removed. The area can then be raked and leveled slightly for staking and location of the supporting posts or piers.

The foundation plan in the working drawings for the house shows all the measurements necessary for construction. The first step in locating the house is to establish a baseline along one side with heavy cord and solidly driven stakes located well outside the end building lines (stakes 1 and 2, fig. 4). This baseline should be at the outer faces of the posts, piers, or foundation walls. When a post foundation is used with an overhang, the post faces will be 13½ inches in from the building line when a 12-inch overhang is used (fig. 5). When masonry piers or wood posts are located at the edge of the foundation, the outer faces are the same as the building line. These details are normally included in the working drawings. A second set of stakes (3 and 4) should now be established parallel to stakes 1 and 2 at the opposite side according to the measurements shown in the foundation plan of the working drawing. When measuring across, be sure that the tape is at right angles to the first baseline. Just as the 1–2 baseline does, this line will locate the outer edge of posts or piers. A third set of stakes, 5 and 6, should then be established at one end of the building line (fig. 4).

A square 90° corner can be established by laying out a distance of 12 feet at line 1–2 and 9 feet along line 5–6. Short cords can be tied to the lines to mark these two locations. Now measure between the two marks and when the diagonal measurement is 15 feet, the two corners at stakes 1–5 and 3–6 are square (fig. 4). The length of the house is now established by the fourth set of stakes, 7 and 8. Finally, the centerline along the length of the house can be marked by stakes 9 and 10. A final check of the alinement for a true rectangular layout is made by measuring the diagonals from one corner to the opposite corner (fig. 4). Both diagonals should be the same length.

In some areas of the country, building regulations might restrict the use of treated wood foundation posts. A masonry foundation fully enclosing the crawl space may be necessary or preferred. For such houses, the details shown in the Appendix to this manual can be used. Details of skirtboard enclosures for post foundations are also included in the appendix.

Footings

The holes for the post or masonry pier *footings* can now be excavated to a depth of about 4 feet or as required by the depth of the *frostline*. They should be spaced as shown in the foundation plan of the working plans and in figure 5. The embedment depth should be enough so that the soil pressure keeps the posts in place.

Place the dirt a good distance away from the holes to prevent its falling back in. Size of the holes for the wood post and the masonry piers should be large enough for the footings. When posts or piers are spaced 8 feet apart in one direction and 12 feet in the other, a 20- by 20-inch or 24-inch-diameter footing is normally sufficient (fig. 5). In softer soils or if greater spacing is used, a 24- by 24-inch footing may be required.

Posts alone without footings of any type, but with good embedment, are being used for pole-type buildings. However, because the area of the bottom end of the pole against the soil determines its load capacity, this method is not normally recommended in the construction of a house where uneven settling could cause problems. A small amount of settling in a pole warehouse or barn would not be serious. Where soil capacities are very high and posts are spaced closely, it is likely that a footing support under the end of the post would not be required. However, because a good, stable foundation is important in any type of house, the use of adequate footings of some type must be considered.

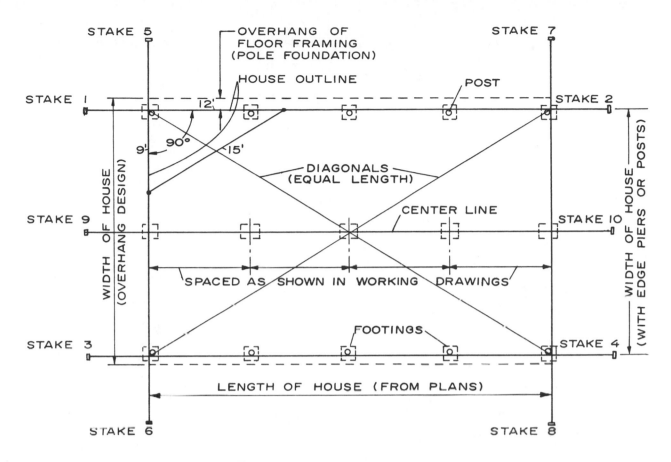

FIGURE 4.—House and footing layout.

Post Foundations—With Side Overhang

After the holes have been dug to the recommended depth and cleared of loose dirt, an 8-inch-thick or thicker concrete footing should be poured (fig. 5). If premixed *concrete* is not available, it can be mixed by hand or by a small on-the-job mixer. Tops of the footings should be leveled by measuring down a constant distance from the level line. A 5- or 6-bag mix (premixed concrete) or a 1 to 2½ to 3½ (cement: sand: gravel) job-mixed concrete should be satisfactory. The footings for the post and pier foundations should be located as shown on the working drawings.

Post Foundations—With Side Overhang

Treatment of the foundation posts should conform to Federal Specification TT–W–571. Penetration of the preservative for foundation posts should be equal to one-half the radius and not less than 90 percent of the sapwood thickness. The selection of posts should be governed also by final finish and appearance. When cleanliness, freedom from odor, or paintability is essential, waterborne preservative-treated posts should be used. The important principle is not to use untreated posts in contact with the soil.

Treated posts having a top diameter as shown on the plans should be selected for each of the footing locations. The length can be determined by setting the layout strings to the level of the top of the beams and posts. Select the corner with the highest ground elevation and move the string on this stake to about 18 to 20 inches above the ground level at this point. The minimum clearance under joists or beams should be 12 inches. However, 18 to 24 inches or more is preferred when accessibility is desired. Then with the aid of a lightweight string or line level (fig. 6), adjust the cord on the other stakes so that the layout strings around the edge of the building and down the center are all truly level and horizontal. To insure accuracy, the line must be tight with no sag and the level located at the center. If available a surveyor's level will serve even better. Thus, the distance from the string to the top of concrete footings will now be the length of the posts needed at each location. A manometer-type level can also be used in establishing a constant elevation for the posts. This type of level consists of a long, clear plastic tube partly filled with water or other liquid. The water level at each end establishes the correct elevation.

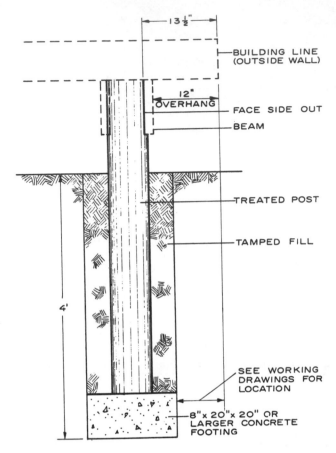

FIGURE 5.—Post embedment and footing alinement (over-hang design.)

Locate the notched and faced posts on the concrete footings using the cord on the stakes as a guide for the faced sides. Place and tamp 8 to 10 inches of dirt around them initially to hold them in place. Posts should be vertical and the faced side alined with the cord from the stakes along each side, the center, and the ends of the house outline. When posts are alined, fill in the remaining dirt. Fill and tamp ro more than 6 inches in the hole at one time to insure good, solid embedment.

Select *beams* the size and length shown in the foundation plan of the working drawings. Moisture content should normally not exceed 19 per-

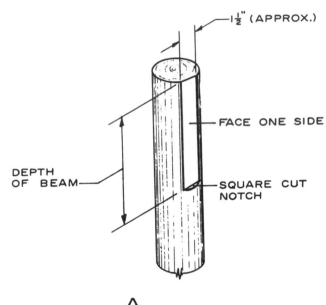

A

If pressure-treated posts are not available in the lengths just determined, use poles more than twice as long as required. Saw them in half and use *with the treated end down.* Now with a saw and a hand ax or drawknife, slightly *notch* and face one side for a distance equal to the depth of the beams (fig. 7A). The four largest diameter posts should be used for the corners and notched on two adjacent sides (fig. 7B). Facing should be about 1½ inches wide, except for corners or when beam joints might be made where 2½ inches is preferable.

Treated 6- by 6-inch or 8- by 8-inch posts can be used in place of the round posts when available. Although they may cost somewhat more and do *not* have the resistance of treated round posts, square posts will reduce on-site labor time.

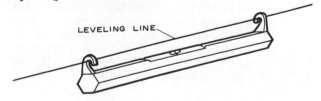

FIGURE 6.—Line level. Locate line level midway between building corners when leveling.

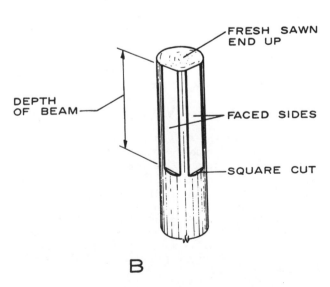

B

FIGURE 7.—Facing posts. *A*, Side or intermediate post; *B*. corner post (use largest).

cent. These beams are usually 2 by 10 or 2 by 12 inches in size and the lengths conform to the spacing of the posts. For example, posts spaced 8 feet apart will require 8- and 16-foot-long beams. The outside beam can now be nailed in place. Starting along one side at the corner, even with the leveling string, nail one beam to the corner post and each crossing post (fig. 8A). Initially, use only one twentypenny nail at the top of each beam, and don't drive it in fully. (The center should be left free to allow for a carriage bolt.) Side beam ends should project beyond the post about 1½ inches or the thickness of the end header (fig. 8B). When all outside beams and those along the center row of posts are erected, all final leveling adjustments should be made. In addition to the leveling cord from the layout stakes, use a carpenter's level and a straightedge to insure that each beam is level, horizontal, and in line with the cord. Final nailing can now be done on the first set of beams. Posts extending over the tops of the beams can now be trimmed flush.

The second set of beams on the opposite sides of the posts should now be installed. Because round posts vary in diameter, the facing on the second side has been delayed until this time. Use a strong cord or string and stretch along the length of the side of the foundation on the inside of the posts and parallel to the outside beams (fig. 9A). This will establish the amount of notching and facing to be done for each post. Use a saw to provide a square notch to support the second beam. All posts can be thus faced and the second beams nailed in place level with, and in the same way as, the outside beams. This facing is usually unnecessary when square posts are used (fig. 9B). However, additional bolts are required when the beam does not bear on a notch. Joints of the *headers* should be made over the center of the posts and stag̲ ̲e̲d. For example, if an 8- and a 16-foot beam are used on the outside of the posts, stagger the joints by first using a 16-foot, then an 8-foot length, on the inside. Only one *joint* should be made at each support.

Drill ½-inch holes through the double beams and posts at the midheight of the beam, and install ½- by 8-, 10-, or 12-inch galvanized carriage bolts with the head on the outside and a large washer under the nut on the inside. Use two bolts for square posts without a notch (fig. 9B). At splices, use two bolts and stagger (fig. 10A). When available at the correct moisture content, single nominal 4-inch-thick beams might be used to replace the two 2-inch members (fig. 10B).

All poles and beams are now installed and the final earth tamping can be done around the poles where required. A final raking and leveling is now in order to insure a good base for the soil cover if required. There is now a solid level framework upon which to erect the floor joists.

Edge Piers—Masonry and Posts

When masonry piers or wood posts are used along the edge of the building line instead of for overhang floor framing as previously described, the 8-inch poured footings are usually the same size as shown in figure 11A. However, check the working drawings for the exact size. For masonry piers, the distance to the bottom of the footings should be governed by the depth of frost penetration. This may vary from 4 feet in the Northern States to less than 1 foot in the Southern areas. The wood posts normally require a 3- to 4-foot-deep hole. The masonry piers or posts should be alined so that the outside edges are flush with the outside of the building line (fig. 11A, B). The foundation plan in the working drawings covers these details further.

Concrete block, brick, or other masonry, or poured concrete piers can now be constructed over the footings. Concrete block piers should be 8 by 16 inches in size, brick or other masonry 12 by 12 inches, and poured concrete 10 by 10 inches. The tops of the piers should all be level and about 12 to 16 inches above the highest corner of the building area. Use any of the previously described leveling methods. Use a 22-gage by 2-inch-wide galvanized perforated or plain anchor strap for nailing into the beams. It should extend through at least two courses, filling the core when hollow masonry is used (fig. 11A). A prepared mortar mix with 3 or 3½ parts sand and ¼ part cement to each part of mortar or other approved mixes should be used in laying up the masonry units. The wood post installation details are shown in figure 11B. Anchor straps are nailed to each post and beam with twelvepenny galvanized nails.

Beams consisting of doubled 2 by 10 or 2 by 12 members (check the foundation plan of the working drawings) can now be assembled. Place them on the posts or piers and make the splices at this location. Make only one splice at each pier. Nominal 2-inch members can be nailed together with tenpenny nails spaced 16 inches apart in two rows. Fabrication details at the corner and intersection with the center beam and fastening of the beam tie (*stringer*) are shown in figure 11C.

Ledgers, used to support the floor joists, should be nailed to the inside of the nailed beams. The sizes are 2 by 2, 2 by 3, or 2 by 4 as indicated in the foundation plan of the working drawings. Ledgers should be spaced so that the top of the joists will be flush with the top of the edge and center beams when bearing on the ledgers (fig. 11A, B). Use sixteenpenny nails spaced 8 inches apart in a staggered row to fasten the ledger to the beam.

The foundation and beams are now in place ready for the assembly of the floor system.

An alternate method of providing footings for the treated wood foundation posts involves the use of temporary braces to position the posts while the

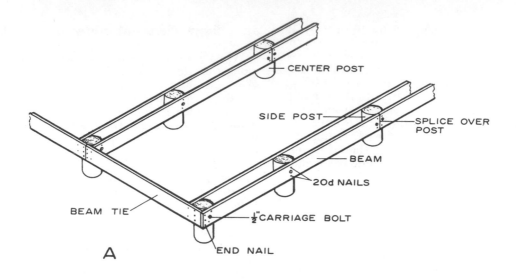

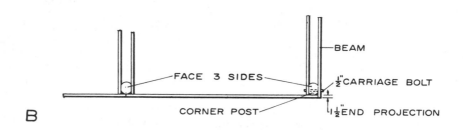

FIGURE 8.—Beam installation. *A*, Overall view; *B*, plan view.

concrete is poured (fig. 12). After the holes are dug, posts of the proper length are positioned and temporary braces nailed to them (fig. 12*A*). A minimum of 8 inches should be allowed for footing depth (fig. 12*B*). When posts are alined and set to the proper elevation, concrete is poured around them. After the concrete has set, holes are filled (*backfilled*) and earth tamped firmly around the posts. Construction of the floor framing can now begin.

This system of setting posts and pouring the concrete footings can also be used when a beam is located on each side of the posts (fig. 8). Posts are placed in the holes, the beams nailed in their proper position, and the beams alined and blocked to the correct elevation. After the concrete has set and the fill has been tamped in place, the beams can be bolted to the posts.

Termite Protection

In areas of the South and in many of the Central and Coastal States, *termite* protection must be considered in construction of crawl-space houses. Pressure-treated lumber or poles are not affected by termites, but these insects build passages to and can damage untreated wood. Perhaps the most common and effective present-day method of protection is by the use of soil poisons. Spraying soil with solutions of approved chemicals such as aldrin, chlordane, dieldrin, and heptachlor using recommended methods will proved protection for 10 or more years.

A physical method of preventing entry of termites to untreated wood is by the use of *termite shields*. These are made of galvanized iron, aluminum, copper, or other metal. They are located over continuous walls (fig. 13*A*) or over or around treated wood foundation posts (fig. 13 *B* and *C*). They are not effective if bent or punctured during or after construction.

Crawl spaces should have sufficient room so that an examination of poles and piers can be made easily each spring. These inspections normally provide safeguards against wood-destroying insects. Termite tubes or water-conducting fungus

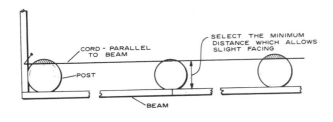

A

should be removed and destroyed and the soil treated with poison. Caution should be used in soil treatment, however, since the effective chemicals are often toxic to animal life and should not be used where individual water systems are present.

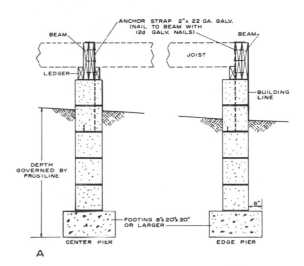

A

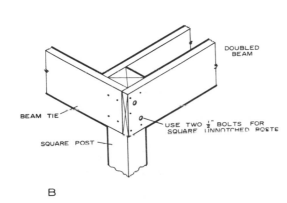

A

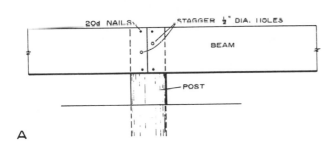

FIGURE 9.—Facing and fastening posts. A, Round posts; B, square post.

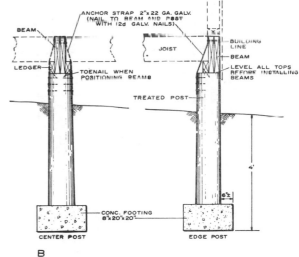

B

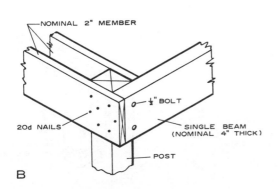

B

FIGURE 10.—Fastening beams to posts. A, Bolting; B, single beam.

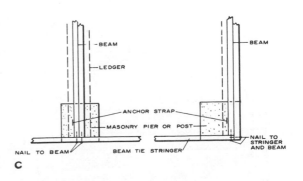

C

FIGURE 11.—Edge posts and masonry piers. A, Masonry edge piers; B, edge post foundation; C, corner and edge framing of beam.

729

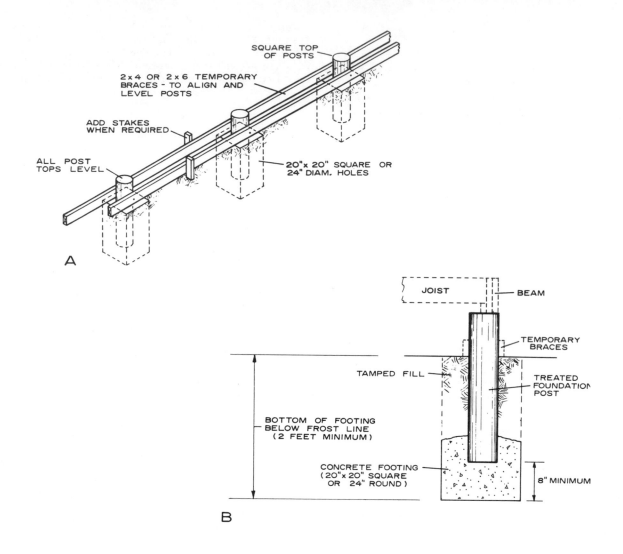

FIGURE 12.—Alternate method of setting edge foundation posts. *A*, Temporary bracing; *B*, footing position.

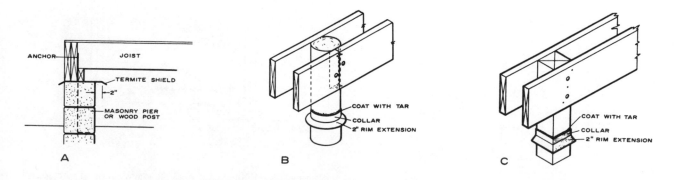

FIGURE 13.—Termite shields. *A*, On top of masonry or wood post; *B*, round posts; *C*, square posts

FOUNDATION ENCLOSURES

A treated wood post foundation is normally constructed with a crawl space having 18- to 24-inch clearance. This access space may be used for placing floor insulation, for installing a vapor barrier soil cover, for examining and treating soil in termite areas, and for other needs. In colder climates, it is often desirable to enclose a crawl space by some low-cost means, even though the floor is insulated. This is commonly done by fastening skirt boards of a long-lasting sheet material to the outside beams or floor framing. Enclosed crawl spaces should have a soil cover and a small amount of ventilation to assure satisfactory performance.

When a treated post foundation is not used, a masonry wall of concrete or concrete block construction will provide a satisfactory enclosure. Unlike the skirtboard enclosure, the masonry wall normally acts as a support for the floor framing system. Ventilation and the use of a soil cover are also required for a masonry foundation. The soil cover can consist of a 4-mil polyethylene or similar material placed over the soil under the house. Its use reduces ground moisture movement to wood members, which could result in excessive moisture and condensation. Ventilation can consist of standard foundation vents made for this purpose.

Masonry House Foundations

Although a treated wood post foundation will reduce the cost of a crawl-space house substantially, local conditions or a personal choice may indicate the use of a full masonry wall. This type of foundation, like the treated post type, will often require only a minimum amount of grading. Concrete block or poured concrete walls and piers with appropriate footings are accepted methods of providing supporting walls for the floor framing. Their use normally eliminates the need for outer beams as the joists are supported by the walls. Only a center flush or drop beam is required. However, the concrete block wall introduces the need for masonry work, which in some areas may be difficult to obtain.

The perimeter layout for the masonry foundation can be made in the same manner as has been outlined for the wood post foundation system (fig. 4). The outside line of the masonry walls will be the same as the outside line of the floor framing. The centerline of the interior load-bearing masonry piers is normally the same as for the posts of the wood foundation. Figure A is the plan of a typical masonry foundation for a crawl-space house. Footings are required for the outside walls as well as for the masonry piers which support the center beam. A soil cover of polyethylene or similar vapor barrier material should be used in all enclosed crawl-space houses. This prevents ground moisture from moving into the crawl-space area. Uncovered ground can result in high moisture content of floor framing, insulation, and other materials in the crawl space. A poured concrete wall requires some type of formwork while concrete block wall and piers are laid up directly on the footings. The information on concrete and the proper mortar mix for the concrete block is outlined in the chapter on "Foundation Systems" in this manual.

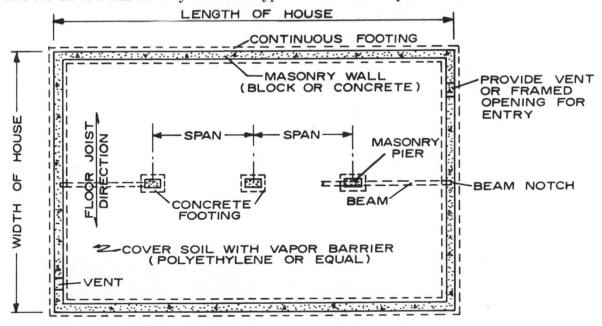

Figure A.—Typical masonry foundation wall (concrete block or poured concrete for crawl-space house).

731

The spacing of the center piers will depend on the size of the center beams. Longer distances between piers will require larger or deeper beams than moderate spans of 8 to 10 feet. These details are shown on the working drawings for each house.

Footings

Footing size is determined by the thickness of the foundation wall. A rule of thumb which is often used for small wood-frame houses under normal soil conditions is: The footing depth should equal the wall thickness and the footing width should be twice the wall thickness. Thus, an 8-inch masonry wall will require a 16- by 8-inch footing (**figure B**). The foundation plan will show the footing size for each house. Unusual soil conditions will often require special footing design.

The bottom of the footings should be located below frost line. This may be 4 feet or more in the Northern States. Local regulations or the footing details of neighboring well-constructed houses will indicate this depth. If the soil is stable, no forms are required for the sides of the footing trench.

One of the important factors in footing construction is to have the top level all around, especially when concrete block construction is to be used. Drive elevation stakes around the perimeter of the footing so that they can be used as guides when pouring the concrete. These elevations can be established by measuring down from the leveling line described in the chapter on "Foundation Systems." Concrete for footings should be poured over undisturbed soil.

A concrete block wall should normally be finished with a 4-inch solid cap block at the top to provide a good bearing surface for the joists, headers, and stringers (**figure B**). Anchor straps for the perimeter joists or headers are desirable in areas where high winds occur. They often consist of perforated or plain 22-gage or heavier galvanized metal straps about 2 inches wide. They are used with a bent "L" shaped base which is placed one or more courses below the cap block. They extend above the top of the wall so that they can be fastened to the edge headers, stringers, or joists. Space them about 8 feet apart and fasten by nailing (**figure B**).

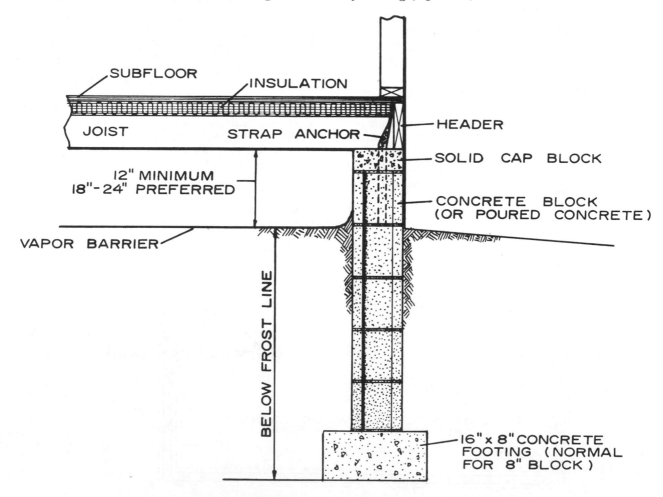

Figure B.—Section through exterior masonry wall (for crawl-space house).

732

Center Beams

The height of the center masonry piers with relation to the wall height is determined by the type of center beam used. The flush beam uses ledgers to support the joists, but when a drop beam is used the joists bear directly on the top surface. Each system requires a different depth notch for the masonry end walls. This is to assure that the top of all joists, headers, stringers, and the flush beam, when used, have the same elevation.

Flush beam.—The flush beam allows for more clearance in the crawl space as only the amount equal to the depth of the ledgers extends below the bottom of the joists (figure C.1). The joists rest on the ledger and are toenailed to the beam. A strap anchor or a bolt should be used to anchor the beam in high wind areas.

A notch must be provided in the end walls for beam support (figure C.2).The depth of this notch is equal to the depth of the ledgers. Bearing on the wall should be at least 4 inches. A clearance of about ½ inch should be allowed at the sides and ends of the beam. This will provide an airway to prevent the beam from retaining moisture. The size of the notch or beam opening in the wall (when 2- by 4-inch ledgers are used and the beam consists of two nominal 2-inch-thick members) would be approximately:

Length (along length of wall).	7 inches
Width _____	4½ inches (4-inch bearing area, plus clearance)
Height (or depth) _____	3½ inches (or width of ledgers)

The top of the beam should be flush with the top of the joists (figure C. 2).

In a concrete block wall, the mason provides the beam notch. When a poured concrete wall is used, a small wood box the size of the notch is fastened to the forms before pouring.

Drop beam.—The drop beam is supported by the masonry piers and the end walls of the foundation. Joists rest directly on the beam (figure D.1).A lap or butt joint can be used for the joists over the beam. In areas of high winds, it is advisable to use a strap anchor or a long bolt to fasten the beam to the piers. One disadvantage of this type of beam is the difference in the amount of wood at the center piers and at the outer walls which can shrink or swell. It is desirable to equalize the amount of wood at both the center and outside walls whenever possible. The flush beam closely approaches this desirable construction feature. The end foundation walls have a notch to provide bearings areas for the ends of the beams (fig. D. 2). This notch should have the same depth as the beam height. Allow clearance at the sides and end for air circulation. Assembly of the beams, the joist arrangement, and other general details are discussed in the chapter on "Floor Systems" in this manual. Specific details on the size, spacing, and location of the beams and joists are included in the floor framing plan of the working drawings for each house.

Masonry Foundation for Entry Steps

The construction of wood entry steps has been discussed in the chapter on "Steps and Stairs." This type of wood stoop provides a satisfactory entry platform and steps at a reasonable cost. However, when masonry walls are used in the foundation of the main house, it may be desirable to also provide a masonry foundation for the entry steps.

Figure E. 1 shows the foundation plan for a typical masonry entry platform and step. The walls are normally of 6-inch concrete blocks or poured concrete. Block units 4 inches thick or a 4-inch poured wall have also been used for the supporting front and sidewalls. The size of the top platform for a main entry step should be a minimum of 5 feet wide and 3½ feet deep. The foundation in figure E. 1 would result in a 6-foot by 3½-foot top with two steps or 6 by about 5 feet when one step is used.

The outer wall and footings are sometimes eliminated in providing support for the concrete steps. Then only the two wing walls are constructed. In such cases, the concrete steps are reinforced with rods to prevent cracking when the soil settles. Use at least two ½-inch diameter rods located about 1 inch above the bottom of the step.

It is important in constructing a masonry entry stoop to tie the wall into the house foundation walls and to have the bottom of the step footings below frost level. A concrete footing should be used for the concrete block wall to establish a level base as well as a bearing for the wall (figure E. 2) The footing should be at least 6 inches wide even though the wall may be less than this. Footings are normally not required for the 6-inch poured wall as the bearing area of the poured wall is usually sufficient. Ties or anchors to the house wall can consist of ½-inch reinforcing rods for the poured wall or standard masonry wall reinforcing for the block wall. These are placed as the house wall is erected. In block construction, both the house foundation wall and the wall for the entry stoop can be erected at the same time and tied together with interlocking blocks.

The finish concrete slab should be reinforced with a wire mesh when fill is used. Concrete is poured over the block wall forming the steps and platform (figure E. 2). Boards at each side of the step and platform and one at each riser provide the desired formwork for the concrete.

Skirtboard Materials for Wood Post Foundation

Sheet materials such as exterior grade plywood, hardboard, or asbestos board are most suitable for enclosing a wood post foundation. They need little

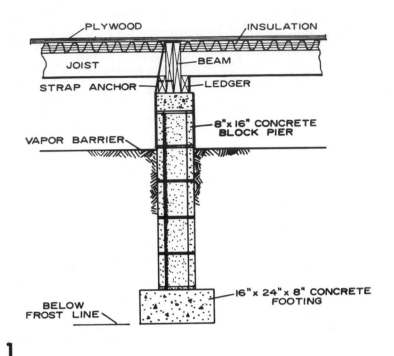

PLYWOOD   INSULATION

JOIST   BEAM

STRAP ANCHOR   LEDGER

8"x 16" CONCRETE BLOCK PIER

VAPOR BARRIER

BELOW FROST LINE

16"x 24"x 8" CONCRETE FOOTING

1

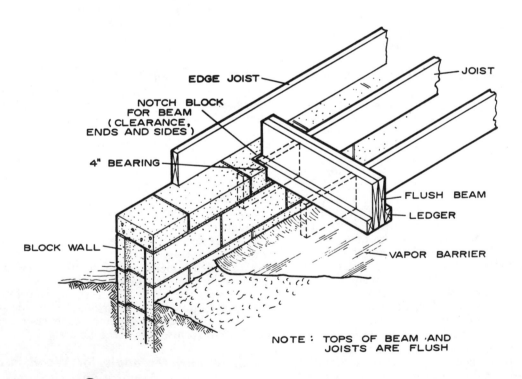

EDGE JOIST   JOIST

NOTCH BLOCK FOR BEAM (CLEARANCE, ENDS AND SIDES)

4" BEARING

FLUSH BEAM

LEDGER

BLOCK WALL

VAPOR BARRIER

NOTE : TOPS OF BEAM AND JOISTS ARE FLUSH

2

Figure C.—Details of concrete block pier for center flush beam. *A*, cross section ; *B*, detail at exterior end wall.

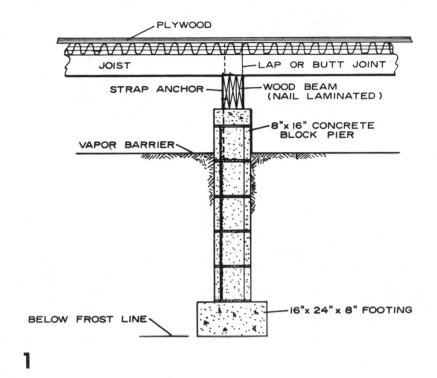

PLYWOOD

JOIST   LAP OR BUTT JOINT

STRAP ANCHOR   WOOD BEAM
(NAIL LAMINATED)

8"x 16" CONCRETE
BLOCK PIER

VAPOR BARRIER

16"x 24"x 8" FOOTING

BELOW FROST LINE

1

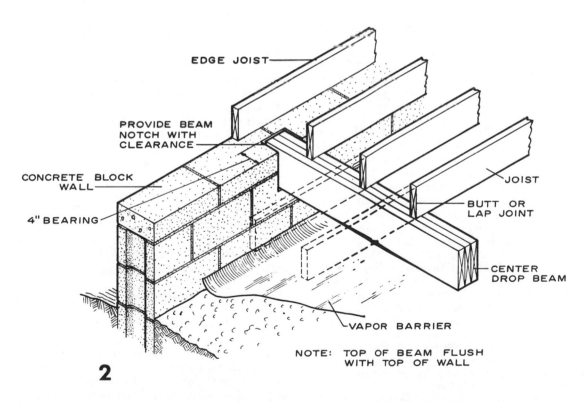

EDGE JOIST

PROVIDE BEAM
NOTCH WITH
CLEARANCE

CONCRETE BLOCK
WALL

4" BEARING

JOIST

BUTT OR
LAP JOINT

CENTER
DROP BEAM

VAPOR BARRIER

NOTE: TOP OF BEAM FLUSH
WITH TOP OF WALL

2

Figure D—Details of block pier for center drop beam. *A*, cross section; *B*, detail at exterior end wall.

if any framing and can be installed during or after construction of the house. When the skirtboard will be in contact with or located near the ground, it is most desirable to use a treated plywood when available. However, a 3-minute dip coat of a penta preservative or a water-repellent preservative along the exposed edges will provide some desirable protection for untreated material. It is also desirable to treat the exposed edges of hardboard with water-repellent preservative. Although asbestos board and metal skirtboards do *not* require preservative treatment, they are not as resistant to impacts as plywood and hardboard. Plywood and hardboard can be easily painted or stained to match the color of the house.

Plywood for skirtboards should be an exterior grade to resist weathering. A standard or sheathing grade with exterior glue is the most economical and can be readily stained. Plywood with rough-sawn or grooved surfaces commonly used for panel siding is also satisfactory.

Tempered hardboard probably provides somewhat better performance than regular-density (Standard) hardboard. However, with paint or similar protective coatings, the lower cost regular-density hardboard should give satisfactory performance. Asbestos board normally requires predrilling to prevent cracking when nailing near the edges.

Skirtboard for House With Edge Beams

Two relatively simple means of fastening the skirtboards to the floor framing can be used. The first method must be employed during construction of the floors and walls. The second can be used either during construction or after the house has been completed.

First method.— **Figure F** illustrates the first method of providing a nailing surface for the skirtboard material. The subfloor and the bottom plate of the walls are extended beyond the edge beam a distance equal to the thickness of the skirtboard (**figure F.1**). The skirtboard is then nailed to the beam and to the foundation posts. Splices should be made at the post when possible. When splices must be made between the posts, use a 2-by 4-inch vertical nailing cleat on the inside. Fourpenny or fivepenny galvanized siding or similar nails can be used for the 1/4-inch hardboard or asbestos board. Space them about 8 inches apart in a staggered pattern. Use sixpenny nails for 3/8-inch plywood and sevenpeny or eightpenny nails for 1/2-inch plywood skirtboards with the same 8-inch spacing. The panel siding is nailed directly over the top portion of the skirtboard. Nailing recommendations for the siding are given in the chapter on "Exterior Wall Coverings."

The detail for the end wall in this type of installation is shown in **figure F. 2**. A 2- by 4-inch cleat should also be used for the skirtboard joints when they occur between the foundation posts.

Some support or a backing may be required for the bottoms of the skirtboard to provide stability and resistance to impacts from the outside especially with thinner materials. Treated posts, treated 2- by 4-inch stakes, or embedded concrete blocks can be used for this purpose. Treated posts or stakes can be driven behind the skirtboard and the concrete blocks can be embedded slightly for added resistance (fig. F. 1 and 2). Space these supports about 4 feet apart or closer if required.

Second method.—A second method of installing the skirtboard can be used either after the house has been completed or during its construction. This system consists of nailing the skirtboard to the inner face of the ledger (fig. G). The skirtboard must be fitted between the posts which are partly exposed. Use the same method of nailing and blocking at the bottom of the boards as previously described. In addition, toenail the ends of the boards into the posts which they abut. **Figure G. 1** shows the details of installation at a sidewall and **figure G. 2** shows the details at an end wall. Use treated posts or stakes or concrete blocks as backers at the bottom of the skirtboard as previously described.

Skirtboard for House With Interior Beam

In those crawl-space houses with interior supporting beams and posts and with side overhang, the application of skirtboards is much the same as when the beams are located under the outside walls. The skirtboard may be nailed to the outside of the joist header or to the inner face (fig. H. 1). When it is fastened to the exterior face, the subfloor and the bottom wall plate are extended beyond the header a distance equal to the thickness of the skirtboard material. The skirtboard can also be fastened to the inside face of the header, but then it must be notched at each joist (fig. H. 1). Use the same type of nails and nailing patterns described for the details shown in **figures F and G**.

Details at the end walls of the house are much the same as those at the sidewalls (**figure H. 2**). When the skirtboard is nailed to the inner surface of the edge joist, no notching is required except at the center and edge beams. Use backers for the bottom of the skirtboard as previously described.

Soil Cover, Ventilation, and Access Door

An enclosed crawl space should not only be protected with a soil cover but should also have a small amount of ventilation. A soil cover of 4-mil polyethylene or similar vapor barrier material placed over the earth in the crawl space will minimize soil moisture movement to floor members and floor insulation. Lap the material 4 to 6 inches at the seams, and carry it up the walls or skirtboards a short distance. Use sections of concrete block or small field stones to keep the material in place.

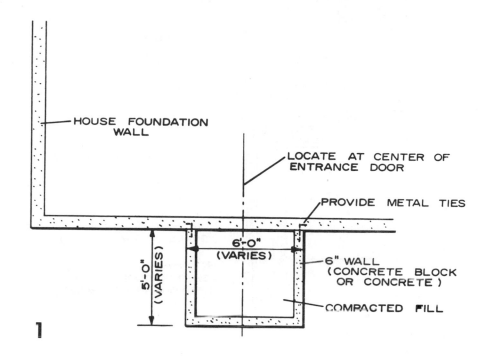

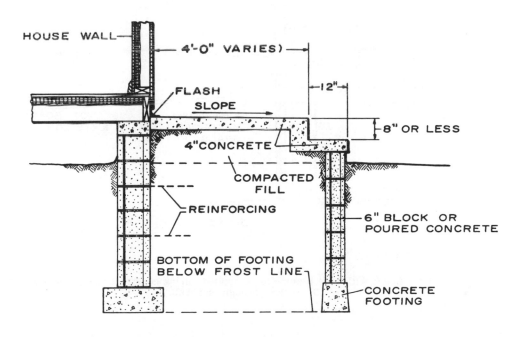

Figure E —Details of typical masonry entrance steps. *A*, plan view ; *B*, section view.

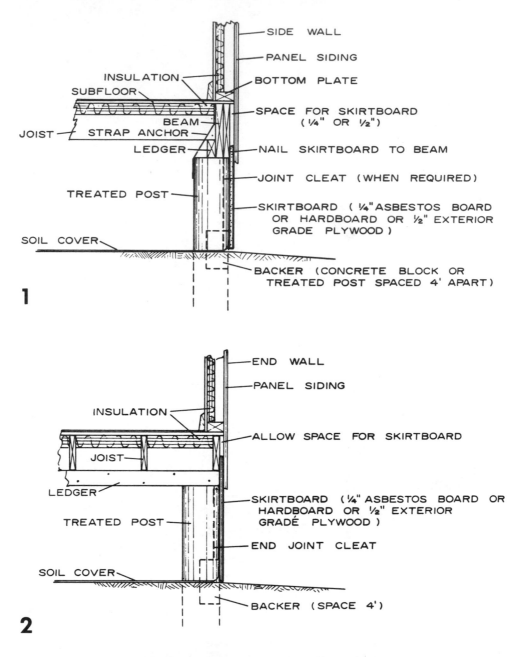

Figure F —Skirtboard for house with edge beams (to be applied during construction). *A*, Section through sidewall; *B*, section through end wall.

When the use of a soil poison is required in termite areas, do not install the soil cover for several days after poisoning or until the soil becomes dry again.

Ventilators should be installed on two opposite walls when practical. Standard 16- by 8-inch foundation vents can be used in concrete block foundations. Usually, in small houses, two screened ventilators, each with a net opening of 30 to 40 square inches, are sufficient when a soil cover is used. Crawl spaces with skirtboard enclosures should also be ventilated with small screened vents.

Access doors should be provided in crawl spaces. With a masonry foundation, they can consist of a 16- by 24-inch or larger frame with a plywood or other removable or hinged panel. Install the frame as the wall is constructed. With skirtboard enclosures, provide a simple removable section. When practical, the access doors should be located at the rear or side of the foundation and at the lowest elevation when a slope is present.

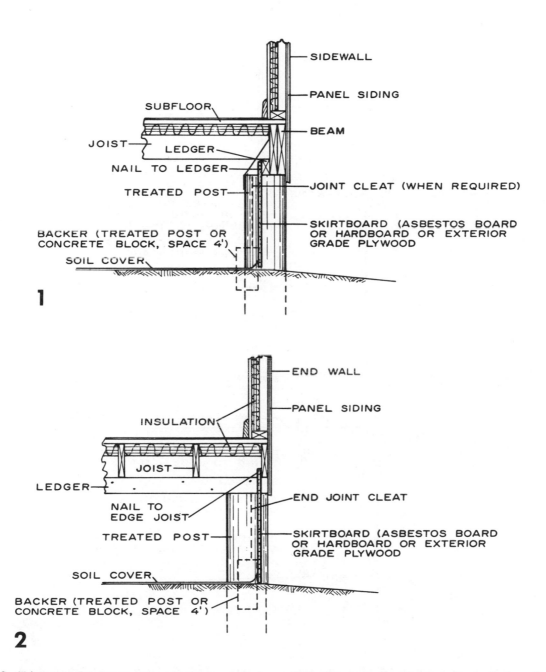

1

SIDEWALL

PANEL SIDING

SUBFLOOR

BEAM

JOIST

LEDGER

NAIL TO LEDGER

JOINT CLEAT (WHEN REQUIRED)

TREATED POST

SKIRTBOARD (ASBESTOS BOARD
OR HARDBOARD OR EXTERIOR
GRADE PLYWOOD

BACKER (TREATED POST OR
CONCRETE BLOCK, SPACE 4')

SOIL COVER

2

END WALL

PANEL SIDING

INSULATION

JOIST

LEDGER

END JOINT CLEAT

NAIL TO
EDGE JOIST

TREATED POST

SKIRTBOARD (ASBESTOS BOARD
OR HARDBOARD OR EXTERIOR
GRADE PLYWOOD

SOIL COVER

BACKER (TREATED POST OR
CONCRETE BLOCK, SPACE 4')

Figure G—Skirtboard for house with edge beams (can be applied after house is constructed). *A*, Section through sidewall; *B*, section through end wall.

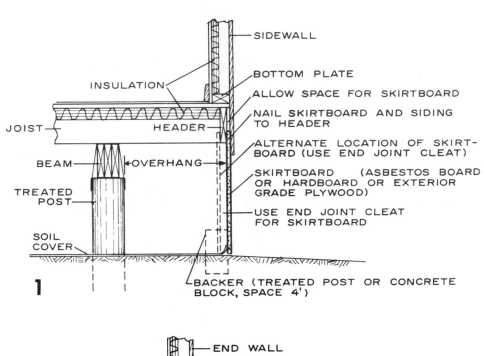

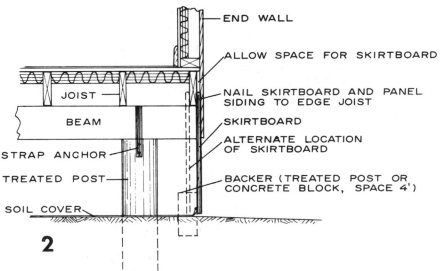

Figure H—Skirtboard for house with interior beam (with overhang). *A*, Section through sidewall; *B*, section through end wall.

JOISTS FOR DIFFERENT FLOOR SYSTEMS

The beams, or beams and ledger strips, and the foundations are now in place and joists can be installed. Size and lengths of the floor joists, as well as the species and spacing, are shown in the floor framing layout in the working plans for the house being built. The joists may vary from nominal 2 by 8 inches in size to 2 by 10 inches or larger where spans are long. Moisture content of floor joists and other floor framing members should not exceed 19 percent when possible. Spacing of joists is normally 16 or 24 inches on center so that 8-foot lengths of plywood for subfloor will span six or four joist spaces.

In low-cost houses, savings can be made by using plywood for subfloor which also serves as a base for resilient tile or other covering. This can be done by specifying tongued-and-grooved edges in a plywood grade of C–C plugged Exterior Douglas-fir, southern pine, or similar species. Regular Interior Underlayment grade with exterior glue is also considered satisfactory. The *matched* edges provide a tight lengthwise joint, and end joints are made over the joists. If tongued-and-grooved plywood is not available, use square edged plywood

and block between joists with 2 by 4's for edge nailing. Plywood subfloor also serves as a tie between joists over the center beam. Insulation should be used in the floor in some manner to provide comfort and reduce heat loss. It is generally used between or over the joists. These details are covered in the working drawings.

Single-floor systems can also include the use of nominal 1- by 4-inch matched finish flooring in species such as southern pine and Douglas-fir and the lower grades of oak, birch, and maple in $25/32$-inch thickness. To prevent air and dust infiltration, joists should first be covered with 15-pound asphalt felt or similar materials. The flooring is then applied over the floor joists and the flood insulation added when the house is enclosed. When this single-floor system is used, however, some surface protection from weather and mechanical damage is required. A full-width sheet of heavy plastic or similar covering can be used, and the walls erected directly over the film. When most of the exterior and interior work is done, the covering can be removed and the floor sanded and finished.

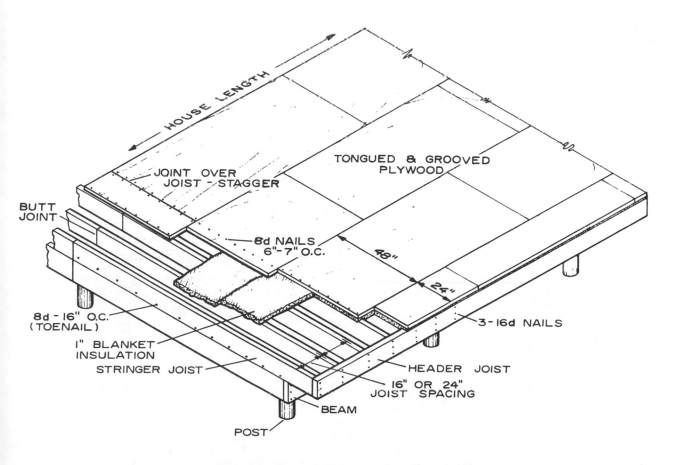

FIGURE 14.—Floor framing (post foundation with side overhang).

741

Post Foundation—With Side Overhang

The joists for a low-cost house are usually the third grade of such species as southern pine or Douglas-fir and are often 2 by 8 inches in size for spans of approximately 12 feet. If an overhang of about 12 inches is used for 12-foot lengths, the joist spacing normally can be 24 inches. Sizes, spacing, and other details are shown in the plans for each individual house.

The joists can now be cut to length, using a *butt joint* over the center beam. Thus, for a 24-foot-wide house, each pair of joists should be cut to a 12-foot length, less the thickness of the end header joist which is usually 1½ inches. The edge or stringer joists should be positioned on the beams with several other joists and the premarked headers nailed to them with one sixteenpenny nail (or just enough to keep them in position). The frame, including the edge (stringer) joists and the header joists, is now the exact outline of the house. Square up this framework by using the equal diagonal method (fig. 4). The overhang beyond the beams should be the same at each side of the house. Now, with eightpenny nails, *toenail* the joists to each beam they cross and the stringer joists to the beam beneath (fig. 14) to hold the framework exactly square. Add the remaining joists and nail the headers into the ends with three sixteenpenny nails. Toenail the remaining joists to the headers with eightpenny nails. When the center of a parallel partition wall is more than 4 inches from the center of the joists, add solid blocking between the joists. The blocking should be the same size as the joists and spaced not more than 3 feet apart. Toe-nail blocking to the joists with two tenpenny nails at each side.

In moderate climates, 1-inch blanket insulation may be sufficient to insulate the floor of crawl-space houses. It is usually placed between the joists in the same way that thicker insulation is normally installed. Another method consists of rolling 24-inch-wide 1-inch insulation across the joists, nailing or stapling it where necessary to keep it stretched with tight edge joints (fig. 14). Insulation of this type should have strong damage-resistant covers. Tenpenny ring-shank nails should be used to fasten the plywood to the joists rather than eightpenny common normally used. This will minimize nail movement or "nail pops" which could occur during moisture changes. The vapor barrier of the insulation should be on the upper side toward the subfloor. Two-inch and thicker blanket or batt insulation is placed between the joists and should be applied any time after the floor is in place, preferably when the house is near completion.

When the house is 20, 24, 28, or 32 feet wide, the first row of tongued-and-grooved plywood sheets should be 24 inches wide, so that the butt-joints of the joists at the center beam are reinforced with a full 48-inch-wide piece (fig. 15). This plywood is usually ⅝- or ¾-inch thick when it serves both as subfloor and underlayment. Rip 4-foot-wide pieces in half and save the other halves for the opposite side. Place the square, sawed edges flush with the header and nail the plywood to each crossing joist and header with eightpenny common nails spaced 6 to 7 inches apart at edges and at intermediate joists (fig. 14). Joints in the next

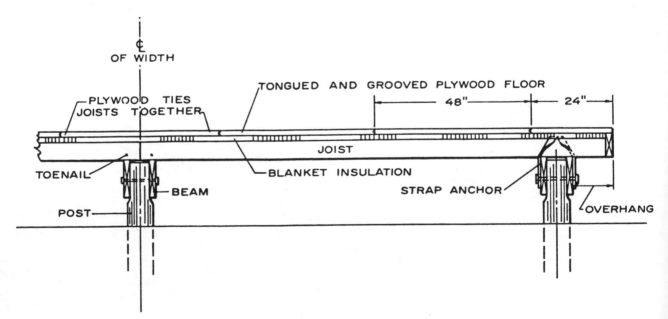

FIGURE 15.—Floor framing details (post foundation).

full 4-foot widths of plywood should be broken by starting at one end with a 4-foot-long piece. End joints will thus be staggered 48 inches. End joints should always be staggered at least one joist space, 16 or 24 inches. Be sure to draw up the tongued-and-grooved edges tightly. A chalked snap-string should be used to mark the position of the joists for nailing.

Edge Foundation-Masonry Pier or Wood Post

When edge piers or wood posts are used with edge support beams, the joists can be cut to length to fit snugly between the center and outside beams

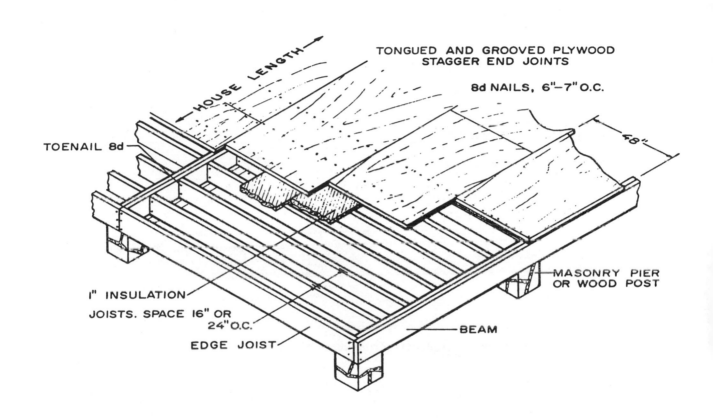

FIGURE 16. — Floor framing for edge foundation (masonry piers or wood posts).

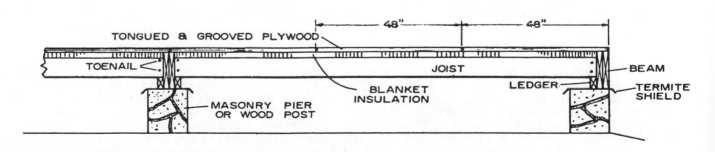

FIGURE 17.—Floor framing details (edge pier foundation).

743

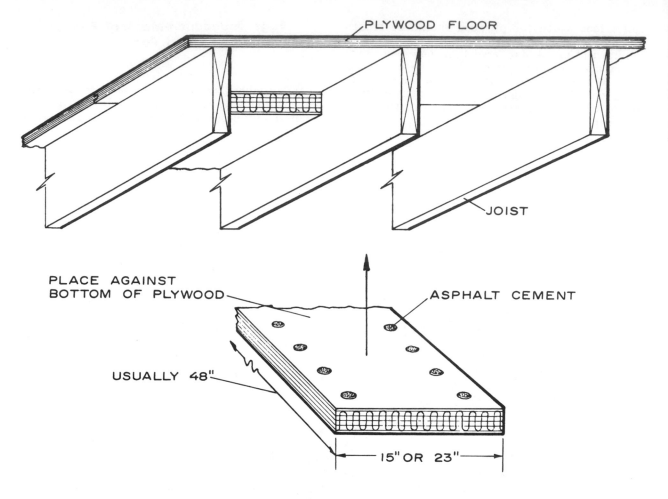

FIGURE 18.—Application of friction-fit batt insulation between floor joists.

so that they rest on the ledger strips. Sizes of headers, joists, and other details are shown on the working drawings for each individual house. Toe-nail each end to the beams with two eightpenny nails on each side (fig. 16). In applying the plywood subfloor to the floor framing, start with full 4-foot-wide sheets rather than the 2-foot-wide pieces used for the side overhang framing (fig. 17). The nail-laminated center beam provides sufficient reinforcing between the ends of joists. Apply the insulation and nail the plywood the same way as outlined in the previous section.

Insulation Between Joists

Thicker floor insulation than the 1-inch blanket is usually required for most houses. This will be indicated in the floor framing details of the working drawings or in the specifications. This type is normally used between the joists. Thus, the subfloor is nailed directly to the joists and the insulation placed between the joists after the subfloor is in place. Friction-fit (or similar insulating batts) 15 inches wide should be used for joists spaced 16

inches on center. Use 23-inch-wide batts for joists spaced 24 inches on center.

Friction-fit batts need little support to keep them in place. Small "dabs" of asphalt roof cement on the upper surfaces when installing against the bottom of the plywood will keep them in place (fig. 18). Standard batts can also be placed in this manner, but somewhat closer spacing of the cement might be required in addition to stapling along the edges. The use of a vapor barrier under the subfloor is important and is described in the section on "Thermal Insulation."

When other types of subfloor are specified, such as diagonal boards, some kind of overlay or finish is usually required. If tongued-and-grooved flooring is applied directly to and across the joists, a tie is normally required at the center butt joints of the floor joists. This is accomplished with a metal strap across the top of the joists or 1- by 4- by 20-inch wood strips (*scabs*) nailed across the faces of each set of joists at the joint with six eightpenny nails. When plywood subfloor is used, the sheets are centered over the center beam and joist ends to provide this tie for overhang floor framing.

Finally, if the plywood is likely to be exposed for any length of time before enclosing the house, a brush coat (or squeegee application) of water-repellent preservative should be used. This will not only repel moisture but will prevent or minimize any surface degradation.

FRAMED WALL SYSTEMS

Exterior sidewalls, and in some designs an interior wall, normally support most of the roof-loads as well as serving as a framework for attaching interior and exterior coverings. When roof trusses spanning the entire width of the house are used, the exterior sidewalls carry both the roof and ceiling loads. Interior partitions then serve mainly as room dividers. When ceiling joists are used, interior partitions usually sustain some of the ceiling loads.

The exterior walls of a wood-frame house normally consist of *studs*, interior and exterior coverings, windows and doors, and insulation. Moisture content of framing members usually should not exceed 19 percent.

The framework for a conventional wall consists of nominal 2- by 4-inch members used as top and bottom *plates*, as studs, and as partial (cripple) studs around openings. Studs are generally cut to lengths for 8-foot walls when subfloor and finish floors are used. This length depends on the thickness and number of wall plates (normally single bottom and double top plates). Studs can often be obtained from lumber dealers in a precut length. Double headers over doors and windows are generally larger than 2 by 4's when the width of the opening is greater than 2½ feet. Two 2- by 6-inch members are used for spans up to 4½ feet and two 2- by 8-inch members for openings from 4½ to about 6½ feet. Headers are normally cut 3 inches (two 1½-inch stud thicknesses) longer than the rough opening width unless the edge of the opening is near a regular spaced stud.

Framing of Sidewalls

The exterior framed walls, when erected, should be flush with the outer edges of the plywood subfloor and floor framing. Thus, the floor can be used both as a layout area and for horizontal assembly of the wall framing. When completed, the entire wall can be raised in place in "tilt-up" fashion, the plates nailed to the floor system, and the wall *plumbed* and braced.

The two exterior sidewalls of the house can be framed first and the exterior end walls later. Cut two sets of plates for the entire length of the house, using 8-, 12-, or 16-foot lengths, staggering the joints. Joints should be made at the centerline of a stud (*on center*). Starting at one end, mark each 16 or 24 inches, depending on the spacing of the studs, and also mark the centerlines for windows, doors, and partitions. These measurements are given on the working drawings. These are cen-

terline (*o.c.*) markings, except for the ends. With a small square, mark the location of each stud with a line about ¾ inch on each side of the centerline mark (fig. 19).

Studs can now be cut to the correct length. When a low-slope roof with wood decking (which also serves as a ceiling finish) is used, the stud length for an 8-foot wall height with single plywood flooring should be 95½ inches, less the thickness of three plates. Thus for plates 1½ inches thick, this will mean a stud length of 91 inches. When ceiling joists or trusses are used with the single plywood floor, this length can be about 92⅛ inches. These measurements are primarily to provide for the vertical use of 8-foot lengths of dry-wall sheet materials for the walls with the ceiling finish in place. Cornerposts can be made up beforehand by nailing two short 2- by 4-inch blocks between two studs (fig. 20). Use two twelvepenny nails at each side of each block.

Begin fabrication of the wall by fastening the bottom plate and the first top plate to the ends of each cornerpost and stud (with two sixteen-penny nails) into each member. As studs are nailed in place, provisions should be made for framing the openings for windows and doors (fig. 20). Studs should be located to form the rough openings, the sizes of which vary with the types of windows selected. The rough openings are framed by studs which support the window and door headers or *lintels*. A full-length stud should be located at each side of these framing studs (fig. 21).

The following allowances are usually made for rough opening widths and heights for doors and windows. Half of the given width should be marked on each side of the centerline of the opening, which has previously been marked on top and bottom plates.

A. *Double-hung window* (*single unit*)
   Rough opening width = glass width *plus* 6 inches
   Rough opening height = total glass height *plus* 10 inches
B. *Casement window* (*two sash*)
   Rough opening width = total glass width *plus* 11¼ inches
   Rough opening height = total glass height *plus* 6⅜ inches
C. *Exterior doors*
   Rough opening width = width of door *plus* 2½ inches
   Rough opening height = height of door *plus* 3 inches

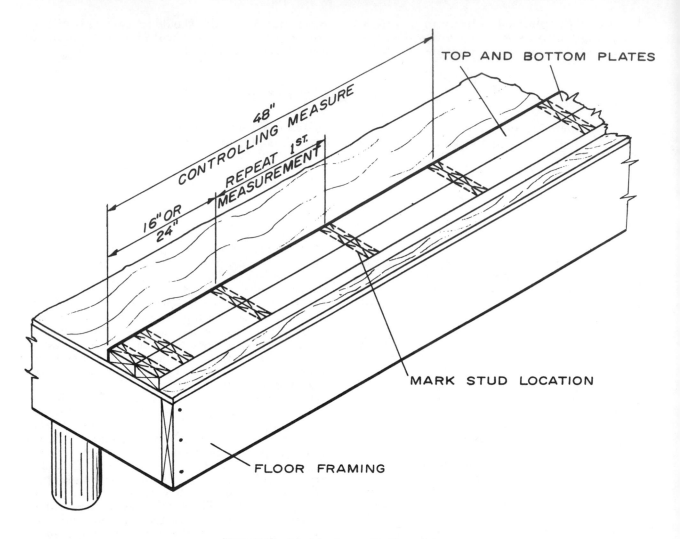

FIGURE 19.—Marking top and bottom plates.

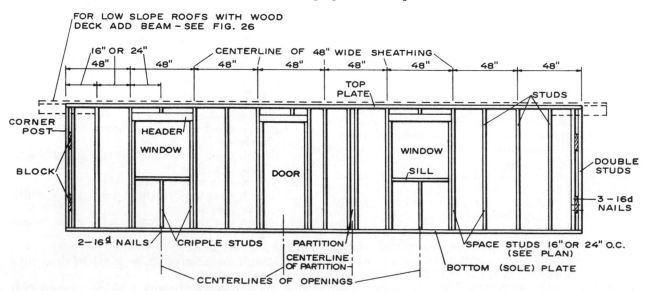

FIGURE 20.—Framing layout of typical wall.

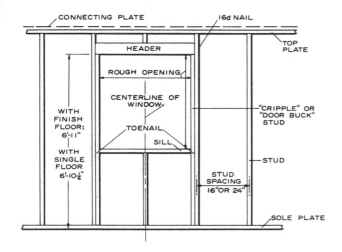

FIGURE 21.—Framing at window opening and height of window and door headers.

Clearances, or rough-opening sizes, for typical double-hung windows, for example, are shown in table 2.

TABLE 2.—*Frame opening sizes for double hung windows*

Window glass size (each sash)		Rough frame opening size	
Width *Inches*	Height *Inches*	Width *Inches*	Height *Inches*
24	16	30	42
28	20	34	50
32	24	38	58
36	24	42	58

The height of the window and door headers above the subfloor when doors are 6 feet 8 inches high and finish floor is used is shown in figure 21. When only resilient title is used over flooring made up of a single layer of material, the framing height for windows and doors should be 6 feet 10¼ inches for 6-foot 8-inch doors. The sizes of the headers should be the same as those previously outlined. Framing is arranged as shown in figure 21. Doubled headers can be fastened in place with two sixteenpenny nails through the stud into each member. Cripple (door *buck*) studs supporting the header on each side of the opening are nailed to the full stud with twelvepenny nails spaced about 16 inches apart and staggered. The sill and other short (cripple) studs are toenailed in place with two eightpenny nails at each side when end-nailing is not possible.

Doubled studs should normally be used on exterior walls where intersecting interior partitions are located. This is often accomplished with spaced studs, (fig. 22A) and provides nailing surfaces for interior covering materials. Blocking with 2-by-

4-inch members placed flatwise between studs spaced 4 to 6 inches apart in the exterior wall might also be used to fasten the first partition stud (fig. 22B). Blocks should be spaced about 32 inches apart. When a low-slope roof with gable overhang is used with wood decking, a beam extension is required at the top plates (figs. 20 and 26).

Erecting Sidewalls

When the sidewalls are completed, they can be raised in place. Nail several short 1- by 6-inch pieces to the outside of the beam to prevent the wall from sliding past the edge. The bottom plate is fastened to the floor framing with sixteenpenny nails spaced 16 inches apart and staggered when practical. The wall can now be plumbed and temporary bracing added to hold it in place in a true vertical position. Bracing may consist of 1- by 6-inch members nailed to one face of a stud and to a 2 by 4 block which has been nailed to the subfloor. Braces should be at about a 45° angle. If the wall framing is squared and braced, the panel siding or exterior covering can be fastened to the studs while the walls are still on the subfloor. In addition, window frames can be installed before erection of the wall. These processes are covered in following sections on "Exterior Wall Coverings" and "Exterior Frames."

End Walls—Moderate-Slope Roof

The exterior end walls for a gable-roofed house may be assembled on the floor in the same general manner as the sidewalls with a bottom plate and single top plate. However, the total length of the wall should be the exact distance between the inside of the exterior sidewalls already erected. Furthermore, only one end stud is used rather than the doubled cornerposts (fig. 23). Window and door openings are framed as outlined for the exterior sidewalls. When 48-inch-wide panel siding is used for the exterior, serving both as sheathing and siding, for example, the stud spacing should conform to the type of covering used. The center of the second stud in this wall should be 16 or 24 inches from the outside of the panel-siding material (fig. 24). This method of spacing should be used from each corner toward the center, and any adjustments required because of sheet-material width should be made at a center window or door.

End walls are erected in the same manner as the sidewalls with the bottom plate fastened to the floor framing. These walls also must be plumbed and braced. The end studs should be nailed at each side to the cornerposts with sixteenpenny nails spaced 16 inches apart. The upper top plate is added and extends across the sidewall plate (fig. 23).

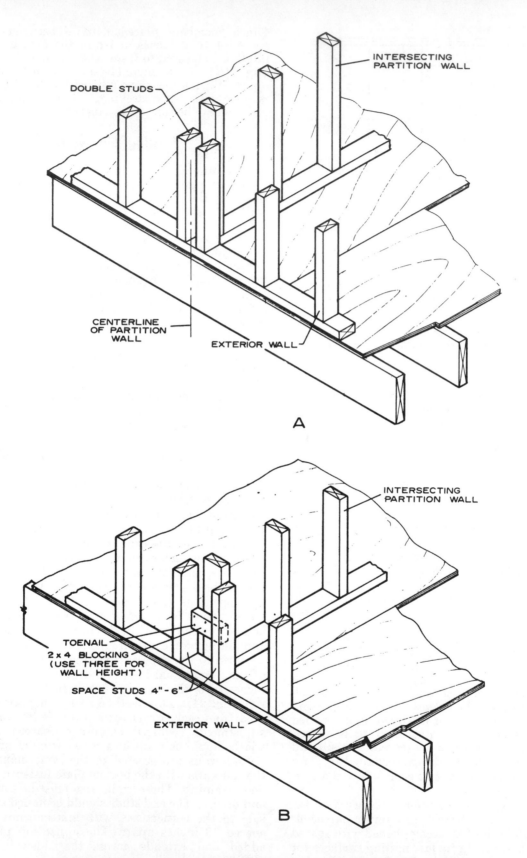

FIGURE 22.—Intersecting walls. *A*, Double studs; *B*, blocking between studs.

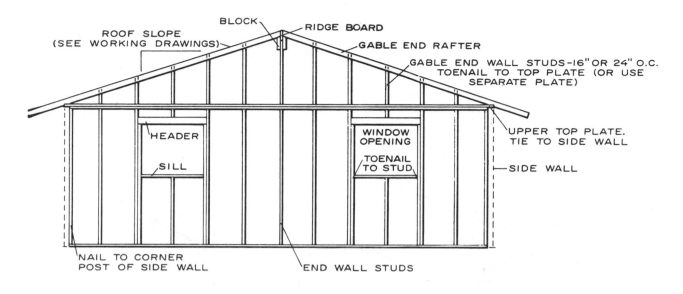

FIGURE 23.—End wall framing for regular slope roof (for trusses or rafter-type).

The framing for the gable-end portion of the wall is often done separately (fig. 23). Studs may be toenailed to the upper top wall plate, or an extra 2- by 6-inch bottom plate can be used, which provides a nailing surface for ceiling material in the rooms below. The top members of the gable wall are not plates; they are rafters which form the slope of the roof. Studs are notched to fit or may be used flatwise. Use the roof slope specified in the working drawings.

End Walls—Low-Slope Roof

End walls for a low-slope roof are normally constructed with *balloon framing*. In this design, the studs are full length from the bottom plate to the top or rafter plates which follow the roof slope (fig. 25). The stud spacing and framing for windows are the same as previously outlined. The top surface of the upper plate of the end wall should be in line with the outer edge of the upper top plate of the sidewall (fig. 26). Thus, when the roof decking is applied, bearing and nailing surfaces are provided at end and sidewalls. A beam extension beyond the end wall is provided to support the wood decking when a *gable*-end overhang is desired. This can be a 4- by 6-inch member which is fastened to the second sidewall stud (fig. 26).

The lower top plate of the end wall is nailed to the end of the studs before the upper top plate is fastened in place. For attaching the upper plate, use sixteenpenny nails spaced 16 inches apart and staggered. Two nails are used over the cornerposts of the sidewalls (fig. 26).

To provide for a center *ridge* beam which supports the wood decking inside the house, the area should be framed (fig. 27). After the beam is in place, twelvepenny nails are used through the stud on each side of the beam. The size of the ridge beam is shown on the working plans for each house which is constructed using this method. When decking is used for a gable overhang, the beam extends beyond the end walls.

Interior Walls

Interior walls in conventional construction (with ceiling joists and rafters) are erected in the same manner and at the same height as the outside walls. In general, assembly of interior stud walls is the same as outlined for exterior walls. The center load-bearing partition should be located so that ceiling joists require little or no wasteful cutting. Cross partitions are usually not load-bearing and can be spaced as required for room sizes. These spacings and other details are covered in the working drawing floor plan.

Studs should be spaced according to the type of interior covering material to be used. When studs are spaced 24 inches on center, the thickness of gypsum board, for example, must be 1/2 inch or greater. For 16-inch stud spacing, a thickness of 3/8 inch or greater can be used. Details of a typical intersection of interior walls are shown in figure 28. Load-bearing partitions should be constructed with nominal 2- by 4-inch studs, but 2- by 3-inch studs may be used for nonload-bearing walls. Doorway openings can also be framed with a single member on each side in nonload-bearing walls (fig. 28). Single top plates are commonly used on nonload-bearing interior partitions.

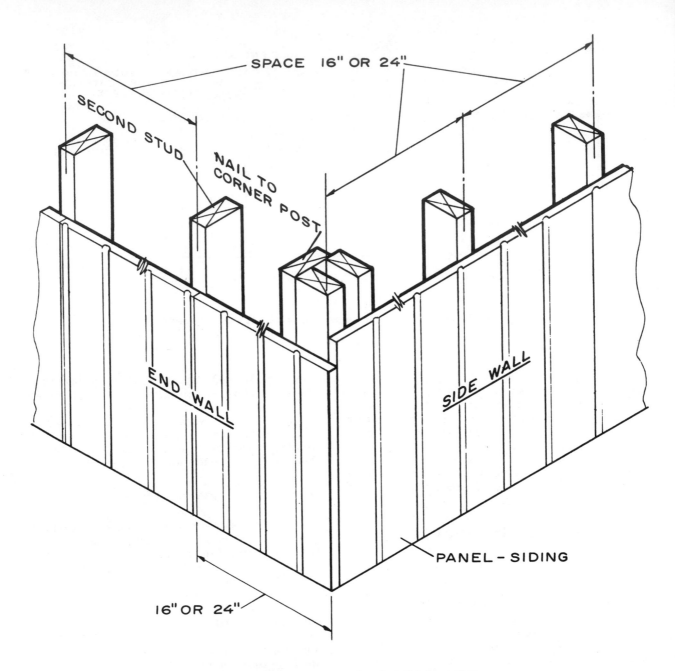

FIGURE 24.—Exterior side and end wall intersection.

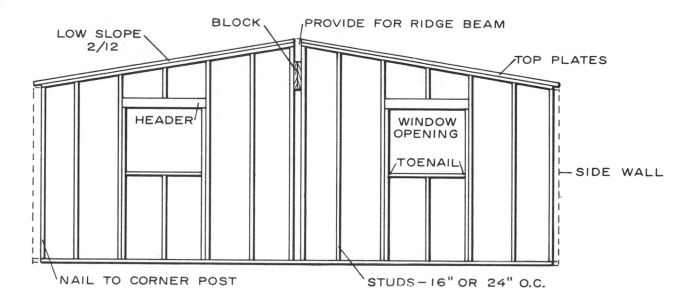

FIGURE 25.—Framing for end wall (low-slope roof).

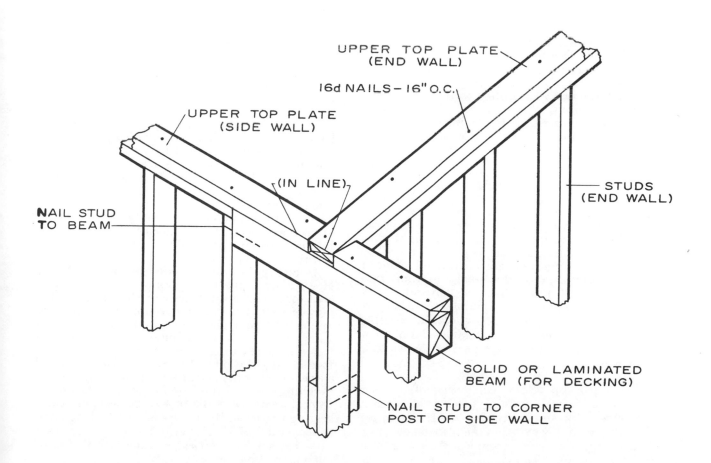

FIGURE 26.—Corner detail for low-slope roof with wood decking.

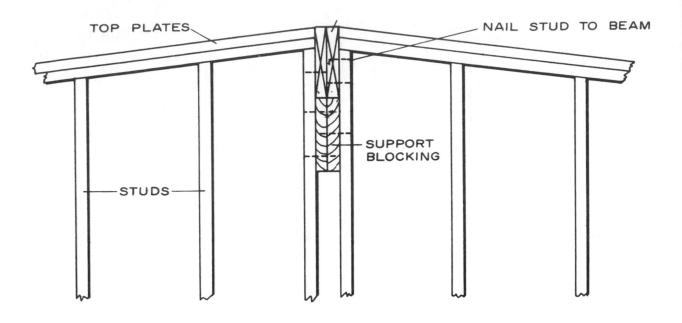

TOP PLATES

NAIL STUD TO BEAM

SUPPORT BLOCKING

STUDS

FIGURE 27.—Framing detail at ridge (end wall for low-slope roof).

ROOF SYSTEMS

Roof trusses require no load-bearing interior partitions, so location of the walls and size and spacing of studs are determined by the house design and by the type of interior finish. The bottom chords of the trusses are often used to tie in with crossing partitions where required. Details of stud location at the intersection of an interior partition with an exterior wall are shown in figure 22 *A* and *B*.

When a low-slope roof is used with wood decking, a full-height wall or a ridge beam is required for support at the center (fig. 29). The ridge beam may span from an interior center partition to an outside wall, forming a clear open area beneath. Cross or intersecting walls are full height with a sloping top plate. In such designs, one method uses the following sequence: (a) Erect exterior walls, center wall, and ridge beam; (b) apply roof decking; and (c) install other partition walls. Size and spacing of the studs in the cross walls are usually based on the thickness of the covering material, as no roof load is imposed on them. These spacings and sizes are part of the working drawings.

The upper top plates (connecting plates) are used to tie the wall framing together at corners, at intersections, and at crossing walls. The upper plate crosses and is nailed to the plate below (figs. 26 and 28). Two sixteenpenny nails are used at each intersection. The remainder of the upper top plate is nailed to the lower top plate with sixteen-

penny nails spaced 16 inches apart in a staggered pattern.

The primary function of a roof is to provide protection to the house in all types of weather with a minimum of maintenance. A second consideration is appearance; a roof should add to the attractiveness of the home, as well as being practical. Happily, a roof with a wide overhang at the cornice and the gable ends not only enhances appearance, but provides protection to side and end walls. Thus, even in lower cost houses, when the style and design permit, wide overhangs are desirable. Though they add slightly to the initial cost, savings in future maintenance usually merit this type of roof extension. Wood members used for roof framing should normally not exceed 19 percent moisture content.

As briefly described in the section on "Major House Parts," the two types of roofs commonly used for houses are (a) the low-slope and (b) the *pitched* roof. The flat or low-slope roof combines ceiling and roof elements as one system, which allows them to serve as interior finish, or as a fastening surface for finish, and as an outer surface for application of the roofing. The structural elements are arranged in several ways by the use of ceiling beams or thick roof decking, which spans from the exterior walls to a ridge beam or center bearing partition. Roof slope is usually designated as some ratio of 12. For example, a "4 in 12" roof slope has a 4-foot vertical rise for each 12 feet of horizontal distance.

The pitched roof, usually in slopes of 4 in 12 and

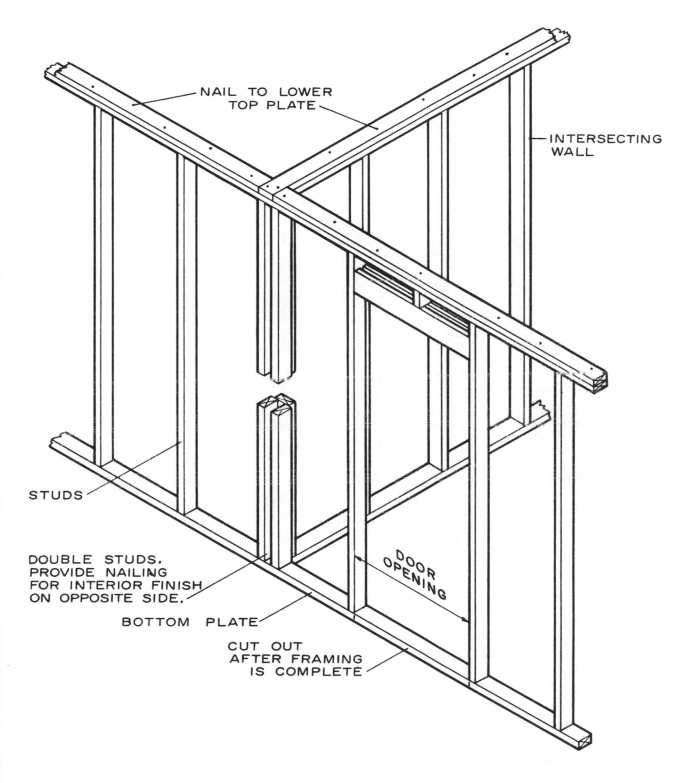

NAIL TO LOWER
TOP PLATE

INTERSECTING
WALL

STUDS

DOUBLE STUDS.
PROVIDE NAILING
FOR INTERIOR FINISH
ON OPPOSITE SIDE.

BOTTOM PLATE

CUT OUT
AFTER FRAMING
IS COMPLETE

DOOR
OPENING

FIGURE 28.—Intersection of interior walls.

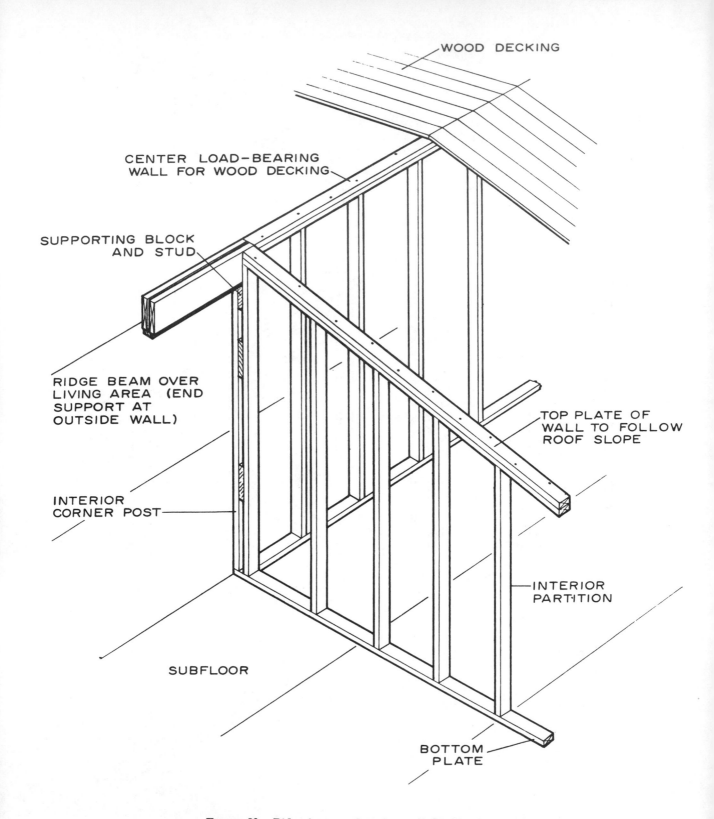

WOOD DECKING

CENTER LOAD—BEARING
WALL FOR WOOD DECKING

SUPPORTING BLOCK
AND STUD

RIDGE BEAM OVER
LIVING AREA (END
SUPPORT AT
OUTSIDE WALL)

TOP PLATE OF
WALL TO FOLLOW
ROOF SLOPE

INTERIOR
CORNER POST

INTERIOR
PARTITION

SUBFLOOR

BOTTOM
PLATE

FIGURE 29.—Ridge beam and center wall for low-slope roof.

greater, has structural elements in the form of (a) *rafters* and joists or (b) trusses (trussed rafters). Both systems require some type of interior ceiling finish, as well as roof sheathing. With slopes of 8 in 12 and greater, it is possible to include several bedrooms on the second floor when provisions are made for floor loads, a stairway, and windows.

Low-Slope Ceiling Beam Roof

One of the framing systems for a low-slope roof consists of spaced rafters (beams or *girders*) which span from the exterior sidewalls to a ridge beam or a center load-bearing wall. The rafter can be doubled, spaced 4 feet apart, and exposed in the room below providing a pleasing beamed ceiling effect. *Dressed and matched* V-groove boards can be used for roof sheathing and exposed to the room below. When plywood or other unfinished sheathing is used, a ceiling tile or other prefinished wallboard can be fastened to the undersurface. Such materials also serve as insulation. Thus, a very attractive ceiling can be provided using a light color for the ceiling and a contrasting stain on the beams. This type of framing can be varied by spacing single rafters on 16- or 24-inch centers. Separate covering materials would normally be used for the roof sheathing and for the ceiling, with flexible insulation between.

The size and spacing details for ceiling beams are shown on the working drawings for each house design. For example, when beams are doubled, spaced 48 inches apart, and the distance from outer wall to interior wall is about 11½ feet, two 2- by 8-inch members are satisfactory for most of the construction species such as Douglas-fir, southern pine, and hemlock. Use of some wood species, such as the soft pines, will require two 2- by 10-inch members for 48-inch spacing. When a spacing of 32 inches is desirable for appearance, two 2- by 6-inch members of the second grade of Douglas-fir or southern pine are satisfactory. In some of the species, such as southern pine and Douglas-fir, a solid 4- by 6-inch member provides sufficient strength for 48-inch spacing over an 11½-foot span.

The details of fastening and anchoring these structural members to the wall elements can vary somewhat. Variations from the details included in this manual are shown on the working drawings for each individual house.

Construction.—We will assume that the details in the working plans specify doubled 2- by 8-inch ceiling beams spaced 48 inches on center. When the beams do not extend beyond the wall, a lookout member is required for the *cornice* overhang. The center wall or ridge beam is in place, so the roof slope, which may vary between 1½ in 12 to 2½ in 12, has been established. As previously outlined in the section on wall systems, the ceiling beams should normally be erected before cross walls are established. Thus, the exterior sidewalls and the load-bearing center wall, all well braced and plumbed, are all that is required to erect the ceiling beams.

There are several methods in which the beams are supported at the center bearing wall or ridge beam: (a) By a 2- by 3-inch block fastened to the stud wall; (b) by a metal joist hanger; and (c) by notching the beam ends and fastening to the stud. The first two may be used for either a stud wall or a ridge beam. Fastening at the outside wall is generally the same for all three methods.

The first or sample beam can now be cut to serve as a pattern in cutting the remainder of the members. Figure 30A shows the location of the ceiling beam with respect to the exterior and load-bearing center walls. When beams themselves do not serve as roof extensions, they can be assembled by nailing a 2- by 6-inch *lookout* (roof extension) at the outer wall and a 2- by 4-inch spacer block at the interior center wall (fig. 30B). Use two twelvepenny nails on each side of the block and twelvepenny nails spaced 6 inches apart for the 2- by 6-inch lookout member. When a block nailed to the stud is used to support the interior beam end, the ends should be notched (fig. 30C).

Details at center wall or beam.—The first system of connecting the inside end of the ceiling beam is most adaptable to ridge-beam construction. It consists of fastening a nominal 2- by 3-inch block to the beam with 4½-inch lag screws (fig. 31A). The 2 by 3 should be the same depth as the ceiling beam. The beam ends are then bolted to the 2- by 3-inch block with ⅜- by 5-inch carriage bolts.

Using joist hangers (fig. 31B) to fasten the inside end of the ceiling beams is another method most adaptable to a ridge-beam system. Hangers are fastened to the ridge-beam face, the ceiling beams dropped in place, and the hangers nailed to the beams. Eight- or tenpenny nails are commonly used for nailing. Hangers will be exposed, but can be painted to match the color of the beams.

A third method which can be used at a center bearing wall includes notching the ends of the ceiling beams (figs. 30C and 31C). When the beam is in place, it is face-nailed to the stud at each side with two twelvepenny nails. Short 2- by 4-inch support blocks are then nailed at each side of the stud with twelvepenny nails (fig. 31C).

When solid 4- by 6-inch or larger members are used as ceiling beams in place of the doubled members, the joist hanger is probably the most suitable method of supporting the inside ends of the beams. If the solid or laminated beams are to be stained, care should be taken to prevent hammer marks.

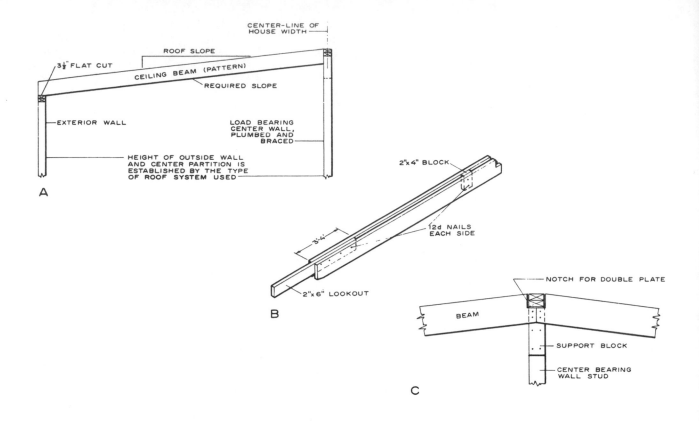

FIGURE 30.—Ceiling beam location and support methods. *A*, Lay out of typical beam; *B*, assembly of beam with roof extension (lookout); *C*, notch for stud support at center bearing wall.

Nailing ceiling beams at exterior walls.—The ceiling beams are normally fastened to the top plate of the outside walls by nailing. In windy areas, some type of strapping or metal bracket is often desirable (fig. 41*B*). The ceiling beams are toenailed to the top of the outside wall with two eightpenny nails at each side and a tenpenny nail at the ends (fig. 32). To provide nailing for panel siding and interior finish, 2- by 4-inch nailing blocks are fastened between the ceiling beams (fig. 32). Toenail with eightpenny nails at each edge and face.

Roof sheathing.—Ceiling beams and roof extensions are now in place and ready for installation of the roof sheathing. Roof sheathing can consist of 1- by 6-inch tongued-and-grooved V-groove) lumber with $25/_{32}$-inch fiberboard nailed over the top for insulation. Use two eightpenny nails for each board at each ceiling beam. The insulation fiberboard can be nailed in place with 1¼-inch roofing nails spaced 10 inches apart in rows 24 inches on center. Cross sections of the completed wall and roof framing are shown in figures 33 *A* and *B*. A nominal 1-inch member about 4¾ inches wide may be used to case the undersides of the beams.

A gable-end extension of 16 inches or less can be supported by extending the dressed and matched V-edge roof boards (fig. 34*A*). A 2- by 2-inch or larger member (*fly rafter*) is nailed to the underside of the boards and serves to fasten the facia board and molding (fig. 34*B*). The V-groove of the underside of the 1 by 6 roof sheathing serves as a decorative surface.

Rafter-Joist Roof

Another type of construction for low-slope roofs similar to the ceiling beam framing is the rafter-joist roof, in which the members are spaced 16 or 24 inches apart and serve both as rafters and ceiling joists (fig. 35 *A* and *B*). Members may be 2 by 8 or 2 by 10 inches in size. Specific sizes are shown on the working drawings for each house plan. The space between joists is insulated, allowing space for a ventilating *airway*. Gypsum board or other types of interior finish can be nailed directly to the bottoms of the joists.

Rafter extensions can serve as nailing surfaces for the *soffit* of a closed cornice (fig. 35*A*). When an open cornice is used, a nailing block is required over the wallplates and between rafters for the siding or *frieze* board.

The inside ends of the rafter-joists bear on an

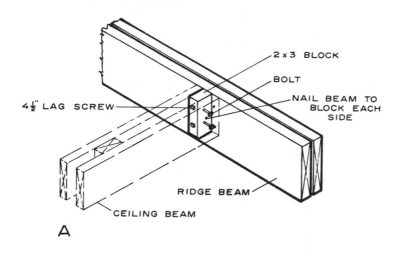

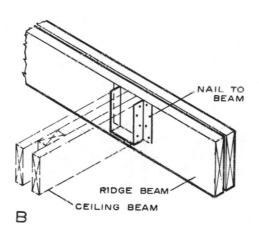

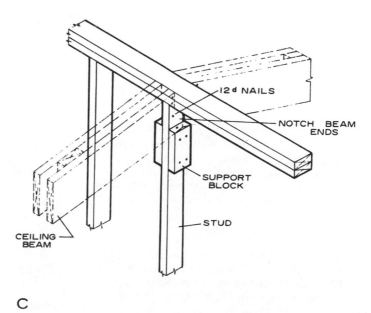

FIGURE 31.—Beam connection to ridge-beam or load-bearing wall. *A*, Block support; *B*, joist hanger; *C*, stud support block.

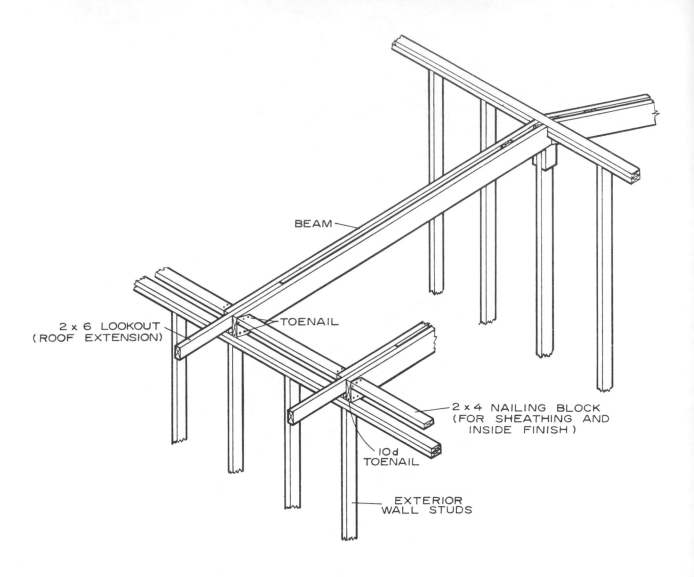

BEAM

2 x 6 LOOKOUT
(ROOF EXTENSION)

TOENAIL

2 x 4 NAILING BLOCK
(FOR SHEATHING AND
INSIDE FINISH)

10d
TOENAIL

EXTERIOR
WALL STUDS

FIGURE 32.—Fastening ceiling beams at exterior walls.

interior load-bearing wall (fig. 35B). Beams are toenailed to the plate with eightpenny nails on each side. A 1- by 4-inch wood or ⅜-inch plywood *scab* is used to connect opposite rafter-joists. This fastens the joists together and serves as a positive tie between the exterior sidewalls.

Low-Slope Wood-Deck Roof

A simple method of covering low-slope roofs is with wood decking. Decking should be strong enough to span from the interior center wall or beam to the exterior wall. Decking can also extend beyond the wall to form an overhang at the *eave* line (fig. 36A). This system requires dressed and matched nominal 2- by 6-inch southern pine or Douglas-fir decking or 3- by 6-inch solid or laminated decking (cedar or similar species) for spans

of about 12 feet. The proper sizes are shown in the working drawings. While this system requires more material than the beam and sheathing system, the labor involved at the building site is usually much less.

When gable-end extension is desired, some type of support is required beyond the end wall line at plate and ridge. This is usually accomplished by the projection of a small beam at the top plate of each sidewall (figs. 20 and 26) and at the ridge. Often the extension of the double top plate of the sidewall is sufficient. Depending on the type and thickness, the decking must sometimes be in one full-length piece without joints unless there are intermediate supports in the form of an interior partition. When such an interior wall is present, a butt joint can be made over its center. The working drawings cover these various details.

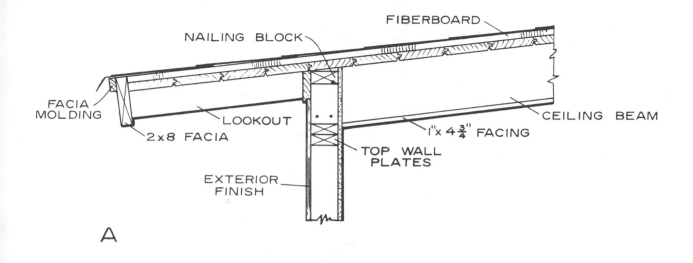

FIBERBOARD

NAILING BLOCK

FACIA MOLDING

LOOKOUT

2x8 FACIA

EXTERIOR FINISH

TOP WALL PLATES

1"x 4¾" FACING

CEILING BEAM

A

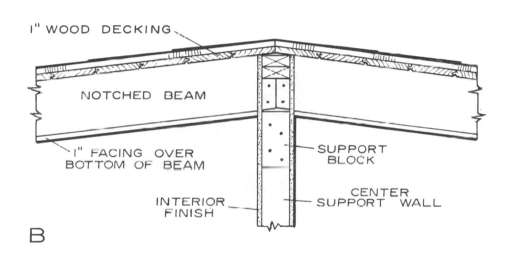

1" WOOD DECKING

NOTCHED BEAM

1" FACING OVER BOTTOM OF BEAM

INTERIOR FINISH

SUPPORT BLOCK

CENTER SUPPORT WALL

B

FIGURE 33.—Cross sections of completed walls and roof framing. *A,* Section through exterior wall; *B,* section through center wall.

Figure 36 shows the method of applying wood decking. This type of wood decking usually has a decorative V-edge face which should be placed down. Often a light-colored stain or other finish is applied to the wood decking before it is installed. Prefinished members can also be obtained. Each 2- by 6-inch decking member is face-nailed to the ridge beam or center wall and to the top plates of the exterior wall with two sixteenpenny ring-shank nails (fig. 36*A*). In addition, sixpenny finish nails should be toenailed along each joint on 2- to 3-foot centers (fig. 36*B*). A 40° angle or less should be used so that the nail point does not penetrate the underside. Nailheads should be driven flush with the surface.

When nominal 3- by 6-inch decking in solid or laminated form is required, it is face-nailed with two twentypenny ring-shank nails at center and outside wall supports. Solid 3- by 6-inch decking usually has a double tongue and groove and is provided with holes between the tongues for horizontal edge nailing (fig. 36*C*). This edge nailing is done with 7- or 8-inch-long ring-shank nails, often furnished by manufacturers of the decking. Laminated decking can be nailed along the lengthwise joints with sixpenny nails through the groove and tongue. Space nails about 24 inches apart.

Decking support at the load-bearing center wall and at the ends of the sidewalls may also be provided by an extension of the top plates (fig. 37).

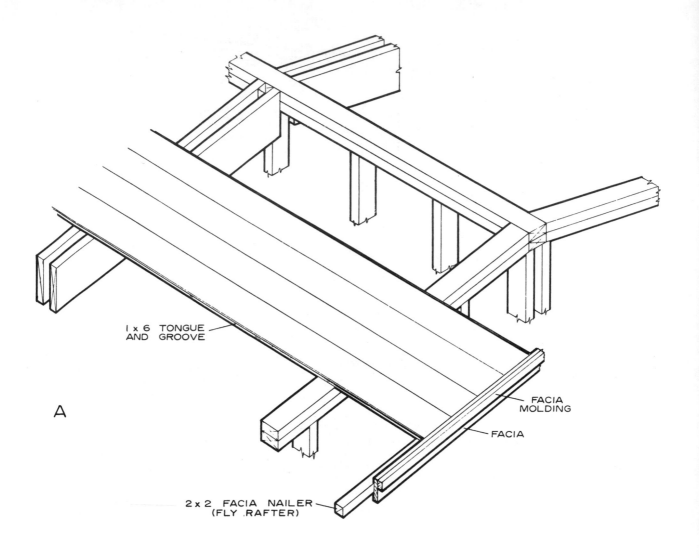

A

1 x 6 TONGUE
AND GROOVE

FACIA
MOLDING

FACIA

2 x 2 FACIA NAILER
(FLY RAFTER)

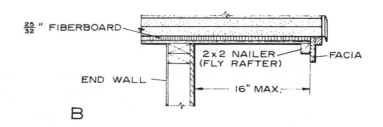

$\frac{25}{32}$ " FIBERBOARD

2 x 2 NAILER
(FLY RAFTER)

FACIA

END WALL

16" MAX.

B

FIGURE 34.—Gable-end extension detail. *A*, Gable extension; *B*, fly rafter.

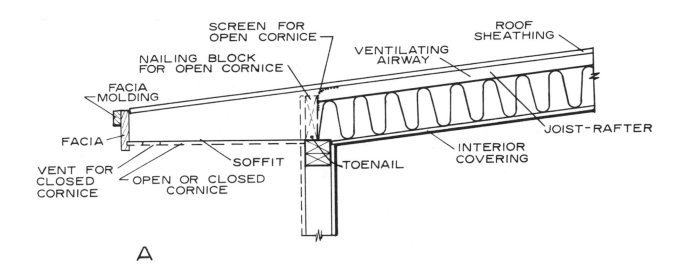

A

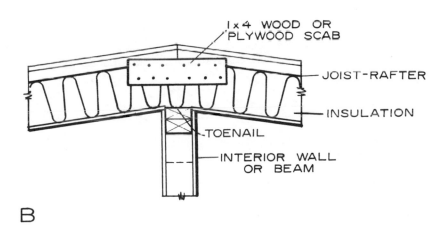

B

FIGURE 35.—Rafter-joist construction. *A*, Detail at exterior wall; *B*, detail at interior wall.

The sides of the members can be faced later, if desired, with the same material used for siding or with 1- by 6- and 1- by 8-inch members.

Insulation for Low-Slope Roofs

It is considerably more difficult and costly to provide the low-slope roof with adequate insulation, except when the rafters-joist system is used with both interior and exterior covering, and even then the joists must have adequate depth to provide space for insulation. Therefore, low-slope roofs are most economical in extremely mild climates or in houses for seasonal use where both heating and air conditioning requirements are quite minimal. Some insulation can be added economically by the use of 25/32-inch insulating board sheathing placed over the wood decking. The use of expanded foam in sheet form as a base for the ½-inch insulation board or tile on the underside of the decking will provide increased resistance to heat loss. Rigid foam insulation can also be used on top of the decking, but ⅜-inch plywood is required over the foam to provide a nailing base for the shingles.

The "U" value of a building component is a measure of resistance to heat loss or gain and the

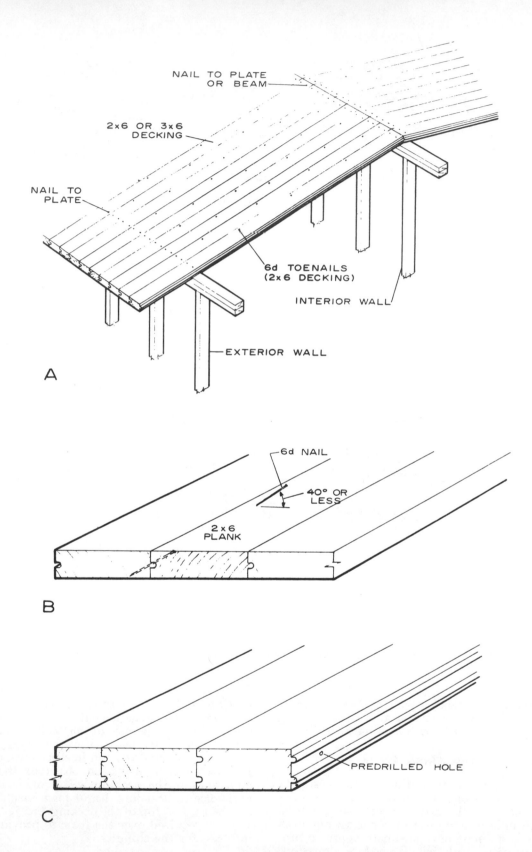

NAIL TO PLATE OR BEAM

2×6 OR 3×6 DECKING

NAIL TO PLATE

6d TOENAILS (2×6 DECKING)

INTERIOR WALL

EXTERIOR WALL

A

6d NAIL

40° OR LESS

2×6 PLANK

B

PREDRILLED HOLE

C

FIGURE 36.—Wood-deck construction. *A,* Installing wood decking; *B,* toenailing horizontal joint; *C,* edge nailing 3- by 6-inch solid decking.

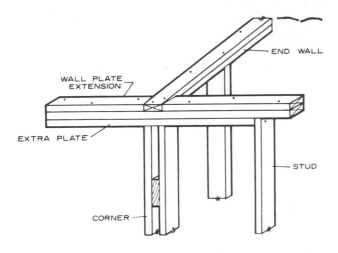

FIGURE 37.—Extension of wallplate for decking support.

lower the number is, the more effective the insulation. Figure 38 shows the value of adding various insulating materials to the basic wood decking.

The method of installing the $^{25}\!/_{32}$-inch insulating sheathing over the wood decking is relatively simple. The 4- by 8-foot sheets should be laid horizontally across the decking. Use 1¼- or 1½-inch roofing nails spaced 10 inches apart in rows 24 inches apart along the length.

Insulating board or tile in ½-inch thickness and the expanded foam insulation are installed with wallboard adhesive designed for these materials. Manufacturers normally recommend the type and method of application. With ½- by 12- by 12-inch tile, for example, a small amount of adhesive in each corner and the use of hand pressure as the tile is placed is one system which is often used. When tile is tongued and grooved, stapling is the usual method of installation. Larger sheets of ½-inch insulating board may require a combination of glue and some nailing. Expanded foam insulation is ordinarily installed with approved adhesives.

When 1-inch wood decking is used over the joist-beam system (fig. 33 A and B), the use of at least $^{25}\!/_{32}$-inch insulating sheathing over the boards and ½-inch insulating board or tile on the inside is normally recommended. A 1-inch thickness of expanded foam insulation under the ½-inch tile would provide even better insulation. When the inner face of the decking is to be covered, lower grade 2- by 6-inch decking is commonly used.

Rigid foam over the deck is usually applied with roofing nails that penetrate the deck a minimum of ¾ inch; however, foams vary in composition, so manufacturer's instructions should be followed. Plywood at least ⅜-inch thick can be applied over the foam as a nailing base for shingles.

Trim For Low-Slope Roofs

Simple trim in the form of facia boards can be used at roof overhangs and at side and end walls. When 2- by 6-inch lookouts are used in the ceiling-beam roof, a 2- by 8-inch *facia* member is usually required to span the 48-inch spacing of the beams (fig. 33 A). In addition, a 1- by 2-inch facia molding may be added. Use two sixteenpenny galvanized nails in the ends of each lookout member. The facia molding may be nailed with six- or sevenpenny galvanized nails on 16-inch centers.

Trim for roofs with nominal 2- or 3-inch-thick wood decking can consist of a 1- by 4- or 1- by 6-inch member with a 1- by 2-inch facia molding at the side and end-wall overhangs (fig. 39 A and B). A 1- by 4-inch member can be used for 2-inch roof decking with or without the $^{25}\!/_{32}$-inch insulating fiberboard. When 3-inch roof decking is used with the fiberboard, a 1- by 6-inch piece is generally required. Nail the facia and molding to the decking with eightpenny galvanized nails spaced about 16 inches apart. The roof deck is now ready for the roofing material.

Pitched Roof

A pitched-roof house is commonly framed by one of two methods: (a) With trussed rafters or (b) with conventional rafter and ceiling joist members. These framing methods are used most often for roof slopes of 4 in 12 and greater. The common W-truss (fig. 40A) for moderate spans requires less material than the joist and rafter system, as the members in the upper and lower chords are usually only 2 by 4 inches in size for spans of 24 to 32 feet. The king-post truss (fig. 40B) for spans of 20 to 24 feet uses even less material than the W-truss, but is perhaps more suitable for light to moderate roof loads. Low-slope roof trusses usually require larger members. In addition to lowering material costs, the truss has the advantage of permitting freedom in location of interior partitions because only the sidewalls carry the ceiling and roof loads.

The roof sheathing, trim, roofing, interior ceiling finish, and type of ceiling insulation used do not vary a great deal between the truss and the conventional roof systems. For plywood or lumber sheathing, 24-inch spacing of trusses and rafters and joists is considered a normal maximum.

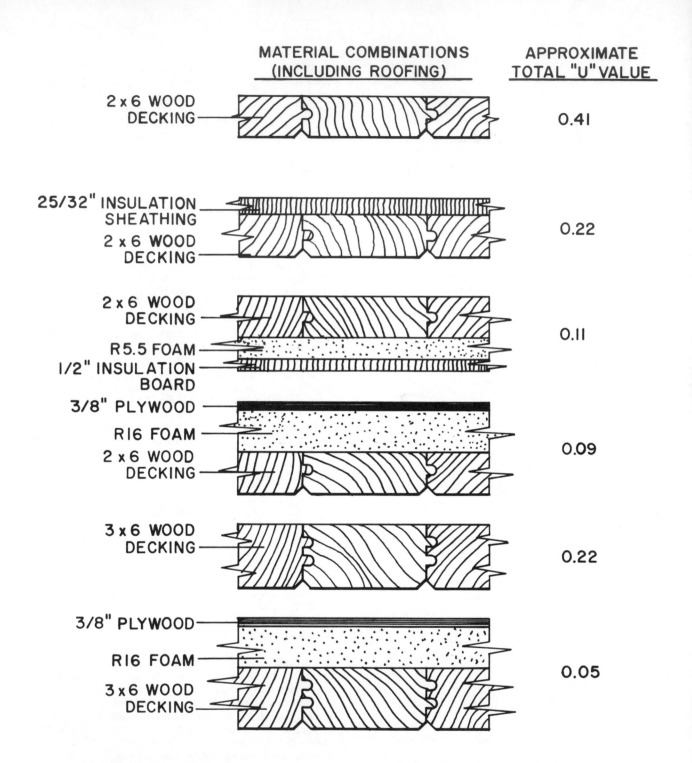

MATERIAL COMBINATIONS (INCLUDING ROOFING)	APPROXIMATE TOTAL "U" VALUE
2 x 6 WOOD DECKING	0.41
25/32" INSULATION SHEATHING / 2 x 6 WOOD DECKING	0.22
2 x 6 WOOD DECKING / R5.5 FOAM / 1/2" INSULATION BOARD	0.11
3/8" PLYWOOD / R16 FOAM / 2 x 6 WOOD DECKING	0.09
3 x 6 WOOD DECKING	0.22
3/8" PLYWOOD / R16 FOAM / 3 x 6 WOOD DECKING	0.05

FIGURE 38.—Insulating values for wood roof decks with various insulating materials added.

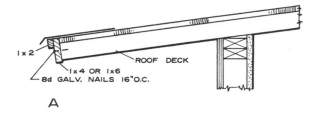

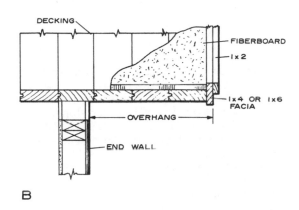

FIGURE 39.—Facia for wood-deck roof. *A*, Sidewall over-hang; *B*, end-wall overhang.

Greater spacing can be used, but it usually requires a thicker roof sheathing and application of wood stripping on the undersides of the ceiling joists and trusses to furnish a support for ceiling finish. Thus, most W-trusses are designed for 24-inch spacing and joist-rafter construction for 24- or 16-inch spacing. Trusses generally require a higher grade dimension material than the joist and rafter

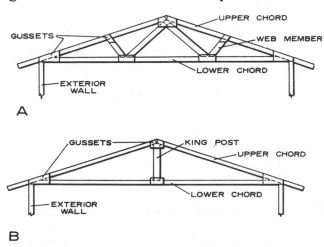

FIGURE 40.—Trussed rafters. *A*, W-type truss; *B*, king-post truss.

roof. However, specific details of the roof construction are covered in the working drawings for each house.

Trussed Roof

The common truss or trussed rafter is most often fabricated in a central shop. While some are constructed at the job side, an enclosed building provides better control for their assembly. These trusses are fabricated in several ways. The three most common methods of fastening members together are with (a) metal truss plates, (b) plywood gussets, and (c) ring connectors.

The metal truss plates, with or without prongs, are fastened in place on each side of member intersections. Some plates are nailed and others have supplemental nail fastening. Metal-plate trusses are usually purchased through a large lumber dealer or manufacturer and are not easily adapted to on-site fabrication. The trusses using fully nailed metal plates can usually be assembled at a small central shop.

The plywood-gusset truss may be a nailed or nailed-glued combination. The nailed-glued combination, with nails supplying the pressure, allows the use of smaller gussets than does the nailed system. However, if on-site fabrication is necessary, the nailed gusset truss and the ring connector truss are probably the best choices. Many adhesives suitable for trusses generally require good temperature control and weather protection not usually available on site. The size of the gussets, the number of nails or other connectors, and other details for this type of roof are included in the working drawings for each house.

Completed trusses can be raised in place with a small mechanical lift on the top plates of exterior sidewalls. They can also be placed by hand over the exterior walls in an inverted position, and then rotated into an upright position. The top plates of the two sidewalls should be marked for the location of each set of trusses. Trusses are fastened to the outside walls and to 1- by 4- or 1-by 6-inch temporary horizontal braces used to space and aline them until the roof sheathing has been applied. Locate these braces near the ridge line.

Trusses can be fastened to the top wallplates by toenailing, but this is not always the most satisfactory method. The heel gusset in a plywood-gusset or metal plate truss is located at the wall-plate and makes toenailing difficult. However, two tenpenny nails at each side of the truss can be used in nailing the lower chord to the plate (fig. 41 *A*). Predrilling may be necessary to prevent splitting. A better system involves the use of a simple metal connector or bracket obtained from local lumber dealers. Brackets should be nailed to the wallplates at sides and top with eightpenny nails and to the lower chords of the truss with sixpenny or 1½-inch roofing nails (fig. 41 *B*) or as recommended by the manufacturer.

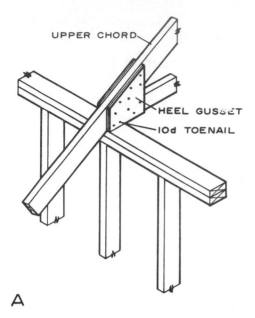

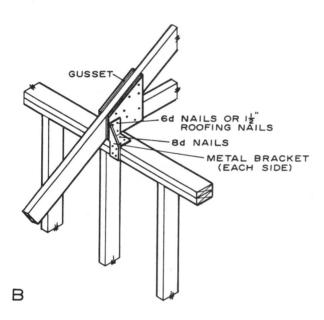

FIGURE 41.—Fastening trusses to wallplate. *A*, Toenailing; *B*, metal bracket connector.

The gable-end walls for a pitched roof utilizing trusses are usually made the same way as those described in the section on "Wall Systems" and shown in figure 23.

Rafter and Ceiling Joist Roof

Conventional roof construction with ceiling joists and rafters (fig. 42) can begin after all load-bearing and other partition walls are in place. The upper top plate of the exterior wall and the load-bearing interior wall serve as a fastening area for ceiling joists and rafters. Ceiling joists are installed along premarket exterior top wallplates and are toenailed to the plate with three eightpenny nails. The first joist is usually located next to the top plate of the end wall (fig. 43). This provides edge-nailing for the ceiling finish. Ceiling joists crossing a center loadbearing wall are face-nailed to each other with three or four sixteenpenny nails. In addition, they are each toenailed to the plate with two eightpenny nails.

Angle cuts for the rafters at the *ridge* and at the exterior walls can be marked with a carpenter's square using a reference table showing the overall rafter lengths for various spans, roof slopes, and joist sizes. These tables can usually be obtained from your lumber dealer. However, if a rafter table is not available, a baseline can be laid out on the subfloor across the width of the house, marking an exact outline of the roof slope, ridge, board and exterior walls. Thus, a rafter pattern can be made, including cuts at the ridge, wall, and the overhang at the eaves.

Rafters are erected in pairs. The *ridge board* is first nailed to one rafter end with three tenpenny nails (fig. 44). The opposing rafter is then nailed to the first with a tenpenny nail at the top and two eightpenny nails toenailed at each side. The outside rafter is located flush with and a part of the gable-end walls (fig. 43).

While the ridge nailing is being done, the rafters should be toenailed to the top plates of the exterior wall with two eightpenny nails (fig. 43). In addition, each rafter is face-nailed to the ceiling joist with three tenpenny nails. The remaining rafters are installed the same way. When the ridge board must be spliced, it should be done at a rafter with nailing at each side.

If gable-end walls have not been erected with the end walls, the gable-end studs can now be cut and nailed in place (fig. 43). Toenail the studs to the plate with eightpenny nails and face-nail to the end rafter from the inside with two tenpenny nails. In addition, the first or edge ceiling joist can be nailed to each gable-end stud with two tenpenny nails. Gable-end studs can also be used flatwise between the end rafter and top plate of the wall.

When the roof has a moderately low slope and the width of the house is 26 feet or greater, it is often desirable to nail a 1- by 6-inch *collar beam* to every second or third rafter (fig. 42) using four eightpenny nails at each end.

Framing for Flush Living-Dining Area Ceiling

A living-dining-kitchen group is often designed as one open area with a flush ceiling throughout. This makes the rooms appear much larger than they actually are. When trusses are used, there

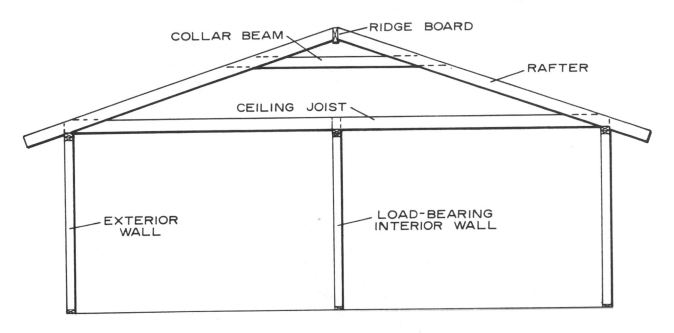

FIGURE 42.—Rafter and ceiling joist roof framing.

is no problem, because they span from one exterior wall to the other. However, if ceiling joists and rafters are used, some type of *beam* is needed to support the interior ends of the ceiling joists. This can be done by using a flush beam, which spans from an interior cross wall to an exterior end wall. Joists are fastened to the beams by means of joist hangers (fig. 45). These hangers are nailed to the beam with eightpenny nails and to the joist with sixpenny nails or 1½-inch roofing nails. Hangers are perhaps most easily fastened by first nailing to the end of the joist before the joist is raised in place.

An alternate method of framing utilizes a wood bracket at each pair of ceiling joists tieing them to a beam which spans the open living-dining area (fig. 46). This beam is blocked up and fastened at each end at a height equal to the depth of the ceiling joists.

Roof Sheathing

Plywood or lumber roof sheathing is most commonly used for pitched roofs. Nominal 1-inch boards no wider than 8 inches can be used for trusses or rafters spaced not more than 24 inches on center. Sheathing (standard) and other grades of plywood are marked for the allowable spacing of the rafters and trusses for each species and thickness used. For example, a "24/0" mark indicates it is satisfactory as roof sheathing for 24-inch spacing of roof members, but not satisfactory for subfloor.

Nominal 1-inch boards should be laid up without spacing and nailed to each rafter with two eightpenny nails. Plywood sheets should be laid across the roof members with staggered end joints. Use sixpenny nails for ⅜-inch and thinner plywood and eightpenny nails for ½-inch and thicker plywood. Space the nails 6 inches apart at the edges and 12 inches at intermediate fastening points.

When gable-end overhangs are used, extend the trim to the plywood or roofing boards when necessary before the 2- by 2- or 2- by 4-inch fly rafter (facia nailer) is nailed in place.

Roof Trim

Roof trim is installed before the roofing or shingles are applied. The cornice and gable (*rake*) trim for a pitched roof can be the same whether trusses or rafter-ceiling joist framing are used. In its simplest form, the trim consists of a facia board, sometimes with molding added. The facia is nailed to the ends of the rafter extensions or to the fly rafters at the gable overhang. With more complete trim, a soffit is usually included at the cornice and gables.

Cornice

The facia board at rafter ends or at the extension of the truss is often a 1- by 4- or 1- by 6-inch member (fig. 47A). The facia should be nailed to the end of each rafter with two eightpenny galvanized nails. Trim rafter ends when necessary for a straight line. Nail 1- by 2-inch facia molding with one eightpenny galvanized nail at rafter locations. In an open cornice, a frieze board is often used between the rafters, serving to terminate siding or siding-sheathing combinations at the rafter line (fig. 47A).

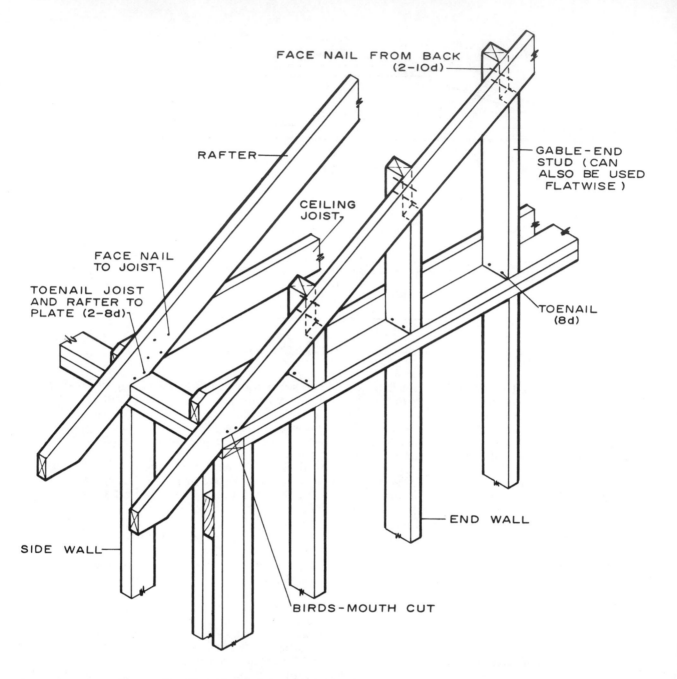

FACE NAIL FROM BACK
(2-10d)

RAFTER

CEILING
JOIST

FACE NAIL
TO JOIST

TOENAIL JOIST
AND RAFTER TO
PLATE (2-8d)

GABLE-END
STUD (CAN
ALSO BE USED
FLATWISE)

TOENAIL
(8d)

SIDE WALL

END WALL

BIRDS-MOUTH CUT

FIGURE 43.—Fastening rafters and ceiling joists to plate and gable-end studs.

A simple closed cornice is shown in figure 47*B*. The soffit of plywood, hardboard, or other material is nailed directly to the underside of the rafter extensions. Blocking may be required between rafters at the wall line to serve as a nailing surface for the soffit. Use small galvanized nails in nailing the soffit to the rafters. When inlet attic ventilation is specified in the plans, it can be provided by a screened slot (fig. 47*B*), or by small separate ventilators.

When a horizontal closed cornice is used, *look-outs* are fastened to the ends of the rafter and to the wall (fig. 47*C*). They are face-nailed to the rafters and face- or toenailed to the studs at the wall. Use twelvepenny nails for the face-nailing and eightpenny nails for the toenailing.

Gable End

The gable-end trim may consist of a fly rafter, a facia board, and facia molding (fig. 48*A*). The 2- by 2- or 2- by 4-inch fly rafter is fastened by

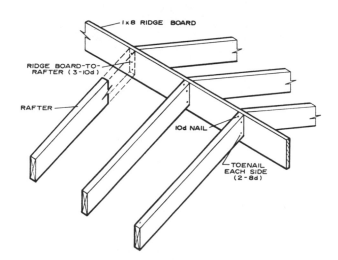

FIGURE 44.—Fastening rafters at the ridge.

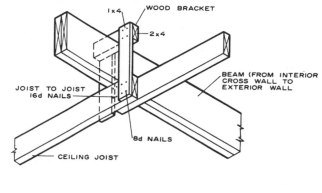

FIGURE 46.—Framing for flush ceiling with wood brackets.

nailing through the roof sheathing. Depending on the thickness of the sheathing, use sixpenny or eightpenny nails spaced 12 inches apart. In this type of gable end, the amount of extension should be governed by the thickness of the roof sheathing. When nominal 1-inch boards or plywood thicker than ⅓ inch is used, the extension should generally be no more than 16 inches. For thinner sheathings, limit the extensions to 12 inches.

A closed gable-end overhang requires nailing surfaces for the soffit. These are furnished by the fly rafter and a nailer or nailing blocks located against the end wall (fig. 48B). An extension of 20 inches might be considered a limit for this type of overhang.

Framing for Chimneys

An inside chimney, whether of masonry or prefabricated, often requires that some type of fram-

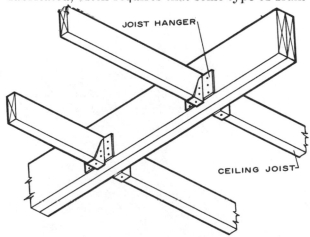

FIGURE 45.—Flush beam with joist hangers.

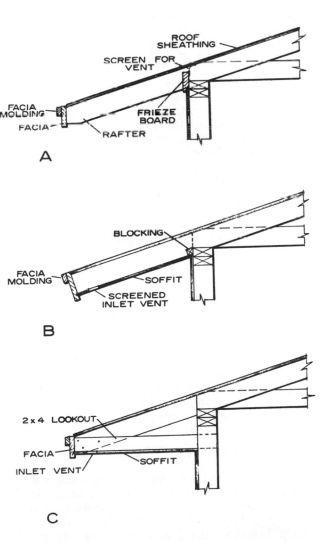

FIGURE 47.—Cornice trim. A, Open cornice; B, sloped closed cornice; C, horizontal closed cornice.

ing be provided, where it extends through the roof. This may consist of simple headers between rafters below and above the chimney location, or require two additional rafter spaces (fig. 49). The chimney should have a 2-inch clearance from the framing members and 1 inch from roof sheathing. When nominal 2- or 3-inch wood decking is used, a small header can be used at each end of the decking at the chimney location for support.

CHIMNEYS

Some type of chimney will be required for the heating unit, whether the home is heated by oil, gas, or solid fuel. It is normally erected before the roofing is laid but also can be installed after. Chimneys, either of masonry or prefabricated, should be structurally safe and provide sufficient draft for the heating unit and other utilities. Local building regulations often dictate the type to be used. A masonry chimney requires a stable foundation below the frostline and construction with acceptable brick or other masonry units. Some type of *flue lining* is included, together with a cleanout door at the base.

The prefabricated chimney may cost less than the full masonry chimney, considering both materials and labor, as well as providing a small saving in space. These chimneys are normally fastened to and supported by the ceiling joists and should be Underwriter Laboratory tested and approved. They are normally adapted to any type of fuel and come complete with roof flashing, cap assembly, mounting panel, piping, and chimney housing.

ROOF COVERINGS

Roof coverings should be installed soon after the cornice and rake trim are in place to provide protection for the remaining interior and exterior work. For the low-cost house, perhaps the most practical roof coverings are roll roofing or asphalt shingles for pitched roofs and roll roofing in double coverage or *built-up roof* for flat or very low-slope roofs. A good maintenance-free roof is important from the standpoints of protection and the additional cost involved in replacing a cheaper roof after only a few years.

Asphalt Shingles

Asphalt *shingles* may be used for roofs with slopes of 2 in 12 to 7 in 12 and steeper under certain conditions of installation. The most common shingle is perhaps the 3 in 1, which is a 3-tab strip, 12 by 36 inches in size. The basic weight may vary somewhat, but the 235-pound (per square of 3 bundles) is now considered minimum. However,

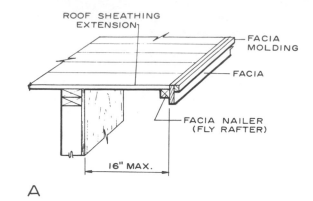

A

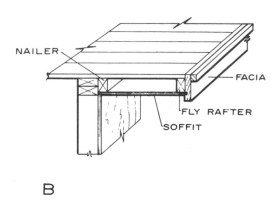

B

FIGURE 48.—Gable-end trim. *A*, Open gable overhang; *B*, closed gable overhang.

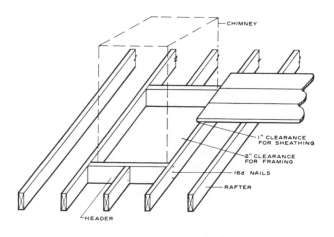

FIGURE 49.—Roof framing at chimney.

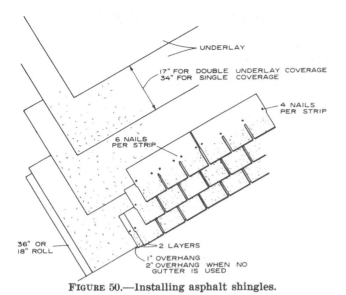

FIGURE 50.—Installing asphalt shingles.

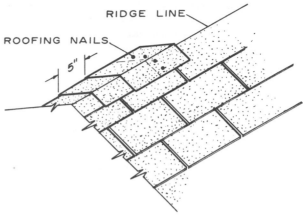

FIGURE 51.—Boston ridge.

many roofs with 210-pound shingles are giving satisfactory service. A small gable roof house uses about 10 squares of shingles, so use of a better shingle would mean about $10 to $20 more per house. Cost of application would be the same.

Installation

Underlay.—A single underlay of 15-pound saturated felt is used under the shingles for roof slopes of 4 in 12 to 7 in 12. A double underlay (double coverage) is required for slopes of 2 in 12 to 4 in 12. Roof slopes over 7 in 12 usually require no underlay. For single underlay, start at the eave line with the 15-pound felt, roll across the roof, and nail or staple the felt in place as required. Allow a 2-inch head lap and install the second strip. This leaves a 34-inch exposure for the standard 36-inch-width rolls. Continue in this manner.

A double underlay can be started with two layers at the eave line, flush with the facia board or molding. The second and remaining strips have 19-inch head laps with 17-inch exposures (fig. 50). Cover the entire roof in this manner, making sure that all surfaces have double coverage. Use only enough staples or roofing nails to hold the underlay in place. Underlay is normally not required for wood shingles.

Shingles.—Asphalt tab shingles are fastened in place with ¾- or ⅞-inch galvanized roofing nails or with staples, using at least four on each strip (fig. 50). Some roofers use six for each strip for greater wind resistance; one at each end and one at each side of each notch. Locate them above the notches so the next course covers them.

A starter strip and one or two layers of shingles are used at the eave line with a 1-inch overhang beyond the facia trim and ½- to ¾-inch extension at the gable end. When no gutters are used, the overhang should be about 2 inches. This will form

a curve during warm weather for a natural drip. Metal edging or flashing is sometimes used at these areas. For slopes of 2 in 12 to 4 in 12, a 5-inch exposure can be used with the double underlay (fig. 50). For slopes of 4 in 12 and over, a 5-inch exposure may also be used with a single underlay.

Ridge

A *Boston ridge* is perhaps the most common method of treating the ridge portion of the roof. This consists of 12- by 12-inch sections cut from the 12- by 36-inch shingle strips. They are bent slightly and used in lap fashion over the ridge with a 5-inch exposure distance (fig. 51). In cold weather, be careful that the sections do not crack in bending. The nails used at each side are covered by the lap of the next section. For a positive seal, use a small spot of asphalt cement under each exposed edge.

Roll Roofing

When cost is a factor in construction of a house, the use of mineral-surfaced roll roofing might be considered. While this type of roofing will not be as attractive as an asphalt shingle roof and perhaps not as durable, it may cost up to 15 percent less for a small house than standard asphalt shingles.

Roll roofing (65 pounds minimum weight in one-half lap rolls with a mineral surface) should be used over a double underlay coverage. Use a starter strip or a half-roll at the eave line with a 1-inch overhang and nail in place 3 to 4 inches above the edge of the facia (fig. 52). When *gutters* are not included initially, use a 2-inch extension to form a drip edge. Space roofing nails about 6 inches apart. Surface nailing can be used when roof slopes are 4 in 12 and greater.

The second (full) roll is now placed along the eave line over a ribbon of asphalt roofing cement or lap-joint material. In low slopes, nailing is done above the lap, cement applied, and the next roll positioned so that the nails are covered. Edge overhang should be about ½ to ¾ inch at the gable ends. When vertical lap joints are required, nail the first edge, then use asphalt adhesive under a minimum 6-inch overlap. Use a sufficient amount of adhesive or lap-joint material to insure a tight joint. On steep slopes, surface nailing along the vertical edge is acceptable. The ridge can be finished with a Boston-type covering or by 12-inch-wide strips of the roll roofing, using at least 6 inches on each side.

Chimney Flashing

Flashing around the chimney at the junction with the roof is perhaps the most important flashed area in a simple gable roof. The Boston ridge over the shingles must be well installed to prevent wind-driven rain from entering, and the flashing around the chimney must also be well done. Prefabricated chimneys are supplied with built-in flashing which slides under the shingles above and over those below. A good calking or asphalt sealing compound around the perimeter completes the installation.

A masonry chimney requires flashing around the perimeter, which is placed as shingle flashing under the shingles at sides and top and extends at right angles up the sides (fig. 53). In addition, counterflashing is used on the base of the chimney over the shingle flashing. This is turned in a masonry joint, wedged in place with lead plugs, and sealed with a calking material. Galvanized sheet metal, aluminum, and terneplate (coated sheet iron or steel) are the most common types used for flashing around the chimney. If they are not rust-resistant, they should be given a coat or two of good metal paint.

EXTERIOR WALL COVERINGS

Exterior coverings used over the wall framing commonly consist of a sheathing material followed by some type of finish siding. However, sheathing-siding materials (panel siding) serve as both sheathing and finish material. These materials are most often plywoods or hardboards. While they are somewhat higher in price than conventional sheathing alone, they make it possible to use only a single exterior covering material. Low-cost sheathing materials can be covered with various types of siding—from spaced vertical boards over plywood sheathing to horizontal bevel siding over fiberboard, plywood, or other types of sheathing. All combinations should be studied so that cost, utility, and appearance are considered in the selection. The working drawings of the house indicate the most suitable siding materials.

Sheathing

In a low-cost house, it is advisable to use a sheathing or a panel-siding material which will provide resistance to racking and thus eliminate the need for diagonal corner bracing on the stud wall. Notching studs and installing bracing can add substantially to labor cost. When siding material does not provide this rigidity and strength, some type of sheathing should be used. Materials which provide resistance to racking are: (a) Diagonal board sheathing, (b) structural insulation board (fiberboard) sheathing in 25/32-inch regular density or ½-inch intermediate fiberboard or nail-base fiberboard sheathing for direct application of shingles, and (c) plywood. The fiberboard and plywood sheathing must be applied vertically in 4- by 8-foot or longer sheets with edge and center nailing to provide the needed racking resistance. Horizontal wood boards may also be used for sheathing but require some type of corner bracing.

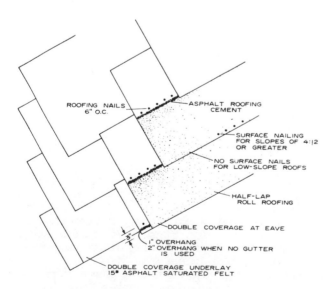

FIGURE 52.—Installing roll roofing.

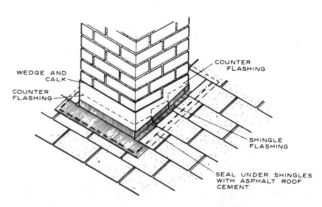

FIGURE 53.—Chimney flashing.

Diagonal Boards

Diagonal wood sheathing should have a nominal thickness of ⅝ inch (resawn). Edges can be square, *shiplapped*, or tongued-and-grooved. Widths up to 10 inches are satisfactory. Sheathing should be applied at as near a 45° angle as possible as shown in figure 54. Use three eightpenny nails for 6- and 8-inch-wide boards and four eightpenny nails for the 10-inch widths. Also provide nailing along the floor framing or beam faces. Butt joints should be made over a stud unless the sheathing is end and side matched. Depending on the type of siding used, sheathing should normally be carried down over the outside floor framing members. This provides an excellent tie between wall and floor framing.

Structural Insulating Board

Structural *insulating board* sheathing (fiberboard type) in 4-foot-wide sheets and in ²⁵⁄₃₂-inch regular-density or ½-inch intermediate fiberboard grades provides the required rigidity without bracing. It must be applied vertically in 8-foot and longer sheets with edge and center nailing (fig. 55). Nails should be spaced 3 inches apart along the edges and 6 inches apart at intermediate supports. Use 1¾-inch roofing nails for the ²⁵⁄₃₂-inch sheathing and 1½-inch nails for the ½-inch

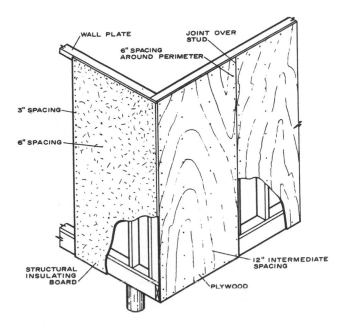

FIGURE 55.—Sheathing with insulating board or plywood (vertical application).

sheathing. Vertical joints should be made over studs. Siding is normally required over this type of sheathing.

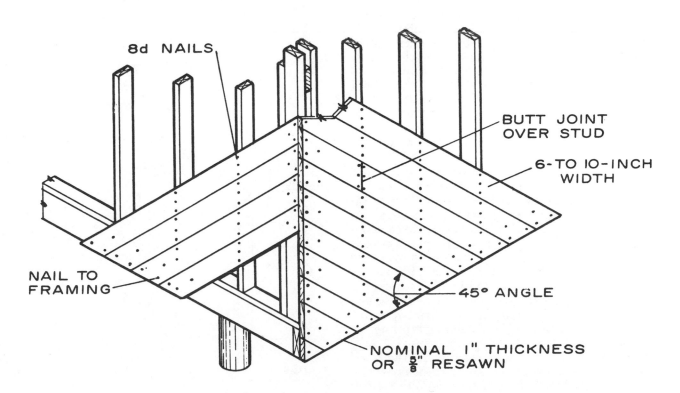

FIGURE 54.—Diagonal board sheathing.

Plywood

Plywood sheathing also requires vertical application of 4-foot-wide by 8-foot or longer sheets (fig. 55). Standard (sheathing) grade plywood is normally used for this purpose. Use 5/16 inch minimum thickness for 16-inch stud spacing and 3/8 inch for 24-inch stud spacing. Nails are spaced 6 inches apart at the edges and 12 inches apart at intermediate studs. Use sixpenny nails for 5/16-inch 3/8-inch-thick plywood. Because the plywood sheathing in 5/16- or 3/8-inch sheets provides the necessary strength and rigidity, almost any type of siding can be applied over it.

Plywood and hardboard are also used as a single covering material without sheathing, but grades, thickness, and types vary from normal sheathing requirements. This phase of wall construction will be covered in the following section.

Sheathing-Siding Materials—Panel Siding

Large sheet materials for exterior coverage (*panel siding*) can be used alone and serve both as sheathing and siding. Plywood, hardboard, and exterior particleboard in their various forms are perhaps the most popular materials used for this purpose. The proper type and size of plywood and hardboard sheets with adequate nailing eliminate the need for bracing. Particleboard requires corner bracing.

These materials are quite reasonable in price, and plywood, for example, can be obtained in grooved, rough-sawn, embossed, and other surface variations as well as in a paper-overlay form. Hardboard can also be obtained in a number of surface variations. The plywood surfaces are most suitable for pigmented stain finishes in various colors. The medium-density, paper-overlay plywoods are an excellent base for exterior paints. Plywoods used for panel siding are normally exterior grades.

The thickness of plywood used for siding varies with the stud spacing. Grooved plywood, such as the "1–11" type, is normally 5/8 inch thick with 3/8- by 1/4-inch-deep grooves spaced 4 or 6 inches apart. This plywood is used when studs are spaced a maximum of 16 inches on center. Ungrooved plywoods should be at least 3/8 inch thick for 16-inch stud spacing and 1/2 inch thick for 24-inch stud spacing. Plywood panel siding should be nailed around the perimeter and at each intermediate stud. Use sixpenny galvanized siding or other rust-resistant nails for the 3/8-inch plywood and eightpenny for 1/2-inch and thicker plywood and space 7 to 8 inches apart. Hardboard must be 1/4 inch thick and used over 16-inch stud spacing. Exterior particleboard with corner bracing should be 5/8 inch thick for 16-inch stud spacing and 3/4 inch thick for 24-inch stud spacing. Space nails 6 inches apart

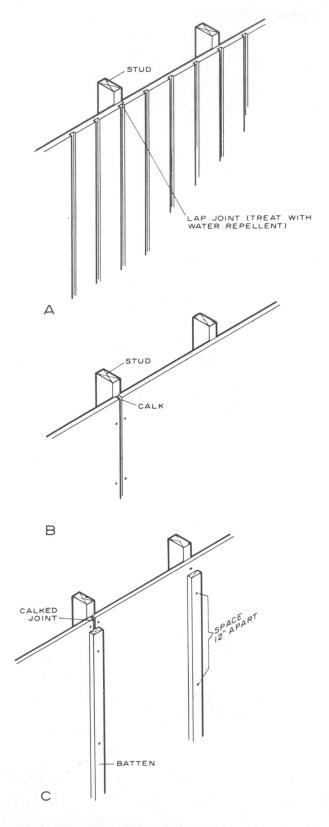

FIGURE 56.—Joint treatment for panel siding. *A*, Lap joint; *B*; calked butt joint; *C*, butt joint with batten.

around the edges and 8 inches apart at intermediate studs.[6]

The vertical joint treatment over the stud may consist of a shiplap joint as in the "1–11" paneling (fig. 56A). This joint is nailed at each side after treating with a water-repellent preservative. When a square-edge butt joint is used, a sealant calk should be used at the joint (fig. 56B).

A square-edge butt joint may be covered with battens, which can also be placed over each stud as a decorative variation (fig. 56C). Joints should be calked and the batten nailed over the joint with eightpenny galvanized nails spaced 12 inches apart. Nominal 1- by 2-inch battens are commonly used.

A good detail for this type of siding at gable ends consists of extending the bottom plate of the gable ⅝ to ¾ inch beyond the top of the wall below (fig. 57). This allows a termination of the panel at the lower wall and a good drip section for the gable-end panel.

---

[6] Federal Housing Administration. Use of Materials Bull. UM–32. Architectural Standards Division. 1961.

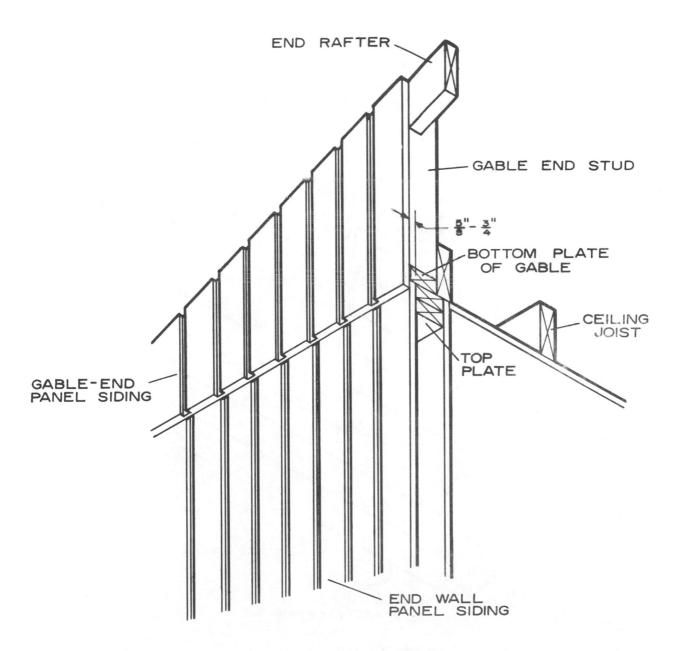

FIGURE 57.—Panel siding at gable end.

Siding—With and Without Sheathing

There are a number of sidings, mainly for horizontal application, which might be suitable for walls with or without sheathing. The types most suitable for use over sheathing are: (a) The lower cost lap sidings of wood or hardboard, (b) wood or other type shingles with single or double coursing, (c) vertical boards, and (d) several nonwood materials. Initial cost and maintenance should be the criteria in the selection. *Drop siding* and nominal 1-inch paneling materials can be used without sheathing under certain conditions. However, such sidings require (a) a rigidly braced wall at each corner and (b) a waterproof paper over the studs before application of the siding.

Application

Bevel siding.—When siding is used over sheathing, window and door frames are normally installed first. This process will be discussed in the next section, "Exterior Frames." The exposed face of sidings such as *bevel siding* in ½- by 6-inch, ½- by 8-inch, or other sizes should be adjusted so that

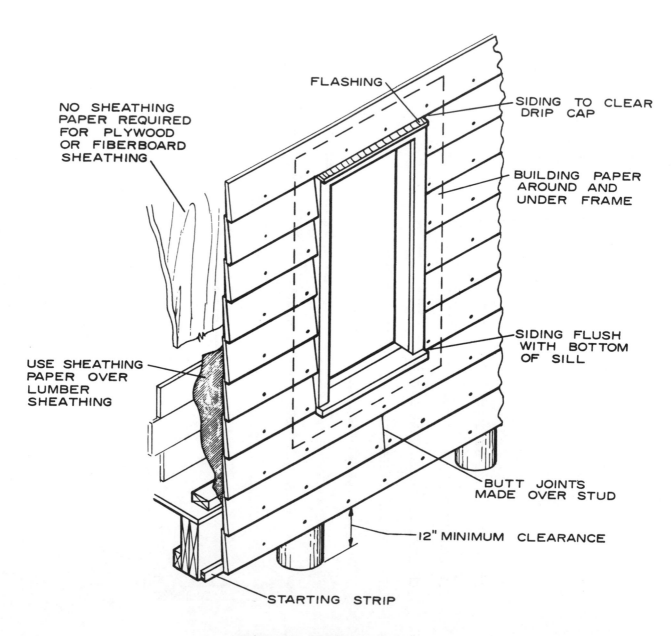

FIGURE 58.—Installing bevel siding.

776

the butt edges coincide with the bottom of the sill and the top of the *drip cap* of window frames (fig. 58). Use a sevenpenny galvanized siding nail or other corrosion-resistant nail at each stud crossing for the ½-in-thick siding. The nail should be located so as to clear the top edge of the siding course below. Butt joints should be made over a stud. Other horizontal sidings over sheathing should be installed in a similar manner. Nonwood sheathings require nailing into each stud.

Interior corners should butt against a 1½- by 1½-inch corner strip. This wood strip is nailed at interior corners before siding is installed. Exterior corners can be mitered, butted against *corner boards*, or covered with metal corners. Perhaps the corner board and metal corner are the most satisfactory for bevel siding.

Vertical siding.—In low-cost house construction, some vertical sidings can be used over stud walls without sheathings; others require some type of sheathing as a backer or nailing base. Matched (tongued-and-grooved) paneling boards can be used directly over the studs under certain conditions. First, some type of corner bracing is required for the stud wall. Second, nailers (blocking) between the studs are required for nailing points, and third, a waterproof paper should be placed over the studs (fig. 59). Galvanized sevenpenny finish nails, which should be spaced no more than 24 inches apart vertically, are blind-nailed through the tongue at each cross nailing block. When boards are nominal 6 inches and wider, an additional eightpenny galvanized nail should be face-nailed (fig. 59). Boards should extend over and be nailed to the headers or stringers of the floor framing.

Rough-sawn vertical boards over a plywood backing provide a very acceptable finish. The plywood should be an exterior grade or sheathing grade (standard) with exterior glue. It should be ½ inch thick or 5⁄16 inch thick with nailing blocks between studs (fig. 60). Rough-sawn boards 4 to 8 inches wide surfaced on one side can be spaced and nailed to the top and bottom wallplates and the floor framing members and to the nailing blocks (fig. 60). Use the surfaced side toward the plywood. A choice in the widths and spacings of boards allows an interesting variation between houses.

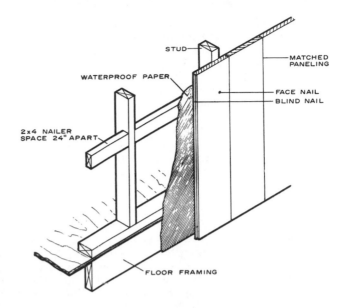

FIGURE 59.—Vertical paneling boards over studs.

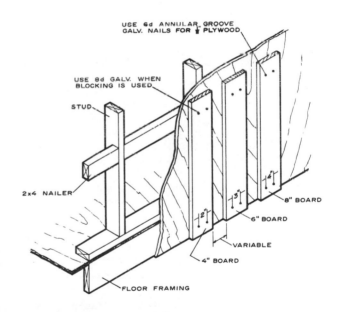

FIGURE 60.—Vertical boards over plywood.

EXTERIOR FRAMES

Windows and Doors

Window and exterior door units, which include the frames as well as the *sash* or doors, are generally assembled in a manufacturing plant and arrive at the building site ready for installation. Doors may require fitting, however. Simple jambs and sill units for awning or hopper window sash can be made in a small shop with the proper equipment, the wood treated with a water-repellent preservative, and sash fitted and prehung. Only the sash would need to be purchased. However, this system of fabrication is practical only for the simplest units and only when a large number of the same type are required. Thus, for double-hung and sill units for awning or hopper window sash desirable to select the lower cost standard-size units and use a smaller number of windows. A substantial saving can be made, for example, by using one large double-hung window rather than two smaller ones.

Window frames are generally made with nominal 1-inch jambs and 2-inch sills. Sash for the most part are 1⅜ inches thick. Exterior door frames are made from 1½- to 1¾-inch stock. Exterior doors are 1¾ inches thick and the most common are the flush and the panel types.

As a general rule, the amount of natural light provided by the glass area in all rooms (except the kitchen) should be about 10 percent of the floor area. The kitchen can have natural or artificial light, but when an operating window is not available, ventilation should be provided. A bathroom usually is subject to the same requirements. From the standpoint of safety, houses should have two exterior doors. Local regulations often specify any variations of these requirements. The main exterior door should be 3 feet wide and at least 6 feet 6 inches high; 6 feet 8 inches is a normal standard height for exterior doors. The service or rear door should be at least 2 feet 6 inches wide; 2 feet 8 inches is the usual width. These details are covered in the working drawings for each house.

Types of Windows and Doors

Perhaps the most common type of window used in houses is the double-hung unit. It can be obtained in a number of sizes, is easily *weatherstripped*, and can be supplied with storms or screens. Frames are usually supplied with prefitted sash and the exterior casing and drip cap in place.

Another type of window, which is quite reasonable in cost and perhaps the one most adaptable to small shop fabrication in a simple form, is the "awning" or "hopper" type.

Other windows, such as the *casement sash* and sliding units, are also available but generally their cost is somewhat greater than the two types described. The fixed or stationary sash may consist of a simple frame with the sash permanently fastened in place. The frame for the awning window would be suitable for this type of sash.

Door frames are also supplied with exterior side and head casing and a hardwood sill or a softwood sill with a reinforced edge. Perhaps the most practical exterior door, considering cost and performance, is the panel type. A number of styles and patterns are available, most of them featuring some type of glazed opening (fig. 64). The solid-core flush door, usually more costly than the panel type, should be used for exteriors in most central and northern areas of the country in preference to the hollow-core type. A hollow-core door is ordinarily for interior use, because it warps

excessively during the heating season when used on the outside. However, it would probably be satisfactory for exterior use in the southern areas.

Installation

Window Frames

Preassembled window frames are easily installed. They are made to be placed in the rough wall openings from the outside and are fastened in place with nails. When a panel siding is used in place of a sheathing and siding combination, the frames are usually installed after the siding is in place. When horizontal siding is used with sheathing, the frames are fastened over the sheathing and the siding applied later.

To insure a water- and windproof installation for a panel-siding exterior, a ribbon of calking sealant (rubber or similar base) is placed over the siding at the location of side and head casing (fig. 65). When a siding material is used over the sheathing, strips of 15-pound asphalt felt should be used around the opening (fig. 58).

The frame is placed in the opening over the calking sealant (preferably with the sash in place to keep it square), and the sill leveled with a carpenter's level. Shims can be used on the inside if necessary. After leveling the sill, check the side casing and jamb with the level and square. Now nail the frame in place using tenpenny galvanized nails through the casing and into the side studs and the header over the window (fig. 66). While nailing, slide the sash up and down to see that they work freely. The nails should be spaced about 12 inches apart and both side and head casing fastened in the same manner. Other types of window units are installed similarly. When a panel siding is used, place a ribbon of calking sealer at the junction of the siding and the sill. Follow this by installing a small molding such as quarter-round.

Door Frames

Door frames are also fastened over panel siding by nailing through the side and head casing. The header and the joists must first be cut and trimmed (fig. 67). Use a ribbon of calking sealer under the casing. The top of the sill should be the same height as the finish floor so that the *threshold* can be installed over the joint. The sill should be shimmed when necessary to have full bearing on the floor framing. A *quarter-round* molding in combination with calking when necessary for a tight windproof joint should be used under the door sill when a panel siding or other single exterior covering is used. When joists are parallel to the plane of the door, headers and a short support member are necessary at the edge of the sill (fig. 67). The threshold is installed after the finish floor has been laid.

A

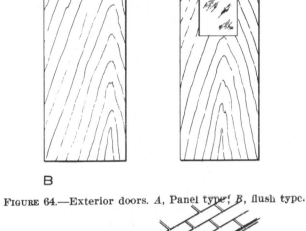

B

FIGURE 64.—Exterior doors. *A*, Panel type; *B*, flush type.

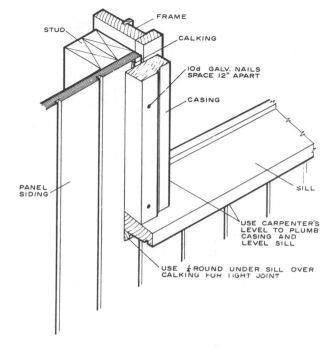

FIGURE 66.—Installation of double-hung window frame

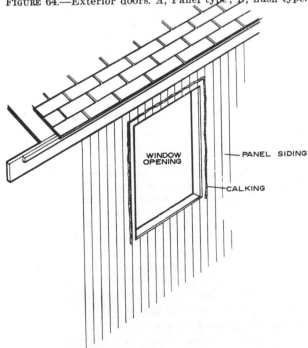

FIGURE 65.—Calking around window opening before installing frame.

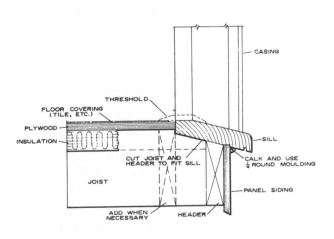

FIGURE 67.—Door installation at sill.

THERMAL INSULATION

Thermal insulation is used in a house to minimize heat loss during the heating season and to reduce the inflow of heat during hot weather. Resistance to the passage of warm air is provided by materials used in wall, ceiling, and floor construction.

In constructing a crawl-space house, other factors must be considered in addition to providing insulation, such as: (a) Use of a *vapor barrier* with the insulation; (b) protection from ground moisture by the use of a vapor barrier ground cover (especially true if the crawl space is enclosed with a full foundation or skirt boards); and (c) use of both attic and crawl-space ventilation when required. Insulation and vapor barriers are discussed here. Ventilation is covered in the next section.

The amount of insulation used in walls, floors, and ceilings usually depends on the geographic location of the house. In all parts of the country, at least 3½-inch thick insulation is essential in walls, and where winters are severe, the addition of foam sheating or the alternate 2 by 6 stud wall with 5½-inch thick insulation is often necessary. Ceilings should have a "U" 0.05 or lower depending on the climate. Some codes allow slightly higher "U" values for roof decks, and requirements are greatly reduced for seasonal homes that need little heating or cooling. Floors also require insulation except for houses over heated basements. Where the basement is heated, walls should be insulated at least down to the frost line. Many States now have minimum insulation requirements and a national mandatory standard for new construction is expected soon. Legal requirements should be checked with local code authorities during the planning stage for any house.

Types of Insulation

Commercial insulating materials which are most practical in the construction of low-cost houses are: (a) Flexible types in blanket and batt forms, (b) loose-fill types, and (c) *rigid insulation* such as building boards or insulation boards. Others include reflective insulations, expanded plastic foams, and the like.

The common types of blanket and batt insulation, as well as loose-fill types, are made from wood and cotton fiber and mineral wool processed from rock, slag, or glass. Insulating or building board may be made of wood or cane fibers, of glass fibers, and of expanded foam.

In comparing the relative insulating values of various materials, a 1-inch thickness of typical blanket insulation is about equivalent to (a) 1½ inches of insulating board, (b) 3 inches of wood, or (c) 18 inches of common brick. Thus, when practical, the use of even a small amount of thermal insulation is good practice.

The insulating values of several types of flexible insulation do not vary a great deal. Most loose-fill insulations have slightly lower insulating values than the same material in flexible form. However, fill insulations, such as *vermiculite*, have less than 60 percent the value of common flexible insulations. Most lower density sheathing or structural insulating boards have better insulating properties than vermiculite.

Vapor Barriers

Vapor barriers are often a part of blanket or batt insulation, but they may also be a separate material which resists the movement of water vapor to cold or exterior surfaces. They should be placed as close as possible to the warm side of all exposed walls, floors, and ceiling. When used as ground covers in crawl spaces, they resist the movement of soil moisture to exposed wood members. In walls, they eliminate or minimize condensation problems which can cause paint peeling. In ceilings they can, with good *attic ventilation*, prevent moisture problems in attic spaces.

Vapor barriers commonly consist of: (a) Papers with a coating or lamination of an asphalt material; (b) plastic films; (c) aluminum or other metal foils; and (d) various paint coatings. Most blanket and batt insulations are supplied with a laminated paper or an aluminum foil vapor barrier. Friction-type batts usually have no vapor barriers. For such insulation, the vapor barrier should be added after the insulation is in place. Vapor barriers should generally be a part of all insulating processes.

Vapor barriers are usually classed by their *"perm"* value, which is a rate of water-vapor movement through a material. The lower this value, the greater the resistance of the barrier. A perm rating of 0.50 or less is considered satisfactory for vapor barriers. Two-mil (0.002-inch-thick) polyethylene film, for example, has a perm rating of about 0.25.

When the crawl space is enclosed during or after construction of the house, or the soil under the house is quite damp, a *soil cover* should be used. This vapor barrier should have a perm value of 1.0 or less and should be laid over the ground, using about a 4-inch lap along edges and ends. Stones or half-sections of brick can be used at the laps and around the perimeter to hold the material in place. The ground should be leveled before placing the cover. Materials such as polyethylene, roll roofing, and asphalt laminated barriers are satisfactory for ground covers.

Where and How to Insulate

Insulation in some form should be used at all exterior walls, floors, and at the ceiling as a sep-

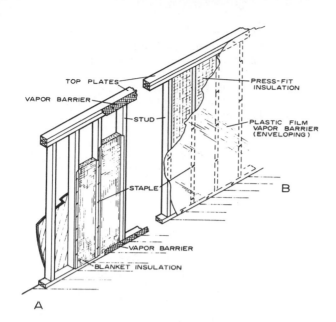

FIGURE 72.—Installing insulating batts in floor.

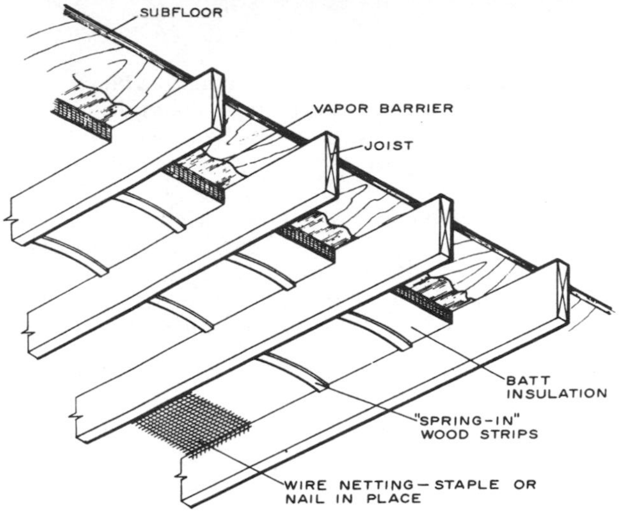

FIGURE 73.—Wall insulation. *A*, Blanket insulation with vapor barrier; *B*, plastic film vapor barrier.

arate material or as a part of the house structure in most climates. The recommended thicknesses for various locations, given at the beginning of this chapter, can be used as a guide in insulating the house.

Floors

Blanket insulation in 1-inch thickness can be installed under a tongued-and-grooved plywood subfloor as previously described and illustrated in figure 14. However, this type is most commonly applied between the joists. In applying insulation, be sure that the vapor barrier faces up, against the bottom of the plywood.

The use of friction or other types of insulating batts between joists has been shown in figure 18. This insulation can be installed any time after the house has been enclosed and roofing installed. When friction-type insulation without a vapor barrier is used, a separate vapor barrier should be placed over the joists before the subfloor is nailed in place. Laminated or foil-backed kraft paper barriers or plastic films can be used. They should be lapped 4 to 6 inches and stapled only enough to hold the barriers in place until the subfloor is installed.

When batts are not the friction type, they usually require some support in addition to the adhesive shown in figure 18. This support can be supplied simply by the use of small, 3/16- by 3/4-inch or similar size, wood strips cut slightly longer than the joist space so that they spring into place (fig. 72). They can be spaced 24 to 36 inches apart or as needed. Wire netting nailed between joists may also be used to hold the batt insulation (fig. 72). The 1-inch-thick or thicker blanket insulation can also be installed in this way if desired.

Wall Insulation

When blanket or batt insulation for the walls contains a vapor barrier, the barrier should be placed toward the room side and the insulation stapled in place (fig. 73*A*). An additional small strip of vapor barrier at the bottom and top plates and around openings will insure good resistance to vapor movements in these critical areas.

When friction or other insulation without a vapor barrier is used in the walls, the entire interior surface should be covered with a vapor barrier. This is often accomplished by the use of wall-height rolls of 2-or 4-mil plastic film (fig. 73*B*). Other types of vapor barriers extending from bottom plate to top plate are also satisfactory. The studs, the window and door headers, and plates should also be covered with the vapor barrier for full protection. The full-height plastic film is usually carried over the entire window opening and cut out only after the dry wall has been installed. Staple or tack just enough to hold the vapor barrier in place until the interior wall finish is installed.

Ceiling Insulation

Loose-fill or batt-type ceiling insulation is often placed during or after the dry-wall ceiling finish is applied, depending on the roof design. In ceilings having no attic or joist space, such as those with wood roof decking, the insulation is normally a part of the roof construction and may include ceiling tile, thick wood decking, and structural insulating board in various combinations. This type of construction has been covered in the section on framing of low-slope roofs and figure 38.

Loose-fill insulation, blown or hand placed, can be used where there is an attic space high enough for easy access. It can be poured in place and leveled off (fig. 74A). A vapor barrier should be used under the insulation.

Batt insulation with attached vapor barrier can be used in most types of roof and ceiling constructions with or without an attic space. The batt insulation is made to fit between ceiling or roof members spaced 16 or 24 inches on center. After the first row of gypsum board sheets has been applied in a level ceiling, the batts (normally 48 inches long) are placed between ceiling joists with the vapor barrier facing down. The next row of gypsum board is applied and the batts added. At the opposite side of the room, the batts should be stapled lightly in place before the final dry-wall sheets are applied. When one set of members serves as both ceiling joists and rafters, an airway should be allowed for ventilation (fig. 74B).

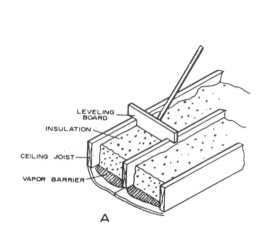

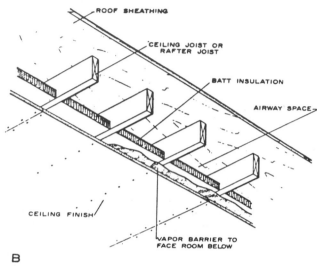

FIGURE 74.—Ceiling insulation. A, Fill type; B, batt type.

INTERIOR WALL AND CEILING FINISH

Some of the most practical low-cost materials for interior finish are the gypsum boards, the hardboards, the fiberboards, and plywood. Hardboard, fiberboard, gypsum board, and plywood are often prefinished and require only fastening to the studs and ceiling joists. The gypsum boards can also be obtained prefinished, but in their most common and lowest cost form, they have a paper facing which requires painting or wallpapering. A plaster finish is usually more costly than the dry-wall materials and perhaps should not be considered for low-cost houses. Furthermore, a plaster finish must be applied by a specialist. Dry-wall finishes can be installed by the semiskilled workman after a minimum of training.

In addition to the four types of sheet materials, an insulating board or ceiling tile can be used for the ceiling. Prefinished tile in 12- by 12-inch and larger sizes usually requires nailing strips fastened to the underside of the ceiling joists or truss members. Tile can also be applied to the underside of roof boards in a beam-type ceiling or on the inner face of the wood decking on a wood-deck roof.

Wood and fiberboard paneling in tongued-and-grooved V-edge pattern in various widths can also be used as an interior finish, especially as an accent wall, for example. When applied vertically, nailers are used between or over the studs.

The type of interior finish materials selected for a low-cost house should primarily be based on cost. These materials may vary from a low of 5 to 6 cents per square foot for 3/8-inch unfinished gypsum board to as much as three times this amount for some of the lower cost prefinished materials. However, consideration should be made of the labor involved in joint treatment and painting of unfinished materials. As a result, in some instances, the use of prefinished materials might be justified. These details and material requirements are included in the working drawings or the accompanying specifications for each house. Before interior wall and ceiling finish is applied, insulation should be in place and wiring, heating *ducts*, and other utilities should be roughed in.

Material Requirements

The thickness of interior covering materials depends on the spacing of the studs or joists and the type of material. These requirements are

782

usually a part of the working drawings or the specifications. However, for convenience, the recommended thicknesses for the various materials are listed in the following tabulation based on their use and on the spacing of the fastening members:

Interior Material Finish Thickness

Finish	Minimum material thickness (inches) when framing is spaced	
	16 inches	24 inches
Gypsum board	⅜	½
Plywood	¼	⅜
Hardboard	¼	------
Fiberboard	½	¾
Wood paneling	⅜	½

For ceilings, when the long direction of the gypsum board sheet is at right angles to the ceiling joists, use ⅜-inch thickness for 16-inch spacing and ½-inch for 24-inch joist spacing. When sheets are parallel, spacing should not exceed 16 inches for ½-inch gypsum board. Fiberboard ceiling tile in ½-inch thickness requires 12-inch spacing of nailing strips.

Gypsum Board

Application

Gypsum board used for dry-wall finish is a sheet material composed of a gypsum filler faced with paper. Sheets are 4 feet wide and 8 feet long or longer. The edges are usually recessed to accommodate taped joints. The ceiling is usually covered before the wall sheets are applied. Start at one wall and proceed across the room. When batt-type ceiling insulation is used, it can be placed as each row of sheets is applied. Use fivepenny (1⅝-inch-long) cooler-type nails for ½-inch gypsum and fourpenny (1⅜-inch) nails for ⅜-inch gypsum board. Ring-shank nails ⅛ inch shorter than these can also be used. Nailheads should be large enough so that they do not penetrate the surface.

Adjoining sheets should have only a light contact with each other. End joints should be staggered and centered on a ceiling joist or bottom chord of a truss. One or two braces slightly longer than the height of the ceiling can be used to aid in installing the gypsum sheets (fig. 78). Nails are spaced 6 to 8 inches apart and should be very lightly dimpled with the hammerhead. Do not break the surface of the paper. Edge or end joints should be double-nailed. Minimum edge nailing distance is ⅜ inch.

Vertical or horizontal application can be used on the walls. Horizontal application of gypsum board is often used when wall-length sheets eliminate vertical joints. The horizontal joint thus requires

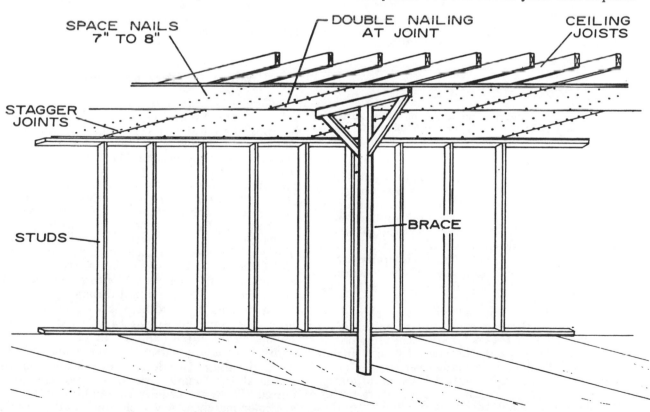

FIGURE 78.—Installing gypsum board on ceiling.

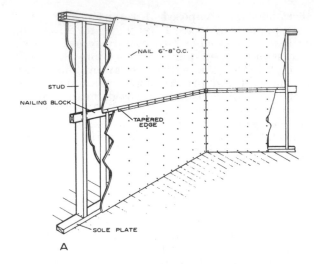

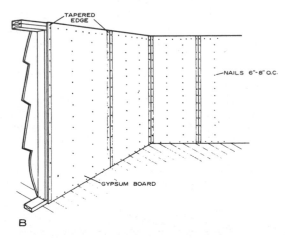

FIGURE 79.—Installing gypsum board on walls. *A*, Horizontal application; *B*, vertical application.

only taping and treating. For normal application, horizontal joint reinforcing is not required. However, nailing blocks may be used between studs for a damage-resistant joint for the thinner gypsum sheets (fig. 79*A*). Horizontal application is also suitable to the laminated system in which ⅜-inch gypsum sheets are nailed vertically and room-length sheets are applied horizontally with a wallboard or contact adhesive. While this results in an excellent wall, it is much more costly than the single application.

Gypsum board applied vertically should be nailed around the perimeter and at intermediate studs with 1⅜- or 1⅝-inch nails, depending on the thickness. Nails should be spaced 6 to 8 inches apart. Joints should be made over the center of a stud with only light contact between adjoining sheets (fig. 79*B*). Another method of fastening the sheets is called the "floating top." In this system, the top horizontal row of nails is eliminated and the top 6 or 8 inches of the sheet are free. This

is said to prevent fracture of the gypsum board when there is movement of the framing members.

Plywood and Hardboard

The application of prefinished 4-foot-wide hardboard and plywood sheets is relatively simple. They are normally used vertically and can be fastened with small finish nails (brads). Use nails 1½ inches long for ¼- or ⅜-inch-thick materials and space about 8 to 10 inches apart at all edges and intermediate studs. Edge spacing should be about ⅜ inch. Set the nails slightly with a nail set. Many prefinished materials are furnished with small nails that require no setting because their heads match the color of the finish.

The use of panel and contact adhesives in applying prefinished sheet materials is becoming more popular and usually eliminates nails, except those used to aline the sheets. Manufacturer's directions should be followed in this method of application.

In applying sheet materials such as hardboard or plywood paneling, it is good practice to insure dry, warm conditions before installing. Furthermore, place the sheets in an upright position against the wall, lined up about as they will be installed, and allow them to take on the condition of the room for at least 24 hours. This is also true for wood or fiberboard paneling covered in the following paragraphs.

Wood or Fiberboard Paneling

Tongued-and-grooved wood or fiberboard (insulating board) paneling in various widths may be applied to walls in full lengths or above the wainscot. Wood paneling should not be too wide (nominal 8-inch) and should be installed at a moisture content of about 8 percent in most areas of the country. However, the moisture content should be about 6 percent in the dry Southwest and about 11 percent in the Southern and Coastal areas of the country. In this type of application, wood strips should be used over the studs or nailing blocks placed between them (fig. 81). Space the nailers not more than 24 inches apart.

For wood paneling, use a 1½- to 2-inch finishing or casing nail and blind-nail through the tongue. For nominal 8-inch widths, a face nail may be used near the opposite edge. Fiberboard paneling (planking) is often supplied in 12- and 16-inch widths and is applied in the same manner as the wood paneling. In addition to the blind nail or staple at the tongue, two face nails may be required. These are usually set slightly unless they are color-matched. A 2-inch finish nail is usually satisfactory, depending on the thickness. Panel and contact adhesives may also be used for this type of interior finish, eliminating the majority of the nails except those at the tongue. On outside walls, use a vapor barrier under the paneling (fig. 81).

Ceiling Tile

Ceiling tile may be installed in several ways, depending on the type of ceiling or roof construction. When a flat-surfaced backing is present, such as between beams of a beamed ceiling in a low-slope roof, the tiles are fastened with adhesive as recommended by the manufacturer. A small spot of a mastic type of construction adhesive at each corner of a 12- by 12-inch tile is usually sufficient. When tile is edge-matched, stapling is also satisfactory.

A suspended ceiling with small metal or wood hangers, which form supports for 24- by 48-inch or smaller drop-in panels, is another system of applying this type of ceiling finish. It is probably more applicable to the remodeling of older homes with high ceilings, however.

A third, and perhaps the most common, method of installing ceiling tile, is with the use of wood strips nailed across the ceiling joists or roof trusses. These are spaced 12 inches on center. A nominal 1- by 3- or 1- by 4-inch wood member can be used for roof or ceiling members spaced not more than 24 inches on center (fig. 82A). A nominal 2- by 2- or 2- by 3-inch member should be satisfactory for truss or ceiling joist spacing of up to 48 inches. Use two sevenpenny or eight-penny nails at each joist for the nominal 1-inch strips and two tenpenny nails for the nominal

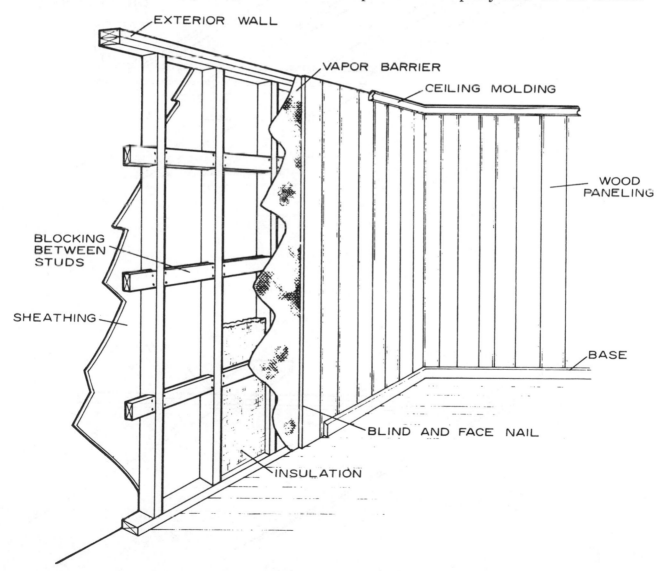

FIGURE 81.—Application of vertical paneling.

2-inch strips. Use a low-density wood, such as the softer pines, as most tile installation is done with staples.

In locating the strips, first measure the width of the room (the distance parallel to the direction of the ceiling joists). If, for example, this is 11 feet 6 inches, use 10 full 12-inch- square tiles and a 9-inch-wide tile at each side edge. Thus, the second wood strips from each side are located so that they center the first row of tiles, which can now be ripped to a width of 9 inches. The last row will also be 9 inches, but do not rip these tiles until the last row is reached so that they fit tightly. The tile can be fitted and arranged the same way for the ends of the room.

Ceiling tiles normally have a tongue on two

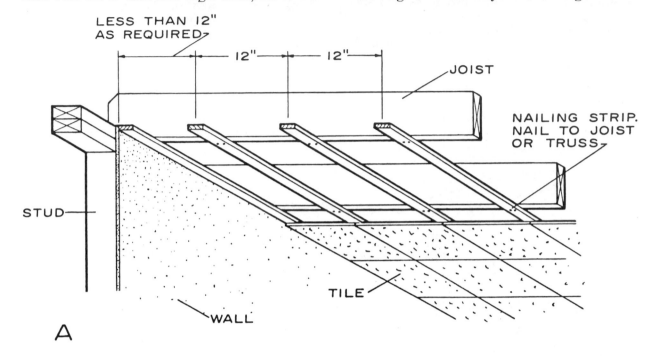

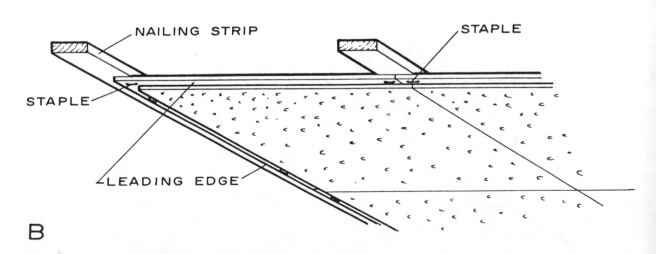

FIGURE 82.—Ceiling tile installation. *A*, Nailing strip location; *B*, stapling.

adjacent sides and a groove on the opposite adjacent sides. Start with the leading edge ahead and to the open side so that they can be stapled to the nailing strips. A small finish nail or adhesive should be used at the edge of the tiles in the first row against the wall. Stapling is done at the leading edge and the side edge of each tile (fig. 82B). Use one staple at each wood strip at the leading edge and two at the open side edge. At the opposite wall, a small finish nail or adhesive must again be used to hold the tile in place.

Most ceiling tile of this type has a factory finish, and painting or finishing is not required after it is in place. Because of this, do not soil the surface as it is installed.

Bathroom Wall Covering

When a complete prefabricated shower stall is used in the bathroom instead of a tub, no special wall finish is required. However, if a tub is used, some type of waterproof wall covering is normally required around it to protect the wall. This may consist of several types of finish from a coated hardboard paneling to various ceramic, plastic, and similar tiles.

In the interest of economy, one of the special plastic-surfaced hardboard materials is perhaps a good choice. These are applied in sheet form and fastened with an adhesive ordinarily supplied by

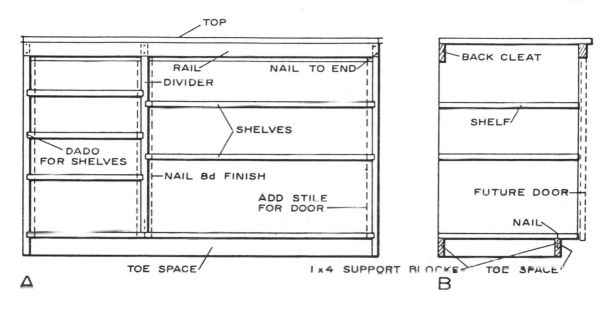

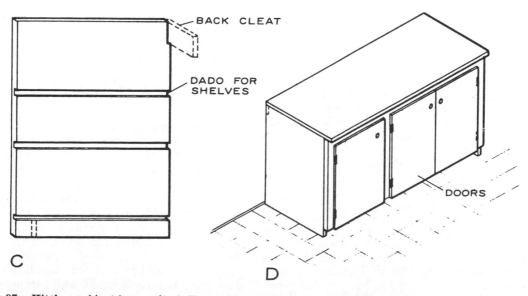

FIGURE 97.—Kitchen cabinet base unit. *A*, Front view; *B*, section; *C*, end from interior; *D*, overall view.

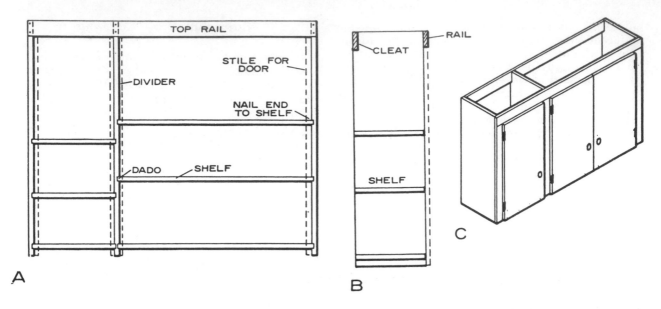

FIGURE 98.—Kitchen cabinet wall unit. *A*, Front view; *B*, section; *C*, overall view.

the manufacturer of the prefinished board. Plastic or other type moldings are used at the inside corners, at tub edges, at the joints, and as end caps. Several types of calking sealants also provide satisfactory joints.

Other finishes such as ceramic, plastic, and metal tile are installed over a special water-resistant type of gypsum board. Adhesive is spread with a serrated trowel and the 4¼- by 4¼-inch or other size tile pressed in place. A joint cement is used in the joints of ceramic tile after the adhesive has set. The plastic, metal, or ceramic type of wall covering around the tub area would usually cost somewhat more than plastic-surfaced hardboard. However, almost any type of wall finish can be added at the convenience of the homeowner at any time after the house is constructed.

Construction of Low-Cost Kitchen Cabinets

Low-cost kitchen cabinets can consist of a series of shelves enclosed with vertical dividers and ends. They can be designed so that flush doors can be added later at little cost and with no alterations. Curtains across the top of the opening can be temporarily used.

Base units.—Figure 97 shows the details of a simple base unit. Any combination of opening widths and shelf spacing can be used. Ripping 4-foot-wide sheets of plywood or particleboard in half provides material for ends, dividers, and shelves. Use ⅝ or ¾-inch thickness in an AC grade when only one side is exposed. When shelf layout is decided, saw slots across the width of the ends and dividers so that shelves will fit them. These *dados* are usually ¼ inch deep. Provide for about a 3½-by 3½-inch toe space at the bottom. When

assembling, nail the ends and dividers to the shelves with eightpenny nails spaced 5 to 6 inches apart. Use finish nails where ends are exposed. A 1- by 3-inch rail is used at the top in front to aid in fastening the top to the cabinet (fig. 97*A*). This will later serve to frame the doors. A 1- by 4-inch cleat is used across the back at the top, serving to fasten the cabinet to the wall studs, as well as for fastening the top (fig. 97*B*). The ends and dividers are notched to receive this cleat (fig. 97*C*). On finished ends, this notch is cut only halfway through. Fasten the cabinet to the wall by nailing or screwing through the back cleat into each stud with eightpenny nails or 2-inch screws.

The top of plywood or particleboard is nailed in place onto the ends, back cleat, and top rail with eightpenny finish nails. When the top has a plastic finish, use small metal angles around the interior and fasten them to the cabinet and top with small screws. When side stiles and doors are added later, a pleasant, utilitarian cabinet will result (fig. 97*D*). Until then, however, curtains can be used across the openings.

Any combination of base cabinets can be constructed. The sink cabinet usually consists only of a pair of doors and a bottom shelf. Vent slots are usually cut in the top cross rail. A cutout is necessary in the top to provide for a self-rimming or rimmed sink.

Wall units.—Wall units are about 30 inches high when installed below a drop ceiling. When carried to the ceiling line, they may be 44 or 45 inches in height. The depth is usually about 12 inches. A 4-foot-wide sheet of plywood or particleboard can be ripped in four pieces if this depth is used. As in base unit construction, the wall cabinet can consist of dadoed ends and one or more vertical dividers nailed to the shelves (fig. 98A). Use eightpenny finish nails when the end is exposed. The cleat is used to fasten the cabinet to the wall (fig. 98B). A top rail ties the sides and dividers together and, when the vertical stiles are added, provides framing for doors (fig. 98C).

Wardrobe-Closet Combinations

Closets are often eliminated in a low-cost house for economic reasons. A closet requires wall framing, interior and exterior covering, a door frame and door, and trim. However, some type of storage area should be included which will be pleasant in appearance, yet low in cost. Practical storage areas, often called wardrobe-closets, consist of wood or plywood sides, shelves, and some type of door or curtain. Figure 99 shows a simple built-in wardrobe that can be initially curtained. A folding door unit can be added later.

Wardrobe-closets are normally built of 5/8- or 3/4-inch plywood or particleboard. Provide dados for the shelf or shelves, as used on the kitchen cabinets. A back cleat at the top provides a member for fastening the unit to the wall. The sides can be trimmed at the floor with a base shoe molding which, with toenailing, keeps the sides in place. Add a simple closet pole. Fasten the unit to the wall at the top cleat and, when in a corner, also to the side wall (fig. 99). Any size or combination of wardrobes of this type can be built in during construction of the house or added later. Shelves or partial shelves can be provided in the bottom section for shoe storage.

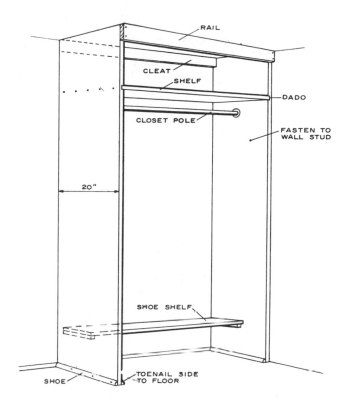

FIGURE 99.—Wardrobe-closet.

PORCHES

Additions such as a *porch* or an attached garage improve the appearance of a small house. Furthermore, in some areas of the country, the porch serves as a gathering place for family and neighbors and becomes almost a necessity.

While it is probably more practical to build a porch at the same time as the house proper, the porch can also be constructed quite readily after the house is completed. The cost of porches can vary a great deal, depending on design. A fully enclosed porch with windows and interior finish would add substantially to the cost of a house. On the other hand, the cost of an open porch with roof, floor, and simple supporting posts and footings might be well within the reach of many home builders. If desired, an open porch could be improved in the future by enclosing all or part of it.

Types of Open Porches

Two locations for porches might be considered for a low-cost house. One is on the end of a house (fig. 100A). Another is at the side of a house (fig. 100B). The porch at the end of a gable-roofed house may have a gable roof similar to the house itself. The slope of the roof should usually be the same as the main roof. When the size of the porch

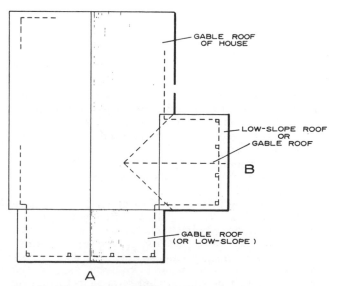

FIGURE 100.—Porch locations. A, End of house; B, side of house.

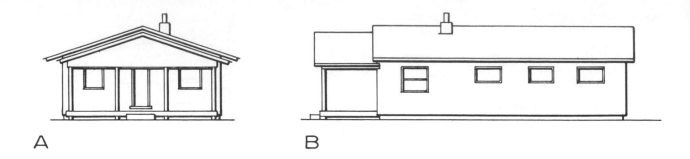

FIGURE 101.—End porch. *A*, End view; *B*, side view.

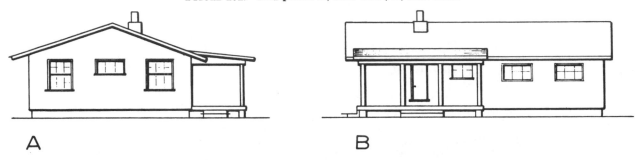

FIGURE 102.—Side porch with low-slope roof. *A*, End view; *B*, side view.

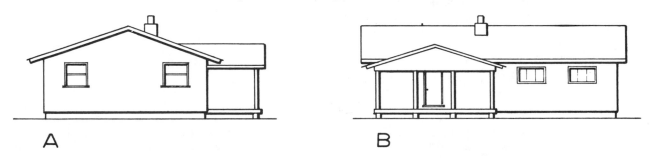

FIGURE 103.—Side porch with gable roof. *A*, End view; *B*, side view.

is somewhat less than the width of the house, the roofline will be slightly lower (fig. 101). When the porch is the same width, the roofline of the house is usually carried over the porch itself. A low-pitch roof in a shed or hip style can also be used for a porch located at the end of a house.

A low-slope roof can also be used for a porch at the side of a gable-roofed house (fig. 102). This system of room construction is usually less expensive than other methods, as the rafters also serve as ceiling joists.

Another method of roofing a porch at the side of the house is with the gable roof (fig. 103). While somewhat more costly than a flat or low-pitch roof, it is probably more pleasing in appearance. Both rafters and ceiling joists are normally used in this system. Thus, if a porch is to be a part of the original house or constructed later, select the style most suited to the design of the house and to the available funds.

Construction of Porches

While many types of foundations might be used for a porch as well as for the house, we will consider only the post or pier type because of its lower cost. A full-masonry foundation wall with proper footings will increase the cost substantially over the post or pier type. With a proper soil cover and a skirtboard of some type, the performance and appearance of such a post or pier foundation will not vary greatly from that of a full-masonry wall.

Floor System

The floor system for the open porch should consist of treated wood posts or masonry piers constructed in the same way as those outlined in the section on foundation systems for the house proper. These posts or piers should bear on footings and support nail-laminated floor beams which are anchored to the posts. These beams, of doubled

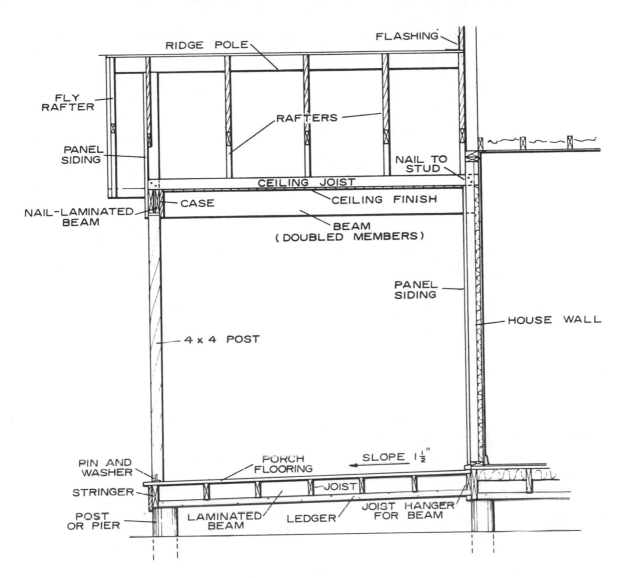

FIGURE 104.—Section through porch with gable roof.

and nailed 2- by 8- or 2- by 10-inch members, serve to support the floor joists by means of *ledgers* (fig. 104). The beams span between the wall line of the house and the posts at the outside edge of the porch. Nail 2- by 3- or 2- by 4-inch ledger members to both sides of the intermediate beams and to one side of single edge beams at the ends of the porch with sixteenpenny nails spaced 8 to 10 inches apart. The beams should be located so that the floor surface of an 8-foot-wide porch slopes outward a total of at least 1½ inches for drainage. Beams are spaced about 10 feet or less apart across the width of the porch, but this depends on the size of joists and the beams. These details are ordinarily shown on the plans.

Joist and beam tables can also be used to determine the correct span-spacing-size relationship. Use a metal joist hanger or angle iron when fastening the beam ends to the floor framing of the house (fig. 104), or allow the ends to bear on a post or

pier. A single header (with ledger) the same size as the beam is used at the ouside edge of the porch.

Now, cut the floor joists to fit between the laminated beams so that they rest on the ledgers. Space them properly according to the details in the working drawings, and toenail the ends to the beam with two eightpenny nails on each side.

Dressed and matched porch flooring, in nominal 1- by 4-inch size, can be applied to the floor joists or installed after the roof framing and roofing are in place. To protect the floor from damage, it is perhaps best to delay this phase of construction. This requires the use of temporary braces for the roof until the flooring and porch posts are installed.

Framing for Gable Porch Roof

A doubled member or nail-laminated beam is required to carry the roof load whether a gable or low-slope roof is used. These beams are made up by using spacers of *lath* or plywood between 2-inch members so that the beam is the same size (3½ inches) as the nominal 4- by 4-inch solid posts used for final support of the roof. End beams made up of doubled 2-inch members are fastened to the outside beam and to the house (fig. 104). Thus, the outline of the porch is now formed by the front and end beams. One method of assembling this roof framing is by nailing it together over the floor framing and raising it in place. Temporary 2- by 4-inch or larger braces are used to support this beam framing while the roof is being constructed. Use enough braces to prevent movement. Use joist hangers or a length of angle iron to fasten one end of the beam to the house framing. The beam ends can also be carried through the wall and supported by auxiliary studs when possible. It is often desirable to provide a 6-foot 8-inch or a 7-foot clearance from the porch floor to the bottom of the beams so that standard units can be used if screening or enclosing is planned in the future.

Ceiling joists are now fastened to the beam at one end and to the studs of the house at the other. They should be spaced 16 or 24 inches on center to provide nailing for a ceiling material.

Rafters, either 2- by 4- or 2- by 6-inch in size depending on the span, are now measured and cut at the ridge and wall line as described in the section on pitched roofs. A 1- by 6-inch ridgepole is used to tie each pair of rafters together. Gable end studs are now cut to fit between the end rafters and the outside beam (fig. 104). Space them to accommodate the panel siding or other finish. Toenail the studs to the outside beam and to the end rafters.

The roof sheathing, the fly rafters, and the roofing are applied as described in the sections on roof systems and coverings. Use flashing at the junction of the roof and the end wall of the house when applying the roofing.

The matched flooring may now be applied to the floor across the joists. Extend it beyond the outer edge (fig. 104). Use sevenpenny or eightpenny flooring nails and blind-nail to each joist. It is good practice to apply a saturating coat of water-repellent preservative on the surface and edges. This will provide protection until the floor is painted. Some decay-resistant wood species require little protection other than this treatment.

The nominal 4- by 4-inch posts can now be installed. When they are cut to length, drill and drive a small ⅜- or ½-inch-diameter pin into the center of one end. A matching hole is drilled into the floor, a mastic calk applied to the area, and the post positioned (fig. 104). Use a large galvanized washer between the post and the floor. This will allow moisture to evaporate and prevent decay. Use toenailing and metal strapping to tie the post to the roof framing. In areas of high winds, use a *bolt* or lag screw instead of the pin from the post to some part of the floor framing.

Matched boards, plywood, or similar covering materials can be used to finish the ceiling surface. The exterior and interior of the laminated beams are normally cased with nominal 1-inch boards except where siding at the gable end can be carried over the exposed face of the beam. A 1- by 4-inch member is normally used under the beam and between the posts. All exposed nailing should be done with galvanized or other rust-resistant nails.

Framing for Low-Slope Porch Roof

Framing a flat or low-slope roof for a porch is relatively simple. A nailed beam, similar to the type used for the gable roof, is required to support the ends of rafter-joists (fig. 105). Single end members, however, are used to tie the beam to wall framing of the house rather than a doubled end beam. Temporary posts or braces are used to hold the beam framing after it is in place. A minimum roof slope of 1 inch in a foot is desirable for drainage.

Cut the rafters and toenail them to the beam with eightpenny nails and face-nail the opposite ends

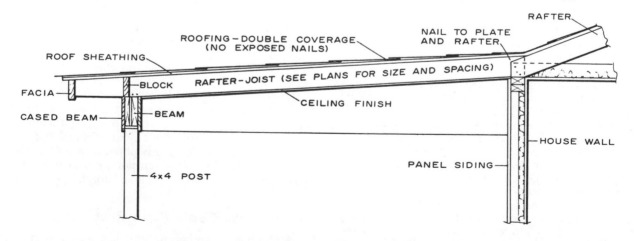

FIGURE 105.--Section through side porch with low-slope roof.

to the rafter or the ceiling joists of the house (fig. 105). When spacing is not the same and members do not join the rafters or ceiling joists of the house, toenail the ends of the porch rafters to the wall plate. Plywood or board sheathing is applied, and when end overhang is desired, a fly rafter is added. When desired, 1- by 6-inch facia boards are applied to the ends of the rafters and to the fly rafter with sevenpenny or eightpenny galvanized siding nails. Roofing is applied as has been recommended for low-slope roofs; underlayment followed by double-coverage surfaced roll roofing. On low slopes, no exposed nails are used. A ribbon of asphalt roof cement or lap seal material is used under the lapped edge. Extend the roofing beyond the facia enough to form a natural drip edge. Ceiling covering and other trim are used in the same manner as for the gable roof.

STEPS AND STAIRS

Outside entry platforms and steps are required for most types of wood-frame houses. Houses constructed over a full basement or crawl space normally require a platform with steps at outside doors. In houses with masonry foundations, these outside entry stoops consist of a concrete perimeter wall with poured concrete steps. However, a well-constructed wood porch will serve as well and normally cost less.

Inside stairs leading to an attic or second-floor bedrooms in a house with a steep roof slope, or to the basement, must be provided for during construction of the house. This includes framing of the floor joists to accommodate the stairway and providing walls and *carriages* for the *treads* and *risers*. Even though the second floor might not be completed immediately, a stairway should be included during construction of such houses.

Outside Stoops

Outside wood stoops, platforms, and open plank stairs should give satisfactory service if these simple rules are followed: (a) All wood in contact with or embedded in soil should be pressure-treated, as outlined in the early section on "Post Foundations"; (b) all untreated wood parts should have a 2-inch minimum clearance above the ground; (c) avoid pockets or areas in the construction where water cannot drain away; (d) if possible, use wood having moderate to good decay resistance; [7] (e) use initial and regular applications of water-repellent preservative to exposed untreated wood surfaces; and (f) use vertical grain members.

Wood Stoop—Low Height

A simple all-wood stoop consists of treated posts embedded in the ground, cross or bearing members, and spaced treads. One such design is shown in figure 106A. Because this type of stoop is low, it can serve as an entry for exterior doors where the floor level of the house is no more than 24 inches above the ground. Railings are not usually required. The platform should be large enough so that the storm door can swing outward freely. An average size is about 3½ feet deep by 5 feet wide.

Use treated posts of 5- to 7-inch diameter and embed them in the soil at least 3 feet. Nail and bolt (with galvanized fasteners) a crossmember (usually a nominal 2- by 4-inch member) to each side of the posts (fig. 106B). The posts should be faced slightly at these areas. For a small, 3½- by 5-foot stoop, four posts are usually sufficient. Posts are spaced about 4 feet apart across the front.

Supports for the tread of the first step are supplied by pairs of 2- by 4-inch members bolted to the forward posts (fig. 106A). The inner ends are blocked with short pieces of 2- by 4-inch members to the upper crosspieces (fig. 106A). Treads consist of 2- by 4- or 2- by 6-inch members, spaced about ¼ inch apart. Use two sixteenpenny galvanized plain or ring-shank nails for each piece at each supporting member.

Some species of wood have natural decay resistance and others are benefited by the application of a water-repellent preservative followed by a good deck paint. If desired, a railing can be added by bolting short upright members to the 2- by 4-inch crossmembers. Horizontal railings can be fastened to the uprights.

Wood Stoop—Medium Height

A wood entry platform requiring more than one or two steps is usually designed with a railing and stair stringers. If the platform is about 3½ by 5 feet, two 2- by 12-inch carriages can be used to support the treads (fig. 107A). In most cases, the bottoms of the carriages are supported by treated posts embedded in the ground or by an embedded treated timber. The upper ends of the carriages are supported by a 2- by 4-inch ledger fastened to posts and are face-nailed to the platform framing with twelvepenny galvanized nails. The carriage at the house side can be supported in the same way when interior posts are used.

When the platform is narrow, a nominal 3- by 4-inch ledger is fastened to the floor framing of the house with fortypenny galvanized spikes or 5-inch lag screws. The 2- by 6-inch floor planks are nailed to the ledger and to the double 2- by 4-inch

[7] Anderson, *ibid.*

beam bolted to the post (fig. 107B). Use two sixteenpenny galvanized plain or ring-shank nails for each tread. When a wide platform is desired, an inside set of posts and doubled 2 by 4 or larger beams should be used.

Railings can be made of 2- by 4-inch uprights bolted or lagscrewed to the outside beams. These members are best fastened with galvanized bolts or lag screws. Horizontal railing in 1- by 4- or 1- by 6-inch size can then be fastened to the uprights. When an enclosed skirting is desired, 1- by 4-inch slats can be nailed to the outside of the beam and an added lower nailing member (fig. 107A). Treat all exposed untreated wood with a heavy application of water-repellent preservative. When a paint finish is desired, use a good deck paint. See section on "Painting and Finishing" for details.

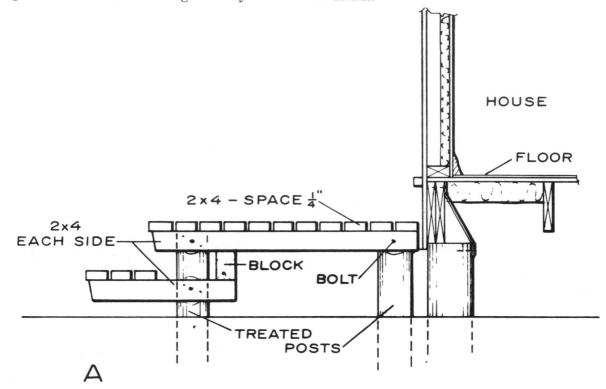

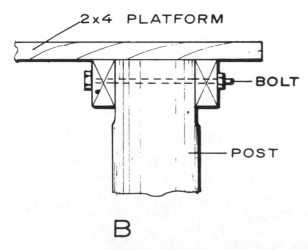

FIGURE 106.—Low-height wood stoop. A, Side elevation; B, connection to post.

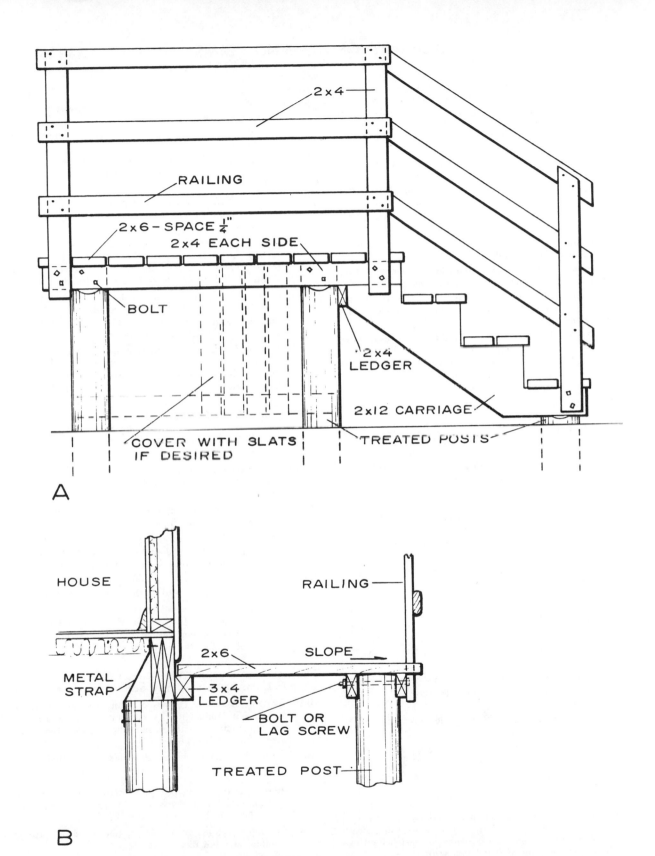

FIGURE 107.—Medium-height wood stoop. *A*, Side elevation; *B*, connection to post.

795

Air-dried lumber. Lumber that has been piled in yards or sheds for any length of time. For the United States as a whole, the minimum moisture content of thoroughly air-dried lumber is 12 to 15 percent and the average is somewhat higher. In the South, air-dried lumber may be no lower than 19 percent.

Airway. A space between roof insulation and roof boards for movement of air.

Alligatoring. Coarse checking pattern characterized by a slipping of the new paint coating over the old coating to the extent that the old coating can be seen through the fissures.

Anchor bolts. Bolts to secure a wooden sill plate to concrete or masonry floor or wall.

Apron. The flat member of the inside trim of a window placed against the wall immediately beneath the stool.

Areaway. An open subsurface space adjacent to a building used to admit light or air or as a means of access to a basement.

Asphalt. Most native asphalt is a residue from evaporated petroleum. It is insoluble in water but soluble in gasoline and melts when heated. Used widely in building for waterproofing roof coverings of many types, exterior wall coverings, flooring tile, and the like.

Astragal. A molding, attached to one of a pair of swinging doors, against which the other door strikes.

Attic ventilators. In houses, screened openings provided to ventilate an attic space. They are located in the soffit area as inlet ventilators and in the gable end or along the ridge as outlet ventilators. They can also consist of power-driven fans used as an exhaust system. (See also **Louver.**)

Backband. A simple molding sometimes used around the outer edge of plain rectangular casing as a decorative feature.

Backfill. The replacement of excavated earth into a trench around and against a basement foundation.

Balusters. Usually small vertical members in a railing used between a top rail and the stair treads or a bottom rail.

Balustrade. A railing made up of balusters, top rail, and sometimes bottom rail, used on the edge of stairs, balconies, and porches.

Barge board. A decorative board covering the projecting rafter (fly rafter) of the gable end. At the cornice, this member is a facia board.

Base or baseboard. A board placed against the wall around a room next to the floor to finish properly between floor and plaster.

Base molding. Molding used to trim the upper edge of interior baseboard.

Base shoe. Molding used next to the floor on interior baseboard. Sometimes called a carpet strip.

Batten. Narrow strips of wood used to cover joints or as decorative vertical members over plywood or wide boards.

Batter board. One of a pair of horizontal boards nailed to posts set at the corners of an excavation, used to indicate the desired level, also as a fastening for stretched strings to indicate outlines of foundation walls.

Bay window. Any window space projecting outward from the walls of a building, either square or polygonal in plan.

Beam. A structural member transversely supporting a load.

Bearing partition. A partition that supports any vertical load in addition to its own weight.

Bearing wall. A wall that supports any vertical load in addition to its own weight.

Bed molding. A molding in an angle, as between the overhanging cornice, or eaves, of a building and the sidewalls.

Blind-nailing. Nailing in such a way that the nailheads are not visible on the face of the work—usually at the tongue of matched boards.

Blind stop. A rectangular molding, usually ¾ by 1-⅜ inches or more in width, used in the assembly of a window frame. Serves as a stop for storm and screen or combination windows and to resist air infiltration.

Blue stain. A bluish or grayish discoloration of the sapwood caused by the growth of certain moldlike fungi on the surface and in the interior of a piece, made possible by the same conditions that favor the growth of other fungi.

Bodied linseed oil. Linseed oil that has been thickened in viscosity by suitable processing with heat or chemicals. Bodied oils are obtainable in a great range in viscosity from a little greater than that of raw oil to just short of a jellied condition.

Boiled linseed oil. Linseed oil in which enough lead, manganese, or cobalt salts have been incorporated to make the oil harden more rapidly when spread in thin coatings.

Bolster. A short horizontal timber or steel beam on top of a column to support and decrease the span of beams or girders.

Boston ridge. A method of applying asphalt or wood shingles at the ridge or at the hips of a roof as a finish.

Brace. An inclined piece of framing lumber applied to wall or floor to stiffen the structure. Often used on walls as temporary bracing until framing has been completed.

Brick veneer. A facing of brick laid against and fastened to sheathing of a frame wall or tile wall construction.

Bridging. Small wood or metal members that are inserted in a diagonal position between the floor joists at midspan to act both as tension and compression members for the purpose of bracing the joists and spreading the action of loads.

Buck. Often used in reference to rough frame opening members. Door bucks used in reference to metal door frame.

Built-up roof. A roofing composed of three to five layers of asphalt felt laminated with coal tar, pitch, or asphalt. The top is finished with crushed slag or gravel. Generally used on flat or low-pitched roofs.

Butt joint. The junction where the ends of two timbers or other members meet in a square-cut joint.

Cant strip. A triangular-shaped piece of lumber used at the junction of a flat deck and a wall to prevent cracking of the roofing which is applied over it.

Cap. The upper member of a column, pilaster, door cornice, molding, and the like.

Casement frames and sash. Frames of wood or metal enclosing part or all of the sash, which may be opened by means of hinges affixed to the vertical edges.

Casing. Molding of various widths and thicknesses used to trim door and window openings at the jambs.

Cement, Keene's. A white finish plaster that produces an extremely durable wall. Because of its density, it excels for use in bathrooms and kitchens and is also used extensively for the finish coat in auditoriums, public buildings, and other places where walls may be subjected to unusually hard wear or abuse.

Checking. Fissures that appear with age in many exterior paint coatings, at first superficial, but which in time may penetrate entirely through the coating.

Checkrails. Meeting rails sufficiently thicker than a window to fill the opening between the top and bottom sash made by the parting stop in the frame of double-hung windows. They are usually beveled.

Collar beam. Nominal 1- or 2-inch-thick members connecting opposite roof rafters. They serve to stiffen the roof structure.

Column. In architecture: A perpendicular supporting member, circular or rectangular in section, usually consisting of a base, shaft, and capital. In engineering: A vertical structural compression member which supports loads acting in the direction of its longitudinal axis.

Combination doors or windows. Combination doors or windows used over regular openings. They provide winter insulation and summer protection and often have self-storing or removable glass and screen inserts. This eliminates the need for handling a different unit each season.

Concrete plain. Concrete either without reinforcement, or reinforced only for shrinkage or temperature changes.

Condensation. In a building: Beads or drops of water (and frequently frost in extremely cold weather) that accumulate on the inside of the exterior covering of a building when warm, moisture-laden air from the interior reaches a point where the temperature no longer permits the air to sustain the moisture it holds. Use of louvers or attic ventilators will reduce moisture condensation in attics. A vapor barrier under the gypsum lath or dry wall on exposed walls will reduce condensation in them.

Conduit, electrical. A pipe, usually metal, in which wire is installed.

Construction dry-wall. A type of construction in which the interior wall finish is applied in a dry condition, generally in the form of sheet materials or wood paneling, as contrasted to plaster.

Construction, frame. A type of construction in which the structural parts are wood or depend upon a wood frame for support. In codes, if masonry veneer is applied to the exterior walls, the classification of this type of construction is usually unchanged.

Coped joint. See **Scribing.**

Corbel out. To build out one or more courses of brick or stone from the face of a wall, to form a support for timbers.

Corner bead. A strip of formed sheet metal, sometimes combined with a strip of metal lath, placed on corners before plastering to reinforce them. Also, a strip of wood finish three-quarters-round or angular placed over a plastered corner for protection.

Corner boards. Used as trim for the external corners of a house or other frame structure against which the ends of the siding are finished.

Corner braces. Diagonal braces at the corners of frame structure to stiffen and strengthen the wall.

Let-in brace. Nominal 1-inch-thick boards applied into notched studs diagonally.

Cut-in brace. Nominal 2-inch-thick members, usually 2 by 4's, cut in between each stud diagonally.

Cornerite. Metal-mesh lath cut into strips and bent to a right angle. Used in interior corners of walls and ceilings on lath to prevent cracks in plastering.

Cornice. Overhang of a pitched roof at the eave line, usually consisting of a facia board, a soffit for a closed cornice, and appropriate moldings.

Cornice return. That portion of the cornice that returns on the gable end of a house.

Counterflashing. A flashing usually used on chimneys at the roofline to cover shingle flashing and to prevent moisture entry.

Cove molding. A molding with a concave face used as trim or to finish interior corners.

Crawl space. A shallow space below the living quarters of a basementless house, normally enclosed by the foundation wall.

Cricket. A small drainage-diverting roof structure of single or double slope placed at the junction of larger surfaces that meet at an angle, such as above a chimney.

Cross-bridging. Diagonal bracing between adjacent floor joists, placed near the center of the joist span to prevent joists from twisting.

Crown molding. A molding used on cornice or wherever an interior angle is to be covered.

d. See **Penny.**

Dado. A rectangular groove across the width of a board or plank. In interior decoration, a special type of wall treatment.

Decay. Disintegration of wood or other substance through the action of fungi.

Deck paint. An enamel with a high degree of resistance to mechanical wear, designed for use on such surfaces as porch floors.

Density. The mass of substance in a unit volume. When expressed in the metric system, it is numerically equal to the specific gravity of the same substance.

Dewpoint. Temperature at which a vapor begins to deposit as a liquid. Applies especially to water in the atmosphere.

Dimension. See **Lumber dimension.**

Direct nailing. To nail perpendicular to the initial surface or to the junction of the pieces joined. Also termed **face nailing.**

Doorjamb, interior. The surrounding case into which and out of which a door closes and opens. It consists of two upright pieces, called side jambs, and a horizontal head jamb.

Dormer. An opening in a sloping roof, the framing of which projects out to form a vertical wall suitable for windows or other openings.

Downspout. A pipe, usually of metal, for carrying rainwater from roof gutters.

Dressed and matched (tongued and grooved). Boards or planks machined in such a maner that there is a groove on one edge and a corresponding tongue on the other.

Drier paint. Usually oil-soluble soaps of such metals as lead, manganese, or cobalt, which, in small proportions, hasten the oxidation and hardening (drying) of the drying oils in paints.

Drip. (a) A member of a cornice or other horizontal exterior-finish course that has a projection beyond the other parts for throwing off water. (b) A groove in the underside of a sill or drip cap to cause water to drop off on the outer edge instead of drawing back and running down the face of the building.

Drip cap. A molding placed on the exterior top side of a door or window frame to cause water to drip beyond the outside of the frame.

Dry-wall. Interior covering material, such as gypsum board or plywood, which is applied in large sheets or panels.

Ducts. In a house, usually round or rectangular metal pipes for distributing warm air from the heating plant to rooms, or air from a conditioning device or as cold air returns. Ducts are also made of asbestos and composition materials.

Eaves. The margin or lower part of a roof projecting over the wall.

Expansion joint. A bituminous fiber strip used to separate blocks or units of concrete to prevent cracking due to expansion as a result of temperature changes. Also used on concrete slabs.

Facia or fascia. A flat board, band, or face, used sometimes by itself but usually in combination with moldings, often located at the outer face of the cornice.

Filler (wood). A heavily pigmented preparation used for filling and leveling off the pores in open-pored woods.

Fire-resistive. In the absence of a specific ruling by the authority having jurisdiction, applies to materials for construction not combustible in the temperatures of ordinary fires and that will withstand such fires without serious impairment of their usefulness for at least 1 hour.

Fire-retardant chemical. A chemical or preparation of chemicals used to reduce flammability or to retard spread of flame.

Fire stop. A solid, tight closure of a concealed space, placed to prevent the spread of fire and smoke through such a space. In a frame wall, this will usually consist of 2 by 4 cross blocking between studs.

Fishplate. A wood or plywood piece used to fasten the ends of two members together at a butt joint with nails or bolts. Sometimes used at the junction of opposite rafters near the ridge line.

Flagstone (flagging or flags). Flat stones, from 1 to 4 inches thick, used for rustic walks, steps, floors, and the like.

Flashing. Sheet metal or other material used in roof and wall construction to protect a building from water seepage.

Flat paint. An interior paint that contains a high proportion of pigment and dries to a flat or lusterless finish.

Flue. The space or passage in a chimney through which smoke, gas, or fumes ascend. Each passage is called a flue, which together with any others and the surrounding masonry make up the chimney.

Flue lining. Fire clay or terra-cotta pipe, round or square, usually made in all ordinary flue sizes and in 2-foot lengths, used for the inner lining of chimneys with the brick or masonry work around the outside. Flue lining in chimneys runs from about a foot below the flue connection to the top of the chimney.

Fly rafters. End rafters of the gable overhang supported by roof sheathing and lookouts.

Footing. A masonry section, usually concrete, in a rectangular form wider than the bottom of the foundation wall or pier it supports.

Foundation. The supporting portion of a structure below the first-floor construction, or below grade, including the footings.

Framing, balloon. A system of framing a building in which all vertical structural elements of the bearing walls and partitions consist of single pieces extending from the top of the foundation sill plate to the roofplate and to which all floor joists are fastened.

Framing, platform. A system of framing a building in which floor joists of each story rest on the top plates of the story below or on the foundation sill for the first story, and the bearing walls and partitions rest on the subfloor of each story.

Frieze. In house construction, a horizontal member connecting the top of the siding with the soffit of the cornice.

Frostline. The depth of frost penetration in soil. This depth varies in different parts of the country. Footings should be placed below this depth to prevent movement.

Fungi, wood. Microscopic plants that live in damp wood and cause mold, stain, and decay.

Fungicide. A chemical that is poisonous to fungi.

Furring. Strips of wood or metal applied to a wall or other surface to even it and normally to serve as a fastening base for finish material.

Gable. In house construction, the portion of the roof above the eave line of a double-sloped roof.

Gable end. An end wall having a gable.

Gloss enamel. A finishing material made of varnish and sufficient pigments to provide opacity and color, but little or no pigment of low opacity. Such an enamel forms a hard coating with maximum smoothness of surface and a high degree of gloss.

Gloss (paint or enamel). A paint or enamel that contains a relatively low proportion of pigment and dries to a sheen or luster.

Girder. A large or principal beam of wood or steel used to support concentrated loads at isolated points along its length.

Grain. The direction, size, arrangement, appearance, or quality of the fibers in wood.

Grain, edge (vertical). Edge-grain lumber has been sawed parallel to the pith of the log and approximately at right angles to the growth rings; i.e., the rings form an angle of 45° or more with the surface of the piece.

Grain, flat. Flat-grain lumber has been sawed parallel to the pith of the log and approximately tangent to the growth rings, i.e., the rings form an angle of less than 45° with the surface of the piece.

Grain, quartersawn. Another term for edge grain.

Grounds. Guides used around openings and at the floorline to strike off plaster. They can consist of narrow strips of wood or of wide subjambs at interior doorways. They provide a level plaster line for installation of casing and other trim.

Grout. Mortar made of such consistency (by adding water) that it will just flow into the joints and cavities of the masonry work and fill them solid.

Gusset. A flat wood, plywood, or similar type member used to provide a connection at intersection of wood members. Most commonly used at joints of wood trusses. They are fastened by nails, screws, bolts, or adhesives.

Gutter or eave trough. A shallow channel or conduit of metal or wood set below and along the eaves of a house to catch and carry off rainwater from the roof.

Gypsum plaster. Gypsum formulated to be used with the addition of sand and water for base-coat plaster.

Header. (a) A beam placed perpendicular to joists and to which joists are nailed in framing for chimney, stairway, or other opening. (b) A wood lintel.

Hearth. The inner or outer floor of a fireplace, usually made of brick, tile, or stone.

Heartwood. The wood extending from the pith to the sapwood, the cells of which no longer participate in the life processes of the tree.

Hip. The external angle formed by the meeting of two sloping sides of a roof.

Hip roof. A roof that rises by inclined planes from all four sides of a building.

Humidifier. A device designed to increase the humidity within a room or a house by means of the discharge of water vapor. They may consist of individual room-size units or larger units attached to the heating plant to condition the entire house.

I-beam. A steel beam with a cross section resembling the letter *I*. It is used for long spans as basement beams or over wide wall openings, such as a double garage door, when wall and roof loads are imposed on the opening.

IIC. A new system utilized in the Federal Housing Administration recommended criteria for impact sound insulation.

INR (Impact Noise Rating). A single figure rating which provides an estimate of the impact sound-insulating performance of a floor-ceiling assembly.

Insulation board, rigid. A structural building board made of coarse wood or cane fiber in ½- and 25/32-inch thicknesses. It can be obtained in various size sheets, in various densities, and with several treatments.

Insulation, thermal. Any material high in resistance to heat transmission that, when placed in the walls, ceiling, or floors of a structure, will reduce the rate of heat flow.

Interior finish. Material used to cover the interior framed areas, or materials of walls and ceilings.

Jack rafter. A rafter that spans the distance from the wallplate to a hip, or from a valley to a ridge.

Jamb. The side and head lining of a doorway, window, or other opening.

Joint. The space between the adjacent surfaces of two members or components joined and held together by nails, glue, cement, mortar, or other means.

Joint cement. A powder that is usually mixed with water and used for joint treatment in gypsum-wallboard finish. Often called "spackle."

Joist. One of a series of parallel beams, usually 2 inches in thickness, used to support floor and ceiling loads, and supported in turn by larger beams, girders, or bearing walls.

Kiln dried lumber. Lumber that has been kiln dried often to a moisture content of 6 to 12 percent. Common varieties of softwood lumber, such as framing lumber are dried to a somewhat higher moisture content.

Knot. In lumber, the portion of a branch or limb of a tree that appears on the edge or face of the piece.

Landing. A platform between flights of stairs or at the termination of a flight of stairs.

Lath. A building material of wood, metal, gypsum, or insulating board that is fastened to the frame of a building to act as a plaster base.

Lattice. A framework of crossed wood or metal strips.

Leader. See **Downspout.**

Ledger strip. A strip of lumber nailed along the bottom of the side of a girder on which joists rest.

Light. Space in a window sash for a single pane of glass. Also, a pane of glass.

Lintel. A horizontal structural member that supports the load over an opening such as a door or window.

Lookout. A short wood bracket or cantilever to support an overhang portion of a roof or the like, usually concealed from view.

Louver. An opening with a series of horizontal slats so arranged as to permit ventilation but to exclude rain, sunlight, or vision. See also **Attic ventilators.**

Lumber. Lumber is the product of the sawmill and planing mill not further manufactured other than by sawing, resawing, and passing lengthwise through a standard planing machine, crosscutting to length, and matching.

Lumber, boards. Yard lumber less than 2 inches thick and 2 or more inches wide.

Lumber, dimension. Yard lumber from 2 inches to, but not including, 5 inches thick and 2 or more inches wide. Includes joists, rafters, studs, plank, and small timbers.

Lumber, dressed size. The dimension of lumber after shrinking from green dimension and after machining to size or pattern.

Lumber, matched. Lumber that is dressed and shaped on one edge in a grooved pattern and on the other in a tongued pattern.

Lumber, shiplap. Lumber that is edge-dressed to make a close rabbeted or lapped joint.

Lumber, timbers. Yard lumber 5 or more inches in least dimension. Includes beams, stringers, posts, caps, sills, girders, and purlins.

Lumber, yard. Lumber of those grades, sizes, and patterns which are generally intended for ordinary construction, such as framework and rough coverage of houses.

Mantel. The shelf above a fireplace. Also used in referring to the decorative trim around a fireplace opening.

Masonry. Stone, brick, concrete, hollow-tile, concrete-block, gypsum-block, or other similar building units or materials or a combination of the same, bonded together with mortar to form a wall, pier, buttress, or similar mass.

Mastic. A pasty material used as a cement (as for setting tile) or a protective coating (as for thermal insulation or waterproofing).

Metal lath. Sheets of metal that are slit and drawn out to form openings. Used as a plaster base for walls and ceilings and as reinforcing over other forms of plaster base.

Millwork. Generally all building materials made of finished wood and manufactured in millwork plants and planing mills are included under the term "millwork." It includes such items as inside and outside doors, window and doorframes, blinds, porchwork, mantels, panelwork, stairways, moldings, and interior trim. It normally does not include flooring, ceiling, or siding.

Miter joint. The joint of two pieces at an angle that bisects the joining angle. For example, the miter joint at the side and head casing at a door opening is made at a 45° angle.

Moisture content of wood. Weight of the water contained in the wood, usually expressed as a percentage of the weight of the ovendry wood.

Molding. A wood strip having a curved or projecting surface used for decorative purposes.

Mortise. A slot cut into a board, plank, or timber, usually edgewise, to receive tenon of another board, plank, or timber to form a joint.

Mullion. A vertical bar or divider in the frame between windows, doors, or other openings.

Muntin. A small member which divides the glass or openings of sash or doors.

Natural finish. A transparent finish which does not seriously alter the original color or grain of the natural wood. Natural finishes are usually provided by sealers, oils, varnishes, water-repellent preservatives, and other similar materials.

Newel. A post to which the end of a stair railing or balustrade is fastened. Also, *any* post to which a railing or balustrade is fastened.

Nonbearing wall. A wall supporting no load other than its own weight.

Nosing. The projecting edge of a molding or drip. Usually applied to the projecting molding on the edge of a stair tread.

Notch. A crosswise rabbet at the end of a board.

O. C., on center. The measurement of spacing for studs, rafters, joists, and the like in a building from the center of one member to the center of the next.

O. G., or ogee. A molding with a profile in the form of a letter *S*; having the outline of a reversed curve.

Outrigger. An extension of a rafter beyond the wall line. Usually a smaller member nailed to a larger rafter to form a cornice or roof overhang.

Paint. A combination of pigments with suitable thinners or oils to provide decorative and protective coatings.

Panel. In house construction, a thin flat piece of wood, plywood, or similar material, framed by stiles and rails as in a door or fitted into grooves of thicker material with molded edges for decorative wall treatment.

Paper, building. A general term for papers, felts, and similar sheet materials used in buildings without reference to their properties or uses.

Paper, sheathing. A building material, generally paper or felt, used in wall and roof construction as a protection against the passage of air and sometimes moisture.

Parting stop or strip. A small wood piece used in the side and head jambs of double-hung windows to separate upper and lower sash.

Partition. A wall that subdivides spaces within any story of a building.

Penny. As applied to nails, it originally indicated the price per hundred. The term now serves as a measure of nail length and is abbreviated by the letter *d*.

Perm. A measure of water vapor movement through a material (grains per square foot per hour per inch of mercury difference in vapor pressure).

Pier. A column of masonry, usually rectangular in horizontal cross section, used to support other structural members.

Pigment. A powdered solid in suitable degree of subdivision for use in paint or enamel.

Pitch. The incline slope of a roof or the ratio of the total rise to the total width of a house, i.e., an 8-foot rise and 24-foot width is a one-third pitch roof. Roof slope is expressed in the inches of rise per foot of run.

Pitch pocket. An opening extending parallel to the annual rings of growth, that usually contains, or has contained, either solid or liquid pitch.

Pith. The small, soft core at the original center of a tree around which wood formation takes place.

Plaster grounds. Strips of wood used as guides or strike-off edges around window and door openings and at base of walls.

Plate. Sill plate: a horizontal member anchored to a masonry wall. Sole plate: bottom horizontal member of a frame wall. Top plate: top horizontal member of a frame wall supporting ceiling joists, rafters, or other members.

Plough. To cut a lengthwise groove in a board or plank.

Plumb. Exactly perpendicular; vertical.

Ply. A term to denote the number of thicknesses or layers of roofing felt, veneer in plywood, or layers in built-up materials, in any finished piece of such material.

Plywood. A piece of wood made of three or more layers of veneer joined with glue, and usually laid with the grain of adjoining plies at right angles. Almost always an odd number of plies are used to provide balanced construction.

Pores. Wood cells of comparatively large diameter that have open ends and are set one above the other to form continuous tubes. The openings of the vessels on the surface of a piece of wood are referred to as pores.

Preservative. Any substance that, for a reasonable length of time, will prevent the action of wood-destroying fungi, borers of various kinds, and similar destructive agents when the wood has been properly coated or impregnated with it.

Primer. The first coat of paint in a paint job that consists of two or more coats; also the paint used for such a first coat.

Putty. A type of cement usually made of whiting and boiled linseed oil, beaten or kneaded to the consistency of dough, and used in sealing glass in sash, filling small holes and crevices in wood, and for similar purposes.

Quarter round. A small molding that has the cross section of a quarter circle.

Rabbet. A rectangular longitudinal groove cut in the corner edge of a board or plank.

Radiant heating. A method of heating, usually consisting of a forced hot water system with pipes placed in the floor, wall, or ceiling; or with electrically heated panels.

Rafter. One of a series of structural members of a roof designed to support roof loads. The rafters of a flat roof are sometimes called roof joists.

Rafter, hip. A rafter that forms the intersection of an external roof angle.

Rafter, valley. A rafter that forms the intersection of an internal roof angle. The valley rafter is normally made of double 2-inch-thick members.

Rail. Cross members of panel doors or of a sash. Also the upper and lower members of a balustrade or staircase extending from one vertical support, such as a post, to another.

Rake. Trim members that run parallel to the roof slope and form the finish between the wall and a gable roof extension.

Raw linseed oil. The crude product processed from flaxseed and usually without much subsequent treatment.

Reflective insulation. Sheet material with one or both surfaces of comparatively low heat emissivity, such as aluminum foil. When used in building construction the surfaces face air spaces, reducing the radiation across the air space.

Reinforcing. Steel rods or metal fabric placed in concrete slabs, beams, or columns to increase their strength.

Relative humidity. The amount of water vapor in the atmosphere, expressed as a percentage of the maximum quantity that could be present at a given temperature. (The actual amount of water vapor that can be held in space increases with the temperature.)

Resorcinol glue. A glue that is high in both wet and dry strength and resistant to high temperatures. It is used for gluing lumber or assembly joints that must withstand severe service conditions.

Ribbon (Girt). Normally a 1- by 4-inch board let into the studs horizontally to support ceiling or second-floor joists.

Ridge. The horizontal line at the junction of the top edges of two sloping roof surfaces.

Ridge board. The board placed on edge at the ridge of the roof into which the upper ends of the rafters are fastened.

Rise. In stairs, the vertical height of a step or flight of stairs.

Riser. Each of the vertical boards closing the spaces between the treads of stairways.

Roll roofing. Roofing material, composed of fiber and saturated with asphalt, that is supplied in 36-inch wide rolls with 108 square feet of material. Weights are generally 45 to 90 pounds per roll.

Roof sheathing. The boards or sheet material fastened to the roof rafters on which the shingle or other roof covering is laid.

Rubber-emulsion paint. Paint, the vehicle of which consists of rubber or synthetic rubber dispersed in fine droplets in water.

Run. In stairs, the net width of a step or the horizontal distance covered by a flight of stairs.

Saddle. Two sloping surfaces meeting in a horizontal ridge, used between the back side of a chimney, or other vertical surface, and a sloping roof.

Sand float finish. Lime mixed with sand, resulting in a textured finish.

Sapwood. The outer zone of wood, next to the bark. In the living tree it contains some living cells (the heartwood contains none), as well as dead and dying cells. In most species, it is lighter colored than the heartwood. In all species, it is lacking in decay resistance.

Sash. A single light frame containing one or more lights of glass.

Sash balance. A device, usually operated by a spring or tensioned weatherstripping designed to counterbalance double-hung window sash.

Saturated felt. A felt which is impregnated with tar or asphalt.

Scratch coat. The first coat of plaster, which is scratched to form a bond for the second coat.

Screed. A small strip of wood, usually the thickness of the plaster coat, used as a guide for plastering.

Scribing. Fitting woodwork to an irregular surface. In moldings, cutting the end of one piece to fit the molded face of the other at an interior angle to replace a miter joint.

Sealer. A finishing material, either clear or pigmented, that is usually applied directly over uncoated wood for the purpose of sealing the surface.

Seasoning. Removing moisture from green wood in order to improve its serviceability.

Semigloss paint or enamel. A paint or enamel made with a slight insufficiency of nonvolatile vehicle so that its coating, when dry, has some luster but is not very glossy.

Shake. A thick handsplit shingle, resawed to form two shakes; usually edge-grained.

Sheathing. The structural covering, usually wood boards or plywood, used over studs or rafters of a structure. Structural building board is normally used only as wall sheathing.

Sheathing paper. See **Paper, sheathing.**

Sheet metal work. All components of a house employing sheet metal, such as flashing, gutters, and downspouts.

Shellac. A transparent coating made by dissolving lac, a resinous secretion of the lac bug (a scale insect that thrives in tropical countries, especially India), in alcohol.

Shingles. Roof covering of asphalt, asbestos, wood, tile, slate, or other material cut to stock lengths, widths, and thicknesses.

Shingles, siding. Various kinds of shingles, such as wood shingles or shakes and nonwood shingles, that are used over sheathing for exterior sidewall covering of a structure.

Shiplap. See **Lumber, shiplap.**

Shutter. Usually lightweight louvered or flush wood or nonwood frames in the form of doors located at each side of a window. Some are made to close over the window for protection; others are fastened to the wall as a decorative device.

Siding. The finish covering of the outside wall of a frame building, whether made of horizontal weatherboards, vertical boards with battens, shingles, or other material.

Siding, bevel (lap siding). Wedge-shaped boards used as horizontal siding in a lapped pattern. This siding varies in butt thickness from ½ to ¾ inch and in widths up to 12 inches. Normally used over some type of sheathing.

Siding, Dolly Varden. Beveled wood siding which is rabbeted on the bottom edge.

Siding, drop. Usually ¾ inch thick and 6 and 8 inches wide with tongued-and-grooved or shiplap edges. Often used as siding without sheathing in secondary buildings.

Sill. The lowest member of the frame of a structure, resting on the foundation and supporting the floor joists or the uprights of the wall. The member forming the lower side of an opening, as a door sill, window sill, etc.

Sleeper. Usually, a wood member embedded in concrete, as in a floor, that serves to support and to fasten subfloor or flooring.

Soffit. Usually the underside of an overhanging cornice.

Soil cover (ground cover). A light covering of plastic film, roll roofing, or similar material used over the soil in crawl spaces of buildings to minimize moisture permeation of the area.

Soil stack. A general term for the vertical main of a system of soil, waste, or vent piping.

Sole or sole plate. See **Plate.**

Solid bridging. A solid member placed between adjacent floor joists near the center of the span to prevent joists from twisting.

Span. The distance between structural supports such as walls, columns, piers, beams, girders, and trusses.

Splash block. A small masonry block laid with the top close to the ground surface to receive roof drainage from downspouts and to carry it away from the building.

Square. A unit of measure—100 square feet—usually applied to roofing material. Sidewall coverings are sometimes packed to cover 100 square feet and are sold on that basis.

Stain, shingle. A form of oil paint, very thin in consistency, intended for coloring wood with rough surfaces, such as shingles, without forming a coating of significant thickness or gloss.

Stair carriage. Supporting member for stair treads. Usually a 2-inch plank notched to receive the treads; sometimes called a "rough horse."

Stair landing. See **Landing.**

Stair rise. See **Rise.**

STC. (Sound Transmission Class). A measure of sound stopping of ordinary noise.

Stile. An upright framing member in a panel door.

Stool. A flat molding fitted over the window sill between jambs and contacting the bottom rail of the lower sash.

Storm sash or storm window. An extra window usually placed on the outside of an existing one as additional protection against cold weather.

Story. That part of a building between any floor and the floor or roof next above.

Strip flooring. Wood flooring consisting of narrow, matched strips.

String, stringer. A timber or other support for cross members in floors or ceilings. In stairs, the support on which the stair treads rest; also stringboard.

Stucco. Most commonly refers to an outside plaster made with Portland cement as its base.

Stud. One of a series of slender wood or metal vertical structural members placed as supporting elements in walls and partitions. (Plural: studs or studding.)

Subfloor. Boards or plywood laid on joists over which a finish floor is to be laid.

Suspended ceiling. A ceiling system supported by hanging it from the overhead structural framing.

Tail beam. A relatively short beam or joist supported in a wall on one end and by a header at the other.

Termites. Insects that superficially resemble ants in size, general appearance, and habit of living in colonies; hence, they are frequently called "white ants." Subterranean termites establish themselves in buildings not by being carried in with lumber, but by entering from ground nests **after** the building has been constructed. If unmolested, they eat out the woodwork, leaving a shell of sound wood to conceal their activities, and damage may proceed so far as to cause collapse of parts of a structure before discovery. There are about 56 species of termites known in the United States; but the two major ones, classified by the manner in which they attack wood, are ground-inhabiting or subterranean termites (the most common) and dry-wood termites, which are found almost exclusively along the extreme southern border and the Gulf of Mexico in the United States.

Termite shield. A shield, usually of noncorrodible metal, placed in or on a foundation wall or other mass of masonry or around pipes to prevent passage of termites.

Terneplate. Sheet iron or steel coated with an alloy of lead and tin.

Threshold. A strip of wood or metal with beveled edges used over the finish floor and the sill of exterior doors.

Toenailing. To drive a nail at a slant with the initial surface in order to permit it to penetrate into a second member.

Tongued and grooved. See **Dressed and matched.**

Tread. The horizontal board in a stairway on which the foot is placed.

Trim. The finish materials in a building, such as moldings, applied around openings (window trim, door trim) or at the floor and ceiling of rooms (baseboard, cornice, and other moldings).

Trimmer. A beam or joist to which a header is nailed in framing for a chimney, stairway, or other opening.

Truss. A frame or jointed structure designed to act as a beam of long span, while each member is usually subjected to longitudinal stress only, either tension or compression.

Turpentine. A volatile oil used as a thinner in paints and as a solvent in varnishes. Chemically, it is a mixture of terpenes.

Undercoat. A coating applied prior to the finishing or top coats of a paint job. It may be the first of two or the second of three coats. In some usage of the word it may become synonymous with priming coat.

Under layment. A material placed under finish coverings, such as flooring, or shingles, to provide a smooth, even surface for applying the finish.

Valley. The internal angle formed by the junction of two sloping sides of a roof.

Vapor barrier. Material used to retard the movement of water vapor into walls and prevent condensation in them. Usually considered as having a perm value of less than 1.0. Applied separately over the warm side of exposed walls or as a part of batt or blanket insulation.

Varnish. A thickened preparation of drying oil or drying oil and resin suitable for spreading on surfaces to form continuous, transparent coatings, or for mixing with pigments to make enamels.

APPENDIX

WOODWORKING TOOLS

Small tools for woodworking can, in general, be used for working the plastic and phenolic materials, and also rubber. As the density of the product increases, carbide-tipped cutting edges become more and more essential for long service and smooth cutting.

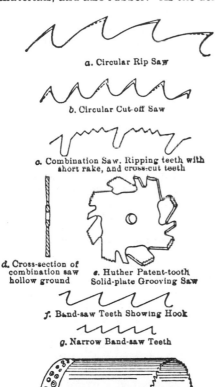

a. Circular Rip Saw

b. Circular Cut-off Saw

c. Combination Saw. Ripping teeth with short rake, and cross-cut teeth

d. Cross-section of combination saw hollow ground

e. Huther Patent-tooth Solid-plate Grooving Saw

f. Band-saw Teeth Showing Hook

g. Narrow Band-saw Teeth

h. Cylinder Saw for Barrel Staves

FIG. 46. Forms of saw teeth.

THEORY OF CUTTING WOOD. Wood is made up of parallel fibers. Cutting lengthwise is simply peeling a layer of fibers off from those below, as with a splitting axe or carpenter's plane. In cutting across the fibers, every fiber must be severed. A knife or chisel driven forcibly across grain is the simplest form of cross cutting. Both examples represent "line" cutting without waste.

Saws divide the fibers, lengthwise or crosswise, by cutting out a swath or *kerf*. In a cross-cut kerf the fiber is first severed at both sides of the swath; the short cut fiber is then removed by a lengthwise cutting at the bottom of the kerf. In ripping a kerf with the grain, the fiber in the swath to be removed is simultaneously cross cut and peeled loose from the sides of the kerf. Cross cutting requires a knife point cut at each side of the kerf. This is made by teeth alternately high at the sides and pointed and beveled on the front. *Raker* teeth may be used to peel or split the severed fiber and remove it. Ripping requires chisel-pointed teeth, set or swaged to the width of the kerf, and with enough hook to sever the fiber with an adzing cut. It is simultaneously split loose at both sides and removed.

Saws are set, or hollow ground, to reduce friction. Surface tension or case hardening of the lumber may cause boards to pinch or close the kerf during a cut. This is one of the principal reasons for wide clearance and wide kerfs. (See Fig. 46*d*.)

With revolving *cutter heads* having knives or bits, the cut generally is lengthwise. It is not exactly a splitting cut, since only at the finish line of straight grain woods is the cut parallel with the grain. At other parts of the cut, the knives are chiseling, or adzing, and cross-cutting the grain.

SAWS have two general classifications. (1) Saws for the *primary conversion* of logs into lumber and timber. Because they are used on live or green logs, there is no danger of overheating and damaging saw temper. Such saws should be designed to cut to approximate dimensions, as subsequent shrinkage necessitates machining to final size and shape. The cutting of tough green wood requires a different tooth from the one used in the cutting of stiffer and stronger, but more brittle, dry wood. (2) Saws for the *secondary conversion* of the board after seasoning, into intermediate or final products. Saws may overheat and lose temper in dry wood, and may scorch or discolor it. Accurate, smooth cuts must be made, to produce final dimensions with maximum smoothness and minimum waste. Teeth must be shaped to cut rapidly without heating, and accurately, with minimum power requirements. The thickness of saws is designated by Stubs gage.

Large saws, chiefly for sawmills in primary production, can be provided with *inserted teeth* of special hard steel, mechanically locked in position, that are interchangeable and can be replaced to compensate for excessive wear due to grinding.

Carbide-tipped teeth are also used on smaller saws in secondary production, brazed electronically to the standard steel blade. Carbide-tipped teeth give long cutting service, on densified wood as well as on casein and resin glue joints, where abrasive characteristics may be encountered. Carbide-tipped teeth have rather blunt cutting edges, cannot be sharpened in the regular way, and require high speeds for smooth cutting.

The two general classifications of primary and secondary conversion may be still further subdivided as below. Figure 46 shows the various forms of saw teeth.

Saws for Primary Production (converting green logs into timber and lumber):
Circular saws.
  Solid disks.
    Parallel-to-grain or rip cut (Fig. 46a).
    Perpendicular-to-grain or cross cut (Fig. 46b).
  Inserted teeth.
  Shingle saws, thin edges with thick centers.
Band saws, continuous or endless (Fig. 46f).
  Parallel-to-grain for cutting logs into lumber and timber.
  Perpendicular-to-grain, drag saws for log lengths.
Band saws, reciprocating, mostly gang resaws.
Special-purpose saws.
  Cylindrical, for cooperage stock (Fig. 46h).
  Concave.

Saws for Secondary Production (converting dry boards into diversified wood products):
Circular saws.
  Ripping (Fig. 46a).
  Cross-cutting (Fig. 46b).
  Combination (Fig. 46c).
  Special-purpose (Fig. 46e).
Band saws, continuous or endless.
  Ripping or resawing, wide (Fig. 46f).
  Curved and special shape, narrow (Fig. 46g).
Band saws, reciprocating.
  Jig and scroll saws.
Special-purpose saws.
  Dadoing (Fig. 46e).
  Tenoning and mortising.

KNIVES AND CUTTERS. The function of a knife or cutter on a planer, molder, or shaper is to reduce the surface or edge of a part to the desired shape and dimensions without marks or scratches, and with a smoothness that can be finished. A roughly sawed surface or edge requires more subsequent machining than one smoothly sawed. Close correlation between the roughness of the saw cut and the depth of the planer cut is necessary. Often, as in jointing plywood cores, a saw can be made to cut smooth enough to eliminate a separate jointing operation. A slightly rough cut is considered better for gluing than one with a very smooth edge.

Following are the basic machines using knives and cutters. (1) Planers or jointers using straight cutting edges to produce plane or level surfaces. They cut continuously along a straight face, not on the side, and parallel or nearly parallel to the grain. (2) Molders or stickers, which are, in effect, multiple planers, usually designed to cut straight or irregular surfaces on all four sides of a piece simultaneously. Cutting may be with either face or side of the knife, but substantially parallel with the grain. (3) Shapers, which are particularly designed to produce irregular and curved shapes or edges both in vertical and horizontal planes. They use a variety of solid, sectional, and laminated cutters, often cutting at an abrupt angle to the grain.

On relatively large diameter wheels, the method of attachment of knives and cutters for planers, molders, and shapers determines the classification. This involves both the provision made for knife attachment, and the corresponding design of head to which knives are attached. Rigid mounting is imperative because of high operating speeds, strong centrifugal forces, and the deep cuts often required.

Knives are distinguished, in general, by thickness. (1) *Thick knives*, usually comprising a tempered steel face and cutting edge, welded to a softer steel back used for clamping. Narrow knives of the chisel type may be entirely of tool steel, tempered principally at the cutting edge. (2) *Thin knives* which require clamp bars and plates, or attachments. Such tool-steel knives seldom are recessed for attachment. They require stiffener bars to provide rigidity and firmness. A knife that chatters will not cut smoothly.

Cutter heads, knives, and router bits can be provided with *tungsten carbide* cutting edges that give long service against very hard species, impregnated, fireproofed, and densified woods, and also resin bonded plywood. The cutting edges of these carbide tips, brazed electronically to a standard steel backing, are rather blunt and cannot be readily sharpened or ground to shape. High linear cutting speeds are necessary for smooth cutting. Solid tungsten carbide cutter heads or knives are seldom practicable, since this alloy lacks the toughness and strength of the standard steels under severe torsional and tension stresses.

Knives and cutters may be classified by machines and by thickness, as follows:

Planers and Jointers (for cutting straight surfaces, single and double, parallel to wood grain):

Thick rigid knives, parallel to axis of head (Fig. 47a).
Square heads.
Two-knife (Fig. 48a).
Four-knife (Fig. 48c).
Face clamping.
Oval holes in knife with bolts adjustable in slotted head (Figs. 47b, 48c).
Slotted holes in knife, with tapped holes in head (Figs. 47a, 48a).
Ribbed-surface knives for rigid clamping (Fig. 47c).
Round heads.
Clamped with countersunk bolt heads.
Internal bolt clamping devices.
Thin Knives, semiflexible, requiring support and clamp bars (Fig. 47d).
Knives parallel to axis of head.
External clamp bars and bolts.
Square heads (Fig. 48d).
Round heads (Fig. 48g).
Internal clamp bars and bolts (Fig. 48e).
Wedge and taper clamps (Fig. 48f).
Knives spiral to axis of head.
External clamp bars and bolts.
Internal clamp bars and bolts.

Molders and Stickers (for cutting multiple straight and irregular surfaces parallel to wood grain [Figs. 50a, 50b, 50c]):

Thick rigid knives (straight, not spiral).
Angular heads.
Two-knife (Fig. 48a).
Three-knife (Fig. 48b).
Four-knife (Fig. 48c).
Face clamping.
Tapped holes in head (Fig. 48a).
Bolt slots in head (Figs. 48b, 48c).
Ribbed-surface knives for rigid clamping (Fig. 47c).
Edge clamping.
External bolts (Figs. 49a, 49b).
Vise or jaw clamps (Figs. 49c, 49d).
Slotted bolts with adjustable cutting angles (Fig. 49c).
Round heads.
Clamped with countersunk bolt heads.
Internal bolt clamping device.
Wedge and taper clamps (Fig. 49f).
Thin knives, semiflexible, requiring support and clamp bars.
Knives, parallel to axis of head.
External clamp plates and bolts (Fig. 48d).
Internal clamp plates and bolts (Fig. 48e).
Wedge and taper clamps (Fig. 48f).
Knives spiral to axis of head.
External clamp plates and bolts.
Internal clamp plates and bolts.

Shapers (for cutting irregular surfaces in two planes and frequently at abrupt angle to wood grain):

Knife blades clamped to head.
Face clamped to head.
Slotted hole, tapped hole in head (Fig. 48a).
Slotted knife and bolt in head (Figs. 48b, 48c).
Edge clamped.
External bolts (Figs. 49a, 49b).
Vise or jaw clamps (Figs. 49d, 49e).
Slotted bolts with adjustable cutting angles (Fig. 49c).
Clamped with pressure bars or plates (thin knives).
External clamped with countersunk bolt heads.
Internal clamped.
Wedge and taper clamps.
Thin knives of slightly irregular shapes (Figs. 48a, 48b, 48c).
Solid knife heads mounted on arbor.
Thick knives of a wide range of shapes (Figs. 48d, 48e).

BORING TOOLS. The principle underlying a boring tool or bit is similar to that of a shaper tool, except that the boring tool advances into the wood. Mechanical means must be provided for this movement, as well as for the removal of chips from the enclosing hole along the spiral shank of the bit. The two basic types of boring tools are these: (1) Boring into *side wood,* where the bit requires side spurs to sever the wood fibers on the circumference of the hole, before the cutting and lifting edge of the bit separates the chips along the fiber line. (2) Boring into *end wood,* where the bit shears off the fibers as it advances, the cutting of the circumference being merely a separation of the fibers. (See Fig. 51.)

Bits. Figure 52 shows four main types of bits. The *double-spur* bit is a general utility tool, usually made with a single-thread screw point. The *extension lip* bit usually has a double-thread point. It is used to produce smoother holes, with less liability to split hard wood. Both types are used to bore holes in side wood. The *taper head drill* is designed for boring end wood. It may be used for side wood if a rough hole is not objectionable. The *spur machine drill* is a modification of the taper head drill. It has spurs and a brad point, to adapt it to better boring service in side wood. It is used in gang boring where holes may be at various angles to the grain.

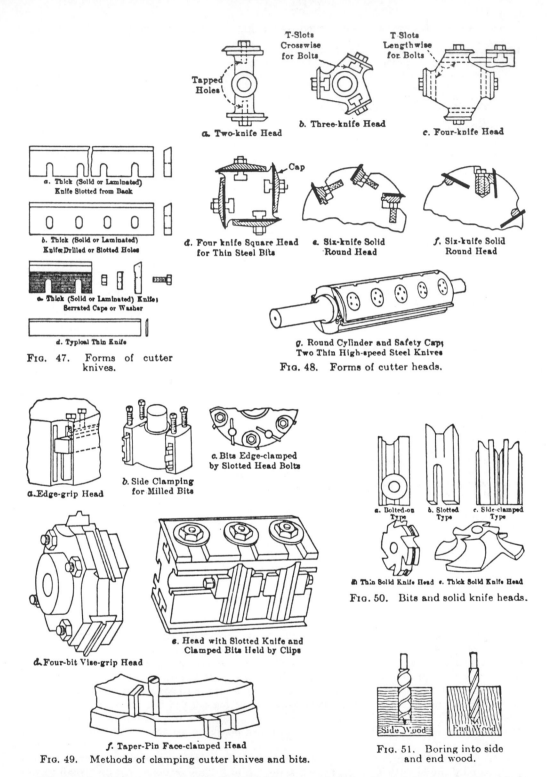

a. Two-knife Head

b. Three-knife Head

c. Four-knife Head

T-Slots Crosswise for Bolts

T-Slots Lengthwise for Bolts

Tapped Holes

Cap

a. Thick (Solid or Laminated) Knife Slotted from Back

b. Thick (Solid or Laminated) Knife Drilled or Slotted Holes

c. Thick (Solid or Laminated) Knife; Serrated Caps or Washer

d. Typical Thin Knife

FIG. 47.   Forms of cutter knives.

d. Four-knife Square Head for Thin Steel Bits

e. Six-knife Solid Round Head

f. Six-knife Solid Round Head

g. Round Cylinder and Safety Cap; Two Thin High-speed Steel Knives

FIG. 48.   Forms of cutter heads.

a. Edge-grip Head

b. Side Clamping for Milled Bits

c. Bits Edge-clamped by Slotted Head Bolts

a. Bolted-on Type

b. Slotted Type

c. Side-clamped Type

d. Four-bit Vise-grip Head

e. Head with Slotted Knife and Clamped Bits Held by Clips

d. Thin Solid Knife Head e. Thick Solid Knife Head

FIG. 50.   Bits and solid knife heads.

f. Taper-Pin Face-clamped Head

FIG. 49.   Methods of clamping cutter knives and bits.

Side Wood   End Wood

FIG. 51.   Boring into side and end wood.

The kind of spiral groove or twist for chip removal depends on bit size, depth of hole, and direction of grain.   Figure 53 shows the more important types.   The *double twists* are usual for medium and large regular boring; left-hand is standard for a bit turning clockwise. Some gang boring machines require right-hand twist bits, rotating counterclockwise.   The *solid center twist*, with a round stem or shank in the center, is used for small bits.   The *ship auger twist* has the largest body of metal on the edges, the center being used for more adequate chip removal.   It is most suitable for deep boring in large timbers.   The *drill twist* has a much more rapidly rising spiral than others, its function being the removal of smaller end wood cuttings, rather than side wood chips. It is ground so that the margin is wide enough to resist wear, thus maintaining the diameter.

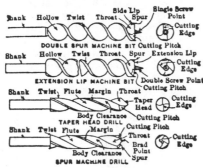

FIG. 52. Types of boring bits.

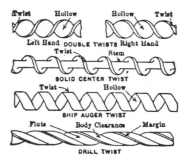

FIG. 53. Types of chip removal grooves.

Figure 54 shows bits for boring concentric holes, sometimes called counterbores. The countersink *a* is for flat-head screws, while recessing, as for washers or bolt heads, is done with one of the forms of counterbore of which *b* and *c* are examples. Many other combinations may be made for both side and end wood.

Large-diameter shallow holes do not require bits with spiral shanks, as chips seldom clog. Figure 55 shows two types, *a* for solid wood and *b*, with saw teeth on the circumference, for plywood. Both have brad points.

Figure 56 shows *plug cutters* with ejector plungers, one of which has inserted high-speed steel teeth for abrasive plywood adhesives, as casein.

Router bits (Fig. 57), while similar in appearance to boring tools, have different functions. A router is chiefly used to cut recessed grooves or to cut out, by means of grooves, regular or irregular outlines. A router bit has a cutting lip throughout its length to produce smooth-edge grooves. It rotates at much greater speeds than a boring bit. The function of side cutting is much more important than end cutting. The advance cut of a spur is not needed, and the fluted clearance slot, for chip removal, may be straight instead of spiral. Router bits may be designed for various shapes, like T slots, ogee grooves, and the like. In this form they may resemble shaper knives and heads.

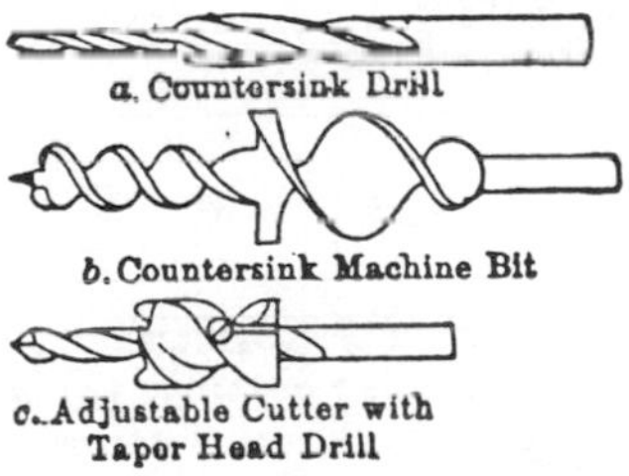

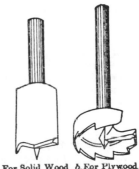

a. For Solid Wood   *b.* For Plywood

FIG. 55. Bits for large diameter shallow holes.

a. Countersink Drill

b. Countersink Machine Bit

c. Adjustable Cutter with Taper Head Drill

FIG. 54. Counterbore bits.

Router bits, for densified types of wood and plywood, are frequently provided with *carbide-tipped* cutting edges, as shown in Fig. 58. Such router bits give long service under severe operating conditions, but must be carefully designed with adequate reinforcing

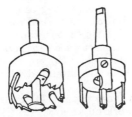

FIG. 56. Plug cutter.

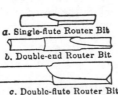

a. Single-flute Router Bit

b. Double-end Router Bit

c. Double-flute Router Bit Made Right and Left Hand

FIG. 57. Router bits.

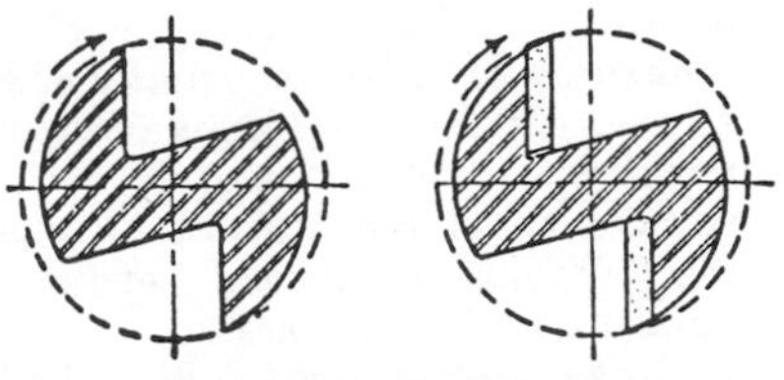

FIG. 58. Sections through router bits without and with carbide-tipped cutting edges on each flute.

backings of tough standard steels to withstand excessive torsional stresses. Carbide-tipped cutting edges are relatively blunt, require high speeds for smooth cutting, and are difficult to sharpen.

MORTISING TOOLS (Fig. 59) are duplex boring and chisel cutting tools. A central standard bit bores a round hole, and, surrounding the bit, a square four-side chisel removes or broaches out the round fillet corners. Chips from both cuts are lifted out by the spiral shank of the bit and discharged through clearance clots in the square chisel shank (one for soft wood, two for hardwood). The slots are located above the wood surface at maximum cutting depth. Chisel points are longer at the corner than at the center, to insure a shearing cut in removing corner fillets. Oblong holes, up to 1 1/2 by 1 dimension ratios, can be made with a rectangular oblong chisel, as in Fig. 59b. If a 2 by 1 ratio is required, and the shape of hole is concealed by shoulders on the tenon, the result can be obtained at a single stroke by the hollow chisel (Fig. 59c). The cutting edges of the chisel are arranged for shear cuts on all four sides. Otherwise, oblong holes are made by multiple strokes of a square mortising tool. Proper clearances must be provided between internal bits and external chisels to avoid friction, heating and burning the wood. In setting mortising tools, the chisel is set first, and then the bit, with the bit cutting lips extending a little beyond the extreme cutting tips of the chisel. Mortising tools are not suited for cutting end wood.

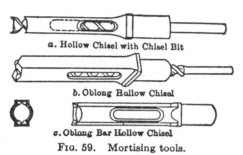

a. Hollow Chisel with Chisel Bit

b. Oblong Hollow Chisel

c. Oblong Bar Hollow Chisel

FIG. 59. Mortising tools.

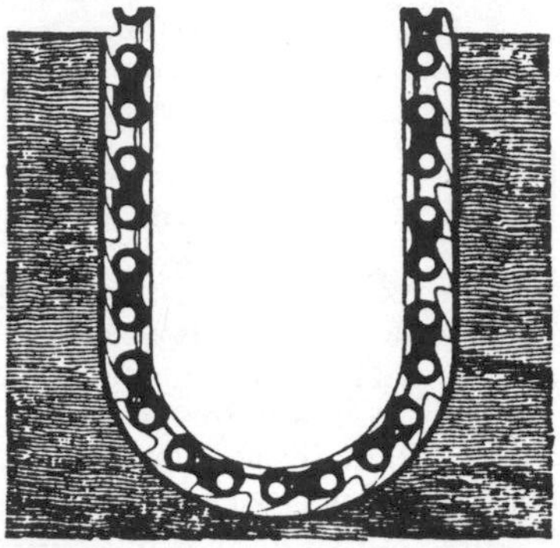

FIG. 60. Mortising chain with saw teeth on outer periphery.

Mortising chains are available for oblong holes, being a series of cutters attached to an endless chain. They are chain saws in principle (Fig. 60), not boring tools, but produce a similar result. The hole has a rough edge, but may be satisfactory for such products as mortise-and-tenon doors, although the joint cannot be very snug.

MACHINE TOOLS FOR WOOD

SAW MACHINES. The five principal types of saw machines are: (1) *rip-saws*, circular cutting; (2) *cross-cut saws*, circular cutting; (3) *variety saw tables*, circular cutting; (4) *band saws*, continuous band cutting; (5) *jig* and *scroll saws*, reciprocating band cutting. In many machines, rip, cross-cut, or combination-tooth saw disks may be used interchangeably, while some are single-purpose machines.

Saw cuts are required on a variety of machines, where the saw cut is auxiliary to the general purpose of the machine. Examples are double-end tenoners, dado heads, or grooving cuts on molders, trim cuts on veneer jointers, or a cut-off on a dowel machine. In general, the best cutting speed for either a circular saw or band saw is approximately 10,000 ft per min.

Rip Saws. Machines designed exclusively for ripping are of two main classes. (1) Single or multiple ripping of full length boards, often the preliminary dimensioning operation in a planing mill, and sometimes in a furniture factory. In gang setups, to avoid a winding cut, saws must be of equal diameter and approximately the same degree of set and sharpness. Saws usually are mounted on the lower arbor, with pressure or feed rollers above. (2) Single ripping of cross-cut stock (often called straight-line rip saws). The saw usually is mounted above, with an endless feed chain below. Such saws often give a sufficiently straight and true edge for gluing. Arbors are sometimes below the work, the saws extending upward between a divided feed chain, above which are mounted idling pressure or feed rollers.

Cross-cutting saws (also called stock-cutting saws) may be of these types. (1) Swing saw (an older type) hung from the ceiling and swinging into the work on an arc. (2) Clip saw, swinging from below, with the motor drive pulley concentric with the pivot. (3) Saws on sliding ways, or other mechanical device, to move them horizontally into the work and out. Cross cutting usually is the first dimensioning operation in a furniture factory.

Variety saw tables may be equipped interchangeably with rip, cross-cut, or combination (rip and cross-cut) saw disks, depending on the work. For grooving or tenoning, dado heads or *wobble saws* may be used. Wobble saws are angle mounted, with wedge-shaped collars, between two perpendicular saws for slotting or rabbeting. Variety saw tables generally are equipped with side gages, back gages, and angle adjustments for miter sawing. For angle cutting, the table may tilt or the arbor may be pivoted. In size, they range from small bench saws up to floor machines. Speed of floor units usually is 3400 to 3600 rpm, requiring 5 hp.

Since all circular saws cut against the feed, the operator should take great care to avoid dangerous "kick backs." Most modern saws have mechanical devices to protect against both kick backs and hand hazards from the saw disks.

Band Saws. Resaws for such operations as ripping special thicknesses usually are about 6 in. wide, with heavy corrugated feed rollers on vertical shafts. Regular band saws may be equipped with a variety of blades, seldom wider than 1 in. for straight work, and down to $1/4$ in. for curved work. Band saw wheels range from 30 to 36 in. in diameter, with direct-connected motor on the lower wheel. Band saw throats take work up to 36 in. wide by 18 in. thick. Motors are 3 to 7 $1/2$ hp; wheel speeds are 600 to 1200 rpm. Small bench models also are made.

Jig and scroll saws are used chiefly for cut-outs that cannot be made on band saws, and are much less costly than band saws for small shops. The development of the high-speed router, producing smoother edges, has greatly lessened the utility of the jig saw.

PLANING MACHINES. The underlying principles of cutting wood surfaces in the planer also are used in the jointer, molder (sticker), and flooring machine. The following explanation of these principles in the planer applies also to these other types of machines.

Cutting Mechanism. Figure 61 shows a section of a typical planer cutting mechanism. The cutting head revolves against the infeed of the stock, and the grooved *infeed rolls*

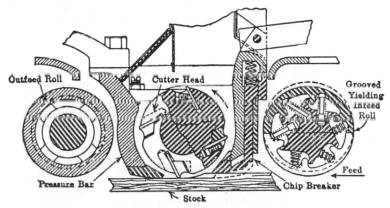

FIG. 61. Planer cutting mechanism.

should be powerful enough to withstand the double load of feeding stock and holding it against the action of the cutters. Infeed rolls often are made in sections to bear more uniformly on thick and thin stock. Large single surfacers and all double surfacers have two pairs of infeed rolls. Pressure of the feed rolls tends to straighten crooked lumber, but the ordinary planer cannot be relied upon to remove wind and warp entirely. Special types are used for this purpose. The *chip breaker* prevents lifting and tearing of the grain, and reduces all chips to a size that will not clog the cutting head or be difficult to move as shavings in the dust collector system. The steel-tipped chip-breaker shoes ride the top surface of the incoming lumber, and are substantially concentric with the cutter head. They are suspended from a sturdy cross bar, as a heavy cut places a severe strain on chip-breaker mechanism. The smooth *outfeed roll* and the *pressure bar* require exact adjustment to the final planed thickness. They must be rigid enough to prevent any chatter or end clipping in the board during the planer cut.

The *cutter head* in Fig. 61 is of the round type with four thin knives, clamped in slots. While it is possible to set the several knives in approximate alignment, it is necessary to joint or grind all knives, at operating speed, to a true cutting line, after they have been set and clamped in the head. This is done preferably with a special motor-driven traveling grinder mounted over the main bearings of the planer. In some cases (as an enclosed lower knife head), the head is partially or wholly removed, with the knives clamped in place and ground in that position.

Planer cutting circles usually are about 6 in. in diameter but vary with the different makers and shaft lengths. Since the standard speed of planer shafts usually is 3600 rpm (7200 when a frequency changer is used), it is important that they be in perfect mechanical and static balance, requiring careful attention to the individual weight of knives and clamping devices. One of the most difficult problems of planer design is to overcome shaft whip or deflection at high speeds, especially in the wider machines, tending to produce uneven thicknesses and rough planing.

Types of Planers. The principal types of planing machines are (1) general-purpose planers; (2) cabinet planers, for high quality work; (3) mill-type planers, for surfacing full length boards.

General-purpose planers, ordinarily used in so-called planing mills and job shops, have a bed (thickness) adjustment in vertical ways by means of two geared screws. With this construction a slight rocking motion is unavoidable, when stock is entering the infeed or leaving the outfeed rolls, which may be unimportant in many rough planing operations that do not require precision. Since the average cost of this type is less than that of the cabinet planer, it usually lacks the better attachments and more accurate adjustments of the latter.

Cabinet planers owe much of their precision to bed adjustment by motor-actuated sliding wedges, giving both accuracy and rigidity. The range of adjustment is less than that of the general-purpose planers, but tendency to rock is entirely eliminated. In cabinet planer design, the point of the chip breaker is placed as close as possible to the actual point of cutting, to reduce the size of chips and lessen the tendency to tear out ahead of the cut. The chip breaker usually is of sectional or flexible construction, to ride equally well on thick, thin, and rough spots in the stock. High-grade planer work demands a large number of knife cuts per inch, which is obtainable either by increasing the number of knives (up to six, rarely more) in the head or stepping up the speed of the shaft, a point on which opinion is divided. Straight knives generally are used, but spiral knives occasionally are used to distribute the load and reduce pull-outs at the back end of the stock.

Motor equipment for planers varies with the service. Standard power ratings for single surface cabinet planers are:

Planer width, in.	26	30	36	40	44	50	60	64
Horsepower	10–15	15–20	20–25	20–25	20–25	20–25	25–30	25–30

Feed mechanism may be driven direct from the motor on the cutter shaft by belt to a constant diameter pulley, or by a variable-speed V-belt drive (between expandable flanged pulleys), thus permitting an adjustment of the feed with relation to the speed of the cutter head. When frequency changers are used on the cutter heads it is an advantage to have separate multiple-speed constant torque motors on the feeding mechanisms.

Speeds usually are up to 60 ft per min, occasionally 100 ft per min, with considerably slower speeds for exacting work.

Production cabinet planers often are fitted with *hopper feed*, the infeed bed or table being equipped with an endless rubber-faced conveyor chain, with slats or side rollers for random lengths. Hopper feeds usually are set on an angle to feed pieces diagonally.

Cabinet planers are made for double surfacing, the lower cutter shaft then being arranged to be pulled out for adjustment and jointing. Double surfacing planers usually are built from 30 to 44 in. wide with larger power requirements than single surfaces.

Special feeding mechanisms are required when the warp and wind in the lumber are beyond the capacity of the infeed rolls to control sufficiently by pressure. These mechanisms, on a single surfacer, called a *facing planer*, have an overhead flexible fingered infeed chain, and produce a true surface on the underside of the board against the cutter head. Two or more surfacings may be required for especially crooked lumber. The rough side is surfaced in a subsequent operation, in the same facing planer or in a standard cabinet planer.

A combination double surfacer, called the *Straitoplane*, has this flexible fingered infeed chain for the lower cutter head to produce an initial true surface on the underside of the board, followed by a standard upper planer head, rigid bed, and outfeed roll for the subsequent facing of the upper side of the board. The second cut can be adjusted for accurate thickness. This machine operates efficiently on stock as short as 10 in., and is particularly adapted to furniture work where most pieces are relatively small. Each cutter head is driven by a 15-hp motor at 3500 rpm. A 5-hp, 2-speed motor drives the feed. The machine is standard at 36 in. wide. Feeds up to 60 ft per min are practicable.

Mill planers are designed primarily for use with full-length lumber. They have been developed, and are used principally, on the Pacific Coast. Feed mechanism comprises endless chains with cross slats of ribbed cast iron, giving a more powerful and positive feed than is possible with rolls, but lacking somewhat in accuracy of cut. The cutter head is above, with the adjustments of a cabinet planer. Bed adjustment is of the sliding wedge type. Capacity is stock 8 in. thick, 30 to 36 in. wide. Feeds of 100 ft per min are practicable.

Jointers are classified as:

Standard Jointers. The cutting mechanism of the jointer is quite similar to that of the planer with a cutter head below the bed, but the feed mechanism is omitted, since edged jointing is the predominant function of the jointer. Jointers are provided with a high *fence* (side gage) to permit them both to square and to true the edge. Portable bench jointers are made and extensively used in preference to the hand plane. Jointers seldom are made over 16 in. wide, and sometimes are hand fed to take the wind out of dimension stock. This is particularly important in producing a true bed surface for molder or sticker work.

Lumber Jointer. The continuous feed lumber jointer, often called a *glue jointer*, has an automatic return feed for edge jointing strips for plywood core stock. This is a duplex or return cycle machine, with an endless feed chain in the center, mounted on vertical shafts. The two cutter heads revolve in opposite directions. A board to be jointed on both edges is placed edgewise at the infeed end against the feed chain, jointed on the lower edge, passed over a tipping mechanism that turns the unjointed edge down and into the reverse feed track. The board travels back to the infeed end, moved by the opposite side of the feed chain, passing over the second cutter head. This machine is suitable only for use on ripped stock, as the cut is insufficient for rough or wany-edge stock, and it will not produce uniform width.

Veneer jointers are highly specialized jointing machines. They operate on a flat *package* of sheets of veneer, aggregating 2 in. thick, which must be clamped firmly during the entire cutting period. The cutter head, usually mounted on a vertical shaft, carries as many as twelve knives to give the smooth edge necessary for matching high-grade veneer faces. In some machines the clamped package travels past the revolving knife head. In others, the clamp is stationary, and the head travels.

Molders or stickers are essentially four-sided planers, with two horizontal and two vertical shaft cutter heads, usually square four-side heads (see Fig. 48c). The feed may be by pressure rolls against a rigid bed or by endless feed chain below. Cuts frequently made are irregular and curved faces on moldings. Cutter heads, so far as possible, have chip breakers and the other auxiliary devices of the planer. Since all cutter shafts are short, knives and cutter heads are bolted on ends of shafts, and it is possible to use high speeds with frequency changers, up to 7200 rpm. Fewer knives are required at high speeds to give the same number of knife cuts per inch. This is a distinct advantage in grinding special knives as most molder heads are four sided, and the uniform setting of four irregularly shaped knives is most difficult. It is usual to operate a molder head at high speed with two opposed knives of the same shape and a correspondingly greater rate of feed. For long runs on standard shapes, solid cutter heads are often used, with knives jointed in a profiling machine.

Modern molders have brackets for *cope heads*, which make re-entrant cuts that are not possible with regular heads. They also often supplement the regular heads by making deep cuts that would require impractical or very costly knives. Molders are close coupled and compact, usually with individual motors for each head and for the feed. A popular size takes stock up to 4 by 4 in., but they are made up to 16 by 6 in. for heavy timbers.

Flooring machines are specially designed molders, the two vertical shafts making the tongue and groove, while the lower horizontal head makes the flooring face, and the upper head cuts the hollow back. A steel lettering wheel often grades and trademarks the back as it leaves the machine. Flooring machines are designed to operate on full-length lumber and at high speeds.

THE SHAPER is one of the oldest types of machines. It still is important in woodworking, although many of its functions have been absorbed by more specialized machines. It is a versatile machine for small shops, as it is more easily set up for short runs than many of the specialized machines that are more efficient for long runs.

The shaper consists essentially of a large flat metal table with a vertical spindle, driven from below, extending up through the center of the table. Revolving cutter heads, mounted on the end of the spindle, accommodate a variety of knives for a wide range of cuts.   Solid heads may be mounted for long runs. A job to which a shaper is best adapted is the shaping of the edges of table tops, especially those of other than rectangular form. The table top blank is firmly mounted, by two or more points, on a jig of the desired shape. A spindle collar is mounted at the machine table level, above which the knives are set and clamped for the required shape, ogee or otherwise. The operator revolves the combined jig and blank, holding it snug against the collar for one complete revolution of the jig, and thus producing the ogee edge around the periphery. Shaper knives cut with the grain rather than against it, as far as possible. Slow feeds are necessary when cutting at right angles to the grain. Blanks may be reversed on the jig to help this condition.

The shaper is especially useful in re-entrant curves, octagons, and other odd contours not suited to machines designed for rectilinear or circular cutting. Small molding, flutings, both straight and curved, and many edge and end cuts can be made on a shaper.

Shapers frequently are equipped with two spindles, usually rotating in opposite directions. All cuts around an irregular table top can then be made in a direction favorable to the grain, and no cuts have to "dig into the grain," leaving a rough and torn edge. The jig, with mounted work, is moved from spindle to spindle according to grain requirements. Automatic shapers are equipped with mechanically operated jigs or revolving tables to which blanks are held with pneumatic clamps. These are suited only to long production runs.

THE ROUTER is a relatively modern machine, achieving its effectiveness with the development of high spindle speeds. It is somewhat the reverse of the shaper, and usually is designed for considerably lighter cuts. The cutting head is above the table, which is provided with an upwardly projecting, jig-guiding round plug of the same diameter as, and directly under, the cutter. The cutter head carries a router bit rotating at speeds of 20,000 rpm or higher. The table may be moved up to the bit, or the bit lowered into the work. The router is admirably adapted to cutting out (with grooved cuts) grills or irregular shapes. It works much better and more quickly than jig saws. Router bits may have shoulders for shaping the edges of cut-outs, or T heads for slotting. Straight cuts on narrow strips can be made by clamping suitable gages to the table. A straight gage, through or near the bit center, permits shaping of straight-edge table tops and similar work.

Router motors, both electric and pneumatic, must be very accurately balanced for excessively high speeds. Brakes are provided to slow down the bits, which otherwise would require a long time to come to rest. Blowers, directed to the exact point of cut, are standard equipment to prevent chips and shaving from clogging the cut and heating the bits. Some routers are equipped with tilting tables. In these the bit is lowered into the work. Motors are 1 to 5 hp for heavy duty. Spindle speeds often are stepped up by V belts.

Portable hand routers are built to operate electrically and pneumatically. Electric machines operate at 10,000 rpm maximum; pneumatic, at 60,000 rpm maximum.

CARVING MACHINES are of two main types, single and multiple.

Single carvers may be rigidly mounted on a floor standard, the work being moved free-hand by the operator. The single carver also is available in the balanced suspension type, hung much like a dentist's drill. The work is held stationary and the drill moved over the pattern by the operator. In both types the mechanism consists of a small high-speed motor with a small spur drill mounted directly on the motor shaft. Single carvers are used for a certain amount of undercutting. Final retouching usually is by hand.

Multiple carvers may carry as many as twenty-four blanks, all to be roughed out uniformly from a master pattern or model, usually of metal. The spur cutters are moved in unison by link levers that are fixed in their relation to each other. Cutters usually are belt driven from one or more motors. Individual motors can be used for extra heavy cuts. Such machines can only rough out the carvings, but their efficiency consists in removing all surplus wood, even on deeply recessed carvings, leaving a minimum of hand carving of finer lines and undercuts. This type of machine has greatly extended the use of carving on classical furniture at very reasonable cost. But one operator is required, who guides the blank cutter over the pattern, carefully watching the direction of the cut to avoid tearing the wood. The hand finishing work reproduces the character of hand carving. This process is displacing pressed and overlaid imitation carved work.

BORING MACHINES range over a wide variety of woodworking machines, from a single bit held in clutch jaws on the end of a motor shaft, to gang or multiple machines 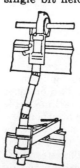 in various arrangements, sometimes equipped for as many as thirty bits. Some bore only vertical holes, others bore horizontally, and still others bore at various angles, depending on the complexity of the work. The maximum number of individual bits seldom exceeds thirty, but special machines have been made for a large number of boring problems. The literature of the industry should be consulted for more specific information.

Vertical multiple borers carry a series of fixed or adjustably positioned spindles, usually arranged in approximately single or double vertical ranks, with a clamp-equipped work table moving up the bits. Spindles may be fixed or *rigid*, i.e., maintain a vertical alignment with the splined drive shaft above, or *adjustable* to points at quite a horizontal distance from the plane of the drive shaft, by a telescopic universal joint (see Fig. 62).

FIG. 62. Boring mill spindle with universal joint.

Horizontal multiple borers differ from vertical borers principally in boring a series of horizontal holes, having a work table that moves in and out. They are used mostly for end boring. Universal-joint spindles are not generally adaptable to horizontal boring. However adjustable multiple or cluster boring heads, to which bits are fixed in position in the head, are available.

Cluster vertical borers (not made in horizontal types) have universal-joint spindles, adjustable over rectangular or circular areas up to 40 in. in diameter, with many combinations of bits that may be attached. Motors are above, and operate the bits through gear trains which terminate in fixed sockets for spindle connections. The universal-joint spindles permit a wide range of adjustment of bits in any horizontal direction within operating limit.

Multiple boring heads are principally (1) fixed cluster (Fig. 63a); (2) adjustable multi-spindle (Fig. 63b). Many other types of heads are made, for use in both vertical and horizontal boring machines.

Standard spindle speeds range from about 1000 to 2000 rpm, varying principally with size of hole, hardness of the wood, and rate of feed.

LATHES differ widely in basic principles, developing by degrees from the plain hand lathe, which is one of the oldest woodworking machines and suited to a wide range of individual work. The principle of the high-speed cutter head has been borrowed from the shaper and adapted to many lathe cuts, especially for duplicating work. Patterns and cams have been adapted from such machines as the carver, and other devices, for various other types of cut. Only the principal types are described below.

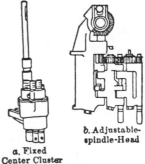

b. Adjustable-spindle-Head

a. Fixed Center Cluster

Fig. 63.   Multiple boring heads.

The copying lathe works entirely from a model or pattern of the exact shape required. It can be used for gun stocks, axe handles, shoe lasts, parts of period furniture, etc. The blank to be turned and the pattern are mounted on slowly rotating gear-driven centers. The cutting member is either a high-speed revolving cutter head or occasionally a heavy saw. Its movement is controlled by a tracing pin, or roller, which traverses all portions of the pattern, producing an exact duplicate. Usually the cutter moves along the length of the blank, but in axe handle lathes the frame, with centers, blank and pattern, moves past the cutter head.

Back knife lathes first turn the blank to a rough round by a ring or die with inside cutting edges, supported on a traveling carriage. The die is moved out of the way, over the dead center, for the turning operation. The long back knife is so set, milled to such a shape, that when in an angling cutting position against the blank, it gradually cuts in, and is stopped when the final turning is of correct size and contour. The back knife never is the exact reverse of the turning, but a mechanically developed extension of it.

The shaping lathe is an exceedingly versatile machine, turning rounds and polygonal sections in almost every conceivable combination. Both lathe centers are "alive," and oscillate slowly into and out of the cutting zone, under the operator's control. Cutter knives, similar to a series of shaper heads on a single shaft, are mounted on a cutter spindle or arbor in the rear, and revolve at high speed. All knives are mounted on the arbor, whether for square, round, or hexagon cuts. As a general rule, where squares and rounds are on one turning, flat surfaces are cut first, the lathe centers turning through a small arc, and keeping the work tangential to the revolving cutters by a cam. The piece to be turned is withdrawn from the cutters, turned through approximately a 90-degree angle (for a square), and the next flat surface is cut. The angle of movement controls the number of sides for squares, hexagons, octagons, etc. After the flats are completed, the rounds are turned, with lathe centers slowly and uniformly rotating. Various modifications may be made in this procedure for complicated types of turning or shaping pieces other than turnings. The setup of the cutter head may be so complex as to demand long production runs for good economy.

Many different types of lathes are built in addition to the major types described above. The literature of the industry should be consulted for detailed information.

MORTISING MACHINES may be classified as (1) hollow chisel mortisers; (2) chain saw mortisers; (3) oscillating bit mortisers.

Hollow chisel mortisers usually provide a vertical movement of the tool into the work. Horizontal types also are available. The tool is attached directly to the motor, the rotating bit being motor driven. The surrounding nonturning chisel is mounted in the chisel carriage on the motor frame, and both are forced together into the work. The tables are equipped with clamps, adjustments, and multiple end-gages, as several mortises may be made in the same piece. Tool strokes may be arranged automatically to repeat cuts for oblong holes, up to about 50 per minute. Horizontal movement of the table also may be automatic for a series of mortises. Motor sizes range up to 4 hp at 3600 rpm. Multiple hollow chisel mortisers always are horizontal stroke, limited to six heads, with a minimum of 4 in. center spacing, and a maximum spacing of 72 in. between outer heads.

Chain saw mortisers operate with cutting teeth mounted on the outside periphery of a link chain. (See Fig. 60.) The minimum size of hole is $3/16$ by $3/4$ in.; the maximum may be $1 1/2$ by 4 in. Longer mortises may be cut at one stroke by using a double roll bar, or by successive cuts. The table moves the work up into the chain cutters. Sprocket motors usually are 3600 rpm, giving the following approximate chain-cutting speeds:

No. of sprocket teeth	5	7	9
Chain speed, ft per min	1330	1930	2400

Multiple-chain saw mortisers are available up to 8 heads, which may be stationary, or adjustable in and out as well as sideways. Chain saw mortisers are not made in horizontal types.

Oscillating bit mortisers make round-end holes, whose depth is limited by the size of the router-type bit. The bit oscillates between stops set for the desired lengths. Oscillations seldom should exceed 200 per min; bit speeds are up to 7200 rpm. Oscillating motor usually is 1 hp; bit motors may be 4 hp. Air clamps hold stock on the tables.

DOUBLE-END TENONERS are highly developed machines that accomplish a wide variety of cuts at one passage of the work. Their functions have been broadened and increased by individual motors on each cutting head, which are more flexible than the belt-driven heads of earlier machines. This machine originally was intended to cut tenons simultaneously on both ends of a piece of stock. Cutting heads of various types were added gradually for a diversity of end and intermediate cuts.

In essential principles, the double-end tenoner consists of a sole (or bed) plate carrying two pedestals. One pedestal is stationary, and carries all operating switches. The other may be moved along the sole plate to accommodate the work, limited in standard machines to about 90 in. Between the two pedestals, and below the work, a pair of endless feed chains, with adjustable lugs, carry the work into the saws and other cutting units. The feed chains are kept relatively close to the pedestals but are adjustable along a spline feed shaft between them. They are fixed in the horizontal plane, all vertical adjustments being made with the cutter heads. Two idling pressure chains hold the work firmly, operating directly above the feed chains.

For double cut-off and double rip cuts, as in plywood, only one saw is used in each pedestal. This cut is made from above. On modern double-end tenoners, the main cutter head can be tilted to make angle cuts. One or more additional horizontal (sometimes tilting) cutter heads are located on each pedestal to make end tenons, grooved, shoulder or shaper cuts on either or both sides of the stock, and at one or both ends. Each pedestal also carries a vertical cutter head. An arbor can be placed between pedestals for intermediate saws or cutters. Cope units, dado heads, relishing cutters, etc., if required, may be attached in various ways. Jump dado cams can be used to enter and withdraw dado cutters near the edge of the stock. Dovetailing attachments may be used for either male or female cuts. Methods of adjustment, attachment of various cutter heads, and other operating features vary considerably in the product of the several machine builders. The variety of work is limited only by the available cutter heads and the extent to which production runs justify elaborate setup.

Single-end tenoners, with corresponding attachments, are available for short runs. Stock is fed on a roller feed carriage, with the cutting units at the left of the operator.

SANDERS. Power-feed drum-sanders are of two general types. (1) Work is forwarded by a rubber-cushioned endless-belt bed. (2) Work is moved ahead by pairs of pressure rollers. Widths range from 31 to 103 in. Generally, the roll-feed type sands one side only, giving thinner and more accurate surface removal as in figured faced veneers on plywood. Two-side types are available, if there is plenty of surface to be removed, and accurate centering between surfaces is unimportant. Endless-bed sanders sand one side only.

Endless-bed Sanders. In a typical machine of this type, the sandpaper drums, usually three, are on top. They are approximately 12 in. in diameter, covered with felt cushion and sandpaper, sometimes wound spirally, but more often attached by clamps in slots. The feed belt, with rubber cushions, is below. Surface speed of the first two drums usually is 3700 ft per min (1200 rpm); that of the third drum is 5600 ft per min (1800 rpm). Drums revolve against, rather than with the work. The work is fed at speeds ranging from 12 to 24 ft per min. The coarsest sandpaper is on the first roll, and the finest on the third roll. Each roller and the feed mechanism are driven by an individual motor. Motor sizes, depending on machine width, are: first and second rollers, 5 to 10 hp; third roller, $7 1/2$ to 15 hp; feed mechanism, usually, 3 hp. A brush roller is frequently installed at the delivery end to remove sandpaper dust. Standard drums oscillate slightly endwise (75 to 100 per min) to simulate the side movement of hand sanding.

Single-surfacing Roll-feed Sanders. The drum arrangement is similar to that of endless-bed sanders, except that drums are below the work, and sand through slots in the feed bed. There is an idler-pressure roller above the work. Pairs of smooth feed rollers are located at the entering end, on both sides of the center drum and at the discharge end. The roll type is not well suited to small dimensioned stock. The sandpaper drums are not so conveniently located for paper changing as in the endless-bed type.

Two-side Roll-feed Sanders. The upper and lower drums are directly over each other in pairs, with the same arrangement of pairs of feed rollers as in the single-surface type.

Belt sanders have a 6- to 8-in. endless sandpaper belt running clockwise on two 18-in. pulleys, 8 to 12 ft apart. In the overhead type the work table, moving on ways toward and away from the operator, is below the lower belt, and the smooth side of the paper is against the pulley. In the underslung type, the table is between the belts, with the sand side of the paper against the pulley. In both types, the work is held on the table by stops at the left and an edge clamp at the right. The operator bears down on the cushion pad, hand operated or mounted on two-way sliding levers. The pad is above and applies controlled pressure through the moving sand belt to sand the stock. Belt sanders are preferred to drum sanders for work that requires a thin cut, as on plywood faces.

Stroke sanders usually are used for polish and finish sanding. They are somewhat similar to belt sanders, except that the pad is mechanically reciprocated under weights. The same type of stroke machine is used for rubbing finish, the sandpaper then being omitted, and a rubbing compound used.

Automatic belt sanders for sanding moldings consist of driven rubber feed rolls which carry the stock under a sandpaper belt, and a flexible shoe shaped to the reverse of the molding contour. Feed is 32 to 50 per min.

Specialized types of sanding machines are belts on vertical pulleys for edge sanding; sand disks on rotating horizontal or vertical tables for drawers and boxes; sand drums of large diameter for curved work; sand arbors for chair legs and irregular pieces, etc. A small portable motor-driven belt sander is available for bench use.

INDEX

mechanical properties, table, 83
Appearance lumber, 139
Apple:
    dimensional change coefficient, table, 261
    moisture content, table, 48
Apron, window, 613
Arches:
    fire resistance of, 267
    glued structural members, 216
Asbestos fiber and paper, thermal conduct-
    ivity, table, 390
Asbestos-cement shingles, siding, 543
Ash:
    characteristics, 5
    characteristics for painting, table, 279
    color and figure, table, 46
    decay resistance, table, 59
    dimensional change coefficient, table, 261
    gluing properties, table, 196
    large pores, 287
    locality of growth, 5
    machining and related properties, table,
      57
    mechanical properties, table, 73
    moisture content, table, 48
    nomenclature, table, 127
    shock resistance, 6
    shrinkage values, table, 51
    uses, 5
    weight, 5
Ash-forming minerals, 42
Aspen, 6
    characteristics for painting, table, 279
    color and figure, table, 46
    decay resistance, table, 59
    dimensional change coefficient, table, 261
    gluing properties, table, 196
    machining and related properties, table,
      57
    mechanical properties, table, 73
    moisture content, table, 48
    nomenclature, table, 127
    shrinkage values, table, 51
    small pores, finishing, 287
Asphalt coatings on plywood, 570
Asphalt shingles, 522
    weight recommended, 520
    with wood sheathing, 505
Asphalt-tile floor, 604
Attic folding stairs, 628
Attic inspection:
    condensation, 680
    maintenance, 681
Attic ventilation, 567, 571, 681
Avodire, 23, 24
    characteristics affecting machining,
      table, 58

    decay resistance, table, 60
    dimensional change coefficient, table, 77
    mechanical properties, table, 58
    shrinkage values, table, 51
Axial stresses, glued structural members,
    217

Bacteria causing decay, 294
Bag molding, plywood, 249
Bagtikan, 24
Balanced construction of plywood, 225, 226
Baldcypress:
    characteristics, 14
    characteristics for painting, table, 279
    color and figure, table, 46
    decay resistance, table, 59
    dimensional change coefficient, table, 261
    gluing properties, table, 196
    locality of growth, 14
    mechanical properties, table, 73
    moisture content, table, 48
    nomenclature, table, 135
    pecky cypress, 14
    resistance to decay, 14
    shrinkage value, table, 51
    uses, 14
Balsa, 24
    decay resistance, table, 60
    dimensional change coefficient, table, 262
    elastic constants, table, 72
    mechanical properties, table, 83
    shrinkage values, table, 53
Balloon-frame construction:
    brick, stone, or stucco houses, 468
    sills, 468
    wall framing, 480
Balustrades, types, 640
Banak, 24
    characteristics affecting machining, table,
      58
    decay resistance, table, 60
    dimensional change coefficient, table, 262
    mechanical properties, table, 83
    shrinkage values, table, 53
Bark, inner and outer, 40
Bark beetles:
    control measures, 296
    damage caused, 296
Basement ceilings, 584
    design to improve fire resistance, 271
Basement floors:
    distance below grade, 449
    drainage, 651
    thickness, 651
Basement, maintenance, 680

installation, 537
nails, 537
plywood, 535, 540
spacing, maximum, 535
storage, 678
treated, 533, 538
types, 533
wood, 85, 533, 540
Sill anchors, depth, spacing, sizes, 456
Sill flashing, extent of, 629
Sill plate:
    balloon-frame construction, use in, 468
    leveling of, 456
Sills, 468
Size factor, for design use, 149
Skin buckling, stressed-skin panels, 221
Skin stresses, stressed-skin panels, 220
Skins, stressed-skin panels, 220
Slash-grained lumber, (see Flat-grained)
Slope of grain:
    determination, 93
    effect on fatigue strength, 104
    in bending stock, 251
    in visual sorting, 143
    strength as related to, table, 93
Smoke shelf, fireplace, 645
Snow dams, protection from, 519
Sod, removal and storage, 449
Soffit, 511
Soft rot, 292
Softwood grading association, 136
Softwood lumber, 130, 132-135
    grading nomenclature, table, 135
    yard lumber sizes, table, 133
Softwood plywood, 226
Softwoods, 3, 4
    bending properties, 250
Soil cover, 575
Soil stack, 546
Solid-bridging, 474
Solvent seasoning of wood, 257
Sound absorption, 580
Sound materials, 582
Sound transmission class ratings, 577
Sound transmission of sandwich panels, 241
Sound-deadening hardboard, 412
Southern hardwoods, list, 4
Southern pines, used for poles, 326
Southern softwoods, list, 4
Soybean glue joints, durability of, 207
Spaced columns, 193
Spaced sheathing, installation, 506
Splash block, 637
Spanish cedar, 36
    characteristics affecting machining,
      table, 58

decay resistance, table, 60
dimensional change coefficient, table, 77
mechanical properties, table, 83
shrinkage values, table, 53
Special densified hardboards, 415
    strength and mechanical properties,
      table, 409
Special drying methods:
    boiling in oil, 257
    high-frequency electrical energy, 257
    infrared, 258
    solvent seasoning, 257
    vacuum drying, 258
Specifications for:
    plywood, 225
    timber piles, 329
Specific gravity:
    coefficient of variability, 54, 88
    definition, 44
    strength as related to, table, 91
    values, table, 73, 82
Specific gravity of wood, 54, 91
Specific heat of wood, 61
Speed of sound, 70
Spikes, 159
Spiral grain, 94
Splits, glued structural members, 215
Springwood, 41
Spruce, Eastern, 21
    characteristics for painting, table, 279
    decay resistance, table, 59
    mechanical properties, table, 73
    nomenclature, table, 135
    shrinkage values, table, 52
Spruce, Engelmann, 21
    characteristics for painting, table, 279
    color and figure, table, 46
    decay resistance, table, 59
    dimensional change coefficient, table,
      261
    mechanical properties, table, 73
    nomenclature, table, 135
    parallel-to-grain tensile strength,
      table, 88
    shrinkage values, table, 52
    toughness values, table, 88
Spruce, Sitka, 21
    characteristics for painting, table, 279
    color and figure, table, 46
    decay resistance, table, 59
    dimensional change coefficient, table, 76
    elastic constants, table, 72
    gluing properties, table, 196
    mechanical properties, table, 73
    moisture content, table, 48
    nomenclature, table, 135
    shrinkage values, table, 52